ENVIRONMENTAL MANAGEMENT
GEO-WATER AND ENGINEERING ASPECTS

PROCEEDINGS OF THE INTERNATIONAL CONFERENCE ON ENVIRONMENTAL MANAGEMENT, GEO-WATER AND ENGINEERING ASPECTS
WOLLONGONG / NEW SOUTH WALES / AUSTRALIA / 8-11 FEBRUARY 1993

Environmental Management Geo-Water and Engineering Aspects

Edited by
ROBIN N.CHOWDHURY & (SIVA) M.SIVAKUMAR
The University of Wollongong, New South Wales, Australia

A.A.BALKEMA / ROTTERDAM / BROOKFIELD / 1993

Cover photo: Val Pola rockslide, Valtellina, Northern Italy
Courtesy of A.Cancelli

The texts of the various papers in this volume were set individually by typists under the supervision of each of the authors concerned.

Published by
A.A. Balkema, P.O. Box 1675, 3000 BR Rotterdam, Netherlands
A.A. Balkema Publishers, Old Post Road, Brookfield, VT 05036, USA

ISBN 90 5410 099 0

Printed in the Netherlands

Environmental Management, Geo-Water & Engineering Aspects, Chowdhury & Sivakumar (eds)
© 1993 Balkema, Rotterdam. ISBN 90 5410 099 0

Table of contents

Waste management

Hydrology and water resources

Soil erosion and sediment transport

Uncertainties, risks and decisions

Natural hazards and environmental impacts

Geotechnology and the environment

Environmental Management, Geo-Water & Engineering Aspects, Chowdhury & Sivakumar (eds)
© 1993 Balkema, Rotterdam. ISBN 90 5410 099 0

Preface

The International Conference on Environmental Management, Geo-Water and Engineering Aspects to be held in Wollongong, Australia from February 8-11, 1993 is a unique, interdisciplinary conference and the first of its kind to be held anywhere. Environmental management has become one of the key priorities in many countries especially during the last two decades and it will continue to represent a dominant influence on planning and development of our physical environment in order to preserve and enhance the quality of life. Scientists, engineers and other professionals must continue to focus on important technical matters along with a keen awareness of wider issues outside the boundaries of any individual discipline. Most conferences encourage contributions which can be, and often are, narrowly specialised. Consequently, there are few opportunities for researchers and professionals from diverse backgrounds and disciplines to share their knowledge and different perspectives and to engage in a meaningful dialogue on a regular and organised basis. In all humility, we have now taken a significant step by providing such an opportunity and we hope that this will encourage a new trend and a new direction for the future.

Any international conference involves a lot of planning and effort but a new type of conference represents a major undertaking and indeed a significant challenge. The organisation of such an international conference was, therefore, not an initiative that we took lightly. Indeed we awaited the response to our call for papers with a mixture of uncertainty and anticipation. In the event we were gratified by an excellent response, having received 230 abstracts from 30 countries. Over one hundred and twenty papers, including twelve keynote papers, have finally been accepted for oral presentation and for inclusion in this proceedings volume. Keynote papers have been grouped together in this volume for the convenience of the reader. Other technical papers have been grouped under nine sub-themes as presented in the contents list. In addition to these papers, 10 papers have been accepted for poster presentation.

The accepted papers represent a wide range of topics and many papers contain significant new material which should be critically examined by interested researchers and professionals worldwide. The opportunity for discussion at this conference will just be a starting point in this regard. The authors of keynote papers have made special efforts to prepare comprehensive contributions on a wide range of environmental management topics. We hope that these papers will encourage further research, development and application of important concepts.

The accepted papers have been edited on the basis of independent reviews. Considerable effort was required to carry out this task within a very limited time. Comprehensive and thorough editing could not be achieved nor was it considered feasible or desirable to achieve a uniform style of linguistic or technical expression. Authors of individual papers bear the sole responsibility for the contents and for following the recommended format of typsetting and presentation. Any errors and omissions are also the responsibility of the authors. The organising committee or the editors do not and can not accept any responsibility for the technical contents or for the claims made by the authors of papers. Our effort as editors has been to deal with significant mistakes in layout and to rectify obvious errors, typographical or otherwise. Occasionally our editing attempted to improve the technical presentation or the syntax and the general flow of expression of a manuscript.

Val McMillan competently carried out her task as the main editorial assistant and was supported by our research students; Xu Dawei, Muralidharan Chinnaiyan and Mahdi Habibi. We wish to thank them for their dedicated work. For general secretarial support we wish to acknowledge Val McMillan as well as Judith Thompson and Pam Burnham. We also thank all the reviewers and members of the organising committee. We appreciate the cooperation of all the authors for providing manuscripts for this conference.

Assistance provided by the Publisher is gratefully acknowledged and extra copies of this proceedings volume can be obtained directly from: A.A.Balkema, P.O. Box 1675, NL 3000 BR Rotterdam, Netherlands.

The organisation of this conference was, to a significant extent, a logical consequence of the formation of the Water Engineering and Geomechanics Research Program within the Department of Civil and Mining Engineering at the University of Wollongong. This was one of the Research Programs formed after the annoucement of the Research Management Strategy by the University in 1988. The Research Program has made considerable progress in achieving its initial aims. The support of the University of Wollongong for academic and research activities is acknowledged. We also appreciate the encouragement and help provided by members of the International Advisory Committee of this conference.

Robin Chowdhury and (Siva) M.Sivakumar
Editors
Wollongong, Australia
October 20, 1992

Keynote papers

Environmental Management, Geo-Water & Engineering Aspects, Chowdhury & Sivakumar (eds)
© 1993 Balkema, Rotterdam. ISBN 90 5410 099 0

Domestic and trade waste disposal in England and Wales

P.B.Attewell
University of Durham, UK

ABSTRACT: Management of waste disposal in England and Wales is now quite rigidly controlled by the Environmental Protection Act of 1990 and associated legislation. The paper briefly notes some of the main points in the legislative framework, indicates the investigation requirements for disposal sites and then discusses some elements of site design.

1. INTRODUCTION

Controls on the disposal of waste in England and Wales are now vested substantially in the Environmental Protection Act of 1990. Under Section 75 Part II of the Act 'waste' is defined as including '(a) any substance which constitutes a scrap material or an effluent or other unwanted surplus substance arising from the application of any process; and (b) any substance or article which requires to be disposed of as being broken, worn out, contaminated or otherwise spoiled; but does not include a substance which is an explosive within the meaning of the Explosives Act 1875'. 'Controlled waste' is defined as 'household, industrial and commercial waste or any such waste'. 'Special waste', under Section 62 of the Act, is 'controlled waste of any kind' that 'is or may be so dangerous or difficult to treat, keep or dispose of that special provision is required for dealing with it'. A decision as to whether a waste is 'special' or not is made by the Secretary of State for the Environment. It may be of a kind that the Secretary of State considers involves or may involve such risk of damage to persons or animals or vegetation that it should be 'disposed of only by him ...' This may be read as being related to a concentration of contaminants which satisfy certain criteria of flammability, carcinogenicity, corrosivity, or toxicity. 'Difficult' could be a liquid or slaughterhouse waste, or wastes difficult to break down, such as, for example, rubber tyres. 'Toxic waste' would be covered by the dangerous waste provision in the Environmental Protection Act, the Dangerous Substances (Notification and Marking of Sites) Regulations 1990 being also relevant.

Although this paper is concerned only with controlled waste, as defined by EPA 1990, it is noted that the Control of Pollution (Special Waste) Regulations 1980, recently revised, define 'hazardous wastes' and meet the obligations of European Community Directive 78/319/EEC. Hitherto, controls on hazardous waste have been insufficiently rigid. The EC seeks to broaden the definition of hazardous waste and improve the management and control of such wastes so that much of the ambiguity and uncertainty is removed from the issue. EC technical directives are to define the types of waste that can be disposed to landfill or incineration, and also the technical standards for landfilling. There will be increasing pressures for pre-treatment prior to landfilling, the percentage of hazardous waste disposed to landfill (currently the least expensive option) will fall, disposal sites will be more strictly controlled (with more severe environmental standards being imposed), and an increasing number of substances will come to be regarded as hazardous, so increasing both volume and cost. The total European market for hazardous waste treatment is estimated to be almost $2.5 billion, rising to $7.6 billion by the end of the century. Special waste may be currently classified by the use of test methods published in a revised edition of Her Majesty's Inspectorate of Pollution Waste Management Paper 23 (HMSO, London).

The European Community has now accepted the concept of co-disposal of controlled and special waste, the requirement in co-disposal being to combine certain difficult wastes with household and similar wastes so that processes operating with a landfill will produce an environmentally benign deposit.

2. LOCAL MANAGEMENT OF WASTE DISPOSAL

Under Section 64 Part II of the Environmental Protection Act 1990, waste regulation authorities in Britain need to maintain registers of licences and applications for licences to deposit 'controlled waste' and 'special waste'. There is provision in the registers for excluding information affecting national security (Section 24 of the Act) and 'confidential information' (Section 22 of the Act). Section 140 Part VIII of EPA concerning the importation, exportation, use and supply of hazardous substances replaces and expands Section 100 of the earlier Control of Pollution Act 1974. Section 142 Part VIII of the Act empowers the Secretary of State to obtain information about potentially hazardous substances. He can impose requirements on manufacturers, importers and suppliers, and extract information from them on such substances. He can prescribe the confidentiality or otherwise of information on the substances and make any public authority responsible for enforcement of the Regulations.

Waste regulation authorities are normally the County Councils in England (not Wales), except in some cases where the District council or a specific waste regulation authority is designated in the legislation. Each waste disposal authority needs to prepare a waste disposal plan for its area. Disposal sites need to be agreed and authorised. Greater powers are given to enforcement bodies, such as Her Majesty's Inspectorate of Pollution (HMIP), the National Rivers Authority (NRA) and the Health and Safety Executive (HSE), and more attention is given both to the various stages in the waste disposal operation and to the care of old disposal sites. A major feature of EPA 1990 is that local authorities in England and Wales will no longer be able to act as both operators and regulators of waste disposal.

The designated waste regulation authorities must exercise direct control over the issuing of waste management licences and operational supervisions. A valid licence is needed for the disposal of waste on land and the use of associated plant. With the exception of special wastes or where a deposit is liable to give rise to an environmental factor there are some exceptions to this licensing requirement as specified in the Collection and Disposal of Waste Regulations 1988.

There are two types of waste management licence: that related to activity on land ('site licences') and that related to treatment or disposal in mobile plant ('mobile plant licences'). A waste management licence authorises the treatment, keeping or disposing of any specified description of controlled waste in or on specified land, or the treatment or disposal of any specified description of controlled waste by means of mobile plant (EPA Section 35(1)). The licence is granted on such terms and subject to such conditions as the authority thinks appropriate (EPA Section 35(3) and continues in force until revoked or surrendered (EPA Section 35(11). A licence cannot be issued until any necessary planning permission is obtained or an established use certificate is in force. The proposal must be referred to the NRA and the HSE. A waste regulation authority may not reject an application unless either they are not satisfied that the applicant is a 'fit and proper person' or that rejection is necessary for the prevention of harm to the environment, harm to human health, or serious detriment to the amenities of the locality (EPA Section 36(3)). The concept of 'fit and proper' is a new one and embodies considerations of technical competence, absence of relevant convictions and financial standing (EPA Section 74).

If planning permission under the Town and County Planning Act 1990 and the Town and Country Planning (Scotland) Act 1972 is required for the use to be made of the land under a site licence then the licence would not be issued unless permission under one or other of those Acts has been granted, or an established use certificate is in force. Planning permission is not generally required for uses of land commenced before 1 July 1948 or for those for which permission has been granted by a special or general development order. Alternative consultative procedures exist for Crown land or land in Crown occupation. Reference may be made to DoE Circular 1/85 'The Use of Conditions in Planning Permission'.

The use of land and the necessary plant thereon may need to be considered separately. Planning permission may not be needed for the former, but may for the latter, and an application could fail on those grounds.

Section 34 of EPA 1990 contains a new duty of care condition in respect of the keeping, control and transfer of waste, the latter involving a scheme for registration of carriers of controlled waste under the Control of Pollution (Amendment) Act 1989 which operates in conjunction with EPA 1990. The Government's Pollution Paper No. 24 (1986) para. 5 states that 'this duty of care should find its main expression in ensuring the integrity of the waste stream on its journey from producer to disposer and beyond this to its safe containment or disposal over time'.

This duty encompasses all those responsible for the generation (excluding householders' domestic waste) or importation of controlled waste and those who have control of such waste at whatever point in the chain from generation to treatment and final disposal. The standard is based on the concept of reasonableness in the light of the measures applicable to the relevant person in his or her particular capacity. 'Reasonable' steps are needed to prevent any other person contravening the law in respect of unauthorised deposit, treatment or disposal of waste, to prevent escape of the waste, to ensure that the waste is transferred only to an authorised person, and to ensure that a sufficient written description of the waste accompanies it on its transfer. Breach of duty of care is an offence punishable by a fine ot £2000 in the magistrates' court and an unlimited fine on conviction on indictment (Section 34 of EPA). Unauthorised or harmful depositing, treatment or disposal of waste is liable on summary conviction to imprisonment for a term not exceeding six months or a fine not exceeding £20 000 or both, and on conviction on indictment to imprisonment for a term not exceeding two years or a fine or both. In the case of special waste, a five year period of imprisonment replaces the two year period (Section 33 of EPA 1990).

It is noted that the European Commission has proposed conditions of strict civil liability for damage caused by waste. Adoption of these would create a parallel liability on a waste producer or controller with a no-fault civil liability in contrast to a criminal liability based on the test of reasonableness. The draft EC Directive (COM(89)282 Final, Sept. 15 (1989) proposes a system of joint and several liability on producers whose wastes are mixed or disposed of together, and cause damage. On the other hand, under Section 34 of EPA liability would arise whether or not damage resulted from the breach. Under the Directive, either personal injury, property damage, or injury to the environment would be a pre-requisite of liability.

In respect of closed landfill sites, every waste authority has a duty under Section 61 of EPA to inspect its area from `time to time', except where a current site licence applies. (County Councils have no formal powers to ensure satisfactory aftercare of leachate and gas on pre-licensed waste disposal sites, and on licensed sites monitoring and control ceases when the licence has expired.) The requirement is to ensure that the site does not cause pollution to the environment or harm human health as a result of the presence of noxious gases or noxious liquids caused by deposits of controlled waste. To satisfy this requirement the authority may enter and inspect any land on which controlled waste has been deposited at any time under a waste management licence or a disposal licence, on land where the authority has reason to believe that controlled waste has been deposited on the land at any time, and land in which there are, or the authority has reason to believe there may be, concentrations or accumulations of noxious gases or noxious liquids. If a problem is discovered, the authority then has a duty to carry out any works or other measures as appear to be reasonable to avoid pollution or harm, and is entitled to recover all or part of the cost from the current owner of the land, unless there was a licence which has been surrendered. The authority is required to take into account in recovering costs any hardship caused by such recovery to the owner of the land. This provision for recovering costs ceases when the authority has accepted the surrender of a licence (Certificate of Completion) under Section 39 of EPA, and so acceptance will be made only rarely and will depend on whether the condition of the land could cause pollution to the environment or harm to human health. The National Rivers

Authority must be consulted before surrender of a licence is accepted. There is a right of appeal procedure against refusal to accept surrender (Section 43(1)(f) or deemed refusal if the authority does not make a decision within 3 months of application to surrender (Section 39(10)).

Local authorities may approve initiatives to develop closed, safe sites, but with the proviso that the developer carries out appropriate protection measures.

3. COSTS OF WASTE DISPOSAL

On the basis of 1990 figures, it costs between about £4 and £20 per tonne to dump rubbish in a UK tip, but it has been estimated that the cost will rise by up to four-times by the year 1995. Another view is that the average cost throughout Britain will probably still be only £20 per tonne by the end of the century, implying that it will still at that time be cheaper to use landfill than to re-cycle suitable waste.

There is an interesting financial implication in the acquisition of private landfill sites. It was decided by the Court of Appeal, upholding Justice Harman's decision in the High Court in favour of the Revenue in the case of Rolfe v Wimpey Waste Management Ltd, that expenditure incurred in acquiring landfill sites (freehold and leasehold) plus the cost of obtaining planning permission and the expense of preparing sites for waste disposal are capital expenditures, not revenue, and are not therefore deductible in computing profits for corporation tax purposes. This decision in the Wimpey case means that a waste management operative cannot treat the costs of air space (landfill site; abandoned quarry) and preparation as a current asset and amortise the expenditure by reference to the utilisation of air space.

It would seem from this case that British taxation law could be amended to provide (a) relief in respect of expenditure on the acquisition of air space on a basis similar to that which exists in relation to acquisition of mineral rights where tax reliefs are given in the form of capital allowances on related capital expenditure, and, (b) relief by way of deduction from profits (or by way of capital allowances, as appropriate) for all expenditure incurred on the preparation, operation, restoration and after-care of landfill sites.

4. LANDFILL GAS

Gas is a form of pollution and claims on the grounds of pollution from gas tend to be largely settled out of court because there seems to be a reluctance to clarify an area where English law remains largely untried.

Under Article 18 of the 1988 Town and Country Planning General Development Order local planning authorities are required to consult waste disposal authorities about applications for planning permission and proposals for building development within 250 metres of waste disposal (landfill) sites in use or which have been used for that purpose within the preceding 30 years As noted in Waste Management Paper 27 (Her Majesty's Inspectorate of Pollution, 1989) responsibility for monitoring, satisfactory control of landfill gas, and general maintenance of environmental protection measures at any site working under a waste management licence rests with the landfill site operator until, as noted above, a certificate of completion is issued by the waste regulation authority (WRA). The WRA then assumes responsibility for the site, and so must be satisfied that it is fully stabilised before issue of the completion certificate. Decades may elapse between completion of filling and issue of a certificate of completion.

It is sometimes necessary to resolve the origin of any gas that is detected. This often reduces to determining whether the gas is from a natural source or has resulted from decomposing organic matter.

The primary gases produced from a landfill site are methane (CH_4 typically about 55% but up to 70%) and carbon dioxide (CO_2 typically up to about 30% but could reach about 40%), although gases such as hydrogen, hydrogen sulphide and other gaseous hydrocarbons such as ethane (C_2H_6) will be present (typically about 2% in total). CH_4 from natural gas can be oxidised by aerobic bacteria in soil to produce CO_2. Similarly ethane and longer chain hydrocarbons are the most easily digested by bacteria, and so any lack of ethane is no absolute guarantee that the methane did not derive from natural gas. If CH_4 and C_2H_6 are present in the ratio of about 20:1, it is fairly certain that natural gas is present. But at higher ratios it is possible that it is still the source of the methane, or it could derive from a mixture of both (landfill gas and natural gas present together). There are also methods involving isotopic content which can be used to help resolve the genesis of a gas when the simpler methods prove to be inconclusive. Hydrogen presence is also a good indicator of potential future methane production. Methane can explode when its concentration with air is between 5% and 15% but the presence of carbon dioxide will modify these explosive limits for methane. CO_2 is an asphyxiant in concentrations greater than 0.5%.

Steady state methane yields tend to be achieved after a fill placement period of one to two years. Typical yields are of the order of 30-300m^3 of methane per tonne of dry refuse compared with a theoretical figure of 470m^3. This 50% or so reduction arises because some of the organic carbon does not contribute to the gases but is removed in the leachate. Further, the actual nature of the fill is unlikely to match the theoretical model. For example, if the refuse is pulverised, then because of the greater area of solid exposed to chemical reaction the initial rate of gas production may be higher than if the refuse is dumped in bales. Gas yield is typically higher in the first few years when the biological activity is at its highest. Expected gas yields are between about 1.5 and 6.0m^3 per tonne dry refuse per annum. Yield is usually low after about 20 years, but small quantities may continue to flow even after 50 years. Of course different refuse material generates different quantities of gas, and under layered placement conditions different layers in the fill will tend to react in different ways at different times.

Methane causes plant death by starving the roots of oxygen. Starvation is by simple dilution of oxygen by methane and also by an oxidising reaction between methane and oxygen under bacterial action.

Grants are available in England and Wales for promoting electrical power generation from local sources of landfill gas. Most new domestic and trade waste sites have active gas extraction systems from their inception.

The potential for gas migration from a landfill site depends to a great extent on the nature of the strata surrounding the fill, together with the fill itself, the style of its containment, and the method of site operation. It is, of course, greater in old sites than in new ones where full containment measures have been implemented. Site development in relatively impermeable clay soil without silty sandy lenses should inhibit lateral migration. If not prevented, substantial migration can occur through sands and gravels and through fissured (discontinuous/jointed) rock strata, as in sandstones and limestones. Many disposal sites are in old quarries, with the fill backing up to the abandoned quarry face. It is then necessary to perform a careful assessment of the discontinuity patterns in the rock to determine the potential for gas migration over substantial distances.

Monitoring the presence of gas takes the form of shallow spike surveys, borehole and gas piezometer monitoring, thermography, false colour infra-red photography, and vegetation studies. It would be industrial waste management general practice to monitor at least four times over a two year period, but in the case of isolated sites with biodegradable waste and sites having high gas yields within 250m of houses the monitoring frequency would be monthly and weekly, respectively. Portable gas monitoring instruments comprise one or more of the following: catalytic oxidation detectors (measuring the concentrations of inflammable gas but inaccurate if the O_2 levels are less than 12%), thermal conductivity meters, flame ionisation detectors (but which can ignite explosive gas mixtures), infra-red analysers and photoacoustic infrared spectroscopic analyses. CO_2 and O_2 levels would also be monitored. In the laboratory, gas samples would be analysed by means of gas chromatography.

It is good practice for continuous testing for methane to be carried out from initial placement until 15 years after the life of the site.

5. ENGINEERING AND GEOLOGICAL INPUT TO A STATEMENT SUPPORTING A PLANNING APPLICATION, ENVIRONMENTAL ASSESSMENT AND OPERATIONAL WORKING PLAN

5.1 *Planning applications and environmental assessments*

- Details of the site investigation.
  The most important factors are the:
  geological regime;
  hydrological regime;
  hydrogeological regime;
  mining history (if any).

  These factors provide the potential for ground instability, water ingress to the waste, and leachate migration. There must be a brief assessment of the potential impact of these features in 'sensitive areas'.

- Details and results of:
  water level and quality from 12 months of monitoring;
  naturally-occurring gas concentrations and the monitoring installations;
  hydrological monitoring of flows, levels and concentrations.

- Outline of engineering works, including:
  type of containment;
  seal construction, including slope stability modelling analyses;
  leachate collection wells;
  landfill gas venting and/or extraction wells;
  capping and after-care.

5.2 *Operational working plan*

This is the final document submitted in support of an application for a waste disposal licence. It must contain a detailed report of the proposed operations. All the above information must be appended and all the detailed designs must be included.

- Slope stability modelling, leading to slope design
- Site investigation data, leading to
  engineering materials;
  engineering details (eg. all seals/drains);
  engineering specification.
- Engineering details and engineering specification, leading to
  quality assurance details (eg. testing of liners, certification)
- All collated SI data, leading to
  cell and operational design details.

6. SITE INVESTIGATION

The fundamental problem in respect of waste disposal sites is water: precipitation, groundwater and leachates. If all three can be controlled, then the problems of waste disposal sites are more or less solved. A site investigation must be designed with a view to identifying the sources of this problem and promoting its solution.

There is an urgent need for site investigation methods used for both contaminated land and waste disposal sites to be standardised. Such standardisation should clearly identify those factors that are essential for overcoming the problems of contamination and pollution, and for ensuring cost effective design of waste disposal facilities. Tendencies to over-investigate must be resisted, but those elements of investigation that are deemed to be essential must be implemented with quality assurance and with full knowledge of British Standard 5750 and ISO 9000.

6.1 *Site assessment*

Sites for waste disposal must be designed and managed in such a manner as to prevent harmful substances penetrating the surrounding ground and groundwater. Reference on this subject may be made to the Technical Guidance document "Geotechnics of Landfills and Contaminated Land" prepared by the European Technical Committee 8 (ETC 8, 1991). There must clearly be an appreciation of those factors that could affect the suitability and safety of the proposed disposal site. The main factors appear to be:

- topography and structure of the area
- type and behaviour of the waste
- geological/hydrogeological setting

The design of the site will take account of these factors and there must be an overall safety plan instituted at the outset. The following features require investigation, the main ones having an asterisk* after them.

General

- morphology
- structure, extent and geological age of the outcropping strata
- tectonic structures*
- deeper sub-soil if it comprises (contains) cavities or soluble rocks
- risk of earthquakes and other natural hazards

Composition and distribution of superficial deposits

- composition, physical and chemical properties and sequence of strata*
- lateral and vertical continuity and distribution of the strata (facies changes)*
- porosity
- permeability (to water and leachate, and to landfill gas)*
- resistance to erosion and washing away of fine particles
- stress deformation behaviour*

Structure and sequence of solid strata
(Underlying bedrock may need to be investigated if the overlying soil cover is thin.)

- type of rock, mineralogical composition and stratigraphy
- state of weathering and weathering resistance*
- solubility in water and leachate or other aggressive solutions
- type and position of geological boundaries*
- extent, degree of separation and widths of individual joints*
- tectonic and petrographical anisotropies in the rock mass
- karstification and risk of subsidence
- deformation behaviour of the rock mass
- permeability to water, leachate, gases and other aggressive solutions (hydrocarbons etc)*

Determination of hydrogeological data

- groundwater regime, direction of flow gradient and rate of flow, including long-term and seasonal fluctuations*
- permeability (horizontal and vertical) or transmissivity of the outcropping strata, with maximum and minimum values*
- distribution, thickness and depth of aquifers, aquicludes and aquitards, including the locations of any springs
- groundwater levels, indicating hydraulic gradients and effective flow velocity in the individual strata components if appropriate, assessed in the context of strata dips*
- groundwater chemistry, including determination of naturally-occurring aggressive substances and groundwater quality
- groundwater protection zones
- groundwater abstraction and its effects*
- groundwater abstraction rights*
- influence of short-term and long-term lowering of the water table, restoration and extraction or augmentation of groundwater in the future
- influence of nearby open waters and their relationship with the groundwater system
- situation in respect of receiving streams, influence of flooding and tides, if appropriate
- effective rainfall, surface runoff, percolation rate, evaporation and groundwater recharge*

Considerations of special factors

- stability of existing slopes if trenches are used
- possibility of sealing off any old tunnels or adits
- potential for subsidence caused by abandoned or existing mine workings (underground or surface workings)
- presence of workable natural materials in the sub-soil*
- presence of geological features or archaeological monuments worthy of protection
- where spoil heaps or landfill are used, survey of the underlying exposed natural sub-soil
- bearing capacity of the soil and/or groundwater*

6.2 *Ground investigation*

Boreholes and trial pits, even shafts and exploratory tunnels in special cases, should be used. At least one borehole per hectare is required for planning purposes. Increasingly the aim should be towards `environmental drilling' whereby drilling equipment and liners are thoroughly (steam) cleaned after each operation (hole) in order to avoid introducing contamination from an earlier hole into a subsequent hole and great care is taken to avoid cross-strata and inter-water-table contamination within boreholes. Conduct of the investigation should follow national recommendations, in the case of the UK, BS 5930: 1981. Colour photographic records should be made of all borehole samples and trial pits. Special emphasis should be placed on the evaluation of groundwater - monitoring of levels and an assessment of the effect of groundwater table fluctuations upon landfill when placed, the nature of any perched water tables, and groundwater chemistry, particularly sulphates and pH. Piezometers would need to installed within the fill area and outside, and should be carefully monitored within the fill area over a period of time before fill placement and outside the fill area both before fill placement and after. (Water from flushing the piezometers should be carefully disposed of according to anti-pollution regulations operative at the time.)

The following would seem to be minimum investigation requirements:

• Soil: Engage in a ground investigation drilling programme, both on the site before any dumping takes place and well beyond the boundaries of the site, the latter for background readings, in order to determine the structure of any permeable lenses in the soil underlying the site and their extent outside the site area. It is unlikely that *in situ* gas permeability tests can be conducted, but it is possible to perform standard packer permeability tests in the boreholes and both water and gas permeability tests on carefully taken borehole cores. Such tests might permit lower bound *in situ* gas permeability values to be estimated by extrapolation. If waste is already dumped on the site, investigation holes may not be put down in the site area, but, if they are, the sampling and testing will be restricted to the soil beneath the tip deposits.

Decisions may then be made as to the permeability quality of the underlying soil. If this is proved to be unsatisfactory as a base material, then attention should be directed to assessing the availability in the proximity of sufficient and suitable clay material for rolling in as a liner/cut-off or as part of a composite clay/geomembrane liner.

• Rock: Use a drilling and coring programme to resolve the structure of the site foundation and adjacent areas. Fracture information will need to be correlated between rock cores, and packer permeability tests must be conducted in the boreholes. Permeability values so derived will relate to the discontinuity structure, and not to the intrinsic pore structure of the rock, so there will be little or no merit in conducting any laboratory gas or water permeability tests. Gas permeabilities will need to be determined from *in situ* water permeability information. If an exposed rock face is available when disposal is in an abandoned quarry, measurements of discontinuity orientation and spatial density distributions will be taken in order to resolve patterns of persistent joints along which gas products from the tip could travel.

Care must be taken to ensure that the equipment used for sampling has no effect on the soil and water samples. Brass, chrome-plated, nickel-plated or galvanised material should not be used if water is being tested for heavy metals. Iron or steel alloys are acceptable when testing for iron and manganese. Samples from dye tests must be kept cool in dark bottles before analysis. Special equipment and care in procedures is needed when sampling volatile organic substances and highly toxic constituents in very low concentrations. Glass containers should generally be used when major ion analyses are proposed. In the case of sampling from several aquifers a sampling borehole should ideally be used for each aquifer to avoid any cross contamination from vertical leakage. If this is not possible then a multi-packer system can be used with control on the pumping rate in particular packer zones. Water samples can be obtained discretely by baling from a borehole or from continuous pumping. Single baling tubes permit rapid sampling at a relatively low degree of accuracy for relatively high ionic concentrations. In the case of pump sampling the recommendation is that pumping should be maintained until the pH, the electrical conductivity and/or temperature have attained constant values. Suitable equipment comprises suction pumps, plunger pumps, submersible electrical pumps and gas-driven diaphragm pumps. Suction pumps are not suitable for determining dissolved gas concentration because of the high degree of induced aeration. Petrol-driven pumps should not be used for sampling low concentrations of hydrocarbons. A sample volume of at least one litre would normally be required.

Samples of soil, rock and water should be tightly sealed immediately they are taken or should be placed in containers which can be sealed in such a way as to minimise the volume of air remaining in the container.

There is a place for geophysical techniques in locating and defining old landfill sites. Electrical resistivity can be used to determine the depth and volume of the landfill and gravity methods can provide an estimate of the density or compaction of the fill. The spontaneous potential method can be used relatively cheaply for delineating the boundaries of an old landfill site and portable magnetometers can prove to be of great assistance in locating buried drums of hazardous waste. New landfill sites near to the coast or to inland waterways pose particular design problems related to liner competence in the presence of possible uplift pressures and to containment of gases and leachates. Potential hydrogeological problems can be identified by the use of geophysics.

It is especially important to check if containment sites have been in the past, or will be in the future, subjected to settlements resulting from shallow mine workings since movements could damage the seal and lead to transmission of gas to the surrounding ground and loss of leachate from the fill.

6.3 *Overall assessment*

The results of the site investigation must be analysed in the context of an overall waste disposal plan, taking account of all safety and environmental impact requirements. This analysis should form part of the geotechnical report. The report should include the following matters as a minimum requirement.

- description in words, and in diagram form if it will increase the clarity, of the geological structure
- presence and suitability (for barrier purposes) of natural low permeability (to both water and gas) strata (thickness, depth, horizontal continuity, permeability, adsorption capacity)
- groundwater regime and permeabilities within the area to be landfilled and its surroundings
- stability of natural and man-made slopes
- bearing capacity and deformation behaviour of the subsoil
- faults, possible subsidence, risk of collapse, earthquake risk and other hazard situations
- comments on any geotechnical measures needed to improve the properties of the sub-soil as a natural barrier to gas and water out-flow

7. SITE TREATMENT

New landfill sites require a cut-off at the base and sides that is virtually impermeable both to leachates which would otherwise pass into the groundwater and to gas. Such a barrier, provided that it remains intact, also prevents any rising groundwater from increasing internal pressures within the waste and thereby increasing the tendency for lateral and upwards de-gassing. Old quarries usually form ideal landfill sites, and in about 50% of cases there is sufficient clay in the quarry area to use for sealing purposes. The 6th Draft of the EC Landfill Directive stipulates liners comprising 3m thickness of clay, having a permeability of 10^{-9}m/s or less, or its equivalent. This naturally-occurring material would be laid in layers, each layer being compacted to specification. In the UK this specification would usefully follow the Department of Transport `Specification for Highway Works', Parts 1-7 (DTp, 1986a; *see also* DTp, 1986b). Compaction by six passes of a vibratory sheepsfoot roller would be in material layers each not exceeding 300mm in thickness.

On-site trials would normally be conducted before beginning the site compaction programme. These provide an opportunity for checks on *in situ* permeability (using rings sealed into the clay with bentonite), *in situ* densities using, for example, a nuclear densitometer (but of no use in clay containing significant quantities of carbonaceous material) or a core-cutter method of sand replacement. There would also be checks on the moisture content (by speedy moisture content equipment), the layer thickness, and also, perhaps, the particle size distribution. The trials provide assessments for the purposes of quality assurance and for confirming that the design values as determined from laboratory test programmes are being achieved.

In addition to its resistance to water penetration the clay liner must possess suitable erosion resistance, a heavy metal absorption capacity related to its clay mineral or organic matter content, and not be unduly responsive to potential swelling and shrinkage conditions. Through its clay content and particle size distribution it must also possess adequate plasticity to be able to accommodate differential settlements caused by differential loadings and by changes in the subsoil properties, and suitable self-healing capabilities. A liquid limit (LL) not greater than 90% and a plasticity index (PI) not greater than 65% would normally be specified for (ideally) an inorganic clay soil of high plasticity.

7.1 *Properties of clay*

For the purposes of design, mechanical tests are required on samples of the clay soil derived from the ground investigation. The most important of these tests are tabulated below.

Tests to be specified on the clay material in order to assess its suitability for use as an engineered low permeability cut-off.

TEST	PARAMETERS	PROPERTY	USE IN DESIGN
Atterberg limits	PL, LL, PI	Physical behaviour	Material suitability
Particle size distribution (wet sieve, hydrometer, pipette)	Percentage clay, silt, sand, gravel	Material composition and variability	Material suitability
Specific gravity	G_S	-	Calculation of percentage air voids
4.5kg compaction	Compaction curve: optimum moisture content and maximum dry density	Remoulded behaviour	Material suitability Engineering specification Quality assurance
Unconsolidated undrained triaxial tests	c and ϕ	Short-term shear strength behaviour	Material suitability Short-term slope design and stability modelling
Consolidated undrained triaxial tests with porewater pressure measurement	c' and ϕ'	Long-term shear strength behaviour	Material suitability Long-term slope design and stability modelling
Permeability (constant head in triaxial cell will often be written into the guidelines)	k coefficient	Remoulded permeability behaviour	Material suitability Engineering specification Quality assurance Rate of leachate migration
Moisture content	Natural moisture content %	Natural moisture content %	Material preparation Material suitability Engineering specification
Slake durability	-	Potential weathering	Material suitability Material preparation

The mineralogical and chemical properties of clay need to be determined in order to assess the effects of leachates upon its permeability and therefore its effectiveness as a seal. Solutions of hydrocarbons may act to increase the clay permeability, the mechanism being a dissolution of the soil binding agents on the clay particles followed by movement of the clay particles out of the barrier. It should also be recognised, and allowance made for the fact, that the field (*in situ*) permeability will usually be greater than the laboratory-determined permeability due to the presence of cracks (enhanced by desiccation) and inter-clod voids. Clay permeability is also very sensitive to placement water content and the degree of compaction actually achieved, and it is also related to differential settlement effects which promote cracking more easily if the clay departs from a plastic state.

In order to specify the grain size of the material the finest fraction (<0.0022mm) must be defined by first dissolving it in, for example, a sodium chloride solution. By treating the samples several times with 0.1 molar solutions of ethylene tetra-acetic acid having a pH value between 4.5 and 8, the Ca and Ca/Mg carbonates occurring as binding agents are gently dissolved and the Ca and Mg ions adsorbed on the clay mineral surfaces are exchanged for Na ions. A sedimentation test can then be conducted on this fraction.

The presence and proportion of swelling clay minerals can be determined by the process of cation exchange. Monovalent or bivalent metal cations or NH_4^+ ions are applied to the clay mineral surfaces of a clay soil and then the fixed cations are quantitatively re-exchanged. The test sample need weigh only a few grammes and is treated several times with an exchange solution such as 3 normal ammonium acetate, with any excess ammonium acetate being removed before analysis by washing several times. Clay minerals, and particularly swelling clay minerals, can be identified and quantified by means of X-ray diffraction analysis. Reference may be made to Klug and Alexander (1954) for full details of the methods that can be used.

It is also necessary to distinguish between the calcitic and dolomitic cementing agents in a clay soil because of the varying chemical stability of the two carbonates. The calcitic and dolomitic content can most simply be determined by titrimetry using a Ca titration and a Ca-Mg cumulative titration (ETA8). It is also possible to detect iron, which, as noted above, may also form a carbonate cementing agent, via the hydrochloric acid solution used for these titration tests. Atomic absorption may be used as an alternative to titration. The calcite/dolomite ratio may also be determined reasonably accurately by the X-ray diffraction method.

7.2 *Design of landfill cells*

In order to minimize the extent of leachate production there is a need to restrict the amount of precipitated water entering the waste. This may be done by segmenting the fill area into `cells' bounded by impermeable bunds. The cell size in terms of its plan area is determined by the water absorptive capacity of the particular type of fill, as derived from tests or by experience. A rather simple waste disposal model is considered in thc Tablc bclow.

Thc figurc of 11 500 tonnes/annum is equivalent to 11 500m^3/annum. Assuming that the density of waste is very approximately equivalent to that of water, and that the average annual rainfall is 1000mm, then

Approximate maximum area of cell

$$= \frac{\text{Estimated annual total absorptive capacity}}{\text{Average annual rainfall}}$$

$$= \frac{11500\text{m}^3}{1000\text{mm}} \times 1000 \quad = \quad 11\,500\text{m}^2.$$

Suppose that the area of the disposal site is 100 000m^2. The operational number of cells will then be

$$\frac{100\,000}{11\,005} \quad = \quad 8.7 \; = \; 9 \text{ in practice if leachate}$$

is to be minimized.

Each cell would be sealed at the top when full of waste in order to inhibit further infiltration.

This system of disposal is satisfactory for flat, low-lying disposal sites. For deep disposal sites in, for example, old quarries, it may be necessary to increase the number of cells as the infilling proceeds upwards, with cells being formed on compacted waste within earlier cells. In this system, if the seal at the top of each cell remains after filling, leachate from a superadjacent cell will be retained. Provision for quite onerous leachate removal operations from each cell must then be incorporated into the design. Alternatively, cell leachates can be allowed to drain downwards in a controlled manner through inserted pipework to one or more take-off points. A further option is to strip the cap from the top of a cell when waste infilling begins above it.

TYPE OF WASTE	EXPECTED DISPOSAL (tonnes equivalent/ annum)	ABSORPTIVE CAPACITY (%)	TOTAL WATER ABSORPTION (tonnes equivalent/ annum)
Domestic	50 000	15	7500
Industrial/commercial	30 000	10	3000
Inert (stones/boulders/ builders' rubble etc)	20 000	5	1000
-	-	TOTAL	11 500

7.3 *Clay sealing at an old quarry face*

Persistent discontinuities `daylighting' at an old quarry face could form channels for gas migration from waste disposal sites. A seal can be achieved by raising a wall of clay progressively as infilling proceeds, the waste providing lateral support for the low shear strength clay. In practice, a tapered wedge (bund) of clay, two or three metres high, will be first placed and the waste then brought up to the top level of the clay. A further bund of clay, tapering upwards, will then be placed on top of the first bund, but also extending over the waste in order to increase the bearing area. As before, infilling will proceed up to the top level of the bund. This sequence of bund construction and waste infilling will proceed until the quarry crest is reached. From a vertical, or near-vertical, face of rock, a composite seal/waste slope of `Christmas tree' configuration, and about 1:2 maximum gradient, can be achieved, such a slope angle being sustainable in soil mechanics terms, with the analysis discounting any lateral support contribution from the waste.

7.4 *Geomembranes*

An alternative or additional choice of sealant is to use geomembranes, usually 0.5 to 2mm thick (but being available in thicknesses of 0.75mm to 5mm ± 10%) and having long-term chemical resistance, combined with geotextile cushion layers and geomesh drainage layers, compositely on top of a 1 metre or thereabouts thick clay formation, comprising 200mm layers compacted with a vibratory roller, for lining before filling with waste. Typically, a 300mm band of sand would then be laid over the liner. This would serve as a protective layer for the geomembrane. The geomembrane would be delivered to site in the form of rolls, usually of width just over 7m or 10m, and would have a permeability of about 10^{-13}m/s. Such a permeability value is within the US EPA guideline of <1 US gallon/acre/day.

Flexible membrane liners are made from several polymer resins:

- chlorinated polyethylene (CPE)
- ethylene co-polymeride bitumen (ECB)
- polyvinylchloride (PVC)
- low density polyethylene (LDPE)
- high density polyethylene (HDPE)
- partial elastomers such as ethylene-propylene-terpolymer rubber (EPDM), nitrobutadiene rubber (NBR), butyl rubber (IIR)

There are several references in the literature to laboratory tests on liner material. The fundamental requirements of a liner are

- Watertightness.
- Toughness resistance to handling and other potential puncturing and tear damage while being laid and when laid.
- Resistant to high tensile forces in one and two directions after being placed and under working conditions. Its tensile strength-elongation (percentage strain) relations must be known for a range of temperature and different liquid environments. This information would be supplied by the manufacturer.
- Adequately plastic to accommodate differential ground movements without its life-long integrity being impaired. (For HDPE the target elongation limit is about 3% and the elongation limit up to which the life-time serviceability of the liner is not destroyed by chemically superimposed stresses is 5% biaxially. For other plastic materials this limit tends to be much lower.)
- Durable and resistant to heat, ultra violet light and stress corrosion. (Wetting agents, solvents, chemicals must not cause early brittle failure under load.)
- Resistant to penetration by roots and fauna to a high standard.
- Easily weldable to a high standard under site conditions.
- Resistant to permeation by hydrocarbons, especially chlorinated hydrocarbons.

Oxidation of polymer is an acknowledged problem. Temperature increases of 20% can lead to 80% reductions in life expectancy, and so assurances on the thermal oxidation stability of the material are required. On the matter of ultra violet radiation, the addition of up to 2% of fine particles of well-dispersed carbon black to the polymer, giving a density of 0.93 to 0.94g/cm^3, is a standard practice for stabilisation. The membrane should not contain plasticisers or additives which could act as foodstuff to biological organism attack. PVC also suffers as a liner in respect of the plasticisers which provide its low modulus of elasticity since these are prone to leaching out.

7.5 *Composite liners*

With increasingly stringent specifications it seems unlikely that any single material alone will be able to meet all these performance requirements. Of all the available polymeric materials HDPE offers the most advantages, and although it is largely resistant to such searching chemicals as chlorinated hydrocarbons a block on their movement to the groundwater table can be achieved by adopting a composite liner system comprising an impermeable mineral layer underneath the plastic liner. Any permeating solvents are thereby blocked by the mineral particles which concentrate the solvents on their surfaces. The addition of organophile bentonites to the mineral layer can improve the blocking mechanism. A composite basal lining system, from the top down, could be

- waste
- transitional layer (if necessary)
- drainage blanket
- protective layer
- geomembrane
- last lift in each case
- 3rd lift
- 2nd lift
- 1st lift
- mineral sealing layer
- subgrade (in the case of embankment or soil replacement)
- subsoil

A simpler system, from the top down, could be

- waste
- drainage layer
- protective layer (0.3m thick)
- flexible membrane liner (2.5mm, or thereabouts, thick)
- mineral layer (>0.6m thick)
- subsoil

The 300mm depth of cover layer required for geomembrane protection should contain a maximum particle size of 20mm, but should comprise a broad distribution of particle size. A cover of at least 1m will be needed if heavy vehicles are to pass over the liner, and waste compactors should not be used until at least 3m of waste, compacted by a track laying vehicle, is in place.

The protective layer, flexible membrane liner and mineral layer comprise the sealing system.

7.6 *Liner placement*

Placement of a liner, taking the form of a natural clay cut-off laid under controlled compaction conditions, or of a geomembrane as part of a composite liner system, will usually be scrutinised by a quality assurance (QA) engineer (BS: 5750/ISO:9000). In the case of a clay seal tests for *in situ* density have often been specified on a 2000m^3 placement basis, but a more rigorous and realistic frequency of 1 test per 20m^3 or 33m^3 is now being adopted.

Man-made liners are supplied on rolls, each roll having a QA certificate attached. Before being covered on site, the liners must be kept free, as far as is possible, from ultra violet light (which can cause embrittlement of the polymer). As a result also of natural light, sheets are prone to stick together (blocking), causing the laminates to peel off. The rolls must also be retained secure from vandals, and must be transported from the site compound to the point of placement suspended by means of slings from any contact with the transporting plant.

HDPE linings, of 2mm minimum thickness and, as noted above, relatively inert to a range of chemical wastes, are used in Germany and Italy, but might lack the necessary flexibility at a site which is prone to large differential settlements. In the UK, HDPE in typically 35m wide strips tend to be used. Reinforcement may be especially useful when uplift gas pressures need to be resisted. Proper supervision and inspection of installation, particularly of jointing, is crucial to the containment efficiency.

Before lining placement, the sub-grade must be compacted to maximum density (for example, according to ASTM D698). Weak spots must be infilled with properly compacted fill having a maximum particle size of 20mm and any standing water removed. There must be a clearly drawn-up sheet layout plan based on a site plan of not less than 1:500.

Geomembrane liners should not be placed in rain or fog or when temperatures are less than 5°C or greater than 35°C. The seams of the liner sheet would normally be laid parallel to the line of the greatest slope within the containment area, the unrolling taking place down-slope, and there would be a 150mm overlap to the adjacent geomembrane panel. During placement, any crinkles on flexible membrane liners (FML's) will need to be smoothed out (or cut out and welded) before the composite is put on top, any tears will need to be suitable patched, and there will need to be assurances as to the quality of the welds generally. Welding trials would normally be conducted, under QA observation, before any landfill site welding begins.

The four basic methods of welding* - hot air and hot (V-) wedge, extrusion overlap and extrusion fillet, together with the methods of weld testing are well documented in manufacturers' literature.

Reinforcement may be needed for a liner when used on steep side slopes of the containment area, a 1:1.5 slope being about the steepest that can be treated in this way. Because sheet polymer is intrinsically smooth, there will always be a tendency on slopes for the sheet, before it is secured, to creep downhill, but this tendency can be overcome to some extent by increasing the liner friction by molding dimples into the sheet. Liner sheets can be held in position on a slope and against wind force by means of sand bags, but should not be fixed at the sharp crest of a rock slope. There will be an anchor trench at the crest, and it will be necessary for the sheet welds to extend substantially into the trench before it is backfilled. Extending down the slope, a `Christmas tree' configuration of laying and support needed, each layer of sheeting being turned over into its own trench and sealed in with clay packing before a superadjacent overhanging sheet is laid.

* Two national standards for welding - American and German - can be quoted: US EPA/530/SW-89/069 Technical Guidance Document 'The Fabrication of Polyethylene FML Field Seams' and 'Richtlinie DVS 2207/26.1 des Deutschen Verbands für Schweisstechnik' (German Committee for Welding Technology). Within the latter, reference may be made to DIN 16726: (Testing of) Flexible Membrane Liners; DIN 16776: Polyethylene Raw Material; DIN 18195: Structural Sealing; DIN 1910: Seaming of Plastics.

If there is space at the quarry crest, slopes may sometimes be re-graded back to gradients of about 1:2.5, so easing the placement of cut-off material.

7.7 *Bitumen/aggregate liners*

Waste tip liners comprising bitumen/aggregate (about 25mm)/sand/ suitable filler are now being used. They are cheaper than composites but more expensive than single liners, and tend to be easier to place and seal at steep inclinations, such as up a quarry face, than high density polyethylene liners.

7.8 *Approximate costs of liner placements in Britain (1992)*

For clay actually on site, the cost of placement, including the necessary implementation of QA procedures, approximates to £2 to £2.5/m^3. The least expensive HDPE liner is about £4/m^2 placed and an HDPE composite liner comprising geotextile and cushion layer would be around £10/m^2. To set these costs in context, the cost of depositing waste on the site is about £4.5/m^3.

8. LEACHATE

Although of limited use for defining the potential chemical effects of leachates on mineral sealing materials the following parameters have been analysed and used:

pH value: a major parameter - acid capacity up to pH 4.3; base capacity up to pH 8.2
electrical conductivity: a major parameter
chloride content: a major parameter
cationic and anionic properties
carbonate hardness
chemical oxygen demand (COD)
biochemical oxygen demand (BOD)
evaporation residue; ignition loss; total organic carbon (TOC)
sulphate
zinc
cadmium
lead
copper
nickel
iron
ammoniacal nitrogen

total oxidised nitrogen
ethanoic acid
propanoic acid
butanoic acid; iso-butanoic acid
pentanoic acid; iso-pentanoic acid
hexanoic acid; iso-hexanoic acid

Factors that may be observed on site are temperature, pH and conductivity as a function of seasonal fluctuations or the age of the waste in the fill. Organic compounds constituting a leachate may be characterised in terms of: extractable organic halogen compounds; organic halogen compounds; highly-volatile to semi-volatile aromatic and aliphatic hydrocarbons; total hydrocarbons; highly volatile halogenated hydrocarbons; phenols.

Leachate production is a function of water ingress to the landfill. Geotextiles can be used to intercept ingress and to drain water away. They may also be used above the protective layer to serve as a drainage blanket. Plastic sheet comprising dimples, corrugations and spacers is wedged between, or sandwich-bonded to, filter fabric geotextiles or a membrane. Leachates drain to the lowest point of the fill and need to be pumped out periodically into suitable tankers or perhaps into a system of lagoons for aeration with activated sludge nutrients. In the case of geosynthetics the material must have a long-term resistance to leachate and have a suitable transmissivity under the imposed mechanical pressures and temperatures likely to develop at the base of the waste infill. HDPE seems to be the only plastic suitable for resisting leachates, but PVC is suitable for any capping that might be applied to the fill. A transitional layer may be needed to prevent the finer-grained fraction of the waste from blocking up the drainage blanket.

Since methane gas generation is a function of the moisture content of the disposed waste, leachate pumped from the base of the tip can be allowed to percolate back into the fill, so seeding it for further gas production.

The National Rivers Authority sets limits to the potential discharge of effluent/leachate from waste disposal sites and contaminated land generally.

9. PERMEABILITY TESTING OF SEALING MATERIALS

As noted above, seals may be achieved either by clay-based natural materials or by man-made materials or by a combination of the two. The properties of man-made materials will have been determined from extensive testing by the manufacturers.

For testing the permeability of fine-grained soils the triaxial test using a constant hydraulic gradient (constant head) is most appropriate. An all-round (isotropic) pressure of about 3 atmospheres above the pore water pressure would be applied to the specimen in the cell. De-mineralised water should be made to flow through the sample from bottom to top in order to achieve maximum saturation, the degree of saturation being determined at the end of the test. Water temperature should be maintained constant, and a suitable correction for temperature made. ETC 8 recommends a hydraulic gradient $i = 30$ for the determination of the coefficient of hydraulic conductivity (permeability). On the other hand if a specific hydraulic gradient can be determined in the field, then this should be used in the laboratory test. A number of tests over different periods of time may need to be conducted before a permeability coefficient is firmly established.

The permeability properties of a seal will be leachate-dependent. Further permeability tests may need to be carried out using leachate from a site containing the same (ideally mono-disposal landfill) material as will be tipped at the target site. If such leachate is not available then the following materials should be used: distilled water (for comparative purposes); strong acids (eg. hydrochloric acid having pH of about 3); strong alkalis (eg. a caustic soda solution having a pH greater than about 11); metallic salt solutions; hexane (to the limit of solubility); halogenated volatile hydrocarbons (to the limit of solubility); acetone (less than or equal to about 10%). Other liquids, such as phenols (100g/l) or benzol (to the limit of solubility) may also be used, but only with care in the laboratory. Tests must be continued until the system is `stable' - the fluid flowing through the sample creates no changes in the sealing material nor in the fluid. Before and after chemical analyses will be required.

10. OLD LANDFILL SITES

In the case of an old landfill site, a cut-off (curtain wall/membrane) may be required to be formed within or just outside the tip material to prevent gas migration. The cut-off must reach down to impermeable ground. There are several possible methods and materials available for constructing a barrier.

10.1 *Vertical barriers*

- A thin steel plate hammered through the fill and into the subsoil and then retracted. The void so formed is then filled with a sealing compound under pressure.

- A diaphragm wall, trench-excavated by back-hoe and supported during excavation by bentonite, to be replaced by tremied concrete. This is a two-phase system. Whenever possible, the wall should be keyed into an underlying low-permeability layer. If this is not practicable, a grout curtain, keyed into such a layer and topped by a wall, might be considered. In order to limit seepage of bentonite into the fill both during and after placement it needs to be stiffened, most simply by cement for self-hardening, but polymer suspensions may also be used. If the excavation can be self-supporting, then there is the possibility of dispensing with the bentonite temporary support and adopting an alternative cut-off using compacted puddle clay, cast-*in situ* concrete, or even pre-cast elements. Back-hoe type tracked excavators normally have a practical dig-depth of about 6 to 7 metres, but specialist diaphragm wall rigs have been used to excavate dam cut-off walls to depths of more than 100m. A wall, raised to maintain level with the contained fill, would be less reliant on self-support if used in an old quarry floor refuse site and constructed close to the quarry face. In a one-phase system, a bentonite-cement mixture would be used. Typical mix proportions of a cement-bentonite wall would be bentonite 20-60kg, cement 100-60kg, water 1000kg Quite detailed information on the composition and manufacture of sealing compounds and the preparation of samples may be found in the ETC 8 manual.

Calcium bentonite is chemically more resistant to leachate attack and provides lower permeabilities than does sodium bentonite. It is, however, more difficult to place. Further, only cement-free bentonite suspensions are resistant to organic solvents or leachates having high organic contents. So, if they are to remain in place they require hardening materials other than cement, or alternatively a plastic membrane inlay as noted below.

Decomposition of organic matter and associated gas generation is accompanied by volume reduction. There is therefore the potential not only for settlement over the site area (and the effect of that upon any surface cover and perhaps any buildings) but also for frictional drawdown on containment walls and any stone chimneys used for venting to atmosphere or collection.

It is generally advisable, in a gas cut-off slurry wall, to start with and maintain a relatively high water content, otherwise the wall becomes gas-permeable. It is important to assess whether the gases themselves act a significant desiccating agents.

Another barrier system comprises patent Bentofix mats lowered to line the outer faces of cut-off trenches filled with stones. Expansion of the bentonite when penetrated by moisture provides a low permeability of about 10^{-10} m/s and the stone provides a preferred venting path towards a gas collection pipe or chimney.

A variant on the slurry wall method is the use of jet grouting to construct the cut-off from the bottom up although, because of the relatively loose nature of the fill, jetting pressures must be carefully chosen so as not to create excessive disturbance.

- A diaphragm wall using a layer of cross orientated polyolefin laminated to 20 micron thick aluminium foil to produce a 1.5mm thick barrier (the Dampseal MB system from Marley Waterproofing). Tests are claimed to have recorded less than 0.2ml/m^2 methane diffusion through the membrane in 24 hours. A problem also arises in the handling of long lengths of the cut-off material, protection of the material from perforation, and in the joining of the individual lengths. An alternative is to use high density polyethylene (HDPE) membranes, a 2mm thick membrane having a permeability of only 10^{-13} m/s. Gas transmission through HDPE is significantly less than through chlorinated polyethylene (CPE), PVC, and low density polyethylene (LDPE).

Operationally, a plastic membrane can be sandwiched between two empty 300mm wide gabion baskets and further protected with an outer line of 750mm wide stone-filled gabion baskets. The outer baskets are then filled with a slurry wall mix. Such a system can provide a venting facility in addition to its barrier function. Alternatively, a plastic membrane can be installed concurrently in-trench with a cement-bentonite slurry. Either a special frame is required for the membrane placement, and a facility for the frame to be withdrawn when the membrane is released, or the membrane can be wound on a drum and then lowered into the bentonite trench using a vibratory mechanism to facilitate penetration. Geomembrane placement tends to decrease the overall *in situ* permeability of a standard cement-bentonite slurry wall by about two orders of magnitude but the overall permeability can decrease by up to 4 to 5 orders of magnitude if high quality jointing can be achieved. There are several patent interlocks designed by different firms on the market, but a major requirement is that the joint must be tight and impermeable. Some devices contain a number of chambers within the interlock, with the possibility of keeping one or more open for monitoring and repair purposes.

- A diaphragm wall comprising a series of linked box piles driven into the fill by vibration, the void created as the piles are withdrawn sequentially being filled with grout. In operation, a line of, say five box piles are inserted. As the first pile is withdrawn it is grout filled bottom-up by pumping from the base of number 2 pile. The number 1 pile just withdrawn is then inserted as number 6 pile in the line. Number three pile is then used to fill number 2 pile from the base, and then number 2 pile becomes number 6 pile in the insertion sequence. In this way a thin contiguous grout-pile wall is created to form the gas barrier. A major advantage of this system is that no waste is excavated and the local compaction of the fill serves to limit loss of grout. The cut-off wall is also much tidier.

10.2 *Horizontal sub-surface barriers*

Possible methods of creating these barriers include jet grouting, chemical grouting, and claquage grouting. Jet grouting can only be used satisfactorily for creating horizontal, contiguous cement-grout layers in fine-grained soils, a situation not encountered in domestic and trade waste tips although if indicated by a suitable site investigation such soils might lie beneath the tip material and so be available for grouting. Unfortunately, continuous coverage could not be guaranteed, and so leakage of gas and leachate would be a possibility. One method uses an auger, with grout injection through the hollow stem as the auger is withdrawn, but there is a problem that frequent passages of the auger through overlying contaminants could contaminate the grout and affect its setting properties. Claquage grouting and chemical grouting are not really appropriate for non-homogeneous soils, the grout seeping away and therefore not forming an impermeable seal. Chemical grouts may also degrade with time and generate added toxicity.

There is really no satisfactory method of post-fill placement sub-surface horizontal containment at the present time.

11. *Surface barriers*

Such barriers have several functions including aesthetic cover, prevention of lightweight fill (paper and dust) from being blown away, reduction of any possible fire hazard through the (partial) exclusion of air (oxygen), reduction of odours, public protection from the fill material toxicity, exclusion of precipitation and any other water inflow which promotes leachate generation and thereby increases the contamination hazard, and containment of gas generated within the fill. Clean soil cover is an obvious choice, but over a long period of time its effectiveness could be reduced through the action of flora and fauna. Also the type of soil will be determined by the primary hazard that it is required to mitigate. For example, any methane hazard would suggest a thick silty sand cover, rich in organic matter capable of oxidising the methane, but such material is less commonly available in the UK than, say, in the Netherlands, and the cost would be high. An alternative would be to create a lower zone of cover, about 600mm thick, between the fill and the top soil by rotovating organic silty sand into the top of the landfill (an example is the use of sewage sludge cake mixed with weathered London Clay to provide a top soil), but checks need to be made to ensure that the cake was not contaminated with phytotoxic metals such as copper, nickel and zinc which would inhibit plant growth. Although unsuitable as a containment material at the base and sides of a fill, the low modulus of elasticity (flexibility) of PVC render it mechanically suitable as a surface capping with a drainage layer above and a fertile soil above the drainage layer. However, PVC is both ultraviolet and methane sensitive. As a sound alternative, membranes produced from linear low density polyethylene resins (co-polymers of ethylene) and higher alpha olefins have a much reduced modulus of elasticity without any significant loss of strength and chemical resistance of conventional high density polyethylenes. These materials are being increasingly specified in varying thicknesses for landfill capping applications.

Over a period of time, waste compacts, with higher void ratios at the top compared with the bottom of the tip, which should be expected to settle overall by about 20% of its depth. In order to avoid surface depressions that could collect water and to promote run-off, the final crest of the tip should be domed to side-slope angles of between about 1:40 to 1:20

It is necessary to ensure that water can drain laterally beneath the cap, otherwise a build-up of water level could lead to buoyancy (and instability) in the top soil and putrefaction of vegetation roots. There must also be an efficient gas-management scheme to avoid gas pressurisation beneath the cap.

In the case of capping liners, thicker membranes are better able to resist stress corrosion and cracking. Stress for a particular load is reduced in proportion to thickness, but since corrosion starts from the outside the time taken for total loss of ductility and sealing capability is greater with the thicker liner. A thickness of 1 to 2mm seems to be satisfactory. Some plastic capping membranes may need `friction-enhancing structures' on either side to increase cap stability. This can be done if the slope of the tip top does not exceed 1:3, the sub-soil is clay and the top soil is a coarse sand for drainage. Should there be high frictional forces, then the friction-enhancing structures need to relax automatically to allow some slippage of the sheet and so avoid tearing. Spray-on structures (bands) having yield strengths below the yield strength of the base sheet allow this movement of the base sheet to take place. Sheets need to be welded together and the weld quality needs to be inspected. This is best done by ultrasonic testing.

Bitumen is another obvious capping material, but at about £8 per m^2 placed it is more expensive than, say, high density polyethylene at about £4 to 5 per m^2. It is important to remember, however, that any prevention of controlled or uncontrolled venting of gas creates a higher potential for lateral migration as gas pressures within the fill increase.

A more exotic hazard mitigation technique would be to use microbes to feed on the methane and convert it into less hazardous gases. The method has been developed on a laboratory scale, but in practice the problem is how to introduce bacteria into the ground. One possibility would be to thread microbes into the vents of a vibropiler as the piles are being driven. Microbiological action can also be adopted to attack the imported waste itself

12. GAS UTILIZATION

Landfill gas has a calorific value 50% that of natural gas. Decisions need to be made as to whether the landfill gas should be collected and exploited (perhaps for power generation) or whether it should be allowed to vent freely to the atmosphere. In 1986, the Department of Energy estimated that Britain's landfills could generate the equivalent of 1.3 million tonnes of coal, of which at least one million tonnes were worth exploiting commercially. In 1989, of the 500 or so disposal sites thought to be suitable for energy production 20 or so were being used and another 20 were at the planning proposal stage. The largest individual project generated 3.5 megawatts, but it was suggested that during the next five years this output could grow to as much as 35 megawatts.

If the gas is to be exploited, then the refuse area may be capped, which will also serve to prevent much rainwater getting into the body of the fill. On the other hand having restricted much of the vertical de-gassing there will then be a tendency, under the greater internal pressures, for lateral migration to occur, particularly if there is an atmospheric pressure drop.

Collection often involves the insertion of stone chimneys by vibropiler, perhaps about 200 millimetres diameter and at about 2 metre intervals, using a retrievable tube the void within which is filled with stone which in turn incorporates a polypropylene slotted pipe of about 50 millimetres diameter. The heads of the pipes are linked by horizontal collector tubes in stone-filled, geotextile-wrapped slit trenches, which in turn are linked to a galvanised steel vent. If the gas is not to be retrieved, similar holes can be inserted around the perimeter of the site and the horizontal collector system omitted.

The gas may be collected passively by expulsion under its own internal pressure or it may be pumped. It is important not to pump out the gas at a faster rate than it is being generated. A lack of pressure within the waste will cause anaerobic bacteria to be killed off and methane production stopped. The production must therefore be monitored.

REFERENCES

Council of the European Communities (1978) Directive on Toxic and Dangerous Wastes, 78/319/EEC, Official Journal L84.

Department of The Environment (1986) Landfilling Wastes, Waste Management Paper No. 26.

Department of the Environment (1979) Landfill Sites: Development Control, DoE Circular 17/89 (also Welsh Office 30/89), 26 July 1989.

Department of Transport (1986a) Specification for Highway Works, Parts 1-7, HMSO, London.

Department of Transport (1986b) Notes for Guidance on the Specification for Highway Works, HMSO, London.

European Technical Committee (1990) Geotechnics of Landfills and ContaminatedLand, Technical Guidance ETF8, Bochum, Germany.

Her Majesty's Inspectorate of Plllution (1989) The Control of Landfill Gas (A Technical memorandum on the monitoring and control of Landfill Gas), Waste Management Paper No. 27, HMSO, London, 56pp.

Klug, H.P. & Alexander, L.E. (1954) X-ray diffraction procedures for polycrystalline and amorphous material, John Wiley, New York.

Environmental Management, Geo-Water & Engineering Aspects, Chowdhury & Sivakumar (eds)
© 1993 Balkema, Rotterdam. ISBN 90 5410 099 0

The tailings dams of Stava (Northern Italy): An analysis of the disaster

R.Genevois
Department of Geological Sciences, University of Bologna, Italy

P.R.Tecca
Institute of Applied Geology, CNR, Padova, Italy

ABSTRACT: On July 19, 1985 the failure of the Upper Dam of a pair of tailings impoundments caused a disastrous flow slide and 269 people died. The paper, after describing the geological conditions of the site and summarizing the geotechnical properties of the tailings materials, analyses the consolidation state of the silty tailings and the drainage conditions of the Upper Dam, both being of paramount significance on the embankment stability. The stability analyses carried out with a finite element computer code show the strong influence of both, together with the dam height, on the stability conditions.

1 INTRODUCTION

On July 19, 1985, liquefaction failure of the two tailings dams of the fluorite Mine of Prestavel, near Stava, North-Eastern Italy,.resulted in a catastrophic flow slide. The villages of Stava and Tesero, situated along the stream channel and at its end respectively, were wiped out or buried. The disaster claimed 269 lives and caused extensive property damages in the valley.

Tailings dams have a long history of failures, resulting in widespread death and destruction. A short review of some of the main disasters is given hereafter.

- Aberfan, South Wales: failure of n. 4 coal refuse dam (1944). Liquefaction failure of n. 7 coal refuse dam; 144 people died (Oct 21, 1966) (Bishop 1966).
- Abercynan, South Wales: liquefaction failure of coal refuse dam (Dec 1939).
- Bafokeng, South Africa: liquefaction failure of a platinum mine tailings dam with a flow of about $3*10^6$ m^3 of liquified silts and clays; 12 people died (Nov 11, 1974) (Blight et al. 1981).
- Buffalo Creek, Sanders, West Virginia: failure of a coal refuse dam due to high water levels and too steep downstream slope; the massive failure of n. 3 dam triggered the progressive overtopping type failure of n. 2 and n. 1 dams; 125 people died (Feb 26, 1972) (Volpe 1978).
- Blackpool and Cholwich, Great Britain: failures of kaolin quarry tailings dams (1967 and 1968, respectively).
- Louisville, Kentucky, USA: failure of a calcium carbide tailings dam due to the freezing of downstream slope (Feb 25, 1963).
- Mochikoshi, Japan: liquefaction failure of a gold and silver mine tailings dam due to earthquake cyclic loading (Jan 14, 1978) (Okusa and Anma 1980).
- Barahona, Chile: liquefaction failure of a copper mine tailings dam due to the Talca earthquake; a flow slide of about $4*10^6$ tons killed 54 people (Dec 15, 1928) (Brawner 1979).
- El Cobre, Chile: seismic induced liquefaction failure with a flow slide of about $2*10^6$ tons that killed 200 people; other 10 tailings dams failed during the 1965 earthquake (1965) (Dobry and Alvarez 1967).

In this paper, after introducing the geological characteristics and the geotechnical properties of the tailings materials, failure conditions .are analyzed and an hypothesis on the triggering mechanism of the failure.is presented.

2 METHODS AND DAMS SAFETY

Tailings dams are the most common disposal system to store the solid by-product of the mining industry.

Frequently, based on economical and practical considerations, the coarser fraction of the tailings themselves is utilized to construct the embankments, wholly or in conjunction with natural soils, while the finer-grained portion is deposited behind the dam. The pulp is transported hydraulically from the mill to the disposal area, where several methods exist to separate the coarse and slime materials.

Segregation process of fine-grained from coarse-grained tailings is one of the key aspects for safe containment of the tailings. It is generally based on natural sedimentation principles, and it is accomplished by different methods.

One of the most common procedures is discharging the slurry through spigots onto the deposits from the dam crest: the coarser particles deposit first, while slimes are transported toward the pond where they settle and from which the water is decanted. The coarse materials can be separated also by means of hydrocyclones, and deposited close to embankment, while finer portion is discharged upstream forming the settling pond.

In any case, particular attention should be paid in avoiding that water could be too close to the outer sandy shell of the dam, where seepage may generate critical stability conditions.

Tailings deposits can be classified as cross-valley, side-hill or diked deposits, depending on whether the embankment blocks a valley or encloses the impoundment partially or wholly by a ring dike (Vick 1983).

The construction of a tailings dam begins with the accomplishment of a starter dike of natural permeable materials. The tailings dam is then raised using the coarser portion of the slurry, as a part of the deposit or as a separate structure behind which the tailings are stored. Depending on whether the dam crest moves upstream or downstream or it is fixed during construction, the embankment is constructed following the so called upstream, downstream or centerline methods. An other, less used, method is the thickened tailings disposal.

The upstream method is the oldest and simplest one and it is particularly cost effective since the volume of the containment embankment is significantly reduced. However, it usually provides low safety factors, because the dried pond surface, constituted by slimes, and the base of each lift represent potential planes of weakness and areas of concentrated seepage. Furthermore, it is quite difficult to control the location of the saturation surface.

With the downstream method the slime layers are not mixed to the sandy tailings to represent planes of weakness, allowing to build safe dams with relevant heights, provided that adequate compaction and seepage control are ensured. However, each successive lift requires more material for construction than the previous one, if the coarse tailings assume their natural angle of repose.

Significant heights can be achieved using centerline method, but the embankment should not be raised too fast, not to have relevant probability of failure of the inside slope.

Since the cost of tailings disposal generates no profit, the techniques employed are aimed to minimize the cost of disposal requiring, therefore, that the most economical design must be employed. On the other hand, design of the tailings embankments must include considerations of safe storage during and after construction and safety in terms of environmental concern. Design criteria involve a detailed assessment of selected site conditions, (climate, topography, geology, seismicity and hydrology); geotechnical properties of foundation soils (ultimate bearing capacity and settlements); physical, mechanical and chemical properties of tailings (dams stability and potential pollution). The method and sequence of dam construction must be selected and design should include an adequate monitoring of the phreatic surface and of the vertical and horizontal displacements. Static and dynamic stability and settlements analyses of the tailings dam must be performed Finally, design requires consideration of a plan of abandonment and reclamation of the selected site (Khon 1972, Campbell and Browner 1971, Kealy 1973, Dept. of Mines 1972, Klohn 1981, U.S. Bureau of Mines 1981, ICOLD 1982, Morgenstern and Kupper 1988).

3 THE TAILINGS DAMS OF STAVA

3.1 Geographic and geological setting

Stava Creek, with a draining area of approximately 21 km^2 is situated in the Dolomite Mountains, North-Eastern Italy, in the upper valley of the Avisio River (Fig. 1).

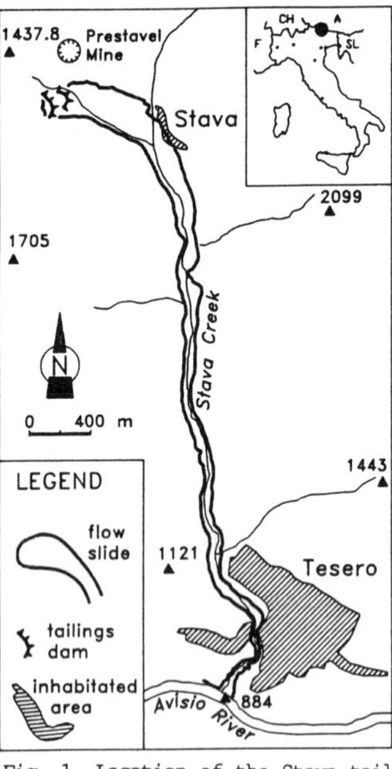

Fig. 1. Location of the Stava tailings dams and the area of the Stava Creek Valley swept off by the flow-slide.

The tailings dams of the fluorite Prestavel Mine, were actually located on the Porcellini Creek, a small tributary to Stava Creek, at an elevation ranging from 1330 to 1380 m above sea level.

The plant area is underlain by a volcanic Permian basement, which comprises massive rhyolitic and quartzolatitic ignimbrites. Subvertical joints and lodes, prevailingly enriched with fluorite, occur in the rock mass. The volcanic basement is overlain by Mesozoic sedimentary deposits.

The morphology of the Dolomitic Region, including Stava Creek area, is characterized by steep cliffs of Triassic carbonates, whereas lower slopes are underlain by older Triassic marly-arenaceous formations. Widespread wurmian morainic deposits and screes occur on the slopes of Stava Valley, whereas glacio-fluvial deposits underlie the valley floor.

The main structural feature of the area is represented by a tectonic line trending approximately WSW-ESE. This line places in contact the volcanic basement, which forms the slope where the Prestavel Mine is located, with the marly- arenaceous formation, where the impoundments are founded on. In addition two other fault systems occur in the Upper Stava Valley, trending NS and NW-SE.

These local tectonic features have an important hydrogeological significance, due to the different permeability of the formations and the subsequent possible existence of confined acquifers.

3.2 History of tailings dams development

The following list summarises the growth history of the two tailings dams (Fig. 2) and the sequence of the more important events which led to the failure on 19th July, 1985.

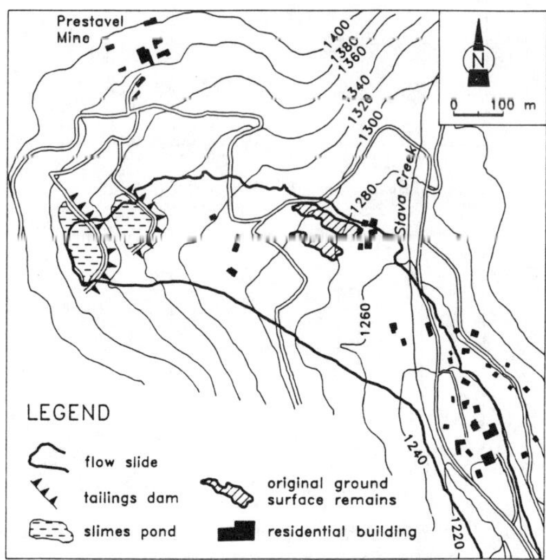

Fig. 2. Stava dams (1983 aerial survey).

1962. Construction of Lower Dam began; the starter dam, built of compacted natural soil, 9 m high and with a downstream slope angle of about 40°, was founded on natural ground, reinforced by concrete soil-nailing.

Late 1960's. Lower Dam, constructed by the upstream method, rose to an ultimate height of approximately 26 m,with an average downstream slope angle of 32°.

Early 1970. Upper Dam construction commenced with a starter dam 5m high, built of natural soil; any foundation pretreatment was used to reinforce foundation soil. Centerline method was used until the dam rose to a height of about 10 m, with a downstream slope angle of approximately 40°, so the downstream toe extended partly over the soft sediments impounded in the Lower Basin. The progressive growth then continued by the upstream method with the same downstream slope angle.

1975. Construction of the Upper Dam continued with an average downstream slope angle of approximately 35°, realizing a 4 m wide berm at a height of 19 m.

1978. Construction of the Upper Dam ceased at a height of 26 m, but natural groundwater continued to flow into the Upper Basin and so high water level was maintained behind the two dams.

1982. Tailings disposal began again; the Upper Dam crest rised to a height of approximately 28 m.

Jan 1985. A small slump occurred on the right-side of the lower portion of the Upper Dam; it was attributed to the blockage by freezing of the decant pipe and to the following leakage.

early Jun 1985. A very large sink hole, 30 m wide and 3-4 m deep, occurred in the Lower Basin, due to the breach of the decant pipe from which silty tailings leaked out down Stava Creek Valley.

July 15 1985. The new decant system was restored; the Lower Basin, drained during repairs, was refilled by flow from the Upper Basin; the Upper Dam crest rose to a final height of approximately 30 m.

12.23 a.m. on July 19 1985. A disastrous failure occurred on the downstream face of the Upper Dam. About 240.000 m^3 of liquified tailings swept down Stava Creek, taking most of the Lower Dam with them, and reached the Avisio Valley where spread laterally (Fig. 1).

4 GEOTECHNICAL PROPERTIES

The geotechnical properties of the Stava Dam were obtained from field and laboratory investigations carried out after the disaster on the remaining tailings materials. Laboratory tests were performed both on undisturbed samples, by the consulting engineers charged by the Court during the legal proceeding, and on recompacted remoulded samples, by the Authors.

4.1 Index properties

It must be emphasized that, as a result of construction method, the dam's material is extremely heterogeneous: the grain size ranges, from the dam shell to the pond area, from medium to fine slightly silty sands, to clayey fine silts, through an area characterized by alternances of sandy and silty materials.

The gradation curves for over 40 samples are summarized on Fig. 3: grading curves envelops are almost completely differentiated, defining materials as silty sands, sometimes clayey, and clayey silts, sometimes sandy.

In terms of plasticity (Fig. 4) the tailings materials are non-plastic (the most sandy) or of intermediate plasticity (the most silty).

The natural water content of the remaining tailings generally increases upwards with mean values of about 18-20% for sandy materials and of about 28-31% for silts; the liquidity index

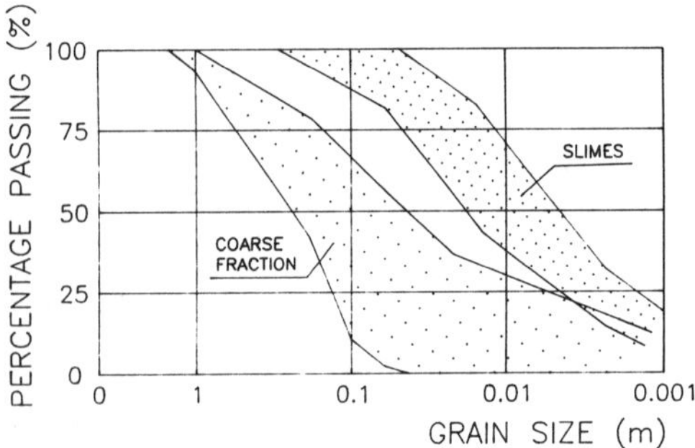

Fig. 3. Grading curves envolopes for tailings.

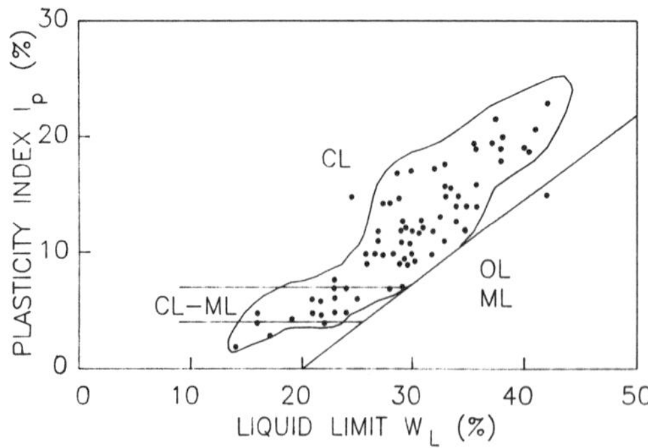

Fig. 4. Plasticity chart.

for the remaining slimes averages 1.20, but it would have been much higher for the material now removed, perhaps higher than 1.40-1.50.

The in- place unit weights of sedimented fine grained materials usually exibit an increase with depth as a result of the consolidation process; but, unlike natural sedimented materials, the depositional process of tailings is intermittent and this brings to density variations within each lift. On the other hand, the unit weight value depends both on grain size of tailings (distribution of fractions, grains shape, uniformity coefficient) and on the flow characteristics of the slurry (discharge quantity and velocity, concentration, drainage conditions) and it is related to the specific gravity of the mined mineral. As a consequence the ranges of bulk and dry unit weights are quite wide (Fig. 5); the statistical analysis showed mean values respectively of 18.6 KN/m^3 and 15.7 KN/m^3, with standard deviations 1.6 and 1.2, for sandy tailings and 19.3 KN/m^3 and 15.0 KN/m^3, with standard deviations 0.9 and 1.0, for silty tailings and silts-sands alternances.

4.2 Relative density

Since the behaviour of sands is sensitive to even small differences in void ratio and the sampling of sands without any change of void ratio is extremely difficult, if not practically impossible, attention has been given to the

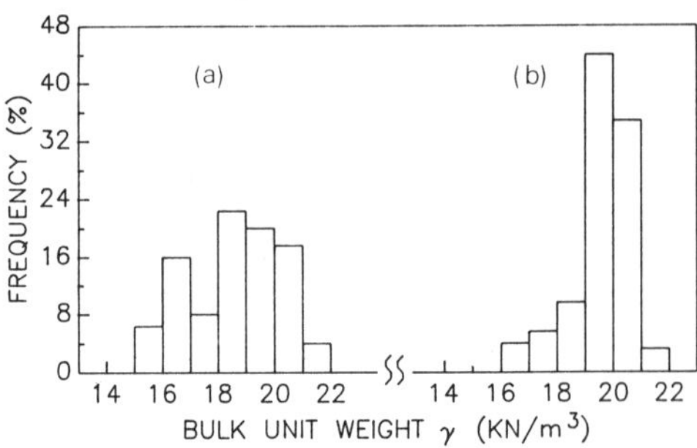

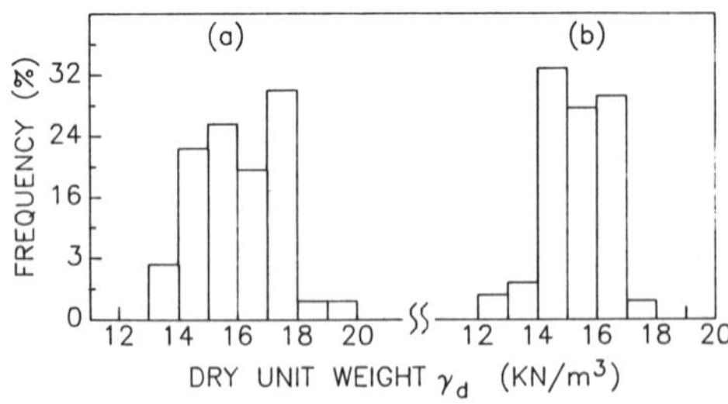

Fig. 5. Frequency distribution of bulk and dry unit weights for sandy tailings (a) and silty tailings and alternances (b).

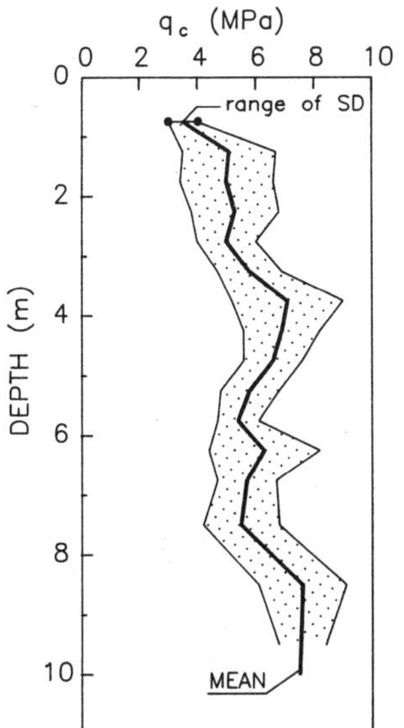

Fig. 6. CPT tip resistance versus depth for all tests.

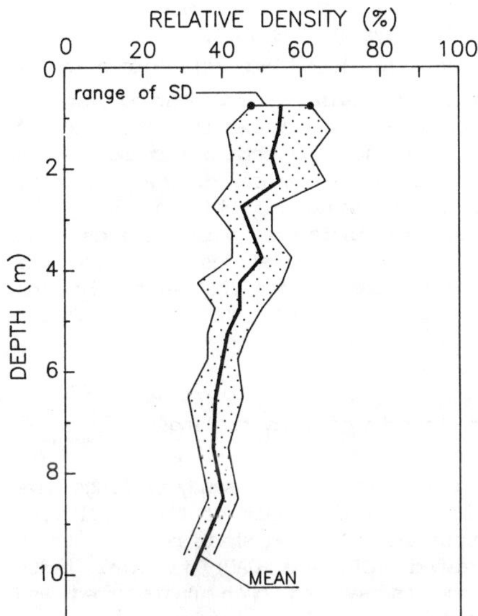

Fig. 7. Relative density from CPT tip resistances versus depth for all tests.

inference of density by means of the cone penetration tests (CPT).

Average CPT tip resistances observed in sandy layers each 50 cm have been then plotted versus depth together with the measured standard deviations (Fig. 6); the relative densities (D_r), inferred on the basis of correlations

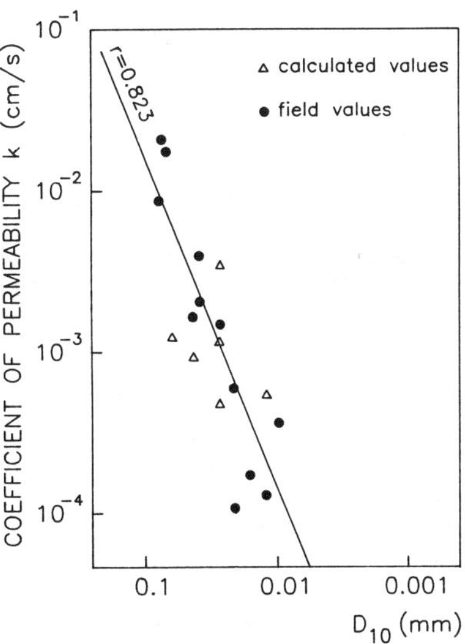

Fig. 8. Permeability versus grain diameter D_{10}.

presented by Baldi et al. (1982) are shown, as average and standard deviations, in Fig. 7.

Examination of data suggests that, while tip resistances slightly increase with depth, average relative density shows a decrease from 50-55% to 40% at depths greater than about 6 m: this is probably due to the fact that the in situ stresses before failure cannot be exactly estimated. In any case, CPT tests show that sandy tailings were in a medium dense state.

4.3 Permeability

The coefficient of permeability of tailings materials can range widely because of the variability in the gradation and in-place density. The differences in the k value have a strong influence on the location of the phreatic surface and on the rate of water flow within the tailings dam. Furthermore, due to the construction method, the tailings deposits have a marked permeability anisotropy that is much more important for the seepage network than the value of k itself. Fig. 8 shows, for the sandy tailings, the variation with the effective diameter D_{10} of k values obtained by field tests: values range from $5*10^{-4}$ to $4*10^{-3}$ cm/s, and so, sandy tailings can be regarded as free draining soils.

The values of k, calculated on the basis of the gradation, show a good agreement, but the range is wider ($k=1*10^{-4}$ - $2*10^{-2}$ cm/s). Laboratory tests on silty tailings, both undisturbed and remoulded, showed values of about three-four orders lesser ($k=1*10^{-6}$ - $5*10^{-7}$ cm/s).

4.4 Compressibility

Consolidation parameters have been evaluated from one-dimensional consolidation tests and from a sedimentation tests. Compression indices come out between 0.15 and 0.28 for undisturbed and between 0.20 and 0.37 for remoulded samples (Fig. 9).

The coefficient of consolidation. at a vertical stress of 200 KPa, showed values generally equal to $2*10^{-2}$ cm^2/s, ranging from

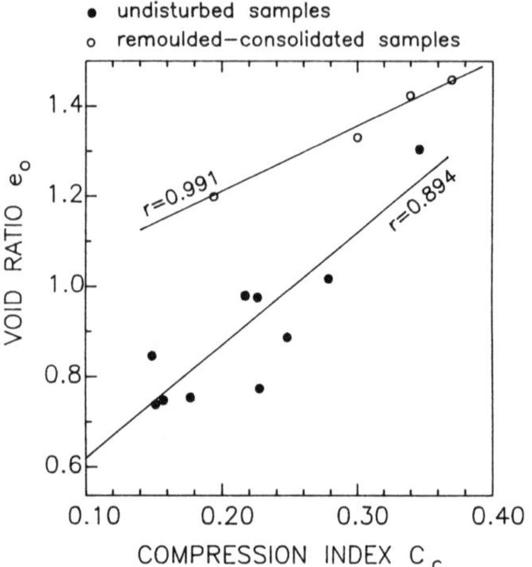

Fig. 9. Compression index versus initial void ratio.

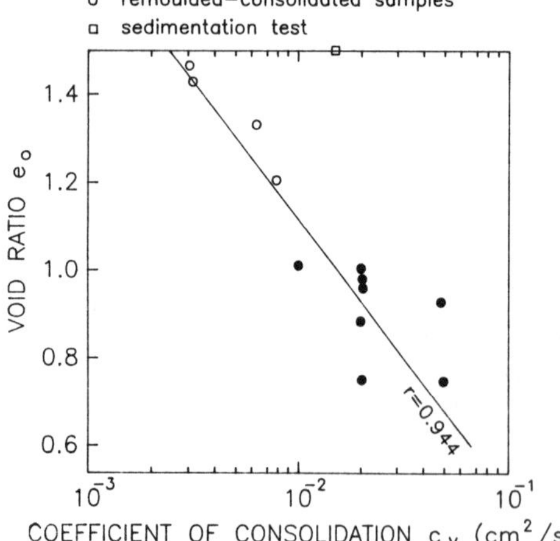

Fig. 10. Coefficient of consolidation versus initial void ratio.

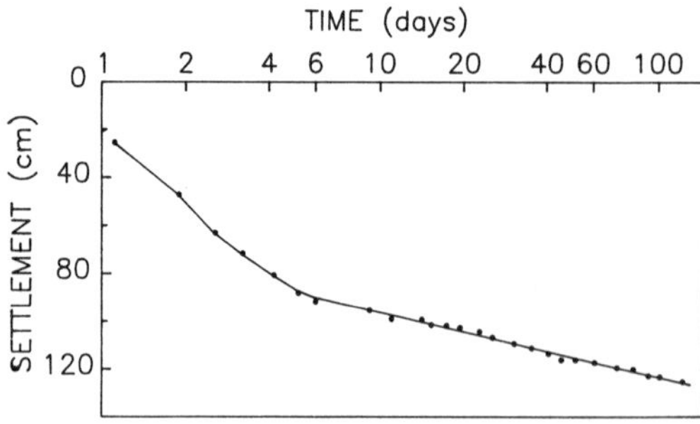

Fig. 11. Sedimentation test.

$1*10^{-2}$ to $5*10^{-2}$ cm^2/s; values slightly lesser have been obtained on remoulded samples (Fig. 10).

The consolidation characteristics of slimes have been besides analysed carrying out a sedimentation tests on a perspex cylinder (height 200 cm; cross area 176 cm^2) with a drainage system at the base. A slurry of silty tailings, with a concentration of 12% by weight, was introduced into the cylinder up to an height of 170 cm and the heights of the segregated and consolidating part have been measured for a period of 120 days (Fig. 11). The coefficient of consolidation ($c_v=1.5*10^{-2} cm^2/s$) agrees quite well with those obtained on undisturbed samples, but it should be outlined that the stress levels are very different. The obtained consolidation parameters values are typical of low plasticity silty materials (Lambe and Whitman 1969): the compressibility indicates that slimes can experience large deformations due to the increase of the dam height; the relatively high value of c_v indicates that consolidation occurred rather fast.

4.5 Shear strength of sandy tailings

The static shear strength of sandy tailings have been evaluated on the basis of the results of drained triaxial (CID) and shear box (SB) tests and undrained (CIU and CAU) triaxial tests carried out on samples both undisturbed and recompacted.

On the whole, considering both drained (ϕ_d) and effective stress (ϕ') friction angles, the values range between 37° and 51° for the undisturbed and between 29° and 40° for the recompacted samples; the statistical analysis shows a mean value and a standard deviation respectively equal to 43.0° and 4.1° (n. 12 tests) and to 34.8° and 3.0° (n. 56 tests).

The variation in the initial density may explain the range of the friction angles (Fig. 12): tests were carried out with initial dry

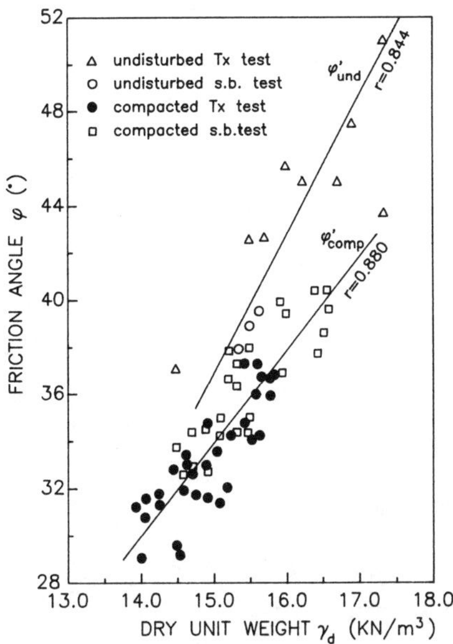

Fig. 12. Variation of friction angle with dry unit weight.

unit weights varying from 15.5 to 17.4 KN/m^3 (D_r=35-60%) for undisturbed samples and from 13.8 to 16.5 KN/m^3 (D_r=20-50%) for recompacted ones. It is noticeable that for undisturbed samples friction angle results are more sensitive to density variations as a consequence of the fabric that sandy tailings develop during sedimentation and dam raising up.

Drained triaxial tests give significantly higher values of the friction angles. Indeed, these tests, carried out on both undisturbed and remoulded samples, give a mean value of the ϕ_d angle equal to 39.1°, that rises to 41.8° at stresses lower than 250 KPa due to the curvature of the failure envelop (Fig. 13), while CIU tests on recompacted samples show a mean friction angle of 33.2° (Fig. 14).

However, if the failure is defined as the point of maximum obliquity instead of the point of peak deviatoric stress, that is if it is considered the steady state condition of loose contractive samples or the post-phase transformation state (Ishihara et al. 1985) of the medium dense dilative samples, the angle ϕ' shows a value of 42.3° (Fig. 14) very similar to the drained angle. A close correspondence between friction angles mobilized at maximum contraction in drained shear and at undrained steady state was postulated by Castro (1969) for sands tested with initially loose states.

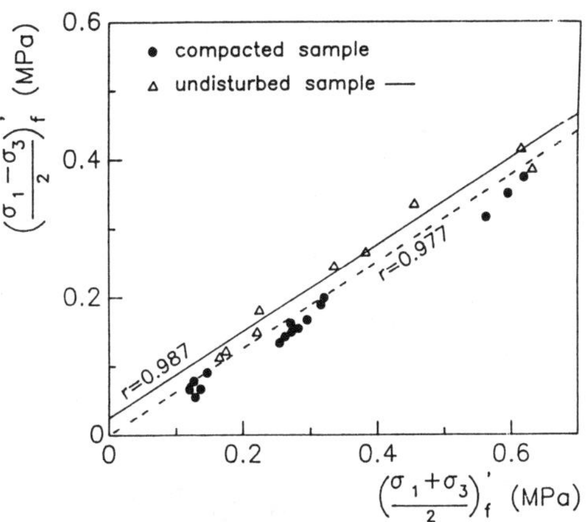

Fig. 13. Triaxial CID tests on recompacted and undisturbed samples. Dashed correlation line for all samples.

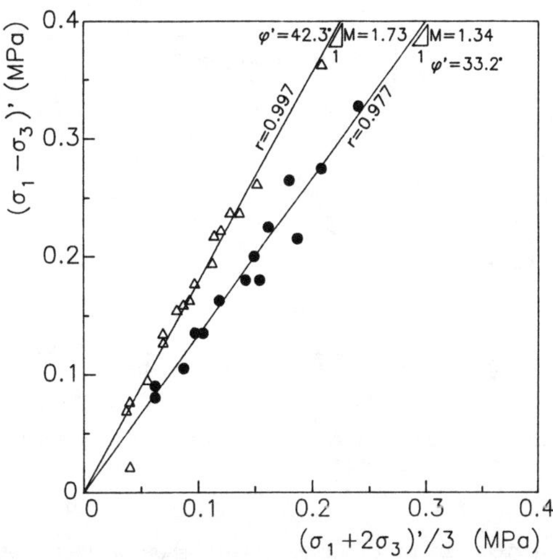

Fig. 14. Triaxial CIU tests on recompacted samples.

4.6 Shear strength of silty tailings

The shear strengths of silty tailings have been evaluated on the basis of field vane and CPT tests and of triaxial UU and CIU tests.

The correlation between field vane strength (s_u) and vertical effective stress existing before failure (Fig. 15) shows values of the ratio s_u/σ'_{vo} ranging from 0.15 to 0.22 with a mean value of 0.19.

It should be noted, however, that the undrained cohesion from triaxial UU tests on undisturbed samples, reconsolidated at the

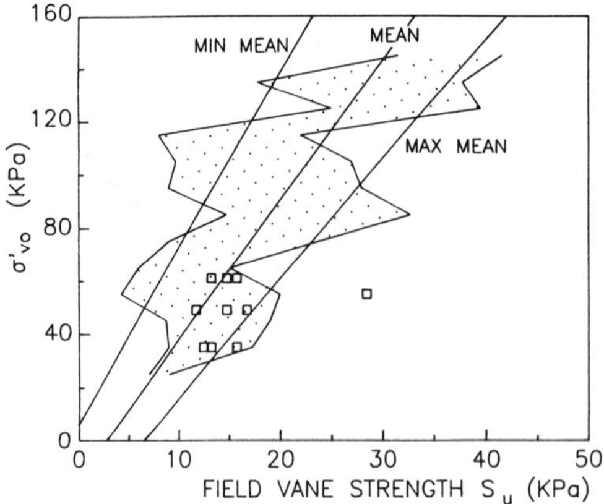

Fig. 15. Variation of field vane strength s_u with the vertical effective stress existing before failure. Small squares: undrained cohesion values from UUTx tests.

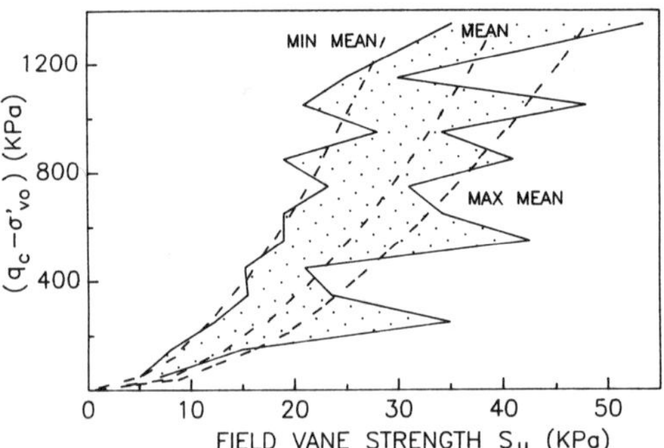

Fig. 16. Correlation between field vane strength s_u and CPT tip resistance.

previous in situ stress level, show values that fit quite well with those ones obtained by the field vane tests. These values are a little smaller than those indicated by Mayne & Mitchell (1987) and Mesri (1988) and this is probably due to the partial remoulding and swelling of slimes as a consequence of both the failure and the subsequent unloading.

CPT tests have been also used to evaluate the undrained strength of silty tailings but, due to the uncertainty on the relation between q_c and undrained shear strength, the value of latter has been evaluated on the base of the correlation between $(q_c - \sigma_{vo})$ and field vane strength s_u (Fig. 16).

The effective stress friction angle (CIU Tx tests) of silty tailings results to be $\phi'=35°-36°$ with a zero cohesion intercept.

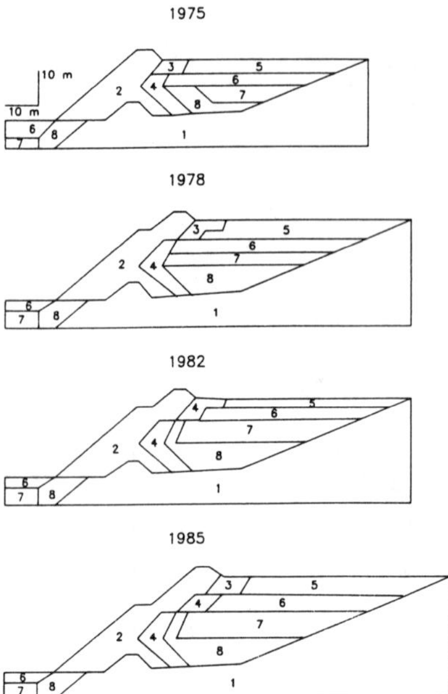

Fig. 17. Cross sections of the Upper basin. Material types: 1= natural soil; 2= sandy tailings; 3= very soft alternances; 4= soft to medium alternances; 5= unconsolidated slimes; 6= very soft slimes; /= soft slimes; 8= medium consolidated slimes.

5 FAILURE ANALYSIS

The Stava tailings dams had not been constructed on the basis of engineered plans: no analyses had been carried out on the geotechnical properties of tailings and of foundations soils; no controls had been made of the location of the phreatic surface; no drawings exist showing the pre-failure embankment configuration as regards particularly the extension of the beach and the distribution of different materials into the embankment.

As a consequence, some hypotheses have been made on the materials distribution considering the construction methods and the rate of raising of the dams. The considered distribution of tailings materials for the Upper Basin is indicated in the four cross sections of Fig. 17.

5.1 Consolidation of the slimes

The consolidation state of the silty tailings is a basic factor in any stability analysis concerning tailings dams. The laboratory tests showed that slimes were normally consolidated or, even, slightly overconsolidated probably due to local essication phenomena; they show, obviously, only the normal consolidation of the remaining silty tailings and not that one of the flowed away slimes.

For the slimes, as a first approximation, it is conceivable to use the relations proposed by Gibson (1958) for a clay layer increasing in thickness with time, to have an idea of the consolidation state. Actually there are some important restrictions to the use of these theories, and mainly: the size of the occurring deformations, considered infinitesimal; the bulk weight of the material, assumed constant; the material properties considered constant during the consolidation process; the existence of an upper layer where sedimentation occurs but where the consolidation theory cannot be applied. In any case, considering that the rate of raising of the Upper Dam has not been constant, the Gibson analysis has been used as a first approach to the knowledge of the consolidation state of the slimes.

In the period from the beginning of the Upper Dam construction (1969-1970) to the suspension (1978), the calculated rate of raising was 2 m/year. With a time t=9 years and a consolidation coefficient of 1.5 cm^2/s we obtain, in the middle of the silty tailings layer, a consolidation degree equal to 83% for a pervious base or to 70% for an impervious base. If the four years break is considered, the remaining excess pore pressures are almost completely dissipated 1n 1982 and the material may be considered normally consolidated.

After the restarting, the rate of raising was 1.75 m/year; with a time t=3.6 years and the same consolidation coefficient, the consolidation degree at the base of silty tailings is not higher than 65% considering that, in this case, the base is made up by the consolidated slimes previously sedimented and, then, practically impervious.

After all, the upper part of the silty tailings has to be considered not completely consolidated: the slimes may not have consolidated as fast as the sandy dam was raising up, as it has happened in other cases (Leon 1979).

In the central part of the tailings pond the consolidation process is generally considered one-dimensional due to the drainage of water in the vertical direction. Nevertheless near the sandy tailings shell the pore water drains mainly in the horizontal direction through the sand dam (Mittal and Morgenstern 1976) and the consolidation process should be considered two-dimensional. These hypotheses have been checked using an explicit finite difference code (FLAC) which may model the flow of ground water through soils in parallel with the mechanical modelling so as to analyse the effects of fluid-soil interaction. The Upper Basin has been then modelled attributing to the tailings materials the permeability coefficients previously indicated, but considering also the horizontal permeability of the sands-silts alternances ten times higher than the vertical one. Besides, a vertical permeability coefficient of $1*10^{-6}$ cm/s and an horizontal one of $1*10^{-5}$ cm/s have been given to natural foundation soils.

Consolidation displacements, occurred after drainage was completed, are shown in Fig. 18:

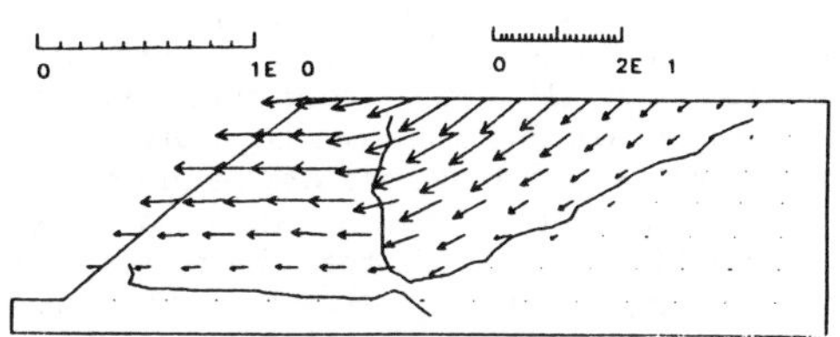

Fig. 18. Displacement vectors and tension region due to consolidation process.

the displacements vectors are really horizontal through sandy tailings and alternances, but in slimes zone they are dipping towards the sandy shell, as a consequence of the relatively low permeability coefficients of the base natural soils.

In conclusion, the seepage flow through the embankment is mainly controlled by the consolidation of the slimes at their transition to the sandy pervious shell; however, a significative part of the stored slimes in the central part of the embankment should have been in an under-consolidated state.

5.2 Phreatic surface within the dam

The location of the phreatic surface is of major interest for the stability analyses but, since the consolidation prior to failure can never be exactly known, it may be only inferred. However, the following boundary conditions can be reasonably established: a low permeability of base natural soils, corresponding to an impervious foundation condition; high ground water levels at the time of failure (July), due to the snow-melt in the spring; the proximity of the ponded water to the crest of the dam.

Nevertheless, the prediction of the equilibrium water content profiles within tailings dams is possible providing the appropriate material properties and distribution and the constitutive relationships for the tailings can be defined.

Due to the poor knowledge of these latter, the computer code FLAC has been used to locate the phreatic surface changing the distribution of tailings materials and the distance of the pond from the dam crest. Four different positions have then been considered for stability analysis purpose.

5.3 Stability analyses

In the case of Stava tailings dams the only threat to slope instability came from static loading and stability can be assessed by carrying out stability analyses using shear strength parameters appropriate to the loading conditions. Drained shear strengths are used to evaluate long term stability condition when all excess pore pressures are fully dissipated, but they can be applied also when no excess pore pressures are developed due to both a slow rate of raising and high values of permeability and of the coefficient of consolidation, as it is the case of the sandy tailings.

On the other hand, the annual raising of the dam for storing the next year's tailings constitutes the most important undrained loading for silty tailings due to their low permeability and coefficient of consolidation. Peak undrained strengths, expressed in terms of effective stress friction angle, can be used in effective stress analyses but it is essential to incorporate the appropriate undrained excess pore pressure due to the field loading. Furthermore, it is important to be aware of the strains required to reach the peak strengths if it is considered to mobilize any strength greater than the steady state strength, but contractive sands and silts show very large differences between strains at peak (Paulos 1988).

The failure of tailings dams containing slimes involves substantial displacements will occur without any significant dissipation of shear induced pore pressures, and the failure is then essentially undrained. In this connection Ladd (1988) emphasizes that "whenever the average shear stress along a potential failure surface reaches the average available undrained shear strength, then an undrained failure will be initiated independent of the prior drainage conditions".

Undrained shear strengths of silty tailings and drained shear strengths of sandy tailings have then been considered for the stability analyses.

A finite element computer code (ZSOIL) has been uilized to calculate safety factors and to analyse deformations.

The used code solves slope stability problems in the framework of a unified nonlinear finite element model. The Drucker-Prager yield criterion is used:non associated flow rule is adopted which imposes a deviatoric plastic flow (no dilatancy) and perfectly plastic behaviour (no hardening).

Table 1. Geotechnical parameters.

soil	γ (KN/m^3)	ϕ' (°)	c_u (KPa)	E (MPa)	ν
1	20.6	40	-	150	0.25
2	18.6	42	-	35	0.30
3	17.2	36	-	15	0.35
4	18.1	37	-	25	0.35
5	13.7	-	7	3	0.35
6	16.2	-	15	6	0.35
7	17.7	-	25	9	0.35
8	18.6	-	35	12	0.35

The geotechnical parameters considered on the analyses are summarized in table 1, where the materials numbers are those ones of Fig. 17. Young's modulus has been estimated considering the empirical correlation with the CPT tip resistances (Mitchel and Gordner 1975).

5.3.1 The influence of phreatic surface location

The four different phreatic surfaces obtained by the flow analysis are expressed by the degree of saturation (S_r %) of the whole sandy shell. In all cases the safety factors show (Fig. 19) an

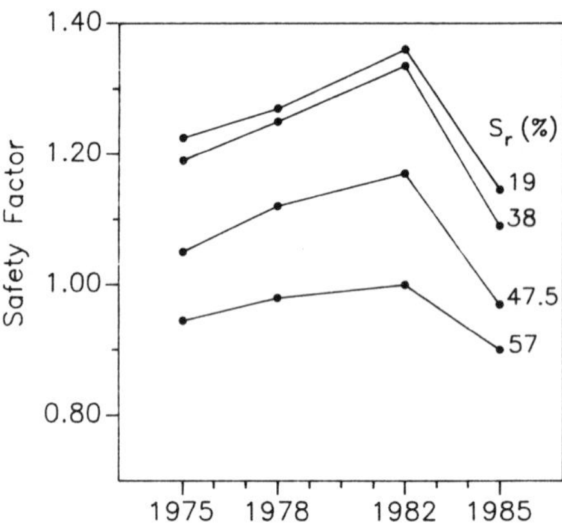

Fig. 19. Variation of safety factor during Upper dam construction as a function of the phreatic surface location expressed by the saturation degree (S_r) of the sandy shell.

initial increase due to the decrease of the downstream slope angle (1975-1978) and to the consolidation of slimes, and even to a probable weak cementation of the sandy tailings during the period of the interruption of the activity (1978-1982).

The decrease of the safety factor as the

construction started again (1982) is very dramatic. The influence of the water table within the dam is significant only for saturation degrees of over the 40%, but only with an S_r of about 45% the dam may experience on instability condition. Higher values of the saturation degree are not actually possible: the dam would have been unstable since 1975. It should be noted that the maximum considered value of S_r (57%) has been obtainbed with a thickness of water in contact with the dam crest equal to 0.5 m, that is a water level in the pond very high and unlike to exist.

5.3.2 The influence of dam height

In loose contractive sands the condition of maximum stress difference ($\sigma'_1-\sigma'_3$) does not coincide with the real failure surface described by the maximum effective stress ratio (σ'_1/σ'_3): the former corresponds then to a condition of minimum stress difference at which instability may develop inside the true failure surface (Kramer & Seed 1988).

Recently, a method has been presented to locate the lower instability boundary that, together with the effective stress line, defines the region of potential instability (Lade 1992). This method has been used to analyze the influence of the dam height on the extent of the potential unstable region. It has been slightly modified to account for the presence of different materials: only toe circles, ending at the limit of the tension region (Fig. 18), have been considered and the stress state in the slope has been calculated by the finite element code FLAC.

A local zone of instability already existed in 1975 and it extended itself more and more upwards and inwards as the dam height increases engaging also the silts sands alternances and even part of the silty tailings due to the particular geometry of the slope resulting from both the berm realisation and the slope angle reduction after 1975 (Fig. 20). In any case, the upper dam was stable as long as drained conditions were secured: any pore pressure build-up, due to an undrained loading, would have result into the enlargement of the unstable region, the engagement of progressively larger volumes of unstable tailings and, finally, into the dam failure by static liquefaction.

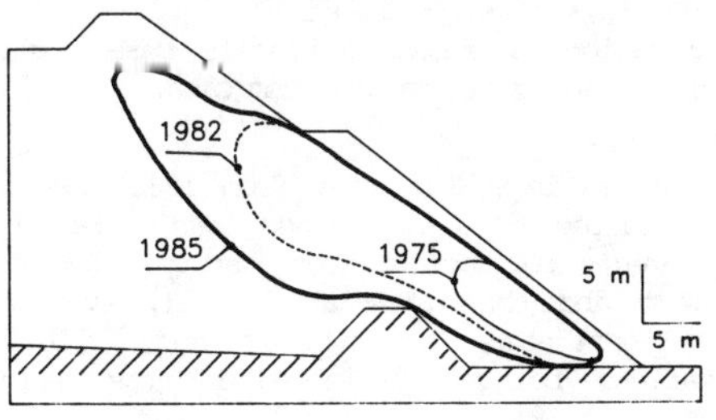

Fig. 20. Progressive extension of the potential instability region during Upper dam construction.

5.3.3 Other possible failure conditions

The stability analyses have been carried out considering two other triggering mechanisms: higher water pressures into the foundation soils; high rate of filling after 1982.

The pressurisation of the foundation soils of the tailings dams must be considered because both the particular geological and hydrogeological conditions of the site and the low permeability of the tailings covering: excess pore pressures may have developed due to the infiltration from the upper pond or high water pressures may be present as a consequence of the likely existence of a confined acquifer in the natural soils. In any case, critical uplift pressures could have developed underneath the toe of the upper dam, which partly rested on the impervious silts of the lower basin. The stability analyses show that the excess pore pressures due to the infiltration from the upper pond doesn't produce any important decrease of the safety factor, at least relatively to the weight of the phreatic surface location into the dam.

Very relevant changes of the stability conditions are showed only if the water pressures in the confined acquifer of the foundation soils are in the order of 100 KPa in excess over the hydrostatic pressures. Safety factor as low as 0.67 has been obtained for such condition with the highest considered water surface into the dam.

Consideration of the rate of fill deposition and likely consolidation of slimes have suggested that excess pore pressures due to construction loading should be very small: the described analyses have then assumed that, prior to failure, no excess pore pressures existed in the failure zone.

The hypothesis has been made, however, that, due to the long period of inactivity, the tailings deposited since 1982 up to the disaster could have acted as an undrained load on the existing tailings embankment. The situation considered for the stability analyses for the 1982 has been the modified applying on the upper surface a distributed load proportional to the final height of the deposited tailings. In such a condition, a safety factor of 1.06 has been obtained having considered the lowest position of the water surface into the dam.

6. DISCUSSION AND FINAL REMARKS

Based on the field investigations and the resulting laboratory studies carried out immediately after the disaster, a number of stability analyses were performed to determine the probable mode or modes of failure, although the complete lack of pre-failure data on tailings distribution and properties and on seepage conditions has been a considerable drawback.

The safety factors, obtained for the considered variations of seepage and loading conditions at four critical times of the Stava Dams history, are summarized in table 2.

Table 2. Safety factors of Upper Stava Dam.

S_r (%)	1975 h_d=19.5	1978 h_d=26	1982 h_d=26	1985 h_d=30	1985a h_d=30	1985b h_d=26
19	1.22	1.27	1.35	1.15	—	1.06
38	1.19	1.24	1.33	1.07	—	—
47.5	1.05	1.12	1.17	0.96	—	—
57	0.94	0.98	1.00	0.91	0.67	—

h_d: dam height (m);
S_r (%): saturation degree of the outer sandy shell;
1985a: hypothesis of pressurization of foundation soils with a water pressure exceeding hydrostatic value by 200 KPa at the base of the sands-silts alternances;
1985b: hypothesis of 1982-1985 tailings thickness acting as an undrained load.

The vector representation of node displacements shows a rotational mechanism of the potentially failing mass, the maximum displacements being vertical and always located immediately upstream the dam crest as a consequence of the material space distribution joined to the used upstream construction method; in the lower part the displacements distribution is controlled by the presence of the more resistant material of the starter dam. As a consequence, the rotational failure surface, after passing through the interlayered sands and silts and skimming the starter dam, intersects the dam slope somewhat above the toe. Examples of the 1975 and 1985 high water level case (S_r=57%) are shown in Fig. 21 a and b.

If the foundation soils pressurization is considered, the failing mass is almost the same, but the movements are quite different: while vertical displacements are always developed in the crest and beach area of the dam, they become parallel to the dam slope below the 1975 berm (Fig. 21b): this is the consequence of the high water level into the dam due to the seepage from the assumed confined acquifer of the foundation soils through the sandy tailings placed upstream of the starter dam by the centerline construction method used at the beginning.

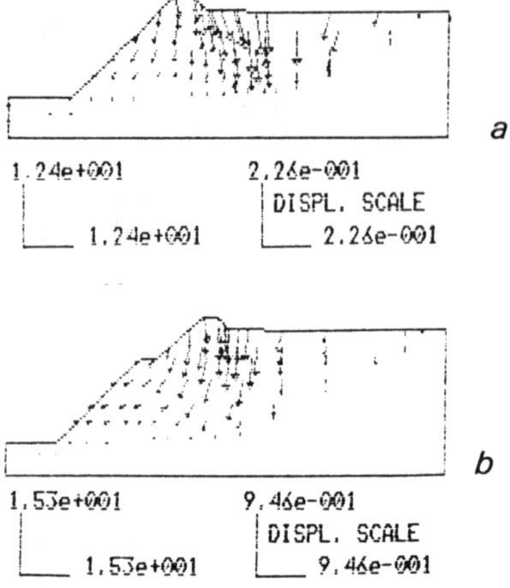

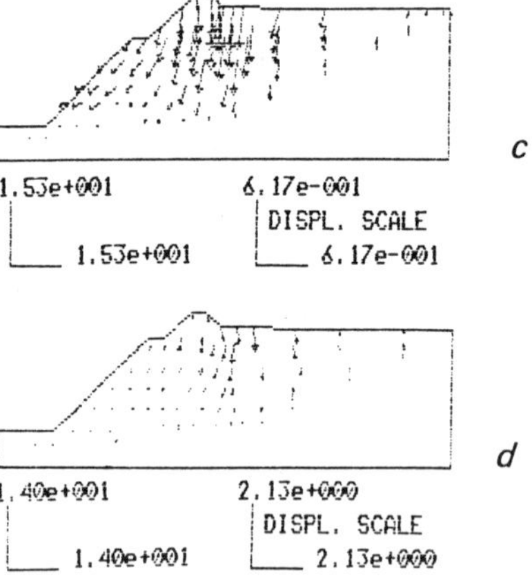

Fig. 21. Stava tailings dams displacements analysis. (a), (b): 1975 and 1985 high groundwater level cases (S_r=57%); (c): 1985 pressurization of foundation soils case; (d): undrained loading of the 1982 dam case.

If the tailings deposited from 1982 onwards are considered as an undrained load, then the displacements are concentrated just upstream the dam crest and they show an initial, even if small, upstream component, very little deformations being experienced by the sandy tailings (Fig. 21c).

In conclusion the stability analysis showed that:

- the unstable mass is always restricted to the sandy-silts alternances, involving the slimes only marginally;

- shear displacements of the sandy outer shell increase with the dam height and the contemporary downstream rotation of the silty tailings displacements vectors mainly due to the consolidation process (Figg. 21 a and b);

- calculated maximum vertical displacements increase from about 16 cm in 1975 to more than 64 cm in 1985, while the horizontal ones range from about 5 cm to more than 18 cm in the same period;

- any failure mode, reasonably conceivable ir the geological, hydrogeological and geotechnical picture of the Stava tailings dams, may lead to the instability of the downstream upper dam slope.

Considering that: no prior investigations of any type was carried out; no design of the dams was ever drawn up; no drainage system was provided beneath the basins; no provision existed to ensure that runoff or near-surface groundwater flow entered the basins; no pore pressure and displacements monitoring instruments were installed; if all this is considered together with both the ever poor geotechnical properties of tailings and the steepness of the downstream upper dam slope, then it can be stated that high failure potential should characterize the upper dam even for relatively small heights.

In July 1985, having an height of 30 m, the upper dam was prone to fail under any small additional effect acting as a trigger, like:

- local static liquefaction of the looser part of the sandy tailings;

- build-up of excess pore-pressure due to additional rapid load on soft sediment;

- critical uplift pressures underneath the dam.

It is quite difficult, if not impossible, to determine the actual trigger of the Stava dams failure, but the height of the upper dam and then the existing state of stresses and strains should be considered of paramount significance from two different point of view:

1- the raising up of the dam produces the progressive enlogement of the potential unstable region engaging so not only the sands and sands-silts alternances but also a small part of the finest silty tailings that may liquefy due to the shear-induced pore-pressure build-up: all the downstream slope may then collapse due to the short shear strengths actually available.

2- The available shear strength is progressively decreasing due to the development of horizontal displacements as a consequence of different phenomena anyhow linked to the dam height: the bidimensional nature of the consolidation of the silty tailings, especially in relation to the upstream construction method; the seismic cyclic loads again and again suffered by the dam during its life, significative but not enough to trigger a cyclic liquefaction; the settlement of the upper dam toe partly laying on the compressible silty tailings of the lower impoundment; the downstream movement of the toe developed on the sloping natural ground to compensate for temporary rises of the phreatic surface and the continuous increase of the weight of the embankment crest (Holubec 1976). Internal movements, monitored during the filling cycle of a tailings dam, resulted of the same order of magnitude of those back-calculated using the finite element code ZSOIL and the significance of these movements to overall stability of tailings dams has been already emphasized (Wahler 1974). The strain-induced internal displacements may then cause the liquefaction failure of the sandy tailings in a loose to medium dense state.

The height of the upper dam reached then in July 1985 a critical value that seems to have produced a too large unstable region or too large deformations or both at the same time: in any case a liquefaction phenomenon, due to excess pore water pressure or strain induced, developed and quickly propagated through the upper sandy shell that suddenly collapsed upon the lower impoundment.

All the existing silty tailings liquefied: those ones of the upper dam due to the undrained unloading and those ones of the lower dam due to the undrained loading.

The following flow-slide killed 269 people: a dreadful cost to be paid.

ACKNOWLEDGEMENTS

The authors are indebted to several individuals for the fruitful discussions and informations, but particularly to Prof. J. Brauns of the Karlsruhe University (Germany) and to the geologist D. Dovera.

LIST OF SYMBOLS

CAUTx: Anisotropically consolidated undrained triaxial test.
CIDTx: Isotropically consolidated drained triaxial test.
CIUTx: Isotropically consolidated undrained triaxial test.
CPT: Cone penetration test.
C_c: Compression index.
c_u: Undrained shear strength.
D_r: Relative density.
D_{10}: Effective size of the soil.
E: Undrained Young's modulus.
e: Void ratio.
SF: Safety factor.
h_d: Tailings dam height.
I_p: Plasticity index.

k: Coefficient of permeability.
c_v: Coefficient of consolidation.
M: q/p' at critical state.
p': Mean principal effective stress.
q: Deviator stress.
q_c: CPT tip resistance.
r: Correlation coefficient.
SB: Shear box test.
SD: Standard deviation.
S_r: Degree of saturation.
s_u: Undrained shear strength from field vane test.
UUTx: Unconsolidated undrained triaxial test.
u: Pore water pressure.
w_L: Liquid limit.
γ: Bulk unit weight.
γ_d: Dry unit weight.
ν: Poisson's coefficient.
σ_1, σ_3: Major and minor principal stresses.
ϕ': Friction angle in terms of effective stresses.
ϕ_d: Drained friction angle.

REFERENCES

Baldi, G., R. Bellotti, V. Ghionna, M. Jamiolkowski & E. Pasqualini 1982. Design parameters for sand from CPT. Proc. 2th ESOPT: 425-432. Amsterdam: Balkema.

Berti, G., F. Villa, D. Dovera, R. Genevois & J. Brauns 1988. The disaster of Stava. Proc. Spec. Conf. on Hydraulic Fill Structures, ASCE Geot. Spec. Pub. 21: 492-509.

Bishop, A.W. 1966. Geotechnical investigation into the causes and circumstances of the disaster of 21 Oct. 1966. London: Her Majesty's Stationery's Office.

Blight, G.E., M.J. Robinson & J.A.C. Diering 1981. The flow of slurry from a breached tailings dam. J.S. African Inst. Min. Metal. 1: 1-8.

Brawner, C.O. 1979. Design, construction and repair of tailings dams for metal mine waste disposal. ASCE Geot. Spec. Pub. : 53-87.

Campbell, D.D. & C.O. Brawner 1971. The tailings dam - an engineered structure. Western Miner.

Castro, G. 1969. Liquefaction of sands. Ph.D. Dissertation, Harvard University, Cambridge.

Dept. of Mines, Energy & Resources 1972. Tentative design guide for mine waste embankments in Canada. Mine Branch Res. Centre, Tech. Bull. TB145.

Dobry, R. & L. Alvarez 1967. Seismic failures of Chilean tailings dams. ASCE J. of the Soil Mech. & Found. Engng. Div.93 (SM6): 237-260.

Gibson, R.E. 1958. The progress of consolidation in a clay layer increasing in thickness with time. Geotechnique 8: 171-182.

Holubec, I. 1976. Geotechnical aspects of coal waste embankments. Can. Geot. J. 13: 27-40.

ICOLD 1982. Manual of tailings dams and dumps. Bull. 45: 1-238.

Ishihara, K., F. Tatsuoka & S. Yasuda 1975. Undrained deformation and liquefaction for sand under cyclic stresses. Soil and Foundation 15: 29-44.

ITASCA Consulting Group, Inc. 1989. Fast Lagrangian Analysis of Continua (FLAC) Manual. Minneapolis.

Kealy, C.D. 1973. Safe design for metal tailings dams. Min. Cong. I. 59 (1): 51-55.

Klohn, E.J. 1972. Design and construction of tailings dams. The Can. Min. Met. Trans, LXXV: 50-66.

Klohn, E.J. 1981. The development of current tailings dam design and construction methods. Design and construction of tailings dams, Colorado School of Mines & Klohn Leonoff Inc.: 1-52.

Kramer, S.L. & H.B. Seed 1988. Initiation of soil liquefaction under static loading conditions. ASCE J. of the Geot. Engng. Div. 114 (4): 412-430.

Ladd, C.C. 1988. Stability evaluation during staged construction. 22th Terzaghi lecture.

Lambe, T.W. & R.V. Whitman 1969. Soil Mechanics. New York: J. Wiley & Sons.

Leon, J.L. 1979. Consolidation of fine grained soils (slimes) in tailings dams built by the upstream method. Proc. 6th Panam. Conf. SMFE, vol. III: 119-131. Lima.

Mayne, P.W. & J.K. Mitchell 1988. Profiling of overconsolidation ratio in clays by field vane. Can. Geot. J. 25: 150-157.

Mesri, G. 1989. A reevaluation of s_u (mob) = 0.22 σ'_p using laboratory shear tests. Can. Geot. J. 26: 162-164.

Mitchel, J.K. & W.S. Gardner 1975. In situ measurement of volume change characteristics. 6th PSC ASCE 2: 279-345.

Mittal, H.K. & N:r: Morgenstern 1976. Seepage control in tailings dams. Can. Geot. J. 13: 277-293.

Morgestern, N.R. & A.A.G. Kupper 1988. Hydraulic fill structures - A perspective. Proc. Spec. Conf.on Hydraulic Fill Structures, ASCE Geot. Spec. Pub. 21: 1-31. Fort Collins, Colorado.

Okusa, S. & S. Anna 1980. Slope failures and tailings dam damage in the 1978 Izu-Oshima-Kinkai earthqake. Engineering Geology 16: 195-224.

Paulos, S.I. 1981. The steady state of deformation. ASCE J. of the Geot. Engng. Div. 107 (GT5): 553-562.

U.S. Bureau of Mines 1981. Mine waste disposal technology. Inf. Circ. 8857: 1-70.

Vick, S.T. 1983. Planning design and analysis of tailings dams. New York: J. Wiley & Sons.

Volpe, R.L. 1978. Engineering aspects of the 1972 Buffalo Creek Dam failure. Proc. 1st Int. Symp. Stability Coal Mining: 349-375. Vancouver.

Wahler, W:A. & Associated 1974. Evaluation of mill tailings disposal practice and potential dam stability problems in southwestern United States. U.S. Dept. of Interior, Bureau of Mines, Vol. 1-5, Usa.

ZACE Services & ZEI Eng. 1989. ZSOIL User Manual. Losanna, Washington.

Environmental Management, Geo-Water & Engineering Aspects, Chowdhury & Sivakumar (eds)
© 1993 Balkema, Rotterdam. ISBN 90 5410 099 0

Environmental geotechnology in the United States: A consultant's perspective

James V. Hamel
Hamel Geotechnical Consultants, Monroeville, Pa. & GTech, Inc., Pittsburgh, Pa., USA

ABSTRACT: Environmental geotechnology, which includes environmental geology and waste geotechnics, is simply a part of civil engineering. The development of environmental geotechnology in the United States over the past two decades is reviewed. Topics and work areas of current interest in the United States are identified. Most of these work areas are expected to remain strong over the next two decades. Anticipated changes in solid waste types and their treatment are noted along with other trends for the next two decades. Some undesirable trends are related to decreased emphasis on geologic training and data collection, limited training in geotechnical practice, and incomplete site characterization which often includes few, if any, geotechnical cross-sections. Other trends and changes, presently unforseen, are anticipated in environmental geotechnology over the next two decades. The future in this field will belong to those who are flexible and adaptive.

1 INTRODUCTION

1.1 Definitions and Philosophy

Environmental geotechnology means different things to different people, e.g., Moh (1977), Sembenelli and Ueshita (1981), Morgenstern (1985), Fang (1986). It is probably impossible to find or propose a general definition of environmental technology with which most people would feel comfortable. The writer favors the definition and related philosophy given by Caldwell and Hobbs (1987). They consider environmental geotechnology to include environmental geology and they define environmental geotechnology and geology as: "a planning tool; a set of problems (and the appropriate solutions) in the relevant disciplines; an attitude of mind; or simply the science and art of the interaction of soils and rocks with their surroundings."

In addition to this definition, Caldwell and Hobbs (1987) present a cogent philosophical overview of environmental geotechnology. Following Caldwell, et al (1986), they "view environmental geotechnology as an attitude of mind that permeates the processes of designing and constructing geotechnical structures... more than a mere tool, a new science, or a categorization scheme for a set of problems... the central aspect of the philosophy that guides professional engineering practice." In this regard, Caldwell and Hobbs (1987) note that environmental geotechnology is really just a part of civil engineering which has all along been concerned with improving and protecting the environment. The writer endorses this viewpoint as does Peck (1983, 1988).

In this paper, the writer will use the term "environmental geotechnology" in the broad sense of Caldwell and Hobbs (1987). To many people, particularly in the United States, this term connotes geotechnical aspects of waste disposal and waste remediation. Herein, the writer will use the term "waste geotechnics" (Gadsby, 1984) for this very important part of environmental geotechnology.

1.2 Perspective

The writer's perspective on environmental geotechnology and waste geotechnics is that of an individual consultant who has practiced in these areas as well as more traditional areas of geotechnical and geological engineering, primarily in the eastern United States, for the past two decades. During this time, he has lived and worked through most of the historic developments in environmental geotechnology and waste geotechnics, modifying his activities and practice according to trends and market conditions. This is the way of the consultant.

1.3 Evolution and Purpose of this Paper

When asked by Robin Chowdhury to present a lecture on environmental geotechnology at this conference, the writer was not particularly busy with consulting activities and the deadline for conference papers was months away. The writer envisioned a case history on a recently completed project that appeared well-suited to the conference theme (Hamel, et al, 1991; Hamel, 1992). Robin, however, had grander plans. He requested "a general section ... on the range of environmental geotechnical problems which are currently of concern and which engineers have to solve in the 1990's and beyond ... at least in the North American context." Along with this, he requested "general comments on the direction in which environmental geotechnics is heading." This was a tall order!

After considering these requests, exchanging letters and faxes with Robin, and reviewing only some of the pertinent literature, time for paper preparation was short and the writer had become exceedingly busy again with consulting activities. Despite some familiarity with environmental geotechnology in Canada (largely through the excellent Canadian Geotechnical Journal and Geotechnical News, the outstanding "Newsletter of the North American Geotechnical Community"), the writer knew little of environmental geotechnology in Mexico and other southern countries of North America. He therefore decided to restrict this paper to environmental geotechnology in the United States, which has a mature industrial society and ever-increasing environmental awareness. Many of the topics treated in this paper are expected, however, to have relevance elsewhere in North America and the world.

Consideration of Robin's requests for "the 1990's and beyond" suggested a need to project forward two decades. Since environmental geotechnology in the United States originated about two decades ago, it seemed appropriate to go back two decades and trace the development of environmental geotechnology up to the present time in order to project two decades forward.

Review of the voluminous literature that environmental geotechnology has spawned in the United States alone over the past two decades was impossible in the time available. The writer therefore decided to trace the development of environmental geotechnology over this period primarily through Geotechnical Specialty Conferences and Civil Engineering magazine of the American Society of Civil Engineers (ASCE). Additional pertinent literature was reviewed as availability and time permitted. Back issues of Geotechnical News which began in 1983 (BiTech Publishers Ltd., Suite 903, 580 Hornby Street, Vancouver, B.C., Canada V6C 3B6), were particularly useful in this regard.

1.4 Scope of this Paper

This paper has two main parts. The first part is a review of the evolution of environmental geotechnology (mainly waste geotechnics) in the United States over the past two decades. This review culminates with lists of the main areas of activity and concern in these fields at present (July 1992). The second part of this paper is a projection of environmental geotechnology in the United States over the next two decades. Major work areas are identified along with some perceived trends.

2 THE PAST TWO DECADES

2.1 Names and Name Changes

Circa 1970, "environmental geotechnology" and "environmental geology" were virtually unknown terms in the United States and probably in most of the rest of the world. Even "geotechnology" and "geotechnics" were little-used terms in the United States at that time (despite the efforts of Krynine and Judd, 1957), though these terms were more widely used in Europe and perhaps elsewhere (Glossop, 1968). When the writer established his first consulting company, Hamel Geotechnical Consultants, in early 1971, he considered the second word of its name to reflect the leading edge of appropriate terminology.

The first use of "environmental geology" that the writer recalls in the United States was in the title of a guidebook for a field trip associated with the 1971 Annual Meeting of the Geological Society of America (Thompson, 1971). This field trip in the vicinity of Pittsburgh, Pennsylvania, probably establishes the

Pittsburgh area as one pioneering in environmental geology.

Until the early 1970's, the "environmental" area of civil engineering was generally called sanitary engineering and the "geotechnical" area of civil engineering was generally called soil mechanics and foundation engineering, at least in the United States. Engineering geology then included most of what is now "environmental geology." Hydrogeology, now a prominent part of "environmental geology," was then considered to be groundwater geology or part of engineering geology.

Environmental awareness developed and environmental issues came to the forefront, with associated legislation and government regulation, in the United States and elsewhere during the late 1960's and early 1970's. Environmental geotechnology and waste geotechnics, as we presently know them, probably date from the early to mid-1970's in the United States.

ASCE has led in these trends. The name of the former Sanitary Engineering Division of ASCE was changed to Environmental Engineering Division in 1972. The former Soil Mechanics and Foundation Division of ASCE became the Geotechnical Engineering Division in 1973. Recently, it was proposed to change the name of this Division's Journal, but not the name of the Division itself, to Geotechnical and Geo-Environmental. It remains to be seen whether this latest name change will occur.

2.2 ASCE Conferences, 1970's to 1990's

The Geotechnical Engineering Division of ASCE has recognized the development of environmental geotechnology in its Specialty Conferences and certain other conferences since the 1970's. Specialty Conferences of particular importance in the development of environmental geotechnology in the United Stated include:

- Geotechnical Practice for Disposal of Solid Waste Materials (1977)
- Geotechnical Practice for Waste Disposal '87 (1987)
- Hydraulic Fill Structures (1988)
- Grouting, Soil Improvement and Geosynthetics (1992)
- Stability and Performance of Slopes and Embankments-II (1992)

Proceedings of these Specialty Conferences and other ASCE publications are available from the American Society of Civil Engineers, 345 East 47th Street, New York, New York 10017-2398, USA.

A "Special Symposium on Environmental Geotech" will be held at the ASCE International Convention and Exposition in New York in September 1992. Sixty papers will be presented in twelve sessions (Section 2.6). No proceedings will be published for this Symposium. Many, if not most, of the papers presented there have probably been published already or they will be published later.

Future ASCE Geotechnical Specialty Conferences relevant to environmental geotechnology include Geotechnical Practice in Dam Rehabilitation to be held at Raleigh, North Carolina, in April 1993 and a conference on Geo-Environmental Engineering to be held in New Orleans, Louisiana, in February 1995. The usual Proceedings can be expected from these Specialty Conferences.

2.3 Work in the Early to Mid-1970's

Much of the work in environmental geotechnology in the 1970's involved disposal of solid and liquid wastes from mining and mineral processing, i.e., mining waste geotechnics. This work resulted in part from world-wide expansion of the mining industry. The coal industry in the United States expanded greatly at that time in response to the oil embargo of 1973. Mining waste work also resulted in part from government regulations following disasters at Aberfan, Wales, in 1966 (Bishop, 1973) and Buffalo Creek, West Virginia, in the United States in 1972 (Davies, et al, 1972). Key references resulting from mining waste geotechnics of the 1970's include Current Geotechnical Practice in Mine Waste Disposal (ASCE, 1979) and Planning, Design, and Analysis of Tailings Dams (Vick, 1983).

In addition to mining and mineral processing wastes, much of the work in the 1970's was related to disposal of municipal refuse and to solid and liquid wastes from industrial processes, coal-fired power generation, and dredging. All of these areas of waste geotechnics are reflected in papers at the 1977 ASCE Specialty Conference Geotechnical Practice for Disposal of Solid Waste Materials. The papers were grouped into four sessions at that Conference:

A. Geotechnical Properties Affecting Disposal and Utilization
B. Utilization in Dams, Embankments, and Structural Fills
C. Environmental Problems: Seepage, Leachate, and Gas
D. Dewatering, Stabilization, and Reclamation

As indicated by these session titles, much of the work at that time was related to rather traditional geotechnical activities, e.g., determination of engineering properties of waste materials and construction of dams, embankments, and fills (Sessions A and B). "Environmental problems" related to leachates and gases were recognized in Session C but most of the papers there actually involved the more traditional geotechnical activities mentioned above. The same comment applies to most of the papers in Session D.

2.4 Work in the Late 1970's and 1980's

Environmental geotechnology broadened to include many non-traditional geotechnical areas in the late 1970's and through the 1980's but the emphasis remained on waste geotechnics. Some of the work during this time involved radioactive waste disposal, studies of which continue to the present time. The number of people working with radioactive wastes has been and continues to be only a small fraction of the number working in other areas of waste geotechnics.

Most of the expansion in waste geotechnics in the United States from the late 1970's to the present has resulted from greatly increased activity related to subsurface disposal of chemicals (including hydrocarbons) and remediation of sites, including surface water features and groundwater systems, contaminated by these materials. This work has been driven primarily by government regulations. Noteworthy in this regard are the Resource Conservation and Recovery Act of 1976 (RCRA); its Hazardous and Solid Waste Amendments of 1984 (HSWA); the Comprehensive Environmental Response, Compensation and Liability Act of 1980 (CERCLA), better known as the first Superfund law; and the Superfund Amendments and Reauthorization Act of 1986 (SARA). Numerous other laws and regulations have been promulgated by the federal government and state governments. Indeed, following and interpreting environmental legislation, regulations, and acronyms has become a full-time activity for some environmental geotechnologists and associated professionals in the United States and at least a part-time activity for many others.

This expanded scope of waste geotechnics was reflected in papers at the 1987 ASCE Specialty Conference Geotechnical Practice for Waste Disposal '87. Papers were grouped into nine sessions at that conference:

1. Regulatory Considerations
2. Site Characterization
3. Geotechnical Properties
4. Containment Structures
5. Landfill Liners
6A. Transport in the Ground
6B. In Situ Barriers
7. Monitoring and QA/QC (Quality Assurance/Quality Control)
8. Insurance and Legal Issues

In addition to these sessions, six workshops on specialized topics were given:

1. Geotechnical Properties of Waste
2. Flow Through Fractured Rock, Clay Fissures, and Discontinuities
3. Permeability of Compacted Clay
4. Slurried Mineral Waste
5. Landfill Site Stabilization
6. Seismic Considerations at Hazardous Waste Sites

All of these session and workshop topics from the 1987 Conference are still of considerable interest today and they are likely to be of interest well into the future.

2.5 Work in the 1990's

Trends and work areas of environmental geotechnology (mainly waste geotechnics) established in the United States by 1987 have generally continued to the present despite the cyclic nature of certain industries, e.g., mining, and the recession which began in 1990. This recession has affected environmental activities, including environmental geotechnology (mainly waste geotechnics), as well as other areas of our economy (Environmental Business Journal, April 1992).

Some new areas of activity have developed, however, again as results of government regulations. In response to the leaking underground storage tank (LUST) problem, the U.S. Environmental Protection Agency issued new standards for underground storage tanks in 1988; these standards will be phased in over ten years. This has created a vast new market for environmental geotechnology (Chedsey, 1988; Fairweather, 1990; Ouelette, 1992). Site assessments for real estate transfers also emerged as a major new work area in the late 1980's (Gerla and Jehn-Dellaport, 1989; Gustin and Neal, 1990) but this area is a "growing liability concern" (Geotechnical News, December 1991).

The "Impact of Environmental Issues on Geotechnical Engineers" circa 1990 was well-described by Collison (1990). Another key paper in this regard is "Are We Ready to Face the Geotechnical

Challenges of the 21st Century?" where Perez (1991) identified the following geotechnical challenges of the 21st century:

1. Restoration and Protection of the Environment
2. More Demanding Performance Criteria for Earth Structures (primarily in regard to waste geotechnics)
3. Better Characterization of Subsurface Conditions (primarily in regard to waste geotechnics)
4. Management and Interpretation of Spiraling Volumes of Data (particularly in regard to waste geotechnics; see, e.g., Duplancic and Buckle, 1989, 1992)
5. Improvement of Aging Infrastructure
6. Dealing with the Predicted Shortage of Engineers

All of these challenges involve environmental geotechnology and most of them involve waste geotechnics.

As noted above, two ASCE Geotechnical Specialty Conferences were held in 1992. Although neither Conference focused specifically on environmental geotechnology, both Conferences had numerous sessions and papers relevant to this field.

Sessions of particular interest at Grouting, Geosynthetics and Soil Improvement included several on grouting and geosynthetics plus one each on in situ waste stabilization and soil modification/environmental geotechnology. Keynote Lectures on "Remaining Technical Barriers to Obtain General Acceptance of Geosynthetics" (Koerner, et al, 1992) and "The Role of Soil Modification in Environmental Engineering Applications" (Mitchell and Van Court, 1992) merit special mention.

Sessions of particular interest at Stability and Performance of Slopes and Embankments-II included those on waste fills and repositories and site improvement/remediation. A large number of Invited Lectures were presented at this Conference. Those by Morgenstern (1992), Galster (1992), Sowers (1992), Mitchell and Mitchell (1992), Byrne, et al (1992), Seed and Bonaparte (1992), and Hausmann (1992) merit special mention.

This brief history of environmental geotechnology in the United States over the past two decades would be incomplete without mention of the major impacts of computers (Geotechnical News, December 1991; Christian, 1991) and geosynthetics (Raymond, 1986; Koerner, 1990; Grouting, Soil Improvement, and Geosynthetics, ASCE, 1992) on geotechnical practice during this time. Computer applications in environmental geotechnology can be expected to grow in the coming years, particularly as more "user friendly" software becomes available in areas such as groundwater modeling. The role of geosynthetics (and other man-made materials and components) can also be expected to grow in the years ahead.

2.6 Areas of Current Interest

Most of the work areas in waste geotechnics reviewed in Sections 2.4 and 2.5 for the period from 1987 to 1992 are still very important today. Session titles from the previously mentioned ASCE "Special Symposium on Environmental Geotech" (New York, September 1992) indicate areas of particular current interest in waste geotechnics:

Earthquake Response of Waste Containment Systems
Stabilized Waste to Construct Engineered Fills
Developments in Groundwater Investigation and Monitoring
In Situ Stabilization of Waste
Field Performance of Waste Containment Systems
Site Remediation: In Situ and On-Site Treatment Processes
Geosynthetics for Waste Containment Systems
Site Remediation: Waste Extraction
Strength and Stability of Solid Waste Landfills
Final Cover Systems
Vertical Containment and Collection Innovations
Geophysical Techniques for Site Characterization

The following areas are also of prime current interest in waste geotechnics:

Underground Storage Tanks
Bioremediation
Groundwater Flow and Contaminant Transport Modeling and Monitoring
Groundwater and Soil Gas Sampling without Well Installation (Chalfant, 1992)
Ground Modification Techniques applied to Waste Site Development and Remediation
Shear Strength of Interface between Geosynthetic Materials
Environmental Risk Assessment and Management

Within the broader area of environmental geotechnology, the following are additional areas of significant current interest in the United States:

Wetlands (see, e.g., Fortell and Johnson, 1992)
Historical and Archaeological Preservation
Natural Hazards

Emergency Response to Disasters, Spills, etc.

Maintenance, Rehabilitation, and Expansion of Existing Facilities and Infrastructure

River Bank, Lakeshore, and Coastal Instability and Erosion

Slope Stabilization and Erosion Control (including Biotechnical Measures; Gray and Leiser, 1982; Gray and Sotir, 1992)

Bridge Scour (see, e.g., Murillo, 1987)

Subsidence from Underground Mining, Fluid Withdrawal, and Karst

Some of these areas (e.g., wetlands, infrastructure rehabilitation, bridge scour) are generating more work than others.

3 THE NEXT TWO DECADES

3.1 General

Chronicling past developments and identifying areas of current interest are much easier than projecting future trends in environmental geotechnology. The writer is neither an oracle nor a soothsayer. He solicited input from a number of colleagues, reviewed some literature (both technical and popular), and attempted to see the future. The following projections are subjective and speculative to a certain extent and certainly not comprehensive. The only thing certain here is that the next two decades will see a great deal of change in environmental geotechnology, perhaps even more change than that experienced over the past two decades (Section 2). The future in environmental geotechnology will belong to those who can adapt swiftly to this change (Stasiowski, 1991).

3.2 Major Work Areas

Waste geotechnics will continue as a major work area over the next two decades and beyond (Environmental Business Journal, April 1992). Work related to development of new waste disposal sites (including waste water treatment facilities as well as facilities for land disposal of solid and liquid wastes) and closure of existing disposal sites will remain strong (Section 3.3). Work related to remediation of presently existing contaminated sites, e.g., Superfund sites in the United States, will eventually diminish. There are three reasons for this:

1. The finite number of presently contaminated sites will eventually be remediated to extents both acceptable and affordable (though monitoring may continue for a long time thereafter).
2. Under existing and future regulations few, if any, new contaminated sites should develop.
3. New standards allowing less than total cleanup or containment will evolve because of limited societal resources and shifting priorities for allocation of these resources.

Maintenance, rehabilitation, and expansion of existing facilities and infrastructure (including dams and reservoirs) will be a major work area in the United States and certain other parts of the world over the next two decades. Historical and archaeological preservation will be a small, but very important, part of this activity.

Environmental awareness and regulations, along with basic engineering requirements, will ensure that environmental geotechnology plays a major role in infrastructure rehabilitation. Evaluation, mitigation, and restoration of wetlands will be but one component of "Ecological Restoration" (Cohen, 1992) associated with some of the work in this area. Traditional geotechnical challenges will result from proximity of existing facilities and structures to those being maintained, rehabilitated, or expanded.

The United Nations established the 1990's as the International Decade for Natural Hazard Reduction and there has been considerable recent interest in both natural and man-made disasters (Morgenstern, 1985; Sowers, 1992; Geotechnical News, March and June 1992). The writer does not, however, expect natural hazard work to be a growth area in environmental geotechnology over the next two decades because of the sheer magnitude of the problem and limited available funding. Disasters will occur from time to time and response will be largely reactive rather than pro-active, despite lessons learned from past disasters (e.g., Voight, 1990).

Some modest amount of work is envisioned in one hazard-related field - risk assessment and establishment of probabilities for failures or malfunctions. Scenarios and probabilities of total or partial failure will be demanded by owners, regulators, and other decision makers as well as society as a whole. In particular, insurers will require these data in order to establish risk categories and insurance costs. Insurance appears the only practicable way to cover risk in monetary terms for both natural and man-made hazards ranging from localized landslide and mine subsidence phenomena to malfunctions of critical waste disposal facilities. Risk assessments for investment as well as insurance purposes are also expected for large scale resource development projects in areas of significant natural hazards, e.g., South Pacific region.

3.3 Solid Wastes in the Future

The types of solid waste disposal sites needed in the future will change with the mix of solid wastes generated. Currently, the largest quantities of such wastes in the United States are mining and mineral processing wastes. Industrial residual wastes are the second largest category. Coal combustion residues are a significant component of the residual waste stream. Municipal and hazardous wastes, though smaller in volume, have received greater publicity.

The future mix of waste will change as a result of current trends. Groundwater problems from past disposal practices have resulted in the ban of landfill disposal for many chemical wastes. Incineration, process changes, and recycling have been substituted. In the case of municipal wastes, recycling has begun and will continue to reduce the volume of material disposed.

Waste sites will be recycled along with waste materials. Mine and mill tailings (including coal wastes) and industrial wastes have been and will continue to be re-assessed for mineral or energy content. Similar re-use of other waste materials is expected. Revenues from re-use of wastes will help fund site remediation and re-use. For example, the writer is currently involved with remediation and closure of a specialty steel slag disposal area where it is anticipated that part of the cost of closure will be paid by specialty steel scrap recovered in the course of remediation. Increased activity in methane recovery, waste-to-energy conversions, and recycling technology suggests that today's municipal landfills will be tomorrow's sources of energy and raw materials. Environmental geotechnology will play a major role in excavation of disposal sites for re-use of waste materials and for replacement of these excavated materials with residuals from re-processing or combustion.

3.4 Coal in the Future

One area of increased future waste volume is the energy area. Past growth trends in global energy production are expected to continue (Starr, et al, 1992). Despite concerns regarding atmospheric emissions and other adverse environmental effects, energy use is projected to increase through the middle of the 21st century. These increases, driven by population and economic growth throughout the world, will occur even with substantial energy conservation.

Coal accounts for more than ninety percent of global fossil energy reserves and its use is expected to increase. As other energy resources (e.g., oil) are depleted, coal prices will increase and reserves that are presently marginal will be used increasingly. Many of these less desirable reserves have remained unexploited because of their higher content of impurities. Use of these lower grade reserves will increase the volumes of residual wastes generated during processing and combustion.

The trend toward development and use of lower grade coals will provide numerous challenges and opportunities in environmental geotechnology. In addition to the problems associated with development, mining, and reclamation are those associated with disposal of increased volumes of processing and combustion wastes. The case study presented by Hamel, et al (1991) and Hamel (1992a) involved a project where new materials and technologies were used to upgrade and expand an existing coal processing waste disposal site. This project is typical of those that will be faced increasingly in the future with regard to waste disposal facilities and other components of our infrastructure.

3.5 Other Trends and Perceptions

Trenchless construction (O'Rourke, et al, 1985) employing microtunneling (Thompson, 1986), pipe jacking, and slip lining (insertion of a new pipe or liner inside an existing pipe) will be used much more extensively in the future, particularly in rehabilitation of infrastructure. Applications of these methods will reduce costs as well as disturbance to the ground surface and existing facilities.

Design-build activities, where one organization acts as both engineer and constructor (Denning, 1992), will increase in the future in waste geotechnics and many other areas (Stasiowski, 1991). Use of independent environmental monitors (Dodds and Sternberger, 1992) will increase for design-build projects as well as more conventional design-bid-build projects. Significant challenges and opportunities in environmental geotechnology are expected to result from both of these trends.

A broad class of problems involving hydraulic and geotechnical interactions related to moving water and their effects on soils is expected to receive more attention in research as well as engineering practice over the next two decades. In addition to the obvious problems related to waste geotechnics, e.g., those associated with contaminated sediments and groundwater, problems in this broad class include:

Earth dam overtopping phenomena, e.g., embankment erosion, armoring for flood overflow, fuse plug spillways

River bank, lake shore, and coastal erosion and instability

Soil erosion from the scale of construction sites to major watersheds

Dredging

Hydraulic transport of slurried solids

Hydraulic fill structures

Debris flows

Bridge scour

Dealing with these problems will require broad perspectives as well as extensive collaboration of geotechnical engineers and engineering geologists with hydraulic engineers, fluvial geomorphologists, soil scientists, and others. As emphasized by Peck (1983), "Nature Ignores Specialties."

3.6 Some Undesirable Recent Trends

In regard to the above-mentioned work areas and problem areas in environmental geotechnology, several undesirable trends have become apparent in the United States over the past few years:

1. Removal of geology from civil engineering curricula
2. Decreased amounts of geologic mapping and hydrologic data collecting by governmental agencies
3. Reduced numbers of geotechnical and geological personnel in review agencies and certain other governmental organizations
4. Limited potential for on-the-job training of the present generation of professionals, especially in the fundamentals of geotechnical practice
5. Lack of appreciation by this present generation of professionals for the overall geotechnical framework of a site or problem and the importance of geotechnical cross-sections in portraying this framework

Despite the admonitions of Peck (1973, 1983), Legget (1979), Galster (1992), Sowers (1979; personal communication, 1992) and others, there has been a tendency to eliminate geology from civil engineering curricula in the United States. This trend should be reversed. Geology is as basic to civil engineering as, e.g., mechanics; most civil engineering structures are constructed in or on earth materials and subjected to geologic processes. Geology is fundamental to geotechnical engineering practice (Hamel, 1983) and crucial in environmental geotechnology. The writer recommends that all civil engineers have at least one course in physical or engineering geology and that geotechnical engineers, including those expecting to work in environmental geotechnology, take as many courses in geology (including field trips) and do as much geologic field work as possible. In most geotechnical work, including environmental geotechnology, the problems and the solutions are both found in the field. The more training in geology, the easier it is to recognize both problems and solutions.

The amount of geologic mapping and hydrologic data collection by the U.S. Geological Survey, state geological surveys, and other agencies has been reduced over the past several years. Funding for these activities has decreased as a result of shifting governmental priorities, budget deficits, and the recent economic recession. The writer recognizes these economic realities but notes that fundamental geologic and hydrologic data are of paramount importance in dealing with most problems of environmental geotechnology.

Also as a result of shifting governmental priorities and reduced funding, there has been a general loss of geotechnical and geological personnel in regulatory agencies and certain other governmental organizations. Personnel who retire or leave (e.g., for higher salaries and greater challenges and opportunties in geo-environmental consulting, operations, or construction) are often not replaced. Few new people are hired, despite greatly increased amounts of review and other work. This makes for an overall lack of geological/geotechnical awareness in many agencies and organizations and extremely lengthy review schedules.

A number of factors have come together to reduce both on-the-job training and overall practical training of the next generation of geotechnical professionals in the United States. Most of these people receive reasonable training on fundamentals in their academic programs, despite the general lack of practical experience of most instructional personnel. In the past, this academic training was built upon through on-the-job training both in the office and in the field. Numerous civil, mining, and geotechnical engineering projects of various sizes and complexities provided opportunity for young engineers to learn the fundamentals of their profession including problem definition (Mirza, 1992), investigation, design, construction, and operation, i.e., how projects evolved and were carried through to completion and productive use.

Because of the demand for personnel and the corresponding high salaries in waste geotechnics, many recent graduates have entered this field. The nature and flow of work in much of waste geotechnics, particularly in remediation of contaminated disposal sites, is much different from that in more traditional engineering areas. Government regulations and review and seemingly endless cycles of RI/FS (remedial investigations and feasibility studies) generate voluminous quantities of paper that often contain little of geotechnical substance. Doing these studies consumes vast amounts of time and defers detailed design and construction, both of which provide the real opportunities to hone geotechnical skills. As a result, many of our younger geotechnical personnel working in waste geotechnics are really not learning much in the way of geotechnical fundamentals.

The training situation is not much better for the lesser numbers of younger personnel working in areas other than waste geotechnics. The large design and construction projects (mining, processing, and industrial facilities; transportation systems; dams; etc.) that formerly provided avenues for training are few and far between these days as a result of economic and other conditions which have evolved over the past decade. Geotechnical work related to rehabilitation of existing facilities provides some training opportunities but these projects often have narrow focus as well as high levels of geotechnical complexity such that they are not well suited for training purposes.

The writer expects that this overall situation will guarantee a demand for services from him and from other members of his generation for as long as they choose to practice geotechnical engineering. The future does not look so favorable, however, for the geotechnical profession as a whole. Without sound training in fundamentals of practice, including geology, construction, and precedents, our successors may spend scarce money in the wrong places and experience other difficulties in many areas of environmental geotechnology including waste geotechnics and infrastructure rehabilitation.

Along with the above-mentioned reduction of training in fundamentals of geotechnical practice is a growing tendency, particularly among younger or newer geotechnical practitioners, for incomplete site characterization. This is particularly common in many areas of waste geotechnics where greater emphasis is often placed on characterization of waste materials and, in some cases, existing contamination than on characterization of the site itself. The increased use of computers for analysis and drafting has contributed to this trend.

The writer has used the term "geotechnical framework" for those components of site geology and project geometry required for site and problem characterization (Hamel, 1983, 1987, 1988, 1992b). With regard to waste geotechnics, the geotechnical framework of a site includes features significant from the standpoint of waste interaction and contaminant migration as well as design, construction, operation, maintenance, and abandonment (Hamel and Ferguson, 1985).

Cross-sections are one of the most useful tools available for characterizing the geotechnical framework of a site or problem. The writer has noticed a general decrease in the quantity and quality of geotechnical cross-sections prepared for projects he has been called upon to review in the United States and elsewhere over the past few years. In fact, some major and complex projects which he has reviewed recently had only one or two geotechnical cross-sections or even none at all. The shortage or absence of geotechnical cross-sections has been most prevalent on projects in waste geotechnics but it has been encountered on all types of projects.

The writer's views on geotechnical cross-sections will be presented in more detail elsewhere. Here he simply notes that, for purposes of development of the geotechnical framework of a site or a problem, cross-sections should be drawn as large as practicable with equal horizontal and vertical scales and they should depict all relevant project features - subsurface, surface, and construction-related. In addition, most projects require development of a considerable number of cross-sections in order to determine which ones are most informative or descriptive, or both, e.g., for inclusion in reports. This is because the "grain" or principal directions(s) of critical site conditions and project features are not always readily apparent or necessarily encountered in the first few cross-sections.

The writer hopes that these undesirable trends will be reversed over the next two decades, but he is not optimistic in this regard.

4 SUMMARY

Environmental geotechnology, a broad field including aspects of geology, geotechnical engineering, and many other disciplines, can and probably should be considered simply a part of the even broader field of civil engineering. Waste geotechnics is presently a major area of environmental geotechnology in the United States and many other countries.

Review of the history and development of environmental geotechnology and waste geotechnics in the United States over the past two decades indicates that these areas have experienced phenomenal growth during this relatively short time interval. This growth was driven by increased environmental awareness and concern and, most particularly, by government regulations.

Topics and work areas of current interest in environmental geotechnology and waste geotechnics in the United States were identified on the basis of recent literature and the writer's professional practice. Most of these work areas, particularly those related to waste geotechnics and infrastructure rehabilitation, are expected to remain strong for the next two decades.

Anticipated changes in solid waste types over the next two decades were reviewed. Wastes from processing and combustion of coal are anticipated to increase over the next two decades as more coal and lower grade coal is used. Disposal of these wastes, plus development, mining and reclamation of these coal reserves will provide numerous challenges and opportunities in environmental geotechnology.

Other trends and perceptions for the next two decades were also reviewed. A broad class of problems involving interactions of soil and moving water was identified and several problems in this area were listed.

Some undesirable recent trends in environmental geotechnology in the United States were also noted. Most of these trends are related to decreased emphasis on geologic training and data collection, reductions in geological/geotechnical staffs in government agencies, limited potential for on-the-job training in geotechnical practice, and incomplete site characterization which often includes few, if any, geotechnical cross-sections.

The history of environmental geotechnology in the United States and elsewhere over the past two decades was one of great change. The only thing certain about environmental geotechnology over the next two decades is that other great changes, many of them presently unforeseen, will occur. The future in this field, as in other fields of business and technology, will belong to those who are flexible and adaptive.

5 ACKNOWLEDGEMENTS

The writer has been fortunate to live and practice for the past two decades near Pittsburgh, Pennsylvania, one of the world's great centers of civil, mining, geotechnical, and environmental activity. During this time, he has benefited from numerous technical society meetings and activities in Pittsburgh and many discussions with members of the large and diverse Pittsburgh technical community. The following members of this community provided oral or written contributions that assisted in preparation of this paper:

William R. Adams, Jr. - Pennsylvania Department of Transportation

Lawrence D. Busack - Pennsylvania Department of Environmental Resources

Ronald L. Frew - Dames & Moore

Donald V. Gaffney - Michael Baker, Jr., Inc.

Richard E. Gray - GAI Consultants, Inc.

Robert J. McLaren - Consulting Engineer

Linda S. Paul - McLaren Hart Environmental

Jack T. Roseman - PSI-Pittsburgh Testing Laboratory

George F. Sowers of Law Engineering, Atlanta, Georgia; Joseph P. Welsh of GKN Hayward Baker, Odenton, Maryland; and Barry Voight of The Pennsylvania State University, University Park, Pennsylvania, provided useful information. The late Harry F. Ferguson (Hamel, 1991) also contributed to this paper in many ways.

The writer's son, Omar, a philosophy student at The Pennsylvania State University, provided philosophical guidance and lexicographic assistance in preparation of this paper. The writer's wife, Betsy, a consultant with Hamel Geotechnical Consultants and GTech, Inc., and an ardent environmentalist, assisted in many ways during preparation of this paper and typed its first draft. Later drafts were typed by Kathleen Donohue and Ruth Konopka of GTech, Inc.

The writer thanks all of the above persons who assisted with this paper and also Robin N. Chowdhury of the University of Wollongong who encouraged its preparation.

REFERENCES

Bishop, A. W. 1973. The Stability of Tips and Spoil Heaps. Quarterly Journal of Engineering Geology 6(3/4):335-376.

Byrne, R.J., Kendall, J. and Brown, S. 1992. Cause and Mechanism of Failure, Kettleman Hills Landfill B-19, Phase IA. Stability and Performance of Slopes and Embankments-II, ASCE, New York: 1188-1215.

Caldwell, J.A. and Hobbs, B.T. 1987. Environmental Geotechnology & Tailings Reclamation. Geotechnical Practice for Waste Disposal '87, ASCE, New York: 362-376.

Caldwell, J.A., Kiel, J. and Thatcher, J. 1986. Environmental Geotechnological Considerations in Design of the Cannon Mine Tailings Impoundment. International Symposium on Environmental Geotechnology, Allentown, Pennsylvania.

Chalfant, C. 1992. Hydropunch and Soil Gas Probes Simplify Groundwater Assessments. The National Environmental Journal 2(3): 24-27.

Chedsey, G.C. 1988. EPA Creates 100,000 to 400,000 New Clients for Consulting Firms. Geotechnical News 6(4):36.

Christian, J.T. 1991. Geotechnical Design and Analysis in the Age of the Modern Computer. Geotechnical Engineering Congress 1991, ASCE, New York: 468-478.

Cohen, T. 1992. Ecological Restoration. Technology Review 95(2):20-21.

Collision, G.H. 1990. Impact of Environmental Issues on Geotechnical Engineers. Geotechnical News 8(3):35-38.

Davies, W.E., Bailey, J.F. and Kelly, D.B. 1972. West Virginia's Buffalo Creek Flood: A Study of the Hydrology and Engineering Geology. U.S. Geological Survey Circular 667.

Denning, J. 1992. Design-Build Goes Public. Civil Engineering 62(7):76-79.

Dodds, P.J. and Sternberger, R.S. 1992. The Evolution of an Environmental Monitor. Civil Engineering 62(6):56-58.

Duplancic, N. and Buckle, G. 1989. Hazardous Data Explosion. Civil Engineering 59(12):6-70.

Duplancic, N. and Buckle, G. 1992. Taming Environmental Data. Civil Engineering 62(8):56-58.

Environmental Business Journal. April 1992. 5(4):1-11.

Fairweather, V. 1990. U.S. Tackles Leaking Tanks. Civil Engineering 60(12):47-49.

Fang, H.Y. 1986. General Report. International Symposium on Environmental Geotechnology, Allentown, Pennsylvania.

Gadsby, J.W. 1984. Waste Geotechnics. Geotechnical News 2(1):18.

Galster, R.W. 1992. The Role of Engineering Geology in Slope and Embankment Stability Analysis. Stability and Performance of Slopes and Embankments-II, ASCE, New York:70-94.

Gerla, P.J. and Jehn-Dellaport, T. 1989. Environmental Impact Assessment for Commercial Real Estate Transfers. Bulletin, Association of Engineering Geologists 26(4):531-540.

Glossop, R. 1968. The Rise of Geotechnology and its Influence on Engineering Practice. Geotechnique 18(2):105-120.

Gray, D.H. and Leiser, A.T. 1982. Biotechnical Slope Protection and Erosion Control. Van Nostrand Reinhold, New York.

Gray, D.H. and Sotir, R.B. 1992. Biotechnical Stabilization of Cut & Fill Slopes. Stability and Performance of Slopes and Embankments-II, ASCE, New York:1395-1410.

Gustin, J.D. and Neal, L.A. 1990. Site Assessments. Civil Engineering 60(8):53-55.

Hamel, J.V. 1983. Fundamentals of Geotechnical Engineering Practice. Geotechnical News 1(2):12-13.

Hamel, J.V. 1987. Geological and Geomorphological Investigations for Cultural Resource Evaluations. Proc. 38th Annual U.S. Highway Geology Symposium, Pittsburgh, Pennsylvania:73-80.

Hamel, J.V. 1988. Geotechnical Aspects of Alluvial Bank Degradation. Proc. 19th Annual Conference International Erosion Control Association, New Orleans, Louisiana:411-419.

Hamel, J.V. 1991. Harry F. Ferguson (1921-1989), Honorary Member Association of Engineering Geologists. Bulletin, Association of Engineering Geologists 28 (1):1-4.

Hamel, J.V. 1992a. Slip Lining a Decant Structure. Coal Prep 92, Maclean Hunter, Aurora, Colorado:200-214.

Hamel, J.V. 1992b. Stability Evaluations for Old Water Supply Dams in Pennsylvania. Stability and Performance of Slopes and Embankments-II, ASCE, New York:1050-1065.

Hamel, J.V., Bartsch, D.L. and Povirk, R.A. 1991. Rehabilitation of Decant Structure for Coal Slurry Impoundment in Ohio. Environmental Management for the 1990's, SME-AIME, Littleton, Colorado:179-188.

Hamel, J.V. and Ferguson, H.F. 1985. Geotechnical Framework of Hazardous Waste Sites. International Symposium on Management of Hazardous Chemical Waste Sites, Winston-Salem, North Carolina, Abstracts & Program:87-88.

Hausmann, M.R. 1992. Slope Remediation. Stability and Performance of Slopes and Embankments-II, ASCE, New York:1274-1317.

Koerner, R.M. 1990. Designing with Geosynthetics, 2nd ed. Prentice Hall, Englewood Cliffs, New Jersey.

Koerner, R.M., Hsuan, Y. and Lord, A.E., Jr. 1992. Remaining Technical Barriers to Obtain General Acceptance of Geosynthetics. Grouting, Soil Improvement and Geosynthetics, ASCE, New York:63-109.

Krynine, D.P. and Judd, W.R. 1957. Principles of Engineering Geology and Geotechnics. McGraw-Hill, New York.

Legget, R.F. 1979. Geology and Geotechnical Engineering. Journal of Geotechnical Engineering Division, ASCE, 105(GT3):342-391.

Mirza, C. 1992. Educating Engineering Students. Geotechnical News 10(1):66-68.

Mitchell, J.K. and Van Court, W.A. 1992. The Role of Soil Modification in Environmental Engineering Applications. Grouting, Soil Improvement and Geosynthetics, ASCE, New York:110-143.

Mitchell, R.A. and Mitchell, J.K. 1992. Stability Evaluation of Waste Landfills. Stability and Performance of Slopes and Embankments-II, ASCE, New York:1152-1187.

Moh, Z.C. 1977. Geotechnical Engineering and Environmental Control. Proc. 9th ICSMFE 4:559-561.

Morgenstern, N.R. 1985. Geotechnical Aspects of Environmental Control. Proc. 11th ICSMFE 1:155-185.

Morgenstern, N.R. 1992. The Evaluation of Slope Stability - A 25-Year Perspective. Stability and Performance of Slopes and Embankments-II, ASCE, New York:1-26.

Murillo, J.A. 1987. The Scourge of Scour. Civil Engineering 57(7):66-69.

O'Rourke, T.D., Flaxman, E.W. and Cooper I. 1985. Pipe Laying Comes Out of the Trenches. Civil Engineering 55(12):48-51.

Ouelette, R.P. 1992. The Underground Storage Tank Market. The National Environmental Journal 3(1):14-17.

Peck, R.B. 1973. The Direction of Our Profession. Proc. 8th ICSMFE 4:156-159.

Peck, R.B. 1983. Nature Ignores Specialties. Geotechnical News 1(1):12-15.

Peck, R.B. 1988. Nature & The Civil Engineer. Geotechnical News 6(2):6-8.

Perez, J-Y. 1991. Are We Ready to Face the Geotechnical Challenges of the 21st Century? Geotechnical Engineering Congress 1991, ASCE, New York:1-13.

Purcell, L.J. and Johnson, T.D. 1992. Creating Wetlands. Civil Engineering 62(8):36-37.

Raymond, G.P. 1986. Geosynthetics - A Short History. Geotechnical News 4(1):35-36.

Seed, R.B. and Bonaparte, R. 1992. Seismic Analysis and Design of Lined Waste Fills: Current Practice. Stability and Performance of Slopes and Embankments-II. ASCE, New York:1521-1545.

Sembenelli, P. and Ueshita, K. 1981. Environmental Geotechnics - State-of-the-Art Report. Proc. 10th ICSMFE 4:335-394.

Sowers, G.F. 1979. Introductory Soil Mechanics and Foundations: Geotechnical Engineering. Macmillan, New York.

Sowers, G.F. 1992. Natural Landslides. Stability and Performance of Slopes and Embankments-II; ASCE, New York:804-833.

Stasiowski, F.A. 1991. 10 Decatrends for the 90's. Professional Services Management Journal, 2 pp.

Starr, D., Searl, M.F. and Alpert, S. 1992. Energy Sources: A Realistic Outlook. Science 256:981-987.

Thompson, J.C. 1986. Microtunneling: No Pipe Dream. Civil Engineering 56(8):56-59.

Thompson, R.D., ed. 1971. Environmental Geology in the Pittsburgh Area. Guidebook for Field Trip No. 6. Annual Meeting of Geological Society of America, 41 pp.

Vick, S.G. 1983. Planning, Design, and Analysis of Tailings Dams. Wiley, New York.

Voight, B. 1990. The 1985 Nevado del Ruiz Volcano Catastrophe: Anatomy and Retrospection. Journal of Volcanology and Geothermal Research 44:349-386.

Environmental Management, Geo-Water & Engineering Aspects, Chowdhury & Sivakumar (eds)
© 1993 Balkema, Rotterdam. ISBN 90 5410 099 0

An approach to seismic microzonation for environmental planning and management

H.G. Poulos
Coffey Partners International Pty Ltd & The University of Sydney, N.S.W., Australia

ABSTRACT: This paper outlines a procedure for seismic microzonation for environmental planning and management, in the broadest sense, to assist in the planning, construction and management of projects. The procedure involves eight steps, each of which is discussed in detail. The steps involve geological, seismological and geotechnical input, and the end-point is the appropriate portrayal of the risk of specific seismic hazards for the selected area. Applications of this approach are illustrated with respect to both a discrete site and to a large regional area.

1 INTRODUCTION

Seismic microzonation may be defined as the assessment of variations in earthquake hazards that result from differing foundation conditions within a larger area that is prone to seismic activity. Earthquake hazards refer to those effects which have the potential to cause harm to people, property or the environment, and may include the following:

1) structural damage arising from additional forces induced by the earthquake
2) liquefaction of sandy soils
3) slope instability
4) failure of earth retaining structures
5) avalanches
6) differential ground settlements.

Seismic microzonation has usually been undertaken to assist in the assessment of risks associated with earthquakes and the mitigation of those risks. It is a valuable adjunct to regional planning and in recent years has been undertaken widely in countries where earthquake risks are relatively high, e.g. Southern Asia (Hattori, 1980), Japan (Sugimura et al, 1984), New Zealand (McCahon et al, 1992; Elder et al, 1992), Italy (Vannuchi, 1991), and Yugoslavia (Mihailov, 1984).

In comparison with many other countries, the risk of serious seismic hazards in Australia is relatively low. However, the 1989 Newcastle earthquake has increased the awareness of engineers and planners of the need to properly assess the possible consequences of earthquakes on both existing and planned construction and environmental projects. Consequently, there appears to be a need to develop consistent procedures for seismic microzonation to assist in the planning, construction and management of projects.

The main objective of this paper is to outline a procedure for seismic microzonation for environmental planning and management in the broadest sense. This procedure involves a combination of both earthquake engineering and geotechnical engineering techniques which are not excessively complicated and which may be implemented with a relatively meagre amount of data. Examples are given of the use of such a procedure with respect to a single site and to a larger regional area.

2 METHODOLOGY

The microzonation process involves a number of component steps, including:

(a) selection of design hazard level
(b) regional geological studies
(c) evaluation of subsurface profiles in the area of interest
(d) estimation of the stiffness and damping characteristics of the subsurface soils
(e) assessment of the characteristics of the design earthquake
(f) analyses of seismic site response at

various locations within the area of interest

(g) assessment of the seismic hazards, including:

i) liquefaction potential

ii) structural damage due to site amplification of earthquake bedrock motions

iii) potential for slope instability

(h) the appropriate portrayal of the above assessment, for example, in the form of seismic hazard maps, damage potential plots, or detailed site-specific information.

Each of these steps is discussed in detail in the following sections.

3 CHOICE OF HAZARD LEVELS

A necessary first step in seismic hazard assessment is to choose an appropriate hazard level. The hazard level will influence the design earthquake and will provide a basis for quantitative risk assessment. This level will depend on the consequences of failure or collapse of the structures or facilities within the area of interest, and as pointed out by Skipp and Ambraseys (1987), this choice is often based on judgement and consensus. A reasonable approach which is described by those authors is to consider two seismic input design earthquakes:

1) an earthquake which has a "reasonable" likelihood of affecting the structure during its lifetime, e.g. the earthquake having a mean return period equal to the selected "life" of the structure. The probability that it will occur during this period is 63% (assuming a Poissonian process). The engineering objective of this analysis is to retain serviceability following the earthquake;

2) a higher level earthquake, with a much lower probability of being exceeded during the life of the structure, e.g. 10% or less. Here, the engineering objective is to avoid collapse, even though the structure may be unrepairable.

Where dangerous industrial processes can expose humans to life threatening consequences, earthquake hazard at a much lower probability level may be appropriate. Skipp and Ambraseys suggest that, for nuclear power plants or liquefied gas storage, appropriate levels would be 10^{-2}/annum exceedance for the "operational" event, and 10^{-4}/annum exceedance for the extreme earthquake.

4 REGIONAL GEOLOGICAL STUDIES

An essential first step of any major geotechnical investigation is a study of the regional geology. With respect to seismic risk assessment, the following geological aspects are of particular importance (Bolt, 1986):

1) regional tectonics and patterns of deformation

2) mapping of significant capable faults within 100-150 km of the site being assessed

3) determination of fault types (e.g. strike-slip, dip-slip)

4) evidence for and against recent displacements along faults

5) location of any landslide, ground settlement, or water-inundation problems which may be attributable to previous seismic activity.

More detailed information on such studies is given in several references, e.g. Wiegel (1970), Elder et al (1991), and Vannuchi (1991).

It is usual to define earthquake source "zones" or "provinces" from which earthquakes may emanate. These are generally based on geological, tectonic and seismological information, and in particular, data on the epicentral locations of recorded earthquakes. An example of this process is described by Rynn (1988) for north-eastern Australia.

5 EVALUATION OF SUBSURFACE PROFILES

The primary source of geotechnical subsurface information is borehole data, although geophysical information can sometimes provide useful supplementary information between borehole locations. Where studies are to be carried out on a single relatively small site, the geotechnical characterisation of the subsoil profile is relatively straightforward, as detailed borehole data will usually adequately cover the area of interest. However, if the studies are of a regional data, a different, and more "broad brush," approach must be adopted. An example of such an approach is as follows:

1) the area of interest is divided into a number of "blocks" of equal area; the size of these will depend to a large extent on the total area to be covered and the geological nature of the region. Typical blocks used in regional studies are 470 m x 370 m, used for a study of Tokyo by Sugimura and Ohkawa (1984), and

500 m x 500 m, used for Christchurch by McCahon et al (1992) and Elder et al (1991);

2) the available borelog data is assembled for the area and the borelog locations plotted on a scale map. If there are large areas in which borehole data are absent, it is desirable to undertake additional drilling if possible;

3) using these borelog data, the soil profile is simplified by classifying the soil types into a discrete number of groups e.g. as adopted by Elder et al (1991). An alternative approach utilising "soil type factors" is described by Sugimura and Ohkawa (1984);

4) a series of depth ranges beneath the ground surface is chosen, and the representative soil type in each depth range is assessed for each borelog. Maps, showing the spatial distribution of representative soil types for each depth range, are then prepared;

5) for each grid, the average or representative soil type is assessed from these maps, so that a simplified geotechnical model is developed to represent each grid.

One of the key elements in developing geotechnical profiles is to select the boundary between soil and firm ground or rock. This decision, which may be somewhat arbitrary or difficult to make, can have a significant influence on the computed seismic response.

In most cases, the primary quantitative information on soil properties will be some form of in-situ penetration data, either Standard Penetration data (SPT) or static Cone Penetration data (CPT). Thus, quantification of relevant engineering properties usually relies on correlation with this data. Such correlations are discussed in the following section.

6 ESTIMATION OF STIFFNESS AND DAMPING CHARACTERISTICS OF SOILS

There are many correlations between dynamic stiffness and damping values of soil, and SPT or CPT data, among these being those of Seed et al (1984), Sun et al (1988), Imai and Tonouchi (1982), and Ohta and Goto (1978). The latter relate initial shear modulus G_0 to N (the SPT value), depth, soil type and geological origin, as shown in Figure 1. Because of the non-linearity of soil behaviour, it is essential to allow for the dependence of shear modulus on cyclic shear strain, and a convenient approach is to use the

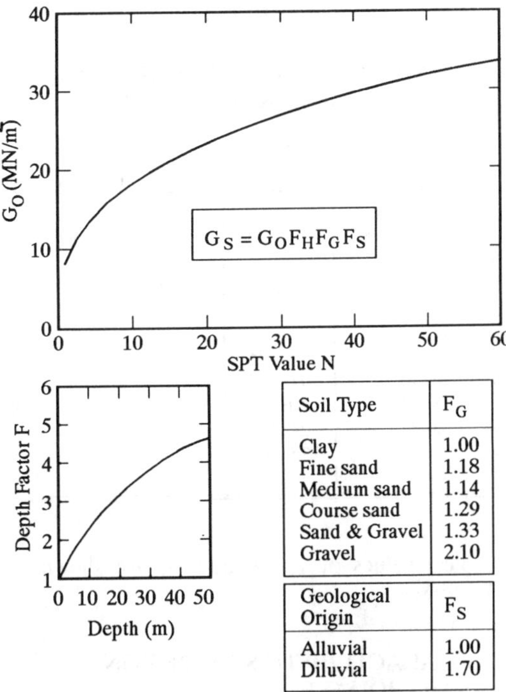

FIG.1 EMPIRICAL CORRELATION FOR INITIAL DYNAMIC SHEAR MODULUS vs SPT
(after Ohta and Goto, 1978)

following relationship developed by Hardin and Drnevich (1972):

$$G = \frac{G_o}{1+\gamma_c/\gamma_r} \qquad ...(1)$$

where G = shear modulus
G_0 = initial shear modulus (at very small strain)
γ_c = cyclic shear strain
γ_r = reference shear strain.

Similarly, the dependence of damping ratio on shear strain must be recognised, and the following relationship has been used (Poulos, 1991):

$$D = D_o + \frac{D_1\gamma_c/\gamma_r}{1+\gamma_c/\gamma_r} \qquad ...(2)$$

where D_0 = damping ratio for very small strains (e.g. $< 10^{-5}$)
D_1 = strain-dependent reference damping component.

TABLE 1

TYPICAL VALUES OF STIFFNESS AND DAMPING PARAMETERS

Soil Type	γ_r	D_0	D_1
Gravelly soils (rel.dens. ≈80%)	1.3x10^{-4}	0	0.260
Quartz sands	3.7x10^{-4}	0.005	0.260
Clays			
PI 5-10	4.0x10^{-4}	0.025	0.265
10-20	7.0x10^{-4}	"	"
20-40	1.1x10^{-3}	"	"
40-80	2.0x10^{-3}	"	"
>80	3.6x10^{-3}	"	"

Typical values of γ_r, D_0 and D_1 are shown in Table 1.

7 CHARACTERISTICS OF DESIGN EARTHQUAKE

The major characteristics of the design earthquake which must be assessed are:

1) the likely Richter magnitude
2) the seismicity rate
3) the maximum bedrock acceleration, and its attenuation with distance from the source fault
4) the duration
5) the predominant period
6) the time-acceleration relationship at bedrock level.

A useful summary of some earlier information on these characteristics has been provided by Seed et al (1969).

7.1 *Earthquake magnitude*

It is generally accepted that the Richter magnitude of an earthquake is related to the length of a fault over which slip occurs. Figure 2 shows an idealised relationship proposed by Housner (1964), a relationship proposed for Australian conditions by McCue (1990), and five points for earthquakes which have occurred in the 20th century. There is generally reasonable agreement among these data, despite the fact that the Australian data relates to intra-plate earthquakes whereas the other data relates to earthquakes along plate boundaries.

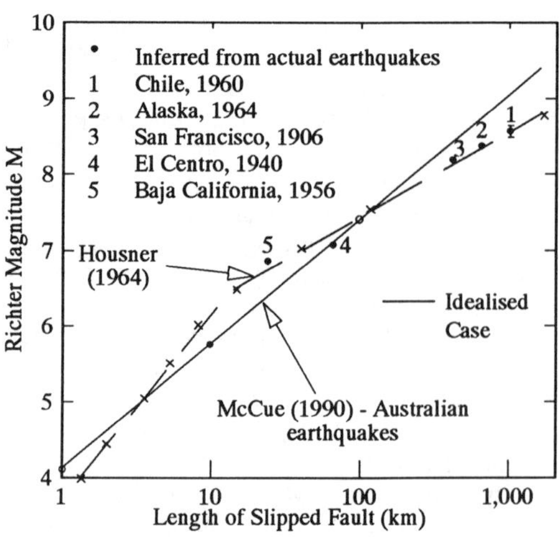

FIG.2 RELATIONSHIPS BETWEEN LENGTH OF FAULT AND RICHTER MAGNITUDE

McCue (1990) has also related the earthquake magnitude to the maximum height of the fault scarp.

7.2 *Seismicity rate*

The seismicity rate refers to the frequency of occurrence of an earthquake of a particular magnitude within the region of interest. It is usually described by the following relationship developed by Gutenberg and Richter (1954):

$$\text{Log } N = A - bM_L \qquad ...(3)$$

where N = number of earthquakes within a time interval with magnitudes greater than or equal to M_L

A = number of earthquakes exceeding magnitude zero

b = recurrence parameter.

Some of the probabilistic implications of this relationship are discussed by Skipp and Ambraseys (1987). Values of A and b for two areas in Australia are shown in Table 2 (Gaull et al, 1990).

TABLE 2

SEISMICITY RATE PARAMETERS
(after Gaull et al, 1990)

Area	A	b
East of Perth, W.A.	3.66	0.94
S.E. Queensland	2.10	0.66

7.3 *Maximum bedrock acceleration*

The maximum bedrock acceleration is related to the magnitude of the earthquake and the distance from the source. A great number of attenuation curves (peak acceleration versus distance) have been proposed, several of which are summarised by Joyner and Boore (1988), Skipp and Ambraseys (1987) and Okamoto (1984). Attenuation curves for Australian conditions have been proposed by Rynn (1986) and Gaull et al (1990), and in each case, the following empirical relationship has been used:

$$a_{max} = c_1 \exp(c_2 M_L)/R^{c_3} \qquad ...(4)$$

where a_{max} = peak ground acceleration (m/s^2)
M_L = Richter magnitude
R = hypocentral distance (km)
= $(\Delta^2 + h^2)^{1/2}$
Δ = epicentral distance (km)
c_1, c_2, c_3 = attenuation constants.

For four Australian/Pacific regions, values of the attenuation constants are shown in Table 3. It should also be noted that Equation (4) can be applied to maximum velocity and ground intensity (in terms of Modified Mercalli units); appropriate attenuation constants for these components of ground motion are presented by Gaull et al (1990).

Contours of peak ground acceleration for Australia have been derived by Gaull et al (1990). These are presented in probabilistic terms, as values which have a 10% chance of being exceeded in a 50 year period.

TABLE 3

ACCELERATION
ATTENUATION COEFFICIENTS
(after Gaull et al, 1990)

Region	**Attenuation Coefficients**		
	c_1	c_2	c_3
W. Australia	0.025	1.10	1.03
S.E. Australia	0.088	1.10	1.20
N.E. Australia	0.06	1.04	1.08
Indonesia (for Indonesian Zone, R>300 km)	37.6	1.11	2.08

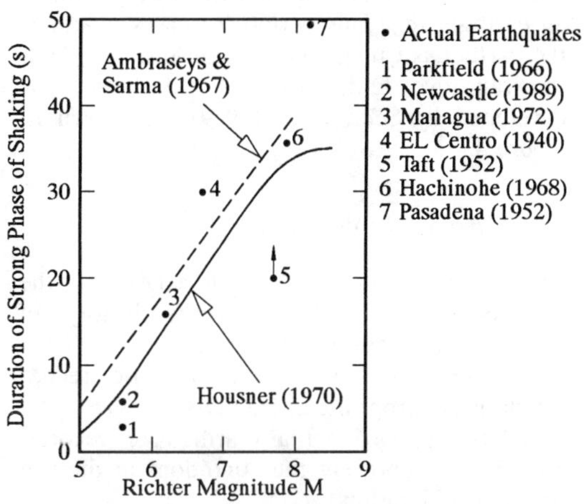

FIG.3 RELATIONSHIP BETWEEN EARTHQUAKE DURATION AND MAGNITUDE

7.4 *Duration*

The duration of an earthquake is related primarily to its magnitude, with longer duration being associated with larger magnitude. Figure 3 summarises two suggested relationships and some selected observations from earthquakes. There is broad agreement between the various sources. It should be noted that the duration shown is that of the strong phase of motion defined by Ambraseys and Sarma (1967) as accelerations which remain above 0.03 g.

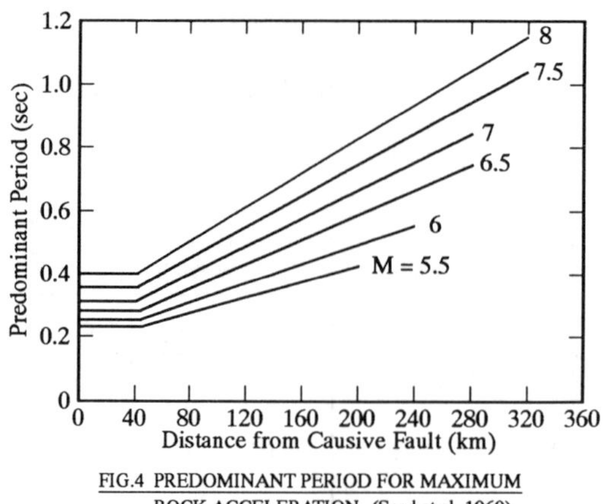

FIG.4 PREDOMINANT PERIOD FOR MAXIMUM ROCK ACCELERATION (Seed et al, 1969)

7.5 *Predominant period*

While there is generally a considerable range of frequency components associated with an earthquake, it is usually possible to define an average predominant or characteristic period of an earthquake. This period increases as either the earthquake magnitude or the distance from the causative fault increases. Relationships developed by Seed et al (1969) are shown in Figure 4.

7.6 *Time-acceleration history*

There are two broad approaches to the development of an appropriate history of bedrock acceleration versus time:

1) the generation of a synthetic record which has appropriate characteristics

2) the use of actual earthquake records, modified to represent the conditions at the site under consideration.

The generation of synthetic records is described by numerous authors (e.g. Clough and Penzien, 1975; Vanmarke, 1976).

The use of existing earthquake records has been discussed by Seed et al (1969), and involves the following steps:

1) selection of an earthquake record with a similar predominant period to that assessed for the site (see Section 7.5)

2) scaling of the record to obtain the appropriate peak acceleration, as estimated from Section 7.3

3) modification of the length of the record to obtain the appropriate duration of strong motion, as estimated from Section 7.4. If this duration (say X seconds) is less than the duration of the selected record, then the first X seconds of the record is used. If the duration (X) is greater than the duration of the selected record, then appropriate parts of the record can be repeated. to give a total of X seconds of significant motions.

8 ANALYSIS OF SEISMIC SITE RESPONSE

The majority of seismic site response analyses consider only the vertical propagation of shear waves from bedrock through the overlying soil. The basis of this type of analysis has been described by Roesset (1970) and several others.

Finite element techniques have been used to analyse two- and three-dimensional wave propagation problems, e.g. soil slopes (Idriss and Seed, 1967) and earth dams (Abdel-Ghaffar and Scott, 1981). However, for routine microzonation studies, such analyses are generally not appropriate and one-dimensional analyses generally provide an adequate basis for assessment of local soil amplification effects.

Several solution methods can be used to solve the one-dimensional wave propagation equation, either in the frequency domain, or in the time domain. For soils having linear stress-strain characteristics, it is possible to develop transfer functions which relate the response spectrum for motion at the surface to the response spectrum for motion at bedrock level. However, for more realistic stress-strain response of the soil, in which both stiffness and damping depend on strain level, it is most common to obtain a solution by numerical integration of the equations of motion in the time domain, with step-by-step linearisation of the stress-strain law for the soil.

An example of the direct time-domain solution is the computer program ERLS, described by Poulos (1991). This program solves the equations of motion by a forward marching finite difference process, using a lumped mass representation of the soil.

The following features are incorporated into the analysis:

1) The shear modulus and damping ratio of each soil layer are dependent on the cyclic shear strain within the layer, as in equations (1) and (2), and are updated continuously throughout the analysis.

2) The bedrock excitation can be specified

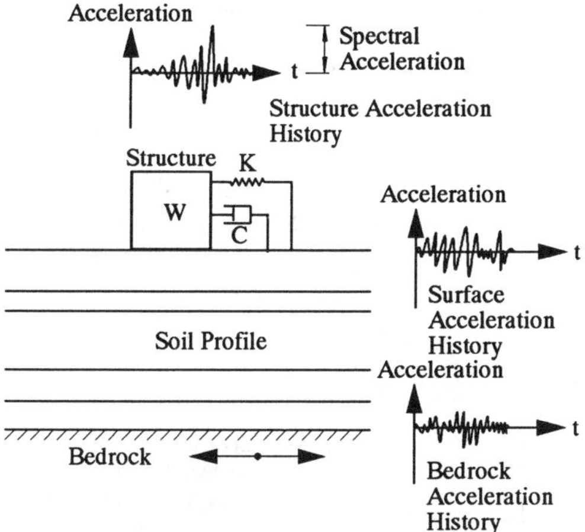

FIG.5 COMPUTATIONAL MODEL

as either a series of harmonic excitations of different amplitudes and periods, or a sequence of "peaks" and "troughs" with the acceleration varying linearly with time between adjacent peak and trough points.

3) Approximate allowance can be made for the radiation damping provided by the bedrock (Roesset, 1977).

4) The distributions of acceleration, velocity, displacement and shear stress with depth may be output at specified times.

5) From the computed surface acceleration versus time response, the program can compute the response spectra for acceleration, velocity and displacement of a single degree of freedom structure having different fundamental periods of vibration.

Figure 5 illustrates the major features of the computational model used in the program ERLS.

9 ASSESSMENT OF SEISMIC HAZARDS

The technique to be used for assessing seismic hazards will depend to a large extent on the area under consideration. For small areas and discrete sites, more detailed and site-specific analyses may be employed. In contrast, where a larger area is being assessed (e.g. in excess of, say, 10 hectares), a more "broad brush" approach may have to be employed. In this Section, both detailed and "broad brush" approaches will be discussed for the assessment of the following earthquake hazards:

a) liquefaction
b) structural damage
c) slope instability.

9.1 *Liquefaction potential*

Liquefaction is a general term describing the unstable behaviour of saturated sands due to increases in pore water pressure leading to a loss of shear strength and development of strains. Extensive reviews of earthquake-induced liquefaction have been published in the last few years. Of particular interest and value are those by Seed and Idriss (1982), CEE (1985), Iwasaki (1986), and Martin (1988).

Most simplified methods of evaluating the liquefaction potential of saturated sand deposits can be classified into one of the following categories:

1) Methods employing data from laboratory tests.

2) Methods using correlations with standard penetration test (SPT) data.

3) Methods using correlations with static cone penetration test (CPT) data.

4) Methods based on some other form of empirical correlation.

Methods falling in the first three categories involve three main steps:

i) estimation of the cyclic shear stress S_s induced at various depths within the soil by the earthquake, and the number of significant stress cycles

ii) estimation of the cyclic shear strength R of the soil, i.e. the cyclic shear stress ratio which is required to cause initial liquefaction of the soil in the specified number of cycles

iii) comparison between the induced cyclic shear stress and the cyclic shear strength; at locations where the induced shear stress exceeds the shear stress required to cause initial liquefaction, there is a potential for liquefaction.

The estimation of the cyclic shear stress induced in the soil by the earthquake can be estimated by one of the following approaches:

i) carrying out a site response analysis, using an input bedrock accelerations versus time history for the bedrock underlying the soil. Such an analysis can be used to identify the maximum cyclic shear stresses at various depths within the soil and to obtain some assessment of the number of significant cycles of loading

ii) using the following approximation

developed by Seed and Idriss (1971) for the cyclic shear stress ratio S_s caused by the earthquake:

$$S_s = \frac{\tau_h}{\sigma_0'} \simeq 0.65\frac{a_{max}}{g} \cdot \frac{\sigma_0}{\sigma_0'} \cdot r_d \qquad ...(5)$$

where a_{max} = peak acceleration at ground level
g = acceleration due to gravity
σ_0 = total overburden pressure at depth under consideration
σ_0' = effective overburden pressure at that depth.

Seed (1979) and Seed et al (1983) have related the cyclic shear strength R to a modified standard penetration resistance and the earthquake magnitude M. Further refinements to standardise the SPT value and allow for the effects of fines in the sand have been presented by Seed et al (1984) and Seed and de Alba (1986).

An estimate of the cyclic stress ratio (R) to cause initial liquefaction may also be obtained from the correlation with a modified cone resistance Q_c presented by Robertson and Campanella (1985).

Sugawara (1989) has examined more closely the influence of fines content on the liquefaction potential. His proposed correlations between R and cone resistance Q_c are shown in Figure 6, which also shows two other correlations (including that of Robertson and Campanella). Sugawara's correlations indicate a higher resistance to liquefaction than the other correlations, and also suggest that increasing fines content leads to a major increase in liquefaction resistance.

Iwasaki et al (1984) have extended the basic approach described above and introduced the concept of a liquefaction potential index to estimate the likely severity of liquefaction at a given site.

They define the liquefaction resistance factor F_L as:

$$F_L = \frac{R}{S_s} \qquad ...(6)$$

where R = in-situ cyclic undrained normalised shear strength of the soil
S_s = cyclic shear stress ratio due to the earthquake

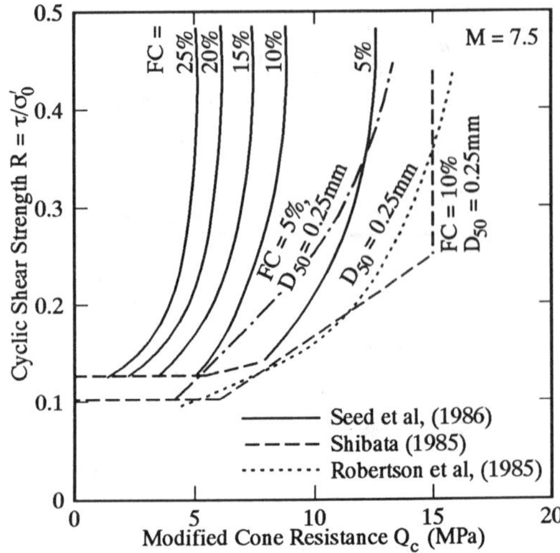

FIG.6 PROPOSED CORRELATION BETWEEN LIQUEFACTION RESISTANCE OF SANDS AND CONE RESISTANCE
(Sugawara, 1989)

The liquefaction potential index I_L is then defined as:

$$I_L = \int_0^{20} W(z)dz \qquad ...(7)$$

where F = $1 - F_L$ for $F_L \leq 1.0$
and F = 0 for $F_L > 1.0$
$W(z)$ = $10 - 0.5z$
z = depth in metres.

Based on field observations, Iwasaki et al propose the following simplified procedure for assessing the risk of liquefaction:

$I_L = 0$
$0 < I_L \leq 5$ low risk
$5 < I_L \leq 15$ high risk
$15 < I_L$ very high risk

The above concept is a very useful extension to the basic approach in that it allows an assessment of the likely significance of a limited zone of the soil being subjected to liquefaction. In many earlier approaches, the existence of even a very limited zone of liquefaction (i.e. a zone where $S_s \geq R$) condemned a site to be classified as liquefiable, whereas the liquefaction potential index gives a much more satisfactory indication of the likely consequences of liquefaction.

9.2 *Potential for structural damage*

Although structural damage may result from several different effects of an earthquake, such as foundation failure due to liquefaction, or foundation displacements associated with fault movements or landslides, the main influence of an earthquake on a structure is the development of additional forces at the base of the structure due to the dynamic ground motions.

Damage or collapse of structures under earthquake loading can arise from several factors, including the following (Williams 1988):

a) inability of the structure to act as a unit due to poor construction

b) inadequate bracing

c) incompatible mix of construction elements

d) existence of eccentric stiffness elements

e) non-uniform distribution of ductile elements, leading to stress concentrations

f) inadequate anchorage to the foundation.

For low-rise buildings in Queensland, Williams (1988) has presented a simplified guide for assessing earthquake-induced damage as a function of earthquake intensity and the general state of repair of the structure; this is reproduced in Figure 7.

Chandler et al (1991) have studied the structural damage caused by the 1989 Newcastle earthquake and concluded that unreinforced brick masonry construction is about twice as vulnerable to damage as reinforced concrete frame buildings, and nearly three times as vulnerable as buildings of timber frame construction.

Detailed analyses of structural response can be carried out by standard structural dynamic analysis techniques. For a particular structure, it is possible to compute the distribution of dynamic forces within the structure and to take account of such factors as structural details, nonlinear material behaviour, and combined lateral and vertical response. Such analyses are essential for major buildings and for environmentally sensitive facilities such as nuclear power plants.

For a "broad brush" approach, it is much easier to treat the structures as equivalent

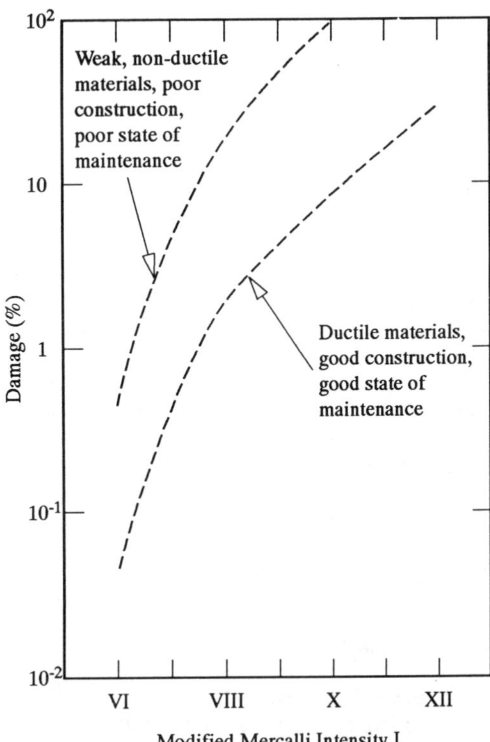

FIG.7 SIMPLIFIED GUIDE TO EARTHQUAKE-INDUCED DAMAGE TO LOW RISE BUILDINGS

linear single degree of freedom systems, and to compute their response spectra for the input ground motion history. Response spectra may be derived readily by solution of the equation of motion for a single degree of freedom system, for a series of natural periods and damping ratios for the structure. For example, the program ERLS uses a simple forward marching procedure, together with the computed surface acceleration versus time history, to obtain the spectral (maximum) acceleration for specified values of natural period and damping of the structure.

For assessment of the potential for structural damage on a regional basis, the following approach may be adopted:

1) the region or area is divided into a number of sub-areas, on a grid as described in Section 5

2) for each sub-area, a seismic site response analysis is carried out to obtain a soil surface acceleration versus time history for the chosen input bedrock time-acceleration history

3) a response spectrum analysis is done for each sub-area, and the peak spectral acceleration is obtained for a series of natural periods of the structure, representing one-storey, two-storey etc. structures

4) for each structure type (i.e. natural period), the areas in which large accelerations may be developed are identified and mapped

5) if in the planning stage of a development, zoning should be carried out to avoid siting structures in areas where they may be subjected to very large accelerations (and hence earthquake-induced forces)

6) if zoning is not possible, then it may be possible to consider some modification of the type of structure or its foundation system so as to reduce the potential acceleration. Either the stiffness may be modified to alter the natural frequency, or else measures taken to increase the amount of structural damping and hence decrease the spectral acceleration.

9.3 *Potential for slope instability*

From the seismic site response analyses, the peak horizontal ground acceleration will be obtained for the site under examination, or the sub-areas comprising the region being studied. The influence of this acceleration on slope stability may be readily calculated from standard limit equilibrium stability analyses, and many computer packages are available to carry out such analyses.

For a "broad brush" approach, a more simplistic analysis may be employed, and the results expressed in terms of the factor of safety F_e (with the earthquake) to the usual factor of safety without the earthquake, F_s. For example, for failure of a clay slope under undrained conditions, Sinclair (1992) derives the following simple expression:

$$\frac{F_e}{F_s} = \frac{1}{(1+\frac{a_{max}}{g}.\cot\theta)} \quad ...(8)$$

where a_{max} = maximum ground acceleration
g = gravitational constant
θ = slope angle.

Thus, for example, with a 30° slope and a_{max}/g = 0.2, F_e/F_s is 0.74, i.e. the factor of safety is reduced by about 26% because of the earthquake.

By using a simple approach such as this, sub-areas in which significant reductions in safety factor occur can be identified, and such areas can then be avoided in development or else be subjected to some form of stabilisation treatment to increase the normal static factor of safety.

10 REPRESENTATION OF SEISMIC HAZARD ASSESSMENTS

The results of a regional seismic hazard assessment are usually presented in the form of a map which identifies the level of risk associated with that hazard. Such maps may be in the form of contours or zones overlying the geographic area, and may take a variety of forms, including the following:

a) maximum computed surface acceleration

b) maximum computed intensity

c) spectral acceleration, for a particular type of structure

d) spectral velocity, for a particular type of structure

e) natural period of soil deposits

f) amplification of bedrock motions

g) liquefaction potential (generally in terms of broad terms such as "high", "medium" or "low", or more specifically in terms of the liquefaction potential index I_L)

h) factor of safety against earthquake-induced slope instability.

A recent example of a regional study is that described by Elder et al (1991, 1992) and McCahon et al (1992), for the city of Christchurch, New Zealand. Elder et al (1992) have identified three effects of the deep alluvium underlying Christchurch:

i) removal of short period spectral accelerations by hysteretic damping

ii) marked increase in peak spectral accelerations across most of the city

iii) shift in spectral shape towards the longer period end of the spectrum, with peak accelerations ocurring at natural periods between 0.5 and 1.5 seconds.

The soil profile at any location, especially that within 20 m of the surface, strongly influences the ground motion and spectral acceleration at that location.

Elder et al (1992) have produced a map showing areas with a likelihood of severe shaking and liquefaction. The areas susceptible to liquefaction have been defined largely on the basis of soil type, rather than site response analyses, while the areas of severe shaking are derived from site response analyses, identifying areas where a spectral acceleration in excess of 80% of the maximum value is likely to occur. Maps such as these are valuable aids in proper planning of future developments and appropriate siting of environmentally sensitive facilities or structures.

In addition to specific geographic maps it is possible to represent the results of seismic hazard analyses in alternative forms. Williams (1988) has presented the results of an assessment of landslides as zones overlain on a family of attenuation curves.

Poulos (1991) has presented the results of response spectrum analyses for the 1989 Newcastle earthquake in the form of a "damage potential plot." This shows contours of spectral acceleration on a plot of soil depth versus number of storeys in the structure (which is related to the natural period of the structure).

This plot is shown in Figure 8, and identifies three zones:

Zone A: where the spectral acceleration exceeds 0.2 g and there is likely to be considerable risk of major damage or collapse

Zone B: spectral accelerations between 0.1 g and 0.2 g, where there may be moderate to severe damage

Zone C: spectral acceleration less than 0.1 g, where only minor damage is likely.

This figure reveals two combinations of soil and structural characteristics which are in Zone A; one- and two-storey structures founded on soil depths of between 5 and 15 m, and three- to five-storey structures on soil deposits between 15 and 28 m deep. This was a post-earthquake investigation, and subsequent comparisons showed that some (but not all) of the expectations from the damage potential plot were realised. Structures which were found to have major damage are plotted on Figure 8, and all lie within either Zone A or Zone B. However, a number of structures falling into Zone A did not collapse or suffer severe damage, especially single storey timber structures. These may have performed better than anticipated because of their greater damping (a relatively low damping ratio of 2.5% was assumed for all structures in the analyses), because the soil conditions at individual sites were more favourable than was assumed in the analysis, or because they were founded on deep foundations, rather than the shallow foundations assumed in the analysis.

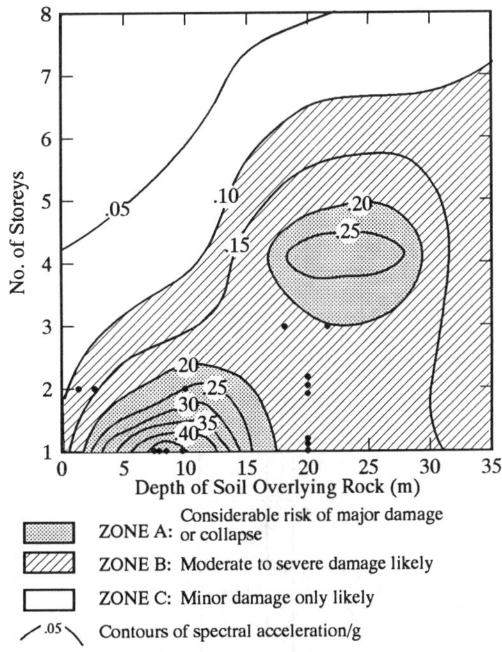

FIG.8 DAMAGE POTENTIAL CURVES FOR 1989 NEWCASTLE EARTHQUAKE

For a relatively small site, it is possible to provide more detailed information on the anticipated seismic response. An example of a detailed analysis is shown in Figure 9 for the Newcastle Technical College site subjected to the 1989 Newcastle earthquake. The shear modulus values were assessed on the basis of SPT data, using the correlations shown in Figure 1, while the damping values were selected from Table 1 on the basis of soil type. The base acceleration-time history was that estimated by Poulos (1991), with a maximum acceleration of 0.05 g and a duration of 6 seconds. Figure 9 shows the computed maximum shear stress in the soil, the input base motion, the response spectrum for this motion, and the computed surface acceleration versus time history and the corresponding response spectrum.

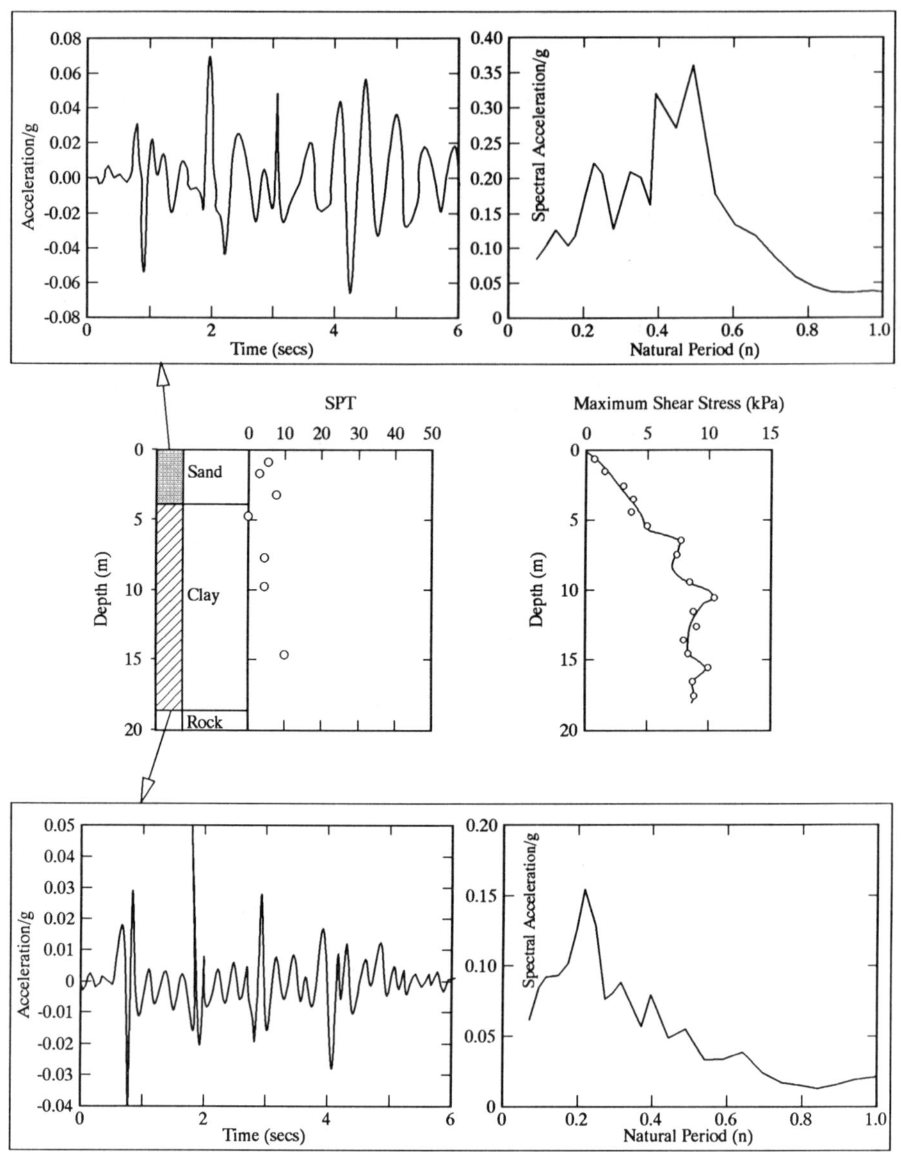

FIG.9 RESULTS OF DETAILED SEISMIC ASSESSMENT OF INDIVIDUAL SITE - NEWCASTLE TECHNICAL COLLEGE

11 CONCLUSIONS

This paper has attempted to set out a step-by-step approach to assessment of seismic hazards and subsequent microzonation. Attention has been focussed on soil liquefaction, structural damage due to soil amplification effects, and slope instability. Any such assessment is carried out in a climate of uncertainty about almost all components, including:

i) the regional tectonic geology
ii) the representation of the design earthquake
iii) the soil profile within the region of interest
iv) the stiffness and damping characteristics of the soil.

Because of these uncertainties, there may be little point in attempting to use sophisticated procedures for seismic hazard assessment, especially if the area is relatively large. Consequently, it is suggested that the following procedures may be appropriate:

a) for liquefaction, correlation of liquefaction potential with penetration test data

b) for structural damage potential assessment, site response analyses using a one-dimensional shear wave propagation analysis

c) for slope instability assessment, a simplified method which relates factors of safety with and without earthquake activity.

The approach outlined herein is essentially deterministic, and requires the application of appropriate engineering judgement. However, risk quantification is possible if the relationship between earthquake magnitude and return period can be established. Interaction between the seismologist, the engineering geologist, and the geotechnical engineer is highly desirable in order to establish this relationship, quantify its implications, and portray the results of the assessment.

REFERENCES

Abdel-Ghaffar AM and Scott RF (1981). "Comparative Study of the Dynamic Response of an Earth Dam." Jnl. Geot. Eng. Divn. ASCE, 107: GT3, 241-269

Ambraseys NN and Sarma SK (1967). "The Response of Earth Dams to Strong Earthquake." Geotechnique, 17, 2, 181-203

Bolt BA (1986). "Earthquakes." New York, WH Freeman and Co

CEE (1985). "Liquefaction of Soils during Earthquakes." Comm. on Earthquake Eng., Nat. Acad. Press, Washingon

Chandler AM, Pappin JW, and Coburn AW (1991). "Vulnerability and Seismic Risk Assessment of Buildings Following the 1989 Newcastle, Australia Earthquake." Bull. NZ. Nat. Soc. for Earthquake Eng., 24, 2, 116-138

Clough RW (1970). "Earthquake Response of Structures." Ch. 12 of Earthquake Engineering, Ed. RL Wiegel, Prentice Hall, Englewood Cliffs

Clough RW and Chopra AK (1966). "Earthquake Stress Analysis in Earth Dams." Jnl. Eng. Mech. Divn., ASCE, 92:EM2, 197-211

Clough RW and Penzien J (1975). "Dynamics of Structures." McGraw Hill, New York

Elder D McG, McCahon IF and Yetton MD (1991). "The Earthquake Hazard in Christchurch. A Detailed Evaluation." Rep. to NZ EQC, ISBN 0-473-01243-X

Elder D McG, McCahon IF and Yetton MD (1992). "Effects of Regional Geology on the Seismic Hazard in Christchurch, New Zealand." Proc. 6th Aust-New Zeal. Conf. on Geomechs., Christchurch, 499-504

Gaull, BA, Michael-Lieba MO and Rynn JMW (1990). "Probabilistic Earthquake Risk Maps of Australia." Aust. Jnl. of Earth Sciences, 37, 169-187

Gutenberg B and Richter CF (1956). "Earthquake Magnitude, Intensity, Energy and Acceleration." Bull. Seism. Society Amer., 46; 105-145

Hardin BO and Drnevich VP (1972). "Shear Modulus and Damping in Soils: Design Equations and Curves." Jnl. Geot. Eng. Divn, ASCE, 98, SM7, 667-692

Hattori S (1980). "Seismic Risk Maps in the Asian Countries......" Int. Conf. on Eng. for Protection from Nat. Disasters, AA Balkema, 491-504

Housner GW (1970). "Strong Ground Motion." Ch. 4 of Earthquake Engineering, Ed. RL Wiegel, Prentice Hall, Englewood Cliffs

Idriss IM and Seed HB (1967). "Response of Earth Banks During Earthquakes." Jnl. Soil Mechs. Foundns. Divn., ASCE, 93: SM3, 61-82

Imai T and Tonouchi K (1982). "Correlations of N Value with S-Wave Velocity and Shear Modulus." Proc. 2nd ESOPT, Amsterdam, 1, 67-72

Iwasaki T (1986). "Soil Liquefaction in Japan: State-of-the-Art." Soil Dyn. and Earthq. Eng., 5:1, 1-68

Iwasaki T, Arakawa T and Tokida K-I (1984). "Simplified Procedures for Assessing Soil Liquefaction During Earthquakes." Soil Dyn. and Earthq. Eng., 3:1, 49-58

Joyner WB and Boore DM (1988). "Measurement, Characterisation and Prediction of Strong Ground Motion." Earthquake Eng. and Soil Dynamics II-Recent Advances in Ground-Motion Evaluation, Geotech. Spec. Pub. No. 20, ASCE, 43-102.

Mihailov V (1984). "Seismic Risk - An Important Parameter for Town Planning in Seismic Regions." Proc. 8th World Conf. on Earthquake Eng., San Francisco, :417-424

Martin GR (1988). "Geotechnical Aspects of

Earthquake Engineering." State-of-the-Art Report No. 2, 5th Aust-New Zeal. Conf. on Geomechs, Sydney. Aust. Geomechs. Special Issue, 42-77

McCahon IF, Elder D McG and Yetton MD (1992). "The Liquefaction Potential in Christchurch." Proc. 6th Aust-New Zeal. Conf. on Geomechs, Christchurch, 526-531

McCue K (1990). "Australia's Large Earthquakes and Recent Fault Scarps." Jnl. Struct. Geol, 12:5/6, 761-766

Ohta Y and Goto N (1978). "Empirical Shear Wave Velocity Equations Determined from Characteristic Soil Indexes." Earthquake Eng. Struct. Dyn, 6:2, 167-187

Okamato S (1984). "Introduction to Earthquake Engineering." Univ. of Tokyo Press, Tokyo, Japan

Poulos HG (1991). "Relationship Between Local Soil Conditions and Structural Damage in the 1989 Newcastle Earthquake." Civ. Eng. Trans. Inst. Engrs. Aust., CE33:3, 181-188

Robertson PK and Campanella RG (1985). "Liquefaction Potential of Sands Using the CPT." Jnl. Geot. Eng., ASCE, 109:11, 1449-1459

Roesset JM (1970). "Fundamentals of Soil Amplification." Seismic Design for Nuclear Power Plants, 183-244, M.I.T. Press, Cambridge, Mass.

Roesset JM (1977). "Soil Amplification of Earthquakes." Ch. 19 of Numerical Methods in Geot. Eng., Ed. CS Desai and JT Christian, 639-682, McGraw Hill, New York

Rynn JMW (1988). "The Assessment of Seismic Risk in North Eastern Australia." Civ. Eng. Trans., Inst. Eng. Aust, CE 30:1. 45-56

Seed HB (1979). "Soil Liquefaction and Cyclic Mobility Evaluation for Level Ground during Earthquakes." Jnl. Geot. Eng. Divn, ASCE, 105:GT2, 210-255

Seed HB and Idriss IM (1971). "Simplified Procedure for Evaluating Soil Liquefaction Potential." Jnl. Soil Mechs. Foundns. Divn, ASCE, 95:SM5, 1199-1218

Seed HB and Idriss IM (1982). "Ground Motions and Soil Liquefaction during Earthquakes." Monograph Series, EERC, Berkeley, Calif

Seed HB, Idriss IM and Arango I (1983). "Evaluation of Liquefaction Using Field Performance Data." Jnl. Geot. Eng., ASCE 109:3, 458-482

Seed HB, Wong RT, Idriss IM and Tokimatsu K (1984). "Modulus and Damping Factors for Dynamic Analyses of Cohesionless Soils." Rep. No. UCB/EERC - 84/14, EERC, Univ. of California, Berkeley

Seed HB and de Alba P (1986). "Use of SPT and CPT Tests for Evaluating the Liquefaction Resistance of Sands." Proc. Conf. on Use of In-Situ Tests, ASCE, Blacksburg, 281-302

Sinclair TJE (1992). "A Simple Method for Assessing Effects of Earthquakes on Slopes." Proc. 6th Aust-New Zeal. Conf. on Geomechs, Christchurch, 551-553

Skipp BO and Ambraseys NN (1987). "Engineering Seismology." Ch. 18 of Ground Engineers Reference Book, Ed. FG Bell, Butterworths, London

Sugawara N (1989). "Empirical Correlation of Liquefaction Potential Using CPT." Proc. 12th Int. Conf. Soil Mechs. and Foundn. Eng., Rio de Janeiro, 1:335-338

Sugimura Y and Ohkawa I (1984). "Seismic Microzonation of Tokyo Area." Proc. 8th World Conf. on Earthq. Eng., San Francisco, :721-728

Sun JI, Golesorkhi R and Seed HB (1988). "Dynamic Moduli and Damping Ratios for Cohesive Soils." Rep No. UCB/EERC-88/15, EERC, Univ. of California, Berkeley

Vanmarke EH (1976). "Structural Response to Earthquakes." Ch. 8 of Seismic Risk and Engineering Decisions, Ed. C Lomnitz and E Rosenblueth, Elsevier, Amsterdam

Vannuchi G (1991). "Seismic Hazard and Site Effects in the Florence Area." Special Volume, 10th Eur. Conf. on Soil Mechs. and Foundn. Eng., Florence.

Wiegel RL (1970). "Earthquake Engineering." Prentice Hall, Englewood Cliffs, NJ

Williams DJ (1988). "Potential Engineering Risks in the Earthquake Hazard to the East Coast of Queensland." Civ. Eng. Trans., Inst. Engrs., Aust., CE 30:4, 307-317

Environmental Management, Geo-Water & Engineering Aspects, Chowdhury & Sivakumar (eds)
© 1993 Balkema, Rotterdam. ISBN 90 5410 099 0

Encapsulation of solid wastes from industrial by-products

Alek Samarin
Boral Resources Limited, Sydney, N.S.W. & University of Wollongong, N.S.W., Australia

ABSTRACT: The paper defines wastes, and highlights the current problem of waste disposal in NSW. Future trends, in the light of world experiences, are evaluated. The theoretical fundamentals of the encapsulation process, and the influence of physical and chemical immobilization on potential leachability of hazardous species are described.

The potential for degradation of cementitious matrix is related to the microstructure and the chemical composition of the encapsulating material. Selected additives, which can enhance the durability of cement matrix are listed. New analytical and modelling techniques (i.e. chemically controlled or diffusion rate controlled chemical attack and quantitative, fractal number assessment of the pore structure intricacy) are proposed.

1 INTRODUCTION

The Oxford Dictionary defines waste as being a "superfluous, a refuse, no longer serving a purpose, left over after use ..." and waste products as being "useless by-products of manufacture or of physiological process".

The term "useless" however, begs the question: "Useless or useful to whom?" Quoting from "Weekend Australian" (24-25 Nov., 1990) - "One man's waste is another's good fortune". It is probably prudent to replace the concept of being "useless" with that of being "unwanted".

The Australian Chemical Industry Council defines waste as "... the unavoidable materials for which there is no economical demand and for which disposal is required". Here again the term unavoidable may be somewhat ambiguous.

Important components of waste management strategies are:- waste prevention, which can be achieved by non-production of the material or product, thus eliminating the process by which particular waste is generated (an extreme case), by substitution of main material or product with one which generates a more "benign" or less voluminous waste, or by optimizing the technology in such a way that the waste is converted into a useful raw material for some other manufacturing process (an ideal case). Thus the term "unavoidable" should probably be best replaced with "not readily avoidable", so that the definition for waste becomes:-

"not readily avoidable by-products, for which there is no economical demand and for which disposal is required".

Now, the South Australian Waste Management Act (1987) puts the main emphasis not on the potential use of the waste, but on the act of disposal, so that the definition for solid waste becomes:-

"any matter, irrespective of value, that is discarded or left over in the course of industrial, commercial, domestic or other activities".

As well as waste prevention, waste management strategies may include reduction of the source of waste, through optimization of the process which is generating a particular waste.

These strategies may involve different methods of recycling waste materials and waste products.

They may utilize waste treatment techniques, such as thermal destruction by incineration, or pyrolysis (the chemical decomposition of a substance by heat in the absence of oxygen).

The treatment may involve chemical destruction, which can take forms of oxidation, reduction or absorption.

The process may also be physical in nature, involving precipitation, filtration, evaporation or condensation.

The treatment may be predominantly biological in essence, such as aerobic and anaerobic (requiring presence or requiring absence of oxygen).

Finally, more effective methods of waste disposal, such as designed systems of landfill or residual repositories, may be used.

In this paper the discussion of waste management techniques will be confined to the specific form of treatment, i.e. encapsulation.

2 ENCAPSULATION OF SOLID WASTE

2.1 Methodology

Waste treatment generally encompasses several objectives, the most apparent of which are:-

1. Improvements in the physical characteristics of waste, so that it can be handled, disposed of, etc., with less difficulty.
2. Reduction of the solubility and leachability of pollutants, which may be present in a particular waste.
3. Detoxification of pollutants, which are contained in the waste.

The process of waste treatment can take a variety of forms. Terms such as solidification, stabilization, fixation, sorption, glassification, self-cementing, etc., are widely used, but often an identical terminology has a different or even conflicting definition, and the interpretation may be misleading.

In the text of this paper the term encapsulation describes the process of waste separation from the environment by a surrounding matrix, which is intended to be as impermeable as practically and economically possible.

The process can be further ranked in accordance with its scale, as either a macroencapsulation, or microencapsulation.

Macroencapsulation is usually a mechanical, physical or physico-chemical process (such as sealing waste in containers or in relatively large blocks or granules).

Microencapsulation, on the other hand, is commonly a chemical or physico-chemical process, such as immobilization of metal ions in a cement gel or silicate gel matrix.

When the process of encapsulation consists of coating the waste with an insoluble cover of inert material, such as polythene or polybutadiene rubber, the costs are usually very high, as the process requires specialised equipment and preliminary drying of the waste.

Ceramic building materials may become a much more affordable vehicle for disposal of putrescibles, or moist organic materials, which in a favourable environment may also break down as a result of the action of microbes. For example, in several developing countries refractories are manufactured using ternary systems of raw materials, consisting of clay, sand and sawdust. Sawdust may be replaced with a wide range of putrescibles, the organic part of which then acts as a fuel, and some of the intractables, such as heavy metals, if present, can be encapsulated and immobilized.

Another economical form of waste encapsulation involves the use of hydraulic cements as a coating material, and from now on the term encapsulation will be used exclusively to describe the process of solidification and stabilization of wastes using these specific types of binders.

2.2 Composition of solid waste

Over one million tonnes of solid waste is produced in Australia annually and approximately half of this waste comes from commercial sources.

The estimated present rate of increase of commercial waste is over ten, and possibly as high as twenty per cent per annum.

Typical composition of solid waste in major Australian cities is given in Table I, and the average is compared with the average for the EEC (European Economic Community) countries, U.K. and the U.S.A. in Table II.

In spite of some variations between the cities and countries (mostly in the ratio

Table I
Solid Waste Composition - per cent total
(in 1980's)

	Major Australian Cities					
	Adelaide	Perth	ACT	Brisbane	Melbourne	Sydney
Putrescible	40	56	44	42	43	48
Paper	27	23	19	30	21	21
Glass	13	7	14	7	16	9
Plastics	9	9	10	12	10	8
Metal	6	4	5	6	6	6
Other	5	2	7	3	4	8

Compiled from data presented in:- "Management & Technologies of Wastes". A Perspective - Australia 1990, by Dr K T Hubick, DITAC, Australia, February 1992.

Table II
Solid Waste Composition - per cent total
Australia and Overseas (in 1980's)

	Average for Australian Cities	European Economic Community	U.K.	U.S.A.
Putrescible	45	40	26	35
Paper	23	25	36	36
Glass	11	10	11	8
Plastics	10	7	8	7
Metal	6	8	11	9
Other	5	10	8	5

Compiled from data presented in:- "Management & Technologies of Wastes". A Perspective - Australia 1990, by Dr K T Hubick, DITAC, Australia, February 1992.

between putrescibles and paper), there is a remarkable overall uniformity in the trends of waste generation in the developed countries.

Approximately 96% of Australian solid waste goes into landfill, about 3% is recycled and the remaining 1% is either incinerated or disposed in some other ways.

The household component breakup for U.S.A. is given in Table III.

Industrial waste composition is much more variable between different countries, or even cities, depending on the peculiarities of local manufacturing, but the wastes from services to the community are inevitably present in each urban area.

Of these, wastes from water supply and sewerage are more consistent in composition, unlike the waste from power generation, which is highly dependent on the type of fuel or the energy source used.

In Australia the cost of waste disposal into landfill is relatively low.

The tipping charges start at $2 per tonne, and rise to about $40 per tonne for some hazardous wastes.

Transfer costs are in the range of $30 to

Table III
Annual Household Waste Generation and Compactability
(U.S.A. - 1977) Ref. Tchobanoglous et al.

Component	Generation per household (kg/yr)	Density (kg/m^3)	Compaction Ratio	Landfill Volume (m^3/yr)
Food wastes	220	288	0.35	0.27
Paper	999	82	0.20	2.45
Carboard	151	50	0.25	0.76
Plastics	42	64	0.15	0.10
Textiles	5	64	0.18	0.01
Rubber	18	128	0.30	0.04
Leather	16	160	0.30	0.03
Garden wastes	332	104	0.25	0.80
Wood	81	240	0.3	0.10
Glass	174	192	0.6	0.54
Tin cans	121	88	0.18	0.25
Non-ferrous metals	35	160	0.18	0.04
Ferrous metals	100	320	0.35	0.11
Dirt, etc.	25	481	0.85	0.05
Total	2,319	-	-	5.55

$35 per tonne for a disposal over a 40 km distance.

Incineration costs in Australia (with energy recovery) are of the order of $60-$90 per tonne.

As reported by the National Solid Wastes Management Association, tipping charges in the United States rose by about 17% on average between 1988 and 1990. Per-ton costs are now close to US$65 (i.e. A$90) in some areas of the densely-populated land-short northeast parts of America.

Thus, approximately 73% of USA's waste goes into landfill, and 14% is incinerated.

In the Netherlands, 14% of the waste is incinerated, but only 57% goes into landfill.

Italy incinerates up to 80% of its waste with only 15% going into landfill.

2.3 Sewerage and water treatment wastes - Sydney experience

The terms sedimentation and clarification are commonly used to describe the process of water purification.

Colloidal particles, which cannot be removed in a practical way by direct settlement, are usually subjected to chemical coagulation in water treatment plants.

The coagulation is achieved by the addition of trivalent metallic salts such as $Al_2(SO_4)_3$ - aluminium sulphate, or Fe Cl_3 - ferric chloride.

The water treatment residuals from these plants in this case are a combination of aluminium hydroxide, alumina, silicon and other organic and inert materials.

Currently water treatment residuals in Australia are mainly disposed of by conventional means, such as landfill.

In one of the largest attempts at privatised construction of municipal drinking water systems, officials in Sydney are soliciting proposals for four filtration plants. At present, water purveyor filters only 5% of water supplied to 4 million residents in the 12,500 square km area of the total Sydney sprawl. Sediment from reservoirs clogs distribution systems and taints the water after heavy rains.

The biggest of the proposed plants, 3.6

billion litres a day capacity Prospect Reservoir will generate approximately 8,000 tonnes per annum of residuals, and disposal by landfill may present a considerable problem. Alternative methods of disposal require urgent consideration.

At the same time, the Water Board for Sydney, Illawarra and the Blue Mountains plans to upgrade 36 sewage treatment plants, including 23 on the Nepean-Hawkesbury River system, from which Sydney draws 97% of its water.

With raw sewage constantly overflowing into Sydney Harbour, the immediate plan calls for the upgrading and most probably enlarging of the three waste water treatment plants on the southwest coast: the 640-million-litre-per day Malabar plant, the 325 m.l.d. North Head plant and the 170 m.l.d. Bondi plant.

A major investigation is currently under way concerning the occasional high levels of nitrogen and phosphorous in the Sydney river systems, as was apparent from the proliferation of blue-green algae during hot, dry weather in summer.

A newly introduced Environmental Planning Policy for New South Wales, SEPP No. 33 detailed definitions of "potentially hazardous industry", "potentially offensive industry", "hazardous industry", "hazardous storage establishment", "offensive industry" and the "offensive storage establishment".

Although SEPP No. 33 has been designed to eliminate some of the unnecessary obstacles to development, its effect on waste-water and sewage sludge disposal methods is not clear. For example, U.S. Environmental Protection Agency (EPA) has a "sludge rule", which dates back to the 1977 Federal Clean Water Act, under revision - it seems that the modified rule will apply to five types of sludge disposal.

- land application
- garden distribution
- sludge-only landfill
- surface disposal and
- incineration

Concurrently U.S. EPA has issued the first set of national rules for the location, design and operation of municipal solid waste landfills.

Under these rules, new landfills or expansions must have composite synthetic liners and groundwater monitoring wells. Landfills consisting of a multi-layered liner incorporating groundwater and leachate collection conduits that feed into access galleries beneath the site have been developed, for example, by Steetly Quarry Products in Ontario, Canada.

It is to be expected that Australian disposal rules will follow similar trends, and the alternatives of waste encapsulation become more and more economically realistic, under these new sets of requirements.

2.4 Ash from coal fired power stations

The coal fired power stations in NSW produce almost 5 million tonnes of fly ash and bottom ash each year, with less than 10% of which being utilized, mostly as a raw feed for building and construction materials and products.

The balance of fly ash was disposed by hydraulic means into the holding ponds - a system which is now considered to be potentially harmful to the environment.

Environmental control agencies in the United States became increasingly concerned with the mobilization of trace elements from the fly ashes disposed by ponding.

Concentrations of Ca, Na, Mg and Fe were determined in water extracts of four lignite fly ashes from three Northern Great Plains mines in the U.S. (Shannon and Fine).

In another investigation (Davison et al) it was found that fly ash contained trace elements As, Sb, Cd, Cr, Pb, Ni, Se, Tl and Zn. Australian coals often differ from their U.S. counterparts, but presence of true metals in some of the leachates was established in our investigations.

The current trend is to convert all NSW plants to dry disposal of ash.

The cost of disposal should be not less than $2 per tonne, with the upper limit influenced by the environmental regulations which are becoming more and more stringent with time.

Depending on the mineralogy of coal, on the firing equipment used and on the power load, the fly ash generated will exhibit various degrees of pozzolanicity.

Pozzolanicity is the reactivity of fly ash

with lime in presence of water under normal temperature conditions. As a result of this reaction, water insoluble composites, similar to calcium silicate hydrates, produced by hydration of Portland cement, are formed.

Thus, fly ash which has good pozzolanic characteristics, can be used as an integral part of hydraulic cement, by blending it with Portland cement.

One of the desirable strategies in minimizing wastes, is the optimization of the coal combustion process, so that the bulk of fly ash produced can have good pozzolanic properties.

3 THE ENCAPSULATION PROCESS

3.1 Definitions and the ingredients used in the process

As already mentioned, encapsulation in this text means the process of waste separation from the environment by a surrounding matrix, which is intended to be as impermeable as practically and economically possible.

Microencapsulation is synonymous with the physico-chemical fixation, by means of which the waste product becomes not only bonded in a solid matrix of low permeability, but, in some cases, the hazardous species are chemically fixed and immobilised concurrently.

The process can employ either organic or inorganic matrix for solidification, but in the content of this article only the inorganic reagent mixtures, i.e. hydraulic cements, will be considered.

Hydraulic cements are inorganic powders capable of setting and hardening under water.

Portland cement is a classical example. Other hydraulic cements include high alumina or magnesium cements, mixtures of lime and pozzolana and blends of the above. Silica fume which is a by-product of silicon alloy manufacturing is an excellent pozzolana. Finally, ground granulated blast-furnace slag has very good pozzolanic properties. When activated with a small amount of Portland cement or lime, it may be considered an hydraulic cement in its own right.

3.2 The immobilization process

Upon hydration, hydraulic cements form hardened paste, if no inert inclusions are present, or mortar, if these inclusions are relatively small in size (such as sand), or concrete is formed, if the inert aggregate is larger in size than sand (say, using Australian standard definitions, coarser than 7mm maximum size). The encapsulating matrix thus can be hardened paste, mortar or concrete.

The physical immobilization of encapsulated waste will depend on the degree of impermeability of this matrix. Overall immobilization will also depend on the chemical reaction and/or chemical adsorption of the aggressive species by the encapsulating matrix.

The evaluation of the impermeability of a matrix may thus involve the analysis of a diffusion process or a mass transfer process and/or a potential chemical reaction.

The diffusion process is generally considered to be governed by Fick's law. The flux, or the amount of species removed over the time and the area of capillaries, according to this law can be expressed thus:-

$$\textit{Flux} = -D \frac{\partial}{\partial x} (\textit{concentration})$$

where D, the diffusion coefficient, is constant, and the variation with time and distance from the surface x is predicted.

If the process is analyzed as a mass transfer, then

$$\textit{Flux} = k\Delta\ (\textit{concentration})$$

where k, a mass transfer coefficient is not constant, variation with time is correlated and the variation with position ignored.

Because of the assumptions made in the development of the above equations, neither the analysis using the diffusion coefficient D, nor that using the mass transfer coefficient k, is always successful.

3.3 Transport process in a porous media

The term "pore structure" can mean different

things in different applications and on the lowest level of information it is simply defined as the "porosity", or the volume fraction of pore space in porous materials (V).

However, the transport process analyses are generally based on the pore network models, and require knowledge of the pore size distribution, of the different shapes and connections between these pores, as well as knowledge of the intricacy of the structure which the network of pores has formed, such as tortuosity.

Many researchers have attempted to establish a reliable relationship between permeability and the pore structure of concrete. The problem is complicated by a number of factors, including the differences of opinion about the morphology of hardened cement pastes and also due to the different experimental techniques used in measurement of pore size distribution.

For the hardened cement paste - a matrix bonding concrete aggregate together, - we can adopt the following classification

Ultra micro - intra crystallite pores -	< 6Å
Micro - inter crystallite pores -	6 to 16Å
Meso - inter gel particle pores -	16 to 1000Å
Macro - entrapped air -	> 1000Å

A differential curve of pore volume distribution gives a relationship between the derivative of pore volume (V) and the radius of pores (r).

The area under the differential curve is equal to the sum total of all pore volumes in a unit volume of hardened cement paste, thus:-

$$V = \int_{m}^{n} \frac{dV}{dr}\, dr$$

$$where\ m = r_{min}\ and\ n = r_{max}$$

The permeability of a porous medium can be expressed as a general coefficient which characterizes the flow of a fluid caused by a pressure head.

In very porous, permeable concretes, not suitable for encapsulation of wastes, an assumption of a laminar fluid flow may not be unreasonable, and in these cases the flow rate, at least in theory, can be calculated using Darcy's equation.

More realistic in application to concrete is the Carman-Kozeny model, which introduces two important factors:- the hydraulic radius, and the tortuosity.

Measurement of these parameters, however, presents significant difficulties, which again hinders the practical application of the equation.

For relatively impermeable concretes, the transport process can be described either as a diffusion (Fick's law) or as a mass transfer, as already mentioned.

However, Fick's law defines the diffusion coefficient of two gases diffusing into one another without hindrance. In concrete the species must diffuse through the pores and the distance to cross the section becomes significantly longer than in unhindered diffusion.

When the pore dimensions are smaller than the mean free path of the gas molecules, Knudson's model would be more appropriate. The model again depends on the measurement of capillary radii.

Gas transport through concrete is also very sensitive to the pore water content.

As the water content of concrete increases, the pores become filled with liquid, starting with the smaller pore sizes and connections, "pore throats" of the "ink-bottle" shaped pores. Since gas diffusion in water is much slower than in air, this formation of clusters of pores cut off from the rest by the "liquid cork" discontinuities can considerably slow down the transport process.

The critical water to cement ratio for complete hydration of Portland cement is approximately 0.24, but unless special chemical admixtures are used, this mix is too stiff for practical purposes, and water to cement ratio of structural concrete can be as high as 0.7 (and in exceptional cases even higher). Hence, the set cement in a hardened concrete consists of a matrix of solids, both crystalline and amorphous, an aqueous phase and pores. Now, depending on the conditions of exposure, almost all structural concrete will absorb moisture from air into the micropores, when the relative humidity is sufficiently high. In the presence of moisture the process of cement hydration in concrete can go on for several years, and as the volume of hydration products is more than double that of unhydrated cement, the voids in this concrete will be gradually filled with calcium-silicate hydrates and lime, the pore volume will decrease with time and the solids-space ratio, and therefore strength of concrete, will increase.

As we can see, the pore volume and pore structure of concrete will generally change with time due to hydration reaction and also generally due to reactions with different chemical species which may be present in the environment interacting with the concrete. Reaction of hydration products with the CO_2 diffusing into concrete from the surrounding air (or carbonation of concrete) is a classical example of this time dependent change.

The time dependent phenomena can include some or all of the following:

- mass transfer process
- diffusion and
- chemical reactions governed usually by the process of chemical kinetics

In modelling chemical reactions two general cases may be considered:-

1. Kinetic model dependent on solution concentration, viz. -

$$S/E = (C/Ci)\ [1 - \exp(-kt)]$$

where

S	=	amount sorbed	mol. kg^{-1}
E	=	sorption maximum	mol. kg^{-1}
C	=	pore solution concentration	mol. L^{-1}
C_i	=	influent concentration	mol. L^{-1}
t	=	time	min

and

2. Kinetic model independent of solution concentration, viz. -

$$dS/dt = -k_{-1}\ S + K_r\ (E - S)$$

3.4 High performance concrete - a pre-requisite for use as an encapsulation matrix

As we have seen, concrete exposed to the environment is a dynamic system, continuously changing its internal structure. Some changes, such as hydration of cement, result in higher strength and impermeability of concrete. Others, such as chemical attack, may lead to its deterioration.

Fig.1 gives a relationship between solid-space ratio, permeability and compressive strength of concrete, manufactured with a specific set of ingredients and tested in accordance with selected standards at some particular age of exposure to a given environment.

Change of any of these parameters can influence the relationship, but the trends would generally retain a large degree of similarity.

From the study of a statistically significant number of relationships of this type, it can be concluded that concretes which are manufactured to reach characteristic strength higher than 100 MPa have the potential for extremely low permeability and thus should be considered for encapsulation of wastes which require a higher degree of physical separation from the environment. The nature of waste will then determine additional requirements for concrete ingredients and composition needed for the immobilization of hazardous species in the matrix of an hydraulic cement.

When physical separation is of lesser significance, concretes of characteristic strengths of between sixty and one hundred MPa may be used.

4 DEGRADATION OF ENCAPSULATING MATRIX AND LEACHABILITY OF STABILIZED WASTES

4.1 Effect of chemical composition on the degradation process

Studies of Portland cement hydration indicate that two types of calcium-silicate-hydrate (C-S-H) phase are formed:- an inner and an outer product.

The first is formed more or less inside the original boundary of cement grains. It has a finely divided pore structure and contains no Portlandite, that is $Ca(OH)_2$ crystals.

The second, an outer product, is deposited in the originally water-filled spaces between the cement grains. This product is formed from hydrated calcium silicate (C-S-H) into a needle-like structure, interspaced with ettringite (C_3A-$3CaSO_4$-32H) and Portlandite, as well as small amounts of other hydrated cement minerals.

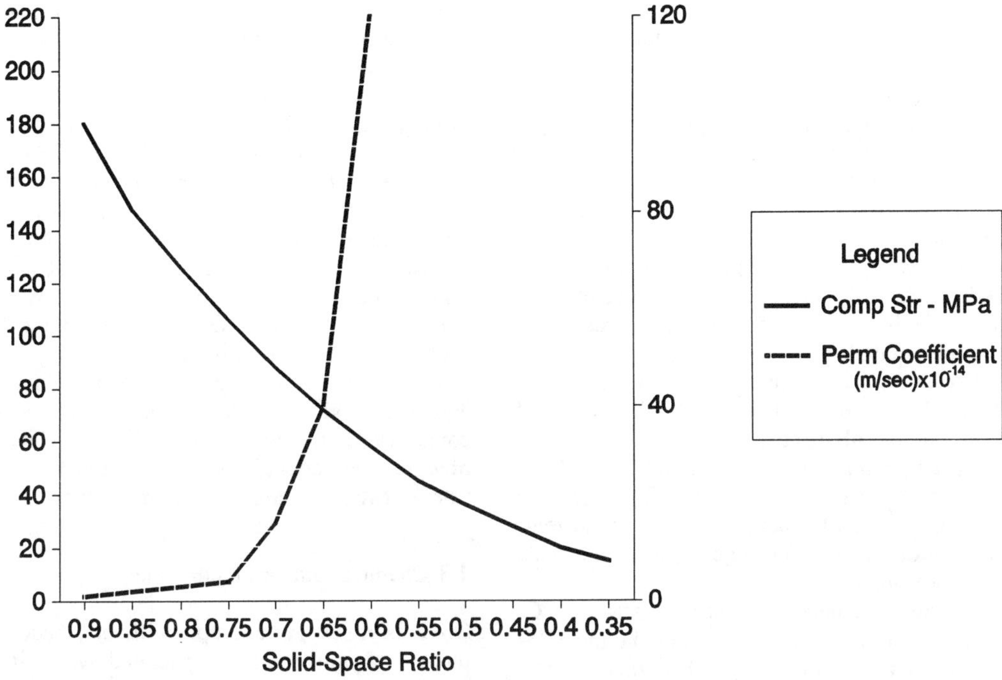

Fig. 1 Relationship between Solid-Space Ratio, Permeability and Compressive Strength of Concrete

It is more porous than the inner product with a porosity directly dependent on the water-cement ratio.

The Portlandite (or calcium hydroxide) formed during hydration of the cement is precipitated as relatively large hexagonal crystals, where space is available within the pore system.

Of all the above phases, it is the Portlandite which is thermodynamically the most stable. As long as calcium hydroxide remains present, the pH in a system is maintained high.

In aqueous solutions pH can be expressed thus:-

$$pH = \log_{10} [H^+]^{-1}$$

where $[H^+]$ is the concentration, or more strictly the activity of the hydrogen ions.

The substances producing H^+ are called acids and substances producing OH^- are called bases. Thus high pH indicates that pore solution in cement paste is highly alkaline.

The stability of a system can also be characterised by its internal redox potential, E_h. Any chemical process in which there is a transfer of electrons, either partial or complete is an oxidation-reduction (or redox) reaction.

Reagents with high values of reduction potential (expressed in volts) are strong oxidizing agents. The internal redox potential, E_h, expresses not only the numerical value, but also a measure of the buffering capacity of the system to resist any change in this value.

Portland cement pastes are generally poorly poised, and their E_h is of the order of +200mV., i.e. only slightly oxidizing.

The E_h however may be affected by supplementary cementitious materials, especially slag which usually contains chemically-reduced sulphur. Slag rich systems develop E_h of the order of -400mV.

Some of the waste components may have similar effect.

Electropositive metals (Zr, Fe, etc.) lower the E_h to that of the hydrogen evolution potential.

The solubility limiting phase for a range of inorganic waste materials depends on both pH and E_h of the cement matrix.

Although predominantly governed by Portlandite, pH of the aqueous phase in cement paste, mortar or concrete is also influenced by the presence of cations of Na and K, (usually present in quantities of less than 1% by weight of cement, expressed as sodium equivalent.)

However, in some blended cements, calcium-silicate-hydrates (C-S-H) will dominate the pH control.

At temperatures of between 10°C and 20°C (approximately) the C-S-H is thermodynamically metastable, but persistent. Natural occurrences on a geological time scale have been recorded.

On the other hand at higher temperatures C-S-H is thermally unsound. Autoclaving C-S-H usually results in its crystallisation.

Although a function of Ca:Si ratio of cement, the autoclaved crystalline phases are much more thermodynamically stable than C-S-H, and hence less soluble in pore solutions, in which pH was reduced either due to soft or acid water attack or carbonation.

4.2 Effect of carbonation and oxygen ingress

When carbon dioxide, which is present at about 0.03% by volume (or 0.0456% by weight) of dry air, penetrates into concrete, it reacts with calcium hydroxide and some other products of hydraulic cement react with water. This process is known as carbonation of concrete.

Oxygen, which is present at about 20.949% by volume (or 23.143% by weight) of dry air also diffuses into concrete, and may cause radiolysis and chemical disruption of water by oxidation of electropositive metals present in the encapsulated wastes.

The porosity of cement matrix in concrete (or mortars) is measurably reduced by carbonation. All pore sizes are affected, but of particular significance to durability is the effect on macropores, especially those with radii in the range of 1 to 0.1mm.

The carbonation of concrete also causes a decrease of the amount of adsorbed water into the hardened cement paste pore system.

Diffusion coefficients of CO_2 and O_2 through carbonated concrete appear to be influenced very little by the equilibrium moisture content in the range of 50% to 90% R.H., unlike uncarbonated concretes, in which the effect of moisture content on the diffusivity over an identical range of R.H. is high. In some cases groundwaters which contain HCO_3^- or CO_3^{2-} species may be sources of carbonation in concrete.

Thus, pore blocking of concrete stored above ground or underground (under or in contact with the water table) due to $CaCO_3$ precipitation should be considered in the safety analysis of a particular repository.

4.3 Chemical attack on concrete

Obviously, every attempt will be made to place repositories of encapsulated waste in a dry environment, which represents a minimum level of chemical attack. However it is doubtful if dry conditions can be guaranteed over the design life of a repository, which is most likely to be of the order of several hundreds of years, rather than under the hundred year cycles, typical of modern civil engineering constructions.

The most likely conditions of exposure are wet-dry cycling, and the progress of degradation can therefore be expressed in terms of water exchange cycles.

The resistance of hardened concrete to chemical attack, which may or may not be combined with physical degradation when exposed to an aggressive environment, is not governed solely by the mechanical strength of the material.

As already stated, chemical composition of hardened cement matrix, and, in addition, the mineralogy of aggregates, may be of reasonable importance in certain types of chemical attack on concrete.

None of the hydrated Portland cement compounds is completely resistant to solutions with a pH value of below 10.

In practice, however, this process of

dissolution is extremely slow, until pH drops to below the value of approximately 6.5.

The calcium then dissolves quite rapidly, and so do the hydroxides.

In the most probable case of corrosion by a rainwater runoff, two mechanisms will influence the degradation kinetics:-

~ ion transfers (diffusion), caused by concentration gradiants between the corrosive water and the pore water of the cement paste;

~ dissolution of portlandite with subsequent dissolution of monosulfoaluminate and ettringite (chemical reaction).

It should be noted that all rainfall, whether polluted or not, is acidic due to carbon dioxide in the atmosphere dissolving in water to form carbonic acid, H_2CO_3. The pH of de-ionised water in equilibrium with the carbon dioxide in the air is 5.6.

In many urban parts of Australia the pH of rain water is well below 5 due to the presence of sulphur dioxide in the air.

In general a chemical attack on concrete involves both:- chemical reaction and diffusion, and the rate of the reaction may be limited by the rate of one of them.

Adopting the previous nomenclature

$$\textit{diffusion rate} = \frac{D}{X}(Ci - C)$$

and the reaction rate = KC.

The second expression assumes first-order kinetics, the rate being taken as proportional to the pore solution concentration and the reaction is in a steady state, i.e. the concentrations are not functions of time. For a steady state pore solution concentrations are equal, and hence

$$C = \frac{(D/X)\ Ci}{K + (D/X)}$$

By substituting C in either of the two previous equations, we obtain

$$rate = \frac{K\ (D/X)\ Ci}{K + (D/X)}$$

It follows, that if $D/X \gg K$, the above expression reduces to

$$rate = KC_i$$

i.e. Chemically controlled while if $D/X \ll K$, it reduces to

$$rate = \frac{D}{X}\ Ci$$

i.e. Diffusion controlled.

If the aqueous pore solution contains n chemical species (or independent components) in solution, for each species there will be only one diffusion equation.

The species H^+ and OH^- are not, strictly speaking, independent components:- their concentrations are determined by the condition of electroneutrality and by the dissociation constant of water.

The n ionic species can give rise to M dissolution-precipitation reactions, so that the number of stoichiometric coefficients of the reactions concerned is greater than one, but less than m - the number of solid phases.

Each solid phase present is associated with a local chemical equilibrium, resulting in the "zoning" of the altered part of the cement, consisting of an assemblage of multi-mineral domains separated by boundaries.

Within each zone, considering transport by diffusion only, the equations of mass conservation can be derived.

Thus, a numerical model of a specific chemical attack on concrete, based on the laws of diffusion and the local equilibrium can be developed.

5 EFFECT OF ADDITIVES ON IMMOBILIZATION OF WASTE

5.1 Supplementary Cementitious Materials, Chemical Admixtures and Adsorbents

The main objective of waste stabilization is to contain the hazardous species and to prevent them from re-entering the environment at a rate that would cause a danger to public health.

For a particular waste, economically achievable physical and chemical properties of optimum encapsulation by the cementitious matrix must be selected.

The leachability of hazardous species from the solidified waste products depends on chemical and physical bonding by the cementitious gel.

The rheological properties of fresh mortars or concretes, and both physical and chemical ability to immobilize the waste in a hardened state is affected by addition to Portland cement of supplementary cementitious materials, chemical admixtures and adsorbents.

Rheology of fresh encapsulating matrix must be optimized for a particular method of placement and compaction to guarantee the minimum water-cementitious ratio, which will translate into the best possible mechanical strength for a particular mix and into the prerequisite for long term durability.

This requirement is usually met by the use of high-range water-reducing admixtures (or "superplasticisers").

Sodium salts of polymerized alkyl naphthalene sulphonic acid and sodium salts of melamine formaldehyde sulphonic acids are two of the best known high-range water-reducers in concrete.

These are generally used in concentrations of between 0.3% and 2.0% by weight of cementitious material to provide good workability of fresh mix at a relatively low water demand.

Pozzolans, such as fly ash, decrease permeability, increase mix fluidity and lower internal heat evolution.

Natural pozzolans also improve sorption of inorganic ions. Zeolites are particularly effective in this respect. However, some of the efficiency can be lost due to their pozzolanic reaction with lime.

Zeolites are a group of complex aluminium silicate minerals containing alkali and alkali-earth metals. The crystallography of the zeolites differs greatly from one to another. Nevertheless, they all have similar open structures that allow certain molecules to pass through.

In encapsulation, process specific zeolites will adsorb certain small molecules, whereas larger ones will be excluded.

In Australia, of the reasonably large zeolite family, Phillipsite, or $(K,Na_2\ Ca)\ (Al_2\ Si_4)\ O_{12}\ .45\ H_2O$ is present.

Silica fume - an industrial pozzolan is also known to increase sorption.

Ground granulated blast furnace slag, although strictly speaking not a true pozzolan but an hydraulic cement, possesses most of the properties of an industrial pozzolan. It is also known to lower internal redox potential, when added to a Portland cement paste.

The leachability of heavy metals from solidified waste products depends mainly on the chemical form in which they exist in the matrix.

Some metals, such as lead and chromium, appear to be bound in the cement gel as silicates, whereas others such as cadmium, appear to be present as hydroxides.

In the presence of pozzolans, hydrates of calcium silicates, calcium aluminates or calcium aluminosilicates are formed from the reaction of calcium in the lime with amorphous aluminosilicates.

It should be noted that aluminoferrite hydrates have a higher themodynamic corrosion resistance than the corresponding aluminate hydrates.

Metal hydroxides will leach from the matrix, if the acidity is sufficiently increased. Silicates, however, remain bound to the cement matrix at much lower pH.

Sodium silicate is often used as an additive to precipitate heavy metals and decrease permeability.

Soluble silicates reduce the leachability of hazardous metal ions by the formation of low-solubility metal oxides-silicates, which encapsulate the metal ions in a silicate or a metal silicate-gel matrix.

Apart from zeolites, adsorbents such as activated carbon, acid clay, etc. have been used in encapsulation of different inorganic wastes with different degrees of success.

5.2 Effect of additives on the microstructure of hardened cement

The physical aspects of immobilization of encapsulated waste are closely related to the transport process in a porous media.

Inclusion of additives in Portland cement paste would generally affect the pore structure.

The pore size distribution, shape of individual pores, as well as the intricacy, complexity and tortuosity of the systems and interconnections of pores can be affected.

All of these changes are reflected in the microstructure of hardened cement.

From a mathematical point of view, the microstructure of hardened cementitious paste is fractal in nature, or, in essence, it is based on the principle of a specific geometrical property which, unlike the classical Euclidian geometry, possesses the scale-invariant self-similarity over a wide range of sizes. In other words, a fractal is a geometrical figure that consists of an identical motif repeating itself on an ever-reducing scale.

The concept of dimension applied to a meandering line or to an uneven surface has various mathematical connotations.

The understanding of fractal dimension requires a sound mathematical basis and the concept was defined differently by several famous mathematicians, including Felix Hausdorff (1886-1942), A.S. Besicovich (1891-1970) and A.H. Kolmogorov (1903-1987).

Results of some of the modern analytical methods, such as small angle X-ray scattering (SAXS), and the somewhat similar method of small angle neutron scattering (SANS) which are applied to study the microstructure of cement paste during hardening and subsequent strength gain, can be interpreted as mass and surface fractal dimensions.

It is apparent that for characterization of the rough structure and porosity of hardened cement pastes, fractal geometry is better suited than the classical, Euclidian description of the internal surface area of the material. An added advantage of these methods is that wet samples can be characterized.

The results of SAXS and SANS analysis can be conveniently expressed as the Hausdorff Fractal Dimensions.

To grasp the concept of a fractal dimension let us consider a problem, the underlying idea of which is to figure out how many small objects or units of size ρ are needed to cover a large object of size P.

For a linear measurement let us assign $P = 10\rho$.

Then for the area measurement:-

$$P^2 = (10\rho)^2 = 100\rho^2$$

and for the volume

$$P^3 = (10\rho)^3 = 1000\rho^3$$

The scale is logarithmic, and the whole process of finding the dimension of the object can be turned into an operation of taking logarithms of the appropriate ratio.

For example, tripling the width of a square, creates a new square, that contains nine of the original squares. Its dimension is calculated by taking logarithms of the size ratio, viz:-

$$\textit{dimension } D = \log^9/\log^3 = \log 3^2/\log^3 =$$

$$= 2 \log^3/\log^3 = 2$$

The result which we know to be correct, as the area is two-dimensional.

In general, for any fractal object size P, constructed of smaller units size ρ, the number N of units that fits into the object is the size ratio raised to a power, and the exponent D is then called the Hausdorff dimension, thus:-

$$N = \left(\frac{P}{\rho}\right)^D$$

or,

$$D = \log N/\log\left(\frac{P}{\rho}\right)$$

Both SAXS and SANS measurements on cement paste result in an isotropic two-dimensional spectrum. After radial averaging, the decay of the scattered intensity I is obtained as a function of the scattering Vector Q, viz.-

$$Q = (4\pi/_{\lambda}) \sin (P/2) \; nm^{-1}$$

The measurement presented as a plot of log I versus log Q would generally show a straight-line relationship over a considerable Q-range.

The slopes of these lines for different values of Q represent a fractal dimension D, in accordance with the Hausdorff definition.

The technique thus is directly applicable to the changes in cement paste microstructure due to incorporation of additives, used in the encapsulation process.

The typical fractal numbers obtained for the internal surface of the cementitious matrix range from 1.9 to 3.0. The higher the number, the more complex, the more intricate is the microstructure.

Some qualitative studies of cement paste pore structures in the past, suggested that the inclusion of supplementary cementitious materials, such as fly ash, resulted in more meandering of pores and in the reduction of permeability for a comparable pore volume.

It will be interesting to see if SAXS and SANS measurements of internal paste microstructure will provide a quantitative verification of these observations, in terms of increasing fractal numbers.

6 CONCLUSIONS

The current situation with Australia's solid waste, approximately 96% of which goes into landfill, is rapidly changing.

Waste management strategies provide ever increasing incentives to implement systems of waste minimization, recycling and treatment, by making landfill disposal an economically unattractive option.

Treatment of waste depends on its composition and on the subsequent handling and methods of disposal.

Of the non-industrial wastes, sewerage sludge, water treatment wastes and ash from power generating plants are common to most of the Australian major cities. Unless treated, all of the above may be adversely affecting some of the existing environments.

There is a potential to encapsulate the above wastes into specially designed concrete-like materials.

High performance, low permeability concretes have the potential to immobilize hazardous chemical species, which may be present in these wastes. Concretes of this type may contain special additives to assist both chemical and physical fixation of potential leachates. These include pozzolans (of which fly ash from power stations is one), adsorbents and chemical admixtures.

If the hazardous species are immobilized effectively and there is little probability of their escape, certain concrete structures, such as foundations, pavements, etc. built in a mild and benign environment can be used as repositories for these wastes.

Theoretical models, assessing the risk of degradation and leachability can be used as a first step in evaluation of this process.

REFERENCES:

Books:

Bear, J. and Corapcioglu, M.Y. 1991. *Transport Processes in Porous Media*, Kluwer Academic Publishers, Printed in Netherlands.

Cussler, E.L. 1984. *Diffusion - Mass transfer on fluid systems*, Cambridge University Press, Cambridge, U.K.

Feder, J. 1988. *Fractals*, Plenum Press, New York.

Hubick, K.T. 1991. *Management and Technologies of Wastes. A perspective - Australia 1990*, Department of Industry, Technology and Commerce, Australia.

Lee, J.D. 1988. *A New Concise Inorganic Chemistry (3rd edition)*, Van Nostrand Reinhold (U.K.) Co. Ltd.

Nigmatulin, R.I. 1991. *Dynamics of Multiphase Media, Volumes 1 and 2*, Hemisphere Publishing Corporation, New York.

O'Gallagher, B. 1990. *Waste Management Technologies. Opportunities for Research and Manufacturing in Australia*, Department of Industry, Technology and Commerce, Australia.

Peavy, H.S., Rowe, D.R., and Tchobanoglous, G. 1985. *Environmental Engineering*, McGraw-Hill Book Co. New York.

Ryan, W.G. and Samarin, A. 1992. *Australian Concrete Technology*, Longman Cheshire, Melbourne, Australia.

Stanley, H.E., and Octrowsky, N. 1986. *On Growth and Form. Fractal and Non-Fractal Patterns in Physics*, Martinus Nijhoff Publishers, Boston.

Tchobanaglous, G., Theisen, H., and Eliassen, R. 1977. *Solid Wastes: Engineering principles and management issues*, McGraw-Hill Book Co., New York, N.Y.

Williams, R.E. 1975. *Waste Production and Disposal in Mining, Milling and*

Metallurgical Industries, Miller Freeman Publications, Inc. U.S.A.

Periodicals:

Adaska, W.S., Tresouthick, S.W. and West, P.B. 1991. *Solidification and Stabilization of Wastes Using Portland Cement*, Portland Cement Association, U.S.A.

Adenot, F. and Buil, M. 1992. *Modelling of the Corrosion of the Cement Paste by Deionized Water*, Cement and Concrete Research, Vol.22 (pp 489-496).

Aldridge, L.P. 1991. *Effective management of hazardous industrial solid waste*, Waste Management, October (pp 32-33).

Brodersen, K and Nilsson, K. 1992. *Pores and Cracks in Cemented Waste and Concrete*, Cement and Concrete Research, Vol.22 (pp 405-417).

Burner, U.R. 1992. *Thermodynamic modelling of cement degradation: Impact of radox conditions on radionuclide release*, Cement and Concrete Research, Vol.22 (pp 465-475).

Cochet, G. 1992. *Process for treatment and embedment of Ion Exchange Resins with a Hydraulic Binder*, Cement and Concrete Research, Vol.22 (pp 287-292).

Cochet, G. and Cariou, B. 1992. *Very High Performance Micro-Concretes for the Confinement of Industrial Waste*, Cement and Concrete Research, Vol.22 (pp 319-324).

Davison, R.L., Natusch, D.F.S. and Wallace, J.R. 1974. *Trace Elements in Fly Ash*, Environmental Science and Technology, Vol.8, No.13 (pp 1107-1113).

Glasser, F.P. 1992. *Progress in the Immobilization of Radioactive Wastes in Cement*, Cement and Concrete Research, Vol.22 (pp 201-216).

Kyle, J.H. 1991. *Stabilization of Hazardous Wastes*, Chemistry in Australia, October (pp 436-438).

Kyle, J.H. 1992. *Chemical fixation solidification*, Waste Management, February (pp 19-21).

Locoge, P, Massat, M., Ollivier, J.P. and Richet, C. 1992. *Ion Diffusion in Microcracked Concrete*, Cement and Concrete Research, Vo.22 (pp 431-438).

Mecholsky, J.J. and Freiman, S.W. 1991. *Relationship between Fractal Geometry and Fractography*, Journal of American Ceramic Society, Vol.74, No.12 (pp 3136-3138).

Mehta, K. 1992. *High-performance Concrete Durability Affected by Many Factors*, Concrete Construction, May, (pp 367-370).

Nishi, T, Matsuda, M., Chino, K. and Kikuchi, M. 1992. *Reduction of Cesium Leachability from Cementitious Resins Forms Using Natural Acid Clay and Zeolite*, Cement and Concrete Research, Vol.22 (pp 387-342).

Olliver, J.P. and Massat, M. 1992. *Permeability and microstructure of concrete: a review of modelling*, Cement and Concrete Research, Vol.22 (pp 563-514).

Revertegat, E., Richet, C. and Gegout, P. 1992. *Effect of pH on the Durability of Cement Pastes*, Cement and Concrete Research, Vol.22 (pp 259-272).

Samarin, A. and Roper, H. 1986. *The Effect of High Replacement Rates of Portland Cement with Granulated Ground Blast Furnace Slag on Concrete Durability*, Proceedings of the International Seminar on Some Aspects of Admixture and Industrial By-Products on the Durability of Concretes, Chalmers University of Technology, Goteberg, Sweden, 28-29 April.

Samarin, A. 1987. *Methodology of Modelling for Concrete Durability*, Katharine and Bryant Mather International Conference on Concrete Durability. Edited by John M Scanlon, ACI SP-100, Vol.2 (pp 1205-1225).

Samarin, A. 1989. *New Cements and Concretes*, Proceeding of the Symposium - Australian Academy of Science - on Application of New Materials, ACT, Australia (pp 16-23).

Samarin, A. 1991. *Advanced Concrete Technologies - Opportunities for the Next Decade Proceedings of the 15th Biennial Conference - Concrete 91*, Concrete Institute of Australia, Sydney (pp 103-112).

Samarin, A. 1992. *Fly Ash from Coal Power Stations - Converting a liability into an asset*, Invited paper at the launch of the Ash Development Association in Australia, Adelaide, S.A.

Sarrot, F.A., Bradbury, M.H., Pondolfo, P. and Spieler, P. 1992. *Diffusion and Adsorption studies on Hardened Cement Paste and the Effect of Carbonatidon on diffusion rates*, Cement and Concrete Research, Vol.22 (pp 439-444).

Shannon, D.J. and Fine, L.O. 1974. *Cation Solubilities of Lignite Fly Ashes*, Environmental Science and Technology, Vol.8, No.12 (pp 1026-1028).

Tamás, F.D., Csetényi, L.J. 1992. *Effect of Adsorbents on the Leachability of Cement Bonded Simulated Low Level Wastes from Nuclear Power Plants*, Cement and Concrete Research, Vol.22 (pp 393-397).

Tamás, F.D., Csetényi, L. and Tritthart, J. 1992. *Effect of adsorbents on the leachability of cement bonded electroplating wastes*, Cement and Concrete Research, Vol.22 (pp 399-404).

Tang Luping and Lars-Olof Nilsson 1992. *A study of the Quantitative Relationship between Permeability and Pore Size Distribution of Hardened Cement Pastes*, Cement and Concrete Research, Vol.22 (pp 541-550).

Teng, S.P. and Lee, C.H. 1992. *Numerical analysis of through-diffusion experimental results*, Cement and Concrete Research, Vol.22 (pp 445-450).

Tritthart, J. 1990. *Pore Solution Composition and Other Factors Influencing the Corrosion Risk of Reinforcement in Concrete* - a chapter in Corrosion of Reinforcement in Concrete edited by Page, Treadaway and Bamforth, Society of Chemical Industry, Elsevier Applied Science, New York (pp 96-106).

Tucker, S.P. and Carson, G.A. 1985. *Deactivation of hazardous chemical wastes* , Environmental Science and Technology, Vol.19, No.3 (pp 215-220).

Warren, C.J. and Dudas, M.J. 1985. *Formation of Secondary Minerals in Artificially Weathered Fly Ash*, Journal of Environmental Quality, Vol.14, No.3 (pp 405-410).

Zamorani, E. 1992. *Deeds and misdeeds of cement composites in waste management*, Cement and Concrete Research, Vol.22 (pp 359-367).

Environmental Management, Geo-Water & Engineering Aspects, Chowdhury & Sivakumar (eds)
© 1993 Balkema, Rotterdam. ISBN 90 5410 099 0

Accuracy of hydrodynamic models of flood-discharge determinations

V.P.Singh & V.Aravamuthan
Department of Civil Engineering, Louisiana State University, Baton Rouge, La., USA

E.S.Joseph
Department of Civil Engineering, Southern University, Baton Rouge, La., USA

ABSTRACT: Hydrodynamic models employed for computing flood discharges are based on the shallow water-wave theory that is described by the St. Venant (SV) equations. These models are derived from either the kinematic-wave (KW) approximation, the diffusion-wave (DW) approximation or the dynamic-wave (DYW) representation of the SW equations. In the studies reported to date, different types of criteria have been established to evaluate the adequacy of the KW and DW approximations, but no explicit relations either in time or in space between these criteria and the errors resulting from these approximations have been derived yet. Furthermore, when doing hydrologic modeling, it is not evident if the KW and DW approximations are valid on one hand for the entire hydrograph or a portion thereof, and on the other hand for the entire channel length or a portion thereof. In other words, most of these criteria take on fixed point-values for a given rainfall-runoff event. This paper attempts to derive, under simplified conditions, error equations for the KW and DW approximations for space-independent as well as for time-independent flows, which provide a continuous description of error in the flow-discharge hydrograph. For space-independent flows, a dimensionless parameter γ is defined which reflects the effect of initial depth of flow, channel-bed slope, lateral inflow, and channel roughness. For time-independent flows, the dimensionless parameter is the product of the kinematic wave number and the square of the Froude number. The kinematic wave, diffusion wave and dynamic wave solutions are parameterized through these parameters. By comparing the kinematic wave and diffusion wave solutions with the dynamic wave solution, equations are derived in terms of these parameters for the error in the kinematic wave and diffusion wave approximations.

1 INTRODUCTION

Physically-based models of overland flow, channel flow, surface irrigation, and many other phenomena involving unsteady, free surface open channel flows are based on the shallow water wave (SWW) theory. These models are based either on the kinematic wave (KW) approximation (Lighthill and Whitham, 1955), diffusion wave approximation (DW), or dynamic wave (DYW) representation. Lighthill and Whitham (1955) showed that at the Froude numbers less than one (appropriate to flood waves) the dynamic waves are rapidly attenuated and the kinematic waves become dominant. Using a dimensionless form of the St. Venant (SV) equations, Woolhiser and Liggett (1967) obtained what is now referred to as the kinematic wave number, K, as a criterion for evaluating the adequacy of the KW approximation. For K greater than 20, the KW approximation was considered to be an accurate representation of the SV equations in modeling of overland flow. However, no relation between K and the error in the KW approximation was suggested. Morris and Woolhiser (1980) modified the above criterion with an explicit inclusion of Froude number, F_0, and showed, based on numerical experimentation, that $F_0^2K \geq 5$ was a better indicator of the adequacy of the KW approximation. A relation between this criterion and the error resulting from the KW approximation was not derived, however.

Using a linear perturbation analysis, Ponce and Simons (1977) derived properties of the KW and DW approximations as well as DYW representations in modeling of open channel flows. They derived a spectrum showing the regions of the validity of the KW and DW approximations. Menendez and Norscini (1982) extended the work of Ponce and Simons by including the phase lag between the depth and velocity of flow. Their results were, however, similar to those of Ponce and Simons (1977). In another but similar study, Ponce, et al. (1978), based on propagation characteristics of sinusoidal perturbation, derived criteria to evaluate the adequacy of the KW and DW approximations. Daluz Viera (1983) compared solutions of the SV equations with those of the KW and DW approximations for a range of F_0 and K, and defined the regions of validity of these approximations in the K-F_0 space.

Fread (1985) developed criteria for defining the range of application of the KW and DW approximations. These were based on an analysis of the magnitude of the normalized errors in the momentum equation due to omission of certain terms. In a comprehensive study, Ferrick (1985) defined a group of dimensionless scaling parameters to establish the spectrum of river waves, with continuous transitions between wave types and subtypes. With the aid of these parameters he was able to discern when the KW and DW approximations would be valid.

In most of these studies, different types of criteria have clearly been established to evaluate the adequacy of the KW and/or DW approximations, but no explicit relations either in time or space between these criteria and the errors resulting from these approximations have been derived yet. Furthermore, when doing hydrologic modeling it is not evident if the KW and DW approximations are valid for the entire flood hydrograph or a portion thereof. In other words, most of these criteria take on fixed point values for a given event. The objective of this study is to derive, under simplified conditions, error equations for the KW and DW approximations for space-independent as well as time-independent flows, which specify errors as a function of time.

2 SHALLOW WATER-WAVE (SWW) THEORY

The SWW theory can be described by some form of the SV equations. For flow over an infiltrating plane subject to uniform rainfall, these equations can be written in one-dimensional form on a unit width basis as

continuity equation,

$$\frac{\partial h}{\partial t} + \frac{\partial}{\partial x}(uh) = q - f \tag{1}$$

momentum equation,

$$\frac{\partial u}{\partial t} + \frac{\partial}{\partial x}\left(\frac{1}{2}u^2 + gh\right) = (g(S_0 - S_f) - \frac{qu}{h} \tag{2}$$

where h is the depth of flow (L), u is local mean velocity (L/T), q is uniform rainfall intensity (L/T), f is uniform infiltration rate (L/T), g is acceleration due to gravity, x is space coordinate in the direction of flow (L), t is time (T), S_0 is bed slope, and S_f is frictional slope. Note Q = uh is discharge (L^3/TL) per unit width. S_f can be approximated as

$$S_f = \beta \frac{u^2}{h} \tag{3}$$

where β is some resistance parameter. If the Chezy relation is used for representing the friction then $\beta = g/C^2$, where C is Chezy's resistance parameter.

The DYW representation employs the full form of equations (1) and (2). The KW approximation is based on equation (1) and equation (2) with the left side omitted,

$$g(S_0 - S_f) - \frac{qu}{h} = 0 \tag{4}$$

The DW approximation uses equation (1) and equation (2) with local and convective acceleration deleted,

$$g\frac{\partial h}{\partial x} = g(S_0 - S_f) - \frac{qu}{h} \tag{5}$$

Analytical solutions of the SV equations or their variants in the KW and DW approximations are tractable only for simple cases. To that end, both space-independent and time-independent cases are considered in this study. In the first case, the water surface is flat.

3 SPACE-INDEPENDENT FLOWS

3.1 Governing Equations

For space-independent (or uniform) flows, equation (1) takes the form

$$\frac{dh}{dt} = q - f \tag{6}$$

and equation (2) becomes

$$\frac{du}{dt} = g(S_0 - S_f) - \frac{qu}{h} \tag{7}$$

Equations (6) and (7) are the governing equations for the DYW representation for spatially uniform flows. The KW approximation is based on equation (6) and equation (7) with the left side dropped,

$$g(S_0 - S_f) - \frac{qu}{h} = 0 \tag{8}$$

The DW approximation uses equation (6) as well as equation (8). Therefore, for spatially uniform flows, the KW approximation is identical to the DW approximation. Equation (8) can also be approximated by neglecting the momentum exchange between lateral inflow and longitudinal channel flow as

$$S_0 = S_f \tag{9}$$

which can be expressed as equation (3). Similarly, equation (7) can be written as

$$\frac{du}{dt} = g(S_0 - S_f) \tag{10}$$

Depending upon the presence or absence of f, equation (6) can also be simplified. If f = 0, then

$$\frac{dh}{dt} = q \tag{11}$$

3.2 Types of Scenarios

Depending upon the presence of lateral inflow and infiltration, four different scenarios can be considered: (1) f = 0, $q = q_0$ = constant; this includes the case q = 0. (2) $q = q_0$ = constant, and $f = f_0$ = constant; and this includes the case q = f = 0. (3) q - f = 0, $q = q_0$ = constant; this includes the case q = f = 0. (4) q = 0, $f = f_0$ = constant; this includes the case f = 0. It may be noted that the scenario with q = 0 in equation (11) or (q - f) = 0 in equation (6) applies to the recession hydrograph. The same applies if (q - f) < 0.

3.3 Initial Conditions

Two types of initial conditions (t = 0) can be assumed:

(1) $h(0) = h_0$, $u(0) = u_0$, (12)

(2) $u(0) = 0$, $h(0) = 0$ (13)

3.4 Scenarios for Determination of Error

Error equations have been derived by Singh (1992a, 1992b) for the KW and DW approximations under the above-mentioned conditions for four different scenarios. Treated together, eighteen cases result and are summarized in Table 1.

4 TIME-INDEPENDENT FLOWS

4.1 Governing Equations

For time-independent flow, equation (1) takes the form

$$\frac{d}{dx}(uh) = q - f \tag{14}$$

and equation (2) becomes

$$\frac{d}{dx}\left(\frac{1}{2}u^2 + gh\right) = g(S_0 - S_f) - \frac{qu}{h} \tag{15}$$

Equations (14) and (15) are the governing equations for the DYW representation for time-independent flows. The KW approximation is based on equation (14) and equation (15), with the left side deleted, leading to equation (8). The DW approximation uses equation (14) and equation (15) with convective acceleration deleted,

$$g\frac{dh}{dx} = g(S_0 - S_f) - \frac{qu}{h} \tag{16}$$

Equations (15) and (16) can also be approximated by neglecting the momentum-exchange between lateral inflow and longitudinal channel flow respectively as

$$\frac{d}{dx}\left(\frac{1}{2}u^2 + gh\right) = g(S_0 - S_f) \tag{17}$$

and

$$\frac{dh}{dx} = S_0 - S_f \tag{18}$$

Depending upon the presence of f, equation (14) can also be simplified. If f = 0, then

$$\frac{d(uh)}{dx} = q \tag{19}$$

4.2 Types of Scenarios

Depending upon the presence of lateral inflow and infiltration, four different scenarios can be considered as in the preceding section. It may be noted that the scenario with q = 0 in equation (19) or q - f = 0 in equation (14) applies to the case of losing flow. This same would apply if q - f < 0.

4.3 Boundary Conditions

Two types of conditions at the upstream boundary (x = 0) can be assumed:

(1) $h(0) = h_0$, $u(0) = u_0$, (20)

(2) $u(0) = 0$, $h(0) = h_0$ (21)

Depending upon the type of flow, the boundary conditions may have to be specified at the upstream boundary as well as the downstream boundary or at the upstream boundary alone.

The upstream boundary condition, given by equation (21), influences both

Table 1. List of cases for error analysis for space-independent flows.

Case No.	Scenario No.	Governing Eqs		Initial Condi-tions	Lateral Inflow q	Lateral Outflow f
		KW/DW Approximations	DYW Representation			
1	1	Eqs. (11) and (8)	Eqs. (11) and (7)	Eq. (12)	q = constant	$f = 0$
2	1	Eqs. (11) and (8)	Eqs. (11) and (7)	Eq. (13)	q = constant	$f = 0$
3	1	Eqs. (11) and (9)	Eqs. (11) and (7)	Eq. (12)	q = constant	$f = 0$
4	1	Eqs. (11) and (9)	Eqs. (11) and (7)	Eq. (13)	q = constant	$f = 0$
5	1	Eqs. (11) and (9)	Eqs. (11) and (10)	Eq. (12)	q = constant	$f = 0$
6	1	Eqs. (11) and (9)	Eqs. (11) and (10)	Eq. (13)	q = constant	$f = 0$
7	2	Eqs. (6) and (8)	Eqs. (6) and (7)	Eq. (12)	q = constant	f = constant
8	2	Eqs. (6) and (8)	Eqs. (6) and (7)	Eq. (13)	q = constant	f = constant
9	2	Eqs. (6) and (9)	Eqs. (6) and (7)	Eq. (12)	q = constant	f = constant
10	2	Eqs. (6) and (9)	Eqs. (6) and (7)	Eq. (13)	q = constant	f = constant
11	2	Eqs. (6) and (9)	Eqs. (6) and (10)	Eq. (12)	q = constant	f = constant
12	2	Eqs. (6) and (9)	Eqs. (6) and (10)	Eq. (13)	q = constant	f = constant
13	3	Eqs. (11) with $q = 0$ and (8)	Eqs. (11) with $q = 0$ and (7)	Eq. (12)	$q = 0$	$f = 0$
14	3	Eqs. (11) with $q = 0$ and (9)	Eqs. (11) with $q = 0$ and (7)	Eq. (12)	$q = 0$	$f = 0$
15	3	Eqs. (11) with $q = 0$ and (9)	Eqs. (11) with $q = 0$ and (10)	Eq. (12)	$q = 0$	$f = 0$
16	4	Eqs. (6) with $q = 0$ and (8)	Eqs. (6) with $q = 0$ and (7)	Eq. (12)	$q = 0$	f = constant
17	4	Eqs. (6) with $q = 0$ and (9)	Eqs. (6) with $q = 0$ and (9)	Eq. (12)	$q = 0$	f = constant
18	4	Eqs. (6) with $q = 0$ and (9)	Eqs. (6) with $q = 0$ and (10)	Eq. (12)	$q = 0$	f = constant

the supercritical and subcritical flow outside zone A of the characteristic solution domain. The downstream boundary condition (x = L) can be of two types:

(1) Critical flow downstream boundary condition:

$$u(L) = [gh(L)]^{0.5} \quad (22)$$

where L is the channel length. This occurs when the channel ends at the steep bank of a river. For supercritical slow

$$u(L) > [h(L)g]^{0.5} \quad (23)$$

(2) Zero-depth gradient downstream boundary condition:

$$\frac{dh(L)}{dx} = 0 \quad (24)$$

4.4 Scenarios for Determination of Error

Error equations have been derived by Singh and Aravamuthan (1992a, 1992b, 1992c) for the KW and DW approximations under the above-mentioned conditions for four different scenarios. Taken together, these will give rise to 21 cases that are summarized in Table 2.

5 DEFINITION OF ERROR

The relative error E is defined as

$$E = \frac{S_K - S_D}{S_D} \quad (25)$$

where S_K is the solution from the KW or DW approximation, and S_D is the solution from the DYW representation. The solution can be either in terms of depth (h), velocity (u), or discharge (Q), i.e., S {u,h,Q}. Subscripts K and D correspond to the KW (or DW) and DYW solutions, respectively. The differential equation of E can be obtained by differentiating equation (25) as

$$\frac{dE}{dx} = \frac{(E+1)}{S_K}\frac{dS_K}{dx} - \frac{(E+1)^2}{S_K}\frac{dS_D}{dx} \quad (26)$$

The differential equation for error can, however, be defined without explicitly knowing S_D, so long as S_K is explicitly known.

6 ERROR EQUATIONS FOR SPACE-INDEPENDENT FLOWS

We consider the case where f = 0, and $q = q_0$.

6.1 Kinematic Wave and Diffusion Wave Solution

Equation (11), subject to equation (12), has the solution:

$$h = h_0 + q_0 t \quad (27)$$

From the kinematic wave approximation,

$$u = \left(\frac{S_0}{\beta}\right)^{0.5} h^{0.5} \quad (28)$$

It is convenient to define a dimensionless time τ as

$$\tau = \frac{h}{h_0} = \frac{h_0 + q_0 t}{h_0}, \quad h_0 \neq 0, \quad \tau \geq 1 \quad (29)$$

Equation (14) can be expressed in dimensionless form as

$$h_* = \frac{h}{h_0} = \tau$$

Equation (34) can be expressed in dimensionless form as

$$v = \frac{u}{U} = \tau^{0.5}, \quad U = \left(\frac{S_0 h_0}{\beta}\right)^{0.5} \quad (30)$$

The kinematic wave (KW) or diffusion wave (DW) solution is given by equations (33) and (34). The velocity increases parabolically with τ.

6.2 Dynamic Wave Solution

Equation (11) has the solution given by equation (27). Equation (2) can be expressed in dimensionless terms as

$$\frac{dv}{d\tau} = \frac{\gamma^{0.5}}{2} - \frac{\gamma^{0.5}}{2}\frac{v^2}{\tau}, \quad \gamma = \frac{4g^2\beta S_0 h_0}{q_0^2} \quad (31)$$

Equation (31) is a special case of the Riccati equation. When $\tau = 1$, v = 1, and $dv/d\tau = 0$. With these initial conditions in hand, equation (31) was solved by using the 4th order Runge-Kutta method. For a fixed γ, v increases with increasing τ, and for a fixed τ, it increases with γ. However, for $\gamma \geq 1.5$, v is not very sensitive to γ.

6.3 Error in KW and DW Approximations

By making use of equations (30) and (31) in equation (26), one obtains the error differential equation:

$$\frac{dE}{d\tau} = C_0(\tau) + C_1(\gamma,\tau)\,E + C_2(\gamma,\tau)\,E^2 \quad (32a)$$

Table 2. List of cases for error analysis for time-independent flows.

Case No.	Scenario No.	Governing Eqs.			Boundary Conditions			Lateral Inflow q	Infiltration f
		KW Approximation	DW Approximation	DYW Approximation	Upstream	Down-stream (critical depth)	Down-stream (zero-depth gradient)		
1	1	Eqs. (14) and (16)	Eqs. (14) and (16)	Eqs. (14) and (15)	Eq. (20)	N.A.	N.A.	q_0	f_0
2	1	Eqs. (14) and (4)	Eqs. (14) and (16)	Eqs. (14) and (15)	Eq. (21)	N.A.	Eq. (24)	q_0	0
3	1	Eqs. (14) and (9)	Eqs. (14) and (16)	Eqs. (14) and (15)	Eq. (21)	Eq. (24)	N.A.	q_0	0
4	2	Eqs. (14) and (9)	Eqs. (14) and (16)	Eqs. (14) and (15)	Eq. (20)	N.A.	N.A.	q_0	0
5	2	Eqs. (14) and (9)	Eqs. (14) and (16)	Eqs. (14) and (15)	Eq. (21)	N.A.	Eq. (24)	q_0	0
6	2	Eqs. (14) and (9)	Eqs. (14) and (16)	Eqs. (14) and (15)	Eq. (21)	Eq. (24)	N.A.	q_0	0
7	3	Eqs. (14) and (9)	Eqs. (14) and (18)	Eqs. (14) and (17)	Eq. (20)	N.A.	N.A.	q_0	0
8	3	Eqs. (14) and (9)	Eqs. (14) and (18)	Eqs. (14) and (17)	Eq. (21)	N.A.	Eq. (24)	q_0	0
9	3	Eqs. (14) and (9)	Eqs. (14) and (18)	Eqs. (14) and (17)	Eq. (21)	Eq. (24)	N.A.	q_0	0
10	4	Eqs. (14) and (4)	Eqs. (14) and (16)	Eqs. (14) and (15)	Eq. (20)	N.A.	N.A.	q_0	f_0
11	4	Eqs. (14) and (4)	Eqs. (14) and (16)	Eqs. (14) and (15)	Eq. (21)	N.A.	Eq. (24)	q_0	f_0
12	4	Eqs. (14) and (4)	Eqs. (14) and (16)	Eqs. (14) and (15)	Eq. (21)	Eq. (24)	N.A.	q_0	f_0
13	5	Eqs. (14) and (4)	Eqs. (14) and (16)	Eqs. (14) and (15)	Eq. (20)	N.A.	N.A.	q_0	f_0
14	5	Eqs. (14) and (9)	Eqs. (14) and (16)	Eqs. (14) and (15)	Eq. (21)	N.A.	Eq. (24)	q_0	f_0
15	5	Eqs. (14) and (9)	Eqs. (14) and (17)	Eqs. (14) and (15)	Eq. (21)	Eq. (22)	N.A.	q_0	f_0
16	6	Eqs. (14) and (9)	Eqs. (14) and (20)	Eqs. (14) and (17)	Eq. (20)	N.A.	N.A.	q_0	f_0
17	6	Eqs. (14) and (9)	Eqs. (14) and (20)	Eqs. (14) and (17)	Eq. (21)	N.A.	Eq. (24)	q_0	f_0
18	6	Eqs. (14) and (9)	Eqs. (14) and (20)	Eqs. (14) and (17)	Eq. (21)	Eq. (22)	N.A.	q_0	f_0
19	7	Eqs. (14) and (4)	Eqs. (14) and (17)	Eqs. (14) and (15)	Eq. (20)	N.A.	N.A.	0	f_0
20	7	Eqs. (14) and (9)	Eqs. (14) and (17)	Eqs. (14) and (15)	Eq. (20)	N.A.	N.A.	0	f_0
21	7	Eqs. (14) and (9)	Eqs. (14) and (17)	Eqs. (14) and (17)	Eq. (20)	N.A.	N.A.	0	f_0

$$E(1) = 0, \ \tau \geq 1 \tag{32b}$$

where

$$C_0(\tau) = \frac{1}{2\tau} \tag{33}$$

$$C_1(\gamma,\tau) = \frac{1}{2\tau}\ (1 - 2\ \Gamma\ \tau^{0.5}) \tag{34}$$

$$C_2(\gamma,\tau) = -\ \frac{\Gamma}{2\tau^{0.5}} \tag{35}$$

Equation (32) is a Riccati equation and has to be solved numerically; it also holds for error in discharge. The distribution of error with τ is shown in Figure 1. The distribution is highly skewed, with a sharp rise and gradual decline over an extended range of τ. For a fixed τ, the error increases with decreasing γ.

7 ERROR EQUATIONS FOR TIME-INDEPENDENT FLOWS: NON-ZERO VELOCITY AT THE UPSTREAM BOUNDARY

7.1 Kinematic Wave Solution

Equation (1) takes the form

$$\frac{d(uh)}{dx} = q_0 \tag{36}$$

which has the solution

$$uh = a + q_0 x \tag{37}$$

where a is constant of integration. The KW approximation is given by equation (28). If

$$H = (\frac{\beta}{S_0})^{1/3}\ (u_0 h_0)^{2/3} \tag{38}$$

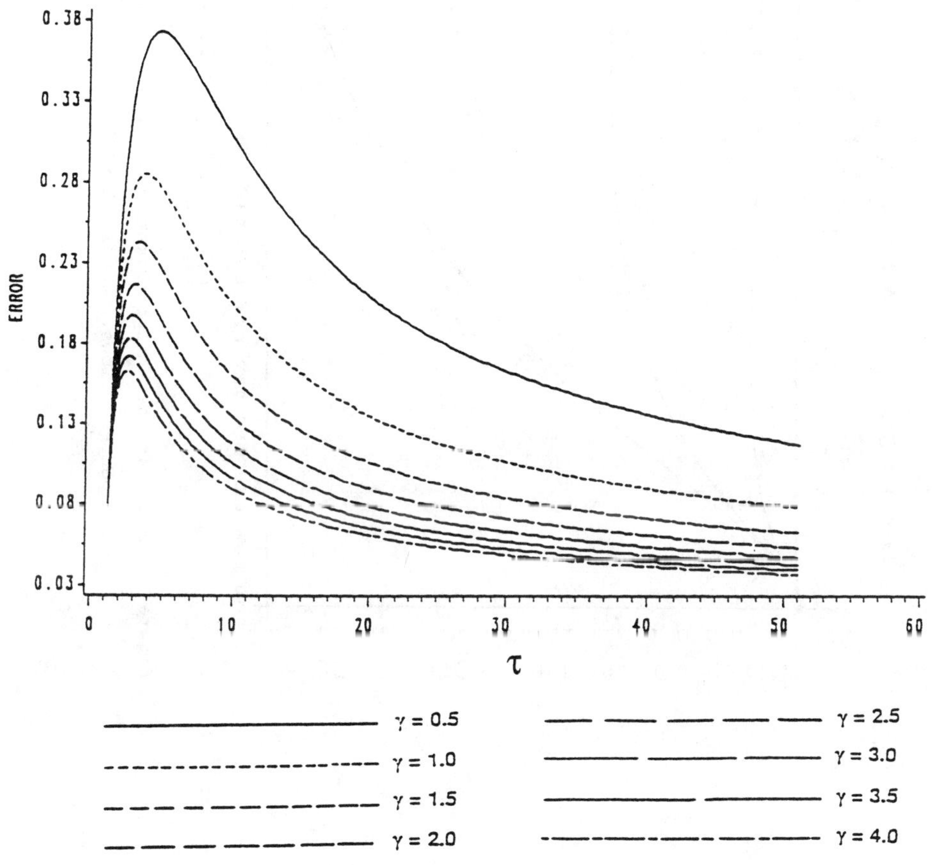

Figure 1. Error in the KW or DW approximation as a function of dimensionless time for space-independent flows.

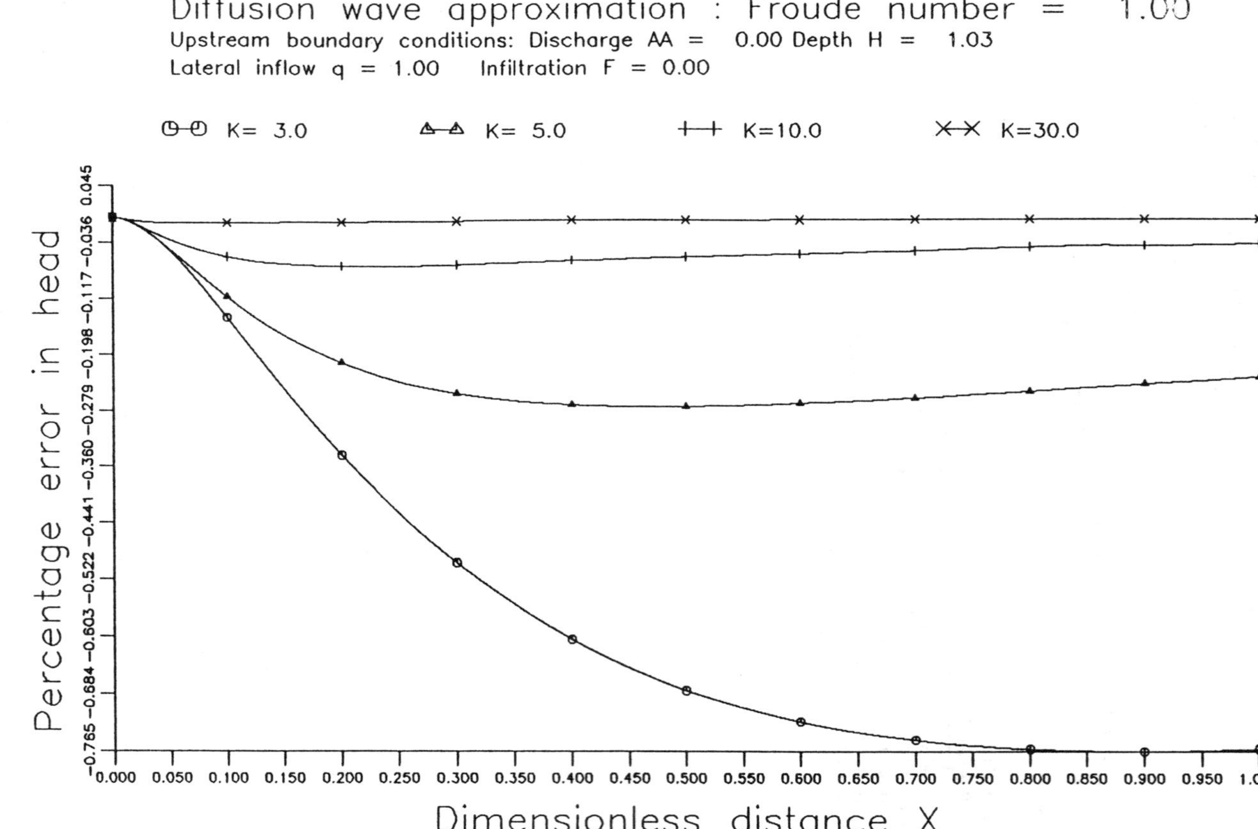

Figure 2. Error in the KW approximation as a function of dimensionless distance for time-independent flows.

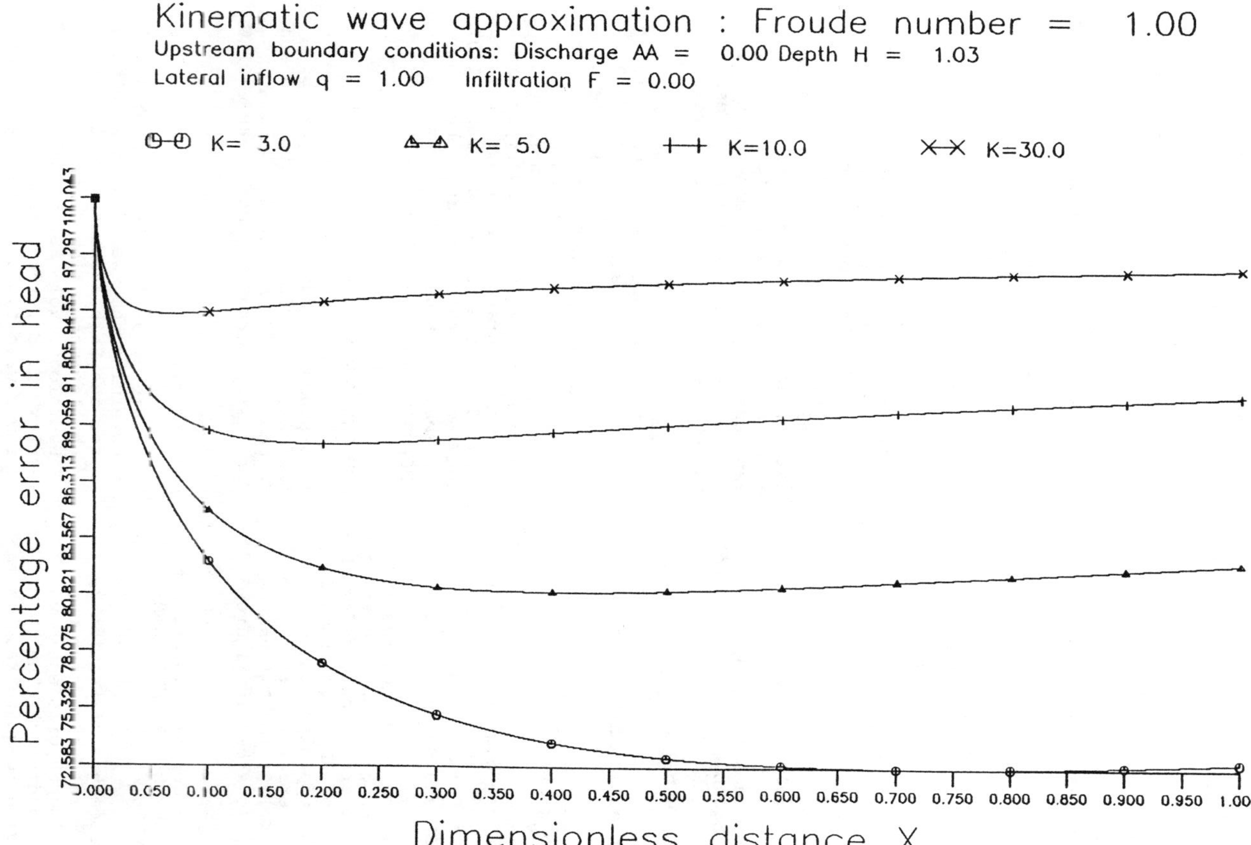

Figure 3. Error in the DW approximation as a function of dimensionless distance for time-independent flows.

then

$$h_* = \frac{h}{H} = \left(\frac{q_0 x + a}{u_0 h_0}\right)^{2/3} \quad (39)$$

7.2 Diffusion Wave Solution

Equation (1) has the solution given by equation (37). Equation (2) reduces to

$$\frac{dh}{dx} = S_0 - \beta \frac{u^2}{h} \quad (40)$$

With the upstream boundary condition given by equation (20), equation (40) can be written in terms of h as

$$\frac{dh}{dx} = S_0 - \beta \frac{(q_0 x + a)^2}{h^3} \quad (41)$$

7.3 Dynamic Wave Solution

Equation (1) has the solution given by equation (44). Equation (2) reduces to

$$\frac{d}{dx}\left(\frac{1}{2} u^2 + gh\right) = gS_0 - g\beta \frac{u^2}{h} \quad (42)$$

which can be expressed in terms of h as

$$\frac{dh}{dx} = \frac{1}{gh^3 - (q_0 x + a)^2} [gS_0 h^3 - q_0 (q_0 x + a) h - \beta (q_0 x + a)^2] \quad (43)$$

7.4 Errors of KW and DW Approximations

The DW approximation gave good results for the entire channel reach with error less than 1%. The error decreased with increasing values of KF_0^2, as shown in Figure 2. The KW approximation gave adequate results (error < 10%) in the region ($0.1 \le x \le 1.0$) for F_0^2 values greater than 30, as shown in Figure 3.

8 ERROR EQUATIONS FOR TIME-INDEPENDENT FLOWS: ZERO DEPTH-GRADIENT DOWNSTREAM BOUNDARY

Following the procedure of Section 7, error equations were derived for the case of the zero-depth gradient downstream boundary condition. The DW approximations gave adequate results in the region ($0.1 \le x \le 1.0$) and for KF_0^2 values greater than 30. For larger values of KF_0^2, the errors were less than 1% in the region ($0.1 \le x \le 1.0$) and increased to about 21% at the upstream end of the channel. The KW approximation gave reasonable results (errors < 10.1) for KF_0^2 greater than 30 and in the region $0.1 \le x \le 1.0$.

9 ERROR EQUATIONS FOR TIME-INDEPENDENT FLOWS: CRITICAL FLOW-DEPTH DOWNSTREAM BOUNDARY CONDITION

The error equations for the critical depth downstream boundary condition were obtained following the procedure of Section 7. For the DW and KW approximations, the errors obtained were almost of the same magnitude as of those obtained using the zero depth gradient downstream condition in the region $0.1 \le x \le 1.0$.

10 CONCLUSIONS

For space-independent flows, the kinematic-wave and diffusion-wave approximations are sufficiently accurate when the dimensionless parameter $\gamma \ge 3$. This parameter reflects the effect of initial flow depth, bed slope, lateral inflow, and channel roughness. The error of these approximations declines exponentially when the dimensionless time exceeds 5. The dimensionless time is obtained with the use of initial depth.

For time-independent flows with a constant depth upstream boundary condition and with parameters ($f = 0.0$, $a = 0.0$, $H = 1.03$, $F_0 = 1.0$), the DW approximation was quite accurate, with errors in depths of less than 1%. The errors decreased with an increase in the value of KF_0^2. The KW approximation was accurate with errors of less than 10% in the region ($0.1 \le x \le 1.0$) and for KF_0^2 values greater than 30. For smaller values of KF_0^2, the KW approximation was not adequate.

ACKNOWLEDGMENT

This study was supported in part by Army Research Office under the project "A Continuum Model for Streamflow Synthesis," Grant No. DAAL03-89-G-0116.

REFERENCES

Daluz Viera, J. H., 1983. Conditions governing the use of approximations for the Saint Venant equations for shallow water flow. Journal of Hydrology, Vol. 60, pp. 43-58.

Ferrick, M. G., 1985. Analysis of river wave types. Water Resources Research, Vol. 21, No. 2, pp. 209-220.

Fread, D. L., 1985. Applicability criteria for kinematic and diffusion routing models. Laboratory of Hydrology, National Weather Service, NOAA, U.S. Department of Commerce, Silver Spring, Maryland.

Lighthill, M. J. and Whitham, G. B., 1955. On kinematic waves: 1. flood movement in long rivers. Proceedings, Royal Society, London, Series A, Vol. 229, pp. 281-316.

Menendez, A. N. and Norscini, R., 1982. Spectrum of shallow water waves: an analysis. Journal of the Hydraulics Division, ASCE, Vol. 108, No. HY1, pp. 75-93.

Morris, E. M. and Woolhiser, D. A., 1980. Unsteady, one dimensional flow over a plane: partial equilibrium and recession hydrographs. Water Resources Research, Vol. 16, No. 2, pp. 355-360.

Ponce, V. M., Li, R. M. and Simons, D. B., 1978. Applicability of kinematic and diffusion models. Journal of the Hydraulics Division, ASCE, Vol. 104, No. HY3, pp. 353-360.

Ponce, V. M. and Simons, D. B., 1977. Shallow wave propagation in open channel flow. Journal of the Hydraulics Division, ASCE, Vol. 103, No. HY12, pp. 1461-1475.

Singh, V. P., 1992a. Errors in kinematic wave and diffusion wave approximations for space-independent flows: 1. Cases 1 to 9. Water Resources Report WRR 17, Department of Civil Engineering, Louisiana State University, Baton Rouge, Louisiana.

Singh, V. P., 1992b. Errors in kinematic wave and diffusion wave approximations for space-independent flows: 2. Cases 10 to 19. Water Resources Report WRR 18, Department of Civil Engineering, Louisiana State University, Baton Rouge, Louisiana.

Singh, V. P. and Aravamuthan, V., 1992a. Errors in kinematic wave and diffusion wave approximations for time-independent flows: Cases 1 to 7. Technical Report WRR20, Department of Civil Engineering, Louisiana State University, Baton Rouge, Louisiana.

Singh, V. P. and Aravamuthan, V., 1992b. Errors in kinematic wave and diffusion wave approximations for time-independent flows: Cases 8 to 14. Technical Report WRR21, Department of Civil Engineering, Louisiana State University, Baton Rouge, Louisiana.

Singh, V. P. and Aravamuthan, V., 1992c. Errors in kinematic wave and diffusion wave approximations for time-independent flows. Cases 15 to 21. Technical Report WRR22, Department of Civil Engineering, Louisiana State University, Baton Rouge, Louisiana.

Woolhiser, D. A. and Liggett, J. A., 1967. Unsteady one-dimensional flow over a plane - the rising hydrograph. Water Resources Research, Vol. 3, No. 3, pp. 753-771.

Environmental Management, Geo-Water & Engineering Aspects, Chowdhury & Sivakumar (eds)
© 1993 Balkema, Rotterdam. ISBN 90 5410 099 0

Application of microfiltration in water and wastewater treatment

S. Vigneswaran & H. Prasanthi
School of Civil Engineering, University of Technology, Sydney, N.S.W., Australia

ABSTRACT: Microfiltration is one of many membrane processes used in various applications due to its low energy requirement and high filtration rate. Crossflow microfiltration (CFMF), wherein the flow is parallel to the membrane surface, thus reducing the clogging, is a very useful technique to treat water and wastewater. In the last two decades much research work has been carried out in improving CFMF to apply more prominently to the water and wastewater treatment systems. This paper reviews the recent efforts in this field to give the reader an overview of the application of CFMF in water and wastewater treatment and its advantages over the conventional processes generally used.

1 INTRODUCTION

Membrane process was initially introduced in 1950s as a part of a research program in the United States concerning the development of sea and brackish water desalination processes. This marked the beginning of the industrial history of membranes. Though many researchers have been working in this field since then, not much progress was achieved until the seventies. In the following years, much improvement was accomplished and as a result of the efforts of many researchers, membrane processes are now being used in variety of applications. Membrane filtration, in recent years, started playing an important role in pollution abatement and control with its wider utilization in both water and wastewater treatment systems.

Membrane processes used in water and wastewater treatment can be categorized into four classes according to the size of particles that can be retained, namely, reverse osmosis (RO), ultrafiltration (UF), microfiltration (MF) and electrodialysis (ED). ED is a proven process for desalting brackish water. RO is also used extensively in desalting applications. It has an added advantage of being able to remove many organic compounds in addition to ionic species and microorganisms. The UF and MF techniques are useful in removing macromolecules, colloids and suspended solids.

Microfiltration (MF) is an important separation process as the permeate flux (or filtration rate) is higher than the other membrane processes and the permeate quality is much better than the conventional separation processes like sedimentation, centrifugation, filtration, flotation, etc. Furthermore, most of the pollutants present in water and wastewater have particle sizes ranging from 0.05 μm to 10 μm and can be removed by MF as they fall within the range of the microfiltration membrane pore size limit.

Microfiltration is a pressure driven process with a microporous membrane as a separating media. In this process, both colloidal and suspended solid particles can be separated from the liquid. The application of high pressure to the feed side of the membrane enables the passage of water through the membrane. This makes bigger-sized particles concentrate on the high pressure side, while the smaller-sized particles infiltrate on both sides of the membrane.

The conventional rapid sand filtration system, where the flow direction is perpendicular to the filter medium, is known as direct filtration or dead-end filtration. In this system, solid particles settle and deposit over the filter medium. Settled solid particles also contribute to the filtration process by retaining other solid particles in the solution to be treated.

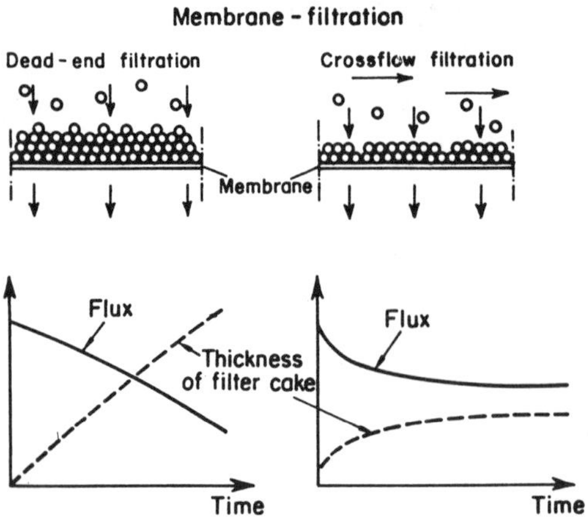

Membrane - filtration

Dead-end filtration | Crossflow filtration | Membrane | Flux | Thickness of filter cake | Time

Fig. 1 Dead-end filtration and crossflow filtration (Ripperger, 1989)

This phenomenon of removal of particles is called "dynamic filtration". Later, when the filtration efficiency reduces due to the excess deposition of particles, the process is stopped for cleaning. But in the case of membrane filtration, membranes, due to their physical and chemical characteristics, themselves are capable of successful separation. So the contribution of retained particles is not essential. Furthermore, if flow direction is perpendicular to the membrane as in the case of conventional filtration, the solid particles retained block the membrane pores and reduce the filtrate flux significantly. This also damages the membrane, thus reducing the life span of the membrane and overall effectiveness of filtration.

To overcome this problem, the crossflow filtration technique is used. This reduces the deposition and increases the efficiency and life span of the membrane. In this technique, since the feed flow is tangential to the membrane i.e., along the membrane surface and perpendicular to the permeate flow direction, the term microfiltration is thus modified to crossflow microfiltration (CFMF). Crossflow generates a shearing force and/or turbulence over the filter medium and limits the thickness of the filter cake, thus maintaining a steady state. Fig. 1 illustrates the schematic of dead-end filtration and crossflow filtration, and the variation of deposit thickness and filtrate flux with time.

Application of MF membranes is well recognized in water and wastewater treatment and also in other domains such as the food and agro industry, chemical industry, metallurgy, biotechnology, paper and pulp industry, pharmaceutical industry etc. Although internal clogging is observed as a drawback in MF systems, because of its higher flux rate, easy washing facilities, flexibility and economy, the application and development rate of microfiltration has quickly progressed during the last decade.

In this review, the applications of microfiltration in water and wastewater treatment are discussed. The results obtained with laboratory-scale, pilot-scale and industrial-scale experiments of microfiltration units are incorporated.

2 APPLICATION OF CFMF IN WATER TREATMENT PROCESSES

Water treatment is possible in conventional systems only by using a number of unit operations like flocculation, sedimentation and filtration. But in the case of the application of CFMF to water treatment, a one-step operation is just enough. CFMF, in water industry, is useful in potable water production, NO_3^- removal, filter backwash wastewater treatment etc.

2.1 CFMF in drinking water supply

Potable water free from microorganisms can be produced by CFMF process using membranes with pore size of 0.2 μm. Thus, a separate disinfection operation is not necessary with this system.

A particular commercial unit, the IMECA microfiltration system, employing inorganic ceramic membranes is used in many potable water production installations in France at varying capacities from 5 to 400 m^3/h. Ceramic asymmetrical membranes of 0.2 μm pore size are used. In addition to the continuous removal of suspended solids, this system can also remove algae, spores, yeast, fungi, parasites, bacteria, amoeba, mineral colloids and ferric compounds.

The system, which incorporates a self-cleaning arrangement, requires only about 0.8 kWh per cubic meter of water treated. The cost is comparable to a conventional system as this MF operates at high filtration velocity of the order of 1 $m^3/m^2.h$.

Wide application of the system for potable

Table 1. Operating parameters used for the choice of membranes

Membrane	Pore size (μm)	Fibre diameter (mm)	Crossflow velocity (m/s)	Operating pressure (bar)
Creaver	0.2	7	0.71	1.2
			1.40	1.2
			0.50	0.5
BRO23	0.1	1	0.75	0.5
			2.50	1.2

Table 2. Bacterial count in the treated water using membranes

Nature of bacteria	Bacterial count
Total coliforms/100 mL	0-140
Fecal coliform/100 mL	0-92
Streptococci fecal/100 mL	0-38
Clostridium/100 mL	0-60
MPN	
at 37°C	<5-108
at 22°C	<5-360

water treatment indicated the following advantages:

* reduction of the usual treatment cycle;
* minimum space and construction cost;
* minimum maintenance (the system is completely automatic and self-cleaning);
* the treated water is crystal clear and without bacterial contamination.

2.2 CFMF in community water supply

CFMF can be used very efficiently for small community water supplies. Since the higher variation of influent turbidity has no significant effect on effluent quality, CFMF is preferable to conventional filtration systems. One notable example of such an application is in Amoncourt, East of France, where treated water is supplied to a small community of 320 people. Previously 180 m^3/d of water was treated by direct filtration and was found unsatisfactory for a higher variation of influent turbidity from 2-5 NTU to 300 NTU. Inhabitants of Amoncourt avoided water supplied for drinking during the highly turbid season and spent 700 French francs/capita/year (approximately A $ 200/capita/year) for the purchase of bottled mineral water. Introduction of CFMF in the supply systems solved this problem. Two different membranes (of pore size 0.1 and 0.2 μm) were tested for their suitability in laboratory-scale experiments. On the basis of easy declogging, acceptable flux (100 L/m^2.h) for a longer period at lower pressure (0.5 bar), and lower energy cost (85 Wh/m^3), organic membranes of type BRO23 were selected. Ceramic membranes needed higher energy (400-550 Wh/m^3) for water production. The effluent was completely free from bacteria and turbidity was less than 0.1 NTU.

Operating parameters were analyzed at a pilot scale unit with BRO23 organic membrane, with 5 m^2 of membrane area (Table 1). The pressure required for the turbidity removal of less than 25 NTU remained almost constant; whereas higher pressure was required for higher influent turbidity.

The BRO23 membrane was used in a real treatment plant in Amoncourt for a maximum production of 10 m^3/h. There were 2 independent lines of 10 modules in parallel in the unit. During the 6 month study, the influent turbidity was observed to vary from its normal value of 2-5 NTU to as high as 100-180 NTU while the effluent turbidity was always less than 0.1 NTU. The membrane was regenerated only twice, after 30 and 130 days of operation. Pressure had to be varied for a constant water production rate. Table 2 presents the bacterial count of the effluent and Table 3 gives the concentration of Al and Fe in the influent and effluent.

2.3 CFMF with in-line flocculation in water treatment

In-line flocculation prior to microfiltration is a good

Table 3. Al and Fe concentration of raw and treated water

Parameter	Raw water	Treated water
pH	7.14-7.46	7.14-7.59
Al (μg/L)	90-730	2-15
Fe (μg/L)	10-140	Below detectable level

Table 4. Influence of in-line flocculation in Garonne river water filtration

Poly-Aluminium Chloride Dose (mg/L)	Steady State Filtration Flux ($m^3/m^2.h$)
0	0.22
60	0.47

method to reduce the internal clogging of membranes. Flocculation is used to achieve two objectives: eliminating the penetration of colloidal particles into the membrane pores and modification of deposit characteristics to increase the filtration flux and to reduce the cleaning requirements.

The mechanism involved in this process is simple. Internal clogging caused by the colloidal particles reduces the permeate flux and life span of the membrane. To avoid this, the membrane can be selected with lower pore size range. By this, the internal clogging phenomena can be controlled. However, higher pressure has to be applied and lower flux will result. To overcome the high energy requirement and low flux problems, prior-flocculation can be used. Prior flocculation of the suspension makes the particle size bigger (microfloc formation) so that CFMF can be used with lower pressure and higher flux.

A study with artificial suspension and poly-aluminium chloride as flocculant indicated that the permeate flux was eight times higher than that without flocculant when the flocculant dose was optimum (Ben Aim et al., 1988). This led to the studies to be conducted with surface water. The laboratory-scale CFMF studies using Garonne river (France) water showed that flocculation clearly improves both permeate flux and effluent quality. Table 4 shows the results obtained in this study. The main factors to be considered in this process are shear or crossflow velocity and flocculant dose.

Peters and Pedersen (1990) conducted a trial test with MEMCOR - Crossflow microfiltration units to treat reservoir water with alum flocculation and gas backwashing. The flocs backflushed from membranes contained enough air to float them to the surface and to form a concentrated sludge layer. This was an added advantage in the system with gas backflushing in large scale applications to reduce sludge volume drastically. Three case studies of problematic surface waters in England and Scotland also were reported by Peters and Pederson (1990). The results are shown in Table 5.

2.4 Natural water treatment using CFMF with backflush cleaning technique

Although CFMF was found to be an effective method for clarification and purification of water for potable purposes, the main problem still encountered is the rapid membrane fouling due to the clogging of membrane pores by colloids, fine particles and dissolved natural organics, with a decrease in the filtrate flux as a consequence. In order to reduce the clogging, many techniques are currently in practice, of which periodic backflushing is the most popular.

In a study reported by Boonthanon et al. (1991), CFMF with backflush technique was used to treat natural water collected from the Chao-Phraya River of Bangkok City. In this study, the membrane used was of ceramic in nature with two layers: a membrane layer comprising several layers of well defined texture of porous ceramic and a support layer composed of macroporous layers of α-alumina. The membrane has a very good chemical resistance (e.g. it can withstand pH range of 0.5-13.5) and at the same time is capable of sustaining mechanical stress (e.g. burst pressure > 30 bars). A single channel membrane with a pore size of 0.2 μm was used. Table 6 gives the characteristics of the river water used in this study.

Although the influent turbidity was low, at the end of the 2-h operation when no backflush was used, the final flux was 0.194 $m^3/m^2/h$. However, when the backflush technique was used, a 21-24% increase in the final flux was obtained, the maximum being 0.240 $m^3/m^2/h$. An improvement of 50-56% in net cumulative permeate volume was also

Table 5. Raw water and permeate qualities of three problematic surface waters, treated by MEMCOR-microfiltration system (Peters and Pederson, 1990)

Water quality parameter	A		B		C	
	Raw	Filtrate	Raw	Filtrate	Raw	Filtrate
E. coli	220	0	18	0	13	0
Coliforms	450	0	18	0	14	0
Turbidity (FTU)	2.6	<0.5	2	0.5	23	0.1
Aluminium (mg/L)	-		0.03	0.02	0.07	0.04
Manganese (mg/L)	0.35	<0.01	0.06	0.02	0.5	<0.01-0.05
Iron total (mg/L)	0.05	<0.01	0.45	0.07	0.85	0.04
Colour (Hazen)	55	0.10	43	11	30	<5

Note:

A South of England: Major river source. System fed direct from river intake pumps with no pre-treatment

B Scotland: Untreated upland reservoir water

C North of England: Upland reservoir. Pre-treatment with 1-2 ppm primary polyelectrolyte to reduce filtrate Hazen levels to less than 1

Table 6. Characteristics of water from Chao-Phraya River, Bangkok, Thailand (Boonthanon et al., 1991)

Parameter	Characteristics
pH	7.26
Conductivity (µS/cm)	275
Turbidity (NTU)	92
TOC (mg/L)	4.6
Suspended solids (mg/L)	73.2

observed with a maximum value of 56% improvement when a T_f (backflush frequency, min) of 5 min and a T_b (backflush duration, s) of 5 s were used. The second best result of 55% improvement in net cumulative permeate volume was obtained when the T_f was 1 min. and the T_b was 2 s. The final flux at this combination was also at its maximum value, i.e. 0.240 $m^3/m^2/h$ (Table 7). The results obtained indicate that the backflush technique improves the flux to some extent (0.194-0.240 $m^3/m^2/h$ which corresponds to a 24% improvement). The lower value of flux obtained when natural water was used as suspension, without backflush, can be attributed to the fact that the fine particles and colloids, whose size range is less than that of the membrane pores, continuously clog the inner membrane surface of the micropores as well as the external surface, thus increasing the membrane fouling to a great extent, which, eventually, reduces the permeate flux. However, when backflush technique was applied, most of the particles and colloids adsorbed/deposited on both the membrane surface and the internal surface of the pores were flushed out. A continuous but gradual decrease in flux with time indicated that the clogging was not removed completely. The decrease in flux may be due to the fine particles present in water.

Table 7. Performance of CFMF with backflush in treating natural water (after 120 min filtration)

Turbidity of suspension (NTU)	Backflush frequency, T_f (min)	Backflush duration, T_b (s)	Final flux ($m^3/m^2/h$)	Percentage improvement in flux	Net cumulative permeate volume (cm^3)	Percentage improvement in net cumulative permeate volume
32	0	0	0.194	0	2600	0
29	1	1	0.234	21	3890	50
29	1	2	0.240	24	4017	55
28	1	5	0.226	17	3636	40
32	2	1	0.225	16	3656	41
24	2	2	0.240	24	3930	51
31	2	5	0.202	4	3462	33
32	5	1	0.231	19	3644	40
33	5	2	0.239	23	3740	44
29	5	5	0.240	24	4064	56

Table 8 Performance of CFMF with backflush in treating filter backwash water (after 120 min filtration)

Turbidity of suspension (NTU)	Backflush frequency, T_f (min)	Backflush duration, T_b (s)	Final flux ($m^3/m^2/h$)	Percentage improvement in flux	Net cumulative permeate volume (cm^3)	Percentage improvement in net cumulative permeate volume
233	0	0	0.235	0	3185	0
223	1	1	1.152	390	12204	50
215	1	2	1.175	400	11753	55
217	1	5	0.772	228	7882	40
232	2	1	1.004	327	11499	41
218	2	2	0.744	216	9151	51
221	2	5	1.127	379	11054	33
231	5	1	0.871	270	8585	40
227	5	2	0.988	320	9676	44
226	5	5	1.010	330	9783	56

2.5 CFMF to treat filter backwash wastewater

CFMF can find its application even in processes other than potable water production. In the conventional water treatment plants, a significant water quantity of 3-6% of water production is used for backwashing the filters and this is generally discharged as a waste. CFMF can be used in treating this wastewater which can be reused for many purposes or recycled back to the filters.

Boonthanon et al. (1991) have used filter backwash wastewater from Bangkhen Water Treatment Plant in Bangkok to investigate the efficient use of CFMF in treating this suspension.

During these studies it was observed that although the influent turbidity of filter backwash water was high, so was the final filtrate flux. The flux was 0.235 $m^3/m^2.h$ even when no backflush was used. The higher flux compared to the case of natural water may be attributed to the fact that the particles in suspension were flocculated (due to the flocculant added prior to rapid sand filters used in the water treatment plant) and were larger in size compared to that of natural water. It was also observed that when backflush technique was used, the flux further increased to a higher value of 1.175 $m^3/m^2.h$ (which is about 400% higher than that without backflush). The optimal T_f and T_b at which the maximum flux was obtained were 1 min and 2s respectively. However, the optimum T_f and T_b having maximum net cumulative permeate volume were 1 min and 1 s, respectively (Table 8).

The permeate turbidity in all the experiments was in the range of 0.1-0.3 NTU.

As a conclusion to the aspect of application of CFMF to water treatment, it can be said that though the initial costs of incorporating CFMF would be high, the benefits of the system are multifold in that it can be used as an efficient tool in obtaining very pure water devoid of bacteria thus eliminating the necessity of disinfection. An integrated system can be useful not only in treating surface water, but for also regenerating the high volume of filter backwash water which otherwise creates further pollution.

3 APPLICATION OF CFMF IN DOMESTIC WASTEWATER TREATMENT

The application of membrane technology in wastewater treatment was first studied by the US Environmental Protection Agency. Because of the presence of suspended solids in the wastewater, membranes with spiral and tubular geometries were suggested as suitable configurations.

In classical biological wastewater treatment, membranes can be inserted at three locations:

- after the primary sedimentation (UF, MF)
- in the activated sludge tank (MF, UF)
- after the secondary sedimentation, in the tertiary treatment (UF, RO, MF, ED) with or without pretreatment

The studies conducted to test the feasibility of using membranes after primary sedimentation (mainly to reduce the variation in solids load to the activated sludge process) were not found to be economical. The cellulose acetate membranes used for this purpose were found to clog rapidly by the organic materials (Witmer, 1974). CFMF (Hydronautics) used for the purpose led to a very low flux (5 $l/m^2.h$) (Sundaram and Santo, 1977).

CFMF is mainly used in domestic wastewater treatment processes in the following ways:

i. on-line flocculation - CFMF process in wastewater treatment
ii. CFMF in bioreactors
iii. CFMF to replace the secondary clarifier
iv. CFMF as a pre-treatment unit in tertiary wastewater treatment

Most of the above options use UF. In view of economy and higher flux, CFMF will gradually replace UF. However, so far the application of CFMF has not been reported much in the literature since this is a relatively new process compared to UF.

3.1 CFMF in place of a secondary clarifier

UF or CFMF can be successfully used as a substitute to the secondary clarifier (Fig. 2) in secondary treatment of sewage. As shown in the figure, the concentrated sludge is returned to the reactor, whereas the permeate is discharged. The use of UF or CFMF in this field is particularly useful in the following cases.

* wastewater from hospitals, where high risk is involved in discharging the treated effluent, as it may contain toxic chemicals

* when stringent standards are enforced for effluents to be discharged

Though a great deal of experimental evidence is not available in the literature to support the use of a CFMF membrane in place of a secondary clarifier,

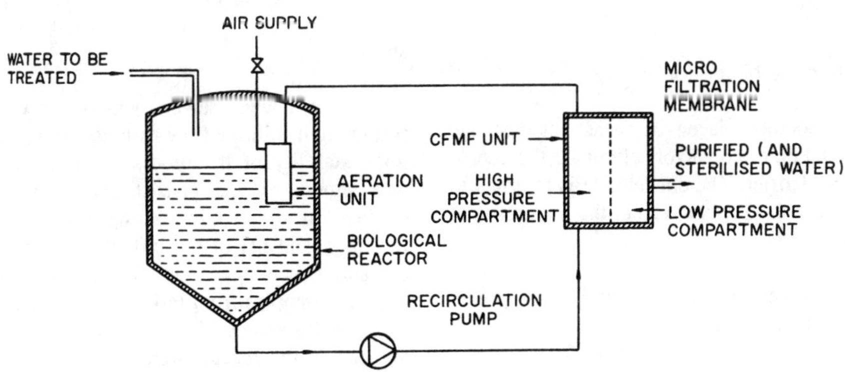

Fig. 2 Biological wastewater treatment with CFMF

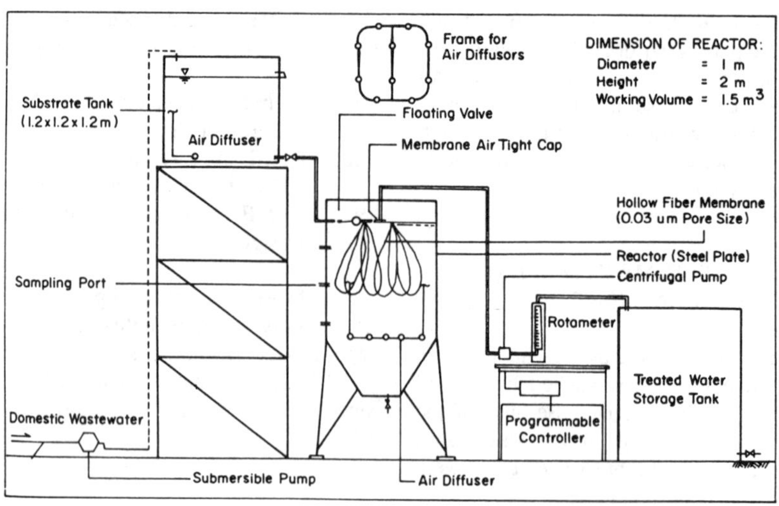

Fig. 3 Schematic of pilot-scale membrane bioreactor (Chiemchaisri, 1990)

a properly selected CFMF membrane (as a secondary clarifier) coupled with an activated sludge process would perform well to produce better effluent at a higher flux rate.

This system has the following advantages:

- can run with varying wastewater loading rates
- produces purified sterilized effluent, free from solids or any other toxic matters
- smaller plant size, since secondary clarifier is replaced by a membrane unit
- smaller aeration tanks since the biomass is recirculated in an effective manner

3.2 CFMF in the bioreactors

Membrane modules placed in an activated sludge reactor (membrane bioreactor) eliminate the use of a secondary clarifier. The concentrate is returned to the reactor and the permeate from the membrane is withdrawn.

Membrane bioreactors have been found attractive in the following applications:

- ■ to treat domestic wastewater which contains toxic compounds (like hospital wastes)
- ■ to treat receiving wastewaters from oysterculture, pisciculture, etc.
- ■ in regions where water is scarce and treated water is recycled for sanitary purposes, lawn watering purposes, etc.
- ■ in treatment plants where large variations in loading and sludge bulking problems occur
- ■ with small treatment units located in building complexes, on-shore activities, etc.

Chiemchaisri et al. (1991) reported their results with experiments conducted using activated sludge process coupled with hollow fibre membrane to treat low strength domestic wastewater. Pilot-scale experimental unit used in this study consisted of 2 modules of 0.03 μm pore size and 9 m^2 surface area each, immersed in the reactor which has a maximum working volume of 1.5 m^3. Fig. 3 gives the schematic of the experimental set up used in this study.

Even when sewage was fed at varying rate following its diurnal flow variation to the bioreactor, good stability of the process was achieved which was shown by the consistency in treated water quality. In this study, lasting sixty days, COD, turbidity, and pH of the influent and effluent were measured once in two days. TKN, NO_2^--N and NO_3^--N of the samples from influent, effluent and reactor were analyzed once in four days. MLSS, MLVSS and pH in the reactor were measured once in two days. The effluent turbidity was observed to be well within 0.5 NTU and it was almost constant

Table 9. Pollutant removal using MEMCO-CFMF with gas backwashing system: Treatment of secondary treated sewage effluent (Peters and Pederson, 1990)

Parameter	Influent	Permeate	Removal efficiency, %
Total coliform (MPN/100 mL)	$8.5x10^4$ - $5.0x10^6$	0	100
Fecal streptococci (MPN/100 mL)	$4.0x10^2$ - $2.4x10^5$	0	100
Enterovirus $TCID_{50}$	10^2 - 10^6	No virus	100
BOD (mg/L)	1-18	<1-7	83
Turbidity (NTU)	6-37	0.6-2	92
Total phosphorus (mg/L)	2.5-9.4	1.8-7.9	22

throughout the run. The results also suggested that the effluent COD concentration maintained a very low value, less than 10 mg/L. It was independent of the variation in the influent COD concentration, 60-200 mg/L, and wastewater loading rate. The COD removal efficiency was in the range of 80-98%.

In a simultaneous study, Chiemchaisri et al. (1991) improved the capacity of the system described above in order to achieve better performance. The improvement was done by introducing a mixing device inside the separation zone and jet aeration device inside the membrane module. Sludge accumulation was found to reduce to a lower value under such conditions, resulting in as much as 4 times of permeate flux improvement. Introduction of intermittent aeration also provided up to 80% of total nitrogen removal by simultaneous nitrification and denitrification resulting in as low as 1-2 mg/L of total nitrogen in the effluent.

3.3 CFMF in tertiary and secondary wastewater treatment

Wastewater effluent is generally disinfected prior to discharge into the water body in order to protect the users downstream. To remove microorganisms and suspended solids, to meet the effluent standards, effluent from secondary clarifier has to be further treated. Disinfection using chlorine would cause formation of THM because of the organic matter left in the effluent. To avoid this, CFMF could be used as a tertiary treatment process.

In Sydney, Australia, large scale studies were conducted with MEMCOR membranes at two treatment plants. The first study was conducted at a sewage treatment plant which provided screening, grit removal, extended aeration, secondary sedimentation, dual media filtration and chlorination. The effluent from the secondary sedimentation was fed to the microfiltration unit (Peters and Pederson, 1990).

A comprehensive microbiological testing program established that all indicator bacteria and viruses were removed by this system operated with gas backwashing. Testing for chemical quality showed significant reduction in BOD, turbidity and oil/grease as well as some reduction for heavy metals and phosphorus, suspended solids being zero.

Consistent filtration rates were achieved over several months without significant biological membrane fouling. The study conclusively showed the capability of the microfiltration system to disinfect and clarify treated sewage.

Table 9 shows the results of secondary treated sewage effluent processes by CFMF with gas backwash.

4 CONCLUSION

Membrane technique, though was in research stages as early as 1950s, is only recently developed for purification and separation of solid-liquid mixtures. Its compactness in size and automatization possibilities, makes it attractive though a few drawbacks such as declining flow rate, membrane deterioration and higher cost exist. New membranes and cleaning techniques discovered overcome these drawbacks.

Microfiltration can successfully be used in water and wastewater treatment, since its separation range falls within the same range as the size of pollutants present in water and wastewater. Most of the existing full scale applications with UF can easily be replaced with MF with small modifications. The use of MF will result in lower cost. It is obvious that more experiments and techno-economical analysis are necessary in order to choose the appropriate membranes and operating conditions.

REFERENCES

Ben Aim, R.; Mietton, P.M.; Vigneswaran, S.; Yamamoto, K. and Boonthanon, S. 1988. Application of Crossflow Microfiltration in Sewage Treatment. In: Advances in Water Pollution Control. Pandward, T.; Polprasert, C. and Yamamoto, K (eds). *Proceedings of the Second IAWPRC Asian Conference of Water Pollution Control*, Pergamon Press, pp. 613-619.

Bersillon, J.L.; Anselme, C.; Mallevialle, J. and Aptel, P. 1989. L'Ultrafiltration appliquee au traitement de l'eau potable: Le cas d'un Petit system, l'eau, l'industrie et les nuisances. pp. 61-64.

Boonthanon, S.; Vigneswaran, S.; Ben Aim and Prasanthi, H. 1991. Use of backflush technique in crossflow microfiltration for treating natural water and filter back wash wastewater in water works, *AQUA*, Vol 40. No. 2, pp. 103-109.

Chiemchaisri, C.C., Yamamoto, K. and Vigneswaran, S. 1991. Household membrane bioreactor in domestic wastewater treatment, *IAWPRC International Conference on Appropriate Waste Management Technologies*, Perth, 27-28 November 1991.

IMECA. 1987. Newsletter on microfiltration, France.

Peters, T.A. and Pederson, F.S. 1990. MEMCOR-Crossflow microfiltration with gas backwashing: design and different applications for a new technical concept. *Proceedings of the 5th World Filtration Congress*, Nice, France, May. pp. 473-478.

Ripperger, S. 1989. Microfiltration. In: *Water, Wastewater and Sludge Filtration.* Vigneswaran, S. and Ben Aim, R. (eds). CRC Press, Florida, USA. pp. 173-190.

Sundaram, T.R. and Santo, J.E. 1977. Removal of suspended and colloidal solids from waste streams by the use of crossflow microfiltration. *ASTM Publication*, ENAS-51.

Vigneswaran, S. and Boonthanon, S. 1992. Crossflow microfiltration with in-line flocculation. *Water, Australian Water and Wastewater Association*, February, pp. 29-31.

Vigneswaran, S., Vigneswaran, B. and Ben Aim, R. 1991. Application of microfiltration for water and wastewater treatment. *Environmental Sanitation Reviews*, Asian Institute of Technology, Bangkok, Thailand.

Witmer, F.E. 1974. The use of semipermeable membrane to filter and renovate sewage effluents. *First World Filtration Congress*, Paris, p F51-1.

Environmental Management, Geo-Water & Engineering Aspects, Chowdhury & Sivakumar (eds)
© 1993 Balkema, Rotterdam. ISBN 90 5410 099 0

Highway transportation and environmental management

L.C.Wadhwa
James Cook University, Townsville, Qld, Australia

ABSTRACT: Two significant developments over the past decades have resulted in profound impacts on issues of environmental management including research focus and policy making. These are the public concern about the environment, quality of life and the greenhouse effect, and demand for greater accountability of public moneys. Highway transportation is responsible for much of urban air pollution and the concern for environment has resulted in number of legislative measures with significant implications for the transportation sector. The requirements of monetary evaluation of all environmental control programs, in response to demand for greater accountability, has meant that techniques for quantifying the impacts on health, vegetation, ecosystem, property, aesthetics etc. to enable the application of benefit-cost analysis, need to be refined. These issues are discussed along with a number of alternative control strategies aimed at reducing the amounts of pollutants released by motor vehicles.

1. INTRODUCTION

Highway transportation has an important role in almost every aspect of the economic and social activity of a nation. There is a strong relationship between a nation's economy and the amount of travel undertaken on its highway system. Highway transportation has a myriad of positive effects on employment, mobility, accessibility, etc. and in stimulating desirable development. However, highway transportation also creates a number of undesirable impacts such as air and noise pollution, depletion of natural resources, undesirable land uses, traffic congestion, accidents, and inefficient space utilization which result in waste or spoilage of natural endowment causing problems, displeasure, anxiety, and grief to society.

Two significant developments over the past decades have resulted in profound impacts on issues of environmental management including research focus and policy making. These are

(a) the public concern about the environment, quality of life and the greenhouse effect, and

(b) demand for greater accountability of public moneys.

In the light of these developments, this paper (i) discuss the environmental impacts of highway transportation, (ii) reviews the methodologies for quantifying air-pollution effects on health, vegetation and ecosystems, property and aesthetics including visibility, and their economic value, (iii) develops a framework for the highway transportation and environmental management studies and evaluating pollution control strategies, and (iv) discusses the application of this approach to regulations concerning motor vehicle emissions.

1.1 Concern for environment

There has been a growing concern relating to the protection of our environment and quality of life during the past 25 years. This "environmental conscience" has affected the planning and development of many major projects including highway transportation projects.

Transportation is a leading consumer of scarce oil resources and a major contributor of the principal greenhouse gas- carbon dioxide. Motor vehicles are also responsible for much of urban air pollution, polluting the atmosphere with carbon monoxide, hydrocarbons, nitrogen oxides, lead, sulphur oxides and particulate matter.

Concern for environment has resulted in the promulgation of stricter emission controls and in global efforts and cooperation on an unprecedented scale in setting goals and standards to reduce emissions of greenhouse gases.

The earliest air pollution legislation in the United States included the Air Pollution Control Act of 1955, a 1959 extension to that act and the Motor Vehicle Exhaust Study Act of 1960. Subsequent legislative measures included:

1963 Clean Air Act; empowered the Department of Health, Education and Welfare (HEW) to convene hearings and conferences on state air pollution; and to pursue polluters in federal courts

1965 Motor Vehicle Control Act; U.S. Congress began to focus on air pollution from the automobile

1966 Department of Transportation Act; provided for protection of publicly owned wildlife refuges, recreation areas, parks and historic sites from highway impacts

1967 Air Quality Act; federal government empowered to designate AQCR's (Air Quality Control Regions), and to specify recommended control strategies

1968 Federal automobile air pollution standards based on the 1967 California standards

1969 Environmental Quality Improvement Act; established Office of Environmental Quality

1969 National Environmental Policy Act; required the preparation of environmental impact statements (EIS), and created Council of Environmental Quality

1970 Clean Air Act Amendments giving federal government the power to set national air quality standards (NAAQS) and to require states to develop implementation plans (SIP's) to meet standards. Federal Motor Vehicle Emissions Control Program (FMVECM) set emission standards for nitrogen oxides, carbon monoxide, and hydrocarbons emitted by new automobiles. These standards were to be tightened successively over a five-year period until emissions from new cars were reduced 90 per cent from their 1970/71 levels. Environmental Protection Agency (EPA) established

1970 Urban Mass Transportation Act; required environmental impact hearings

1975 Endangered Species Act; prohibited federal activity including highway construction which may further jeopardize the existence of a plant or animal on the endangered species list

1975 Federal air quality standards compliance date. EPA authorized to carry out federal air policy and set uniform standards for new sources of pollution.

1977 Clean Air Act Amendments which included a PSD (Prevention of Significant Deterioration) policy. It also required the reexamination of primary and secondary ambient standards for all criteria pollutants every five years.

1990 Clean Air Act Amendments with major impacts on transportation planning and project development in areas not meeting the National Ambient Air Quality Standards for ozone and carbon monoxide.

1.2 Demand for accountability

The second concern has resulted in requirements of monetary evaluation of all environmental control programs. Serious questions are being asked about whether major environmental control programs are worth the cost. The societal resources are limited and many infrastructural and welfare programs compete for these resources. It is imperative that proper evaluations of all programs be carried out if societal resources are to be allocated optimally.

Historical data on pollutant emissions and ambient pollution levels undoubtedly show an improvement in the level of environmental quality in the U.S. These have unarguably resulted in substantial and valuable national benefits in the form of improved human health, recreational opportunities, visibility, and general environment integrity. An ideal comparison of the costs and benefits of pollution control would require that these benefits be identified, quantified , and monetized. This is an extremely difficult task. Total annualized costs for all pollution control activities have been estimated at $115 billion dollars representing just over two per cent of Gross National Product (EPA, 1990). The Environmental Protection Agency has set out, as a priority, to ensure that the resources devoted to achieving the national environmental goals are used as efficiently and effectively as possible.

2. EMISSIONS FROM TRANSPORTATION ACTIVITIES

Transportation activities account for a considerable fraction of the recognized pollutants discharged to the atmosphere in metropolitan areas. Among the principal transportation-related pollutants are carbon monoxide (CO), reactive and non-reactive hydrocarbons, and oxides of nitrogen (NO_x).

2.1 Mobile emission sources

Mobile emission sources can be separated into two major categories:

(i) highway vehicles
- (a) passenger cars and light trucks
- (b) heavy duty trucks and buses, and
- (c) motor cycles

(ii) other sources
- (a) off-highway mobile equipment
- (b) aircrafts

Emissions from highway vehicles form the major part of highway transportation emissions. Emissions from a motor vehicle can differ greatly because of differences in duty cycles and in engine and fuel characteristics. In estimating the emissions from a vehicle population, the class, model, year, mileage and emission factors for each vehicle must be considered.

Off-highway sources include a wide variety of construction equipment, railway locomotives, and marine equipment. These represent concentrated sources of emission with local-area effects. However, the use of diesel engines reduces the amount of emissions. Most railway locomotives are electric with minimal emissions.

Aircrafts are responsible for relatively high yields of CO_x during engine idling and aircraft taxing, and of NO_x during take-off and climb out. Only emissions during the lower altitude (say up to 1000 meters) are assumed to affect surface air quality.

Large-scale land use complexes associated with transportation-related or impacted facilities may be treated as indirect sources for air quality management purposes. Examples of these complexes include shopping centers, airports, sports complexes, and other major activity areas where large number of vehicle trips originate or terminate. These sources represent a large number of cold starts, and distinct pattern of in- and out- traffic flow patterns. In addition, other stationary source emissions (e.g. from space heating and release of volatile organic vapors) may also contribute to the emissions from these complexes. In case of airports, emissions from aircraft landing, taxing, take-offs, and climb outs may be determined from the air traffic measurements and flight path information. Other emission sources at airports include fuel handling, ground service vehicles, etc.

2.2 Variables affecting transportation emissions

With continuous changes in emission standards and technology, the progressive changes in vehicle emissions characteristics have to be properly recognized and duly considered.

Emissions from highway vehicles fluctuate during the vehicle duty cycle - i.e., the sequence of starting, engine idling periods, accelerations, and different travelling speeds. This is recognized by using separate test cycles representing the driving patterns in downtown metropolitan areas and the freeway driving conditions.

(i) Temperature is a critical variable in the production of CO and HC emissions. Vehicles in the cold start phase of operation have higher emissions and lower fuel economy than warmed-up vehicles. Most violations of CO standard occur during winter-time, at temperatures below 10°C.

(ii) Vehicle speed is another significant factor affecting vehicular emissions. CO and HC emissions are higher at slower speeds, especially below 30 kph, which are typical speeds on urban streets and congested highways. The relationship between vehicle speed and emissions is non-linear. At speeds below 30 kph, emission rates increase at a greater rate as speeds continue to decrease. However, NO_x emissions decrease with reduction in speed.

(iii) Vehicle age and maintenance can have considerable effect on the rate of emissions from vehicles. Wear and deterioration of engine parts and emission controls with age can cause emissions to increase. Proper maintenance can significantly reduce this increase in emissions.

2.3 Transportation-produced pollutants

In its normal operations, a motor vehicle burns fuel and emits the by-products of the combustion process into its surrounding environment. The bulk of the byproducts - water vapor and carbon dioxide - are considered relatively harmless, although the effect of carbon dioxide as a greenhouse gas is currently a serious and emotional issue. However, some other byproducts are considered to have serious consequences for the surrounding environment.

Motor vehicle emissions constitute a large portion of three of the five major pollutants:

(a) Carbon monoxide (CO): Most of CO emissions (75 to 95%) come from motor vehicles.

(b) Hydrocarbons (HC): Between 70 to 85% of hydrocarbons are emitted from motor vehicles.

(c) Nitrogen oxides (NO_x): About 50 - 65% of NO_x are contributed by motor vehicle emission with the other major source being the fuel combustion from stationary sources.

(d) Sulfur oxides (SO_x): Primarily contributed by stationary sources as a product of fuel combustion. Transportation sources contribute to just about 10% of SO_x emissions.

(e) Particulates: Primarily contributed by stationary sources with only 5 - 10% from transportation sources.

The principal transportation-related pollutants - carbon monoxide (CO), reactive and non-reactive hydrocarbons, and oxides of nitrogen (NO_x) are now subject to control when emitted by road vehicles. Other pollutants, including several categories of partial oxidation products, particulates, and combustion derivatives of fuels, fuel treatment additives and fuel impurities, should also be recognized in any inventory of source emissions. The use of catalysts to control CO and unburned HC emissions may result in the further oxidation of SO_2 to SO_3 in the exhaust system. Lead emissions resulting from the use of lead tetraethyl as an anti-knock compound in motor fuel have been almost eliminated with the use of lead-free gasoline.

2.4 Air quality and emission standards

Between 1968 (uncontrolled pre-1968) and 1985, the Federal exhaust emission standards required substantial reduction in the amount of pollutants emitted by road vehicles. The first standards were set for HC and CO for automobiles and light-duty trucks in 1968; EPA subsequently set tighter standards in 1970 for HC and CO, added NO_x to the list of controlled pollutants, and also set standards for heavy-duty trucks. Particulate emission standards were set in 1981. Table 1 shows the standards set for various road vehicles.

Table 1. Exhaust (tailpipe) emission standards (grams per mile)

Pollutant	Pre-control	1985-	% reduction
Automobiles (1968-)			
HC	8.2	0.41	95
CO	90.0	3.4	96
NO_x	4.0	1.0	75
Particulates	1.0	0.2	80
Light-duty Trucks (1968-)			
HC	8.2	0.8	90
CO	90.0	10.0	89
NO_x	4.0	1.0	75
Particulates	1.0	0.26	74
Heavy-duty Trucks (1970-)			
HC	20.4	2.0	90
CO	261.0	26.0	90
NO_x	11.4	2.8	75
Particulates	2.0	0.6	70
Motorcycles (1978-)			
HC	12.3	5.0	59
CO	36.7	12.0	67

The 1990 Clean Air Act automobile tailpipe standards are shown in Table 2.

Table 2. 1990 CAA automobile tailpipe standards (gms/mile)

Pollutant	Standard
HC	0.25
CO	3.4
NO_x	0.4

As can be seen from Tables 1 and 2, the CO standard has been retained at 3.4 grams/mile, the

HC standard has been reduced from 0.41 to 0.25 grams/mile, and NO_x standard has also been reduced from 1.0 to 0.4 grams/mile. A new cold temperature CO standard of 10 grams/mile (at 20°F) has been defined beginning with 1994 model.

The new tailpipe standards are to be phased in over time. At least 40% of the 1994 and 80% of 1995 model cars sold must comply with the new standards. By 1996 and all cars sold must meet the lower emission standards.

The automotive emission targets set by the EPA in the 1970's were considered by most to be too ambitious and beyond the capabilities of the existing technological capabilities of automobile manufacturers. The targets were specifically designed to be "technology forcing". The bargaining activity followed and target dates for achieving standards had to be delayed.

3. EMISSIONS AND AMBIENT AIR QUALITY

Three types of basic atmospheric processes provide a link between emissions and ambient air quality.

(a) Transport and diffusion processes determine the movement and dispersion of pollutants in the atmosphere,

(b) Chemical processes convert certain pollutants to other forms, and

(c) Removal phenomenon associated with cleansing of the atmosphere.

These processes are very complex involving many interactions some with antagonistic effects while some pollutants may interact in a biological synergism. Factors influencing these processes include topography and meteorology.

3.1 Transport and dispersion processes

These processes depend primarily on three meteorological factors: wind speed, wind direction, and turbulence. Wind blowing over the surface of the earth and the vertical temperature gradient create a turbulent, well-mixed layer in the lower atmosphere. The height of this mixing layer is usually of the order of about a kilometer during the daytime but can be much less at night. However, if the temperature pattern is inverted so that it increases with height, the pollutant concentration near the ground level will be higher.

3.2 Chemical processes

Once pollutants enter the atmosphere, they may be subjected to a variety of chemical reactions. These may be thermally induced or activated by sunlight. Sunlight has been found to initiate a complicated chain of chemical reactions involving air, nitrogen oxides, and hydrocarbon vapors which yield photochemical smog, also called oxidant, which is primarily composed of ozone. The relationship between oxidant and its precursors is non-linear.

The atmosphere also contains a wide variety of suspended particles (aerosols) that create the visibility reducing haze prevalent in many urban areas. Most of this haze is the result of chemical reactions in the atmosphere with the formation of particulate sulfate and nitrate.

3.3 Removal processes

The atmosphere provides natural processes that remove certain pollutants within a few days of their emission. These processes include the formation of aerosols and their subsequent fallout from the atmosphere; water cloud formation and precipitation; and absorption at the earth's surface into buildings, soil, vegetation, and the seas. The residence times of air-borne pollutants depend on many factors including water solubility and chemical reactivity.

3.4 Relationship between emissions and ambient concentrations

The relationship between emissions and average air quality (ambient concentration) is not one of simple proportionality. For example, a 44% reduction in particulate emissions was found to result in 12-17% reduction in particulate levels. However, an 8% reduction in the emission of sulfur compounds resulted in 15-23% decrease in the ambient concentration of sulfur dioxide (EPA, 1978).

The transportation, dispersion, chemical transformation and removal of pollutants are important mechanisms that govern the ambient air quality resulting from emission of pollutants into the atmosphere. However, our understanding of these processes and interactions is deficient.

3.5 Ambient air quality standards

The National Ambient Air Quality Standards (NAAQS) established for six pollutants - carbon monoxide, sulfur dioxide, nitrogen dioxide, ozone, particulate matter and lead -, under the U.S. Clean Air Act Amendments of 1970 are summarized in Table 3.

Table 3. U.S. national ambient air quality standards (microgram per cubic meter)

Pollutant	Aver. period	Primary	Secondary
CO	8-hr	10,000 (9)	10,000
	1-hr	40,000 (35)	40,000
SO_2	Annual	80 (0.03)	
	24-hr	365 (0.14)	
	3-hr	1,300 (0.5)	
NO_2	Annual	100 (0.05)	
O_3	1-hr		240 (0.12)
PM	Annual	50	
	24-hr	150	
Lead	3-month	1.5	

PM -- particulate matter
Values in parentheses are equivalent ppm

The primary pollutants are emitted directly into the air by a source while secondary pollutants are formed in the atmosphere by a chemical reaction between various pollutants.

4. ENVIRONMENTAL CONSEQUENCES OF TRANSPORTATION

The effects of motor vehicle emissions on the environment, humans, vegetation and ecosystem, exposed materials and aesthetics including visibility are significant. The effects of individual pollutants and their derivatives are shown in Figure 1.

4.1 Effect on the environment

Motor vehicles are the dominant source of many air pollutants such as CO, NO_x, HC, particulates lead, and sulfur compounds.

Carbon monoxide is a colorless, odorless gas produced through the incomplete combustion of motor fuel. It combines with the haemoglobin in the blood reducing the ability of blood to carry oxygen. At concentrations found in urban air, CO can aggravate cardiovascular disease and impair mental function. At high enough concentrations, CO can be fatal to humans. Currently 41 urban areas in the U.S. exceed the air quality standard for CO.

NO_2 is a brownish gas with pungent smell. It is a pulmonary irritant and short exposures to it may increase susceptibility to acute respiratory diseases. School-age children have been found to develop sore throat, coughing and running nose. NO_x can react chemically in the air to form nitric acid and is a contributor to acid rain. It can harm vegetation and a wide variety of materials such as fibre, plastics and rubber.

Although hydrocarbons are not directly harmful, the non-methane hydrocarbons (NMHC) participate in chemical reactions that form NO_2 and ozone. These are also called "reactive hydrocarbons". Hydrocarbons also refer to volatile organic compounds (VOC) such as aldehydes and alcohols. Benzene and formaldehyde are toxic pollutants contributed by transportation.

Transportation contributes to the production of particulate matter through diesel fuel exhaust, and includes unburned hydrocarbons, sulfur dioxide and sulfuric acid. Fine particles small enough to enter the lungs can impair their function and can cause respiratory cancer. Particulate matter can impair visibility, cause corrosion, and soiling of exposed materials.

Although lead is a poisonous heavy metal with grave health effects, transportation sector emissions of lead are declining rapidly due to the use of unleaded fuel.

Ozone, also known as photochemical smog is produced by the photochemical reaction of hydrocarbons and NO_x emissions. It is, therefore, primarily a transport-related pollutant. Ozone is a strong pulmonary irritant which can affect lung functions, cause chest discomfort and can affect people in good health. It cause eye irritation, is toxic to plants and damages many materials. Ozone can travel long distances. About one-third of all American currently reside in areas which do not meet the air quality standard for ozone.

Acid rain is the phenomenon of rain interacting with polluted air having greater acid content than natural rain. The deposition of natural rain has adverse effects on aquatic system, crops, forests, materials, human health, and visibility.

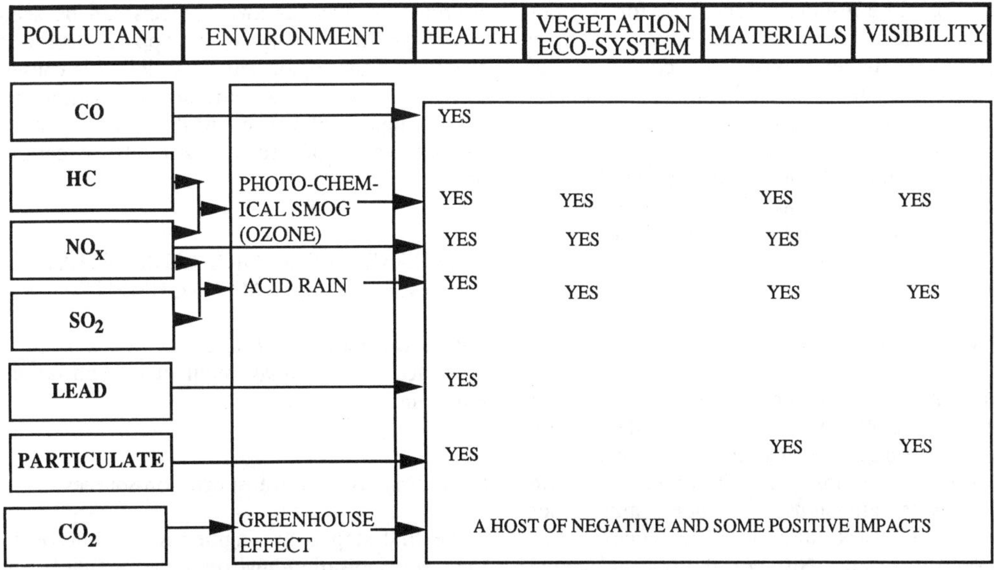

Figure 1. The environmental, physical and physiological consequences of motor vehicle emissions

Transportation emissions which are considered precursors for acid rain formation include VOCs, NO_x, and SO_2.

There has been a dramatic increase in the amount of CO_2 with transportation being a significant contributor. Scientists believe that increases in the level of CO_2 and other greenhouse gases could change the earth's climate. CO_2 allows solar radiations to enter the earth's atmosphere but then traps it preventing heat from escaping into space. A slight increase in earth's mean temperature could affect rainfall patterns, producing floods and draught. Possible melting of polar ice caps, rise in sea level and potential flooding in numerous cities and regions are seen as potential consequences.

4.2 Effect on human health

The effects of air pollution on human health can range from
(i) slight increase of pollutant in the body (body burden) through
(ii) physiologic changes of uncertain consequences to
(iii) pathologic changes of illness and death.

Health effects commonly associated with air pollution include

(a) eye irritation
(b) tightness in the chest
(c) scratchy throat
(d) aggravation of asthma
(e) increased respiratory illness
(f) reduced tolerance of exercise
(g) reduced mental ability
(h) nausea
(i) anemia
(j) heart disease
(k) kidney damage
(l) leukemia
(m) cancer
(n) other ailments.

Methods of estimating health effects include aggregate and dis-aggregate epidemiological studies which use health interview surveys and regression analysis techniques.

4.3 Effects on vegetation and ecosystems

For agricultural crops and livestock, the effects of pollutants are the reduction in the quantity and quality of marketable products. For gardens and flowers, losses take the form of reduction in aesthetic value caused by visible injury to the plant.

A number of studies have been undertaken to

establish dose-response functions for various crops in selected areas. These studies have been undertaken under controlled conditions of temperature, light intensity, humidity and rainfall, and soil nutrients so that the effects of increasing pollutant concentration can be isolated.

Most agricultural crops have been divided into sensitivity classes for each pollutant based on the air-pollution susceptibility of the plant and its economic importance.

4.4 Effects on materials

Many materials of economic value can be damaged by air pollution. The most common form of damage is corrosion of exterior surfaces by acidic conditions, primarily originating from sulfur-oxide emissions. Rubber and other elastomers crack and lose resiliency when exposed to ozone. Nitrogen dioxide and ozone accelerate the fading of fabric dyes, and sulfur-dioxide exposure can reduce the strength of textile fibre.

The rate of corrosion in metals, the loss of paint film thickness through erosion, breaking of the paint film due to poor adhesion or a loss of film flexibility, fading and loss of fibre strength, etc. have been studied as function of increasing pollutant concentration to develop dose-response functions.

4.5 Effect on aesthetic values

Although significantly difficult to estimate, the economic loss from damage to aesthetic values may actually be larger than the losses from damage to vegetation and materials. Aesthetic damage is caused by

(a) odors associated with a variety of pollutants
(b) increased soiling
(c) reduced visibility from suspended particulate matter, and
(d) loss of property values

The degree of an individual's annoyance with an odor depends both on personal reactions to the character and intensity of odor and on the frequency and duration of odor episodes.

Exposed materials may pick up soot and dirt from the air. The character of suspended particulates may vary significantly from place to place being blacker and stickier in some areas.

Reduced visibility may be noticed both as a decrease in how far away objects can be seen clearly and as a change in the apparent color of the sky. This impairment in visibility is caused primarily by very small particles that scatter and absorb light. Photo-chemical smog may be a primary cause of reduced visibility in certain areas.

5. A FRAMEWORK FOR TRANSPORTATION -ENVIRONMENT MANAGEMENT

Figure 2 shows the framework for a systems approach to the management of transportation-environment systems.

5.1 Emissions from transportation sources

The first step is an engineering dimension to the transportation-environment management system.

It involves the estimation of the amount of pollutants generated by conducting an economic activity, in our case the movement of people and goods by road.

Factors influencing the quantity of emissions include the total amount of travel undertaken on the highway system, the emission characteristics of the vehicle fleet (age, size, and model distribution), traffic flow conditions (speed, acceleration, braking, stops, idling etc.), weather conditions etc. This is shown in Figure 3.

Mathematically,

$$Q = f_1(\text{VMT, fleet char., traffic char., weather})$$

where Q is a vector representing the amount of various pollutants generated by motor vehicles, and VMT are the vehicle-miles travelled. An appropriate transportation planning model will be able to provide a reasonable estimate of VMT as a function of land use patterns, socio-economic variables, transportation supply and transportation systems management measures. The emissions generated by other transportation sources must be added to the emissions from motor vehicles. The latter are, however, predominant.

Tonnes of pollutants by type (HC, CO, NO_x, SO_2, etc.) emitted in time and space are obtained at this step.

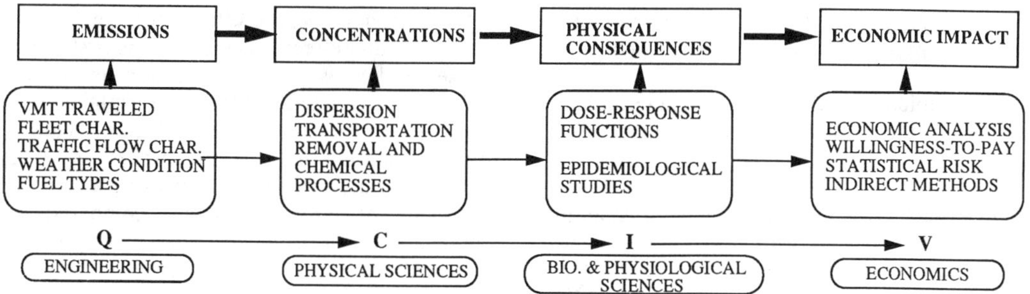

Figure 2. A framework for transportation-environment systems management

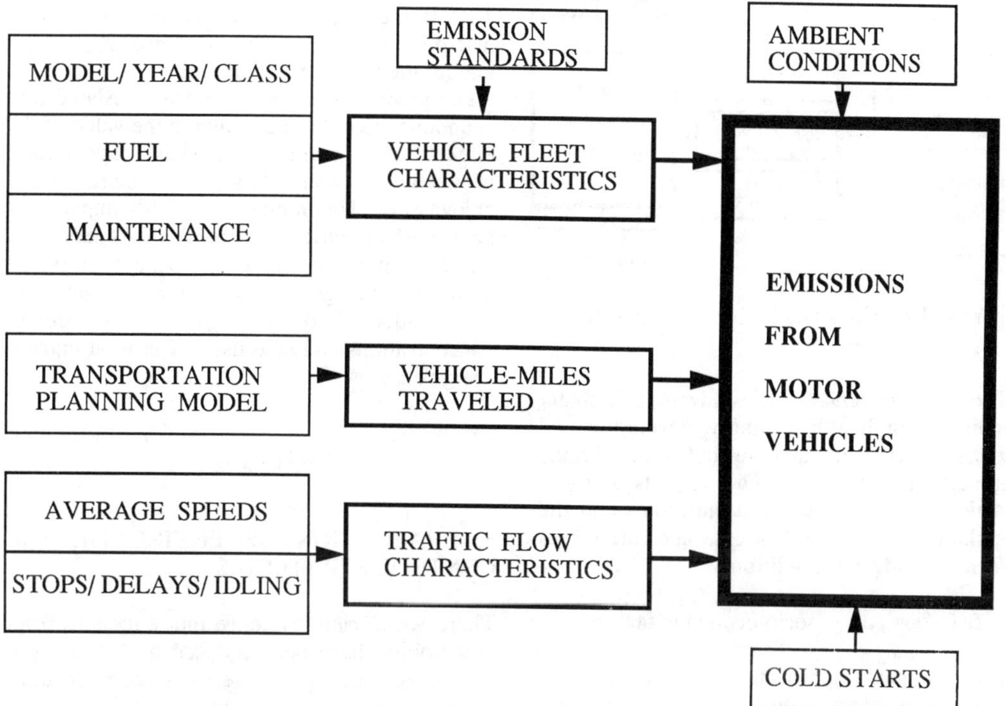

Figure 3. Factors affecting motor vehicle emissions

5.2 Ambient concentration

This step represents the natural science dimension of transportation-environment systems management.

As discussed in section 3 above, transportation of pollutants, dispersion, chemical transformation and removal processes govern the ambient air quality resulting from the emission of pollutants (Figure 4). The ambient conditions resulting from a given quantity of emissions depend on the absorptive capacity of the environment and the various interactions some with antagonistic effects and some with biological synergism. It must be remembered that the environmental impacts are largely a function of ambient concentrations and not the quantity of emissions. Thus the ambient condition vector C is given by

$C = f_2(Q$, topographical char., meteorological factors, environment assimilating capacity).

The ambient concentrations are determined by models of the natural processes discussed above.

5.3 Consequences of adverse ambient conditions

This step represents the physical sciences and physiological dimension of transportation-environment systems management.

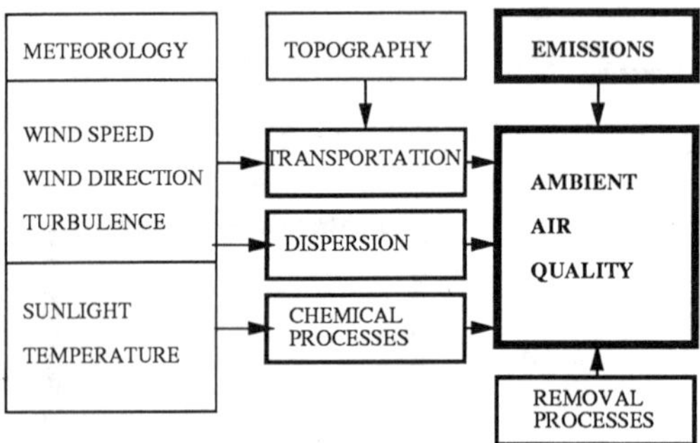

Figure 4. Emissions and ambient air quality

The consequences of adverse ambient conditions on health, property, vegetation and ecosystem, visibility and aesthetics are briefly outlined in section 4. The impacts I are a function of the ambient concentrations and the population, agricultural areas, and properties affected by adverse conditions.

$I = f_3(C$, pop., agri., socio-economic factors)

The common approach to estimating the consequences of air quality is by developing dose-response functions from statistical analysis of available data and by carrying out special surveys. There is a substantial evidence that the response to a pollutant is often related to the total quantity of the pollutant received over the time period of exposure (dose). The amount of response to an increasing dose of pollutant is described by a dose response function. The common forms of dose response functions are linear, threshold (hockey-stick) and sigmoid.

Studies of pollutant effects involve more factors than the concentration of the pollutants and the duration of exposure. In general, air pollutants do not produce unique effects but aggravate or increase other effects. Therefore, the first step in these studies is the identification of all the variables that may be relevant in explaining the changes in impacts. Either control experimentation may be undertaken in which situations which are identical in all respects other than air pollution levels are compared, or multivariate regression techniques may be used to separate the effects of the pollutants from the effects of the other possible influences.

5.4 Estimation of economic impacts

This is the economic dimension of the multi-disciplinary approach to transportation-environment systems management.

The physical consequences are translated into economic effects by determining the value of the effects on health, materials, agricultural productivity, and loss of environmental enjoyment. The value, V_{ijk}, of ith impact (i= health effect, agricultural productivity loss, etc.) due to jth pollutant (j= CO, NO_x, etc.) in region k (the nation's geographical area is assumed to be divided into k regions with similar concentrations and land-use). The total impacts are given by

$$V = \Sigma_i \Sigma_j \Sigma_k V_{ijk}$$

6. EVALUATION OF ENVIRONMENTAL CONTROL STRATEGIES

There is sufficient evidence that emissions from automobiles have been reduced by U.S. revised standards. However, little work seems to have been undertaken to demonstrate the cost effectiveness of these reductions. On the contrary, many believe that these reductions could have been achieved at lower costs.

Actual emissions as a percentage of estimated emissions using 1970 levels of control are stated by EPA (1990) in a report to the U.S. Congress are given in Table 4.

However, the commitments embodied in Clean Air Act and its various amendments involve very large costs for the society. With billions of dollars at issue, the question of whether it is all worth is certainly well taken.

Table 4. Actual emission as a % of estimated emissions using 1970 levels of control

Pollutant	1984	1988
Particulate matter	33	30
Nitrogen oxides	82	72
Sulfur dioxide	71	58
VOC	60	58
Carbon monoxide	56	43
Lead	19	3

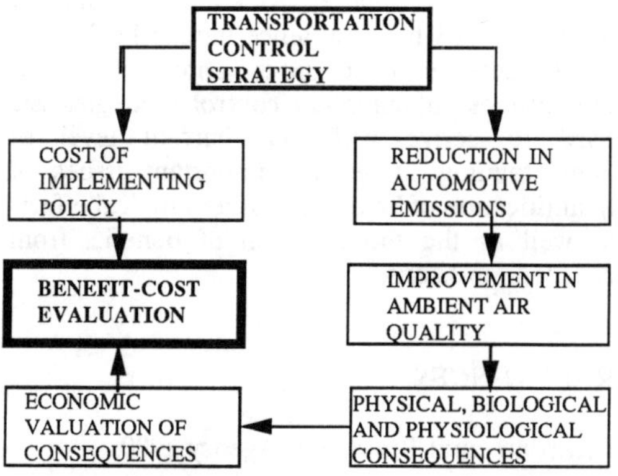

Figure 5. A systematic procedure for evaluating emission control strategies

A systematic procedure for evaluating emission control strategies is shown in Figure 5.

6.1 Complexities in evaluating emission control strategies

The complexities in undertaking the benefit-cost evaluation of emission control strategies include the ever changing, new and additional information concerning
(a) the physical, biological, and medical effects of pollutants (consequences)
(b) development of new emission-reducing technologies (costs)
(c) valuation of the physical, biological, and medical consequences (benefits)
(d) the effect of emission-reducing technologies on ambient air quality (cure for one type of emission problem may actually worsen other types of air pollution)

6.2 Strategies to reduce highway transportation emissions

Emission from motor vehicles may be reduced by
(i) controlling the amount of pollutants released by each vehicle per unit of travel (technological solutions including Australian Design Rules emission controls),
Engine modifications
Treating the exhaust gases
Use of oxygenated fuels
Regular vehicle maintenance to ensure efficient operation of exhaust control equipment.
Some federal programs aimed at reducing emissions include tailpipe standards, evaporative controls, and fuel volatility.
(ii) by reducing the amount of travel by motor vehicles (TDM - transportation demand management). This would include
High occupancy vehicle lanes
Bicycle and pedestrian facilities and lanes
Employer-based transportation plans
Areawide ridesharing incentives
(iii) improving the traffic flow through urban areas (TCM - traffic control measures). Congestion has a profound impact on vehicular emissions. It has been predicted that by 2005, there will be a 300-400% increase in congestion on the U.S. highways resulting in significant increases in aggravation, delays and emissions. Some salient programs include
Improving public transit
Restricting vehicle use in downtown areas
Park-and-Ride/Fringe parking
Flexible work schedules
Transportation systems management
The new Clean Air Act has identified 16 major TCMs and the EPA is preparing information regarding the formulation and emission reduction potential of these measures.

6.3 Estimating the benefits of control strategies

The dose response functions quantify the extent to which an improvement in air quality improves health, conserves materials, reduces structural maintenance costs, improves agricultural productivity or contributes to environmental enjoyment.

Well-structured surveys, control experimentation and reliable data and information sources should be used in conjunction with appropriate statistical tools to obtain reliable dose response functions.

Although results of several studies have appeared in the literature, very few models of physical, physiological, and eco-system impacts have applicability beyond the time and space over which they have been developed.

Estimation of the economic effects of improvement in air quality involves answering questions such as

What is the dollar value of the health, materials, agricultural or other benefits attributable to clean air, both in the aggregate and at the margin?

What is the current cost of the harmful effects of polluted air?

How much do the total costs fall as air quality is improved?

Unfortunately answers to these questions can not be determined with any degree of precision. Most of the benefits of cleaner environment are public goods are can only be valued in contingent markets. The indirect approaches such as statistical value of life, surveys and bidding games, willingness-to-pay and willingness-to-accept etc. yield widely varying results. Wadhwa (1992) has suggested a hybrid approach for valuing human life and injuries which incorporates the direct costs associated with mortality and morbidity and relies on indirect valuations for pain, suffering and reduced quality of life only.

7. DISCUSSION

There has been a growing concern relating to the protection of our environment and quality of life in the past 25 years. This "environmental conscience" has affected many areas including highway transportation. The society demands environmental protection in all facets of life. Highway professionals must address a variety of environmental issues in planning, developing, constructing, and implementing transportation programs.

The effects of highway transportation on health, materials, vegetation, aesthetics, etc. are substantial. However, the need to protect public health is the primary reason for strict regulations relating to motor vehicle emissions since 1965. These programs have undoubtedly achieved substantial reductions in air pollutant emissions from motor vehicles which has resulted in considerable social benefits.

These programs have not been subjected to the scrutiny of cost-effectiveness partly as the techniques of benefit-cost analysis have been found to be inadequate and partly on the conviction that clean air is an absolute good and that the costs involved in achieving this goal generally do not matter.

A few studies (such as White, 1982) have found that the current standards are generally too stringent and the programs have not been cost effective. The credibility of these findings can be no more than that of the approaches used to estimate costs and benefits of these programs. For example, in addition to enormous difficulties with valuing public goods, the long-term health effects of pollutants emitted by motor vehicles have not been isolated from the multitude of other factors which influence human health. In short, definite conclusions about the cost-effectiveness of emission control strategies are hard to derive with our current level of understanding of several important issues of quantification of health and other physical effects as well as the monetization of benefits from reduced emissions.

REFERENCES

Environmental Protection Agency 1990. *Environmental investments: the cost of a clean environment*. Report to the Congress of the United States. Washington D.C. U.S.A.

Environmental Protection Agency 1991. *Transportation control measures*. Draft report. Cambridge Systematics Inc. Cambridge MA.

Environmental Protection Agency 1978. *National air quality and emissions trend report*. Research Triangle Park.

Jackson, C.J. et al. 1976. *Benefit-cost analysis of automobile emission reductions*. Report no. GMR 2265 G.M.C. Research Laboratories, Warren MI U.S.A.

National Academy of Sciences/National Academy of Engineering 1974. *The costs and benefits of automobile emission control*. vol. 4 of Air quality and automobile emission control series. Washington, D.C. U.S.A.

Wadhwa, L.C. 1992. The cost of road crashes: A review and analysis. *ITE compendium of technical papers*. Washington, D.C. U.S.A.

White, Lawrence J. 1982. *The regulation of air pollutant emissions from motor vehicles*. American Enterprise Institute for Public Policy Research. Washington D.C. U.S.A.

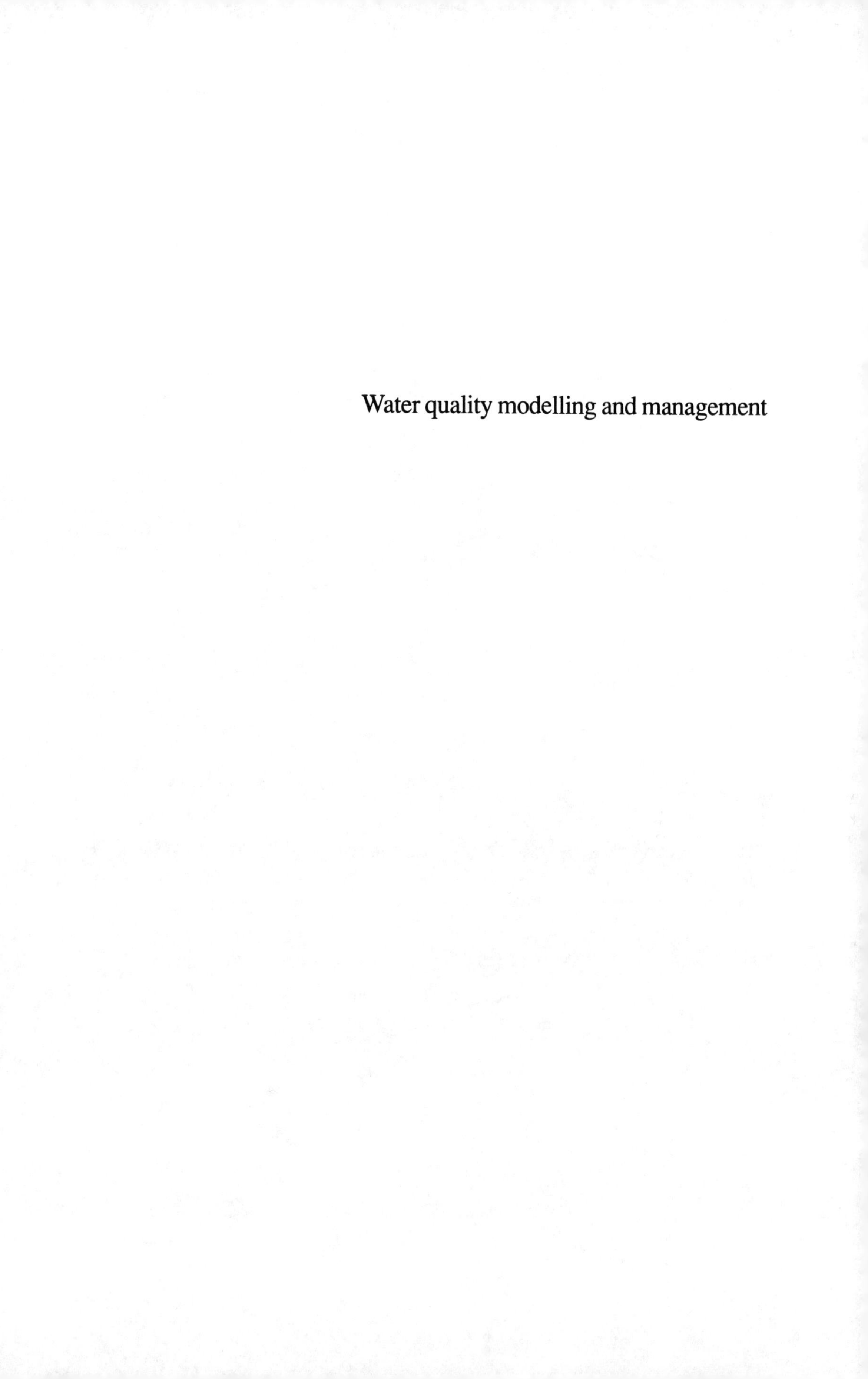

Water quality modelling and management

Environmental Management, Geo-Water & Engineering Aspects, Chowdhury & Sivakumar (eds)
© 1993 Balkema, Rotterdam. ISBN 90 5410 099 0

Performance study of a new membrane anaerobic digester configuration

Fakhru'l-Razi Ahmadun
Department of Civil and Environmental Engineering, Universiti Pertanian Malaysia, Serdang, Malaysia

ABSTRACT: One of the major difficulties faced in the anaerobic treatment of wastewaters is the ability of the process to retain a sufficient quantity of active biomass in the reactor. Present designs of anaerobic treatment processes allows for very limited control of the solids retention time (SRT). This paper discusses the development of a new anaerobic process utilising membranes both for biomass retention and for the production of a clear final effluent. The performance of the new process configuration was evaluated by operating the reactor over a range of hydraulic retention times and organic loading rates. Six steady states were attained over a range of 20,000 to 26,000 mg/l of mixed liquor suspended solids. The maximum organic loading rate applied was 9.54 kg COD/m³/d. The methane content of the gas produced ranged from 68.1 to 74.4 %. The percentage of COD removal was above 98% throughout the study. The results indicated that the new anaerobic process configuration is capable of higher loading rates and had not reached it's maximum treatment capability. The membranes used have practically prevented any biomass loss through the effluent or permeate indicating that positive control of the biomass concentrations and SRT are feasible with this configuration.

1 INTRODUCTION

Membrane separation technology have been applied for biomass recycling in biotechnology (Strathman, 1985) and wastewater treatment (Fakhru'l-Razi, 1991). To predict membrane flux, Fane et al (1980) developed a computer model of a combined ultrafiltration and activated sludge wastewater treatment system. The major resistance to flux was provided by the suspended solids. Various membrane technologies such as reverse osmosis (RO), ultrafiltration (UF) and microfiltration (MF) have been used successfully for various water and wastewater treatment applications. These technologies require no phase change, little energy and can be cost effective.

A major feature in anaerobic digestion systems is the difficulty of retaining a sufficient quantity of active biomass in the reactor to ensure good digester performance. By preventing anaerobic bacteria from leaving the effluent, the digestion process becomes eventually independent of growth rate. This will make it possible to attain high populations of bacteria resulting in higher rates of digestion inspite of very low growth rates. A new membrane anaerobic wastewater treatment system was tested and its performance is presented.

2 EXPERIMENTAL SET UP

The system set-up consists of a crossflow UF membrane unit, a pump and an anaerobic reactor (120 l) treating high strength wastewater as shown in Figure 1. The feed velocity and pressure were controlled by adjusting the flow and pressure regulators. The temperature control was provided by a heat exchanger in the reactor. The UF membrane used has a molecular weight cut-off of 10,000, pressure range of 2-15 bar, pH range of 2-12 and maximum operating temperature of 70°C. The study was conducted at 35°C and a pH range of 6.5-7.2. After an earlier study (Fakhru'l-Razi, 1992) was completed, no deliberate wastage of digester sludge was done and the biomass concentration was allowed to attain a mixed liquor suspended solids (MLSS) of approximately 20,000 mg/l. Deliberate wastage was resumed and six steady states were obtained over a mixed liquor suspended solids (MLSS) range of 20,000 to 26,000

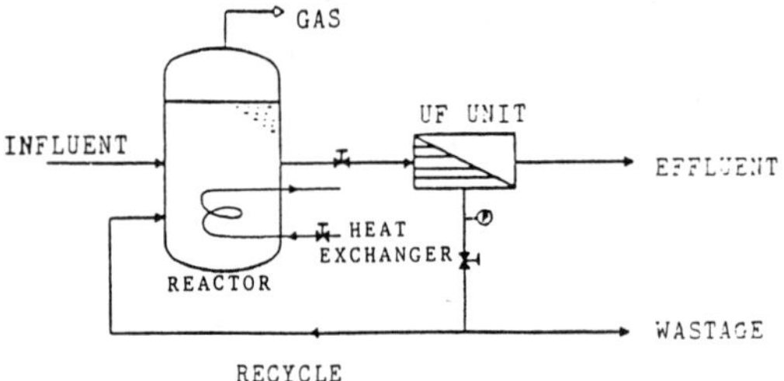

Fig. 1 System Set-up

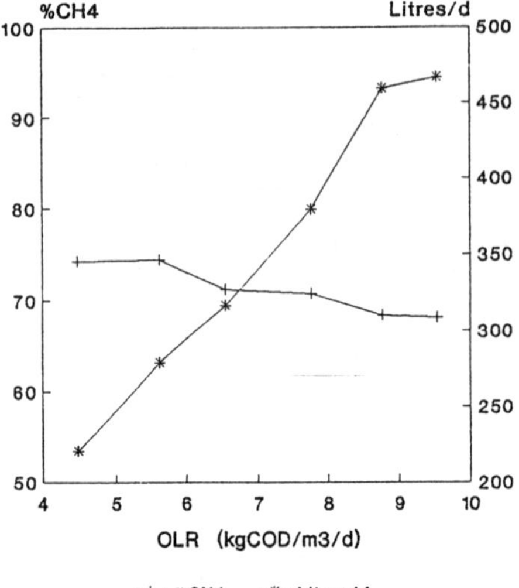

Fig. 2 Gas production

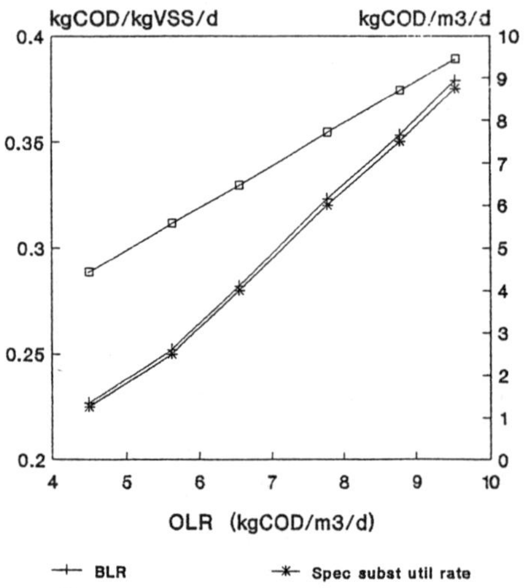

Fig. 3 Membrane reactor performance

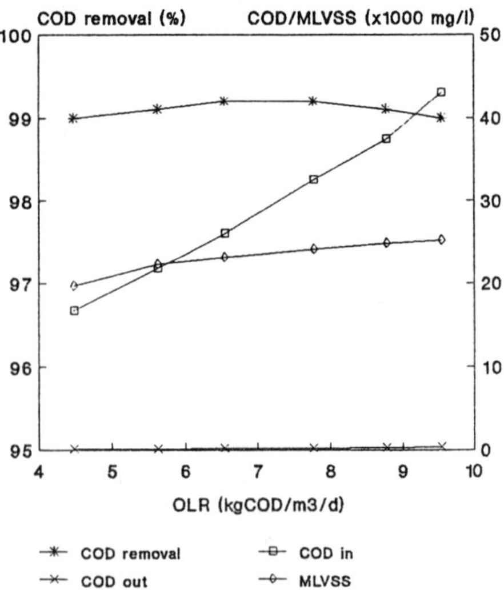

Fig. 4 Treatment efficiency

mg/l by varying the organic loading rates (OLR) and solids retention times (SRT) of the reactor.

3 RESULTS AND DISCUSSION

After the reactor has reached its first steady state the feed COD and sludge wastage rate was then increased and the system was allowed to achieve a new steady state. This took about 12 to 15 days for every new operating conditions. A total of six steady states were achieved. Table 1 summarizes the results obtained during the study. The maximum OLR applied was 9.54 kgCOD/m^3/d and the feed COD was 43,040 mg/l. The hydraulic retention times (HRT) was between 3.65-4.30 days which were largely determined by the membrane flux rates. Increase in total volatile acids concentration in the effluent and reduction in gas production were taken as indicators of any impending failure.

Figure 2 illustrates the variation in gas production and methane percentages of the reactor gas at different OLRs. The methane yield ranged from 0.28-0.31 lCH$_4$/gCOD utilised, at standard temperature (273 K) and pressure (1 atm). The gas production rate increases with increasing OLR while the gas methane content ranged from 68.1-74.4%. There was no sudden upsurge in volatile acids concentration with increasing OLR indicating that the reactor was performing well. In Figure 3, the membrane reactor performance is presented. The substrate utilisation rate, the specific substrate utilisation rate and the biological loading

TABLE 1 Summary of results

PARAMETER	STEADY STATE 1	STEADY STATE 2	STEADY STATE 3	STEADY STATE 4	STEADY STATE 5	STEADY STATE 6
COD in, mg/l	16,680	21,765	25,970	32,505	37,460	43,040
COD out, mg/l	160	185	210	265	320	410
Vol acids, mg/l	120	170	195	215	290	320
Gas prod., l/d $\%CH_4$	220.4 74.3	278.7 74.4	316.8 71.2	379.4 70.7	458.9 68.3	466.0 68.1
$l.CH_4/gCOD$	0.31	0.31	0.29	0.29	0.30	0.28
MLSS, mg/l	19,900	23,100	24,000	24,750	25,100	25,750
MLVSS, mg/l	19,700	22,300	23,150	24,100	24,850	25,200
Wastage, l/d	0.72	0.84	1.01	1.13	1.25	1.28
Feed (Q), l/d	32.2	31.0	30.2	28.7	28.1	26.6
HRT, d	3.65	3.77	3.84	4.02	4.09	4.30
SRT, d	167	143	119	106	96	94
OLR $kgCOD/m^3/d$	4.48	5.62	6.54	7.77	8.77	9.54
BLR kgCOD/kgVSS/d	0.227	0.252	0.282	0.323	0.353	0.379
Subst util rate $kgCOD/m^3/d$	4.43	5.57	6.48	7.71	8.70	9.45
Sp sub ut rate kgCOD/kgVSS/d	0.225	0.250	0.280	0.320	0.350	0.375
COD removal, %	99.0	99.1	99.2	99.2	99.1	99.0

rate (BLR) all increased proportionally with increasing OLR. There was also a corresponding increase in biomass concentration in the reactor. This shows that the loading rate applied was the limiting factor for biomass production and this is in agreement with an earlier study (Fakhru'l-Razi, 1992). As shown in Figure 4, the treatment efficiency of the system shows a consistent COD removal of 90% and above. The mixed liquor volatile suspended solids (MLVSS) increases steadily with increasing OLR indicating that the bacterial population had increased with higher COD inputs.

The results obtained showed that the system is capable of much higher loading rates and had not attained it's maximum treatment capability. The system also demonstrated that positive control of biomass concentrations in the reactor is possible by manipulating the OLR and deliberate sludge wastage from the reactor.

4 CONCLUSION

The treatment efficiency of the system indicates it's potential use for treating high strength wastewaters. The UF membranes used was capable of efficient biomass/effluent separation and produced a clear final effluent. The UF membranes had practically prevented any biomass loss through the effluent or permeate. This had resulted in very high operating SRTs making the system tolerant to variations in influent characteristics. The membrane anaerobic system was subjected to a maximum OLR of 9.54 kg $COD/m^3/d$ and a HRT of 4.3 days. At this loading rate the system was capable of achieving an overall COD removal of 99.0%.

This system has demonstrated it's ability to positively control SRT and HRT which had always been one of the major problems in anaerobic processes. The bacterial growth rate in anaerobic processes are slower than in aerobic systems thus requiring a longer SRT in order to accomodate the slower net growth rate.

REFERENCES

Fakhru'l-Razi, A. 1991. The retention and control of biomass concentrations in anaerobic reactors by UF membranes. Journal of the Institution of Engineers, Malaysia 48: 23-30.

Fakhru'l-Razi, A. 1992. Membrane anaerobic system for wastewater treatment. Proceedings of the Seminar on The Role of the ASAIHL in Combating Health Hazards of Environmental Pollution, Hong Kong University.

Fane, A.G., Fell, C.G.D., Nor, M.T. 1980. UF/AS system - Development of a predictive model. UF Membranes and Applications, Editor: R. Cooper. Plenum Press, London-New York.

Strathman, H. 1985. Membranes and membrane processes in biotechnology. Proceedings of the 39th Industrial Waste Conference, Purdue University: 627-636.

Environmental Management, Geo-Water & Engineering Aspects, Chowdhury & Sivakumar (eds)
 ISBN 90 5410 099 0

Effects of leachates on variable selection in water quality monitoring network design

N. Alpaslan & N. B. Harmancioglu
Dokuz Eylul University, Turkey

V. P. Singh
Louisiana State University, La., USA

ABSTRACT: The paper presented investigates the water quality monitoring network design problem in relation with the presence of solid waste disposal areas within sampled sites. Such areas constitute significant polluting sources as wastes are carried in the form of "leachate" to groundwater and surface waters. With respect to monitoring network design, the basic aspect where the effects of leachates are considered is the selection of variables to be sampled. It is emphasized that monitoring programs must be modified to infer on the significance and the rate of leaching effects. The presented paper analyzes how such a modification can be realized in variable selection to represent the landfill and leachate characteristics of sampled sites. The scope of the study is limited to the monitoring of surface waters since the pollution of groundwater by leachates is a rather different problem requiring the analysis of some other factors.

1 INTRODUCTION

The two current issues in water resources and environmental engineering, which have recently received considerable attention from researchers and practitioners, are the design of water quality monitoring networks and the handling of leachates from waste disposal sites and landfills. Both are highly significant with respect to water pollution control; yet they are subject to numerous uncertainties which complicate their analysis. These issues are often analyzed separately by different disciplines so that the relation between the two are vaguely developed.

Water quality assessment and management is a complicated process due to several uncertainties involved. In this respect, monitoring appears to be the only effective means of dealing with water quality related uncertainties as there is no other way of being informed about the phenomenon. We need to monitor not only the prevailing conditions of water quality but also the results of every decision made or every action taken.

On the other hand, design and operation of water quality monitoring networks is still a controversial issue so that monitoring practices also suffer from various uncertainties. The major difficulty relates to delineation of monitoring objectives for a precise description of what is expected from sampling. Second, the technical design of a network is also a multifaceted problem, where the basic issues awaiting sound solutions relate to the selection of: (a) variables to be sampled, (b) sampling sites, (c) sampling frequencies, and (d) duration of sampling. Various design methods are proposed and used for the determination of (b) and (c), the last issue still remaining unresolved. The selection of variables is probably the most difficult problem as there are several hundreds of water quality variables to choose from. No definite criterion has yet been developed to identify the variables to be sampled although it is stressed that such a selection depends upon the objectives of monitoring.

Despite all uncertainties in network design, there are some basic guidelines established with respect to both objective definition and technical design features. However, many river basins reflect specific characteristics that preclude the use of conventional planning guidelines. Among such characteristics, the most typical one may be the presence of solid waste landfill or dumping areas within the basin. These areas are where domestic and/or industrial solid wastes are regularly (landfill) or randomly (dumping) stored. As a result of certain biochemical reactions and liquefactions, such wastes produce a highly polluted liquid called "leachate". The precipitation infiltering through the layers of landfills and dumps increase the amount of leachate produced so that the by-product constitutes a highly significant source of pollution. The problem is enhanced when the solid waste sites contain hazardous wastes.

If left uncontrolled, leachate is infiltrated through the soil to pollute groundwater. It may also reach surface waters via subsurface or groundwater transport mechanisms. The occurrence of precipitation, however, speeds up the process so that surface waters are more likely to become polluted by leachates from landfill and/or dumping areas.

The assessment of surface water pollution

by leachates requires sampling or monitoring to infer on both the significance and the rate of the leaching effect. In this case, the design and operation of a water quality monitoring network have to be modified to account for this polluting source. The basic modification must be made in "variable selection"; that is, the monitoring program must be extended to include new variables that represent the landfill and leachate characteristics of the sampled site.

The presented paper develops the relation between the two most significant water quality management problems: (a) design of water quality monitoring networks, and (b) handling of leachates from solid waste disposal sites and landfills. It is emphasized here that monitoring programs must be implemented so as to include variables that represent pollutant wastes carried to surface waters by leachates. In the following sections, current problems associated with design of water quality monitoring networks is discussed first. Next, characteristics of solid waste landfill and dumping areas are summarized to eventually infer on their incorporation into the monitoring program. Specific features of the problem are also reviewed as they relate to different practices of solid waste management in developed and developing countries.

2 CURRENT METHODS OF WATER QUALITY MONITORING NETWORK DESIGN

2.1 Review of the general approach

The basic approach in initiating water quality observations has been to collect data at potential sites for pollution problems. Consequently, the early water quality monitoring practices were often restricted to what may be called "problem areas", covering limited periods of time and limited number of variables to be observed.

Recently, however, water quality-related problems have intensified so that the information expectations to assess the quality of surface waters have also increased. The result has been an expansion of monitoring activities to include more observational sites and larger number of variables to be sampled at smaller time intervals. These efforts have indeed produced plenty of data; yet they have also raised the question whether one "really" needs "all" these data to meet the information requirements.

The above considerations have eventually led to the realization that a more systematic approach to monitoring is required. Following up on this need, monitoring agencies and researchers have proposed and used various network design procedures either to set up a network or to evaluate and revise an existing one.

Current methods of water quality monitoring network design basically cover two steps: first, the description of design considerations, and second, the actual design process itself. Researchers emphasize the proper delineation of design considerations as an essential step before attempting the technical design of the network. This step is to provide answers to the questions of why we monitor and what information we expect from sampling water quality. In other words, objectives of monitoring and information expectations for each objective must be specified first. Various objectives or goals for monitoring have been proposed up to date by different researchers, i.e. assessment of trends, delineation of water quality characteristics for water use, assessment of compliance, evaluation of water quality control measures, etc. (Whitfield, 1988; Ward and Loftis, 1986; Tirsch and Male, 1984; Sanders et al., 1983; Langbein, 1979). In practice, the definition of objectives is not an easy task since it requires the consideration of several factors, including social, legal, economic, political, administrative and operational aspects of monitoring goals and practices. Therefore, the delineation of design considerations, inevitably includes assumptions and subjective views of the designers and decision-makers no matter how objectively the problem is approached. In this case, design considerations are often presented as general guidelines, rather than fixed rules to be pursued in the second step of actual design process (Sanders, et al., 1983).

The technical design of monitoring networks relates to the determination of:
(a) sampling sites,
(b) sampling frequencies,
(c) variables to be sampled, and
(d) the period or duration of sampling.

It is only at this actual design phase that fixed rules or methods are proposed. Current literature provides considerable amount of research carried out so far on the above-mentioned four aspects of the design problem. One may refer to Sanders et al. (1983), Tirsch and Male (1984), or to Whitfield (1988) for a rather thorough survey of research results and practices on the establishment of sampling strategies with respect to these factors.

Basically, designers and researchers recognize water quality monitoring as a statistical procedure and address the design problem by means of statistical methods. Ward and Loftis (1986) stress that information expectations from a monitoring system must be defined in statistical terms and that these "expectations are to be in line with the monitoring system's statistical ability to produce the expected information". This implies that one can infer on the types of data needed to perform the statistical methods which, in turn, will eventually lead to the expected information. Then, the selection of sampling strategies (sampling sites, variables, frequencies, and duration) can be realized by starting off with such a statistical approach (Ward and Loftis, 1986; Sanders, et al., 1983).

Statistical analyses based on regression theory as well as decision theory and optimization techniques are used to select the spatial and temporal design features of a network. Although none of these methods are widely accepted, they serve at least to assess the effectiveness of design decisions and the efficiency of an existing network. The problem is much more difficult in case of variable selection as there are no methods established yet for defining objective selection criteria.

2.2 Variable selection in network design

Selection of variables to be sampled depends basically on the objectives and economics of monitoring. It is a highly complicated issue since there are several variables to choose from in representing surface water quality. Some of the selection procedures stress water uses as the major criterion to be pursued; some define levels of monitoring efforts (e.g., surveillance, intensive control, or project-oriented programs) with different groups of variables included at each level (UNESCO-WMO, 1972).

There are also studies which apply quantitative statistical techniques in selection of variables to be sampled. These techniques are basically regression-type methods to investigate the relationships between water quantity and water quality variables or between water quality variables themselves (Harmancioglu et al., 1986). The purpose of such analyses is to reduce the number of variables to be observed.

Sanders et al. (1983) suggest ranking of water quantity and quality variables among which information may be transferred. In this ranking, water quantity appears as the basic variable followed by "associated quality variables of aggregated effects" (often regularly observed) and then by "quality variables that produce aggregated effects" (often unobserved or observed sporadically). If information transfer between the first and the second group of variables is possible, then the required number of variables to be observed may be reduced as long as there is no doubt as to the reliability of information transfer.

A fairly recent study (Mazlum, 1992) uses an objective-based criterion in the selection of variables. Two basic objectives are defined for a monitoring network:
(a) water use (e.g., irrigation, domestic water supply, recreation, power generation, fish and wildlife preservation, etc.), and
(b) impact assessment for evaluating point and nonpoint polluting sources.
Consideration of water use purposes results in a list of "gross variables" to be monitored at every station in the network. Further selection criteria, such as cost and ease of measurement, statistical properties (i.e., statistical variability), and common occurrence in more than one water use objective, are proposed to reduce the long list of gross variables. Further reduction in the number of variables to be sampled may be attained by analyzing information transfer between the already selected variables. With respect to the second monitoring objective, i.e., impact assessment, a group of "specific variables" may be defined to assess point and/or nonpoint pollution by domestic, industrial and agricultural effluents. As such, the list of specific variables is often shorter than that of gross variables. The most significant feature of this group is that they include station-specific or station-based variables which do not need to be observed at every station in the network. The above approach proposed by Mazlum (1992) intends to clarify the complex problem of variable selection by proposing definite selection criteria. It has produced reasonable results in the sense that it pinpoints the most relevant variables with respect to a specific monitoring objective.

3 SIGNIFICANCE OF LEACHATES

Solid wastes arising from man's domestic, social and industrial activities are eventually disposed to the land with or without any prior processing. Many types of solid wastes, including the majority of industrial, construction and demolition wastes are suitable only for disposal to land. Domestic and other similar wastes can be land-disposed without prior treatment provided that environmental pollution is controlled. By far, the largest proportion of solid wastes has been disposed of by storage in landfills and dumping areas. Despite the development of new techniques of waste treatment and disposal (e.g., recovery of certain components of the waste for re-use), the current practice of land storage is likely to continue for many years as the most practical and economic method (Knox, 1985)

On the other hand, landfills and/or dumping areas have a number of disadvantages, the most significant one being the production of "leachate". Leachate is the contaminated water resulting from the percolation of rain water through wastes stored in a landfill or a dumping area. It is an important environmental problem since it constitutes a significant polluting source not only for groundwater but also for subsurface and surface waters. In some cases, serious harm can result from inadequate control of highly polluted leachate (Straub and Lynch, 1982).

Currently, the increased concern about environmental pollution caused by leachates has led to significant efforts for mitigation of this contaminating source. However, these efforts are often hindered by the uncertainties in leachate producing mechanisms. It is highly difficult to estimate both the quantity and the quality of water leaking from a landfill so that control and handling of leachates are not easy tasks. The containment, treatment and disposal of leachates are often costly processes. Thus, landfill sites are selected so as to permit slow transport and natural degradation of the output leachate by physical, chemical and biological processes in the ground. Some researchers (Knox, 1985) claim that by intensive control measures, careful operation and thorough investigations, leachate problem can be minimized; however all these efforts are often more expensive than treating the leachate. In fact, such a minimization can not be realized at all sites (Knox, 1985).

It follows from the preceding discussion that not only groundwater but also surface water quality is always threatened by polluted water leaking from a solid waste disposal site nearby. This is true despite the development of new technologies for treatment and disposal of leachates. Furthermore, the problem is worse when a landfill site also stores hazardous wastes. In addition, the output sludge from domestic or industrial wastewater treatment plants are often disposed to solid waste sites with or without prior processing (dewatering). Sludge waste is a concentrated mass of all pollutants in wastewater treated by the plant. When placed in a land disposal site, such a mass becomes a significant component of the solid waste composition. Release of interfacial water from the sludge eventually contributes to leachate characteristics.

In the above cases, the need for identifying the nature of the resulting leachate is much more significant. Apparently, the required control measures are also more costly and difficult.

To gain an insight into the leachate generation and transport mechanism at a site, it is necessary to understand and quantify all factors which affect leachate production. These factors can be best examined through the landfill hydrologic cycle, which is a local subsystem of the overall hydrologic cycle. Such a subsystem may be highly complex with different inputs to and outputs from the landfill storage system, as schematically shown in Figure 1.

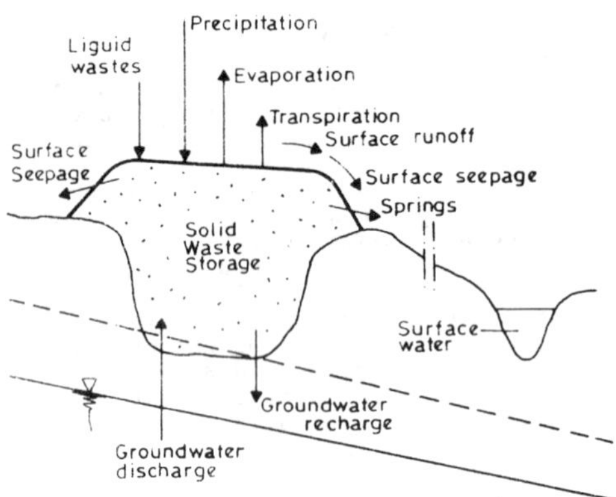

Figure 1. Hydrologic cycle at the landfill site

Within the framework of this hydrologic cycle, the variables that are of interest to the water quality monitoring network may be surface runoff, surface seepage, groundwater recharge, and springs. In addition, the hydrology of the region can have a direct and significant effect upon the nature and the size of the leachate problem within the landfill site. The above discussion is valid for developed countries where solid wastes are orderly and regularly disposed to landfills. Despite all controls, it is observed that landfills and leachate production are still significantly threatening pollution sources for groundwater and surface water as well. For example, in the 70's surface waters near Byron, Illinois were reported to be contaminated with cyanide, heavy metals, phenols and several other toxic materials leaking from a solid waste disposal site nearby. Similarly, in Kentucky, EPA has identified about 200 organic chemicals and 30 metals in soil and surface waters due to the same reason (EPA, 1985). As a result of such occurrences, continuous monitoring of groundwater or surface waters nearby are advised (WHO, 1981).

In developing countries, however, it is rarely possible to find properly designed and operated landfills so that the problem is much worse. Solid wastes in such countries are often disposed to uncontrolled open dumps selected by municipalities at empty or unused sites. The problem here is that leachate production is also left uncontrolled when the dumping area lacks control. Such a practice in developing countries is likely to continue as it is for years to come since establishment of new technologies either to convert dumps to landfills or to control the disposal site is often too expensive to be afforded by such countries (WHO, 1981).

On the other hand, authorities in developing countries are in efforts of initiating and expanding water quality monitoring networks. If controlled land disposal of solid wastes is not affordable, then the work is up to these authorities to be aware of the leachate problem and its potential threat to surface waters. The best approach they can use will be to include in their monitoring programs the variables that represent leachate produced near sampled sites.

4 VARIABLE SELECTION IN THE PRESENCE OF LEACHATE

As discussed in section 2.2., basically two groups of variables may be specified for a water quality monitoring network: (a) gross variables which are water-use based and which need to be monitored at every station; and (b) specific variables that are site-specific variables to reflect the effects of particular point or nonpoint sources. This classification indicates that, in the presence of landfills and/or dumping areas, the "specific variable" list has to be expanded to include those variables that represent possible effects of leachates. The "gross variable" list does not need to be changed so that only particular stations will have to monitor the consequences of the leaching effect.

The next question will be what to monitor if there exists a leachate problem. In this case, the most reasonable approach would be to identify those variables that constitute significant components of leachate. That is, characteristics of leachate must be specified first to select variables to be monitored. Yet, this is not an easy task as enormous variations occur in leachate quality.

Management of landfill areas and identification of the leachate problem had been pretty random in the past. Recently, however, significant research efforts have been put into the investigation of leachate characteristics generated in domestic and/or industrial landfills. More is identified now with respect to the underlying physical, chemical and biological processes that occur in landfills and that partially determine the widely varying nature of the resulting leachate (Knox, 1985).

Leachate characteristics are spatially and temporally variant; that is they not only vary from one landfill site to another but also change with the age of the landfill. For example, young landfills may constitute a more severe polluting source as their leachate is more aggressive and strong than that of old landfills, where pollution levels are relatively lower.

Despite all these variations, some researchers have specified leachate characteristics with respect to their sources. Knox (1985) identifies leachate characteristics from domestic landfills in different phases. For instance, in early years after emplacement, he identifies high concentrations of volatile fatty acids, acidic pH, high BOD, high BOD/COD ratio, several hundred mg/l of ammonia and nitrogen, high concentrations of iron and manganese, moderately high concentrations of some heavy metals like zinc, and high concentrations of inorganics like Na, K, and Cl. In the following period, he notes that these characteristics change within the bacterial phase. Industrial landfills reflect a wide spectrum of hazardous wastes including mercury, lead, and phenols. Leachate composition becomes much more complex when wastewater treatment plant sludges are also included. Knox (1985) states that these characteristics also change through different phases and that seasonal variations are also likely. Similar observations are presented by other researchers (Tchobanoglous et al., 1977; Chian and DeWalle, 1976).

In view of significant variations in leachate quality, continuous monitoring of landfill sites is proposed for close control of their polluting effects (Knox, 1985; WHO, 1981). Such monitoring must disclose not only leachate characteristics but also factors that affect the leachate generating mechanism. These factors include the nature of the input waste, hydrology and hydrogeology of the landfill site, operational features, meteorological conditions and the like. Knox (1985) suggest that different levels of monitoring can be performed to observe physical, chemical and biological characteristics of leachate from domestic and/or industrial landfills. For example, chemical characteristics must be checked for sanitary analyses, inorganic components, growth inhibiting pollutants, and special chemical characteristics. Biological features to be monitored include toxicity, mutagenicity and microbiological characteristics.

Coming back to water quality monitoring network design, one may state that if adequate leachate monitoring is performed at or around the landfill site, then the surface water quality monitoring network has to monitor only for the purpose of a final check-up. That is, what may have escaped control at the landfill site will then be detected by monitoring along the surface water. In this case, the majority of the work applies to monitoring practice at the landfill site so that relatively less burden is to be carried by the network. Then, specific variables to be observed by a station will include only the most significant constituents of leachate so that their number can be kept at a minimum. Apparently, the above practice is valid particularly for developed countries where the most sophisticated technologies can be used for a strict control of landfill site. In developing countries, however, storage of solid wastes is randomly realized in uncontrolled dumping areas which are equally or even more dangerous than landfill areas as polluting sources. This means that leachate generation remains as an uncertain process, the consequences of which are also unknown. Then, the responsibility of pollution control is left to the water quality monitoring network to detect the adverse effects of leachate. This means that there is a greater need in such countries to expand their monitoring programs for inclusion of a larger number of specific variables.

5 CONCLUSION

The design and operation of surface water quality monitoring networks have to be modified to account for leachate if landfill and/or dumping areas exist in the vicinity.This modification applies particularly to the variable selection problem where *specific variables* to be monitored must include the most significant components of leachate. In the presence of controlled landfills, continuous leachate monitoring at the storage site may help to reduce the number of specific variables to be sampled by the network. In this case, a network station serves to detect what may have escaped control at the landfill site. If the storage area is an uncontrolled dumping site, as the case is with most developing countries, the network design has to foresee a more significant modification in variable selection. In this case, the number of specific variables to be sampled will increase to control more closely the polluting effects of leachate.

REFERENCES

Chian, E.S.K. & F.B. DeWalle 1976. Sanitary landfill leachates and their treatment. *Journal of the Environmental Engineering Division, ASCE,* 102 EE2, 411-431.

Harmancioglu, N.B., V. Yevjevich and J.T.B. Obeysekera, 1986. Information Transfer Between Variables. *Proceedings of Fourth International Hydrology Symposium-Multivariate Analysis of Hydrologic Processes,*(ed.H.W.Shen, et.al.),pp.481-499.

Knox, K. 1985. Leachate production, control and treatment. In: *Hazardous Waste Management Handbook* (ed. by A. Porteous),

Butterworth & Co (Publishers) Ltd., 98-145.
Mazlum, S. 1992. Variable selection in design of water quality monitoring networks. *Dokuz Eylul University, Institute of Science and Technology,* Master Thesis (adv. N. Alpaslan), Izmir,Turkey, 85p.
UNESCO-WMO, 1972. Hydrologic information systems: Studies and reports in hydrology, no.14, (ed.by G.W. Whetstone, and J.J. Grigoriev), *prepared by the Panel on SAPHYDATA,* 74p.
Sanders, T.G., R.C. Ward, J.C. Loftis, T.D. Steele, D.D. Adrian & V. Yevjevich 1983. Design of networks for monitoring water quality. Littleton CO:Water Res. Publ.,328p.
Straub, W.A. & D.R. Lynch 1981. Models of landfill leaching: Moisture flow and inorganic strength. *Journal of the Environmental Engineering Division, ASCE,* 108 EE2, 231-249.
Tchobanoglous, G., H. Theisen & R. Eliassen 1977. *Solid Wastes, Engineering Principles and Management Issues.* McGraw-Hill Book Company, 621p.
Tirsch, F.S. & J.W. Male 1984. River basin water quality monitoring network design: Options for Reaching Water Quality Goals, (ed. by T.M. Schad), *Proceedings of Twentieth Annual Conference of American Water Resources Associations,* AWRA Publications, 149-156.
Ward, R.C. & J.C. Loftis 1986. Establishing statistical design criteria for water quality monitoring systems: Review and synthesis. *Water Resources Bulletin,* AWRA, 22 (5), 759-767.
Whitfield, P.H. 1988. Goals and data collection designs for water quality monitoring. *Water Resources Bulletin,* AWRA, 24 (4), 775-780.
WHO, 1981. *Hazardous Waste Management.* Report on a Working Group Garmisch-Partenkirchen, ICP/RCE 402(1) 4856K, 15p.

Environmental Management, Geo-Water & Engineering Aspects, Chowdhury & Sivakumar (eds)
© 1993 Balkema, Rotterdam. ISBN 90 5410 099 0

Aquifer vulnerability and groundwater resource management in the Po and Venetian Plain (Northern Italy)

M.Civita
Dipartimento di Georisorse, Turin Polytechnic, Italy

G.Giuliano
IRSA CNR (National Council of Research), Rome, Italy

M.Pellegrini
Dipartimento di Scienze della Terra, Modena University, Italy

ABSTRACT: This paper describes the hydrogeological characteristics and the problems induced by groundwater exploitation and aquifer vulnerability of the largest Italian plain, where the most intense productive systems and the 37% of the total Italian population are concentrated. The aquifers are made up of prevalently alluvial thick sedimentary sequences defined as a monostratum system although large portions of it appear in many areas to be subdivided into several compartmented levels. At the foot of the mountains where the aquifers crop out and are characterised by a high permeability, several cases of pollution, deriving mainly from nitrates, have taken place. Indeed, in these areas the distribution of the richest groundwater resources coincides with the presence of the most intense industrial and farming activities. Also the groundwater protection and safeguard measures have been so far largely neglected by the responsible public boards which in most cases have simply increased the potability limits fixed by national and European laws, without enterprising adequate upgrading initiatives. Only in recent times the government has given to the Universities and other research Boards the task of investigating the hydrogeological characteristics of the Po and Venetian Plain in order to produce groundwater vulnerability maps, whose methods of elaboration are briefly illustrated.

1 INTRODUCTION

Northern Italy is characterised by a vast plain formed by alluvial sediments, deposited on an area of over 69,000 km^2 (23% of the whole Italy), stretching continuously from the Po Plain to the west (corresponding to the River Po and its tributaries) to the Venetian Plain to the east. To the north and west it is surrounded by the Alps, to the south by the Apennines and to the east by the Adriatic Sea (Fig. 1). Its maximum altitude is 600 m (western extremity) whilst its average height is less than 50 m. The humid-temperate climate is characterised by rainfall ranging from 600 mm along the southern margin to 2,000 mm on the north-eastern sector. Precipitations are prevalently concentrated in the spring and autumn. The summer months are the driest with a rainless period that can last up to three or even four months, with a consequent sharp increase, compared with the average amount, in the water demand for civil and farming purposes.

With relation to the favourable climatic and morphological conditions of this vast plain, which is the third largest in Europe, the highest population concentration, as well as the farming and industrial activities of the whole of Italy are here found (over 21 million inhabitants corresponding to circa 37% of the total national population, with a density of 310 people/km^2, compared with a national average of 186). The residing population is mostly concentrated in the numerous towns (urban population up to 90% in over 2900 municipalities) but many rural houses scattered in the countryside, where each farm has an average surface of c. 60 ha, are still inhabited.

The utilised farming surface is c. 40,000 km^2, with over 12 million heads of farm animals, and corresponds to 25% of the national farming surface. On it, though, about 50% of the total cattle farming and 70% of the pig farming is concentrated, while 3.6 million people work in the industrial sector (61% of the national industrial workers).

It is obvious that within this context of demographic and economic development the hydric demand is particularly high, both for water-supply purposes and for farming and industrial use. Nearly everywhere on this plain, the main hydric supplies are constituted by groundwaters although during the last decade widespread and serious cases of pollution have occurred, undermining the water quality and posing new problems, which are dealt with in this article, for a correct exploitation and management of this important resource.

2 HYDROLOGICAL SCHEME

The Po and Venetian plain is formed by the Quaternary alluvial deposits of the River Po, of its main and minor tributaries and other watercourses flowing in the eastern sector of the plain. The

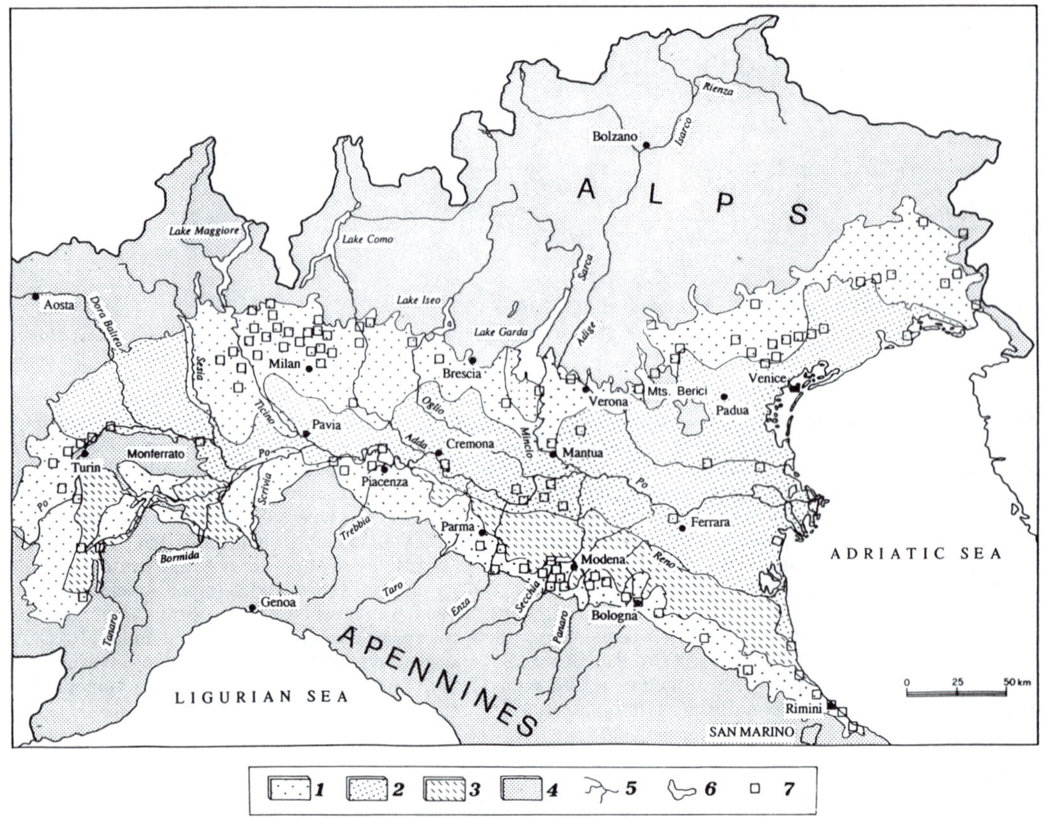

Fig. 1 - Schematic hydrogeological map of the Po Valley and surrounding areas. Legend: 1) gravels and coarse sands with high productivity aquifers; 2) sands and subordinate gravels and silts with high to low productivity aquifers; 3) silts, clays and subordinates sands with low to very low productivity aquifers; 4) Alpine and Apennine bedrocks; 5) main water courses; 6) lakes; 7) main wells for water-supply purposes.

sedimentary basin is strongly subsiding, although in a markedly differentiated way. In fact the base of the Pliocene which shows a very articulated surface, conditioned by the tectonic structures of the substratum, reaches in some points a depth of about 6,000 m, while the average thickness of the Pliocene-Quaternary sequence is c. 3,500 m of which over half are attributed to the Quaternary. Continuous and generalised marine sedimentation started in the lower Pliocene and went on until the end of the lower Pleistocene, even if the conclusion of this sedimentary cycle was not synchronous in all the area since several ingression and regression cycles are recorded in concomitance with different glacial episodes (in the Quaternary Europe was affected by at least four ice ages). Also the overlying continental deposits show an extremely variable thickness (up to 300 m and beyond) still controlled by the deep substratum structures. From the particle-size point of view, the sediments are made up of gravels and sands at the outlet of the watercourses into the plain, whilst in the central part and toward the coast silts and clays prevail.

The base of the aquifer does not correspond to a well defined physical limit but rather to the fresh water-salt water interface, whose trend is conditioned by the buried tectonical structures which determine differential squeezing phenomena of the coeval waters. In some areas (culmination of buried anticlines) located also in the centre of the plain, the thickness of the aquifer saturated by fresh waters is reduced to less than 50 m, whilst elsewhere, usually in correspondence with synclines, it can reach 700 m. One can estimate that the average aquifer thickness is about 200 m, although, as discussed later, the exploitable level is much thinner than that.

From a regional and structural viewpoint, the aquifer makes up a monostratum system (Castany & Margat, 1977), although in large portions in the central part of the plain and along the coast-line, it clearly appears to be subdivided into several layers.

At the foot of the Alpine and Apennine chains the gravels and sands of the fluvial and glacio-fluvial depcsits form an indifferentiated stratum of unconfined groundwaters often without a low-permeability cover. Towards the centre of the plain it is subdivided into more layers ("multi-compartmental monostratum") characterised by the presence of thick covers of fine sediments and confined groundwaters.

Considering the particle-size distribution of the sediments, the aquifers' feeding takes place in the upper plain, at the margins with the hills, where the aquifers outcrop. The effective infiltration of meteoric waters, determined also by means of theorical models and direct measurement, is up to 25-30% and depends mostly on the soil permeability and local meteo-climatic characteristics. As for the relationships between surrounding mountains and watercourses, the situation appears to be extremely differentiated in the various sectors.

Hydric exchanges with the aquifers located along the hill margins are possible and have been found only in the upper part of the north-eastern plain at the foot of the Alps, east of Milan as far as Trieste, due to the presence of fissured carbonatic formations often of Karst-type (southern calcareous Alps). In other sectors (i.e. in southern Piedmont), these exchanges take place only locally and occasionally or are totally absent (southern plain, near the Apennine margin). In the Piedmont and Lombardy plain (north-western sector) the presence of terraced alluvial deposits with surfaces in some cases considerably high compared to the valley bottoms (up to 50 m) is quite common. They sometimes constitute important superficial phreatic aquifers with groundwaters often drained away by the underlying rivers. In their turn, these aquifers receive water also from the surface irriguous network, water reservoirs and storage basins connected with the cultivation of rice (Piedmont). In the Venetian plain (eastern part) and in the southern sector bordering the Apennines the watercourses together with the irriguous canals play a major rôle for the underground hydric balance and the feeding of the aquifers, owing to the different morphological conditions. Indeed, they contribute to at least 30% of the groundwater reserves as witnessed by some experimental measurements carried out on some rivers of the Venetian plain (Antonelli & Dal Prà, 1980): for each river investigated the annual flow-rates dispersed underground are evaluated as 2 to 3 m^3/s per kilometer of dispersing water-bed.

Due to these hydrogeological conditions (outcropping or nearly outcropping aquifers and feeding characteristics), all the high plain located at the foot of the mountain chains appears to be affected by a very high degree of vulnerability to pollutants. In these portions of the plain the highest transmissivity values are recorded and, when pollution phenomena are absent, the best water is pumped out of these aquifers, from the qualitative viewpoint. Instead, in the central part of the plain and along the sea-coast where the aquifers are confined, extremely long times of permanence of the waters underground are recorded (even more than 25,000 years, cfr. Venturini *et al.*, 1990). Therefore low or negative redox potentials are established with groundwaters showing NH_4, Fe and Mn concentrations much higher than the limits fixed for drinkable waters by the World Health Organization. Unfortunately, at least 30% of the Po Plain groundwater resources is characterised by these unfavourable chemical conditions.

3 EXPLOITED HYDRIC RESOURCES

Since ancient times (sometimes already in the Roman period) in all the Venetian and Po Plain groundwaters have constituted, apart from few exceptions, the most important or exclusive hydric resource for water-supply purposes and, in more recent times, for industrial and farming activities as well. Due to their hydrogeological conditions, to their quality and high transmissivity, the most important wells for water-supply uses are located in the upper plain sectors, at the foot of the mountain chains, where also the main urban centres are found together with the most important industrial and farming activities. A coincidence is therefore established between the areas characterised by the best hydrogeological conditions for groundwater supplying purposes and the production settlements, the latter being the main hazard sources for water quality maintenance.

Within the project of a complete restructuring of the water-supply sources, the problem of a consistent water demand in the northern sector could be in the long term overcome by turning to a more efficient exploitation of the mountains' groundwaters (Alps) and of the main watercourses, already characterised by the presence of numerous dams and reservoirs. Moreover, these rivers show a rather constant flow also in the summer while the great Alpine lakes (such as Garda, etc.) could constitute potentially exploitable extra storage basins. Instead in the southern sector, south of the River Po, the groundwaters are an irreplaceable resource since in the summer the watercourses are characterised by extremely low or next to nil flow rates. Even the geological characteristics represent an obstacle for the construction of water reservoirs, owing to the stability problems given by the widespread flysch and clayey soils which determine also a high solid transport of the rivers.

The pumpage and distribution of the groundwaters takes place mainly through private wells (with a density of up to 10 wells per square kilometer) since the water-supply networks are extremely fragmentary and managed only at a local scale for civil and urban purposes. Only in very few cases centralised systems of distribution are present at a district level. In some urban centres, with populations up to 5,000 inhabitants, there are no water-supply networks and each household relies on its own well. Inevitably, in these situations there are no adequate controls on the drinking water quality. The water pumped from the subsoil and used for urban supply is subjected to treatment with sodium hypochlorite, for safety reasons. Instead, the groundwater rich in iron, manganese and ammonia is subjected to hypochlorite break-point treatment (induced advanced oxidation) for the same reason.

From a managemental point of view, the main

problem is given by the fact that the pumpage centres, being very scattered and located in intensely populated areas, present serious difficulties for a correct safeguard and management of groundwaters. In nearly all the developed countries the safeguard of the water-supply sources is guaranteed by the definition of protection zones which are combined with specific monitoring systems. In this situation, however, these protection techniques would be ineffective and inapplicable since they imply such rigorous restriction measurements that no other use or development of the territory would be possible. In fact, in many large towns the pumpage points which originally were located in peripheral areas, are now well inside the urban centres, making the application of any protection norm impossible. The situation looks even more serious if compared with the aquifer high vulnerability to pollutants, since the development of vast industrial and urban areas has inevitably increased the proliferation of many kinds of real and potential groundwater pollution sources.

4 THE QUALITY OF GROUNDWATERS

During the last 40 years an intense urban and industrial development has occurred, in most cases without adequate and rational planning norms which could have prevented enivironmental degrading and pollution, thus determining a progressive decline of the quality characteristics of groundwaters. In nearly all the situations, the disposal of industrial waste has taken place in an uncontrolled way, directly on the soil or subsoil, whilst agricultural and farming procedures have used more and more chemical substances such as fertilizers, pesticides and weedkillers.

In the Po and Venetian Plain at least 60% of all national high hazard industrial plants is concentrated, as well as an undetermined amount of liquid and solid waste, while it has been assessed that agricultural and farming activities scatter into the environment 400 Mkg/y N equivalent and 80 Mkg/y of chemicals. Among the factors contributing to the deterioration of the groundwater quality one should mention the thousands of poorly bored and managed wells which link together the most superficial and often polluted aquifers with the deeper ones. Also quarrying activities of soil materials have noticeably increased the pollution risks by digging deep trenches as far as the aquifers' top, thus depriving them of the natural self-depuration processes that take place through the ground and the unsaturated soil layer.

The most characteristic problems of man-induced pollution in the territory of this area can be ascribed to three kinds of substances: organic-halogenic compounds, nitrates and weedkillers. The pollution cases referred to the first group of substances (tri- and tetra- ethylene chloride, chloroform, etc.) have been recorded in Lombardy since the early 1970s; in this region the groundwaters are used for supplying over 2 million people. Similar problems have also occurred in Piedmont and Veneto, with over 1,5 million people involved. The aquifers' natural high vulnerability to pollutants and the presence of well developed mechanical, textile and galvanic industries have sometimes determined in the groundwaters maximum concentrations up to 500 mg/l. The origin of pollution due to industrial waste disposal is also confirmed by the frequent association of organic-halogenate substances with toxic metals, in particular chromium. In some cases the extension and intensity of these pollutants has caused the total elimination of water-supplying sources and the resort to alternative resources. Otherwise, where possible, the waters have been chemically treated in order to make them suitable for human consumption. In some other situations provisional exceptions to the national and EC norms regulating the quality of drinkable waters have been granted.

The nitrate contamination deriving from fertilising and farming activities, as well as from leaks in the sewage pipes, reaches sometimes extremely high levels up to and beyond 150 mg/l. To the areas with the highest pollution values, still relatively limited, is nearly always superimposed a widespread constant level of pollutants of 5 to 15 mg/l (according to national and EC norms the recommanded values are 5 mg/l while the maximum admitted concentration is 50 mg/l).

As for weedkillers, the first recordings of pollutants go back to the 1980s, following the more reliable and detailed analyses performed in this period, and concern the rice and maize cultivations where atrazine, symazine and bentazone have been detected in the groundwaters. Generally the contamination of groundwaters has an extensive character, with peak concentrations only at a local level, affecting mainly the northern sector of the plain (Piedmont, Lombardy and Veneto regions), where pollutants can easily be transmitted from the superficial aquifers into the deeper ones through the hydric wells which intercept several aquifers arranged one above the other.

5 CONCLUSIONS: MANAGEMENT AND CONTROL TRENDS

As mentioned before, groundwaters make up one of the most important elements for the developing activities of the Po and Venetian Plain but, nevertheless, during the past ten years not much has been done to adequately contrast the rapid decline of these vital hydric resources. Rather than enterprise effective policies of control and safeguard, both on a national and local scale, up to now the government initiatives have just been directed to granting disputable exceptions to the potability limits fixed by national and European laws. In other cases, alternative hydric resources have been sought after or water has been conveyed from nearby areas. In any case, a global approach to the correct management and exploitation of groundwaters has always been lacking.

Even without the support of properly organised national and local geological and hydrographical boards, the Government Departments in charge of the protection and management of groundwaters have recently promoted, with the collaboration of the Italian National Council of Research and some University Institutes, an accurate and multidisciplinary analysis of the hydrogeological situation of the Po and Venetian Plain. The research has been carried out through specific methods of investigation, representing on appriopriate maps the intrinsic vulnerability characteristics of groundwaters to pollutants, the classification of the contamination sources in relation with the hydrogeological features and the water-supplying centres, the impact evaluation given by the extraction of fluids from the subsoil (land subsidence) and the depletion of the groundwater resources. This research programme ("VAZAR" = Evaluation of Aquifers Vulnerability in High Hazardous Areas) has been developed along several lines: lay out of conservation and safeguard techniques of the water catchment equipments and determination of appropriate protection zones defined by means of hydraulic and hydrogeological parameters (Francani & Civita, 1988), employment of effectual upgrading methods and elaboration of specific mapping of the aquifers' vulnerability to pollutants (Civita, 1990 a,b). Fig. 2 shows the first maps so far produced in the Po and Venetian Plain in areas representative of the different hydrogeological and man-induced impact situations. The elaboration of such maps will be progressively extended to the remaining areas as a means to aim for a correct management of the groundwaters and territorial planning and development (Muratori, 1990).

The evaluation of the integrated aquifer vulnerability to pollutants is deduced from specific maps which emphasize two different kinds of information. The first aspect to be pinpointed concerns the intrinsic vulnerability, that is all the hydrogeological elements which define the specific susceptibility to pollutants of an aquifer system in all its different components and geometrical and hydrodynamic situations. Thus the intrinsic characteristics of an aquifer to absorb, propagate and mitigate in space and time the effects of a pollutant on the natural groundwater quality is better assessed. The second kind of information contains instead the description of the hydric supplying sources, the pollution producers, the upgrading structures (such as waste liquids depurators, controlled dumping grounds, etc.) and the physical and man-made elements spreading pollution (karst landforms, leaking wells, soil pits and quarries, mining activities, etc.).

In the vulnerability maps, around the public water-supplying wells, zones of protection have been traced. They have been defined adopting a time-based criterion, that is choosing a dimension to attribute to the protection zones corresponding to the time necessary for the hydric flow to run through a certain distance (safety times at 60 and 365 days). Such times and therefore the perimetration of the different zones have been chosen through the determination of the hydrogeological parameters resulting from the elaboration of pumping tests. The

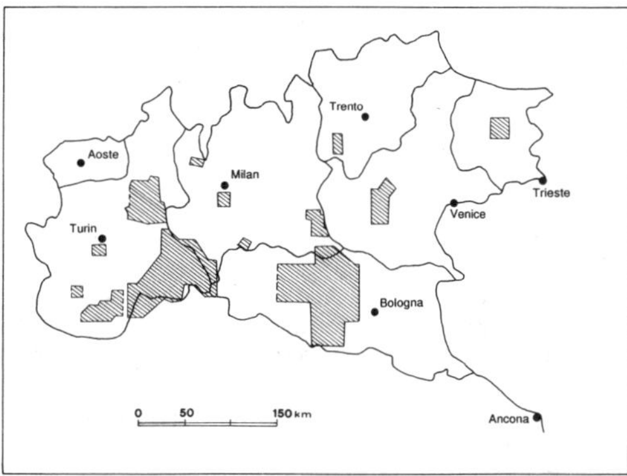

Fig. 2 - Vulnerability maps elaborated during the 1987-1992 period.

reconstruction of the piezometric surface has been determined with a permanent regimen condition and the time calculation procedures carried out applying Bear's (1979) expression with the finite differences method. Thus three areas have been defined (Francani & Civita eds., 1988):

"A" zone of immediate protection (10 to 20 m radius from the axis of each groundwater catchment).

"B" zone of limited protection, determined on the basis of the distance run in 60 days from a possible outer pollutant to reach the tray-well; that is its perimeter corresponds to the 60 day isochronal line. In this area the activities susceptible of altering the feeding trends and the groundwater natural flow are forbidden. It must also be guaranteed that no bacterial activity takes place, that non-biodegradable substance cannot reach groundwaters and the biodegradable ones do not constitute a supplementary biological load (oxygen consumption). Finally, in case of accident a sufficient time should be left in order to prepare adequate upgrading measures. In this zone it is forbidden to dig ditches, sewage works, canals, to use chemical substances on the ground such as most kinds of fertilizers, pesticides, etc.

"C" zone, defined adopting a distance corresponding to a 365 day safety time. The potentially polluting activities are here regulated and controlled limiting the number of industrial plants. A sufficient dilution of groundwaters should also be allowed before the pollutants can reach the water catchments. Theoretically this area should coincide with all the feeding area of the groundwater catchment but it is also allowed to consider administrative elements, such as municipal borders, and the presence of transport infrastructures (roads, highways, railroads, etc.).

The integrated vulnerability maps, thus elaborated, allow a reliable forecast of the aquifer system patterns of behaviour at a low cost and with simple investigation methods, performable also by local management offices. At the same time, these maps constitute a useful instrument for identifying priorities and strategies of intervention and subdividing the Po and Venetian Plain into areas characterised by different degrees of aquifer vulnerability to pollutants.

REFERENCES

Antonelli R. & Dal Prà A. 1980. Carta dei deflussi freatici dell' alta pianura veneta con note illustrative. *Quad. Ist. Ric. Sulle Acq. C.N.R.*, 51(I), Rome.

Bear J. 1979. Hydraulics of Groundwater. McGraw-Hill series in: *Water Resource and Environmental Engineering*, New York.

Bortolami G., Braga G., Colombetti A., Dal Prà A., Francani V., Francavilla F., Giuliano G., Manfredini M., Pellegrini M., Petrucci F., Pozzi R., Stefanini S. 1978. Hydrogeological features of the Po Valley (Northern Italy). Proc. of the Budapest Conf. May /June 1986, IAH-IAHS, *Annales Instituti Geologici Publici Hungarici*, Budapest.

Castany G. & Margat J. 1977. Dictionnaire français d'hydrogéologie. B.R.G.M., Orléans.

Civita M. 1990 a. Unified Legend for the aquifer pollution Vulnerability Maps. *Studi sulla vulnerabilità degli acquiferi (1), Quad. Tecn. Protez. Ambientale*, Pitagora Ed., Bologna.

Civita M. 1990 b. La valutazione della vulnerabilità degli acquiferi all' inquinamento. Proc. "*I° Conv. Naz. Protez. e Gestione delle Acque Sotterr.*", Marano s/P, Modena.

Francani V. & Civita M. eds. (1988) - Proposta di normativa per l' istituzione delle fasce di rispetto delle opere di captazione di acque sotterranee. Geo-Graph, Segrate, Milan.

Muratori A. (1990) - Le carte di vulnerabilità: possibilità di impiego in sede operativa e di pianificazione. *Quad. Tecniche Protez. Ambient., 12*, Pitagora Ed., Bologna.

Paltrinieri N., Pellegrini M. & Zavatti A. eds. 1990. Studi sulla vulnerabilità degli acquiferi. 2 - Alta e media pianura modenese. *Quad. Tecniche Protez. Ambient.*, 12, Pitagora Ed., Bologna.

Venturini L., Zuppi G.M. & Bertoni W. 1990. Studio idrogeologico e idrochimico delle acque sotterranee della pianura ravennate. *Studi sulla vulnerabilità degli acquiferi (1), Quad. Tecn. Protez. Ambientale,* 11, Pitagora Ed., Bologna.

Environmental Management, Geo-Water & Engineering Aspects, Chowdhury & Sivakumar (eds)
© 1993 Balkema, Rotterdam. ISBN 90 5410 099 0

Pollution studies on surface waters of the Thika area, Kenya

T.C. Davies
University of Nairobi, Kenya

ABSTRACT: The high level of industrialisation in the Thika area has kept its waterways relatively polluted. In order to identify the nature, sources and extent of this pollution, various physical and chemical parameters of the surface waters were determined in samples collected during three field campaigns mounted from October, 1990 to August, 1992. A comparison is made between these results, world averages for river-water and international water quality standards for drinking water. Copper, lead, zinc, cadmium and chromium are found in concentrations within the safety limits for drinking water, whereas iron and manganese as well as several other chemical parameters are well in excess; for these parameters, values above threshold are obtained at loci co-incident with effluent discharge points for factories and the municipal sewage treatment works in the area. The maximum levels for some of these parameters at effluent discharge points along the main rivers are: biochemical oxygen demand (BOD), 990 mg $1^{-1}0_2$; chemical oxygen demand (COD), 1088 mg 1^{-1} 0_2; total suspended solids (TSS), 46 mg 1^{-1}; total dissolved solids (TDS), 390 mg 1^{-1} and electrical conductivity (EC), 4420 μS cm^{-1}. It is therefore concluded that most of the pollution of the rivers in the area is due to untreated or partially treated effluents discharged into them from adjacent factories, sewage treatment plants solid waste dumps and run-off from agricultural lands. The present trophic status of the rivers is also evaluated in the light of the collected data.

1 INTRODUCTION

The Thika area in Kenya ranks among the most industrialised, populous and agricultural regions of the country. Its industrial orientation is the result of the influence of a set of physical and human factors, among which are the topography, availability of industrial water resources and raw materials, and the influence of the infra-structure.

For sometime now a lot of popular concern has been expressed on the quality of river-water in this region for drinking purposes. There is however a dearth of information on the hydrogeochemistry, of these waters, which should form a basis for the identification of pollutants, their sources and levels. Such information would help municipal authorities in formulating guidelines and control measures regulating water use by industry and the population in general. An intensification of research effort in these areas is therefore desirable. The aim of this paper is to provide quantitative information on some aspects of water quality for rivers in the Thika area.

2 PHYSIOGRAPHY AND GEOHYDROLOGY

The study area is located about 40 km north-east of Nairobi. It is characterised by a flat volcanic plain at an altitude of about 1524 m. The average annual rainfall is about 900 mm and the level of agricultural activity is relatively high. The most important river is the Thika, which receives its waters throughout the year from streams rising further west on the high ground at the edge of the Rift Valley. Natural waters of the area are in contact with the Kapiti Phonolite (Miocene) and a range of Tertiary pyroclastic rocks (trachytic tuffs, agglomerates) with thin basalts, and their weathered derivatives.

3 SAMPLING AND ANALYTICAL PROCEDURES

Surface water samples and effluents were collected from 18 stations along waterways and drains as indicated in Fig. 1. To detect any seasonal variation in the water quality parameters, each locality was sampled four times - once each in October 1990, February 1991, February 1992 and August 1992. Samples were obtained using a polyethylene bucket to which was attached a 10 metre-long nylon rope for immersing it below the river surface. The bucket was washed five times with the river water or effluent at a particular locality before a sample is collected. The samples were then stored in acid-cleaned linear polyethylene bottles. Aliquots for the determination of heavy metals were acidified to pH <2 with purified nitric acid.

Variables such as temperature, pH and electrical conductivity were measured in the field. Dissolved cadmium, chromium, copper, manganese, lead and zinc were analysed by flame atomic absorption spectrometry after filtration of the sample (Whatman No. 1 filter paper) followed by evaporation and treatment with 1:1 purified nitric acid. Iron was determined by the Lovibond Comparator and Disc method after treatment of the sample with dilute HCl, potassium permanganate, ammonium thiocyanate and amyl acetate - alcohol solution.

Other parameters such as biochemical oxygen demand, total hardness, total dissolved solids, etc. were determined by standard techniques in the untreated samples as soon as they were brought to the laboratory.

4 RESULTS AND DISCUSSION

Table 1 gives the range and mean values of water quality parameters determined in this study, along with international averages for unpolluted freshwater, and maximum permissible levels in drinking water. The results show that surface waters of the Thika area are in general, highly oxygen demanding. The highest values for BOD, COD, TDS and electrical conductivity are obtained at industrial effluent discharge points mainly along the Thika river, and at the point of discharge of sewage wastewater at the Komu river (Fig. 1). Wastewaters produced by the tanning, coffee and sisal factories (localities 3, 8, 14 and 15; Fig. 1) give BOD and COD values close to 1000 mg $l^{-1}O_2$. A smaller but significant contribution of organic pollutants comes from diffuse sources such as the leaching of solid waste dumps and runoff from agricultural lands (eg. the Del Monte pineapple plantations). At the main municipal sewage works near Ndarugu, chemical treatment to remove nutrients is not included in the routine treatment procedure and BOD values of the final effluent into the Komu river can be as high as 40 mg $l^{-1}O_2$. As a result the Komu river teems with microbial life and is coloured green most of the year owing to algal bloom.

The dissolved content of the heavy metals examined, except iron and manganese, are within the acceptable levels for drinking water. It must be emphasised that these values represent only dissolved concentrations obtained by analysis of the filtered water samples. Metals in natural waters can exist in truly dissolved, colloidal and suspended forms (Chapman and Kimstach, 1992). The binding of pollutants by colloidal and sediment phases generally diminishes their toxic properties to organisms (Spitzy and Leenheer, 1991). Since the dissolved fraction of the metal is the part relevant to toxicology (Reuter and Perdue, 1977; Spitzy and Leenheer, 1991; etc.) the waters studied can be considered largely free from metal pollution on account of the low dissolved concentrations found.

Iron and manganese are however found in concentrations several times above the international averages for freshwater and is the only case where a relationship between bedrock geology and water chemistry appears significant. The occurrence in the drainage areas of alkaline volcanic rocks (phonolite, trachytic tuff, etc.) relatively rich in pyroxenes, amphiboles and iron ore, accounts for the release of iron (and manganese) through weathering (high mean temperature and moderate precipitation) and soil leaching. The ferrous form of iron (Fe^{2+}) can persist in waters devoid of dissolved oxygen (Rail, 1989). Like iron, manganese occurs in the divalent and trivalent forms and its chlorides, nitrates, and sulphates are highly soluble in water (McKee and Wolf, 1972). In natural waters that are subject to reducing conditions, manganese is leached from rocks and soils and occurs in high concentrations in the water. Manganese in such waters accompanies iron, although they normally occur together (Rail, 1989). The anaerobic conditions (reducing) set up by organic pollutants in waters of the study area are responsible for holding iron and manganese in their reduced forms (Fe^{2+}, Mn^{2+}), once they are released by weathering and leaching.

High iron and manganese levels are objectionable in water supplies used for domestic or industrial purposes. They cause unpleasant tastes, stain laundry and plumbing fixtures, and in the case of manganese, fosters the growth of certain micro-organisms in distribution systems (McKee and Wolf, 1972). The daily requirement for iron in human

Table 1. Selected hydrochemical data for surface waters of the Thika area, recorded during the period Oct. 1990 to Aug. 1992.

Parameter	Range	Arith. mean	International average for fresh water	Maximum permissible level in drinking water
pH	5.5-9.5	7.6	6.5-8.5(R)	6.5-8.5 (R)
EC ($\mu S/cm$)	70-4420	526	10-1000(R)	——
Hardness ($mg\ l^{-1} CaCO_3$)	20-48	32	——	500
TDS ($mg\ l^{-1}$)	105-390	205	<100	1000
TSS ($mg\ l^{-1}$)	1-46	22	150	——
BOD($mg\ l^{-1} O_2$)	5-990	177	2.0	3.0
COD($mg\ l^{-1} O_2$)	8-1088	194	<200	——
Cd ($\mu g\ l^{-1}$)	1-4	2	.001	5
Cr ($\mu g\ l^{-1}$)	1-265	20	.1	50
Cu ($\mu g\ l^{-1}$)	1-57	10	1.4	1000
Fe ($\mu g\ l^{-1}$)	400-2500	883	50	300
Mn ($\mu g\ l^{-1}$)	1-2976	442	10	100
Pb ($\mu g\ l^{-1}$)	1-26	7	.04	50
Zn ($\mu g\ l^{-1}$)	10-163	56	.2	5000

R = Range

Sources: WHO, 1984 (a); CCREM, 1987; CEC, 1980; Sayre, 1988; Meybeck, 1988; Schiller and Boyle, 1985,1987; Meybeck and Helmer, 1989, and Chapman and Kimstach, 1992.

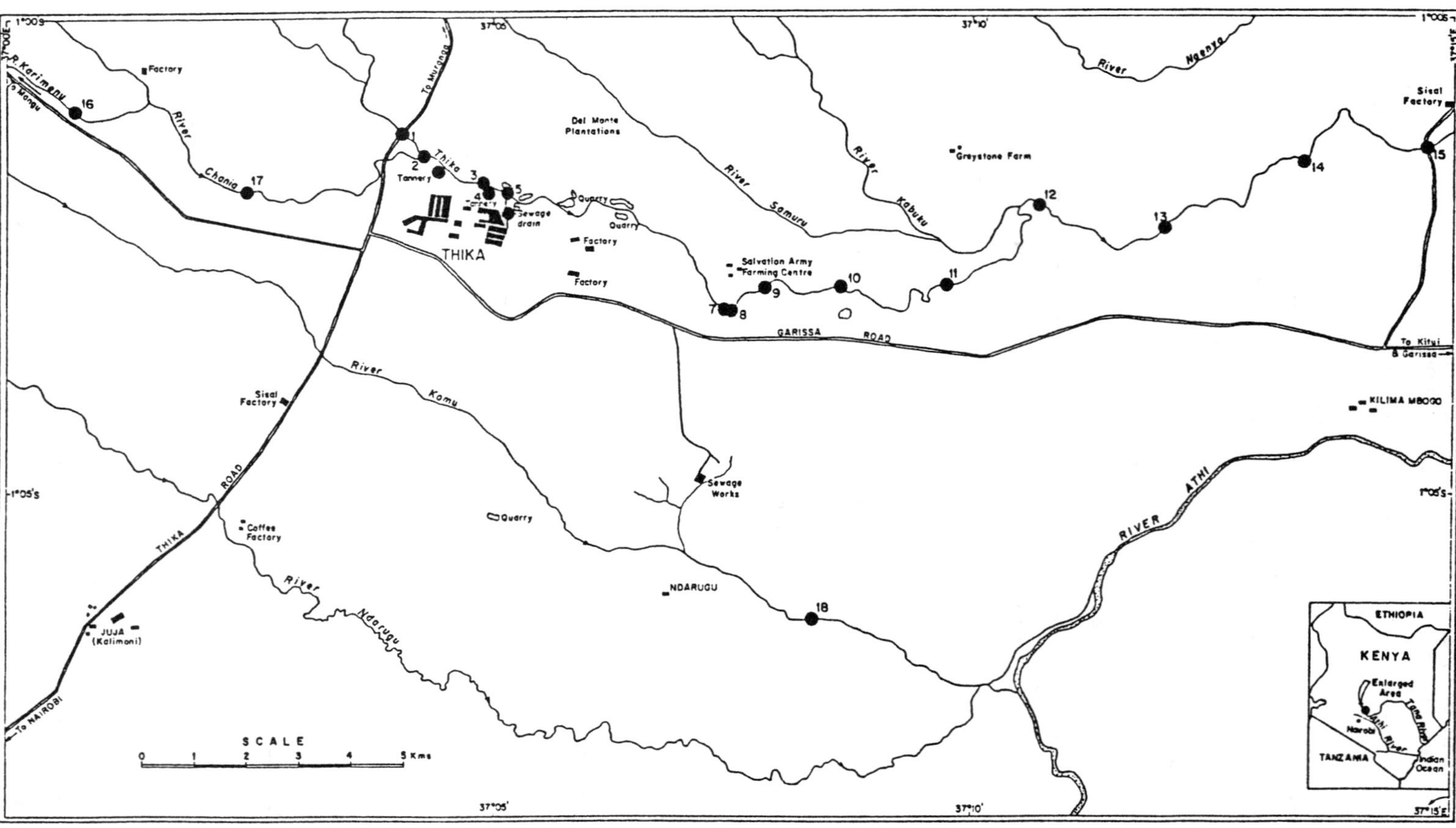

Fig. 1 Map of the Thika area showing sample locations

individuals is 1-2 mg, but the intake of larger quantities could be required if poor absorption occurs (Sollman, 1957). The iron standard given for drinking or domestic water (300 $\mu g\ l^{-1}$) is meant to prevent objectionable tastes or laundry staining and is of aesthetic rather than of toxicological significance, although iron at high concentrations is toxic to livestock by interfering with the metabolism of phosphorus (NAS, 1974). In the case of manganese, the WHO 1984 guideline value for drinking water is 100 $\mu g\ l^{-1}$(WHO, 1984 a). The daily intake of manganese required for a normal human diet is 10 mg (Rail, 1989). The element is !absorbed slightly and accumulates in liver and kidney tissues.

5. CONCLUSIONS

Water quality data collected for surface waters of the Thika area indicate that although it is an industrially influenced environment, the dissolved heavy metal contents in its rivers are generally within the acceptable limits for drinking water. Iron and manganese are however found in concentrations several times above their global averages for freshwater, but the undesirability of such high levels of these metals is for aesthetic rather than toxicological reasons.

The most important group of pollutants in rivers of the study area are organic, entering the drainage system as point sources directly from industrial effluent discharge drains and sewers, and also as non-point sources from the leaching of solid waste dumps, and agricultural runoff. These have rendered the rivers quite eutrophic (mean TDS, 205 mg l^{-1}; mean BOD, 177 mg $l^{-1} O_2$); this situation is brought to an extreme in the Komu river which receives the final effluent from the main sewage works in the municipality.

Relationship between bedrock geology and chemistry of the waters is quite weak and reflects the extent of anthropogenic chemical perturbations.

The enforcement of regulations governing the way industries operate and dispose of their wastes, should be made a priority by the relevant mjnicipal authorities. The need for reducing the use of artificially manufactured chemicals in agriculture should be stressed. Further steps aimed at removing nutrients (especially phosphorus and nitrogen) should be incorporated into the treatment procedure at the main sewage works to ensure that the final effluent discharged into the Komu river is of acceptable quality.

ACKNOWLEDGEMENT

This work was funded by the Deans' Committee, University of Nairobi.

REFERENCES

CCREM (Canadian Council of Resource and Environment Ministers). 1987. Canadian Water Quality Guidelines. Environment Canada, Ottawa.

CEC (Commission of European Communities). 1978. Council Directive of 18 July 1978 on the quality of fresh waters needing protection or improvement in order to support fish life, (78/659/EEC). Official Journal, L/222, 1-10.

CEC (Commission of European Communities). 1980. Council Directive of 15 July 1980 relating to the quality of water intended for human consumption, (80/778/EEC). Official Journal, L/229, 23.

Chapman, D. and V. Kimstach. 1992. The selection of water quality variables. In: D. Chapman (Ed.) Water Quality Assessments, Chapter 3. Chapman & Hall, London, 585 pp.

McKee, J.E. and H.E. Wolf. 1972. Water Quality Criteria. California State Water Resources Board, Sacramento California, Publ. 3-A.

Meybeck, M. 1988. How to establish and use world budgets of riverine materials. In: A. Lerman and M. Meybeck (Eds.). Physical and Chemical Weathering in Geochemical Cycles. Kluver, Dordrecht: 247-272.

Meybeck, M. and R. Helmer. 1989. The quality of rivers: from pristine stage to global pollution. Palaeogeogr., Palaeoclimatrol., Palaeoecol. (Global Planet. Change Sect.), 75, 283-309.

National Academy of Sciences. 1974. Water Quality Criteria. National Academy of Engineering, Washington, D.C.. U.S. Government Printing Office.

Rail, C.D. 1989. Groundwater Contamination: Sources Control and Preventive Measures. Technomic Publ. Co. Inc., Lancaster: 132 pp.

Reuter, J.H. and E.M. Perdue. 1977. Importance of heavy metal-organic matter interactions in natural waters. Geochim. et Cosmochim. Acta, 41:325-334.

Sayre, I.M. 1988. International Standards for drinking water. Journal AWWA, January, 53-60.

Schiller, A.M. and E.M. Boyle. 1985. Dissolved zinc in rivers. Nature, 317: 49-52.

Schiller, A.M. and E.M. Boyle. 1987. Variability of dissolved trace metals in the Mississippi. Geochim. et Cosmochim Acta, 51: 3273-3277.

Spitzy, A. and J. Leenheer. 1990. Dissolved organic carbon in rivers. In: E. Degens, S. Kempe and J. Richey (Eds.) Biogeochemistry of Major World Rivers. SCOPE Report 42, John Wiley and Sons, Chichester, pp. 213-32.

WHO. 1984 a. Guidelines for Drinking-Water Quality. Volume 1. Recommendations. World Health Organisation, Geneva, 130

Environmental Management, Geo-Water & Engineering Aspects, Chowdhury & Sivakumar (eds)
© 1993 Balkema, Rotterdam. ISBN 90 5410 099 0

Environmental and health impact of irrigation and drainage system in Egypt

A. El-Zawahry
Civil Envineering Department, Sultan Qaboos University, Sultanate of Oman

I. El-Assiouti, E. Emam & S. El-Diddi
Irrigation & Hydraulics Department, Cairo University, Egypt

ABSTRACT: No doubt that the introduction of the Irrigation and Drainage (I & D) system in Egypt has had a significant impact on the environment setting in the region. To have an irrigated agriculture in the country during the whole year round, an extensive network of canals and drains were constructed along with their control structures. Moreover, control structures in the form of dams and barrages were built across the Nile to allow better control and use of its water. All these activities and projects have resulted in a drastic impact on the physical, chemical, biological, cultural and socio-economic setting in the area. The study, covers the following aspects: (1) Assessment of chemical and biological pollution of the Nile below Aswan and the irrigation and drainage system; (2) Assessment of the occurrence of water borne diseases and pathogens in the I & D system; and (3) Recommendation for further actions.

1 INTRODUCTION

In Egypt, the River Nile is the main source of water supply for drinking, industrial, and agricultural purposes, Fig. 1. Therefore it has been under investigation to define its water quality changes. For thousand of years the Nile has behaved in the same way. One of the most important development schemes is the construction of dams. No dam can be built and no lake can be created without environmental costs and benefits of some kind. The environmental side effects of dam construction are generally divided into categories: (a) the local effects and (b) the effects resulting from the change of the hydraulic regime. Both categories have their physical, biological and socio-economic elements . Water impoundment may drastically affect the water quality. The physical, chemical and biological properties of water leaving a lake may differ significantly from that of water entering the lake. In Egypt, the construction of the Aswan High Dam has created a new impact so far as river water quality is concerned. Water impoundment in the lake Nasser exerts significant impact on the downstream water conditions and multipurpose water use. Furthermore, downstream changes in water quality are primarily due to river flow control by the different barrages, river navigation for transport purposes, and pollution both from point as well as non-point sources.

Surveying has shown that there are industrial installations along the river. Most of these installations are not connected with a regional sewerage system and do not provide any treatment before discharging their waste water in the river. Others discharge their wastes into agricultural drains, which in turn flow into the river especially in Upper and Middle Egypt. So it is important to have an assessment of the chemical and biological pollution in the River Nile including factors affecting water quality.

The total length of the irrigation and drainage networks in Egypt is about 47,807 kilometers and serve more than six million feddans. To control such huge system, thousands of water structures (regulators, intakes, weirs, escapes, bridges, ... etc.) were built. The water is transferred from the Nile to the field through a system of main Canals and Rayahs, Secondary Canals, third order Canals and Meskas. The length of the irrigation system is about 30,309 km. The cultivated area is covered with a network of drains starting from the collectors and ending with the main drains. Due to the increase of the soil salinity and ground water table a massive subsurface drainage network is introduced. The surface drains in upper Egypt are allowed to discharge their water to the Nile. The lower Egypt drains are designed to discharge to the Northern Lakes or directly to the sea. Due to the expansion in the cultivated land, drainage water is pumped back to some canals to be reused.

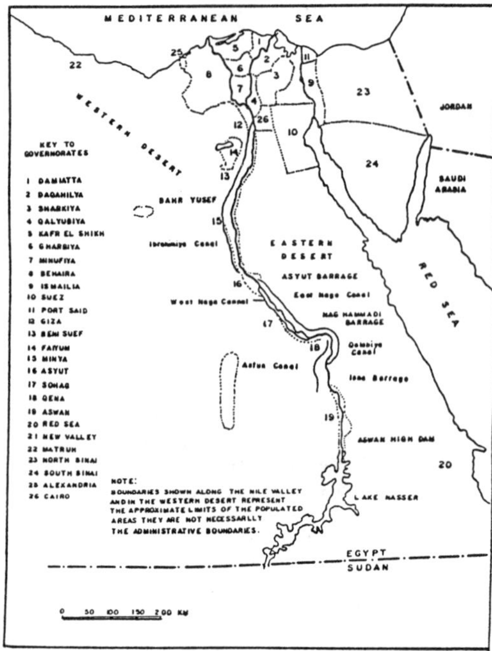

Fig. 1 River Nile and Lake Nasser

Hence, the irrigation and drainage systems are cross connected, environmental problems are created.

2 ASSESSMENT OF CHEMICAL POLLUTION

Before the construction of the High Aswan Dam (HAD) there were seasonal changes in certain water quality constituents of the Nile water. During the flood period the water was muddy. During this periods the Nile water has minimum dissolved solids but maximum suspended solids, whereas during the rest of the year dissolved solids increased and suspended solids decreased. Since the construction of the dam the water released from the reservoir is practically silt-free, and the maximum discharge is about a quarter of the earlier flood discharge. Both the population and the industrialization of the country have increased significantly during the past two decades. It is thus necessary to assess the quality of the Nile water for domestic, agricultural and industrial uses.

2.1 Salinity

Prior to the construction of the HAD, the concentration of dissolved salts varied with season; being low during flood and high during drought period. Results of chemical analysis of Nile water samples taken at Cairo (averaged for the period 1919-1927) indicated that dissolved salts reached their highest value of 232 ppm during April and May and their lowest value of 128 ppm during September. Same trend and almost values were also reported for samples of Nile water at Cairo in 1963 (prior to the High Aswan Dam). As for changes in Nile water quality after the HAD, analyses of water samples taken at Cairo during 1972 indicated a high dissolved salts (DS) concentration of 207 ppm during April and May, and a low of 177 ppm during August. It can be seen from the figures that HAD has damped out the fluctuations in DS during the impoundment and mixing in the lake. However, it has retained higher DS levels for longer periods than prior to the HAD.

2.2 Turbidity

Although silt concentration in the Nile water is vital for increasing land fertility, yet, it is viewed differently from water quality viewpoint, high silt concentration means turbid water which has a certain biological state at equilibrium. On the other hand, clear water develops a different biological states of equilibrium. Table 1 presents a comparison between average concentrations at El-Gaafra, downstream of Aswan, before and after impoundment behind the HAD, for the periods 1958-1963, and 1966-1976 respectively.

Table 1. Impact of HAD on Silt Concentration in the River Nile

PERIOD	Silt Concentration in the Nile at Gaafra (ppm)											
	J	F	M	A	M	J	J	A	S	O	N	D
BEFORE HAD (1958-1963)	64	50	45	42	43	85	647	2702	2422	925	124	77
AFTER HAD (1968-1976)	44	47	45	50	51	49	48	45	44	43	48	47

2.3 Eutrophication

The HAD reservoir can be visualized as a huge biological reactor which releases a large number of planktonic organisms. The main algal species in the Nile belong to the Cyanophyceae (blue-green Algae), Chlorophycease (green algae) and Bacellariophyceae (Diatoms). The change from lacustrine to riverine conditions as the Nile water leaves the reservoir is accompanied by significant changes in algal distribution. Typically the Cyanophyceae will decrease and Diatoms and the Chlorophyceae will increase. Further changes downstream reflect local flow control patterns and nutrient inputs.

After the initial decrease in Upper Egypt,

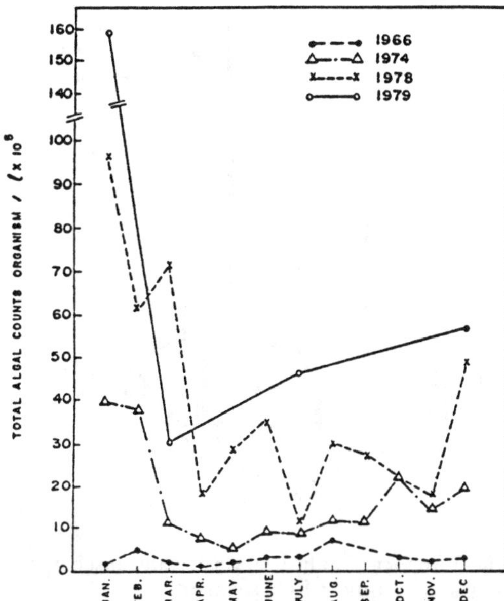

Fig. 2 Average Phytoplankton (1966-1977)

Cynophyceae reappear in large numbers during the summer at the end of the two delta branches, Rosetta and Damietta. This is an indication of stagnant river conditions. In addition to causing taste and odor problems, these undesirable organisms are toxic to certain types of fish.

Fig. 2 shows average Phytoplankton fluctuation between 1966 (during construction of HAD) and 1974, 1978, and 1979 (after impoundment). The distinct rise in Phytoplankton density is a prevailing feature of the Nile water after the construction of the HAD. This is attributed to the low level of turbidity after impoundment, in addition to the sharp reduction in flow rate of the Nile downstream HAD.

2.4 Pesticides

The average consumption of pesticides in Egypt during 1962-1982 is about 25,000 ton/year. More than 1000 pesticides belonging to more than 30 classes of chemical compounds exist. A considerable number containing chlorine and phosphorous pesticides may enter a body of water : (1)as a result of wash off from plants and soil; (2) as aerial and surface spraying; (3) with agricultural drainage; (4) with drainage water of pesticides manufacturing.

Table 2 summarizes the statistical presentation of DDT residue survey (1979-1983). The number of samples collected in the irrigation and drainage waters during the five years is only

Table 2. Statistical Presentation, DDT Residue survey (1979-1983)

YEAR	No. per year	No. of samples		Substrate									
				Foliage		Fruit		Soil		Irrigation Water		Drainage Water	
		ND	D	ND	D	ND	D	ND	D	ND	D	ND	D
1979	5	2	3	2	-	-	-	-	3	-	-	-	-
1980	34	23	11	16	2	7	-	-	3	-	4	-	2
1981	4	4	-	2	-	2	-	-	-	-	-	-	-
1982	27	16	11	9	10	7	1	-	-	-	-	-	-
1983	36	27	9	13	5	10	1	-	-	4	-	-	3
Total	106	72	34	42	17	26	2	-	6	4	4	-	5

D = Detected
ND = Not Detected

thirteen. Additionally, pesticides and herbicides are used recently on a large scale for the protection of crops, pest control, and weeds control on bank sides of water bodies. The amount of pesticides used in the Egyptian environment over the last 30 years is about 620,000 tons of about 200 different types. The pesticides and herbicides residues pass through the soil by the irrigation water into drains and then to the River Nile. It is worth noting that there is a limited amount of data, work and research in this field by some universities and research centers.

2.5 Fertilizers

Inefficient use of Nitrogen fertilizers results in losses of nitrates through leaching out from the soil contaminating surface and ground water. The increase of the average concentrations of Nitrogen compounds in drains (irrigation return flow) was found to be roughly corresponding with the growth of application rates of Nitrogen fertilizers.

Before the Aswan High Dam was constructed, 24,000,000 tons/year of solids were deposited on the Egyptian flood plains. After the High Dam construction, this amount decreased to 2,100,000 tons/year. The remainder settled out in lake Nasser. The loss of sediment N and soluble P is estimated to be 12,000 and 6,000 tons/year respectively. This required a significant increase in fertilizers, which caused an increase in nutrients and phosphorus in soil, canals and drains.

2.6 Industrial sources of pollution

There are many sources of industrial pollution in the River Nile. At Aswan segment, the survey showed that El-Sail drain is the main source of pollution in this area. It receives sewage, agricultural drainage water and wastewater from Kima factory.

In Qina segment, the most significant and potentially dangerous pollution effect is due to wastewater from the sugar factories waste effluents which are discharged into the river without any kind of treatment.

In Sohag segment, the hydrogenated oil and onion drying factories discharge their wastewater directly to the river causing a zone of pollution of about 200 m in length.

In Assiut segment, the super phosphate factory discharges its waste water into the river. The waste is highly acidic with a pH of around 2.0 and high total solids content (1854 mg/1).

In Cairo segment, the river receives wastes from a number of factories. The most polluting wastes are those discharged from the sugar factories at El-Hawamdia and the mixture of wastes discharged from the Iron and Steel, Coke and fertilizer factories into a runoff which in turn reaches the river in front of the Iron and Steel factory.

Deterioration of water quality occurs at localized areas near the river bank at the sites of wastewater discharge. Mediterranean conditions are usually not affected.

In the case of Rosetta segment, there are two main sources of pollution. The first is El-Rahawy drain which carries the effluent of El-Zenin sewage treatment plant. The second source is due to the industrial wastes from the three factories at Kafr El-Zayat, namely, the pesticides, salt and soda, and the super phosphate factories. Water samples collected along Damietta segment showed that the most polluting source along this segment is Talka factory.

Figures 3 to 5 show the impact of some industrial sources on the River Nile water quality.

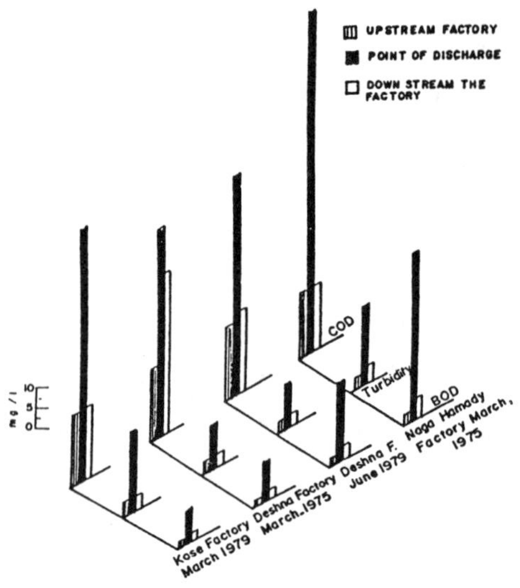

Fig. 3 Impact of industrial wastewater discharge on water quality at Kina segment

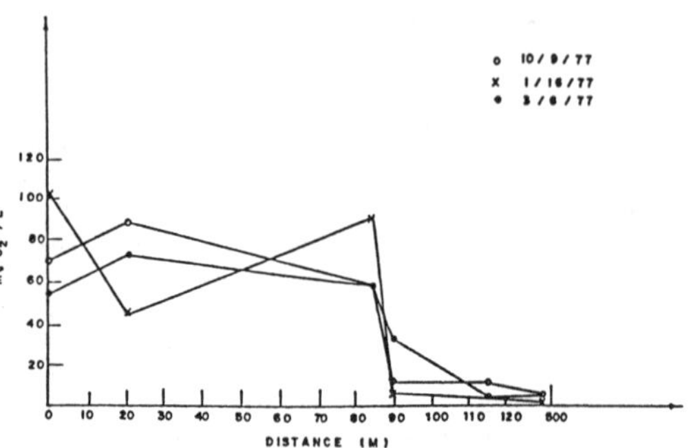

Fig. 4 Variation of BOD values from the point of discharge to mid-way of the River Nile at the iron and steel factory (Helwan)

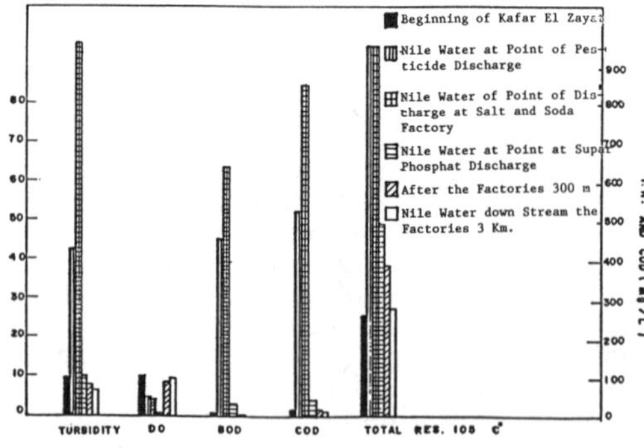

Fig.5 Impact of industrial wastewater discharge on the water quality at Kafr El-Zayat.

3 ASSESSMENT OF BIOLOGICAL POLLUTION

The health environment in rural Egypt has been affected by the biological pollution of the water of the irrigation and drainage system. Water related diseases can be classified into three main categories: (a) water-borne, (b) water-hygiene, and (c) water-habitat diseases.

3.1 Water-borne diseases

Water-borne diseases include enteric diseases due to pathogens in drinking water which are of faecal origin. The occurrence of water-borne diseases depends on the existence or non-existence of water supply system. With the government policy of supplying the Egyptian villages with potable water, these diseases are reduced.

3.2 Water-hygiene diseases

Water-hygiene diseases are the consequence of the inadequate use of water to maintain personal cleanliness. Water quantities appear to be of higher priority than quality in this category.

It appears that the existence of the irrigation canals may have had a positive impact in this category. Obviously personal hygiene still governs the spread of these diseases.

3.3 Water-habitat diseases

Water-habitat diseases are vector-borne and are the most important group of diseases related to the introduction of irrigation. Three different types of vectors are involved in diseases transmission: (a) snail vectors such as Schistosomiasis (Bilharizia), (b) mosquito vectors, and (c) fly vectors.

No doubt that the existence of irrigation canals has created favorable conditions for the life cycle of Bilharizia snail. Based on a sample of 16,295 persons selected from several places, the results indicated that: (1) the diseases rates tend to decrease from north to south, (2) the peak infection is usually in young age (15 to 19 years age), (3) for any specific age the prevalence is higher in males than females. Table 3 presents the overall percent prevalence of Schistosomiasis in different study areas.

Table 3. Overall percent prevalence in different areas

Study area	Schistosomiasis percentage
Nile Delta	42 (22-67)
Upper-Middle Egypt	27 (17-37)
Upper Egypt	
Aswan desert	4 (1-7)
Agricultural villages	25
Kom Imbo (New Nubia)	--

-The lowest and highest prevalence figures are recorded in parenthesis

4 CONCLUSIONS AND RECOMMENDATIONS

Water quality in the Nile prior to the introduction of the irrigation and drainage system is not known. But with the construction of control works and the expansion of irrigated areas, water quality is definitely suffering.

The existence of the network of canals and drains has affected the social and health environment in Egypt in many ways, both positively and negatively. Schistosomiasis, has always threatened Egypt due to the contact between the people and the polluted water.

To minimize the negative effects of the Irrigations and drainage system, four groups of recommendation can be made. The first group is related to the improvement of water quality such as: (1) domestic and industrial wastewater must be treated before discharging to drains; (2) education about personal hygiene; (3) the types, rates and loads of agricultural chemical should be reviewed from an environmental point of view; and (4) the present program for Bilharizia control by combating snails should be enforced.

The second group aims at better identification of environmental setting through a well planned monitoring system. All water bodies should be monitored with respect to pertinent water quality parameters. Short and long term planning is greatly needed to provide new facilities and upgrade existing facilities. All discharge points are to be carefully examined, and permissible flows, concentration, and provided treatment should be shown on the permits. It is recommended to establish a National Data Bank for water resources, and their quality related information.

Thirdly, several studies are recommended to aid decision makers in managing the water resources. It is suggested to develop appropriate mathematical models to simulate difficult water quality issues. Priority given to a specific study should reflect the importance of the issue to be studied. Typical studies suggested include: (a) Impact of domestic and industrial wastewaters on the water resources; (2) study the environmental impact of major irrigation and drainage projects (storage of fresh water in Northern Lakes, Reuse of drainage water, conversion of Northern Lakes to agricultural land, transport of Nile water to Sinai, and integrated development of the region surrounding the High Aswan Dam reservoir).

Last group of recommendations deals with training and education in the area of environmental management of water resources. This includes training programs for farmers and villagers, personnel of agencies in charge of sampling and analysis, and personnel who are in charge of granting discharge permits and also detecting violation.

REFERENCES

Atta, A.A. 1978. Egypt and the Nile after construction of the Aswan High Dam, Ministry of Irrigation, Cairo.

El-Gohary, F. 1981. Water quality changes in the River Nile and impacts of waste discharges, proceedings of international conference on water resources management in Egypt, Cairo.

Miller, F.D., Hussein, M., and Mancy, K. H. 1978. Aspects of environmental health impacts of the Aswan High Dam on rural population of Egypt, Progress Water Technology, Pergamon Press, 11,173.

Volker, A., and Henry, J.C. 1988. Side effects of water resources management, IAHS Pub. No.172, IAHS press, Oxfordshire.

Water Master Plan 1981. Report on fisheries, ecology, and health. Technical report 13,UNDP-EGY/73/024.

Environmental Management, Geo-Water & Engineering Aspects, Chowdhury & Sivakumar (eds)
© 1993 Balkema, Rotterdam. ISBN 90 5410 099 0

On the estimation of river reaeration coefficient with the soluble solids probe

M. F. Giorgetti
School of Engineering at São Carlos, University of São Paulo, Brazil

ABSTRACT: Results of experimental studies and data correlation analyses are presented for a proposed method for the indirect determination of gas transfer coefficients at gas-liquid interfaces — particularly reoxygenation of water — based on information given by the dissolution of floating solid probes.

1. INTRODUCTION

Natural stream reaeration or reoxygenation is the physical absorption of oxygen from the atmosphere by a flowing stream. It is the fundamental process by which a stream replenishes the oxygen consumed in the biodegradation of organic wastes. The primary use of reaeration coefficients is to quantify the process of reaeration in dissolved-oxygen water-quality models. The reaeration coefficient is the rate constant for the absorption of oxygen from the atmosphere. These models, which simulate the exchange of dissolved oxygen, are used to calculate waste-load allocations for the stream so that dissolved oxygen concentration standards are not violated.

Prior to 1965, reaeration rate coefficients could be estimated only by indirect methods, which were subject to significant errors. Tsivoglou and others (1965) developed a direct method of measuring stream reaeration coefficients employing the radioactive tracer gas krypton-85. Tsivoglou's method is based on the principle that a gas tracer injected into a stream would be desorbed from the stream similary to how oxygen would be absorbed. This provided the first direct means of measuring stream reaeration capacity. In the early 1970's, Rathbun and others (1975) modified the technique: hydorcarbon gas such as ethylene or propane was substituted for radioactive krypton as the gas tracer. Theoretically, this method does not differ substantially from the krypton method; the only differences are in performance techniques. The modified method is largely used as a standard method in many countries.

A third kind of technique is presented and discussed in this paper; a method based on the rate of dissolution of a floating solid probe, drifting down the stream. The physical fundamental for the new proposal is the fact that both, the gas transfer at the water surface and the mass transfer from the solid probe to the water are affected by the water turbulence and by the larger-scale motions responsible for the rate of water surface renewal and, thus, for the rate of oxygen exchange. The costs and the difficulties of the new proposal are minimal as compared to the demands of the other existing methods. The basic idea involved is very simple. A solid cylindrical disk of the soluble solid, with only one of its flat surfaces exposed to the water, is placed in contact with it, near the free surface, hanging from a wire support set on styrofoam floats; details are illustrated in figures 1 and 2. Velocities of dissolution V_s are obtained from the initial and final weight of the probe and the time of duration of each experiment.

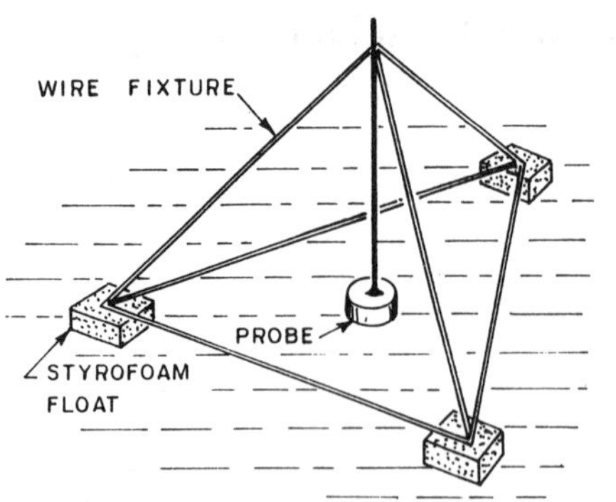

Fig. 1. General assembly of floating probe

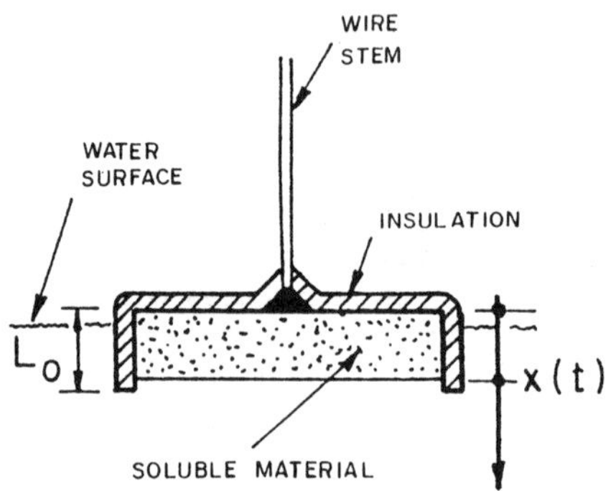

Fig. 2. Detail of probe

2. EXPERIMENTS

Experiments were carried out in a 61 cm diameter cylindrical tank with approximately 100 liters of water volume, and in a 20.0 m long recirculating channel with a rectangular cross section with 17.5 cm of depth and 35.0 cm of width. The tank was baffled to suppress the main circulation. Agitation was provided by a four-blade propeller with 14.6 cm diameter rotating with speed N. The assembly duplicated the one described by Rainwater and Holley (1983).

The channel was specially designed to operate in a closed circuit without interruption of the free water surface in order to permit long times of continuous operation of the floats. Near the end of each 10 - meter stretch a part of the flowing water is diverted and subsequentely re-injected as a fast bottom wall jet into the beginning of the other stretch. The momentum transfer from the jets to the main flow keeps the water in motion. Five different bottom roughnesses were used in combination with three different flow rates (four in two cases) to produce a set of 17 distinct experimental situations varied within the limits of 16.9 and 29.1 liters per second; artificial roughness at the bottom was produced by gluing quasi-homogeneous materials (sand or chipping) with mean diameters of 2.38, 4.76, 6.35, and 9.52 mm. The fifth condition was that of smooth painted steel with surface equivalent sand roughness of 0.97 mm.

The material chosen for the soluble probes was the oxalic acid, with chemical formulation $HOOC - COOH - 2H_2O$, in function of its adequate sensitivity to the velocities of dissolution observed at the hydraulic conditions described above. The probes were manufactured from the typical powdered material used for chemical analysis, by dry compression in a metal cylindrical die with a load of 50,000 N. The final dimensions of the oxalic acid probes were 28.5 mm diameter, 5.0 mm thickness, and 5.0 g initial mass. One flat surface and the cylindrical surface were covered with an impervious material prior to the final assembly; the probes were completely saturated in water before utilization.

3. EXPERIMENTAL RESULTS AND ANALYSIS FOR THE TANK EXPERIMENTS

Analyses of experimental data are based upon equations (1) and (2), which model the two processes. Equation (1) is a classical model, and equation (2) is derived in a paper by Schulz and Giorgetti (1991).

$$C = C_s - (C_s - C_o) \exp(-K_L t/H) \qquad (1)$$

in which C = dissolved oxygen concentration, C_s = saturation DO concentration, C_o = initial DO concentration, K_L = liquid phase transfer coefficient, H = ratio of water volume to surface area, and t = time.

$$\frac{M}{M_o} = 1 - \frac{V_s t}{L_o} \qquad (2)$$

in which M_o = initial mass of solid probe,

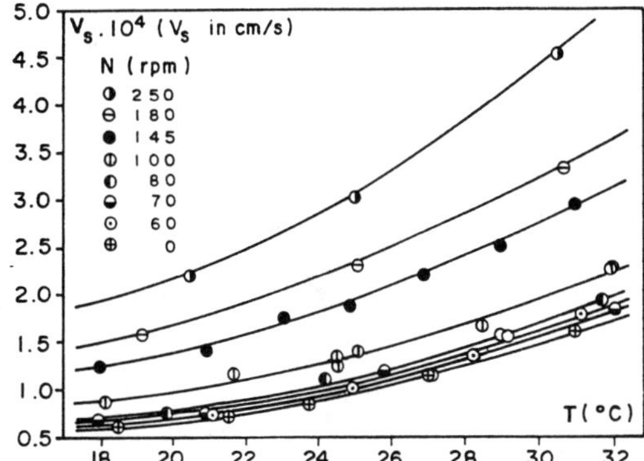

Fig. 3. Velocity of dissolution as a function temperature and agitation

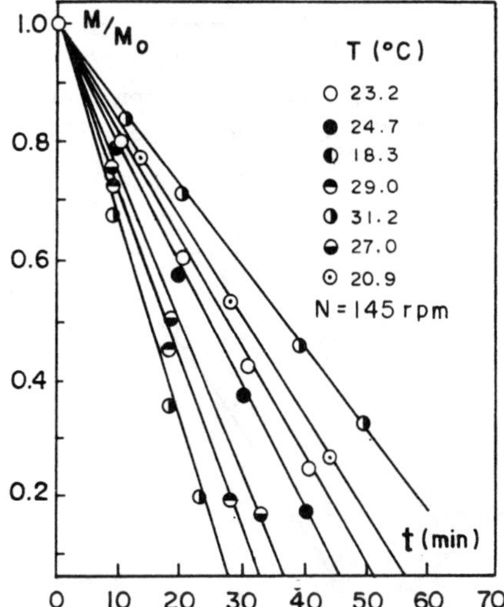

Fig. 4. Typical experimental results obtained in the tank with agitation

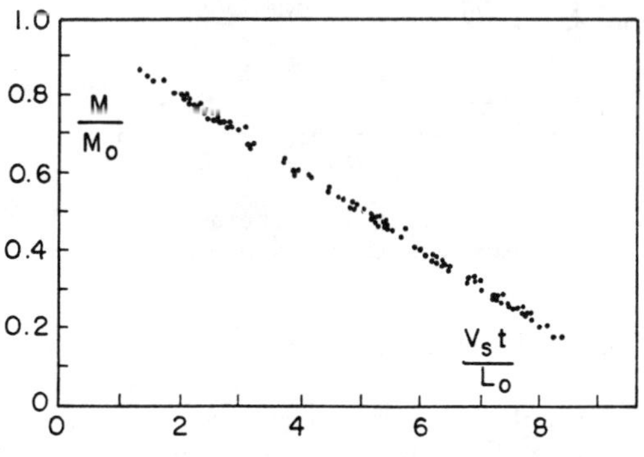

Fig. 5. Time variation of solid mass

V_S = velocity of dissolution, L_O = initial thickness of probe, t = time, and M = mass at time t.

Figure 3 illustrates values of V_S obtained at the tank from tests with 40 oxalic acid probes at different temperatures and different states of agitation characterized by the rotational speed N of the propeller. The curve for zero rotation is a lower bound for the whole set; curves for different rotational speeds are distinct, showing that the oxalic acid is a material adequately sensitive to variations in the levels of mixing as examined in the experiments. Figure 4 ilustrates a typical experimental result obtained with N = 145 rpm; notice the linearity of the process for all temperatures. The value of V_S for each temperature is obtained from the slope of the straight line of best correlation, and is used to plot figures 3 and 5. Figure 5 shows the relationship between the probe mass ratio and the non-dimensional parameter $V_S t/L_O$, for all experimental data.

The same tank had been used for the measurement of reaeration coefficients of water under conditions similar to the ones described for the measurement of velocities of dissolution of solid probes; results were published by Barbosa Jr. and Giorgetti (1990). Figure 6 ilustrates in a log-log plot the dependence of K_L (at 25°C) upon the speed of stirring N. For the range covered by the experiments the stirring Reynolds number is sufficiently high to render the power number a constant. Therefore, N is proportional to the cubic root of the power input, and as in every experiment the quantity of water was constant, N^3 is also proportional to the power dissipation per unit mass of liquid, ε.

Schulz and others (1991) have shown that of data such as those in figure 6 it may be expected a dependence of K_L upon ε raised to the power 1/3 for lower rates of power dissipation, and to the power 1/2 for higher rates of power dissipation. The two straight lines shown are best fits to partial sub-sets of the data, and are denoted as $K_{LO}(N)$ and $K_{L\infty}(N)$:

$$K_{LO}(N) = A_O N \quad (3)$$

$$K_{L\infty}(N) = A_\infty N^{3/2} \quad (4)$$

A good continuous correlation, asymptotically, adhering to the two curves can be obtained as proposed by Churchill and Usagi (1974), and is given by

$$K_L(N) = A_\infty N\left[\left(\frac{A_o}{A_\infty}\right)^n + N^{n/2}\right]^{1/n} \quad (5)$$

in which n is a positive number whose value regulates the closeness of the adherance to the two curves. The solid curve in figure 6 is drawn with $n = 8$; $A_o = 1.654 \times 10^{-5}$; and $A_\infty = 1.579 \times 10^{-6}$.

A general description for the data in this paper, including the effect of temperature is:

$$K_L(N, T) = AN\left[B^n + N^{n/2}\right]^{1/n}\left[1 + \beta\right]^{(T - T_r)} \quad (6)$$

with $A = A_\infty = 1.579 \times 10^{-6}$; $B = 10.475$; $\beta = 0.024$; $T_r = 25°C$; N in rpm; T in °C; K_L in cm/s.

Data for V_s show a tendency toward saturation both for very low and very high rates of agitation. A form of description found adequate to these trends, for any temperature, is:

$$V_s(N) = V_{so} + (V_{s\infty} - V_{so})\, e^{-N_r/N} \quad (7)$$

where $V_{so} = 9.81 \times 10^{-5}$ cm/s; $V_{s\infty} = 5.95 \times 10^{-4}$ cm/s; and $N_r = 270.1$ rpm.

A general description, including the effect of temperature is:

$$V_s(N, T) = V_{so} + (V_{s\infty} - V_{so})\, e^{-N_r/N}\left[1 + \alpha\right]^{(T - T_r)} \quad (8)$$

where $\alpha = 0.076$; $T_r = 25°C$; N in rpm; T in °C; and V_s in cm/s.

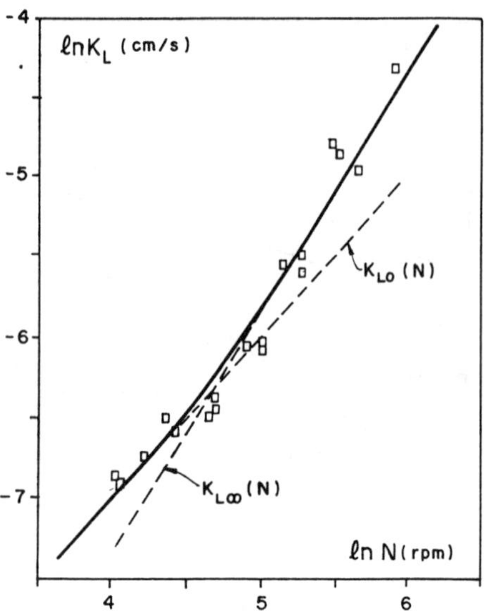

Fig. 6. Liquid phase oxygen transfer coefficient as a function of agitation at 25°celsius

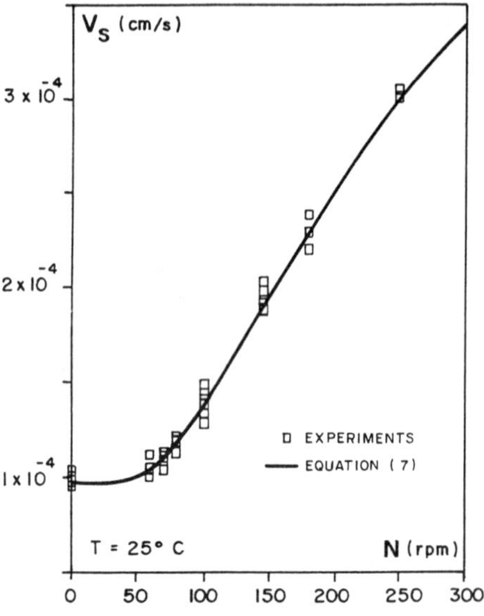

Fig. 7. Velocity of dissolution at 25°celsius as a function of agitation

Data and this correlation are shown in figure 7.

As a concluding result equations (6) and (8) can be used to generate corresponding values for K_L and V_s at 25 °C by varying

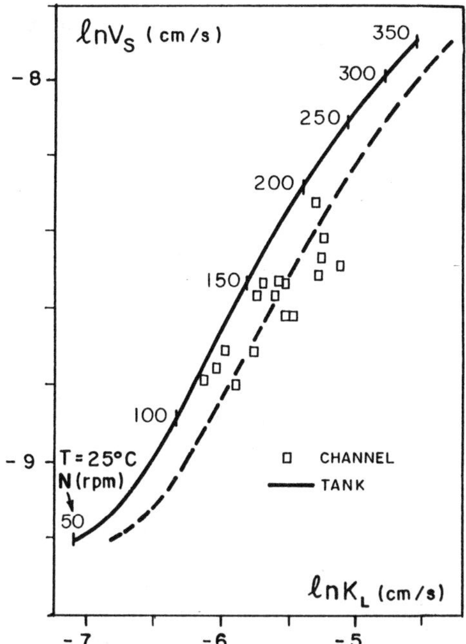

Fig. 8. Oxygen transfer coefficient vs. velocity of dissolution at 25° C

the parameter N. Results are illustrated by the solid line in figure 8 on which the markers identify values of N.

3. EXPERIMENTAL RESULTS AND ANALYSIS FOR THE CHANNEL EXPERIMENTS

Data for K_L and corresponding V_s, simultaneously measured in the channel, are shown in figure 8, also reduced to 25 °C. The general trend appears to be very similar to the solid-line curve except for a translation of it to the right or down, a feature theoretically anticipated in Schulz and Giorgetti (1991). Either translation means that when one moves from the tank to the channel an increase in the relative value of K_L/V_s is observed. Some speculation is possible at this point. It appears that with the increase in the physical scale and the consequent shift in the structure of turbulence and in the flow macro-structure towards larger scales (eddies and vortices), the two transfer phenomena suffered different influences. This is natural to expect, since the geometry of the water body changed very much, while the geometry of the solid probe did not change.

If this is what is expected to happen when one goes from the laboratory to a river for measurement of K_L, then the method is useless unless calibration is done for the specific stretch of the river. But then, another speculation is possible, and an answer to it resides in further research efforts already planned for: what happens if the size (diameter of the contact face) of the solid probe is "conveniently" increased? Would it be possible to correctly scale-up the probe in such a way that the K_L vs. V_s data would plot the same for different flow geometries? If this is possible, then the rules for the scaling-up would probably permit the scaling of data, not of probes, for measurement in any flow geometry with a probe calibrated by a single standard laboratory procedure. An important parameter for this consideration should be the rate of energy dissipation per unit mass. Unfortunately only very crude estimates are available now, and they do not permit a reliable analysis. Its limits for the channel experiments, for instance, are estimated as 0.84 and 6.40 W/m^2, with an average of 3.62 W/m^2. If this value is used for comparison with the tank data with the power number assumed as 0.3, a reference value for N results as 158 rpm. When this point is located on the solid line one sees that it corresponds to approximately the same value of V_s as for the mean of the ensemble of points obtained in the channel experiments. Does it suggest a horizontal translation? Nothing can be said at this point, but if in further experiments accurate measurements of the rate of power dissipation are made, it would help in the better understanding of this question. The dashed line in figure 8 represents, tentatively, an interpolation upon a translation done under these hypotheses.

4. ACKNOWLEDGEMENTS

This research effort was made possible by the generous support of CNPq - Conselho Nacional de Desenvolvimento Científico e Tecnológico, and FAPESP - Fundação de Amparo à Pesquisa do Estado de São Paulo, in the form of research grants and graduate students scholarships.

REFERENCES

Barbosa Jr. A.R. and Giorgetti, M.F. 1990 Determination of Reaeration Coefficients of Water Bodies with the Use of Ethylene as a Gas Tracer", in Portuguese, Proceedings of III ENCIT, Associação Brasileira de Ciências Mecânicas: 1085-1090.

Churchill, S.W. and Usagi, R. 1974. A Standardized Procedure for the Production of Correlations in the Form of a Common Empirical Equation. Ind.Eng. Chem.Fundam., Vol. 13, nº 1.

Rainwater, K.A. and Holley, E.R. 1983. Laboratory Studies on the Hydrocarbon Gas Tracer Technique for Reaeration Measurement. Center for Research in Water Resources. University of Texas at Austin, 114 p.

Rathbun, R.E., Schultz, D.J. and Stephens, D.W. 1975. Preliminary Experiments with a Modified Tracer Technique for Measuring Stream Reaeration Coefficients, U.S. Geological Survey Open File Report. 75-256.

Schulz, H.E.; Bicudo, J.R.; Barbosa Jr., A. R., and Giorgetti, M.F. 1991. Turbulent Water Aeration: Analytical Approach and Experimental Data. Air-Water Mass Transfer, Selected Papers from the Second International Symposium on Gas Transfer at Water Surfaces, ASCE, Editors Steven C. Wilhelms and John S. Gulliver, pp. 142-155.

Schulz, H.E. and Giorgetti, M.F. 1991. Measurement of Reaeration Coefficient with the Soluble Solids Probe. Air-Water Mass Transfer, Selected Papers from the Second International Symposium on Gas Transfer at Water Surfaces, ASCE, Editors Steven C. Wilhelms and John S. Gulliver, pp. 278-293.

Tsivoglou, E.C.; O'Connel, R.L.; Walter, C. M.; Godsil, P.J. and Logsdon, G.S. 1965. Tracer Measurements of Atmospheric Reaeration I, Laboratory Studies. Journal of Water Pollution Control Federation. Vol. 37, nº 10, 1343-1362.

Environmental Management, Geo-Water & Engineering Aspects, Chowdhury & Sivakumar (eds)
© 1993 Balkema, Rotterdam. ISBN 90 5410 099 0

Entropy theory to identify water quality violators

I.C.Goulter & A.Kusmulyono
University of Central Queensland, Qld, Australia

ABSTRACT: A new procedure for estimating the most likely locations of changes in water quality levels in the upstream reaches of a river basin given an observed change in the mean of the water quality at a downstream monitoring location is proposed. The procedure is based upon fundamental entropy theory and requires knowledge of the prior (previously determined) distribution of water quality level at the upstream and downstream locations. The procedure is demonstrated by application to an example problem. The impacts of a range of restrictions on the probability distributions of water quality at the upstream and downstream stations are investigated for this example problem.

INTRODUCTION

The cost of collecting data is one of the more difficult problems facing authorities involved in monitoring and managing water quality in river systems. The greater the number of stations in a monitoring network for a river system, the greater the information available about the system. However, establishing and operating water quality stations is costly and, in most cases, budgets available for water quality monitoring are limited. Furthermore in the environment of decreasing overall budgets being experienced in many countries, the budgets allocated to water quality monitoring are also being reduced and the networks 'rationalised'.

Many studies have been conducted previously to optimise the design of water quality monitoring networks and monitoring programs, e.g., Loftis and Ward (1980), Palmer and Mackenzie (1985), Dunnette (1980), and Harmancioglu (1984). All of these studies were directed at identifying procedures or models which were able to identify network design and/or monitoring programs which gave the best information within specified budget limits.

This study addresses one aspect of the problem of network design, namely, how to predict water quality changes at an upstream location after the station has been discontinued following a period of data collection sufficient to establish the base–line distribution of the water quality at each station. The water quality levels at the discontinued stations are then predicted with an approach based on entropy/information theory using measured data at a downstream station.

BACKGROUND

The entropy concept has been introduced to water resources only recently. Some of the applications of the entropy concept in water resources are :

- parameter estimation and derivation of frequency distribution (Sonuga, 1972).
- derivation of functional rainfall–runoff relationship (Sonuga, 1976).
- measuring the information content of random process (Harmancioglu, 1981)
- evaluation of information transfer between hydrologic processes (Harmancioglu and Yevjevich, 1987).
- assessment of recharge systems for a river basin (Harmancioglu and Baran, 1989)
- assessment of model performance (Amorocho and Espildora, 1973).
- velocity distribution in open channels (Chiu 1987, 1988, 1989, 1991; Chiu and Chiou 1986)
- redundancy measures in water distribution network design (Awumah, et al.,1991)

The application of entropy examined in this paper embodies the use the Principle of Maximum Entropy (POME) to develop updated probability distributions of water quality levels in tributaries, given a new distribution of the water quality levels observed at the downstream location and knowing the previously established probability distribution of the water quality levels at the tributary stations. Once the updated probability distributions have been identified the likely location of 'violators' causing changes in water quality at the downstream station can be identified.

THEORETICAL BACKGROUND

Shannon's Measure of Information.

The basic principle of the procedure is the interpretation of entropy as expressed by Shannon's measure of information (Shannon, 1948). This entropy expression can be applied as a measure of uncertainty. Let the probabilities of n possible outcomes $A_1, A_2, \ldots, A_n$ of an experiment be $p_1, p_2, \ldots, p_n$ respectively.

Shannon's formulation can now be written mathematically as :

$$H_n(p_1, p_2, \ldots p_n) = -\sum_{i=1}^{n} p_i \ln p_i \qquad (1)$$

where:

$$\sum_{i=1}^{n} p_i = 1 \qquad (2)$$

to ensure development of a complete probability distribution

The important characteristics of Equation 1 are:

- H_n takes on its maximum when all events have the same probability or uncertainty, i.e, $p_i = 1/n$.
- H_n take on its minimum value (equal to 0), when there is a certainty among the events.
- any random probabilities will give the value of H_n between these two extremes.

The results obtained from maximising H_n (Equation 1) by assigning values to p_i has been discussed extensively in earlier papers, e.g. Jaynes (1983) and Sonuga (1972) and is known as the Principle of Maximum Entropy (POME), also known as the most unbiased estimator for its characteristics described. Maximising the value of H_n in this manner also develops the most unbiased estimates for p_i for any condition defined in constraints on the values of p_i.

One such formulation involving constraints is as follows :

$$\text{Max } H_n = -\sum_{i=1} p_i \ln p_i \qquad (3)$$

subject to :

$$\sum_{i=1}^{n} p_i = 1 \qquad (4)$$

$$\sum_{i=1}^{n} p_i x_i = \mu \qquad (5)$$

$$\sum_{i=1}^{n} p_i x_i^2 = \mu^2 + \sigma^2 \qquad (6)$$

where:

p_i = probability of event x_i

$$0 \leq p_i \leq 1 \qquad \forall_i \qquad (7)$$

μ = mean of the outcomes of events x_i for all i

σ = standard deviation of the outcomes of events x_i for all i

Solving the above formulation for the unknown values of p_i results in a normal distribution with the mean of μ and the standard deviation of σ. Solving the same formulation but with the constraints defined in Equations 5 and 6 removed results in the specification of uniform distribution of x_i. The principle of maximising H_n subject to a range of constraints on the values of p_i is adapted in the following approach to predicting water quality levels at upstream locations in a river system.

PROBLEM FORMULATION

Consider a river basin in which sufficient water quality data have been gathered to define probability distributions of the water quality on the major tributaries and in the downstream reaches of the main channel of the river. For simplicity of explanation at this time consider the case of a stream with two tributaries as shown in Figure 1.

Define events as the possible values of the

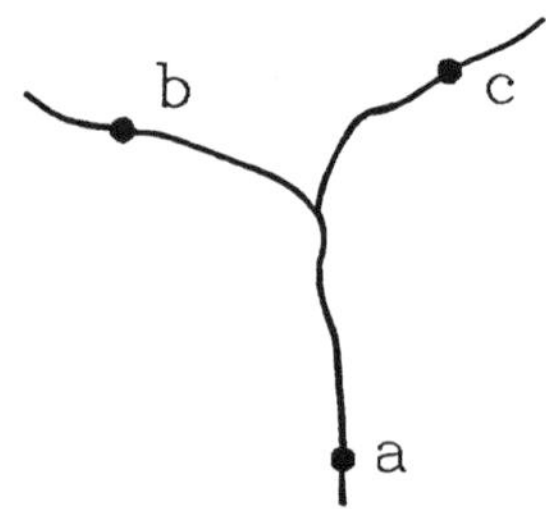

Figure 1. Schematic of explanatory example.

water quality levels at each sampling station on the tributaries and the mainstem of the stream. To simplify the problem at this stage of study, assume that the total water quality level at the downstream sampling location is a simple function of water quality levels measured at the upstream locations. For the situation shown in Figure 1, the water quality level at any point downstream of the confluence of all tributaries can be expressed mathematically as :

$$X_a = f(X_b + X_c) \tag{8}$$

where X_a, X_b, and X_c are water quality levels at stations a (downstream, main channel), b (upstream, tributary) and c (upstream, tributary) respectively.

In other words, the concentration of a pollutant at the downstream location a is defined by the contribution of the pollutant from each of the tributaries b and c.

When the total water quality level at the downstream main channel station a is not in the expected range, it is likely that extreme values of water quality levels are occurring at one or more of the upstream tributaries. The POME is proposed here as a means of identifying, without bias, in which tributary the extreme water quality condition is the most likely to occur.

The knowledge of the prior (previously determined) probability distribution of the possible water quality level at each of the upstream tributary stations can be introduced in the formulation through Kullback–Leibler's Principle of Minimum Discrimination Information (MDI) (Kapur, 1989). The MDI principle is used in this study to assign the probabilities to all of the possible values of the water quality levels at each upstream contributing locations, knowing only the observed mean of water quality at the downstream sampling station and the prior distribution of water quality at each upstream station. Knowing the probabilities of each of the possible levels of water quality at each upstream location, the distribution, and therefore the new mean, of the water quality at each of the upstream locations, can be estimated. Application of the MDI principle in this manner ensures that the allocation of the probabilities to each of the upstream water quality levels is unbiased. This process also ensures that the probability distribution of water quality levels arising from the initial allocation of probability is also an unbiased estimate. The upstream tributary location that process estimates as being the most likely location at which the water quality distribution has changed the most can then be identified. The changes in the observed mean of the water quality at the downstream main channel location are attributed to this location in the first instance and procedures can be undertaken to determine if water quality standards are actually being violated at that location.

Procedure

As noted in the above discussion the procedure is based upon knowledge of the prior probability distributions of water quality levels at all sampling locations in the basin. This knowledge is based upon sufficient previous measurements to permit an adequate description of those distributions. Once these distributions are defined the range of possible water quality levels at each station is identified, e.g, four standard deviations either side of the mean. This range is then discretized. The existing or 'prior' probability of each possible interval of water quality (as discretized in the previous step) at each station is then determined.

Given a new observed mean at the downstream station, the MDI principle is then modified to assign new probabilities to each of the water quality levels at the upstream stations as follows:

$$\text{Max } H_n = -\sum_{i=1}^{n}\sum_{j=1}^{m} p_{ij} \ln\frac{p_{ij}}{q_{ij}} \tag{9}$$

subject to :

$$\sum_{i=1}^{n}\sum_{j=1}^{m} p_{ij} = 1 \tag{10}$$

$$\frac{\sum_{i=1}^{n} p_{ij}\, x_{ij}}{\sum_{i=1}^{n} p_{ij}} = \mu_j \qquad (j = 1, 2, \ldots, m) \tag{11}$$

$$\frac{\sum_{i=1}^{n} p_{ij}\, x_{ij}^2}{\sum_{i=1}^{n} p_{ij}} = \mu_j^2 + \sigma_j^2 \qquad (j = 1, 2, \ldots, m) \tag{12}$$

$$\sum_{j=1}^{m} \mu_j = \mu' \tag{13}$$

$$0 \le p_{ij} \le 1 \qquad \forall i, \forall j \tag{14}$$

$$0 \leq q_{ij} \leq 1 \qquad \forall i, \forall j \qquad (15)$$

$$\mu_j \geq 0 \qquad \forall j \qquad (16)$$

$$\sigma_j \geq 0 \qquad \forall j \qquad (17)$$

where:

x_{ij} = i^{th} possible water quality level at upstream station j (known)
q_{ij} = prior probability of event x_{ij} (known)
μ_j = mean of the water quality level at upstream station j (unknown)
σ_j = standard deviation of the water quality level at station j based on the prior distribution (known)
μ' = a function of the new observed mean level of the water quality at the main channel downstream (measured)
p_{ij} = probability to be assigned to water quality level x_{ij} knowing the mean of downstream water quality level (μ).

Solving this formulation results in the assignment of probabilities (p_{ij}) to every event (water quality level) at every station j. The value of the mean at every station can then be derived.

The following variations of the above formulation were examined in this study. Firstly, the variance of the water quality at every upstream station was maintained at the value associated with previously determined probability distributions. Secondly, the variances of water quality levels at the upstream locations were bounded only by requirement that their sum equal a function of the variance of the water quality values at the downstream station. (The variance of the values at the downstream station was assumed to be constant at the level of the prior distribution at the downstream station). In this case the MDI model determines the value of the variance at each upstream station. In the last case, the prior probability is modified to accommodate the possibility that there are some known changes in the environment around the upstream tributaries.

In this initial work, the water quality data from a section of the Citarum River Basin in West Java, Indonesia were used to demonstrate how the MDI describes the situation. The schematic location of the stations is shown in Figure 2. It is recognized that the information about how the data were collected is very limited therefore it is only presumed that the data are reliable. Furthermore, there are only 7 months of records in 1984 and 6 months of records in 1986 available. Hence, the results are for demonstration purposes only. A summary of these data which are normally distributed according to the Saphiro–Wilk W test is given in Table 1.

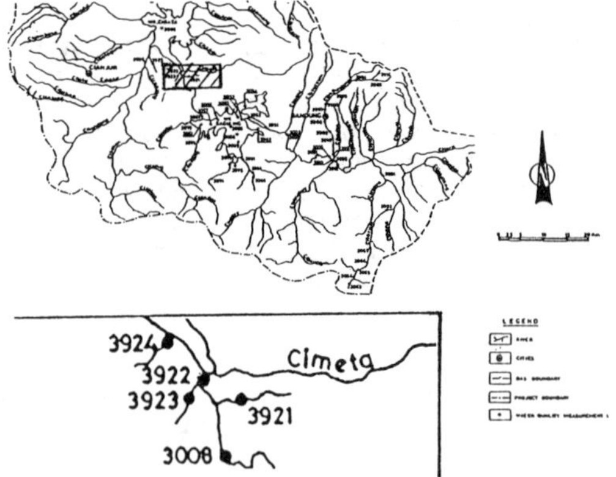

Figure 2. Location of monitoring stations.

Table 1. Dissolved Oxygen Levels (mg/l).

YEAR	MONTH	Station 3921	Station 3923	Station 3922
1984	6	7.22	8.44	8.03
	7	6.98	7.30	6.82
	8	8.12	7.14	7.30
	9	7.79	8.12	8.12
	10	8.44	8.76	9.09
	11	8.59	8.43	8.59
	12	6.32	6.06	6.44
	MEAN	7.64	7.75	7.77
	STD	0.8316	0.9597	0.9572

YEAR	MONTH	Station 3921	Station 3923	Station 3922
1986	6	8.44	9.01	7.79
	7	8.40	7.80	8.10
	8	--	--	--
	9	7.30	7.30	7.11
	10	7.95	7.14	5.19
	11	4.87	5.19	9.41
	12	8.60	6.65	4.87
	MEAN	7.59	7.18	7.08
	STD	1.4150	1.2641	1.7567

In the analysis, the data from 1984 is used as the known/prior probability distribution and the data from 1986 is the one that is to be assigned. Based on a preliminary analysis of the data the relationship between the mean water quality level at the downstream station and the water quality levels at the upstream tributary station was found to be:

$$\mu = 0.505\,\mu' \qquad (18)$$

where:

μ = mean of the observed value of the water quality level at the downstream station.
μ' = sum of the mean values of the water quality level in the tributaries.

The formulation was solved using a nonlinear optimization package program the GRG2. Table 2 shows the examples of values achieved for the three variations of the basic model.

Table 2. Values of Dissolved Oxygen (mg/l) Assigned by the Procedure

(a) Original Distribution with Fixed Variances

OBSERVED MEAN at Station 3922 μ	μ'	VALUE ASSIGNED TO MEAN		Hn	$(X-\mu)/\sigma_m$	
		Station 3921 (x)	Station 3923 (x)		Station 3921	Station 3923
7.77	15.39	7.64	7.75	0.693	0.0	0.0
7.51	14.88	7.42	7.46	0.646	-1.183	-1.351
7.15	14.16	7.15	7.01	0.436	-2.635	-3.448
7.08	14.02	7.08	6.94	0.382	-3.011	-3.775
6.58	13.02	6.58	6.44	-0.242	-5.700	-6.105

(b) Original Distribution with Modified Variances

OBSERVED MEAN at Station 3922 μ	μ'	VALUE ASSIGNED TO MEAN		Hn	$(X-\mu)/\sigma_m$		VARIANCE	
		Station 3921 (x)	Station 3923 (x)		Station 3921	Station 3923	Station 3921	Station 3923
7.77	15.39	7.64	7.75	0.693	0.0	0.0	0.691	0.921
7.51	14.88	7.44	7.44	0.646	-1.076	-1.444	0.707	0.905
7.15	14.16	7.14	7.02	0.441	-2.689	-3.402	0.693	0.919
7.08	14.02	7.08	6.94	0.380	-3.011	-3.775	0.648	0.964
6.58	13.02	6.64	6.38	-0.204	-5.377	-6.384	0.709	0.902

(c) Modified Distribution at Station 3921 with Fixed Variances

OBSERVED MEAN at Station 3922 μ	μ'	VALUE ASSIGNED TO MEAN		Hn	$(X-\mu)/\sigma_m$	
		Station 3921 (x)	Station 3923 (x)		Station 3921	Station 3923
7.77	15.39	7.35	8.03	0.645	1.129	1.305
7.57	14.88	7.14	7.75	0.693	0.0	0.0
7.15	14.16	6.84	7.32	0.584	-1.613	-2.004
7.08	14.02	6.77	7.25	0.535	-1.990	-2.330
6.58	13.02	6.37	6.65	0.526	-4.140	-5.126

The values of $\frac{X-\mu}{\sigma_m}$ in each of Tables 2(a), 2(b) and 2 (c) show the deviation of the values of the means assigned to the upstream stations in terms of the standard deviations of the means of the prior distribution at that station from the original mean of the distribution at that station.

The value of σ_m used in the application of this formula is $\sigma/\sqrt{n}$ where n is the number of samples used to get the observed value at the downstream station. At this time n is set to 20.

DISCUSSION

The results presented in Table 2(a), indicate that the water quality at the station with the greater variance will tend to deviate further from the mean of the prior distribution when the observed value of the water quality downstream changes. Therefore, when there is a significant change of the mean of water quality level at the observed downstream station, the technique will identify that the tributary with the greater variance is the most likely to be the one where distribution of water quality has changed significantly.

Table 2(b) shows the results when the sum of the variances upstream is bounded by a function of the variance at the downstream site. It can be seen that the means have only slightly been shifted from those shown in Table 2(a).

The results in Table 2(c) show the effect of accommodating the knowledge of a change in the prior distribution of water quality levels specified in the formulation. Such a change might occur as a result of development changes within the tributary basin or the location of a new industry on the river. In this case the change in the prior distribution caused by knowledge of existing conditions is a decrease in the mean value of the water quality at station 3921. This change has shifted the result considerably. In particular it can be seen that the value of the mean at Station 3923 does not drop as much for the modified low value of μ'.

CONCLUSION

The formulation presented in this study is a new approach to estimating the most likely water quality levels at upstream locations in a river system, knowing the prior (previously determined) probability of water quality at those locations and at a downstream location. The procedure is based upon using a variation of the entropy principle to assign new probabilities to upstream water quality levels in an unbiased fashion when a change is observed in the mean of water quality levels downstream.

Demonstration of the procedure by a simple example shows it to be capable of making sensible inferences about water quality conditions.

The quantity and quality of the data used for the demonstration was very inadequate. Therefore, further study with more reliable data is essential for a more rigorous evaluation of the technique.

In the study, it was assumed that the data are independent, which is not the usual case in the water quality issue. Therefore, it would be interesting in future study to examine the case considering the dependency among the data. Further work is also required to investigate how prior knowledge of changes at the upstream station can be incorporated into the procedure on a conditional probability basis.

REFERENCES

Amorocho, J. and Espildora, B., 1973, "Entropy in the Assessment of Uncertainty of Hydrologic Systems and Models", *Water Resources Research*, 9(6), pp 1515–1522.

Awumah, K., Goulter, I. and Bhatt, S., 1991, "Entropy Based Redundancy Measures in Water Distribution Network Design", *Journal of Hydraulic Engineering, ASCE*, 117(10), pp 595–614.

Chui, Chao–Lin, 1987, "Entropy and Probability Concepts in Hydraulics", *Journal of Hydraulic Engineering, ASCE*, 113(5), pp 583–600.

Chui, Chao–Lin, 1988, "Entropy and 2–D Velocity Distribution in Open Channels", *Journal of Hydraulic Engineering, ASCE*, 114(7), pp 738–756.

Chui, Chao–Lin, 1989, "Velocity Distribution in Open Channel Flow", *Journal of Hydraulic Engineering, ASCE*, 115(5), pp 576–594.

Chui, Chao–Lin, 1991, "Application of Entropy Concept in Open–Channel Flow Study", *Journal of Hydraulic Engineering, ASCE*, 117(5), pp 615–628.

Chiu, Chao–Lin and Chiou, J.D., 1986, "Structures of 3–D Flow in Rectangular Open Channels", *Journal of Hydraulic Engineering, ASCE*, 112(11), pp 1050–1068.

Dunnette, D.A., 1980, "Sampling Frequency Optimization Using a Water Quality Index", *Journal of Water Pollution Control Federation*, 52(11), pp. 2807–2811.

Harmancioglu, N.B., 1981, "Measuring the Information Content of Hydrological Processes by the Entropy Concept", Centennial of Ataturk's Birth, *Journal of Civil Eng.*, Faculty of Ege Univ., pp 13–38.

Harmancioglu, N.B., 1984, "Entropy Concept as used in Determination of Optimum Sampling Intervals", *Proceedings of Hydrosoft '84, International Conference on Hydraulic Engineering Software, Portoroz, Yugoslavia*, pp 6–99.

Harmancioglu, N.B and Baran, T., 1989, "Effects of Recharge System on Hydrologic Information Transfer Along Rivers", *IAHS, Proc. of the Third Scientific Assembly of the International Associates of Hydrologic Science – New Direction for Surface Water Modelling*, IAHS Publ. 181, pp.223–233.

Harmancioglu, N.B and Yevjevich, V., 1987, "Transfer of Hydrologic Information Among River Points", *Journal of Hydrology*, Vol. 91, pp.103–118.

Jaynes, E.T., 1983. "*Papers on Probability, Statistics and Statistical Physics*", D.Reidel Publishing Company, Dordrecht, Holland.

Kapur, J.N., 1989, "*Maximum Entropy Models in Science and Engineering*", Wiley Eastern Ltd., New Delhi, India.

Lasdon, L.S. and Waren, A.D., 1984, "*GRG2 User's Guide*", Department of General Business Administration, The University of Texas at Austin, Austin, Texas, 60 p.

Loftis, J.C. and Ward, R.C., 1980, "Water Quality Monitoring – Some Practical Sampling Frequency Consideration", *Environmental Management*, 4 (6), pp. 521–526.

Palmer, R.N. and Mackenzie, M.C., 1985, "Optimization of Water Quality Monitoring Networks", *Journal of Water Resources Planning and Management*, 111(4), pp. 478–493.

Shannon, C.E., (1948), "A Mathematical Theory of Communication", *Bell System Technical Journal*, 27(3), pp 379–423, 623–659.

Sonuga, J.O., 1972, "Principle of Maximum Entropy in Hydrologic Frequency Analysis", *Journal of Hydrology*, Vol. 17, pp. 177–191.

Sonuga, J.O., 1976 "Entropy Principle Applied to the Rainfall–Runoff Process", *Journal of Hydrology*, Vol. 30, pp. 81–94.

Environmental Management, Geo-Water & Engineering Aspects, Chowdhury & Sivakumar (eds)
© 1993 Balkema, Rotterdam. ISBN 90 5410 099 0

A two-dimensional unsteady water quality model

H.S.Jin & Y.M.Zheng
Hohai University, People's Republic of China

J.Ye
Nanjing Hydraulic Research Institute, People's Republic of China

ABSTRACT: A planar 2-D unsteady water quality model for simulating pollutant transport in a natural water environment is reported, in which the boundary-fitted orthogonal curvilinear coordinate system and the grid "block" technique are used to eliminate the obstacle resulted from irregularity and moveableness of the domain boundary. The model is employed to simulate the COD concentration field in a tidal river. Comparison of the computational results to the measurement data and the error analysis show that the results are reasonable.

1 INTRODUCTION

In pace with economic development, environmental pollution, including river pollution, has become one of the most important problems within all the world. In China, especially some coastal cities in south east China, in which industry is more developed and population is more gathered, water pollution in river is one of the key factors by which economical development can be restrained.

For effectively controlling and treating the water pollution, it is very necessary to study the law of pollutant transport in water body. During the past decades, some theoretical and semi–empirical methods which can be used to predict pollutant transport in water body, e.g., cumulative discharge method, imaginative source method and so on (Yotsukura 1976, Fischer 1979, Zhang 1988), have been reported. However, natural water body is usually very complicated, especially in river, its boundaries are irregular, it may also include some bends, shallows, sandbars, etc, furthermore, position and shape of the bank boundaries will gradually vary with fluctuating of tide or with flood, i.e., they are moving boundaries, analytical and simple method cannot well describe the complete process of the flow. In the past, field or laboratory experiment is almost sole method to investigate the pollutant transport in complex environment. Recent decades, many researchers have done a lot of works on numerical simulating of the pollutant transport (Rodi 1980).

In tidal river, the fluid flow moves reciprocately because of the tide and the pollutant oscillates with the flow. With the result, pollution occurs not only downstream but also upstream of the sewage outlet, e.g., the water pollution in Huangpu River which is a main water resource in the key industry city—Shanghai, China. In this paper, based upon a boundary–fitted orthogonal curvilinear coordinate system and a grid "block" technique, a water quality mathematical model for simulating planar 2–D unsteady pollutant transport in tidal river is reported.

2 BOUNDARY–FITTED ORTHOGONAL COORDINATE SYSTEM AND GRID "BLOCK" TECHNIQUE

Boundary–fitted coordinate system is one which coordinate axes completely coincide with the body boundaries and are orthogonal curves. It is generated by numerically solving the two elliptic equations about the physical coordinate variables (Cartesian coordinate system, x and y) and the boundary–fitted coordinate variables (ξ, η) (Thompson 1985, Jin 1992). In physical plane (x, y system), the domain is irregular and the coordinate axes are curves, and in transformed plane (ξ, η system), the domain is rectangle and the axes are straight lines. By means of transformation, the correspondent relation between the point (x, y) in physical plane and the point (ξ, η) in transformed plane is determined, discretizing and solving of the govern equations can be conducted with regular grids in the rectangle domain.

In spite of that, it is very expensive in the problems involved with moving boundary that every new domain is transformed in order to adapt the variation of the boundaries and obtain a relevant boundary–fitted orthogonal coordinate system. In the present model, a grid "block" technique is introduced so that not only the computation can be carried out in a rectangle domain but also the computational cost is low.

Grid "block" means that no water will flow outfrom or into the grid and just as the grid is blocked. If river bottom at the water surface level node is above the water surface, i.e., the river bottom elevation is greater than the water surface level, the grid of which the node is at center is considered as a emergent one from water surface and a special measure (Jin 1992) will be taken to assure that the velocities at the grid periphery calculated by the momentum equations will approximately equal zero, i.e., no water will flow outfrom or into the grid. Hence, the grid at which the river bottom is emergent from the water surface can be included in the computational domain and the numericalsolutioncan be carried out in an unvaried area. Although the computational load will be increased, the benefit obtained from the technique is distinct and the computational cost is still cheaper than that by some other methods.

3 WATER QUALITY MATHEMATICAL MODEL

3.1 Governing equation

The equations which govern the planar 2–D transport feature of pollutant in water environment consist of a depth–averaged continuity equation, two momentum equations and a pollutant transport equation. In the orthogonal curvilinear coordinate system, the equations can be expressed as follows (supposing that the pollutant simulated is neutral substance, the interaction between the pollutant and the flow is ignored, i.e., flow movement equations and concentration equation don't link each other):

$$\frac{\partial \zeta}{\partial t}+\frac{1}{J}\left[\frac{\partial(u^{*}hg_{22})}{\partial \xi}+\frac{\partial(v^{*}hg_{11})}{\partial \eta}\right]=0 \tag{1}$$

$$\frac{\partial u^{*}}{\partial t}+\frac{u^{*}}{g_{11}}\frac{\partial u^{*}}{\partial \xi}+\frac{v^{*}}{g_{22}}\frac{\partial u^{*}}{\partial \eta}+\frac{u^{*}v^{*}}{J}\frac{\partial g_{11}}{\partial \eta}-\frac{v^{*2}}{J}\frac{\partial g_{22}}{\partial \xi}=-\frac{g}{g_{11}}\frac{\partial \zeta}{\partial \xi}-\frac{gu^{*}\sqrt{u^{*2}+v^{*2}}}{C_z^{2}h}+\frac{\varepsilon_t}{h}\left(\frac{1}{g_{11}}\frac{\partial A}{\partial \xi}-\frac{1}{g_{22}}\frac{\partial B}{\partial \eta}\right)+fv^{*} \tag{2}$$

$$\frac{\partial v^{*}}{\partial t}+\frac{u^{*}}{g_{11}}\frac{\partial v^{*}}{\partial \xi}+\frac{v^{*}}{g_{22}}\frac{\partial v^{*}}{\partial \eta}+\frac{u^{*}v^{*}}{J}\frac{\partial g_{22}}{\partial \xi}-\frac{u^{*2}}{J}\frac{\partial g_{11}}{\partial \eta}=-\frac{g}{g_{22}}\frac{\partial \zeta}{\partial \eta}-\frac{gv^{*}\sqrt{u^{*2}+v^{*2}}}{C_z^{2}h}+\frac{\varepsilon_t}{h}\left(\frac{1}{g_{22}}\frac{\partial A}{\partial \eta}+\frac{1}{g_{11}}\frac{\partial B}{\partial \xi}\right)-fu^{*} \tag{3}$$

$$\frac{\partial C}{\partial t}+\frac{u^{*}}{g_{11}}\frac{\partial C}{\partial \xi}+\frac{v^{*}}{g_{22}}\frac{\partial C}{\partial \eta}=\frac{1}{hJ}\left[\frac{\partial}{\partial \xi}\left(\frac{hg_{22}}{g_{11}}\varepsilon_{\xi}\frac{\partial C}{\partial \xi}\right)+\frac{\partial}{\partial \eta}\left(\frac{hg_{11}}{g_{22}}\varepsilon_{\eta}\frac{\partial C}{\partial \eta}\right)\right]-kC+r \tag{4}$$

in which,

$$A=h\left(\frac{1}{g_{11}}\frac{\partial u^{*}}{\partial \xi}+\frac{1}{g_{22}}\frac{\partial v^{*}}{\partial \eta}\right),$$

$$B=h\left(\frac{1}{g_{11}}\frac{\partial v^{*}}{\partial \xi}-\frac{1}{g_{22}}\frac{\partial u^{*}}{\partial \eta}\right),$$

u^{*} (and v^{*}) is depth–averaged velocity along ξ (and η) direction in transformed plane,

$$u^{*}=(ux_{\xi}+vy_{\xi})/g_{11},$$
$$v^{*}=(ux_{\eta}+vy_{\eta})/g_{22},$$

u (and v) is depth–averaged velocity along x (and y) direction in the physical plane, C is depth–averaged pollutant concentration, ζ and h are respectively water surface level and water depth, $h=\zeta-Z_b$, Z_b is river bottom elevation, ε_t is horizontal coefficient of eddy and molecular diffusivity, ε_{ξ} and ε_{η} are pollutant diffusion coefficients along ξ and η directions, they respectively correspond the longitudinal and the transverse diffusion coefficients, g and f are respectively the gravitational acceralation and the Coriolis coefficient, C_z is Chezy resistance coefficient, $C_z=\frac{1}{n}h^{1/6}$, k is a general pollutant attenuation coefficient, r is pollution source (or sink), g_{11}, g_{22} and J are relations of the orthogonal coordinate transformation between (x, y) and (ξ, η),

$$g_{11}=\sqrt{x_{\xi}^{2}+y_{\xi}^{2}},$$
$$g_{22}=\sqrt{x_{\eta}^{2}+y_{\eta}^{2}},$$
$$J=x_{\xi}y_{\eta}-x_{\eta}y_{\xi}=g_{11}g_{22},$$
$$x_{\xi}=\frac{\partial x}{\partial \xi},\quad x_{\eta}=\frac{\partial x}{\partial \eta},\quad y_{\xi}=\frac{\partial y}{\partial \xi},\quad y_{\eta}=\frac{\partial y}{\partial \eta}.$$

3.2 Boundary condition and initial condition

On the upstream and the downstream boundaries of the computational domain, time varying water surface level and pollutant concentration are specified.

On the bank boundaries, unpenetrable condition, i.e., normal velocity equals zero, is employed, and normal gradient of pollutant concentration is taken as zero, however, if any tributary joins with the river, the velocities at the joint will be calculated according to discharge and geometry feature of the tributary and the pollutant concentration equals that at the tributary exit.

The initial conditions for the model are approximately specified, in which water surface level and pollutant concentration at interior node are computed by interpolating between the water surface level and the concentration on the upstream boundary and those on the downstream boundary at time t=0, velocities at interior nodes are always taken as zero. In the model, some special measures are taken to eliminate the effect of intial condition on the output. The results show that it is successful.

3.3 Numerical scheme

According to the feature of the movement equations (eqs.(1)~ (3)) and the pollutant transport equation (eqn.(4)), a "staggered" grid system (as Fig.1), in which the nodes for velocities (u^*, v^*) and water surface level (ζ) do not coincide with each other and the node for pollutant concentration (C) overlaps that for water surface level, is set up in the computational domain.

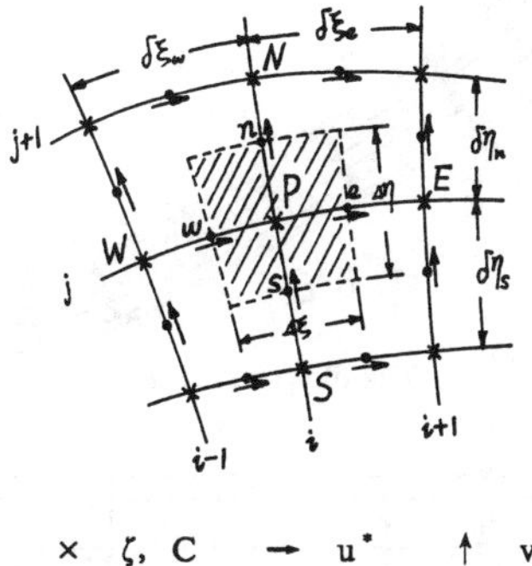

× ζ, C → u* ↑ v*

Fig.1 "Staggered" grid and control volume (shade area)

In orthogonal curvilinear coordinate system, numerical methods for partial differential convection-diffusion equation well applied at present can be adopted conveniently, on condition that relevant geometry feature resulted from coordinate transformation is considered.

Because the movement equations and the pollutant transport equation do not link each other, they can be solved separately. Within a computational time step (Δt), eqs.(1)~(3) are firstly solved to achieve the water surface level and the velocities at each node, then eqn.(4) is solved to obtain the pollutant concentration field.

Eqs.(1)~ (3) are more complicated, a numerical scheme based on a time-splitting method, with 3 steps: advection, diffusion and propagation along with source terms, is used (Jin, 1992) Eqn.(4) is discretized with a control volume method, i.e., eqn.(4) is integrated in a control volume for C (as the shade area on Fig.1), in which the variation of concentration between two adjacent nodes is assumed to obey power function form (Patankar 1980). We can get the following linear algebraic equation set of five nodes interrelation, i.e., the concentration at central node is directly related to the concentrations at four adjacent nodes, which are respectively located at the forward, the backward, the left and the right,

$$a_p C_{ij} = a_e C_{i+1j} + a_w C_{i-1j} + a_n C_{ij+1} + a_s C_{ij-1} + b \tag{5}$$

in which,

$$a_e = \frac{1}{J_{ij}}\{D_e max[0,\ (1-0.1|P_e|)^5] + max[-F_e,\ 0]\},$$

$$a_w = \frac{1}{J_{ij}}\{D_w max[0,\ (1-0.1|P_w|)^5] + max[F_w,\ 0]\},$$

$$a_n = \frac{1}{J_{ij}}\{D_n max[0,\ (1-0.1|P_n|)^5] + max[-F_n,\ 0]\},$$

$$a_s = \frac{1}{J_{ij}}\{D_s max[0,\ (1-0.1|P_s|)^5] + max[F_s,\ 0]\},$$

$$b = \{max[-k,\ 0]C_{ij}{}^0 + r_{ij}\}h_{ij}\Delta\xi\Delta\eta + a_p{}^0 C_{ij}{}^0,$$

$$a_p = a_e + a_w + a_n + a_s + a_p{}^0 + max[k,\ 0]h_{ij}\Delta\xi\Delta\eta,$$

$$a_p{}^0 = \frac{h_{ij}{}^0\Delta\xi\Delta\eta}{\Delta t};$$

the source term in eqn.(4), $S_c = -kC + r$, has been expressed as:

$$S_c = \{max[-k,\ 0]C_{ij}{}^0 + r_{ij}\} - max[k,\ 0]C_{ij};$$

the upper index "0" means initial value within a time step (Δt) and the notation "max[a_1, a_2]" means to take the maximum one in the brackets; (D_e, F_e, P_e), (D_w, F_w, P_w), (D_n, F_n, P_n) and (D_s, F_s, P_s) are respectively diffusion intensity, convection intensity, Peclet number at correspondent control face, and they can be expressed as below:

$$D_e = \left(hg_{22}\frac{\varepsilon_\xi}{g_{11}}\right)_e \Delta\eta / \delta\xi_e,$$

$$F_e = (hu^* g_{22})_e \Delta\eta,\ P_e = \frac{F_e}{D_e},$$

$$D_w = \left(hg_{22}\frac{\varepsilon_\xi}{g_{11}}\right)_w \Delta\eta / \delta\xi_w,$$

$$F_w = (hu^* g_{22})_w \Delta\eta,\ P_w = \frac{F_w}{D_w},$$

$$D_n = \left(hg_{11}\frac{\varepsilon_\eta}{g_{22}}\right)_n \Delta\xi / \delta\eta_n,$$

$$F_n = (hv^* g_{11})_n \Delta\xi,\ P_n = \frac{F_n}{D_n},$$

$$D_s = \left(hg_{11}\frac{\varepsilon_\eta}{g_{22}}\right)_s \Delta\xi / \delta\eta_s,$$

$$F_s = (hv^* g_{11})_s \Delta\xi,\ P_s = \frac{F_s}{D_s}.$$

The above algebraic equation set is solved with a line by line TDMA iterative method.

3.4 Computational results and discussion

The above model is applied to the water quality simulation in Huangpu River, China, which is a tidal river. The water quality variable is the depth–averaged concentration of COD (Chemical Oxygen Demand), i.e., the pollutant is COD. The computational domain is the length 23.8km and the average width about 400m, as Fig.2.

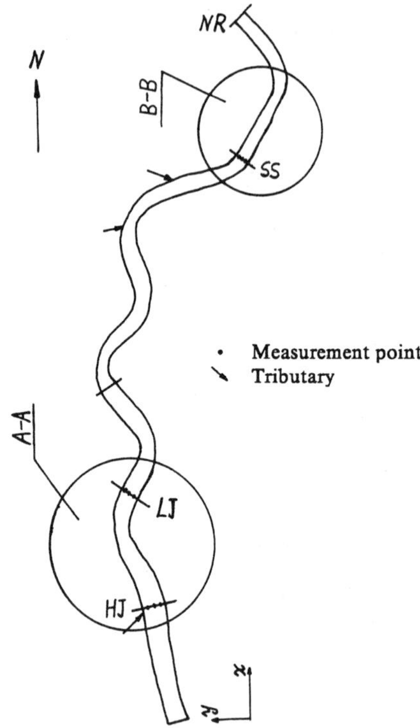

Fig.2 The computational domain

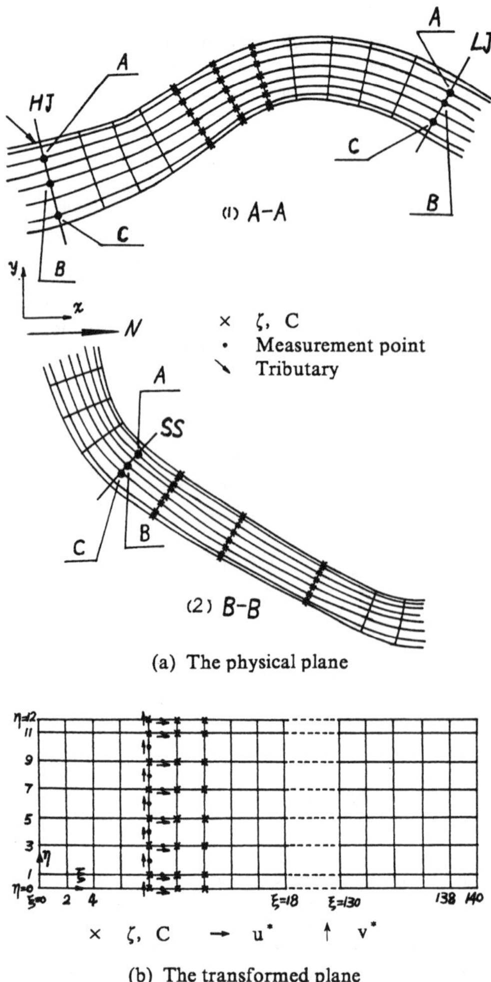

Fig.3 Part of the grids and the nodes

According to the river pattern, a set of nonuniform grids with 72 × 8 computational nodes, maximum grid size along longitudinal direction is greater than 1000m and the minimum along transverse direction is about 20m, is set up within the computational domain. Part of the grids and the nodes is displayed on Fig.3.

According to numerical experiment based on the requirement of calculational accuracy, the computational time step (Δt) is selected to equal 120s. The computational results are compared with field measurement data. Those of flow variables agree fairly with the field data (Jin 1992). Fig.4 describes the comparison of time varying COD concentration computed by the present model to field data at 5 locations which are shown on Fig.2 and Fig.3. It shows that the computational results of COD are reasonable.

Analysis of the relative error (defined as percentage of absolute difference between calculational value and measurement datum to the measured) indicates that in all 180 data of 9 measuring points which are situated on three cross–sections as Fig.2, the maximum error equals 38.66% and the error of 85.56% data is less than 20% . However, in water quality monitoring, measurement error of two samples is allowed to reach about 20%, it is thus clear that the present water quality model can well simulate the planar 2–D unsteady COD concentration field.

4 CONCLUSION

Nowadays, water quality model is one of the most important tools for predicting water quality. Usually, the mixing of sewage to environmental water is of two–dimensional nature, i.e., the distribution of concentration along depth is almost uniform. However,

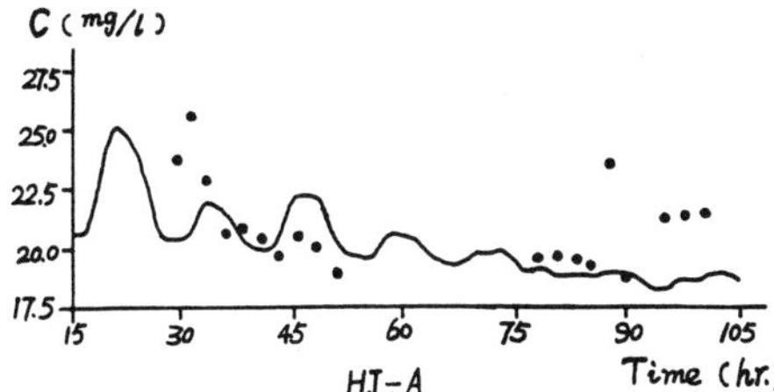

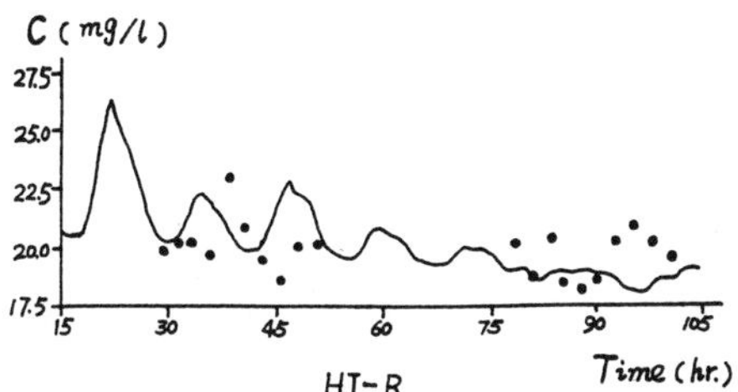

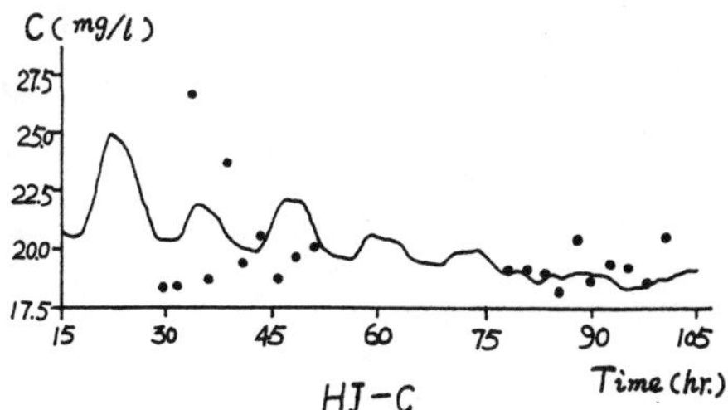

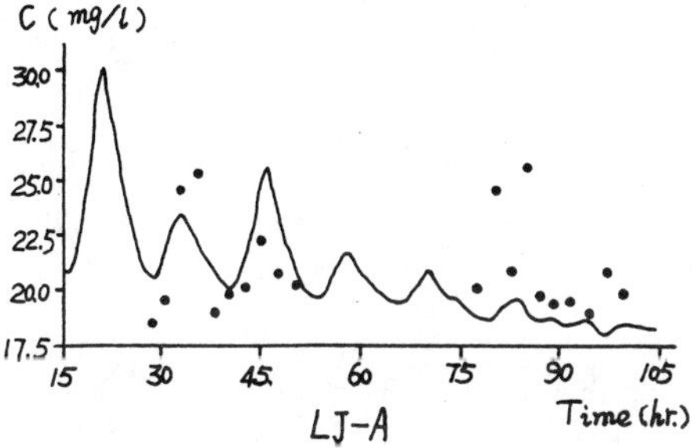

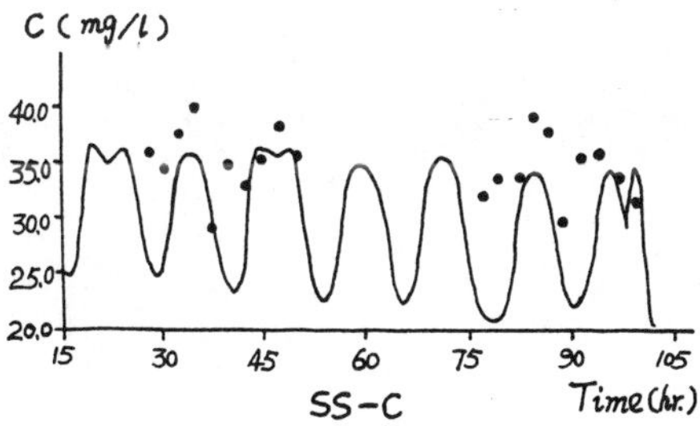

Fig.4 Comparison of COD concentration computed to the field data measured

(• Measurement, — Computation)

planar 2–D unsteady water quality simulation in nature environment is difficult because its boundaries usually vary with the fluctuating of water surface level. Application of the boundary–fitted orthogonal curvilinear coordinate system and the grid "block" technique can eliminate these difficulties. Based on these, a general mathematical model for simulating planar 2–D unsteady water quality concentration is presented. The COD concentration simulation in Huangpu River, China shows that the agreement of the results computed by the present model to the field observations is good.

It can be believed that the present model is of good adaptability and can be conveniently applied to simulation of the other water quality variable and (or) in the other type of water environment.

REFERENCES

Yotsukura, N. and Sayre, W.W. 1976. Transverse mixing in natural channels. Water Resource Research, 12: 695–704.

Fischer, H.B., List, E.J., et al. 1979. Mixing in inland and coastal waters. Academic Press, New York: 120–122.

Zhang, S. N. 1988. Environmental hydrodynamics. Hohai University Press: 120–134(in Chinese).

Rodi, W.1980. Turbulence models for environmental problems. Prediction methods for turbulent flow, edited by W. Kollmann. Hemisphere Publishing Corporation: 259–349.

Thompson, Joe F., Warsi, Z.U.A., et al. 1985. Numerical grid generation, foundations and applications. Elsevier Science Publishing Co., Inc.New York: 331–347.

Jin, H.S., Zheng, Y.M., et al. 1992. A general mathematical model of tidal current in natural rivel. Proceedings of the Second International Conference on Hydraulic and Environmental Modelling of Coastal, Estuarine and River Waters. Bradford, England.

Patankar, S.V.1980. Numerical heat transfer and fluid flow. Hemisphere Publishing Corporation and McGraw Hill Book Company.

Environmental Management, Geo-Water & Engineering Aspects, Chowdhury & Sivakumar (eds)
© 1993 Balkema, Rotterdam. ISBN 90 5410 099 0

Water management aspects of coal fired thermal power plant in India

Somnath Mukherjee
Indian Institute of Technology, Kharagpur, India

N.C. Das
Bangladesh Institute of Technology, Dhaka, Bangladesh

ABSTRACT : An overview of water management techniques for coal-fired thermal power plants in India is presented in this paper. Topics discussed include Environmental regulations, water requirements, wastewater generated and its treatment principles and concepts of zero discharge for total reuse. As an example, a water management plan for typical 4*210 MW power generation station is also presented adopting once-through condenser cooling system. The total contents of the paper wraps up useful information regarding Environmental Impact Assessment (EIA) study to get consent order from Government for implimentation of the project.

1 INTRODUCTION

In India, the thermal power station utilize coal, oil, and natural gas as fossil fuels, amongst which coal is the most widely used fuel to produce electric power. In the early days, wastewater produced by electric power generation was not seen as having adverse environmental impacts, as a result, waste water treatment was either minimal or non existent. In India, even today, water pollution caused by thermal power plants is not considered to be significant and takes a second place compared to air pollution. As a matter of fact, the power generation industry uses large quantities of water which approximately 70% of the total water utilized by total industrial water demand. Owing to Minimal National Standards (MINAS) adopted by Central Board for prevention & control of water pollution, who is supreme agency responsible for formulating effluent regulations for various industries, all new industries are required to obtained a "Consent Order" from the concerned state Board prior to commencing operation. This enactment, yielded a very effective result amongst industrialists for carrying out Environmental Impact Assessment (EIA) study. As a part of EIA programme, water management aspect is considered to be essential requisite. Topics discussed in the present paper include effluent regulation, water requirement, waste water generated, treatment requirements and technologies for total reuse. Water management aspect for a typical 4*210 MW generating station is also presented.

2 EFFLUENT REGULATION

The standard laid down by the central Board for the prevention and control of water pollution for wastewaters discharged from condenser cooling, cooling tower blow down, ash pond effluent and boiler blow down for thermal power plants are reproduced as Table 1.

3 WATER REQUIREMENTS

The typical generating station selected for the purpose consists 4*210 MW coal field units. The station is situated near adequate source of a perennial river and utilizes a once-through condenser cooling system. The water balance diagram for the reference station is presented as figure 1.

Table 1.

parameters	Limits
a) Condenser cooling water	
(1) once-through cooling	
i) pH	6.5-8.5
ii) Temp.	Not more than 5^0C higher than intake water.
iii) Free avialable chlorine	0.5 mg/l
(2) Cooling tower blowdown	
i) Total chromium	0.2 mg/l
ii) Zinc	1.0 mg/l
iii) Free available chlorine	0.5 mg/l
iv) Phosphate	5.0 mg/l
b) Ash pond effluent	
i) pH	6.5-8.5
ii) Suspended solids	100 mg/l
iii) Oil & grease	20 mg/l
c) Boiler blowdown	
i) Suspended solids	100 mg/l
ii) Oil & grease	20 mg/l
iii) Copper	1 mg/l
iv) Iron	1 mg/l

3.1 Condenser cooling

The condenser cooling water flow is dependent on the heat dissipation required and design temperature rise in the effluent stream. With a 10^0 C temperature rise the total circulating water requirements for the reference station is 126,400 m^3/hr. This includes the water requirement for auxiliary cooling also. Normally the cooling water is withdrawn from the surface water source and the heated water is discharged into the surface water.

3.2 Demineralized water system

In todays boilers, it is critical that the feed water be of the highest quality. Concentration of total solids in the feed water is less than 0.15 mg/l. In order to maintain boiler water quality, demineralizer trains are utilized for the feed water. A small quantity of makeup feed water is required to compensate for boiler blowdwon and other losses. In addition deminearlized water is used for initial filling and periodic chemical cleaning of the boiler. The average water requirements for the demineralizers are 160 m^3/hr including regeneration water requirements for the reference plant.

3.3 Ash transport

A significant amount of water is required at coal field power plant for ash transport. Fly ash is collected at electrostatic precipitators and bottom ash is collected at the bottom ash hoppers. The ash conveyed hydraulically to the ash slurry pump house and then to the ash disposal area in slurry form. Average water required for the reference station for ash transport is 3200 m^3/hr.

3.4 Miscellaneous

Other requirements at a power plant are for pump bearings and sealing, air conditioning and ventilation, coal dust suppression, service and drinking water etc. Normally surface water sources have varying amounts of suspended solids and are not suitable for the use of above requirements without treatment. Pretreatment in the form of flocculation and clarification is usually provided. Average estimation of such requirements for pretreatment unit for the subject station is about 895m^3/hr inclusive of D.M. and potable water requirement.

4 WASTEWATER MANAGEMENT

Wastewater generated from various sources in the power plant are shown in figure 1. The treatment required for each waste stream are discussed in the following sections

4.1 Cooling water

Once-through condenser cooling is preferable then a recirculating type cooling system wherever feasible because of greater operational efficiency. While condensers are normally designed for a temperature increase of between $10^0$and 15^0C in the effluent, the stipulated regulation says that this should not be

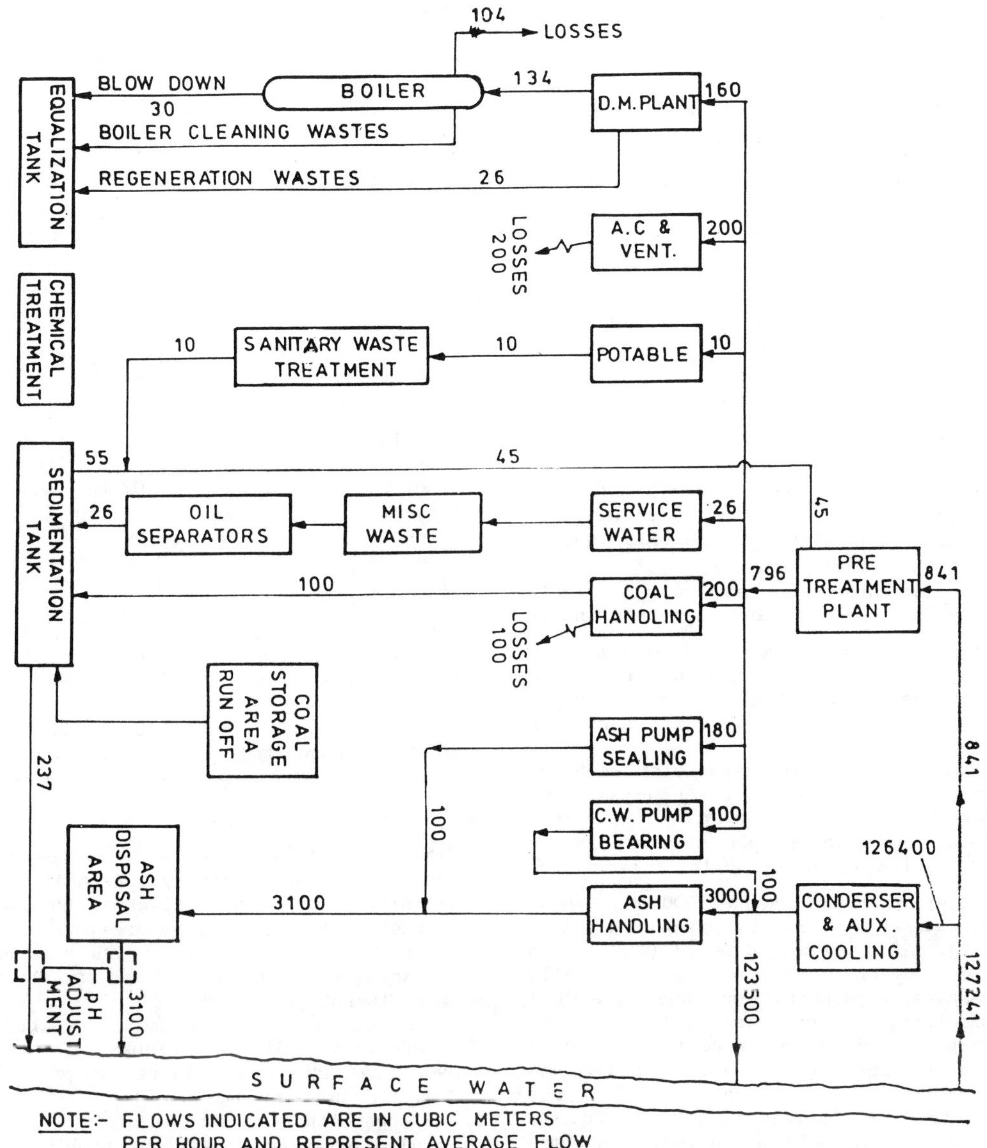

FIG. 1. WATER BALANCE DIAGRAM

more then 5^0C above the intake water temperature. The major problem with once-thorough cooling is thermal pollution. Large quantities of hot water discharged into a natural water body (river, lake or sea) affect the physical, chemical and biological characteristics of the receiving water bodies. Changes also occurs in metabolism, reproduction and development rate of many organisms. These changes result in a change in the structure of aquatic ecosystem. Another possible impact of the thermal discharges on the aquatic community is due to thermal shocks, that is a rapid change in temperature caused during start up and shutdown of the stations. Sometimes the heated discharges may prove to be beneficial to certain commercial species. It is therefore all the more important to be able to predict the impacts of the heated discharges on the natural water bodies. Impact assessments studies are normally conducted to predict the impacts.

In case the adoption of once - through cooling systems, site of the project is selected in such a way that impacts on the environment will be minimal. For doing so, adequate flow in the river would be necessary. In some cases discharge channels are properly designed to reduce the temperature of discharges from the cooling system.

4.2 Deminearlization stream waste

Discharges from the demineralized water system (D.M.) are boiler blow down, regeneration waste and periodic boiler cleaning wastes.

Boiler blow down is usually highly alkaline in nature and less in quantity and average of 30 m^3/hr is produced as blow down waste for the reference stations. The blow down requires neutralization prior to discharge or reuse.

Boiler cleansing waste water emanates very occasionally. Frequencies vary from twice a year to once in five years. Total waste volume per cleaning period is in the order of 2 to 3 times one boiler fill or around 1000 m^3. These wastes contain high in suspended solids (SS), total dissolved solid (TDS), iron and sometimes copper and normally require sedimentation and chemical precipitation of dissolved iron and copper. D.M. plant regeneration waste are usually high in S.S. and TDS and exhibits wide variation in pH. A degree of self neutralization is achieved by detention in holding basins. Sometimes chemicals are dosed for correction of pH.

4.3 Ash handling

About 3100 m^3/hr of waste water from ash pond is emanated. Ash transport water normally picks up dissolved solids from the ash. Wide variations in the pH of the ash transport water has also been observed. Typical characteristics of ash pond effluent on yearly average for Indian condition is given in Table 2.

Table 2. Ash pond effluent water quality data

parameters	range, mg/l
Chlorides(Cl)	3.00 to 28.00
Fluorides(F)	0.14 to 2.10
Nitrates(N)	0.18 to 0.27
Sulphate(SO_4)	9.00 to 14.00
Calcium(Ca)	16.00 to 25.00
Iron(Fe)	0.05 tO 35.60
Magnesium(Mg)	7.20 to 18.20
Potassium(K)	0.99 to 6.00
Sodium(Na)	6.00 to 16.30
Arsenic(As)	0.002 to 0.05
Cadmium(Cd)	0.006 to 0.01
Chromium (Cr^{6+})	0.005 to 0.05
Copper(Cu)	0.0005 to 0.1
Lead(Pb)	0.02 to 0.1
Manganese	0.003 to 0.6
Mercury(Hg)	0.001
Selenium(Se)	0.005
Zinc(Zn)	0.01 to 0.14
Cyanides(Cn)	0.003
Detergents(As MBAS)	0.06 to 0.8
Phenolic compounds	0.001
Hardness (As $CaCO_3$)	76.0 to 116.0
TDS	120.0 to 196.0
pH (Standard units)	7.4 to 11.0

Site for ash disposal is selected as low lying areas preferably barren in the vicinity of the plant. Natural depression are utilized where possible, otherwise dykes are built surrounding the area. Recirculation of ash transport water is not practiced.

The ash pond is normally designed carefully to maintain a blanket of water over the ash at all times to prevent fugitive dust emission. The impact of the supernatant on surface water bodies is also assessed. A serious problem in ash pond is leachate generation and contamination of ground water thereon. Leachate generation is controlled by the use of locally available impervious clay lines. Nowadays, synthetic mat/geotextile lining is also being thought of for preventing leachate generation.

As an alternative it is now being considered to develop indigenous technique for dry disposal of fly ash adopting pneumatic transport systems. Water is recycled either from the ash pond or from ash dewatering bins.

4.4 Miscellaneous

4.4.1 Sanitary wastes

Sanitary wastes are normally treated in small extended aeration package plants.

4.4.2 Floor and area drains

Wastewater flows through plants outlet normally contain SS, oil and grease, trace metals etc, which receive discharges from laboratory drain, routine cleaning and maintenance of equipment, floor washing etc. Normally these wastes are collected and routed through oil separators and sedimentation basins.

5 ZERO DISCHARGE CONCEPT

For reducing waste discharge load, as well as minimization of withdrawal of surface water, it is now advised for a rational approach towards zero discharge. Naturally, the first step towards zero discharge is the introduction of a recirculating condenser cooling system. Significant amount in reduction of blow down is achieved through high cycles of concentration.

Ash pond overflow may be collected and reused for transporting fly ash towards ash disposal area and it can be reused.

If cost permits with the introduction of adsorption tower and reverse osmosis units in the wastewater treatment systems, removal of dissolved solids is possible. Under such circumstances wastewater can be reused as potable water.

6 WATER QUALITY MONITORING PROGRAM

It is of great importance to assess the status of the environment during plant operation so that proper mitigative measure can be taken whenever necessary. This helps to build up a data base and provides a comparison between the predicted impacts and actual status of the environment.

Sample collected from plant outlet are monitored at regular interval. Effluent from ash pond outlet analyzed on monthly basis since water quality does not change much. A well equipped laboratory is needed within the plant for conducting the above mentioned test.

REFERENCES

Edinger, J.E., Brady, D.K., and Geyer, C.J. 1974. Heat exchange and electric power research institute, Palo Alto Calif.

Havens and Emerson Ltd. 1971. Feasibility study for wastewater management programme. A report to the Buffalo District, U.S. Army Corps of Engineers.

Neeri, Nagpur, India. 1980. Proceedings on Indo-US workshop on Environmnental Impact Analysis and Assessment held on October 27-31.

Rau, J.G. and Wooten, D.C. 1980. Environmental Impact Analysis Handbook. McGraw Hill.

Environmental Management, Geo-Water & Engineering Aspects, Chowdhury & Sivakumar (eds)
© 1993 Balkema, Rotterdam. ISBN 90 5410 099 0

Role of surface drain on quality and nutrient levels in groundwater and surface water, near Lake Thomson, Jandakot, Western Australia

C.E.Olsen
Groundwater Technology Australia Pty Ltd, Perth, W.A., Australia

A.Q.Rathur
Curtin University of Technology, Perth, W.A., Australia

ABSTRACT: Analysis of surface water and groundwater 2.5 kilometres north east of Lake Thomson at Jandakot, Western Australia revealed 'high' to 'extreme' concentrations of Total Phosphorus(TP) and Total Kjeldahl Nitrogen (TKN). Sandy soils in the area have built up a phosphorus store in the soil profile due to applications of fertilizers from past agricultural practices. High winter rainfall is responsible for leaching of excess nutrients from the shallow subsurface into open drains which flow into Lake Thomson. This enables the nutrient-responsive, free-floating fern Azolla sp to rapidly colonise the drains, completely covering the drain water surface. Large stands of bulrushes also grow within the drain indicating high levels of phosphorus and nitrogen.

1 INTRODUCTION

The project area is located 20 kilometres south of Perth and lies with a declared Water Authority of Western Australia (W.A.W.A.) public water supply and underground water pollution control area (Cargeeg et al 1987). The project site is situated on the south west crest of the Jandakot Mound groundwater flow system which covers an area of approximately 500 km^2 immediately to the south of the Swan River, and is one of the two major groundwater resources being utilised for the Perth Metropolitan Municipal Water Supply (Fig: 1). The Jandakot Mound is located on the Bassendean dune system with extremely infertile, extensive areas with high watertable and generally scattered and low intensity agricultural activity.

The area has a typical Mediterranean climate characterised by hot dry summers and cool wet winters. The average annual rainfall for Perth is 870 mm of which 60% falls in the winter months.

2 LOCAL GEOLOGY AND HYDRO-GEOLOGY

The project site is situated within the Perth Basin which extends north-south for some 1000 kilometres along the western margin of the Darling Scarp. The Swan Coastal Plain lies within the Perth Basin and is a low lying, gently undulating area covered largely by Holocene and Pleistocene coastal dune and shoreline deposits. The project site is located on the Swan Coastal Plain and is immediately underlain by the Superficial Formations. Here they comprise a sequence of calcareous quartz sands, shelly limestones and carbonate silts of the Jandakot beds, unconformably overlain by quartz sands, silts and clays of the (Guildford Formation and Bassendean Sands. Unconfined shallow groundwater occurs in the Superficial Formations where the water table forms a small mound (Jandakot Mound) rising to about 25.0 metres above mean sea level (Table 1).

The highly permeable Bassendean Sands which are almost pure silica contain very little organic matter. A shallow water table at about 1.3 metres depth is present within the Bassendean Sands during winter months. Summer groundwater levels drop significantly to 3.5 metres depth.

3 TECHNIQUES

Monitoring of surface water from two drains located east of Lake Thomson was carried out

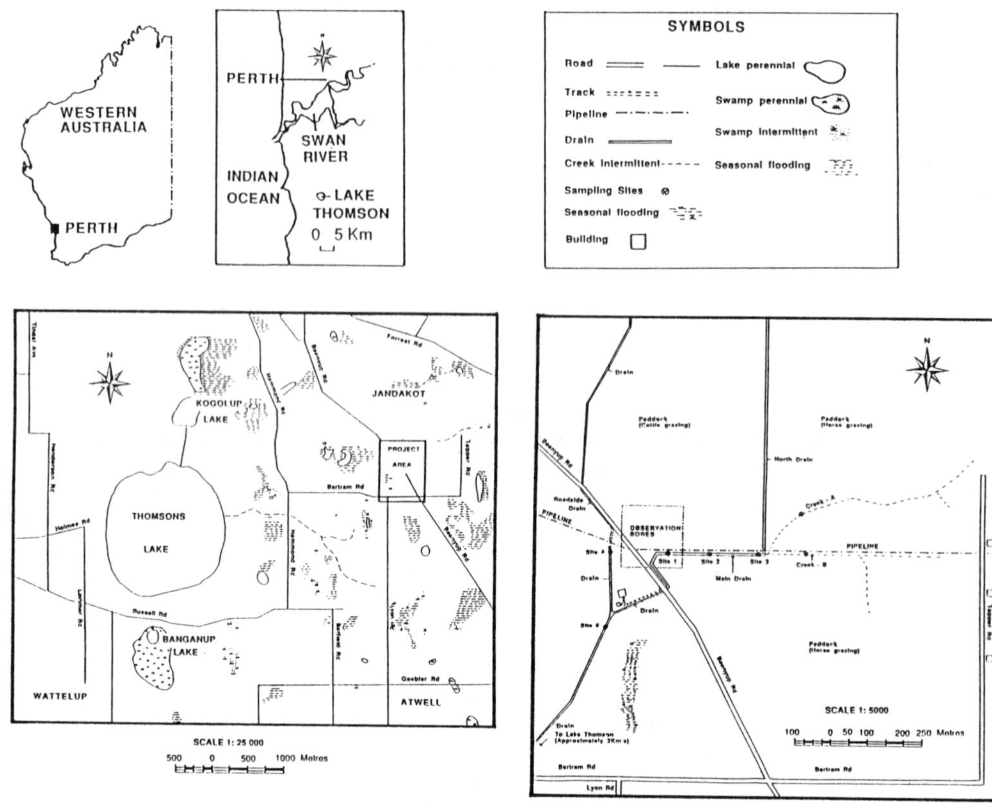

Fig. 1. Location map Jandakot Western Australia

Table. 1. Statigraphic sequence of the project area,(adaptedfrom Martin and Baddock,1989)

AGE	FORMATION	UNIT	MAX. THICK.	DEPTH (m)	LITHOLOGY	REMARKS
Quaternary	Superficial Formations	Bassendean sand	5m	0-5	Fine to course quartz sand, sub rounded,poor sorting.	Thickness uncertain, holds water table
					UNCONFORMITY	
		Guilford Formation	18m	5-23	Fine to coarse quartz sand. Sub-rounded to rounded, poorly sorted, trace of heavy minerals. Ferruginized sand layer is present near the top of the unit. Minor Clay and silt sometimes occurs near the base.	Variable thickness and lithology
					UNCONFORMITY	
Late Tertiary		Jandakot Beds	21m	23-44	Fine sand to gravel, sub-rounded to rounded, poorly sorted. The unit is highly fossiliferous with minor clay and abundant calcareous silt. Variably cemented calcarenite is also present.	Variable thickness and lithology. Calcareous component of the sediments is silt sized.
					UNCONFORMITY	
Cretaceous	Osborne Formation		130m	44-174	Gluaconitic siltstone, clay and minor sand.	Major confining layer, forms base of aquifer.

in order to evaluate the quality and nutrient levels in these drains and their effect on the groundwater. These drains eventually discharge into Lake Thomson thus affecting the water quality of the lake.

Monitoring of surface water and groundwater was carried out by collecting and analyzing water samples from 3 sites in these drains and some piezometers placed in the vicinity of these drains and penetrating the shallow unconfined aquifer. This was done at infrequent intervals - at shorter time intervals of about two weeks during wet winter months and longer time intervals during dry summer months. Main emphasis of this work was to measure Total Phosphorus (TP) and Total Kjeldahl Nitrogen (TKN) concentrations using normal chemical laboratory techniques. Another objective of this work was to evaluate the environmental implications of TP and TKN concentrations in these waters.

4 RESULTS AND DISCUSSION

Analytical results show increase in TP and TKN levels in the drains in early May at the start of rainy season. Bott (1990) has tabulated TP and TKN concentrations for classification of these waters (Table 2a and b). The "first flush" of surface water through the drain occurred during mid-May. Drainwater analysed indicated 'extreme' levels (greater than 3.0 mg/L) of TP and TKN. These excessive levels were unable to be flushed away quickly due to the shallow gradient of the drain and low drainwater volumes.

Increasing precipitation during winter months resulted in substantial increases in drain flow. A significant reduction in nutrients occurred as drain flow increased, although levels of TP and TKN still remained 'moderate' to 'very high' over the wet winter period.(Fig: 2 a and b).

Shortly after the initial first flow of surface water in May, the drain was rapidly colonised by a nutrient-responsive, free-floating fern Azolla sp which harbours colonies of blue-green Anabaena, capable of fixing and storing atmospheric nitrogen. The Azolla sp fern

Table 2a. Guide to interpreting drainage phosphorus concentrations (after Bott 1990)

APPROXIMATE RANGE OF PHOSPHORUS IN mg/L	STATUS	COMMENTS
< 0.05	pristine	Bush catchment
0.05 - 0.15	low	Partially cleared catchment (> 70% bushland)
0.15 - 0.25	moderate	30% - 70% catchment cleared
0.25 - 0.40	high	70% - 90% catchment cleared
0.40 - 3.00	very high	Approximately 95% catchment cleared and/or a proportion of sandy soils
> 3.00	extreme	Point sources of nutrients (e.g. piggeries)

Table 2b Guide to interpreting drainage nitrogen concentrations (after Bott 1990)

APPROXIMATE RANGE OF NITROGEN IN (mg/L)	STATUS	COMMENTS
< 1.0	pristine	Largely bush catchment or low flow (< 30% bush)
1.0 - 1.4	low	Partially cleared catchment (> 70% bushland)
1.5 - 2.0	moderate	60% - 85% catchment cleared
2.0 - 3.0	high	> 85% catchment cleared
3.0 - 6.0	very high	"First-flush", flood conditions (high organic N content) or stock access near sampling site (high nitrate content)
6.0 - 100	extreme	Point sources of nutrients (e.g. piggeries)

These tables were prepared by Bott, 1990 using data obtained from the Swan Coastal Plain and south coast of Western Australia.

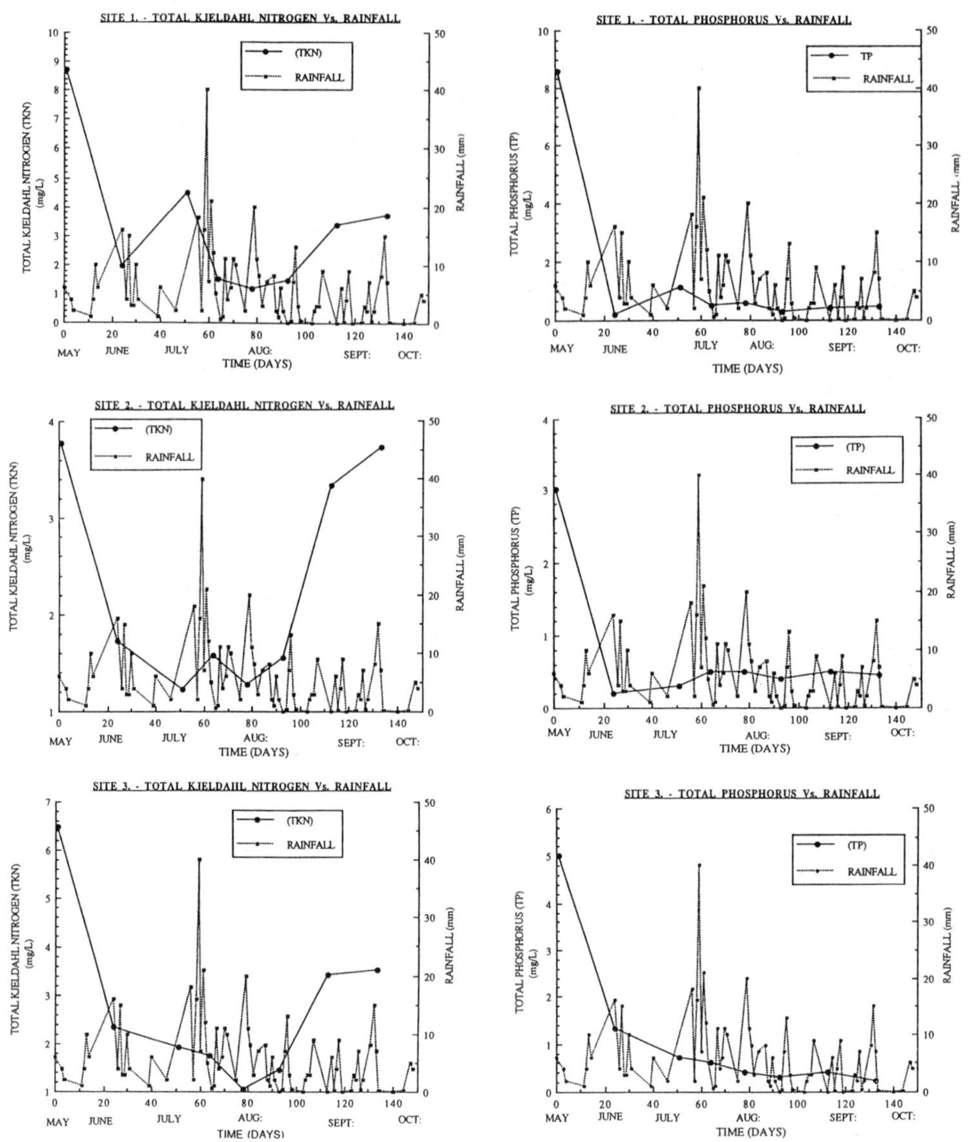

Fig 2. Total Phosphorus and TKN concentrations vs rainfall over 150 days period for site 1, 2 and 3 on the main drain.

completely covered the surface of the drain water in many areas. Large stands of bulrushes also grew and thrived within the drain which gave an indication that phosphorus and nitrogen levels were relatively high.

Groundwater analysed for TP and TKN concentrations in two shallow monitoring wells showed 'high' to 'very high' levels. Groundwater in four other deeper wells generally showed 'low' to 'moderate' levels of TP and TKN. Organic ferruginized sand, silt and clay layers underlying the Bassendean Sands appear to act as nutrient filters, removing a large proportion of excess nutrient leached down into deeper aquifers.

5 CONCLUSIONS

Agricultural sources are responsible for excess nutrients detected in the surface water flowing in open surface drains. Over past years a phosphorus store or 'super bank' appears to have built up within the soil profile of the project area. This has resulted from the application of fertilizers to the sandy soils to sustain the growth of oat crops.

Although the farming of agricultural crops has ceased in the area, continued leaching of residual phosphorus contained within the sandy soil profile may account for the 'excessive' phosphorus levels detected in surface water and groundwater.

Other sources of nutrients within the project area include animal faeces from grazing stock and plant material, especially from Azolla sp and bulrushes decaying within the drains during summer months. This is likely to be the main source of the 'extreme' TP and TKN levels measured from drain water during the "first flush". Biologically fixed nitrogen from the floating Azolla sp also contributes towards the elevated concentrations of TKN detected in the drain water.

The shallow water table underlying the project area appears polluted with excess phosphorus and nitrogen. Surface water runoff and groundwater seeping into the open drain eventually discharges into Lake Thomson thus affecting the water quality of the Lake.

6 REFERENCES

Bott, G. 1990. Guide to interpreting drainage phosphorus and nitrogen concentrations from Swan Coastal Plain and south coast of Western Australia. Environmental Protection Authority of Western Australia (Per.Comm:).

Careeg, G.C., G.N.Boughton, L.R.Townley, G.R.Smith, S.J.Appleyard and R.A.Smith 1987. Perth urban water balance study. Volumes 1 and 2. Water Authority of Western Australia.

Martin, W.M. and L.J.Baddock 1989. Jandakot Bore J270P pumping test. Hydrogeology Report No: 1989/41. Geological Survey of Western Australia. (unpublished).

Environmental Management, Geo-Water & Engineering Aspects, Chowdhury & Sivakumar (eds)
© 1993 Balkema, Rotterdam. ISBN 90 5410 099 0

Water quality management issues for Lake Tinaroo

Grant Sadler & Wojciech Poplawski
DPI, Water Resources Commission, Qld, Australia

ABSTRACT: The lake and the catchment are described. A brief outline of the water quality monitoring program is given. Statistical methods used for trend analysis are presented. Emerging issues including water quality changes and management strategies are discussed.

INTRODUCTION

Tinaroo Falls Dam is a major water conservation project on the Atherton Tablelands near Cairns in Far North Queensland (see Fig. 1).

The dam was originally constructed in 1958 to provide water for irrigation in the Mareeba Dimbulah Irrigation Area. Supply is currently available from the dam for the following.

. Irrigation of over 15,000 ha of mixed crops in the irrigation area.

. Urban water supplies for the towns of Mareeba, Atherton, Tinaroo, Yungaburra, Walkamin, Dimbulah and Mutchilba.

. Power generation using surplus water at the Barron Falls Hydro Electric Power Station.

In addition to the above, clients in the Irrigation Area use water from the irrigation system for domestic purposes. Lake Tinaroo has significant environmental values including scenic amenity, fisheries, wildlife habitats and aquatic ecosystems. The lake is a major regional resource for recreation with up to three quarters of a million people estimated to visit the storage each year.

The level of utilisation of the storage as measured by water allocated is currently around 60%. Strategies are in place to increase the level of utilisation and thus maximise the economic return to the community from the State investment in the scheme.

Figure 1. Locality map - N.E. Australia

The WRC has a responsibility to ensure that water quality in Lake Tinaroo is not degraded and is suitable for all purposes including protection of environmental values.

Table 1. Land use change, Tinaroo catchment

	1952 ha	1952 %	1978 ha	1978 %	1988 ha	1988 %
Forest Areas	26791	49	23713	43	23137	42
Crop Areas	7429	14	5814	11	4947	9
Pasture Areas	18089	33	19079	35	20549	38
Lakes and Streams	2387	4	6062	11	6062	11
Total	54696	100	54695	100	54695	100

Water quality monitoring of the lake has been undertaken with the following water quality objectives in mind:

. To prevent deterioration of the storage ecosystem (e.g. no significant algal blooms that affect the use of the storage).

. To ensure acceptable water quality for the following uses:

. irrigation

. town water supply

. contact recreation

There have been a number of occasions in recent years where significant growth of water weeds has been noted. Such increase in the presence of weeds is an indication of an increase in availability of nutrients in the water body. Other factors such as temperature and sunlight are also important.

If weed growth on the storage were to continue unchecked the Commission's concern was that a general deterioration in water quality was likely.

It was the presence of weed growth which prompted the Commission in 1987 to commence a more comprehensive water quality monitoring program with a view to identifying water quality trends and assessing the condition of the storage.

1. CATCHMENT LAND USE

Tinaroo Falls Dam has a catchment area of 545 sq km. Land use within the catchment comprises approximately 40% Forest and National Park, 40% grazing and agriculture with the remaining 20% including the lake, rural residential and urban land uses.

Based on aerial photography the University of New England (Hlaing 1991) has estimated the land use changes as shown in Table 1.

This table shows a decrease in forest areas, decline in cropping areas and an increase in pasture. It is possible that the reduction in forested areas and the subsequent increase in the use of agrochemicals may have created more adverse effects on water quality in the system.

Table 2. Fertiliser applied, 1988

	kg/yr
Nitrogen	905000
Phosphorus	242000
Potassium	243000

SOURCE: Valentine, 1990.

In a 1990 report James Cook University have estimated the elemental nutrients applied in the catchment. Their findings are detailed in Table 2. It is considered that these nutrients impact adversely on waterways although fertilisers are not the only source of nutrient input to the system.

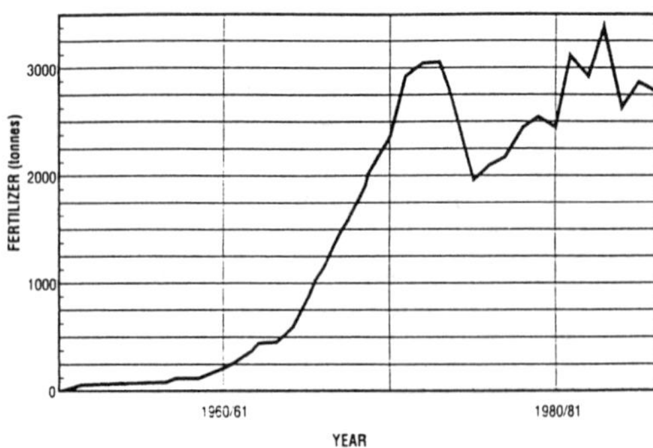

Figure 2. Atherton Shire total fertiliser use

As a guide to the historical pattern of fertiliser use in the catchment, figures for Atherton Shire (source - Aust. Bureau of Statistics) are shown in Figure 2. This figure shows a dramatic increase in fertiliser application during the 60's and 70's.

The application of both nitrogen and phosphorus have increased with phosphorus now applied in the more soluble form of fertiliser rather than as superphosphate.

While agriculture and grazing are still the dominant land uses in the catchment, the population of the catchment is also growing. Land for urban and rural residential uses and recreational activities will further increase pressure on waterways.

2. WATER QUALITY MONITORING PROGRAM & RESULTS

As a result of concern over possible future water quality problems, in 1987 the Commission established an interdepartmental advisory committee to steer a study to assess the trophic state of the lake and to identify nutrient loads and the limiting nutrient in the lake.

A water quality monitoring program was initiated comprising measurement of nutrient and chlorophyll-a levels at 6 sites on the lake and measurement of nutrients at 8 sites in the catchment. Furthermore bacteria levels are monitored at 12 sites on and around the lake.

The Commission has also recently constructed 5 bores in the area for the purpose of assessing groundwater. Sediments in the lake adjacent to major streams have been sampled to determine nutrient levels in sediment.

Time series analyses were used for water quality data evaluation. The period of observation (2.5 years) is not long enough to make a definite assessment of trends however preliminary estimates of trends can be made. The Kendall-τ Test and the Wilcoxon Signed Rank Test (median analysis) were used for trend detection (Rayment and Poplawski, 1992).

The results indicate a positive trend for nitrates (particularly in the bottom layer, $0.13 mgL^{-1}$ per yr) and chlorophyll-a (Figure 3). However phosphates during this period remained static.

It is suspected that unavailability of phosphorus in the water column is the reason that

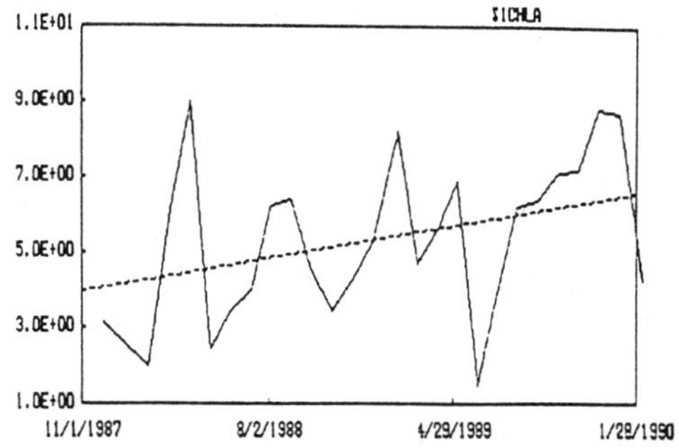

Fig. 3 Lake Tinaroo. Station 1.
Kendall-τ Test for Chlorophyll-a (μgL^{-1})

Lake Tinaroo is still mesotrophic. However there is a potential of the lake becoming eutrophic because it is likely that sediments may have reached a maximum P fixing capacity.

Results of the excursion analysis for bacteria tend to indicate that an area near Yungaburra township may become unfit for recreational uses. At one station, the E. Coli count exceeded 50 colonies for some 75% of the time and was higher than 200 during 30%. Groundwater in this area is also high in bacteria.

The results of the water quality monitoring indicate that the lake is under considerable stress. Further monitoring of nutrients, bacteria and other parameters is warranted. The Commission is now establishing an enlarged monitoring network.

3. MANAGEMENT STRATEGIES

The catchment area of Tinaroo Falls Dam contains prime agricultural and grazing lands which support the economy of the upper tablelands area. Furthermore the lake has proven to be a popular tourist destination and an attractive area in which to live.

It has been recognised for some time that all land uses in the catchment are contributing to the water quality status of the lake. In 1973 a Declared Catchment Area was constituted under the Water Act. This gave the Water Resources Commission powers to control close subdivision or any land use which could cause pollution of the lake.

More recently the Commission has been

working closely with the Atherton and Eacham Shire Councils and the Department of Housing and Local Government to prepare specific Development Control Plans which provide objectives and implementation standards that will ensure the long term protection of water quality in Lake Tinaroo. These plans may include buffer zones, construction of a treatment plant for Yungaburra and some restrictions on development.

In conjunction with the Local Authorities in the catchment, the Department of Primary Industries is also undertaking a study of the issues related to the management of the natural resources of the Barron River Catchment.

High on the list of issues to be addressed are the protection and enhancement of water quality. This study and the community participation process proposed are expected to result in improvements in the way the catchment is currently used and to reduce the threat of the storage becoming eutrophic. This process will no doubt result in further improvements in our knowledge which can be incorporated in subsequent control plans for the catchment.

Water quality objectives will need to be agreed for the storage so that performance of strategies against these objectives can be continually monitored.

CONCLUSION

The lessons learnt from the experience gained at Lake Tinaroo has prompted the Commission to look more closely at its other storages and projects.

Similar monitoring networks are being designed for irrigation areas in Emerald and St. George (cotton) and Barker - Barambah where increased salinity may be a problem.

It is also obvious that protection of Lake Tinaroo against pollution can only be achieved through the concerted effort of all involved. Therefore integrated management of this catchment with participation of the State and Local Governments, land users, and general public is so important.

ACKNOWLEDGMENT

The authors acknowledge with thanks the permission of the Commissioner of Water Resources to publish the paper and the assistance of members of the Commission staff in preparing the paper. The conclusions reached and comments made do not necessarily reflect those of the Water Resources Commission.

REFERENCES

Hlaing, C. 1991. Watershed - a key issue in tropical resource management. University of New England (unpublished).

Rayment, G.E. & W.A. Poplawski 1992. Training Notes on Sampling for Water Quality Monitoring. Condamine - Balonne Water Committee, Dalby, Australia.

Valentine, P.S. 1990. A land use study to investigate potential agricultural chemical inputs to marine environments in the Northern Great Barrier Reef Region. James Cook University, Townsville.

Environmental Management, Geo-Water & Engineering Aspects, Chowdhury & Sivakumar (eds)
© 1993 Balkema, Rotterdam. ISBN 90 5410 099 0

Possible water resource for the Nabire area, Irian Jaya, Indonesia

Djoko Santoso & Sigit Sukmono
Department of Geology, Bandung Institute of Technology, Indonesia

Joko Supriyanto, Haryadi Permana & Eko Soebowo
Institute of Geotechnology, Indonesia Institute of Science, Indonesia

ABSTRACT: Irian Jaya is the most rare populated province in Indonesia. There are some new settlement area being constructed in the Nabire region, Irian Jaya. Nabire is situated around 0 - 600 m above mean sea level. Previously the water for irrigation purpose would be supplied from Wadio river and for domestic purpose from Nabire river. However, based on some spot investigations and considering the number of people and land use planning, it seem that the water quantity from Kalibumi river whose catchment area is wider should be taken into account. Consequently, the management of water resources for the region should be reviewed, i.e. the Kalibumi river for covering irrigation purpose, Nabire river and shallow groundwater for covering the domestic purpose.

1 INTRODUCTION

1.1 General

Irian Jaya is the most rare populated province in Indonesia. Under the transmigration program, inhabitants from Java Island as the most populated region in Indonesia have been moved partly to Nabire area, Paniai District, Irian Jaya Province. Their activities are in food production or paddy field agriculture (Wanggar, Kalibumi and Kimi area) and cultivation (Topo area). Therefore water is very important for domestic and irrigation purposes. Water consumption for a person in Indonesia is calculated approximately 80 litre/day in rural area and 300 litre/day in the city area (Simoen, 1986). Accordingly the water resources in Nabire and surrounding area should be properly managed.

1.2 Physical condition of the studied area

The studied area is a catchment area of Kalibumi river of 994,400 hectares wide. Morphologically it can be divided into five units (Figure 1). The coastal plain area with slope 0 - 2%, the alluvium plain area with slope 0 - 2%, intrahilly plain with slope 3 - 13%, hilly area with medium to steep slope or 14 - 20%, and mountainous area of steep slope 21%. Geologically, the area is underlaid by rock formation from Cretacous/Mesozoic to recent age (Figure 2) volcanic rocks, conglomerate, mudstone and alluvium.

1.3 Objective

Previously the water for irrigation and domestic purposes would be supplied from Wadio and Nabire river, but it can not satisfy the requirements. Therefore the objective of this study is to evaluate the water resources management for Nabire and surrounding area.

1.4 Investigation methodology

The aerial photo was the most important tool in this study. It was panchromatic, black and white of 1 : 120,000 scale. The photographs were taken on 1976. Some climatic data were also used in this study.

The airphoto interpretation gave the geological map, geomorphological map, catchment area map, and landuse map.

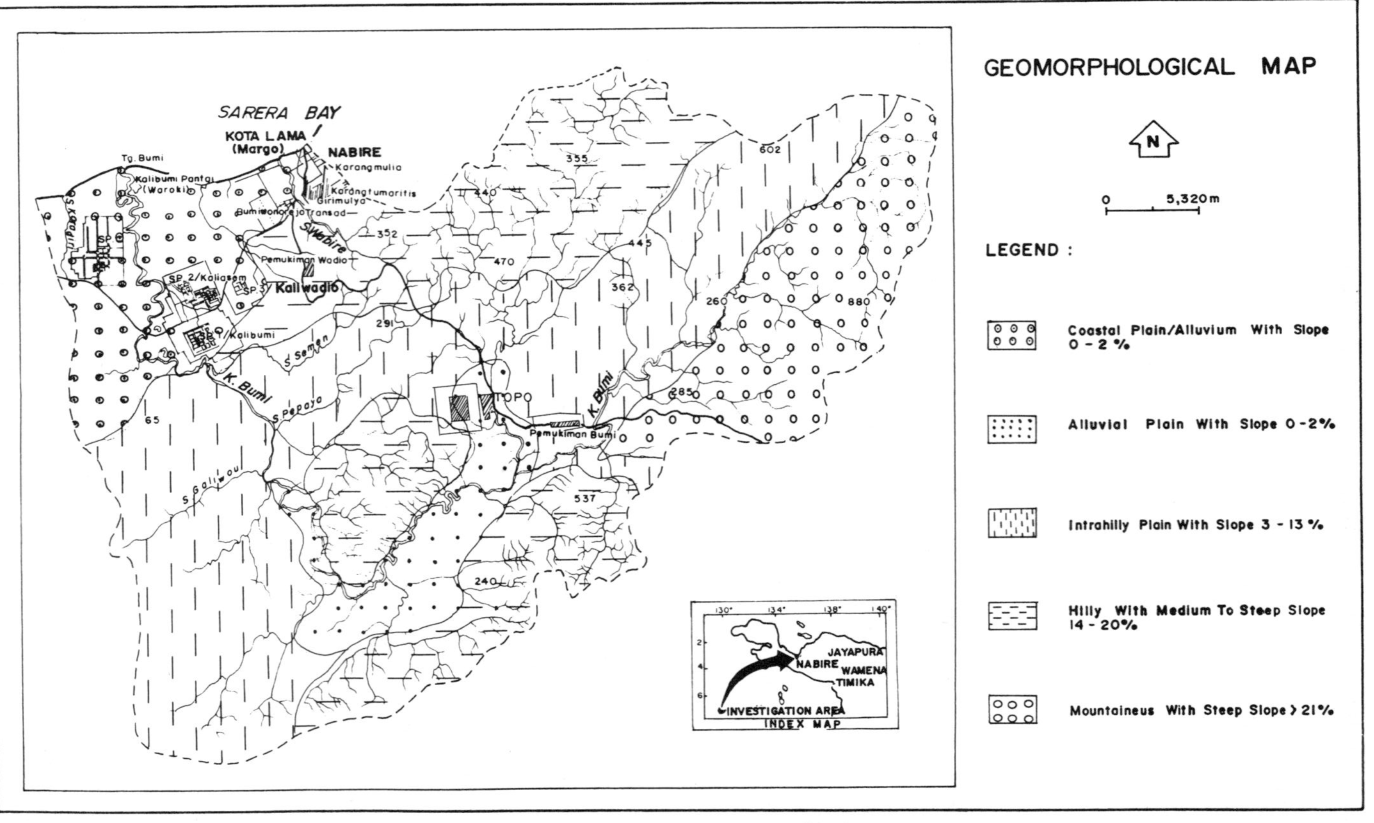

Figure 1. Geomorphological Map of Kali Bumi

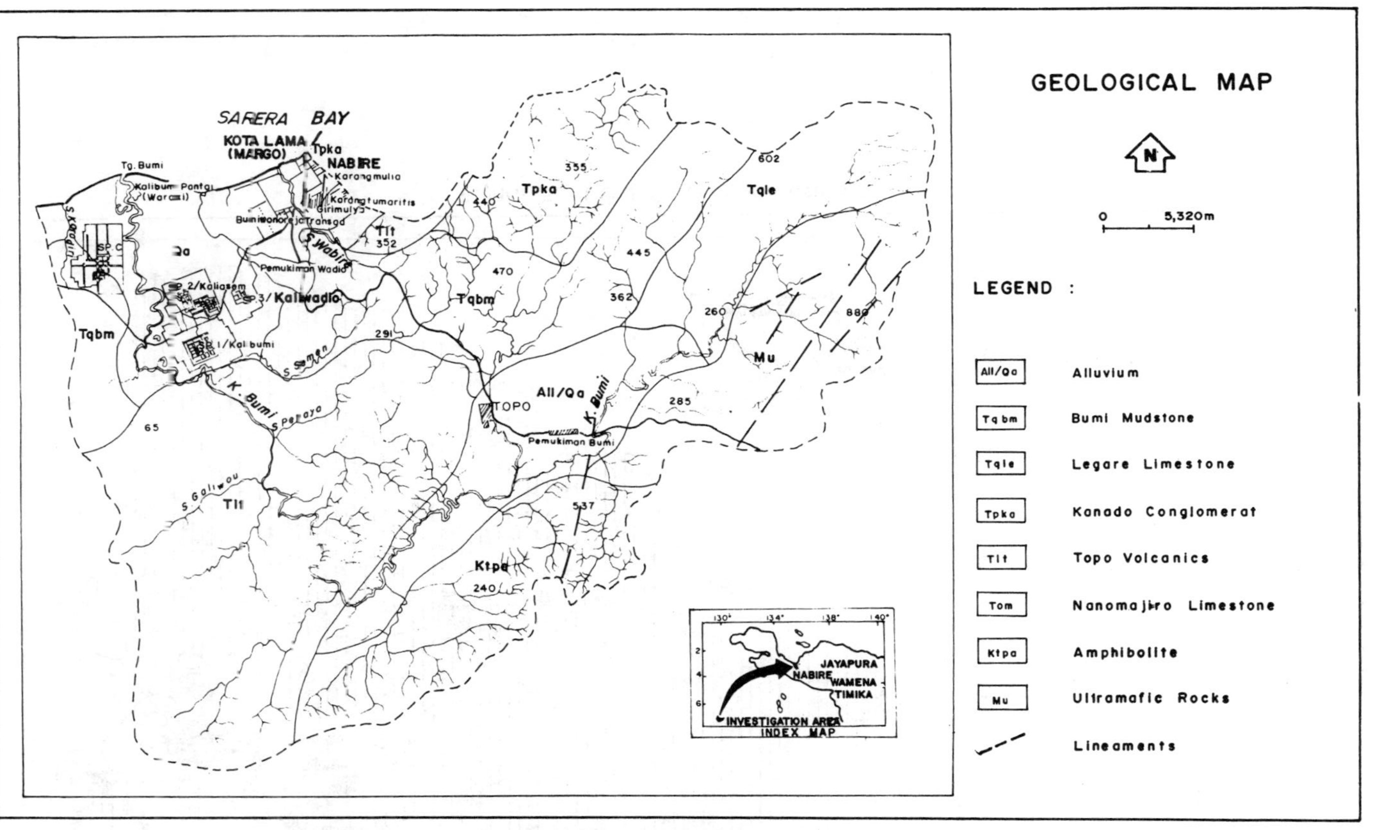

Figure 2. Geological Map of Kali Bumi Catchment Are

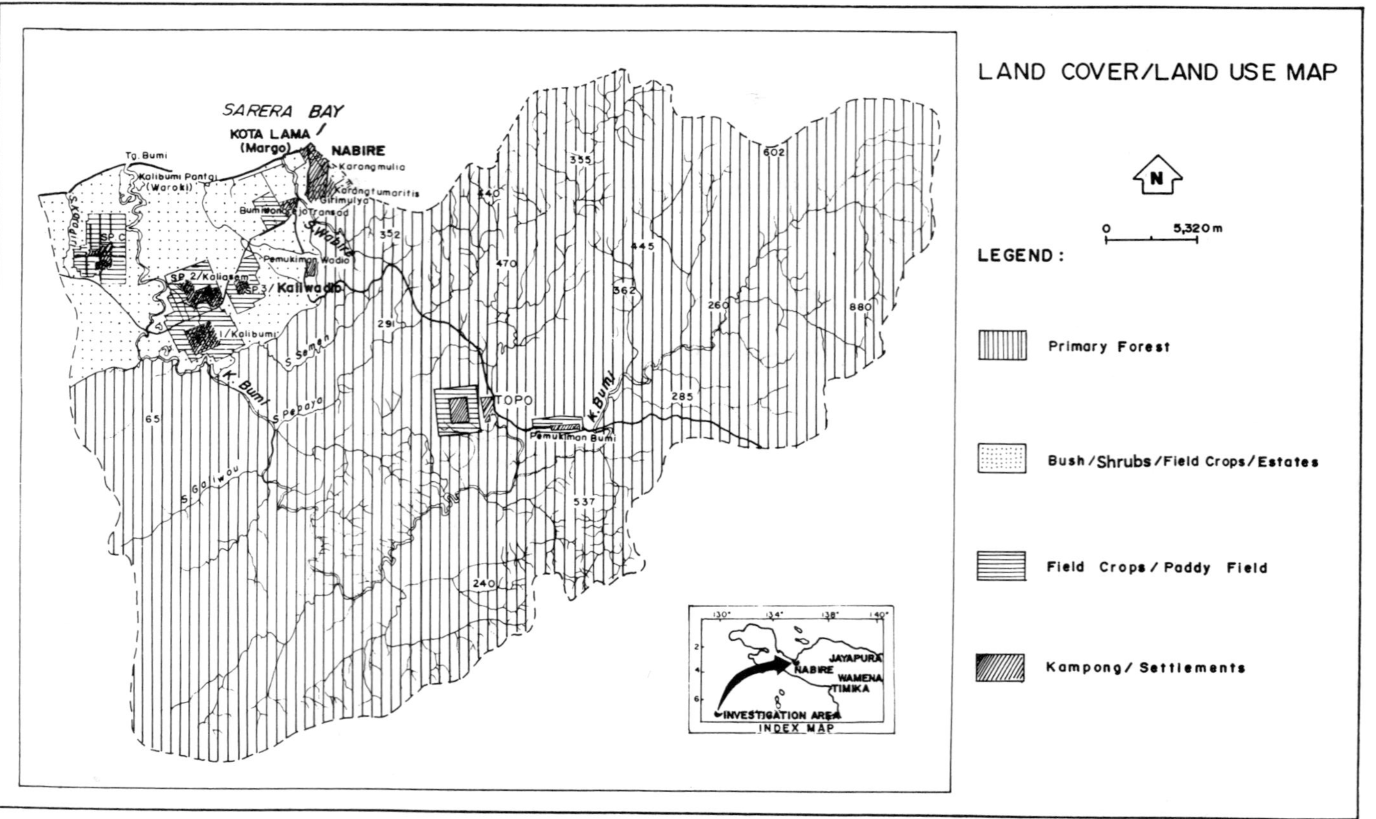

Figure 3. Land Cover/ Land use Map

2 WATER RESOURCES

2.1 Surface water

The surface water condition is much influenced by the land use of the area. The land use in the area, which is also the Kalibumi catchment area (994,400 hectares), can be divided into 4 units (Figure 3). They are primary forest (85%, 848,200 hectares) which some have been exploited, bush/shrubs/field crops/estate which cultivated by coffee, clove and cocoa (11%), field crops/paddy field (2.5%) and settlement area (1.5%).

The rainfall data is shown in Table 1. It shows that the highest rainfall is on January and the lowest on November. The calculation of evapotranspiration gave Monthly Head Index of 50.99 and Evapotranspiration Potential of 2,053.6 mm/year.

Table 1. Rainfall data of 1981-1990 from Geophysics & Meteorological Agency of Paniai District.

Month	Mean rainfall (mm)	Mean Temperature (°C)	Evapotrans-piration (mm)
January	449,2	27,3	177,6
February	441,2	27,2	158,4
March	416,0	27,3	174,2
April	373,3	27,4	168,6
May	320,0	27,5	173,4
June	270,5	27,2	165,5
July	295,9	26,9	168,0
August	309,0	27,0	170,2
September	319,8	27,1	165,8
October	327,3	27,3	176,6
November	260,7	27,5	174,8
December	282,0	27,6	180,5

The flow rate of the Kalibumi river (Figure 4) was measured as 4 cubic metre/sec, and for Nabire river was 4 litre/sec with 22,500 hectare catchment area. Therefore Kalibumi river water can be used for paddy field irrigation and Nabire river water for domestic purpose. Wadio river itself is only an intermittent river, therefore it can not be used for irrigation purpose.

2.2 Groundwater

Some water samples from shallow groundwater in Nabire town were taken for laboratory analysis. The depth of shallow groundwater was varied from 4.5 meters to 7.0 meters. For comparison the downstream water sample of Nabire river were also taken and analyzed. The result is shown in Table 2.

Table 2. Water quality analysis.

Location/ depth	K	Na	Ca (mg/litre)	Mg	CI	SO4
Kotalama/ 4.5 m	2.45	10.40	77.50	17.08	10.18	11.80
Oyehe/5 m	3.32	14.12	33.70	6.00	16.97	9.20
Karang-mulya/7 m	2.97	10.40	63.74	6.98	6.79	10.20
W.Anggrek/ 4.5 m	2.97	14.74	41.56	12.45	6.76	1.65
Down stream Nabire	1.40	10.40	49.48	24.59	6.79	12.00

2.3 Analysis

Those limited data were used for calculating the Tentative Water Balance in order to get the effective flood. The calculations are as follows :

In = P x catchment area (1)
O = Ep x catchment area (2)

Where:
In = Influx water
O = Evapotranspirated water
P = Mean monthly rainfall in mm
Ep = Monthly evapotranspiration in mm

Because infiltration rate is unknown, then Maximum Water Losses (MWL) is assumed to be 30% In.

Therefore:
Effective flood = (In)-(0)-MWL (3)

The calculated result can be seen in Table 3. It shows that the highest flood is 0.136 litre/month in January and the lowest 0.0076 litre/month in November. This value is a small amount of effective flood, therefore there is no flood season or hazard in the area.

The flow rate of Kalibumi river of 4 m^3/ sec can be used for irrigation purpose. On the other hand the water quality of Nabire river and shallow groundwater are possible to be consumed for domestic purpose. The water quality is also influenced by sea water, as shown by the Cl content, while in Oyeke (150m from the shore line) was only is 16.97 mg/litre,

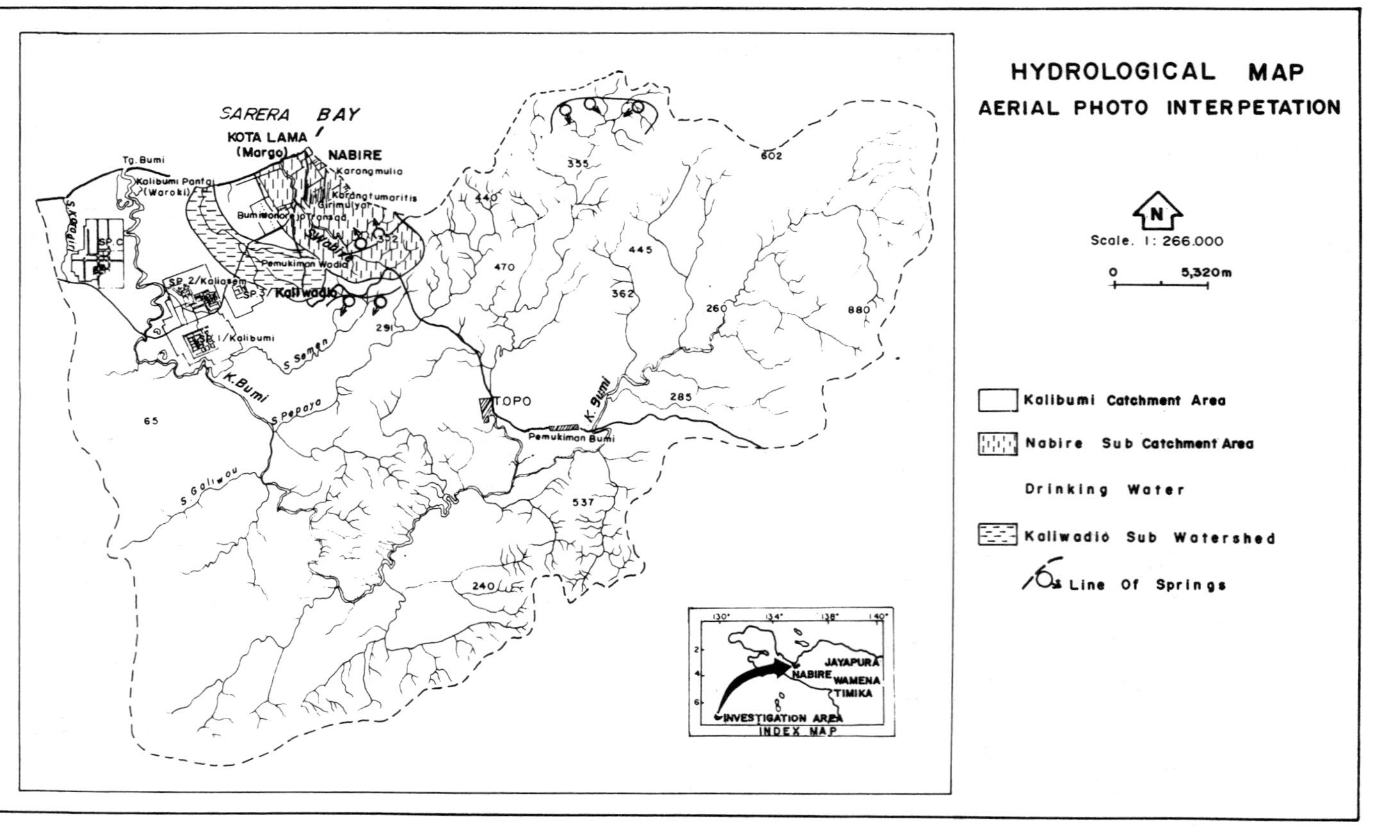

Figure 4. Hydrological Map of Kali Bumi Catchment

Table 3. Effective flood calculation.

Month	In ($x10^3$ mm)	O ($x10^3$ mm)	MWL ($x10^3$ mm)	Effective Flood ($x10^3$ mm)
Jan.	4467.5	176.6	134.0	136.1
Feb.	4089.6	157.5	122.6	128.7
Mar.	4137.5	173.2	124.1	116.4
Apr.	3712.7	167.6	111.3	92.2
May	3182.0	172.4	95.4	50.3
Jun.	2690.2	164.5	80.7	23.7
Jul.	2942.4	167.0	88.2	38.9
Aug.	3072.6	169.2	92.1	45.8
Sep.	3180.2	164.8	95.4	57.7
Oct.	3258.9	175.6	97.7	52.5
Nov.	2592.6	173.8	77.7	7.6
Dec.	2805.1	179.4	84.1	16.8

on the contrary in Kotalama (600 m from the shore line) was 10.18 mg/litre.

3 CONCLUSION

It can be concluded that :

1. Water resources for irrigation purpose can be supplied from Kalibumi river whose flow rate approximately 4 cu.m/sec.
2. No flood hazard will be encountered, as the highest effective flood only 0.316 litre/month in January.
3. Water resources for domestic purpose can be supplied from Nabire river whose flow rate 4 litre/sec and shallow groundwater. This water resource can also be used for drinking water.

REFERENCES

Simoen, S. 1986. The need of water for Indonesia population of the year 2000, Seminar of Geographical Approach for Regional Planning toward 2000, Faculty of Geography UGM, Yogyakarta.

Environmental Management, Geo-Water & Engineering Aspects, Chowdhury & Sivakumar (eds)
© 1993 Balkema, Rotterdam. ISBN 90 5410 099 0

Runoff source areas and water quality

M.G.Sharpin
Gutteridge Haskins and Davey, Sydney, N.S.W., Australia

ABSTRACT: An investigation was undertaken into the influence of stormwater runoff-producing source areas on the characteristics of nutrient export from two rural catchments. It was found that the catchments contained variable extent source areas and exhibited the expected correlation between area size and the proportion of rainfall appearing as runoff. However, the predicted influences of source areas upon the behaviour of nutrients were not observed.

1 INTRODUCTION

The influence of saturated soil zones (or source areas) upon the hydrology of non-urban catchments in humid regions is becoming increasingly recognised. The influence that these areas have upon water quality has, however, been the subject of less investigation. This paper presents a discussion of this issue and describes the results of a study into the influence of source areas on the water quality of two rural catchments near Canberra, Australia.

2 SOURCE AREAS AND HYDROLOGY

Early hydrological concepts of overland flow generation (eg Horton's method) related the soil infiltration capacity to the rainfall intensity during a storm event. However, this proposition has been criticised as failing to account for the overland flow observed under conditions where the theoretical infiltration capacity exceeds the rainfall depth.

It was contended by Kirkby and Chorley (1967) that this overland flow form predominates in arid regions with thin soil and little vegetative cover. At the other end of the spectrum, under humid conditions with substantive soil and vegetation cover, a throughflow (or interflow) model was considered more appropriate. This model accounted for the ability of soils to store rainfall as soil moisture. This throughflow concept was based upon the down-slope movement of infiltrated water through the upper horizons of the soil profile, generally above a less permeable layer. This flow was heavily influenced by the soil moisture conditions within a catchment and their spatial and temporal distribution. This moisture increased in a downslope direction, reaching a maximum adjacent to the stream channel.

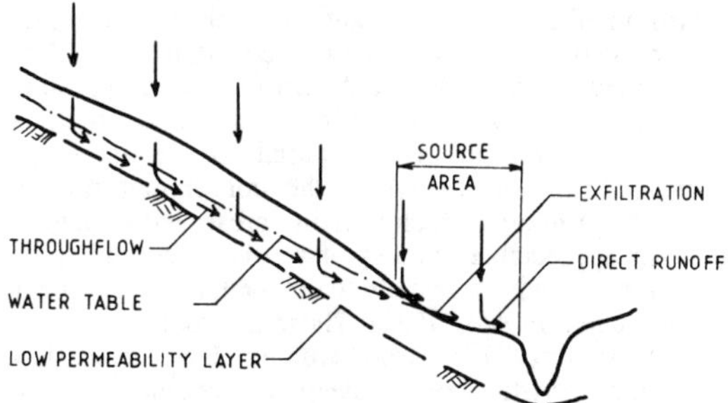

Fig 1: Throughflow model schematic

Under this model, overland flow occurs when the throughflow encounters saturated soil and is forced to the surface (exfiltration) or by runoff from rain falling directly on to the saturated soil. This area of saturated soil was predicted to move up-slope during an extended storm event by temporary storage of throughflow, increasing the proportion of rainfall appearing as runoff. This approach resulted in runoff being produced from a portion of the catchment under conditions whereby the theoretical infiltration capacity of the catchment (as predicted by the Horton model) has not been attained (Figure 1).

The loss of soil moisture from these saturation zones following the passage of a storm event was predicted to occur by throughflow. Soil moisture was reduced in a downslope direction, with the throughflow generating a baseflow from the catchment which may endure for many weeks. Due to the supply of throughflow from up-slope areas, soils adjacent to a watercourse were often found to remain at or near saturation for extended periods.

Another important difference between the two models is in their prediction of flow velocities.

Overland flow velocities tend to be two to three orders of magnitude higher than throughflow velocities. Theoretically, a catchment will therefore reach steady state conditions under overland flow in a considerably shorter time than under throughflow conditions (although equilibrium is rarely attained under the latter conditions).

This early work also recognised a number of catchment characteristics that influence the nature of the throughflow process. These were contour curvature (convergence or divergence of flow), gradient variation (throughflow being highest in concave rather than convex topography), soil thickness (thin soils reaching saturation earlier than thick soils) and proximity to a perennial stream (were saturated soil conditions are common). Algorithms were derived to account for the influence of these parameters on the throughflow.

Since publication of the above paper, numerous researchers have further investigated the throughflow process, although the basic process outlined by Kirkby and Chorley (1967) has been found to be essentially valid. While confirming and expanding upon this process, Dunne et al (1975) also noted that the daily extent of source areas could be related to the baseflow (where baseflow could be regarded as the flow in a stream following cessation of overland flow). Difficulties were also noted in defining variation in the spatial extent of source areas during storm events the consequent rate of runoff, problems which remain unresolved to date.

An appreciable number of source area studies on pastoral and forested catchments have been undertaken in Australasia. These include:

- Mosely (1979), who studied the streamflow characteristics of a steep catchment in New Zealand's (NZ) South Island;
- O'Loughlin (1981), who derived equations for predicting the existence and spatial extent of source areas, and illustrated their applicability for three catchments in the west of Canberra;
- another study in the same area by O'Loughlin et al (1982) who investigated the hydrological response of a catchment to fire;
- an investigation of the hydrological impact of fire and logging four catchments near Eden, New South Wales by Mackay and Cornish (1982);
- Topalidis and Curtis (1982), who investigated the runoff regime in a catchment east of Canberra;
- characterisation of source areas in a catchment near Wellington, NZ, by McColl et al (1985);
- development of a computer model by O'Loughlin (1986) to define source areas in catchments north of Melbourne and west of Canberra;
- Cooke et al (1988), who investigated the streamflow process in a pastoral catchment near Hamilton, NZ.

These studies all confirmed the presence of source areas, which were most notable in flatter catchments. These areas covered up to 100% of the catchments area during storm events on wet catchments, with volumetric runoff coefficients ranging between 0.04 and 0.99.

3 SOURCE AREAS AND WATER QUALITY

Source areas can influence water quality by two primary mechanisms, namely the flooding of soils and generation of overland flow. For the macro-nutrients nitrogen and phosphorus, these mechanisms relate to the adsorption/desorption process.The adsorption of these macronutrients in their common anionic forms (NO_3^-, PO_4^{3-}) can be via non-specific adsorption or ligand exchange (White, 1987). The former can be illustrated by the adsorption of nitrate to an aluminium clay mineral:

$Al(s)\text{-}OH_2 + NO_3^-(aq) \rightarrow Al(s)\text{-}OH_2NO_3(s)$

The latter form relates to the displacement of OH or OH_2^+ ions on a minerals surface by the O in an oxyanion, as illustrated by the adsorption of orthophosphate onto an aluminium clay mineral:

$Al(s)\text{-}OH_2^+ + H_2PO_4^-(aq) \rightarrow Al\text{-}PO_2(OH)_3(s)$

In addition to these adsorption mechanisms, orthophosphate can form insoluble salts with Fe, Al and Ca. This is illustrated for Fe by:

$Fe^{3+}(s) + PO_4^{3-}(aq) \rightarrow FePO_4(s)$

These reactions are dependent upon the soil pH, and can be reversed following changes to this pH. The redox reactions which occur in soil also influence these reactions. During the long term flooding of a soil, anaerobic soil conditions develop and substances other than oxygen accept the free electrons generate by respiration (the decomposition of organic matter by microorganisms). Relevant redox reactions, in order of descending equilibrium redox potential, include:

$O_2 + 4e^- + 4H^+ \rightarrow 2H_2O$

$2NO_3^-(aq) + 12H^+ + 10e^- \rightarrow N_2(g) + 6H_2O$

$MnO_2(s) + 4H^+ + 2e^- \rightarrow Mn^{2+}(aq) + 2H_2O$

$Fe(OH)_3(s) + 3H^+ + e^- \rightarrow Fe^{2+}(aq) + 2H_2O$

The first of these reactions lowers the soil pH, while the second lowers nitrate levels in the soil. The third and fourth reactions cause the gleying of soils, resulting in a mottled (caused by the incomplete reduction of Mn and Fe through the soil profile due to variation in the oxygen condition).

Source areas have a strong influence on the spatial distribution of overland-flow dependant erosion within a catchment. Erosion was considered by Kirkby and Chorley (1967) to be dependant upon soil and topographic conditions, and indirectly by the density of the drainage system. This contrasts with the predicted dominanace of the overland flow path length in the erosion process according to the Horton model. It was noted that Hortonian overland flow (and hence sheet erosion) may occur on catchments dominated by throughflow processes during severe but infrequent storm events, where the soil infiltration capacity is exceeded.

Hence assuming an equal degree of vegetative cover across a catchment, suspended solids entering a watercourse will be derived from this source. From a stream load perspective, this process is further complicated by in-stream sources of suspended solids (from bed or bank erosion and the of particles deposited during the falling limb of the previous storm's hydrograph). Further, erosion can be generated from areas where land use disturbances have occurred (eg earthworks, cultivation) and the dislodgement of particles by the kinetic energy of rainfall becomes important.

Nutrients adsorbed to this particulate matter can be transported by overland flow generated by source areas. Whether the nutrients reach the watercourse relates to the length of the overland flow phase relative to the distance to the stream. However, phosphorus concentrations of this particulate matter can be influenced by whether the overland flow is generated from long or short turn source areas (although the boundary between these areas is indistinct). In long term source areas, the reduction of Fe^{3+} to soluble Fe^{2+} will have reduced the iron concentration in the soil. This in term releases precipitated phosphorus to the soil solution, and hence potentially reduces the amount of particulate phosphorus exported by erosion. In short term source areas, the reduction process will be less important and higher export of particulate phosphorus could be expected.

The filterable (dissolved) phosphorus contained in the soil solution can be readily transported to the stream by throughflow. Phosphorus in this form can, however, be adsorbed by particulate matter being conveyed by the stream or on the stream bed/banks. In isolated source areas resulting from thin soil cover or topographic convergence, such phosphorus may be re-adsorbed by downslope soils during the throughflow process.

Nitrate in soils can be readily leached by throughflow, as the non-specific adsorption is considerably weaker than ligand exchange. It could therefore be expected that the nitrate concentration will be low in long term source areas, due to leaching and the reduction of nitrate to nitrogen gas.

Based upon the above theoretical considerations, it could be expected that the water quality of catchments containing source areas could exhibit the following characteristics:

- nutrient loads from an event occurring on a catchment with a small source area would be lower than those from an equivalent event falling on the catchment with a large source area (as less overland flow will occur);
- dissolved phosphorus loads (concentrations) under base flow conditions and during small events (not generating appreciable overland flow) would form a relatively high proportion of the total phosphorus load. During large events (generating extensive overland flow), particulate phosphorus would comprise a large proportion of total phosphorus loads.
- the percentage of total nitrogen represented by nitrate would be high during baseflows and small events with limited overland flow where only throughflow occurs. This may reverse during large events, when particulate (organic) nitrogen would be transported by overland flow.

However, when concentrations and flows are measured at the outlet of a catchment, these processes may be masked by factors such as:

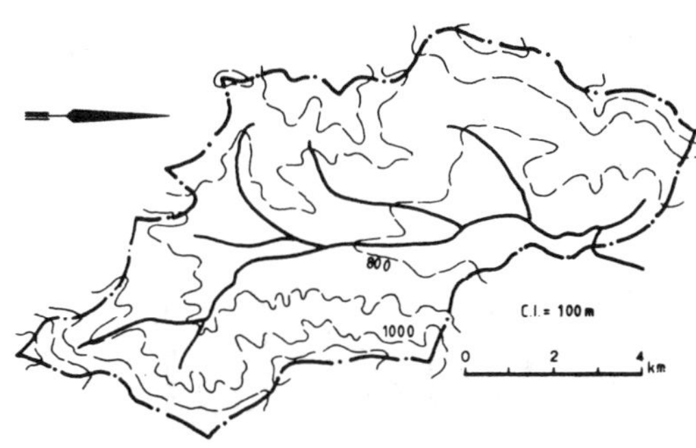

Figure 2: Burra Creek catchment

- spatial/temporal variations in soil geochemistry;
- in-stream biogeochemical processes such as biological uptake, adsorption-desorption, sedimentation-resuspension and erosion;
- land use changes within the catchment, including cyclical agriculture;
- vegetation growth cycles (both terrestial and aquatic), from rapid growth to decay and the influence of climatic conditions on these cycles;
- the availability of nutrients and the efficiency of the overland and stream flow processes to convey the nutrients to the catchment outlet;
- the nature of atmospheric pollutant sources (precipitation and dustfall).

While the importance of source areas in non-point source pollution export has been recognised by a number of reporters, (eg Cullen and O'Loughlin, 1982), only a small number of Australasian case studies have been reported:

- Farmer (1987) found that no long term increase in phosphorus loads occurred when fertiliser was applied to non-source areas in a forested catchment west of Canberra;
- Cooke and Cooper (1988) noted that the majority of phosphorus entering the stream from a pasture catchment near Hamilton, NZ, was from surface runoff derived from source areas;
- Cooke and Dons (1988) found that the majority of the total kjeldahl nitrogen from the above catchment entered the stream by the same mechanism, while nitrate was derived from subsurface flow.

Laboratory studies on the phosphate adsorption characteristics of rice soils were reported by Willett and Higgins (1978). They noted that adsorption was influenced by clay content in non-flooded soils, and the iron oxide content in long term flooded soils. In a later study, Willet (1989) found that the reduction of Fe^{3+} was the dominant cause of phosphorus release in flooded soils.

4 STUDY CATCHMENT DESCRIPTIONS

The catchment of Burra Creek to Burra Rd has an area of 71 km^2, and is located 40 km SW of

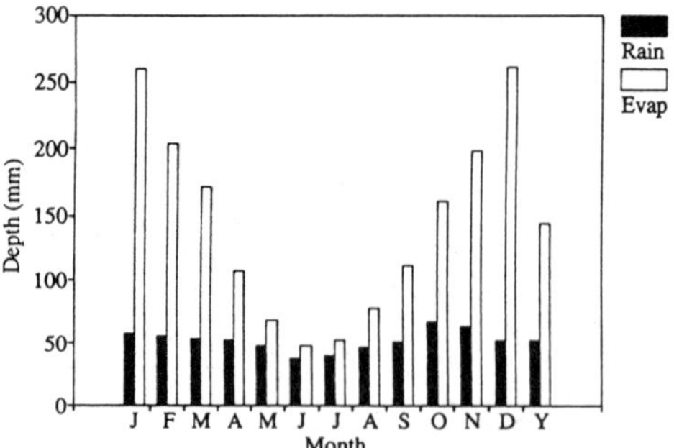

Fig 3: Avg rainfall & pan evap at Canberra Airport

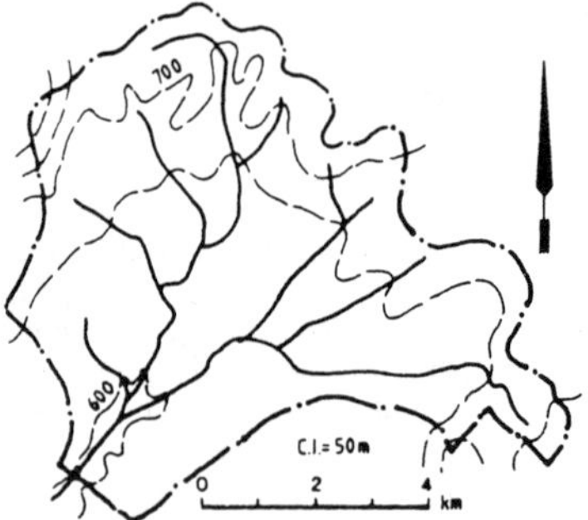

Fig 4: Ginninderra Creek catchment

Canberra. The topography is relatively steep (refer to Figure 2), with an average stream grade of 12% and mid-catchment side slopes of 14%.

Land use west of the creek is pastoral with native eucalypt forest covering the area east of the creek. Soils in the lower portion of the catchment are texture-contrasting red duplex soils and the higher areas are uniform texture earthy loams (Gunn et al, 1969). The average annual rainfall in Canberra is 630 mm and is uniformly distributed throughout the year (Figure 3). Pan evaporation considerably exceeds rainfall in summer but drops appreciably in winter, averaging 1720 mm per annum.

The Ginninderra Creek catchment (to the Barton Highway), located 15 km north of Canberra, has an area of 52 km^2. The topography is flat, with a main channel slope of 1% and mid-catchment valley slopes of 2% (refer to Figure 4).

Land use within the catchment is predominantly pastoral, with some eucalypt forest fringing the northern catchment boundary. The catchment is dominated by yellow duplex soils, with pockets of earthy loams and red duplex soils (Gunn et al, 1969). The presence of gleying in the soils adjacent to watercourses was noted by NCDC (1988), as saturated conditions persist for much of the year due to a relatively impermeable B horizon.

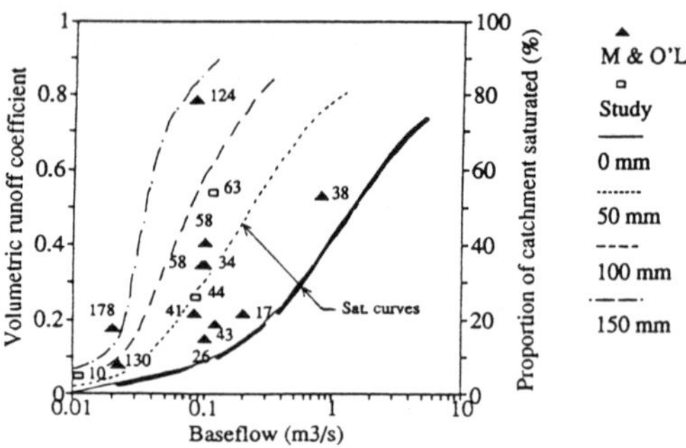

Fig 5: Runoff coefficients for Burra Ck

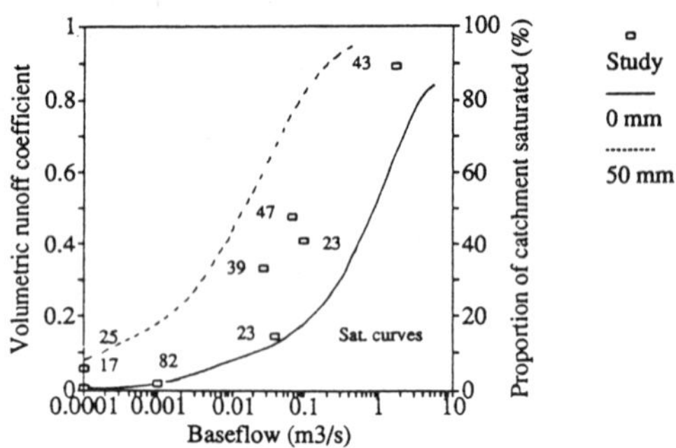

Fig 6: Runoff coefficients for Ginninderra Ck

5 HYDROLOGICAL REGIMES

Stream gauging stations have operated at the outlet of the Burra and Ginninderra Creek catchments for 13 and 7 years respectively. A pluviograph is located at the outlet of Burra Creek and two are located in the Ginninderra Creek catchment (centrally and at the outlet).

The influence of source areas on the runoff characteristics of Burra Creek have been previously reported by Mein and O'Loughlin (1989). They derived relationships between pre-storm baseflow (a surrogate measure of source area size), total rainfall depth and volumetric runoff coefficient. Curves showing the proportion of the catchment covered by source areas for a given rainfall depth were also estimated as a function of baseflow. The resulting graph, with three additional study events included and indicating assumed saturation curves (percentage of catchment covered by source areas) for various rainfall depths, is presented as Figure 5.

A similar analysis was undertaken for eight study events on Ginninderra Creek, the results of which are presented as Figure 6.

For both of these catchments, there appears to be a strong correlation between the size of the source area and the volumetric runoff coefficient, although some minor anomalies do exist (possibly due to the spatial and temporal distribution of the rainfall and source areas). Hence the source area philosophy is

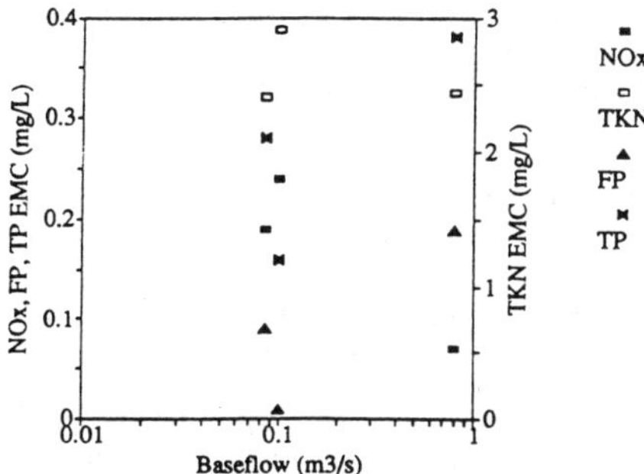

Fig 7: EMC and baseflow - Burra Ck

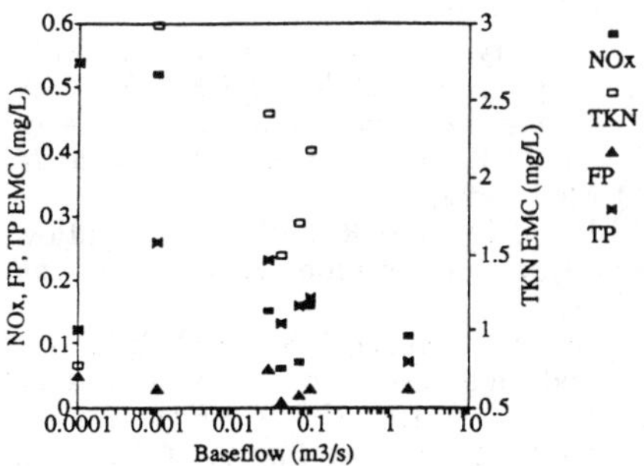

Fig 8: EMC and baseflow - Ginninderra Ck

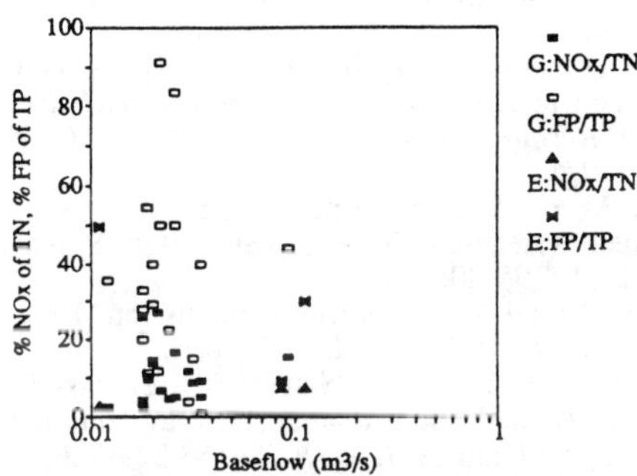

Fig 9: Concentration ratios and baseflow - Burra Ck

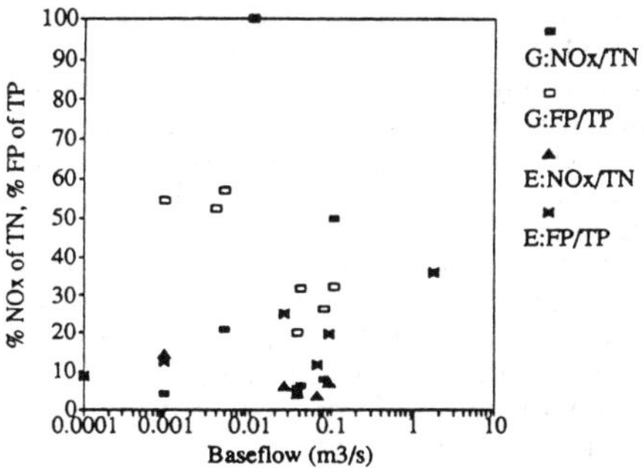

Fig 10: Concentration ratios and baseflow - Ginninderra Ck

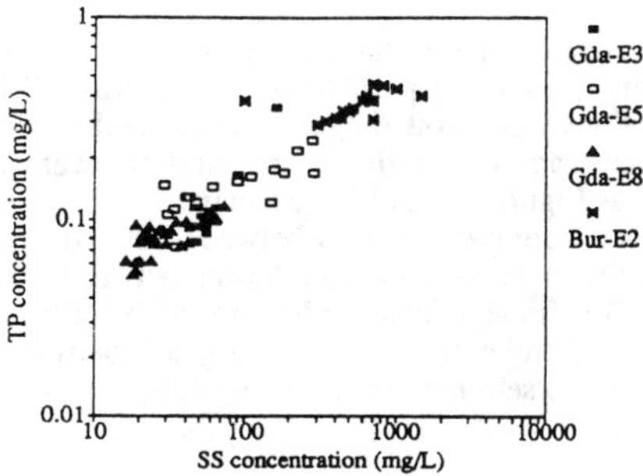

Fig 11: TP and SS concentration

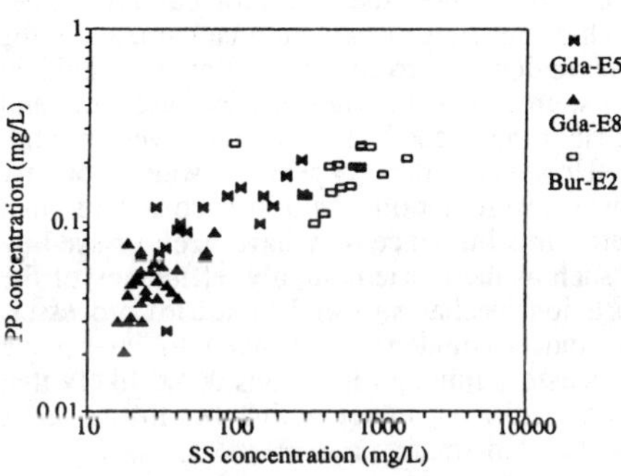

Fig 12: PP and SS concentration

apparently applicable to these catchments and a correlation between source area geochemical characteristics and stream water quality could be expected.

6 NUTRIENT CHARACTERISTICS

Continuous monitoring of nutrients is available for three storm events in Burra Creek and eight events in Gininderra Creek, in the forms of nitrate and nitrite (NO_x), total kjeldahl nitrogen (TKN), filterable phosphorus (FP) and total phosphorus (TP). Total nitrogen can be calculated as the sum of NO_x and TKN. Suspended solids was monitored for four of these events. Additional grab sample results are available for baseflow conditions.

To investigate the influence of source area size on normalised nutrient loads, the event mean concentrations (EMC: total event load divided by total runoff volume) were plotted against baseflow. Figures 7 and 8 are the resulting graphs for Burra and Ginninderra Creeks respectively.

Insufficient data exists in Burra Creek to enable

firm conclusions to be drawn, although no trend is apparent. Ginninderra Creek exhibits a relatively wide scatter, although there is a tendency for the EMC of TKN and TP to decrease with increasing baseflow and for NO_x and FP EMCs to be relatively constant. This is the reverse of the expected relationship.

An evaluation was made of the proportion of dissolved nutrient concentrations occurring under base flow grab sampling (G) and storm events (E) conditions (using EMCs). This is indicated in Figure 9 for Burra Creek and Figure 10 for Ginninderra Creek, plotted against baseflow.

Both graphs exhibit a wide degree of scatter and overlap occurs between the event and grab sample ratios. There is, however, no evident trend for the baseflow nutrients to be in a predominantly dissolved form and for the reverse to apply during storm events.

To investigate the influence of the adsorption phenomenon, total and particulate phosphorus (PP) concentrations were plotted against suspended solids (SS) concentrations for the four common events, presented as Figures 11 and 12 respectively.

There is a strong correlation between TP and SS, and a surprisingly weaker correlation between PP and SS. Insufficient data exists to allow for a meaningful correlation to be investigated between phosphorus and sediment loads during storm events.

7 CONCLUSIONS

An investigation was carried out into the influence of source areas upon the hydrological and water quality characteristics of two rural catchments in the Canberra region. A strong correlation was evident between source area size (defined by baseflow) and volumetric runoff coefficient for a given rainfall depth. However, the expected water quality relationships were not observed. It is considered that any source area influence may have been masked by factors such as the nutrient supply, efficiency of the transportation mechanisms and in-stream processes. Further studies could be undertaken to investigate these processes, although it is considered likely that difficulties will exist in separating the influence of catchment and in-stream processes.

ACKNOWLEDGEMENTS

The data used for this study was obtained from ACT Electricity and Water and the assistance of Ross Knee, Greg Long, Gary Newton and Tony Falkland is gratefully acknowledged. The comments from Ian Jollife on a draft of this paper are also appreciated.

REFERENCES

Cooke, J.G. 1988. Sources and sinks of nutrients in a New Zealand hill pasture catchment, II. Phosphorus, *Hydrol Processes*, 2:123-133.

Cooke, J.G. and Cooper, A.B. 1988. Sources and sinks of nutrients in a New Zealand hill pasture catchment, III. Nitrogen, *Hydrol Processes*, 2:135-149.

Cooke, J.G. and Dons, T. 1988. Sources and sinks of nutrients in a New Zealand hill pasture catchment, I. Streamflow generation, *Hydrol Processes*, 2:109-122.

Cullen P. and O'Loughlin, E.M. 1982. Non-point sources of pollution, *Prediction in Water Quality*, Aust Acad of Sci.

Dunne, T., Moore, T.R. and Taylor, C.H. 1975. Recognition and prediction of runoff-producing zones in humid regions, *Hydrol Sci Bull*, 20(3):305-327.

Farmer, N.R. 1987. *The impact of fertiliser application on phosphorus exports from a radiata pine forested catchment*, MAppSc thesis, Canberra Coll Advan Ed.

Gunn, R.H., Story, R., Galloway, R.W., Duffy, P.J.B., Yapp, G.A. and McAlpine, J.R. 1969. *Lands of the Queanbeyan-Shoalhaven Area, A.C.T. and N.S.W.*, Land Research Series No 24, CSIRO, Melbourne.

Kirkby, M.J. and Chorley, R.J. 1967. Throughflow, overland flow and erosion, *Hydrol Sci Bull*, 12(3):5-21.

Mackay, S.M. and Cornish, P.M. 1982. Effects of wildfire and logging on the hydrology of small catchments near Eden, N.S.W., *Proc 1st Nat Symp Forest Hydrol*, Inst Eng Aust Nat Conf Publn 82/6.

McColl, R.H.S., McQueen, D.J., Gibson, A.R. and Heine, J.C. 1985. Source areas of storm runoff in a pasture catchment, *Jour Hydrol New Zealand*, 24:1-19.

Mein, R.G. and O'Loughlin, E.M. 1991. A new approach to flood forecasting, *Proc Intl Hydrol Water Resour Symp*, Inst Eng Aust Nat Conf Publn 91/4.

Mosley, M.A. 1979. Streamflow generatiom in a forested watershed, New Zealand. *Wat Resour Res*, 15(4):795-806.

National Capital Development Commission 1988. *Gunghalin Draft EIS*, NCDC, Canberra.

O'Loughlin, E.M. 1981. Saturation regions in catchments and their relationships to soil and topographic features, *Jour Hydrol*, 53:229-246.

O'Loughlin, E.M., Cheney, N.P. and Burns, J. 1982. The Bushrangers experiment: hydrological response of a eucalypt catchment to fire, *Proc 1st Nat Symp Forest Hydrol*, Inst Eng Aust Nat Conf Publn 82/6.

O'Loughlin, E.M. 1986. Prediction of surface saturation zones in natural catchments by topographic analysis, *Wat Resour Res*, 22(5):794-804.

Topalidis, S. and Curtis, A.A. 1982. The effect of antecedent soil water conditions and rainfall variations on runoff generation in a small eucalypt catchment, *Proc 1st Nat Symp Forest Hydrol*, Inst Eng Aust Nat Conf Publn 82/6.

White, R.E. 1987. Introduction to the principles and practice of soil science, 2nd edition, Blackwell

Scientific Publications, Oxford.
Willet, I.R. 1989. Causes and prediction of changes in extractable phosphorus during flooding, *Aust J Soil Res*, 27:45-54.
Willet, I.R. and Higgins, M.L. 1978. Phosphate sorption by reduced and reoxidised rice soils, *Aust J Soil Res*, 16: 319-326.

Environmental Management, Geo-Water & Engineering Aspects, Chowdhury & Sivakumar (eds)
© 1993 Balkema, Rotterdam. ISBN 90 5410 099 0

Reservoir as a tool for brackish water management: A case study in southern Iran

K. Shiati
Yekom Consulting Engineers, Tehran, Iran

ABSTRACT: The model DYRESM was used to investigate the response of the planned Jarreh Reservoir (Iran) to various management strategies. Among various strategies investigated, the diversion of the most saline part of the inflow will result in significant water quality improvements of both offtake and reservoir. The adopted management policy (diversion of inflows with a salinity greater than 3000 ppm) reduced the total salt accumulation in the reservoir by $72*10^3$ kg which is 8.5% improvement on the " no policy " results.

1 INTRODUCTION

In several arid and semi-arid countries, the main water resources for regional development are from reservoirs constructed on brackish sediment-laden rivers. Such salinity is caused by different processes, and results in a rise in salinity of the water stored. The presence of saline inflows requires a careful management of such reservoirs.

The models for a stratified reservoir, can be used to predict the effects of various measures to be taken. Such simulation studies have been used to find appropriate strategies for the future Jarreh Reservoir in Southern Iran. This storage dam with a capacity of 470 Mm^3 on the brackish Shapur river will be constructed to irrigate 13000 ha of a coastal plain. The catchment of Shapur river is suffering from salinity caused by natural processes (Shiati, 1989). Due to this, approximately $1.2*10^6$ tons of salt are brought into the reservoir in an average year. The river discharge and salinity of the Shapur river at the damsite (Jarreh Station) are shown for the period 1975-1990 in Fig. 1. It shows that the salinity (TDS) of the river is strongly influenced by both long term and seasonal variations in stream flow. For example, the high annual inflows of 1982 (discharge 1127 Mm^3, 2.12 times the median) averaged 1500 mg/l TDS whereas the inflows of 1984, a dry year (discharge 280 Mm^3, 0.54 times the median) averaged 2180 mg/l in salinity. On a seasonal basis, the flood flows (winter period) are generally less saline. Low flows (summer period), however, are in part due to groundwater flows and remain highly saline.

In reservoirs, the temporal variation of salt content as shown in Fig. 1 is partially alleviated, and the water quality is mainly controlled by stratification phenomena and retention time. It is the aim of this study to show how appropriate management strategies can improve the quality of the water to be withdrawn and at the same time minimize the risk of a salinity build-up in the reservoir. Such strategies might be developed by studying the interaction between inflow characteristics, withdrawal policy and stratification in the reservoir.

The dynamic reservoir simulation model DYRESM as developed by Imberger et al (1978) is used to investigate the response of the Jarreh Reservoir to various management strategies. Furthermore, it is used to simulate the salt build-up in the reservoir over a long time span. For the former purpose, a five-year period (1982-1986) of daily inflow and meteorological data and for the latter a fifteen-year period (1975-1990) of daily inflow data and an average year of meteorological data (1986) are used.

2 STRATEGIES AND MEASURES

Among the factors which influence the reservoir stratification (meterological, inflow and outflow) only inflows and outflows can be controlled. As a result, only those management options are investigated that are based on the manipulation of both inflow and outflow. Inflow manipulation is achieved by diverting the most saline river inflows (summer inflows) before they reach the body of the reservoir ("by-pass policy"). Outflow manipulation

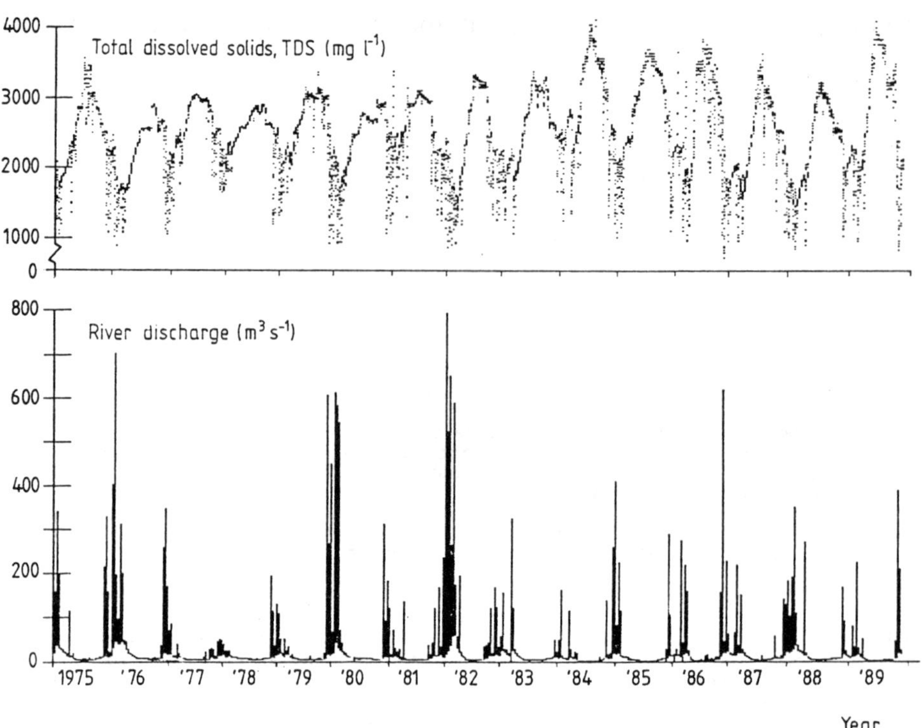

Fig. 1 Distribution of river discharge and salinity of the Shapur river at Jarreh Station (Iran).

includes release of excess water (subject to safe yield constraints) from the most saline part of the reservoir at appropriate outlets and appropriate times ("scouring policy").

Different strategies of course depend on the management objective; solely to improve the offtake salinity, and/or to improve the average salinity in the reservoir. The objective for Jarreh Reservoir is primarily to maintain the status quo or, if possible, minimize the salinity build-up in the reservoir. The quality of the irrigation supply is not a primary objective because on one hand any improvement in supply salinity is not going to be significant and on the other hand it will be of little importance for the moderately salt tolerant crops that are already being cultivated in the study area.

A general conclusion on the aspects of reservoir management cannot be given; every reservoir of this kind will require its specific management. For example, the results of adopted options for the Jarreh Reservoir show some similarities but also differences with Wellington Reservoir, W. Australia (Patterson et al., 1978; Fischer et al., 1979; Imberger, 1981).

3 DETERMINATION OF EXCESS WATER FOR BY-PASSING OR SCOURING

The amount of water that can be diverted or released should be consistent with the demand, inflow and the storage volume. Long-term simulation may lead to the development of operation rules to minimize the amount of spillage and maximize the available water for scouring or diversion (e.g. Loh and Hewer, 1977; Yeh, 1985; He, 1987). Long-term simulation might also lead to improved water quality at the expense of supply quantity.

It should be noted that the simulation techniques used to derive reservoir operating rules should include quality criteria. At present little work has been done to include water quality in the optimization procedure. Instead the state of reservoir system is simply described in terms of a single variable, water volume. The above-mentioned problem i.e. how to couple an optimization algorithm with a simulation model to assess the water quality and account for multiple objectives, are the core of a research is currently undertaken for the Jarreh Reservoir, Iran (Bogardi, 1990).

At present in the Jarreh Reservoir,

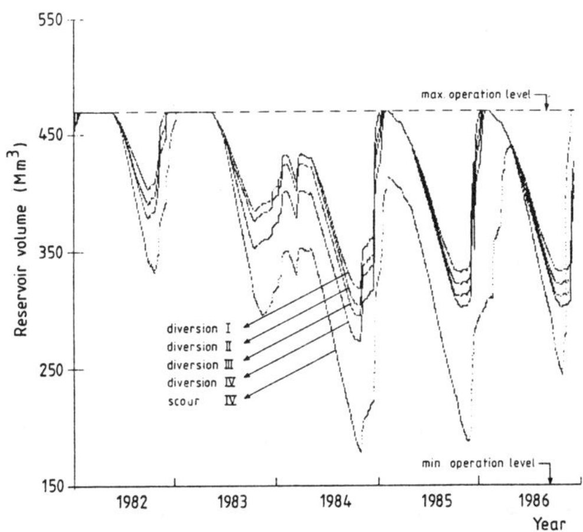

Fig. 2 Simulated Jarreh Reservoir volume corresponding to different diversion volumes.

for the period 1982/1986, different values for diversion or scouring are examined, using a simulation program, RELEAS (Shiati and Torfs, see Fig.2).

Various strategies were investigated with this model, to wit:

- DIVERSION I to IV: bypassing of summer flows with high salinity to a point downstream of the last irrigation intakes.

DIVERSION I: diversion of flows with salinity of 3600 ppm TDS and over;
DIVERSION II: same, with cutoff level 3400 ppm;
DIVERSION III: same, with cutoff level 3200 ppm;
DIVERSION IV: same, with cutoff level 3000 ppm.

- SCOUR I to IV: selective withdrawal of water from the reservoir.

SCOUR I: in summer, through midlevel outlet at height 52 m;
SCOUR II: in summer, through midlevel outlet at height 67 m;
SCOUR IIa: in late summer to early autumn, at height 67 m;
SCOUR III: in summer, through top outlet at height 78 m;
SCOUR IV: combination of SCOUR IIa and DIVERSION IV.

For the amounts and times of diversion and scouring see Table 1.

4 THE EFFECT OF BY-PASSING POLICY

The simulated reservoir volume corresponding to diverting all inflows with a salinity greater than 3000, 3200, 3400, and 3600 ppm is shown in Fig. 2, Diversion I-IV. As can be observed, for the simulated period (1982/1986) an amount of 158 Mm^3 (corresponding with cutoff salinity 3000 ppm) can be diverted without adversely affecting the reliability of supply. The corresponding volumes to be diverted, the period of diversion and the simulation results for some specific days of the simulation period are shown in Table 1. Fig. 4 shows the salt content of the reservoir (kg) and the salinity (TDS) of the withdraw water, respectively. These results show that a diversion policy will have a major effect on the salinity of the withdraw water as well as on the overall salinity of the reservoir. When a cutoff salinity of 3000 mg/l is examined (diversion IV) the average salinity of the reservoir on day 84295 is reduced by 7.3% (equivalent to 172 mg/l TDS) and the salinity of released water by 4.6% (equivalent to 100 mg/l TDS). The maximum reduction in supplied salinity (18 %) occurs on day 853-54, with also 8.7% reduction in average salinity of the reservoir equivalent to 387 and 198 mg/l TDS, respectively. At the end of this simulation the total salt accumulation in the reservoir is reduced by $72*10^3$ tons which is 8.5% improvement on the "no policy" results.

The price paid for this reduction in salinity is a diversion of $158*10^6$ m^3 of the river flow during the periods shown in Table 1.

5 THE EFFECT OF SCOURING POLICY

Apart from diversion, advantage can be taken of strong summer stratification in Jarreh Reservoir. In summer, a thin layer of warm and brackish water will form at the top, above the level of the thermocline. This brackish water layer persists up to late autumn, when a turnover occurs and the water in the reservoir will be mixed. The development of this strong stratification may be used to our advantage by selective withdrawal of this layer. In this respect, much will depend on the flexibility of offtake level selection.

Based on the stratification and a strategy of selective withdrawal, some of the most saline part of the reservoir water can be removed in summer (scoured) before internal mixing distributes it over the remainder of the storage.

A number of options are examined. Furthermore, the policy is tested for the effects of the timing of the scour and the level at which withdrawal take place.

The simulation results are also summarized in Table 1. The results are also compared with the simulation results of "no policy" option and the percentage of improvement is cited. For all type of poli-

Table 1 Effect of management policies in the Jarreh Reservoir, southern Iran

	Policy	Diversion/ Scouring Period	Diversion Volume 10^6 m3	Scouring Rate m^3/d/1000	(82001) S.S. PPM	(82001) R.S.C. 10^8 kg	(82001) A.R.S. PPM	(82275) S.S. PPM	(82275) R.S.C. 10^8 kg	(82275) A.R.S. PPM	(83295) S.S. PPM	(83295) R.S.C. 10^8 kg	(83295) A.R.S. PPM	(84295) S.S. PPM	(84295) R.S.C. 10^8 kg	(84295) A.R.S. PPM	(85354) S.S. PPM	(85354) R.S.C. 10^8 kg	(85354) A.R.S. PPM	(86365) S.S. PPM	(86365) R.S.C. 10^8 kg	(86365) A.R.S. PPM
	No Policy	-	-	-	1864	842	1859	1470	683	1768	1810	781	2093	2176	765	2381	2349	867	2273	1804	850	.1808
DIVERSION POLICY	Diversion I	84177-235 85184-231 86185-190 86204-218	7.9 7.7 0.8 2.1	- - - -	1864	842	1859	1470	683	1768	1810	781	2093	2176	735	2345 (1.5)	2036 (13.3)	836	2238 (1.5)	1776 (1.5)	840	1788 (1.1)
DIVERSION POLICY	Diversion II	84147-283 85163-265 86149-190 86185-236	22.8 18.8 2.3 6.1	- - - -	1864	842	1859	1470	683	1768	1810	781	2093	2176	683	2287 (4)	1997 (15)	790	2177 (4.3)	1738 (3.6)	817	1739 (3.8)
DIVERSION POLICY	Diversion III	82181-231 83184-210 83266-281 84139-283 85149-294 86142-276	12.9 5.5 4.0 24.9 29.1 24.7	- - - - - -	1864	842	1859	1470	641	1716	1788 (1.2)	744	2045 (2.4)	2142 (1.5)	648	2253 (5.4)	1984 (15.5)	750	2181 (4.1)	1701 (5.7)	798	1699 (6.1)
DIVERSION POLICY	Diversion IV	82173-265 83181-290 84139-283 85137-315 86142-294	24.9 32.3 24.9 39.4 37.3	- - - - -	1864	842	1859	1471	603	1669 (5.6)	1763 (2.6)	669	1960 (6.4)	2076 (4.6)	589	2209 (7.3)	1962 (18)	711	2075 (8.7)	1662 (8)	778	1656 (8.5)
SCOURING POLICY	Scour I	82230-320 83260-350 84270-360 85270-360 86250-340	- - - - -	277 359 277 438 414	1864	842	1859	1471	664	1778	1817	760	2108	2193	685	2416	2362	782	2270	1763 (2.2)	831	1768 (2.2)
SCOURING POLICY	Scour II	82230-320 83260-350 84270-360 85220-360 86250-340	- - - - -	277 359 277 438 414	1864	842	1859	1459	662	1773	1796	757	2100	2177	679	2398	2036 (13.3)	774	2249 (1.1)	1758 (2.6)	828	1762 (2.6)
SCOURING POLICY	Scour IIa	82275-320 83300-345 84280-325 85300-345 86290-335	- - - - -	554 718 555 876 829	1864	842	1859	1460	680	1764	1812	780	2090	2160	672	2389	2033 (13.4)	763	2237 (1.6)	1751 (3)	824	1753 (3.1)
SCOURING POLICY	Scour III	82180-270 83200-290 84170-280 85200-300 86200-300	- - - - -	273 359 277 394 373	1864	842	1859	1477	638	1765	1824	711	2063 (1.5)	2145 (1.5)	636	2308 (3.1)	2068 (12)	765	2195 (3.4)	1720 (4.6)	824	1724 (4.6)
SCOURING POLICY	Scour IV		24.9 32.3 24.9 39.4 37.3	554 718 555 876 829	1864	842	1859	1537	601	1728 (2.3)	1829	661	2036 (2.9)	2129 (2.2)	502	2290 (3.8)	2220 (5.5)	557	2095 (7.8)	1579 (12.4)	733	1560 (13.7)

1) S.S. - Supply Salinity R.S.L. - Reservoir Salt Content A.R.S. - Average Reservoir Salinity
2) In brackets are percentage improvements

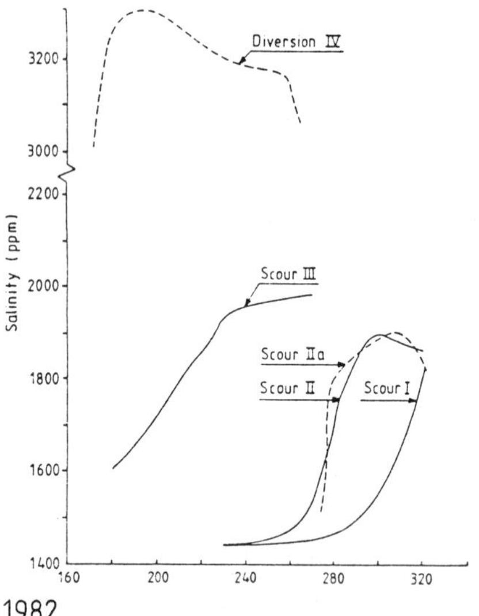

1982

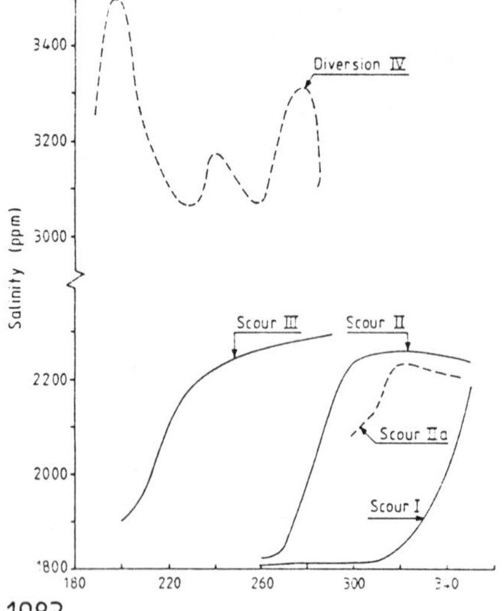

1983

cies tested, the irrigation water is supplied from the bottom outlet (height 32 m) and an amount of $153*10^6$ m^3 of water (the same amount as diversion IV) is scoured.

The first class of policies tested, SCOUR I and SCOUR II, is to remove the top brackish water layer during mid summer up to mid autumn in 90 days, before the occurrence of turnover in the reservoir. In SCOUR I, scour takes place from the midle-

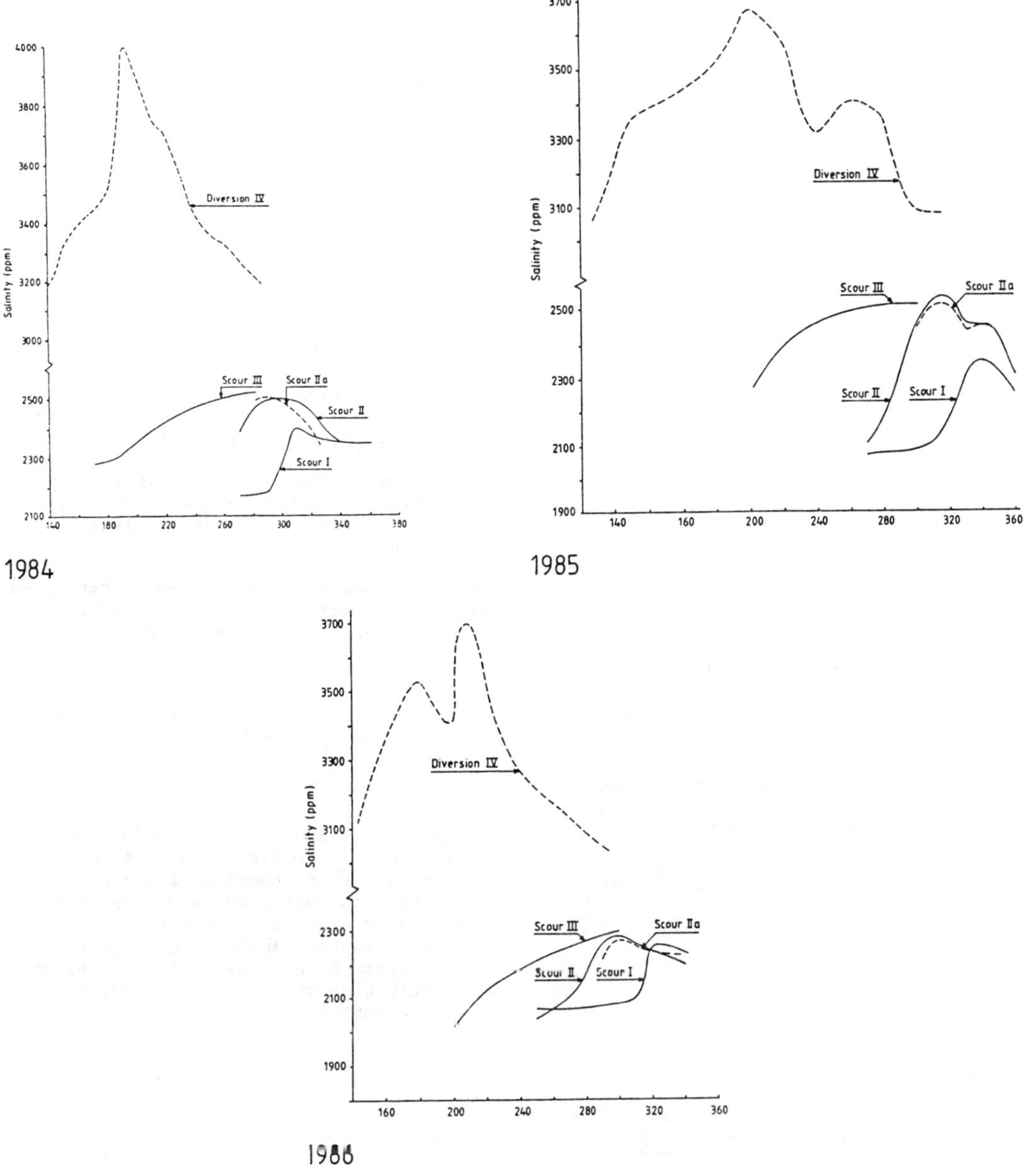

Fig. 3 Scouring policy, the salinity of released water.

vel outlet at height 52 m, as was envisaged in the feasibility design of the Jarreh dam. However, the final design of the dam fixed the midlevel outlet at a height of 67 m above the base of the dam. SCOUR II simulates scouring at this level. SCOUR IIa shows the effect of the timing of the operation: it was delayed to 45 days after all salt wedges have been fully inserted and after the depth of the brackish layer has deepened with the cooling at the surface and by the increasing wind effect.

The second class of policies tested, SCOUR III, involves the removal of the most saline wedges at the top (about 78 m from the base, see Fig. 3), before they participate in mixing. SCOUR IV examines the combined effect of diverting the inflows with salinity greater than 3000 ppm (Diversion IV) and SCOUR IIa. In this case

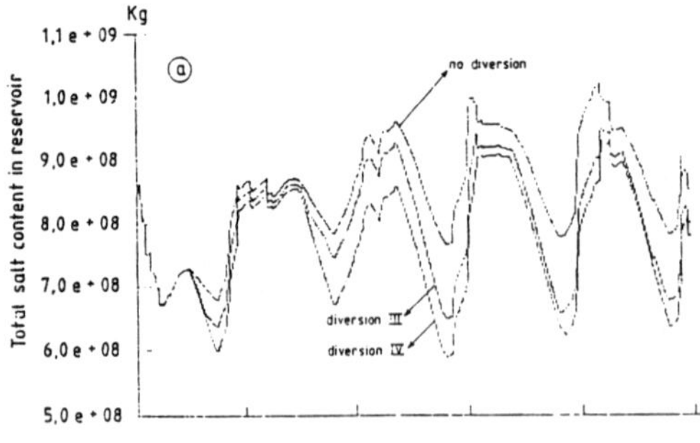

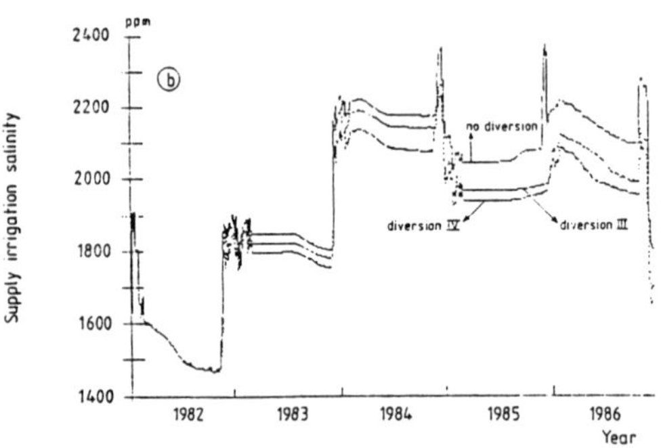

Fig. 4 Simulated results of diversion policy; (a) total salt content of reservoir (kg), (b) supply irrigation salinity (ppm).

Table 2 Water and salt balance for Jarreh reservoir, simulation period 1975-1990 (withdrawal from bottom outlet)

	water (10^6 * m^3)	salt concentration (ppm)	salt content (10^6 *kg)
Reservoir on day 0	450	1850	832
Inflow	8404		14820
Spill	3787		6740
Irrigation	4022		80204
Evaporation excess	574		
Reservoir on day 89365	470	Ave. salinity =1887	887

a quantity of 316 $*10^6$ m^3 is removed, which leads to intolerably low reservoir levels (Fig. 2, SCOUR IV).

The effectiveness of the scouring policy is shown in Fig. 3, where the salinity of the scoured water is illustrated. It is evident that this salinity is consistently less than the salinity of the diverted water, indicating that the policy is not as effective as diversion. The maximum salinity of the scoured water is 2540 mg/l compared with 4000 mg/l of diverted flows.

6 SALINITY TREND IN THE RESERVOIR

It is important to study the behaviour of a salt-affected reservoir on a long-term basis. For this purpose, a salt balance calculation of the reservoir is necessary.

- the water balance in reservoir reads:

$$Q_i = Q_u + B + dV/dt \qquad (6.1)$$

- the salt balance in reservoir reads:

$$Q_i\ C_i = Q_u\ C_u + dZ/dt \qquad (6.2)$$

where
Q_i = inflow (m^3/s)
Q_u = outflow (m^3/s)
B = evaporative excess (m^3/s)
V = volume of water in reservoir (m^3)
C = salt concentration in the reservoir (kg/m^3)
C_i = salt concentration of inflow (kg/m^3)
C_u = salt concentration of outflow(kg/m^3)
Z = salt stored in reservoir (kg)
t = time (s)

For a completely mixed reservoir ($C=C_u$), an analytical solution can be obtained from eqs. (6.1) and (6.2) (see van der Molen, 1980, eq. 8):

$$C=\frac{Q_iC_i}{Q_i-B}+[C_0-\frac{Q_iC_i}{Q_i-B}]\cdot[\frac{V_0}{V_0+(Q_i-B-Q_u)\,t}]^{\frac{Q_i-B}{Q_i-B-Q_u}} \qquad (6.3)$$

However, for a stratified reservoir eq. (6.3) cannot be applied because of $C \neq C_u$. For such a case a numerical solution is often used. In this study a module is added to DYRESM to calculate the total mass content, salt content and average salinity of reservoir as well as outlets water at each daily time step. The following scheme is employed.

$$\text{Total water mass} = \sum_1^{NS} (V_i*\rho_i) \qquad (6.4)$$

$$\text{Total salt content} = \sum_1^{NS} (S_i*V_i*\rho_i) \qquad (6.5)$$

$$\text{Average salinity} = \frac{\sum_1^{NS} (S_i*V_i*\rho_i)}{\sum_1^{NS} (V_i*\rho_i)} \qquad (6.6)$$

where

V_i = volume of the i-layer (m^3)
ρ_i = density of the i-layer (kg/m^3)
S_i = salinity of the i-layer (ppm), and i= 1, NS

The salinity trend in the Jarreh Reservoir is obtained by using the 15-year daily inflow data. The results of this simulation are shown as total salt content and average salinity of the reservoir, Fig. 5, and water and salt balance at the end of the simulation period, Table 2. It is evident from Figs. 5 and 1 that the salt content of the reservoir is primarily determined by the annual variability of inflow. The high and fresh flows of the wet years 1975-76, 1979-80 and 1987-88 flushed out the reservoir and reduced the salt content to its lowest level of about $7.5 * 10^8$ kg and the average salinity to 1550 ppm TDS. On the other hand, the low and saline flows of the dry years 1977-78, 1978-79 and 1983-1984 increased the salt content of the reservoir up to about $1.1 * 10^9$ kg and an average salinity to 2600 ppm.

7 CONCLUSIONS

It is clear that the diversion of the most saline part of the inflow will result in significant water quality improvements of both offtake and reservoir. Implementing such a policy for the Jarreh Reservoir does not involve severe engineering constraints.

Control by scouring is less effective at the presently planned offtake levels. However, an alteration of the outlet structures, or use of a multi-level offtake structure would result in more effective implementation of this policy. Nevertheless, this policy will remain always less effective than the diversion policy. Scouring the salt wedges in autumn, is a standard procedure in the management of salinity in Wellington Reservoir, W. Australia (Fischer et al., 1979; Imberger, 1981). This is because for the Australian case the high salinity inflow water is usually cold. Hence saline water lodges in the deeper places and can be easily scoured through the bottom outlet. The above mentioned condition is not to be expected for Jarreh Reservoir, where the most saline inflows remain near the surface at the level of thermocline.

In Fig. 6 the average salinity of the reservoir operating policies (with and without implementing the management policy) are compared with the original river salinity for the period 1982-1986. It is evident that the reservoir has alleviated the consequences of extreme river salinity (up to 4000 mg l^{-1}) to an extent acceptable for moderately salt tolerant crops (up to 2400 mg l^{-1}). In this period the reservoir salinity varies in a range between 1450 and 2415 mg l^{-1}. The adapted management policy (diversion of inflows with a salinity greater than 3000 ppm) reduces this figure to about 1430 to 2200 mg l^{-1}.

The salinity trend in Jarreh Reservoir is

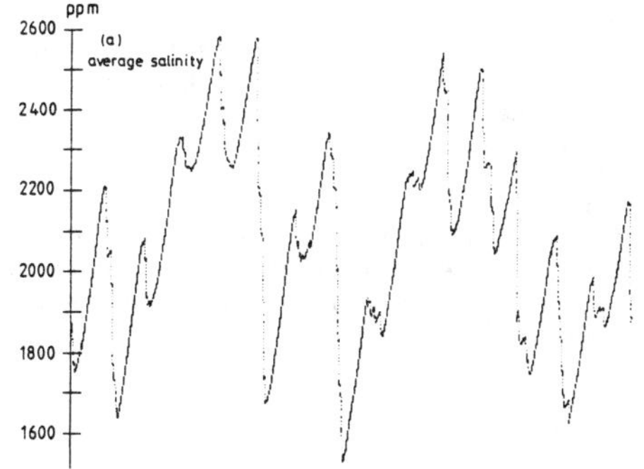

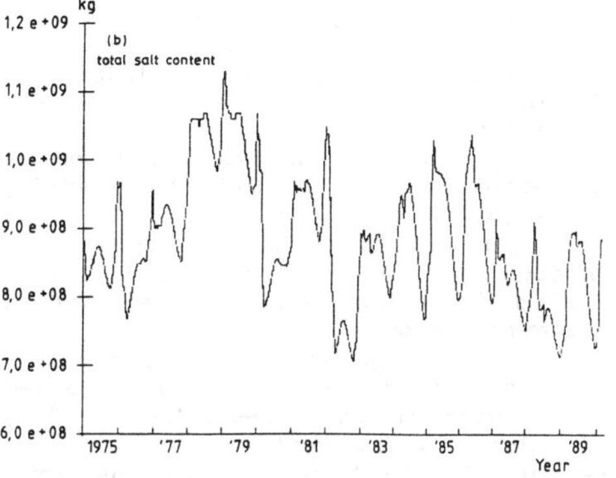

Fig. 5 Salinity trend in Jarreh R.; (a) average reservoir salinity (ppm), (b) total salt content (kg).

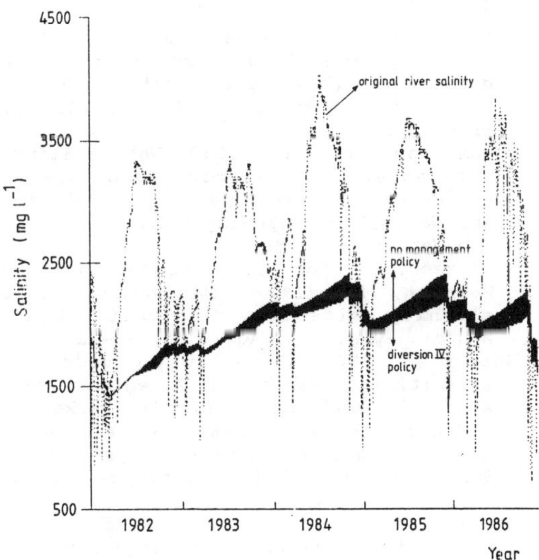

Fig. 6 Comparison between the original river salinity and average reservoir salinity with and without management policy.

determined mostly by annual variability in river discharge. The high and fresh flows of the wet years flushed out the reservoir, whereas the low and salty flows of the dry years deteriorate the quality. This indicates that an equilibrium condition hardly will be acheived, but that the reservoir salinity will vary within a range. For the period 1975-1990 the average reservoir salinity, without management policies varies between 1500-2600 ppm.

REFERENCES

Bogardi, J. J., 1990. Reservoir management under consideration of stratification and hydraulic phenomena. Research Proposal to PhD Program, Dept. of Hydrology, Soil Physics and Hydrology, Wageningen Agric. Univ., The Netherlands.

Crank, J., 1967. The mathematics of diffusion. vi+347 pp. Oxford University Press (p. 31, eq. 3.13).

Fischer, H. B., List, E. J., Koh, R. C., Imberger, J. and Brooks, N. H. 1979. Mixing in inland and coastal water. Academic Press .

He, Q., 1987. Operation policies for a three-reservoir system in Tianjin area, North China. M. Eng. Thesis No. WA-87-7, AIT, Bangkok, Thailand.

Imberger, J., Patterson, J.C., Hebbert, R.H.B., and Loh, I., 1978. Dynamics of reservoir of medium size. J. Hydraul. Div., ASCE, 104:725-743.

Imberger, J., 1981. The influence of stream salinity on reservoir water quality. Ag ric. Wat. Manage., 4:255-273.

Loh, I. C., and Hewer, R. A., 1977. Salinity and flow simulation of a catchment reservoir system. Proc. Hydrology Symposium, I. E. Aust., Brisbane.

Molen, van der W. H., 1980. Waterquality: influence of transport and mixing processes. Lecture Notes, Dept. of Land and Water Use, Wageningen Agric. Univ., The Netherlands, 88p.

Patterson, J., Loh, I., Imberger, J., and Hebbert, B., 1978. Management of a salinity affected reservoir. Proc. Hydrol. Symp. I.E. Aust. pp. 18-22.

Shiati, K., 1989. The origin and source of salinity in river basin: A regional case study in southern Iran. In: S.Ragone (Editor). Regional Characterization of Water Quality. Proceedings of a Symposium, May 1989, Baltimore, IAHS, 182:201-210.

Shiati, K., Torfs, P. J. J. F. 1990. Simulation of reservoir storage level by model RELEAS. Internal Note, Dept. of Hydrology, Soil Physics and Hydraulics, Wageningen Agric. Univ., The Netherlands.

Volker, A., 1942. Voorloopige resultaten en conclusies van een onderzoek naar den invloed van grondwaterstroomen op de chloorgehalten van het grondwater in de IJsselmeerkom. Unpublished Report, Dienst der Zuiderzeewerken (in Dutch).

Yeh, W. W-G., 1985. Reservoir management and operations models: A state-of-the-art review. Water Resour. Res. 12:1797-1818.

Environmental Management, Geo-Water & Engineering Aspects, Chowdhury & Sivakumar (eds)
© 1993 Balkema, Rotterdam. ISBN 90 5410 099 0

Toxic substances model of Tuen Mun River, Hong Kong

Y.S.Sin
Department of Civil & Structural Engineering, Hong Kong Polytechnic, Hong Kong

ABSTRACT: Along the stretch of the Tuen Mun River in Hong Kong, industrial pollution is evident by the elevated levels of heavy metals such as Boron, Manganese and Iron. A tidally averaged segment box model for the analysis of the heavy metals dispersion in the Tuen Mun River is developed. The river is assumed to be one-dimensional and is divided into a number of segments. Each segment is assumed to be completely mixed. It is based on the conservation of the pollutant mass over a segment, and the pollutant transport due to convection, tidal dispersion , decay, bottom diffusion, sources and sinks are considered in the analysis. The dispersion coefficients are obtained by measuring the averaged salinities of the river segments. For validation purposes, dissolved oxygen and BOD concentrations are also predicted based on industrial as well as municipal organic carbon loads. Predicted in-stream dissolved oxygen and BOD levels are compared to measured field data resulting in successful model validation. The validated models are then used to predict the heavy metals concentrations in the Tuen Mun River. It is concluded that the model developed is a simple and useful tool for long-term planning purposes of the water quality of a one-dimensional tidal river.

1. INTRODUCTION

The main channel of Tuen Mun River is about 4 km long. It flows through the Tuen Mun New Town and drains into the Castle Peak Bay. Livestock waste and domestic sewage are the major pollution sources. Along the stretch of the main river channel, industrial pollution was also evident by the elevated high levels of heavy metals such as Boron, Manganese and Iron (EPD 1989). Since the Tuen Mun River is subjected to the tidal flushing effect by the Castle Peak Bay, the values of the hydraulic parameters, sources and sinks of the constituents as well as the boundary conditions may vary within a tidal cycle. A finite difference method using explicit scheme (Sin 1991 & 1992) can be adopted for solving the advection-diffusion equation. The tidally varying water quality parameters are determined by assuming that a quasi-steady state condition is reached. However, a large amount of information on the intra-tidal variation of water quality must be obtained for the validation of the model. Due to the presence of limited field data in this study, a tidally-averaged model is attempted to describe the tidally averaged water quality of the Tuen Mun River. Some of the rate coefficients are obtained by tuning the model parameters with the field data.

2. MODELLING EQUATIONS

The methodology used by Officer (1980) for the investigation of nonconservative quantities in a one-dimensional estuary is adopted for this study. The estuary is divided into segments within each of which the water quality is assumed to be well mixed. The nomenclature and box enumeration system is given in Fig. 1. The boundaries of the boxes are the estuary bottom configuration, the water surface, and vertical sections normal to the net advective flows. All observed quantities are taken as tidal and volume averaged values within each box. It is assumed that both advective and nonadvective exchanges occur only at the boundaries of the boxes. The required inputs are salinity, estuary geometry and river flows, which are often known quantities.

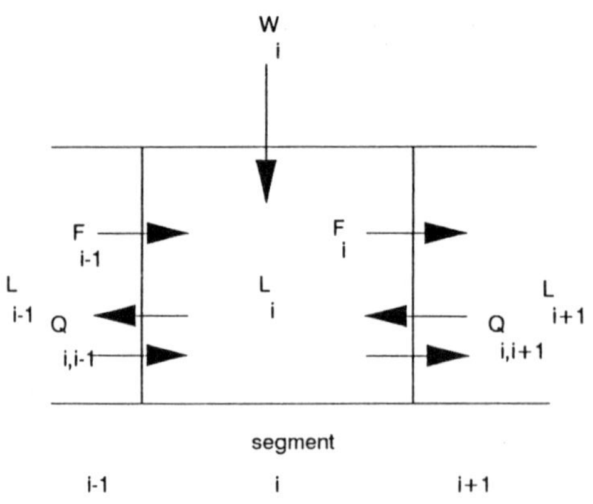

Fig. 1 One dimensional box model nomenclature

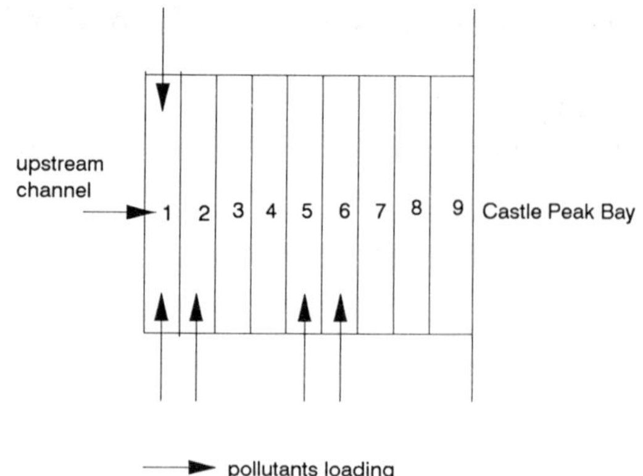

Fig. 2 Model schematization of Tuen Mun River

Let C_i^k and L_i^k be the dissolved oxygen (DO) and BOD_5 concentration in the segment i of tidal cycle k. Considering a mass balance to segment i will yield:

$$L_i^{k+1} = L_i^k + \underbrace{\left(\frac{W_i^k \Delta t}{V_i^k}\right)}_{\text{BOD load}} + \underbrace{\frac{Q_{i,i+1}^k \Delta t}{V_i^k}\left[L_{i+1}^k - L_i^k\right]}_{\text{tidal exchange}}$$

$$+ \underbrace{\frac{Q_{i,i-1}^k \Delta t}{V_i^k}\left[L_{i-1}^k - L_i^k\right]}_{\text{tidal exchange}} + \underbrace{\left(\frac{F_{i-1}^k \Delta t}{V_i^k}\right) L_{i-1}^k}_{\text{freshwater}}$$

$$- \underbrace{\left(\frac{F_i^k \Delta t}{V_i^k}\right) L_i^k}_{\text{flushing}} + \underbrace{R_i^k \Delta t}_{\text{resuspension}} - \underbrace{(1 - e^{-k_1 \Delta t})\, L_i^k}_{\text{BOD decay}} \qquad (1)$$

$$C_i^{k+1} = C_i^k - \underbrace{\beta\,(1 - e^{-k_1 \Delta t})\, L_i^k}_{\text{BOD decay}}$$

$$+ \underbrace{\frac{Q_{i,i+1}^k \Delta t}{V_i^k}\left[C_{i+1}^k - C_i^k\right] + \frac{Q_{i,i-1}^k \Delta t}{V_i^k}\left[C_{i-1}^k - C_i^k\right]}_{\text{tidal exchange}}$$

$$+ \underbrace{\frac{F_{i-1}^k \Delta t}{V_i^k} C_{i-1}^k - \frac{F_i^k \Delta t}{V_i^k} C_i^k}_{\text{freshwater flushing}} - \underbrace{\frac{B_i^k \Delta t}{H_i^k}}_{\text{benthal demand}}$$

$$+ \underbrace{(1 - e^{-k_2 \Delta t})\,(C_{s,i}^k - C_i^k)}_{\text{reaeration}} \qquad (2)$$

where

V_i^k = mean volume of segment i during tidal cycle k

H_i^k = mean depth of segment i

F_i^k = freshwater flow across transect i

W_i^k = BOD loading

R_i^k = resuspension of bottom organic material

B_i^k = benthal oxygen demand by bottom deposits on overlying water

β = ratio of ultimate to 5-day BOD

$Q_{i,i+1}^k$ = tidal exchange flows between adjacent segments

k_1 = deoxygenation coefficient

k_2 = reaeration coefficient

$C_{s,i}^k$ = saturated dissolved oxygen at segment i

Δt = tidal period

The average ammoniacal nitrogen concentration in the river is small and hence the nitrification process is unimportant in the modelling equations. For modelling the heavy metals concentrations in the river, only Eq. (1) is used and L_i^k will represent the heavy

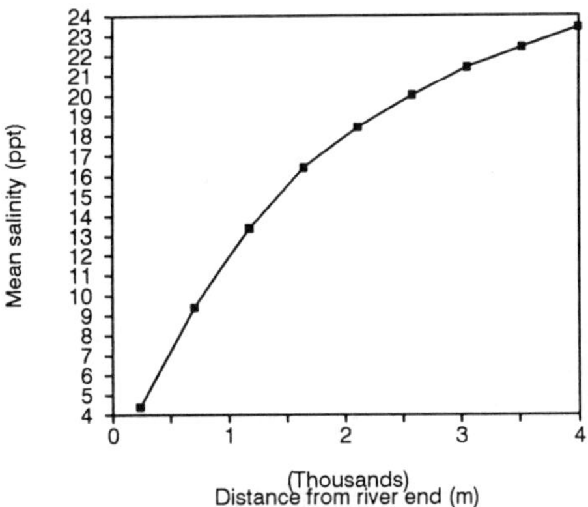

Fig. 3 Measured salinity for Tuen Mun River

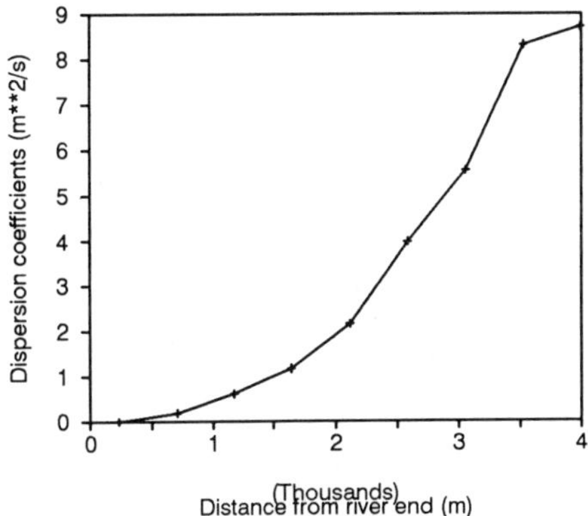

Fig. 4 Computed dispersion coefficients for Tuen Mun River

metals concentrations. k_1 will then describe the metals decay coefficient.

For a steady state condition, $L_i^k = L_i^{k+1}$ and $C_i^k = C_i^{k+1}$, and equations (1) and (2) can be further simplified to a system of non-linear algebraic equations. Several approaches can be used to estimate the tidal dispersion coefficients. Firstly, the estuarine salinity is used as a tracer of tidal mixing. Secondly, dye is discharged as tracer for tidal river freshwater portions of estuaries. Thirdly, hydrodynamic theory incorporating velocity shear and salt diffusion mechanisms is used. In this study, the first approach is used to calculate the dispersion coefficients by balancing the freshwater flushing and upstream salt transport. This is also called "the fraction of freshwater method".

$$Q_{i,i+1} = \frac{F_i S_i}{S_{i+1} - S_i} \tag{3}$$

where S_i is the observed salinity at segment i. The dispersion coefficient $E_{i,i+1}$ is then related by:

$$Q_{i,i+1} = \frac{E_{i,i+1} A_{i,i+1}}{0.5(x_i + x_{i+1})} \tag{4}$$

where:

$A_{i,i+1}$ = midtide cross-sectional area of boundary

x_i = length of segment i

3. RESULTS AND DISCUSSION

The Tuen Mun River can be approximated by a rectangular channel. Only the length of the river subjected to tidal effect which is about 4 km is considered in this study. The river is divided into 9 segments. The major pollutant loadings come from the upstream channel and the storm sewer flows due to illegal discharges from the factories (Fig. 2). The tide in the Castle Peak Bay is mainly mixed and semi-diurnal and hence a tidal period of 12.42 hours is assumed. From the field measurements, it is clear that the observed salinity (Fig. 3) is decreasing away from the mouth of the river. The tidal exchange flows are then computed from the salinity data. It is shown that the tidal effect is the greatest near to the mouth and is diminished towards the end of the river. The dispersion coefficients computed from eq. (4) range from 1-10 m^2/s (Fig. 4).

Theoretically, an infinite time is required for the steady state BOD and DO conditions to develop from any given initial condition. However, an absolute maximum difference of 0.01 mg/l of the water quality parameters at each segment between two consecutive tidal cycles is defined to be the steady state condition. In this study, the downstream boundary condition is specified as $L_9 = \alpha L_8$, and $(C_s - C_9) = \alpha(C_s - C_8)$, where α is the tidal exchange ratio. A general value of 0.5 used by Lee and Choi (1986) is adopted in the calculations. The numerical computation is started from a zero BOD concentration and a saturated DO. An iteration procedure using Jacobi method is

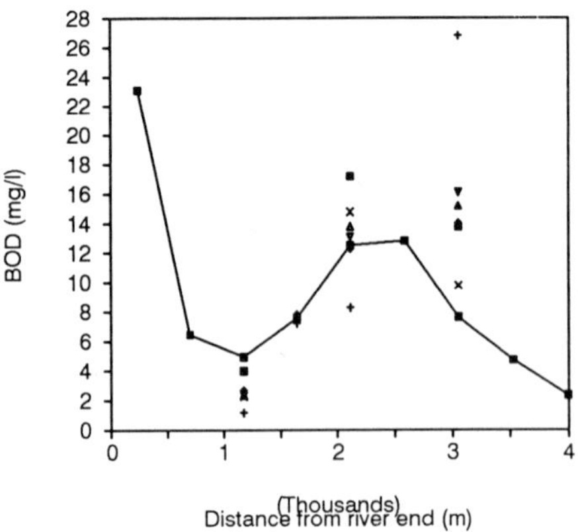

Fig. 5 Comparison of measured and computed BOD concentrations

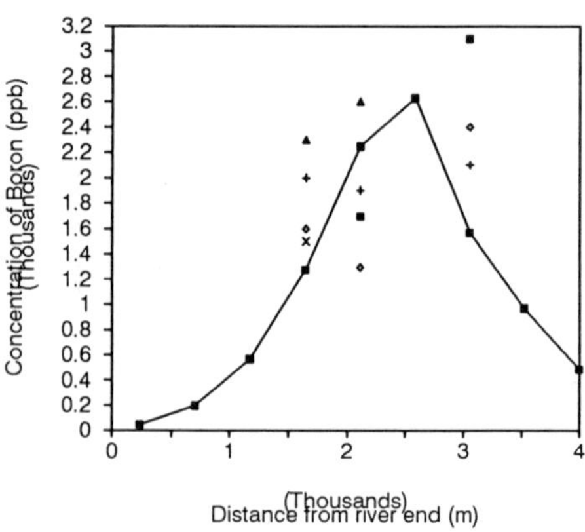

Fig. 7 Comparison of measured and computed Boron concentrations

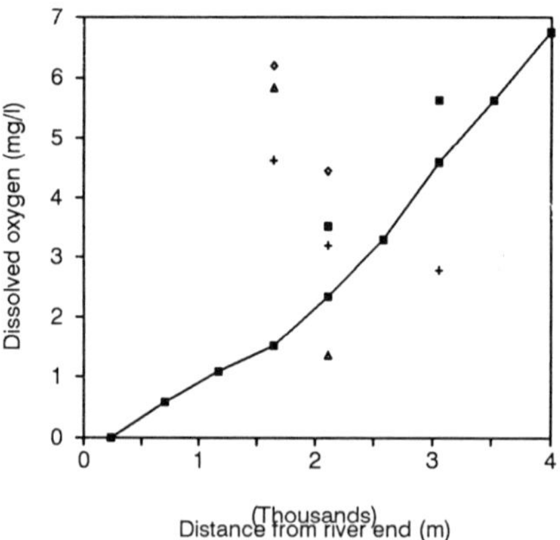

Fig. 6 Comparison of measured and computed DO concentrations

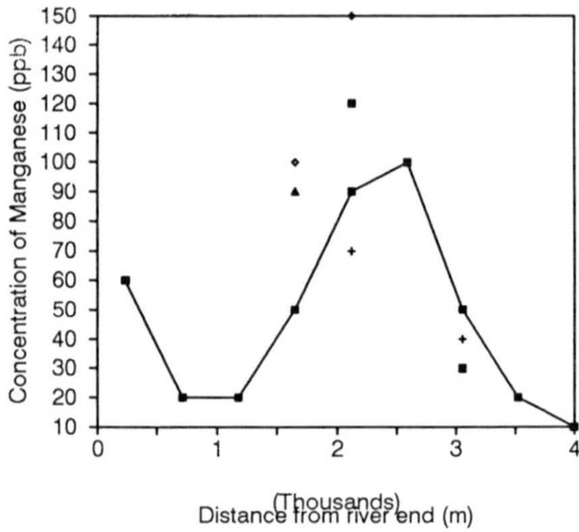

Fig. 8 Comparison of measured and computed Manganese concentrations

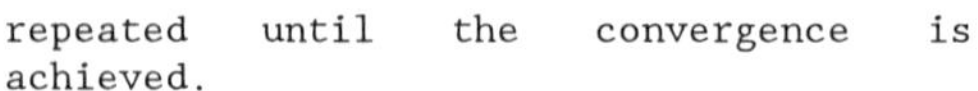

repeated until the convergence is achieved.

The model developed is then used to predict the water quality of the river at the dry seasons. Water samples are collected from November to March of a year. After the calibration of some of the model parameters, it can be shown from Fig. 5 and 6 that the trends of the BOD and DO are modelled.

The validated model is then used to predict the heavy metals concentrations along the river channel during the dry seasons. The numerical model is started from a zero metals concentrations, and the procedure is repeated as above until the convergence is achieved. The trends of the heavy metals concentrations such as Boron, Manganese and Iron are shown in Fig. 7, 8 and 9.

The model developed can also be modified to include the prediction for the metals concentrations in the river sediments by one additional mass conservation equation for the sediment. However, this type of model will become more complex and require more field data of the sediments for the validation purpose.

It is noted that the gazetted Water Quality Objectives for Tuen Mun River are such that DO $\geq$ 4 mg/l and 5-day BOD $\leq$ 5 mg/l. If f is taken to be the fraction of the existing pollutants discharges, then it can be shown from Fig. 10 & 11 that about 80 % removal of the loadings are required in order to attain the Water Quality Objectives for the whole river channel. For heavy metals concentrations, the maximum allowable concentration is about 4.5 μg/l in order to provide a

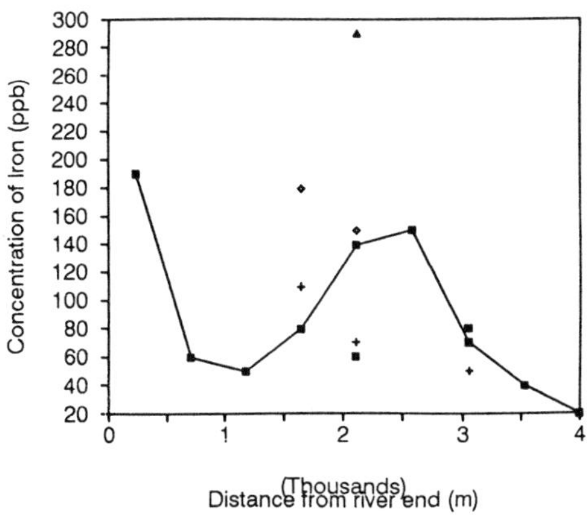

Fig. 9 Comparison of measured and computed Iron concentrations

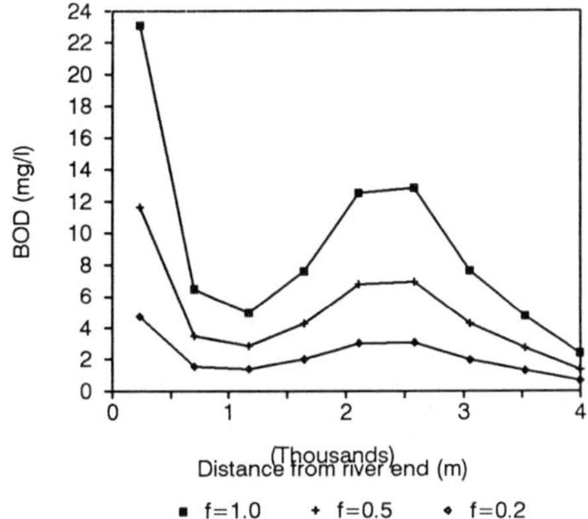

Fig. 10 Prediction of BOD concentrations under various pollution loadings

suitable environment for the freshwater aquatic life. More than 95 % of the industrial discharges have to be removed in order to reduce the heavy metals concentration levels to be acceptable to the aquatic life.

4. CONCLUSION

It is concluded that the tidally averaged segmented box model discussed is capable to predict the general trend of the heavy metals concentration in a one-dimensional tidally river after the validation of the model results against the field BOD and DO data. The model developed is especially suitable for long-term water quality planning of a tidal river.

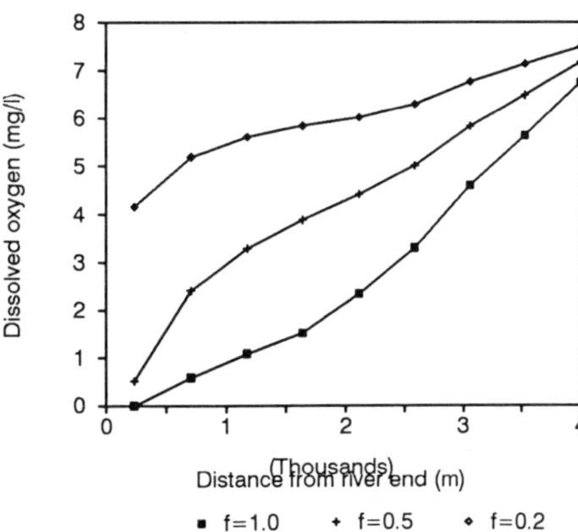

Fig. 11 Prediction of DO concentrations under various pollution loadings

ACKNOWLEDGEMENTS

The author wish to thank my colleague, Mr. J. Fung, for providing the water quality monitoring data for the Tuen Mun River.

REFERENCES:

Environmental Protection Department, Hong Kong (EPD) 1988. River water quality in Hong Kong.

Lee, J.H.W. & Choi, K.W. 1986. Slack tide oxygen balance model. J. of Environmental Engg., 112, No. 5, Oct., pp. 985-991.

Officer, C.B. 1980. Box models revisited in Estuarine and Wetland Processes. P. Hamilton and K.B. Macdonald (ed.), Plenum, pp. 65-114.

Sin, Y.S. 1991. Water quality modelling study of Shatin Shing Mun River, Hong Kong. Extended Abstracts of The Fourth International Conference on Computing in Civil & Building Engineering, 1991, pp. 188.

Sin, Y.S. 1992. Finite difference modelling for a one-dimensional rectangular estuary. Water Science and Technology, Vol. 26, No. 9-11, 1992, pp. 2591-2594.

Environmental Management, Geo-Water & Engineering Aspects, Chowdhury & Sivakumar (eds)
© 1993 Balkema, Rotterdam. ISBN 90 5410 099 0

Urban storm water quality modelling of NO_x and TP pollutant loads

M. Sivakumar & S. Boroumand-Nasab
Department of Civil & Mining Engineering, University of Wollongong, N.S.W., Australia

ABSTRACT : A semi-empirical equation is presented for the prediction of urban stormwater based on Nakamura concept of pollutant transport. The predicability of this equation is tested for two Australian urban catchments for the transport of total oxidised nitrogen (NOx) and total phosphorous (TP) pollutant loads. The proposed method is based on the assumption that the transport rate (washoff pollutant rate) is proportional to the distribution and availability of pollutants. This method is simple and can be readily applied. The results show good agreement between measured and predicted data.

1 INTRODUCTION

Urban runoff quality and quantity problems are both historic and current. A growing population, industrialisation and industrial productions all increase pollutants in an urban environment. These pollutants are carried by runoff during storm events. The risk to public health increases when unacceptable concentrations of pollutants are present in storm waters. Sediment transport in urban areas is another problem particularly due to construction activities. Rapid rate of sediment inflow in lakes and streams causes infilling and in some cases degradation of ecosystems. The nutrients such as nitrogen and phosphorous exported from urban areas also cause eutrophication in receiving waters.

The review of existing formulations for pollutant washoff from urban areas indicated that washoff equations are more empirical rather than process oriented. The newly proposed method is derived to develop process oriented semi-empirical algorithms for modelling urban runoff quality. Using this method total washoff pollutant load for each event can be predicted.

2 THEORETICAL BASIS OF THE PROPOSED EQUATION

The pollutant generation and wash off processes in urban areas are diffuse in nature. Hence it can be assumed that the washoff pollutant rate is proportional to the distribution and availability of pollutants. The force which causes pollutants to leave the surface is due to shear stress. It can be assumed, therefore, that washoff rate of pollutant transport(-dP/dt) is proportional to shear stress acting on surface.

$$-\frac{dP}{dt} \propto \tau_o \qquad (1)$$

where,
P = pollutant load remaining at time t, kg
τ_o = bed shear stress, Nm^{-2}

If the storm runoff regime may be assumed to be steady and uniform, the shear stress acting on the bottom surface of the flow is given by :

$$\tau_o = \gamma_s \, r \, S_o \qquad (2)$$

in which,
γ_s = specific weight of storm runoff, Nm^{-3}
r = hydraulic radius, m
S_o = slope of the surface

If the pollutant available for transport in an urban catchment can be assumed to be proportional to the specific weight of storm runoff (Nakamura, 1984) then,

$$\gamma_s \propto P \qquad (3)$$

By combining equation (1), (2) and (3) the washoff rate of pollutants from the surface can be shown to be:

$$\frac{dP}{dt} = - k' r P S_o \tag{4}$$

in which k' is a proportionality coefficient. The solution to equation (4) is:

$$P = P_o e^{-\int_0^t k' S_o r \, dt} \tag{5}$$

where, P_o = initial pollutant load, kg.

The expression $\int_0^t k' S_o r \, dt$ is impossible to determine in practice. For urban runoff, the hydraulic radius r can be assumed to be the depth of runoff. The depth of runoff is non linearly related to the runoff rate. When the runoff rate is integrated with respect to time, it gives a value of cumulative runoff depth. Hence the following expression is suggested which incorporates a readily measurable parameter namely the depth of cumulative runoff.

$$\int_0^t k' S_o r \, dt = k v^a \tag{6}$$

in which "v" is the cumulative runoff depth in mm, "k" is a modified washoff coefficient and "a" is an exponent. Hence equation (5) reduces to

$$P = P_o e^{-k v^a} \tag{7}$$

and the washoff equation is :

$$P_o - P = P_o (1 - e^{-kv^a}) \tag{8}$$

where, $P_o - P$ = pollutant washoff, in kg.

Equation (7) shows a functional relationship between the amount of pollutant available for transport with the parameters Po, k, v and a. Nakamura(1984) has observed from field experiments that the plot of ln(P/Po) versus depth of cumulative runoff was non-linear. This observation supports the functional relationship as shown in equation (7) where "v" is shown to have a non-linear relationship with ln(P/Po).

3 THE STUDY CATCHMENT AND DATA AVAILABILITY

Two independent data sets from Giralang and Jamison Park catchments(Figure 1) were selected to test the above method. The Giralang catchment (National Capital Development Commission, 1980) in the ACT is an urban catchment which is located approximately 10 km from Canberra city centre and has an area of 94 hectares with 27% impervious surface. The land use includes 69.4 hectares of medium density residential and 24.6 hectares rural and pasture. The Jamison Park is an urban catchment in Penrith 40 km from Sydney which has a 20.6 hectare, fully residential catchment. The catchment contains 35% impervious surface.

General deficiencies in the data set for example non availability of data for complete events and timing errors between hyetograph, hydrograph and pollutograph have been identified. Nine events from Giralang catchment and eleven events from Jamison Park catchment were selected for testing the new method. The data of the events have been identified on the basis of the date of the event. For example the storm of 30 September 1977 is numbered as 770930.

Fig. 1 Giralang & Jamison Park Catchments

Table 1. Dust and dirt buildup parameters for rural and residential land use

Parameters	Residential	Rural
Limit (kg/ha)	90	80
Power	0.2	0.12

Table 2. Pollutant as a fraction of dust and dirt (g/kg) for residential and rural land use

Pollutant	Residential	Rural
NO_x	4	5
TP	1	1.3

Table 3. Calibration results for NO_x

Event No.	Runoff depth mm	k	a	Pollutant washoff (kg)		Ratio r
				measured	predicted	
770218	1.18	0.024	1.08	0.82	0.75	1.09
770912	1.30	0.069	0.84	1.43	1.47	0.97
780510	1.81	0.018	1.24	1.05	1.10	0.95
770930	3.36	0.018	1.17	2.47	2.56	0.96
770812	1.91	0.028	1.26	1.97	2.02	0.97
770308	2.68	0.034	0.84	2.63	2.76	0.95

Table 4. Calibration results for TP

Event No.	Runoff depth mm	k	a	Pollutant washoff (kg)		Ratio r
				measured	predicted	
770912	1.30	0.113	1.16	0.60	0.60	1.01
780510	1.81	0.057	0.76	0.74	0.75	0.99
770930	3.36	0.061	0.93	1.42	1.39	1.02
770812	1.91	0.088	0.73	1.06	1.06	1.00
770308	2.68	0.213	0.63	2.72	2.85	0.95

4 POLLUTANT BUILDUP

The pollutant accumulation is derived by the dust and dirt method(Huber et al 1989). An exponential form is used to predict the dust and dirt build up. The initial NO_x loads and total phosphorous(TP) loads are determined from the quantities of dust and dirt collected by specifying the fraction of pollutant present per unit mass of dust and dirt. Table 1 gives the parameter values used to estimate pollutant build up rate whereas Table 2 provides the fraction of dust and dirt contributing to NO_x and TP from each land use (Sriananthakumar and Codner, 1992).

5 GIRALANG CATCHMENT

5.1 Calibration results

The "k" and "a" parameters were determined by calibration using stormwater data from Giralang catchment.

The calibration results for each event for total NO_x load and TP load are given in Table 3 and 4 respectively. The values of "k" and "a" are the best fit for each event, as computed by nonlinear regression option of the statistical analysis package SAS.

A discrepancy ratio "r" is defined as the ratio between measured and predicted washoff pollutant load for the purpose of comparison. A value of "r" near unity gives the best prediction.

The values of "r" for NO_x and TP were found

to vary between 0.95 to 1.03 and 0.95 to 1.02 respectively. The overall values of "r" for total NO_x and TP are 0.98 and 0.97 respectively for each event which are indeed very close to unity.

5.2 Verification results

Equation 8 was verified after calibration for the Giralang catchment using three independent data sets. The calibrated pollutant average washoff parameters for oxidised nitrogen and total phosphorous are given in Table 5. The measured and predicted results for total NO_x and TP are shown in Table 6 and 7 respectively. The value of "r" varies between 0.80 to 0.94 and 0.51 to 1.62 for NO_x and TP respectively. The average values of "r" are 0.90 and 0.88 for NO_x and TP respectively and show reasonable prediction. It also appears that the NO_x pollutant load prediction is better than the prediction for TP pollutant load.

Table 5. Calibrated washoff parameters values for NOx and TP

Pollutant	k	a
NO_x	0.030	1.07
TP	0.106	0.84

Table 6. Verification results for NO_x

Event No.	Runoff depth mm	Pollutant washoff (kg)		Ratio r
		measured	predicted	
770113	0.93	0.85	0.91	0.93
780409	2.23	1.18	1.46	0.81
790303	3.16	3.13	3.33	0.94

6 JAMISON PARK CATCHMENT

The proposed model has shown good prediction for the Giralang catchment. However it should be tested with data from other catchments. Hence Jamison Park urban catchment was chosen for this purpose.

Table 7. Verification results for TP

Event No.	Runoff depth mm	Pollutant washoff (kg)		Ratio r
		measured	predicted	
780115	2.55	1.88	1.16	1.62
780409	2.23	0.77	1.01	0.76
790303	3.16	1.01	1.96	0.51

6.1 Calibration results

The parameters of "k" and "a" in equation 8 are calibrated for the Jamison Park catchment using the statistical package SAS for 4 events for NO_x and 5 events for TP. The calibration results for each event for total NO_x load and TP load are shown in Tables 8 and 9 respectively. The calibration results show better agreement between measured and predicted values for NO_x load in comparison to TP load. The value of "r" varies between 1.00 to 0.98 and 0.95 to 1.50 for NO_x and TP respectively. The average values of "r" are 1.00 for NO_x and 1.08 for TP indicating very encouraging calibrations.

6.2 Verification results

For wider applicability the parameters obtained from statistical techniques need to be verified from independent data sets. Table 10 provides the calibration parameters "k" and "a" for NO_x and TP which have been determined by four and five independent events respectively. The results indicate that the values of "k" and "a" for Giralang catchment are different from Jamison Park catchment.

Based on the calibrated parameters, the equation 8 is verified and the results are given in Tables 11 and 12 for NO_x and TP respectively. The variation of discrepancy ratios for NO_x and TP are between 0.94 to 4.08 and 0.42 to 1.78 respectively.

The overall values of "r" for NO_x and TP are 1.35 and 0.71 respectively. The discrepancy ratios for NO_x for events 881109 and 881115 are 4.08 and 2.67 which are too large and hence not shown in Figure 2.

The values of "r" for Jamison Park are found to vary more in comparison to those for the Giralang catchment for both pollutant loads. This may be due to the fact that the dust and dirt build up parameters for Jamison Park catchment were assumed to be the same as for

Table 8. Calibration results for NO_x

Event No.	Runoff depth mm	k	a	Pollutant washoff (kg) measured	Pollutant washoff (kg) predicted	Ratio r
840215	8.44	0.014	1.34	1.16	1.16	1.00
840409	1.60	0.026	1.06	0.09	0.09	1.00
840620	7.62	0.008	1.14	0.43	0.42	0.98
840717	2.47	0.015	0.7	0.18	0.18	1.00

Table 9. Calibration results for TP

Event No.	Runoff depth mm	k	a	Pollutant washoff (kg) measured	Pollutant washoff (kg) predicted	Ratio r
840215	8.44	0.102	0.09	0.90	0.89	1.01
840409	1.60	0.146	1.42	0.06	0.06	1.00
840620	7.62	0.158	0.76	0.85	0.83	1.02
840717	2.47	0.046	0.91	0.19	0.20	0.95
841108	8.01	0.108	1.70	0.51	0.34	1.50

Table 10. Calibrated washoff parameters values for NO_x and TP

Pollutant	k	a
NO_x	0.0157	1.06
TP	0.117	1.14

Table 11. Verification results for NO_x

Event No.	Runoff depth mm	Pollutant washoff (kg) measured	Pollutant washoff (kg) predicted	Ratio r
840704	10.7	1.28	1.25	1.02
881109	1.0	0.49	0.12	4.08
881115	1.5	0.32	0.12	2.76
830907	1.5	0.17	0.18	0.94
831128	1.1	0.13	0.10	1.30

Table 12. Verification results for TP

Event No.	Runoff depth mm	Pollutant washoff (kg) measured	Pollutant washoff (kg) predicted	Ratio r
840704	10.7	0.62	1.47	0.47
871109	14.2	0.66	0.37	1.78
840921	1.51	0.10	0.12	0.83

the Giralang catchment. The data on dust and dirt for Jamison Park catchment is not available to the authors.

7 DISCUSSION

Figure 2 shows the ratio between measured and predicted NO_x pollutant washoff load for the two catchments. The calibration and subsequent prediction with independant data for NO_x washoff loads are very good for Giralang catchment.

However for the Jamison Park catchment, the ratio "r" varies widely and it is possible that "k" and "a" values need fine tuning with more data including data on dust and dirt buildup.

Figure 3 shows the ratio "r" for TP load for both catchments. As observed from Tables 7 and 12, the TP pollutant load predictions are variable. The "r" values predominantly fall within $0.5 < r < 2$. By comparing Figures 2 and 3, it is clear that the proposed model predicts NO_x load better than TP load.

The calibration and predictions have been done in this paper only for low runoff events. However the procedure should be checked for high runoff events where the cumulative pollutant loads can be very significant.

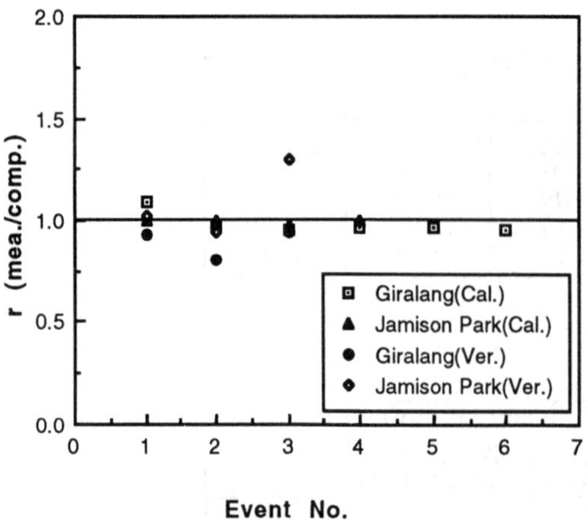

Fig. 2 Calibration and Verification Results for total NO_X washoff (Giralang & Jamison Park)

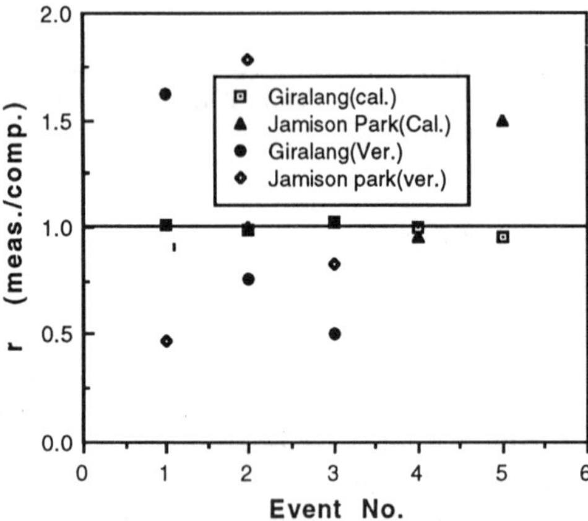

Fig. 3 Calibration and Verification Results for TP washoff (Giralang & Jamison Park)

8 CONCLUSION

A new semi-empirical method is developed to predict pollutant washoff loads from urban catchments based on the assumption that transport rate(washoff pollutant rate) is proportional to the distribution and availability of pollutants. The proposed washoff equation was calibrated and verified for two Australian catchments. The calibrated results for total oxidised nitrogen (NO_X) and total phosphorous (TP) pollutant washoff loads show good agreement between measured and predicted data for low runoff events. The verification results for NO_X load are better than TP load. However the new method should be tested by a large number of reliable urban water quality data for both high and low runoff events before further conclusions can be arrived.

ACKNOWLEDGMENT

This work is part of the activities of the Water Engineering and Geomechanics Research Program of the Department of Civil and Mining Engineering of the University of Wollongong and the support of the University is acknowledged. The ACT Electricity and Water as well as the Environmental Protection Authority of NSW are acknowledged for providing data. The second author wishes to thank the Iranian Government for providing a postgraduate research scholarship.

REFERENCES

Huber, W.C. & Dickinson, R.E 1989. Stormwater Management Model, Version 4: User's Manual. *U.S. Environmental Protection Agency,* Athens, Georgia 30613, U.S.A.

Nakamura, E. 1984. Factor affecting the removal of street surface contaminants by overland flow. *Journal of Research,* Public Works Research Institute, Japan, Vol. 24, 1-49.

National Capital Development Commission, 1980. Monitoring stormwater flow and water quality in paired rural and urban catchment in ACT, *Technical paper No. 29.*

Sriananthakumar, K. and Codner, G.P. 1992. Factors affecting pollutant washoff decay coefficient in urban catchments. *Proc. International Symposium on Urban Stormwater Management,* IAHR-IEAust., Nat. Conf. Pub. No. 92/1, Sydney, 390-396.

Environmental Management, Geo-Water & Engineering Aspects, Chowdhury & Sivakumar (eds)
© 1993 Balkema, Rotterdam. ISBN 90 5410 099 0

Geochemical and geophysical characteristics of an unconfined coastal aquifer at Tomago, NSW, Australia

Michael J.Thom
D.J.Douglas & Partners Pty Ltd, Sydney, N.S.W., Australia

Stephen R.Jones
D.J.Douglas & Partners Pty Ltd, Newcastle, N.S.W., Australia

ABSTRACT: Sand mining has taken place within the coastal unconfined aquifer at Tomago for 20 years. During this time, a programme of groundwater chemical tests has been conducted to monitor the effects of mining on groundwater quality, resulting in a substantial database which traces the changing groundwater quality with time. Changes in groundwater quality can be related to mining activity. Geophysical testing has been carried out using a Geopulse electrical resistivity meter to provide electrical resistivity profiles in various areas of the catchment. The results of the electrical resistivity measurements are compared to the groundwater quality test data. This paper presents a review of the geology and hydrogeology of the aquifer together with a review of the results and the likely causes of increases in iron concentration. It also outlines directions for future research.

1 INTRODUCTION

The Tomago Sandbed aquifer is located on the central portion of the east coast of New South Wales, about 15 kilometres north of the industrial City of Newcastle (see Fig 1). Control of the aquifer and all activities within the catchment is vested in the Hunter Water Corporation (HWC) whose main function is to provide water and sewage services to the people of the Hunter region.

The sandbeds cover an area of approximately 100 square kilometres (Hartwell & Viswanathan, 1983) and have a saturated thickness of about 15 m. At an effective porosity of 20%, the amount of water stored in the aquifer is about $3x10^8$ cubic metres, or 300,000 megalitres. This is an important regional resource which supplies up to about 20% of Newcastle's water requirements in summer.

The sandbeds at Tomago have been used since 1939 as a source of water supply. Mineral sand mining within the aquifer commenced in 1972 (within catchment) using dredges to extract the mineral. This conjunctive use of the sandbeds has required strict guidelines for monitoring of groundwater quality which takes place from numerous spearpoints located roughly on a 300 m grid over the catchment. The testing of groundwater quality is under-

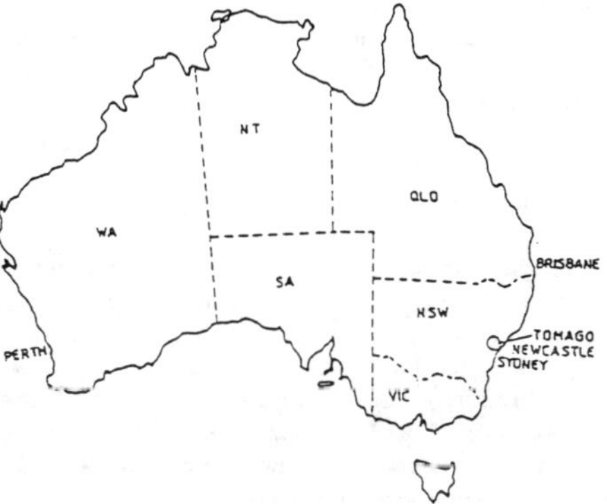

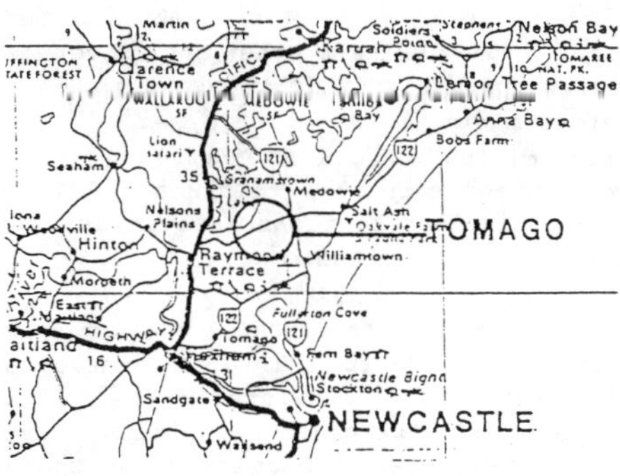

Fig.1 Locality plan

taken by the mining company who sample the groundwater at three monthly intervals and test principally for pH and iron concentration.

2 GEOLOGY

The geomorphological processes which have lead to the formation of the NSW coastal lakes and aquifer systems have been studied intensively (Chapman et al, 1982). These processes have resulted in unconsolidated sediments up to 80 m thick (Packham, 1969) but the depths around Tomago are typically 20 m (Soil Mechanics Ltd, 1971).

The Tomago Sandbeds stratigraphy has been described as follows (SML, 1971)

. UPPER LIGHT SAND (ULS) - medium dense light brown medium grained sand typically 2 - 5 m thick

. DARK SAND (DS) - medium dense dark brown or black medium grained cemented sand typically 2 - 20 m thick

. LOWER LIGHT SAND (LLS) - medium dense light brown fine to medium grained sand; 2 - 8 m thick

. GREY SAND (GS) - medium dense light grey fine to medium grained sand with some silt; 3 - 10 m thick

The above is a generalised profile and not all layers are recorded at all borehole locations.

3 HYDROGEOLOGY

The Tomago Sandbeds are an unconfined aquifer system, with groundwater flow generally from north to south (see Fig 2). Recharge is predominantly from precipitation although some water enters the aquifer via Moffat's Swamp and from very minor flows from Grahamstown Reservoir (Douglas & Partners Pty Ltd, 1991).

Numerous pump tests over the aquifer and surrounding areas (SML,1971; Valentine and Herzog, 1970) indicate that the average hydraulic conductivity of the sand is

. before mining 2×10^{-4} m/sec

. after mining 3×10^{-4} m/sec

For an average saturated thickness of 15m, this is equivalent to transmissivities of

. before mining 260 m^2/day

. after mining 390 m^2/day

Hydraulic gradients within the aquifer vary from 1 in 500 to 1 in 1500 averaging about 1 in 1000. Calculated actual flow velocities (not Darcian velocities) are about 6 - 40 m/year (Douglas & Partners Pty Ltd, 1986) although local flows around mining areas may be much higher because of elevated water tables resulting from increased rainfall infiltration.

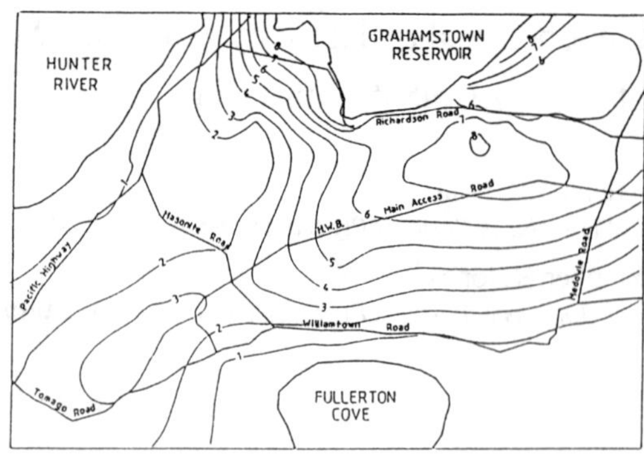

Fig.2 Groundwater Contours

4 HYDROGEOCHEMISTRY

The hydrogeochemistry of the Tomago Sandbeds varies from point to point because of different conditions created by

- the depositional environment
- topography
- groundwater movement
- degree of cementing
- leaching
- type of bacteria
- type of vegetation
- stratigraphy
- geomorphology

Added to the natural phenomena is the homogenising effect of mining and it is not difficult to realise that a single hydrogeochemical model for the entire aquifer is not possible.Indeed substantial variations in groundwater quality are sometimes seen in closely spaced spearpoints.

The Tomago Sandbeds aquifer consists of four distinct strata, ie upper light sand, dark sand, lower light sand and grey sand. The results of chemical analysis of samples of these strata (Soil Mechanics, 1971) are shown in Figure 3. They indicate that the highest mean values of iron, aluminium and carbonate occur in the Grey Sand (GS). Also organic content of the GS is lowest of all the strata. It is concluded that the GS is significantly different from the overlying three strata with respect to organics, iron, aluminium and carbonate and is probably of a different origin from the other strata.

Iron concentrations within the groundwater at Tomago are a function of

- prevailing soil conditions
- oxidation/reduction potential around

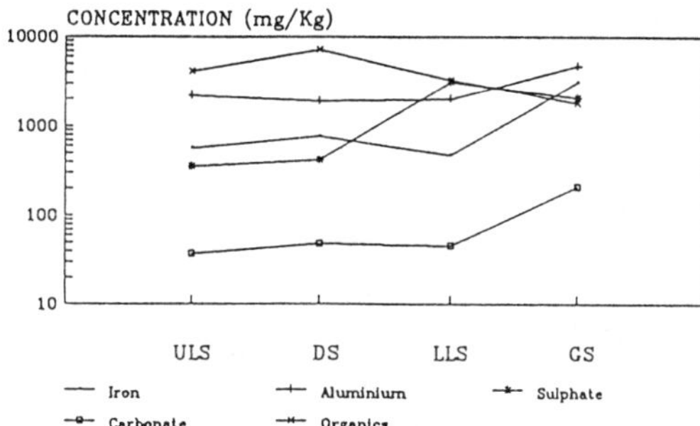

Fig.3 Results of soil chemical analysis

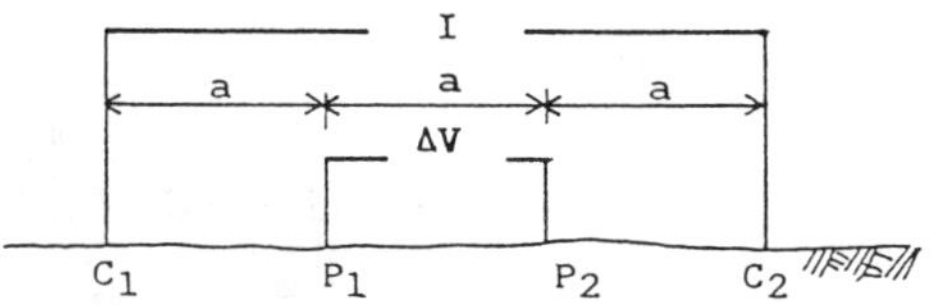

Fig.4 Schematic layout of resistivity profiling

the sampling spearpoint

- the degree to which sulphide minerals are oxidised by fluctuating water tables
- bacterial activity brought about by changes from anaerobic to aerobic conditions
- leaching of organic acids from topsoil spread on the mining path
- rainfall patterns and hence water levels

With such a diverse list of factors contributing to fluctuations in iron concentrations, it is little wonder that conflicting trends may be apparent from individual spears. To overcome this difficulty, statistical analysis of the overall trends in iron concentrations is considered the only appropriate method of assessing long term trends.

Analysis of data from 300 spearpoints across the catchment, indicates that yearly average iron concentrations are

- before mining 2 - 5 mg/l
- after mining 8 - 13 mg/l
- after remining 20 - 40 mg/l
- after deep mining > 100 mg/l

Remining has taken place in some locations to similar depths to the original mining (ie 6 - 8 m). Deep mining has recently commenced to the base of the aquifer at depths of around 20 m.

5 GEOPHYSICAL SURVEY

Geophysical testing in the aquifer comprised electrical resistivity profiling using a Geopulse resistivity meter manufactured by Campus Geophysical Instruments Ltd of Birmingham, UK.

Testing took place at four areas as follows

- Pump Station 12 (PS12) (east and west) in unmined ground where water quality is good
- Elephant Walk in mined ground where slime from the mining process was deposited on, and mixed with, the tailings
- Plant 4 where slime was left at the pond bottom
- Deep mining area where mining has been to the bottom of the aquifer

These four areas were selected because they represent different mining operations for which differences in electrical properties should be identified by geophysical testing.

The profiling utilised twenty-four 1 m long steel stakes driven 0.9 m into the sand at 5 m intervals and connected to a 24 core cable used for recording resistances.

Geopulse profiling was carried out at all four locations between 25 and 27 September 1991. Repeat tests were performed on part of the Elephant Walk on 30 November, 12 December, 1991 and 16 February 1992, to check the integrity of the data and ascertain any diurnal or other time dependent variations.

Resistivity profiling, using an equidistant electrode configuration similar to that adopted for Wenner soundings, involves passing a current through two outer electrodes (C_1 and C_2) and measuring the potential difference between two inner electrodes (P_1 or P_2). Figure 4 shows details of the arrays used.

The apparent resistivity can be calculated from the measured resistances by

$$\rho a = \frac{\Delta V}{I} \cdot K$$

where ΔV = voltage drop
I = current
K = geometric factor
= $2\pi a$ for the four electrode, equal spaced array

To enable comparison with conductivity tests on groundwater samples and cone conductivity tests (not reported here), the results of the resistivity testing have been reported as apparent electrical conductivity.

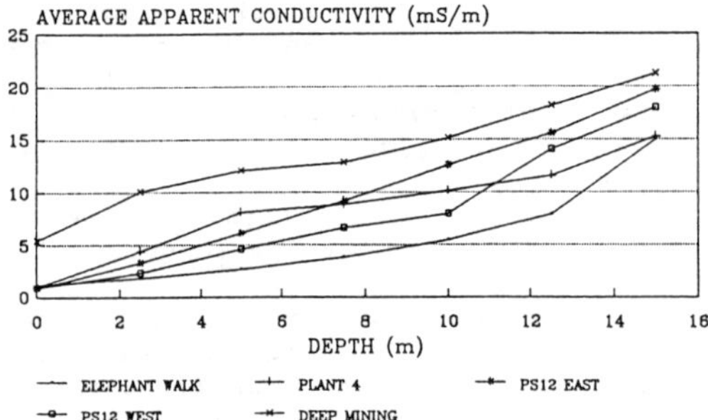

Fig.5 Average conductivity

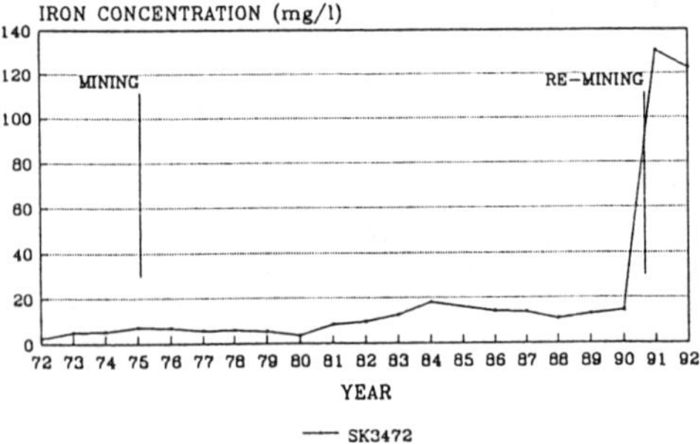

Fig.7 Iron concentrations at SK 3472

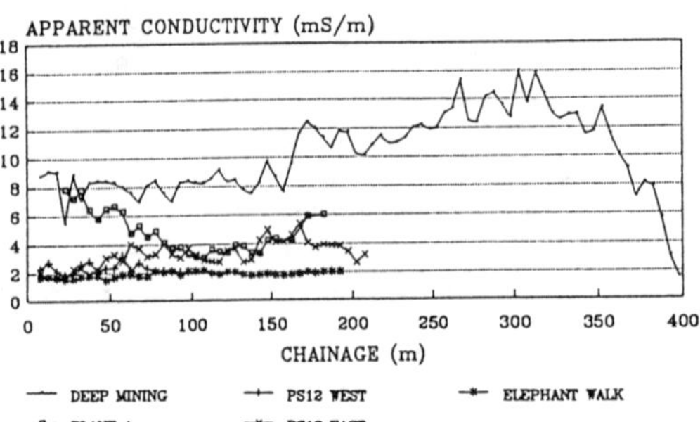

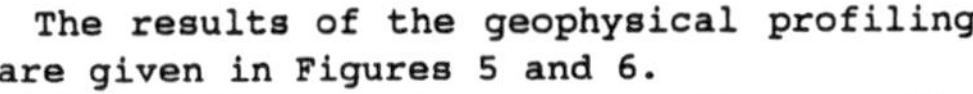

Fig. 6 Conductivity at depths of 2.5 m

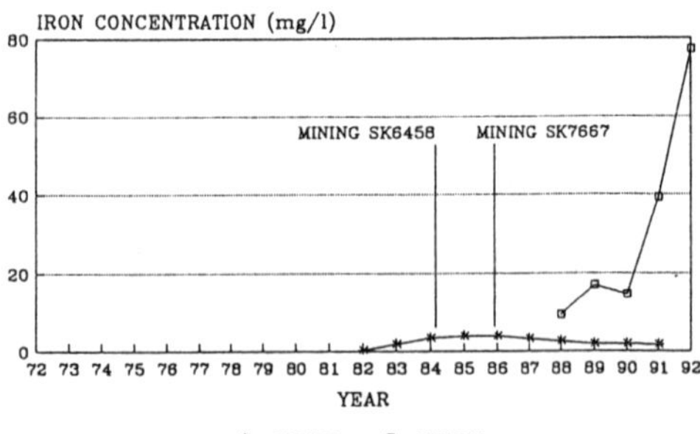

Fig.8 Iron concentrations at SK 6458 and SK 7667

The results of the geophysical profiling are given in Figures 5 and 6.

The difference in the results for the individual areas are easily explained in terms of groundwater chemistry and mining procedure as follows

. Deep mining within the aquifer has resulted in greater than normal increases in soluble iron concentrations (see Fig.7). The electrical profiling has shown much higher apparent conductivities due to the greater concentration of dissolved salts.

. The apparent electrical conductivity of Plant 4 is about twice that at the Elephant Walk due to greater iron concentrations in the groundwater (see Fig. 8 for comparative results) and a slime layer deliberately deposited on the pond bottom at Plant 4.

. Deep mining has disturbed the grey sand (GS) near the base of the aquifer, previously undisturbed by shallow mining. The GS has abundant iron (see Fig. 3) which is brought into solution as a result of disturbance which causes oxidation of sulphide minerals to sulphate with an accompanying drop in pH.

Several interesting points arise from close inspection of the bulk electrical conductivities. These are:

. The electrical conductivity at Elephant Walk (mined) is much lower than at PS12 (unmined) even though the groundwater iron concentrations are similar (see Figs. 9 and 10).

This indicates that electrical conductivity changes due to mining are not entirely explained by increases in iron concentrations.

. The electrical conductivity at Plant 4 changes along the line of testing due to an increase in depth to the water table because of elevation changes (see Fig. 11). Interpretation of conductivity data must therefore take into account the degree of saturation of the soils.

. The formation factor calculated from the bulk electrical conductivity and fluid conductivity from groundwater samples ranged generally from 2.1 to 5.9 with a mean value of 3.2. This is much lower than expected for a sand with a porosity of 35 - 40% (see Fig.12), due possibly to the high bulk conductivity caused by the presence of electrically conductive slime.

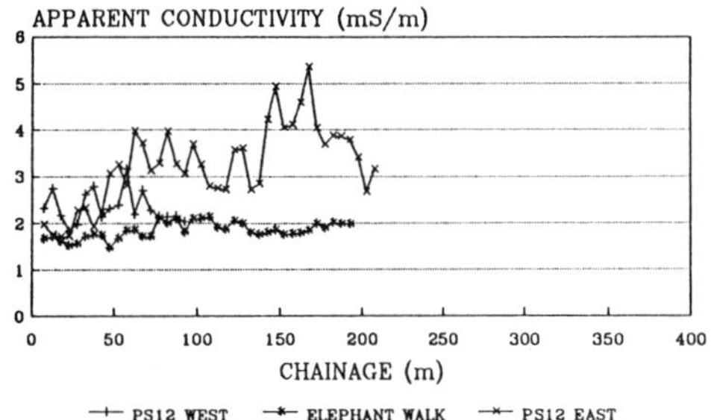

Fig.9 Conductivity at Elephant Walk and PS12

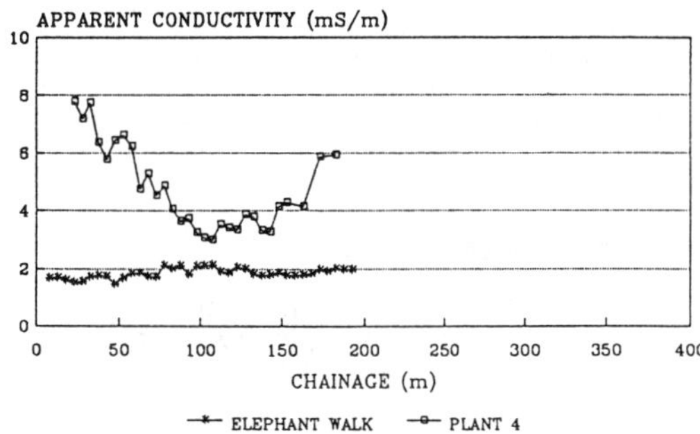

Fig.11 Electrical conductivity at Plant 4

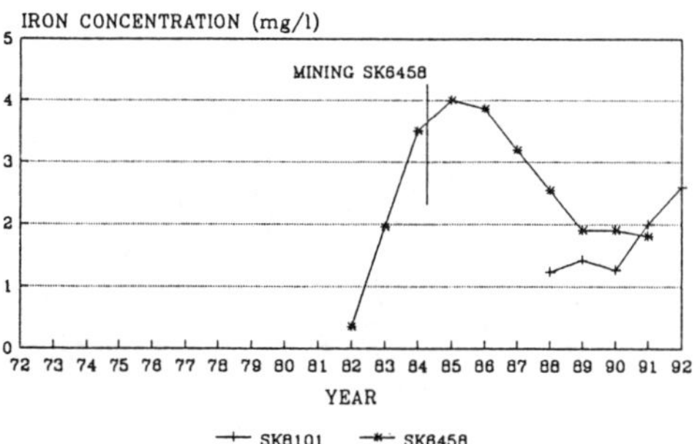

Fig.10 Iron concentrations SK 8101 and SK 6458

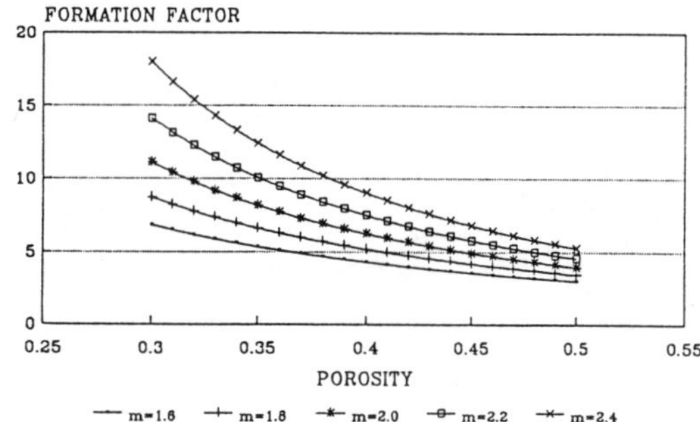

Fig.12 Porosity-Formation factor relationship

6 CONCLUSIONS

The conclusion that can be drawn from the data presented above can be summarised as follows (Thom, 1992)

. Electrical resistivity profiling can be used to measure changes in groundwater quality that occur due to mining and deep mining. Highly conductive plumes should be easily tracked if, in fact, they move out of the deep mining area.

. The data indicates that electrical conductivity in the deep mining area is 2 to 5 times that where deep mining has not occurred.

. Slime layers within the saturated soil profile leads to a doubling of electrical conductivity.

. Electrical conductivity decreases as the depth to the water table increases. In this respect, the electrical conductivity profile is an inverse measurement of the water table depth. Plant 4 profiling took place over a hill which is directly reflected by the drop in conductivity (see Fig. 5).

. Formation factors are lower than expected due to high bulk conductivity caused by highly conductive slime.

REFERENCES

Chapman, D M, Geary, M, Roy, D S & Thom, B G, 1982. Coastal evolution and coastal erosion in NSW., Coastal Council of NSW.

D J Douglas & Partners Pty Ltd. Annual review of mining and monitoring. February 1986. Unpublished report to Hunter Water Corporation.

D J Douglas & Partners Pty Ltd. Hydrochemical study, Grahamstown Storage Reservoir, Raymond Terrace, 1991. Unpublished report for Hunter Water Corporation.

Hartwell, D J & Viswanathan, M N, 1983. Mineral sands mining in a developed aquifer at Tomago, New South Wales. Int. Conf. on Groundwater & Man, Sydney, 1983.

Packham, G H (Ed), 1969. The geology of NSW. Geol. Society of NSW.

Soil Mechanics Ltd, (1971). Tomago aquifer study - consideration for mineral sand mining. Unpublished report No 5532/12, Hunter Water Corporation.

Thom, Michael J. The geochemical and geophysical characteristics of an unconfined coastal aquifer at Tomago, NSW. M.App.Sc Thesis, Centre for Hydrogeology & Groundwater Management, UNSW, unpublished.

Valentine, H R & Herzog, A. Assessment of the effects of rutile mining on the yield of an aquifer near Tomago, NSW. Progress Report No 1 to Hunter Water Corporation, August 1970. Unpublished.

ACKNOWLEDGEMENT

The authors wish to thank Mr W Down of the Hunter Water Corporation for his permission to publish this paper.

Environmental Management, Geo-Water & Engineering Aspects, Chowdhury & Sivakumar (eds)
© 1993 Balkema, Rotterdam. ISBN 90 5410 099 0

Groundwater quality for irrigation and human consumption

Vikram P. Kumar, I. Mehrotra & R. P. Mathur
Department of Civil Engineering, University of Roorkee, India

ABSTRACT: The quality of ground water was assessed to determine its suitability or otherwise for irrigation and domestic use. The samples drawn from different sources exhibited a wide range of spatial variation. In general, a sampling station close to river yielded water of high salinity. The salinity potential, residual alkalinity, sodium absorption ratio, magnesium hazard were computed to determine the irrigation potential of water.

1 INTRODUCTION

Agra, the city of Taj Mahal is a tourist city with scarce water supply from river Yamuna to meet the daily requirement. In order to balance water deficit the aquifers are tapped to install personal as well as community hand pumps, tubewells and open wells. However, for a developmental work and organised usage of water it is rather imperative to assess the quantity and quality of ground water in the aquifer. The purpose of this article is two fold : (1) to describe the quality of ground water in an area of about 2x2 Sq Km in Agra, India infront of Taj Mahal on eastern bank of river Yamuna and (ii) to assess the suitability of water for irrigation and human consumption. This study forms a part of a developmental activity under the framework of the 'Agra Heritage Project' supported by US park service. It is proposed to develop Taj National Park. The core has key historic structures, the Taj Mahal, the Agra Fort and the area across Yamuna River designated as Mumtaz Bagh. Around that core are buffers, the green spaces protecting views and offering visitor facilities. Surrounding the buffers are satellite sites (e.g. Fatehpur Sikri, Itmad-ud-Daulah, Sikandra) so that when unified, they may provide a far richer experience than any of the individual sites. Since ground water in the core and buffer areas of the proposed park will be used for irrigation and human consumption, this site was selected for evaluating ground water quality.

The ground water quality in space and time is a much dependent on the recharge and draft as on geological formations comprising the aquifer and human activities within the recharge area. The chemical composition of ground water is determined by a series of complex physical, chemical and biological processes occuring as water moves through the soil, subsoil, unsaturated and saturated zones of the aquifer. For the entire aquifer vertical and areal difference in water-sediment interactions are affected by residence time of water as well as soil type, vegetation, lithology and minerology of the aquifer material and groundwater flow pattern. The observed water quality may reveal the existing as well as the forthcoming geochemical processes due to the proposed development of land and water based activities. Therefore in the first phase the quality parameters of the ground water from the site were determined.

2 METHODOLOGY

A reconnaisance survey was done to identify sampling locations. Sixteen stations identified for sample collections are located in cluster of villages across Taj Mahal on the eastern bank of River Yamuna (Fig. 1). Samples were drawn from open wells, tube wells,

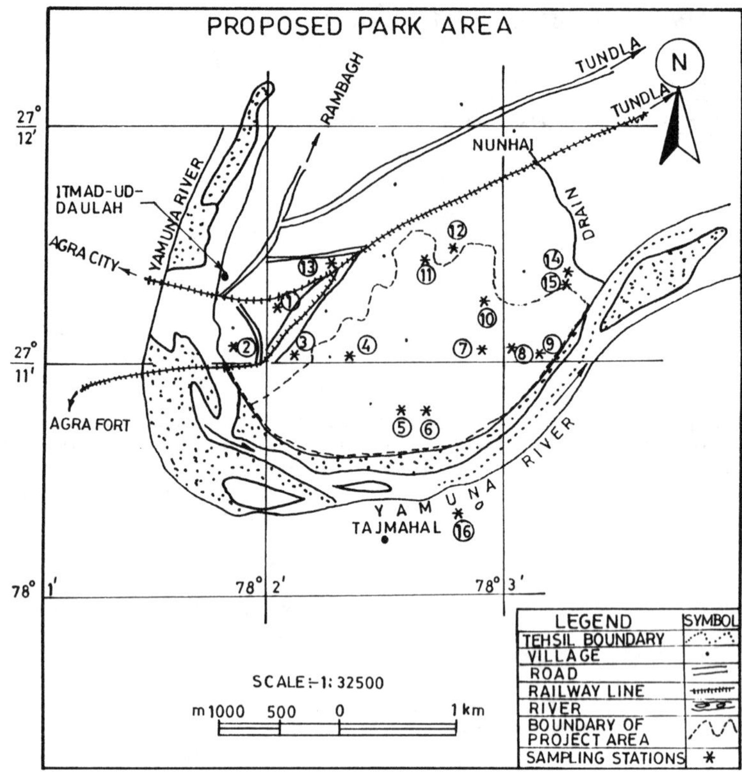

FIG. 1-SAMPLING LOCATIONS

Table 1. Details of sampling stations

Sample station No.	Location	Depth of Ground water (approx.)(m)
1.	Yamuna Bridge	16.0
2.	Moti Mahal	33.0
3.	Nagla Devjit	20.0
4.	- do -	17.0
5.	Kachhpura	33.0
6.	- do -	14.0
7.	Garhi Chandni	25.0
8.	- do -	18.0
9.	- do -	18.0
10.	Sushil Nagar	33.0
11.	Gautam Nagar	18.0
12.	Railway Colony	18.0
13.	Yamuna Bridge St.	15.0
14.	Prakash Nagar	33.0
15.	- do -	9.0
16.	Taj Mahal East	7.0

1,4,7,8,12 - Private handpumps
2,5,10,14 - India Mark-II (IM-II)
3,6,11,15,16 - Open wells
9,13 - Tube wells

shallow and deep hand pumps (designated as India Mark II). (Table 1). Three water samples were drawn at one month interval in October, November and December, 1990. Detailed quality assurance practices were followed for collection of samples in field, measurement of field parameters and for the analysis of these samples in the laboratory. The parametric estimations were carried according to standard procedures (Standard Methods 1985). Average values for each sample were determined by considering all the three or at least two close observations.

3 RESULTS AND DISCUSSION

The average major ion concentration is presented diagrammatically in Fig. 2 and the summary of solids., conductivity and pH are given in Table 2. From these wide range of spatial variation in water quality is noticed. The charge-balance error in fifteen out of sixteen samples was less than 5%. In one of

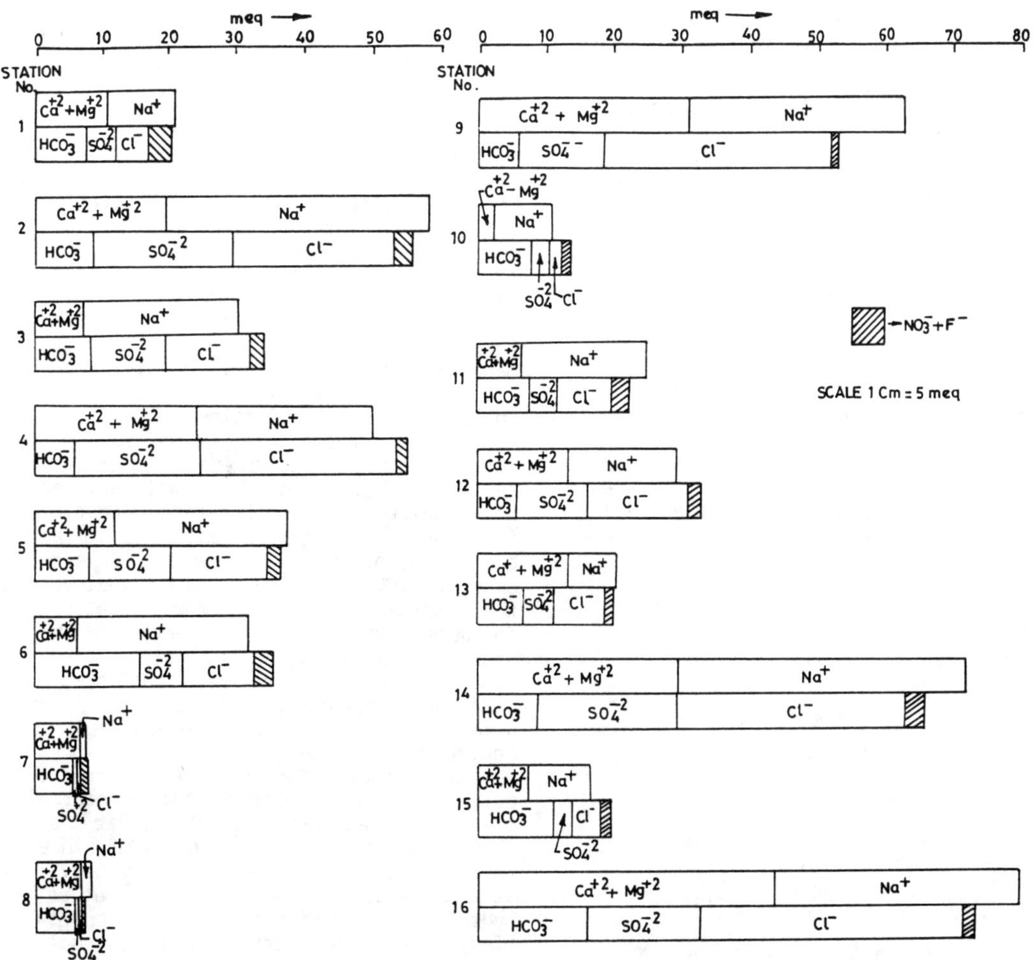

FIG. 2 – IONIC COMPOSITION OF GROUNDWATER SAMPLES

the samples the error was around 11%. Often, relatively large charge-balance error (30%) is used as cut-off for water with low dissolved solid concentration (Katz and Chiquette, 1991).

3.1 Ionic composition

Perusal of data given in Fig. 2 reveal that among the cations sodium is the dominating cating in most cases excepting for samples 7, 8 and 13 which are rich in calcium and magnesium. Chloride happens to be the dominating anion particularly in the samples with relatively higher TDS (samples 2,4,9,14,16). In general, the sodium is either equal to or greater than the chloride. The imbalance between the equivalent amounts of the chloride and the alkali ions may due to the cationic exchange and breakdown of silicates which releases more alkali than chloride ions. The occurance of calcium and magnesium together with bicarbonates in samples 7 and 8 indicate that these samples have probably been drawn from the zone of active flushing (Freeze & Cherry, 1979). The samples at station 10 with relatively low ionic content has been collected from a deeper source (IM-II). It has sodium along with biocarbonate as the major cation and the anion respectively.

The major ion composition of water samples is shown in a hydrochemical facies diagram (Fig. 3) originally presented by Piper (1944). From

Table 2. Water quality parameters (mean values)

Sample No.	TDS (mg/l)	Conductivity (u mhos) (cm^{-1})	pH range
1.	1162	2234	7.4 - 8.3
2.	4010	5411	7.0 - 7.8
3.	1975	5838	7.5 - 8.5
4.	3086	4323	7.4 - 8.2
5.	2110	4607	7.3 - 8.1
6.	1964	4403	8.0 - 8.8
7.	499	795	7.5 - 8.5
8.	316	783	7.3 - 8.6
9.	3649	5062	7.0 - 8.0
10.	707	1129	7.9 - 8.7
11.	1438	1393	7.7 - 8.5
12.	1805	4876	7.5 - 8.3
13.	1224	2317	7.3 - 8.2
14.	4164	6501	7.3 - 8.2
15.	1046	1955	7.7 - 8.6
16.	4984	6388	7.2 - 8.1

Temperature - 23.5°C - 30°C
TDS ~ 95% of total solid

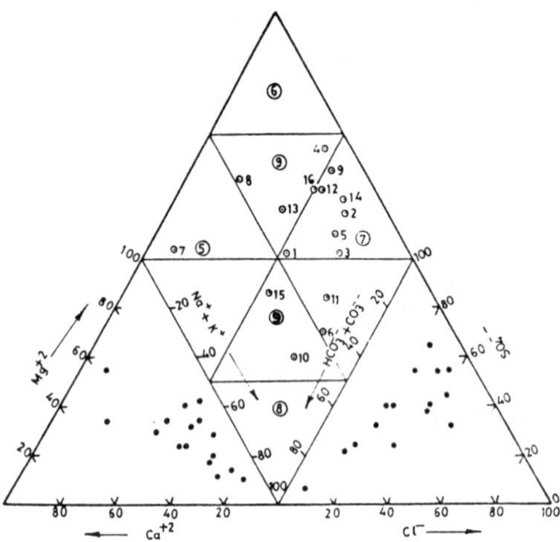

FIG. 3-HILL PIPER TRILINEAR DIAGRAM

this it is clear that there is no distinct water type. Samples are scattered in zones 5,7 and 9. Zone-5 is characterized by a composition where the carbonate hardness exceeds 50 percent i.e. chemical properties of water are dominated by alkaline earths with weak acids. The samples from stations 4,10,13,15 and 16 show a mixed character. In these samples, no single cation anion pair exceeds 50% of the total ionic strength. The remaining samples which fall in zone-7 have dominance of alkalies and strong acids over alkali earth and weak acids. In these samples, primary salinity (non carbonate alkalinity) exceeds 50 percent.

3.2 Total dissolved solid (TDS)

One might expect that the concentration of selected constituents in or overall composition of ground water is related to the depth of the well sampled. This assumption is based on previous studies suggesting the subdivision of aquifer into distinct upper active zone, intermediate zone and lower passive zone (Freeze & Cherry, 1979). In order to obtain such a correlation if existed, a visual conventional comparison was attempted by three dimensional presentation of TDS, depth of the well sampled and sample from different location (Fig. 4). The samples arranged in an increasing order of TDS exhibit a large spatial variation. These results indicate that although there are some variations in water chemistry with depth in the aquifer for parts of the study area, concentration of these constituents does not consistently increase or decrease with depth from the variations in chemical composition, that are observed near the water table. Sample from station 16, collected from the most shallow source (7m deep), adjacent to the river has a maximum concentration of TDS since this location is recharged by river water, water flowing under natural hydraulic gradient picks various impurities. Thus the matter such as soil rock, nature or recharge water, transport mechanism and eventually conversion processes at this location are different from those prevailing at other stations.

The samples collected from 33m depth from India Mark-II handpumps do not have uniform TDS level. The sample from station 10 has much lower value as compared to the values of the samples 2 and 14. Thus the TDS in water sample depends on both the sampling location and the depth from which the sample has been taken. The samples from the stations located in the vicinity of the river bed have a higher TDS content. The sampling stations 14 and 15 are close to each other (approximately 3m apart) as well as to the river. However, the TDS in the sample 14 is about four times the value for the sample 15. This may be attributed to the differences in the depth of the source. The hand pump IM-II at station 14 is close to the river bed

(TDS 4000 mg/l) whereas the hand pump IM-II at station 10 is away from the river bed (TDS 700 mg/l). It is also seen that the shallow sources (station 7 and 8) which are away from the river have still lower values of the TDS (300 mg/l). The sources 11 and 12 are shallow and away from the river but have relatively higher TDS values (1438 and 1804 mg/l). These sources may be under the influence of human activity being located in the inhabited Nunhai area. It may therefore be inferred that the apparent randomness in data may be due to the location of the sampling station with respect to the river, depth of the source and human influences (Engle, 1981).

Trace quantity of aluminium, copper, iron, maganese, nickel, zinc and silicon were detected in all the samples. The occurance of metals in trace quantities is not usual in water samples having high concentration of primary and secondary alkalinity. Copper concentration were in slight excess to limit the use of this water for irrigation of fine textured soils of pH 6-7-8.0 (FAO Soil Bulletin Rome, 1979). The concentration of other metals, however, is less than the recommended value of water usage, continuously for all types of soils.

3.3 Bacteriological quality

Water from the stations already in domestic use, viz. from station 1,3,6,7,8, 10,13 and 15 were selected for the evaluation of bacteriological quality. The records of the total count and the MPN at these eight representative stations are given in Table 3. The stations 1,3,10 have superior bacteriological quality as compared to that of water from stations 6,7. These wells are located at sites which have cleaner sanitary surroundings. The higher counts at other stations (6,7,8,13 and 15) may be due to solid waste accumulation and its leachate, prevailing unsanitary practices in buffer areas, manuring in agriculture and open defaction.

3.4 Water for irrigation use

The parameters determining irrigation potential of groundwater alongwith samples satisfying suitability criteria are presented in Table 4. TDS in samples, given in Table 3 is also a general indication of the overall suitability

Table 3. Bacteriological quality of water samples

Sample station no.	Total count	MPN
1.	227	0
3	60	39
6	-	460
7	Numerous	2400^+
8	Numerous	2400^+
10	281	9
13	Numerous	2400^+
15	Numerous	150

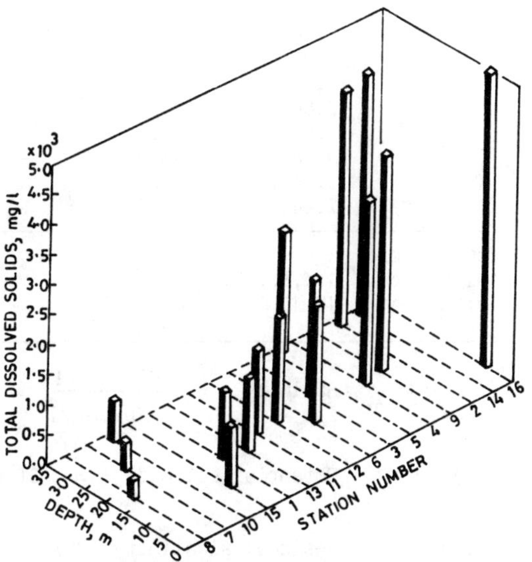

FIG. 4 — 3D-REPRESENTATION OF TOTAL DISSOLVED SOLIDS

of water for many types of uses. Water having too much dissolved minerals alter the soil structure, permeability and aeration. It is generally accepted that water containing around 1000 mg/l TDS mainly in the form of bicarbonate and carbonate is unsuitable for irrigation purpose. According to the salinity criteria, only waters with less than 0.5 salinity are suitable for irrigation. A moderately saline water has a salinity of 0.5 to 1.5. Water samples from eight stations (1,3,6,7,8,10,11 and 13) satisfy these criteria. Water from station 9 has maximum salinity and should not be used for irrigation of crops. However, during survey and sampling trips agriculture fields were noticed only around station 9 whose water was being used

Table 4. Water quality parameters evaluated for irrigation use

Sample No.	Salinity	Potential salinity (meq/l)	Residual alkalinity (meq/l) (RA)	SAR	Magnesium hazard	Permeability index
	(I)	(II)	(III)	(IV)	(V)	(VI)
1.	0.67	7.4	- 2.8	4.6	49.9	63.3
2.	2.76	33.7	-11.0	12.2	58.6	33.9
3.	1.47	17.9	1.2	23.1	47.9	85.6
4.	4.83	38.2	-18.2	7.3	75.4	74.8
5.	1.75	20.3	3.7	8.4	52.0	14.5
6.	0.66	13.7	9.4	17.2	60.4	94.2
7.	0.08	-	- 1.0	0.6	-	42.6
8.	0.08	0.4	- 0.5	1.0	42.8	50.7
9.	5.58	40.2	2.8	8.0	73.6	54.3
10.	0.22	3.1	5.6	7.8	59.7	103.4
11.	1.00	10.1	- 0.6	10.2	58.7	85.2
12.	2.50	20.0	- 7.6	6.2	65.1	62.5
13.	1.07	9.6	- 6.4	2.8	58.6	47.8
14.	3.78	43.9	-20.5	11.1	74.1	63.2
15.	3.64	5.7	3.5	4.7	50.4	74.4
16.	2.44	47.3	-27.6	7.8	83.7	50.5
Sample suitable for irrigation	1,3,6,7,8, 10,11,13	7,8,10 class-I type	-	1,7,8,10 11,15	1,3,8	10 - LP* 3,6,11- MP 4,15- HP

(I) Salinity $= \dfrac{Cl^- \text{ meq/l}}{HCO_3^- \text{ meq/l}}$

(II) Potential salinity $= Cl^-(\text{meq/l}) + 0.5\ SO_4^{--}$ meq/l

(III) Residual alkalinity (RA) $= (CO_3^- + HCO_3^-) - (Ca^{++} + Mg^{++})$

(IV) SAR (Sodium Absorption Ratio) $= \dfrac{Na^+}{\sqrt{(Ca^{++} + Mg^{++}) / 2}}$

(V) Magnesium hazard (MH) $= \dfrac{Mg^{++}}{Ca^{++} + Mg^{++}} \times 100$

(VI) Permeability index $= \dfrac{Na^+ + HCO_3^-}{Ca^{++} + Mg^{++} + Na^+}$

* - LP, MP, HP - Low, medium and high permeability; all terms in meq/l.

to grow wheat crop. Although it has not been possible to determine yield etc. from this land, the owner of the agriculture field and the tubewell appeared to be quite satisfied. The crop was grown on slanting fields. Following classification procedures outlined by Donnen (1966) and Ayers and

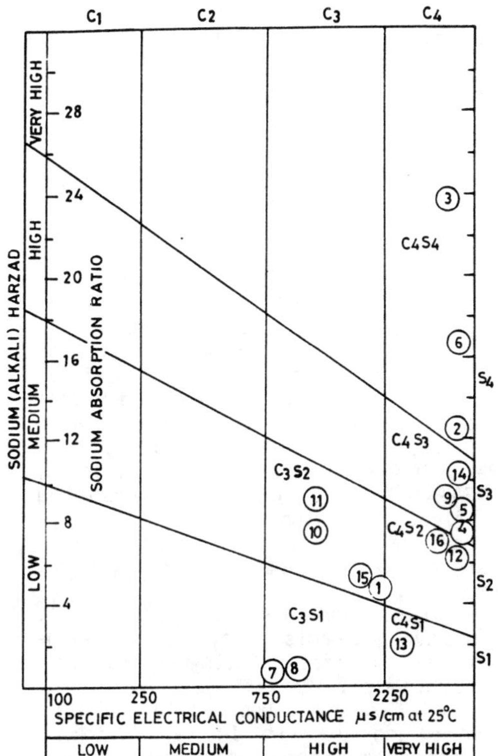

FIG. 5 - U.S SALINITY LABORATORY DIAGRAM

Westcot (1976) water samples from station 7,8 and 10 belong to class-I category & pose on potential salinity hazards.

Sodium adsorption ratio (SAR) used to evaluate the sodicity hazard ranges from 0.6 to 23.0. The SAR values in combination with conductivity shown in Fig. 5 indicates that these samples are scattered in C_3S_1, C_3S_2, C_4S_1, C_4S_2, C_4S_3 and C_4S_4. According to this, water belonging to C_1S_2, C_2S_3, C_3S_1 and C_3S_2 of bad quality. Water from stations 7,8 (C_3S_1), 1,10,11 and 15 (C_3S_2) may be used for irrigation.

Water with more than 2.5 meq/l of residual alkalinity are not considered suitable for irrigation use, those containing 1.25 - 2.5 are only marginally suitable, whereas those containing less than 1.25 meq/l RA are probably safe. The values of residual alkalinity calculated for samples under considerations is slightly misleading (Table 4). A large number of samples have negative value. A negative value indeed is less than 1.25 meq/l, but it reflects high chloride concentration and poses chloride toxicity problem. Chloride upto 4 meq/l in water generally does not lead to any problem.

If the irrigation water contains calcium and magnesium ions in a quantity that equals or exceeds the quantity of sodium, a sufficient concentration of magnesium will be retained on clay particles of the soil to maintain its permeability. Such waters serve irrigation well even though the total mineral content may be quite high. However, an excess of magnesium in water poses magnesium hazard. Amounts greater than 50% are not desirable for irrigation. Samples 1,3,8 do not pose magnesium hazard.

The permeability index of samples analysed varied between 14.5-103.4. The classification, however, is dependent on soil permeability. The class-I type water from station 10 is suitable for soil of low permeability; from stations 3,6,11 for soil of medium permeability, and from 4 & 15 for soil of high permeability.

The other impurities such as boron, fluoride & trace metals are well below the toxic levels.

3.5 Water for human consumption

During the sampling surveys it was found that except for two sources (stations 9 and 14) water from other stations was in general public domestic use without any treatment. The water characteristics do not conform to the guidelines suggested by several authorities determining the consumptive use of water. Developed countries have more stringent regulations in this regard compared to regulatory provisions in developing countries. (WHO, 1984) Local population seems to have adapted itself to general use of this water showing no observable health effects. However, it is necessary to abide by the guidelines especially for the safety of overseas tourists (NRC, 1980; EPA, USA, 1972). In many arid regions human beings have known to consume groundwater with much higher than 1000 mg/l TDS content without any obvious harmful effects. In semi-arid regions at times, the accepted upper limits of sulphates and chlorides have been prescribed as 2400 mg/l and 2800 mg/l respectively. A biocarbonate content in the range 31-610 mg/l (0.5-10 meq/l) has been suggested by the Department of National

Health and Welfare, Canada (1969) to alleviate corrosive or encrusting effects and to eliminate gastrointestinal problems, irritation and other health related problems. The guidelines for limiting calcium and magnesium contents are not based on health considerations but on minimising soap wastage. Acceptable limits of sodium (270 mg/l) in potable water are determined by taste threshold considerations and not by any toxic effects.

In general, the samples collected from the sixteen stations in the Agra Heritage Area pose no problem of toxicity due to heavy metal contents, nitrates and fluorides. All samples have been found aesthetically acceptable as regards considerations of colour and turbidity. Except for samples from stations 7,8,10 other samples are nonpotable on account of undesirable TDS, salinity and hardness. The chemical parameters are only indicative of selective use but the tourists may find these waters undesirable as judged by taste alone.

Considering bacteriological quality water from station 10 is suitable for domestic consumption, as it has zero MPN. However, it would still be desirable to resort to chlorination/ozonation to ward off any possible dangers associated with gastric disorders and other health hazards.

4 CONCLUSIONS AND RECOMMENDATIONS

Water from three stations (7,8, and 10) is better suited for domestic use. Chlorination/ozonation to sterilise the water before use is, however, necessary. Suitability of water for irrigation may be summarized as follows.

Parameter	Acceptable	Limited acceptability	Unacceptable
TDS	St. 7,8,10	St.1,11, 13,15	St.2 to 6,9, 12,14,16
Salinity	-do-	St.1,3,6, 11,13	St.2,4,5,9, 12,14,15,16
Potential salinity	-do-	St.1,11, 13,15	St. 2 to 6,9, 12,14,16
SAR/EC	-	St.1,7,8,	St.2 to 6,9, 12 to 14,16

It is thus seen that quality of ground water fit for human consumption and irrigation may be drawn from new Garhi Chandni and Sushil Nagar. In order to fruitfully use water from other sources it is necessary to consider a proper selection of crop type, soil characteristics and appropriate irrigation method.

REFERENCES

Ayers, R.S. and D.W. Westcot 1976. Water quality for agriculture. Irrigation and Drainage paper FAO Rome, 29:1-97.

Doneen, L.D. 1966. Water quality requirement for agriculture. Proc. National Symposium on Quality Standards for Natural Waters, School of Public Health, University of Michigan, 213-218.

Englen, G.B. 1981. A systems approach to ground water quality. The Science of Total Environment, 21:1-15.

EPA, USA 1972. Water quality criteria, EPA-R3-73-033, US Govt. Printing Office, Washington, D.C. 592.

Freeze, R.A. and J.A. Cherry 1979, Ground water, Prentice Hall, Inc. NJ, USA.

Katz, B.G. and A.F. Choquette 1991. Aqueous geochemistry of the sand-and-gravel aquifer, Northwest Florida, Ground water, 29:47-55.

National Research Council 1980. Drinking water and health in safe drinking water committee. Washington D.C., 3:415.

Piper, A.M. 1944. A graphic procedure in the geochemical interpretation of water analyses, Trans. Am. Geophys. Union, 25:914-928.

Shainberg, I. and J.D. Oster 1978. Quality of irrigation water. III C Publication No. 2, Ottawa, Canada.

Standard Methods for the Examination of Water and Wastewater 1985. 16th Edition APHA/AWWA/WPC, Washington D.C.

WHO 1984. Guidelines for drinking water quality, Recommendations, WHO, Geneva, 1:130.

Environmental Management, Geo-Water & Engineering Aspects, Chowdhury & Sivakumar (eds)
© 1993 Balkema, Rotterdam. ISBN 90 5410 099 0

Efficient solutions for solute transport in hillside seepage

R.E.Volker
Department of Civil & Systems Engineering, James Cook University, Townsville, Qld, Australia

W.W.Read
Department of Mathematics & Statistics, James Cook University, Townsville, Qld, Australia

C.Demetriou
Australian Centre for Tropical Freshwater Research, James Cook University, Townsville, Qld, Australia

ABSTRACT: For gently sloping hillsides, the large length to depth ratios for the saturated permeable layer lead to difficulties in accurately simulating seepage and solute transport processes. A method is presented which takes advantage of analytical solutions for the velocity field allowing it to be determined continuously throughout the entire saturated domain. This has a consequential benefit in solution of the transport equation because it facilitates the introduction of a fractional step algorithm so that optimum solution techniques can be applied to each of the relevant transport processes, advection and dispersion. The finite element mesh for solution of the transport equation is generated to take advantage of the fact that streamlines are precisely known and velocities can be determined at any arbitrary set of mesh points. Solutions are presented to show the effect of a range of parameter values on the rate of progression of solute through the hillsides.

1 INTRODUCTION

An understanding of seepage processes in hillsides is important for management of both the landscape itself and streams receiving flow contributions from this seepage. Changes in land use often result in increases in groundwater recharge beneath slopes due to factors such as clearing of large trees and the introduction of irrigated agriculture. Deterioration in water quality frequently accompanies the increased recharge due to mobilisation of salt stored in the soil profile at levels above the previous water table elevation. Pollutants can also be introduced at the soil surface due to agricultural activities and carried with deep drainage through the groundwater in the slope.

An ability to predict the movement of solutes under these landscapes is therefore important in determining appropriate management strategies. Unfortunately the geometry of the permeable zone involves large length to depth ratios (often of the order 50 to 1 or 100 to 1) and this leads to high computational costs in obtaining numerical solutions of the governing equations. There are two basic equations, the flow equation which determines the seepage velocities, and the mass transport equation from which solute concentrations are obtained.

The mass transport equation is notoriously difficult to solve numerically and even computationally expensive solutions may be inaccurate due to numerical dispersion. Although analytical solutions are available for contaminant transport in uniform seepage (Hunt, 1983) these are of limited use in practice where nonuniform flow is common. The validity of the solute concentrations is also dependent on the accuracy of the velocity field and, if using non-stationary meshes for the transport equation, the velocity field must be recalculated for each mesh location.

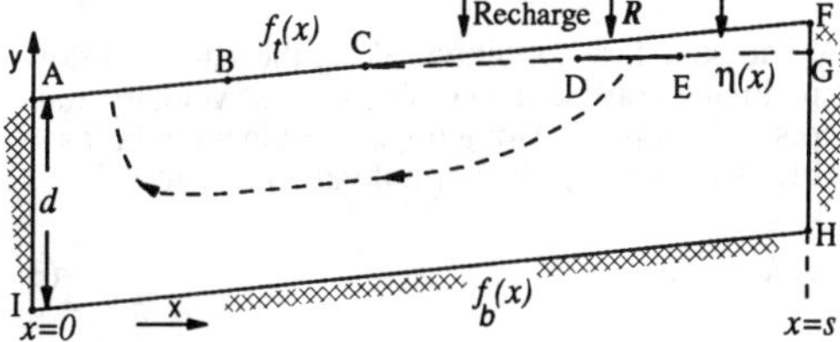

Fig.1 Simplified hillside section

The simplified geometry for a general hillside seepage problem is shown in Figure 1. For a modification of this case to include a horizontal bottom boundary, the velocity field can be determined continuously in the saturated domain, by using a series expansion of the velocity potential (Powers et al, 1967; Powell and Kirkham, 1976). Until recently, however, series solutions have been severely restricted by the necessity for a horizontal lower boundary. Fortunately, by applying least squares methodology to the problem, series expansions for arbitrary bottom geometry can now be obtained (Volker and Read, 1990; Read and Volker, 1990). Given the velocity field throughout the flow domain, grid refinements for the transport equation can be readily made, and consequently the effects of numerical diffusion can be minimised.

2 FORMAL PROBLEM DESCRIPTION

Determination of the rate of solute transport in groundwater requires the solution of two equations, the flow equation for seepage velocities, and the mass transport equation for solute concentrations. For each equation boundary conditions must be imposed on all elements of the boundary of the saturated domain shown in Figure 1.

2.1 *The flow equation*

Seepage through saturated, permeable soil is governed by Darcy's Law. Assuming the soil is homogeneous and isotropic, the velocities u,v in the x,y directions are given by

$$u(x,y) = -K\ \partial\phi/\partial x, \quad v(x,y) = -K\ \partial\phi/\partial y \tag{1}$$

where $\phi(x,y)$ is the hydraulic potential and K is the constant hydraulic conductivity. For an incompressible fluid, application of the continuity condition leads to Laplace's equation, in the fully saturated zone (Figure 1):

$$\partial^2\phi/\partial x^2 + \partial^2\phi/\partial y^2 = 0 \tag{2}$$

The boundary conditions along the saturated flow perimeter consist of a combination of velocity and potential forms. Along impermeable boundaries FHIA (Figure 1), the normal velocity is zero:

$$\partial\phi/\partial m = 0 \tag{3}$$

where m is normal to the boundary. When the impermeable boundary is vertical (FH and IA), this condition reduces to:

$$\partial\phi/\partial x = 0 \tag{4}$$

The upper saturated flow boundary $y_t(x)$ is permeable, and consists of a seepage face $f_t(x)$ on AC and water table $\eta(x)$ on CG. For steady recharge along the water table, the surface potential along the seepage face ($y_t(x)$, $0 \leq x \leq x_C$) and the water table ($y_t(x)$, $x_C < x \leq s$) is equal to the elevation above an arbitrary datum:

$$\phi_t(x) = y_t(x) = \begin{cases} f_t(x), & 0 \leq x < x_C \\ \eta(x), & x_C \leq x \leq s \end{cases} \tag{5}$$

For recharge R along the water table, conservation of mass stipulates that

$$R = K\,[\partial\phi/\partial y - \mathrm{d}\eta/\mathrm{d}x . \partial\phi/\partial x\,] \tag{6}$$

2.2 *The solute transport equation*

The equation for solute transport in an incompressible fluid can be expressed as (Hunt, 1983):

$$\frac{\partial}{\partial x}[D_{xx}\frac{\partial C}{\partial x} + D_{xy}\frac{\partial C}{\partial y}] + \frac{\partial}{\partial y}[D_{yx}\frac{\partial C}{\partial x} + D_{yy}\frac{\partial C}{\partial y}] - u\frac{\partial C}{\partial x} - v\frac{\partial C}{\partial y} + \frac{\partial C}{\partial t} = 0 \tag{7}$$

where C is the concentration of solute (mass per unit volume) and D_{xx}, D_{xy}, D_{yx}, D_{yy} are the components of the dispersion tensor D which is given by

$$\begin{aligned} D_{xx} &= (\alpha_I u^2 + \alpha_{II} v^2)/w + D_m T_{xx} \\ Dxy = Dyx &= (\alpha_I - \alpha_{II})uv/w \\ D_{yy} &= (\alpha_I v^2 + \alpha_{II} u^2)/w + D_m T_{yy} \end{aligned} \tag{8}$$

where α_I is the longitudinal dispersivity in the direction of fluid velocity, α_{II} is the lateral dispersivity normal to the direction of the fluid velocity, D_m is the molecular diffusion coefficient, T_{xx} and T_{yy} are components of the tortuosity tensor, and w is the magnitude of the velocity:

$$w = (u^2 + v^2)^{1/2} \tag{9}$$

Note that the dispersivities α_I and α_{II} are constant with respect to time and position.

The boundary conditions for solute transport need to be described along the impermeable boundaries, seepage face and water table. Referring again to Figure 1, along impermeable boundaries FHIA, the boundary condition for $C(x,y,t)$ is (Bear, 1979)

$$m_x[D_{xx}\frac{\partial C}{\partial x} + D_{xy}\frac{\partial C}{\partial y}] + m_y[D_{yx}\frac{\partial C}{\partial x} + D_{yy}\frac{\partial C}{\partial y}] = 0 \tag{10}$$

where m_x, m_y are the direction cosines of the normal to the boundary.

Along the water table (CG), the concentration is specified as zero except for a section (DE) where it is C_0 to indicate the pollution source. The length of the water table (CG) varies for different cases tested.

The boundary condition on the seepage surface ABC is set so that the concentration is governed by the solute advected and diffused to the surface. In the saturated zone immediately below BC solute can reach the surface only by diffusion since there is a very small advective component into the soil. Along AB both advection and diffusion will transport solute upwards toward the surface and the imposed boundary condition requires that there be no deterrent to the passage of the solute through the soil surface.

3 SERIES SOLUTIONS FOR SEEPAGE VELOCITY

An analytical solution for the hydraulic potential and hence the seepage velocity can be obtained by applying the method of separation of variables. The appropriate truncated form of the series expansion, for the given boundary conditions is:

$$\phi(x,y) = c_0 + \sum_{n=1}^{N} \left(a_n e^{\frac{n\pi y}{s}} + b_n e^{\frac{-n\pi y}{s}}\right) \cos \frac{n\pi x}{s} \qquad (11)$$

The bottom boundary condition equation (3) determines the relationship between a_n and b_n. This boundary condition can be expressed concisely as

$$\overline{\phi}_b(x) = 0.c_0 + \sum_{n=1}^{N} \overline{u}_n^b(x) a_n + \sum_{n=1}^{N} \overline{v}_n^b(x) b_n = 0 \qquad (12)$$

where

$$\overline{u}_n^b(x) = \frac{\pi}{s} n\left(\cos \frac{n\pi x}{s} + \mathrm{d}f_b/\mathrm{d}x \,.\, \sin \frac{n\pi x}{s}\right) e^{\frac{n\pi\, \mathrm{d}f_b/\mathrm{d}x}{s}}$$

and (13)

$$\overline{v}_n^b(x) = -\frac{\pi}{s} n\left(\cos \frac{n\pi x}{s} - \mathrm{d}f_b/\mathrm{d}x \,.\, \sin \frac{n\pi x}{s}\right) e^{\frac{-n\pi\, \mathrm{d}f_b/\mathrm{d}x}{s}}$$

The two sets of basis functions $\overline{u}_n^b(x)$ and $\overline{v}_n^b(x)$ form a linearly dependent spanning set; after removing the linear dependence the squared error ε_b in the bottom boundary can be expressed as

$$\varepsilon_b = \int_0^s \left(\sum_{n=1}^{N} [\overline{u}_n^b(x) - \sum_{k=1}^{N} h_{nk} \overline{v}_k^b(x)]\, a_n\right)^2 \mathrm{d}x \qquad (14)$$

Minimisation of ε_b with respect to h_{ij} leads, after some simplification, to N^2 equations of the following form:

$$\int_0^s \overline{u}_i^b(x) \overline{v}_j^b(x) dx = \sum_{k=1}^{N} h_{ik} \int_0^s \overline{v}_k^b(x) \overline{v}_j^b(x)\, \mathrm{d}x \qquad (15)$$

This set of N^2 simultaneous equations can be solved for the h_{ij}.

The series expansion for the hydraulic potential can now be expressed concisely as

$$\phi(x,y) = \sum_{n=0}^{N} c_n w_n(x,y) \qquad (16)$$

where $w_0(x,y) = 1$,

and (17)

$$w_n(x,y) = \left(e^{\frac{n\pi y}{s}} - \sum_{k=1}^{N} h_{nk} e^{\frac{-n\pi y}{s}}\right) \cos \frac{n\pi x}{s}$$

for n = 1,...,N.

The least squares technique can again be used to evaluate the coefficients c_n, by applying the top boundary condition for $y_t(x)$, equation (5). After minimising the squared error in the top boundary, and after some simplification the following set of equations is obtained for the coefficients c_n:

$$\int_0^s \phi_t(x) w_j^t(x)\, dx = \sum_{n=0}^{N} c_n \int_0^s w_n^t(x) w_j^t(x) \mathrm{d}x \qquad (18)$$

where the sub- and super-script t represents values on the top boundary of the saturated flow domain (ACF in Figure 1) This set of (N+1) simultaneous equations can be solved for the c_n, after a slight modification to the surface potential.

The surface potential $\phi_t(x)$ has a small discontinuity at the downstream impermeable boundary as discussed by Klute et al (1965). It can be removed by using a cubic spline to take the surface potential smoothly to zero gradient at $x = 0$ without appreciably affecting the surface potential.

The solution technique for the series coefficients c_n depends on the initially unknown water table location $\eta(x)$. The standard approach for solution of the water table location in numerical methods such as boundary integral and finite element methods is to assume an initial position and to improve upon it iteratively by applying the relevant boundary conditions. A similar approach is used here since the boundary condition, equation (6), that determines $\eta(x)$ is both nonlinear and implicit, precluding any exact representation for the water table. However, the water table can be approximated by a cubic spline, and the approximation iteratively improved using equation (6).

4 FINITE ELEMENT SOLUTIONS FOR CONTAMINANT TRANSPORT

The solute transport equation (7) is not easily solved analytically for the boundary conditions of Figure 1 and a numerical solution is therefore employed. High computational costs are incurred in obtaining solutions over large spatial and long time domains because most numerical techniques are prone to numerical dispersion. In the work presented here, a fractional step algorithm (Sobey, 1983) is used so that the advective and dispersive steps are solved separately. Equation (7) is thus divided into two components:

$$\frac{\partial C}{\partial t} = (L_1 + L_2)[C] \qquad (19)$$

with

$$L_1[C] = -u\frac{\partial C}{\partial x} - v\frac{\partial C}{\partial y} \qquad (20)$$

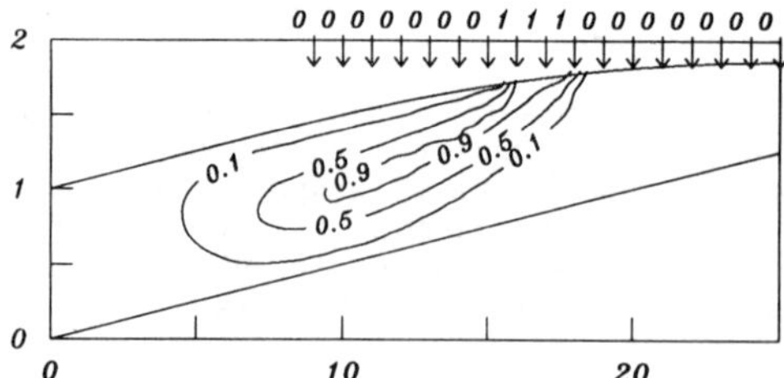

Fig. 2 Solute concentrations for case 1 at $t_{nd} = 50$ (90 years).

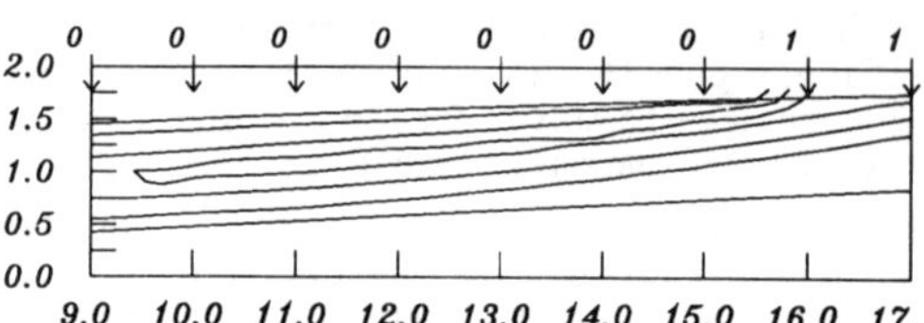

Fig. 3. Solute concentrations for case 1 at t_{nd} = 50 (90 years) plotted with undistorted length scale.

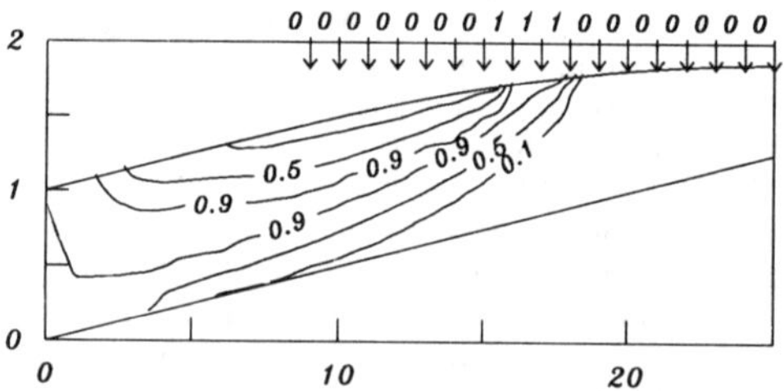

Fig. 4. Solute concentrations for case 1 at t_{nd} = 100 (180 years).

and

$$L_2[C] = \frac{\partial}{\partial x}\left[D_{xx}\frac{\partial C}{\partial x} + D_{xy}\frac{\partial C}{\partial y}\right] + \frac{\partial}{\partial y}\left[D_{yx}\frac{\partial C}{\partial x} + D_{yy}\frac{\partial C}{\partial y}\right] \qquad (21)$$

Equation (20) can be solved by a Runge-Kutta method to determine the advection of solute over any time step while equation (21) is solved using a finite element method (FEM). The advective step is accomplished by tracking the location of points carrying solute from the source as they move with the seepage flow. Solution of the dispersive component (21) requires a mesh of nodes for the FEM and this is formed by establishing a triangular mesh through points along the streamlines. The availability of the continuous solution for the stream function and velocities enables a set of points to be generated along an appropriately spaced set of streamlines from the water table to the seepage face. These points represent the successive locations (at the end of equal time intervals) of points commencing from the inflow surface at an instant of time. By adopting this approach only one mesh has to be generated through a non-uniformly spaced set of points and the need for interpolating solute concentrations to a fixed mesh is avoided.

5 RESULTS AND DISCUSSION

Solutions for solute transport have been obtained for hillslope geometry of the general form shown in Figure 1, consisting of sloping, linear top and bottom boundaries, combined with a steady water table. A section DE of the water table CG (Figure 1) is subjected to input of solute with a concentration C_0; the recharge rate R is uniform over the entire water table including the solute inflow section. The solution method could equally be applied if the source of solute were located within the saturated slope rather than at the surface. Results are presented in the form of contours of dimensionless concentration (relative to the source concentration C_0) for a range of parameter values. Referring to Figure 1, Figure 2 shows results for $s/d=25$, $R/K=3\times10^{-3}$, at a dimensionles time of 50.

A note about the presentation of results using distorted length scales is necessary here. The length to depth (s/d) ratio is so extreme for relevant slopes that the horizontal dimension must be scaled down to a much larger extent than the vertical dimension in order to fit the diagram on a page but still show meaningful detail in the vertical direction. Plots of lines of equal concentration such as those shown in Figure 2 must then be interpreted with care to avoid drawing incorrect conclusions about the pattern of solute movement. To illustrate the point, Figure 3 shows a short segment of the flow domain of Figure 2 at an undistorted scale. The section shown is the entire solute source segment and a short segment immediately down slope from it. It will be noted that the perceptions of the shape of the solute zone are substantially different from the two figures.

Results for several hillside dimensions and flow and aquifer parameters are presented in Figures 2 to 10. Details of the cases and the corresponding parameter values are given in Table 1. The results are in nondimensional form. Concentration is normalised by the source concentration, length dimensions are scaled relative to the depth d (Figure 1), and recharge is scaled relative to the hydraulic conductivity K (equation 1). A non-dimensional time is defined as:

$$t_{nd} = \frac{td}{nR} \qquad (22)$$

where n is the porosity of the soil.

The dimensionless times quoted are therefore only comparable in a relative sense when other parameters

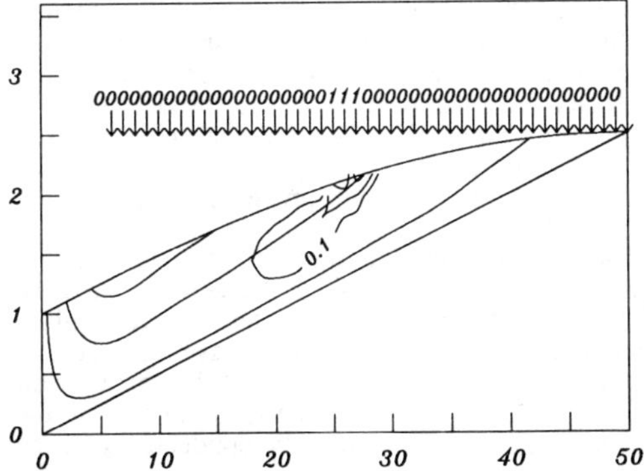

Fig. 5. Solute concentrations for case 2 at t_{nd} = 50 (270 years).

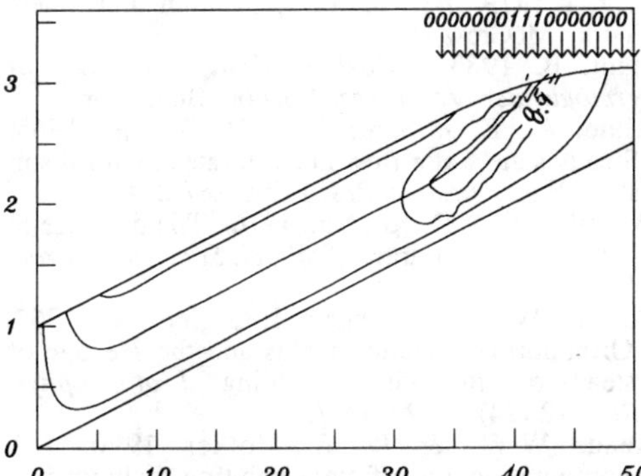

Fig. 6. Solute concentrations for case 3 at t_{nd} = 50 (90 years).

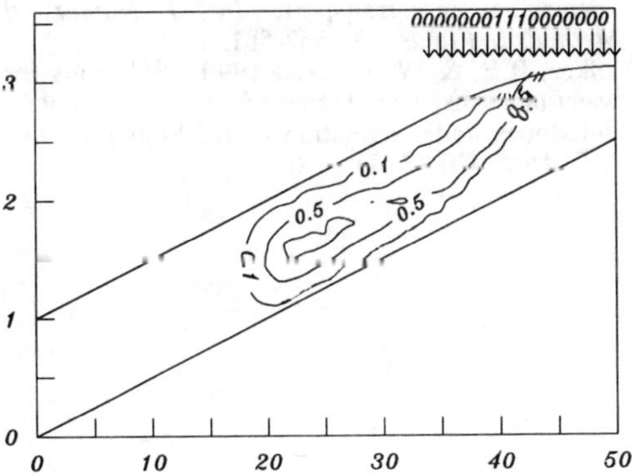

Fig. 7. Solute concentrations for case 3 at t_{nd} = 100 (180 years).

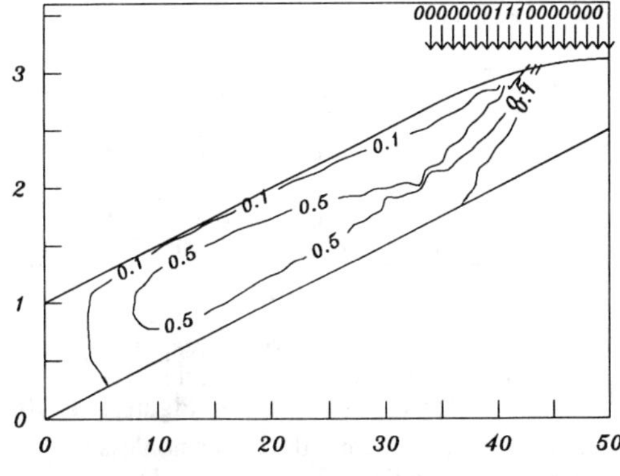

Fig. 8. Solute concentrations for case 3 at t_{nd} = 150 (270 years).

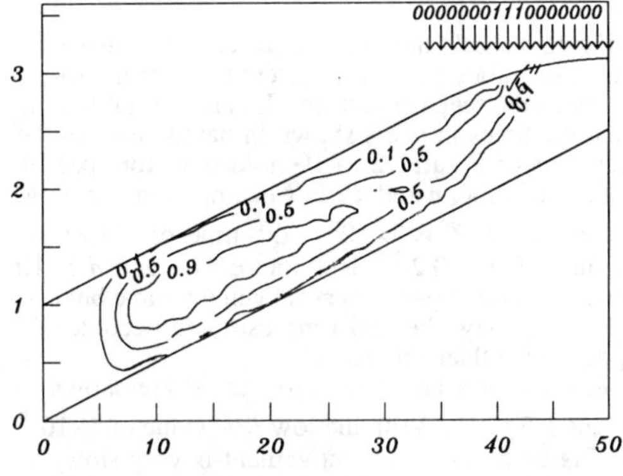

Fig. 9. Solute concentrations for case 4 at t_{nd} = 150 (270 years).

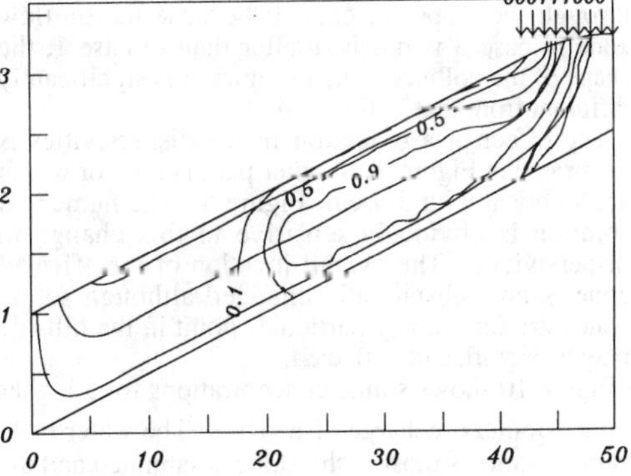

Fig. 10. Solute concentrations for case 5 at t_{nd} = 100 (90 years).

Table 1 Details of flow cases

Case	s/d	$10^3 R/K$	α_I/d	α_{II}/d
1	25	3.0	1.00	0.010
2	50	1.0	1.00	0.010
3	50	3.0	1.00	0.010
4	50	3.0	0.10	0.001
5	50	6.0	1.00	0.010

are constant. Determination of corresponding real times requires a knowledge of R, d and n. In all cases given in Table 1 and shown in Figures 2 to 10 the slope θ is 5 percent and the dimensionless length of the solute source DE is 2.

The extent of the water table is indicated by arrows in Figures 2 to 10. The fresh water inflow zone is indicated by zero values above the arrows while the extent of the solute source is shown by values of unity.

Figure 4 shows the shape of the contaminated zone for case 1 after there is sufficient time for the solute to reach the seepage surface. It is important to note that the times in years shown in parentheses in the captions of Figures 2 to 10 hold only for specific values of R, K, n and d. For example, in Figure 4, given that R/K is 3×10^{-3}, the time of 270 years applies if n is 0.2, K is 1 metre/day, and d is 10 metres. If other parameters remain the same but K is 10 metres/day, the real time estimate becomes 27 years rather than 270 years.

Results for a larger s/d ratio of 50 are shown in Figures 5 to 10. With the low R/K value of 1×10^{-3} in Figure 5, the solute movement is very slow. A threefold increase in R/K (Figure 6) causes a marked reduction in the length of the water table but, as expected, the solute reaches approximately the same location as for Figure 5 in one third of the time. Figures 7 and 8 show the progression of the solute through the slope for case 3; because the outflow zone in case 3 is much smaller than in case 1, the shape of the polluted zone in Figure 8 is significantly different from that in Figure 4.

The effect of a reduction in the dispersivities is illustrated in Figure 9, all other parameters for which are identical with those of Figure 8. The numerical solution is obviously sensitive to this change in dispersivities. The overall location of the affected zone is not substantially changed although solute concentrations at any particular point in the hillside may be significantly affected.

Figure 10 shows solute concentrations for a higher dimensionless recharge of 6×10^{-3}. The water table is quite short as most of the slope is saturated and the solute movement is faster in response to the higher recharge.

6 SUMMARY

Series solutions of the velocity field for hillside seepage have been combined with numerical solutions of the mass transport equation for solute movement. The continuous description of the velocity field has been shown to provide an ideal foundation for an efficient numerical solution based on a fractional step algorithm, which allows the advective and dispersive processes to be treated separately.

Application of the method to long shallow permeable layers in hillsides has demonstrated the influence of various parameters on the solute movement. While dispersivities have a discernible effect on the shape and location of the contaminant zone, the movement of solute is dominated by the recharge rate and the hydraulic conductivity of the permeable zone.

REFERENCES

Bear, J. 1979. *Hydraulics of groundwater.* New York: McGraw-Hill.

Hunt, B. 1983. *Mathematical analysis of groundwater resources.* London: Butterworths

Klute, A., E. J. Scott & F. D. Whisler 1965. Steady state water flow in a saturated inclined soil slab. *Water Resour. Res.* 1 (2): 287-294.

Powell, N.L. & D. Kirkham 1976. Tile drainage in bedded soil or a draw. *Soil Sci. Soc. Amer. Proc.* 40: 625-630.

Powers, W.L., D. Kirkham & G. Snowden 1967. Orthonormal function tables and the seepage of steady rain through soil bedding. *J. of Geophys. Res.* 72 (24): 6225-6237.

Read, W.W. & R. E. Volker 1990. A computationally efficient solution technique to Laplacian flow problems with mixed boundary conditions. *Computational Techniques & Applications: CTAC '89:* 707-714. New York: Hemisphere

Sobey, R.J. 1983. Fractional step algorithm for estuarine mass transport. *Int. J. Numerical Methods in Fluids* . 3: 567-581.

Volker, R.E. & W.W. Read 1990. Solutions for water table shape and seepage face location on long hillslopes under irrigation. *Civ. Engng Trans., I. E. Aust.* CE 32 (1): 1-6.

Environmental Management, Geo-Water & Engineering Aspects, Chowdhury & Sivakumar (eds)
© 1993 Balkema, Rotterdam. ISBN 90 5410 099 0

Water quality simulations for Chao Phraya River, Thailand

L.C.Wadhwa
James Cook University, Townsville, Qld, Australia

S.Supatanasinkasem
National Environment Board, Bangkok, Thailand

ABSTRACT: The Chao Phraya river which passes through the heart of Bangkok frequently gets flooded causing significant damages. Plans to reduce the effect of flooding have detrimental effect on water quality since the river also carries most of the residential and industrial waste generated by the metropolis of well over six million people. A computer model is presented along with the simulation of water quality under various options of waste treatment and a specific flood protection plan. It has been found that the river's water quality in year 2002 will be unsatisfactory for about nine months in a year unless sewage is treated before discharging into the river. Various operating procedures for the flood protection plan have been evaluated and recommendations have been made for the optimal operating procedure.

1. INTRODUCTION

Bangkok, the capital of Thailand and a growing metropolis with over six million people, has a number of canals which were once used as a mean of transportation. They are now used as outfall sewers which carry domestic and industrial wastes to the Chao Phraya river. The 380km. long river is a union of 4 merging rivers and has a catchment area of about 162,000 sq.km. It originates from Nakorn Sawan province in northern Thailand and passes through the heart of Bangkok before reaching the Gulf of Thailand (Fig. 1). It is the country's most important river and plays a major role in the city's drainage, sewage, water supply as well as economic systems.

Presently, the river suffers from two major conflicting problems of flooding and water quality. Bangkok is located in a very flat terrain near the mouth of the Chao Phraya river. It receives an average annual rainfall of about 1,300 mm. primarily from May to October. During this wet season the city is frequently flooded due to rising of water level in the Chao Phraya river. The major flood in 1983 was estimated to have caused damage to the city of about Baht 6600 million (about $250 million). The frequency of flooding has increased due to continuing land subsidence in the Bangkok area. The present water quality is unfortunately also not very satisfactory. Dissolved Oxygen (DO), essential for aquatic life, is heavily used in the process of decomposition of organic matters which are present in the wastes. The DO concentration is gradually decreasing particularly in the densely populated areas. In the dry season, when flow rate is low, the DO level may reach zero.

The flood protection plan proposed in 1984 by the Thai-Austrian Consortium, ACE-CONSULTCO-CAE, aims to protect the entire Bangkok and its vicinity from flooding. The plan is entitled "Integrated Flood Relief Plan of the West Bank", also known as "the Chao Phraya 2". The reduced flow in the river may, however, have serious implications for the water quality as the waste assimilative capacity of the river partially depends on its flow rate.

This paper describes a model which can simulate water quality due to growth in population and industry, estimate the need for waste treatment to maintain desired water quality level and evaluate the effect of proposed flood protection plan on river's water quality. The model is also used for designing an alternative operational policy for maintaining water quality in the Chao Phraya river.

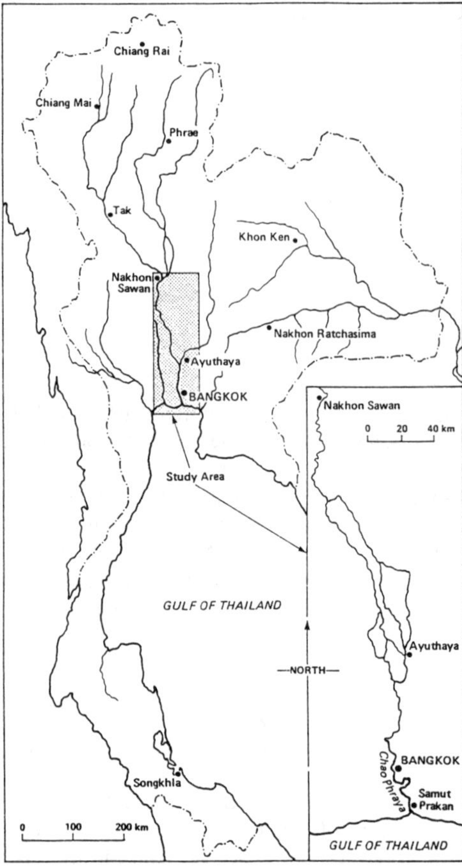

Figure 1. The study area

2. WATER QUALITY MODELLING

The concern for water quality and the consequent desire to understand the physical, chemical and biological activities in streams, which are the major source of water supply, has resulted in the development of a number of water quality mathematical models. The modelling of water quality allows the evaluation of various pollution loading and treatment policies. One of the first mathematical models which is still being widely referred to is the model developed by Streeter and Phelps (1925). The model involves the analytical prediction of dissolved oxygen deficit as a function of time. Following their work, several more advanced water quality models have been developed. Examples include models developed by Beck and Young (1975), Whitehead et al. (1981), Gowda (1983), Finney (1982), Dewey (1984). The studies of the water quality of Chao Phrya river vary from salinity (Tosukhowong, 1968), pollution loading (Bernadino, 1974), heavy metals (Hungspreuggs ,1984), etc. Leffel (1968) used a mathematical model to study the DO sag curve of the Chao Phraya river. This study investigated the effect of the mixing of oxygen-saturated sea water in the lower part of the river which increases DO concentration. Extensive studies and monitoring of the river's water quality have been carried out since the establishment of the Office of the National Environment Board (ONEB) in 1975.

3. THE MODEL STRUCTURE

The study area encompasses the city of Bangkok and part of Samut Prakarn and Nonthaburi provinces. A diversion channel is proposed to be located in the west bank of the river where population density and land subsidence are less compared to the area on the east bank. The channel and the river will be occasionally joined through canals. The study area, apart from being high in population density, has a large number of industrial plants. Most of the plants are located in the lower part of the study area close to the river mouth.

The model used in this study comprises of a pollution source submodel (population and industries), a BOD loading submodel and a water quality submodel. A brief description follows:

3.1 Pollution source submodel

The area has been divided into 8 zones for the purpose of allocating BOD generated by domestic and industrial sources to various canals. Each zone is bounded by districts for which the population data are known. Canals in each zone are assumed to carry the zonal BOD generated both by domestic and industrial sectors for eventual discharge in the Chao Phraya river. These zones are shown in Fig. 2.

The prediction of population growth made by Thai-Austrian Consortium for each 5-year period until year 2002 are used in this model.

Industrial plants are located mostly along the lower part of the river. Wastes generated by these plants, though partly treated, still account for a large amount of BOD being discharged into the Chao Phraya river.

3.2 BOD loading model

The BOD loading model calculates the BOD generated in each zone by the resident population and industries in the zone. The BOD generated is transported through the canals in the area to the river. The BOD loading model determines the location and amount of BOD discharged along the river within the study area.

The estimate of per capita domestic waste generation of 36 gms/day by Japan International Cooperative Agency (JICA) study (1982) is used in this study. The BOD loading from industry is obtained from the AIT study (1978,1980).

3.3 Water quality model

The water quality model is based on the material balance formulation. It can be considered in terms of mass flow in and out of a given element in which kinetic reactions occur as shown in Figure 3. Inputs include flow rate, initial water quality, and the addition of BOD in the element. Reactions occur within the element such as reaeration, deoxygenation, etc.

Based on the Streeter-Phelps model (1925), the BOD concentration in the upreaches of the river is given by

$$\frac{dL}{dt} = -(K_1+K_3)L$$

and the differential equation for the rate of change of dissolved oxygen deficit can be written as

$$\frac{dD}{dt} = (K_1+K_3)L - K_2D$$

where L = BOD demand, mg/l, D = oxygen deficit i.e. the difference between the DO saturation concentration and the concentration at time t, K_1, K_2, and K_3 are deoxygenation rate coefficient, reaeration rate coefficient and the BOD sedimentation rate constant, respectively.

The equations for the dispersion of BOD and DO, considering the BOD added from the bottom sediments and the oxygen produced by photosynthesis, respectively become

$$\frac{\partial L}{\partial t} = E\frac{\partial^2 L}{\partial x^2} - U\frac{\partial L}{\partial x} + (K_1+K_3)L - L_a = 0$$

and

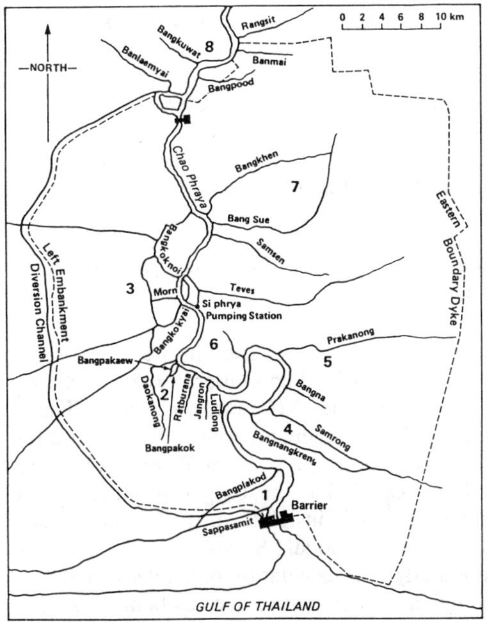

Figure 2. Zones in the study area

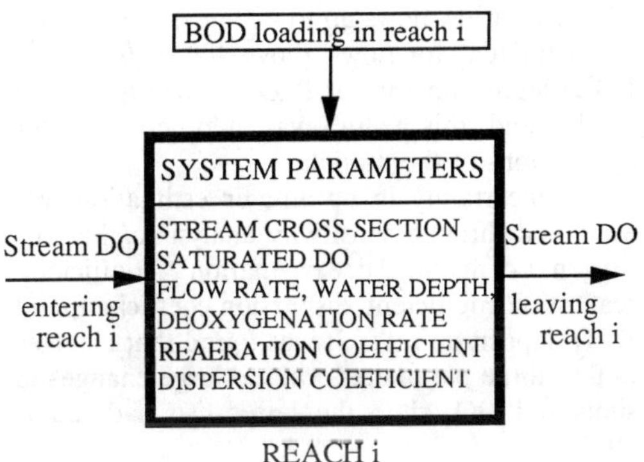

Figure 3. Stream dynamics

$$\frac{\partial C}{\partial t} = E\frac{\partial^2 C}{\partial x^2} - U\frac{\partial C}{\partial x} - K_2(C_s-C) + (K_1+K_3)L - L_a = 0$$

where L_a = addition of BOD introduced by bottom deposit, E is the longitudinal dispersion coefficient, C_s = DO saturation concentration, C = DO concentration and U = average stream velocity.

The finite element model used in this study is adapted from AIT (1978,1980). The Galerkin's weighted residual method is used to obtain the weak form and the discretized form of substance balance equation. A vertically averaged two dimensional substance balance equation incorporating time derivative, advection, dispersion, decay and the sources is used in calculating DO distribution in the Chao Phraya river due to BOD discharged.

3.4 Coefficient estimation for the Chao Phraya River

Most of the parameters used in the model were estimated from field and laboratory experiments conducted in the AIT study (1980). These included the deoxygenation coefficient, reaeration coefficient, and dispersion coefficient. The field study indicated that the oxygen produced from photosynthesis is insignificant and hence is not included in the model. Samples from 17 stations were used in the calculation of the deoxygenation coefficient. Parameters such as benthic oxygen demand (BEN) and BOD deposition rate were determined in the process of model calibration using the field measurements undertaken in 1983. Based on this, a value of BEN at 6 $g/m^2/day$ was adopted for low flows up to 400 m^3/s and a value of 8 $g/m^2/day$ for flows above 400 m^3/s. The BOD deposition rate at 0.05 per day gave best results and this value was adopted in later applications of the model.

The uncertainty in parameter estimation was examined through sensitivity analysis of benthic oxygen demand, deoxygenation coefficient, reaeration coefficient, dispersion coefficient, and BOD deposition rate. It was found that changes in first three parameters caused large changes in simulated DO while the latter two had minor effect.

4. MODEL APPLICATIONS

4.1 Criteria for interpreting model output

The output of model simulation is a set of values of DO for each reach of the river. A meaningful interpretation of the effectiveness of a policy is obtained by determining
(a) the proportion of reaches with DO below the minimum value of 2.0 mg/l during dry months (flow of 200m^3/s) which is a critical indicator of water quality.
(b) the minimum value of the river flow which will result in at least 80% of the reaches with DO above 2.0 mg/l, and
(c) the number of months for which the water quality is likely to be unsatisfactory.

These statistics correspond to the definition of the water quality adopted by the National Environment Board, Thailand (NEB 1983/1984). During periods of high river discharge, (flows above 1000 m^3/s), DO in the river well exceeds 2.0 mg/l. The criteria become significant at low level of flow when many reaches of the river may have DO of less than 2.0 mg/l.

4.2 Model validation

Having established the value of all parameters used in the water quality model by field observations and model calibration, simulation was carried out to establish if the model output reproduced the 1978 field measurements.

Three flow rates were used in the experiment, 225, 475, and 865 m^3/s. The goodness of fit test confirmed that the model parameters are realistic representations of the physical and biological processes in the Chao Phraya river.

4.3 Predicting future water quality in the Chao Phraya River

The effect of continued growth in population and industry on the water quality in the Chao Phraya river was simulated.

It was observed that the critical water quality in the river occurs during the extreme case of low flow. It is shown in Table 1 that 61.45 % of reaches would have DO less than 2.0 mg/l. The river's water quality gradually deteriorates with increase in population. Critical DO in the river occurs from about 50 km. down to the river mouth. The DO concentration could be maintained above the minimum required level only during the wet months of September to November.

4.4 Evaluating waste treatment strategies

To determine the effect of establishing waste treatment plants to reduce BOD loading in the river, it is assumed that four treatment plants will be built in the most populated zones of the Bangkok Metropolitan area (zones 3 and 6) and these plants will operate at an efficiency of (a) 65% and (b) 89%

Table 1. Summary of model simulation runs

Policy/Scenario	% of reaches with DO < 2.0	Min. flow for maint. water quality (m^3/s)	Months/yr. of unsat. quality
1. Base case 1983	53.01	490	7
2. Future year 2002	61.45	560	9
3. Sewage treatment @ 65% efficiency	50.96	300	6
4. Sewage treatment @ 89% efficiency	24.10	235	4
5. Flood protection plan	53.01	335	6
6. Recommended flood Protection plan	--	162	3

It was found that the 65% treatment of the two most populated zones cannot maintain the river water quality during periods of very low flow. The increase in the treatment efficiency to 89%, however, reduces the percentages of reaches that had DO less than 2.0 mg/l from 50 to about 24. The results are shown above in Table 1.

4.5 Evaluating the proposed Chao Phraya 2 plan

The proposed flood protection scheme involves the construction of:
(i) a diversion channel which will carry excess water, by-pass the city, and discharge to the sea. The channel is designed to carry water at a maximum rate of 2,000 m^3/s.
(ii) a diversion dam at Pak Kred about 70 km. from the river mouth to control the diversion of run off.
(iii) sea-barriers at the river mouth and the end of the channel to protect sea water intrusion. Water pumps will be installed at the barrier where they will be used in case of closure of the barrier.

The plan is expected to prevent an area of 14, 000 $km.^2$ from flooding. The effect of the proposed diversion plan on the water quality in the Chao Phraya river with operating procedures suggested by the plan was simulated on the assumptions that the plan will be operated in the year 2002.

From the river mouth up to about 50 km., DO reached zero mg/l for most of the reaches when flow rate was 200 m^3/s. At 380 m^3/s, DO concentration increased but near the river mouth it was still lower than 2 mg/l. A flow rate of 600 m^3/s however did not produce undesired DO level in the river. It was concluded that if the flow rate was reduced according to the plan, the operation could affect the river water's quality particularly during the time when the river is recovering from low DO due to low flow in a dry period. Certain measures are required in order to improve the water quality of the Chao Phraya river.

4.6 Designing an improved operating policy for the proposed Chao Phraya 2 plan

An alternative operating policy designed to improve the water quality in the river without increasing the risk of flooding was simulated. The treatment plants could withhold the treated effluent during the period of closure of the barrier. This would bring the required flow in the Chao Phraya river to the lowest acceptable level. The rates of diverting of water to the channel were varied to reflect the proposed policy.

By relating the DO to the current stream standard and the proposed operating scheme, it was found that the required flow during the closure and the opening of the barrier were 335 and 560 m^3/s respectively. With the presence of treatment plants, required flow rate was further reduced to 185 m^3/s at 89% treatment efficiency. This flow is reasonably low when compared to the flow proposed by the plan and reduces the risk of flooding.

5. SUMMARY AND CONCLUSIONS

The model was calibrated to obtain certain parameters which were also examined for their sensitivity. After validating against field measurements, the model was used to predict future water quality in the river and also to evaluate various policies.

The model was used to predict water quality in the future with anticipated growth in population and industry. It was observed that the proportion of elements with DO of less than 2.0 mg/l increased from 53% in 1983 to 61% in 2002 at low flow rate. In case of high flow, no reach showed poor water quality. It was found that over 80% of elements would have acceptable water quality at a flow rate of 820 m^3/s. These results signify that the desired water quality is unlikely to be maintained except during the months of September to November.

The water quality model was simulated with four waste treatment plants in the most densely populated areas of Bangkok. It was found that with 65% efficiency, the river water quality could not be maintained during periods of very low flow. The increase in the treatment efficiency to 89%, however, reduced the percentages of reaches that had DO less than 2.0 mg/l from 50 to about 24.

The flood control plan which involves diverting part of the upstream flow to the diversion channel was found to create undesirable DO concentration throughout the year. With the closing of canals along the west bank of the river, more than 50% of the reaches will have DO less than 2.0 mg/l. When the barrier is open, the number of such reaches increases.

An alternative operating policy to improve the water quality in the river without increasing the risk of flooding was simulated. It is recommended that the flow in the Chao Phraya River should be at least 560 m^3/s as far as possible. When the barrier is closed and waste discharge from the west canals is temporarily withheld, flow rate can be reduced to 335 m^3/s. If these flows are too large and could affect the flood relief plan, treatment plants should be provided in zones 3 and 6. These could reduce the required flow to 185 m^3/s at the maximum treatment efficiency of 89%. Further flow reduction may be possible by holding treated effluent from the treatment plants.

REFERENCES

Asian Institute of Technology 1978,1980. *Mathematical optimization model for regional water quality management : A case study of Chao Phraya river (Phase I).* The Office of the National Environmental Board Thailand, Pub. 1981-005.

Beck, M. B. & Young, P. C. 1975. A dynamic model for DO-BOD relationships in non-tidal stream. *Water Research.*, 9 : 769-776.

Bernadino, A. 1974. Pollution survey of Chao Phraya River in Phathumthani Province, M.Eng. *Thesis No. 629*, AIT, Bangkok.

Dewey, R.J. 1984. Application of stochastic dissolved oxygen model. *Env. Eng. Div. ASCE.*, 110 (2) : 412-429.

Finney, B.A. 1982. Random differential equations in river water quality modelling. *Water Resource Research.*, 18 (1) : 122-134.

Gowda ,T.P. Halappa 1983. Modelling nitrification effects on the dissolved oxygen regime of the speed river. *Water Research.* 17 (12):1917-1927.

Hungspreuggs, M. et.al. 1984. Heavy metals and polycyclic hydrocarbon compounds in benthic organisms of the Upper Gulf of Thailand. *Mar. Pol. Bull.*, 15 (6):213-218.

Japan International Cooperation Agency, 1982. *Feasibility study of Bangkok sewage system project in Kingdom of Thailand.*

Leffel. E. R. 1968. Pollution of the Chao Phraya River estuary. *J. San. Eng. Div. ASCE.*, 94 (SA2): 295-306.

National Environment Board 1982-84. *A study of the lower part of the Chao Phraya 's water quality.* The Office of the National Environment Board Thailand, Pub. 1982-003 (in Thai).

Streeter, H.W. & E.B. Phelps 1925. A study of the purification of the Ohio river, Ill. Factors concerned in the phenomena of oxidation and reaeration. *US Public Health Service*, Washington DC Bulletin 146, 75pp.

Tosukhowong, S. 1968. Oxygen demand of bottom muds in Bangkok. M.Eng. *Thesis No. 220*, AIT Bangkok.

Whitehead, P., B. Beck & E.O' Connell 1981. A system model of stream flow and water quality in the Bedford - Ouse River-II. *Water Research.* 15:1157-1171.

Environmental Management, Geo-Water & Engineering Aspects, Chowdhury & Sivakumar (eds)
© 1993 Balkema, Rotterdam. ISBN 90 5410 099 0

Electro-kinetic barrier to contaminant transport

Albert T. Yeung
Texas A&M University, Tex., USA

ABSTRACT: The United States Environmental Protection Agency mandates the use of an engineered double liner containment system for hazardous waste. However, the long-term integrity and effectiveness of the synthetic liner as physical barriers to contaminant transport in such a harsh service environment are always in question. The compacted clay liner would always allow some seepage in time because of its porous nature. An electro-kinetic barrier created by the continuous or periodic application of an electrical gradient across a compacted clay liner may inhibit migration of hazardous constituents into the environment. This paper presents the theoretical formulation of the complex coupled transport processes of fluid, electricity and contaminants through such a barrier under the combined influences of hydraulic, electrical and chemical gradients, and the results obtained from an experimental program performed to evaluate the validity of the developed theory on compacted clay.

1 CURRENT SITUATION

The industrial commitment required to support the needs of our ever-increasing population and ever-improving living standard is producing more and more waste products and by-products. For the protection of public health and the minimization of environmental damage to present and future generations, these wastes must be destroyed, treated or stored properly somewhere in the environment. Techniques being practiced to achieve these aims include chemical, physical and biological treatment, resource recovery, land treatment or land spreading, incineration, detonation, solidification and stabilization, deep well injection, ocean dumping and land disposal.

Though ground burial of waste is not an ideal solution, not necessarily even a good solution, it may be the only feasible and economical solution in most cases. The hazardous nature of the waste remains unchanged, but it is isolated from immediate human contact. However, indiscriminate or careless dumping of materials ranging from household sewage to toxic metals to lethal organic compounds will definitely cause environmental damage to present and future generations. The problem then is to design, construct and operate the containment system so that contaminant migration into the environment is prevented, or at least limited to an acceptable level.

Caps and liners are used in conjunction with each other to meet this end. Regulations, performance standards and technological guidelines for the design, construction and operation of landfills and surface impoundments have been issued by the U. S. Environmental Protection Agency (U. S. EPA 1985) in compliance with the Hazardous and Solid Waste Amendments (HSWA) which was signed into law by President Ronald Reagan on November 8, 1984.

In brief, the Agency recommends engineered double liner containment systems with leachate collection systems above and between liners. These liners may be flexible membrane liners (FML), compacted clay layers or a composite composed of both components. All these liners should be designed, constructed and operated to minimize the migration of any hazardous constituents into the environment during the operation period and a 30-year post-closure monitoring period of the facility. The synthetic flexible membrane liners are supposed to be designed, constructed and operated to prevent the migra-

Materials	Dimensions and specifications	Nomenclature
		Solid waste
Graded granular filter medium	Recommended thickness > 6 in.	Filter medium
Granular drain material	Recommended thickness > 12 in. Hydraulic conductivity > 1×10^{-2} cm/s	Primary leachate collection and removal system
Flexible membrane liner (FML)	Recommended thickness of FML > 30 mils	Top liner (FML)
Granular drain material	Recommended thickness > 12 in. Hydraulic conductivity > 1×10^{-2} cm/s	Secondary leachate collection and removal system
Flexible membrane liner (FML)	Recommended thickness of FML > 30 mils	Compression connection (contact) between soil and FML
Low permeability soil, compacted in lifts	Recommended thickness > 36 in. Recommended hydraulic conductivity $< 1 \times 10^{-7}$ cm/s Prepared in 6 in. lifts, surface scarified between lifts	Bottom liner (composite FML and compacted low permeability soil)
In-situ soil	Unsaturated zone Saturated zone	Native soil foundation/sub-base

Figure 1 Schematic of a FML/composite double liner system for a landfill

tion of any hazardous constituents through the liner system. The function of the low hydraulic conductivity compacted clay liner is to ensure there will not be any significant quantity of waste leachate penetration into the environment for at least thirty years after closure, even if the flexible membrane liner fails. A schematic of the FML/composite double liner system for a landfill is illustrated in Fig. 1.

There are always questions of the chemical compatibility; i.e., the long term integrity and effectiveness of the synthetic liner in harsh service environments such as landfills. Construction flaws always aggravate the problem. Bass et al. (1985) surveyed twenty-seven lined facilities and found twelve failures at ten sites resulting in four or five cases of groundwater contamination. Thus, the important role of the compacted clay liner in the protection of the environment is evident. No matter how well it is designed, constructed and operated, however, it is still a porous medium. There will be some seepage through it in time if there is a sustained hydraulic head. Moreover, hazardous constituents of the leachate can migrate from the impoundment into the groundwater system by hydrodynamic dispersion in addition to advection. Thus low hydraulic conductivity compacted clay liners cannot prevent contaminant transport completely. The capacity of a soil to attenuate contaminants by adsorption is limited, and adsorption mechanisms are not fully understood. Hence, it cannot be relied upon for the protection of the groundwater system.

This paper presents the theoretical and experimental results of a study to investigate the viability of electro-kinetic counterflow to create a leakage proof barrier to contaminant transport through compacted clay.

2 WHAT IS AN ELECTRO-KINETIC BARRIER ?

The concept of an electro-kinetic barrier to contaminant transport through a compacted clay liner is depicted in Fig. 2. v_h and v_d are the migration velocities of the contaminants due to advection and hydrodynamic dispersion, respectively. v_e is the electro-osmotic counterflow velocity induced by the imposed electrical gradient. v_c and v_a are the effective ionic migration velocities of the cation and the anion relative to the motion of fluid induced by the electrical gradient, respectively. u_c and u_a are the effective ionic mobilities of the cation and the anion in soil, respectively. The in-place saturated hydraulic conductivity k_h of the compacted clay liner for a hazardous waste landfill has to be 1×10^{-9} m/s or less to fulfill the mandatory requirement of the HSWA of 1984 (U. S. EPA 1985). The coefficient of electro-osmotic permeability k_e of most soils is generally in the range of 1×10^{-9} to 10×10^{-9} m²/Vs (Casagrande 1983; Mitchell 1976). Consider a fine-grained soil with the maximum acceptable hydraulic conductivity for a compacted clay liner k_h of 1×10^{-9} m/s and a typical electro-osmotic permeability k_e of 5×10^{-9} m²/Vs. For equal flow rate per unit cross-sectional area,

$$k_h i_h = k_e i_e \qquad (1)$$

Rearranging

$$i_h = \frac{k_e}{k_h} i_e \qquad (2)$$

where i_h and i_e are the imposed hydraulic and electrical gradients, respectively. If an electrical gradient of 100 V/m is used, then

$$i_h = \frac{5\times10^{-9}}{1\times10^{-9}} \times 100 = 500 \qquad (3)$$

Thus, an electrical gradient of 100 V/m or 1 V/cm can move water as effectively as a hydraulic gradient of 500. Hence, the hydraulic flow rate induced by a small electrical gradient applied in the direction shown in Fig. 2 can stop the advection component of contaminant migration.

The migration of contaminants by hydrodynamic dispersion may also be hindered by an electrical gradient in excess of that needed to balance the hydraulic gradient. In fine-grained soils typical for liner construction, molecular diffusion can be expected to be the dominant component of hydrodynamic dispersion. Ionic contaminants will be migrated relative to the hydraulic flow under the influence of an electrical gradient. However, the direction of migration depends on the sign of charges and the effective ionic mobilities of the cation and the anion. In physical chemistry, the ionic mobility of an ion is defined to be the velocity of the ion in free solution under the influence of a unit electric field. However, as the paths for ionic migration in soils are much longer and more torturous than in aqueous solutions, the effective ionic mobilities of ions in soil are considerably less than the ion mobilities in free solution (Mitchell 1991). Fig. 3 illustrates the migrations of cations and anions under the influences

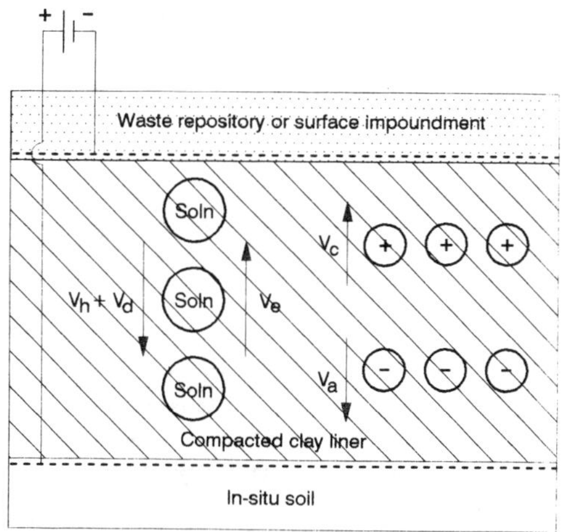

Figure 2 Concept of an electro-kinetic barrier to contaminant transport

of hydraulic, chemical and electrical gradients in terms of concentration against distance at a certain time.

3 THEORETICAL FORMULATION

Because of the length limit on the paper, only the results of the theoretical formulation are presented here. The complete development is given elsewhere (Yeung 1990; Yeung & Mitchell 1992). The contaminant transport through an electro-kinetic barrier involves the coupled flows of fluid, electricity and contaminants under simultaneously imposed hydraulic, electrical and chemical gradients. These complex transport processes are formulated by the formalism of non-equilibrium thermodynamics to give a macroscopic quantitative description of the phenomenological coefficients.

By the assumptions of non-equilibrium thermodynamics, the flows or fluxes J_i are linear, homogeneous functions of the gradients or driving forces X_i. Thus, any flow or flux is related to the gradients or driving forces by

$$J_i = \sum_{j=1}^{n} L_{ij} X_j \qquad (i = 1,2,\ldots,n) \qquad (4)$$

where the phenomenological coefficients L_{ij} are independent of the driving forces. Moreover, the matrix of phenomenological coefficients is symmetric, i.e.,

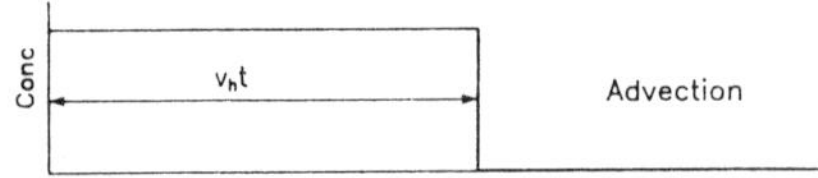

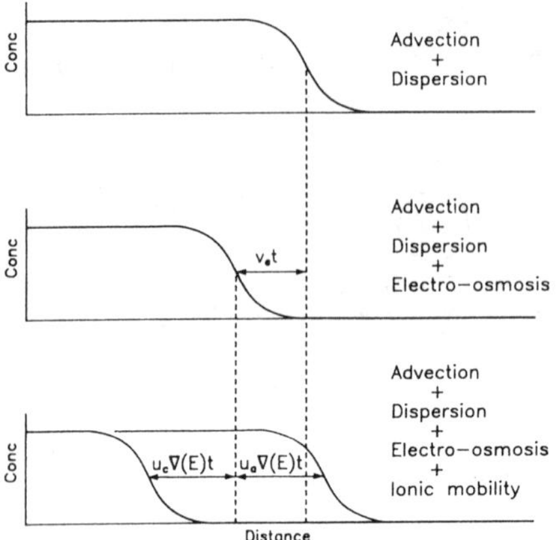

Figure 3 Movements of cations and anions under the influences of hydraulic, electrical and chemical gradients

$$L_{ij} = L_{ji} \qquad (i,j = 1,2,\ldots,n) \tag{5}$$

provided a proper choice is made for the fluxes and driving forces (DeGroot & Mazur 1962).

Assuming the contaminant consists of an ionic salt that dissociates into one cation and one anion, the driving forces are identified to be the hydraulic gradient $\nabla(-P)$, the electrical gradient $\nabla(-E)$, the concentration dependent parts of the chemical gradients of the cation $\nabla(-\mu_c^c)$ and of the anion $\nabla(-\mu_a^c)$. The fluxes are the volume flow rate of the solution per unit area J_v, the electric current density I, the diffusional flow rate of the cation J_c^d and of the anion J_a^d per unit area relative to the flow of fluid. The diffusional flows are related to the absolute flows by

$$J_i = J_i^d + \frac{c_i}{c_f} J_f \tag{6}$$

where c_i is the concentration of ion i and c_f is the concentration of fluid. Hence the set of phenomenological equations relating the four flows and the four driving forces is

$$J_v = L_{11}\nabla(-P) + L_{12}\nabla(-E) + L_{13}\nabla(-\mu_c^c) + L_{14}\nabla(-\mu_a^c) \tag{7a}$$

$$I = L_{21}\nabla(-P) + L_{22}\nabla(-E) + L_{23}\nabla(-\mu_c^c) + L_{24}\nabla(-\mu_a^c) \tag{7b}$$

$$J_c^d = L_{31}\nabla(-P) + L_{32}\nabla(-E) + L_{33}\nabla(-\mu_c^c) + L_{34}\nabla(-\mu_a^c) \tag{7c}$$

$$J_a^d = L_{41}\nabla(-P) + L_{42}\nabla(-E) + L_{43}\nabla(-\mu_c^c) + L_{44}\nabla(-\mu_a^c) \tag{7d}$$

Hence there are ten independent coefficients characterizing the system. Considering the testing conditions that are used for the measurements of hydraulic conductivity, coefficient of electro-osmotic permeability, electrical conductivity, osmotic efficiency, effective diffusivities and the effective ionic mobilities and assuming the solution is dilute and there is no interaction between the cation and the anion, the L_{ij}'s were determined to be (Yeung 1990):

$$L_{11} = \frac{k_h}{\gamma_f n} + \frac{L_{12}L_{21}}{L_{22}} \tag{8a}$$

$$L_{12} = L_{21} = \frac{k_e}{n} \tag{8b}$$

$$L_{13} = L_{31} = \frac{-\omega c_c k_h}{\gamma_f n} + \frac{L_{12}L_{23}}{L_{22}} \tag{8c}$$

$$L_{14} = L_{41} = \frac{-\omega c_a k_h}{\gamma_f n} + \frac{L_{12}L_{24}}{L_{22}} \tag{8d}$$

$$L_{22} = \frac{\kappa}{n} \tag{8e}$$

$$L_{23} = L_{32} = c_c u_c \tag{8f}$$

$$L_{24} = L_{42} = -c_a u_a \tag{8g}$$

$$L_{33} = \frac{D_c c_c}{RT} \tag{8h}$$

$$L_{34} = L_{43} = 0 \tag{8i}$$

$$L_{44} = \frac{D_a c_a}{RT} \tag{8j}$$

where

k_h is the hydraulic conductivity
k_e is the electro-osmotic permeability
κ is the bulk electrical conductivity of soil

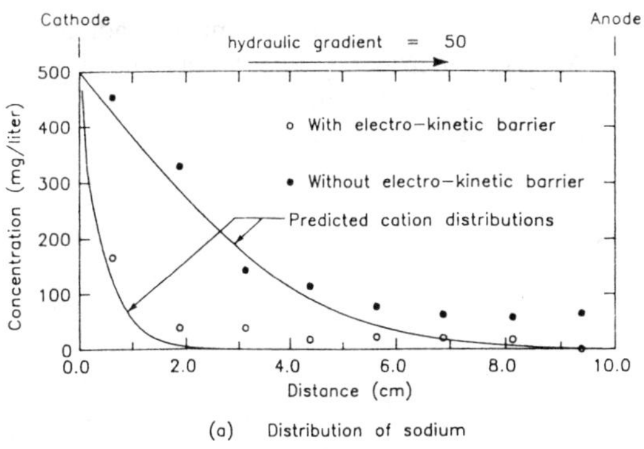

(a) Distribution of sodium

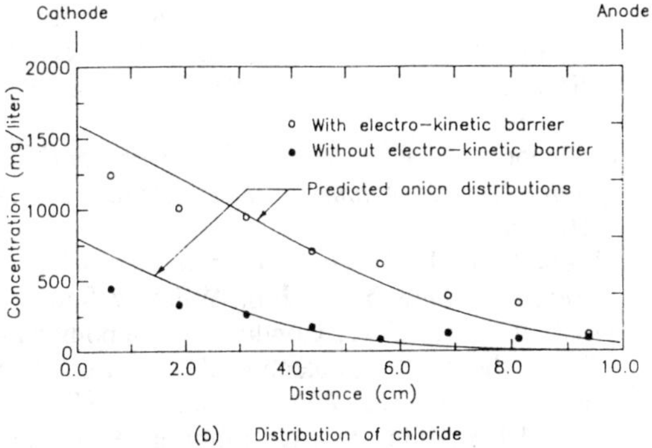

(b) Distribution of chloride

Figure 4 Distribution of contaminants after 20 days

ω is the coefficient of osmotic efficiency
γ_f is the unit weight of fluid
c_c is the concentration of cation
c_a is the concentration of anion
u_c is the effective ionic mobility of cation
u_a is the effective ionic mobility of anion
D_c is the effective diffusivity of cation
D_a is the effective diffusivity of anion
n is the porosity of soil
R is the universal gas constant
T is the absolute temperature

The contaminant flows relative to the soil are of more significance than the diffusional flows relative to water. Combining Eqs. (6) and (7) gives the flow equations for cations and anions under the influences of hydraulic, electrical and chemical gradients,

$$\begin{aligned} J_c \;=\; & (L_{31} + c_c L_{11})\gamma_f \nabla(-h) \\ & + (L_{32} + c_c L_{12})\nabla(-E) \\ & + (L_{33} + c_c L_{13})\frac{RT}{c_c}\nabla(-c_c) \\ & + (L_{34} + c_c L_{14})\frac{RT}{c_a}\nabla(-c_a) \end{aligned} \tag{9a}$$

$$\begin{aligned} J_a \;=\; & (L_{41} + c_a L_{11})\gamma_f \nabla(-h) \\ & + (L_{42} + c_a L_{12})\nabla(-E) \\ & + (L_{43} + c_a L_{13})\frac{RT}{c_c}\nabla(-c_c) \\ & + (L_{44} + c_a L_{14})\frac{RT}{c_a}\nabla(-c_a) \end{aligned} \tag{9b}$$

where $\nabla(-h)$ is the hydraulic gradient (dimensionless). Applying the principle of conservation of mass for steady flow, the governing equation for the concentration of species i is

$$\frac{\partial c_i}{\partial t} = -\nabla \cdot J_i - G_i \tag{10}$$

where G_i is a source-sink term denoting the removal rate of species i per unit volume. Solution of this equation gives concentration of species i as a function of time and position.

4 EXPERIMENTAL EVALUATION OF THE THEORY

The soil used in this laboratory test program is a gray brown silty clay of moderate plasticity - Unified Soil Classification CH. The maximum dry density is 110 pcf and the optimum water content is 17.4% as determined by the modified Proctor compaction test. The liquid limit, plastic limit and plasticity index of the soil are 52%, 27% and 25%, respectively.

Ten uniform replicate samples of 35.6 mm in diameter and 101.6 mm in length were compacted wet of optimum to 90% relative compaction. They were divided into two groups of five samples each. All the samples were fully saturated with tap water and then permeated with sodium chloride solution under identical hydraulic gradients. A periodic electrical gradient of one volt per centimeter was applied for one hour per day to one group of samples. One sample from each group was dismantled at intervals of 5 days. Each sample was then sectioned into eight pieces and complete chemical analyses were performed on the pore fluid extracted from each piece.

The hydraulic conductivities of these samples were measured to be in the order of 1×10^{-11} m/s. The coefficients of electro-osmotic permeability were measured to be 2×10^{-9} m^2/Vs. The diffusion coefficient of NaCl was measured in a separate experiment on similar samples to be 3×10^{-10} m^2/s under identical conditions. The periodic application of an electro-kinetic flow barrier to the compacted clay samples indicated that the migration of sodium ions was halted, whereas the migration of chloride ions was accelerated. In this case, the effective ionic mobility of the chloride ions exceeded the electro-osmotic counterflow resulting in a net migration of chloride ions towards the anode. However, the electrical gradient provided a very effective flow barrier to the migration of the cation.

The distributions of contaminants (NaCl) after 20 days of permeation is presented in Figure 4. The curves give the computer model predictions using the measured parameters as input. It can be observed that the measured concentration profiles are in good agreement with the predicted profiles given by the theory.

5 CONCLUSIONS

1. The results from this research indicate that electro-kinetics can stop advective and dispersive flows through compacted clay. When applied to a landfill liner, the cathode would be installed on the top of the compacted clay layer. By maintaining a small net inward flow, i.e. a greater inward electro-osmotic flow than outward hydraulic flow, transport of contaminants by these mechanisms would be prevented. In typical cases, the required sustained DC voltage across a compacted clay landfill liner would be quite low, of the order of a few tenths of a volt. Thus problems resulting from gas and heat generation and cracking that are often associated with electro-kinetic treatment of soil may be minor. As both the voltage and current density would be low, the power costs should be small.
2. However, the results show also that ionic transport in the electric field would cause migration of anions to the anode underneath the liner. Should these anions pose a risk to the environment, a collection system would be required for their removal.
3. The developed coupled flow theory reasonably predicted the migration of the contaminant ions under the combined influences of the hydraulic, electrical, and chemical gradients.
4. An electrical gradient may move some inorganic species in fine-grained soil much more effectively than a hydraulic gradient.

REFERENCES

Bass, J. M., W. J. Lyman & J. P. Tratnyek 1985. *Assessment of synthetic membrane successes and failures at waste storage and disposal sites*. Report No EPA-600/2-85/100, Cincinnati: U. S. EPA.

Casagrande, L. 1983. Stabilization of soils by means of electro-osmosis - state-of-the-art. *J. Boston Soc. Civ. Eng. Sect.*, ASCE, 69: 255-302.

DeGroot, S. R. & P. Mazur 1962. *Non-equilibrium Thermodynamics*, Amsterdam: North Holland.

Mitchell, J. K. 1976. *Fundamentals of Soil Behavior*, New York: John Wiley & Sons.

Mitchell, J. K. 1991. Conduction phenomena - from theory to geotechnical practice: 31st Rankine Lecture. *Géotechnique* 41: 299-340.

U. S. EPA 1985. Draft minimum technology guidance on double liner systems for landfills and surface impoundments - Design, construction, and operation (2nd version). Report No. EPA/530/SW-85/014, Cincinnati.

Yeung, A. T. 1990. Coupled flow equations for water, electricity, and ionic contaminants through clayey soils under hydraulic, electrical and chemical gradients. *J. Non-Equilib. Thermodyn.* 15: 247-267.

Yeung, A. T. & J. K. Mitchell 1992. Coupled fluid, electrical and chemical flows in soil, accepted for publication in *Géotechnique*.

Yeung, A. T., S. M. Sadek & J. K. Mitchell 1992. A new apparatus for the evaluation of electro-kinetic processes in hazardous waste management. *Geotech. Test. J.*, ASTM, 15(3).

Environmental Management, Geo-Water & Engineering Aspects, Chowdhury & Sivakumar (eds)
© 1993 Balkema, Rotterdam. ISBN 90 5410 099 0

Temporal and spatial variations of the total dissolved solids, Williams River, Western Australia

B.Yu
Griffith University, Brisbane, Qld, Australia

D.T.Neil
University of Queensland, Brisbane, Qld, Australia

ABSTRACT: The Williams River has one of the highest documented rates of stream salinity increase in south-west Western Australia. Investigation of long-term variation of salinity shows that this increase has occurred during a period of decreasing rainfall. Furthermore, the period of most intensive stream water sampling (1975-80) coincided with a period when the streamflow was the lowest on record. The increase in salinity was not uniform across the range of streamflows and was highest for intermediate flows. On a shorter time scale, a single-peaked flood was usually associated with a clockwise hysteresis. For multi-peaked floods, on the other hand, counter-clockwise hysteresis may occur when the first peak discharge was larger than those of the subsequent floods. Spatially, salinity of the Williams River peaks about three fourths of the way from the Hotham River confluence in spite of the fact that rainfall is lowest in the upper catchment. Salinity of stream water along the main stem is greater than that of tributary streams, suggesting that the Williams River valley is the primary source of salt input.

1 INTRODUCTION

Williams River catchment has an area of approximately 1,500 km^2. It lies about 150 km south-west of Perth in Western Australia (Fig. 1). The river flows westward, mostly through agricultural land. The Williams and Hotham Rivers are the two major tributaries of the Murray River. Mean annual rainfall at the Hotham River confluence is about 800 mm and gradually decreases to about 500 mm in the headwater area. In south-west Western Australia winter rainfall dominates. At Williams, rainfall in June to August constitutes more than half of the annual rainfall total, and summer rainfall (December to February) contributes no more than 10 % of the annual total. The Williams River has one of the highest rates of increase in stream salinity (95 mg/L/yr, for the period 1966-1986) in south-west Western Australia (Schofield *et al.*, 1988). With an average salinity of about 3,000 mg/L there is little development potential for municipal water supply.

The increase in the stream salinity throughout south-west Western Australia has largely been attributed to human activities (Schofield *et al.*, 1988). With forests cleared for agriculture, percolation to groundwater increases. As a consequence groundwater levels rise until they intersect the valley surface. Stored salt in the soil is mobilised, resulting in increased stream salinity. Stream salinity increase has been most dramatic in low rainfall areas (< 600 mm per annum) where forest has been cleared for agriculture, whereas in areas of high annual rainfall salinity has remained low irrespective of whether the catchment has been cleared or not.

2 DATA

Shortly after the 1991 winter season, an intensive survey of the surface water quality was made to provide a 'snapshot' of the

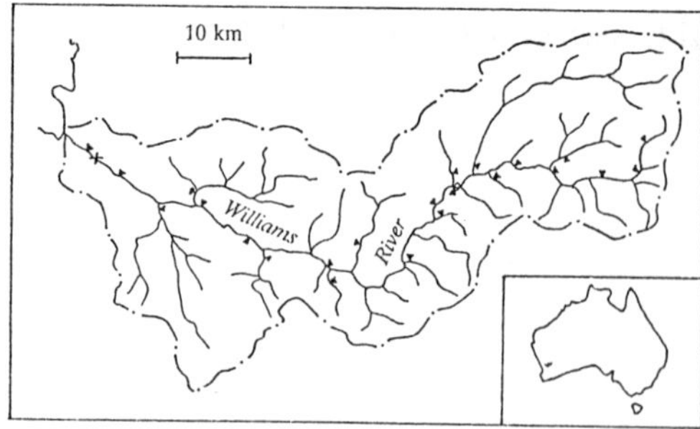

Fig. 1 Williams River catchment in south-west Western Australia and sampling locations (solid triangle) during the survey in 1991 (open circle: rainfall station; cross: stream gauging station).

salinity pattern of the Williams River catchment. 40 measurements were made in four days during which there was little change in the flow condition. These data were supplemented by instantaneous salinity and discharge measurements and monthly streamflows data for the period 1967-1990, courtesy of Surface Water Branch, Water Authority of Western Australia. The gauging station (GS614196) is located at Saddleback Road Bridge on the Williams River with a catchment area of 1,437 km^2. Also available to us was a set of spatial salinity data collected during repeated surveys of the Murray River in 1972 (Collins, 1974).

3 TEMPORAL VARIATION

To put the stream salinity issue in perspective, annual rainfalls over the past eighty years (1911-1990) at Williams are shown in Fig. 2.a. It is clear that salinity has been monitored during a period of decreasing rainfall. The impact of long-term decreasing rainfall on stream salinity is largely unknown and probably difficult to evaluate because, for most streams in south-west Western Australia, salinity of stream waters was not monitored until the 1960s.

Annual mean streamflows during the period of water quality monitoring for the Williams River are shown in Fig. 2.b. Within the period of water quality measurement, the sampling frequency is by no means uniform. As shown in Table I, measurements of water quality were most intensive between 1975 and 1980 when annual streamflows were among the lowest for the last 26 years (Table II). In fact, 1976-80 and 1975-80 had the lowest streamflow averages for five and six year periods, respectively. Thus, monitoring of stream water quality has been concentrated in the driest period within a decreasing rainfall regime.

A good experimental design requires that most measurements be taken at both the lower and upper end of the range for each independent variable. With respect to the temporal variation of stream salinity for the

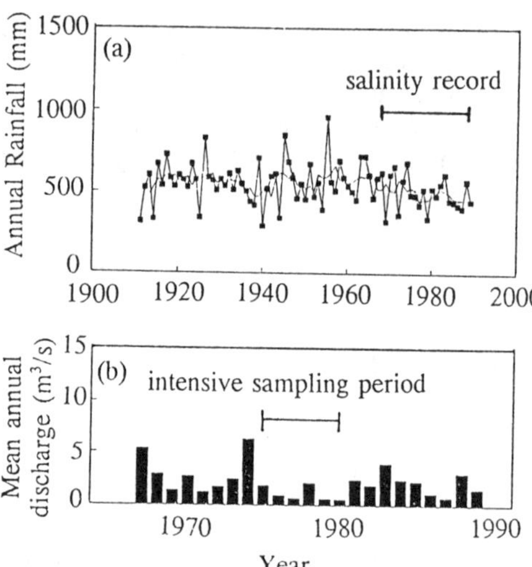

Fig. 2 (a) Annual rainfall and 5-year running mean at Williams (1911-1990); (b) Annual streamflow, Williams River at Saddleback Road Bridge (1966-1990).

Table I Williams River sampling intensity at Saddleback Road Bridge.

Period	Measurements per month
5/1966 - 7/1975	1.6
7/1975 - 11/1980	25.0
12/1980 - 11/1990	1.8

Williams River, most salinity measurements should ideally be made in the beginning and near the end of a period in order to detect the trend in stream salinity for the period. This is certainly not the case for the Williams River during the period of salinity monitoring for the last 26 years. Another important factor to consider is the effect of stream discharge. Generally speaking, salinity decreases when the discharge is increased. Thus the effect of discharge should be removed or accounted for when detecting trend in stream salinity. Because of these considerations, we examined two periods of roughly the same streamflow conditions and equal number of measurements (n=110). The average flow rate was 1.856 m^3/s for the first period (1969-73), and 1.863 m^3/s for the second (1984-88), and the two periods are 15 years apart (between mid-points). Under similar hydrological conditions, the difference in the rating curves for different periods would be more appropriate to quantify the temporal change in salinity levels for a given flow rate.

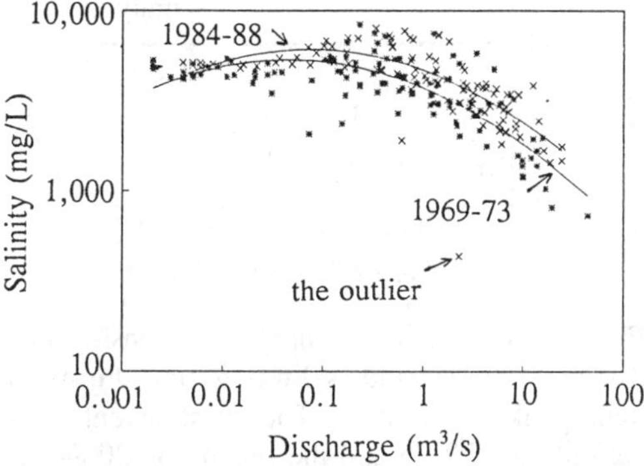

Fig. 3 Rating curves of stream salinity 1969-1973 and 1984-88.

Table II Average (Q) and peak discharge (Q_p) during the intensive salinity sampling period and their rankings in ascending order (lowest = 1).

Year	Q (m^3/s)	Rank	Q_p (m^3/s)	Rank
1975	1.79	12	56.5	14
1976	0.85	5	23.6	7
1977	0.637	3	16.8	4
1978	2.07	14	89.9	19
1979	0.54	2	8.10	2
1980	0.532	1	5.27	1
25 yr mean	2.13		67.4	

A scatter plot of salinity versus discharge showed that a simple log-linear model was inadequate to describe their relationship for either period. A second order term was introduced to fit the data. One discharge measurement was zero in the first period, and it was removed from the data set. For the second period, there is an outlier (Fig. 3). It was excluded from the final analysis. Its inclusion would reduce the contrast between these two periods. Fig. 3 shows both the scatter plot and fitted quadratic function for the two contrasting periods. It can be seen that the maximum concentrations occurred during intermediate flows and the difference between the fitted curves is smaller than the amount of scatter around each one of them. The rate of salinity increase was estimated using the difference in the rating curves for the two periods. The maximum rate of salinity increase is 68 mg/L/yr which occurs at discharge of 0.5 m^3/s. For discharge of 0.01 m^3/s the rate of increase is 24 mg/L/yr and for discharge of 10 m^3/s the rate of increase is 41 mg/L/yr.

Frequent measurement of water quality between July 1975 and November 1980 permits detailed examination of stream salinity during a flood event. In most streams, concentration of total dissolved solids (TDS) for a given discharge is lower during the falling stage of the hydrograph, resulting in a clockwise salinity rating curve. For example, during the 1977 winter season, a peak discharge of 16.7 m^3/s was reached on 11 August 1977. The salinity decreased as the discharge increased (Fig. 4) and during the streamflow recession the salinity recovered but to a lower level (Fig. 4). The rating curve becomes complicated when multiple events occur during the same rainy season, especially when the second peak is smaller in magnitude. The rating curve associated with multiple flood

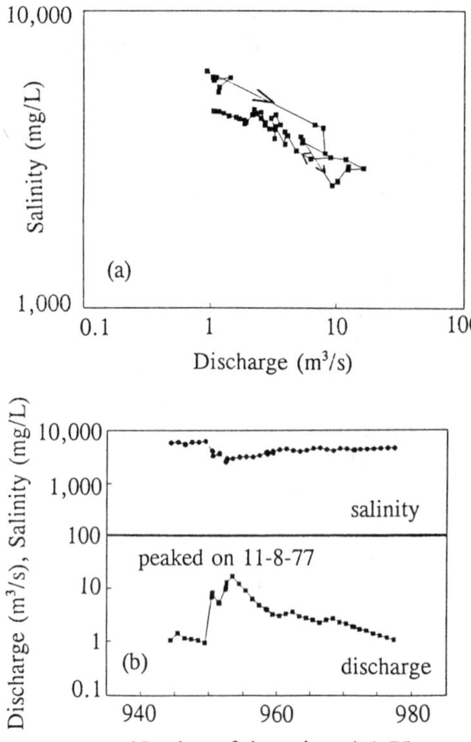

Fig. 4 (a) Salinity-discharge relationship and (b) hydrograph and salinity-graph during a single-peaked event.

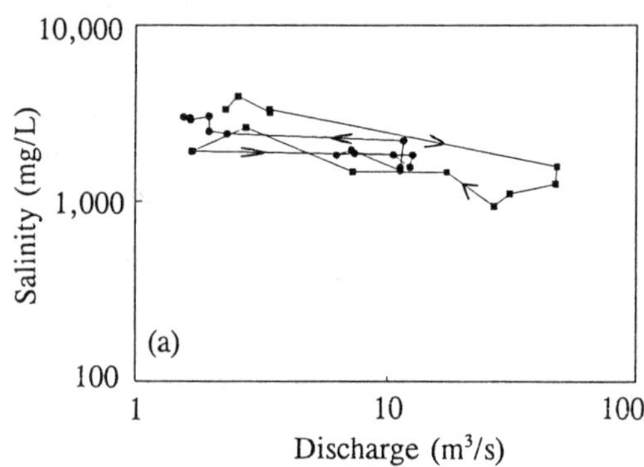

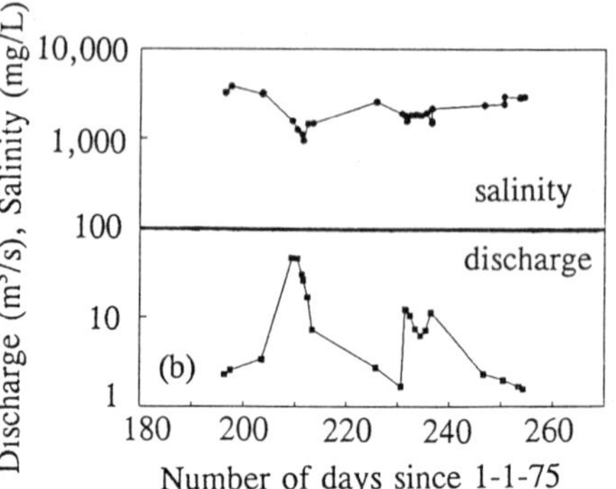

Fig. 5 (a) Salinity-discharge relationship and (b) hydrograph and salinity-graph during a multi-peaked event. The first event was peaked on 27-7-75 and the second on 20-8-75.

events may show counter clockwise hysteresis (Fig. 5).

4 SPATIAL VARIATION

Between 28 and 30 September 1991, 40 conductivity measurements were made along the main stem of the Williams River and its tributaries to identify sources of high salinity. Water samples were retained for laboratory determination of major ion concentrations.

An empirical relationship between the measured concentrations of TDS and field conductivity readings was very good (r^2 = 0.997) and was used to estimate TDS for each site. No rainfall had occurred since 19 September 1991 (daily records, Williams Post Office) and repeated measurements at selected sites in the beginning and at the end of the survey showed nearly identical results. This survey has provided a 'snapshot' of the spatial distribution of salinity around the catchment near the end of the 1991 winter season. Survey results are plotted in Fig. 6.a as a function of AMTD (Adopted Middle Thread Distance) upstream from the Hotham River confluence. Only the measurement of salinity closest to the Williams River confluence is plotted for each tributary surveyed. Results of a similar survey in September 1972 (Collins, 1974) are shown in Fig 6.b. Both surveys have quite similar spatial patterns. Near the headwaters of the Williams River the salinity was low in spite of the low mean annual rainfall and high proportion of land cleared. Salinity peaked around AMTD 70 km (about 10 km upstream from Williams) and then steadily decreased downstream due to dilution by the tributary streams. None of the tributaries had a higher salinity than that of the adjacent Williams River. It seems that the broad valley associated with the Williams River main stem may be the principal source of stream salinity.

5 DISCUSSION AND CONCLUSION

In south-west Western Australia mean annual regional rainfall has decreased for the last forty to fifty years (Pittock, 1983). Given this regional rainfall trend and the consequent increase in salinity in the Williams River, it seems likely that the pattern of long-term salinity variation documented for other streams in the region may also be strongly influenced by variation in streamflow as well as intrinsic trends in salinity. If the decreasing trend in regional rainfall continues, as suggested by recent results of GCM models under double CO_2 conditions (Pittock and Allan, 1989), continued stream salinity increase is a possible response.

Trends in salinity should be evaluated in the context of changing climate (particularly rainfall) regimes in order to discriminate between, for example, system equilibration and increased rainfall with a decreasing salinity trend, and between continued response to land clearing and decreased rainfall with an increasing salinity trend.

The dramatic increase in Williams River salinity occurred during a period of low rainfall and streamflow. The increase in salinity between periods of similar hydrological regimes is most pronounced for intermediate flows (up to 68 mg/L/yr), with much lower rates of increase for both high and low streamflows. Repeated surveys have shown that salinity of the Williams River is higher along the main stem than in tributary streams, suggesting that the valley alluvium is the main source of salts.

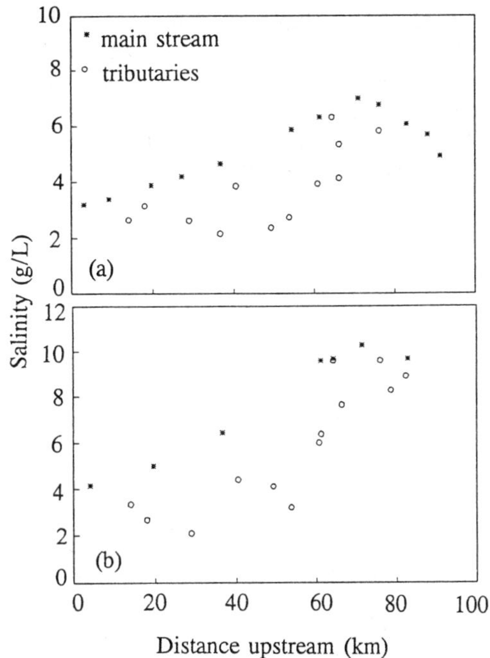

Fig. 6 Longitudinal salinity profile of the William River and its tributaries for (a) September 1991 and (b) September 1972.

ACKNOWLEDGMENTS

We thank the Department of Biogeography and Geomorphology, Australian National University for financial and technical support and Mr Greg May of Water Authority of Western Australia for providing streamflow and salinity data.

REFERENCES

Collins, P.D.K. 1974. Murray River basin surface water resources survey. Technical Note No. 45, Water Resources Section, Planning, Design and Investigation Branch, Public Works Department of Western Australia, 54 p.

Pittock, A.B. 1984. Recent climatic change in Australia: Implications for a CO_2-warmed earth. *Climatic Change*, 5: 321-340.

Pittock, A.B. & Allan, R.J. (eds.), 1990. The Greenhouse Effect: Regional implications for Western Australia. 1st Interim Report, Climate Impact Group, CSIRO Division of Atmospheric Research, Western Australia E.P.A., 65 p.

Schofield, N.J., Ruprecht, J.K & Loh, I.C. 1988. The impact of agricultural development on the salinity of surface water resources of south-west Western Australia. Water Authority of Western Australia, Report No. WS 27, 69 p.

Slope stability and landslide management

Environmental Management, Geo-Water & Engineering Aspects, Chowdhury & Sivakumar (eds)
© 1993 Balkema, Rotterdam. ISBN 90 5410 099 0

Landslide hazard zonation (LHZ) mapping of a part of Doon valley, Garhwal Himalaya, India

R.Anbalagan, Luv Sharma & Sushil Tyagi
University of Roorkee, India

ABSTRACT: The landslide hazard zonation (LHZ) maps are useful for selecting suitable locations to implement development schemes in mountainous terrains as well as for adopting appropriate mitigation measures in the unstable hazard prone areas. A LHZ map of Maldeota - Sahastradhara area in Garhwal Himalaya has been prepared using landslide hazrd evaluation factor (LHEF) rating scheme, a quantitative approach based on major inherent causative facators of slope instability.

1. INTRODUCTION

A landslide hazard zonation (LHZ) map depicts division of land surface into zones of varying degree of stability based on the estimated significance of the causative factors in inducing instability. The LHZ maps are useful in planning and implementation of development schemes in mountainous terrains. They help the planners to take into account the existing instabilities, while implementing the development schemes so that the environmental hazards due to construction activities are kept to the minimum. The LHZ mapping is based on a quantitative approach called landslide hazard evaluation factor (LHEF) rating scheme. This technique is useful during the preliminary stages of geotechnical investigations.

2. LANDSLIDE HAZARD ZONATION (LHZ) MAPPING BASED ON LHEF RATING SCHEME

The LHEF rating scheme is based on the major inherent causative factors of slope instability such as geology, slope morphometry, relative relief, land use and land cover and ground water condi-

Table 1. Proposed maximum LHEF rating for different contributory factors for macro-zonation.

Contributory Factor	Max. LHEF Rating
Lithology	2
Relationship of structural discontinuities with slope	2
Slope morphometry	2
Relative relief	1
Land use and land cover	2
Groundwater conditions	1
Total	10

Table 3. Land hazard zonation on the basis of total estimated hazard (TEHD)

Zone	TEHD Value	Description of zone
I	3.5	Very low hazard(VLH) zone
II	3.5-5.0	Low hazard (LH) zone
III	5.1-6.0	Moderate hazard (MH) zone
IV	6.1-7.5	High hazard (HH)zone
V	7.5	Very high hazard (VHH) zone

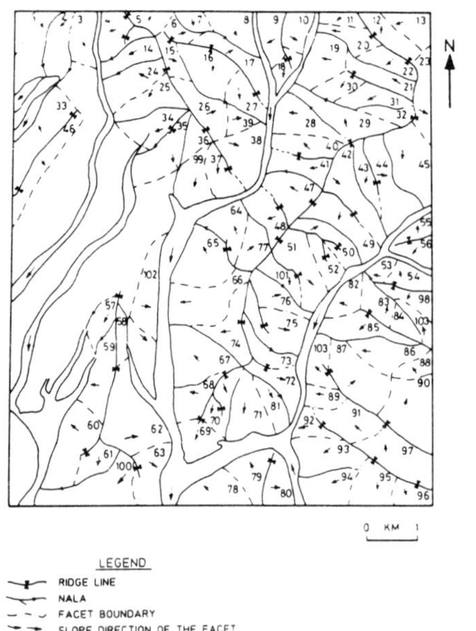

Fig. 1: Slope facet map

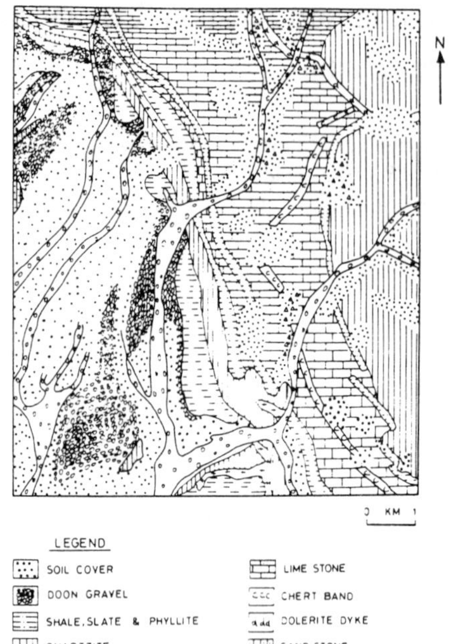

Fig. 2. Lithological map

tions. The maximum LHEF ratings for different categories are determined on the basis of the estimated significance in causing instability (Table 1). A total of 10 indicates maximum value of total estimated hazard (THED).

The hill slopes of the area of study are divided into a number of facets mostly delimited by ridges, spurs, gullies and rivers. A facet is a part of hill slope which has more or less similar characters of slope showing consistent slope direction and inclination. A detailed LHEF rating scheme showing ratings for a variety of subcategories for individual causative factors (Anbalagan, 1992) has been given in Table 2.

The total estimated hazard (TEHD) indicating the net probability of instability is calculated facet wise. The TEHD of an individual facet is obtained by adding the ratings of the individual causative factors obtained from LHEF rating scheme. On the basis of TEHD 5 categories of landslide hazard zones have been identified (Table 3) namely very low hazard (VLH), low hazard (LH), moderate hazard (MH), high hazard (HH) and very high hazard (VHH).

3. LHZ MAPPING OF MALDEOTA - SAHASTRADHARA AREA

The Maldeota-Sahastradhara area falls in a part of Doon valley in Garhwal Himalaya, between longitude 78° 5' to 78° 10' E and latitude 30° 20' to 30° 25' N covering an area of 75 sq km. Physiographically the south-western part of the area is flat. Moderate hills are present with the broad valleys in the central and southeastern parts and the high hills with narrow valleys are present in the north and northeastern part of the area. The LHZ map of this area has been prepared using the LHEF rating scheme. For that purpose maps such as slope facet map (Fig. 1), lithological map (Fig. 2), structural map (Fig.3) slope morphometry map (Fig. 4), relative relief map (Fig. 5), land use and land cover map (Fig. 6) and hydrogeological map (Fig. 7) have been prepared covering all the slope facets. The LHZ map has been prepared on the basis of TEHD of facets calculated using LHEF rating schemes (Fig. 8).

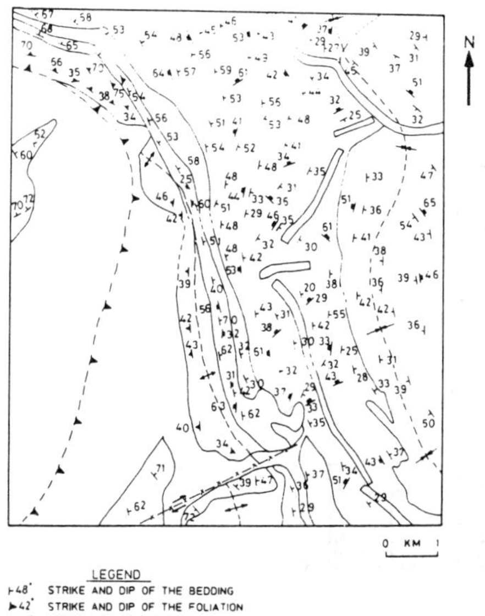

Fig. 3. Structural map

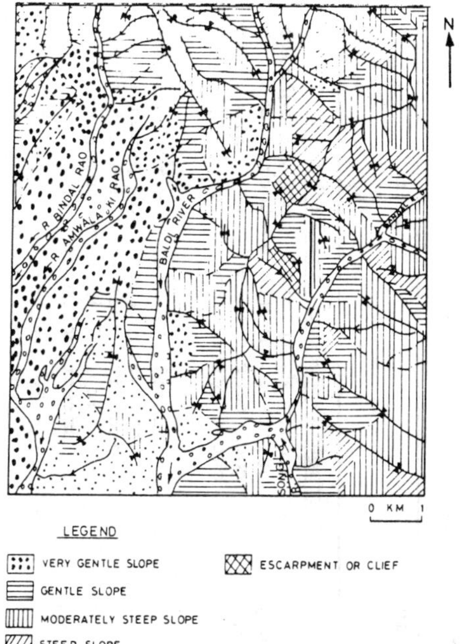

Fig. 4. Slope morphometry map.

The important rocks exposed in the area are limestone, quartzite shale, slate and doon gravel (Fig.2). The physical properties of these rocks vary due to different environments of deposition and extent of weathering. The rocks have been gently folded into broad open folds. A number of joints have been observed to traverse through the rocks. The geological discontinuities considered for LHEF rating purposes include bedding, joints, foliations and faults (Fig. 3).

The slope morphometry map shows that the central and the northeastern parts have steep to escarpment slopes while very gentle to gentle slopes are present on the western and southwestern parts. The southeastern part has mostly moderate to gentle slopes (Fig. 4). The area generally has high relative relief in the central, northern and northeastern parts. The slopes have low to medium relative relief on the western and southwestern parts of the area (Fig. 5).

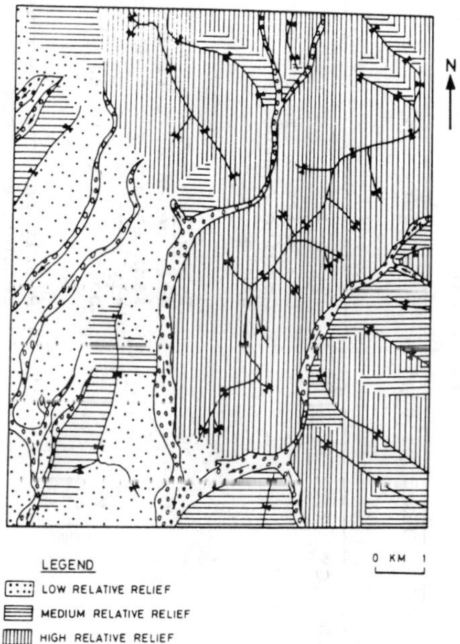

Fig. 5. Relative relief map.

Major portions of western northwestern and southwestern parts have cultivated land with good concentration of human population. Dense and moderately dense forest cover comparatively less areas in the northern, northeastern and

TABLE. 2

Land hazard evaluation factor (LHEF) rating scheme

Contributory Factor (1)	Description (2)	Category (3)	Rating (4)	Remarks (5)
A. LITHOLOGY	Rock Type	Type-I		Correction Factor for Weathering
		Quartzite & Limestone	0.2	i) Highly Weathered - rock discoloured, joints open with weathered products, rock fabric altered to a large extent - Correction Factor C_1
		Granite & Gabbro	0.3	
		gneiss	0.4	
		Type-II Well cemented terrigenous sedimentary rocks dominently sandstone with minor beds of clay stone	1.0	ii) Moderately Weathered - rock discoloured with fresh rock patches, weathering more around joint planes, but rock in-tact in nature - Correction Factor C_2
		Poorly cemented terrigenous sedimentary rock dominently sand rock with minor clay shale beds.	1.3	iii) Slightly Weathered - rock slightly discoloured along joint planes, which may be moderately tight to open, in-tact rock - Correction Factor C_3
		Type-III		**The Correction Factor for weathering to be multiplied with the fresh rock rating.**
		Slate & Phyllite	1.2	For Rock Type-I
		Schist	1.3	
		Shale with interbedded clayey and nonclayey rocks	1.8	$C_1 = 4$, $C_2 = 3$, $C_4 = 2$
		Highly weathered shale, phyllite & schist	2.0	For Rock Type-II $C_1 = 1.5$, $C_2 = 1.25$, $C_3 = 1.0$
	Soil Type	Older well compacted fluvial fill material	0.8	
		Clayey soil with maturally formed surface	1.0	

(1)				
			Sandy soil with naturally formed surface(Alluvial)	1.4
			Debris comprising mostly rock pieces mixed with clayey/ sandy soil(Colluvial)	
			-Older well compacted	1.2
			-Younger loose material	2.0
B.STRUCTURE	Relationship of Structural Discontinuity with slope			
i)	Relationship of parallelism between the slope and the discontinuity*	I	>30°	0.20
		II	21°-30°	0.25
		III	11°-20°	0.30
	Planar ($_j$- $_s$)	IV	6°-10°	0.40
	Wedge ($_i$- $_s$)	V	<5°	0.50
ii)	Relationship of dip of discontinuity* and inclination of slope	I	>10°	0.3
		II	0°-10°	0.5
		III	0°	0.7
	Planar (B_j-B_s)	IV	0°-(-10°)	0.8
	Wedge (B_i-B_s)	V	>(-10°)	1.0
iii)	Dip of discontinuity*	I	<15°	0.20
		II	16°-25°	0.25
	Planar-B_j	III	26°-35°	0.30
	Wedge-B_i	IV	36°-45°	0.40
		V	>45°	0.50
	Depth of Soil cover		<5m	0.65
			6-10m	0.85
			11-15m	1.30
			16-20m	2.0
			>20m	1.20

*Discontinuity refers to the planar discontinuity or the line of intersection of two planar discontinuities whichever is important from the point of view of instability.

$_j$ = Dip direction of joint — B_j = Dip of joint

$_i$ = Direction of line of intersection of two discontinuities — B_i = Plunge of line intersection of two discontinuities

$_s$ = Direction of slope inclination — B_s = Inclination of slope

Category I = Very favourable, II = favourable, III = fair, IV = unfavourable, V = very unfavourable

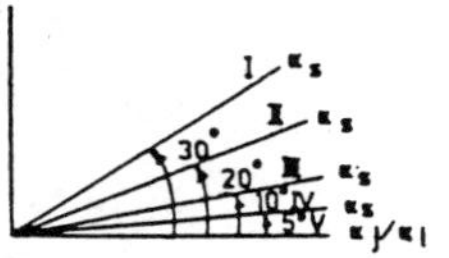

PARALLELISM BETWEEN THE SLOPE AND THE DISCONTINUITY ($\alpha_j/\alpha_i - \alpha_s$)

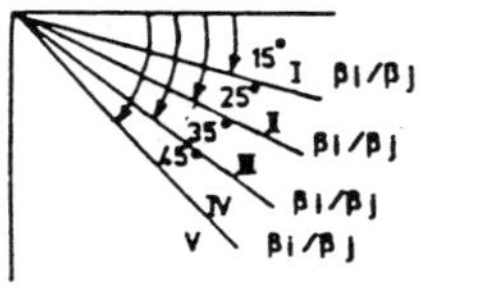

DIP OF DISCONTINUITY (β_i/β_j)

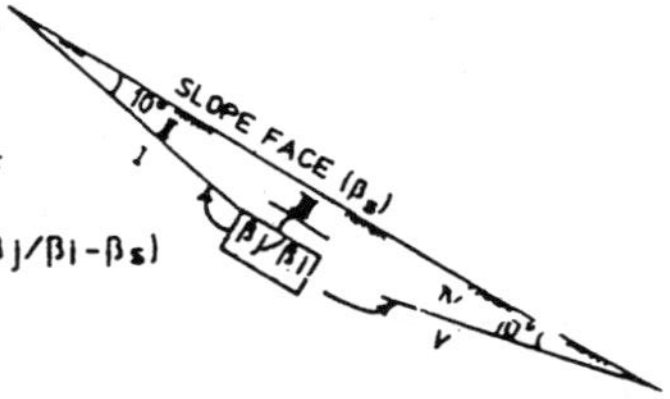

RELATIONSHIP OF DIP OF DISCONTINUITY AND THE INCLINATION OF SLOPE ($\beta_j/\beta_i-\beta_s$)

(1)	(2)	(3)	(4)	(5)	
C.SLOPE MORPHOMETRY	Escarpment/ cliff	>45°	2.0	No. of contour lines over one cm length (1:500000)	Slope angle
	Steep slope	36°-45°	1.7		
	Moderately steep slope	26°-35°	1.2	>35	>45°
	Gentle slope	16°-25°	0.8	19-25	36°-45
	Very gentle slope	<15°	0.5	13-18	26°-35°
				8-12	16°-25°
D.RELATIVE RELIEF				<7	<15
	Low	<100m	0.3		
	Medium	101-300	0.6		
	High	>300m	1.0		
E.LAND USE AND LAND COV	Agricultural land/ populated flat land		0.6		
	Thickly vegetated forest area		0.80		
	Moderately vegetated area		1.2		
	Sparsely vegetated area with lesser ground cover		1.5		
	Barren land		2.0		
F.GROUND WATER CONDITIONS	Flowing		1.0		
	Dripping		0.8		
	Wet		0.5		
	Damp		0.2		
	Dry		0.0		

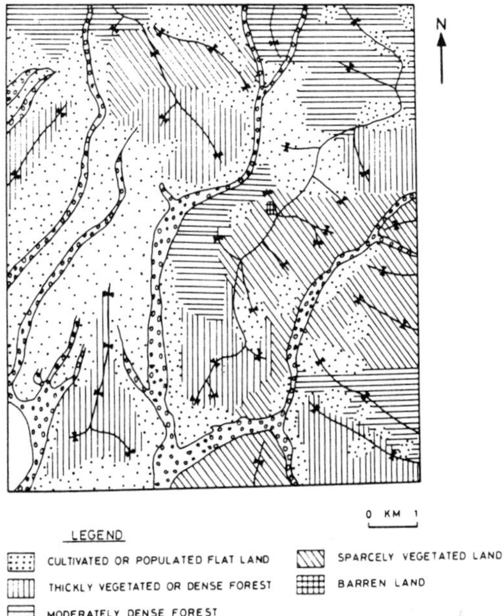

Fig. 6. Land use and land cover map

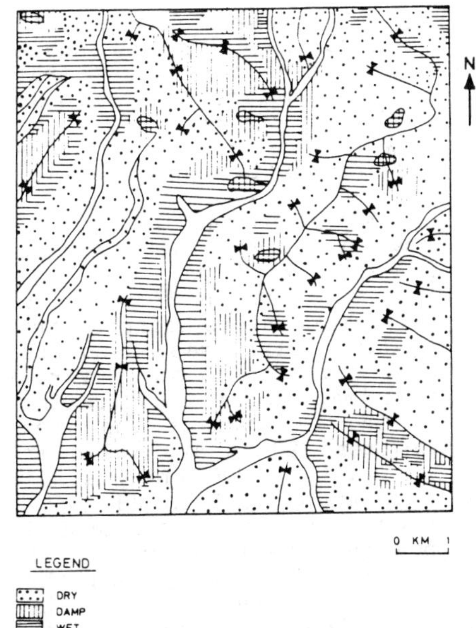

Fig. 7. Hydrogeological map.

central parts. The southern and southeastern parts generally have sparsely vegetated land with small patches of moderately dense to dense forests (Fig. 6).

The hydrogeological map shows that the northeastern part of the area is generally dry with some patches showing damp to wet conditions. The central part generally has damp to wet conditions. The western and southern parts generally have dry conditions (Fig. 7)

The LHZ map has been prepared by calculating TEHD for individual facets using LHEF rating scheme (Fig. 8). The map shows very low hazard (VHH) to high hazard (HH) zones. The moderate hazard (MH) zones contain some minor unstable slopes locally causing rock slides and rock falls. The HH zones need careful attention while planning excavations, as these areas are likely to pose maximum stability hazards during implementation of the development schemes.

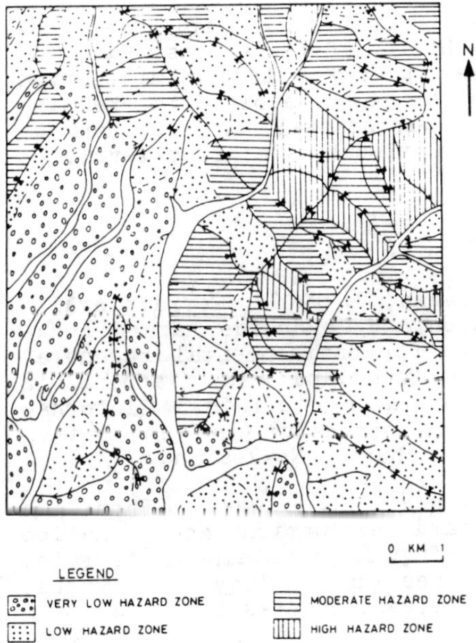

Fig. 8. Landslide hazard zonation map

4. CONCLUSIONS

For planning of development schemes in mountainous terrains, the existing instabilities of the slopes should be taken into consideration so that the schemes may be executed systematically with minimum disturbance to the environment of the area. For that purpose the landslide hazard zonation (LHZ) maps are much useful as they show the nature and distribution of instability in the area.

The LHZ map of Maldeota-Sahastradhra area in Garhwal Himalaya has been prepared with the help of derivative maps such as lithological structural map, slope morphometry map, relative relief map, land use and land cover map and hydrogeological map using LHEF rating scheme. The LHZ map of the area indicates that 20.2% are falls in VLH zone, 48.36% in LH zone, 21.17% in MH zone and 10.27% in HH zone. The areas falling in HH zone should be mapped in detail on scales 1: 1000 to 1:5000 depending upon the total area of the facet. The factor of safety of these area may be calculated by preparing geological cross sections across the slope, determination of engineering properties for various lithologies and determining the causative factors responsible for the instability of the slopes. Based on these studies suitable mitigation measures can be evolved for keeping the environmental hazards, due to construction to a minimum level.

REFERENCES

Anbalagan, R. 1992. Terrain evaluation and landslide hazard zonation for environmental regeneration and land use planning in mountainous terrain. International symposium on landslides, Christchurch, Newzealand.

Anbalagan, R. 1992. Landslide hazard evaluation and zonation mapping in mountainous terrain. Engineering Geology, Elseviers, Amsterdam, in press.

Environmental Management, Geo-Water & Engineering Aspects, Chowdhury & Sivakumar (eds)
© 1993 Balkema, Rotterdam. ISBN 90 5410 099 0

Landslide management in historical towns in Italy: The case of Magliano in Tuscany

R. Bertocci, P. Canuti, C.A. Garzonio, V. Menestrina & P. Vannocci
Earth Sciences Department, Florence, Italy

ABSTRACT: Magliano in Toscana is a typical example of the phenomenon of unstable towns lying in the Italian hills. The main geological, geomorphological and geotechnical features of the terrains composing the relief on which Magliano stands and the effects the phenomena had on the construction are described in this note. The research enabled us to identify the landslide mechanisms and to comment on the remedial works carried out by the landslide management authorities.

FOREWORDS

In Italy there are many towns built on hilly reliefs affected by slope instability, particularly in the central-northern regions. These villages and towns are of considerable importance for their history, monuments and works of art and are consequently great tourist attractions. However they often pose problems, at times difficult to solve, from a landslide management point of view. We can name Orvieto, Todi and Assisi in Umbria, S.Leo and Ancona in Le Marche and Pienza and S.Miniato in Tuscany, as some of the most famous examples of this situation.

Magliano in Tuscany is a typical example of the phenomenon of unstable towns lying in the Italian hills. Its instability is well known historically, but the recent intensification of these phenomena has made it necessary to intervene in order to stabilize these gravitational phenomena once and for all and to consolidate and restore the urban area.

The main geological, geomorphological and geotechnical features of the terrains composing the relief on which Magliano stands are described in this note. The effects the phenomena have on the constructions are also analysed. An evaluation of the morphological dynamics and the developing trends of the movements, together with the interpretation of previous geotechnical surveys, enabled us to identify the landslide mechanisms and consequently establish the correct criteria to be adopted for the planning of remedial works and to draw a comparison with the works carried out by the landslide management authorities. Particular attention is paid to the description of the results of historical research which enabled us to reconstruct the evolution of the failures on the ancient buildings. The research also interprets the anomalous presence and architectural typology of some constructions (take for example the case of the castle keep) as probably due to their being modified as a result of this landslide morphology.

GEOMORPHOLOGICAL AND GEOLOGICAL FEATURES

The town is situated in the south of Tuscany (Fig. 1) and lies on a gentle relief (about 130 m asl) composed of a top calcarenite plate (Upper Pliocene) with maximum thickness 25-30 m which is fractured and faulted in places and is superimposed on clayey deposits (Lower-Middle Pliocene). The morphology is therefore characterized by a slightly sloping wide bench bordered by gentle slopes with the exception of

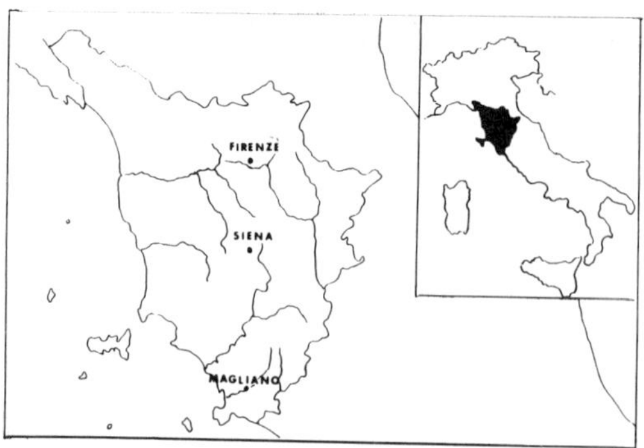

Fig. 1 - Location of Magliano in Tuscany

the east side where there is an escarpment which reaches an elevation of over 20 m in some places. The instability phenomena occur in fact in this latter part.

The top limestones (Fig. 2) are characterized by an extreme lithological variability: large stone banks (2m thick), often presenting weathering and microcavities linked to microkarst phenomena, alternate with 20-30 cm thick layers composed solely of shells and rather friable layers with a powdery consistency. The underlying grey-blue clays are over 30 m thick (Middle-Upper Pliocene) and are lightly overconsolidated. They are superimposed in discordance on the calcareous flysch of the "Argille a Palombini".

From a tectonic point of view the area corresponds to a high structure connected to a recent disjunctive phase with fault systems which have caused the fragmentation of the "Horst" of Magliano. One fault in particular, which divides the built up area in two parts, can be identified. This fault, together with the sometimes intense jointing connected to it, plays a decisive role in the evolution of the gravitational phenomena (consecutive opening fractures, tension cracks, etc.).

Fig. 2 gives the main geomorphological evidence, describing above all the gravity caused phenomena, as well as showing the principal geological units outcropping. These complex phenomena are composed of falls from the calcareous escarpment and multiple rotational slips in the underlying clayey materials, which evolve into mud-slides and flows. Note must be taken of the large thick debris sheet, with large calcareous elements in places, which is to be found on the unstable slope in the east side of the hill. This testifies to the gravitational evolution of the slope even before man settled there.

HISTORICAL DETAILS

The historical town of Magliano is particularly important for its enclosing walls, only partly in ruins , which constitute one of the most interesting and complete fortifications to be found in Tuscany. They were built by the Republic of Siena and most of them date back to the 15th century. One part (SE side) can be dated to the Middle Ages (completed in 1323). It is believed the the early walls initially surrounded the parish church of S. Martino(11th century) and later the church of S. Giovanni (13th century).

The large medieval tower of S.Martino and the course of the walls in this area (East side) are particularly important elements for the evaluation of the interaction between the geomorphological dynamics active on the slope and the evolution of the urban order of the historical town. At present only a few ruins of the large tower remain and they lie in the debris strip caused by the failure which affects this side of the town.

On a map dating back to the 16th century the large tower is given the name "Cassero" (Castle Keep) and is shown as separate from the rest of the walls. From the map in Fig. 3, which shows the plans for further fortification works which were never carried out, this sector is also shown as being separate from the rest of the circuit.

In substance we can formulate two hypotheses to explain this irregularity. The first attributes the separation of the building from the rest of the walls to a progressive detatchment of this portion of the stronghold as result of a landslide. The second interprets this separation as being an element of the building plan and considers the tower a demilune. As such it was a defence outpost which was built separate from the rest of the fortifications and was connected to them by a drawbridge. There are a few elements which favour the first hypotesis as opposed to the second. Firstly the demilune does not belong to

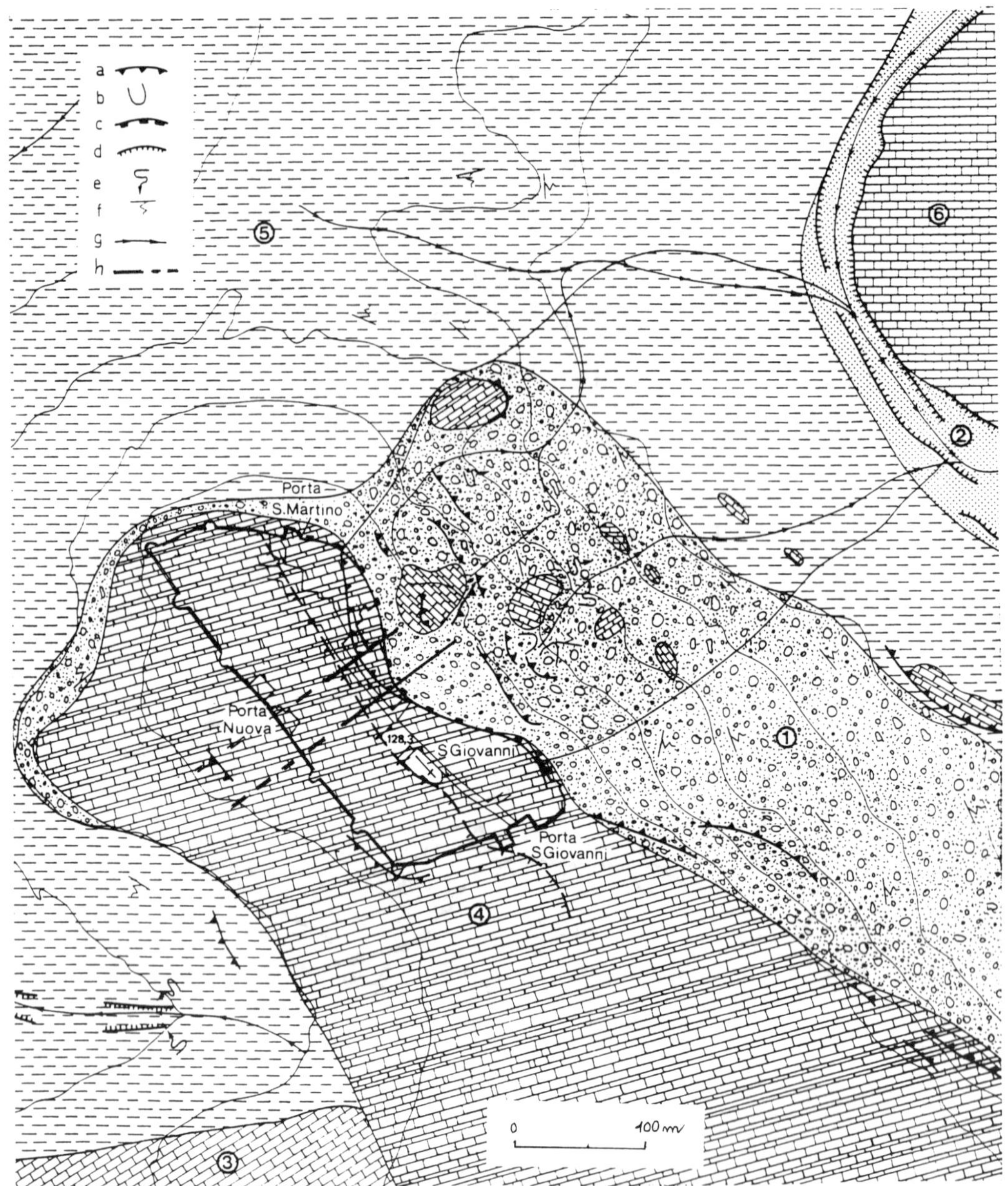

Fig. 2 - Geological and geomorphological map of Magliano area. 1) Landslide and slope debris; 2) Alluvial deposits; 3) Lacustrine limestone (Pleistocene); 4) Magliano Calcarenite (Upper Pliocene); 5) Grey-blue clays (Lower and Middle Pliocene); 6) Argille a Palombini (Cretaceous?). a) crown; b) landslide body; c) scarp (h > 5 m); d) scarp (h < 5 m); e solifluction area; f) rill erosion; g) gully and fluvial erosion; h) fault.

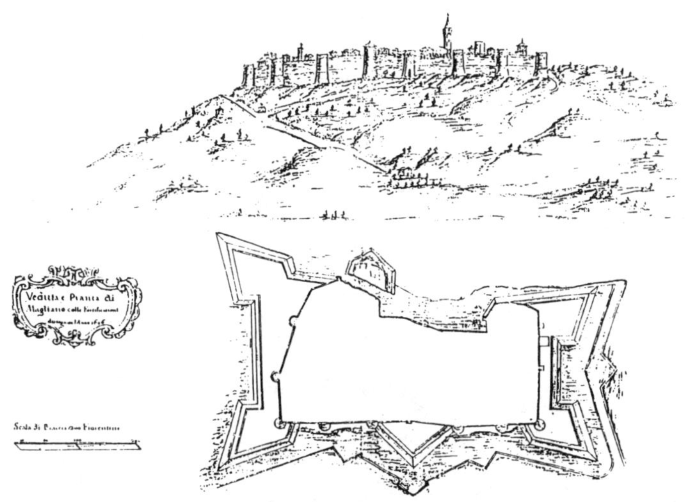

Fig. 3 - Ancient map of Magliano (1646).

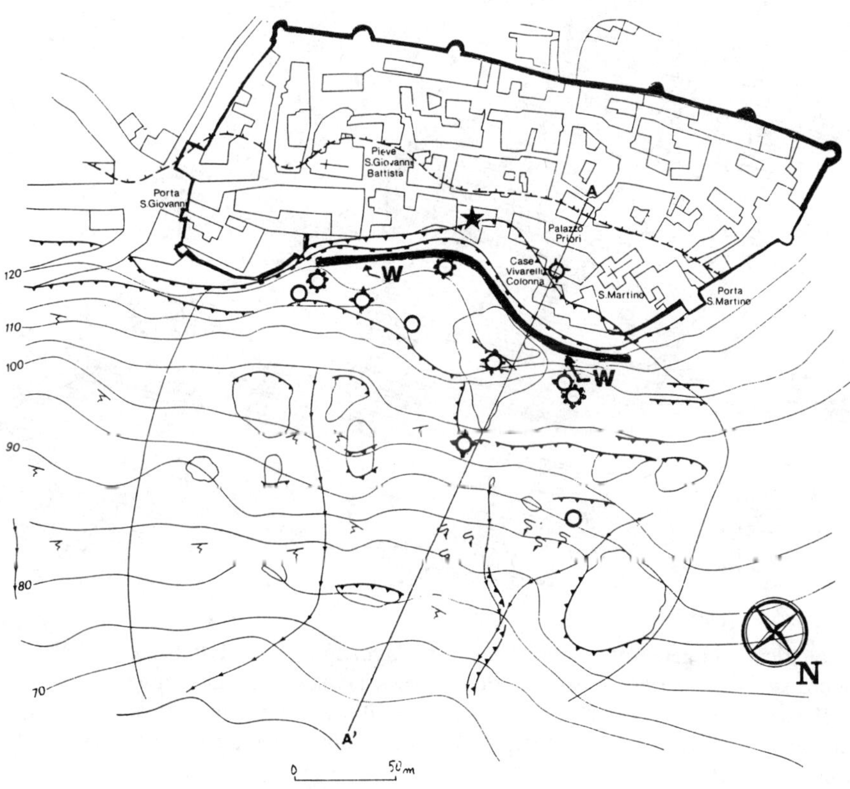

Fig. 4 - Planimetry of the unstable area. ✪: drainage wells; O drillings with piezometer; ✧: inclinometer; ★: destroyed buildings; W: reinforced wall with tie rods. All the buildings lying close to the crown are severely damaged.

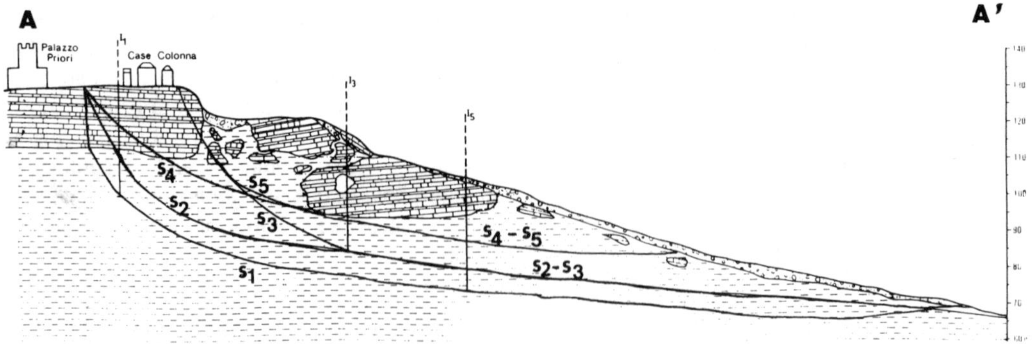

Fig. 5 - Geological section of the unstable slope (geological symbols as in Fig. 2). Slip surfaces located on the basis of geognostic data (S_3, S_4, S_5) and of geomorphological evidence (S_1, S_2, S_3).

the military architecture of this area. Another element is the irregularity of the walls in this same part of the stronghold which is in contrast with the linearity of the other sectors of the fortifications. This seems to point to a morphological influence which can easily be linked to the presence of a back moving landslide crown in this area.

ANALYSIS OF THE LANDSLIDE

Fig. 4 shows a planimetry of the unstable area and indicates the sites of the drilling and the piezometers and inclinometers set up by the Council Administration at various times and stages. A landslide model (see section in Fig. 5) was drawn up on the basis of a detailed field survey, the interpretation of the stratigraphic columns and inclinometric-piezometric and geotechnical data relating to studies, having different methods and objectives, carried out over a period of time.

The phenomenon in question is therefore a complex one, composed mainly of multiple rotational movements, favoured by the presence of a groundwater table in the permeable limestones and by the consequent hydraulic overpressure on the underlying overconsolidated clays presenting fissures in places. The important fractures in the rocky mass and the inferred presence of discontinuity among some clayey samples (unfortunately they cannot be correlated), together with identified block slide movements of large calcareous elements -one of these can be connected to the failure of the castle keep- would seem to suggest the presence of ancient slip surfaces which are deeper than the present more active ones.

GEOTECHNICAL ANALYSIS

Tab. 1 show the main geotechnical data of the clayey materials. The table was drawn up on the basis of data obtained from previous studies and from the analysis we carried out on samples taken from the drainage wells (disturbed) or from excavation walls (undisturbed) during the remedial works.

Table 1. Geotechnical data from clayey samples.

	γ kN/m3	W %	Wl %	Ip %	C kPa	φ' deg	φres deg
A	18.7	31	53	25.5	28	22.5	14.35
B	19.3	31	52	23.7	29	23.9	15.75

A: Surface level, disturbed and/or weathered (15-20 m); B: Deeper level, partially disturbed by ancient movements (10-30 m).

The inclinometric data, even if only partial, showed more active slip surfaces (measurement period 1988-89), displacements in the range of 6-8 cm, broken pipes. Tab. 2 gives the results of the stability tests carryed out taking different slip surfaces (Fig. 5) -that have been located on the basis of the inclinometric data and of

geomorphological evidence- and hydraulic conditions into account (Janbu method).

Tab. 2 - Main results of the stability tests on different slip surfaces.

	Dmax m	Lmax m	D/L	ϕ_{res} deg		ru 0.5	ru 0.3	ru 0.25	ru 0
S1	35.5	345	0.1	15.7	Fs	0.75	0.98*	1.03	1.50
S2	30	350	0.09	15.7	"	0.77	1.03*	1.06	1.60
S3	28	341	0.08	15.7	"	0.70	0.9*	0.93+	1.40
S4#	25	180	0.14	15.7	"	0.82	1.06*	1.11+	1.60
S5#	19	140	0.135	13.3	"	0.69	0.83	0.89*	1.38+

*Actual and recurrent hydraulic conditions; #Slip surface considered by the works; +New hydraulic conditions due to the effects of drainage wells. Dmax: maximum thikness of landslide; Lmax: maximum length of landslide; ru: pore pressure ratio; Fs: factor of safety;

LANDSLIDE MANAGEMENT

Fig. 4 also shows the works performed to reclaim the slope (work carred out 1985-1990). The philosophy of the work programme was to lower the level of the water table with drainage wells having structural function in order to reduce the instability caused by the pore overpressure in the clayey materials. It was also aimed at holding back the large calcareous blocks on which the built up area stands by means of tie rod walls placed along the top escarpment.

Even if the project carried out took account of the need to intervene on the causes (e.g. the changes to the watertable) and to mitigate the effects of the phenomenon (retrogressive evolution of the movement of the fractured blocks) it did not however take into correct consideration the landslide model and the identification of the real landslide mechanisms. Only qualitative checks of the efficaciousness of the programme carried out are currently underway (checking for failure in the buildings). In fact in the case of Magliano, previous studies, partial as they were, were not taken into consideration in the project planning phase. The geognostic field study results are incomplete and contradictory, the monitoring data is inadequate (inclinometric pipes and piezometers were abandoned). No historical research on the evolution of the failure was carried out.

The reason for all this is the incorrect logic behind the policy which public financing has adopted in the consolidation of built up areas up to now. The policy is for works to be carried out immediately and only very rarely financing is allocated to studies and geological, geomorphological and geotechnical analyses. Land management works on the logic of the emergency and hardly ever that of prevention and forecasting. As a result, a programme for the checking of the efficaciousness of works in the future and financing for the maintenance of works are often completely lacking on the part of local governments (Regional Councils, Provincial Councils, Town Councils).

CONCLUSIONS

This paper has described a typical instability situation in a built up area in Tuscany. We wanted to highlight the need to define a methodology for detailed geological-geomorphological analysis in order to optimize these interventions and thus obtain correct landslide management. These analyses serve as a guide for the subsequent geotechnical surveys and tests then used to define the landslide model. As we are dealing with ancient built up areas we must also add adequate historical research to these other analyses.

As a result of what we have just said, it is clear that all kinds of public financing, be it from the Civil Defence of The Monuments and Fine Arts Service etc., should always be subject to detailed geological studies which should, in turn, provide complete landslide and failure models.

REFERENCES

Bettelli G. (1985) - Geologia delle alte valli dei F. Albegna e Fiora. Geol.Rom., XXIV,147-188.

Mancini F. (1960) - Sulla geologia della Piana di Albegna. Boll. Soc. Geol.79.

Environmental Management, Geo-Water & Engineering Aspects, Chowdhury & Sivakumar (eds)
© 1993 Balkema, Rotterdam. ISBN 90 5410 099 0

The natural and induced instability of colluvium and other weak rock during urban development in Taiwan

Hongey Chen
Department of Geology, National Taiwan University, Taipei, Taiwan

ABSTRACT: The stability of colluvial and weak rock slopes were investigated by comparing and contrasting data obtained from a number of examples in Taiwan. These data included both an analysis of the history of slope processes and the mechanical behaviour of their constituent materials in each case. The history of each example is discussed and geomorphological mapping is used to eluçidate the characteristics of the activity in a number of examples. Analysis of mechanical behaviour is demonstrated under laboratory conditions where obvious contrasts between samples from different weak rock and colluvium slopes were recorded. Results of the testing programme indicate that the colluvium has a greater shear strength than the weak rock does. The transient colluvium slope may temporarily have a higher safety factor than the weak rock slope. The differences in their mechanical behaviour is related to the fact that the colluvium consists of more fine grain size fractions (created by gradual weathering) and the lower shear strength.

Clearly, the changeable grain size fractions in the colluvium material which induced the mechanical behaviour will have an unavoidable effect on the slope stability--and thereby the safety factor.

1 INTRODUCTION

Taiwan is an island that has more than 75% of its land surface covered by steep forested mountains and their foothills. Around 25% of the total land area could be classified as plain. The greatest part of the 20 million occupants of the island live on the western coastal plain (about 75%), and a further 7% live within the city of Kaohsiung and 10% in the city of Taipei. To satisfy the necessity for more residents and more homes in the vicinity of cities, urban development has been pressed to take in aggressive engineering plans to build on the steep slopes of the hilly areas.

The successful advance of urban developement requires that wise and safe use be made of many slopes. Colluvial and weak rock materials have produced a number of problems during the early stages of urban developement. This study examined engineering sites in the area of Chitan (a suburb of Taipei) and Chisan (a suburb of Kaohsiung). The results of this study clearly indicate that the colluvial and weak rock materials on these sites may need to be excavated from the steep hillsides for maximum, perhaps essential, safety.

This study analysed in detail all mechanical behaviour aspects of both colluvium and weak rock, which might influence the stability of slopes containing these materials. This investigation was divided into two parts. The first part included surveying of geomorphological features and the second part included experimental tests on the materials. These data have been collected and repeatedly analysed--particulary with respect to safety factors of the particular slope materials studied.

2 GEOLOGY AND GEOMORPHOLOGY

The geology of both study areas is of sedimentary faces in the Tertiary Age (Ho, 1986). The rock types of the Chitan area belong to the earlier Miocene Taliao Formation. The Taliao Formation, Chitan area, has a sequence of thick and massive sandstones alternated with shale and interbedded sandstone with shale. The sandstone is light grey to light bluish grey and fine to medium-grained. The shale is dark grey or greyish black. The interbedded unit comprises dark grey shale with grey, fine-grained sandstone. The rock types of the Chisan area belong to the Kuanmiao Formation late on Pliocene and earlier Pleistocene (Ho, 1986). The Kuanmiao Formation, Chisan area, is composed of thick conglomerate, fine-grained sandstone, sandy shale, and subordinated mudstone. The conglomeratic formation disappeared locally several years ago (due to excavation and erosion) in this study area. The sandstone of the Chisan area is fine to coarse grained and contains light-grey quartz and mica minerals. The sandy shale is dark grey and has fully developed exfoliation. The bluish-grey mudstone is of very fine-grained particle composition and its cementation has been partial.

The strike of bedding of Chitan area is N68° E and the dip is 29° S. There are 3 sets of joints. Their respective orientations are N25° E, dip 72°N; N34°E, dip 76°S and N84° W, dip 80°S. The strike of bedding of Chisan area is N22°E and the dip is 62°S. There also have 3 sets of joints. Their respective orientations are N63°w, dip 86°N; N48° E, dip 47°S and N37°E, dip 72°N.

The geomorphological mapping of the study area potrays the slope direction, the average slope angle, convex and concave shapes of the slope forms, and the nature of the drainage system on the ground surface. The characterisation of the Chitan area depicted its cuesta topography of the landform (Chen, 1989). The orientation of the hinge is from north-east to south-west, and dip direction is from north-west down to the south-east (Figure 1). In general, the distribution of slope angle is from 6 to 40 degree. The elevation of the study area is between 280 metres to 450 metres. The converged groundwater flowed to the lower valley from the study area and entered the Hsing-Dien stream finally. The height difference of the valley is around 1 to 3 metres. Most construction was done on the colluvium slope and located on its ground surface.

In the Chisan area, the elevation is between 100 metres and 300 metres. The inclination of the hill is mostly from the west down to the east. The distribution of the inclinations are regular except in some local areas. In general, the inclinations of this area are more

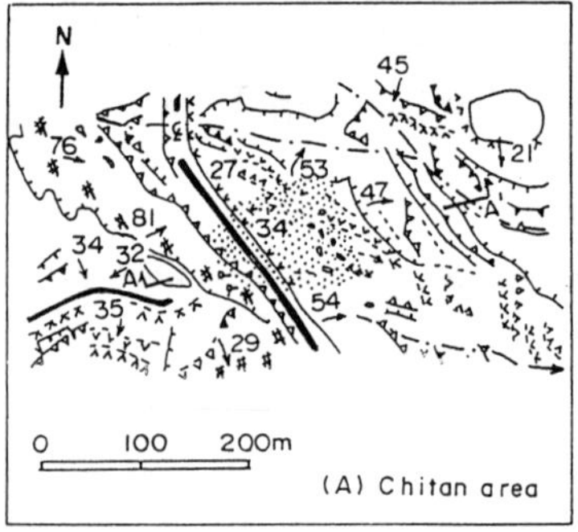

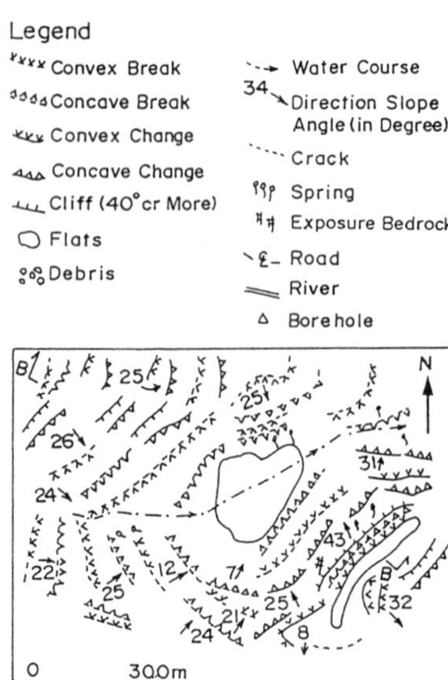

Figure 1 Geomorphological map of study areas

smooth than those in the Chitan area and its slope angle is around 30%. Cracks appear in the old depoists and extend to the other slope surfaces of this study area. The natural valley is not distinct and the differences in height are of around 1 metre. The Chisan area water flow converges and flows mostly into the convex of the slope area (Figure 1). The average of these slope angles is less than 60%.

3 COLLUVIUM AND WEAK ROCK

Colluvium is defined as a regolith material and a chaotic mixture of rock fragments and soil materials. It developed on the upper slope of the weathering weak rocks transport material and spread to the downward slope and formed a transient deposit.

The colluvium consists of a poorly sorted gap-graded assemblage of grain particles with parametres in size from sand grade down to the clay fractions. The material exhibits a highly variable grain size and contains 1 - 2 metres boulders with pebbles which are set in a matrix of sand and clay fraction. The distribution of the grading curve for the colluvium normally reveals around 30% clay fraction material (Chen, 1992). The original weak rock material was evidently separated into several small parts by the rolling and collapsing factors during the sliding activity of slope failure. Therefore, the fine fragments of the rock material were probably created in this colluviation process. The percentage of the larger grain sized materials evidently decreased, as a portion of the overall material, during the transportation process of the geological agent.

The weak rocks of the study areas were mainly composed of the sedimentary deposits. The Chitan area has a main composition of sandstone, shale and interbedded sandstone and shale. These rock materials have three major resovable components which contain quartz, feldspar and mica minerals. Clay minerals and organic matter were also present in the study area. The clay minerals consist of illite, kaolinite and calcium montomorillonite. Each rock formation is of various thicknesses from 5 to 30 metres. The Chisan area normally consists of sandstone, shale and mudstone. These rock materials also contain three major minerals which include quartz, feldspar, mica and some of the clay minerals. The clay minerals are mostly made up of chlorite, illite and kaolinite.

4 SHEAR STRENGTH

This study has concentrated on material strength characteristics because they are the most important factors relevant to an analysis of unstable slopes. The shear behaviours of the colluvium are nearly identical to the shear behaviours of its recombined fraction. In this paper, the shear strength parameters of the materials were obtained by conducting laboratory experiments using the direct shear test apparatus. In adopting this test procedure, it is realised that the shear strengths obtained are for a remoulded material. In this case, this may be appropriate because the colluvium is continually being naturally remoulded as it is being transported downhill. The physical properties of colluvium on both areas are summarized in Table 1. In Chitan area, the average in-situ water content is 31%. The unit weight of colluvium is 2.0 T/m^3 . The void ratio is 0.54 and the plastic index is 8.3. In Chisan area, the average in-situ water content is 22%. The unit weight of colluvium is 2.1 T/m^3. The void ratio is 0.48 and the plastic index is 9.4.

This experimental analysis method used on materials in an undrained

Table 1 The physical properties of colluvium on both study areas

	Wn	Rt	e	PI
Chitan area	31%	2.0	0.54	8.3
Chisan area	22%	2.1	0.48	9.4

Wn: water content, T/m^3, Rt: unit weight, e: void ratio, PI: plastic index.

condition was used because failure of the colluvium always occurs under a high piezometric head during the typhoon season. The high piezometric levels observed during the typhoon season were taken to justify the choice of undrained experiments to obtain shear strength data. Only this method allows the laboratory procedures to approximate actual conditions in the field of materials experiencing relevent water forces during the typhoon season. It is widely recognised that changes in water content and water pressure will lead to increased stability or instability of the subject material (Morgenstern, 1980). The transport of the colluvium material could be due to sliding along a well-defined slip surface with the deformation rate increasing in the wet season when the pore pressure increases. The values of shearing resistance determined from the remoulded colluvium and weak rock using the direct shear test on saturated specimens are summarized in Table 2.

Table 2 The shear strength of colluvium and weak rock materials on both study areas

	Material	Rt	C'	Ø'
Chitan area	Co	2.0	2.3	29
	SS	2.3	0	27
	SH	2.5	0	22
	SS & SH	2.5	0	24
Chisan area	Co	2.1	6.0	29
	SS	2.4	0	31
	SH	2.4	0	22
	MS	2.5	0	27

Co: colluvium, SS: sandstone, SH: shale, SS & SH: sandstone and shale interbedded, MS: mudstone.

In the Chitan area, the unit weight of the colluvium is 2.0 T/m^3. The effective friction angle of the colluvium is 29^0 and its cohesion is 2.3 KN/m^2. The unit weight of the sandstone is 2.3 T/m^3. The effective friction angle of the sandstone is 27^0 and no cohesion. The unit weight of the shale is 2.5 T/m^3. The effective friction angle of the shale is 22^0 and no cohesion. The unit weight of the sandstone and shale interbedded is 2.5 T/m^3. The effective friction angle of the sandstone and shale interbedded is 24^0 and no cohesion as well. In the Chisan area, the unit weight of the colluvium is 2.1 T/m^3. The effective friction angle of the colluvium is 30^0 and its cohesion is 6.0 KN/m^2. The unit weight of the sandstone is 2.4 T/m^3. The effective friction angle of the sandstone is 31^0 and no cohesion. The unit weight of the shale is 2.4 T/m^3. The effective friction angle of the shale is 22^0 and its effective cohesion is 0.5 KN/m^2. The unit weight of the mudstone is 2.5 T/m^3. The effective friction angle of the mudstone is 27^0 and no cohesion as well.

The experimental data indicate that between 40% and 50% of the colluvium matrix is composed of fine and coarse sand on both areas. The coarser fractions of the colluvium matrix are themselves matrix supported. The ultimate strength of the colluvium will lie in its finest grain fractions (silt 20% and clay 30% of the boulder supporting matrix). The boulder supporting matrix contains between 40 and 50% sand. The sand sized fractions in the boulder supporting matrix of the colluvium are in a continuous state of change. The deformation is dominated by grain size reduction which leads to a significant increase in the fine fractions as the rock fails in a shear event. The shear behaviour of the colluvium is therefore nearly identical to the shear behaviour of its finest fractions.

5 SLOPE STABILITY

The purpose of the stability analyses is to verify the safety factors for the potential slip surfaces within the slope study areas. A major part of the slope stability analysis should include a field investigation which also involves the aspects of geomorphological mapping (Hutchinson, 1977). The history of the slope approach is organized and correlated with both laboratory investigation and stability analysis in the study areas of each example. The stability analyses in this paper were conducted to understand the behaviour of the colluvium and weak rock slopes in the two study areas.

In this paper, Janbu's method (1973) was selected for the stability analysis of the colluvium slopes. The most active slips on both areas appear in the cracks on the slope surface. These cracks are bounded to the slip margins. The data of the safety factor presented in Table 3 indicate that some parts of the observed slip surfaces could be in critical stable condition.

Table 3 The results of safety factor of the slope stability on both study areas

		Safety Factor	
		F. Sa.	M. Sa.
Chitan area	A-1	1.38	2.03
	A-2	0.92	1.70
	A-3	1.12	1.85
Chisan area	B-1	1.14	1.78
	B-2	1.21	1.80

F.Sa.: fully saturated
M.Sa.: mean saturated

In the Chitan area, three slip surfaces were subjected to the analysis and these are shown in Figure 2. Slip surface A-1 is defined by fractures and scarps at its top, sides and toe and by the recognition of a slickensided failure surfaces in the core samples. Slip surface A-2 is defined by outcropping fractures and a small rear scarp. A-3 is defined by outcropping fissure of the weak rock at the top of A-1 slip surface. In the Chisan area, two slip surfaces were subjected to the analysis and these are shown in Figure 3. Slip surface B-1 is

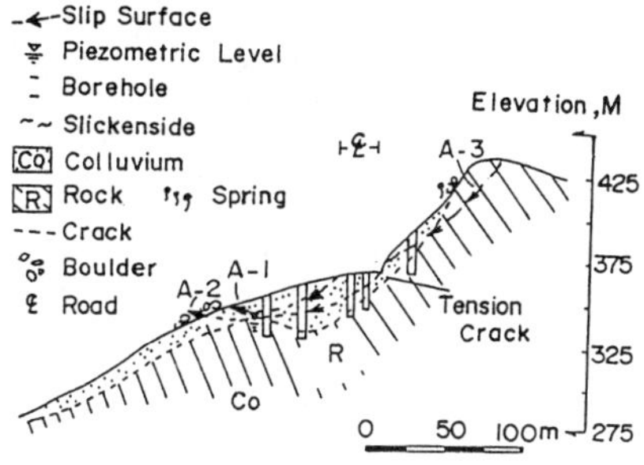

Figure 2 A-A Section of Chitan area

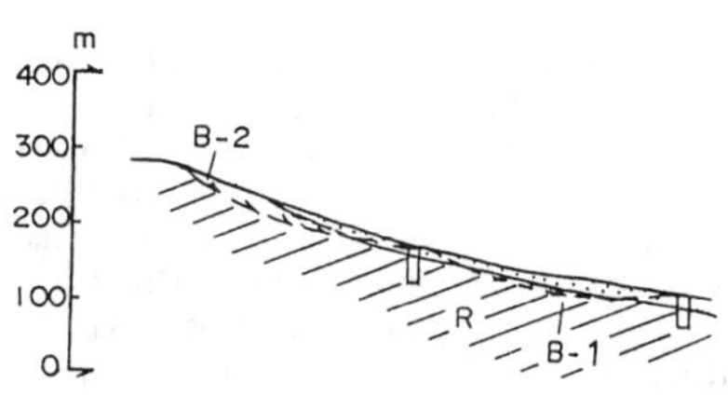

Figure 3 B-B section of Chisan area

defined by fractures at the middle of the colluvium slope and by recognition of slickensided failure surfaces in the cores. Slip surface B-2 is defined by outcropping fissures of the weak rock on the top of B-1 slip surface.

The back analyses method of investigating slope stability should be seriously re-examined in the light of promoted or improved historical and morphological data (Chandler, 1977). These data will be examined in a new back analysis process when the data become available. Table 4 gives the results of back analysis of the potential failure surface for the condition of F=1.0 under a fully or mean saturated condition. These indicate, what are taken to be, realistic friction angles for the colluvium and weak rock slopes, and show the very large friction angle needed to maintain the stability on the slope of the study areas. The results given in Table 4 indicate that the friction angle for the colluvium slope should be less than 29.1° . The friction angle for the weak rock slope should be less than

Table 4 The results of back analysis (F=1.0) on both study areas

		Friction angle	
		F. Sa.	M. Sa.
Chitan area	A-1	29.1	18.6
	A-2	41.6	10.2
	A-3	33.1	20.2
Chisan area	B-1	32.8	17.5
	B-2	31.5	19.2

33.1° at Chitan. In Chisan area, the friction angle for the colluvium slope should be less than 32.8°, for the weak rock slope should be less than 31.5° . Construction developement on the colluvium and weak rock slopes of the Chitan and Chisan areas and in similar areas around Taipei and Kaohsiung is a response to the increasing pressure of urban development in the northern and southern parts of Taiwan. This resulting development presents a hazard of serious proportions because of the numerous potentially severe slope failures.

6 CONCLUSIONS

The results of all the analyses of the studied colluvium and weak rock materials reveal a number of important factors worthy further consideration.

1. The comparison of shear strength data obtained from direct shear tests and the triaxial tests conducted on remoulded samples of the colluvium and weak rock materials adds significantly to our understanding of the mechanical behaviours of such materials in the types of contexts studied.
2. Back analyses of slopes that are known to be stable except when fully saturated--enables us to define a range of friction angles. The shear strength results of the back analyses seem to lie between those determined for saturated and those determined for unsaturated colluvium and weak rock.
3. The colluvium matrix is composed of shale fragments of different grain sizes. It is the behaviour of this shale (particularly the finer fractions) that dominates the mechanical behaviour of the colluvium.

7 ACKNOWLEDGEMENTS

The author wish to express his appreciation to Mr. K.L. Wu of Shyh-Chyi Corporation for supplying the borehole and piezometric data for this atudy in both of Chitan and Chisan area. Thanks are also given to National Science Council, Republic of China, for their funding support in this research project.

REFERENCES

Chandler,R.J. (1977) Back analysis techniques for slope stabilisation works: a case record. Geotechnique 27, 479-495.

Chen,H. (1989) A case study of the morphology on the unstableslopes. 6th Intl. IAEG Congress, A.A. Balkema, P1603-1608.

Chen,H. (1992) The stability and mechanical behaviour of colluvium slopes.Jl. Geol. Soc. of China, Vol.35, No.1,P95-114.

Ho,C.S. (1986) An introduction to the geology of Taiwan:Explanatory of the geologic map of Taiwan. The Ministry of Economic Affairs, Republic of China.

Hutchinson,J.N. (1977) Assessment of the effectiveness of corrective measures in relation to geological conditions and types of slope movement. General Report, Theme 3, Bull., Inter. Ass. Engng. Geology, No.16, 131- 155.

Janbu,N. (1973) Slope stability computations. In: Embankment dam engineering, Casagrande Memorial Volume (R.C. Hirschfield and S.J. Poulos, Ed.). John Wiley & Sons, New York, 47-86.

Morgenstern, N. R. (1980) Factors affecting the selection of shear strength parameters in slope stability analysis. Pro. Int. Symp. on Landslides, Vol. 1, New Dehli, 83-93.

Environmental Management, Geo-Water & Engineering Aspects, Chowdhury & Sivakumar (eds)
© 1993 Balkema, Rotterdam. ISBN 90 5410 099 0

Lateral spreading in rock formations: Mechanism, analysis, hazard assessment and control measures

A.Cancelli, N.Chinaglia & D.Mazzoccola
Department of Earth Sciences, University of Milan, Italy

ABSTRACT: The origin and subsequent evolution of large extent landslide areas, of complex type but with prevalent lateral spreading mechanism, are controlled by particular geological, lithological, structural and geotechnical conditions. Some typical situations, recently recognized in the Northern Apennines (Italy) are described and analyzed in this paper.

1. INTRODUCTION

Lateral spreading represents one of the most important landslide mechanisms in rock masses. Such forms of mass movement are mostly associated to the presence of a slab of competent rocks, overlying a substratum formed by ductile rocks (e.g. clayshales).

Numerous examples from many parts of the World are reported in the geological and geotechnical literature, where both superficial and deep-seated failures are considered. Large extensions are generally involved in the slope movements; the mechanism is complex, including not only lateral spreading. The rate of movement is generally low, according to the characters of creep deformations, with temporary accelerations in part of the landslide area.

For such complex landslide areas, recognition, geomorphological surveying and hazard evaluation can found unexpected difficulties; many situations have been misinterpreted in the past, and attributed to purely tectonic causes (faulting) or, alternatively, to other geomorphic processes (e.g. karstic). In Italy, more and more cases have been recently put in evidence, and are presently reconsidered; the paper reports the preliminary results of geological, geomorphological and geotechnical investigations for an area situated in the Northwestern Apennines (Northern Italy).

2. INSTABILITY FACTORS, FAILURE MECHANISM AND EVOLUTION

Most of lateral spreadings are associated to a substratum formed by structurally (and lithologically) complex formations. Landslide mechanism results from a complex interaction of multifarious instability factors, as assessed by some evolutive models that are found in the most recent literature (among all, Pasek 1974; Radbruch-Hall et al. 1976; Cancelli & Pellegrini 1987).

According to the first author, when a competent rock slab, cut by subvertical fractures of tectonic origin, is subject to erosion by surface waters, in the zones of more intense fracturation an erosion channel can deepen progressively and reach the clayey substratum; due to the stress relief, bending of clay layers and swelling processes take place on the bottom and the flanks of the valley, so favouring the opening of vertical fractures, their filling with debris and the starting of horizontal movements for the blocks of competent rock situated at the edge of the slab; at this time, a further deepening and enlarging of the valley induces new complex movements of the blocks, also by tilting or by sliding (both translational and rotational); therefore, the detatched blocks continue to move along the hillslope, dividing into smaller and smaller portions; mudslides and earthflows are the final stages of the slope instability process.

Alternatively, a slightly different starting mechanism was also proposed, not requiring any initial slope gradient (Cancelli & Pellegrini 1987). In the same lithological and structural situation, the competent rock slab behaves as a superficial raft foundation on compressible medium; the differential settlement from the central to the external part of the slab can induce the first subdivision into vertical prisms and lateral bulging of the unconfined shaly substratum; then, a first rotational sliding surface can form, followed by the movement of external rock prisms; successively, further movements can take place, by sliding and/or by tilting, giving origin to a complex landslide mass, as in the mechanism previously described.

In any case, the intrinsic causes of the instability are the superimposition of massive rocks over weak or plastic clayey materials, the presence of shear or tension fractures in the former, the high

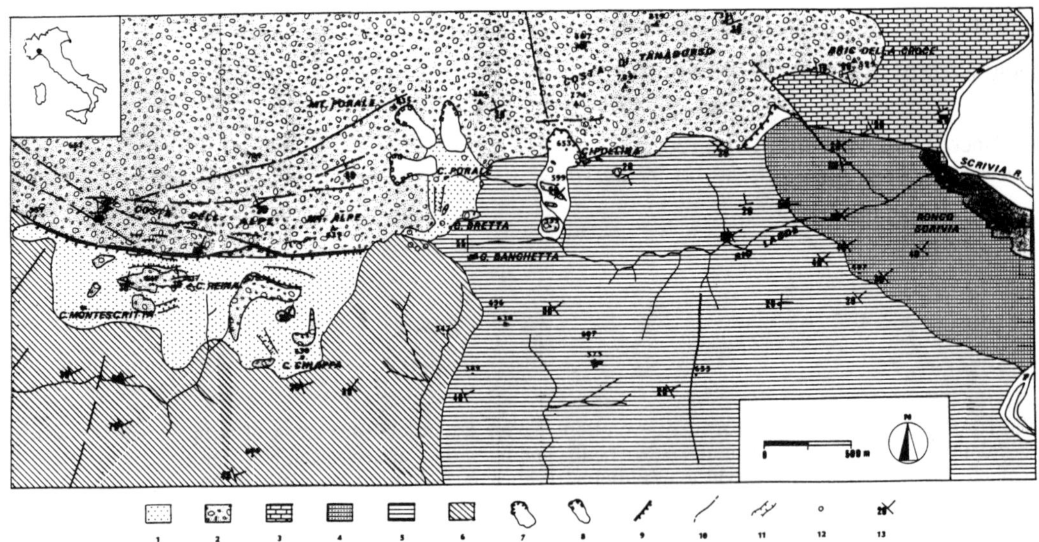

Fig. 1. Detailed map of landslides along the Southern edge of the conglomerate slab. Legend: 1) slope debris; 2) "Formazione di Molare"; 3) "Calcari del M. Antola"; 4) "Formazione di Ronco"; 5) "Argilliti di Montanesi"; 6) "Argilliti di Mignanego"; 7) landslide areas (lateral spreads prevailing); 8) rockfalls; 9) fault cliffs; 10) faults; 11) trenches; 12) springs; 13) strike and dip of layers.

deformability and low shear strength of the latter.

Among external causes, natural erosion by rivers and minor streams is probably the most effective on long-term instability; progressive creep and softening of clays also play an important role.

Pore pressure insurgence by rainfall or snowmelt, lateral water pressures in open vertical cracks, seepage pressures of percolating groundwater are generally recognized as triggering causes of accelerated slope movements.

3. LATERAL SPREADINGS IN THE NORTHERN APENNINES

A large number of block-type slope movements, prevalently showing lateral spreading deformations, have been recognized in the Northern Italian Apennines (Cancelli & Pellegrini 1987); researches are still in progress in many areas. This paper reports the results of recent investigations in the extreme NW part of the chain (Ligurian Apennines), where Oligocene conglomerates are superimposed to Cretaceous turbiditic formations (clayshales, with calcareous or arenaceous interlayers).

The former belong to a terrigenous formation known as "Formazione di Molare", mainly constituted by coarse grained conglomerates and breccias (with some lenses and interlayers of sandstones, siltstones and marls), gently dipping (10÷20°) to NNW. They form a slab with an overall thickness varying from 200 m up to 500÷600 m, limited by a cliff abruptly rising from the clayey substratum: a significant part of the border area is represented in Fig. 1.

The Cretaceous sequence includes several formations (partly heteropic each to other); within the limits represented in Fig. 1 we can recognize (from top to bottom and outcropping from E to W in the map):

- "Calcari del M. Antola" (Upper Cretaceous): marly limestones, somewhere with interlayered marls and sandstones;
- "Formazione di Ronco": rhythmically interbedded marls and sandstones (somewhere with clayey interlayers);
- "Argilliti di Montanesi": essentially clayshales, with rare interlayers of fine grained sandstones and siltstones;
- "Argilliti di Mignanego" (Lower Cretaceous): clayshales (largely prevailing), interlayered with fine grained sandstones and/or calcarenites.

From a lithological point of view, Montanesi and Mignanego formations are characterized by high clayshale/competent rock ratios and show a typical ductile behaviour.

Following the paroxysms of the alpine orogeny, all the Cretaceous sequence was subjected to intense folding and faulting; as a general trend, within the examined area the layers dip to ENE with variable slope angle (20÷70°), but minor folds, break thrusts and shear zones furtherly complicate the structural path.

As a consequence of their depositional and tectonic stress history, these geological units are typical examples of both lithologically and structurally complex formations; evident schistosity and tendency to be subdivided into scales and laminae are common to all argillaceous lithotypes and influence the shear strength behaviour of the bedrock.

The distribution of landslides along the edge of the conglomeratic slab is controlled by a tardo-orogenetic system of subparallel faults (W-E directed), increasing the degree of freedom of large competent rock masses; this is demonstrated by the general W-E trend of most landslide crowns and trenches downhill. An example is clearly visible in the left part of Fig. 1: the repetition of the same failure mechanism, for successive conglomerate masses, along the same geological structure and the coalescence of adjacent landslides have led to the formation of a unique, large landslide body (about 50 Mm^3), tectonically controlled, characterized by anomalous width/length ratio (about 4/1).

Landslide areas at different grade of evolution are also observed. Due to the initial, gently counterslope setting of the conglomerate slab, the most common starting mechanism consists in rotational sliding failures, involving the external conglomerate blocks and the underlying clayshales (in turn affected by softening processes). Successively, translational movements with prevalent horizontal component (i.e. lateral spreads) take place, favouring the opening and progressive widening of trenches and reducing the lateral constraint on adjacent conglomerate blocks; the subsequent evolution of the slope includes a more or less complicate alternance and combination of different kinematisms: tilting, translational and rotational sliding of the blocks, together with rotational sliding within the clayey substratum. Finally, as conglomerate blocks move farther and farther along the hillslope, their average dimensions progressively decrease, by comminution and erosional processes, giving origin to a colluvial mantle, including heterometric conglomerate debris in abundant clayey matrix: mudslides and earthflows represent the final stage of most landslide areas.

The landslide involving some houses of the small Cipollina hamlet is a representative example of landsliding hillslope at this stage of evolution (Figs. 1 and 2-a,b). This complex landslide is 670-m long and 180-m wide; the sliding mass should have a thickness of 130÷150 m in the upper part and less than 50 m at the foot; an overall volume of about 5÷6 Mm^3 has been preliminarly evaluated.

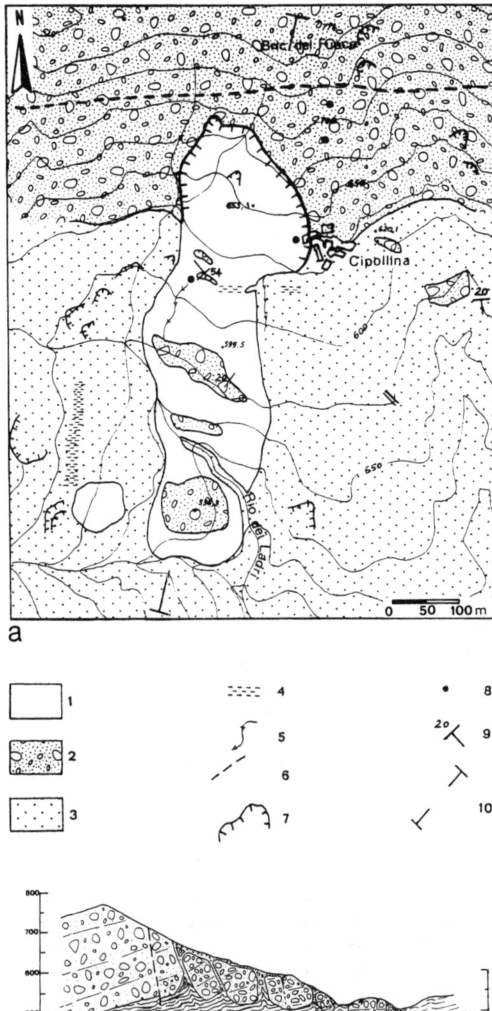

Fig. 2. Cipollina landslide: a) plan; b) interpreted section. Legend: 1) landslide debris; 2) conglomerates ("Formazione di Molare"); 3) clay shales ("Argilliti di Montanesi"); 4) wet ground; 5) surface water flow; 6) supposed faults; 7) landslide scarps; 8) springs; 9) strike and dip; 10) trace of section.

4. MECHANICAL PROPERTIES OF CLAYSHALES

The mechanical behaviour of these rock masses is essentially governed by the argillitic component. The degree of diagenesis and lithification is typical of a "rock-like shale", but the presence of shear bands, characterized by scaly structure, and the small size of single scales (1 ÷ 10 mm) could justify the routinary approach to slope stability problems in terms of Soil Mechanics (at least for shallow slope movements). A laboratory disaggregation technique, consisting in alternate wetting-drying cycles followed by remoulding, was adopted, in order to test the clay-shaly component as a "soil-like material"; this technique allowed to determine the classification properties (Figs. 3 and 4) and to prepare specimens for shear resistance testing.

For this study, clayshales samples were taken from both Montanesi and Mignanego formations

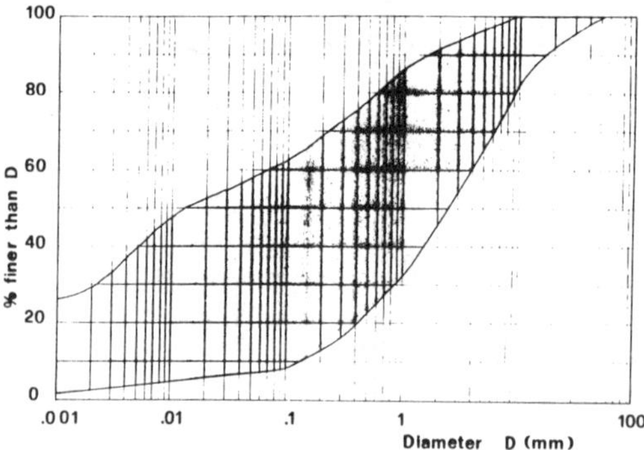

Fig. 3. Grain size distribution of clayshales (after disaggregation by wetting-drying cycles).

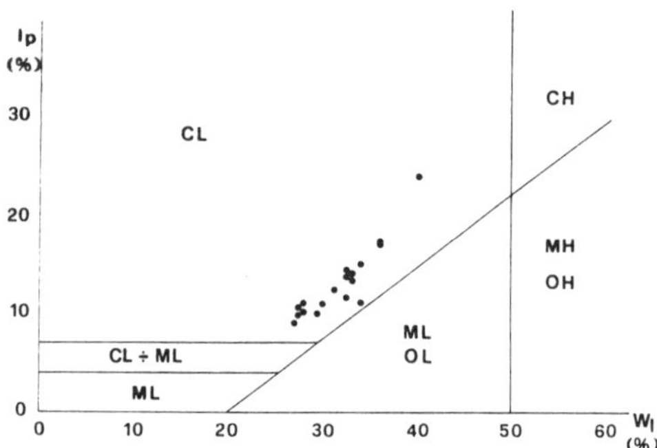

Fig. 4. Plasticity chart.

(not showing any substantial differences beteewn the respective clayey components). According to classification tests, the clay fraction is characterized by low plasticity (plasticity index I_p = 9÷24 %) and is composed by clay minerals of moderate activity index (A_c = 0.6÷0.8).

The shear resistance has been determined on both undisturbed and remoulded samples, by means of triaxial, direct shear and torsional shear tests; different values of the shear strength have been measured (Fig. 5) and the results can be summarized as in the following:

a) peak resistance (undisturbed specimens): fairly scattered values, interpretable by an angle of shear strength = 22÷25° and by an intercept cohesion varying from 0 to about 60 kPa (according to the prevalent orientation of the scales with respect to the imposed failure plane);
b) fully softened state: values from both DS-PR and TS-PR tests, fairly aligned with the origin and giving an average angle of 22° (in good agreement with the lowest peak value found for undisturbed specimens);
c) residual resistance: highly scattered values (see dashed area in Fig. 5), indicating residual angles of shear strength ranging from 11° to 17°; the lowest values were determined by direct shear tests along a soil/rock contact and (at least in this case) they seem too conservative and unrealistic for most stability problems in natural slopes; in accordance with the well known correlations with index properties (Lupini et alii 1981), a limited range of 15÷17° was adopetd for stability analyses.

5. RESULTS OF STABILITY ANALYSES

Results of back analyses reported in the geotechnical literature show that in fissured argillitic rocks the shear strength parameters at failure are always higher than the residual ones, as obtained in the laboratory. For some Italian sheared clayshales ("Argille scagliose" of the Apennines), this behaviour has been justified with the presence of irregularly distributed fragments of more competent rock (D'Elia et alii, 1986).

Stability analyses were also performed for some landslides developing in the same Montanesi and Mignanego formations (and also in similar clayshales formations in the neighbour areas - Cancelli 1986); some general considerations can be drawn.

In a few cases of first-time failures, along the edge of the conglomerate slab, the mobilized shear strength at failure resulted lower than the measured peak strength, but higher than the fully softened one, still including some cohesion (not yet completely annihilated by natural softening processes). Furtherly, the preferential orientation of scales should be considered for a best evaluation of shear strength parameters to be introduced into stability analyses.

For landslides subjected to periodical acceleration of movements (e.g. the same Cipollina landslide shown in Fig. 2), the mobilized strength was always higher than the residual shear strength.

Of course, all stability analyses (reported both in the literature and in this paper) are related to simple failure mechanisms and to shallow slip surfaces (up to a few tens meters deep). Conventional analyses (in terms of limit equilibrium of "rigid" bodies) can lead to non negligible errors if applied to complex failure mechanisms; in addition, the results of tests performed according to routinary Soil Mechanics criteria, devices and procedures cannot be simply extrapolated to landslides involving thick rock masses and stress levels higher than 2÷3 MPa. In all these cases, a thorough evaluation of both strength and stress-strain relationships of the geological medium and numerical analyses (by finite elements or distinct elements methods) are required.

6. ASSESSMENT OF THE HAZARD LEVEL

Hazard assessment requires detailed topographical, geological, geomorphological, hydrogeological and geomechanical surveying. The superimposition of competent rocks on clayey substratum, and the

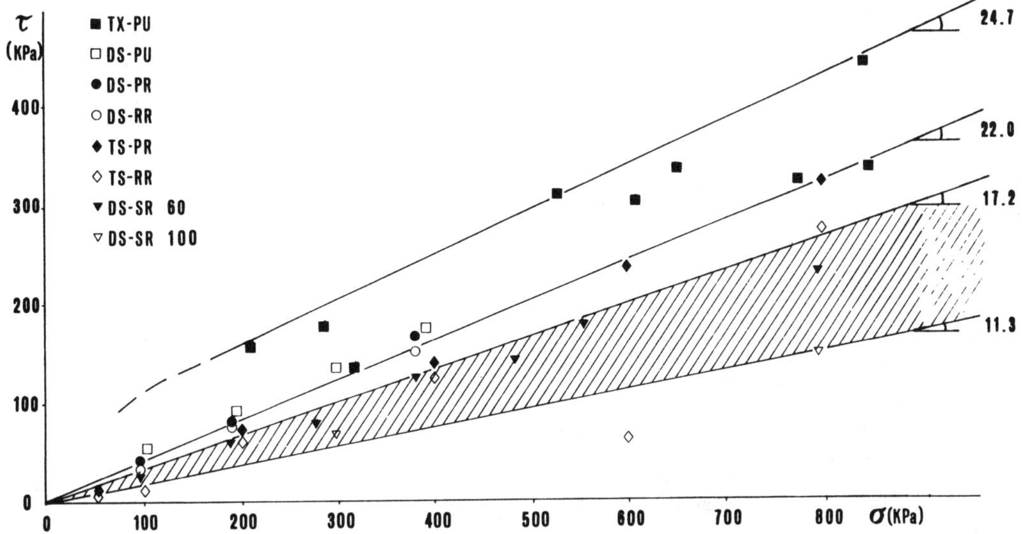

Fig. 5. Results of shear resistance tests. Legend: TX) triaxial compression; DS) direct shear; TS) torsional shear; PU) peak value, undisturbed specimen; PR) peak value, remoulded specimen; RR) residual value, remoulded specimen; SR) soil/rock contact, remoulded specimen; 60) circular box, diameter in mm; 100) square box, width in mm.

presence of colluvial material or squeezed clay between blocks greatly increase the difficulties for subsurface, geophysical and geotechnical investigations.

As said above, due to the complexity of failure mechanisms, the results of stability analyses in the most unfavourable conditions can be affected by a large degree of uncertainty.

The rate of movement is generally very low (from "slow" to extremely slow", according to Varnes' 1978 classification); consequently, the impact of these slope movements on human lives, structures and economic activities is relatively low, so that their "damage potential" can be classified "low" to "moderate" (Hungr 1981; Morgenstern 1985). In any case, the monitoring of movements should be mainly based on long term geodetic and topographic surveys; in this view, the application of a modern global positioning system (GPS) can be helpful, and a long term, field monitoring campaign is now in progress for the same area representeed in Fig. 1.

Also due to the generally low rate of movement, high precision, long time in use and maintenance are essential requisites for every instrument (e.g. crack gauges, tiltmeters, extensometers, in-place inclinometers, etc.), installed to monitoring surface and subsurface displacements.

Periodical accelerations of landslide movements are generally governed by the piezometric response of phreatic or confined groundwater to the meteorological conditions of the area (rainfall duration and intensity, snowmwlt). Monitoring piezometric levels can be necessary for high risk, particular situations, provided that piezometers be carefully selected and properly installed in the most significant zones (to be selected on the basis of thorough hydrogeological investigations).

7. CONTROL MEASURES IN UNSTABLE AREAS

For many reasons (coexistence of different mechanisms within the same unstable slope, uncertainties related to the subsurface geological model, low rate of movements and degree of damage potential), the true effectiveness of geotechnical stabilization measures is difficult to be evaluated.

The use of stabilization measures is generally limited to a few local situations, where slope movements could damage inhabited areas (e.g. villages or towns on the top of a competent rock slab) or life lines (roads; railways; water, oil and gas pipelines). Such measures include most methods and techniques of common use for slope stabilization, as:

- modifications in the slope geometry;
- surface and subsurface drainage;
- retaining structures.

The lack of coordination between different agencies (each one looking at its particular problem) and the consequent spatial limitation of the respective stabilization works are typical defects of this "active" approach to the problem and can make the measures ineffective (with a loss of time and money). For most situations, "passive" control measures, consisting in limitations to the land use, should be preferred; in any case, the large extent and depth of unstable masses should be taken into account, much more than in the common engineering practice.

ACKNOWLEDGEMENTS

The research has been sponsored partly by CNR (National Research Council) and partly by MURST (Ministry for the University and Scientific and Technological Research).

REFERENCES

Cancelli, A. 1986. The determination of shear strength parameters for sheared clay shales. Proc. Int. Symp. of Engineering in Complex Rock Formations, Beijing: 297-303.

Cancelli, A. & M. Pellegrini 1987. Deep-seated gravitational deformations in the Northern Apennines, Italy. Proc. 5th Int. Conf. & Field Workshop on Landslides, Australia & New Zealand: 171-178.

D'Elia, B., D. Di Stefano, F. Esu & G. Federico 1986. Slope movements in structurally complex formations. Proc. Int. Symp. of Engineering in Complex Rock Formations, Beijing: 430-436.

Hungr, O. 1981. Dynamics of rock avalanches and other types of slope movements. Ph.D. Thesis, Univ. of Alberta.

Lupini, J.K., A.E. Skinner & P.R. Vaughan 1981. The drained residual strength of cohesive soils. Geotechnique 31:181-213.

Morgenstern, N.R. 1985. Geotechnical aspects of environmental control. Proc. 11th Int. Conf. Soil Mechanics & Foundation Engineering, San Francisco 1:155-185.

Pasek, J. 1974. Gravitational block-type slope movements. Proc. 2nd Int. Conf. IAEG, Sao Paulo, 2 Th. V-PC-1.

Radbruch-Hall, D.H., D.J. Varnes & W.Z. Savage 1976. Gravitational spreading of steep-sided ridges ("sackung") in Western United States. Bull. IAEG 14:23-35.

Varnes, D.J. 1978. Slope movements types and processes. Landslides Analysis and Control (R.L. Schuster & R.J. Krizek, eds.), Nat. Acad. of Sciences, Washington, Spec. Rep. 176:11-33.

Environmental Management, Geo-Water & Engineering Aspects, Chowdhury & Sivakumar (eds)
© 1993 Balkema, Rotterdam. ISBN 90 5410 099 0

Some cases of landslides affecting urban areas in the Marche, Central Italy

S.Conti & G.Tosatti
Department of Earth Sciences, University of Modena, Italy

ABSTRACT: The valley of the River Marecchia (Northern Apennines, central Italy) is characterised by the presence of competent fractured formations outcropping in isolated and dismembered slabs overlying shaly clayey soils with a prevalent plastic-viscous behaviour. This geological superimposition has favoured the construction of castles and villages but, at the same time, has determined the activation of several mass movements, which have for the first time been recognised as deep-seated gravitational deformations. The cases of Pennabilli, San Leo and other minor centres of the R. Marecchia Valley are being considered because of their historical importance and representation of different structural situations. The relationships between the evolution of the landslides, the structural characteristics and the mineralogical and geotechnical properties of the substratum have been investigated. On these sites, monitoring systems have also been installed and consolidation and upgrading works have been enterprised. Notwithstanding this, the interventions carried out so far have not solved the problems related to slope movements, that by now involve a large number of buildings, since they were not based on a correct and thorough interpretation of the instability causes.

1 INTRODUCTION

Widespread landslide-prone conditions characterise all the valley of the River Marecchia (Marche and Romagna regions), owing to the superimposition of Miocenic competent formations belonging to the Epiligurian Sequence (semi-allochthonous) on mostly argillaceous soils (Cretaceous-Eocene Ligurian allochthonous units). The latter correspond to dark grey, greenish and reddish overconsolidated fissured clays and clay shales, within which bedding surfaces and overturned and compressed folds are often recognizable.

In this article the stability conditions of some of the historical centres of the valley are taken into account, since most of them were built on top of the limestone cliffs outcropping all over the valley, owing to their privileged positions that turned these settlements into practically impregnable strongholds. Unfortunately the geological and mechanical characteristics of these competent slabs and their clayey substratum have always favoured the formation of widespread and complex gravitational movements, in great part still active, which directly threaten the conservation of many ancient and artistic buildings. In particular, the cases of Pennabilli and San Leo are described, important historical villages which exert a strong touristic appeal thanks to the richness of their well preserved monuments. Previous interpretations tended to consider the landslides affecting these urban areas as due mainly to superficial movements taking place only along the margins of the slabs (Persi, 1973; Soccodato,1980; Dramis *et al.*,1987; Ribacchi & Tommasi, 1988). Instead, according to the failure model here proposed, the mechanism originating most of these landslides and their related features, which include karst-type landforms such as trenches, gullies, and doline-shaped depressions, should be ascribed to deep-seated deformations, due to the slabs' load stresses transmitted to the compressible substratum, with differential settlements more accentuated in the inner part (Conti & Tosatti, 1991). The differential settlements are induced by plastic-viscous flows in the clay shales towards the less confined external boundaries of the slab. This phenomenon is accompanied by bulging in the clays outcropping along the margins and formation of large rotational slides, thus undermining the stability of the overlying rocks which are subject to block-tilting and collapse, being afterwards involved in the superficial earth flows. The further fragmentation of the rock masses is favoured by the numerous tectonic fractures found all over the slabs, along which the lateral spreading of the blocks begins, with opening of gulls that tend to be filled from the top with colluvial material and, in some cases, also from the bottom with clay shales, following a diapiric-type mechanism (Cancelli & Pellegrini, 1987).

The incomplete identification of the real causes responsible for the long-term instability has implied a lack of satisfactory results in the remedial measures carried out so far. Other examples of landslides affecting ancient sites will also be considered, although to a lesser extent.

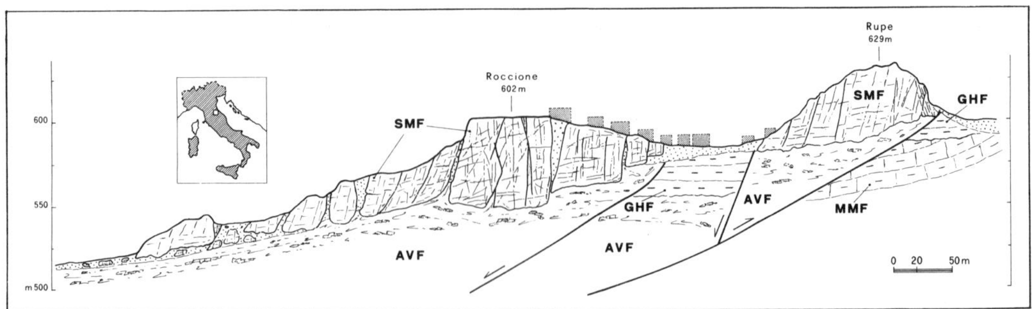

Fig. 1 - Location of the area studied and geological section of the slabs of Pennabilli (modified from M. Gabrielli): GHF = Ghioli di Letto Formation; SMF = San Marino Formation; MMF = Monte Morello Formation; AVF = Argille varicolori Formation.

2 THE CASE OF PENNABILLI

The area where the village of Pennabilli rises is formed by two intensely fractured rocky slabs, called Roccione and Rupe, separated by a central lowered part (Fig. 1). From the lithological point of view, these rocks are made up of organogenous limestones belonging to the San Marino Formation (Lower-Middle Miocene), with sub-horizontal strata or slightly inclined bedding to SW, which in the southern part overlie directly the Argille varicolori Formation, while in the northern sector they overthrust the Upper Miocene pelites of the Ghioli di Letto Formation. The fracture systems affecting the two slabs are ascribable to three main groups: the first with a N-S direction, the second with a WSW-ENE direction and the third with a dominant NW-SE direction. The first system is more evident in the Roccione, whilst the third one is predominant in the Rupe. These fracture systems derive from two tectonical phases that accompanied the emplacement of the allochthonous sheet and their different orientations can be explained by the different degree of translation, also with a rotational component, which affected these tabular slabs in the Upper Miocene and Lower Pliocene.

The distribution and intensity of the mass movements affecting the village of Pennabilli are conditioned by the presence of a substratum characterised by two clayey formations that show different mechanical properties, as resulted from the geotechnical and mineralogical investigations. The highest concentration of landslides is found along the southern boundary of the two slabs, where the Argille varicolori substratum determines more accentuated deformations, with particularly intense phenomena of flowing and plasticisation at the solid state. These deformations affect also the overlying calcareous rocks which, deprived of their support, are subjected to extensive falls and topples. The areas where these phenomena have occurred more frequently are morphologically represented by ancient detachment scars, always linked to the fracture systems. These features are more evident in the Roccione, which has undergone a more advanced dismembering than the Rupe, as also witnessed by a considerably lowered trench in correspondence with the old historical buildings, most of which are affected by tension cracks, only partially equipped with glass tale-tellers, fissure meters and strain meters.

Along the southern flank of Roccione, many cavities, ascribable to recent-formation rock cracks, are present. They have been subjected to width variations up to a few dm during the last 30 years and bear a clear witness of the progressive displacement and collapse of the calcarenitic blocks that make up this slab. Roccione can in fact be considered as formed by a set of independent blocks, separated by the previously described fracture surfaces. The largest boulders of the San Marino limestones, scattered from the southern margin of the slab down to the river bed of the torrent Messa, correspond to the final stage of the degradation and movement processes. An intermediate stage of the mass-movement concerning Pennabilli's Roccione is put well into evidence by the sloping down terrace-morphology of the slab's middle part, where the blocks are moving away one from the other, although in some points they are not completely separated. The initial stage of the dismembering process is observable only in a reduced area located at the top of Roccione, where the fractures are closed or, at the most, show very limited openings.

The different evolution of the landslides affecting the two slabs of Pennabilli is also deriving from the unequal extension of the contact with the highly deformable Argille varicolori which, according to the field data, seem to extend on a wider area in correspondence with Roccione, while the Rupe slab lies mostly on the Ghioli di Letto pelites, which show a more limited deformability and plasticisation than the clay shales of the Argille varicolori Formation.

The origin of the numerous slope movements affecting the area of Pennabilli is therefore ascribable to the long-term plastic-viscous deformations affecting the substratum, not only along the outer margins but at a deep-seated level, owing to the load stress induced by the overlying competent rocks. The dismemberment of the slabs, which shows a more

Fig. 2 - Air view of the southern cliff of San Leo (Buga aerial photograph no. 675/70).

Fig. 3 - The slab of San Leo after a water colour by Francesco Mingucci da Pesaro, 1626 (Vatican Library Codex no. 4434).

advanced stage compared with the case of San Leo, is furthermore accentuated by the erosive action of the meteoric waters, only partially collected and drained.

3 THE CASE OF SAN LEO

The slab where the village of San Leo lies is made up, from the bottom to the top, of the San Marino Formation limestones (Lower-Middle Miocene) and the Mt. Fumaiolo glauconitic sandstones (Middle Miocene), characterised by cross-bedding and thin marly intercalations. The substratum of this competent slab is homogeneous in all its extension, being made up of shaly clays belonging to the Argille varicolori Formation (Cretaceous-Eocene). From a structural viewpoint, this slab is characterised by an arcuate arrangement of the cliff, with convexity facing ENE and a fan-shaped system of sub-vertical fractures. Other important groupings of fractures have been recognised in San Leo: the first with a N-S direction, the second with a ENE-WSW direction. They are both considered as parts of a single conjugated fracture system. These structural discontinuities subdivide the slabs into several blocks, progressively lowered to the west (Fig. 2). A third fracture system, with a sub-horizontal attitude, plays a major rôle in the movement of the blocks towards the outer margins. All the fracture systems investigated clearly result from a compressive tectonical origin, following the overthrust movements which have accompanied the emplacement of the Marecchia valley allochthonous sheet.

In San Leo the landslides are particularly widespread and affect both the calcareous-arenaceous rocks, with rock falls and topples, and the clayey substratum, with slumps and earth flows. The most intense movements are ascertained along the outer margin of the arcuate structure, where the intersection of the radial system with the conjugated one determines the detachment of wedge-shaped rock boulders, even of considerable dimensions. Indeed, the rock falls are accentuated by the presence of radial fractures which tend to open toward the convex side of the arc. The mass movements localised along the slab boundaries are nevertheless only favoured by the structural conditions, since their main cause should be traced back to the long-term differential settlements of the clay substratum, determined by the load stresses transmitted by the overlying competent rocks. The central part of the slab, where the settlements are more intense shows a wide trench, laterally limited by antithetical fractures, which emphasize its wedge-shaped form. Another trench, even if smaller, is located below the southern flank of the castle.

Rotational slides in the less confined clays along the margins of San Leo follow the deformations in the shaly substratum, involving in their movement also the superimposed calcareous and arenaceous rocks, thus giving origin to complex landslides. Sudden rock falls have occurred several times in the past: particularly noticeable was the detachment and successive dismemberment of a large rock block which collapsed from the cliff underneath the castle late in the XVIIth century. Before this date the access to the stronghold was situated on the northern side. Other landslides which have progressively reduced the size of the slab's northern margin took place also in the following centuries. The landslide-controlled evolution of this cliff is well documented by numerous paintings and engravings depicting the village of San Leo in various epochs, among which particularly significant is the picture shown in Fig. 3, where a large calcarenitic boulder is observable on the left, just under the castle. At present, this rock is completely collapsed and dismembered into several smaller blocks, most of which have been carried away by the earth flows affecting the clayey substratum. Also the main access road to the stronghold, which in the Middle Ages ran along this flank, has been totally obliterated by the repeated slope movements of the following centuries. The latest movements, although more reduced, affected the cliff's northern side in 1938 and 1952 and the southern one in 1949.

Fig. 4 - Geological sketch of Rocca Pratiffi: GLF = Ghioli di Letto Formation; SMF = San Marino Formation; MAF = Marnoso-arenacea Formation; RD = Rock detritus (thick arrows show main displacement directions).

4 OTHER EXAMPLES OF LANDSLIDES

Among the other types of mass movements which affect the historical sites of the Marecchia Valley, worthy of note is the case of Rocca Pratiffi. This old village rises on top of two small limestone cliffs (San Marino Formation) separated by a central lower part (trench) made up of Miocene pelites (Ghioli di Letto Formation). These limestones slid together with levels of Argille varicolori within the sedimentary basin of the Ghioli di Letto, during the Upper Miocene. At present the active landslides are essentially linked to the further dismemberment and slide of the two cliffs towards the less confined outer areas (Fig. 4), with occasional rock topples and falls which in the past have determined the damage and destruction of several buildings. The extent of these movements is nevertheless more limited than in the cases previously discussed, due to the lower degree of deformability of the substratum. Nowadays the few houses still inhabited do not seem to be threatened by these still active movements.

The ancient village of Maiolo (Fig. 5), situated on the SW flank of a poorly cemented sandstone-conglomerate hill (Lower Pliocene) with a characteristic cone-shaped form, was struck by a sudden landslide in the spring of the year 1700, after particularly intense rainfall. Over 100 people were killed and the rate of destruction was such that it was decided to abandon the place and establish a new settlement on a more stable area downhill. The failed slope, on which the village was placed, coincided with a bedding surface dipping towards the valley (Fig. 6). The type of movement is complex: earth flows and erosion processes in the substratum, particularly intense during the rainy periods, led to a lack of support for the overlying competent formation that was subjected to vast rock block slides along weaker layers and rock and debris falls.

5 GEOTECHNICAL AND MINERALOGICAL PROPERTIES

The stability conditions of the areas examined depend essentially on the geotechnical properties of the shaly argillaceous soils which underlie the calcareous-arenaceous competent formations and which show different degrees of deformability according to the varying content of clay minerals. The Argille varicolori Formation is characterised by extremely variable water contents, ranging from 14% (dry conditions) to 70% (completely softened conditions in active landslides). In all the situations examined, the clay fraction does not show particularly marked variations, with CF value of 60% to 75%. Quite different results have been found in the determination of the Atterberg limits: in San Leo the Argille varicolori give very high values of the liquid limit and the plasticity index (w_L>110% and I_P>80%) while in Pennabilli and Rocca Pratiffi they are more limited (w_L~60% and I_P~35%). This fact seems to derive from the marked variations of the clay minerals which characterise the whole Argille varicolori Formation. In fact, the diffrattometric analyses have shown that smectite reaches the highest concentration in the San Leo samples, although its presence is found to a lesser extent also in the Pennabilli and

Fig. 5 - The ancient village of Maiolo and the stronghold of Maioletto before the destruction of May 29, 1700 (after a geographical map of the XVIIth century).

Fig. 6 - The cliff of Maioletto today, showing the ruined towers of the castle and the dip-downstream bedding surface along which the landslide occured.

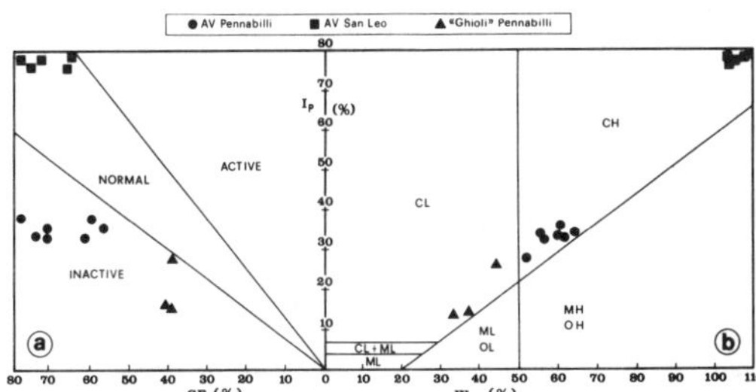

Fig. 7 - Activity chart of Clay Fraction (a) and Plasticity chart (b).

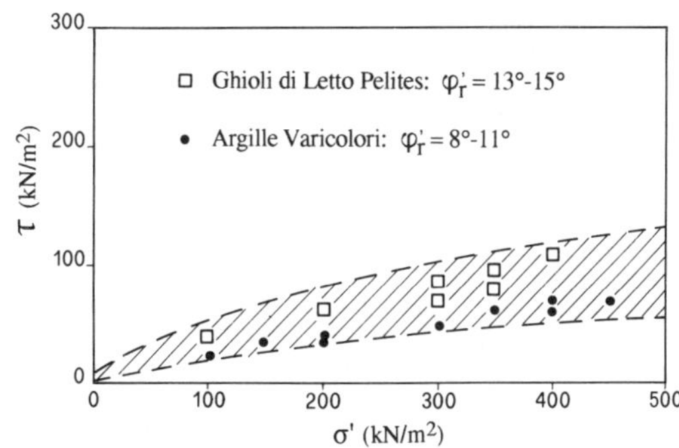

Fig. 8 - Distribution of the residual shear resistance values determined on remoulded samples.

Rocca Pratiffi samples. The very high values of the liquid limit and the plasticity index in the San Leo clay shales derives from a prevalent content of calcium and sodium smectite, which is characterised by high cation exchange capacity and plasticity. Instead, the pelites of the Ghioli di Letto Formation show lower values of the Atterberg limits than those found in the Argille varicolori, in agreement with a reduced percentage of the clay minerals (CF~40%), and a geotechnical behaviour corresponding to low-middle plasticity soils subject to a more limited deformability (Fig. 7).

As for the mechanical characteristics, it should be emphasized that since the substratum argillaceous soils are essentially made up of overconsolidated fissured clays, the long-term shear strength is governed by the residual friction angle φ'_r. The widespread diffusion of fissures in these soils determines a progressive decrease of the shear strength parameters, due mainly to decompression and absorption of meteoric waters accompanied by swelling and softening, thus giving origin to vast earth flows which drag in their movement a large amount of dismembered blocks and boulders fallen from the overhanging cliffs. Reversal shear box tests have been performed on remoulded samples, considering that the ultimate residual shear resistance is unaffected by the initial structure of the soil (Bishop *et al.*, 1971; Lupini *et al.*, 1981), giving φ'_r values of 8° to 11° for the Argille varicolori and 13° to 15° for the Ghioli di Letto pelites (Fig. 8). The lowest values of φ'_r have been found with samples containing large amounts of smectite, in agreement with the fact that the residual shear strength of natural soils is primarily dependent on mineral composition (Kenney, 1967). These results are also consistent with other researches carried out in different areas of the Northern Apennines (Cancelli *et al.*, 1987; Gori & Vannucci, 1987; Conti & Tosatti, 1991) and are sufficiently low to justify the persistent slope instability which has been affecting these villages since ancient times.

6 REMEDIAL MEASURES AND CONCLUSIONS

Geognostic investigations consisting of borings and monitoring surveys of the tension cracks on the buildings were performed in Pennabilli between 1979 and 1984. The data collected have confirmed that the instability phenomena affect in all their extension the two rocky slabs on which the village is built. The mass movements are particularly active along the western and southern side, in agreement with other geological and structural investigations recently carried out (Conti & Tosatti, 1991). The interventions, consisting in the execution of a diaphragm wall with bored piles anchored to the rock and trench drains at shallow depth, have proved to be insufficient to noticeably reduce the entity of the movement, as witnessed by the fact that most of the fissure and strain meters, installed in the 1981 surveying and monitoring campaign, have undergone such a rate of displacement that they are at present unusable, due also to years of neglect. Many buildings which only a few years ago were marginally affected by tension cracks, today appear to be severly damaged. The continuous movements affecting the historical centre of Pennabilli could also induce displacements and bursts in the water-supply and sewage pipes and the resulting leaks of water in the subsoil would inevitably reduce the stability conditions. Considering the type and mechanism of the landslides, involving both the calcareous sandstones and the underlying clayey substratum, the remedial interventions should mainly consist of draining wells throughout the thickness of the competent rocks, together with widespread horizontal drains, in order to catch most of the water circulating within the fractured rocks. More superficial structures such as retaining walls and anchors in the dismembered boulders should also be considered, since they would effectively contribute to a general stabilisation of the southern slopes of the Pennabilli slabs.

The remedial measures in San Leo have been carried out prevalently along the NE margin, underneath the stronghold, and in correspondence with the central lowered area, stretching from the castle to the cathedral (Fig. 2). This part, where the village historical centre is concentrated, shows serious problems concerning the buildings' stability. The differential movements which affect also this central area of the slab are in fact witnessed by the numerous cracks present in several buildings, which entail continuous restoration works also on the most important monuments, such as the two Romanesque churches and the town hall. Nevertheless, the most immediate landslide risks are concentrated on the NE corner of the cliff, where the control of the earth flows has been attempted by means of concrete weirs and gabions of metal wire filled with pebbles, constructed in correspondence with the most active movements. These retaining structures have been partially "overflowed" by the earth flows that they were supposed to contain, since they have been placed too low down along the slopes. On the southern part of the cliff the stabilisation interventions have been essentially directed to the setting up of anchors with load distribution plates and low-pressure grouting. Also drainage holes provided with waterproofing have been drilled into the fractured rocks, aiming at the collection and discharge of the percolating water. Moreover, drainage systems for the collection of meteoric and sewage water have been set up at the base of the cliff (Soccodato, 1980; Diamanti & Soccodato, 1981; Ribacchi & Tommasi, 1988). The consolidation measures so far performed are nevertheless inadequate, especially for what concerns the hydrogeological boundary conditions. Also in the case of San Leo a more effective reduction of the porewater pressures in the underlying materials could be better reached by means of horizontal drains set up deep at the rock/soil contact, in order to reduce the adsorption by the clayey soils, since their high smectite content makes them particularly prone to plasticisation.

A final aspect that should not be underestimated in the area studied is given by the middle-high degree of seismicity (S=9). In the Marecchia valley a high recurrence of earthquakes has in fact been recorded, with maximum intensity up to the IXth degree of the MCS scale (Ferrari *et al.*, 1980). From the viewpoint of landslide-prone conditions, a high frequency of earthquakes tends to further reduce the soils' residual strength, thus accentuating the slopes' instability, as documented also by previous investigations in the Northern Apennines (Pellegrini & Tosatti, 1982), which recognised that seismic events, even of reduced intensity, can suddenly trigger mass movements of considerable dimensions.

ACKNOWLEDGMENTS

The research has been financially supported by the Italian Ministry for the University and Scientific and Technological Research (M.U.R.S.T. 60%, 1991).

REFERENCES

Bishop A.W., Green G.E., Garga V.K., Andresen A. & Brown J.D. 1971. A new ring shear apparatus and its application to the measurement of residual strength. *Géotechnique* 21, No.4, 273-328, London.

Cancelli A. and Pellegrini M. 1987. Deep-seated gravitational deformations in the Northern Apennines, Italy. *5th Internat. Conf. & Field Workshop on Landslides,* Australia & New Zealand.

Cancelli A., Pellegrini M. and Tosatti G. 1987. Alcuni esempi di deformazioni gravitative profonde nell'Appennino settentrionale. *Mem. Soc. Geol. It.*, 39, 447-466, Rome.

Conti S. and Tosatti G. 1991. Le "placche" di San Leo e Pennabilli (Val Marecchia): rapporti fra gli elementi strutturali e le deformazioni gravitative profonde. *Atti I° Conv. Naz. Giovani Ricercatori in Geologia Applicata,* 57-66, Milan.

Diamanti L. and Soccodato C. 1981. Consolidation of the historical cities of San Leo and Orvieto. *Proc. X ICSMFE*, 3, 75-82, Rome.

Dramis F., Garzonio C.A., Nanni T. & Principi L. 1987. Franosità e dissesti dei centri abitati delle Marche. Primi risultati del censimento e dello studio delle situazioni a rischio. *Mem. Soc. Geol. It.*, 37, 105-116, Rome.

Ferrari G., Gasperini P. & Postpischl D. 1980. Catalogo dei terremoti della Regione Emilia-Romagna. R.E.R.-C.N.R. *Collana di orientam. geomorf. e agronom. forestali,* Pitagora, Bologna.

Gori U. and Vannucci S. 1987. Argille alloctone e parautoctone della Val Marecchia. Relazioni fra caratteri petrografico-fisici e stabilità. *Mem. Soc. Geol. It.*, 37, 277-286, Rome.

Kenney T.C. 1967. The influence of mineral composition on the residual strength of natural soils. *Proc. Conf. on Shear Strength*, vol.1, 123-129, Oslo.

Lupini J.F., Skinner A.E. & Vaughan P.R. 1981. The drained residual strength of cohesive soils. *Géotechnique* 31, No.2, 181-213, London.

Pellegrini M. & Tosatti G. 1982. Alcuni esempi di frane determinate da sismi nell'alto Appennino modenese e reggiano. *Atti Soc. Nat. Mat. di Modena,* vol.113, Modena.

Persi P. 1973. Il "monte" di San Leo nel Montefeltro: geomorfologia e riflessi sulla stabilità dell'abitato. *Atti XXI Congr. Geogr. Ital.*, 3, De Agostini, Novara.

Ribacchi R. and Tommasi P. 1988. Preservation and protection of the historical town of San Leo (Italy). *Eng. Geol. of Ancient Works, Mon. and Hist. Sites*, Marinos & Koukis eds., A.A.Balkema, Rotterdam.

Soccodato C. 1980. Consolidamento dell'antico centro abitato di San Leo (PS). *A.G.I., Atti XIV Conv. Naz. di Geotecnica,* Florence.

Environmental Management, Geo-Water & Engineering Aspects, Chowdhury & Sivakumar (eds)
© 1993 Balkema, Rotterdam. ISBN 90 5410 099 0

Analysis of a mass movement in structurally and geotechnically complex arenaceous marly formation

V.Cotecchia & L.Monterisi
Institute of Engineering Geology and Geotechnics, University of Bari, Italy

A.Salvemini
Department of Structures, Geotechnics and Engineering Geology, University of Basilicata, Potenza, Italy

ABSTRACT: The following describes the stability conditions on the right shoulder of the Campolattaro Dam (Benevento, Southern Italy) which is under construction. The shoulder is affected by an old landslide, a consequence of the geomorphological evolutionary context, which is present on both slopes of middle valley of River Tammaro. This mass movement restarted during the dam building works although it was not caused by them. The geological formation on which the dam is based must be considered "structurally complex". The soils involved in the landslide consist in the upper layer (10-15 m) of old landslide detritic cover formed from arenaceous or marly arenaceous particles included in remoulded silts and clays. The lower layer of the landslide (5-10 m) consists of marly clayey arenaceous sets, that are fairly remoulded and de-compressed. The landslide has been monitored for several months period of time. The influence on the slope stability of several factors in different seasons has been examined. Finally, the main landslide causes and possibile stabilizing interventions have been pointed out.

1 INFORMATION

The area covered by this study is located in the central part of the valley of Tammaro river, between Campolattaro (BN) and the earth dam which has not yet been completed. The morphological features of the Tammaro valley in general (Cencvoic et al.1982) and the area of the dam in particular (Melidoro 1967) are typical of an Appennine environment. The surface presence of "structurally complex" formations, the irregular hydrological structure that often characterize these areas, their mainly clayey lithologic nature and the high risk of earthquakes in this part of the Southern Appennine Chain have combined to define an extremely delicate and complex geomorphological picture, that is characterized by both diffused and strong mass movements. These appear as multiple, rotating and progressive landslides, or plastic gravitation pours, or, as creep phenomena of greater or lesser depth.

2 GEOLOGY

As far as its structural geological aspect is concerned, the area examined is included in the Molisano-Sannitica Depression, one of the geological structures that bears a regional importance, is part of the Southern Appennine Chain, and is located on the north east sector of the F°173 "Benevento" according to the 1:100.000 State Geological Map. Other geological studies carried out by other Authors on the river Tammaro have pointed out a remarkable difference among some lithological complexes which should belong to different paleogeographical domains. There have always been problems in attributing variocoloured clays in the Valley of River Tammar either to the "Flysch Rosso" of the Lagonegresi Units or to the unit of Variocoloured Clays of the structural-geological scheme which has been defined by OGNIBEN (1969) for the Calabrian Lucanian Border. In this study, because of the limited size of the area of interest, we will consider the examined soils as "informal lithological complexes" without referring to codified formations or geostructural units. According to the above mentioned point of view, on the barrage area near the dam and on a hill near there, the following lithological complexes have been identified (Fig. 1):

- Variocoloured Clay Complex. In the barrage area it is only on the right of River Tammaro, at levels well above that of the crest of the dam. From the lithological point of view the complex is composed of red, green, violaceous and greyish clays and silty clays interbedded, with calcilutites and calcarenites. The structure of the clay lithotypes is scaly, whereas the stony layers are fractured and disjointed. Micropaleontological analysis carried out on this complex (Melidoro 1967) dates it back to the Superior Cretaceous--Oligocene. In the examined area the above mentioned complex is tectonically superimposed on the "Marly-Arenaceous" one.
- Marly-Arenaceous Complex. As far as the studied area is concerned, it represents the basal unit, being the bed-rock of the foundation of the dam and its supporting works. It is a sedimentary flyscioid sequence, made up of clayey marls, marly clays and silty sandstones. This formation is characterized by numerous systems of primary and secondary discontinuity which, by intersecting, reduce the rock to a mass of polyhedrons of different size. The primary structures depend on strata, level flat or convolute lamination, slumpings, etc. Olisthostromes of Variocoloured clays are often found and probably derive from submarine slides in the Miocene sedimentary basin. As a whole the Marly-Arenaceous Complex has a monoclinal stratification, where strata have a 20°N-40°E orientation and a NW dip of 30° to 80°. This lithological complex seems to date back to the Upper Miocene.
- Molassic Sandstone Formation. The rocks of this formation outcrop in discontinuous plates, left by erosion, on both sides of the valley near the dam supports. The most representative rocks are either massive or thick slightly stratified sandstones with a yellowish quartz-micaceous-feldspathic composition and are characterized either by calcareous cement or calcareous-clay cement. This mass, lying on the Marly-Arenaceous Complex, seems to result sometimes from an advance of the waters and sometimes from tectonic activity. Therefore, the detachment of the molassic sandstones from the underlying formation was probably caused by tectonic movements during the Miocene, Pliocene and Quaternary periods.
- Alluvial deposits. These outcrop along the bottom of the valley of the River Tammaro and are often disarranged by frequent slides which have reached the river bed. Lithologically the alluvial deposits, are made up, in the lower part, of polygenic and heteromerous gravel and, on the surface, of sand and silty sand.
- Detritic cover. Such cover is due partly to the erosion and alteration of the shallower soil outcropping over the slopes, partly to the diffuse and intense geomorphological activity that up to now has concerned the same. From the lithological point of view, such a complex is represented by two fundamental lithotypes: the first is derived from the decay of the Marly-Arenaceous Complex, is made up of blocks of an arenaceous and marly nature, immersed in a Silty-clay matrix. The second is made up of a diverse mixture of soils deriving from the erosion of the above lithological complex, like that of the "Variocoloured Clays". Both the "Lithotypes" widely cover the slopes of the River Tammaro, comprising the valley section of the dam zone.

3 GEOMORPHOLOGY

In the final design of the dam (1976) on the right bank of the River Tammaro, at the proposed site of the dam, a large area in which considerable mass movement was taking place was observed. This area starts at the bottom of the valley, where the most recent slides had covered the alluvial deposits, and reaches up to 370 m above mean sea level, that is to say a level higher than the crest of the dam. The feeding basin for the displaced soil

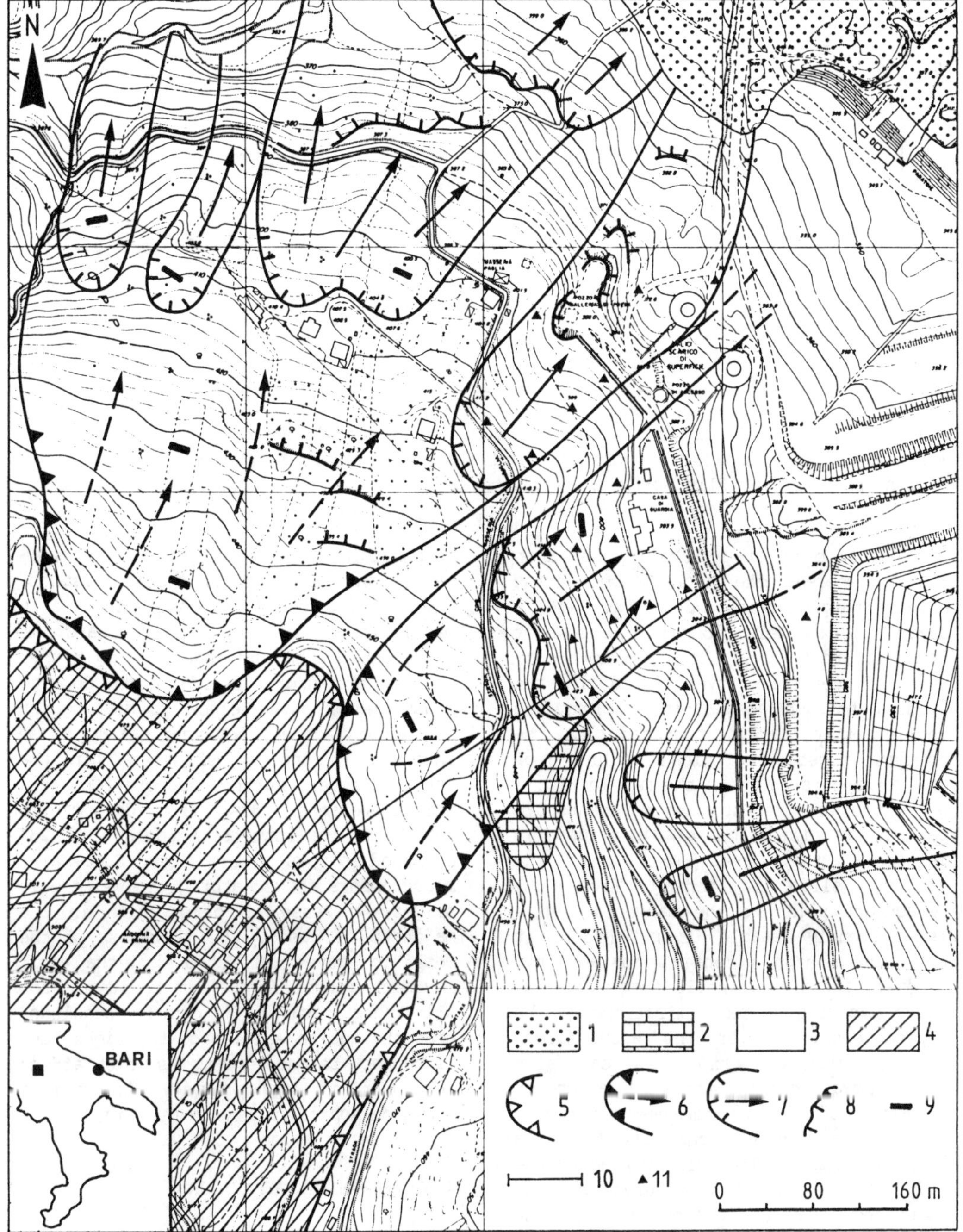

Fig.1 Geomorphological Map. 1) Alluvial deposits; 2) Molassic Sandstone Complex; 3) Marly Arenaceous Complex; 4) Variocoloured Clays Complex; 5) Overthrust; 6) Old landslides; 7) Recent and actual landslides; 8) Morphological slope; 9) Morphological depression; 10) Geological section; 11) Boreholes.

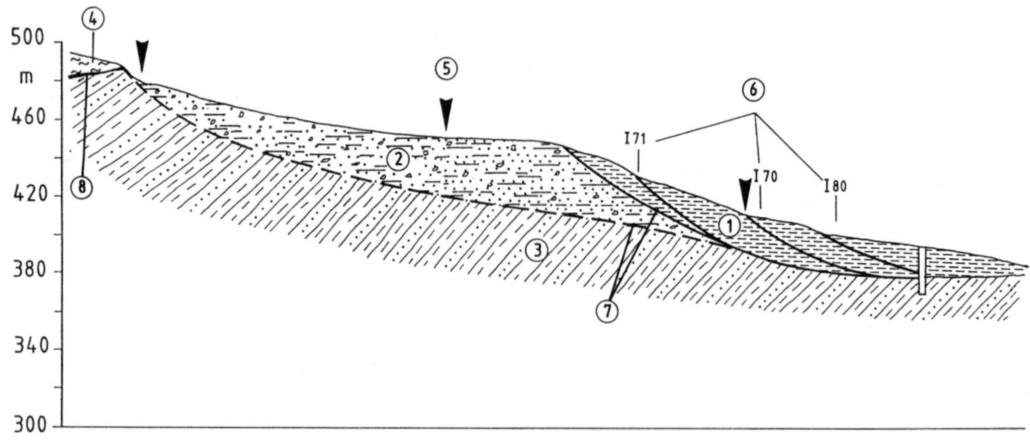

Fig. 2 Geological section. 1) Recent and actual landslides; 2) Old landslide; 3) Marly Arenaceous Complex; 4) Variocoloured Clays Complex; 5) Morphological depression; 6) Inclinometers; 7) Slip surface; 8) Overthrust.

was found to be at a level higher than 370 m, whereas the highest detachment points ranged from 440 m to 450 m a.m.s.l., extremely close to the town of Campolattaro (Fig. 1). The maximum thickness of detritic cover found in the slides, through geognostic, nuclear and inclinometric surveys, averaged 20 m. The continuing geomorphological evolution of the versant is shown in the map in Fig. 1 where the most recent mass movements are partial remobilizations of old slides. In particular the area uphill from the ancient slide stretching from the detachment point to the Masseria Paglia is currently stable even though it exhibits a high potential instability. However, the area downhill from the slide shows a series of remobilizations of the ancient detritic cover. The recent slides are "multiple and retrogressive rototranslational movements". The zone at the foot of the same movement directly affects the Control House and the area downhill from it. The soils making up the body of the slides are very soft and saturated plastic clayey silts. The saturation is due to the irregular hydrogeological conditions of the valleyside which, following periods of rain, takes the piezometric level of the numerous water tables above ground level. The dimensions of the recent slide have been defined by making reference to morphologic, lithostratigraphic, inclinometric and geotechnical characteristics. The geotechnical measurements in particular clearly show shears at different depths and zones having higher weathering and/or plastification. On site measurements have helped to distinguish zones of recent movement from those involved in ancient slides within the detritic cover. Further support for these interpretations comes from deformation tests on inclinometric tubes (N. 23) arranged in rows on the slope. Successive measurements showed a progressive, albeit slight, increase in deformation in the majority of cases. The thickness of the detritic cover affected by the soil movement ranges from 5 m to 10 m. From all these results it can be concluded that the entire section of the right bank is subject to mass movements which affect the detritic cover down to a maximum depth of approximately 10 m. (Fig. 2).

4 GEOMECHANICAL BEHAVIOUR

The base lithological formation of the area under study, i.e. the Marly-Arenaceous Complex, must be considered as a structurally complex formation according to the most recent classifications. The detritic cover should also be considered as a structurally complex formation in relation to the soils from which it is made up. With reference to the classification proposals formulated by AGI (1979, 1985) the Marly-Arenaceous Complex can be included in the class B1, the original sedimentary structure having been well conserved. The detritic cover, on the other hand, can be classified as B3. Since, following repeated remouldings caused by slides, the original structure is no longer recognisable apart from a few integral blocks within the formation. The values of certain geotechnical parameters are reported in table 1 and figs. 3, 4.

The difference between values measured in the laboratory and those measured on site is attributable on one hand to slight remoulding during sampling and on the other hand to the fact that the undisturbed samples are representative of general conditions of the formation. Particularly interesting are the on site measurements of the density and natural water content since they provide a diagram of the above mentioned parameters along significant vertical lines (Fig. 5). From the study of these diagrams, correlated with lithostratigraphical observations it was possible to locate fairly accurately the slip surface and the areas of great breakdown and/or weathering of the formation. When considering the detritic cover, the values of geotechnical parameters reported in Tab. 1 show a high heterogeneity both lithologically and in the consistency of the materials constituting the detritic cover. The shear strength of the material forming the detritic cover was analysed during triaxial compression tests, CU so obtaining values of the angle of shearing resistance varying from 16° to 28° in the absence of cohesion (Fig. 6). Values not dissimilar to these were obtained in direct CD shear tests. These tests have also made possible the determination of the residual shearing resistance through reversal shear tests. The results obtained were ϕ' = 16°-18°. These are values that the authors consider over estimations because the clayey mass contains stoney elements which have

Table 1. Geotechnical properties of soils.

	Cover	M.A. Complex
γ_t (KN/m^3)	18,9-22,5	20,0-22,7
γ_d (KN/m^3)	17,6-20,7	17,8-20,8
	(14,0-18,0)(*)	(19,0-21,0)(*)
W	8%-36%	10%-26%
	(14%-32%)(*)	(8%-18%)(*)
e	0,32-0,84	0,33-0,57
s	68%-100%	75%-87%
CF	8%-49%	25%-46%
WL	30%-50%	35%-56%
PI	9%-33%	18%-37%

(*) by in situ tests

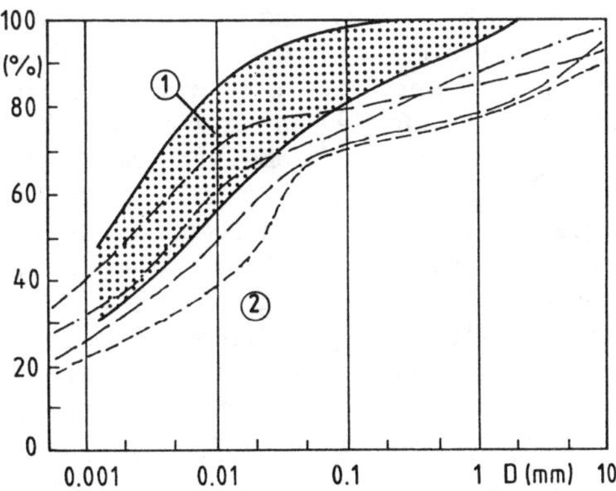

Fig.3 Soil grading. 1) Marly Arenaceous Complex; 2) Detritic Cover.

certainly exagerated the results of the tests. As far as the base marly arenaceous formation is concerned, the variation intervals of the principle geotechnical parameters are affected in a way by the lithologic variability that characterises a complex formation. The shear strength of the clayey and marly levels of this formation was studied during triaxial CU compression tests and direct CD shear tests. The results of TRX CU tests are not very indicative of mechanical properties of the soils under study because of the fissured state of the material, accounted for by the scale of the sampling, as well as by the orientation of the particles. The results obtained from reversal shear tests are, however, more reliable and give values of residual resistance expressed by the parameter ϕ' = 16°-21° in the absence of cohesion.

5 PIEZOMETRIC LEVELS

The piezometric system of the numerous water tables observed during the course of the geognostic surveys was reconstructed through measurements taken with Casagrande cells introduced at different levels during specific geognostic surveys. Measurements were made every 15 days for more than a year and provided a large amount of data, from a large number of measurement points, which allows some reliable conclusions to be drawn concerning the circulation of underground waters (Fig. 7). By observing the measurements taken it can be seen that piezometric levels vary from point to point along the same vertical; the shallower piezometric cells generally registered a higher piezometric level than the deeper ones. Such a phenomenon probably depends on the fact that the slope is affected by a filtration motion with lines of flow approximately parallel to the slope with a piezometric surface not very far from the ground level.

6 COMPARISON BETWEEN LABORATORY DATA AND RESISTANCE PARAMETERS MEASURED ON SITE

The recurrence of slides took place along shear surfaces within the detritic cover. The causes can be traced back to the successive interventions on the slope and the increase in interstitial pressure. The influence of this last factor has been studied in particular through back analysis of the movement. The analysis was carried out according to a shearing mechanism where circular shear surfaces corresponded with potential shearing surfaces or at least surfaces with less resistance as shown by geognostic surveys. The method adopted was that of BISHOP, by means of a specific computer program. The model of slope calculation and the results of the stability analysis are reported in fig. 8. The shearing resistance measured on the slope, in

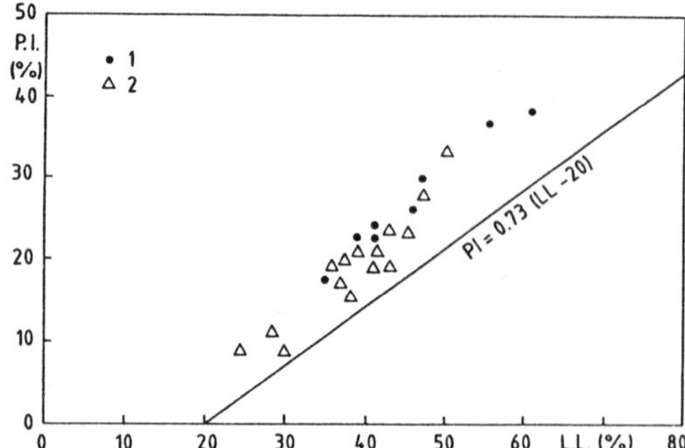

Fig.4 Plasticity chart. 1) Marly Arenaceous Complex; 2) Detritic Cover.

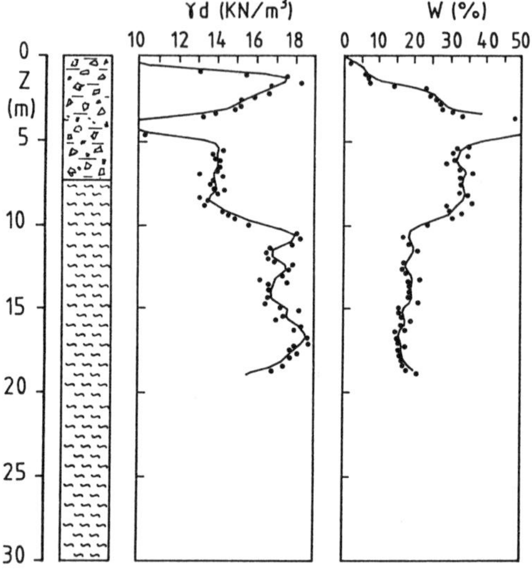

Fig.5 In situ tests. Dry density and natural water content.

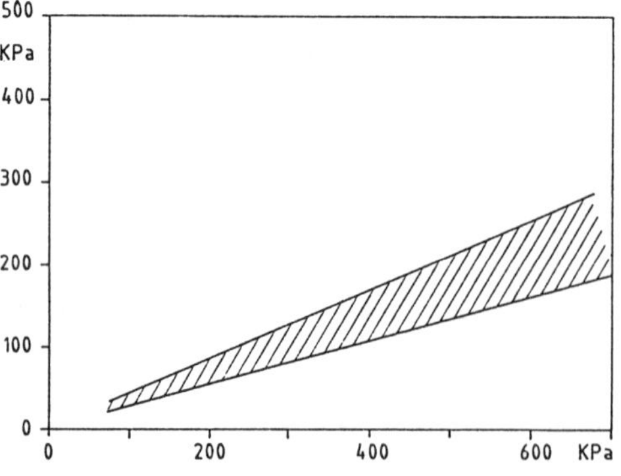

Fig.6 Triaxial CIU tests.

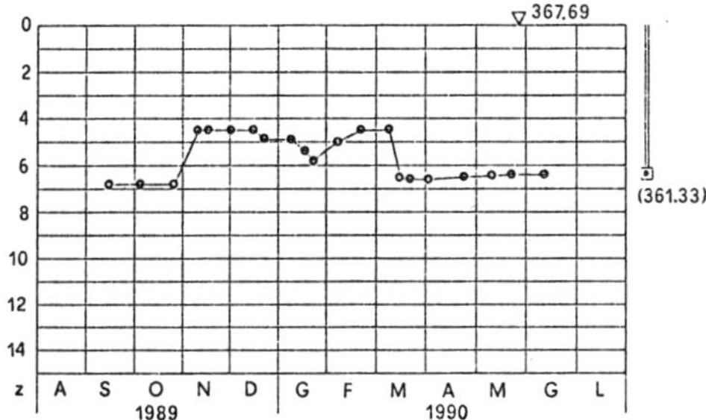

Fig.7 Piezometric measurements.

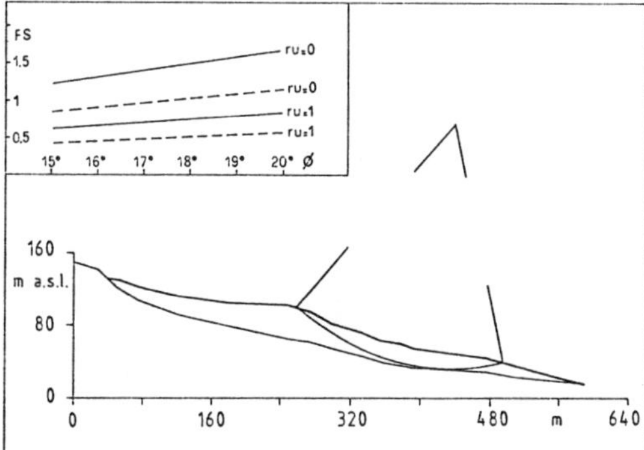

Fig.8 Back Analysis.

particular in relation to the piezometric levels measured, was found to be 18° which is in fairly good agreement with the resistance values measured by direct shear tests. Within the limits of the approximations introduced in the calculation, the influence of the variations in interstitial pressures on the safety factor is also evident. Similar results were also obtained on an instability model of the infinite slope type involving the entire thickness of the detritic cover.

7 STABILIZATION WORKS

The surveys carried out have clearly shown that the geotechnical parameter that most influences the slope stability is given by interstitial pressures. The hydrogeological irregularity of the slope together with the chaotic structure of the detritic cover and the variation in the flow pathways caused by the slow deformation of the valley side certainly have a great influence on this interstitial pressure system. The most suitable stabilization work seemed to be intensive drainage of the valley side to both break down the piezometric surface and encourage the consolidation of the soil of the cover. The operation in question involved the drilling of a series of large diameter drainage wells (1500 mm) down to the marly-arenaceous formation. The wells were placed in rows approximately 40 m apart. Within each row the wells were spaced at 10 m intervals. To enhance draining efficiency slightly inclined radial microdrains were constructed. The wells are linked at their bases by a system of interconnected drainage pipes. A series of large diameter multianchored piles were sunk at the foot of the landslide. These served to hold up the area in front of the control house. Other improvement works were carried out at individual points on the slope involving the digging of drainage trenches and gutters and the planting of trees.

REFERENCES

A.G.I. 1979. Some italian experiences on the mechanical characterization of structurally complex formations. Int. Congr. on Rock Mechanics. Montreux.

Airò Farulla, C. and Nocilla N. 1984. Assetto strutturale e resistenza di campioni di argille a scaglie. Rivista Italiana di Geotecnica, n. 3, 148-158.

Bilotta, E. 1984. Contributo allo studio della resistenza a taglio di argille a scaglie con prove di laboratorio. Atti del Dipartimento di Difesa del Suolo. Cosenza.

Budetta, P., Corniello A., de' Medici G.B., De Riso R., Lucini P. and P. Nicotera. 1979. Il bacino del F. Tammaro (Campania): geologia, geomorfologia, idrologia, risorse idriche. Mem. e Note Ist. Geol. Appl. Napoli. Vol. XV.

Corniello, A., De Riso R. and P. Lucini. 1979. La franosità potenziale del bacino del F. Tammaro (Campania). Memorie e Note Ist. Geol. Appl. Napoli. Vol. XV.

Cotecchia, V., Monterisi L., Salvemini A., Spilotro G. and G. Trisorio Liuzzi. 1977. Geolithological, structural and geotechnical aspects of some arenaceous-marly formations cropping out in central-southern Apennines. Int. Symp. The Geotechnics of structurally complex formations, Vol. II, Capri.

Esu, F. 1985. Geotechnical properties and slope stability in structurally complex clays soils. A.G.I.

Genevois, R. and A. Prestininzi 1982. Deformazioni e movimenti di massa indotti dal sisma del 23.11.1980 nella media valle del F. Tammaro (BN). Geol. Appl. e Idrogeol., Vol. XVII.

Melidoro, G. 1967. Geologia e geomorfologia applicata allo studio di una diga di ritenuta sul fiume Tammaro (Sannio). Geol. Appl. e Idrogeol., Vol. II, Bari.

Oddone, E. 1930. Studio sul terremoto avvenuto il 23 luglio 1930 nell'Irpinia. La Meteorologia Pratica, 16-86, Roma.

Ogniben, L. 1969. Schema introduttivo alla geologia del confine Calabro-Lucano. Mem. Soc. Geol. It., Vol. 8, fasc.4, Roma.

Pescatore, T. 1965. Ricerche geologiche sulla depressione molisano-sannitica. Atti Acc. Sc.Fis. Mat., Sez. 3,4,99-164.

Servizio Geologico D'Italia. Note illustrative della Carta Geologica d'Italia. F. 173 "Benevento".

Vari, V. 1930. Il terremoto dell'Alta Irpinia. Boll. Soc. Sism. It., 29, 181-196.

Environmental Management, Geo-Water & Engineering Aspects, Chowdhury & Sivakumar (eds)
© 1993 Balkema, Rotterdam. ISBN 90 5410 099 0

A complex rockfall-debris slump failure on a rockfall prone cliff

G.Crosta
Dipartimento Scienze della Terra, Milano, Italy

S.Agostoni & M.Ceriani
Servizio Geologico, Regione Lombardia, Milano, Italy

ABSTRACT: A rockfall prone cliff, in Valtellina (Italy), has been subject to 3 different failure events in the last 60 years (1931, 1977 and 1991). The two most recent events are compared and particular attention is given to the stability analysis and movement evolution of the 1991 events, probably occurred in consequence of a 4.4 Richter's magnitude earthquakes.

1 INTRODUCTION

This paper treats about an alpine rockfall prone cliff (Valtellina, Italy), that underwent two important consecutive failure phenomena in a 14 years time interval (1977-1991). The landslide phenomena, because of the high relief energy and the geological and morphological settings of the area, can be placed at the boundary between the rockfall and the complex landslides category (Varnes, 1978). In particular, we can observe that the 1991 landslide, occurred few days after a medium intensity earthquake, may be considered, for some of its features, as a sort of aborted rock avalanche. In any case, the relative complexity of the sliding phenomenon is cause by itself of great interest. The landslide mass dammed the valley bottom, destroying the road, reconstructed after the 1977 event, 10 mountain huts, and causing extended damages to 4 quarry plants, electrical and acqueduct networks, without any casualties or injuries.

2 LOCATION AND GEOLOGICAL SETTINGS

The investigated area is located in the Central Alps (Italy), in the Masino valley (fig. 1), a tributary valley of the well known Valtellina, especially after the 1987 Val Pola landslide, close to the Italy and Switzerland state borderline. The Masino valley is a NE-SW oriented hanging valley and the landslides occurred on the steep valley flanks of a ENE-WSW secondary valley stem

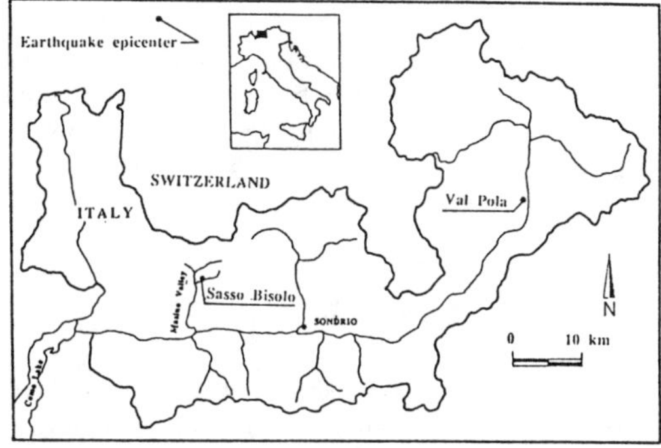

Fig. 1 - Location map (Valtellina, Italy)

(Sasso Bisolo valley), where the minimum and maximum elevation are 800 and 3678 m (Mt. Disgrazia), respectively.

From a geological point of view, this area belong to a Tertiary intrusive massif, called "Masino - Bregaglia Massif", constituted by granodioritic (locally called "Ghiandone") and tonalitic ("Serizzo") rocks. In particular, the unstable cliff is located within a narrow border, around the main granodioritic massif, made by tonalitic rock, commonly characterised by evident iso-oriented minerals (horneblende phenocristals) that result in a regularly foliated texture.

Foliation dips toward the valley bottom (160°/66°) with a constant 60° to 65° inclination controlling the entire right-hand side of the valley. Besides the foliation, 5 main joint sets have been recognised in the rock mass (K1: 070°/80°, K2: 217°/53°, K3: 250°/47°, K4: 355°/80°, K5: 000°/55°) through structural field survey. These joint sets, easily recognizable also from the aerial photographs taken in different years (1970, 1977, 1991) (see fig. 2a, b, c), are shown in fig. 3 where the heavy line represents the foliation and the cliff slope at the same time. Joints are generally characterised by high spacing intervals, between 1 and 30 m (the average is about 10 m) while the foliation spacing ranges between 0.5 and 5 m, increasing up to 10 m moving uphill along the cliff wall toward the crown area. These high values of spacing are at the origin of the large rock unit volumes that can be found inside the talus, at the foot of the entire rock cliff, and within the landslide material of the various events.

A structural feature, that can be considered of first importance for the consequences on slope stability, is represented by the presence of a fault system, with weak milonitic layers, mainly with ENE-WSW strike direction, parallel to the foliation and to the valley bottom. These milonitic layers, with thickness ranging between 1 to 15 centimeters, can be described as milonitic gneiss, variable in colour from green to black.

3 HISTORIC LANDSLIDES

The history of these alpine valleys is full of reports about landsliding and especially about rockfalls (Comunità Montana Valtellina, 1981; Govi, 1988). For this particular area three main rockfall events are known in historic times. The first event, occurred on January 1931, the second one on March 22nd, 1977 and the last one on November 24th, 1991.

Few data exist about the 1931 event, except for the damages to the small village (Valbiore) that has been repeatedly striken by the following events, and eventually destroyed. Looking at fig. 2a, it may be possibile to recognise the 1931 landslide debris, partially revegetated and cut by the road. More information exist about the 1977 rockfall and figures 2b and 3 (n° 1 and thin line) show the complete scenario, with total destruction of two quarries and partial destruction (60%) of the Valbiore village. At that time the volume of rock that slided down along the foliation and the K1 joint set, was estimated quite in a broad range: 0.04 Mm^3 (Govi, 1988), 0.2 Mm^3 and 0.5 Mm^3 (Comunità Montana Valtellina, 1981). We think that the rockfall volume has been a little under- and over-estimated respectively, as from calculations and field and aerial photoes observation, and that 0.15÷0.2 Mm^3 could represent a good upper limit value. No external action was supposed at the origin of that lanslide, but just the rockfalls and rockslides prone cliff characters, even if someone suggested an antropical origin as a consequence of need in new quarring material by some quarrymen. No talus material was remobilized at that time.

Quarrying activities continued up to november 24th, 1991 when, few days after the Piz Platta earthquake (11-20-1991; 2:54:37 AM) a rock wedge of an estimated 0.19÷0.2 Mm^3 volume detached and fell from the cliff, impacting against the scree slope, originated by ancient failure phenomena, and remobilizing in some way a huge slump inside the coarse material (average size is 1 m).

The 4.4 Richter's magnitude earthquake (6th Modified Mercalli Intensity scale) was localized 35 km away from the Sasso Bisolo valley, within the Swiss territory (between Piz Platta and Tiefen Castle, 46°40' lat. N, 09°24' long. E, see fig. 1), and the shaking, besides for the occurrence of some damages in Switzerland, was felt all along the Valtellina area and neighbouring regions. During the following days up to the evening of the 24th, 27 small aftershocks earthquakes were recorded with decreasing frequency (15 aftershocks on Nov. 20th and 2, 5, 3, 2 for the following days up to Nov. 24th).

This time the Valbiore village was completely covered by blocks and a debris tongue, of remobilized talus material, with thickness ranging between 10 and 20 m and with huge blocks up to 2000÷3000 m^3 of volume. One of these blocks has been considered as fundamental for the reconstruction of the landslide mechanism. This block, visible in fig. 2b (1977) (middle low sector, close to the valley creek), is included between the lower U-turn of the road and the main accumulation, while in fig. 2c and 3 it is placed at the tip of the debris tongue. Such a huge boulder was recognised in consequence of its volume, shape and of bad quarring quality that caused it to be discarded by previous quarring activities.

a b c

Fig. 2 - a) 1971 aerophoto b) 1977 aerophoto c) 1991 aerophoto: 1) 1977 landslide scar; 2) 1991 landslide scar with water leakage; 3) location of impact point (about 1450 m a.s.l.) - North is upward

4 1991 LANDSLIDE EVOLUTION

On November 24th, 1991, at about 4:00 AM, no witness was, fortunately, at the sliding site and so we inferred the landslide evolution through different steps from successive studies of the scar areas and accumulation zone. In the following, we try to present all the different features considered important to better understand the instability conditions and the evolution of the movement. As a consequence three main phases have been individuated: detachment, fall and bouncing, impact and debris slumping.

4.1 Detachment

The first step of sliding was the failure of the huge 0.2 Mm^3 wedge (see fig. 2c and 3) from elevation 1870 m and with toe downward at 1625 m. The detachment occurred along three major planes: the foliation and the K2 (217°/53°), K3 (250°/47°) joint sets. From a field survey in the crown area, the authors observed the smooth (joint roughness coefficient, JRC = 4÷6) and flat surface of the foliation plane, with almost no rock bridges ruptures visible anywhere, suggesting the contribution to stability along this surface only by friction, excepted for the lower stepped surface section. Furthermore, localized water leakages were identified and are still visible on the sliding surface also in fig. 2c as a dark stripe.

The head of the hanging wedge developed with a stepped profile for a total height of about 8÷10 m (measured perpendicular to the sliding surface) and a total width of 80 m, mainly along the K1, K3 and K4 joint sets, and part of the lateral dihedral face (K2-K3). The dihedral observed in consequence of the failure has a maximum width of about 35 m and length of 150 m. The various head steps, especially the lower one, were characterised differently from the main basal surface by the presence of fresh conchoidal rupture surfaces further than joint planes and intersections.

In figure 3, one of the performed graphical stability analysis is shown (stable area is hatched) and clearly it is a first evidence of the instability conditions of the rock wedge and of the potentially unstable areas all along the cliff and the valley flank.

Such a condition close to instability has been verified also conducting a series of pseudo-static stability analysis by Sarma's method (1979), which permits the introduction of inclined slice boundaries to simulate main rock joint set directions. The adopted strength values were a 40° friction angle and cohesion only for the uppermost slice of the discretized wedge to introduce the strength mobilized along the conchoidal step surfaces. The safety factor for static conditions resulted slightly more than the unit (FS about 1.15÷1.2), while the critical accelerations necessary to reach limit equilibrium resulted in the range of 0.1÷0.2 times the gravitational acceleration value (g). Also this analysis points out the precarious wedge stability and gives us a good connection with the recorded seismic activity.

4.2 Fall and bouncing

The intermediate path followed by the landslide can be further subdivided in two more steps. The first one characterised by the rapid sliding and free falling of the rock mass, along the remaining cliff wall, for a total length of about 470 m, up to the more important impact on the talus slope and partially on the cliff foot. After the impact a second bouncing phase started up to the complete stoppage of the blocky and then dispersed mass.

A check of these phases has been carried out through two methods: a rockfall simulation and the energy line technique (Crosta, 1992). Adoption of the first analysis method was suggested by the observed distribution of main blocks along the scree slope where the average inclination ranges between 35° and 40° (see fig. 3). Such distribution is comparable to that for classic rockfall situations except for the central sector (between dashed lines in fig. 3) where a large part of the talus debris has been remobilized and transported downward, resulting in a long scar and an elongated tongue deposit (see fig. 3, 4).

The simulation demonstrated, as observed on the field that few blocks, of different shapes, could overcome the valley bottom of few tens of meters. The energy line technique was considered for easiness of its approach, for the possibility of occurence of a rock avalanche like phenomenon, even if in presence of small rock volumes, and the necessity to evaluate maximum impact velocity at an elevation of about 1450 m (see fig. 3). This elevation was in fact the more probable impact point, as from field surveing, because of an aeroplanning effect from the cliff foot where low inclination buttresses are present and where opened tension cracks, as a consequence of a violent impact, have been observed. Impact velocity, as computed by means of these different techniques, was found to be in the range of 27 up to 45 m/sec.

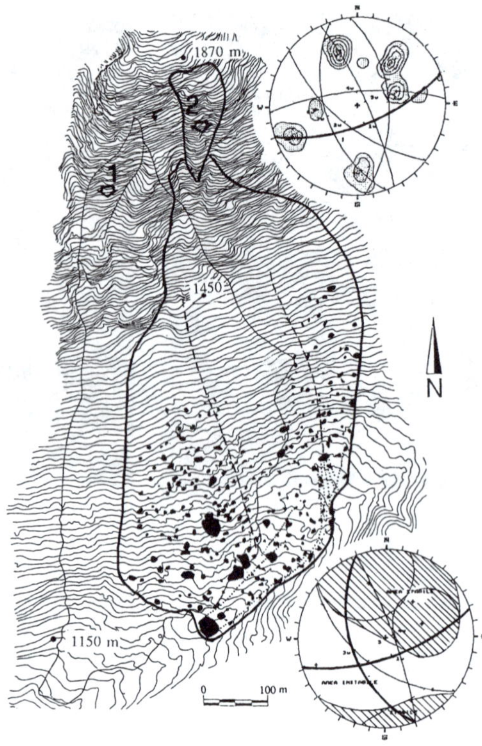

Fig. 3 - Map of: 1) 1977 and 2) 1991 landslides with structural data and stability analysis plot. Equidistance: 5 m. Dotted lines: main lateral levees; dashed lines: scar boundary inside the talus slope.

4.3 Debris slumping

The conversion through impact of the kinetic energy of the falling mass could have been the cause of debris remobilization along the scree slope (Crosta, 1992). For the kinetic energy computation we have to consider: the subdivision of the entire falling mass in two main directions along the cliff, one straigth downhill (N160°) and another to the left (looking at fig. 3) along the dihedral direction (N250°). Together with the subdivision of the mass, the number of impacts must be distributed within a certain time span.

The debris slump was about 300 m long and 150 m wide for a maximum 15 to 20 m depth and an approximate total volume of 0.20÷0.25 Mm^3, to be added to the original 0.2 Mm^3 of the rock wedge. The debris, mainly blocks with almost no fine matrix, ran downward for about 300 m, after a 90° bend, along the valley bottom and

Fig. 4 - Downhill view of the accumulation area. Debris slump turns 90° toward the right

finally produced a debris tongue, with the above mentioned huge block at its front. Under such an accumulation the creek continued to flow for the high permeability of the material, starting to cause some lateral erosion and undercutting of the opposite valley flank only some months later. Along the debris tongue evident lateral levees (dotted lines in fig. 3), particularly the more external one, and blocks embrication in the movement direction have also been observed. A series of ridges in a fan like assemblage are recognizable just uphill of the 90° bend (figs. 3 and 4) and could be explained as due to different shear velocity zones, increasing toward the flow centerline, or to the bending itself of the flow.

The position of the huge frontal block seems to prove the debris slump model discarding that of a "turbulent" rock avalanche scouring the talus slope during its run downhill. The slump movement could be assimilated to a low level inertial granular flow (Pierson & Costa, 1987), because of the estimated velocity and of the almost complete absence of matrix and water. Furthermore, the field survey showed an important lateral shearing and scouring effect. In fact, in corrispondence of the multiple sharp contacts between the landslide and the road (see fig. 2c) and far of the main scar within the talus, we observed besides numerous impacts and bouncing marks a shear movement and sliding action developed for at least 3÷4 m in depth. This seems to be a further demonstration of the transition from the granular flow to the slumping type of movement and of the strong lateral decrease in the velocity of sliding, probably along few discrete shear surfaces.

From a rough estimation of the debris superelevation at the 90° bend, in correspondence of the valley bottom (see figs. 3 and 4) we found the tilt angle (α) of the debris surface and the average radius of curvature (R). Appling Morris' formula (1963; adopted also to debris flow modelling by Johnson, 1970) for subcritical flow :

$$V_{average} = \sqrt{R * g * \tan\alpha}$$

it has been calculated for the study case a mean velocity of flow of about 30 km/h, that can be corrected to a lower bound of 21 km/h (0.71 * $v_{average}$) according to Ikeya and Vehara (1982). These values are in good agreement with the mean velocity for granular flow as proposed by Pierson and Costa (1987).

A pseudo-static analysis performed on the talus material, before of slump occurence and for a cohesionless material with internal friction angle of 40°, gives a static safety factor of 1.5 and a coefficient of critical acceleration equal to 0.75 g. This value gives the constant force required for the debris remobilization and it could be compared with the acceleration and impulse energy transmitted through the instantaneous impact. Such a comparison could be performed assuming a certain value of elastic moduli or stiffness and a coefficient of restitution, for the impact, and reconstructing a complete energy balance for the phenomenon.

5 CONCLUSIONS

The study of stability conditions of the rock mass and debris slope, involved in the 1991 rockfall, together with observations about morphology and geometry of the valley as of the debris accumulation permit the reconstruction through different steps of the landsliding scenario. Particularly, the analysis discusses and shows the main rockfall triggering factors both for antecedent and actual conditions existing on the cliff.

Furthermore, it has been shown how field surveying and applications of different modelling techniques can be adopted to prove the evolution of complex landslides. Worth of mention is the recognition of a sliding and falling phase followed by the remobilization, through the mass impact, of the debris slope. This remobilization occurred in two different way: like a granular flow for the main body, with sustained velocity, and in form of sliding along the lateral bounding areas.

In particular, the application of a rockfall simulation and of the energy line technique, used to estimate impact velocity of the falling mass, have shown comparable results useful for the energy balance evaluation. The possibility of an energy balance reconstruction is suggested in order to compare critical acceleration values, evaluated from pseudo-static stability analysis methods and necessary to mobilize the debris, with the ones induced by impact and instantaneous conversion of kinetic energy.

As a final consideration, instability has to be considered an explicit problem because of the orientation and the geological and structural features of such a cliff. Risk from the same kind of failure mechanisms similar to the studied ones might be expected.

Furthermore, from a designing point of view it is not possible to stabilize the entire cliff and other zones along the same valley flank. At the same time, considering the mechanisms of formation of the debris material and the recurrence interval of the events (14 ÷ 20 years), any type of passive protection work (embankments, trenches or artificial tunnels, etc.) seems to be unreliable on this valley flank. The easiest solution for the planners could be to consider the reconstruction of the main road on the opposite valley flank and the monitoring of particular areas in order to permit new quarrying activities.

ACKNOWLEDGEMENTS

The first author thanks prof. A. Cancelli, Univ. Milano, for the critical review of the paper and E. Crosta for his thoughtful advices. The authors are also indebted to the Geological Survey of the Regione Lombardia to make available some of the information.

REFERENCES

Comunità Montana Valtellina, 1981. Indagine geologico strutturale sul pendio roccioso coinvolto nel movimento franoso di Sasso Bisolo (Val Masino). Technical report.

Crosta, G. 1992. Meccanismi di movimento e messa in posto di grandi frane in roccia. Gli esempi della Val Pola e Sasso Bisolo. Atti 2° Convegno dei Giovani Ricercatori di Geol. Applicata, Viterbo, Oct. 28th-31st, 1992 (submitted for print)

Govi, M. 1988. Processi d'instabilità naturale: tipologie, distribuzione, frequenza e pericolosità. 2° Ciclo Conf. Meccanica e Ingegneria delle Rocce, Nov. 28th - Dec. 1st, MIR '88, Politecnico Torino, COREP, Torino, 1, 10 pp.

Ikeya, H., S. Vehara 1982. Debris flow in S-shaped channel curves. Japanese Engng. J., 24, 645-650 (in japanese).

Johnson, A.M. 1970. Physical processes in geology, Freeman & Cooper Co., 577 pp.

Morris, H.M. 1963. Applied Hydraulics in Engineering. Ronald Press, New York, 455 pp.

Pierson, T.C., J.E. Costa, 1987. A rheologic classification of subaerial sediment-water flows. In "Debris flow-avalanches: Process, recognition and mitigation", J.E. Costa and G.F. Wieczoreck Eds., Geol. soc. Am., Reviews in Engineering Geology, vol. VII, pp. 1-12.

Sarma, S.R. 1979. Stability analysis of embankments and slopes. J. Geotech. Engng. Div. Am. Soc. Civ. Engrs., 105, GT12, 1511-1524.

Varnes, D.J. 1978. Slope movement types and processes. In "Landslides Analysis and Control", R.L. Schuster and R.J. Krizek, Eds., Special Report 126, Nat. Res. Council, Washington D.C., p. 12-33.

Environmental Management, Geo-Water & Engineering Aspects, Chowdhury & Sivakumar (eds)
© 1993 Balkema, Rotterdam. ISBN 90 5410 099 0

Large landslides in flysch formations in the Northern Apennines, Italy: Analysis and comparison of the geomorphological features and geomechanical behaviour

C.Elmi, A.Fini, R.Francia, A.Lizzani & R.Genevois
Department of Geological Sciences, University of Bologna, Italy

ABSTRACT:
In the Northern Apennines, where more than 60% of the area is covered by flyschoid formations, 23 large landslides (block slide, rock slide or rock slump) ranging from 1 to 3 km in length and from 1 x 10^6 to 1 x 10^7 m^3 have been examined in their geomorphological and geomechanical aspects. The statistical evaluation of the geomechanical behaviour of those formations performed on a wide area and on a large number of events has been carried out in view of an assessment and a more accurate forecasting of slope stability in presently stable areas, showing comparable geomorphological conditions.

1 GEOLOGICAL AND GEOMORPHOLOGICAL SETTING

On the Adriatic side of the Northern Apennines more than 60% of the area is covered by large flyschoid units: the Monghidoro Formation (upper Cretaceous), the Macigno fm. (lower to middle Miocene), the Marnoso-arenacea fm. (middle to upper Miocene) (fig. 1). They show similar sedimentological characters, i.e. regular alternance of marls and sandy shales rhythmically interbedded with graywackes and

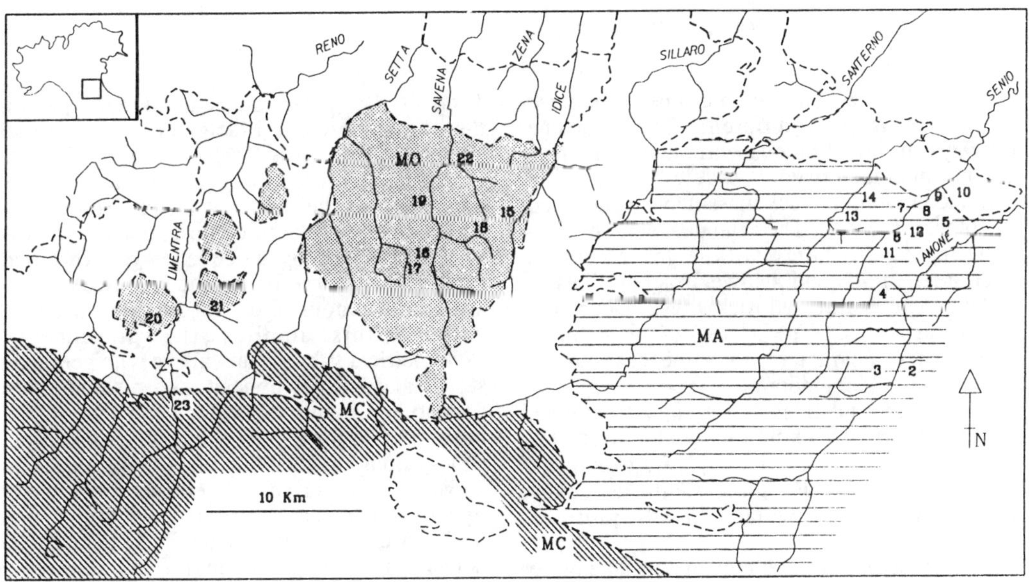

Fig. 1 - Location of landslides. M.A : Marnoso-arenacea formation; MO: Monghidoro fm; MC: Macigno fm.; 1-23 nos.: landslides (see table 1 and 2).

Table 1. Types and dimensions of landslides

N.	Locality (age)	Form. (1)	Type (2)	Slope h. (m)	Length (m)	Width min (m)	Width max (m)	Volume (10^3 m^3)
1	S. Eufemia (1939)	M.A.	j2-q1	445	1900	70	750	12500
2	Casoni di Tura	M.A.	j3	400	725		900	10300
3	M.Romano (1690)	M.A.	j3	400	1000	650	1000	33750
4	Purocielo (1690)	M.A.	j2	340	1000		1300	29250
5	Torretta (1978)	M.A.	j3	125	340	175	240	850
6	Cavinella	M.A.	j2-g	200	600	100	360	2800
7	Zattaglia	M.A.	j2	180	780	70	230	620
8	S. Stefano in Z.(1500)	M.A.	j2-q4	175	1460	140	180	1850
9	Lame	M.A.	j2-g	210	1050	600	1350	24600
10	Rio Bobbo	M.A-PL	g-q1	195	800	80	330	4800
11	Rio Pozza	M.A.	j2	195	620	100	340	4300
12	S. Croce	M.A.	j2	175	780	200	300	5500
13	M. Castellaro R.	M.A.	j2	171	850	100	210	480
14	Ca' d. Antonio	M.A.	j2	230	1230	310	700	6800
15	Campeggio (1890)	MO	g-q1	430	1800	40	320	4250
16	Castel d Alpi 1 (1951)	MO	g	225	820	190	500	8400
17	Castel d.Alpi 2 (1951)	MO	g	100	510	180	370	2600
18	Fosso d.Macchie	MO	j2	162	1100	100	550	6600
19	Balzo d. Cigni (<1935)	MO	g	281	630		300	5300
20	Castel d. Casio (1874)	MO	j3	480	2570	160	610	28700
21	Camugnano (<1930)	MO-AS	g-m2	225	1000	500	1150	22900
22	Rio Lognola	MO	j2	160	1100	190	550	6700
23	S. Giorgio	MC	g	690	1760	280	600	18900

(1) M.A. = Marnoso-arenacea formation (middle to upper Miocene); M.A.-PL = Marnoso-arenacea fm. and Pliocene blue clays; MO = Monghidoro formation (upper Cretaceous); MO-AS = Monghidoro fm. and liguride nappe (argille scagliose); MC = Macigno formation (upper Oligocene-lower Miocene)
(2) According to D.J. Varnes (1978)

coarse sandstones. The tectonic style of the three formations is quite different . The older one overlays an allochtonous complex (Liguride nappe) and underwent severe tectonization and fracturation: the rock mass may be classified as "blocky rock". The younger one shows a more regular structure, with large, sometimes overturned folds and a lower fracture density ("slabby" to "buttressed rock" type, Duncan & Goodman 1968). The Macigno fm. exhibits intermediate characters, in some cases more similar to the Monghidoro fm.

Some large landslides (fig. 1 and table 1), ranging from 1 to 3 km in length and from 1 x 10^6 to 1 x 10^7 m^3 in volume, were examined. They hit in the last three centuries, even repeatedly, several villages, roads, railroads and rivers.

Different types of rock movements take place in the three formations. Rock block slides or rock slides are almost the only types of movements in the Marnoso-arenacea fm. (table 1, no. 1-14); rock slumps are more frequent in the Monghidoro and Macigno fm, even if rock slides are common (table 1, no. 15-23). As mentioned before, the type of movement is connected to the different tectonic conditions and positions of the three formations. The M.A. is characterized by a sequence of overturned folds and by large monoclines, dissected by a structurally controlled rectangular drainage pattern: this implies the presence of asymmetrical subsequent valleys with gentle dip slopes and steeper opposite scarp slopes . The formation presents a low degree of jointing and only translational movements (rock block slides) along the bedding planes take place (fig. 2, 3). Circular surface movements can develop only in isolated cases (no. 10), as for example in intensely deformed zones. Ultimately the landslides develop only on dipslope sides, even if the slopes have low inclination, sometimes less than 10°. Complex landslides can be also observed, in the shape of rock slide-rock slump

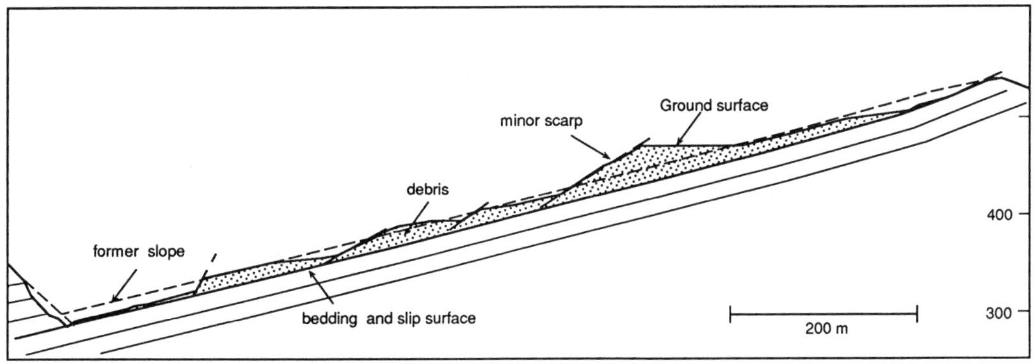

Fig. 2 - The Purocielo 1690 landslide (Marnoso-arenacea fm.) (no.4 of table 1).

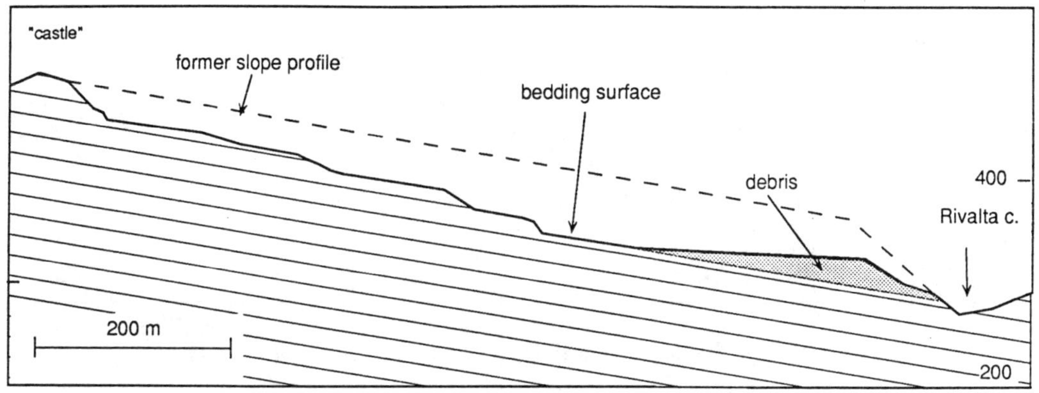

Fig. 3 - The M. Castellaro -Renzuno landslide (Marnoso-arenacea fm.) (no 13 of table 1).

(no 6 and no 9), or landslides evolving to debris flow (no.1, 8 and 10).

Because of the different tectonic style the Monghidoro fm. and Macigno fm. present a higher degree of jointing. The general geomorphological set-up is however similar to M.A. fm., with rectangular drainage pattern and highly asymmotrical monoclinal valleys. Rock block slides and rock slides on dip slope sides can be seen (table 1, no. 18, 20 and 22). But rock slumps on scarp slope sides are more frequently observed (fig. 4, 5; table 1 no. 15, 16, 17, 19, 21 and 23). In the first case (rock block slides and rock slides) the landslides develop on surfaces with an inclination of about 10°, that means similar values to M.A. fm. In the second case (rock slumps), the slope angles vary from 15° to 24°. Sometimes the movement is complex and includes lateral spreading (no. 21) or flows (no.15).

2 BACK ANALYSIS OF LANDSLIDES

Back analysis was possible in most cases: in some slides the presence of clear geomorphological features made it possible to reconstruct the former slopes. For the most recent slides, some detailed maps or aerial photographs preceeding the movement were really helpful. Due to the complexity, dimensions and heterogeneity of movements, neither in situ nor laboratory shear test were performed: some isolated tests on the silty marls of the M.A. formation are available. The shear strength data were obtained from back analysis. We assumed that:

- only frictional resistance (c'=0) should be considered both in planar and rotational failures; otherwise unusually low friction angles and/or high uplift water forces are required.
- the ground water level was near or at the ground surface in the most of cases
- in some failures the piezometric head tends

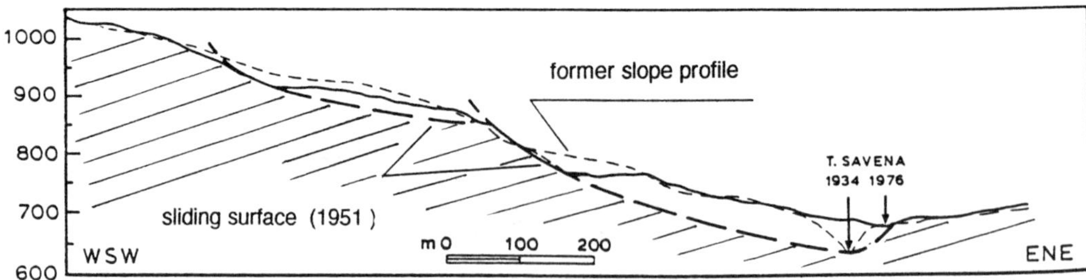

Fig. 4 The 1951 landslide of Castel dell'Alpi (Monghidoro fm.) (no. 16 -17 of table 1).

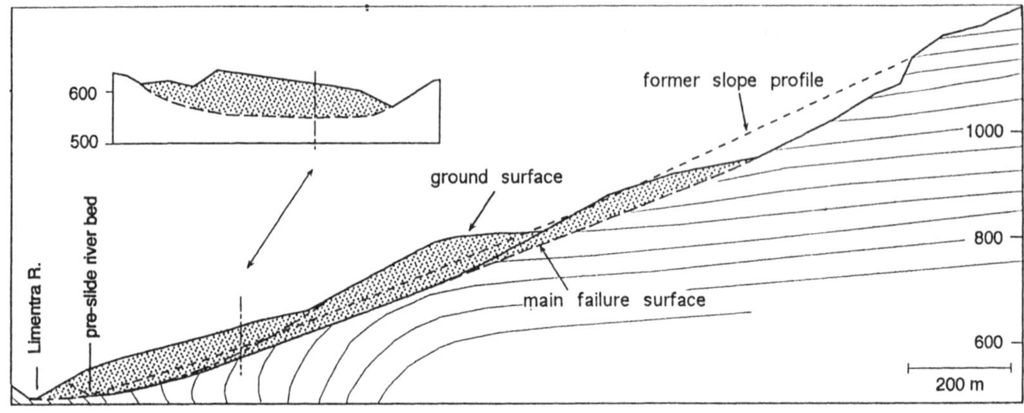

Fig. 5 - The S. Giorgio landslide (Macigno fm.) (no 23 of table 1).

to 'geostatic' value (r_u tends to 1)

- the presence of highly permeable sandy beds underlying less permeable clayey marls and of inhibited drainage at the toe causes the rising of piezometric surface
- the pore-water force u_f generally increases from the slope crest to the slope toe;

The geometrical and geomechanical parameters are listed in table 2.

3. DISCUSSION AND CONCLUSIONS

From the data obtained, some comparisons and remarks can be made.

3.1 *Rock slides*

The movement usually develops on a single bedding surface (rock block slide), extended to the entire slope (fig. 2), followed by successive slump of the displaced mass. In other conditions (fig. 3) the sliding occurs along different and gradually higher bedding planes, forming a typical terraced profile. At the slope crest an isolated slab, locally named "castle" often remains , where the uplift water force did not reach values high enough to start the movement. The peak velocities are generally low, less than 1 m/day. The slide debris is rapidly disrupted and eroded, particularly in the younger parts of M.A. formation, where the degree of cementation is lower. The debris causes a progressive displacement of the water courses at the toe and the formation on the opposite slope of active faceted spurs. Among the various flyschoid formations no difference in rock slides was observed.

Statistical analysis of data shows that:

1. the maximum frequency of the slip surface dip angle (β) for the M.A. fm. (fig. 6) is found in the 8°-10° class, but more than 60% of the cases considered shows values of β between 8° and 13°. The frequency distribution is very similar if all cases are considered, including the Monghidoro fm.
2. In more than the 75% of both M.A fm. and all

Table 2 - φ'_{max} and φ'_{min} : back calculated friction angles with water table at ground surface (z_w = 0) and measured z_w > 0 respectively; β_{min}, β_{max}: slide surface dip; i_{min}, i_{max}: slope angles.

No.	φ'_{max}	φ'_{min}	β_{max}	β_{min}	i_{max}	i_{min}
1	27		12		13	
2	23		15	10	10	
3	21		20		16	15
4	23		19	8	19	
5		20	10		35	10
6		20	12		13	
7	18		10		10	
8	19		8		10	
9		17	11		10	
10	24		15		14	
11	27	20	12		15	
12	23	18	18		13	
13	23	18	10		9	
14	27	22	10		10	
15	22		14		15	
16	24	14	18	10	15	
17		18	22	5	18	
18	25	19	10	9	9	
19	41	25	50	17	24	
20	24	19	10		11	
21		18	14	13	15	
22	32	18	18		8	
23	32	20	26	13	22	18

cases (fig. 7, 8) the friction angle φ', obtained by back analyses carried out with a water table at the ground surface, was between 18° and 23°, with a minimum of 17° and a maximum of 32°, mean 21°45': those values are generally smaller than the peak friction angles obtained on the marl samples (φ' = 35°).

3. The correlation between friction angles and slip surface dip (fig. 9), is linear as it should be due to the back analysis technique used, but friction angles seem to be related to the stratigraphic position of the slide: the more ancient are the rocks, the higher the friction angle (fig. 10).

3.2 *Rock slumps*

As said before, these landslides have been observed only in the MO and MC fm., on the scarp slope sides, and only when the rock mass shows a very high degree of tectonization. Generally, landsliding is of the retrogressive type and more than one slip surface or minor scarps can be observed (fig. 4 , 5). The slide

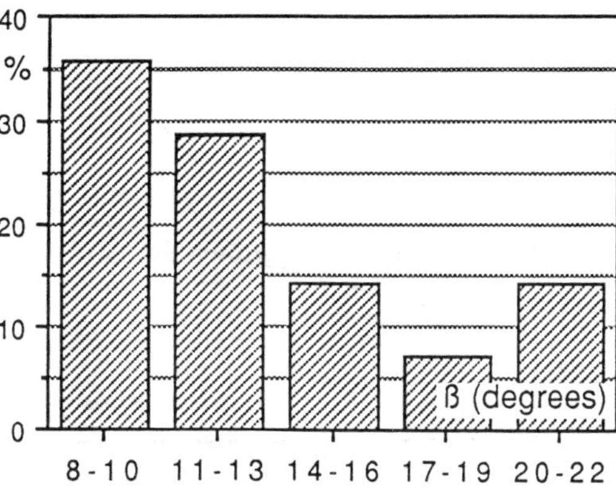

Fig. 6 Frequency classes of slide surface inclination (M.A. formation, 14 landslides)

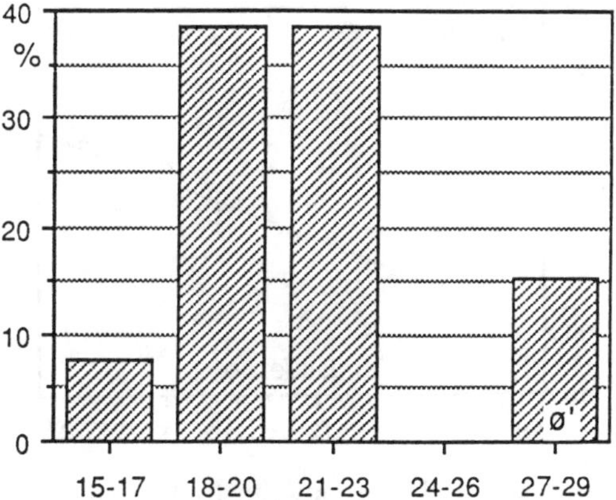

Fig. 7 Frequency classes of friction angle (M.A. formation, 14 landslides)

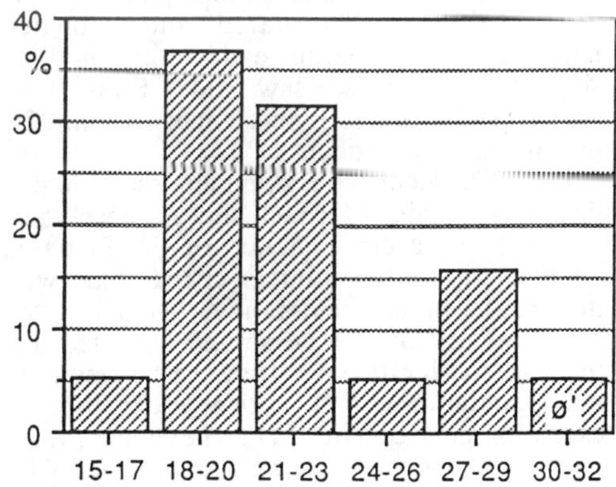

Fig. 8 Frequency classes of friction angle (all cases)

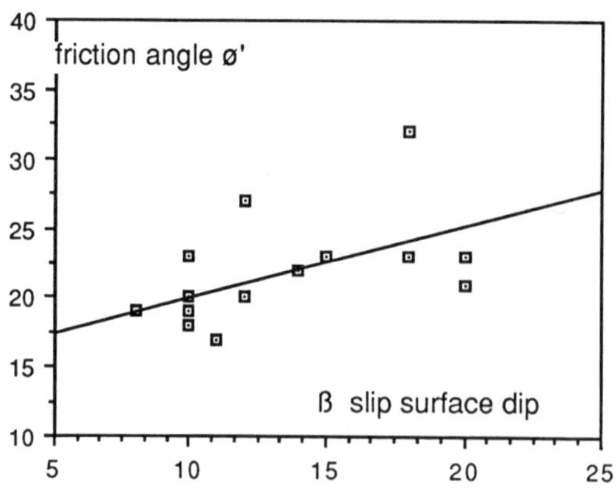

Fig. 9 Friction angle vs slip surface dip (M.A. formation, 14 landslides)

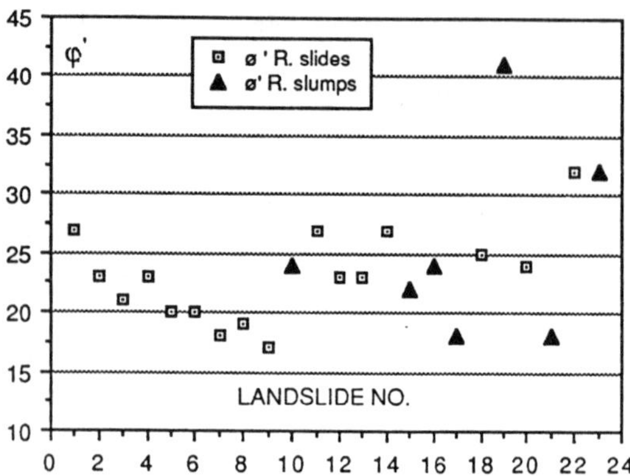

Fig. 11 Values of friction angles of all cases

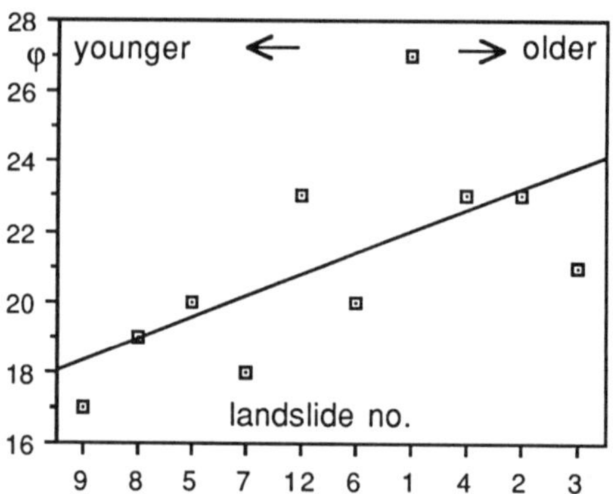

Fig. 10 Relation between friction angle and relative age of rocks (M.A. fm.)

mass causes in most cases the deviation of the river bed and the erosion of the opposite slope: a condition of alternated and repeated landsliding may therefore develops on both slopes (no. 15, 16, "see-saw" slides, Elmi 1988). The slope angles are higher (mean value 22°) than in rock slides (about 13°).

The back calculated friction angles are higher than those obtained for rock slides (mean value 25° 36') as a consequence of the fact that circular slip surfaces intersect materials with different shear resistances (marls and sandstones). However, it should be noted that there is no significant difference between the two populations (rock slides, rock slumps), except in one case (fig. 11) . The highest value of the back-calculated friction angle ($\varphi' = 41°$) was obtained for the Balzo dei Cigni landslide (no. 19); but the analysis carried out without pore-water pressures showed anyway the same friction angle value ($\varphi' = 25°$). It is then possible that in rock slumps the failure mechanism was mainly rotational, but large parts of the slip surface were planar, thus developing the shear resistances of the marls-sandstones contacts as happens in the rock slides.

In any case the back-calculated friction angle values are smaller than the peak ones, obtained both by laboratory tests and by related papers, showing that in both cases (rock slides and slumps) progressive failure phenomena have to be considered.

REFERENCES

Duncan , J.N. & Goodman, R. E. 1968 . Finite element analyses of slopes in jointed rocks. U.S. Corps of Engineers, Contract report S-63-3.

Elmi, C. 1990. I movimenti franosi di Castel dell'Alpi. CNR, GNDCI, Previsione e prevenzione di eventi franosi a grande rischio. Convegno su Cartografia e monitoraggio dei movimenti franosi, Bologna, 10-11 Nov. 1988.

Varnes, D.J. 1978. Slope movements types and processes. In: Landslides: Analysis and control. Transportation Research Board, Nat. Ac. of Sc., Special report 176.

Zecchi , R. & Montanari, E. 1990. Franosità nei bacini montani della Provincia di Bologna. Acta Naturalia At. Parmense 26, 1/2, 51-68.

Environmental Management, Geo-Water & Engineering Aspects, Chowdhury & Sivakumar (eds)
© 1993 Balkema, Rotterdam. ISBN 90 5410 099 0

Remobilization phenomena of a landslide body occurred in a structurally complex formation and in an area already affected by deep-seated gravitational slope deformations (Basento River, Southern Italy)

Damiano Grassi & Francesco Sdao
Department of Structures, Soil Mechanics, Engineering Geology, Basilicata University, Potenza, Italy
Luigi Monterisi
Institute of Engineering Geology and Soil Mechanics, Politechnics of Bari, Italy

ABSTRACT: This paper reports predisposing and determining factors associated with recurrent remobilization that an landslide body undergoes bringing about serious damages to an important railway-line passing through it. Among predisposing factors there are: the flysch nature of soils; the orogenic transportation; failures and deformations; the poor mechanical features of soils. The main determining factor is represented by the periodical heavy rainfalls occurring in this area.

1 INTRODUCTION

The examined landslide is an ancient multiple compound slide; as a matter of fact, its foot is hid by recent and present alluvial deposits of the Basento river being scantly deep. The most important railway-line of the region cuts transversely the middle-lower portion of the landslide body; so it frequently undergoes serious damages because of remobilizations of the landslide which still nowadays take place sometime in a moderate way sometime remarkably. The specific aim of this study is to detect the causes of the non-stop remobilizations . For, the fact that the landslide has not found stability over the time, despite its attitude with dipping inside the slope strata, and that its foot is any longer disturbed by the erosive action of the river, clearly indicates that several predisposing and determining factors take part in its evolution and they still do so.

2 MAIN FEATURES OF THE INVESTIGATED LANDSLIDE

It is a multiple compound slide with a roughly distinct well cut scarp wherein, at a slightly lower altitudes, two more distinct scarps, located side by side, may be noted (Fig. 1). Lesser scarps are found along the landslide body whose average slope is of about 25° and 15° respectively above and below the railway-line.

The main sliding surface is located in

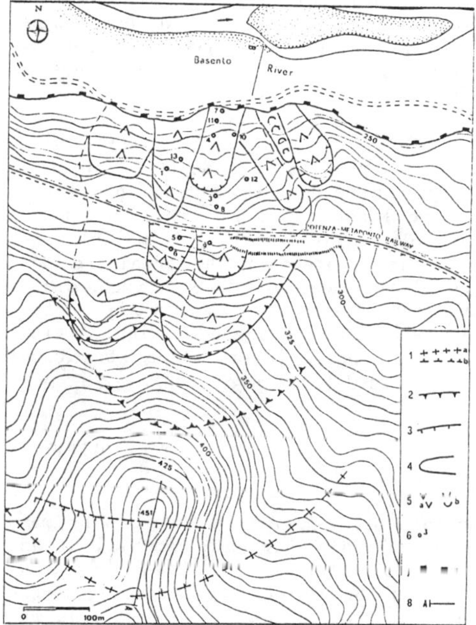

Fig.1 Geomorphological map of the examined landslide. 1) Possible rupture surface (a) and edge of the trench due to deep-seated gravitational slope deformations. 2) main scarp of ancient and recent landslide; when degraded are marked by a dashed line; 3) scarp of present landslide; 4) limit of the landslide body; uncertain limits are marked by a dashed line; 5) compound slide (a) and earthflow (b); 6) boreholes; 7) edge of the river bed . 8) section line

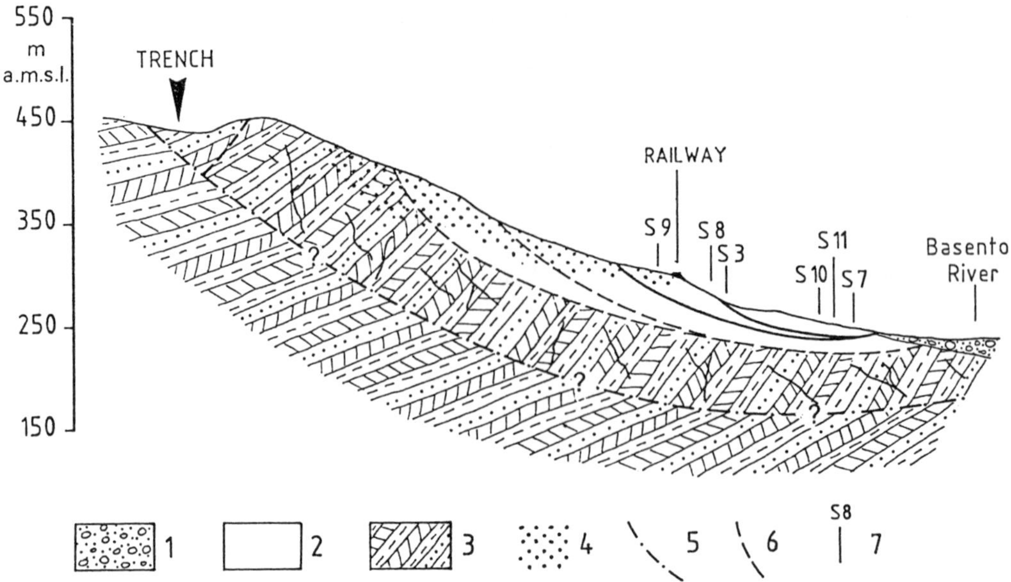

Fig. 2 Geomorphological section. 1) alluvial deposits; 2) landslide debris; 3) Serra Palazzo Flysch likely to be affected by D.G.S.D.; 4) swelling and deformation phenomena caused by D.G.S.D.; 5) likely deep-seated failure surface trend; 6) sliding surface; continuous line represents failure surfaces on which the May 1985 movements occurred; 7) boreholes.

marly-clayey soils at a depth of about 40 m . At various depths other failure surfaces are found (Fig. 2). The top of the landslide body (first 10 - 12 m) is formed by detrital material; the remaining portion of the landslide body, though dislocated and showing sometime a rather detrital state, still has the original attitude, with dipping inside slope strata but more inclined (even 50° - 60°).

3 PREDISPOSING FACTORS OF THE INVESTIGATED LANDSLIDE

The main predisposing factors of the mass movement and recurrent remobilizations are: nature and morphostructural setting of soils; the hydrogeological and geotechnical features of soils; deformations and failures caused by ancient and inactive deep-seated gravitational slope deformations (D.G.S.D.) (Fig. 3).

Soils belong to the Serra Palazzo Flysch (Miocene) (Boenzi et al., 1968) (Fig. 3b), representing one of the complex formations from the structural and geotechnical standpoint. Such a Flysch is sometime almost entirely marly-clayey, as the area where studied landslides occurred, sometime rich in sandstones arranged in layers and banks. Owing to tectonic events which took place and the imposing gravitational movements triggered by the last up-lifts, resistance parameters were often significantly decreased in magnitude. In fact, the formation underwent a considerable orogenic transportation which has upset strata and caused large and small tectonic nappes to overthrust. The investigated area is located near the overthrusting front of the Appennine tectonic units on the Plio - Pleistocene formations of the Bradanica foredeep (Fig. 3). Moreover, the landslide in question is located at the foot of an imposing deep-seated gravitational slope deformations (D.G.S.D.) of compound slide type, covering an area of about 4.5 sq.Km. This last is located at the northern edge of a 20 sq.Km wide area being massively deformed and upset by D.G.S.D. whose morphology has already undergone tectonic events, and is characterized by : considerable scarps; trenches (which add to large deep open fractures and intensively fractured zones); splitting of ridges, morphological depressions, often undrained; etc. D.G.S.D., which produced these morphological effects, are mostly of compound slide and lateral spreads type of previously dislocated masses; sackungs are also present locally. Said remarkable mass movements have been triggered following the last tectonic

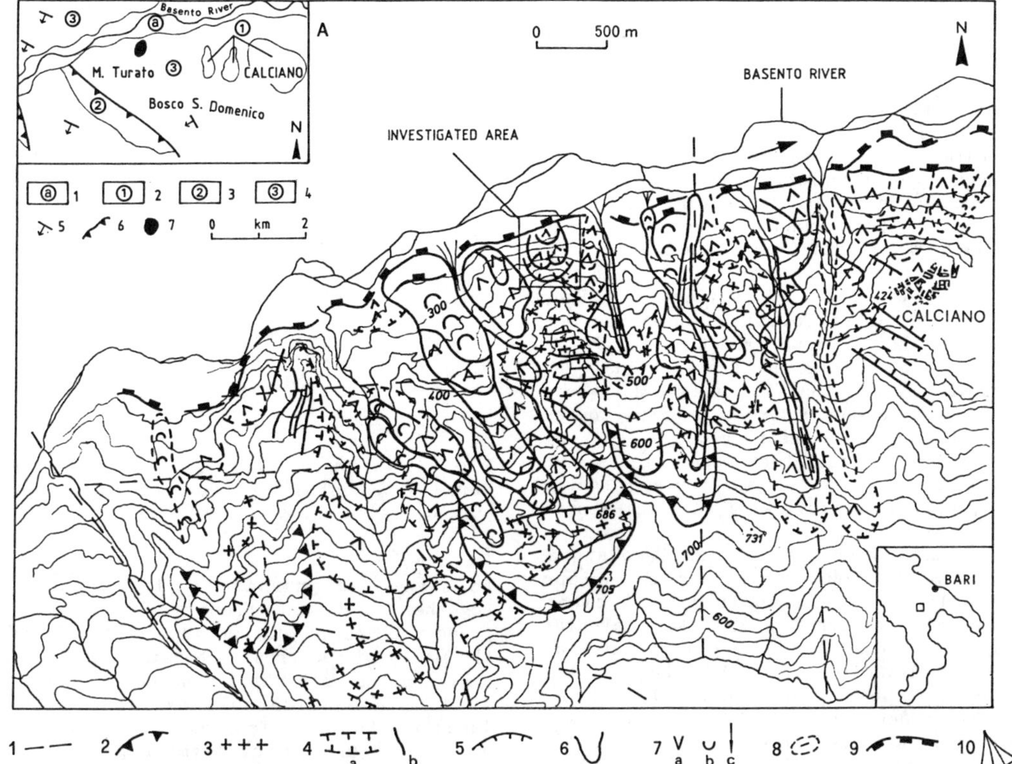

Fig. 3 Deep-seated gravitational slope deformations and landslides in the left slope of the Basento river valley near Calciano village. 1) Important fault or fracture; 2) main scarp of D.G.S.D.; when degraded are marked by a dashed line; 3) lesser failure surfaces due to D.G.S.D.; 4) trench (a) and large deep open fracture (b); 5) scarp of landslide; when degraded are marked by a dashed line; 6) limit of landslide body; uncertain limit are marked by a dashed line; 7) rotational or compound slides (a), earthflows (b) and debris flows (c); 8) morphological depression caused by landslide; 9) edge of the fluvial terraces; 10) Alluvial fans. Geological skecht (a): 1) Alluvial deposits ; 2) Sandy - clayey deposits (Pleistocene - Pliocene) 3) Numidian Flysch (Langhian - Upper Oligocene); 4) Serra Palazzo Flysch (Langhian - Tortonian); 5) attitude of strata; 6) overthrusts; 7) location of examined landslide;

uplifts bringing about a roughly sudden deepening of the River Basento Valley (more than 200 m from the Mindel age) (Cotecchia and Del Prete, 1986), and, then, mass imbalance, loosening of tectonic stress, remarkable increase of relief energy (450 m). Landslides, occurring during or following to D.G.S.D., have been favoured by the high seismicity characterizing such an area (seisms of even VII° - IX° MCS) and by heavy rains (1 every 6 years).

The D.G.S.D., housing the landslide in question at its foot, is located in the middle portion of the whole area disturbed by D.G.S.D. (Fig. 3). Its scarp runs along the ridge at 700 m of altitude. Several lesser failures appear at different altitude of the slope causing frequent splitting of ridges, intensively fractured zones and trenches. The main trench is located immediately below the main scarp being 1 km long and 200-300m wide. In the middle-lower portion of the slope (around the main scarp of the investigated landslide) the stresses caused by the above deformations originate continuous swellings of the slope (Fig. 2) and then a roughly continuous state of stress which is enhanced by rainfalls.

The Serra Palazzo Flysch, when it is not much disturbed, has a quite good mechanical resistance, with friction angle values likely to achieve 30°; since, in this case, the flysch is entirely marly-clayey, in general, it is mostly impervious.

Where the flysh underwent both tectonic failures and effects of a great deal of D.G.S.D. and very ancient landslides, its resistance often turns out to be poor. Furthermore, its acquired permeability by fracturing allows a deep seepage of rains and originates a sort of underground water circulation occurring through (scant and sometime transitory) overlapping water levels, whose piezometric heads may deeply vary one from the other.

Owing to such structural features, referring to recent classifications (Esu, 1977), the above formation sometime belongs to the B1 group, sometime to the B2 group, according to the gravitational and tectonic disturbance it underwent. Where the formation has become a detrital material it belongs to the B3 group.

In situ and laboratory measurements and tests, carried out on marly-clayey material, proved the following phisical-mechanical features .

In situ dry density and soil moisture measurements, obtained by radioisotope borehole surveys, resulted in a clear difference between the detrital material and still intact marly - clayey material even if dislocated (Fig. 4). The detrital material shows a dry density,with no visible order, from 11 KN/m3 to 18 KN/m3; the natural water content ranges from 45% to 5%. The underlying material has a dry density tending to increase with the depth from a minimum value of 18 KN/m3 to a maximum one of 24 KN/m3; of course the equivalent natural water content tends to decrease with the depth ranging from max values of about 30% to minimum values of about 5%.

Figure 5, representing grain-size distribution, plasticity chart and the activity diagram , shows that soils influenced by landslides are characterized by marly-clay or slightly sandy clayey - silts, with an average plasticity and low activity.

Direct CD shear tests on the clayey - marly material proved that: the peak friction angle ranges from 20° to 25°; the peak cohesion varies from 20 - 60 KPa; the residual friction angle ranges from 10° to 14°, this conforms to the results which may be derived from Kanji correlation (Kanji, 1974) between $\phi'r$ and plasticity index. It is very likely that the detrital material has a residual resistance of the same order.

4 DETERMINING FACTORS OF LANDSLIDE REMOBILIZATION PHENOMENA

Such resistance characteristics that the material has acquired with the time, owing to several disturbances it underwent, together with said predisposing factors, prove that rains play an important role as the main and recurrent determining factor.

Even rains with quite a normal intensity lead to a considerable overloading of landslide bodies which make up the slope, and the occurrence of harmful pore water pressure in them.

As a matter of fact, measurements proved that piezometric heads undergo sudden and remarkable changes (even of 2 - 3 m) during or shortly after rains characterized by a moderate occurrence period. Then, in some places and often, near the landslide bodies, the piezometric surface is the same as the topographic one.

On the basis of the last remobilizations occurred in May 1985 and in winter-time of 1991 supported by appropriate records, it may be said that: the upper portion of the landslide body (up to 10 - 12 m depth from the ground level) where the soil is mostly formed by detrital material, is the most subject to mobilizations, with a horizontal displacement of a few metres. The underlying portion, as a shown by some inclinometric measures, moves on pre-existing surfaces located at different depth; its displacement is, at most, a few centimetres.

In order to verify a clear correlation between rains and landslide remobilizations, the mass movement which took place in May 1985 has been investigated. This last affected only the middle-lower portion of the pre-existing landslide body, involving , for a maximum thickness of about 20 m (Fig. 2), both a detrital material and part of the already disloca-

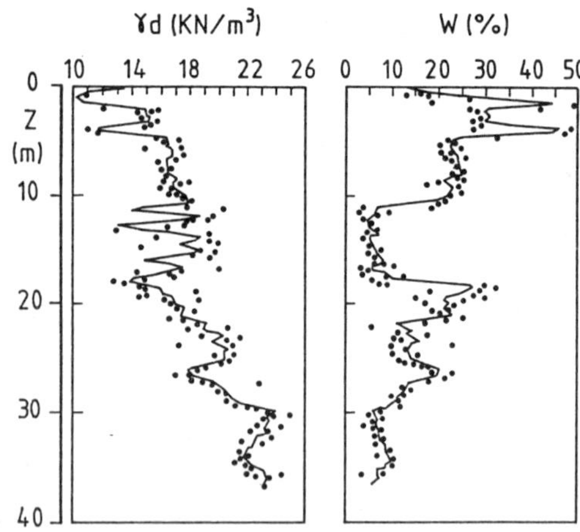

Fig. 4 Natural water content and dry density , obtained from radioosotope borehole surveys.

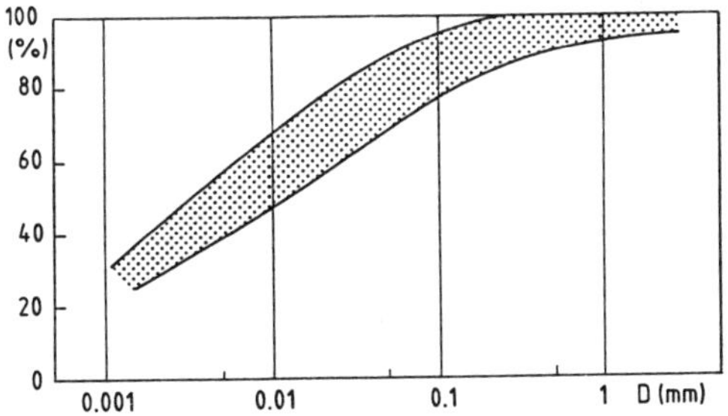

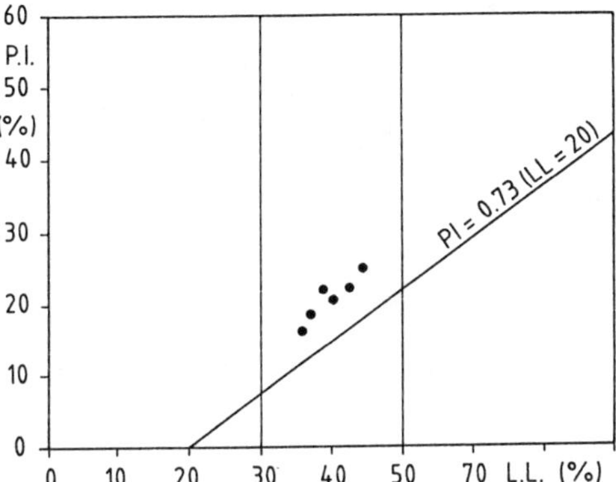

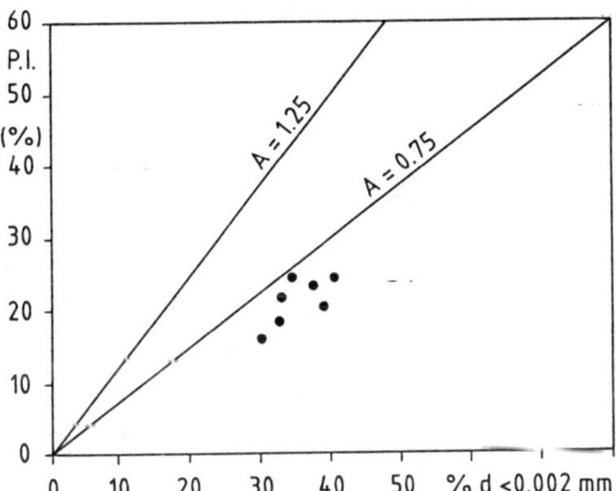

Fig. 5 Grain-size distribution, plasticity chart and activity diagram of examined soils.

ted marly-clayey material. The mass movement occurred according to two different failure surfaces at various depths (Fig. 2) ; the deeper one being a pre-existing shear surface, located a maximum depth of about 20 m from the ground level; the more superficial one originated during the remobilization along the contact between detrital material and the marly-clayey material which certainly is less disturbed by previous landslides and is sited at a maximum depth of about 12 m from the ground level.

Few days prior to the remobilitazion (from 16th to 19th of April) 154.3 mm of rain fell on the landslide area (such a value stands for 25 % of the total yearly rainfall). Moreover, the exceptional aspect of rains may be related to the fact that during 150 days before the movement 849 mm of rain fell down, this value has a occurrence period of about 20 years (Lazzari et al., 1990).

In order to verify if said rains had played a role in causing the landslide, a back stability analysis has been carried out resorting to the Bishop's simplified method and taking into account a failure surface sited at a depth of about 20 m from the ground level. Fig. 6 shows the calculation diagram and results of the stability analysis, carried out using the pore pressure adimensional parameter ru, both under seismic and static conditions. Such an analysis shows that: without the groundwater the equilibrium of the slope is granted by resistance values slightly higher than the residual resistance ones. During heavy rains, as the above mentioned ones, the increase of the piezometric surface requires values higher the residual ones, close to the peak resistance.

As above said, following to rains with a moderate occurrence period the change of the piezometric head have been even of 2 - 3 m; hence, the assumption that because of rains with considerable occurrence periods, as the ones of April 1985, the piezometric level has more significant changes comparable with those assumed during the stability analysis. Therefore, it is likely to assume that rains stand for the main predisposing factor which caused the investigated landslide remobilizations.

REFERENCES

Boenzi, F., Ciaranfi, N. and Pieri P. 19[illegible]. Osservazioni geologiche nei dintorni di Accettura e di Oliveto Lucano. Mem. Soc. Geol. It. 7 : 379 - 392

Cotecchia, V. and Del Prete, M. 1986. Some observations on stability of old landslides in the historic centre of Grassano after the earthquake of 23rd November 1980. Proc. of Inter. Symp. on Engineering geology problems in seismic areas. IV : 155 - 167.

Esu, F. 1977. Behaviour of slopes in structurally complex formations. Int. Symp. on the Geotechnics of Structurally Complex Formations, Gen Rep. v. 2: 292 - 304, Capri.

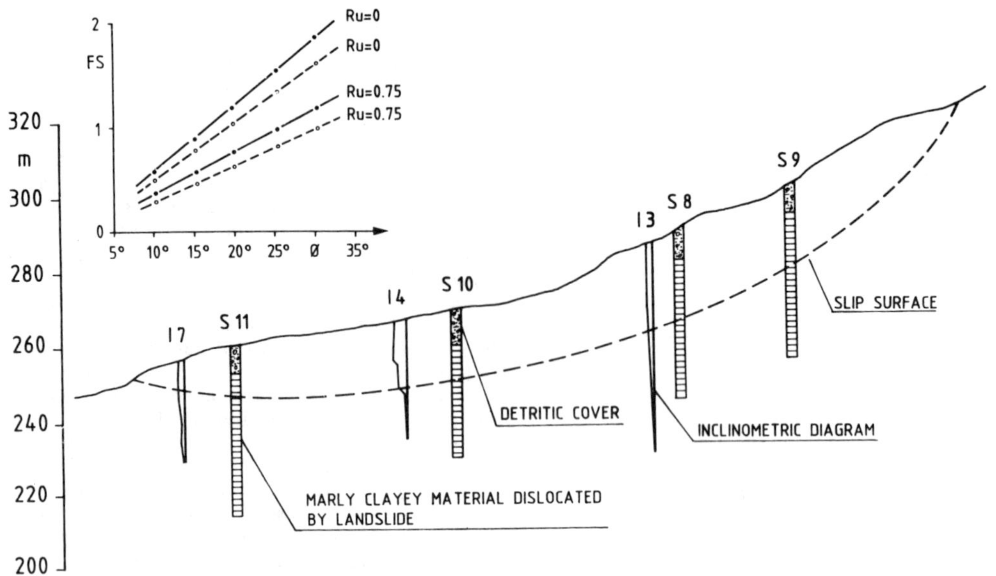

Fig. 6 Back analysis stability of the 1985 landslide remobilization phenomenon.

Kanji, M.A. 1974. The relationship between drained friction angles and Attemberg limits of natural soils. Geotechnique, v. 24, n.4: 671 - 674.

Lazzari, S., Caloiero, D., Gabriele, S., Lambiase S., Mercuri, T., Pesce, F., Tramutoli, M., Versace, P. 1990. Effetti prodotti in Basilicata dagli eventi alluvionali verificatisi nel periodo Dic. 1984 - Apr. 1985. GEODATA n. 37 (1990) C.N.R. I.R.P.I..

Environmental Management, Geo-Water & Engineering Aspects, Chowdhury & Sivakumar (eds)
© 1993 Balkema, Rotterdam. ISBN 90 5410 099 0

Deep-seated gravitational slope deformations and landslides located near some Southern Appennine chain areas (Southern Italy)

Damiano Grassi & Francesco Sdao
Department of Structure, Soil Mechanics, Engineering Geology, Basilicata University, Potenza, Italy

ABSTRACT : Deep - seated gravitational slope deformations (D.G.S.D.) in the Southern Appennine chain (Southern Italy) are further widespread than it is presently assumed, and some of them are still active. Experience proved that the chronic and spread proneness to landslides of certain areas is because said areas are located within or at the edge of D.G.S.D. First this correlation deserves particular attention owing to the fact that often the proneness to landslides relates to urban areas and/or their infrastructures. Cases described later deal with proneness to landslides caused by D.G.S.D. in the proximity of urban areas. D.G.S.D. are likely to both act as predisposing factors (when they are inactive) and as determining factors (when they are active) for landslides.

1 GENERAL REMARKS

Since 1988 the authors pointed out there was a direct or indirect relationship between the proneness to landslides of certain areas and the local presence of D.G.S.D. (Grassi et al., 1988). The investigation carried out later proved the validity of the previous assumptions and verified that the phenomenon was much more widely spread (Grassi et al., 1989; Cotecchia et al., 1989). The main problem is that D.G.S.D. geomorphological and geomechanical effects cannot always be easily detected, above all dealing with very erodable loose soils which may show plastic deformations and soil creep phenomena. In some cases D.G.S.D.have been discovered by chance during excavations or geognostic survey. In rocky soils failures and deformations caused by D.G.S.D. are often confused with tectonic failures. The extension of D.S.G.D. phenomenon is about 30 - 40 sq.Km . The determination of a relationship between a sliding occurrence and D.S.G.D. is an all-important factor to produce an appropriate analysis of the slope stability and for a good choice of type and practice for strengthening measures.

D.G.S.D. act as the main landslide predisposing factors when they are inactive, and as determining factors of proneness to landslides when they are active.
The determination of the activity or inactivity of such a phenomenon stands for the first difficulty the researcher has to cope with. This is because it takes place by very slow movements whose pick phases are likely to occur after decades or centuries. Hence, the need of a historiographical investigation.

As predisposing factors the effects caused by D.G.S.D. play a role much more important than those belonging to the slope itself (morpho-structural profile and hydrogeological physicomechanical features of soils). The main effects are represented by: mass imbalance, and anomalous state of stress; failures and various horizontal and vertical deformations; contour effects; soil permeability increase and origin, also in soils mostly anhydrous or impervious, of numerous groundwaters being under pressure or overlapping; possibility of soils to receive sudden and considerable increase in pore pressure, even on the occasion of rather ordinary rainfall; reduction down to residual values, or nearly so, of soil resistance parameters (especially with clayey interbedded layers); acceleration of the erosive action on rocks and start and / or acceleration of selective eroson phases.

In the case of inactive D.G.S.D., the slope is characterized by the proneness to landslides, postponed in time, tending to reach an equilibrium based on the new environmental geomorphological and geomechanical status D.G.S.D. created. Once the

Fig. 1 Location of mass movements presented in this paper

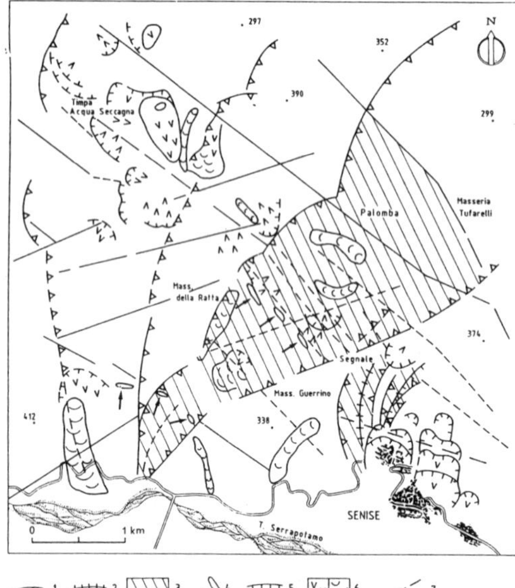

Fig. 2 Territory of Senise Village. 1) Main edge of D.G.S.D.: trench axis; 3) trench area showing a complex gravitational morphology produced by sackung, lateral spread, block slides and landslides; 4) hill ridge dislocated by D.G.S.D.; 5) landslide scarp; 6) rotational slides (a), earthflow (b); 7) main fault

new equilibrium has been attained, by means of individual and autonomous sliding movements, the versant turns out to be stable and undergoes the ordinary geomorphological evolution.

Conversely, where D.G.S.D. are still active it comes about that: during standstill periods of the phenomenon, the evolution of versants takes place as indicated above, by individual geomorphological movements; during active periods, all landslide bodies undergo a new simultaneous mobilization including those which already reached the equilibrium, thus trigging new landslides.

Experience proved that even a decimetre displacement is likely to produce considerable effects for the whole area. Shortly before D.S.G.D. remobilization, old and new failures become visible as well as the remobilization and acceleration of movements of some sliding bodies which did not reach stability yet.

All D.G.S.D. examined so far (Fig. 1) come under highly seismic areas (III° thru X° M.K.S.) or subject to frequent earthquakes.

2 Examples of landslides triggered by D.G.S.D.

2.1 Territory of Senise village (Basilicata Region).

Four fifths of the urban installations of Senise, realized in the last two decades, suffer from serious and continuous lack of the state of stability. This is due to the fact that they have been carelessly or unknowingly sited on level areas which hid the danger of a genesis characterized by ancient slides (Fig.2). Of the main landslides occurred in the urban area, the most recent (1986) is the Timpone hill one, which knocked down all buildings and killed eight people. Such a proneness to landslides, which mainly takes place through retrogressive both translational and compound slides, is only partly justified by the stratigraphic - structural features of the territory. For, this area is characterized by a stratified sandy Pleistocene formation with muddy-clayey interbedded layers dipping down - slope , laid on a clayey substratum. The main predisposing factors responsible for the chronic spread and remarkable proneness to landslides are likely to be ascribed to the above specified destabilizing effect triggered by inactive D.G.S.D.. These last took place through lateral spread and sackung processes; the ravaged area is larger than 25 sq. Km. The most remarkable effect which is still well evident, though massive erosive processes occurred, is a trench exceeding 5 Km in length and 300 m deep; lesser trenches may be seen close to the town - area (Fig.2). All over the area, especially inside the main trench there are lesser phenomena such as: sackungs, block slides, a great deal of the landslides of various type and size. Following to the above D.G.S.D. the slope is characterized by: lack of mass stability; large deep open fractures (even some metres wide) which are often filled with debris; reduction of mechanical parameters of soils; under pressure, overlapping, multiple aquifers (in indisturbed conditions the sandy formation is strictly anhydrous). Such effects stand for the main predisposing factors of the recent and present proneness to landslides of the urban areas and surroundings.

2.2 Territory of Canolo village (Calabria Region)

Most of the urban area of Canolo (Fig. 3) is affected by D.S.G.D. and landslides. The village is sited against a Mesozoic

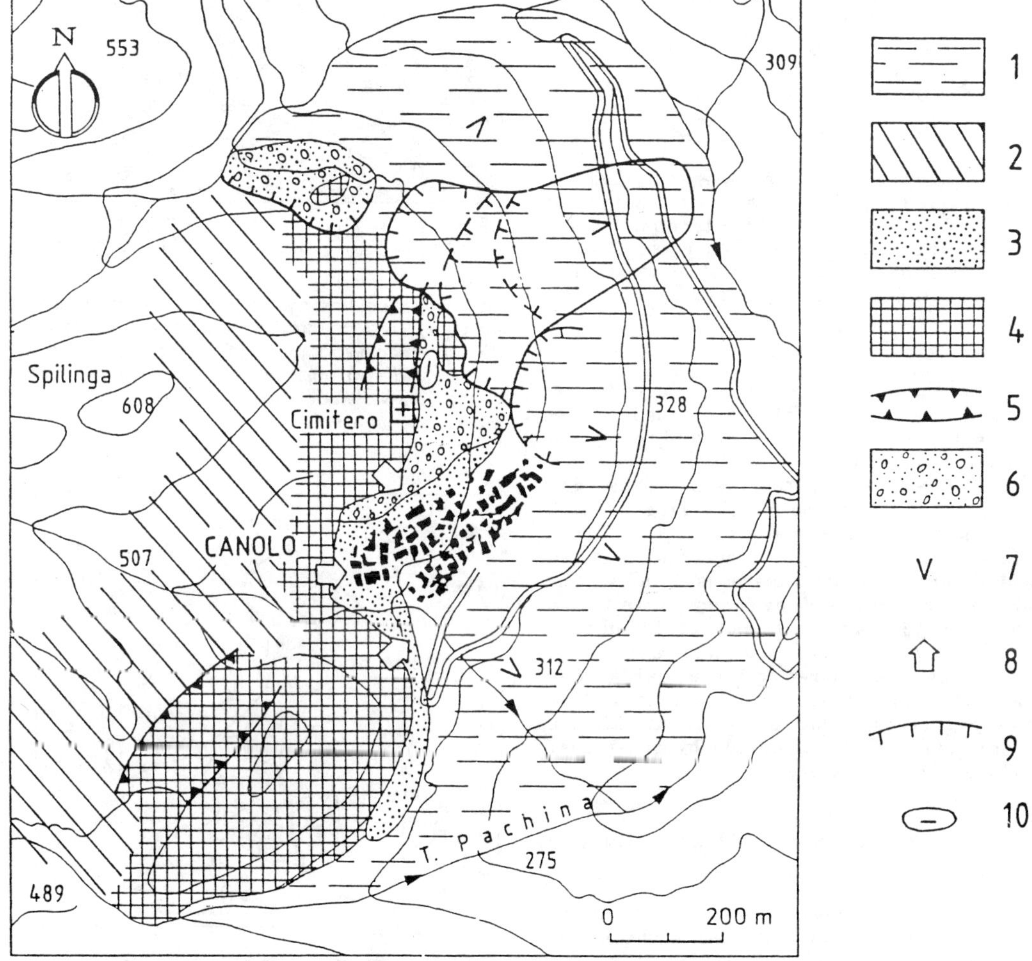

Fig. 3 Territory of Canolo Village. 1) Palaeozoic phyllites; 2) calcareous rocks; 3) landslide and/or talus debris; 4) Area disturbed by lateral spread and sackung; 5) trench caused by D.G.S.D.; 6) area with calcareous slides blocks; 7) phyllites rocks disturbed by block slides and rotational slides; 8) Area interested by falls; 9) well preserved landslide scarp; 10) morphological depression caused by landslide

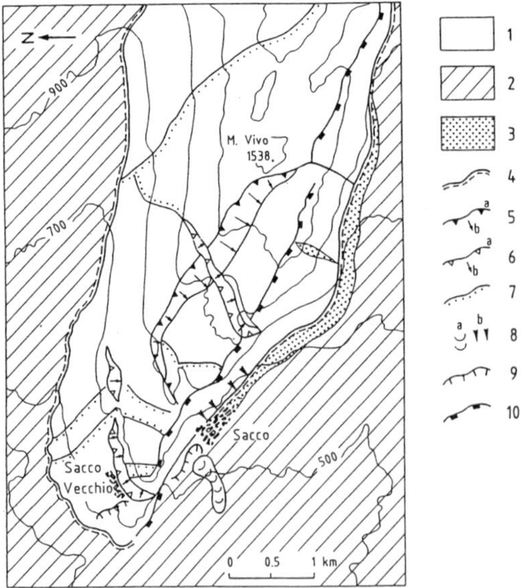

Fig. 4 Territory of Sacco Village. 1) Mesozoic limestones; 2) varicoloured clays; 3) landslide and/or talus debris; 4) fault; 5) edge (a) and failure surface (b) of main D.G.S.D.; 6) edge and failure surface of gravitational deep failures after the sackung; 7) edge of minor deep seated failure; 8) earthflow (a) and fall (b); 9) landslide scarp; 10) cliff edge

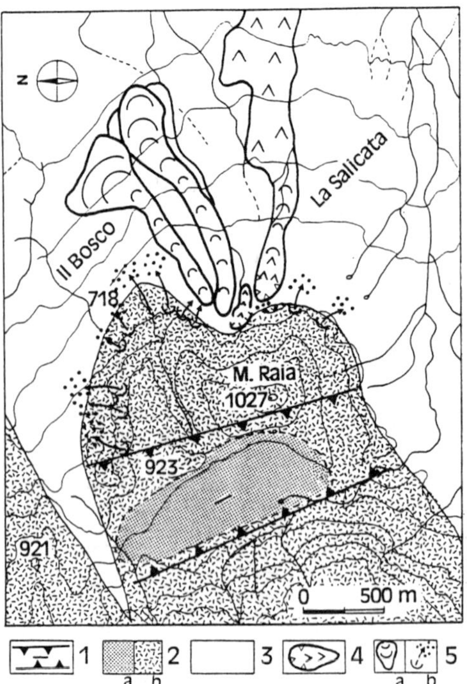

Fig. 5 Territory of Raia mount. 1) Edge and depression of trench; 2) fluvio-lacustrine deposits (a), carbonate rocks (b); 3) varicoloured clays; 4) compound slides; 5) earthflow (a) and rock fall (b); the phenomena induced by the 23rd November 1980 earthquake are indicated in bold face type

carbonate mass based on Paleozoic phyllite rocks, sometimes on disarticulated calcareous blocks sometimes on talus or/and landslide debris.

The carbonate mass comprises : sackungs, lateral spreading and block slides, which originated trenches (even 200 - 300 m wide), large deep open fractures, rotated and disarticulated calcareous blocks. The whole area ravaged by these events is characterized by past and still active block slides, falls and topples; these phenomena are shared by the calcareous wall the village is placed against. Also the phyllites along the versant of the town-centre is characterized by rotational slides and block slides.

2.3 Territory of Sacco village (Campania Region).

The urban area rises at the foot of Mount Vivo - Mount Motola versant made up of Mesozoic carbonate rocks (Fig. 4a). For quite a time the versant is subject to falls and topples of large calcareous blocks threatening the village. Landslides originate at the edge of a wide area (exceeding 13 sq.Km) characterized by complex and inactive D.S.G.D. being mainly

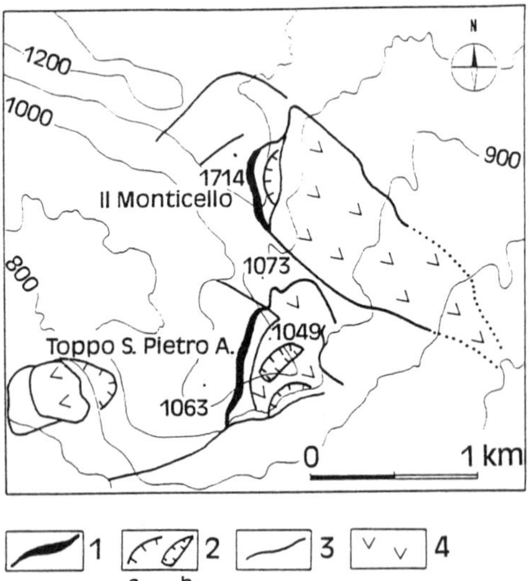

Fig. 6 Territory of Toppo San Pietro Hill. 1) Main trench caused of D.G.S.D.; 2) landslide scarp (a), minor trench (b); 3) main fracture; 4) landslide area

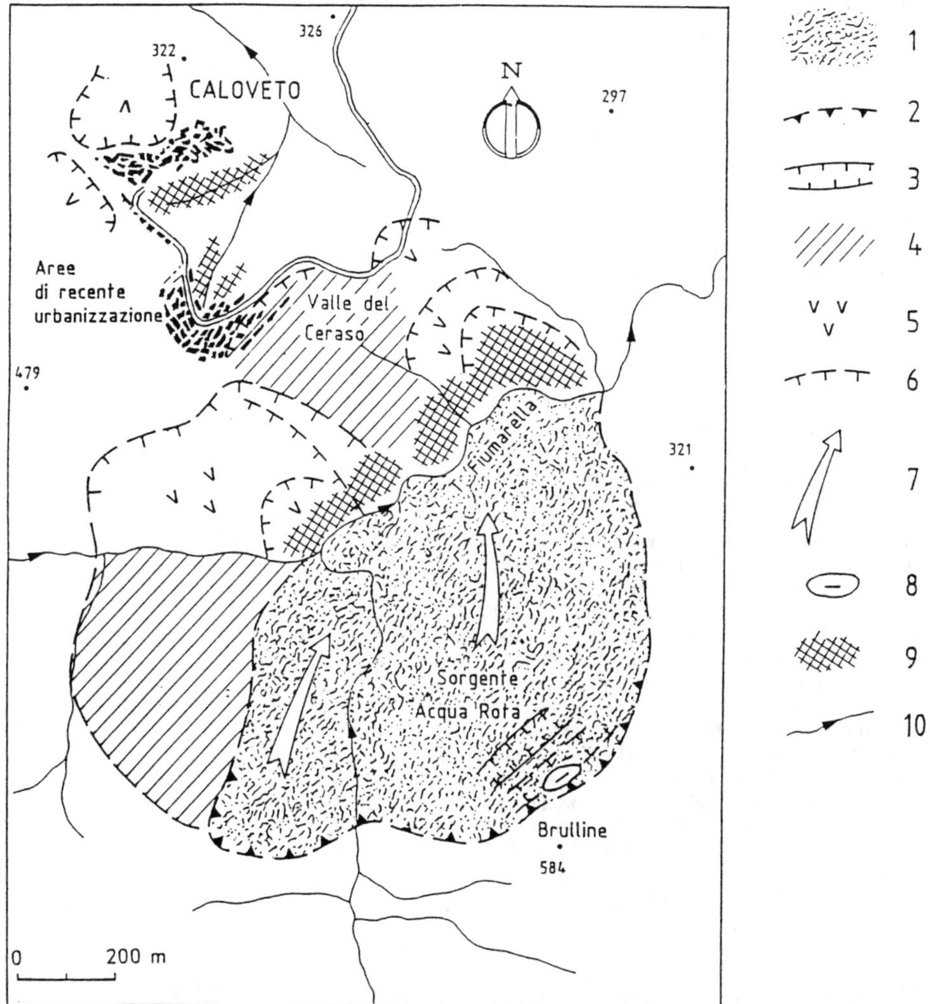

Fig. 7 Territory of Caloveto Village. 1) Area interested by lateral spread, sackung and block slides; 2) main edge of D.G.S.D.; 3) trench caused by D.G.S.D. 4) Area disturbed by block slides and rotational slides; 5) superficial rotational slides; 6) scarp of landslide; 7) main direction of deep mass movement; 8) swallow - hole; 9) area disturbed by erosion; 10) surface drainage

ascribable to the sackung of a large calcareous wedge (covering an area of 2.5 sq.km) (Fig.4) and to lateral spreads. Within said area especially along the versant the Sacco village is placed against, there are: lesser sackungs, deep subvertical open fractures, disarticulated and rotated calcareous blocks.

2.4 Territory of Raia Mount (Campania Region)

Raia Mount is made up of Mesozoic calcareous rocks surrounded with Varicoloured Clays Formation (Cretaceous - Miocene) (Fig.5). The whole carbonate mass shows that it underwent a remarkable lateral spreading phenomenon originating a large and deep trench partly filled with lacustrine deposits. This D.G.S.D., by conferring the calcareous mass a massive fracturing, is responsible for the considerable and repeated rockfall phenomena along the eastern wall of the carbonate mass. Moreover, earthflow phenomena which are in progress, originate from the bottom of the mountain where Varicoloured Clays formation outcrops.

2.5 Territory of Toppo San Pietro Aquilone hill (Basilicata Region)

The territory of Toppo San Pietro Aquilone hill, characterized by Mesozoic limestone laid on the marly - clayey formation of the Galestrino Flysch (Cretaceous), is widely affected by D.G.S.D. (Fig. 6). These last characterize an area of about 4.5 sq.Km and may be ascribed to sackung and lateral spread phenomena. The whole area includes : trenches (often filled with debris), deep open fractures, double ridges.
Landslides take place by rotational slides and falls and they are subject to recurrent mobilization during seismic events.

2.6 Territory of Caloveto village (Calabria Region)

The lack of stability characterizing the left versant of the Fiumarella torrent, housing Caloveto village, is mostly due to the erosive action of the torrent itself (Fig.7). In its turn, the erosive action is due to the migration to the left bank, that the active riverbed undergoes every time it is invaded by landslide bodies coming from the opposite versant. In this last versant D.G.S.D., being still active, affects the outcropping Mesozoic limestones and Paleozoic phyllites. It is an active lateral spread forming trenches and deformations, and it contributes to develop considerable rotational and translational slides whose activity is still under way.

REFERENCES

Cotecchia, V., Grassi D. & Sdao F. 1989. Evoluzione geomorfologica e movimenti di massa del versante vallivo impegnato dall'abitato di Senise (Basilicata). Geologia Applicata e Idrogeologia 25:1 - 25.

Cotecchia, V., Grassi D., Merenda L. & Sdao F. 1989. Movimenti di massa connessi a deformazioni gravitative profonde di versante accertate nell'ambito di aree urbane dell'Appennino meridionale. Proceedings del Convegno Centri Abitati Instabili, Ancona (Italy), S.C.A.I. National Research Council of Italy : 155 - 164.

Grassi D., Merenda L. & Sdao F. 1988. Esempi di fenomeni gravitativi di diverso tipo nell'Appennino dell'Italia meridionale. Proceedings of 74° Congr. Soc Geol. It. Sorrento, vol A: 313 - 319.

Grassi D. & Sdao F. 1992. Landslides caused by deep-seated gravitational slope deformations affecting also urban areas of Central-Southern Italy. Proceedings 29th Intern. Geol. Congr. Kyoto (Japan).

Environmental Management, Geo-Water & Engineering Aspects, Chowdhury & Sivakumar (eds)
© 1993 Balkema, Rotterdam. ISBN 90 5410 099 0

Some landslides along the Nabire-Epomani highway, Irian Jaya, Indonesia

Djoko Santoso & Sigit Sukmono
Department of Geology, Bandung Institute of Technology, Indonesia

Muhamad Ruslan & Yugo Ruslan
Institute of Geotechnology, Indonesia Institute of Science, Indonesia

ABSTRACT: Irian Jaya is one of the most sparsely populated province in Indonesia. In recent years, the highway path between Nabire - Epomani is under construction. The geologic map along the path had been prepared by air photo and some field-checks. During the construction some landslides occurred in the residual soil or weak rocks. They were thick clayey-sandy soils (5-15 meters) of the dioritic rock origin. The type of mass movement was debri slide. The sizes of landslides were approximately 25 - 30 meters long and 15 - 30 m high. These landslides were controlled by the high precipitation, high angle of cut slope, shallow ground water as indicated by numerous springs and very thick soil or debris. Some earthflows were also recognised. The sizes of earthflows were similar to the first type of landslide, with lower slope angle.

1 INTRODUCTION

Irian Jaya is the most eastern Indonesian province. Population in the area is very sparse. At present some development projects have been initiated. One of the project is the Nabire - Epomani National highway (Figure 1). In the future it will continue to Enarotali - Ilaga toward South. At the end of 1991 the road construction had reached Bedodipa at the 103 + 00 Km (KM 103 + 00) distance from Nabire (Figure 1). Along the path some mass movements were recognised. They were in KM 57 + 00 - KM 59 + 00, KM 75 + 400 - KM 75 + 600 and KM 97 + 00 - KM 97 + 500. The investigation were carried out by using air photo followed by field observation. Air photo had the scale of 1 : 130,000, on the other hand the available base map was the topographic map of 1 : 100,000 scale. Some rocks and soil samples were taken and tested in the laboratory for identification and understanding of the material behaviour. This paper discusses mainly the understanding of landslide occurrences in the area.

2 GEOLOGY

The area along the road path are composed of numerous rock formations. Therefore it is reflected by some geomorphic expressions.

2.1 Geomorphology

Geomorphic analysis was done by using air photo supported by some field observations Accordingly, the area between Nabire-Bedodipa can be divided into four units, as follows (Figure 2):

1. Flat low lying area. This area situated in the beach and surrounding area. Town of Nabire is located in this unit. The region has an altitude of 50 - 100 m above mean sea level. The slope angle is in the range of 0°- 2°.

2. Flat high lying area. This region is located between 290-450 m above mean sea level. It is situated in the intra mountain region of Tobo to Bumi settlement villages. The region is underlain by unconsolidated materials or river alluvium and colluvium.

3. Undulating hilly area, situated in the south of Nabire to north of Topo village, and the road path from the east of Bumi settlement to Bedodipa village toward south The landscape is wavy terrain with high angle slope (8°- 16°). It is underlaid by sedimentary, igneous, volcanic and metamorphic rocks.

4. Mountainous area, situated in the north of road path from east of Bumi settlement to Bedodipa village. The region has an altitude of 700 - 3200 m above mean sea level. It is a part of Middle Mountain range. This unit has high slope angle of cliff and valley (20 %).

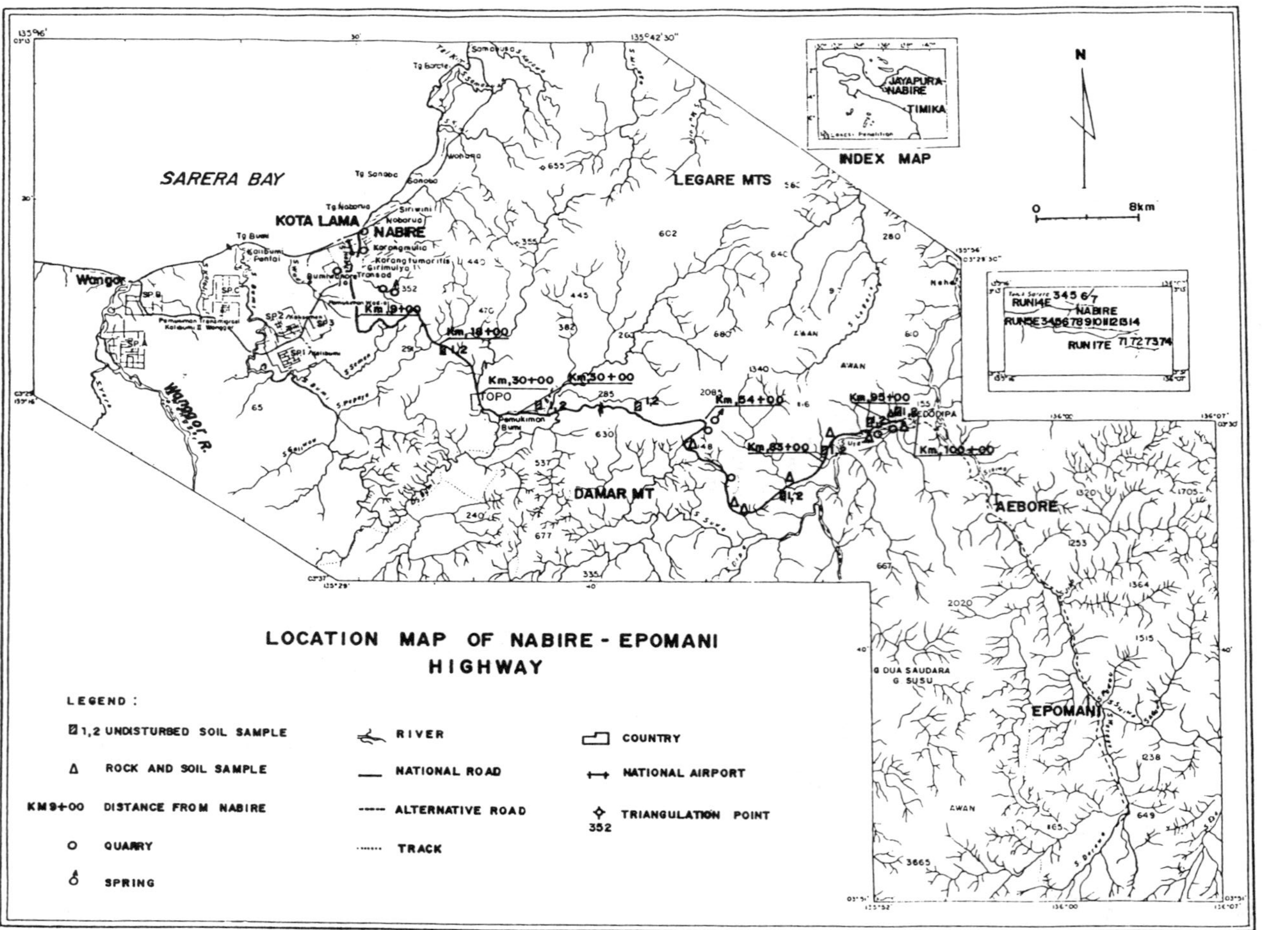

Figure 1. Location Map of Nabire - Epomani Highway

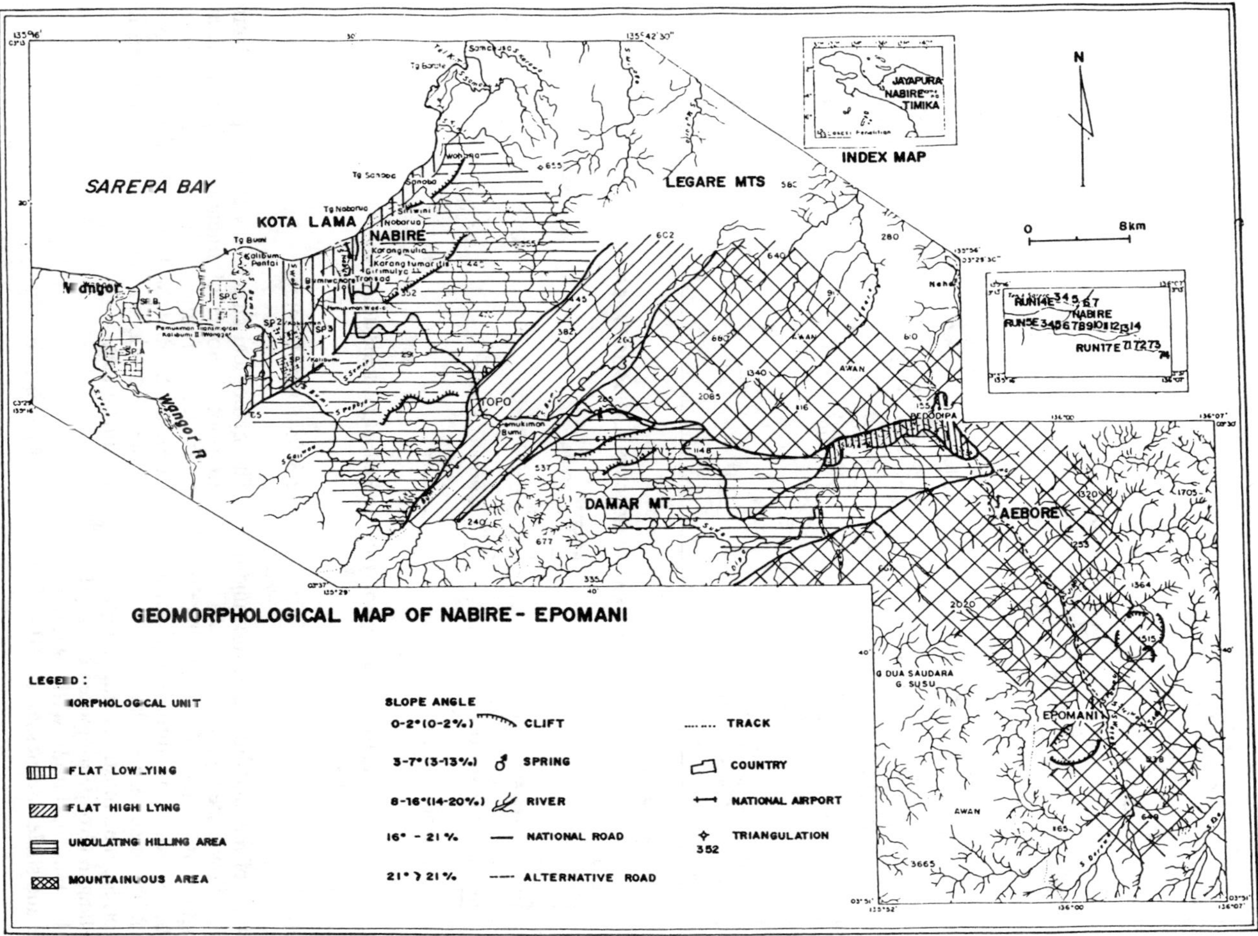

Figure 2. Geomorphological Map of Nabire - Epomani

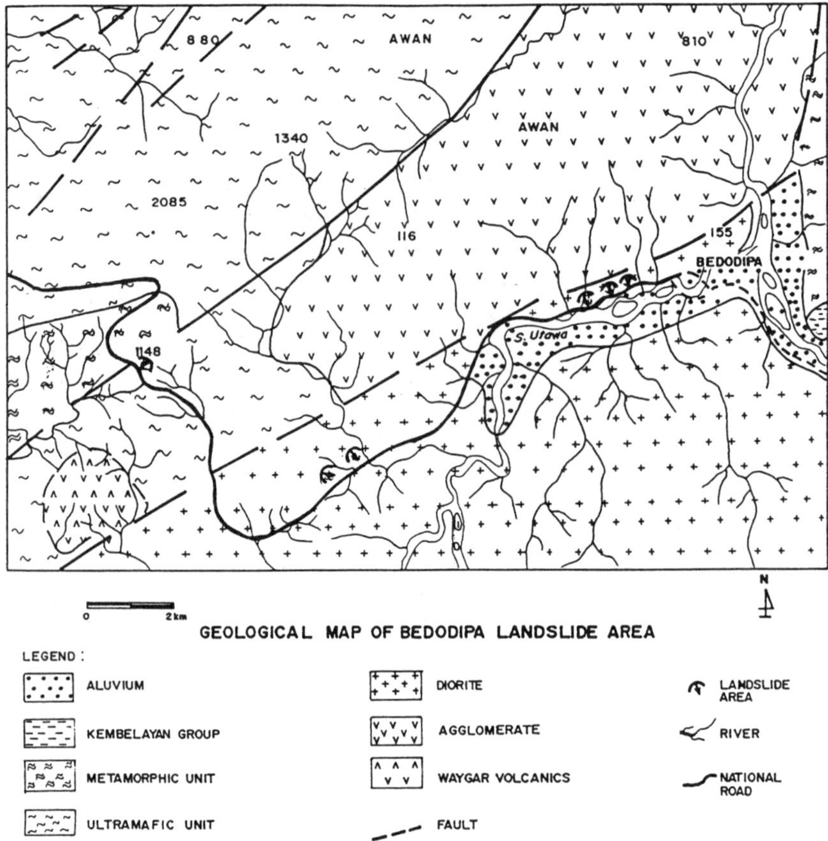

Figure 3. Geological Map of Bedodipa Landslide Area

2.2 Lithology

The lithology of the landslides and surrounding area are as follows (see Figure 3) :
1. Ultramafic unit, consist of serpentinite, found in KM 65 + 00. It is intensively jointed, therefore it becomes very weak.
2. Metamorphic unit, mostly is amphibolite.
3. Kembelangan Group, composed of claystone and shale, intercalated with sandstone. Claystone and shale thicknesses are 10 - 65 cm, weak.
4. Alluvium, this unit consists of unconsolidated material, i.e. clay, sand and gravel. It is found around Utama river at Bedodipa and surrounding area.
The volcanic rocks are :
1. Diorite as igneous rock. In the surface mostly as highly weathered rock to soil. The thickness of soil is more than 10 meters.
2. Agglomerate, it was found between KM 84 + 00 to KM 85 + 00. This unit consists of agglomerate and dacitic tuff. Mostly are strong.

2.3 Structure Geology

The geological structure was interpreted from air photos. Some of lineaments indicated as fault were found with the direction of Northeast-Southwest. It was identified that one of the landslide was approximately located in that lineament.

3 LANDSLIDE

Some mass movements were found in three locations, they were around KM 57 + 00, KM 75 + 400 - KM 78 + 800, KM 90 + 00 - KM 97 + 500.

3.1 Landslide in KM 57 + 00 - KM 59 + 00

Landslide in this area mostly are "rock fall" (according to Varnes (1978) classification) type. As indicated before, the ultramafic rocks

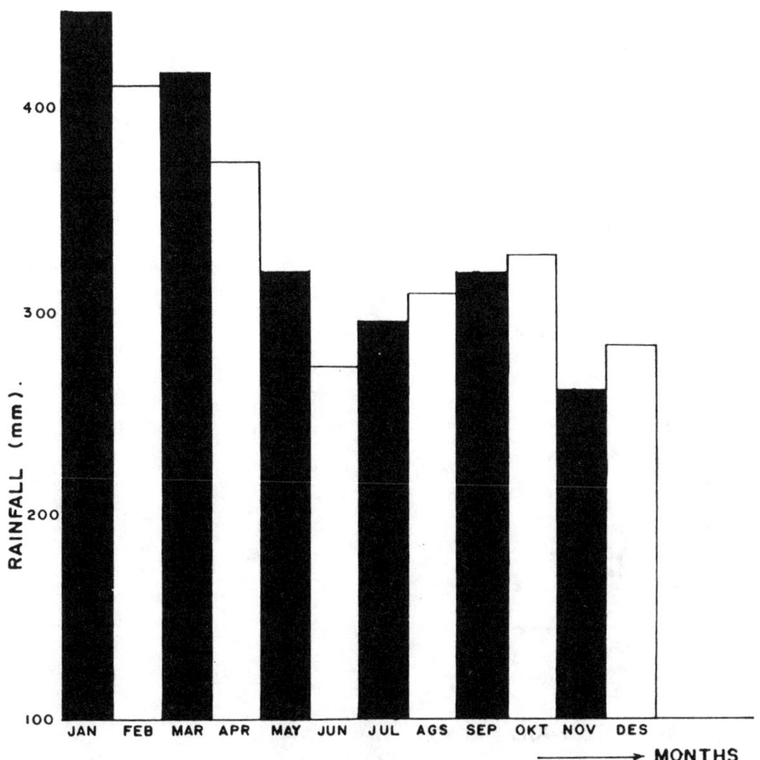

Figure 4. High Rainfall in Nabire and Surrounding Area

unit is highly jointed or fractured. This landslide could be interpreted as occurred due to highly jointed rocks or jointed, high angle slope of cut off (45%), and high rainfall (Figure 4). According to geological map one of the fault was run through this location. Due to intensive weathering, the "debris slide" were also recognised. The soil mostly has red colour, porous, thickness 2 - 5 m, and according to laboratory analysis has plasticity index 9%, water content 43%, and no cohesion.

3.2 Landslide in KM 75 + 400 - KM 75 + 60C

Landslide in this area mostly occurred in the residual soil of diorite rocks. The soil identification shows clayey-sandy soil, yellow to pink, porous, 10 - 25 m thick. According to laboratory data they show plasticity index of 3 - 20% and water content 35 - 50%, grain size analysis gave 60% fine-medium sand. This type of soil will become very weak due to the high precipitation in the area, therefore some high angle cut slopes 70 - 80%, failed and some landslides occurred. Most of landslides along this traverse occurred in the diorite residual soil. The schematic diagram showing the landslide in the area is shown in Figure 5. These landslides are "debris slide" type. Another type of landslide is "debris flow" which occurred also in the diorite residual soil with medium angle cut slope (40 - 50%)

4 CONCLUSION

From the above discussion, it can be concluded as follows :

1. The rock fall type of landslides only occurred in the two lithology's, i.e. ultramafic and diorite rocks.
2. Debris slide type landslides occurred in the residual soil. The main parameters which

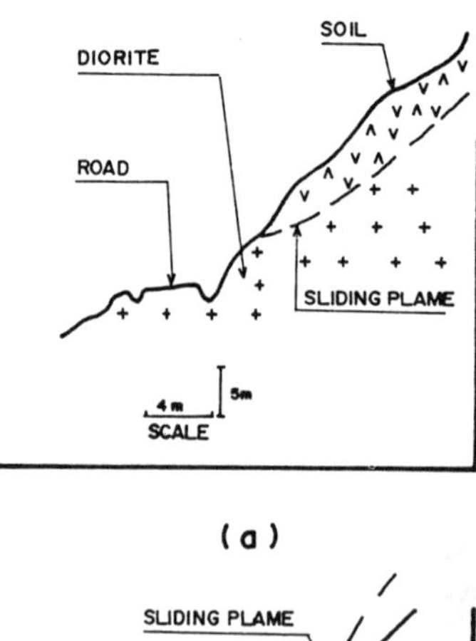

(a)

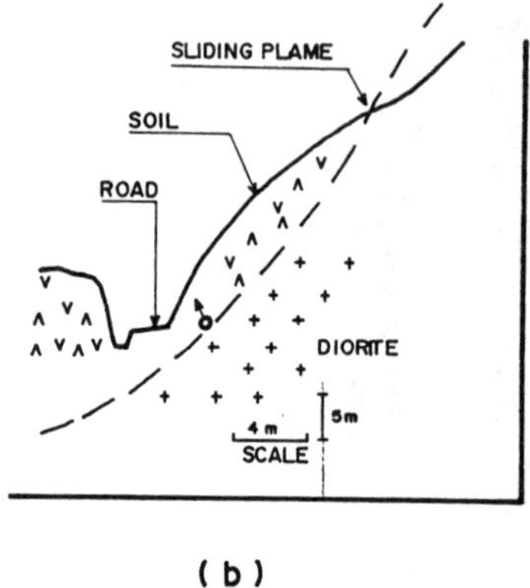

(b)

Figure 5. Landslide in KM 75+400 and KM 75+600

controlled mass movement were thick soil, porous, low plasticity index, high rainfall and high slope angle.

3. Debris flow type occurred with similar parameters but lower slope angle.

REFERENCES

Varnes, D., 1978. Slope Movement Types a Process in Landslide Analysis and Control Ed. by Schuster, R. and Krizek, R., Nat. Acad. of Sci, Spec. Rep. 176, pp 11 - 33.

Environmental Management, Geo-Water & Engineering Aspects, Chowdhury & Sivakumar (eds)
© 1993 Balkema, Rotterdam. ISBN 90 5410 099 0

Causes and mechanism of landslides in Sri Lanka

K.S.Senanayake
Kyowa Engineering Consultants Co. Ltd, Tokyo, Japan (Formerly: National Building Research Organisation, Colombo, Sri Lanka)

ABSTRACT: Occurrence of landslides in Sri Lanka has increased recently both in number and frequency. Over 200 past landslides in the island were studied to examine causes and mechanism of instability. Failure of weak rocks due to multiple fracture systems and feldspar rich compositions, kaolin lubricators, weathered quartzites and crystalline limestone are found to be the major engineering geological factors that affect instability of slopes. Human activities, such as, excavations and earth cuttings for buildings and roads, impounding of water on hill slopes in reservoirs and ponds, poorly maintained irrigation channels and contour drains excavated on the slopes, double cropping and resulting perennial irrigation of slopes, denudation of forest cover and intensive cultivation of tobacco and vegetables, poor land management, gem mining and improper quarrying are found to be the major man made causative and contributory factors. Most important triggering factor of landslides in Sri Lanka is the rainfall received during the monsoons.

1. INTRODUCTION

The hill country in the central and south western parts of Sri Lanka is prone to Landslides. The tropical forests that once covered the hill slopes have been gradually replaced with commercial plantations introduced by the colonial rulers. Estate centered human settlements began to develop and population began to congregate on the hill slopes. Haphazard development activities that followed resulted in environmental degradation leading to soil erosion and landslides. Landslides have become a frequent phenomena in the island particularly during the past decade taking many lives and causing immense damage to property. As one step towards arresting the landslide problem in Sri Lanka, the government initiated a study on landslides. One of the objectives of this study undertaken by the National Building Research Organisation (NBRO) was to examine the causes and mechanism of landslides in Sri Lanka.

2. NATURAL CONDITIONS

2.1 *Physiography*

Physiographically, Sri Lanka broadly consists of a central mountainous mass stepping down towards the sea in three distinct peneplains or levels of erosion (Cooray 1967). In the highest peneplain which has a general elevation of 1500~1800m, a few peaks rise to 2000~2500m above sea level. The altitude of the middle peneplain varies from 400~750m while the lowest peneplain has general levels less than 30m although some hills rise to about 100m above sea level. Landslides occur mostly in the highest peneplain and somewhat less in the middle peneplain.

2.2 *Geology*

Two main types of rocks are found in the hill country; (a) metamorphosed sediments like garnet-sillimonite schists and gneisses, quartz and quartz schists, quartz-feldspar granulates and garnetiferous gneisses, crystalline limestones (marbles) and calc granulates and graphitiferous schists and (b) charnockites.

The tropical rainfall and temperature intensify the weathering of these rocks. A complex distribution of lineaments some of which may be currently active, and geological evidence of sheared zones have been observed (Vithange 1983,1986).

2.3 *Soils*

Residual soils, colluvial soils and mixed hill slopes are the three common types of earth mantles observed in the hill country.

Residual soils are formed as a result of in situ weathering of metamorphic rocks. The upper layers of residual soil have undergone weathering to a greater degree and contain soil minerals such as quartz, mica, iron oxide and clay minerals like kaolinite and illite. Foliation joints, joint patterns and fault patterns remain as relict joints in the partially weathered lower layer. Finer material formed by advanced weathering are found among these relict joints. Micaceous materials which do not weather readily as some other minerals are found within the soil mass or along the foliation planes. Such material found in relict joints and planes are likely to differ from the rest of the soil mass in texture, shear strength and permeability characteristics.

Colluvial terrain which occur at foot of rock scarps or steep mixed hill slopes generally contain mixtures of fallen or rolled down rock boulders and soil masses. Transported residual soil may occur as colluvium within which gravel, cobbles and boulders may present, but the soil mass having properties similar to the in situ residual soils.

In the mixed hill slopes, earthy mantles co-exist with outcrops of bedrock. Earthy slopes where stones and boulders are enclosed in the soil or laid on its surface may also fall into this type.

2.4 *Rainfall*

Sri Lanka receives rains mainly during the southwest and northeast monsoons and to less extent during the intermonsoon periods. Tropical cyclones that pass through the island bring unusually heavy rains. Many recent cyclones have cut across the flat lowlands in the northern part of the country, but some others passed through the hill country causing disastrous landslides.

Based on the average annual rainfall, Sri Lanka is divided into three zones; dry zone, wet zone and intermediate zone. Major part of the hill country fall within the wet zone.

Annual rainfall could be as low as 750mm in some parts of the dry zone but has exceeded 5000mm at some locations in the wet zone. The annual precipitation in the hill country is generally above 1700mm, but in the south western parts it is about 3750mm. The highest annual rainfall recorded so far is 5403 mm at Watawala. However it is not rarely that consecutive two-day rainfall has exceeded 500mm.

3. DISTRIBUTION OF LANDSLIDES

The Geological Survey Department of Sri Lanka had been involved with the investigation of landslides and unstable slopes often at the request of the Government Agents of the landslide affected districts. Records of past landslides had been maintained also by a few other agencies who had conducted ad hoc studies. However, systematic documentation of such landslides were not available when the NBRO undertook the present study. As the first step of the study, information on over 200 past landslides was collected from the limited records available and through field reconnaissance. Moreover 64 past landslides were selected for detailed study to examine the causes and mechanism of slope instability.

The distribution of reported or known landslide occurrences over the past five decades is shown in Fig.1. It is seen that past landslides are scattered over ten administrative districts However, considering the environment and the impact of landslides on the society at present, only seven districts were identified as the landslide prone areas. It could be observed that the areas of past landslides generally fall mostly within the wet zone and partly in the intermediate zone indicating the close relationship between landslides and rainfall. It is also noticed that landslides on the north eastern hill slopes occur mostly during the north east monsoon and similarly on the south western hill slopes during the south west monsoon.

4. CAUSES AND MECHANISM OF LANDSLIDES

4.1 *General*

Landslides studied in Sri Lanka, like elsewhere, were often influenced by a number of contributory or causative factors an it is difficult to isolate one as the main factor. Contribution of natural, but gradual and ever changing environmental conditions of the terrain and its geology towards instability of slopes is further increased by the more vigorous environmental changes caused by adverse human activities. As almost all the landslides in Sri Lanka have occurred during or soon after heavy rains, rainfall is considered the key triggering factor.

4.2. *Environmental Factors*

In about 40% of the landslides studied, traces of old landslides were observed. A slope which has reached a state of equilibrium after a landslide can be easily destabilized by subsequent environmental and soil moisture changes.

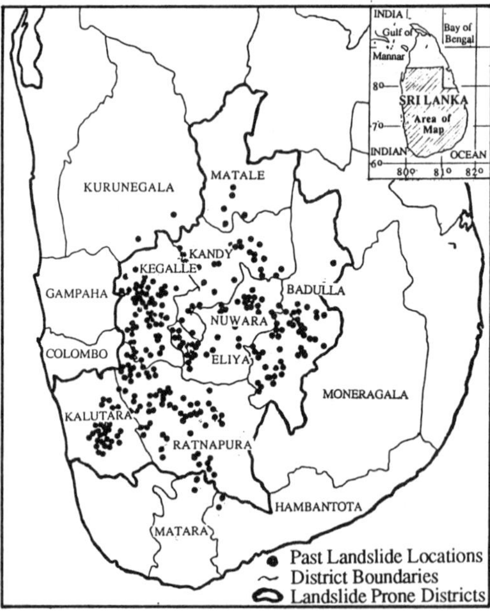

Fig. 1 Distribution of Landslides in Sri Lanka

Majority of the landslides were on mid slopes while the others were on bench in slopes or in natural terraces. Nearly two thirds of the slopes where failure occurred were between 16° to 30°. About 20% were in steeper slopes in the 31° to 45° range and the remaining was in gentle slopes less than 15°. Rock and earth falls were observed mostly in steep slopes whereas creep was common on gentle slopes which were rain fed or otherwise irrigated almost perennially for rice or vegetable cultivation.

Nearly one third of the landslides were on dip slopes, about 20% on scarp slopes and the rest in intermediate slopes. Sliding and subsidence were common where the plane of the rock bedding was nearly parallel to the natural slope. Rock falls and scarp slides were observed where the dip of rock bedding and the dip of surface slope are opposite.

Quartzite or quartzitic gneiss was present as the main rock type at 57% of the locations and as the secondary at 51% of the locations. Charnockites or charnockitic gneisses were present as the main rock type in 19% of the slides and biotite gneiss in the rest.

At least one predominant joint plane was evident at all the locations, though two to three joint planes were obvious in some. Joint planes were mostly vertical or near vertical. Spacing between two joints was less than 1m in 77% of the locations and even less than 30cm in 24% of them. The depth to bedrock was less than 5m in most cases and rarely exceeded 10m.

Profuse joints often in three directions were observed in quartzitic rocks. Joints which are closely spaced, about 2cm to 5cm, in this metamorphic rock allow easy passage of water and deposition of clays by the weathering of feldspar present in the rock thus making the slopes vulnerable to rock falls and landslides. On advanced weathering, such jointed rocks leave a coarse grained soil mass which tend to crumble easily. A few landslides were associated with steep road cuttings or building excavations in weathered quartzite.

Crystalline limestone or marble was found among other metamorphic rocks at several landslide locations. Solution caverns and cavities are formed in these soluble rocks that increase the propensity to retain water and provide subsurface passage downslope for emission as springs or marshes.

Seams or layers of clays and micaceous minerals were observed near the failure plane at most of the locations. With ingress of water, shear strength of clays tend to decrease while micaceous layers provide passages for water and become slippery.

4.3 *Human Activities*

The hill country of Sri Lanka was once covered with tropical jungle and forests with homesteads developed in the valleys and lower parts of the hill slopes. Large extents of this natural vegetation was replaced with commercial plantations like coffee, cocoa, tea and rubber. Natural forests today stand at less than 20 % of the total land area of the seven landslide prone districts.

Majority of the landslides have occurred in the plantation areas on slopes, with 35% in tea, 20% in rubber and 10% in coconut. A reasonable number of landslides were on slopes which are kept irrigated almost perennially, with 13% in rice terraces and 8% in vegetable cultivations.

Land management in the plantations has gradually deteriorated with little consideration given to soil erosion measures. Particularly after some of the large estates were redistributed as small holdings under the land reform act, land management has been badly neglected by many of the new owners. Overflowing of poorly maintained drainage and irrigation channels was a major cause in 10 landslides.

As the population increased gradually, homesteads and villages began to extend higher and higher up the hill slopes. Where shift cultivation is practiced by the peasants, entire vegetation is burned at the end of dry spell in preparing the land for the next season. The result is heavy soil erosion of the denudated slope when the rains come. Tobacco which became a lucrative crop at one time was intensively grown in the hill country and became so attractive that it was illegally grown even in the state forest reserves in the higher slopes leading to denudation of the forest. It was found that vegetation in 51% of the landslide locations was low bush, tobacco or crops of shift cultivation.

In rubber plantations, old trees are uprooted using heavy machines to remove entirely the stumps of roots which otherwise on decaying could be harmful to the young trees that will be newly planted. This procedure not only let the soil cover loosen, but also leave large pits which when kept open through the rainy period provide room for heavy infiltration into the ground.

Overflowing of waterways and bank erosion of streams are closely associated with the landslides. At 57% of the locations streams ran through or adjacent to the landslide. Water is collected on the hill slopes mainly for irrigation and domestic purposes often in impounding reservoirs, ponds and pits which are not properly designed. Overflowing or seepage of water into the ground or breaching of bunds was found to be one of the causes in 30% of the landslides studied.

Gem mining is practiced in some parts of the hill country. Diversion of drainage paths through deep pits and subsurface ducts and toe removal of slopes caused by mining activities have contributed to a few landslides.

With rising pressure for land throughout the country, expansion of human settlements on hill slopes cannot be avoided and is difficult to be controlled. Hardly any building guides are available to house builders and as a result earth cutting and filling is done haphazardly on the slopes. Combined with lack of proper drainage measures, such construction activities lead to instability of the slopes.

In the last decade, many large scale development activities have been undertaken in the hill country of Sri Lanka, for the construction of several large dams, roads and highways, housing schemes and other infrastructure facilities. Rock quarrying had been done by heavy blasting with explosives. Vibration of rock blasting is alleged to be a cause of at least three landslides studied. It is reported that minor tremors have been felt by the people living in the vicinity of some landslide affected areas.

The vibrations caused by a landslide can contribute to the instability of an adjacent slope. A series of landslides that occurred in the same vicinity in Kalutara district is said to have taken place almost simultaneously as reported by the inhabitants. Jayawardene (1986) does not rule out the role of earth tremors as a triggering factor even though Sri Lanka is supposedly situated in an earthquake free zone.

4.4 *Rainfall*

Landslides in Sri Lanka are closely associated with rainfall. It was found that 80% of the landslides have occurred when the antecedent three day cumulative rainfall has exceeded 200mm. However, in a few cases where creep was the mode of failure, the three day total was as low as 50mm, and the site has previously experienced much higher precipitations. In some others, failure has occurred only after the three day rainfall has reached 600mm or even more. From the observation of rainfall pattern and intensities, a thumb rule for early warning of landslides was derived:- if more than 200mm rainfall had occurred on the hill slopes in a period of three days, and if wet weather was continuing, then the possibility of landslides looms ahead.

5.0 MITIGATION OF LANDSLIDE HAZARDS

The recent increase in the occurrence of landslides in Sri Lanka can be attributed to the gradual deterioration of the environment caused by human activities which are associated partly with haphazard development and partly due to ignorance of their adverse effects. Living with landslides had been taken for granted by the developers and the inhabitants in the hill country and no serious effort was taken to mitigate this hazard. Financial and technical constraints did not permit the administrators to do more than attending to the post disaster matters.

As the outputs of the study of landslides in Sri Lanka, several activities were initiated to promote landslide disaster mitigation.

District Landslide Monitoring Units were set up in each of the landslide prone districts to attend to all landslide related matters at district level. In addition to the training of the members of these units, awareness programmes were organised to discuss problems and mitigation of landslides both at district and village levels. Vigilance committees of villagers for regular monitoring of the hills were set up in some districts. Awareness was created among the inhabitants of the landslide prone hill slopes on the symptoms of possible slope instability, activities that are harmful to the environment, proper land use and the measures to be taken for better land management on hill slopes.

In a developing country like Sri Lanka, socioeconomic conditions pose serious constraints to disaster mitigation. Relocating of people from hazardous areas is not easy due to scarcity of safer land and the reluctance of the relocated people to live in a new area because it affects living patterns and livelihood they are accustomed to. Identification of the hazardous areas, itself is a major task. To assist the landslides monitoring units, guidelines were prepared for the selection of geotechnically suitable lands for relocation purposes. The government has now launched a project for landslide hazard mapping in the two districts Badulla and Nuwara Eliya as proposed based on a pilot study.

Large civil construction works for landslide control are very expensive from Sri Lankan standards and these are adopted only at selected strategic locations. Control of adverse human activities, stabilization of slopes by vegetative and drainage measures, vigilance through continuous monitoring, early warning and taking timely mitigatory action such as evacuation of people from endangered areas etc., remain as some actions that can be readily taken to mitigate landslide hazards until more systematic approach is worked.

ACKNOWLEDGEMENT

Author is grateful to the Director General of NBRO for permission to present this paper and to the Engineering Adviser and the NBRO Research Team involved in this study. He is indebted to the Government of Sri Lanka and the United Nations Development Programme with whose joint sponsorship and assistance the Study of Landslides in Sri Lanka was conducted.

REFERENCES

Cooray, P.G. 1967. *An introduction to the geology of Ceylon.* National Museum of Ceylon .

Jayawardene, M.P.J. 1986. Landslides in Pasdun Koralaya, Kalutara District, Sri Lanka, May 22, 1984. *Proc. Asian regional symposium on geotechnical problems and practices in Foundation Engineering.* Colombo: NBRO:

NBRO 1990. *A study of landslides in Sri Lanka.* Colombo: NBRO

Vithanage,P.W. 1983 *A study of geomorphology and morphotectonics in Ceylon.* ECAFE/ UNESCO Doc. GSM 2/89.

Vithanage,P.W. 1986. *Lineaments and tectonics around Victoria and Kotmale Reservoirs and engineering geological implications* . Proc. Mahaweli Seminar, Sri Lanka.

Environmental Management, Geo-Water & Engineering Aspects, Chowdhury & Sivakumar (eds)
© 1993 Balkema, Rotterdam. ISBN 90 5410 099 0

Landslide management in Sri Lanka – Issues and strategies

K.S. Senanayake
Kyowa Engineering Consultants Co., Ltd, Tokyo, Japan (Formerly: National Building Research Organisation, Colombo, Sri Lanka)

ABSTRACT: Sri Lanka frequently experiences natural disasters such as cyclones, floods, landslides, coastal erosion and drought. With abundant monsoon rainfall received in the hill country and with poorly managed hill slopes hosting a high population, landslides have become a recurrent phenomena that has taken many lives and caused damage to property to the tune of several million dollars over the past decade. Confronted with natural disasters, attention of the government and the people is rapidly growing on management of environment and sustainable development.

A study undertaken by the National Building Research Organisation paved the way to develop several strategies for mitigation of landslide hazards in the island and to bring together the concerned agencies to make a concentrated effort towards this goal. District landslide monitoring units were sets up in landslide prone districts for efficient management of landslide related activities at district level down to grass root level. Through public seminars, training courses and workshops, and dialogue with the inhabitants a district landslide management plan was formulated. Steps were taken to create awareness among the communities in the landslide prone areas on; the identification of symptoms of a possible landslide so that timely mitigatory action could be taken, the activities that should be prohibited or avoided on hill slopes which lead to slope instability and the activities that should be effected or promoted for land improvement on hill slopes. Guidelines were prepared to select alternate lands for relocating and resettling the victims. A project was formulated for landslide hazard mapping in two districts with a total area of 4525 km^2 as a major outcome of the study.

1. BACKGROUND

Twenty five percent of the land area of Sri Lanka is mountainous as remarkably represented by the Central Highlands and the South-western Hill Range where a few peaks rise above 2000m to 2500m. Hill country receives high rainfall exceeding 5000mm/year at some locations mainly during the monsoons. Metamorphosed sedimentary rocks and charnockites are commonly found as weathered and weak rocks.

Agriculture had been the core of development activities with rice being the staple food of the people for centuries. In the past agriculture had flourished in the generally flat dry zone with the help of gigantic irrigation schemes. But pressure on land with gradual growth of population, and intermittent foreign invasions forced migration of the people in to the hilly terrains. Habitation began to concentrate on the gentle hill slopes and along river valleys. For sometime, self sustained homesteads with mixed vegetation continued to provide a good canopy over the lower slopes while the hardly accessible upper slopes remained as rich tropical montane forests.

European invasions since 16th century and the colonization of coastal areas and subsequently the entire island under their administration brought further pressure on land forcing the people to find refuge in the hill slopes higher and higher. Agricultural practices began to change with the forests being cleared of vegetation and exploited for Chena (shift) cultivation. Paddy fields began to appear even on higher slopes with water being retained in ponds and reservoirs built on unstable slopes for irrigation.

Degradation of hill slopes increased with the introduction of Land Ordinance of 1840 by the British colonial administration which paved the way to forcible acquisition of land which has been utilized by peasants, along with the waste lands, unoccupied or underutilized lands having being declared as Crown land. Large extents of virgin forest land were denuded for the cultivation of coffee, tea, rubber and the spices. Displaced peasants found new habitations within the limited marginal lands which gradually became densely populated leaving no control on proper land management.

Attention of the government had been drawn to the occurrence of earthslips in the hill country as far back as 1880's and the various committees who studied this subject had come up with worthy recommendations which unfortunately have never received the attention they deserved. Negative effects of the mass exploitation of the hilly terrain were soon realized and attempts had been made, though rather too late, to introduce improved land policies and release of land to the farmers. Village expansion under the Land Development Ordinance introduced in 1935 helped relaxing congestion in villages, but the deterioration of hill slopes continued.

When a large number of people were rendered homeless due to landslides in Kotmale valley and the adjacent catchments in 1947, the political interest generated was sufficient to introduce the Soil Conservation Act in 1951. Under this act, mitigatory and control measures could be taken against soil erosion and earthslips. However, it did not remain a useful tool to curb soil erosion and mass wasting as it received less attention with time, overpowered by other socio-economic factors.

Beginning in the 1960's, land policy took a different turn with the nationalization of plantations. Many of the well maintained tea and rubber plantations began to suffer from lack of proper maintenance and land management. Some of them when split up into small holdings or homestead allotments with the introduction of the subsequent Land Reform Act, suffered further as land management was almost totally neglected by the new occupants.

Development activities in the hill country gained momentum in the late 1970's with the construction of multi-purpose large dams and reservoirs, tunnels and canals, housing schemes and new roads in the hilly areas.While development activities increased on one side, occurrence of landslides also increased in number and frequency during the past two decades. Landslides that occurred in Nuwara Eliya and Badulla districts in January 1986 alone took 52 lives. Misery struck again in the districts of Kegalla, Ratnapura and Nuwara Eliya in May-June 1989 killing nearly 300 persons and causing tremendous damage to property. As a result of landslides, thousands of families were rendered homeless. It is often the poorest of the poor peasants who become the victims of landslide disasters and plunged into further misery pausing new socio-economic problems. Burdened with the increasing environmental and socio-economic problems, time has come once again for the government and the public to draw their attention on the landslide problem. People in Sri Lanka have begun to witness and realize the direct and indirect, favourable and adverse impact of haphazard development efforts. Fortunately enough, many quarters have now realized the need of sustainable development and have begun to rethink on matters concerned with land conservation and environmental protection from this point of view. Since mid 1980s, several agencies have taken interest to study the landslide problem.

2. LANDSLIDE STUDIES IN SRI LANKA

Hooker in 1873 and Triner in 1880's had warned of the possible adverse consequences of the clearing of forests in the hill country, when devastating impact of the colonial land policy on the stability of hill slopes as well as the general ecological balance began to be noticed. Reference is made on the subject of earthslips in a report by the Indian Forest Service, which was largely instrumental in the enactment of Forest Ordinance of 1885.

Baker (1855) commenting on the mechanism of earthslips that occurred in the hills above 1500 ft. elevation, where the forest had been cleared to plant coffee, explains the incapability of clayey soil overburden in infiltrating the rain water as a cause of instability.

During the colonial era, landslide problem had been treated under the broad perspective of Soil Erosion. The Commission on Soil Denudation in the Kelani Valley had recommended in 1904 that no land above 50% gradient should be cleared and this had helped to preserve parts of the hill forests in the Upper Kelani Valley. A Land Commission was appointed in 1927 and the land policy was revised. Soil Erosion Committee in 1931 had identified poor land practices as the cause of land degradation and suggested a five year education programme.

With the onset of unprecedented rainfall of about 450mm in two days on 14th and 15th, August 1947, a vast area in around Kotmale Valley was affected by landslides causing severe damages to several villages. Gorrie & Sirimanne (1954) identified removal of forest cover on the upper slopes of catchments and from talus debris at the foot of scarp faces, insufficient drainage of steep slopes under plantation, and ponding of water for paddy cultivation on terraces above unstable slopes as the main causes of human intervention which had increased the potentialities for slope failures inherent in the area. They recommended remedial measures for a) amelioration of the latter set of conditions and arresting of further deterioration of steep hill slopes by deforestation, chena firing etc., b) adoption of proper land use and soil conservation methods for stabilizing the already slipped area and the potentially unstable slopes.

Kotmale slides of 1947 being instrumental, further studies on landslides had been conducted for several years by the committees appointed for this purpose. During this period, the Soil Conservation act of 1951 was passed in the Parliament while the political interest was high. Kandyan Peasants Commission of 1951 had stressed on the need for scientific investigations on landslide prone areas, and effective control of deforestation, reforesting stream banks and afforesting patana lands.

Analysing the causes and the types of earthslip movements Gorrie (1951) described the cures classifying them in to three categories, viz; preventive, palliative and remedial measures.

With landslides becoming more frequent and with population congregating on the hill slopes, their disastrous consequences also began to grow. The government responded to them by entrusting the Mineralogical Survey Department and its successor, the Geological Survey Department to study the problem, mainly from the geotechnical point of view. Many studies had been undertaken by them, though mostly on an ad-hoc basis, made on specific requests from the Government Agents. The records of these studies with recommendations by the department provide a valuable data base of past landslides in the country.

Some typical landslides are discussed by Cooray (1967) giving geological background and the causes of instability. In a broadcast talk given to create awareness on landslides and floods, Sithamparapillai (1970) discusses the causes and simple preventive and mitigatory measures that could be easily taken by the residents. If soil conservation is not practised-if excessive land clearing is done-if drainage lines are choked up, he says that these are the breeding places for landslides.

The interest to study the geotechnical aspects of landslides in Sri Lanka is seen in the work of Dissanayake (1973), in which the geological and soil mechanics characteristics in thirteen landslides have been discussed. Further interest on landslide research was developed among the scholars in mid 1980's after several devastating landslides occurred in various parts of the hill country taking a toll of over 135 human lives between 1977 and 1985. Particularly, the Pathulupana landslide of 1982 which killed 13 persons and a series of landslides in Kalutara district which took 29 lives in separate incidents were sufficiently influential to call the attention of the authorities and the scholars as well, who had to gear themselves to find solutions.

Several groups and individuals have made independent studies of the landslides in Kalutara district . Jayawardena (1986) does not rule out earth tremors as a possible triggering factor of the simultaneous occurrence of a large number of landslides in the same region although rainfall too had been excessively high for two consecutive days just before the event. Mampitiyarachchi (1986) suggests high rainfall as the triggering factor while inherent poor geological conditions have been a contributive factor. For the first time in Sri Lanka, a simple early warning device was developed and installed at the landslide site in Agalawatta to monitor ground movements and alert the inhabitants in the event of a recurring instability.

The need for a systematic study of landslides was soon realized by the government in an environment where independent studies were being conducted by different agencies or individuals without any obvious coordinated effort for the mitigation of landslide hazards. With the initiative taken by the National Building Research Organisation (NBRO, 1990), under the Ministry of Local Government, Housing and Construction, a research project was launched in late 1985 with the assistance of the United Nations Development Programme (UNDP). The broad objectives of this project titled "Study of Landslides in Sri Lanka" were to examine the causes and mechanism of landslides in Sri Lanka, to set up guidelines for identifying landslide prone areas and to evolve criteria for selecting land for development in the hilly areas, and to develop suitable early warning systems in unstable slopes. Apart from the above noted outputs, this project which was implemented with close association and participation of the many relevant agencies has paved the way to formulate several constructive programmes aimed at the mitigation of landslides.

Heavy rains, exceeding 1000mm in five days, poured over the north eastern parts of the central hill range in the first week of January 1986 caused disastrous landslides. Reportedly, in the two districts of Badulla and Nuwara Eliya alone, over 20,000 people had to be housed in refugee camps and the death toll exceeded fifty.

The Prime Minister who promptly visited the areas affected by the landslides and floods pointed out that the nature and extent of the damage caused is such that it will be necessary to bring about a coordinated approach for speedy reconstruction of the damaged infrastructure and housing and the rehabilitation of the affected people. An Inter-ministerial Coordinating Committee was set up and the committee was charged with the following responsibilities; to assess the extent of damage, to determine alternative locations for relocation of housing and the infrastructural facilities and services, to identify the sources of funding, to obtain approval for allocation of additional funds to conduct necessary geological and other survey to establish the causes of earthslips, and to recommend ways and means of minimizing damage and destruction to private and public property in the future.

Further, a team of scientists and engineers from several relevant agencies were rushed to the affected areas to conduct a survey with a view to identify the causes of the disasters and to recommend immediate and long-term remedial measures. In the report (CEA 1986) submitted by this team the possible causative factors are discussed along with the recommended short-term strategies for implementation before the next monsoon rains. Evacuation of people from the potential danger areas and their resettlement in areas known to be safe and stable, acquisition of high risk areas, monitoring and warning of landslides and training of local personnel for this purpose, remote sensing with low altitude aerial surveys etc., were among the short-term strategies suggested. Stabilization of landslide affected areas through vegetative, mechanical and drainage measures, preparation of landslide hazard maps, stressing on developers and planners to work closely with the Department of Geological Survey when planning development activities in highland areas, monitoring of micro-seismic activities and rainfall intensity, promoting post-graduate studies on landslides, setting up a permanent multi-disciplinary investigation committee to arrest the land degradation phenomena and appointing a coordinating agency to conduct and monitor the implementation of the short-term and medium-term strategies, were recommended as the mid-term strategies.

The report suggested unfavorable topography with steep hill slopes, weak rocks underlain by thin soil cover, poor and unscientific land management subsequent to deforestation, poor drainage and farming activities including tobacco and vegetable cultivation, earth cutting for roads etc. as some of the causative and contributory factors besides identifying excessive rainfall as the main triggering factor.

Meanwhile, a Joint Landslide Research Group was set up at the Institute of Fundamental Studies (IFS), in 1985. This Study group consisting of representatives from most of the relevant agencies presented its recommendations for a plan of action to the government in June 1986. Public seminars were organized also by a few other groups with a view to create public awareness on landslides where the geological, geotechnical, socioeconomic and land policy aspects of landslides, causes and control & mitigatory measures were discussed.

Based on field reconnaissance of many reported or known past landslides, NBRO compiled a report documenting past landslides which were considered high risk areas and therefore need urgent attention. Highlighting the reported damage and future risks, NBRO suggested the remedial measures to be taken and the relevant agencies who could initiate such action. Evacuation of occupants from houses in imminent danger, restabilizing natural drainage regime on the affected slopes, terracing the slopes to safer gradients where possible, prevention of soil erosion through planting of fast growing trees and special varieties of grass, avoiding cultivation of rice, chena practices etc., on high and steep slopes and resettling the cultivators on low risk areas are some of the suggestions made case by case. The necessity of detailed investigations by specialist agencies in certain locations to determine the long-term stabilization measures has also been pointed out in this report.

NBRO's study of the past landslides have shown that much of the damage that occurred in the past could have been avoided had the high risk areas under habitation been identified in time and remedial measures and, in extreme cases, evacuation initiated before the occurrence of these landslides. Lack of information about the stability of the different parts of the hill slopes has been the main obstacle in taking timely preventive action in this matter. NBRO also recommended the mapping of landslide hazards to be undertaken initially in the two most affected districts Badulla and Nuwara Eliya on an urgent basis in order that future landslide dangers can be minimized through proper land use planning.

As it was of utmost importance that the district administration and other concerned agencies initiate the suggested remedial measures with the least possible delay to limit future damages to a minimum, NBRO noted that the concerned Government Agents must be given the necessary technical support and additional funds for execution of the remedial measures. Further, formation of a monitoring committee in each district, with representatives from all concerned agencies, to watch over the progress of the remedial measures was also suggested. These recommendations subsequently formed the basis of a Cabinet Paper.

3. SOME STRATEGIES AND ACTION FOR LANDSLIDE MITIGATION

3.1 *Inter-ministerial Committees*

On a Cabinet decision an Inter-ministerial Steering Committee was set up with Secretaries of Ministries handling development activities. Later the subject of landslides was transferred to the Inter-ministerial Committee on Land Use Policy Planning. Meetings of this committee held once in 3 months helped to keep the authorities and personnel concerned constantly interested and involved in landslide disaster management related matters.

3.2 *District Landslide Monitoring Units*

District Landslide Monitoring Units (DLMU) were set up in seven landslide prone districts and the members of these committees, who are the senior district level officials, were imparted training in landslide management and control and on selection of lands for relocation and resettlement of persons affected by landslides. Key functions of the DLMUs were identified as follows; a) carrying out technical investigations in landslide prone areas b) continuous monitoring of landslide prone areas within the district, c) attending to immediate physical remedial measures in affected areas and d) advising the Government Agent on evacuation of affected people. Availability of DLMU members on the spot supported by monitoring units at divisional and village levels will greatly helps in providing early warning to the people likely to be affected which can make the all important difference in the event of a landslide.

3.3 *District Landslide Management Plan*

Based on the proceedings of seminars and workshops held with the members of DLMUs, NBRO formulated a District Landslide Management Plan. The head of district administration assisted by the DLMU is responsible for the implementation of the plan. The responsibilities of the district level, divisional level and village level monitoring units at the different stages of; awareness, preparedness, warning, threat, emergency, restoration & rehabilitation and reconstruction are stipulated in the plan for its smooth and effective implementation.

3.4 *Guidelines for selection of lands suitable for development and human settlements.*

Realizing the need for rational use of land, policy measures are being framed to achieve this objective. Land use maps to the scale of 1:10,000 are prepared based on topography, slope, soil texture and depths and drainage. Land units are demarcated based on above characteristics and the appropriate land use for each unit is recommended taking into consideration of socioeconomic and political environment. The consequence may be that people living in a landslide prone area will be completely evacuated and relocated in safe areas elsewhere. Again, human activities, including agriculture, construction and development activities may be controlled or prohibited in a certain area and the land may be afforested. A major problem faced by the district administration each time a landslide disaster occurs is the identification of lands for relocating landslide victims. It is a prerequisite that the lands identified for resettling of evacuees should be geotechnically safe and stable. lands but also that they should be educated to manage the new lands properly. Therefore NBRO prepared guidelines for the selection of land suitable from geotechnical point of view to assist the DLMUs in identifying and selecting lands in the hilly areas for relocation resettlement purposes.

3.5 *Guidelines for proper land management on hill slopes*

Even if the land selected for resettlement purposes are found to be geotechnically satisfactory at the time of selection, future activities of the settlers may affect the stability of the slopes. For the resettlement efforts to be effective, the occupants of the land should be appropriately educated and guided to manage the new lands properly. For this purpose, NBRO prepared guidelines describing the Do's and Don'ts that should be observed in the new settlements.

3.6 *Guidelines for detection of potential landslides and slope instability*

Moreover, for the benefit of the inhabitants of hazardous areas, guidelines were prepared for the detection of potential landslides and slope instability. Warnings were issued to the residents living on hill slopes facing the hazards of landslides advising them to be observant and vigilant on the signs of possible instability which are given in check list form. They are also advised to observe precautionary measures at all times by refraining from adverse activities that promote destabilization of the slopes and by maintaining good land management practices. Vacating the residence as soon as possible, at least temporarily, on observing serious symptoms is encouraged.

3.7 *Early Warning of Landslides*

Timely warning of landslides is very helpful in saving lives from a disaster. However, to issue such timely warning, the authorities themselves should be geared with correct information so as not to create panic by false warnings. Warnings may be issued by monitoring ground movements and effecting an alarm when the displacement exceed a threshold value. However, establishing criteria for threshold values for extent and rate of ground movement itself is difficult as they depend on numerous factors such as geologic, geotechnical and hydrological characteristics, mode of failure etc. NBRO developed an extensometer type mechanical early warning system for monitoring ground movements at Beragala landslide. Durability and maintenance of the equipment, vandalism and human intervention that affect reliability of monitoring and results etc. are some serious problems confronted be establishing threshold values. Much research may be needed before such a system becomes practical within the socio economic framework and technology in Sri Lanka.

Compared to such mechanical early warning systems, timely transfer of information on contributory and triggering factors of instability becomes the basis of the information type early warning systems considered by NBRO. Human brain and other informatory organs are its key components and proper and effective use of human resources in the communication system becomes a vital factor in successful operation of this system. Knowledge, attitude, responsibilities, skills etc. of the personnel involved need to be upgraded through systematic training in order to avoid catastrophic failure of the early warning system. In this system it is always not necessary to wait and issue last moment warnings based on triggering factors immediately prior to a possible disaster, but measures could be taken to issue very early warnings based on contributory factors of instability, particularly the man made ones which could be controlled well ahead to eliminate or minimize the risks.

Rainfall is identified as the main triggering factor of landslides in Sri Lanka. Therefore, intensity and duration of rainfall may be considered as a criteria for early warning. Based on the information collected from past landslides studied, a thumb rule was set up as follows; if cumulative rainfall for consecutive three days exceeds 200mm and if the rains continue, landslides are likely to occur. However, there are many pitfalls in predicting landslides based on rainfall alone (Bhandari et al, 1992). Rainfall pattern examined for past landslides indicated that some landslides have occurred long before the cumulative three day rainfall reached 200mm whereas elsewhere failure had taken place after the rainfall has exceeded 500 mm.

However, as far as rainfall is considered a major triggering factor of landslides in Sri Lanka, vigilance of the pattern and intensity of rainfall will definitely be helpful in mitigating much of the disastrous effects of landslides. This concept is nothing new. The residents of the hilly areas have been often watchful about the rains in their own way and make crude judgments for their own protection. Nevertheless, a systematic and scientific approach is necessary in this direction.

3.8 *Landslide hazard mapping*

An important outcome of the study is the project proposal for hazard mapping of landslides in two districts with a total area of 4525 km^2. Under this project which is now being implemented with technical assistance from the United Nations Development Programme, a complete set of graded hazard maps to the scale 1:10,000, particularly for the urban, homesteads and marginal lands will be produced to serve as a useful planning tool and to assist in the formulation of a landslide hazard mitigation strategy. As some socioeconomic conditions in the landslide prone areas pose a major constraint in the hazard mitigation process, a socioeconomic study is being carried out to identify the anticipated problems and possibilities associated with relocation of affected communities to safe places. Another aspect of the hazard mapping project is the training of resident communities, non-governmental organizations, public and private sector development agencies in landslide prone areas, and creating awareness about the adverse effects of improper landuse on environment.

3.9 *Disaster preparedness and management*

Response to natural disasters had been very poor and often limited to post disaster activities covering mostly the emergency stage and rarely extending to restoration & rehabilitation and reconstruction stages. Senanayake (1986,1991) points out that a national scale disaster management plan is essential to mitigate the disasters which are experienced from time to time in Sri Lanka, the essential elements of such a plan would be to identify the research needs, infrastructural requirements and the training needs. However, the major obstruction to the implementation of measures related to disaster preparedness and management appears to be the lack of organization, infrastructure, finance and facilities and above all a coordinated effort. Awareness of the problems at all levels, is essential for effective implementation of such measures. Therefore a powerful agency at national level would be required to plan and coordinate implementation. But rather than setting up a new institution it would be more logical to coordinate among the existing organizations at the highest level and to make a concerted effort for their maximum interaction, essentially because the country has limited resources in qualified and trained personnel, besides the large sums of funds that may be required for institution building. As research and development are important part of disaster management and can be continued by the agencies where relevant expertise and know-how, though scarce, are already available. The structure for a national set up for disaster preparedness, mitigation and rehabilitation activities has also been proposed.

ACKNOWLEDGEMENT

Author is grateful to the Director General of NBRO for permission to present this paper and to the Engineering Adviser and the NBRO Research Team involved in the Study of Landslides in Sri Lanka. He is indebted to the Government of Sri Lanka and the United Nations Development Programme with whose sponsorship and assistance the Study was conducted.

REFERENCES

Bhandari,R.K, Senanayake, K.S. and Thayalan, N. 1992. Pitfalls in the prediction on landslides through rainfall data. *Landslides, Bell(ed).* Rotterdam: Balkema.

Cooray, P.G. 1967. *An introduction to the geology of Ceylon.* National Museum of Ceylon .

Dissanayake, J.B. 1973. Some aspects of landslides in Sri Lanka. *M.Sc. Thesis. Univ. Peradeniya,* Sri Lanka.

Gorrie, R.M.1951. Landslides and soil creep in the Ceylon Hills. *Transactions of the Engineering Association of Ceylon.*

Gorrie,R.M. and Sirimanne, C.H.L. 1954. Report on Kotmale landslips, *Ceylon Sessional Papers.*

Jayawardene, M.P.J. 1986. Landslides in Pasdun Koralaya, Kalutara District, Sri Lanka, May 22, 1984. *Proc. Asian regional symposium on geotechnical problems and practices in Foundation Engineering.* Colombo: NBRO

Mampitiyaarachchi, D.K. 1986 Earth movements in natural slopes. *Proc. Asian regional symposium on geotechnical problems and practices in Foundation Engineering.* Colombo: NBRO

NBRO 1990. *A study of landslides in Sri Lanka.* Colombo: NBRO

Senanayake, K.S. 1986. Natural disaster mitigation planning and some aspects of landslide disasters in Sri Lanka. *International Seminar on Regional Development Planning for Disaster Prevention,* Japan.

Senanayake,K.S. 1990. Landslide disaster mitigation in Sri Lanka. *ESCAP UNDRO Regional Symposium on IDNDR,* Bangkok, Thailand.

Sittamparapillai, P.M. 1970. Landslides and Floods. *No.8CPS/8/p/7/70, Ceylon Association for the Advancement of Science.*

Environmental Management, Geo-Water & Engineering Aspects, Chowdhury & Sivakumar (eds)
© 1993 Balkema, Rotterdam. ISBN 90 5410 099 0

Tectonic and geomechanic aspects in relation to the stability of Colle Giove's slopes (Santarcangelo di Romagna, Italy)

G.Toni & R.Genevois
Department of Geological Sciences, University of Bologna, Italy

ABSTRACT. The historical town of Santarcangelo di Romagna rises on a hill (Colle Giove) made of sandstones, gravels and subordinately clays of the last levels of the Plio-Pleistocenic marine sediments. In the area two tectonic situations are present that result in anti and synclinalic axes and faults system with different orientations.

The field survey and analyses have shown the presence of continous micromovements that generates cracks visible in a great number of structures in the historic centre of the town.

All the orientation data have shown that the most recurring systems are nearly orthogonal among themselves not having therefore any interpretative significance on the deep tensional condition. However there are other discontinuity planes that fit almost exactly the Riedel's model, if we consider the existence of the following Apenninic compressive stress, still active nowadays with a N 30° direction, that is able to determine in depth a right wrench tectonic regime.

The study of the mechanical behaviour of the materials has been made on undisturbed samples and the geomechanical parameters obtained have been used for the finite element stability analysis of the north-eastern slope deeping as the stratification.

The hydraulic conditions considered are those theoretically critical due to periods of prolonged and intense precipitations. The results of such analysis show high safety factors (Fs = 1.6-1.7) but considerable strains, particularly in vertical direction.

The geological structural and geomechanical studies have made possible to develop two different hypothesis. The first one brings back cracks to the presence of an active compressive stress that would have formed a right wrench system with a northerly direction and additional systems according to the Riedel's pattern.

The second hypothesis calls instead for the development of deformations due to stress fluctuations induced by the variation of water table levels as a consequence of the hydrometeorics phenomena: such hypothesis is based on the consideration that the soil materials may show time-dependent deformations due to the application of stresses even lower than those who entail the short-term failure.

1 INTRODUCTION

The historic centre of Santarcangelo di Romagna rises on Colle Giove (Figure 1) made up of the last spur of the Plio-Pleistocenic arenaceous, gravelly and subordinately argillaceous lithotypes that submerges below the alluvial plain of the Marecchia and Uso rivers.

In the area two different tectonic situations are present: the first westerly one with anti and synclinalic axes in a N-NW direction and the second easterly one with axes having a NE direction. Moreover two parallel faults systems NW-SE oriented have been observed, within which some systems due to Apenninic thrust (SW-NE), have been developed.

The field investigations have highlightened the presence of continous micromovements generating a number of cracks visible in many structures of the historic centre of the town. The observed cracks and the lineations detected by the examination of aerial photographs can be accounted either to the tectonic stresses or to the actual mechanical behaviour of various lithotypes.

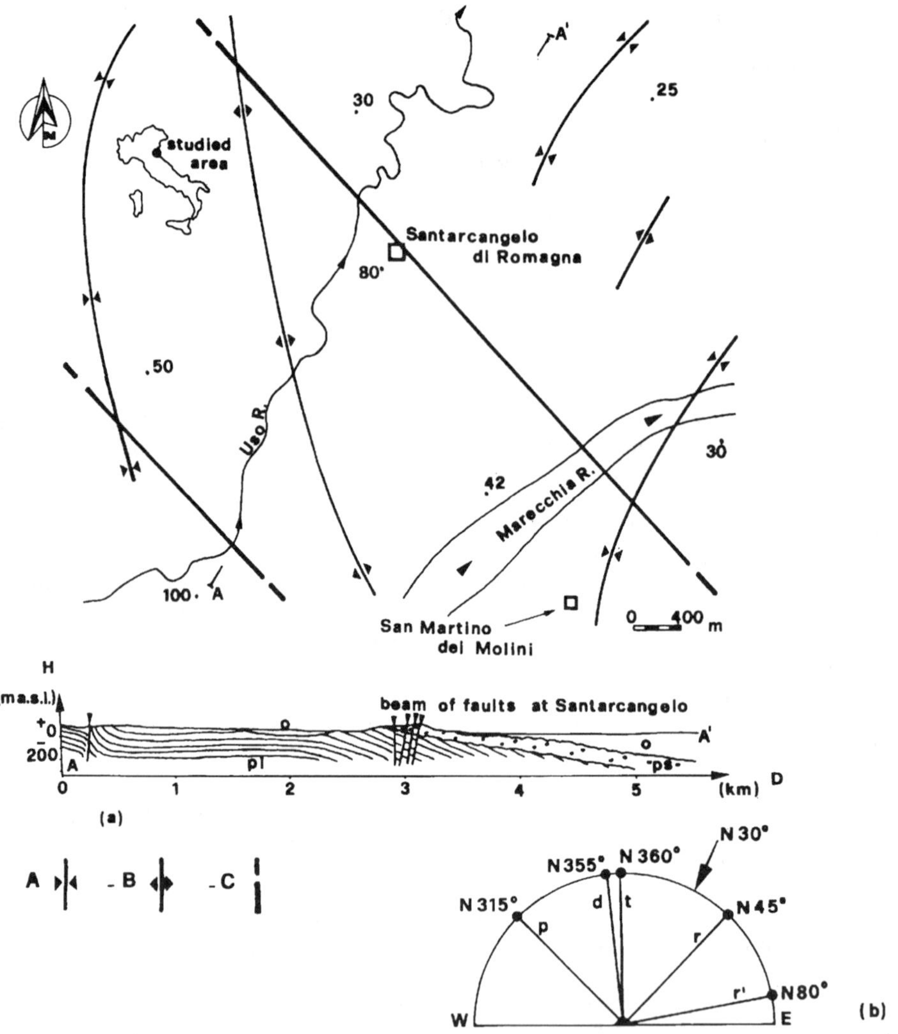

FIG. 1: (a) Geostructural schetch of the examined area and geological cross section (A-A'): syncline axes (A) and anticline axes (B); faults systems (C). (b) Orientation of principal systems of observed fractures in relation to the Apenninic compressive stress.

2 STRUCTURAL SURVEYS

All data (Figure 1b) including those obtained inside over 100 tunnels existing in the Colle Giove (Figure 2), have shown that the most recurring systems are nearly orthogonal (N 315°-N45°), not having therefore any interpretative significance on the deep tensional condition (Mattauer 1973).

However there are other discontinuity planes that fit exactly the Riedel's model (Riedel 1929): considering the Apenninic compressive stress, still active nowadays, with a N 30° direction, the orientation of the resulting planes will be N 80°-355°-360°.

3 LABORATORY RESULTS

The mechanical behaviour of materials has been analysed with triaxial C.I.U. tests on undisturbed soil samples and with unconfined compression tests on samples with lithic behaviour (Figure 3).

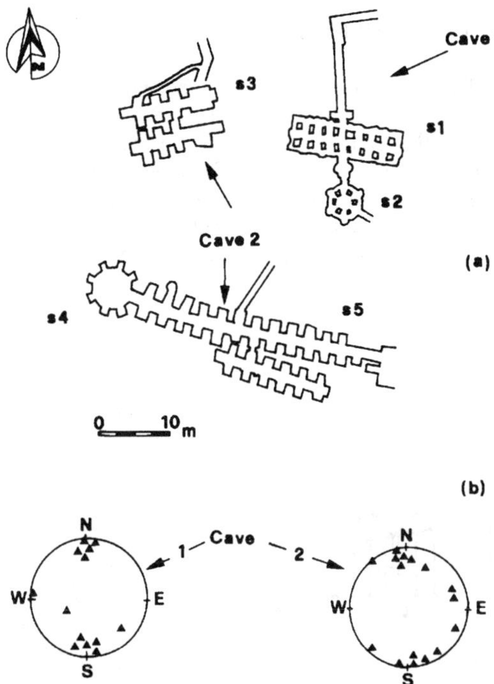

Fig. 2: (a) Planimetric survey of the two principales caves of Giove Hill with location of undisturbed samples. (b) Polar equiareal projection of lithoclases.

Elastic parameters have been determined on site with plate load tests (Toni 1988).

The average geotechnical characteristics of outcropping formations as resulted from the analyses carried out on 5 samples are the following: effective stress coesion (100 kpa); effective stress friction angle (30°); bulk density ($1.9\ g.cm^3$); pore pressure ratio on failure ($\Lambda = 0.3$); Young's modulus about (5000-20000 kpa) due to the alteration grade of arenaceous level actually tested. These parameters have been used for stability analysis of the NE slope where the dip of the stratification coincides with that one of the slope.

4 STABILITY ANALYSIS

The analysis has been carried out utilizing a finite elements programme (Z-soil) that allowes to analyse also the deformative state of the slope and that is based on the plasticity theory (simplified Drucker-Prager model) (Zienkiewicz 1977).

The hydraulic conditions considered are those

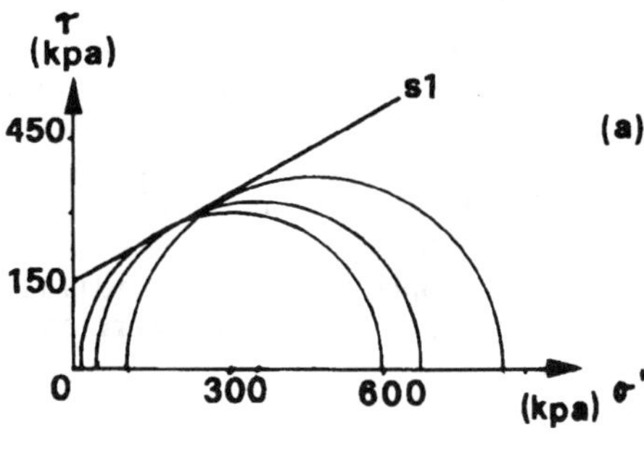

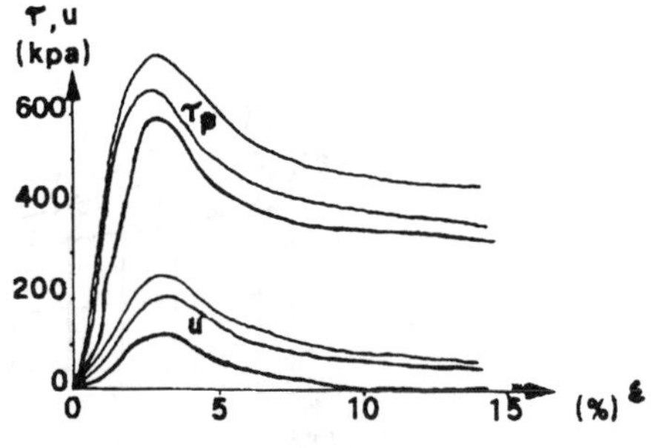

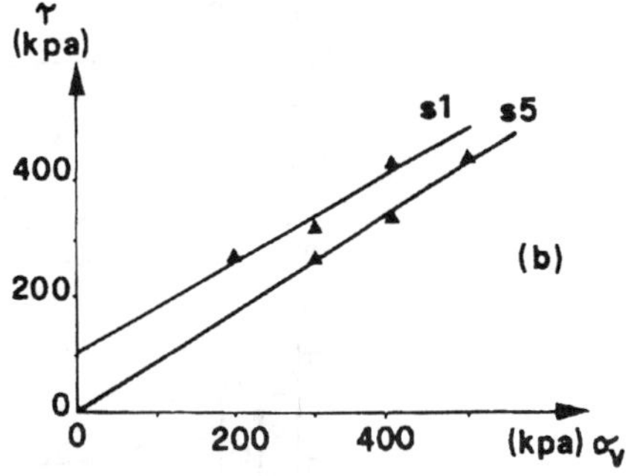

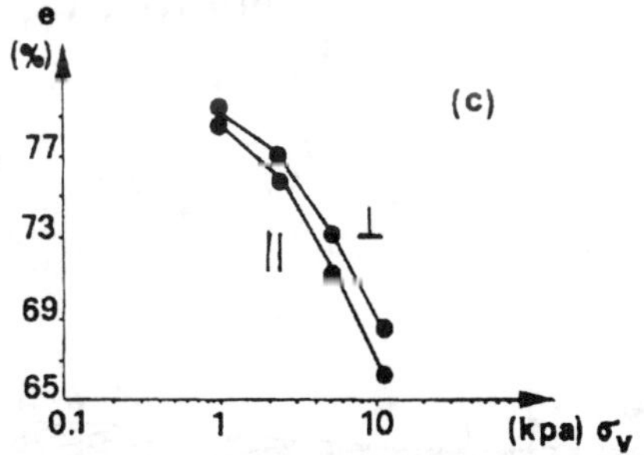

Fig. 3: Laboratory results of some examined samples: (a) triaxial C.I.U. tests: effective stresses envelope; shear stresses strains and pore pressure strains diagrams. (b) direct shear tests. (c) oedometer tests on samples parallel or normal to the stratification.

theoretically corresponding to the critical one's due to periods of prolonged and intense precipitations. The results of such analysis (Figure 4) show that, also the safety factors are quite high (Fs = 1.6-1.7), considerable strains, particularly in the vertical direction are developed: the maximum value of vertical displacement is about 50 cm in front and about 10cm for horizontal displacement.

5 CONCLUSIONS

The geological, structural, and geomechanical analyses of the Santarcangelo di Romagna's area have allowed to develop two different hypotheses that can justify the cracks observed both in the buildings of the historical centre and in the caves existing below the centre itself.

The first hypothesis brings back these cracks to

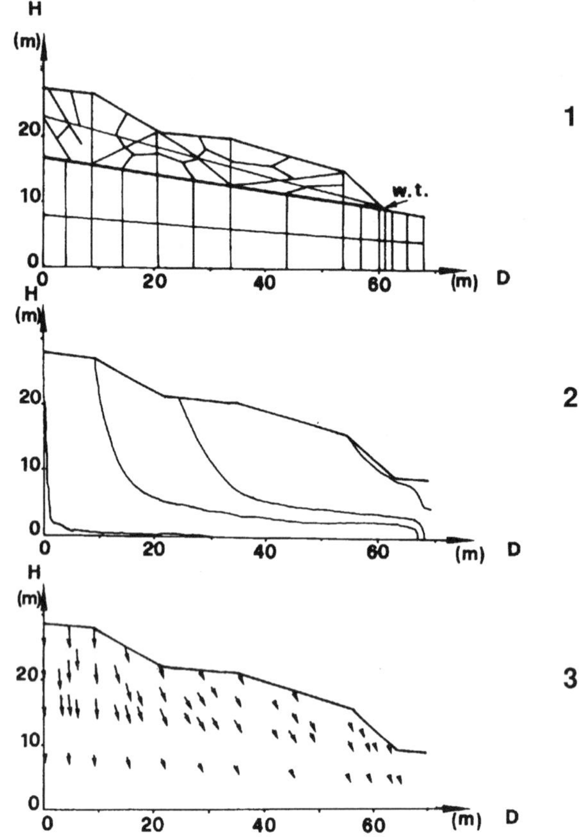

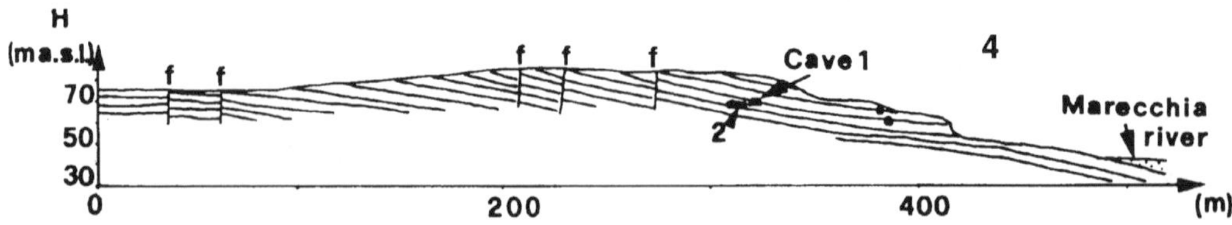

Fig. 4: Nodal disoretization of soil medium (1); stream lines (2); vector displacements in actual conditions (3) and transverse section of Giove Hill (4).

the presence of a still active compressive stress, N 30° oriented, that would have formed both a right northerly wrench system and additional fracture's systems according to Riedel's pattern.

It is conceivable infact that natural slopes, where considerable strains have developed due to the tectonic rotation of principal stresses axes, could be subjected to deformative phenomena even without a generalized failure, as in our particular case, that affect mainly the building structures, whose behaviour is certainly more rigid than that of foundation's soils.

The second hypothesis calls instead for the development of deformations due to stress fluctuations induced by variations of water table levels as a consequence of the hydrometeorics phenomena. Such hypothesis is based on the consideration that the soil materials may show time-dependent deformations due to the application of stresses even lower than those that entail the short-term failure of the slope (Wittke 1990).

REFERENCES

Angelier, J. and Mechler, P. 1977. Sur une méthode graphique de recherche des contraintes principales également utilisable en tectonique et en seismologie: la méthode des diedres droits. Bull. Soc. Geol. France (7), 19 n. 6: 1309-1318.

Barton, N. - Lien, R. and Lunde, J. 1974. Engineering classification of rock masses for the design of tunnel support. Rock Mechanics 6. Verlag: 189-236.

Hancock, P.L. 1985. Brittle microtectonics: principles and practice. Journal of structural Geology, vol. 7/3-4. Bristol: 437-457.

Mattauer, M. 1973. Les déformations des matériaux de écorce terrestre. Hermann. Paris: 1-493.

Pellegrini, M. and Vezzani, L. 1978. Faglie attive in superficie nella pianura padana presso Correggio (RE) e Massa Finalese (MO). Geogr. Fis. Dinam. Quat., 1: 141-149.

Ramsay, J.C. 1967. Folding and fracturing in rocks. McGraw Hill. New York: 1-568.

Riedel, W. 1929. Zur mechanik geologischer brucherscheinnungen. Zentral blatt Min. Geol. Pal. B: 354-369.

Toni, G. and Zaghini, M. 1968. Prime osservazioni geologico-tecniche del colle Giove su cui sorge il centro storico dell'abitato di Santarcangelo di Romagna (FO). CNR-SCAI. Bologna: 1-18.

Zienkiewicz, O.C. 1977. The finite element method. McGraw Hill. London.

Wittke, W. 1990. Rock Mechanics. Springer Verlag. New York: 1-1076.

Environmental Management, Geo-Water & Engineering Aspects, Chowdhury & Sivakumar (eds)
 ISBN 90 5410 099 0

Consideration on prediction method of slope failure based on critical amount of rainfall

Ryuichi Yatabe, Norio Yagi & Meiketsu Enoki
Ehime University, Japan

ABSTRACT: The traffic control of national highway is generally conducted on the basis of a critical amount of rainfall. This critical amount of rainfall has been empirically determined by investigating the actual data gathered on the past slope failures, but is not based on the geometry or the soil properties of slopes in the district. The critical amount of rainfall for three districts, where slope failures are frequently caused by rainfall, are estimated in this by theoretical and numerical analyses considering the geometries and soil properties of slopes in the districts. The results are then compared with those obtained by the past data of actual slope failures.

1 Introduction

In Japan, slope failures of sandy soils caused by the rainfall are one of the most serious disasters for human's lives and properties. It is then an urgent necessity to establish a prediction method of these slope failures, especially a prediction method of the time of occurrence of failure, since we can have sufficient time before the slope failure to avoid the damages.

The present methods of predicting the slope failures are classified into following two groups;

1) predict the failure by premonitory symptoms such as movement of slope surface
2) predict the failure by the amount of rainfall

Based on the second prediction method, some critical amounts of rainfall have been empirically determined by district by many public offices, and the inhabitants in the district are required to evacuate their home when the total amount of a series of rainfalls is over this critical amount, and the traffic control of national highway is also conducted by some critical amounts of rainfall. This amount is called "critical amount of rainfall" or "limit amount of rainfall". As this method of predicting the slope failure is simple and practical, it is useful to improve the reliability of this method.

In the past, this critical amount of rainfall for the object district has been empirically determined investigating the actual data gathered on the past slope failures, and neither the geometry nor the soil property of slopes in the district has been introduced in the determination. However, as clarified by the authors, this critical amount of rainfall can be obtained by theoretical and numerical analyses of seepage and stability of the slopes.

In this paper, the present situation and controversial point of traffic control by the continuous rainfall which has been empirically determined, is described. The critical amounts of rainfall for three districts where slope failures are frequently caused by rainfall, Kure district in Hiroshima prefecture, Ehime prefecture and Kagoshima prefecture, will be obtained by theoretical and numerical analyses considering the geometries and soil properties of slopes in the districts. Then, they will be compared with those obtained by the past data of actual slope failures.

2 The present situation and controversial point of traffic control in Japan

Japan is divided into nine administrative blocks, Hokkaido, Tohoku, Kanto, Hokuriku, Chubu, Kinki, Chugoku, Shikoku and Kyushu, as shown in Fig.1.

Fig. 1 The administrative blocks

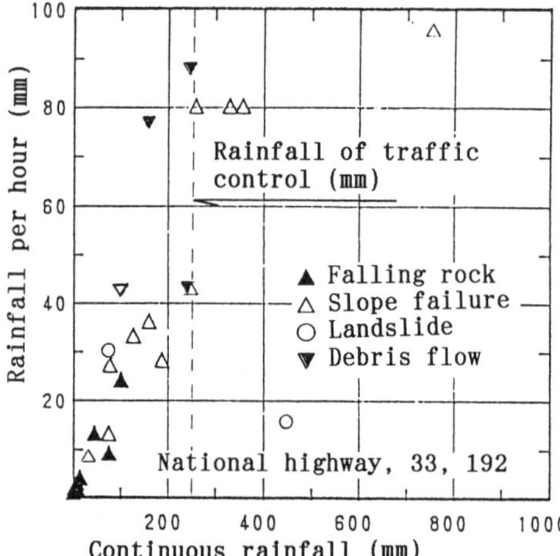

Fig. 2 Actual of the slope failure in Shikoku district

The traffic control of national highway in a heavy rain is issued when the continuous rainfall exceeds some standard value. The length of national highway is 19288 km in the whole of Japan and the length of traffic controlled section is 1154 km.

The representative standard of amount of continuous rainfall for traffic control of national highway is shown in Table 1 by blocks. The amount of rainfall for traffic control vary with the district, for instance, 100 mm in Hokkaido district, 150 mm in Kanto district and 250-300 mm in Shikoku district. This depends on the difference of topography, geology, the nature of soils and weather conditions. The number of times of traffic control is shown in Table 2. The traffic control is frequently issued in Hokkaido district. It is because the critical rainfall amount for traffic control is too small. The value for

Table 1 Standard value of amount of rainfall for traffic control

Block	Length of highway(km)	Continuous rainfall(mm) -100	-150	-200	-250	-300
Hokkaido	5818	182km				
Tohoku	2465		59km	19km		
Kanto	2012		87	12		
Hokuriku	976	6	54	35		
Chubu	1546		99	39	26km	23km
Kinki	1635		9	105	86	1
Chugoku	1529		12	22	24	
Shikoku	1158		9		129	18
Kyushu	2077			59	38	

Table 2 Actual results of traffic control (1981-1990)

Block	Number of times	Hours Total	Average
Hokkaido	393	11450	29.1
Tohoku	156	1597	10.2
Kanto	185	2736	14.8
Hokuriku	177	2511	14.2
Chubu	151	1644	10.9
Kinki	48	1420	29.6
Chugoku	15	498	33.2
Shikoku	75	1894	25.3
Kyushu	103	1745	16.9

Hokkaido district must be re-examined. The relation of rainfall per hour with the continuous rainfall which caused the slope failure, debris flow, falling rock or landslide in Shikoku district, is shown in Fig.2. The standard of rainfall for traffic control is 250 mm in Shikoku district. Many slope failures have occurred under 250 mm. This results indicate that the prediction of the slope failures by critical amount of rainfall are very difficult.

3 Theoretical analysis for obtaining critical amount of rainfall

3.1 Analysis method

The problem of slope failure caused by rainfall can be treated as a complex problem of seepage flow and stability of slope.

Unsteady saturated and unsaturated seepage flow in a slope can be analyzed numerically using the finite element method.

Stability of the slope in which pore pressure is distributed can be analyzed by several methods, and the authors adopted the Fellenius method. By these seepage and stability analyses the geometry and the soil property of slopes can be introduced in determining the critical amount of rainfall. The details of this analysis method in which the seepage analysis is combined with the stability analysis have been demonstrated in the papers (Yagi, Yatabe, & Yamamoto 1983, Yagi & Yatabe 1987) together with the validity of the method.

3.2 Slope model, soil properties and conditions used in the analyses

In this theoretical analysis, the model slope shown in Fig.3 is used. The thickness H of the slope is 0.5 m or 1.0 m, and the inclination of model slope is 35°.

The soil of surface layer of kure district and Ehime prefecture is composed of weathered granite, called Masado, and that of Kagoshima prefecture is composed of volcanic ash, called Shirasu.

Strength parameters are obtained by triaxial tests on the saturated samples under consolidated drained condition. ϕ_d is about 30° for the corresponding density to the loose surface layer ($\rho_d <$ 1.5kN/m^3). c_d becomes zero by saturation.

Seepage properties are obtained by the permeability test and pF test during dewatering process using ceramic plate. Relative permeability Kr ($Kr=k/k_s$ k:permeability coefficient of unsaturated sample, k_s:permeability coefficient of saturated sample) is obtained numerically using Irmay method.

In order to obtain the critical amount of rainfall for a slope to fail by theoretical analysis, the initial degree of saturation of soil is required. The authors had long-term observation of the suction in a slope. Typical change of the suction in the slope soil is demonstrated in Fig.4. This figure suggests that the suction decreases during a rainfall and increases after the rainfall, and that during non-rainy days the suction of the shallow part becomes higher than that of the deep part.

Based on this observation, a relation of the suction (pF) with N_d is obtained and demonstrated in Fig.5, where N_d expresses continuous non-rainy days.

In the following theoretical analysis of critical rainfall, ϕ_d of Masado and Shirasu are assumed to be 35°, and c_d is assumed to be zero for Masado and 1.96 kPa for Shirasu. ρ_d is assumed to be 1.4 kN/m^3 for Masado and 1.0 kN/m^3 for Shirasu. k_s is assumed to be 3.0x10^{-4} cm/s for Masado of Ehime, 3.0x10^{-3} cm/s for Masado of Kure and 3.0x10^{-2} cm/s for Shirasu. The initial suction of slope soil is assumed to be 2.94 kPa for zero non-rainy day, 9.8 kPa

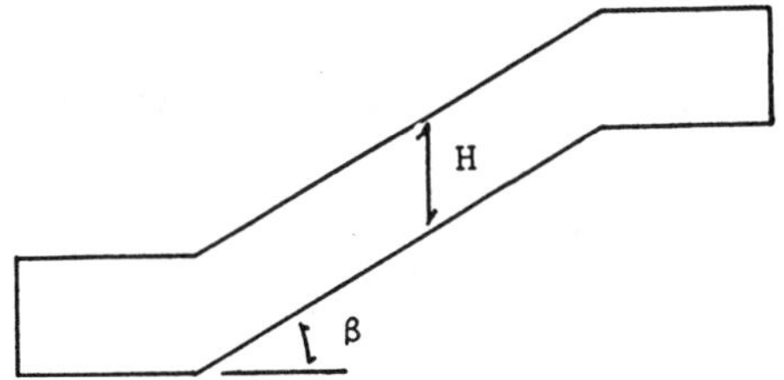

Fig. 3 Slope model

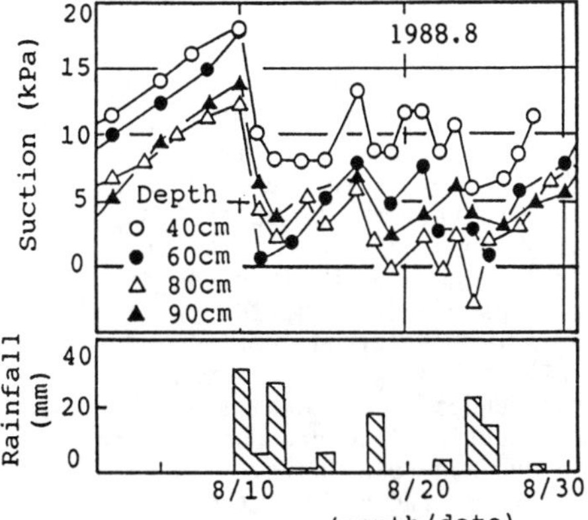

Fig. 4 Typical change of the suction in a slope of weathered granite soil (Yatabe, Yagi & Enoki 1987b)

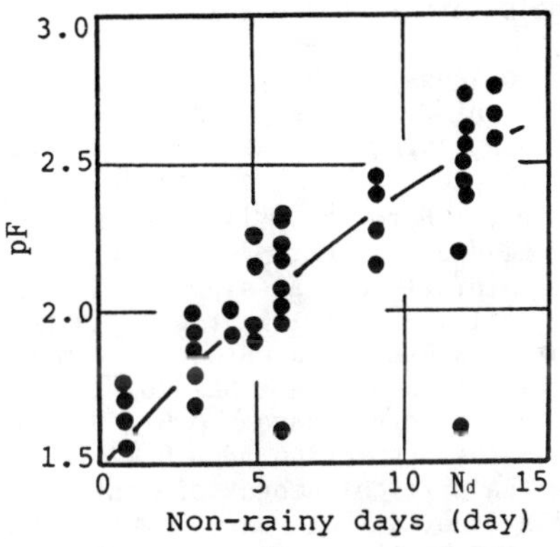

Fig. 5 Relation of suction with non-rainy days

for continuous 5 non-rainy days, 18.6 kPa for continuous 10 non-rainy days and 39.2 kPa for continuous 14 non-rainy days. The pF, K_r-θ relations are obtained experimentally and numerically. The intensity of rainfall is assumed 20 mm/h, which corresponds to the critical intensity above which most of slope failures have been observed. In addition, only for the Shirasu slope, influence of permeability of base rock on critical amount of rainfall is investigated changing the permeability from impermeable to 3.0×10^{-3} cm/s. The details of this soil properties and inithial conditions have been demonstrated in the papers. (Yagi, Yatabe & Enoki 1985)(Yatabe, Yagi & Enoki 1986)(Yagi, Yatabe & Enoki 1990)

3.3 Critical amounts of rainfall in various regions

As the results of the theoretical analyses, relations of R_c with N_d for the slopes of 0.5 m and 1.0 m thick are shown in Fig. 6 and 7, respectively, where Rc is the total amount of rainfall required for the slope to fail and N_d is the continuous non-rainy days prior to the rainfall. These relations are obtained as follows;

1)Pore pressure distribution within the slope after a given period of heavy rainfall is obtained by the seepage analysis under the condition that the initial suction is corresponding to the continuous non-rainy days.

2)The safety factor of the slope is obtained by the slope stability analysis using this pore pressure distribution.

3)If the safety factor is larger than 1.0, the period of heavy rainfall is increased and the procedure 1) and 2) is repeated until the safety factor becomes less than 1.0.

Before discussing about the critical amount of rainfall thus obtained, some problems concerning the influence of the assumptions on the analyses are investigated, here. As is shown by comparing Fig. 6 with Fig. 7, the critical amount of rainfall for a slope of 1.0 m thick to fail is about 1.5 times larger than that of 0.5 m thick. Since the most of actual slopes in Masado and Shirasu districts have 0.5-1.0 m thickness, the assumption of 0.5 m thick may give the critical amount of rainfall less than the actual one, but the assumption of 1.0 m thick may give the critical amount larger than the actual one. As shown in Fig. 7, the assumption of c_d=1.96 kPa for Shirasu gives the critical amount of rainfall 1.4-1.5 times larger than the assumption of c_d=0 does, and this suggests that the cohesion c_d has significant influence on the obtained critical amount. In Fig. 8, the critical amounts of rainfall for Shirasu obtained on the two assumptions that the base rock of the slope has the permeability one order less than that of the surface layer soil and the base rock is impermeable, are

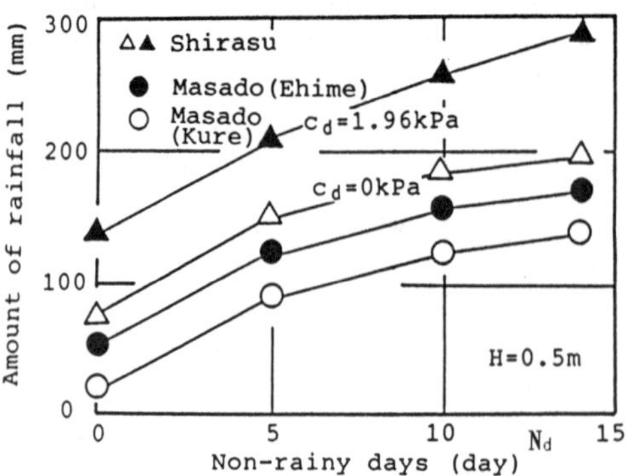

Fig. 6 Relation of critical amount of rainfall with non-rainy days (H=0.5 m)

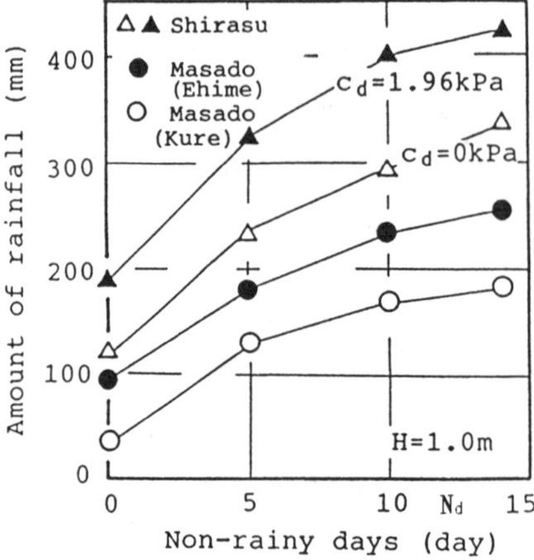

Fig. 7 Relation of critical amount of rainfall with non-rainy days (H=1.0 m)

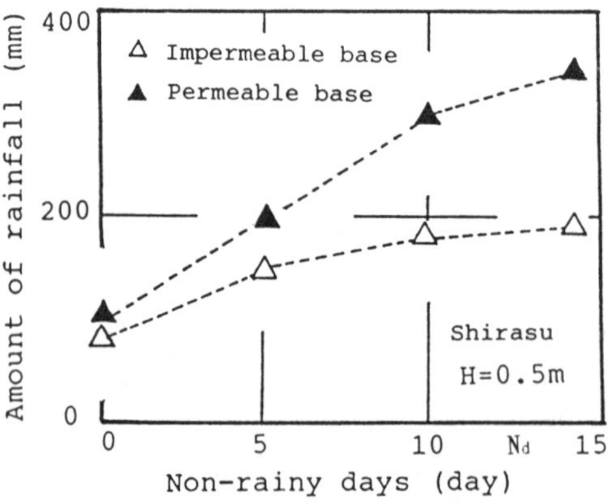

Fig. 8 Influence of assumptions on the critical amount of rainfall

compared with one another. The former assumption gives 1.2-1.7 times larger than the latter assumption does. These investigations on the assumptions indicate that the critical amount of rain fall for a slope to fail is influenced by the assumptions on the thickness of surface soil, cohesion and the permeability of the base rock. In the case where 0.5 m thick, $c_d=0$ and impermeable base rock are assumed, a minimum amount of rainfall will be obtained.

The critical amounts of rainfall R_c for the slopes of Shirasu, Masado (Ehime) and Masado (Kure) obtained by the theoretical analyses are in the following order of amount; Shirasu, Masado (Ehime) and Masado (Kure). This result suggests that the slopes of Masado in Kure are more likely to fail due to rainfall than those in Ehime. This may be caused by the difference of the permeabilities of the two districts, since the other parameters are approximately common with these districts. The result also suggests that the slopes of Masado are more likely to fail than those of Shirasu. This may be caused mainly by the difference of cohesion parameters (in saturated condition, $c_d=0$ for Masado and $c_d=1.96$ kPa for Shirasu) , since the difference in permeability and the difference in preservation capacity of water may be canceled out.

On the other hand, the relation of R7 with R1 is obtained by the data gathered on the past actual slope failures and shown in Fig. 9, where R7 is the total amount of rainfall for a week prior to the day of occurrence of slope failure, and R1 is the amount of rainfall for 24 hours prior to the time of occurrence of slope failure. In most cases many slope failures are caused in a district during a series of rainfalls. In Fig. 9, R7 and R1 at the time of the first slope failure are plotted. So, Fig. 9 implies the minimum of the critical amount of rainfall for the district.

For the slopes of Masado in Ehime and Kure, the relations of R7 with R1 can be represented by the curves drawn in the figure, but for the slopes of Shirasu the representative curve cannot be obtained according to the scattering of the points. This scattering may implies that the slope geometry and the slope soil property differs much from each slope.

Here, the critical amount of rainfall obtained by the theoretical analysis is compared with that obtained by the records of actual slope failures. Note that R7+R1 required for a slope to fail in Fig.9 has

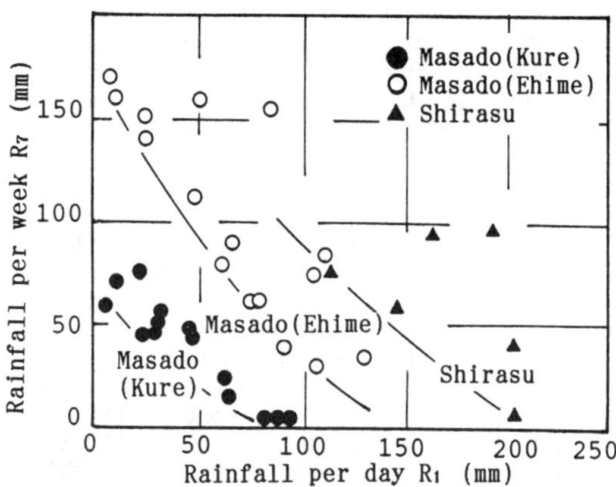

Fig. 9 Relation of the total amount of rainfall for week before the day of slope failure and the amount of rainfall for 24 hours before the time of slope failure

the same characteristic with R_c in Figs. 6 and 7.

In Fig.9, R7+R1 is 50-100 mm for Masado district in Kure, 100-200 mm for Masado district in Ehime, and over 200 mm for Shirasu district. On the other hand, in Figs. 6 and 7 obtained by the theoretical analyses, R_c is 50-180 mm (H=1 m) or 20-130 mm (H=0.5 m) for Masado district in Kure, 100-250 mm (H=1 m) or 50-170 mm (H=0.5 m) for Masado district in Ehime, and 120-330 mm (H=1 m, $c_d=0$) or 80-200 mm (H=0.5 m, $c_d=0$) for Shirasu district. As the parameter of non-rainy days appears in Figs. 6 and 7, the theoretical analyses differ a little in the characteristic from the parameter R7 which appears in Fig. 9. Thus a direct comparison of Figs. 6 and 7 with Fig. 9 is difficult. However, note that the case where non-rainy days exceed 7 days in Figs. 6 and 7 corresponds to R7=0 in Fig. 9. These comparisons indicate that the critical amount of rainfall for slopes in a district obtained by the theoretical analysis agrees well with that obtained by the records of actual slope failures.

4 CONCLUSIONS

The present situation and the controversial point of the traffic control in Japan have been clarified. It seems that the critical amount of rainfall must be decided for a particular district considering geology, nature of soils and topography.

In the past, the critical amount of

rainfall for a particular district has been empirically determined on the basis of the recorded number of slope failures. However, it has been clarified in this paper that it can be theoretically obtained by the seepage and slope stability analyses considering the geometry and the soil property of slopes. Though there are various slopes in a district and the dangerousness of slope failure differs from each other, the critical amount of rainfall is used as the representative value for an object district. If the representative geometry and soil property of slopes in the district are available, the critical amount of rainfall can be theoretically obtained. The application of this predicting method of slope failure to many regions will be useful to relieve the loss of human's lives and properties.

ACKNOWLEDGMENTS

The authors thank to Mr. O. Futagami for his preparing the apparatus of experiments, and also thank to the staffs of Ehime Univ. Forest for their arranging the site of the field experiment.

REFERENCES

Yagi, N., Yatabe, R. & K.Yamamoto 1983. Slope failure mechanism due to seepage of rain water. Proc. 7th ARCSMFE, Vol.1. 382-386.

Yagi, N. & R.Yatabe 1987. Prediction method of slope failure in sandy soil due to rainfall. Proc. 8th ARCSMFE, Vol.1. 217-220.

Yagi, N. & R.Yatabe 1985. Shear Strength property of undisturbed masado. JSCE, No.364. 133-141 (in Japanese)

Yatabe, R., Yagi, N., and M.Enoki 1986. Prediction of occurrence time of slope failure caused by rainfall. JSCE, No.376. 297-305 (in Japanese).

Yagi, N., Yatabe, R. & M.Enoki 1990. Prediction of slope failure based on amount of rainfall. JSCE, No.418. 65-73 (in Japanease)

Yatabe, R., Yagi, N. and M. Enoki 1987b. Measurement of suction for prediction in a slope failure. Proc. Symp. on Unsaturated Soil. 301-306 (in Japanese)

Waste management

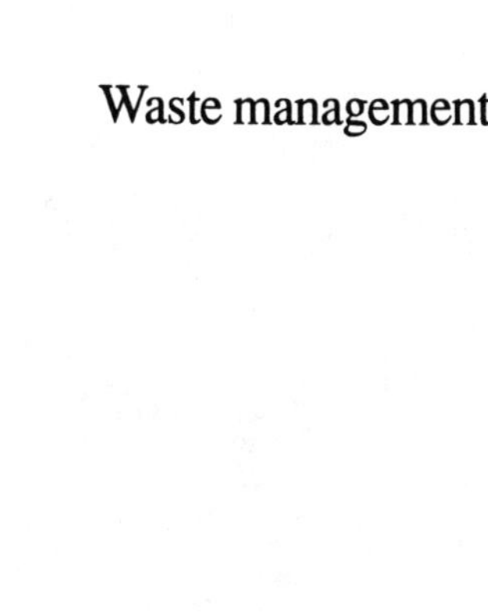

Environmental Management, Geo-Water & Engineering Aspects, Chowdhury & Sivakumar (eds) © *1993 Balkema, Rotterdam.*
ISBN 90 5410 099 0

Migration of methane and leachate from landfill

S.Abaci & J.S.Edwards
Department of Mining Engineering, University of Nottingham, UK
B.N.Whittaker
Department of Mining and Mineral Engineering, University of Leeds, UK

ABSTRACT : Although landfill constitutes a major component of the solid waste disposal system, it can cause serious environmental pollution. In recent years, hazardous products of landfill have caused some danger to the public and environment. To prevent or minimize environmental pollution it is necessary to recognize and understand landfill decomposition processes, migration of gas and leachate production.

1 INTRODUCTION

Solid waste disposal is one of the most important and expensive problems facing all communities in the world. The volume of waste material generated by domestic, commercial and industrial sources continues to increase whilst land availability, especially in urban areas, is decreasing. In addition, an increasing awareness of environmental issues puts increasing pressure on Waste Disposal Authorities.

A landfill offers an acceptable and economic solution to the disposal of solid wastes. Despite the apparent simplicity, the landfilling of waste materials may give rise to environmental pollution and pose risks to human health.

The major environmental impacts from a disposal site originate from the flux of waste material differing in composition and/or concentration from the surrounding strata. The main problems are: the generation of hazardous, potentially toxic or explosive, landfill gases (mainly methane and carbon dioxide) which are the by - products of decomposition of wastes by anaerobic bacteria; and leachate, which is a potential contaminant liquid (Esmaili 1975). These hazardous products can, under suitable conditions, migrate from the landfill site through any permeable materials to adjacent formations where they can cause unacceptable changes in the quality of the surrounding environment. The main problems associated with migrating methane and leachate can be summarized as:

1. Oxygen depletion in soils: Methane migrating to the aerobic root zones of plants may displace oxygen, or methane-oxidizing bacteria may use all the available oxygen in the root zone, resulting in the death of vegetation.

2. The build - up of explosive concentrations of methane within constructions: Methane is a flammable gas and forms an explosive mixture with air at between 5 % and 15 % by volume. The gas can migrate from the landfill and may accumulate in enclosed spaces in explosive concentration.

3. Asphyxiation: The addition of any contaminant gas will reduce the oxygen content in the atmosphere (Card and Roche 1991). If the proportion of oxygen in the air decreases from the average 21% by volume to below 17 % volume there is a threat of anoxia. Carbon dioxide is a respiratory and nervous system stimulant and may also produce unconsciousness and death at very high concentrations (Edwards 1989).

4. Water pollution by leachate: Leachate which contains organic and inorganic compounds and suspended solids cause extensive pollution of aquifers, streams and water wells etc.

To prevent or minimize environmental pollution arising from landfills, it is essential to understand the processes of landfill gas and leachate generation and also the factors which control their migration through soils and sediments.

2 FORMATION OF LANDFILL GAS AND LEACHATE

The organic and inorganic constituents of landfill

waste undergo decomposition due to physical, chemical and microbial processes resulting in the release of gaseous and dissolved compounds (Findikakis and Leckie 1979). The refuse will, soon after disposal, become anaerobic, and a consortium of bacteria which are named methanogens will start degrading the organic material to carbon dioxide and methane.

A waste disposal site in a region of heavy rainfall will accumulate water which percolates through from the surface, and which tends to be absorbed by the waste, until field capacity is reached. In time, the infiltrating water reacts with the solids in the waste and dissolves some into solution. This polluted wastewater, named leachate, contains high concentrations of several organic and inorganic pollutants. The first stage of anaerobic degradation, a hydrolysis process, is a very important one in the landfill since the solid organic refuse present must be solubilized before the micro-organisms can convert it. In this first anaerobic biodegradation process the fermenters convert dissolved organic compounds from the leachate to primarily volatile fatty acids, hydrogen and carbon dioxide. The hydrolytic stage is followed by acidogenesis in which an acidogenetic group of bacteria produce acetic acid, hydrogen and carbon dioxide from the first stage of products. The third stage starts with the slow growth of methanogenetic bacteria. This phase mainly comprises two reactions; acetophilic bacteria convert acetic acid to methane and carbon dioxide or hydrogenophilic bacteria convert hydrogen and carbon dioxide to methane. According to these reactions hydrogen and carbon dioxide concentrations in the gas decrease while the methane concentration increases.

Landfill gas compositions produced in these three stages is shown in table 1.

Table 1. Landfill Gas Composition

Phases	Composition by volume %	
	CH_4	CO_2
Acidogenic	-	20 - 70
Acetogenic	5 - 50	70 - 40
Methanogenic	50 - 65	40 - 30

The production of landfill gas and leachate can be affected by many factors. The major factors which may vary from site to site, are moisture content and movement, pH level, the compaction of the waste and temperature (McEntee and Jelley 1991).

3 MIGRATION OF METHANE

Methane may exist in the subsurface in a gaseous form or dissolved in ground water and when pressure is reduced to atmospheric methane comes out of solution and forms a separate gaseous phase. All fluids (gases and liquids) may migrate from one place to another through porous media at a rate controlled by the permeability of the medium and the driving force. Gases can move through permeable consolidated rocks by movement through interconnected porosity or discontinuities such as joints, fractures and fault zones. In unconsolidated materials such as sands and gravels gas may flow through pore spaces between individual sediment grains. Permeability characteristics of soils and rocks can vary by many orders of magnitude, depending on the texture and structure of the geologic formation. For instance, in unconsolidated materials the permeability can be controlled by particle size and particle size distribution. Another important impact affecting the permeability is the saturation level of the medium. In the other words,the permeability of the medium to one of the fluids (gas or water) is dependent on the saturation of the other fluid. If the material is completely saturated with water, gas may be transported only by slow diffusion through water (Williams and Aitkenhead 1991). In practise, partially saturated conditions may exist such that the rocks and soils are saturated with varying proportions of gas and water. Multiphase flow is valid if the media is partially saturated. Only when the medium become completely unsaturated with water will the gas permeability be at the maximum value.

The another dominant factor in controlling migration of gas from the landfill is a driving force, which could be a combination of:

1. concentration gradient, and
2. pressure gradient.

Gases may migrate through porous media by a number of mechanisms: molecular diffusion due to concentration gradient, or viscous flow (or mass flow) due to pressure gradient (Williams 1989).

3.1 Diffusion

Diffusion is a process in which matter is transported from one part of a system to another by molecular motion. The substances will be transported via molecular diffusion since there is a gradient in concentration, although the hydraulic gradient is zero. In this case, an advective flow will not occur. It can be said that the factors controlling diffusion are the diffusion coefficient and concentration gradient. Diffusion is assumed to occur in accord with Fick's first law which states that the rate of gas diffusion through a unit cross - sectional area of a given porous medium is dependent on the gradient of concentration measured normal to the section (Williams and Hitchman 1989). Fick's first law states that one dimensional flow can be expressed as:

$$F = - Do \frac{\partial c}{\partial x} \quad (1)$$

where F is the rate of transfer per unit area (normal to the direction of transport), Do is the diffusion coefficient, c is the concentration of the diffusing solute and x is the direction in which diffusion is occurring.

Solutes will not diffuse as quickly in soil as they will in free air. For soil, equation (1) can be rewritten as:

$$F = - (t\, Do)\, q \frac{\partial c}{\partial x} \quad (2)$$

or

$$F = - De\, q \frac{\partial c}{\partial x} \quad (3)$$

$$\text{and } q = n\, Sr \quad (4)$$

where t is the tortuosity factor (dimensionless), q is the volumetric moisture content (dimensionless), De is the effective diffusion coefficient, n is the porosity and Sr is the degree of saturation.

Fick's first law describes steady - state flow, Fick's second law applies for unsteady - state of flow :

$$\frac{\partial c}{\partial x} - De \frac{\partial^2 c}{\partial x^2} \quad (5)$$

The differential equation (5) is defined for initial and boundary conditions to obtain a description of changes in concentration with respect to time and distance. For initial boundary conditions, gas distribution can be determined from the solution to equation (5) which assumes that concentration is constant with time :

$$\frac{c}{co} = \operatorname{erfc}\left[\frac{x}{2\sqrt{De\, t}}\right] \quad (6)$$

where co is the initial concentration of gas at x = 0 and erfc is the complementary error function.

3.2 Mass or Viscous Flow

Viscous or mass flow of gas occurs in response to a pressure gradient which arises from gas production in a confined medium, or from changes in ground water level and/or in atmospheric pressure. Mass flow through a porous medium is assumed to occur in accordance with Darcy's law in which

$$Q = K\, A \frac{\Delta P}{L} \quad (7)$$

where Q is the volumetric flow rate, K is the effective permeability of the medium, A is the cross - sectional area, Δp is the differential pressure and L is the length.

3.3 Factors affecting gas migration

The rate and extent of gas migration within and/or from landfill will depend upon various site conditions and those pertaining within the surrounding strata. According to conditions, those factors (environmental, climatic and geophysical) which significantly affect gas migration potential can be summarized from two aspects: influence in the landfill waste itself and influences within the surrounding environment.

Climatic conditions have major impacts on gas emission via landfill site surfaces, while sudden changes in atmospheric pressure affect the quantities of gas released. Temperature and moisture content, on the other hand, control microbial activity and gas production rate. Within the landfill, compaction, settlement of wastes and types of cover materials will also have impacts on migration pathways.

The second main aspect, which is important with respect to gas migration beyond landfill site boundaries, is the nature of the strata beneath and around the site and their permeability characteristics.

4 CONTROL OF LEAKAGE

A waste disposal site is usually sealed with an impervious lining, both to prevent gas migration and to prevent leachate from seeping into adjacent strata and into ground water. These linings can either be a natural feature of the site such as clays or an impermeable strata around the site, or an artificial material such as synthetic membrane liners, asphalt, or concrete. The choice of these liners depends on expense and on the quantity of gas collected. A liner design requires a

Table 2. Types of Lining Systems Used in Landfill

Liners / Type of Liners	Single Liners		Composite L.	
	S. Clay	Synthetic	Single	Double
	Number of Layers			
Compacted Clay	1	-	1	2
Synthetic Liner	-	1	1	2

detailed site investigation, especially the stratigraphic, hydrogeologic situation and climate in and/or around site.

A liner system which is chosen to prevent leakage must perform well in the long - term. Criteria for determining the long - term performance of liners are the permeability, leakage resistance and chemical resistance. Synthetic liner materials have very low permeabilities but they can easily be damaged during landfill operation. The permeability of soils can vary greatly. In the case of natural liner materials, layer thickness and compaction are the other important characteristics. In some cases,both natural and artificial liner materials can be used in one landfill. Table 2 shows types of liner system using in landfills.

5 CONCLUSION

This paper describes the generation and migration of landfill gases and leachate and the factors which control leakage. Although landfill is an integral part of existing strategies for solid waste disposal there are a number of environmental considerations that need to be taken into account. Hazardous products from the degradation of waste materials by anaerobic bacteria can migrate through permeable materials to the surrounding environment by molecular diffusion and / or mass flow and may thus pose a danger. For the prevention of leakage, natural impermeable strata and/or synthetic liners are used in landfill operation.

In the future, the main role of landfill is to ensure that waste disposal is carried out with minimal environment impact. For this reason, it is essential to carry out pre - treatment steps before landfilling and adopt high quality standards for landfill design. The quantity of wastes to be landfilled can be decreased if recycling policies are adopted and the quality of those wastes may be changed.

REFERENCES

Card, G.B. & Roche, D.P. 1991. The design of gas control measures for development affected by landfill gas. Symposium on Methane - facing the problems, 26 - 28 March, Nottingham, England.

Edwards, J.S. 1989. Gases - Their basic properties. Symposium on Methane-facing the problems, 26-28 September, Nottingham, England

Esmaili, H. 1975. Control of gas flow from sanitary landfills. ASCE, Journal of the Environmental Engineering Division. Vol. 101. No. EE4: 555 - 566.

Findikakis, A.N. & Leckie, J.O. 1979. Numerical simulation of gas flow in sanitary landfill. ASCE, Journal of the Environmental Engineering Division. Vol. 105, No.EE5: 927 - 945.

McEntee J.M. & Jelley J.C. 1991. Legal implications of methane generation from old landfill sites. Symposium on Methane- facing the problems, 26 - 28 March, Nottingham, England.

Williams, G.M. 1989. Case study - The gas explosion at Loscoe, Derbyshire. Symposium on Methane-facing the problems, 26-28 September, Nottingham, England

Williams, G.M. & Hitchman, S.P. 1989. The generation and migration of gases in the subsurface. Symposium on Methane- facing the problems, 26 - 28 September, Nottingham, England.

Williams, G.M. & Aitkenhead, N. 1991. Lessons from Loscoe: the uncontrolled migration of landfill gas. Quarterly Journal of Engineering Geology. 24: 191- 207.

Environmental Management, Geo-Water & Engineering Aspects, Chowdhury & Sivakumar (eds)
© 1993 Balkema, Rotterdam. ISBN 90 5410 099 0

Anaerobic biotreatment of soils contaminated with toxic organic compounds and heavy metals

S.K. Bhattacharya & M.R. Haghighi-Podeh
Tulane University, USA

ABSTRACT: A literature search showed that there is a lack of quantitative information on anaerobic biotreatment of contaminated soils. The objective of this research was to study the treatability of leachate containing selected heavy metals and organics. This paper includes results on cadmium and nitrophenols. The results showed a good correlation between free cadmium and toxicity indicating the importance of knowing the concentrations of free species of the heavy metals. The inhibition models worked well to predict both cadmium and nitrophenol toxicity in anaerobic systems. Among the nitrophenols studied, 4-nitrophenol and 2-4 dinitrophenol were most toxic.

1 INTRODUCTION

Treatment of soils contaminated with toxic organics and heavy metals could be very expensive. A good option is in-situ treatment which reduces the cost of excavation. In-situ vitrification is becoming more popular in the U.S. The other option is to excavate the soil and treat it using physical, chemical, or biological processes. It is generally accepted that biological treatment is a cheap option but it may not be reliable. Such skepticism about biotreatment generally prevails among environmental engineers. A literature search shows that there is a lack of quantitative information on anaerobic biotreatment of contaminated soils. The goal of this research was to study the toxic effects of selected contaminants on biotreatment using an Inhibition Coefficient Model. To determine the extent of biodegradation, various fate mechanisms such as adsorption, chemical transformation, and volatilization were also studied using model equations and experimental data. The selected organic compounds were nitrophenols, chloro-phenols, nitrobenzene, chlorobenzene, dichloroethane and pentachlorophenol. The heavy metals studied were cadmium, zinc, and nickel. In this paper the results from the studies on nitrophenols and cadmium have been included. The focus of this research was on the hazardous leachate that could be gen erated from soil contaminated with these toxic substances. Unless the leachate is collected and treated, groundwater contamination could occur.

2 BACKGROUND

Nitrophenols are among the most important and versatile industrial organic compounds. These compounds are widely used in the manufacture of explosives, pharmaceuticals, pesticides, pigments, dyes and rubber chemicals. Nitrophenols may be inadvertently produced by microbial or photodegradation of pesticides that contain the nitrophenol moiety (EPA, 1980). Approximately 6,800 metric tons of 2-nitrophenol and 19,000 metric tons of 4-nitrophenol are produced in U.S.A. Although production data for 3-nitrophenol are not available, the amount is probably <500 metric tons. Nitrophenol production is high and increasing in many industrialized countries. Of the six possible isomers of dinitrophenols, 2,4-dinitrophenol is commercially the most important. It is used as an intermediate in the production of explosives, dyes, wood preservatives, and photochemicals. Current output now exceeds 500 metric tons per year. Among the mononitrophenols, 4-nitrophenol is probably the most important in terms of quantities used and potential environmental contaminations. Demand for 4-nitrophenol was 16,000 tons in 1976 and production in 1980 was increased to 19,000 tons (Haghighi-Podeh, 1991).

Table 1. Summary of analytical technique.

Volatile Suspended Solids-(Method 423 Standard Method)
Total Suspended Solids-(Method 209C Standard Method)
Total Volatile Acids-(Titration-Jenkins et al., 1983)
Acetate-(Gas Chromatography (GC))
Methane and Carbon Dioxide-(GC)
Gas production-(Gas Measurement Device)
Free Cadmium-(Dialysis/Ion Exchange)
Heavy Metals-(Atomic Absorption Spectrometer)
Nitrophenols-(HPLC)

Human exposure to the nitrophenols or dinitrophenols has not been monitored. Unspecified amounts of 4-nitrophenol have been detected in samples of urban ambient particulate matter. 4-nitrophenol has been detected in rainwater in Japan. Available data indicated that the general public may be exposed to nitrophenols in the atmosphere where severe photochemical fog conditions develop. Quantitative estimates of such exposures are not possible at the present time.

4-nitrophenol has been detected in the urine of 1 percent of general population at levels as high as 0.1 mg/l. However, these urine levels are not believed to be the result from direct exposure to 4-nitrophenol. A number of widely used pesticides, including parathion, are readily metabolized to 4-nitrophenol in the human body and are believed to be the source of 4-nitrophenol residues in human urine (EPA, 1980). The literature survey shows that there is a lack of quantitative information on nitrophenol toxicity in anaerobic treatment systems. More information is required on the mechanism of toxicity by these organic compounds. This research was undertaken to provide such information.

Mosey et al. (1971) stated that heavy metals cause digestion failure only when concentrations of their free ions which are in soluble form exceed a certain threshold concentration. In addition to the relative affinity of the free metal ion for various biological functional groups, it has become increasingly recognized that toxicity greatly depends on the form or species of metal to which an organism is exposed. The valency or oxidation state of a metal also has considerable influence on its bioavailability. Methylated species of mercury are even more toxic than mercuric ion (Florence, 1983). Other metals which can be biomethylated by bacteria or fungi include cadmium, lead, selenium and arsenic (Brown and Lester, 1973). Yetis and Gokcay (1989) showed that the toxic effects of metals are exerted when they are present in the dissociated or ionic form. Resistance of biological treatment system to metal toxicity may be enhanced greatly by proper acclimatization. During acclimatization, either resistant organisms are selected and/or microbes are adapted to the metal environment metabolically (Chang et al., 1986). Research with a wide range of solid retention times (SRTs) indicted that heavy metal toxicity decreases at longer SRTs (Sujarittanontal and Sherrard, 1981). Organonmetallic complexation is believed to be a major mechanism responsible for association of heavy metals with sludge solids (Gould and Genetelli, 1978). The biological effect of such complexation is that the toxicity

Table 2. Effect of cadmium on acetate enrichment systems.

Spiked Concentration	TSS (mg/l)	Free Metal (mg/l)	Soluble Metal (mg/l)	Effects on systems
Control	1,012	0.018	0.04	N
Control	1,040	0.016	0.05	N
20 PPM	1,068	0.03	1.9	N
20 PPM	1,050	0.07	2.4	N
20 PPM	1,324	0.01	1.8	N
40 PPM	1,006	0.10	3.1	N
40 PPM	1,024	0.08	1.9	N
60 PPM	1,016	0.13	3.0	F
60 PPM	1,070	0.12	0.87	F
80 PPM	1,480	0.07	0.9	N
80 PPM	1,754	0.03	0.4	N
80 PPM	1,320	0.38	3.1	F

of several transition metals to microorganisms had been found to decrease indicating that the free metal ion is the toxic form (Allen et al., 1980). The diffusion transport of metal species across a dialysis membrane provide a useful simulation of the biological process (Morrison, 1987). In addition, the membrane itself can be used to simulate the cell membrane of microorganisms. Dissolved metal species, particularly those that are free and weakly complexed are potentially available to organisms. Morrison (1987) determined bioavailability of zinc, lead and copper using dialysis with receiving resins.

The metabolisms of cadmium is closely related to zinc metabolism. Cadmium seems to displace zinc in many vital enzymatic reactions. Cadmium and its compounds have found increasing application in a variety of industrial products and operations, causing a marked increase in direct production. Cadmium is used extensively in industries such as electroplating and coating, battery and plastic manufacturing (Stephenson and Lester, 1983). The literature survey shows that there is a lack of quantitative information on cadmium toxicity in anaerobic treatment systems. More information is required on the mechanism of toxicity by cadmium. This research was undertaken to provide such information.

3 OBJECTIVES

The specific objectives of this research project were:

1. to determine the toxic effects of nitrophenols (2-nitrophenols, 3-nitrophenols, 4-nitrophenols, and 2-4-dinitrophenols) on methanogenic bacteria.
2. to determine a correlation between free bioavailable cadmium concentration and toxicity in anaerobic systems, and
3. to quantify the effects of 4-nitrophenols and cadmium using an inhibition model.

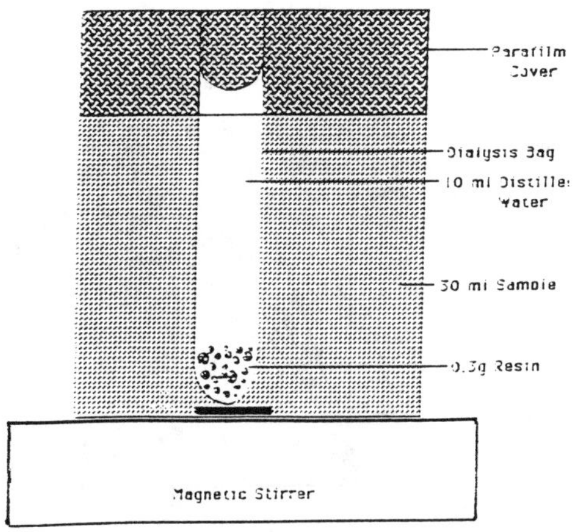

Figure 1. Dialysis/ion exchange method.

4 MATERIALS AND METHODS

Stock Culture: The stock acetate enrichment culture was developed from anaerobically digested sludge using acetic acid as the sole source of organic carbon. The culture was developed in 20-L plastic carboys in a constant temperature room ($35^{o}C$).

Anaerobic Toxicity Assay (ATA): Serum

Table 3. Effect of nitrophenol on methanogenic bacteria.

Compound(mg/l)	Recovery(day)	Compound(mg/l)	Recovery(day)
2-Nitrophenol		**4-Nitrophenol**	
1	0	1	0
10	0	5	0
20	3	10	4
30	5	20	4
40	7	40	5
3-Nitrophenol		80	Failed
1	0	**2,4-Dinitrophenol**	
10	0	1	0
20	2	5	0
40	5	10	3
		15	Failed
		20	Failed

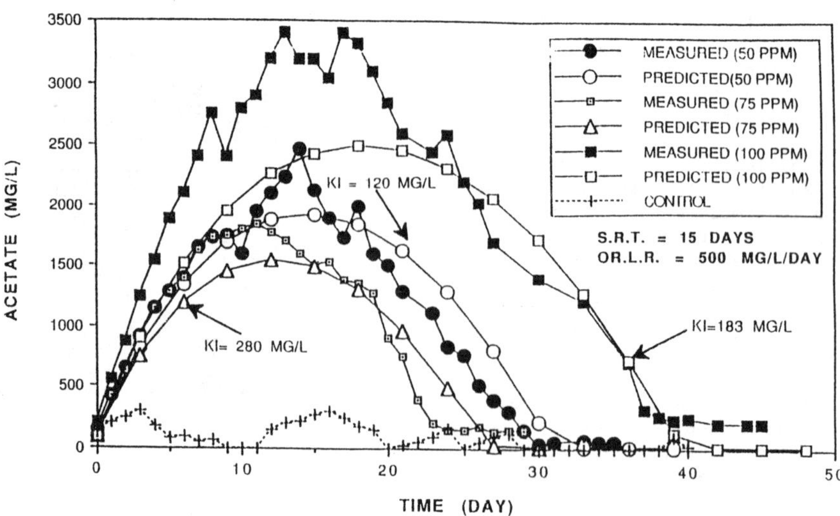

Figure 2. Comparison of predicted and observed data with slug additions of cadmium to acetate enrichment systems.

bottles (150 ml) were seeded anaerobically from the stock cultures. Clean serum bottles were first filled with tap water which was then displaced with N_2 gas. The bottles were then capped with rubber stoppers to maintain the anaerobic conditions. In each prepared bottle, 45 ml of acetate enrichment culture, 3 ml of nutrients, 2 ml of yeast extract (stock concentration of 50 g/l) were anaerobically transferred with syringes. For each experiment, triplicate controls and test bottles were maintained. All experiments were performed at 35°C in a constant temperature room. Total gas production was measured daily by using a gas displacement device. After measuring the gas production, 50μL of acetate acid was added to each serum bottle to compensate for previously consumed acid and to bring the acetate level back to 1050 mg/l.

When the serum bottles reached steady state, different concentrations of cadmium ($CdCl_2 . 2.5H_2O$) and selected organic compounds were added to the respective bottles. The serum bottles were maintained until the systems failed or there were no effects for several days.

Chemostats: Chemostats were used to model the toxic effects of cadmium and 4-nitrophenol. The chemostats were glass bottles (2-L) containing 1.5-L of acetate enrichment culture. Solids retention time/Hydraulic retention time (SRT/HRT) was maintained at 15 days. Having no recycling of sludge, SRT equals HRT in chemostats. Cassette pumps (Manostat, New York, N.Y.), were used to maintain the continuous feed. All chemostats were seeded anaerobically from the stock cultures.

Analytical Techniques: Methods used for measuring the various parameters are listed in Table 1.

A combination of dialysis and ion exchange methods were used to determine the concentrations of free cadmium in the serum bottles and chemostats. Dialysis bag (MW cut off of 10,000; Spectrum, Inc., Los Angeles, California) was used to separate high MW complexes. DOWEX SOW-X8, Na^+ form, 20-50 mesh resin (0.3g), and 10 ml deionized water were placed in dialysis bags. The dialysis bags were placed in glass bottles containing the samples which were mixed with magnetic stirrers (Figure 1). The mixing was continued for 20 hours to ensure equilibrium. Samples from both inside and outside the dialysis bags were analyzed for cadmium after 24 hours. The metal associated with resins was eluted with 1:1 HNO_3. The acid solution was then analyzed for free metal content. All samples for metal analysis were filtered (Whatman 934-AH glass microfiber filter), and the clear supernatants were used for the metal analysis. A Perklin-Elmer Model 5000 Atomic Absorption Spectrometer(AA) following Method 303A (Standard Method, 1989) with an air/acetylene flame was used to determine cadmium concentrations. Samples were centrifuged and clear supernatants were used for cadmium analysis.

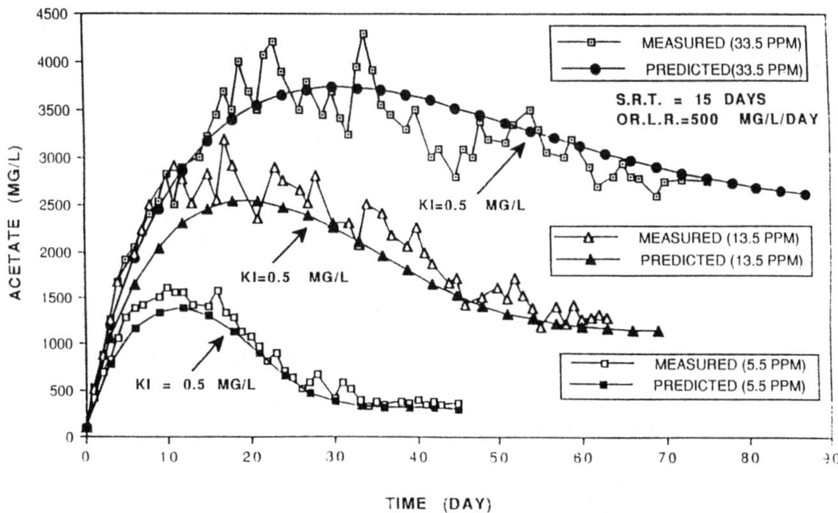

Figure 3. Comparison of predicted and observed data with combination additions of 4-nitrophenols to acetate enrichment systems.

Absorbance was measured at wavelengths of 228.8 nm and 240.7 nm for cadmium and cobalt, respectively.

Methane and carbon dioxide were measured by gas chromatography (GC) (Shimadzu, GC-8) with a thermal conductivity detector. The GC column was 6 ft x 1/8 inch stainless steel column, packed with 60/80 Chromosorb 102. The column was held at a constant temperature of 60°C. The injector and detector temperatures were held at 160°C.

5 INHIBITION MODELS

The uncompetitive and competitive inhibition models had been used in earlier reseach for toxicants such as ammonia, nickel, formaldehyde, and methylene chloride (Bhattacharya and Parkin, 1986). In this paper only the application of the models for cadmium and 4-nitrophenol has been described.

6 RESULTS AND DISCUSSION

The results from the experiments with cadmium showed a good correlation between free cadmium and toxicity in anaerobic systems. Table 2 summarizes these results. With free cadmium concentration up to 0.1 mg/L, the anaerobic systems were not adversely affected. Free cadmium of 0.12 mg/L or higher showed system failures. The soluble cadmium concentration, on the other hand, did not show any correlation with adverse effects on the anaerobic organisms (Table 2). Figure 2 shows the measured concentrations of acetate in the systems. Accumulation of acetate indicated toxic effects. Such accumulation of acetate was predicted using the uncompetitive inhibition model. This model includes the term K_I, the inhibition coefficient, which is a measure of resistance of the organisms to toxicants. Figure 2 shows that the model can predict and quantify the effects of cadmium on anaerobic systems.

Table 3 summarizes the effects of nitrophenols on methanogenic systems. Up to 10 ppm of 2- and 3-nitrophenols and up to 5 ppm of 4-nitrophenols and 2-4 dinitrophenol did not cause any toxicity. Higher concentrations of these toxicants either caused a reversible inhibition (with recovery after a few days) or an irreversible inhibition leading to system failures. Figure 3 shows that the competitive inhibition model was successfully used to predict the toxic effects of 4-nitrophenol on acetate-utilizing methanogens. The inhibition coefficient, K_I, was determined to be 0.5 mg/L. This information will be useful to predict 4-nitrophenol toxicity in other biosystems.

7 CONCLUSIONS

1. The Inhibition Models are useful to

predict toxicity in anaerobic biosystems.
2. Free cadmium concentrations show a good correlation with anaerobic toxicity.
3. Among the nitrophenols studied, 4-nitrophenol and 2-4 dinitrophenol are most toxic.

8 ACKNOWLEDGEMENTS

This research was funded by the Center for Bioenvironmental Research at Tulane University.

9 REFERENCES

Allen, H.E., Hall, R.H. and Brisbin, T.D. 1980. Metal Speciation Effects on Aquatic Toxicity. Environmental Science and Technology, 14, pp. 441-443.

Bhattacharya, S.K. and Parkin, G.F. 1988. Fate and Effect of Methylene Chloride and Formaldehyde in Methane Fermentation Systems. Jour. of Water Pollution Control Fed., 60, 531.

Brown, M.J. and Lester, J.N. 1973. Role of Bacterial Culture and Activated Sludge. Water Res., 16, 1539.

Change, M.H., Patterson, J.W., and Minear, R.A.J. 1986. Heavy Metals Uptake by Activated Sludge. Jour. Water Pollution Control Fed., 47, 2, 362.

Environmental Protection Agency 1980. Ambient Water Quality Criteria for Nitrophenols. EPA 440/S-80-063.

Florence, T.M. 1983. Trace Element Speciation and Aquatic Toxicity. Trends Anal. Chem., 2, 162.

Gould, M.S. and Genetell, E.J. 1978. The Effect of Methylation and Hydrogen ion Concentration on Heavy Metal Binding by Anaerobically Digested Sludges. Water Research, 12, 889.

Haghighi-Podeh, M.R. 1991. Fate and Effect of Cadimum, Cobalt, and Nitrophenols as Anaerobic Systems. Ph.D. Dissertation, Tulane University, New Orleans.

Jenkins, S.R., Morgan J.M., and Sawyer, C.L. 1983. Measuring Anaerobic Sludge Digestion and Growth by a Simple Alkalimetric Titration. Jour. Water Pollution Control Fed., 55, 448.

Morrison, G.M.P. 1987. Bioavailable Metal Uptake Rate Determination on Polluted Waters by Dialysis with Receiving Resins. Environmental Technology, 8, 393.

Mosey, F.E., Swanwick, J.D., and Hughes, D.A. 1971. Factors Affecting the Availability of Heavy Metals to Inhibit Anaerobic Digestion. Jour. Water Pollution Control Fed., 7, 668.

Standard Methods of Examination of Water and Wastewater 1989. 17th ed. American Public Health Association, Washington, D.C.

Stephenson, T. and Lester J.N. 1983. Heavy Metal Behavior During the Activated Sludge Process. Sci. Total Environmen., 63, 199.

Sujarittanonta, S. and Sherrard, J.H. 1981. Activated Sludge Nickel Toxicity Studies. Jour. Water Pollution Control Fed., 53, pp. 1314-1322.

Yetis, U. and Gokcay, C.F. 1989. Effects of Nickel(II) on Activated Sludge. Water Res., 23, 8, pp. 1003-1007.

Environmental Management, Geo-Water & Engineering Aspects, Chowdhury & Sivakumar (eds)
© 1993 Balkema, Rotterdam. ISBN 90 5410 099 0

Clean up operations in contaminated land

E.E. Hellawell & C. Savvidou
University of Cambridge, UK

J.R. Booker
University of Sydney, N.S.W., Australia

ABSTRACT: A solution is derived for the hydrodynamic clean up of a contaminated site. Clean up using flow reversal is investigated as a means of containing the contaminant plume whilst reducing its concentration. A closed form solution is derived for pumping with and without reversal which is then used to calibrate a finite difference formulation of the problem. A parametric study is performed to evaluate the influence of permeability, velocity and hydrodynamic dispersion upon two clean up techniques.

1 INTRODUCTION

Present techniques for the clean up of contaminated ground are often considered time consuming, expensive and unreliable. A popular method involves the hydrodynamic withdrawal of contaminants from the ground followed by surface treatment. Water is pumped through the site, flushing out the contaminant and then the contaminated solute extracted. An investigation into the clean up of Price's Landfill (Gray and Hoffman 1983) revealled the difficulty of capturing all the pollutant and highlighted the fact that the plume front may move past the extraction wells. Such data from full scale insitu operations are rare due to lack of extensive fully monitored activities and also because of confidentiality. Using a simple model an analysis and parametric studies can be obtained which can give valuable guidance in the planning of remedial actions.

This paper investigates flow reversal as a method of containing the plume during clean up and compares this technique to the present method of flushing in one direction. The closed form solution is derived for one-dimensional hydrodynamic clean up. The model considers the migration mechanisms advection and dispersion/diffusion, but assumes a non sorbing contaminant.

2 THEORY

2.1 *Governing Equations*

The one-dimensional advection-dispersion equation based on the approximation of a Fickian-type dispersion law (Gillham and Cherry, 1982) has the form:

$$D_{hL}\frac{\partial^2 c}{\partial x^2} - u\frac{\partial c}{\partial x} = \frac{\partial c}{\partial t} \qquad [1]$$

in which the first term on the left represents dispersive transport and the second term advective where u : average pore fluid velocity c : concentration (mass/unit volume of fluid), D_{hL}: longitudinal hydrodynamic dispersion coefficient

2.2 *Non-Dimensional representation*

The variables in the the two models can be non-dimensionalised. This enables comparisons to be made and the results to be adapted to similar situations. Thus the one dimensional advection-dispersion equation for a non reacting solute becomes

$$\frac{\partial^2 C}{\partial X^2} - U\frac{\partial C}{\partial X} = \frac{\partial C}{\partial T} \qquad [2]$$

The non-dimensional groups are

$$C = \frac{c}{c_o} \qquad X = \frac{x}{L}$$

$$U = \frac{uL}{D_{hL}} \qquad T = \frac{D_{hL}t}{L^2}$$

where c_o is the initial concentration of the contaminant and L is the overall length.

2.3 *Boundary Conditions*

Consider a layer of contaminated soil of length L, bounded above and below by impermeable soil. A well is constructed at each end. The concentration of solute in each well is assumed to be that of clean water. (Figure 1)
Thus

$$C(0,T) = C(L,T) = 0 \qquad [3]$$

2.4 *Initial condition*

Initially the site is assumed to have a contaminant concentration distribution F(X). The site is then flushed in one direction until time T" when G(X) describes the new distribution. Thus

$$C(X,0) = F(X)$$
$$C(X,T") = G(X)$$

2.5 *General solution*

Equation [2] may be solved, satisfying the boundary conditions by the method of separation of variables, to give

$$C(X,T) = \sum_{n=1}^{\infty} A_n\ \phi_n(X)\, e^{-\lambda_n[T-T']} \qquad [4]$$

where

$$\phi_n(X) = \Delta_n\, e^{\mu X} \sin(n\pi X) \qquad [5]$$

$$\mu = \frac{U}{2}$$

$$\lambda_n = \frac{U^2}{4} + n^2\pi^2$$

A_n are constants to be determined

It is also useful to define the associated set of orthogonal functions $\phi_n(1\text{-}X) = \psi_n(X)$

Introducing the initial condition into equation [4] gives

$$f(X) = \sum_{n=1}^{\infty} A_n\ \phi_n(X) \qquad [6]$$

Δ_n can then be conveniently chosen such that

$$\int_0^1 \phi_n(X)\ \psi_n(X)\ dX = 1 \qquad [7]$$

so that

$$\Delta_n = 2\,(-1)^{n+1}\, e^{-\mu} \qquad [8]$$

thus enabling A_n to be evaluated

$$A_n = \int_0^1 f(X)\ \psi_n(X)\ dX \qquad [9]$$

2.6 *Flow reversal*

At time T" the concentration profile in the site is g(X). The direction of flow is then reversed. The governing equation becomes

$$\frac{\partial^2 C}{\partial X^2} + U\frac{\partial C}{\partial X} = \frac{\partial C}{\partial T} \qquad [10]$$

The solution to this, satisfying the initial and boundary conditions is

$$C(X,T") = \sum_{n=1}^{\infty} B_n\ \psi_n(X)\, e^{-\lambda_n(T-T")} \qquad [11]$$

It may then be shown (analogous to equations 5-7) that

$$B_n = \int_0^1 g(X)\ \phi_n(X)\ dX \qquad [12]$$

This solution may be obtained directly or by introducing a system of matrices as outlined below. These two methods enable cross checking of the analysis and computer programs to ensure accuracy.

2.7 *Alternative solution using matrices*

$\underline{\Phi}$ and $\underline{\Psi}$ are defined as matrices containing the functions $\phi_n(x)$ and $\psi_n(x)$ respectively. These can be related by introducing a third matrix $\underline{P}$, thus

$$\underline{\Phi} = \underline{P}\,\underline{\Psi} \qquad [15]$$

It can be shown using equation 4, in the first phase of flow at time T

$$C(X,T) = \underline{A}^T\,\underline{D}\,(T-T')\,\underline{\Phi} \qquad [16]$$

where

$$\underline{A}^T = [A_1\ A_2\ A_3\ \ldots\ \ldots]$$

$$\underline{D}(T-T') = \begin{bmatrix} e^{-\lambda_1(T-T')} & 0 & 0 & \ldots \\ 0 & e^{-\lambda_2(T-T')} & 0 & \ldots \\ 0 & 0 & e^{-\lambda_3(T-T')} & \ldots \\ \ldots & \ldots & \ldots & \ldots \end{bmatrix}$$

Thus for equal reversal periods of $\bar{T}$, it is possible to show that

$$C(X,0) = \underline{A}^T\,\underline{\Phi} \qquad C(X,\bar{T}) = \underline{A}^T\,\underline{Z}\,\underline{\Psi}$$

$$C(X,2\bar{T}) = \underline{A}^T\,\underline{Z}^2\,\underline{\Phi} \qquad C(X,3\bar{T}) = \underline{A}^T\,\underline{Z}^3\,\underline{\Psi} \qquad [17]$$

where $\underline{Z} = \underline{D}\,(\bar{T})\,\underline{P}$

2.8 *Finite difference formulation*

Equation [2] may be solved numerically using a forward difference formulation.

The solutions to equation [2] obtained using the general analytical equation, matrices and finite differences were then compared. Excellent agreement was obtained except in highly advective flow when instabilities occurred.

3 RESULTS

The solutions developed can be used to derive predictions and compare clean up operations

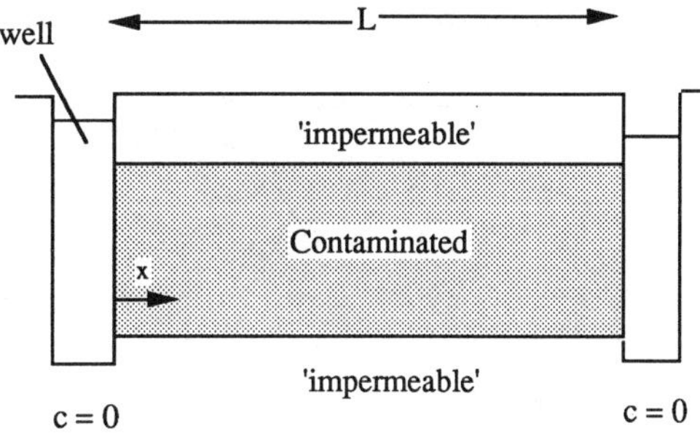

Fig. 1 One dimensional model of a contaminated site

using flow reversal to ones that use flushing only in one direction. For simplicity a 1m site is considered and an initial uniform contaminant concentration is assumed. For each soil type hydrodynamic dispersion was evaluated from the Peclet number and the diffusion coefficient of migration of a chloride ion (Bear 1972). A parametric study is performed by varying the soil and flow parameters (hydraulic permeability k, u, D_{hL}) for the two clean up techniques.

The clean up of contaminated sites for three different permeability soils, typical of sand, silt and clay were compared. Pumping was considered to establish the same head across each site. Figure 2a shows the variation in contaminant concentration in the sand site with lateral position after 0.5, 1 and 1.5 hours using the two clean up techniques. Similar plots for the clean up of silty and clayey soils are shown in figures 2b and 2c respectively. Advection is seen to be the dominant transport mechanism in the highly permeable sand. Diffusion becomes more important as permeability decreases, occurring at the site edges due to the large concentration gradients. Hydrodynamic clean up is shown to be unsuitable for low permeability soils. Clean up without using flow reversal is shown to be more effective at reducing the maximum and average contaminant concentrations. The reversal technique however is effective in keeping the maximum concentration more central and preventing spreading of the contaminant.

The time to reduce the average concentration in the site by 75% for a range of pore velocities is shown in figure 3. Dispersion was assumed

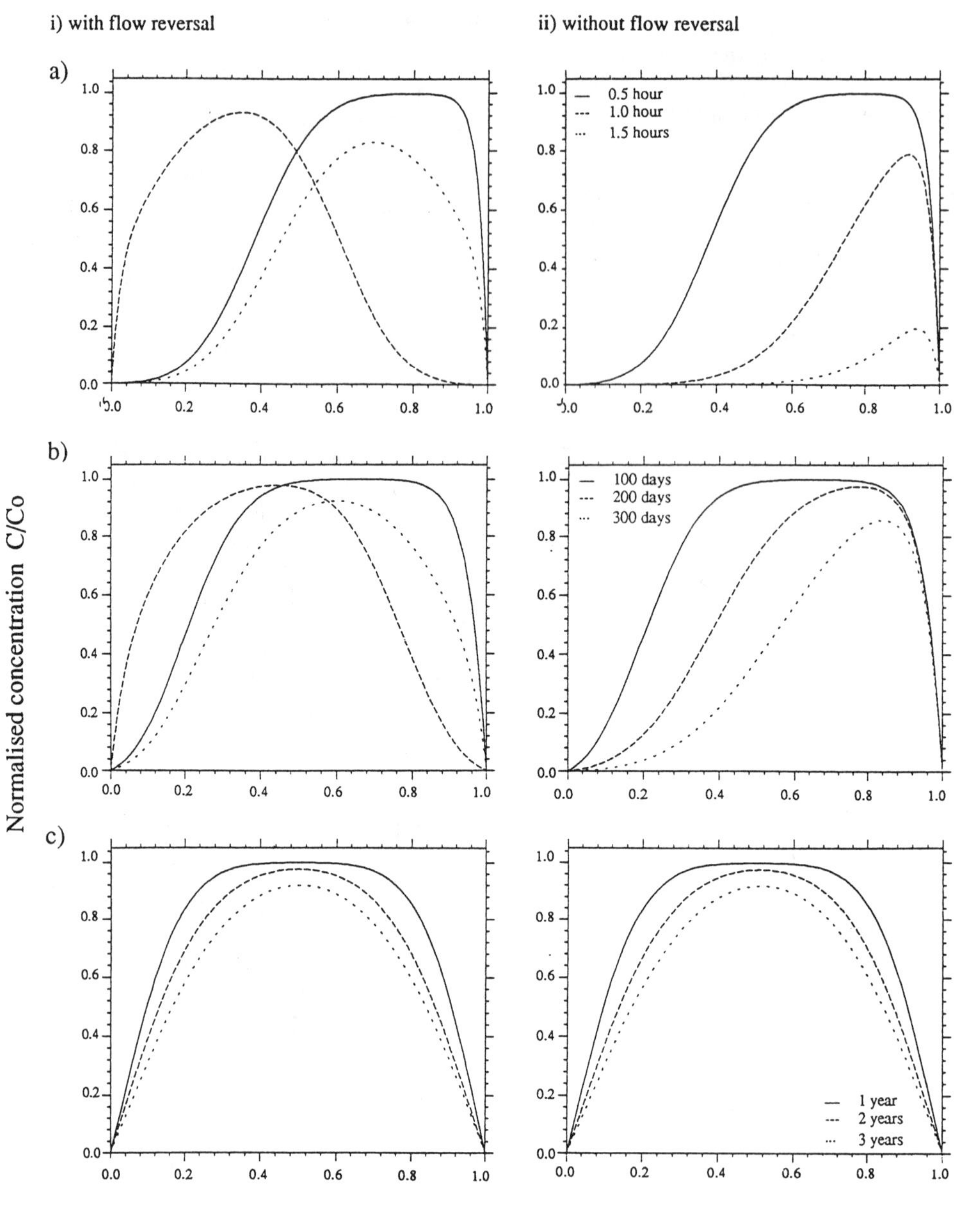

Figure 2 - Variation in contaminant concentration with distance during clean up of a) sandy soil, b) silty soil and c) clayey soil

to be a function of the pore velocity. It is clear that the flow reversal technique takes longer to reduce contaminant concentrations.

The time to reduce the maximum concentration of a site by 75% for a range of dispersions at constant pore velocity is shown in figure 4. A decrease in the hydrodynamic dispersion has a great effect on clean up times particularly for the flow reversal technique.

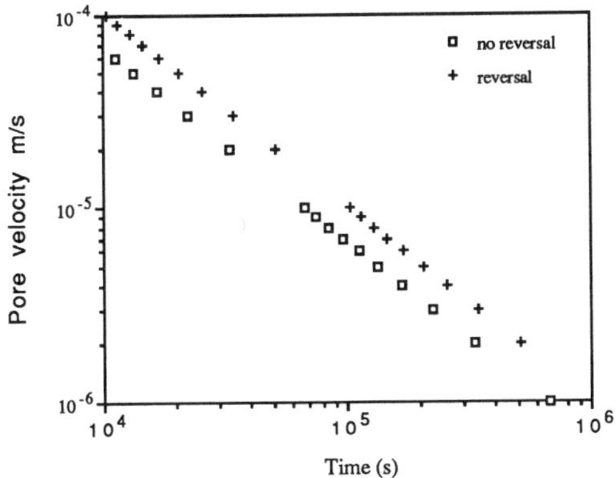

Figure 3 - Effect of pore velocity on the time to reduce the contaminant concentration by 75%

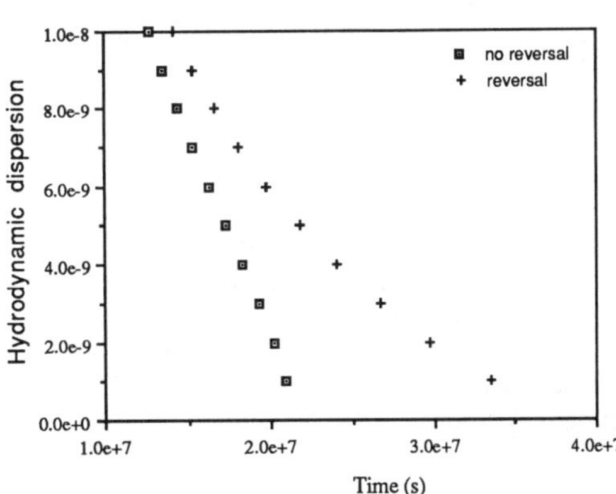

Figure 4 - Effect of disperson on the time to reduce the peak contaminant concentration by 75%

a) with flow reversal

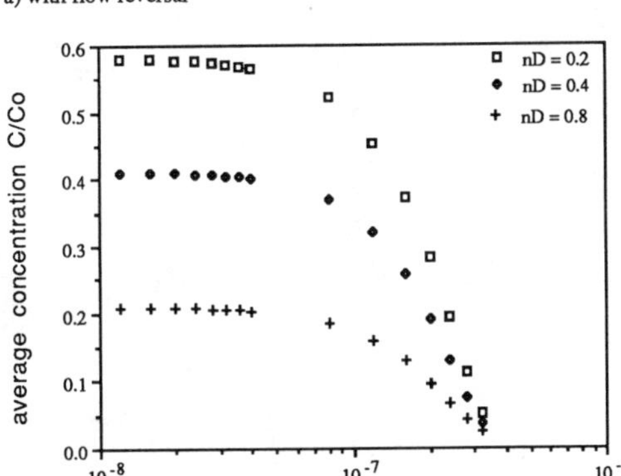

b) without flow reversal

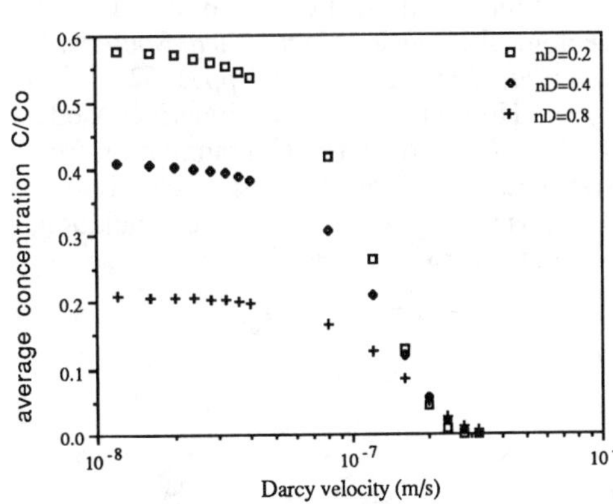

Figure 5 - The effect of Darcy velocity on the average contaminant concentration of a site after 25 days

The variation in maximum contaminant concentration after 25 days of clean up with Darcy velocity is shown in figure 5 for different values of nD. Dispersion is shown to be important at low values of velocity.

The variation of maximum concentration with time for the two techniques is shown in figure 6. As with the previous graphs, flow reversal is the less effective technique at reducing the maximum concentration.

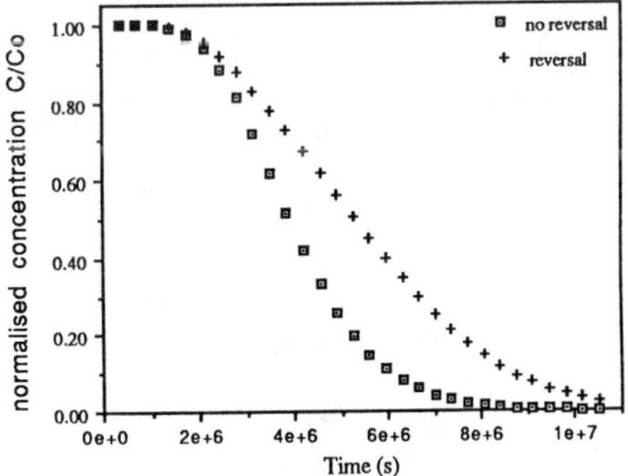

Figure 6 - Variation in peak concentration with time during clean up for a site of contaminated silt

4 CONCLUSIONS

A closed form solution of the one-dimensional hydrodynamic clean up of contaminated land with and without reversal has been obtained. It has been shown that:

Clean up without reversing the flow is a more effective technique at reducing the maximum and average contaminant concentration but that the peak concentration is more centralised if flow reversal is used. The latter technique may be more effective in avoiding the spreading of the pollutant.

In the case of clayey soils other remedial action techniques should be applied.

REFERENCES

Bear, J (1972) *Dynamics of Fluids in Porous Media,* American Elsevier, New York

Gilham and Cherry (1982) Contaminant Migration in Saturated Unconsolidated Geological Deposits, *Geophysical Society of America,* Special Paper 189 pp31-62

Gray and Hoffman (1983) A Numerical Model Study of Ground Water Contamination from Price's Landfill, New Jersey II- Sensitivity Analysis and Contaminant Plume simulation, *Groundwater* vol. 21 No. 1

Environmental Management, Geo-Water & Engineering Aspects, Chowdhury & Sivakumar (eds)
© 1993 Balkema, Rotterdam. ISBN 90 5410 099 0

Engineering environmental safeguards for disposal of coal washery reject

D.R.Jephcott
Austen & Butta Limited, Australia

ABSTRACT: During the past 25 or so years in Australia the safe and economical disposal of coal washery waste has become increasingly more important. During this period disposal of the waste by landfill emplacement has been the traditional method employed by many companies including Austen & Butta Limited at its South Bulli Colliery. On the narrow Illawarra coastal plain competition for land is intense and it is of real importance that the final land use for any area of extensive land fill is one that is appreciated by as wide a cross section of the community as possible. At the planning stages it is essential that wherever possible potential environmental hazards are engineered away. Correct drainage is of particular importance. This paper primarily explains how the development of a golf course using coal waste assists in maintaining the viability of local industry and at the same time, provides a popular recreational resource. Reference is also made to the use of coal washery waste for a widening range of purposes.

1 INTRODUCTION

Austen & Butta Limited owns and operates South Bulli Colliery which has been located at Russell Vale on the South Coast of New South Wales, Australia, for over 100 years. During the last 30 years the colliery has been a major supplier of coking coal to the South East Asian and European steel mills and more recently has supplied fuel coal to a range of markets. The combined effect of market quality requirements, modern mining methods and coal preparation techniques has, as a matter of course, led to the production of substantial quantities of reject material.

Traditionally, the satisfactory disposal of this reject material has been based upon land fill techniques. Many local sporting facilities have resulted from land fill operations over the years, and these include playing fields, ovals, a racetrack and general recreation areas. Inert cover for garbage disposal areas is another use for the reject material. All of these land fill projects have had to comply with the requirements of the many interested Local and State Government authorities which in essence mean that the finished work must be stable, environmentally acceptable and aesthetically pleasing.

In recent years the shortage of land in the Illawarra District for the disposal of industrial wastes generally and coal washery wastes in particular has led to the coal industry more closely examining its waste disposal techniques. The options for localised long term landfill projects have been increasingly more limited and wherever possible a more innovative approach has been required if the washery in question was to remain in the district.

At South Bulli the waste disposal problem has been attacked in the following ways:

1. Where deep (10 metres) landfill disposal is appropriate the final land use plan has been one of creating a more worthwhile facility than previously, e.g. from land suitable for grazing cattle to a golf course.
2. Where shallower (0.5 to 3 metres) landfill disposal is appropriate areas have been developed for industrial, residential and commercial uses apart from sporting facilities mentioned earlier.
3. Fine (sand sized) tailings is sought after by plant nurseries for use

in potting mix and is also popular for lawn topdressing and backfilling under concrete slabs and around houses.

4. Commercial brick manufacturing is a proposed use currently under investigation.

5. Road and highway construction continues to be an avenue for the use of coal washery waste as a suitable sub-grade alternative to other traditional materials.

In all of the above cases 1. to 5. and with most other processes, good quality control of the raw material is essential to facilitate sound end results. In this regard South Bulli has tightened its control over the quality of coal washery waste in recent years to ensure that the "market" for the waste remains firm. Currently, demand far exceeds supply for good clean coarse reject and sand sized fine tailings. Clay slimes may be mixed with coarse reject for general fill purposes where quality control requirements are less stringent but still important.

In this paper I will expand upon the environmental safeguards employed for deep filling during construction of the golf course. The principles applied here are also basic to those required for successful emplacement of coal waste in shallower areas which are to be used for industrial, residential, commercial and roadway construction purposes.

2 DEVELOPMENT APPROVAL

As with most large scale emplacement proposals approval from Local and/or State Government authorities is required.

In 1977 a development application was made to Wollongong City Council by South Bulli Colliery for the purpose of emplacement of coal washery reject material in an area of its own freehold land immediately north of the coal preparation plant. The overall proposal was that an eighteen hole championship golf course would be built on land already being filled by the Council and operated as the Northern Suburbs Waste Disposal Depot, plus land to be filled by the Company. Conditional approval was granted in 1978 to initially fill a gully adjacent to Council's Waste Disposal Depot. This first area identified as the Northern Gully, was to provide sufficient space for reject material to be placed for eight years. The ultimate landform proposal incorporating all areas amounted to a total area of 60 hectares.

During the period 1978 to 1984 filling progressed according to plan and as final levels were attained and grass and trees planted, moves were made by Council to develop as an interim measure a nine hole par 27 course. Council established the trees, fairways and greens during 1985 and a section of filled area was opened to the public as a golf course in March 1986. The course was subsequently extended by an additional nine holes and an eighteen hole par 54 course was in use by mid 1990. This facility will serve the community until the total area becomes available for the par 72 eighteen hole course about five years from now.

In early 1986 the Minister for Environment and Planning granted approval for the second stage of the project to continue in the Southern Gully. As with the other areas of the project detailed attention was given to all environmental aspects and these were submitted in the form of an Environmental Impact Statement when the development application was first lodged.

3 CONSTRUCTION OF THE GOLF COURSE

The different processes involved in creating the golf course can be identified as follows:

3.1 The preparation of the coal for sale

The coal preparation process which brings about saleable products also causes large quantities of reject material to be produced.

At South Bulli approximately 85% of the reject is "coarse", this is in the size range 125 x 0.5mm with the balance as "fines" or minus 0.5mm. These materials are trucked to the emplacement area separately.

3.2 The emplacement process

The preparation of the area for disposal of the reject material involved:

1. Site selection - as discussed in paragraph 2 above.
2. Site approval - also discussed in paragraph 2 above.
3. Site preparation - this involved the removal of timber from the site, construction of diversion drains to control clean surface water run off and the stripping and stockpiling of topsoil and removal and stockpiling of selected subgrade material.
4. The identification of wet areas and installation of sub-surface drain-

age pipes and construction of filter dams for treating drainage from "dirty" areas.

5. Construction of a coarse reject drainage blanket as a first layer over the floor of the area being filled.

6. Construction of coarse reject walls to form cells for mixing the fine and coarse reject material. The size of these cells average 30x30 metres square with a wall height of 2.0 metres. This size of cell will hold approximately 1000 tonnes of fine material containing approximately 40% water by volume.

3.3 Placement of the fines material (tailings)

The fines are partially dewatered in two solid bowl centrifuges and a band press filter as part of the normal coal preparation plant operation. A flocculant is added prior to mechanical dewatering to ensure that the fines material produced is handleable to the extent that it can be safely trucked. The haul road to the current emplacement area is only 200 metres from the preparation plant. Nevertheless it remains an important aspect of the operation that the fines are as viscous as possible to prevent the risk of spillage. This is important as the haul road crosses the main access road to the colliery pit top. Particular attention is paid to sealing the tailgate of the fines truck and an additional safeguard is that the tailgate is hydraulically locked into position.

3.4 Mixing coarse reject with the fines

This operation is achieved by backfilling with coarse reject across the surface of the cells containing the semi fluid fines. A caterpillar D7 dozer is used initially to work the coarse reject into the soft fines. The aim is to produce a homogenous mix of the coarse and fine material to provide overall strength and therefore stability of the emplaced material. Once the dozer has worked the area a vibrating roller is used to "puddle up" the fines to the surface through the coarse reject. Further compaction is achieved as fully laden trucks carrying coarse and fine material use the surface of the cell as a roadway during construction of the next layer of cells. Experience has shown that as the fines and coarse material mix dries out excellent compaction is achieved. Dewatering of the mix is facilitated by the coarse reject cell walls which readily drain water away and are located by design at each level through the entire emplacement area. Thus these walls provide good strength and drainage capabilities and so overall stability.

3.5 Monitoring

A general monitoring programme has always been part of the emplacement operation and involves various inspections, sampling, recording and reporting before, during and after emplacement. In particular the following aspects are considered:

1. Before emplacement - as part of the Environmental Impact Statement, background readings relating to air quality, noise levels and water quality were determined. Drainage features, apart from creek lines, such as springs and associated slip areas were identified.

2. During emplacement - parameters which are monitored at set intervals include:

2.1 Water quality of the local creek and emplacement leachate. The discharge of leachate from the emplacement area is licensed by the Environment Protection Authority. Dry weather flows have proved to be of very acceptable quality and one main reason for this is that the leachate which drains from the tailings is clear of solids because of the flocculant mentioned in paragraph 3.3 above. Wet weather run off from disturbed areas on and around the emplacement area constitute a potential pollution hazard. As already mentioned in paragraph 3.2 clean water is diverted wherever possible. However the remaining contaminated stormwater is collected in settling dams. Overflow from these dams discharges into filter beds which further ensure removal of contaminants.

2.2 Air quality - dust deposition gauges are located at selected positions to record fallout. A range of dust control measures are employed to minimise this hazard and these include spillage avoidance as already referred to in paragraph 3.3 above, use of mobile water tankers during and after production periods and the addition of agglomerating agent to bind the fine surface dust particles. This chemical is mixed into the water being sprayed by the mobile tanker.

2.3 Compaction testing - at various levels as filling progresses compaction testing is carried out. Results which are generally excellent are recorded on a plan of the emplacement area and along with other results are forwarded to Council as required.

2.4 Temperature and water table level monitoring - these parameters are determined via cased boreholes which are drilled down through the emplaced material once final levels have been achieved.

2.5 Vertical and horizontal movement - when final levels are reached the area is surveyed to detect any ground movement. Results show that in areas filled to a depth of up to 20 metres movement is measured in only a few millimetres. The maximum readings since commencement of the emplacement project in 1978 are 33mm vertical and 15mm horizontal. Such stability is encouraging for the long term view of establishment of expensive greens over the surface.

3. After emplacement - monitoring for surface movement, leachate quality, water table levels and temperature changes will continue for some years after emplacement until consistent results show the area is stable.

Vegetation growth in terms of grass, trees and shrubs will need to be monitored to ensure the successful rehabilitation of the area.

If bare patches occur then these will require topsoiling and with regular fertilising good growth will continue. In view of the proposed final land use, good maintenance of the grass, trees and shrubs will be an essential part of the management of the area in any case.

Surface scour due to failure of drainage facilities is an area of particular importance, however again due to the proposed use of the land such potential problems should be well monitored by good golf course management.

4. Overburden replacement - overburden, previously won and stockpiled, is spread, using a dozer or grader, to a depth of 300mm. This material serves the purpose of insulating the emplaced material against entry of fires and also provides a moisture and nutrient bank for the grass.

Where trees and shrubs are to be grown between fairways, an additional layer of overburden will be provided to give an improved growing medium.

5. Topsoiling - Topsoil, either freshly won or previously stockpiled, is spread over the surface to a depth of 150mm. Kikuyu runners are incorporated into the topsoil wherever possible.

6. Revegetation - the practice to date of using kikuyu runners as a quick growing grass cover has generally been very successful. During winter months or if kikuyu runners are unavailable, seed is sown. The mixture includes White Clover and Perennial Rye. "Grower II" fertilizer has been spread at the rate of 250kg per hectare over the finished surfaces to provide additional growth.

Generally speaking, revegetation is relatively simple on the coast because of the mild climate and good rainfall. In some areas where minor surface scour occurred during heavy storms prior to a complete grass cover being achieved, topsoil has been used to repair the damage.

Control of dust during revegetation periods has also been improved by the addition of surface agglomerating agents otherwise used as a dust control measure on the coal stockpiles. Also, unstable topsoil stockpiles are treated in this way.

7. Green construction - an important aspect of green construction is correct drainage and coarse coal washery reject has proved to be an excellent material for this purpose. As required, the areas designated as greens were brought to level using the coarse reject to overcome side slope and to provide the desired shape of the green and its immediate surrounds.

Hybrid bent grass has been established in the specially selected sandy topsoil which overlays the coarse reject to a depth of approximately 150mm. As expected, the establishment of this grass required large quantities of fresh water and the excellent results can be seen from the photographs.

8. Fairway and tee construction - in areas such as grassed ramps, tees and other selected sections of fairways which receive heavy pedestrian and/or some vehicular traffic, good drainage again was of particular importance. Accordingly, wherever possible selected coarse reject was used to achieve these requirements. Where a significant drainage problem needed to be solved, reject from the rotary breaker at the washery was used to form a rubble drain. This material is harder than most of the other reject material and is of a coarser size, ie 100 to 200mm diameter.

Some sections of the golf course, particularly close to the club house and car park, have been paved. As a foundation material, coarse reject was used to ensure good drainage and stability. At the golf course and at South Bulli Colliery experience has shown that coal reject material, when placed correctly, is an excellent subgrade material for vehicular traffic both light and heavy.

4 CONCLUSIONS

The Illawarra region of New South Wales is featured geographically as a narrow coastal plain limited in area by the Pacific Ocean to the east and the Illawarra Escarpment to the west. Industrial and urban expansion is thus severely limited now after more than 150 years of development. Competition for land use requires that planners at all levels in both private and public sectors need to ensure the best utilisation of all available land.

The decision to construct an eighteen hole championship golf course at Russell Vale north of Wollongong came about through the need for a suitable site of more permanent nature and in close proximity to the colliery to dispose of coal washery reject material. The site was originally considered as a possible filling area in 1970 and when finally selected in 1978 it had the twin advantage of being adjacent to the colliery and adjacent to an area in use by Wollongong City Council for disposal of household waste.

Filling in both sites has progressed to a point where an 18 hole par 54 course has been operating for several years. A small club house is in use and a professional golfer with two assistants has been appointed.

To ensure sound development of the project, close environmental monitoring has been a feature throughout. Parameters monitored include air quality, leachate quality, noise levels, horizontal and vertical movement, compaction testing and water table and temperature levels. Sound rehabilitation and drainage principles have been followed also. This work ensures that the finished project is one which will provide a stable, aesthetically pleasing and valuable community resource for future generations whilst fulfilling current needs of industry and local residents for disposal of waste materials.

Although sporting facilities are important, the demand for coarse reject for other purposes has increased recently. In particular Wollongong City Council now specify that coarse reject from South Bulli Colliery is suitable as backfill around building sites. Accordingly the building industry in and around the Wollongong Central Business District during recent years has seen significant quantities of coarse coal reject used. Projects such as the new grain terminal at Port Kembla Harbour have used large quantities of the material also. In August 1987 the Department of Main Roads commenced using reject as subgrade material for the extension of the Northern Distributor in Wollongong. Further, Wollongong City Council have a major subdivision being filled at Bellambi. This area will be for industrial and residential development purposes. Also, in early 1987 an application for financial assistance through The National Energy Research Development and Demonstration Council was approved to investigate the use of coal washery reject as a suitable material for the manufacture of bricks. Results to date are very encouraging and work is continuing on that project.

REFERENCES

Ellis, J. 1986. Monitoring of effluents from landfill sites at Whytes Gully and Kembla Grange; paper presented on 11th July 1986 at the 12th Symposium of the Water Research Foundation of Australia's Illawarra Regional Committee.

Foreman, I.K. 1981. The suitability of coarse coal washery refuse for the various stages of pavement construction. Thesis for the award of Master of Engineering from The University of Wollongong.

Jephcott, D.R. 1986. From black waste to greens. Paper given AMIC Conference Launceston 1986.

Johnstone Environmental Technology Pty Limited. 1989. Environmental Impact Statement. Coal Washery Reject Emplacement in the Southern Gully Russell Vale NSW Stage 3. Report prepared for Austen & Butta Limited.

Longworth & McKenzie Pty Limited. 1984. Environmental assessment for extension of the coal washery reject emplacement in the northern gully Russell Vale, NSW. Report prepared for Austen & Butta Limited.

Longworth & McKenzie Pty Limited. 1985. Environmental Impact Statement. Coal Washery Reject Emplacement in the Southern Gully Russell Vale, NSW. Report prepared for Austen & Butta Limited.

Environmental Management, Geo-Water & Engineering Aspects, Chowdhury & Sivakumar (eds)
 ISBN 90 5410 099 0

Solid waste treatment: A steel industry perspective

D.E. Jones
BHP Steel Slab & Plate Products Division, Wollongong, N.S.W., Australia

SYNOPSIS: An inevitable result of any manufacturing process is the production of other materials. These associated products often represent additional business opportunities, for the generator or some other party. In an integrated steelworks, the tonnage of associated products is about equal to the tonnage of primary steel produced. Many of these materials are energy or ferrous based, with the other significant volume being useful in the construction industry. A significant quantity of these associated products have markets of their own either within the steelmaking process as raw materials or as products for other industries.

This paper examines some of the work undertaken by BHP Steel Slab & Plate Products Division over the past 30 years in recycling and the development of markets for associated products. During this period,there has been an increasing recognition of the value of some associated products and a range of management and marketing techniques.In recent times there have been some significant changes in the way the handling and marketing of some of the major associated products is carried out . A move to focus on core steel making activities has been one of the driving forces. Environmental and cost pressures are causing industry to focus on waste management as a matter of urgency.These programmes and the impact of our efforts to date on the business are discussed.

INTRODUCTION

Modern society relies heavily on the products of industry for the very building blocks of our civilisation. Steel, concrete, glass and plastic surround and support every aspect of modern living. None of these products is present in final usable form in nature, having to be manufactured from ores and minerals which are present in the earth as oxides, silicates sulphides or minerals such as oil or coal. To extract metals in particular, there is usually a process (often pyrometallurgical) which splits the ore into metal and an associated product, containing the silicates aluminates and oxides and other materials present in the ores, fluxes and fuels used. The quantity of associated products generated will depend on the concentration of the metal in the ore body.

Steel, as used by the customer, is the principal end product resulting from the refining of iron ore in an integrated steelworks. There exists a series of linked processes, all necessary to produce high quality steel to exacting customer requirements. At BHP Steel's Slab and Plate Products Division at Port Kembla (the largest integrated Steelworks in Australia), a number of refinement processes are necessary to produce finished steel from iron ore and scrap. At each processing step, a primary product and one or more associated products result. Each of the processes, their purpose, products and associated products will be described in this paper. For each of the associated products (by products) their current utilisation status is also described.In round figures, one tonne of associated product is generated for each tonne of raw steel production. This means the production of some 4.1 million tonne of associated products, worth more than $132 million each year to the Slab and Plate Products Division.

As will be shown below, many of these associated products are metallic or energy based and as such are recycled within the plant. Products not used internally are marketed as raw materials to other industrial production or construction processes. Some associated products do not as yet have a productive use and are placed in landfill sites. While these products may at present be "wastes", at some future trial they may become useful products.

The following definition is currently being used as part of the Port Kembla Steel Works waste management plan:

"A waste is any material (or part thereof) for which, at the current time, there is not the technology or market to enable its use.

The distinction between solid and liquid wastes needs to be made. For this paper, solid waste refers to any waste which has a spadeable consistency. This includes sludges which, because of their physical characteristics may have quite high moisture contents (even as high as 50%). Further, while there are many liquid and some gaseous associated products of the steel industry, only those considered to be solid will be considered for this paper.

The underlying principle applied to the solid associated products generated from steel production and manufacture, is to market them externally or internally to achieve the best value to the SPPD business. In some cases this involves SPPD taking some value added step to achieve it. More recently, as in the case of blast furnace and steel furnace slag, the principle has been to contract these non core SPPD activities to others for whom such business is core. This is giving rise to the generation of satellite businesses developing around the core steel business, based on further value adding associated products of steel production and manufacture. Minimisation of Waste is now a key issue for cost effective and environmentally aware management of large enterprises such as the steel industry.

SUMMARY OF PROCESSES AND PRODUCTS AT SPPD PORT KEMBLA.

To produce 4 million tonne of steel each year at the Port Kembla Steelworks involves the bringing together of approximately 5 million tonne of iron ore, 1.3 million tonne of limestone and other fluxes, and 5 million tonne of coal, with significant amounts of natural gas, electricity, paper, oxygen, refractories, acids and a myriad of other materials. Some of these materials are ready for use, like sulphuric acid paper and refractories, while others like iron ore fines, coal and limestone require some primary processing even before they can be used in the basic metal separation and refining processes. The primary and associated products from each of the processes involved in iron and steel making are shown in Table I.

Coalwash

This product contains the shales from the process of washing coal, to reduce its ash content .Each year, approximately 1.3 million tonne of coalwash is produced. Of this total, 21.5 % is used for construction fill applications. Such applications included the establishment of screening mounds for landscaping and afforestation around industrial and light industrial establishments, the development of playing fields, construction fill applications, land reclamation and covering municipal garbage. Other uses include production of BHP soilmix,

Ammonium Sulphate

Ammonium Sulphate is the product resulting from the extraction of ammonia from raw coke ovens gas. All of the Ammonium Sulphate produced is used in the fertiliser industry as a direct application product or in blending with other products, to produce combination fertilisers. The Ammonium Sulphate from Port Kembla is sold at market rates on both the domestic and overseas markets.

Blast Furnace Slag

Blast Furnace Slag is cast from the furnace along with the molten iron. It is separated from the iron by gravity, in the trough adjacent to the taphole. Once separated, the molten slag can be granulated, pelletised or allowed to air cool. While some holding stockpiles of blast furnace slag may be held from time to time, very little of it is irrevocably dumped as waste. BHP Steel has entered into a long term contract with service company, Australian Steel Mill Services (ASMS), to receive molten slag at the blast furnace, solidify, produce and market the products required by the market place and to develop new markets for slag products.

Blast furnace dust

Blast furnace dust is extracted from the gas and collected. It is then recycled to the sinter plant, via the ore blending beds as a source of iron units in sinter production.

Thickener slurry

Thickener slurry is the fine dust fraction which is wet scrubbed from blast furnace gas. It is collected in the thickener and mixed with a similar dust fraction from the BOS. The low Zinc fraction is cycloned out, dried and recycled through the sinter plant. The high Zinc (up to 2% Zn) fraction is landfilled, pending a process to economically retrieve the zinc and iron units.

Steel furnace Slag

Steel furnace slag is decanted from each ladle of steel and solidified by pouring it into bays adjacent to the BOS building.Since the slag arises as an associated product with steel, it contains a considerable amount of free metallics. Once cooled and dug from the slag bays, steel slag is recycled through a metallic separation process. Here around 20% of the material is removed as metallics, further refined and returned to the steelworks via the sinter plant as metallic sinter fines

Summary of processes and products at SPPD Port Kembla.

TABLE I

Process	Primary Product	Associated Product(s)
Coal	Clean Coal	Coalwash
Washery	Middlings	
Coke		Coke Ovens Gas
Ovens	BTX	
	Naphthalene	
	Tar	
	Ammonium Sulphate	
Sinter	Sinter	Sinter Fines
Plant		Recovered Heat
Blast	Molten Iron	Blast Furnace Slag
Furnace	Blast Furnace Gas	
	Blast Furnace Dust	
	Thickener Slurry	
	Electricity	
BOS	Molten Steel	Steel Furnace Slag
	BOS Dust	
	Flue Gas	
	Kish	
Slab	Steel Slabs	Tundish Skulls
Caster	Crop Ends	
	Caster Scale	
Rolling	Rolled Steel	Steel Offcuts
Mills	Mill Scale	
Tin	Tinplate	Scrap Offcuts
Mill	Strapping	Recycled Tin Cans
	Spent Palm Oil	
	Spent Pickle Liquor	

and the BOS as scrap and coolant. These make a valuable contribution to the steel industry. Some of the slag remaining after the metallic separation is screened to -63 + 26 mm for rail ballast. This product is used exclusively on the steelworks extensive rail network. Fine slag - 26 mm is finding use as road pavement material on the internal road system. A screened fraction of this product is also used in the majority of asphaltic concrete placed in the works. Pavement material made with steel slag has a high stability factor, very suitable for heavy duty roads.Steel furnace slag is also handled by the ASMS service contract.

BOS Dust
BOS Dust is the product which is wet scrubbed from the exhaust gas of the BOS process. It is taken in slurry form and mixed with the slurry from the blast furnace thickener.

Caster Scale
This is fine iron oxide which flakes off the surface of the steel. It is recyclable to the sinter plant as a source of iron units. Caster scale is also useful in exothermic powders.

Tundish Skulls
Tundish Skulls are the solidified steel residues in the receiving trough of the caster. This butt of steel is cut up and becomes part of the scrap feed for the BOS or for export to overseas steelmakers.

Crop Ends
Crop Ends are the beginnings and ends of slabs. Since they are all steel, they can be recycled to the BOS. Larger sections are able to be re rolled by smaller steel producers into steel products. Such sections are usually exported.

Scrap offcuts
Scrap offcuts are first de-tinned and then recycled as steel scrap back into the BOS or exported as scrap.

Tin Cans:
Steel cans once de-tinned are used as part of the scrap feed to the BOS or exported as scrap.

Recovered Tin:
Tin removed from cans or scrap offcuts is again recycled after refinement. Tin is still the preferred coating for steel used to package food.

Plant Refuse
Plant Refuse is all of the miscellaneous packaging (ie drums, boxes, cardboard, plastic) that accumulates from purchases of some raw materials and consumables. Additionally, there is a considerable quantity of clean-up material some excavation material and miscellaneous dusts and similar materials which are included in this category.

There still remains however, a considerable volume of waste material for disposal. At Port Kembla Steel Works, this is around 1.3 million tonne annually. Much of this is coalwash (approximately 1 million tonne) and the remainder plant refuse and a mixture of sludges from the iron and steelmaking gas cleaning systems.

TECHNOLOGIES FOR DEALING WITH WASTE MATERIALS

As can be seen from the above list, many of the associated products from the steelworks are energy or iron unit based. For these, the logical process for reuse is to turn them back into the iron or steel making processes. Progressively this has happened, with captured gases from the blast furnaces and coke ovens, providing over 60 % of the purchased energy needs of the plant; making a significant contribution to cost reduction and improvement of the environment.

Around half of the associated product output however, is in the form of coalwash or slag. These products are not energy or iron based, being predominantly the clays, shales, silicates lime and aluminates arising from the raw materials or slag formation materials. As such, these products are more appropriate for use in the construction industry. It is in this area that considerable development and marketing effort has been expended over the last two or so decades.

Iron Blast Furnace Slag

In molten form, blast furnace slag comes from the furnace at around 1500 degrees centigrade. After gravity separation from the molten iron, the potential exists to produce any one of a range of products, depending on the cooling regime chosen.Typical chemical analysis of blast furnace slag is shown in Table II.The five analysis results shown, represent a month's production from one furnace (in this case about 28000 tonne). This shows the low variability of analysis, due to the fact that slag is produced as a co product with the molten iron, from the same selected raw materials. Each of the following Iron Blast Furnace Slag products have the same chemistry, from the same molten slag source. The difference, which provides a range of materials and properties, comes from the method of solidification chosen for the molten slag.

Allowed to cool slowly, in the absence of significant quantities of water, a tough vesicular rock material will be produced,with a crystalline structure. Properties of the rock slag are influenced by the rate of gas evolution, rate of cooling, presence or absence of moisture and initial slag temperature and chemistry. Once solidified, rock slag is won by front end loader or excavator and processed in crushing and screening plant typical of that used in the Quarry Industry. Resulting processed products compete in most of the same markets as quarry products. In the area of product development, the blending of various forms of blast furnace slag and also steel furnace slag is producing a new range of road and aircraft pavement materials of superior performance. It is also possible to form a light weight or foamed blast furnace slag by control pouring of molten slag into a bay,the base of which has been moistened to facilitate the foaming action. Another form of light weight (pelletised) slag can be produced, using a part water, part mechanical system. The resultant product is useful as lightweight aggregate and has been used in Australia for both lightweight concrete and lightweight / fire rated masonry blocks.

Iron Blast Furnace Granulated Slag is formed by passing of molten blast furnace slag through high volume, high pressure water sprays. Water to slag volumes range from 6:1 to around 20:1 v/v. This is necessary to quench the slag to a glass, preventing the formation of crystalline material. As determined by optical microscopy, the glass content of iron blast furnace granulated slag in Australia is 95%. Lee (1974) states that "granulated slag is regarded as a supercooled liquid and that it is a characteristic of silicate melts, that on rapid cooling, they tend to form a glass." The resultant product is a coarse sand like material.

Grinding granulated slag, liberates the hydraulic energy, with reactions very similar to that of Portland Cement. Most granulated slag in Australia is either separately ground or interground with cement clinker, in a conventional cement grinding ball mill. The latest slag milling plant in Victoria, commenced full production in 1991, it is a vertical roller mill. Separately ground granulated slag is generally post blended by the cement company before supply to the customer. Unlike North America, separately ground slag is not generally available on the Australian market at present. Fineness of grind affects the rate of hydraulic reaction, similarly to Portland Cement. Ground granulated slag is now covered by Australian Standard AS 3582.2 (1991).

Rod Milled Granulated Slag is coarser ground than that produced by ball or roller mills. It is essentially a hydraulic sand. Though more reactive than unprocessed granulated slag, rod milled granulated slag reacts at a much slower rate than product ground to around 350 M^2/Kg.

BOS Steelmaking Slag

To remove materials such as phosphorus from steel,a slag is formed, by the addition of lime to the molten bath, before blowing with oxygen. This slag is then decanted from the molten steel into a large slag pot, transported molten, by rubber tyred vehicles to nearby slag bays, poured in layers, allowed to solidify dug by conventional large scale excavation equipment and transported to a processing plant for metallics removal. After metallic separation, the resulting steel slag becomes a useful construction material. Chemistry of the product is quite different to blast furnace slag, the principal constituent being CaO, with SiO_2 and FeO / Fe_2O_3 being the next most important constituents. Structurally, it tends to be a solid solution of oxides, rather than the mineral formations typical of blast furnace slag. By comparison with blast furnace slag, steel furnace slag is higher in density. Indicative chemical analysis is shown in Table II.

Electric Arc Furnace Slag

Another method of producing steel, is directly from the remelting of recycled scrap. In this case, the scrap is placed into a smaller (50 to 150 tonne) refractory lined vessel. Once the charge is loaded, large carbon electrodes are lowered into contact with the scrap. Melting is effected by high voltage electric current. Again, lime is added to form a slag. Once the steel refining is complete, the slag is decanted similarly to the BOS slag, solidified and metallically separated. The cooling regime of Electric Arc Furnace slag is similar to BOS slag.Its porosity and density are also similar.

BHP SLAB AND PLATE PRODUCTS WASTE MANAGEMENT PROGRAMME

In all manufacturing processes, the preoccupation has been with the prime product. Other materials have traditionally been discarded, consistent with the low cost of raw materials and landfill in Australia. Most of the major industrial plants, have access to a landfill site used to take all of the residues from the production process. Increasing urbanisation around many of these sites, together with higher disposal costs and increasing environmental pressures has brought about a greater focus on waste management by most of the major industrial processes in Australia.

At the Port Kembla Steel Works, the emphasis in waste management is consistent with the oft cited waste management hierarchy of:

Reduce
Reuse
Recycle
Disposal

Contracting the Processing and Marketing of 2 Million Tonne Per annum of Slag

Steel companies have long recognised that some associated products have value and have taken many different paths to the management of this part of their businesses (Jones 1989). Many of these tend to be large volume low value products. For this reason, blast furnace slag received a great deal of attention over the past 30 years at the steel

TABLE II

Typical Analysis of Blast Furnace Slag

	S1	S2	S3	%S4	S5	Mean	SD
FeO	.53	.51	.34	.37	.39	.43	.008
SiO2	35.6	35.3	35.8	35.8	35.5	35.6	.19
Al2O3	14.8	15.2	15.3	15.0	14.4	14.9	.32
CaO	41.0	41.4	40.7	40.9	41.5	41.1	.30
MnO	.48	.43	.43	.48	.45	.45	.02
MgO	5.4	5.6	5.8	5.6	6.0	5.7	.20
K2O	.50	.48	.49	.49	.42	.48	.03
S	.55	.60	.57	.54	.52	.56	.03
Na2O	.36	.33	.29	.32	.31	.32	.02
TiO2	.59	.58	.60	.60	.59	.59	.007

(Representing the total production (28 000t) of slag from one blast furnace over a one month period)

TABLE III

Indicative Chemical Analysis of BOS Steel Furnace Slag

	%
Fe	19.8
MgO	5.0
CaO	41.2
SiO2	15.7
A12O3	.96
TiO2	.51
MnO	7.0
S	.01
P	.76

works. Commencing in 1971, the first slag contract for the processing and marketing of rock slag was let. This was paralled by a significant research and development programme carried out by steelworks researchers. From this work came the development in Australia of granulated slag and its progression into the cement market. This research and development work was carried out by the Steel Company, which also established the market. The processing of granulated blast furnace slag for cement represents a potential $80 million per annum industry.

In 1989, a new long term slag contract was commenced. Unlike the earlier contracts, total handling processing and marketing of slag (both blast furnace and steel furnace - some 2 million tonne per annum) is the responsibility of one contractor. Slag is the core business of this new company,(Australian Steel Mill Services) which can devote its whole effort to developing markets for slag products.

Additionally, in 1990, BHP joined with its slag contractors and significant slag users, in the formation of the Australasian Slag Association.

Landfill Site for Coalwash

As indicated above, annual production of coalwash at Port Kembla Steel Works is 1.35 million tonne. Of this, around 300,000 tonne is used for construction fill material,building mounds for landscaping and BHP Soilmix. This leaves approximately 1 million tonne per year to be landfilled. Although landfilling of waste materials is not new, the terms and conditions for industrial landfill have changed significantly over the past two decades. The coalwash landfill site for Port Kembla Steel Works is a significant, controlled engineering structure, located 19 km away from the works. Site approval was granted by the Minister for Environment and Planning, after a public examination of the issues. There are some 26 conditions applying to the site and its operation.

BHP Soilmix

There has been an ongoing research programme since 1982, at the Port Kembla Steel Works with our Newcastle Research Laboratories, using steel plant waste products as plant growing media. Coalwash became the base for soilmix. The mix is produced by blending and mixing fine coalwash with sewage sludge and granulated slag. BHP Soilmix is used in all of the Division's re afforestation work, providing another significant use for coalwash and eliminating the need to purchase topsoil for this work.

Fluidised Bed Combustion for Coalwash

Coalwash still contains some quantity of combustible coal. This is due to the coal washing process and the fact that coal seams contain materials which are part coal and part shale at the margins. Fluidised bed combustion has been investigated on many occasions as a means of disposal for coalwash. It is reported in Business Review Weekly (1991) that a fluidised bed combustor burning slimes from a coal washery will supply 80 MW of electricity to the power grid in the Hunter valley. This project is a direct follow on from the early CSIRO pilot programmes in the 1970's and 1980's. The plant is located adjacent to the Wambo mine near Singleton. While it is technically feasible to produce electricity from fluidised bed combustion of coalwash, burning of the coalwash, only takes away around 20 % of the total volume of the material and also leaves residues such as fly ash and bottom ash.

Coalwash for Pavements

Practical experience in the Illawarra Region, has demonstrated the potential to build road sub pavements with coalwash. Coalwash stabilised with lime has been successfully used in this manner for a major road construction North of Wollongong. This work assisted in having coalwash listed as a potential fill material for construction of the Sydney Airport Third Runway.

Treatment of Iron Bearing Dusts and Sludges

As identified earlier, the Steelworks generates a considerable amount of iron bearing dusts and sludges (around 100,00 tonne per annum for Port Kembla). The limitation for recycling these materials is the level of Zinc present (average 3% for an integrated Steelworks such as Port Kembla). Zinc accumulates as a recirculating load in the integrated system and creates significant operational difficulties if allowed to accumulate. This is a problem common to many integrated steel plants around the world, particularly those who buy in merchant scrap or have zinc bearing ores to refine. At present the iron bearing dusts and sludges are accumulated in landfill emplacements, awaiting the technology to recycle them.

The most usual approach to zinc (and from some steel plants around the world, lead) removal is to use a pyrometallurgical process such as plasma arc or rotary kiln. These processes work best on the iron bearing dusts from electric arc furnaces which make steel from the melting of scrap. Here the zinc level is usually greater than 15 %, making the zinc recovery process an important economic contributor to the process. This is given added impetus by USA EPA legislation, requiring such dusts with 15% zinc content to be treated in this way and those with lower zinc levels, to be chemically stabilised, usually by making a concrete from the dusts prior to landfilling because the iron bearing dusts from Australia's integrated Steelworks' produce low zinc dusts, the pyrometallurgical processes are not economic at present. Further, there is no evidence here that low zinc dusts constitute a significant health hazard. Australian environmental regulations do tend to follow the USA EPA.

Within the Australian Steel Industry, the technology to recycle the Iron bearing (zinc containing) dusts under the principle of Best Available Technology, Economically Achievable does not exist. However considerable effort is going into research on this subject at present.

Fly Ash Utilisation

The most regular use for Fly Ash is as part of the cementitious binder in concrete. Its spherical particle shape aids the workability of concrete made with it and reduces the amount of Portland Cement required. For much of the concrete (around 1 million m3) placed at the Port Kembla Steelworks from 1966 to 1986, the cementitious binder was 40% Portland Cement, 40% Ground Granulated Slag and 20% Fly Ash. The Fly Ash improved workability of the mix and substituted for some of the fine aggregate. It is also used as a filler in asphaltic concrete. In NSW the market for Fly Ash is approximately 340 000 tonne per annum, with the Australian market 700 000 tonne per annum.

With Fly Ash production well in excess of market requirements for concrete, new uses have been developed. One of the larger volume potential uses for Fly Ash is as mine backfill. Fly Ash is mixed with a small quantity of lime and pumped into old mine sections to prevent subsidence and allow structures such as roads to be constructed in these areas. Research work at the University of Wollongong has produced "biofly" bricks from flyash and sewage sludge.

TECHNOLOGY AND CHANGE IN WASTE MANAGEMENT.

There is no doubt that Technology has had and will continue to have a significant role in the development of solutions to this nation's waste management problems. However, to rely on hardware alone, is to adopt an approach which will undoubtedly be expensive and ignores the basic issues of waste generation in the first case. This requires a critical review of all of the activities of a company that lead to waste generation, and a reassessment of each material and process step. This route may be described as a soft technology or systems approach.

This soft technology approach, is not new and embraces the key elements of reduce-reuse-recycle-dispose philosophy. However, it does require a significant questioning of the current practices for any individual process. Often raw materials are taken as given as are process considerations and styles of management. Attitudes of process operators are important. If the process operators continue to treat a particular stream as waste, it will remain so. Secondary or associated product streams require the same style of thought given to primary products, to prevent them becoming part of the waste stream.

One of the support systems for this approach in our industry is the full embrace of total quality control processes. Under this philosophy, all parts of the process are monitored and improved. Often the associated products from a process are made under the same conditions as the primary process, such as the production of blast furnace slag simultaneously with Iron in a blast furnace. Proper handling and processing of this molten slag is the mechanism for turning a waste stream into a product. Some of these steps are as simple as avoiding contamination and having quality control/assurance procedures which apply to the associated product. Using this logic some significant associated product streams have been diverted from the waste stream and become specifiable products, often at significantly reduced costs or increased value.

Environmental review of processes are a regular part of todays business management. BHP Steel SPPD is currently developing an Environment Compliance Manual to cover all of its processes. Part of the development of the manual is the identification and quantification of all of the products associated with each process. The process and its inputs and outputs are subject to review to enable completion of the manual and appropriate management of the associated products. Our division is committed to the reduce reuse recycle management hierarchy (Moore and Worral, 1991). This current review and subsequent audit process will ensure maximum financial and environmental benefit to the company and the community of which it is part. If a product is unavoidable or has a market, the aim is to maximise its end use value. For products which are to be disposed of, the aim is to minimise disposal and environment cost.

Another potentially overlooked aspect of waste management technology is the existence of a market for the products of the technology. The technology to granulate blast furnace slag has been in existence for many decades and commenced in Australia in 1966. However, it was not until the mid 1980's that this product made any real impact on the Australian market, despite overseas acceptance. Such changes are generally hard won, in proving products technically and changing the perceptions of specifiers and users of construction material. This has resulted in slag products being specified in nationally significant construction projects such as the North West Shelf Off shore Gas Platforms and Infrastructure and the Sydney Harbour Tunnel submersed tube units.

Formation of the Australasian Slag Association is a means of keeping pressure on this process of change.

MANAGING THE RECYCLING PROCESS

Over the past 30 years, significant effort has been put into the use of the major associated products from the steel manufacturing process. However, there remains the issue of dealing with the smaller streams and improving the returns from streams currently being recycled.

BHP Slab and Plate Products Division has been involved over the past decade in developing Total Quality Management System. In applying the system to raw material purchases, and all of the operating process. This has impacted on the associated products equally, since many of these are produced simultaneously with the respective primary product. Hence there is greater predictability in quality and quantity making the marketing or re use of these products possible. This involves significant pressure on generators of waste to minimise it and applying TQC principles, each waste material is to be evaluated periodically to revise its treatment or disposal. A typical process of evaluation is given in FIG 1.

Another important feature of industrial waste management is to ensure that the full disposal loop is closed. In the SPPD context, a supply agreement covering generator, waste process operator, planner and technical / research embodies not only disposal, but the notion that the waste stream will be under regular review. Removal from the waste stream is a key objective. This may come by improvements in the product which allow it to be marketed or the elimination of that product. The relationship between these elements are shown in FIG 2

CONCLUSION

With such large volumes and extensive range of by products SPPD has always been mindful of the costs.However in more recent years, the emphasis has firmly shifted to the point of considering each of the by products to have potential to add value to the business.The generation of by products is inevitable the opportunity is always being sought to find the highest value market for all of our by products. Greater effort is being taken to ensure that these products are consistent in quality to enable them to be sold as raw materials for other industries, maximising their value as raw materials for other industries, or by recycling within the steel industry.

PROCESSES IN WASTE MANAGEMENT

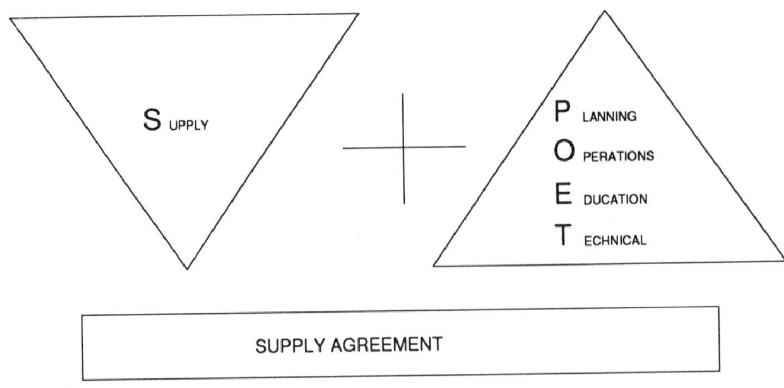

Figure 1

DETERMINING PROCESS OPTIONS
FOR ASSOCIATED PRODUCTS

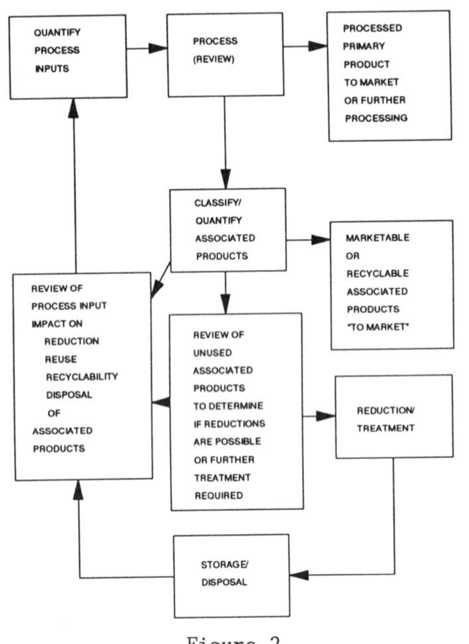

Figure 2

REFERENCES

Lee, A.R., Production Properties and Uses Blast Furnace and Steel Furnace Slag, Halsted Press Book John Wiley & Sons Ltd NY, 1974.

Australian Standard AS 3582.2 1991 Ground Iron Blast Furnace Slag for Use with Portland Cement

Jones, D. E., A study of the Current Approaches to the Management of Non Core Business in The Australian Steel Industry - with reference to overseas experience., MBA Thesis - University of Wollongong, Wollongong November 1989

"Energy from Coal Washery Slimes" Business Review Weekly, November 1991

Moore, S.J. and Worral, M., Waste Management Plans for Major Industries IRR Waste Management & Technology Conference Sydney, February 1991

Environmental Management, Geo-Water & Engineering Aspects, Chowdhury & Sivakumar (eds)
© 1993 Balkema, Rotterdam. ISBN 90 5410 099 0

Engineering uses of steel slag – A by-product material

D.G. Montgomery
University of Wollongong, N.S.W., Australia

G.Wang
JEGEL, Downsview, Ont., Canada

ABSTRACT: In Australia Basic Oxygen Steelmaking (BOS) slag has to date received little attention with regard to useful purposes, with the accepted tradition being to merely dump the material as land-fill. This approach is wasteful and, in many cases, environmentally unacceptable. Some limited uses, chiefly as railroad ballast and skid resistant aggregate for bituminous concrete, have been achieved. However, material properties, in particular the tendency to swell due to extensive free lime content, have often precluded the material from consideration for several applications. Much of the current opinion as to material behaviour is based on overseas experience which may not relate directly to local BOS slags. Research currently being carried out into the properties and potential uses of BOS slags produced in New South Wales is discussed. Material properties considered include the volumetric stability and potential expansion of the slag when used in applications such as in road construction. BOS slag, as produced in NSW, can be successfully applied for this purpose without detrimental effects, thereby increasing the utilization of a valuable resource material with a consequent reduction in an environmental problem.

1 INTRODUCTION

Slags are the main by-products produced in steelworks, with steel slag comprising approximately thirty-five percent of total slag output. In the broad sense, the comprehensive utilization of steel slag will have significant benefits in three major ways.

Firstly, there will be a substantial reduction in environmental pollution due to changes in current practice, whereby the existing material is disposed off by dumping or stockpiling. Secondly, the use of such material will supplement, or replace, the need for using natural materials, thereby resulting in protection of natural, non-renewable resources and a reduction in energy requirements associated with the winning of natural materials. Thirdly, there exists the possibility of altering, or modifying, physical and chemical properties of the basic materials to produce various engineering materials which can be utilised for specific applications.

The comprehensive, rational utilization of steel slag should comprise these three aspects.

2 PRODUCTION OF STEEL SLAG AND ENVIRONMENTAL CONSIDERATIONS

More than 150 million tonnes of steel slag were discharged in the world in 1988 (Xiao and Feng 1990). In Australia, three million tonnes of blast furnace slag and BOS slag are produced each year (Spencer 1991).

There are three steel production centres in Australia, which are, in order of steelmaking capacity, Port Kembla (4.0 Mt/a), Newcastle (1.8 Mt/a) and Whyalla (1.2 Mt/a) (Jones 1990). Production of around 7 Mt/a of steel gives rise to the generation of some 3.2 Mt/a total of iron and steelmaking slags. Usually, about 35% of the slags are BOS slag. Therefore, more than one million tonnes of BOS slag are discharged each year in Australia.

Unlike normal materials, which are supplied on demand, slag is generated daily. The traditional disposal methods of filling swamps and valleys are no longer socially acceptable. Modern society demands efficient and environmentally acceptable usage of any by-product (Hanley 1990)

In some steelworks, steel slag heaps have a height of up to 28 metres and large areas of usable land are occupied by slag dumps. On average, 10 million tonnes of steel slag will require about one square kilometre of land. Steel slag dumping can also interfere with the normal steel production schedule in some steelworks. In addition, leaching of steel slag to the underground water systems has sometimes occurred or dumping has resulted in changes to water courses or rivers.

The use of steel slag to replace natural aggregate in concrete is initially based on considerations of natural resources and the good engineering characteristics of steel slag. In nature, the resources of natural mineral aggregates of high quality which can be used ultimately will become exhausted. In Japan, the proportion of natural mineral aggregates in concrete is diminishing each year and is being replaced by artificial aggregates and industrial by-product

Table 1. Chemical composition of steel slags from various steelmaking processes.

Steelmaking Furnace		FeO	MnO	P_2O_5	SiO_2	CaO	Al_2O_3	MgO	CaF
Thomas Furnace		12-17	3.5-5.5	16-22	3.5-6.0	42-50	1.5-2.5	3-4	-
BOS Furnace		10-25	5.0-15	0.5-3.0	12-17	35-45	-	3-15	-
Open Hearth Furnace		5-25	2.0-10	0.5-3.0	15-20	35-50	3-10	5-20	-
Electric Furnace	oxidising	12-25	5.0-10	0.5-2.0	10-20	40-50	1-3	5-10	-
	reducing	< 0.5-0.6	< 0.4	-	15-20	50-55	2-3	< 10	5-8

aggregates (Japanese Civil Engineering Society 1980). In recent years, shortages of high quality natural aggregates have been experienced in several areas of the United States (Lewis 1982a).

3. BASIC PROPERTIES AND RESEARCH METHODOLOGY

3.1 Basic properties

3.1.1 Chemical composition

The chemical composition of steel slags from different steelmaking processes have been reported in the literature (Jones 1988, Kuwayama et al 1989). The comparison of chemical composition for different steel slags can be summarized as Table 1. Some steel slags contain traces of V_2O_5 and TiO_5 which are not included in the table.

3.1.2 Mineral Composition of Steel Slag

The mineral composition of hardened (cooled) steel slag is related to the forming process and chemical composition. Basic steel slag is composed of $2CaO{\cdot}SiO_2$, $3CaO{\cdot}SiO_2$ and mixed-crystals of MgO, FeO and MnO (i.e. MgO·MnO·FeO) which is expressed as RO. CaO can also enter the RO phase. In addition $2CaO{\cdot}Fe_2O_3$, $CaO{\cdot}Fe_2O_3$, $CaO{\cdot}RO{\cdot}SiO_2$, $3CaO{\cdot}RO{\cdot}2SiO_2$, $7CaO{\cdot}P_2O_3{\cdot}2SiO_2$ and some other oxides exist in steel slag.

Kubodera et al (1979) reported that the X-ray diffraction pattern of steel slag is close to that of OPC clinker. In fact, steel slag is called low grade portland cement clinker by some researchers.

3.1.3 Physical Properties

Compared to air-cooled blast-furnace slags, the steelmaking slags are much heavier, harder, denser, and less vesicular in nature (Lewis 1982b). They have high resistance to polishing and wear in pavement surfaces.

Solid steel slag exhibits both block shape and honeycomb shape. The former possesses lustre, the latter being non lustreous and with reduced brittleness. The specific gravity of steel slag is dependent on viscosity, surface tension of the liquid steel slag, amount of dioxide and ferrous material contained and porosity. Moisture content of steel slag is 0.2-2%, specific gravity 3.2-3.6, compressive strength 169-300 MPa and Mohs' scale number 5-7. Grindability of steel slag is less than that of blast furnace slag. Hardness and specific gravity are greater than those of blast furnace slag.

3.2 Research methodology

The final aim of research on steel slag utilization is to open up avenues of rational use for steel slags with various characteristics.

The following three essential points are important for conducting research effectively and form the basis of research on slag utilization.

Firstly, a sound understanding of the comprehensive properties of steel slag, especially the expansion-related properties of steel slag, is necessary. The properties include mineral composition and the characteristics of the minerals, chemical, physical and mechanical properties and comparison of these properties with those of other related materials.

Secondly, a sound, comprehensive understanding of the properties, production, design, construction and specification of matrix materials in, or with, which steel slag will possibly be put into use is also important in order to achieve rational, optimum utilization. The matrix materials could be inorganic non-metallic construction materials, highway materials, or other silicate materials. Their properties, processing procedures, related specified construction methods and spheres of application should be considered. The optimum use of steel slag in construction can only be achieved as a result of comprehensive knowledge in these areas.

Thirdly, a realisation that steel slag is a distinctive, unique material, being different from any other natural or man-made mineral material. Steel slag, like any other waste or by-product, is a type of special "raw" material with its own characteristics or properties. Like other waste materials, steel slag is not a raw material which can be used without additional processing. Steel slag can not be utilized for its properties according to the conventional standards and specifications relating to similar natural raw materials or mineral resources and it has to be pre-processed before being used as raw material or mineral resource. This pre-processing procedure, along with other factors, has been the topic of research into its intrinsic properties to investigate how it can be rationally utilized. It is because of the difference between steel slag and any other natural mineral resource (e.g. natural limestone or basalt) or any other man-made end product or semi end product (e.g. OPC clinker), that the aim of utilization will be different. For example, steel slag is usually inter-used with other materials to reduce the effect of expansion.

The technical specification for steel slag should also be different. For instance, in some applications, the amount of steel slag used has to be restricted within limits, suggesting that the technical requirements for steel slag use should be different from those for other materials or replacement materials. In research, however, on the one hand, consideration of the similarity of steel slag to other materials (comparison of properties) should be determined while, on the other hand, confusion of steel slag with other materials should be avoided. For example, in the study of the mechanism of expansion of steel slag, individual properties of steel slag and OPC may give rise to differing effects. For example, transformation of β-C_2S to γ-C_2S and existing $C_{12}A_7$ should be avoided in the production of OPC, because they will lower its hydraulicity. However, in steel slag, consideration of β-C_2S to γ-C_2S transformation and existing $C_{12}A_7$ are not generally the consequence of expansion of the steel slag, although they may still result in lower hydraulicity.

Utilization of steel slag is an involved process which includes several stages from production of steel slag to final end products. Successful utilization is not expected to be fulfilled in one stage. The overall process can be expressed as Fig.1. There are eight links or areas in this whole process. Any individual link in the process might affect the final use of the slag in engineering practice. These links, or areas, include: the pre- and post-treatment of slag; the chemical, physical and expansion properties and their affecting factors; the evaluation and measurement of the expansion. The comprehensive utilization can be related to three main stages: treating and processing, intrinsic properties of steel slag and properties of end products. The utilization shown in Fig.1 is in three areas, i.e. road engineering, concreting and cement manufacture. In addition to the engineering applications, there could be other areas for utilization, such as use in agriculture as a fertiliser or soil additive.

The comprehensive utilization of steel slag is a topic which requires a knowledge of materials and a sound understanding of the properties of both steel slag and the target usage is essential. That means, on the one hand, the intrinsic properties of steel slag have to be investigated before finding out its potential use, while on the other hand, the requirements of target use should be well understood. Without the study of the both aspects, utilization will be blind, inactive or low grade. Usually, in the utilization, steel slag comprises a composite with other materials or matrix, and therefore it is necessary to understand the matrix properties. This principle is also suitable for other waste materials utilization.

In the utilization of steel slag, expansion of the steel slag is the main factor likely to affect its successful application as a construction material. Expansion is dependent on the chemical and mineral composition and affects the end use of the steel slag. There are three relationships to be considered for the utilization of steel slag, these being shown as Fig. 2. They are: (i) the relationship between chemical composition and expansion properties; (ii) the relationship between the expansion properties and the properties of the matrix containing the slags and (iii) the rational use of slag with different properties to ensure optimum use of every type of BOS slag.

To effectively use steel slag, it is necessary to know how the chemical and mineral composition affect the expansion properties, i.e. volume expansion and expansion stress (relationship 1) and how the expansion properties affect the properties of the composite end materials (relationship 2). Setting up of relationship 3 if dependant on relationship 1 and relationship 2. Relationship 1 determines the treatment and modification of the properties of slag. Relationship 2 is essential to enable the steel slag, with known properties, to be put into use in construction engineering. Once relationship 2 is known quantitatively, the use can become reality. Once relationship 1 is known, suitable treatment methods can be chosen and relationship 3 can be set up based on the other two relationships. Relationship 3 has the same effect as relationship 2. However, because elaborate measurement of expansion is omitted, relationship 3 is more convenient for determining suitable utilization.

Steel slag is an energy-containing material. In determining the utilization, proper attention should be paid to the optimum use of potential energy. This is one of the three aims of waste utilization, i.e. protecting environment, full use of all waste resource/energy and technical beneficiation. The term "utilization of steel slag" can be defined as the effective use of steel slag to achieve all of the three aims. Under this definition, land-fill is not utilization.

Any use of hydraulicity and strength of steel slag could be considered as the use of energy. For example, use as a skid-resistant material involves the use of its hardness, where hardness is a strength related property and therefore an energy. Use as an additive in steel slag blended cement (SSBC) is a use of its hydraulicity (chemical energy), use as road base material or as an aggregate in concrete also depends on hydraulicity or strength attainment and is therefore

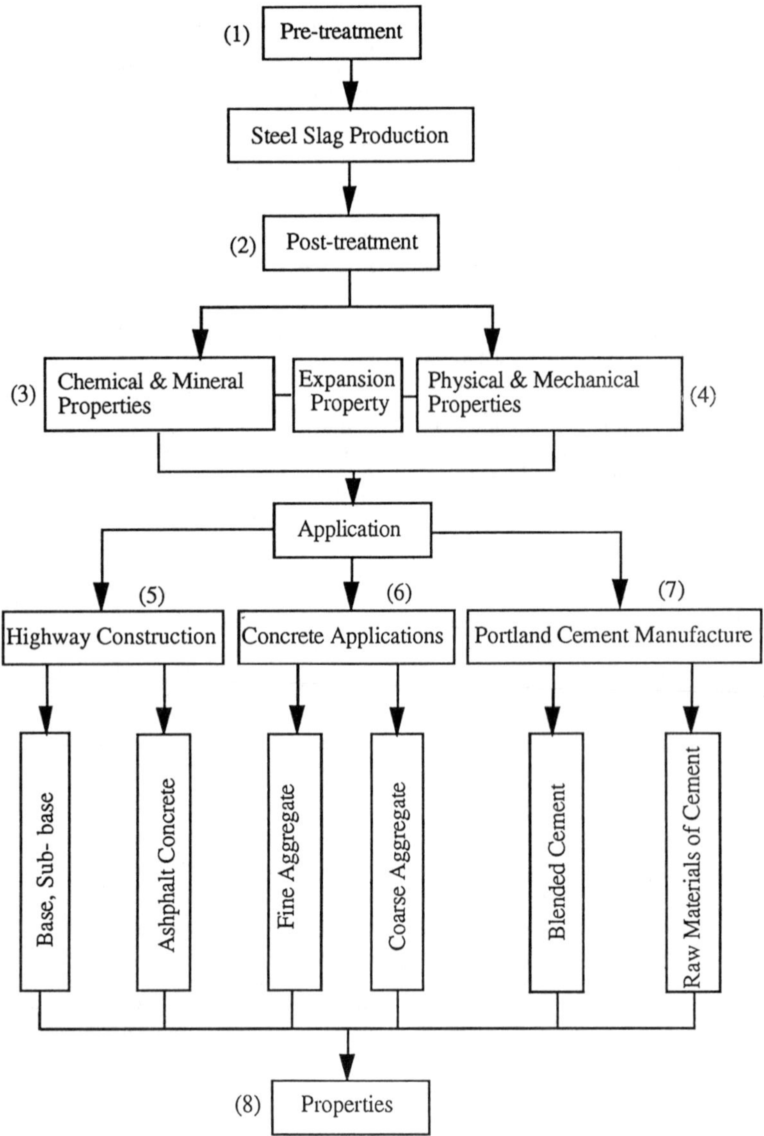

Fig. 1 The Process of Steel Slag Utilization - Eight Areas of Research

the use of energy. Although any use of strength can be considered as energy use, the question is how to use this energy to its highest extent.

The higher the requirement for energy in the use of steel slag, the more quantitative work is needed. This is shown as Fig.3, in which, from left to right, more potential chemical energy is used, necessitating the need for higher requirements of stability and more rigorous quantitative work to quantify properties.

The reason why steel slag is not fully utilized is considered to be due to a general lack of quantification work on the relationships between expansion properties of steel slag and that of the matrix containing the steel slag or of the steel slag composite material. Opinion unsupported by thorough testing is not sufficient to encourage the use of steel slag without misgiving or concern.

Quantification work is therefore the setting up of the relationships shown in Fig. 2, which can be divided into two aspects: quantification of the properties of steel slag itself and quantification of the relationships between properties of steel slag and matrix materials or end products. These relationships can be based on

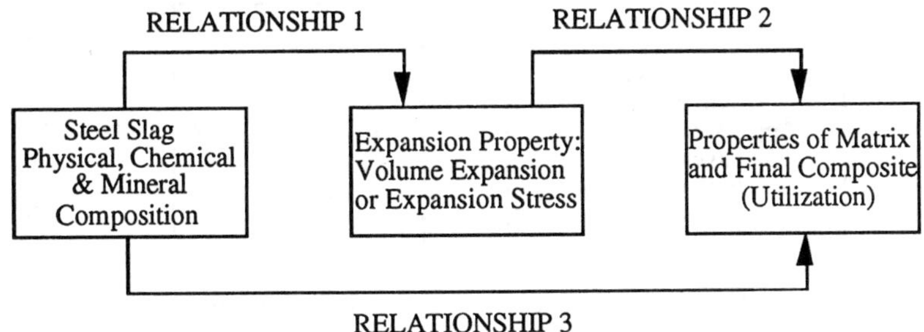

Fig. 2 Three Important Relationships for Steel Slag Utilization

both experimental work and theoretical work.

The properties of steel slag, and in particular the expansion property, constitute the main parameters affecting the use of steel slag as a construction material. However, the degree to which these play a role varies depending on different end usage. For the different uses, different emphasis should be placed on their importance to achieve rational, optimized use.

Usability criteria and properties of composites made with steel slag can be determined based on studies of the basic properties of steel slag and their respective expansion assessment index.

The relevant quantification work which is necessary includes: laboratory testing and calculation of the volume expansion of steel slag; rational mix proportioning for road base considering particle interference and the effect of particle size and grading on expansion. For use in cement concrete as an aggregate, necessary work includes: assessment of expansion property; feasibility of determination of expansion stress; setting up of relationships between the expansion stress and the maximum plane expansion stress and allowable stress. For use in blended cement, necessary work includes: assessment of grindability properties of the slag (main aspect of concern in SSBC manufacture), and determination of dosage rates of steel slag in relation to stability of SSBC. Details of theoretical and experimental work pertaining to the latter two areas, i.e. cement concrete aggregate and blended cement, will not be presented in this paper due to space restrictions.

All of the above work is essential to establish certain usability criteria to increase the use of BOS slag in highway construction, concrete applications and blended portland cement manufacture.

Research into the utilization of steel slag should be oriented and developed in the following directions:

1. from inert use to active use (use energy at higher level).
2. from low-grade use to high-grade use (achieve the three aims of steel slag utilization).
3. from single use to multi-use (use steel slag to fullest extent).

These factors should then determine the direction and trends for future use and research.

4 THEORETICAL AND EXPERIMENTAL RESULTS

4.1 Volume expansion testing of steel slag

Some 24 materials were tested for volume expansion during the study. The materials included BOS slag and GBF slag from Newcastle Steelworks and Port Kembla Steelworks, road base containing BOS slags and materials proportioned by the authors.

The aims and emphasis of the volume expansion experiments were: to determine the volume expansion of Australian BOS slags; to compare the volume expansion of BOS and GBF slag for both Australian BOS slag and overseas BOS slag; to determine volume expansion of proportioned road base materials containing BOS slag; to derive relationships between the volume expansion of steel slag and surcharges and to establish relationships between the volume expansion of steel slag and particle size.

The experimental apparatus used was adapted from the water soaking test method (Emery 1977, PMOT 1978, ASTM 1988), with water temperature of 74 $^{o}C \pm 3^{o}C$. The steel slag sample was compacted in a cylindrical steel mould with perforated base plates to allow for moisture movement during the immersion period. The mould containing the sample was fully immersed in a water bath and a dial gauge was installed in the top of the mould to record expansion at daily intervals.

Maximum values of expansion ranged from zero (BOS slag, 16-76mm particle size) to 2.25% (BOS slag, 0-16mm particle size). In general, the maximum volume expansions of the Australian steel slags tested were less than 0.5%, these values being relatively low compared with published results for some overseas slags, although some variability in potential expansion must be expected due to variations in materials and testing methods.

4.2 Volume expansion criterion

From the expansion of F-CaO, an equation for estimating the potential volume expansion of steel slag containing F-CaO can be deduced.

For a given volume of steel slag, the volume

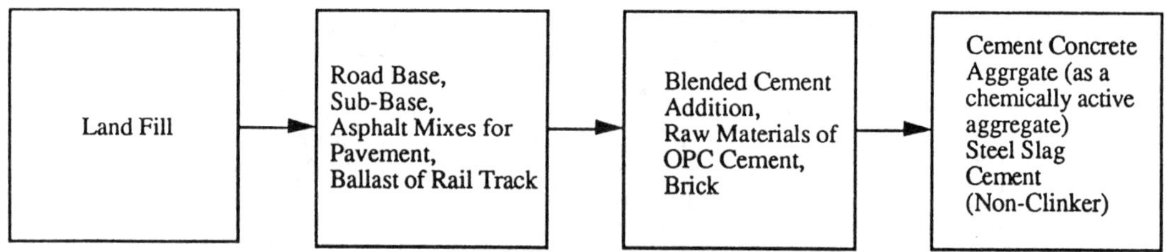

The requirement for stability of steel slag in use (lenient ➔ rigorous)

The degree of use of chemical energy in steel slag (nil ➔ full)

Fig. 3 Four Different Utilizations of Steel Slag

expansion of the steel slag, Es, can be expressed as:

$$E_s = \frac{1}{V_o D} \times \frac{V_o D \gamma_s F}{\gamma_l} \times \frac{E_l}{100} \qquad (1)$$

where E_l is the volume expansion of lime (%), E_s is the volume expansion of steel slag (%), V_o is the apparent total volume of steel slag (cm^3), D is the denseness of steel slag (γ_o / γ_s), F is the content of F-CaO (%), γ_o and γ_s are the bulk density and compacted density of steel slag (g/cm^3) and γ_l is the compacted density of lime (g/cm^3).

Substituting values of $\gamma_l = 3.34$ g/cm^3 and $E_l = 127.78\%$ (theoretical value) into (1) gives

$$E_s = 0.38 \times \gamma_s F \qquad (2)$$

From equation (2), it can be seen that the volume expansion of steel slag is related to the compacted density γ_s and the F-CaO content of steel slag. However, γ_s is a physical constant for a particular slag. Therefore, the volume expansion is directly related to F-CaO content.

In the above equation the carbonation of free lime is not considered because of the low proportion of CO_2 in the atmosphere (about 0.03%) and the lower expansion rate from $Ca(OH)_2$ to $CaCO_3$ (about 7.6%) compared with the expansion rate of free lime (127.78%).

Equation (2) can be treated as one of the category 1 relationships and it is easy to estimate the potential volume expansion of steel slag if the content of free lime is known.

4.3 Expansion Criterion

When the slag is used as a geomaterial such as in road base or sub base, it must be treated as an entirety. Voids will exist in the steel slag mass and the actual D will not generally be equal to 100% (although it could be 100% theoretically), i.e. the voids volume of steel slag will not be equal to zero. It must be considered as to whether the voids in steel slag can 'absorb' part of the volume expansion of the steel slag.

In the analysis it is assumed that, for road base construction, the allowable volume expansion is zero, i.e. no apparent volume expansion is allowed.

For consideration of category 2 relationships, comparative expansion tests, with and without surcharges, of steel slag as road base materials were carried out.

The test method used was as previously described. Two types of steel slag were tested in the mould under conditions with and without 4560 g surcharges to give a surcharge per unit area of 25 g/cm^2.

From the results, for the two slags tested, if the slag is acted upon by a 25g/cm^2 surcharge, 7-8% of the expansion can be absorbed in the interior of the voids volume of the steel slag. This means that if the potential pre-estimated expansion (equation 2) is larger than an estimated voids content of 7-8% (say 7.5% shown below) and the material is acted upon by a 25 g/cm^2 surcharge, actual expansion volume of the steel slag will occur. If the expansion is less than this value of voids content, then the expansion can be absorbed, even though actual volume expansion occurred. In other words, if

$$E_s < 7.5\% \, (1\text{-}D) \qquad (3)$$

or

$$0.38 \, \gamma_s F < 7.5\% \, (1 - \gamma_o/\gamma_s) \qquad (4)$$

expansion will not occur, indicating that a 7.5% void content is sufficient to fully absorb any actual expansion which will occur in the slag particles.

Equation (4) can be rewritten in the form shown in equation (5),

$$F < \frac{0.075 \, (\gamma_s - \gamma_o)}{0.38 \, \gamma_s^2} \times 100\% \qquad (5)$$

i.e. if F-CaO in steel slag is less than the right hand term, the steel slag will not expand macroscopically, or the expansion resulting from free lime given by equation (5) can be 'absorbed' by the voids volume of steel slag itself under a pressure of 25 g/cm^2 (i.e.

overall expansion will not occur if this condition is met). Things worthy of note are that equation (5) does not infer that the content of free lime has an intrinsic relationship with the compacted density and bulk density of steel slag (both of these are physical constants within some range). It simply provides a convenient, referential estimation for a given steel slag with a known content of free lime and particle properties; in this case equation (5) has been determined experimentally for steel slag exerted upon by a pressure of 25 g/cm^2. Under this condition, about 7-8% (7.5% used in the deduction) of the voids can be filled by increased volume of steel slag. If the surcharge pressure exerted on the slag, or the strength of slag, is very high, the allowed free lime content could be considerably increased irrespective of its relation to the particle properties of the steel slag.

Some of the literature specifies a limitation of free lime in steel slag used for road construction. The suggested value is about 4% or so, although others suggest the limitation of 4% can be extended. Equation (5) may be modified dependent upon the in-situ internal strength of the steel slag in the road base material and the actual surcharge in practice on the road materials (up to the 25 g/cm^2 which was applied in the laboratory). Results of the two slags tested gave limit values of $F < 2.53\%$ and $F < 3.03\%$ respectively. These are close to the suggested free lime limitation given in the literature for steel slag to be used in road engineering.

5 CONCLUSIONS

The dominant cause of expansion in steel slag is F-CaO. This has been proven experimentally from the expansion test, F-CaO test and the theoretical calculation of F-CaO.

Steel slag produced in Australia possesses comparitively stable properties, such as lower F-CaO and lower expansion, compared with some overseas steel slags. From the mineral composition, all Australian steel slags can be classified as active slag in terms of basicity.

The calculation of volume expansion based on F-CaO is reliable and practical for estimating the volume expansion and is useful for determining rational and optimum use of steel slag.

Research concerned with utilization of steel slag should be oriented and developed along the following lines: from inert use to active use (use energy at higher level); from low-grade use to high-grade use (achieve the three aims of steel slag utilization); from one use to multi-use(use steel slag to fullest extent).

The total content of F-CaO should be considered for evaluating the expansion property. The volumetric methods for determination of F-CaO are effective. The content of F-CaO in Australian steel slag from both Newcastle Steelworks and Port Kembla Steelworks are comparative low and usually range from 0.08 -1.2% for steel slags with particle size between 16 and 76 mm.

The volume expansion of Australian steel slag is relative low compared with some overseas steel slag, although test methods vary in different countries (some tests use water at 81 °C in the accelerated test). The apparent volume expansion of the steel slag should be considered when it is used as a road construction material. The equations established in relation to volume expansion give results in keeping with the suggested empirical F-CaO content limitation. This confirms that the dominant factor contributing to volume expansion of steel slag is F-CaO and that the calculations and criteria based on the content of F-CaO are reliable.

Particle size, gradation and particle interference of steel slag should be taken into consideration in order to minimise the potential volume expansion of road base mixes containing steel slag and blast furnace slag or other inert mineral aggregate.

REFERENCES

ASTM Designation: D 4792-88. 1988. Standard test method for potential expansion of aggregates from hydration reactions. *Annual Book of ASTM Standards* 566-567.

Emery, J.J. 1977. Steel slag application in highway construction. *Silicates Industriels* 4-5: 209-218.

Hanley, P.J. 1990. Slag-towards 2000 Australian trends. *Concrete for the Nineties:* Leura, Sept.: 11p.

Japanese Civil Engineering Society. 1980. *Handbook of Civil Engineering-Concrete.* 7.

Jones, D.E. 1988. Utilization of ground granulated slag in Australia. *Proc. of Concrete Workshop 88:* Sydney, July: 307-329.

Jones, D.E. 1990. Development in the use of granulated blast furnace slag for concrete and stabilization. *Concrete for the Nineties:* Leura, Sept.: 22p.

Kubodera, S., Koyama, T., Ando, R. and Kondo, R. 1979. An approach to the full utilization of LD slag. *Trans. Iron and Steel Inst. Japan:* 19: 419-427.

Kuwayama, T., Mise, T., Yamada, M. and Honda, A. 1989. Properties of electric slags geomaterials. *Proc. Japan Congress Materials Research:* 32: 213-218.

Lewis, D.W. 1982a. Resource conservation by use of iron and steel slags. *ASTM Tech. Pub..* 774.1. 31-42.

Lewis, D.W. 1982b. Properties and uses of iron and steel slags. *Symposium on Slag, Nat. Inst. Transport and Road Res., South Africa:* Feb.: 8p.

PMOT. 1978 (revised). Method of test for evaluation of potential expansion of steel slag: 6 p.

Spencer, K.W. 1991. An overview of slag usage in North America and Australia. *Proc. Int. Seminar on Slag-The Materials of Choice:* Port Kembla, July: 1-13.

Xiao, L. and Feng, D. 1990. Comprehensive utilization of steel slag. *Iron and Steel:* 25:3: 66-69.

Environmental Management, Geo-Water & Engineering Aspects, Chowdhury & Sivakumar (eds)
© 1993 Balkema, Rotterdam. ISBN 90 5410 099 0

Textile residue management – A case study

N.S. Rathore & A.N. Mathur
College of Technology and Agricultural Engineering, Udaipur, Rajasthan, India

ABSTRACT: Cotton waste or willow-dust is one of the rich organic matter available as by-product after the processing and spinning. This is a potential alternate feed-stock for biogas generation. The Renewable Energy Centre has designed and developed a willow-dust based pilot biogas plant.

1 INTRODUCTION

In India, cotton processing is associated with production of 30,000 to 3,000 mt waste and residues annually. This willow dust consists of cellulosic material containing suitable substances which, if placed anaerobically, than can favour generation of biogas (Balasubramanya 1986). As such the available waste is associated with the pollution, and safe disposal of these material is necessary. Keeping in view the potential and problem associated with waste and for broadening the scope of biogas there is a need to develop a technology for anaerobic digestion of willow dust. Such an approach has been initiated in Udaipur Cotton Mills, Udaipur. The design of the plant is developed on the basis of extensive laboratory study.

1.1 Laboratory studies

The laboratory studies were carried out to identify the potential of biogas generation from willow dust. The material were analysed for their physical and chemical characteristics. The average values are shown in Table -1. The moisture content of willow dust varies from about 245% - 40%, also the C/N ratio is 45-50 and cellulosic content is 28%, both of which seems to be suitable for anaerobic digestion since optimum C/N ratio reported is 25-30. Hence, in our case 1% of urea may be added initially to bring down the optimum C/N ratio.

Table - Physico-chemical Characteristics of Willow-dust(Mathur,1987)

S. No.	Particulars	Willowdust
1.	Physical Characteristics	
	(a) Moisture	24%wb
	(b) Total Solids	76%wb
	(c) Volatile Solids	89%db
	(d) Ash	10-11%
	(e) Density	0.18gm/cc
2.	Chemical Characteristics	
	(a) Carbon	55%
	(b) Nitrogen	1.2%
	(c) C/N ratio	45-50
	(d) Cellulose	28%
	(e) Hemicellulose	15%
	(d) Lignin	16%

After bio-chemical analysis of willow dust the gas production was evaluated through lab set up. It has been observed that about 6-8 litre of gas can be produced through 1 kg. of willow-dust per day, which reveals that gas production through willow-dust is more in comparison to cattle waste. Based on above observations a pilot batch plant of 25 cum capacity has been designed and commissioned. (Rathore, 1989)

2. DESIGN OF 25 CUM BATCH TYPE PILOT PLANT

The plant has been constructed in Udaipur Cotton Mills premises. Where daily avilability of willow-dust is about 110-150 kg. The block diagram of the plant is illus trated in Fig. 1. The salient features are summarized as follows:

1. The plant is essentially a fixed dome type having rectangular digester.

2. The recycling is divided in three stages namely pre-digestion for partial aerobic treatment, digestion chamber for anaerobic digestion of waste and slurry digestion tank.

3. In pre-digestion chamber the willow dust remain in an open tank of 10.5 x 2.7 x 1.0m, which can accommodate 2.5 tonnes of willow dust. Here the material is treated with 1% urea in order to make up optimum C/N ratio. The pre-digestion under partial aerobic/anaerobic condition is recommended for 6-10 days depending on atmospheric temperature.

4. The main digester for anaerobic digestion has one opening at the top of feed pre-digested willow dust and a gate at the bottom for removing the spent slurry. The main digester is having three batch section configuration having volume of 50 m^3 , which are designed on 40 days retention period, In fact, the idea of keeping three batch is for getting continuous gas production by rotation through individual batch. Each batch is charged after every 40 days duration.

5. Third section of plant is post digestion chamber, where the water is being separated through screen from digested slurry. This separated water in-turn is being recirculated in the main digested chamber in the spray through a pump which help in agitation of slurry in ther digester. The separated water can also be used as inoculum for fresh loading.

6. In order to enhance the gas production through activating bacteria in the main digester chamber, a provision has been made to recirculate biogas in the digester chamber. It also helps in agitating the slurry to ehance gas production.

7. The produced gas is stored in a separate gas holder of capacity of 15 cum. , which floats in a water/oil jacketed circular tank. The outer peripheral channel (water jacket) is provided for movement of the gas holder to reduce corrosion. The gas holder has two outlets. One goes to kitchen for use, where as other goes to the compressor, for recirculation in the main digester.

8. The plant is equipped with self-loading and unloading mechanism. The input is provided through a masonary wall of pre-digester near the top of the main digester and output is from the bottom to facilitate the feeding of fresh and floating material and discharging of digested slurry.

3 MONITORING OF THE BIOGAS PLANT

The plant was tested for leakage by filling the digester with water and compressed air. There after the pre-digested 2.5 tonnes willow dust, 1% urea and 17.5 Kl water was charged in one digester. A rotation of 15 days was adopted for charging the other two digester.

3.1 Measurement of gas production

The gas through storage gas holder is measured daily through a wet type gas flow meter of capacity 1 lit. in 3 revolutions at an accuracy of ± 0.5% at cm of water pressure. The gas from each batch is stored in gas storage tank.

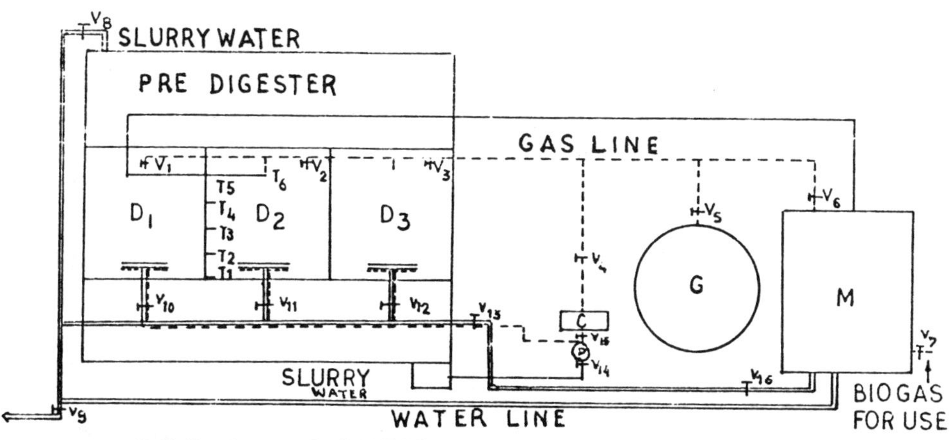

$D_1 D_2 D_3$: MAIN DIGESTER
C: COMPRESSOR
G: GAS COLLECTOR
M: MONITORING
P: PUMP
T_1 T_6: TEMP. SENSORS
V_1 V_{16}: VALVES

FIG. 1-FLOW CHART OF WILLOW DUST BASED BIO GAS PLANT

3.2 Measurement of temperature

The slurry temperature in digestion is being continuously measured at five vertical locations once in a day. For this a temperature indicator with RTD's is being employed. The RTD's are placed in first batch at a distance of 50 cm apart starting from bottom of digester, the Sixth RTD's is placed at the top central portion of digester for measuring gas temperature.

3.3 Measurement of slurry content, pH and gas analysis.

The pH of slurry is measured on weekly basis in pre-digester and slurry tank taking sample and using a digital pH meter. The analysis of gas in terms of CH4 and CO2 is also done twice a month using a gas chromotograph. The chemical analysis of influent & effluent consists of measuring total solids, volatile solids, C/N ratio, and nutritional value in terms of nitrogen, phosphorous and potassium (N.P.K.), has been made using the standard procedures.

4. CONCLUSIONS

The plant hs been running well for ten months producing 22 cum gas per day. Such biogas plant will be useful for the textile industries and other similar industries based on jute or yarn-residues. Installation of such plants will not only help to solve the problem of waste disposal which unnecessarily causes foul odour and fly menance but also provide a fuel and other non-quantifiable benefits like increasing the fertilizer value of the stabilised waste.

REFERENCES

Balasubramanya, R.H., H.V., Gangas V.H. Khandeparkar and Sundaram 1986. Production of biogas from willow dust. Proc. Seminar on Microbial ecology.

Mathur A.N.1987. Recycling of Textile Residues. Proc. IV BES Convention, 229-230.

Rathore N.S. and Mathur A.N. 1989 A new approach for recycling of Textile Residues. Proc. ISAE Hissar Convention, 166-171.

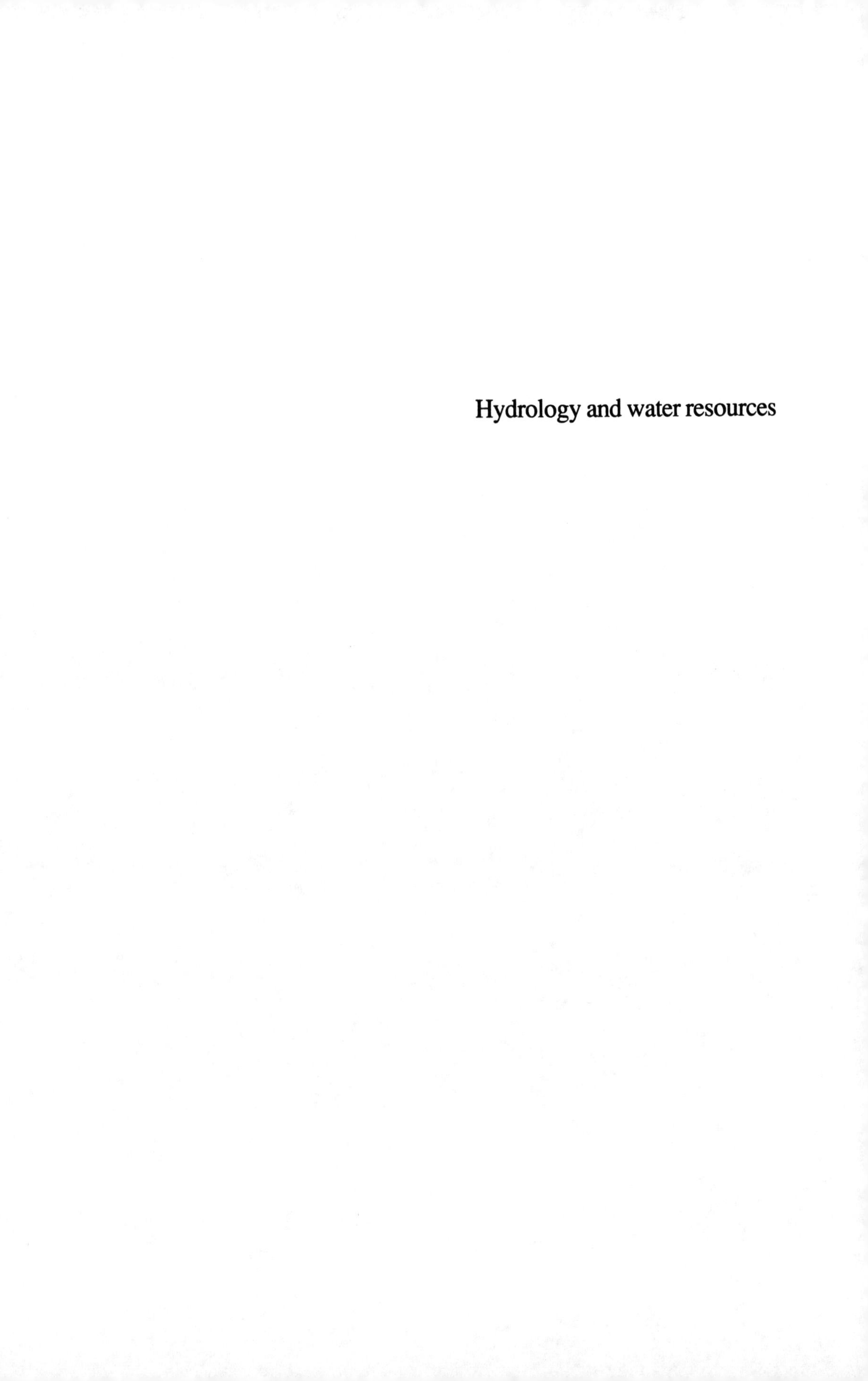

Hydrology and water resources

Environmental Management, Geo-Water & Engineering Aspects, Chowdhury & Sivakumar (eds)
© 1993 Balkema, Rotterdam. ISBN 90 5410 099 0

Does onsite detention storage reduce total catchment flooding?

M.J. Boyd
Department of Civil and Mining Engineering, University of Wollongong, N.S.W., Australia

ABSTRACT: Onsite detention storage is frequently used to reduce flood peaks from a developed site to be no greater than the pre-development values. The question often asked is, does OSD reduce flood peaks further downstream in the catchment or can it in some cases increase downstream flooding? A review of published literature found little evidence of increased downstream flooding, and this finding is supported by calculations for a range of site and catchment sizes.

1. INTRODUCTION

Provision of onsite detention storage (OSD) to reduce urban stormwater runoff peak discharges is becoming widespread and is required by many drainage authorities (ASCE, 1982; Boenisch and Ogle, 1989; Tomkins and Cooper, 1989; Lees and Lynch, 1992). Since the OSD accepts water from a small site (typically having an area of some hundreds of m^2) within the larger catchment, volumes of OSD are generally small (typically tens of m^3). However if OSD is provided at many sites throughout the catchment, total volumes can be significant.

Design of OSD to handle flows from the site is relatively straightforward, and can be based on flood hydrographs and volumes (Boyd 1981; Culp, 1948). It has been suggested however that OSD should be designed considering its effect on total catchment flows (Phillips, 1987). This case is considerably more complicated, requiring details of impervious areas and the drainage system for the complete catchment, and in practice blanket requirements are often used. For example the Upper Parramatta River Catchment Trust (Lees and Lynch, 1992) recommends OSD of $470 m^3$ and permissible site discharges of 80 litres/s per hectare of developed area.

Many papers (for example McCuen, 1979; Lakatos and Kropp, 1982) have commented on possible adverse effects of detention storage, but convincing evidence of these effects is hard to find. For example McCuen uses a storage which is actually under designed for the site in question, since the site discharge remains 12 % greater than the pre-development value, and despite this, downstream flood discharges are increased by less than 7%. These papers are mainly concerned with large detention storages on relatively large sub-catchments of the main catchment. McCuen (1979) considers that a large detention storage is not a good substitute for the natural storage which is distributed over the catchment, and suggests that several small on-site detention storages, as used in the present study, would be better.

The question considered in this paper is, does OSD designed to reduce flooding from a site within the larger catchment, also reduce total catchment flooding? If it does, then OSD can be designed considering the site alone.

The approach taken is to use design storms of selected recurrence interval and critical storm duration for the site and for the complete catchment. Flood hydrographs are calculated for the site and the catchment for the pre-development condition. The site is developed, and increases in site and catchment flooding are noted. OSD is then designed for the site to reduce site post-development discharges to the pre-development value. Finally, the effect of this OSD provided at the site, on total catchment discharges is examined.

2. DESIGN OF OSD TO REDUCE FLOOD PEAK DISCHARGES

Figure 1 shows concepts involved in design of detention storage to reduce flooding

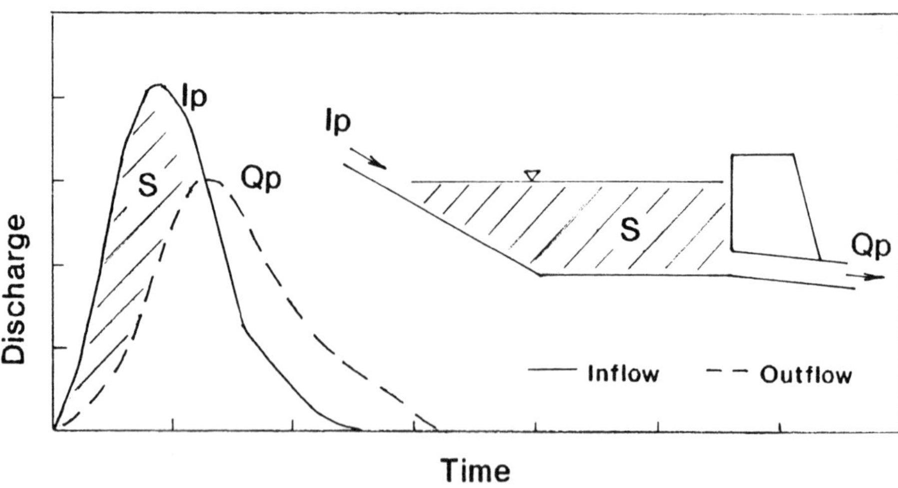

Figure 1. Reduction of flooding with detention storage.

from a site. The flood routing effect of the detention storage reduces the peak discharge of the inflow hydrograph from the site into the storage Ip, to a lower value Qp. The amount of reduction depends on the volume of storage S provided in the OSD, which in turn depends on it's geometry and the size of the OSD outlet. A range of storm durations should be tried and the one requiring the largest volume S adopted as the critical one.

Computer programs are available to calculate OSD requirements, including SITEDET from Swinburne University of Technology and ONSITE from the University of Wollongong.

3. EFFECT OF DEVELOPMENT ON STORM RUNOFF

Development of a site from less intensive to more intensive land use affects runoff in two ways. Firstly, because impervious surfaces increase, the volume of runoff also increases. We could calculate a weighted runoff proportion for the pervious and impervious surfaces. Alternatively, because we are dealing with small site areas, we could use Rational method runoff coefficients as a guideline. Australian Rainfall and Runoff (1987), Ch. 14 for example, indicates that an increase in fraction impervious from 0.3 to 0.7 on a typical site, increases the runoff coefficient from 0.43 to 0.64.

The second effect is for flow travel times to decrease as flow paths are improved. Studies of lag parameters on urban catchments (Rao et al, 1972; Aitken, 1975; NERC, 1975) have shown that conversion from natural to 100% urban reduces lag parameters to approximately 25% of the natural value, while 50% urbanisation reduces them to 50%. Similar reductions in the time parameter are obtained using kinematic wave models of overland flow on plane surfaces. For example a decrease in surface roughness from 0.17 to 0.011 reduces the time to equilibrium to 20% of the less developed value. This decrease in the time parameter makes the site more responsive to rainfall and therefore produces higher peak discharges. It also reduces the critical storm duration for the site, thus increasing the design rainfall intensity, adding to the increase in flows.

It is generally accepted that effects of development on flooding become less pronounced as the average recurrence interval ARI increases (Espey and Winslow, 1974; ASCE Task Committee, 1975). In the present study, 100 year ARI storms were used with post-development time parameters reduced to 50% of pre-development values, and 100% runoff for both pre and post-development. Five year ARI storms were used with the time parameter reduced to 25%, and runoff increased from 43% to 64% for the post-development case.

4. RESULTS OF CALCULATIONS

4.1 General

The effect of providing OSD at a number of sites within a catchment is to distribute storages, which give a flood routing effect, over the catchment. This was modelled in the study using the distributed runoff routing model WBNM

Table 1. Development of One Site (11% of catchment)

ARI (years)	Storm Duration (minutes)	4	8	12	16	20	24	28	32	36
100	Site (pre)	43*	54	56	55	53	50	47	45	42
	Site (post, no OSD)	71	76	70	63	56	53	49	46	43
	Site (post, OSD = $17m^3$)		51	56	56	56	51	48	46	43
	Catchment (pre)		208	255	287	313	325	335	341	339
	Catchment (post, no OSD)	141	214	262	294	322	332	342	347	344
	Catchment (post, OSD=$17m^3$)		205	252	284	312	322	334	340	339
5	Site (pre)		13	14	13	13	12	12	11	10
	Site (post, no OSD)	37	32	27	24	22	19	18	17	16
	Site (post, OSD=$14m^3$)		11	13	14	14	14	14	14	14
	Catchment (pre)		50	62	70	78	80	84	85	84
	Catchment (post, no OSD)		56	69	77	85	87	91	91	90
	Catchment (post, OSD=$14m^3$)		50	61	69	77	79	83	84	84

*Peak flows in litres/second

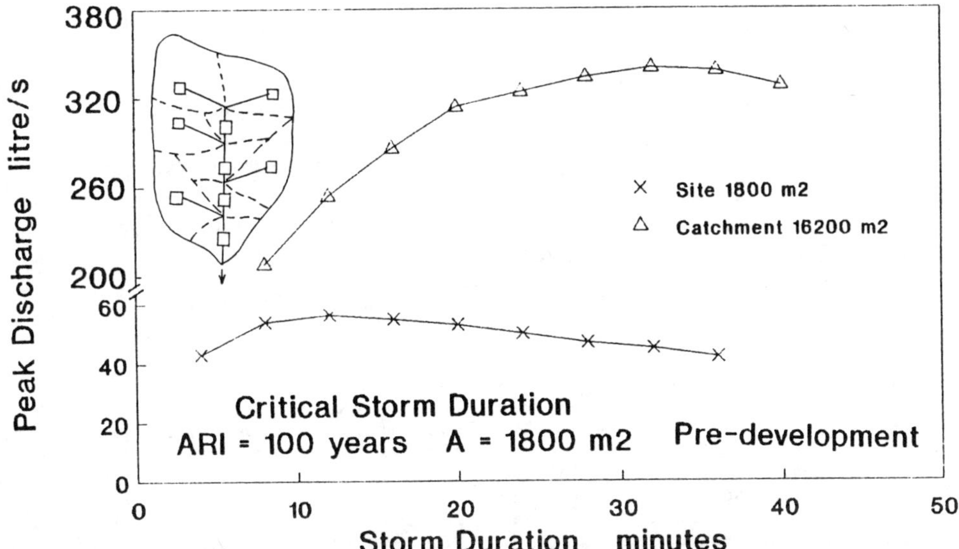

Figure 2. Critical storm durations .

which calculates realistic hydrographs and routes them through detention basins. (Boyd et al, 1987). The site area was adopted as $1800m^2$ and the catchment consisted of 9 sites (total area $16,200m^2$). A site travel time or lag time of 8.8 minutes was adopted for the pre-development condition.

Two recurrence interval storms were used. One had ARI = 100 years, post-development travel times of 4.4 minutes, and zero rainfall losses for pre and post-development conditions. The other storm had ARI = 5 years, post-development travel time of 2.2 minutes, and pre and post-development runoff coefficients of 0.43 and 0.64 respectively.

Design rainfall intensity - frequency - duration data for the Sydney region were used.

4.2 Development of one site within the catchment

The first case examined is for one site occupying 11% of the total catchment area of $16200m^2$. Results are summarised in Table 1. For ARI = 100 years and the pre-development condition, the critical duration for the site of $1800m^2$ is 12

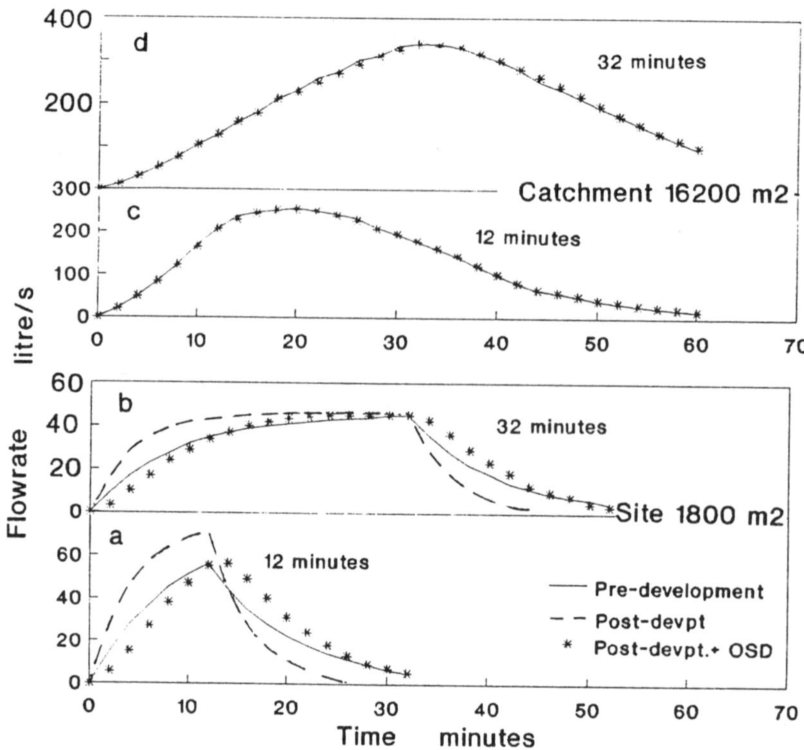

Figure 3. Effect of onsite detention on flood hydrographs.

minutes with a peak site discharge of 56 litres/s, while the critical duration for the total catchment of 16200m^2 is 32 minutes with peak discharge 341 l/s. (These results are also shown in Figures 2 and 3). Post-development, the shorter site response time increases peak discharges and also reduces the critical storm duration to 8 minutes, giving a peak discharge of 76 l/s. Approximately 17m^3 of OSD is required to reduce the post-development site discharge to the pre-development value of 56 l/s (Figure 3a). Note that the presence of OSD has shifted the critical duration back to the slightly longer value of 12 minutes.

Development of the site causes significant increases in site peak discharge for storm durations near to its critical duration of 8 to 12 minutes, and OSD is required to reduce these down to the pre-development value of 56 l/s. For longer duration storms near to 30 minutes however, equilibrium has been reached for both pre and post-development conditions (Figure 3b) and post-development peaks are only marginally higher than pre-development values. Since equilibrium has been reached, OSD has little effect in reducing peak flows, but this does not matter since the increase is only marginal. For the catchment's critical storm duration of 32 minutes, development of the site increases catchment peak discharge from 341 to 347 l/s. Provision of 17m^3 of OSD at the site reduces this back to 340 l/s, so that the OSD designed to reduce site peak discharges also acts to reduce catchment peak discharges to pre-development values (Figure 3d). Note that the OSD provided at the site reduces catchment peaks below the pre-development value for all durations (Figure 3c and Table 1).

Essentially the same results were obtained for ARI = 5 years. Because of the greater difference in pre and post-development runoff proportion and travel times, post-development site discharges increased significantly (from 14 to 37 l/s), and a considerable amount of OSD (14m^3) was needed to reduce these back to pre-development values, even though the small ARI meant smaller volumes of runoff. Despite this difference, provision of OSD designed for the site was found to reduce total catchment discharges back to the pre-development values across the whole range of storm durations, in the same way as for ARI = 100 years.

Table 2 Development of Larger Proportions of the Catchment

ARI (years)	Area of developed site (m^2)	Site			Catchment		
		Pre	Post	Post ($OSDm^3$)	Pre	Post	Post ($OSDm^3$)
100	1800 (11%)	56	76	56 (17)	341	347	340 (17)
		56	76	56 (17)	341	346	341 (17)
		56	76	56 (17)	341	345	342 (17)
		56	76	56 (17)	341	343	342 (17)
	5400 (33%)	145	200	145 (65)	341	365	336 (65)
		145	200	145 (65)	341	382	363 (65)
	9000 (55%)	218	309	223 (120)	341	389	334(120)
		218	309	223 (120)	341	423	359(120)
	12600 (77%)	283	400	283 (183)	341	433	333(184)
5	1800 (11%)	14	37	14 (14)	85	91	84 (14)
		14	37	14 (14)	85	91	85 (14)
		14	37	14 (14)	85	91	86 (14)
		14	37	14 (14)	85	91	87 (14)
	5400 (33%)	36	96	36 (49)	85	109	84 (49)
		36	96	36 (49)	85	117	92 (49)
	9000 (55%)	54	151	54 (95)	85	136	81 (95)
		54	151	54 (95)	85	146	88 (95)
	12600 (77%)	71	201	71 (138)	85	169	83 (138)

The results shown in Table 1 were for the developed site located at the top end of the catchment. Calculations were repeated for the developed site in the middle and near the outlet of the catchment. In each case catchment discharges were reduced to pre-development values, with slightly greater reduction for the site located further from the outlet.

Overall, the results show that when a site occupying 11% of the catchment is developed and OSD is provided at this point to reduce site discharges to pre-development levels, the OSD also reduces total catchment discharges to very close to pre-development values.

4.3 Development of larger sites in the catchment

Table 2 summarises results when one or more sites were combined and treatment as a larger site occupying 11 33, 55 and 77% of the total catchment area of 16,200m^2. Where several rows of results are shown, the first row is for the developed site near to the top end of the catchment, and successive rows for the site located closer to the catchment outlet.

All results are for the critical storm duration applying to that case. In each case OSD was provided at the outlet of the developed site, and the volume required to reduce post-development peak flows from the site back to the pre-development values was calculated.

The volume of OSD obviously is larger when the developed site makes up a larger proportion of the catchment. Post development catchment flows are also larger, but are reduced by the large OSD to be close to the pre-development catchment values.

Overall the results show that if a small proportion of the catchment is developed, catchment discharges increase by only a small amount, and the small volume of OSD is sufficient to reduce them to catchment pre development values. Conversely, if a large proportion of the catchment is developed, total catchment discharges increase significantly and the large volume of OSD required to control site discharges is also sufficient to reduce total catchment discharges to pre-development levels. In only a few cases, when the developed site is located near to the catchment outlet, was the catchment peak greater than the pre-development value.

4.4 Effects of OSD at several sites within the catchment

In sections 4.2 and 4.3 the size of the developed site ranged from 11% to 77% of the total catchment and one OSD storage was provided at the outlet of each of

Table 3 Effect of OSD at Several Sites within the catchment

ARI years	Number of Developed Sites	Percent of total catchment area	Peak Discharge at Catchment Outlet (litre/s)		
			Pre	Post	Post with OSD
100	1	11	341	347	340
	1	11	341	343	342
	2	22	341	353	339
	2	22	341	347	343
	3	33	341	358	339
	3	33	341	352	342
	4	44	341	362	339
	4	44	341	358	311
	5	55	341	364	341
5	1	11	85	91	84
	1	11	85	91	87
	2	22	85	98	84
	2	22	85	98	88
	3	33	85	105	86
	3	33	85	105	90
	4	44	85	112	88
	4	44	85	112	68
	5	55	85	119	91

these sites. In this section the effect of providing a small volume of OSD at several small sites within the larger catchment is investigated. Each site area was adopted as $1800m^2$ and OSD designed to reduce post-development flows from the site to pre-development values. The total catchment area was $16{,}200m^2$. For ARI = 100 pre-development, post-development flows and required OSD were 56 l/s, 76 l/s and $17m^3$ respectively; for ARI = 5 the values were 14 l/s, 37 l/s and $14m^3$ respectively.

Table 3 summarises results and, as in section 4.3, where more than one row of results is shown the first row is for the site or sites located at near the top of the catchment and the next rows are for sites closer to the outlet. All results are shown for the particular critical storm duration. If only a small number of sites are developed, then total catchment flows increase by only a small amount and the OSD provided at each site keeps catchment discharges close to the pre-development values. When many sites within the catchment are developed, total catchment flows increase more, but the OSD provided at this larger number of sites is still sufficient to reduce total catchment flows to be close to pre-development values.

As in section 4.2, OSD gives a slightly greater reduction in catchment peak discharge when the site is located further from the catchment outlet.

5 CONCLUSIONS

From the calculations in this study, it is clear that OSD designed to reduce flood peaks from a developed site within a larger catchment is also effective in reducing flood peaks from the total catchment. If the site is located further away from the catchment outlet, the reduction in catchment peak flow is greater than if the site is located near to the catchment outlet. In a few cases when the site was located near to the catchment outlet, catchment discharges were slightly greater than the pre-development values.

If the site is small relative to the total catchment, development of the site causes insignificant increases in catchment peaks, and the small volume of OSD provided at the site is sufficient to reduce them to pre-development values. If the site is large relative to the total catchment, development causes significant increases in catchment peaks, and the large volume of OSD required to control site discharges is again sufficient to similarly control catchment discharges.

If several small sites each have OSD individually designed to control peaks from the site, they are together effective in reducing peaks from the total catchment to near to pre-development values.

Concern is sometimes expressed that provision of detention storage can increase total catchment flooding because the delayed outflow from the storage may co-incide with peak flows from the remainder of the catchment. However for the long storm durations which are critical for the catchment both pre and post-development site discharges approach equilibrium. The resulting site hydrographs have a broad flat peak so that time shifts have little impact, and the possibility of a time shifted peak coinciding with other flows to give a greater peak is unlikely. The only time when this may occur is when the site is large relative to the remainder of the catchment, and even then the increases are not large (Table 2).

The results presented here are for a particular model of the catchment and a range of drainage networks, using design storms and assumptions about the effects of development on flood discharges. It should be seen as a preliminary study of the problem. It would be desirable to study one or more real urban catchment drainage systems using a specific urban drainage model such as ILSAX, SWMM or MOUSE. Some calculations of this kind have been carried out (O'Loughlin, 1989). ILSAX (a variant of ILLUDAS) was applied to a 2.9 hectare residential area. One site of 0.46 ha. was re-developed from housing to home units and OSD was provided at the site. The OSD reduced the site peak discharge from 120 litre/s to the pre-development value of 58 l/s and also reduced total catchment flooding from 580 l/s to the pre-development value of 555 l/s. These results are in agreement with those of the present study.

A review of case studies by Boyd (1993) found very few cases in which downstream flooding was increased by detention storage, consistent with these results.

REFERENCES:

Aitken A.P., 1975. Hydrologic investigation and design of urban stormwater drainage systems. Australian Water Resources Council Technical Paper 10.

American Society of Civil Engineers Task Committee, 1975. Aspects of hydrological effects of urbanisation. Proc. ASCE, 101, HY : 449-468.

American Society of Civil Engineers, 1982. Proc. Confce. on Stormwater Detention Facilities, New Hampshire.

Boenisch, M.E., and A. Ogle, 1989. On-site detention basins - Wollongong, I.E. Aust. Seminar on On-Site Stormwater Detention Systems, Sydney.

Boyd, M.J., B.C. Bates, D.H. Pilgrim and I. Cordery, 1979. A storage routing model based on catchment geomorphology. Journal of Hydrology, 42: 209-230.

Boyd, M.J., 1981. Preliminary design procedures for detention basins. Second Intl. Confce. on Urban Storm Drainage, Urbana: 370-378.

Boyd, M.J., 1991. Critical storm durations for detention basins. Instn. Engineers Australia, Sixth National Local Govt. Engg. Confce., Hobart : 169-173.

Boyd, M.J. (1993). Effect on detention storage on total catchment flooding. 6th Intl. Cont. Urban Storm Drainage, Niagara Falls (in press).

Culp, M.M., 1948. The effect of spillway storage on the design of upstream reservoirs. Agricultural Engg., 29:344-346.

Espey, W.H., and D.E. Winslow, 1974. Urban flood frequency characteristics. Proc. ASCE. 101, HY2 : 279-293.

Institution of Engineers Australia, 1987. Australian Rainfall and Runoff, 2 volumes.

Lakatos, D.F., and R.H. Kropp, 1982. Stsormwater detention - downstream effects on peak flowrates. Proc. ASCE Confce. on Stormwater Detention Facilities, New Hampshire : 105-120.

Lees, S.J., and S.J. Lynch, 1992. Development of a catchment on-site stormwater detention policy. Intl. Symposium on Urban Stormwater Management, Sydney Sydney : 343-349.

McCuen, R.H., 1979. Downstream effects of stormwater management basins. Proc. ASCE. 105, HY11 : 1343-1356.

Natural Environment Research Council UK, 1975. Flood Studies Report, Vol. 1 Hydrological Studies.

O'Loughlin, G.G., 1989. Successful implementation of on-site detention policies. I.E. Aust. Seminar on On-Site Stormwater Detention Systems, Sydney.

Phillips, D.I., 1987. On-site stormwater detention for small urban re-development projects. Fourth Intl. Confce. on Urban Storm Drainage, Lausanne : 361-366.

Rao, A.R., J.W. Delleur and P.B.S. Sarma, 1972. Conceptual hydrologic models for urbanising basins. Proc. ASCE, 98, HY7 : 1205-1220.

Tomkins, N. and C. Cooper, 1989. On-site detention in Kuringai - 10 years on. I.E. Aust. Seminar on On-Site Stormwater Detention Systems, Sydney.

Environmental Management, Geo-Water & Engineering Aspects, Chowdhury & Sivakumar (eds)
© 1993 Balkema, Rotterdam. ISBN 90 5410 099 0

Large scale catchment simulation using the MIKE-SHE model: 1. Process simulation of an irrigation district

R.S.Carr
Lawson and Treloar Pty Ltd, Sydney, N.S.W., Australia

J.F.Punthakey & R.Cooke
New South Wales Department of Water Resources, Sydney, N.S.W., Australia

B.R.Storm
Danish Hydraulic Institute, Horsholm, Denmark

ABSTRACT: Description and conceptualisation of hydrologic processes in an irrigation district with rising watertables is discussed. Application of a distributed deterministic numerical model with variable spatial and temporal scales to a large heterogeneous catchment has lead to an understanding of the utility and limitations of such models.

1 INTRODUCTION

The Berriquin irrigation district in south-western New South Wales has experienced rising water tables for several years. The irrigation district covers an area of approximately 320,000 Ha, with 1300km of unlined supply channels and 700km of drainage channels and natural streams. The terrain is flat, with average slopes of 0.0005. Rice, annual pasture and perennial pasture are the main crop types, the average farm area being of the order of 250Ha (2.5km^2) in area. Rice makes up approximately 20% of the total crop on average, the average rice bay being of approximately 50Ha (0.5km^2). A detailed description of the catchment and data is given in Punthakey et. al. (1993). The New South Wales Department of Water Resources has assembled a multi-disciplinary team of Departmental staff and specialist consultants to undertake a study of the irrigation district with a view to defining the impact of management alternatives. In order to evaluate these options, a numerical model was required which should be able to describe the impacts of farm-scale alternatives on catchment-scale behaviour, and to be able to address the impacts of proposed management strategies on salinity levels once the initial flow studies were complete.

Because of the large scale of the catchment and the heterogeneity of parameters, a distributed model was required with appropriate temporal and spatial scales to describe the hydrologic processes. The Department had a number of separate surface and groundwater models available for application to the irrigation district, but difficulties with simulating the close interaction that occurs when watertables are close to the surface meant that an integrated modelling system was necessary. After consideration of an in-house development program and a review of public domain and commercially available models, the Department chose the MIKE-SHE system (Danish Hydraulic Institute 1991) as being the most cost-effective modelling platform. The broad range of process facilities available in the MIKE-SHE were considered sufficient to conceptualise the operation of the irrigation district and proposed management strategies. The paper describes those facilities and the way the irrigation district has been conceptualised.

2 DESCRIPTION OF THE MIKE-SHE

The SHE (Abbott et. al. 1986a & 1986b) is a generalised mathematical modelling system originally developed by a consortium of European agencies in the 1980's as a research project. The Danish Hydraulic Institute (DHI) have further developed and enhanced the basic code, included pre- and post-processors, graphics systems and manuals to the extent that it has been made available for commercial release and named the MIKE-SHE to signify its difference from the original system. This paper will refer to the model as SHE where features are common with the original system, and MIKE-SHE where the package represents a departure from the SHE. A number of additional modules have been added to the basic SHE, including three-dimensional groundwater, soil erosion, solute transport, and salinity/chemical (ion exchange, complexation, precipitation/dissolution) process description in both the unsaturated and saturated zones.

The SHE represents the deterministic view of catchment hydrology, containing no empirically based process descriptions. The model has an ability to take into account a large range of spatial and temporal data which was previously seen as requiring excessive effort when the results in terms of calibration to existing conditions was no better than simple lumped models.

Lumped conceptual models are efficient for applications where significant data are lacking, or where catchment changes are of a broad nature that can be reasonably lumped. Unfortunately, lumped conceptual models lack the ability to define the impact of specific changes, i.e. the deepening of a specific drainage channel by 1 metre, or the laser grading of certain fields. Within a modern Environmental Impact

Statement, relative impacts of detailed measures are often required. Advances in hydrographic data collection, storage and retrieval over the past decade, as well as the development of GIS systems has made that data far more accessible than before. As computer hardware advances continue to provide last decades mainframes on a desktop (a trend that shows no sign of slowing), computational speed issues will be solved by hardware manufacturers. The present study is using workstation hardware which provides reasonable run times for the calibration period.

The SHE was originally developed with a view to describing the entire land phase of the hydrologic cycle in a given catchment with a level of detail sufficiently fine to be able to claim a "physically-based" concept (Abbott et. al. 1986a). The equations used in the model are well-known and tested, so the fundamental scale-dependency of parameters remain unsolved. Hence the model is able to represent fine details in the flow of water around the surface and groundwater system, but the model parameters are still scale-dependant. The pre-processing systems provided are designed to be able to change model scales with a few keystrokes, so the impact of scale effects on model results and conceptualisation are easily obtainable. This was seen as an important capability in the Berriquin study so that the impacts of scale dependency could be tested.

In the SHE, all process descriptions operate at time steps consistent with their own temporal scales. For example, the unsaturated zone solution cannot be expected to accurately represent the movement of wetting fronts if forced to operate at time steps of days or weeks which are appropriate for groundwater modelling. Hence, each process is simulated at different time steps (taking into account the level of stress in the system), and coupled (with the appropriate accumulated mass) with physically adjoining processes when their time steps coincide.

The catchment can be discretised by a number of layers of horizontal finite difference meshes for the groundwater system, and a similar single-layer mesh for overland flow. These mesh systems are linked by vertical unsaturated zone columns at each the centre of each grid square as shown in figure 1.

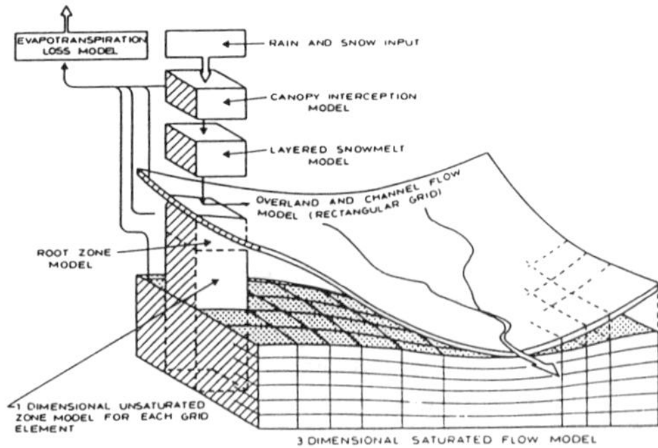

Figure 1: Structure of MIKE-SHE

Flows around the catchment are modelled by a finite difference solution of the partial differential equations describing overland and channel flow and unsaturated and saturated subsurface flow. Interception, evapotranspiration, root zone development and snowmelt are simulated by individual process models or external inputs.

Each grid square in the finite difference mesh is able to have differing parameters for rainfall, potential evapotranspiration, soil properties, leaf area index, root zone depth, etc. Each process description in the water movement part of the MIKE-SHE, with modifications as used in the Berriquin study are briefly described below:

2.1 Overland and channel flow

Flow in the channels and streams in the SHE is simulated using a branched and looped one-dimensional diffusive wave approximation of the St Venant equation. It is solved in a grid system which runs between the overland and groundwater cells (i.e. along the grid lines) and so has a rectangular form in plan. If the discretisation is fine enough, this mesh provides a satisfactory approximation to the actual system.

Figure 2 shows a part of the Berriquin drainage system as digitised, and represented by 500 metre, 1km and 2km grids. The 1km grid model also shows the digitised channel locations for direct comparison. As the mesh spacing reduces, the difference between the digitised and modelled channels is reduced. One impact of this discretisation error is that as the mesh size is reduced, more bends and twists in the channel system are reproduced, and the stream system becomes longer. This in turn increases the stream leakage to groundwater by the same proportion as the stream length increases.

The model is able to simulate weirs and other structures in the channel system, so the impact of various structural measures in the supply and drainage system can be quantified. Each river link is defined by a cross-section, so the impacts of channel deepening may also be addressed.

Overland flow occurs either because net rainfall exceeds the infiltration rate, or as a result of the groundwater table rising to the ground surface and causing discharge zones. Overland flow is solved by a two-dimensional diffusive wave approximation of the St Venant equation which allows backwater effects on the ground surface. Hence water can be stored in surface depressions (if the soil is sufficiently impermeable) to be removed by evaporation, or infiltrate into the unsaturated zone.

2.2 Unsaturated zone component.

The unsaturated zone processes are described by the one-dimensional Richard's equation solved by a finite difference scheme in the vertical. Node spacings in the finite difference mesh are typically in the order of a few centimetres near the soil surface, gradually increasing

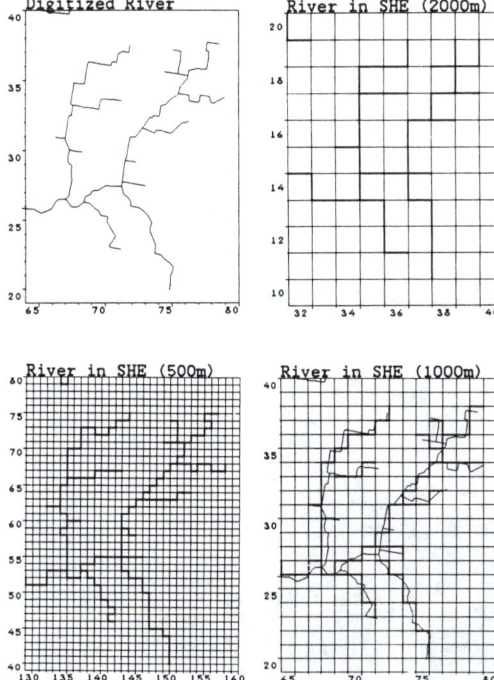

Figure 2: Comparison of digitised river at three grid scales.

in spacing with depth. Soil properties are described by saturated hydraulic conductivity, an exponent to describe hydraulic conductivity as a function of saturation, and a retention curve. Transpiration and soil evaporation moisture losses are incorporated as sink terms at the node points in the root zone. The infiltration rate in each column is determined by the upper boundary condition which may be a flux condition when net rainfall is less than capacity, or as a head boundary when net rainfall exceeds soil capacity. The solution technique has an advanced feature which estimates the exchange of water between the saturated zone and the unsaturated zone based on the retention curve, and adjusts the phreatic surface accordingly (Storm, 1988). Several horizons of different soils can be incorporated into each unsaturated column. The unsaturated zone may disappear if the water table intercepts the ground surface.

2.3 Evapotranspiration/Root Zone

The root zone model used in the present MIKE-SHE version 4.3 is based on a non-linear description of root zone shape (Kristensen & Jensen 1975). The crop parameters (root zone depth and shape, leaf area index) are varied over the growing season to represent root zone development for each crop type. The shape and depth of the root zone determines the extraction function of soil moisture from the unsaturated zone.

The estimation of potential evapotranspiration rates was undertaken using a quasi two-layered resistance energy balance model (Punthakey 1988; Punthakey & Gurner 1986). The model is based on a modified Penman-Monteith combination method which incorporates aerodynamic and surface resistance as controls on the evaporation process, and utilises a numerical procedure to evaluate the energy balance of a composite crop-soil surface. The modified equation eliminates the canopy resistance term allowing estimation of transpiration from measured meteorological data and prescribed crop and soil parameters. A stomatal function has been incorporated into the Penman-Monteith equation. The stomatal function accounts for the effect of mean soil water potential and relative humidity on the transpiration rate and thus provides a mechanism for influencing the energy balance equation when moisture in the root zone becomes limiting.

2.4 Saturated zone component

The saturated groundwater flow system is modelled by using an implicit finite-difference solution of the groundwater equation. The upper layer is described by the non-linear unconfined equation, while lower layers are described by the linear confined equation. The groundwater system may be two or three-dimensional. Stream-aquifer exchange is estimated using a leakage layer using driving heads based on the difference in water level between the river and groundwater table.

2.5 Graphical Presentation

One difficulty encountered in the use of a model as detailed as MIKE-SHE has been the understanding of complex interactions between processes during simulation. Development of advanced graphical systems in recent years has resulted in the ability to present highly complex information in a simple form that can be understood in a physical sense. Figure 3 shows a profile through a soil column with time, unfortunately reproduced in black and white which does not reflect the powerful communication tool of colour. The figure shows moisture content with depth and time in an unsaturated column, potential and actual evapotranspiration rates and rainfall for a two-month test period of data from Berriquin. The figure shows recharge events which penetrate to the groundwater table, and those which do not on account of low initial soil moisture, or high evapotranspiration rates transmitted through an active crop. A close inspection of the results of this type of presentation, in conjunction with recorded data, gives physical insight into the performance of the model and how calibration can be improved.

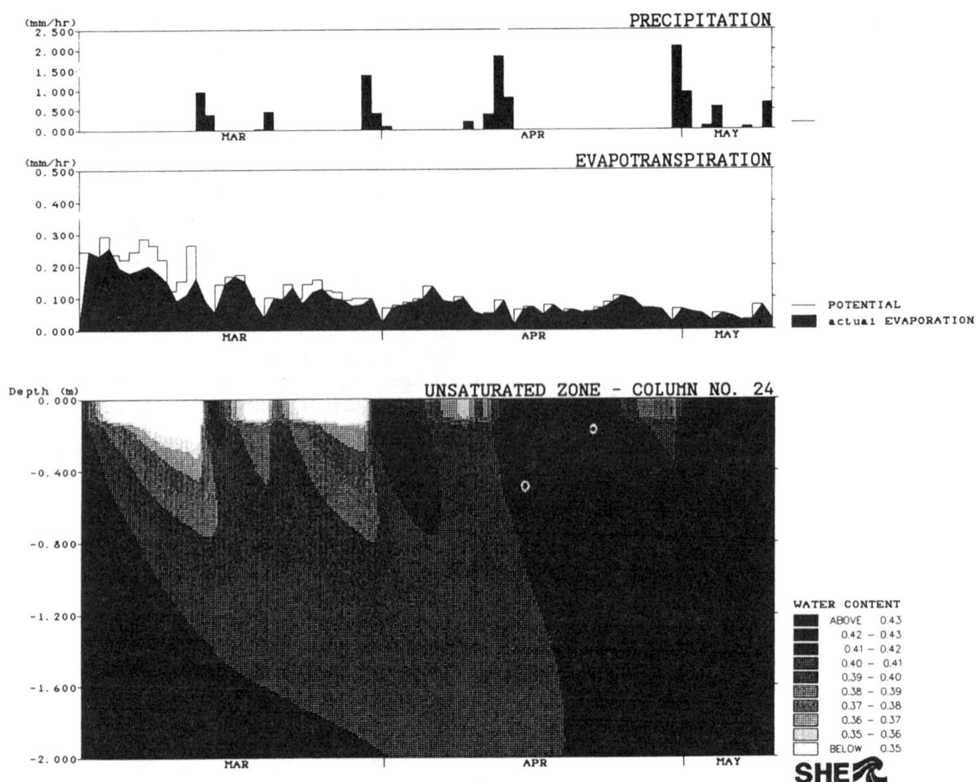

Figure 3: Graphical presentation of rainfall and unsaturated zone dynamics.

3 PROCESS DESCRIPTION

The fundamental hydrological regime to be addressed in the model is continued application of irrigation water to an area with watertables close to the surface. Hence, rainfall, stream leakage and irrigation water application provide the basic driving forces, with evapotranspiration (linked to potential ET and crop types), groundwater flow and surface drainage being exporting processes.

The model calibration procedure revolves around achieving the correct balance between rainfall, runoff and infiltration at the soil surface. The correct proportion of runoff from rainfall is indicated by streamflow data. If the model over-estimates streamflow, then there is clearly insufficient infiltration (provided rainfall volumes are correct). If the groundwater subsequently rises to quickly, then there is either insufficient evapotranspiration (which can be tested by setting actual ET equal to potential ET with a deep crop root zone), or the soil and groundwater parameters have too little storage. Many of these processes can be addressed at a larger grid scale because, if the unsaturated zone parameters are unchanged, the spatial scale can only affect the timing of runoff and stream leakage rates (proportional to model stream length).

If the spatial and temporal variation in the rainfall and irrigation inputs can be distributed onto a model in which the spatial variation in topography, soils, drainage system and groundwater system are well defined, then there is a good physical basis for representing the flow of water. If the rainfall and irrigation rates can also be defined spatially (not always possible due to data inadequacies), then the model has a good chance of calibration. Rainfall can be described on a daily basis, and potential ET rates estimated from climatic data. The drainage system can be described by digitising available mapping and design drawings. Soil conductivity and saturated moisture content information can often be estimated from the literature and available data. If laboratory data are not available, retention curves can be estimated from the literature and particle size distribution. Crop development can be described by drawing on the large body of lysimiter data and field studies in the literature.

3.1 Supply System

One driving force is the supply system which provides water to the irrigators through a system of unlined channels.The supply system is more extensive than the drainage system and is thought to lose water to the groundwater system, thereby contributing to rising groundwater system, thereby contributing to rising groundwater levels. The SHE can only have one surface channel system, which has been defined as the drainage system. The supply channels are formed by mounding earth at the surface, so are unlikely to experience connection of the water table to the channel bed.

The supply system was therefore conceptualised as a completely separate channel system with one inflow point to the district. Total leakage to the district could be estimated by holding the water level in the first river link constant, then the discharge through that link must equal the total leakage. The leakage rate was variable over the district because of the soil heterogeneity. Total leakage in segments of the channel system could be obtained from the difference in discharge between the start and end of the segment.

Once the supply system had reached steady-state, the leakage was input into the total district model by setting up a series of lines of groundwater injection bores corresponding to the supply channel. This process reduced the amount of data manipulation required, since the MIKE-SHE post-processors and one custom-written service program could be used to automatically generate the necessary files.

3.2 Surface scale considerations

The issue of spatial scales becomes important not only from the point of view of the sensitivity of the model to parameters, but also from the perspective of having enough detail to address necessary management issues. The fundamental issues are the spatial variability of irrigation application rates over a heterogeneous soil regime, modified by channel leakage and crop water usage. Clearly, rice irrigation over sandy soils will result in large leakages to groundwater, but since the irrigator would see a large water bill result, the rice may be moved to another part of the farm. Since the average rice area per farm is of the order of 50Ha, we would require grid spacing of that order (500-750 metre grid) to simulate details of rice rotation.

If grid spacings of 2km are considered, then the rotation of crops becomes unresolvable and the simulation of average gross application rates over 1-2 farms will be included in each grid square. Analysis of satellite imagery is expected to provide details of rice and other crop densities across the study area, from which cropping patterns can be established for the calibration period. Sensitivity analysis will then be undertaken to grid spacing to determine the appropriate scale for modelling work. Preliminary runs have indicated that a 2km grid is sufficient to define the rainfall-runoff and groundwater processes, and a finer grid will only be required to account for irrigation water variation between rice and pasture. The range of likely grid spacings are therefor expected to be between 500 metres and 2km.

4 CONCLUSIONS

The range of processes operating in an irrigation district have been conceptualised in a distributed deterministic catchment model. The likely range of grid scales necessary to define the hydrologic performance of the district have been identified, and further calibration and sensitivity work will be undertaken to finalise the grid spacing. The model is then expected to be able to provide sufficient detail for a comprehensive land management study involving small-scale changes to the infrastructure of the irrigation works and on-farm practices.

REFERENCES

Abbott, M.B., Bathurst, J.C., Cunge, J.A., O'Connell,P.E. & Rasmussen, J. 1986: An Introduction to the European Hydrological System - Systeme Hydrologique Europeen, 'SHE', 1: History and Philosophy of a Physically-Based, Distributed Modelling System. *J. Hydrol. 87:45-59.*

Abbott, M.B., Bathurst, J.C., Cunge, J.A., O'Connell, P.E. and Rasmussen, J. 1986: An Introduction to the European Hydrological System - Systeme Hydrologique Europeen, 'SHE', 2: Structure of a Physically-Based, Distributed Modelling System. *J. Hydrol. 87:61-77.*

Danish Hydraulic Institute 1991 SHE - European Hydrological System Version 4.3 Users Guide and Technical Reference. Horsholm, Denmark.

Kristensen, K.J. & Jensen, S.E. 1975: A Model for Estimating Actual Evapotranspiration from Potential Evapotranspiration. *Nordic Hydrol. 6:70-88.*

Punthakey, J.F. 1988: Actual Evapotranspiration and Energy Balance for Irrigated Horticultural Crops in the Riverland of South Australia. *Dept of Agric. South Australia, Tech Paper No 20, 134p.*

Punthakey J.F. & Gurner, E.K. 1986: Predicting Actual Evapotranspiration by Modelling the Crop and Energy Balance. *Hydrology and Water Resources Symposium, Institution of Engineers, Australia pp 33-37.*

Punthakey, J.F., Cooke, R., Carr, R.S., Somaratne, N. & Watson, K.K. Large scale catchment simulation using the MIKE-SHE model: 2. Modelling the Berriquin irrigation district, 1993. Int. Conf. on Environ. Management. Wollongong, Australia.

Storm, B. 1988: Deterministic Modelling of Saturated Flow and the Coupling of Surface and Subsurface Flow. In Recent Advances in the Modelling of Hydrologic Systems, NATO Advanced Study Institute, Portugal.

Environmental Management, Geo-Water & Engineering Aspects, Chowdhury & Sivakumar (eds)
© 1993 Balkema, Rotterdam. ISBN 90 5410 099 0

About the dynamic parameter of flood models in urban drainage networks

V.A.Copertino, M.Fiorentino, B.Molino & V.Telesca
Department of Environmental Engineering and Physics, University of Basilicata, Potenza, Italy

ABSTRACT: Relationships between the dynamic parameter of flood models and the potential energy dissipation characteristics of the unsteady flow were analyzed. A new equation relating the average flood wave celerity to some hydraulic characteristics such as mean depth and width of the flow and to a shape factor of the hydrograph was derived. The assessment of the proposed equation was carried out by using real world data recorded in a urban experimental catchment, located at Malvaccaro (Potenza,Italy).

1. INTRODUCTION

Most physically based models used for the simulation of the flood hydrographs in a urban drainage network require, other than the characterization of the drainage surface, the determination of the dynamic parameters needed to define the phenomenon of flood routing through the drain network.

In the last years with regard to the capacity of the models to outline correctly the topology of the network great developments have occured . However, a more reliable application is still related to the possibility of better outlining the dynamic characteristics of the flow and, at least for basins of small dimensions, the runoff production at the hillslope scale. It appears clear, therefore, how the detailed knowledge about soil properties and ground typology and shape, can improve the capacity to describe the complex response of the basin to rainfall events. A systematic approach to verify theoretical hypotheses of the hydrologic response at the hillslope scale is possible using Geographical Information Systems (GIS), that allow us to treat a large quantity of spatial data in the framework of a distributed hydrological model. However, this aspect, on which authors are now working, still requires further research.

In this paper one is concerned with the possibility to outline, on a theoretical basis, more consistently, the dynamic characteristics of the flow in the network, when topological flood models are used.

Topological models are based on the fact that the arrival times of water particles are proportional to the topological distances of the entry point into the network from the outlet; for topological distances of a point we state the number of entries between the outlet and the point itself. Furthermore, the constant of proportionality is given by the ratio between real mean distances among entries and flow velocity of the water particles.

Agnese & La Loggia (1987) suggested this velocity to be estimated as a function of velocity of the full pipe discharge. Such an estimate finds one of its comparisons in the design of networks, but it may result inadequate in the estimate of the flood peak and volumes for a set of events that do not fill completely the pipes themselves, events that are, also, of notable interest for management problems .

The aim of the present work is the research of a relationship that clarifies the modality with which the mean celerity of the flood flow depends on the hydraulic quantities of a clear physical meaning and on hydrological parameters of easy interpretation. Such a relation is derived, with regard to the outlet of the drainage network, by using energy equations which allow us to relate the celerity to the rate of potential energy dissipation of the flow.

The assessment of the proposed equation is carried out by using thirty flood events with peak discharge greater than 100 l/sec, recorded at the experimental basin of Malvaccaro (Southern Italy) during 1989-1991. In a previous work by Fiorentino et al. (1992) the proposed model was also applied to three basins making part of the international database UDM (Maksimovic & Radojkovic, 1986).

2. THE DYNAMIC PARAMETER OF THE TOPOLOGIC MODEL

The hydrologic response of a drainage network, that is the frequency distribution f(t) of arrival times of the water particles, can be represented as the Instantaneous Unit Hydrograph (IUH). The topological IUH, proposed by Agnese & La Loggia (1987), who transferred to the urban drainage case the approach followed by Rodriguez-Iturbe & Valdès (1979) and Gupta et al. (1980) for river basins, gives:

$$f(t)=\sum_{i=L}^{1}\left(Xi\Big/Z\sum_{j=1}^{2}Aj/A\,EXP(-(t-iT)/tmj)\right) \quad (1)$$

where:

i = topological level of the entry, or topological distance of the entry;
L = maximum topological distance;
Xi = number of network links that belong to the level i;
Z = number of sources (external nodes);
A = total area of the drainage basin;
Aj = pervious or impervious area (j=1,2);
tmj = holding time in Aj;
T = holding time in the links at level i .

To use equation (1) it is necessary to determine, other than topologic characteristics of the network, which are well synthesized through the width function (Troutman & Karlinger, 1984), the travel times of the water particles from the point they fall to the outlet. These times are given once dynamic characteristics of the flow are known, with regard to both hillslope and network.

For the holding times tmj in pervious and impervious areas, one can refer to what has been described in literature (e.g. Fair et al., 1966).

The flood wave celerity c , i.e. the dynamic parameter of the model, is here derived by using the equations of De Saint Venant , which are, with straightforward meaning of notation:

$$\frac{\partial Q}{\partial x}+\frac{\partial A}{\partial t}=0 \quad (2)$$

$$\frac{\partial h}{\partial x}+\frac{V}{g}\frac{\partial V}{\partial x}+\frac{1}{g}\frac{\partial V}{\partial t}+s_f=0 \quad (3)$$

A simplification of the system formed by equations (2) and (3) is obtained by introducing the Boussinesq hypotheses , according to which the flood routing phenomenon can be approximate by a succession of almost uniform states. The dynamic equation (3) then reduces, in this case, to a relationship between the discharge Q and the cross area A, given by:

$$Q=kA^m \quad (2')$$

with k and m constants. Thus, celerity c, being the derivative of Q with respect to A, is given by:

$$c=\frac{dQ}{dA}=m(\frac{Q}{A})=c(Q)\,. \quad (4)$$

As

$$\frac{\partial Q}{\partial t}=\frac{dQ}{dA}\frac{\partial A}{\partial t}=c(Q)\frac{\partial A}{\partial t}\,, \quad (5)$$

substituting $\frac{\partial A}{\partial t}$,obtained by equation (5), into equation (2), one obtains:

$$\frac{1}{c}\frac{\partial Q}{\partial t}+\frac{\partial Q}{\partial x}=0\,. \quad (6)$$

Equation (6) is the differential equation of the kinematic model in which c is the celerity of a given discharge value Q.

For a rectangular section, having width w, equation (4) becomes:

$$c=\frac{dQ}{dA}=(\frac{1}{w})\frac{dQ}{dh} \quad (7)$$

from which, after simple mathematics one obtains:

$$\frac{dh}{dt}=(\frac{1}{cw})\frac{dQ}{dt}|_{x\ fixed} \tag{8}$$

where h represents the flow depth. Equation (8) gives the instantaneous unit stream power of the flow. With regard to the elementary volume Qdt, the stream power is given from $\rho g(\frac{dh}{dt})Qdt$. Assuming that at time t=0 one has Q(0) = 0, the stream power for t >0, is given by:

$$P=\int_0^t \rho g\ (\frac{dh}{dt})\,Q(t)\,dt=\int_0^t (\rho g/cw)(\frac{dQ}{dt})\,Q(t)\,dt \tag{9}$$

that, by means a variable transform gives:

$$P=\int_0^Q (\rho g/cw)\,q\,dq. \tag{10}$$

Which, for w and c constants gives:

$$P=(\frac{\rho g}{2cw})Q^2 \tag{11}$$

Integrating with respect to time, one obtains the total potential energy expenditure in the interval (0,t), that is also equal to the work done by the flow in that given section:

$$W=(\frac{\rho g}{2cw})\int_0^t Q^2(\tau)d\tau. \tag{12}$$

Then, the work per unit weight results in:

$$E=(\frac{1}{2cw})\frac{\int_0^t Q^2(\tau)d\tau}{\int_0^t Q(\tau)d\tau} \tag{13}$$

Equation (13) can also be written as:

$$E=(\frac{1}{2cw})Qm(1+Cv^2) \tag{14}$$

in which Qm is the average discharge of the hydrograph and Cv represents the coefficient of variation of the flood event.

Equation (14) shows how the average potential energy E dissipated by the flow during the flood event is a function of the average celerity c, of the mean discharge Qm and of a shape factor of the hydrographs expressed by Cv.

Moreover, equation (13) shows the way with which c is linked to the average potential energy dissipated by the flow; in particular, one has :

$$c=(\frac{1}{2wE})Qm(1+Cv^2) \tag{15}$$

Under the hypotheses that the flood event begins and ends at Q=0 and that, the term E, is the mean value y of the flow depth, equation (15) can be written as:

$$c=(\frac{1}{2wy})Qm(1+Cv^2) \tag{16}$$

In general , the knowledge of the main parameters of the streamflow process (Qm and Cv) allows, along with the knowledge of the rating curve of the channel section, the estimation of y and therefore of the dynamic parameter c.

3. THE EXPERIMENTAL BASIN OF MALVACCARO

Urban drainage basin of Malvaccaro is located near the town of Potenza in southern Italy. The catchment has an extension of 9 ha, is characterized by a maximum elevation of 870 m above sea level, a minimum elevation of 770 m above sea level, an average slope of 7%.

The urban basin consists of one urbanized zone that has 15% of pervious areas (planting, flower bed and private gardens) which are not directly linked to the drainage network and 85% of impervious areas, as sloped roofs, roads, etc. This last zone is served by a sewer network of a separate type; the one which drains the precipitation is composed by two principal PVC pipes 500 mm in diameter, characterized by lengths of 310 m and 409 m, and slopes of 8.7% and 4.4%, which drain two sub-basins of 3.8 ha and 4 ha respectively.

The excess rainfall drains into 38 road intakes, placed at a distance of 20 m between each other. The rainfall which falls on roofs reaches straightforward the network, because the 110 zinc drainpipes located just below the sloped roofs are connected to the network itself.

Type and number of equipment that are used to measure rainfall and discharges were accurately selected. For rainfall we adopted a rain gauge CAE PMB2, with a resolution of 0.2 mm. The measure of the discharge is carried out through three electromagnetic sensors Toshiba (TOSMAC 335/372). Such sensors are linked to an electronic gearcase C.A.E. model SP200, 64 K RAM memory, for data acquisition, using time-steps of one minute.

It is necessary to emphasize that electromagnetic sensors, in order to be able to make exact measures, requires that the measuring section must be full of water, then a siphon in measuring sections was placed. Such siphons, with a diameter of 500 mm, are characterized by a convergent part, upstream the measuring equipment, and a divergent one; they are equipped with waste device for periodic purging of accumulated sediments; the siphon-sensor system is put in a concrete pit.

4. DATA PROCESSING

On the basis of the morphological data, the topology of the drainage network of the Malvaccaro basin was outlined and the topological width function evaluated (fig.1).

The travel times of water on pervious and impervious areas were selected as 5 minutes in both cases. Travel times within the network pipes were evaluated according to the procedure indicated in section 2.

Using equation (6) the value of the average celerity for each event was estimated, evaluating the average flow depth y, as well as the average flow width w.

Table 1 and figures 2 to 7 show the results. It may be noted the good behavior of the topological model adopted. In fact, the mean deviation between the observed and simulated discharge is equal to 13.8%.

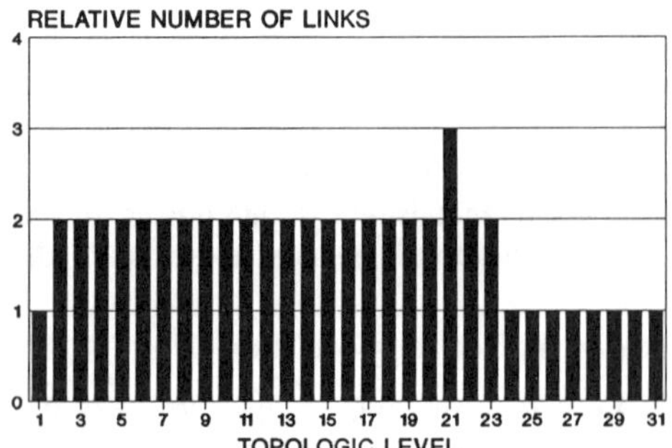

Fig. 1. Width function

Table 1. Comparison between observed and simulated peak discharges.

Event	Q observed (l/sec)	Q simulated (l/sec) topologic model	Error %
26/02/89	123.61	87.00	-29.62
04/03/89	136.00	138.70	1.98
04/03/89	103.30	135.10	30.78
15/05/89	598.00	627.40	4.91
26/07/89	354.00	326.50	-7.77
26/07/89	657.00	843.10	28.33
26/07/89	255.00	293.60	15.15
28/07/89	155.60	126.40	-18.72
23/09/89	195.00	214.40	9.92
28/09/89	161.00	157.70	-2.06
07/10/89	115.80	105.50	-8.92
10/10/89	103.90	120.00	15.46
09/04/90	276.10	218.40	-20.91
20/05/90	421.90	444.20	5.29
26/10/90	101.40	111.00	9.43
16/11/90	107.80	91.20	-15.37
17/11/90	120.60	115.50	-4.23
28/11/90	118.10	141.10	19.47
20/12/90	268.10	304.60	13.60
26/04/91	101.10	124.50	23.16
29/07/91	145.00	167.80	15.71
13/08/91	236.70	244.30	3.22
AVERAGE ERROR %			13.82

5. CONCLUSIONS AND FUTURE DEVELOPMENTS

The authors paid their attention to the estimation of the dynamic parameter of flood models analyzing the existing link between the flood wave celerity and the potential energy expenditure. An equation relating the average celerity to the flow characteristics, such as depth

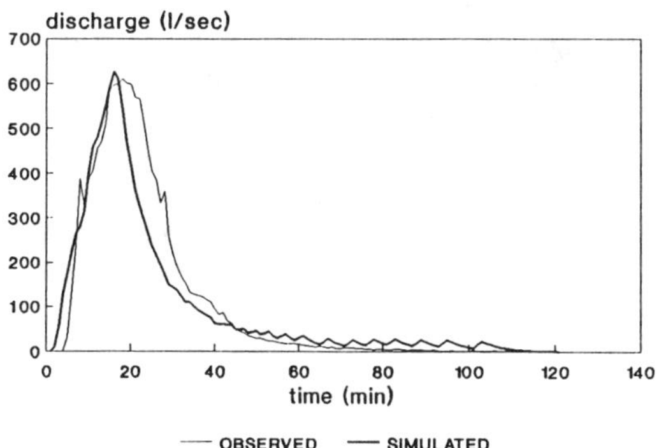

Fig. 2. Observed and simulated hydrographs

EVENT OF 10 OCTOBER 1989

discharge (l/sec)

time (min)

OBSERVED SIMULATED

Fig. 5. Observed and simulated hydrographs

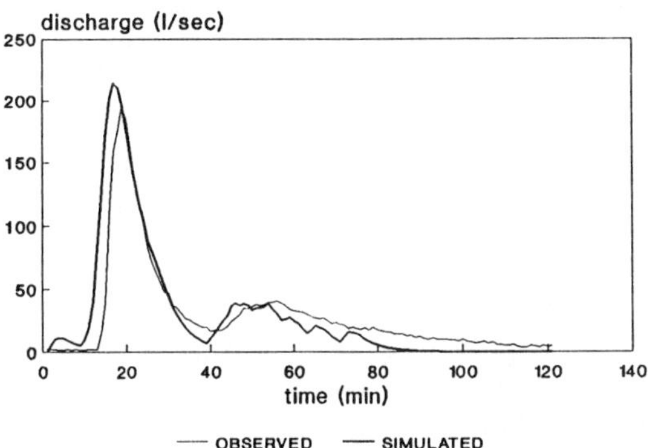

Fig. 3. Observed and simulated hydrographs

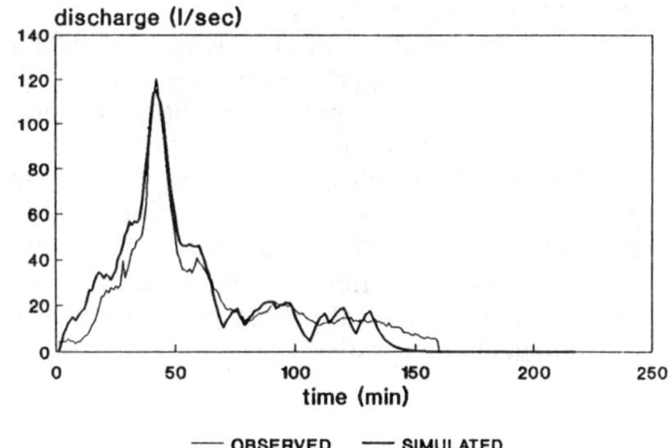

Fig. 6. Observed and simulated hydrographs

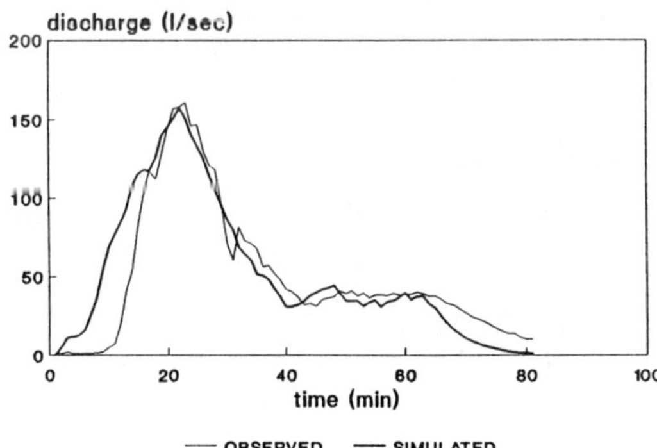

Fig. 4. Observed and simulated hydrographs

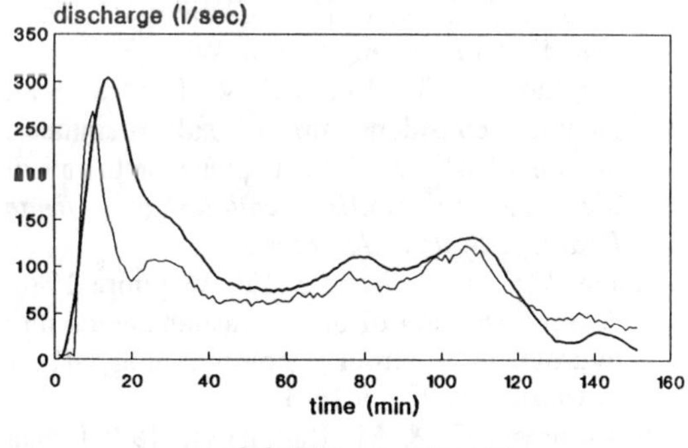

Fig. 7. Observed and simulated hydrographs

and width and to a hydrograph shape coefficient, has been derived. This equation was used to estimate the discharge travel time in the topological model given by equation (1).

The computations carried out showed a good agreement between the observed hydrographs and the model itself. That is, the percentage mean error detected with reference to a set of observed events (max discharge greater then 100 l/sec) was equal to 13.8%, consistent with the results given by other classical models. These results seem to provide a good behavior of the model. Fiorentino, Molino & Telesca (1992) obtained comparable results, although data recorded in a shorter period were used.

The paper brings a contribution for the estimation of dynamic parameter of the topological model. Some more developments are needed for the estimation of the parameters related to the overland flow.

A new contribution in that way, should be given by the detailed terrain analysis with consequent local scale parameterization, leading to a physically based estimation of the process characteristic variables. As stated above, the parameterization can be also obtained using the GIS analysis techniques able to give more contributions in this field as well as in the other ones involving physical processes with intrinsic distributed characteristics.

REFERENCES

Agnese, C. & G. La Loggia 1987. Testing a morphology based runoff model for urban drainage networks. *XXII Meeting AIRH*.Losanna.

Fair, G., J.C. Geyer & D.A. Okun 1966. *Water and Wastewater Engineering*.Wiley.

Fiorentino, M., B. Molino & V. Telesca 1992. Alcune considerazioni sul parametro dinamico dei modelli di piena nelle reti di drenaggio urbano.*III Seminar on Urban Drainage Systems*.Ancona Italy.

Gupta, V.K., C.T. Wang & Ed. Waymire 1980. A representation of an instantaneous unit hydrograph from geomorphology.*Water Resources Research* 16(5).

Maksimovic, C. & M. Radojkovic 1986.Urban Drainage Modeling. *Proceeding of the International Symposium on Comparison of Urban Drainage Models with Real Catchment Data UDM'86*.Dubrovnik-Yugoslavia.

Rodriguez-Iturbe,I. & J.B. Valdès 1979.The geomorphologic structure of the hydrologic response.*Water Resour.Res.* 15(6).

Troutman, B.M. & M.R. Karlinger 1984. On the expected width function for topologically random channel networks. *Journ. Appl. Probab.* 21.

Environmental Management, Geo-Water & Engineering Aspects, Chowdhury & Sivakumar (eds)
 ISBN 90 5410 099 0

High pressure artesian wells to tap Torbido Spring, Italy

V.Cotecchia
Istituto di Geologia Applicata e Geotecnica, Politecnico di Bari, UO 4.14 del GNDCI CNR, Italy

G.D'Ecclesiis
Geologist, UO 4.23 of the GNDCI CNR, Italy

M.Polemio
CNR CSATAI Bari, UO 4.14 of the GNDCI CNR, Italy

ABSTRACT: Torbido Spring is located at the foot of Mount Sirino (Basilicata, Italy). From the bottom upwards the geological formations include: Flinty Limestones with high secondary permeability, forming the spring aquifer; Siliceous Schists consisting of radiolarites and multicoloured jaspers; and Galestrino Flysch, formed of argillites and marls. The Spring is singular in that the artesian waters from the limestones which are at a depth of 80 m here, come to the surface because of the marked fissuring of the Siliceous Schists. The headworks consist of a system of unpumped wells. The groundwaters are intercepted before they leak away in the detrital surface cover and before they lose their hydrostatic head (more than 4 bar above ground level). The maximum discharge which can be abstracted and handled by the wellfield and appurtenant works is around 300 l/s. This permits peak demand to be met without any waste of resources and guarantees the annual hydrologic balance between recharge and discharge. These objectives are attained by continuous monitoring of the significant hydrogeological and hydraulic parameters.

1 INTRODUCTION

Torbido Spring provides an example of the role played by geological phenomena which can result in groundwater flow being preferentially vertical. Such conditions create hydrogeological and technological problems for the rational use and conservation of groundwater resources.

Mount Sirino is a tectonic window. Lagonegro I Unit which forms the mountain is bounded by the Lagonegro II Unit to the north, the Liguride Unit and Carbonate Platform Units to the east, the Liguride Unit to the south and Carbonate Platform Units to the west (Fig. 1).

The mountain itself is characterised by outcrops of a succession belonging to Lagonegro I Unit, originated from the deformation of Lagonegro Basin sediments during the tectonic phases which led to the building of the Apennines during the Tertiary (Scandone 1972).

From the bottom upwards Lagonegro I Unit consists of:

- Flinty Limestone Formation: dolomitic limestones with flint bands and nodules (Upper Triassic);
- Siliceous Schists Formation: multicoloured radiolarian jaspers, siltites and red and green marls (Upper Jurassic-Upper Triassic);
- Galestrino Flysch: argillites, marls and dark-grey siliceous limestones (Lower Cretaceous-Upper Jurassic).

The succession forms a brachyanticline with Flinty Limestones in the core; the southern limb of the structure connects up to the south with a synform fold. The structure is truncated by faults at Il Vallone gorge.

There are axial culminations and depressions on the N-S axis of the structure.

The Spring is located where the Torbido Gorge - cut into the Siliceous Schists - emerges from a deep ravine.

For many years the Torbido Spring was held to be a barrier spring due to an impervious sill. However, field investigations and schematic flow models have revealed a number of incongruent points.

The jaspers possess exceptional mechanical properties, primary permeability being absolutely negligible, while secondary permeability is very variable, depending on the degree of fracturing.

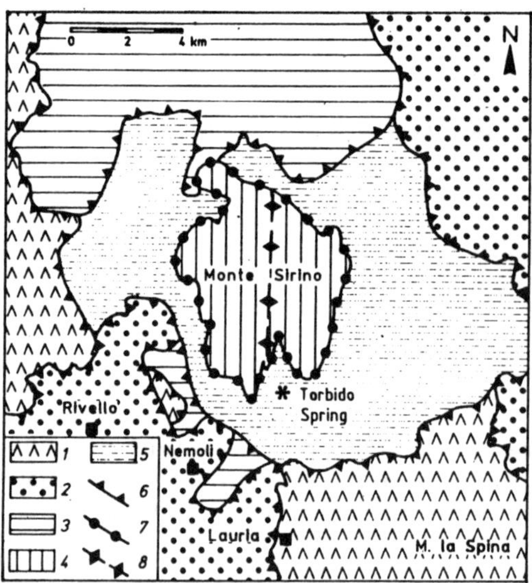

Fig. 1 - Structural schematic. 1) Platform Unit; 2) Liguridi Unit; 3) Lagonegro II Unit; Lagonegro I Unit; 4) Flinty Limestones and Siliceous Schists; 5) Galestrino Flysch; 6) Tectonic contact; 7) Limit of water-bearing formations; 8) Main axis of structure.

Despite the fact that the bottom of the gorge reaches the elevation of the resurgence, it is dry upstream of that point. Downstream, instead, water oozes from the clayey-marls of the Galestrino Flysch in many places, giving rise to permanent widespread surface runoff.

Piezometers drilled upstream of the Spring encountered the Siliceous Schists for the most part, as well as a thin shallow water table. As it was originally thought that these shallow groundwaters fed the Torbido Spring, a system of drainage galleries was built; these provide an average discharge of about 100 l/s.

Despite the construction of these galleries, however, a considerable amount of spring water continued to flow freely, the average discharge being estimated to be around 315 l/s, a figure which could not be justified on the basis of the presumed size of the aquifer and the poor hydrogeological properties of the Siliceous Schists.

The hydrogeological model of the Torbido Spring thus had to be reconsidered and the idea was explored of the existence of a substantial vertical flow from the much more extensive limestone aquifer (with greater transmissivity) to the ground surface.

2 GEOLOGICAL AND STRUCTURAL CHARACTERISTICS OF THE SPRING

Argillites and siliceous marls belonging to the Galestrino Flysch outcrop in this area, while jaspers and radiolarites belonging to the Siliceous Schists Formation occur right in the highly-incised channel of the Torbido.

At between 96 and 116 m below ground level, underlying the jaspers there are grey dolomitic limestones with bands and nodules of black flint.

Locally the formations occur as an asymmetric fold striking roughly NNE-SSW, the western limb having the steepest dip (20 to 30°) (Figs 2 and 3). In the spring area there is an axial culmination on the axis of the fold which induces a decidedly southerly dip.

Here the Siliceous Schists dip below the Galestrino Flysch and come to the surface again about 3 km south near Il Vallone, where the structure is interrupted by normal faults which bring it into contact with the Lagonegro II Unit and with the Liguride Unit.

The jasper beds, which outcrop along the steep slopes upstream of the Spring, are visibly affected by a system of subvertical tension fractures that markedly increase the permeability of the rocks along the hinge of the fold, into which the gorge has cut its bed. This effect reaches a maximum in the Spring area where another system of fractures is superimposed on the first, at the axial culmination.

3 HYDROLOGIC AND CLIMATIC CHARACTERISTICS

Torbido Spring lies at 905 m a.s.l, while the highest point of the catchment basin is 1907 m a.s.l. (Mount Sirino). Rainfall and temperature data are available from a station at about 650 m a.s.l. (LL.PP. 1922-1987). Precipitation is of the maritime type, so it is influenced by the vicinity of the Tyrrhenian Sea and the mountain (Fig. 4). Mean annual rainfall is 1832 mm.

There is generally only one minimum temperature and one maximum temperature during the year. The climate is moderate, the mean annual temperature being 13.4 °C.

According to the Thornthwaite-Mather and Turc methods, evapotranspiration is 590 mm, so an average of 1242 mm of rainfall is available for infiltration and runoff (Thornthwaite, Mather, 1957) (Turc 1954).

Fig. 2 - Geological map. 1) Detritus; 2) Liguride Unit: argillites and siliceous marls; 3) Platform Unit: limestones and dolomitic limestones, subordinately marly clays; 4) Lagonegro II Unit: clayey marls, siltites and sandstones, subordinately limestones with bands of flint; Lagonegro I Unit; 5) Galestrino Flysch: argillites and siliceous marls; 6) Siliceous Schists Formation with multicoloured jaspers and radiolarites; 7) Flinty Limestone Formation: dolomitic limestone with flint bands and nodules; 8) Overthrust; 9) Fault; 10) Attitude and dip of strata: (a) 0-10°, (b) 11-45°, (c) 46-80°, (d) 80-90°, (e) overturned; 11) axis of secondary antiform structure; 12) axis of secondary synform structure; 13) axial culmination; 14) spring.

The presumed catchment basin of the spring measures 8 km^2, which gives an annual effective rainfall volume of 9.9 Mm^3, equivalent to an average discharge of 315 l/s, which is exactly that of the average from the Spring, determined through a series of measurements.

Since the available rainfall and temperature data were recorded at a station with a lower elevation than the average of the basin, it can be taken that the effective rainfall is underestimated. Regional-scale studies on temperature and rainfall variability put this underestimate at not more than 25%, which is more or less the portion of runoff in such environments.

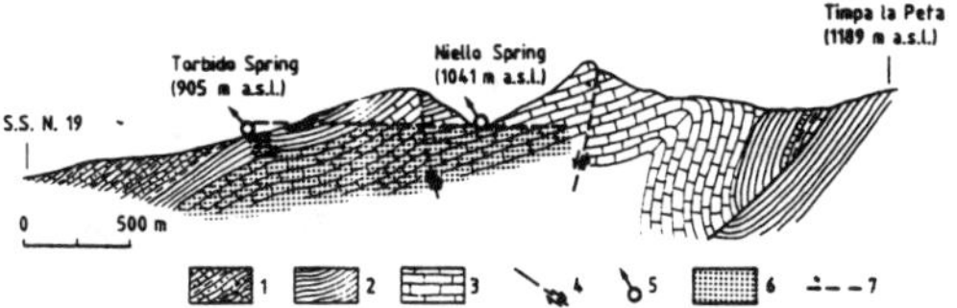

Fig. 3 - Hydrogeological section. 1) Galestrino Flysch; 2) Siliceous Schists Formation; 3) Flinty Limestones Formation; 4) Fault; 5) Spring; 6) Aquifer; 7) Piezometric line.

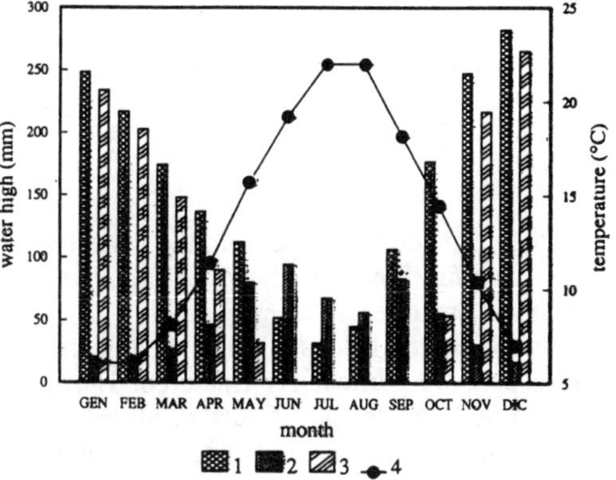

Fig. 4 - Hydrological regime. 1) Mean monthly rainfall; 2) Real evapotranspiration; 3) Surplus water; 4) Mean monthly temperature.

4 HYDROGEOLOGICAL CHARACTERISTICS

Mount Sirino contains an aquifer formed mainly of the Flinty Limestones with secondary permeability. The Siliceous Schists and the Galestrino Flysch for an impervious belt around the aquifer, rendering the system closed and independent (D'Ecclesiis, Grassi, Sdao, Tadolini, 1990)

Because of the brittle behaviour and the thickness (about 70 m) of the Siliceous Schists, they are so closely fractured in stress zones

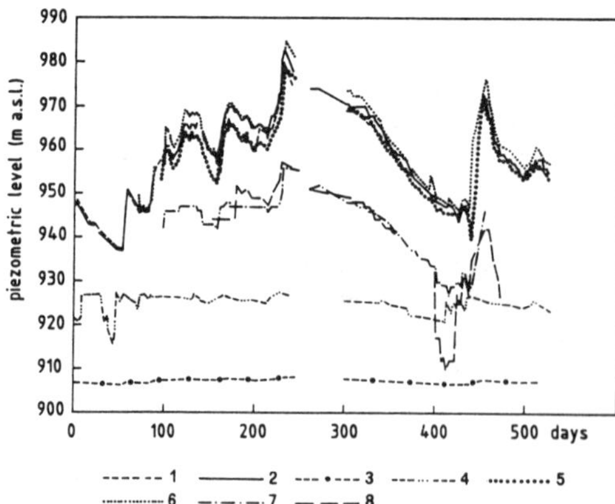

Fig. 5 - Piezometric Hydrograms. Piezometers considered are: 1) S1; 2) S2; 3) S3; 4) S4; 5) S4b; 6) S5; 7) S6; 8) S6b (For locations see Fig. 6).

that they provide an outlet from the underlying water-bearing limestones, which allows the groundwaters to reach the surface.

Torbido Spring is located in an area in which the jaspers are particularly fractured. Here, in fact, in addition to the effects resulting from the tension exerted on the hinge of the fold with a NW-SE axis, along which runs the line of the Torbido gorge, there are those present in the axial culmination of the fold where the spring lies.

The piezometric surveys indicate that it is very likely that the groundwaters seep through the fractured Siliceous Schists, losing a considerable amount of head, and come to the surface at the Spring.

The piezometric surveys lasted about 525 days (Fig. 5). A few days after the beginning of the observations the preexisting well P2 completed in the Siliceous Schists started to be used, the intention being to abstract 45 l/s. However, this discharge rapidly emptied the well, creating a depression of over 10 m in the 80-m deep piezometers S1 and S2, while also resulting in a delayed piezometric drop in S4, which is only 30 m deep (Fig. 6).

As P2 is close to S2 and around 100 m from S1 and S4, this substantial difference in behaviour is bound up with the depth of S4 which does not penetrate through the slightly-fractured or fractured strata that occur at depths of over 50 m (RQD>80%).

The use of P2 was suspended and S4b was completed at a depth of about 100 m close to S4.

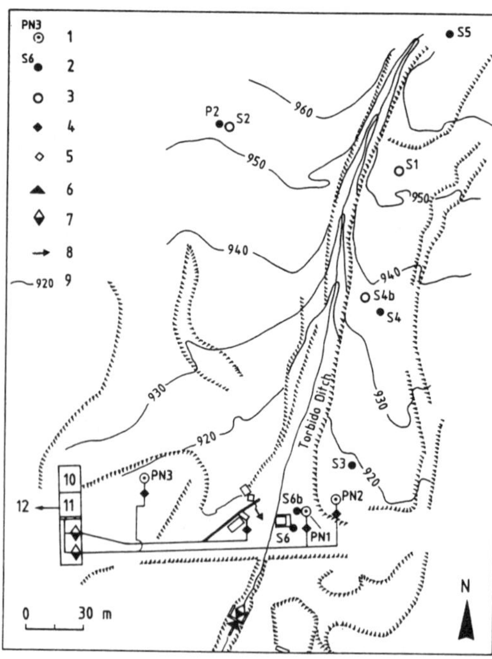

Fig. 6 - Spring area and monitoring works. 1) Production well; 2) piezometer; 3) Piezometer with pressure and temperature measurements; 4) Measurements and safeguards on wells; from upstream to down: pressure measurement, dissipator, discharge and pressure measurements; 5) discharge measurement on pressure line; 6) discharge measurement on gorge; 7) multiparameter measurements: pH, dissolved O_2, conductivity, temperature; 8) secondary users; 9) contour (m a.s.l.); 10) data logging; 11) collection tank; 12) to water supply line.

The levels revealed by that borehole stabilized at 26 m above that of piezometer S4, indicating artesian flow conditions. This result adds weight to the hypotheses postulated by the authors, though some workers are of the idea that two aquifers are involved, one shallow (unconfined) and the other deep (confined).

However, during the drilling of S5, the farthest from the Spring (340 m) no waters were encountered that could be tied in with a presumed shallow aquifer. On the other hand, maximum piezometric levels were recorded which tie in well with those in the piezometers completed in the presumed deep aquifer.

The small piezometric fluctuations of the two shortest piezometers (S3 and S4) would appear to indicate that this temporary aquifer in the Siliceous Schists is fed regularly from below during the year and that it reaches piezometric levels which depend markedly on the boundary conditions, namely the geomorphology of the Spring area, being influenced to only a secondary extent by the hydrostatic heads found in the true carbonate aquifer.

The start-up, after four hundred days, of the first hydrological well designed - PN1 - , influenced the 80-m deep piezometers S6 and S6B.

5 DESIGN AND CONSTRUCTION OF PRODUCTION WELLS

From pumping tests it was estimated that the permeability of the limestones close to the Spring is around 10^{-5} m/s and that the radius of influence varies between 10 and 20 m. The usual formulae were adopted to study various solutions for a wellfield containing from 3 to 6 wells at four different spacings and completed at three different depths in the limestones.

When seeking the optimum solution it was always assumed that at least 20 of the available 40 to 50 m of hydrostatic head at the wellhead would not be used. This assumption was adopted to ensure that the aquifer would not be tapped too energetically and to guarantee that it would be invulnerable for an extensive area around the Spring.

It was finally decided to drill three wells (Fig. 7) using a cable-tool rig and reverse circulation. To prevent the leakage of rising waters under high lateral pressure during boring and to ensure there was no lateral diffusion through the highly-fractured mass, drilling was interrupted several times to permit the squeeze cementing of stretches of hole affected by intense fracturing.

The wells reached the water-bearing formation at different depths, due to their diverse position vis-a-vis the hinge zone of the fold. Near the roof of the aquifer, every increase in drilling rate due to the increase in fracturing, was accompanied by an increase in the discharge tapped. The maximum discharge of 150 l/s was provided by the well which penetrates farthest into the aquifer. The hydrostatic head of the three wells varies between 4 and 5 bar.

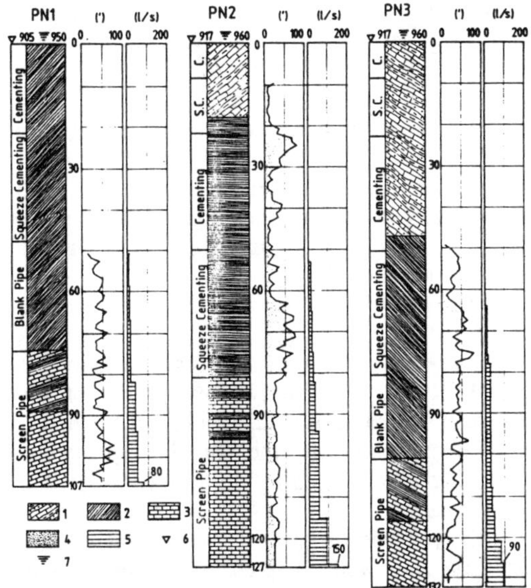

Fig. 7 - Characteristics of hydrological wells. 1) argillites, siliceous marls and blackish siltites; 2) multicoloured jaspers and radiolarites; 3) dolomitic limestones with flint nodules; 4) time required to drill one metre; 5) cumulative discharge tapped; 6) wellhead elevation (m. asl); 7) piezometric elevation with well closed (m a.s.l.).

6 MONITORING AND MANAGEMENT OF THE ABSTRACTION

A hydrogeological and hydraulic monitoring programme (Fig. 6) has been implemented to ascertain the effects of tapping the Torbido Spring via artesian wells. This information is required for a variety of reasons, namely, to conserve this precious resource, minimise the risk of pollution and optimise management of the installation so as to satisfy demand without the need for balancing reservoirs and without any waste, while at the same time not overexploiting the groundwaters.

The programme provides for a series of observation points represented by production wells, piezometers, and hydraulic structures for branching and control, complete with sensors for monitoring parameters of interest, all connected to a data logging and storage system.

The piezometric heads vary over the year, sometimes rising above ground level. On three piezometers it is planned to install two pairs of water temperature and pressure sensors.

The vertical position of these sensors is

established on the basis of stratigraphic knowledge and comparison with the gamma-logs and the RQD, so as to cover the most significant water-flow levels.

The purpose of these measuring stations is to ascertain variation in the vertical flow components during natural conditions and during well operation. It is planned to install electric piezometers as well as Pt-type temperature sensors.

To safeguard the groundwater resources some of the water-flow energy in the production wells is absorbed by fixed dissipators consisting of gravel filters immediately downstream of the wellhead. These devices also filter out fragments of rock transported by the fast-flowing waters, protecting the measuring equipment installed downstream.

Each well has its own station for measuring discharge, as well as pressure upstream and downstream of the filter. The discharge is measured by devices of the magnetic-inductive type, while pressure is measured by common transducers.

There are small springs around the operating artesian wells. Other springs in the area feed the Torbido Gorge. It is planned to measure the variations in the discharges of these minor occurrences.

To bring out any eventual significant differences in the routes or the water velocities of the main resurgences, it is planned to install three multiparameter sensors, which will provide real-time indications of pollution.

The data measured by the monitoring network are logged and stored every sixty minutes in a 30-day logger. This unit is complete with a thermostating facility and thermal insulation because the winter temperatures can fall to below zero Celsius. Regarding the sensors at the monitoring points, simple thermal insulation of the electronic circuits is all that is necessary because their proximity to the groundwaters elminates any risk of freezing.

Water demand varies from season to season, because of the number of tourists present in the summer, when all available water is consumed. The design of the headworks permits the aquifer to be used as a seasonal balancing reservoir.

The continuous logging of data on recharge, piezometric levels, water quality and especially the amount of offtake ensures control of the system in such a way that there is no waste, thus safeguarding groundwater resources.

7 CONCLUSIONS

It was indicated the role played by the Siliceous Schists which are impervious elsewhere, but are so fissured here that they provide the route for the resurgence of waters from the carbonate aquifer. Tapping the spring waters by means of artesian wells produces a discharge in excess of 300 l/s, with a head of more than 4 bar. The installation of fixed dissipators on the wellheads ensures that all the hydrostatic head is not used. This precaution prevents the risk of pollution and anyway guarantees that the aquifer is invulnerable for an extensive area around the Spring.The production wells have been incorporated in a system for monitoring and managing the water-supply/spring complex in such a manner as to permit the aquifer to be utilized as a "seasonal balancing tank", preventing waste in the months when demand is low, while assuring equilibrium between groundwater availability and abstraction thereof.

REFERENCES

D'Ecclesiis G., Grassi D., Sdao F., Tadolini T. (1990). *Potenzialità e vulnerabilità delle risorse idriche sotterranee del Monte Sirino (Basilicata).* Geologia Applicata e Idrogeologia Vol. XXV, Bari.

Ministero dei LL. PP. (1922-1987). *Annali Idrologici.* Roma: Servizio Idrografico - Sezione Idrografica di Catanzaro.

Scandone P. (1972). *Studi di geologia lucana: carta dei terreni della serie calcareo-silico-marnosa e note illustrative.* Bollettino della Società dei Naturalisti., 81, Napoli.

Thornthwaite C. W. and Mather J. R. (1957). *Instructions and tables for computing potential evapotranspiration and the water balance.* Centerton: Drexel Inst. of Climat. 10.

Turc L. (1954). *Le bilan d'eau des sols. relations entre les precipitations, l'evaporation et l'ecoulement.* Paris: Ann. agron., 1954-1955.

Environmental Management, Geo-Water & Engineering Aspects, Chowdhury & Sivakumar (eds)
© 1993 Balkema, Rotterdam. ISBN 90 5410 099 0

Water balance analysis of natural and rehabilitated areas at Weipa

M.R.Crees & R.E.Volker
Australian Centre for Tropical Freshwater Research, James Cook University, Townsville, Qld, Australia

ABSTRACT: The water balance of a seasonally recharged shallow aquifer at Weipa, north Queensland has been analysed and the effect of bauxite mining has been considered. Neutron Moisture Meter measurements have shown a considerable dependence on the type of material, and a system using differences in moisture content rather than absolute values has been adopted. The analysis has provided indications of water use by vegetation including comparisons between different types of vegetation, but the close proximity of the water table to the surface leads to difficulties in differentiating between water extractions from above and below the water table.

1 INTRODUCTION

All activities on the Weipa peninsula in north Queensland, Australia, depend on water from an aquifer located a few metres below the surface. This aquifer is recharged annually during the summer wet season when an average of about 2 m of rainfall causes groundwater levels to rise rapidly to within two meters of the surface over most of the peninsula. During the remainder of the year when there is virtually no rain, water levels fall gradually as water is removed from the aquifer by evapotranspiration, drainage to the sea, and extraction for industrial and residential uses. Bauxite mining is the main industry of Weipa. The mining operation involves clearing the native vegetation, moving the topsoil to areas already mined, removing the deposits of bauxite and then rehabilitating the mine site by deep ripping the mine floor, spreading topsoil and seeding with native species of vegetation. This leaves behind a landscape which is approximately 2 m lower than the natural surface, and for many years after mining, has vegetation of considerably different species and size distributions to undisturbed areas. This disturbance has a number of consequences for the water resource. Firstly, recharge is affected by the reduced size of the soil water store above the aquifer, by the altered amount of water intercepted in the unsaturated zone by vegetation, and by possible changes to the surface and subsurface infiltration characteristics. Another effect manifests itself during the wet season when groundwater levels more frequently rise above the lowered ground levels of old mine sites, and water that would normally remain within the soil profile is lost via surface drainage. Throughout the remainder of the year, water loss from the aquifer is altered by the changed water use characteristics of the vegetation, and by the closer proximity of the surface, and hence vegetation, to the water table.

The effect of landscape disturbance due to bauxite mining on the recharge of the shallow aquifer system of the Weipa peninsula has been the subject of an extensive study sponsored by Comalco. The goal of this study has been to develop an understanding of the recharge mechanisms and then transfer this understanding into a process based computer model to provide well founded estimates of recharge to groundwater models.

This paper considers one of the aspects of this study in detail, namely the use of Neutron Moisture Meter measurements of soil moisture in analysing the water balance of the system. Particular emphasis was placed on quantifying the water use by vegetation, and on comparison of water use by re-established vegetation of different ages. A common water balance approach to assess evapotranspiration is to define the lower limit of the root zone and to calculate water movement across this boundary based on measured soil moisture profiles and soil hydraulic characteristics. In this case however the water table rises close to the surface and inundates most of the root zone. Also since the water table is seldom deeper than 10 m it is likely that roots from well established vegetation would always be able to draw water more or less directly from the saturated zone.

2 STUDY SITES AND INSTRUMENTATION

Two pairs of study sites were established late in 1989 at the locations shown in Figure 1. Each pair of sites was chosen to give an unmined reference site in close proximity to a mined and revegetated site. At each site a water level observation bore was

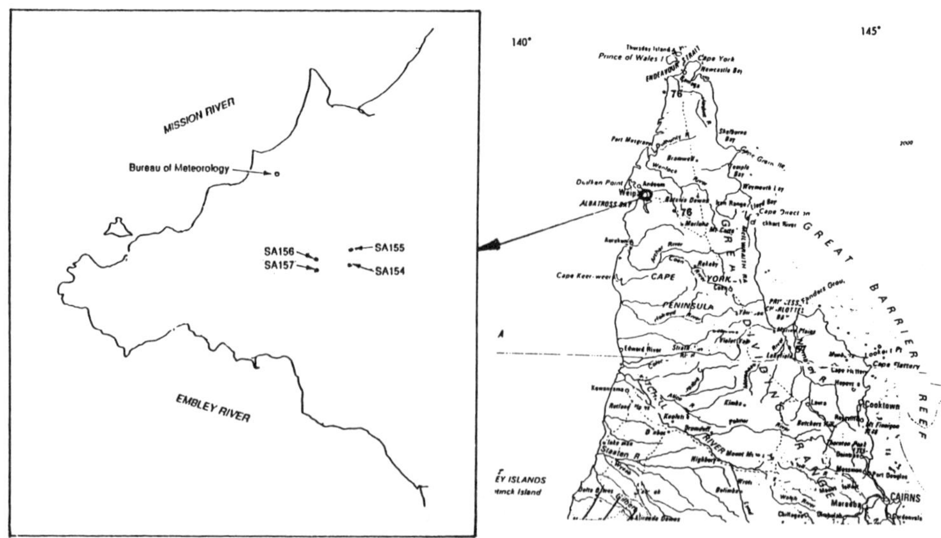

Figure 1 Water levels and rainfall over the study period

installed to a depth of about 10 m and a nest of five Neutron Moisture Meter (NMM) access tubes were installed to depths in the order of 8 m. The observation bores were instrumented with an electronic water level recording device, while a tipping bucket rain gauge and electronic data logger were installed at a convenient location adjacent to each pair of sites. The sites are referenced using the code assigned to the observation bore at each site.

Site SA154 is an undisturbed site paired with a relatively well established revegetated site at SA155. This pair provided the most complete set of data over the monitoring period. Site SA156 was the undisturbed site paired with an area of relatively recent revegetation at site SA157. During the monitoring period data collection ceased at the SA156 site when it was required for mining.

Instrumentation was installed and data collection commenced in late 1989 and continued until early 1991. Data loggers were downloaded on a nominal fortnightly cycle, at which time manual measurements of water levels were also made and the volume of water that had flowed through each tipping bucket rain gauge was recorded. The measurement of soil moisture profiles was commenced on a weekly cycle which was extended to a fortnightly cycle after the first wet season.

Data recovery from the electronic instrumentation was generally disappointing. Problems were experienced with all the types of equipment in use, and operator error and accident also contributed to data loss. As a result, water level records were limited to manual measurements during substantial periods, and a continuous record of rainfall was not obtained from the instruments installed. However the manual checks on rainfall indicated that daily rainfalls from a nearby Bureau of Meteorology station were applicable to the study site. Figure 2 shows groundwater levels measured at sites SA154, SA155 and SA157 during the study period, and daily rainfalls recorded by the Bureau of Meteorology.

3 NEUTRON MOISTURE METER RESULTS

The interpretation of Neutron Moisture Meter (NMM) results was not straightforward. Examination of the NMM readings showed considerable variations in apparent moisture content where such variations were not expected, thus suggesting that differences in material or differences in access tube installation through the profile had a significant influence on the NMM readings.

The material encountered in the profiles was divided into three broad categories, topsoil, bauxite ore, and ironstone. Laboratory calibrations were performed using samples of each of these materials. Each sample was reconstituted in a drum to a representative bulk density and a NMM access tube was located in the centre of the drum to replicate field installation conditions. NMM readings were taken in each material at moisture contents of 0% (oven dried), approximately 10%, and at saturation. Calibration equations of the form

$$\theta = b\,n + a \qquad (1)$$

where

θ = volumetric moisture content
n = NMM count rate
b = calibration coefficient

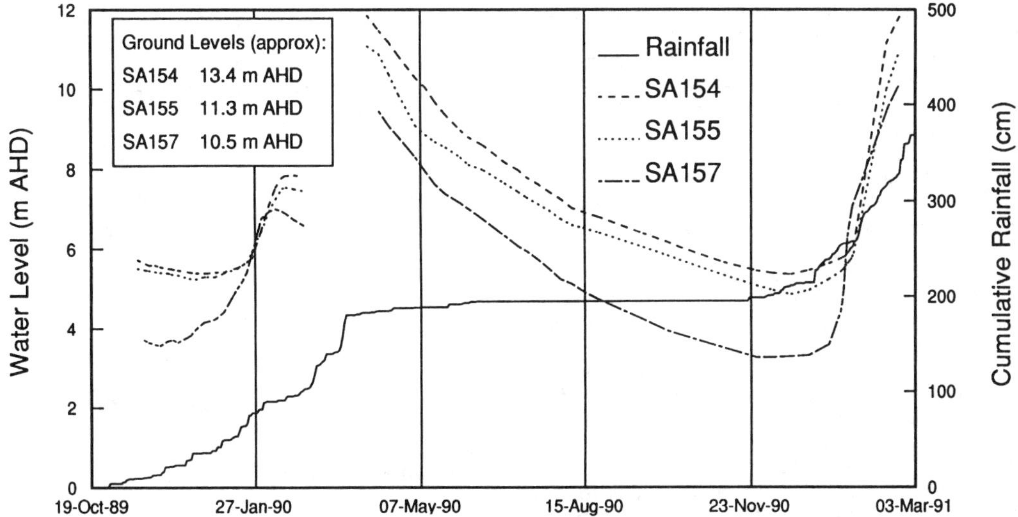

Figure 2 Water levels and rainfall over the study period

a = calibration intercept

were derived for each type of material. The results of these calibrations are shown in Figure 3 which also includes calibration points obtained during installation of the SA157 NMM access tubes.

The type of material in which the NMM is being used has an obvious effect on the intercept of the calibration line with variations of about 20% moisture content between the topsoil and bauxite calibrations, but there is much less effect on the slope of the calibration line. The calibration data from SA157 is consistent with the drum calibrations.

Given the difficulty in determining the correct intercept term for each measurement location, absolute moisture contents were not used, and an approach utilising differences in moisture content was adopted instead. A time of low moisture conditions was chosen as the 'dry reference' time, and for each sampling point the moisture content at this time was recorded as the 'dry reference' moisture content. Changes from the 'dry reference' moisture content were calculated using an equation of the form

$$\theta = b\,(n - n_d) + a \tag{2}$$

where

n_d = 'dry reference' NMM count rate

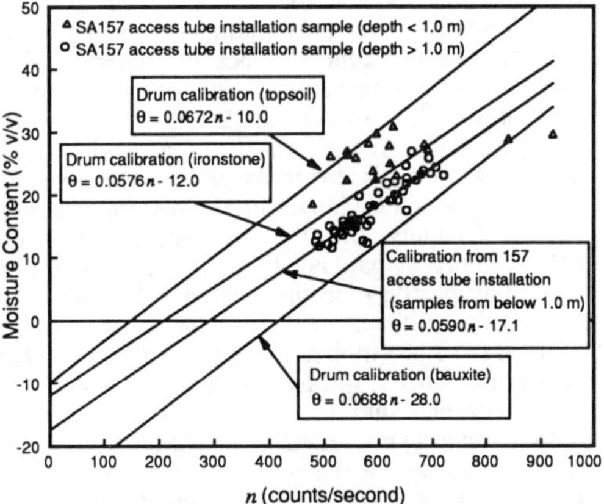

Figure 3 Neutron Moisture Meter calibrations

4 WATER BALANCE

4.1 *Method*

A water balance of a column of soil extending from the surface of the ground to an arbitrary depth below the ground can be expressed by

$$P - \Delta S = ET + R + D \tag{3}$$

where

P = precipitation
ΔS = change in water content of profile
ET = evapotranspiration
R = surface runoff
D = drainage component

The terms on the right hand side of this equation represent the unknown components of water loss from the profile. Precipitation is known from measured values of rainfall, and the NMM readings allow the change in moisture content of the profile to be calculated.

The calculation of profile moisture contents from the NMM readings has been complicated by the method of installation of the access tubes. A jetting technique was used to install the lower portion of the tubes into saturated aquifer sands. This left the lower ends of the tubes unsealed and water levels could rise and fall within the tubes in response to changes in groundwater level. As NMM readings could not be obtained below the water level in the access tube, moisture contents corresponding to saturated conditions could not be measured and an approximate method has been employed to quantify the change in water content associated with movement of the water table.

It has been assumed in this analysis that for any given measuring location, the change in moisture content from the 'dry reference' moisture content to saturated conditions is a constant value of $\Delta\theta_f$. A first approximation for $\Delta\theta_f$ has been estimated from a limited number of measurements of saturated conditions and from information on the hydraulic properties of the general types of material. The effect of the chosen value of $\Delta\theta_f$ on the results of the analysis has been evaluated.

Figure 4 is a representation of a NMM access tube, showing the NMM measurement locations, the zones over which each measured moisture content is assumed to apply, and the two categories of water table locations that need to be considered in calculating profile moisture content. In both cases the portion of the zone below the water level is assumed to be at a moisture content $\Delta\theta_f$ greater than the 'dry reference'. In case (a) where the water level has encroached into the zone of a measurement location but has not prevented a NMM reading being taken, the portion of the zone above the water level is assumed to be at the moisture content indicated by the NMM reading. In case (b) where the water level has risen above the measuring point and prevented a reading being taken, the portion above the water level is assumed to be at the same moisture content as the zone immediately above.

The change in profile moisture content from the 'dry reference' value has been calculated for each access tube by summing the changes in moisture content for each of the zones associated with NMM measurement points. A value for the site has then been found by averaging all the tubes at the site.

4.2 *Results*

Figure 5 shows values of the right hand side of Equation 3 calculated by the above method for sites SA154, SA155 and SA157. These values have been expressed as an average rate of water loss (cm of water per day) between the preceding NMM measurement time and the current measurement time.

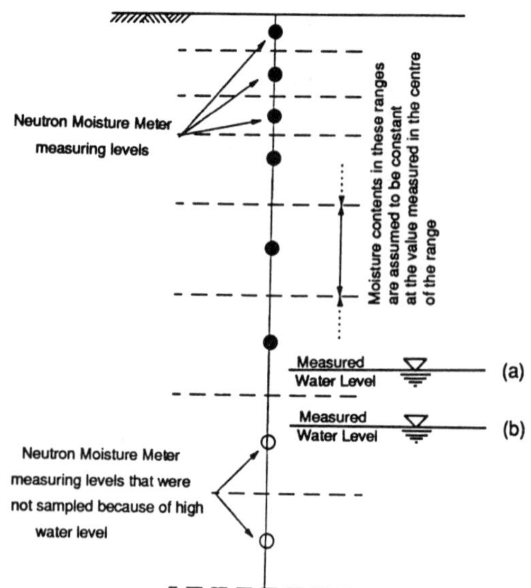

Figure 4 Schematic of Neutron Moisture Meter measurement locations within profile

Rainfall and pan evaporation expressed similarly, are also shown. The values in Figure 5 are calculated with $\Delta\theta_f$ equal to 15 cm/m (15%). The break in continuity during March and April 1990 is due to high groundwater levels which prevented NMM readings being taken. A five point moving average has been used to smooth the calculated values of water loss.

Overall the water loss from the profiles at SA154 and SA155 is very similar throughout the entire monitoring period, and the losses from SA157 are close in magnitude to the other two but show some notable differences. The results of Figure 5 will be considered by dividing the monitoring period into a number of discrete intervals with distinct characteristics.

The period from early November 1989 to mid January 1990 is the early stage of the 1989/1990 wet season. The calculated losses at SA154 and SA155 follow similar trends and are of similar magnitude, with the undisturbed SA154 in general having the slightly higher losses. Losses at SA157 also follow similar trends but are generally much lower and in some cases approach zero or become negative.

From the end of January 1990 to early March 1990, water levels rise sharply in response to wet season rainfall. Loss rates at SA154 and SA155 drop rapidly to significant negative values and then rise sharply to high positive values. Losses at SA157 do not follow the fall shown at the other sites, but do rise to a high value during the same period for which the rise at SA154/SA155 occurs.

The negative losses observed during this period are also observed during the early stages of the 1990/1991 wet season and are unusual in that they

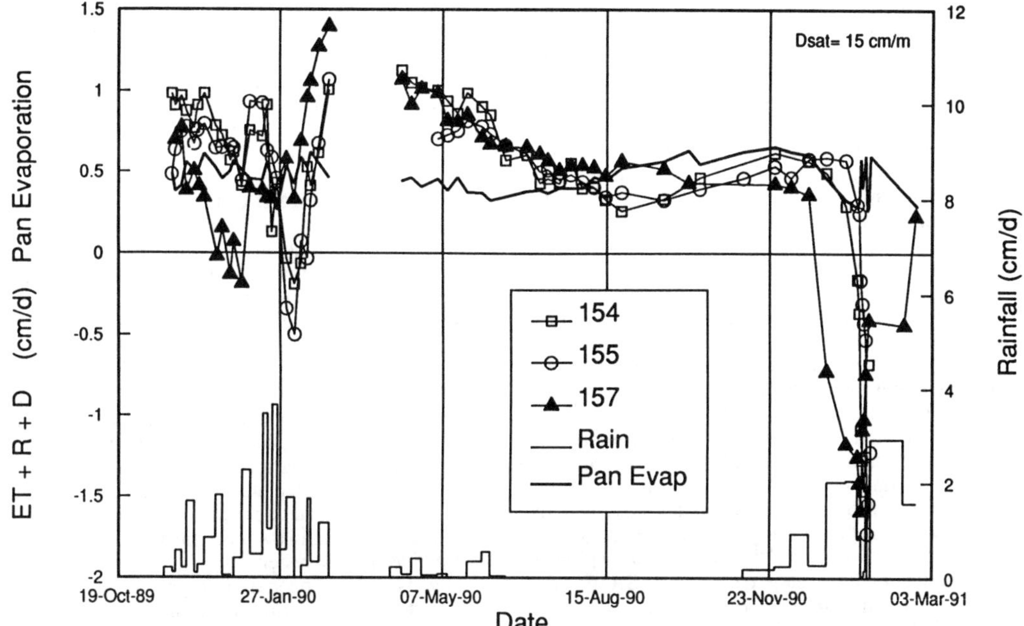

Figure 5 Results of water balance calculations

indicate that water from a source not accounted for in Equation 3 is being measured in the profile. It is possible that the region being studied is the recipient of a net inflow of water from the saturated zone, making the drainage term a gain of water rather than a loss. However there is no evidence to suggest that this is the case. Another explanation is that complete saturation below the observed water level does not occur as assumed, and the calculation consequently gives too much water in the profile and hence a negative loss. This situation could arise in the field when an apparent water table is detected but the measured level is actually due to water rising rapidly within a series of macropores while the rate of rise is too rapid to allow the material between the macropores to become saturated. Thus the measured water level in this case does not represent a smooth surface of the top of the saturated zone but rather the peaks reached temporarily by water rising through macropores. As can be seen by considering Figures 2 and 5, negative losses occur in the early periods of rapid water level rise which is consistent with this explanation. Negative losses are not observed later in the wet season, possibly because the soil matrix is relatively wet by that time and occurrences of non-saturated conditions would not result in such noticeable errors in water content. The next period of consideration begins immediately after the wet season in April 1990 when all sites show high losses of virtually the same magnitude. This is a period of high groundwater levels when saturated zone flow from the aquifer is at its highest, and when water availability to plants is also high. There was no runoff from the surface of the study sites during this period, but a base flow in creeks a few kilometres away indicated that high groundwaters were being drained at these points rather than remaining as subsurface flow to the sea. Losses at all sites fall steadily at almost the same rate until mid August 1990.

From mid August through to the first rains of the 1990/1991 wet season in late November 1990 is a period of zero rainfall. At sites SA154 and SA155 losses rose steadily during this period, almost in parallel with the increases in pan evaporation. Losses at SA157 did not fall to a low in mid August as they did at SA154 and SA155, but instead gradually declined until the wet season rains.

The increase in losses at SA154 and SA155 in this period can only be attributed to increased evapotranspiration since a decrease in drainage through the saturated zone would be expected with falling water levels. It therefore seems that evapotranspiration at these sites is at least partially controlled by potential evaporation, suggesting that both the undisturbed and the rehabilitated vegetation are able to obtain adequate water during these times of low water level. At SA157 where no increase is observed, it appears that the availability of water to the plants rather than the potential evaporation is the control, thus indicating a lesser ability of this younger vegetation to obtain water when the water table is low.

The period from the beginning of December 1990 to the end of January 1991 is the beginning of the 1990/1991 wet season. The onset of negative losses corresponding to the rapid rise in water levels as discussed above is observed at all sites.

Differing values of $\Delta\theta_f$ have been evaluated and shown to have primarily a scaling effect on the losses. For example, values of profile losses calculated with $\Delta\theta_f$ = 25 cm/m follow the same pattern of loss and maintain the same relationship with losses at other sites as those calculated with $\Delta\theta_f$ = 15 cm/m, with only the magnitude of the losses being changed.

4.3 *Components of water loss*

During the periods for which water loss has been calculated there was no widespread surface runoff from the study sites, although it is possible that there may have been some localised runoff during high intensity rainfall.

The calculated losses are therefore a combination of the evapotranspiration and the saturated zone drainage components. Water loss from various depths of the profiles has been examined and the results confirm that during the dry months (e.g. October and November), the implied transpiration component of the water balance at SA154 and SA155 must come from close to or below the water table. This analysis cannot readily distinguish between the components although some indication of the appropriate division can be obtained from the results at SA157. At SA157 during the period from 15 August 1990 to 23 November 1990, losses from the profile between the surface and 4.5 m depth vary in the range 1.2 mm/d to 0 mm/d in synchronisation with the total loss from the profile. This leaves a relatively constant 5 mm/d loss of water from below a depth of 4.5 m. Since it appears that the vegetation use at SA157 during this period is not enough to cause potential evaporation to be a control, a major portion of the 5 mm/d can be attributed to saturated zone drainage. At the other two sites where the evapotranspiration is assumed to be higher, however, total losses reach a minimum of about 4 mm/d which suggests that saturated zone losses would be significantly less than 4 mm/d. There is no obvious explanation for this apparent difference in saturated zone drainage.

4.4 *Uncertainty in results*

The values of profile water loss calculated in this analysis show a good deal of scatter before smoothing. One reason relates to the interval of time over which losses are averaged. Intervals that include substantial rainfall generally have higher losses than those with little or no rain, but because the losses depend to some degree on the distribution of the rainfall, averages over periods that contain different numbers and types of rainfall events will produce scattered results. However the integral of these results over time will not be in error because of this scatter.

The most significant cause of uncertainty in these analyses was the inability to obtain NMM readings below the water table. Moisture contents in the profile below the water table had to be estimated, with a consequent loss of accuracy. In addition, vital information on the substantial changes in moisture content that occur just above the water table was also not available.

5 DISCUSSION AND CONCLUSIONS

This analysis has shown qualitative differences in the losses of water from an area of relatively recent revegetation compared with areas of more mature revegetation and undisturbed forest. It has indicated that the more mature vegetation obtains enough water during the dry times for potential evaporation to be a controlling influence, whereas the recent revegetation does not obtain this amount of water. The results place upper limits on the amount of water used by the vegetation, but do not provide a definite value. Calculated negative values of water loss early in the wet season support the suggestion that preferred pathway mechanisms for water flow are active in this system.

The NMM has been used to provide an account of water stored in the soil profile, but the full potential of the approach was not realised because of an inability to measure moisture contents below the water table. Future programs should ensure that NMM access tubes are sealed at the bottom so that moisture content measurements can be made over the full profile depth at all times. Measuring locations should be no greater than 0.25 m apart for 1 m above and below the observed water table. Greater separation is acceptable in regions of the profile where moisture contents are not changing rapidly.

A monitoring program incorporating an appropriate unvegetated site would provide water loss data that could help define the evapotranspiration loss from the vicinity of the water table, and could be combined with the results presented here to establish confident estimates of evapotranspiration and saturated zone drainage.

ACKNOWLEDGEMENTS

The work reported in this paper forms part of a larger study of groundwater recharge on the Weipa peninsula that has been funded by Comalco Aluminium Limited. The majority of the data on which this work is based was collected by Comalco personnel, and their diligent efforts are gratefully acknowledged. Dr John Williams of CSIRO Division of Soils provided valuable assistance with the design and installation of the NMM access tubes.

BIBLIOGRAPHY

Greacen,E.L. (Editor) 1981, *Soil water assessment by the neutron method*, CSIRO Division of Soils, Australia

Environmental Management, Geo-Water & Engineering Aspects, Chowdhury & Sivakumar (eds)
© 1993 Balkema, Rotterdam. ISBN 90 5410 099 0

Managing the Great Artesian Basin

N.M.Eigeland
Department of Conservation and Land Management, Inverell, N.S.W., Australia

M.Jones
Department of Resource Engineering, University of New England, Armidale, N.S.W., Australia

ABSTRACT: Bores in the Great Artesian Basin of Australia are declining in pressures and flows. Piping improves water use efficiency and, by saving water in the aquifer, replenishes pressures and flows. It provides the opportunity to reverse land degradation, introduce new landuse management and improve productivity. This paper briefly outlines the present condition of the Great Artesian Basin and suggests several measures to improve water and land management.

1 INTRODUCTION

The Great Artesian Basin is one of the largest artesian basins in the world and a major Australian water resource. In many parts of arid and semi arid inland Australia, it is the only reliable source of water for towns, agricultural enterprises and industry.

Since the first bore was drilled in 1878, pressures and flows have declined throughout the basin, sheep and cattle grazing has intensified, and land degradation has become widespread. The survival of a large number of enterprises, especially grazing, is threatened.

2 THE GREAT ARTESIAN BASIN

The Great Artesian Basin (Basin) underlies twenty two percent of the Australian continent. Over 20 000 bores have been drilled into the Basin. About 3 000 free flowing bores discharge approximately 1 500 megalitres per day (Habermehl, 1980).

Water temperatures range from 40 to above 60 degrees centigrade. Some bores in central Queensland exceed 80 degrees. Water quality is generally good, between 500 and 1 000 milligrams of dissolved solids per litre.

The main recharge areas consist of exposed aquifers on the western slopes of the Great Dividing Range, with some recharge occurring in the west.

Pastoral enterprises, mining and towns rely heavily on artesian water in the arid and semi arid parts of the Basin. In the 1988/89 financial year, $2.5 million, or eleven percent, of the total value of Australia's agricultural commodities was produced in the Basin (Eigeland,1991). Income and employment from other industries such as mining and tourism is substantial.

2.1 Pressures and Flows

The widespread decline of pressures and flows in artesian bores is well documented (Qld. Public Works, 1945; Habermehl, 1980; Eigeland, 1992). Decreasing flows and pressures are mainly attributed to the release of elastic storage and aquifer subsidence.

2.2 Bore Condition

Deterioration in aged bore casings is generally caused by corrosion, encrustation or collapse. In steel casings, internal corrosion is mainly caused by dissolved carbonates and sulphates. External corrosion is generally from shallow saline aquifers. In some parts of Queensland, bores without protection from corrosion can disintegrate within 6 to 12 months (Pittman et al.,1912).

Corrosion can be controlled by pressure cementing the outside of the bore casing or by using inert materials such as Poly Vinyl Chloride (PVC), Fibreglass Reinforced Plastic (FRP), Acrylonitrile Butadiene Styrene (ABS), or Polyethylene.

Many bores have become uncontrollable because of corrosion of the headworks. Intrusions of saline water into failed bores may contaminate artesian aquifers.

About 2 000 bores, or ten per cent, require reconditioning. The loss of water from deteriorated bores is estimated at about 140 megalitres per day, or ten percent of present Basin discharge (A.W.R.C., 1987).

2.3 Open Bore Drains

Water is conveyed by gravity through a total of about 32 000 kilometres of open earth drains.

The drains are inefficient, requiring constant maintenance to keep them flowing. Over ninety per cent of water reaching the surface, or 1 350 megalitres per day, is lost by seepage and evaporation in open bore drains (A.W.R.C., 1987).

In Queensland an average of 30 megalitres per kilometre of drain is required to maintain flows and about 80 megalitres is required in South Australia (A.W.R.C., 1987).

Less than ten percent of the water in drains is used by live-stock.

Landholders experience a number of problems with bore drains which include blockages caused by weeds, drying up in summer months, deteriorating water quality, restricted access to paddocks and several forms of land degradation such as erosion, soil structure decline and the spread of woody weeds.

2.4 Administration

The Basin is managed by state government departments. Each sets standards of bore construction , rehabilitation and administers policy.

South Australia alone has a policy of piping rehabilitated bores (A.W.R.C., 1987).

The Federal Water Resources Assistance Program (F.W.R.A.P.) offers financial assistance for bore rehabilitation. Each state makes provision for financial assistance to landholders for piping.

2.5 Future Demand

Increasing demand, especially for high volume extraction, is coming from several activities including feed lots, irrigation, town water supply and mining.

Thirty two new bores were drilled in the Basin during 1988 (Interstate Working Committee, 1989). Drawdown, from high flow extractions of water, can influence pressures and flows at neighbouring bores and result in adverse changes in water quality (Habermehl, 1980).

3 ANALYSIS

3.1 Total Basin Discharge

Vertical leakage and losses to other basins amounts to about thirty percent of the Basin dis-charge (Habermehl, 1991). Almost sixty percent of Basin discharge is lost through the combination of deteriorated bores and antiquated, inefficient bore drains. Total Basin discharge is estimated at 2 500 megalitres per day. Only about six percent of total basin discharge is actually used (Eigeland, 1991).

3.2 Equilibrium

Equilibrium, by definition, occurs when total discharge equals recharge (Habermehl, 1980). Many bores will cease to flow before total equilibrium is reached. If demand increases, equilibrium may never be reached and even more bores will stop flowing.

3.3 Modelling

An outflow target for each hydraulic region can be set by the GABHYD Model (Seidel, 1980). This will help address the imbalance caused by declining flows and pressures. The use of optimisation models can further improve management.

3.4 Integrated Water and Land Management

Piping provides the opportunity to introduce new land management techniques. Setting low stock ceilings, adjusting fencelines and strategically placing water points can lead to more sustainable landuse, increase plant species cover, and maintain income levels to the producer (Waite and Nicholson, 1984; Morrisey and O'Connor, 1988).

3.5 Costs and Benefits

Between ninety and ninety eight percent of water reaching the surface, nearly half a million megalitres per year, is lost in highly inefficient bore drains. This costs the community $16.4 million per year, based on a market price of $30 per megalitre (Eigeland, 1991).

The maximum benefit from investment in the conservation of artesian water will come from maximising piping and ensuring that all rehabilitated bores are piped (Eigeland, 1991).

The average benefit to cost ratio for piping the Milroy Bore was 1.3 to 1 (Christiansen, 1991).

There are a number of other benefits that result from piping artesian water including less maintenance, the same quality water at the bore head to any point on the property, reliable water supply, lower stock losses from bogging, increased stock yields, better feral animal control, and improved stock and pasture management.

Uncosted benefits include the provision of good quality water for domestic use, increased property values and conservation of the waters and land of the Basin.

4 MANAGEMENT

The pattern of water use in the Great Artesian Basin has resulted in the decline of pressures and flows.

The most obvious way of improving water use efficiency in the Basin is by replacing open bore drains with pipelines.

For each megalitre of water saved, piping has a higher benefit to cost ratio than bore rehabilitation and should have priority over rehabilitation, except where recharge and discharge targets, pollution of artesian aquifers, land management, wildlife management or other environmental and social aspects arise (Eigeland, 1991). Incentives available for bore rehabilitation such as financial assistance, should also apply to piping.

4.1 Pricing

The most controversial artesian water management strategy, particularly for water users, is pricing. The prime objective of pricing is to increase efficiency and achieve equity. Additional revenue could fund new programs and assist the conversion to pipelines.

Other ways of increasing water use efficiency and equity include metering and education.

4.2 Options

I Existing Management

The present management, by the states and the Interstate Working Committee, adopts limited controls on discharge without the use of market forces. It does not address social, economic, land management or environmental issues in any detail and results in the least beneficial use of the resource (Eigeland, 1991).

II Unregulated Extraction

Unregulated extraction has economic benefits but social and environmental impacts are high. Subject to free market forces, some individuals may suffer extreme hardship (Pasqual and Rocabert, 1989).

III Planned Extractions

Avoiding conflicts in water use by careful planning has economic benefits and minimises social and environmental impacts (Custodio, 1989). Competition for water improves water use efficiency and equity.

5 GREAT ARTESIAN BASIN AUTHORITY

A planned extraction policy with water marketing (Option III) yields a higher benefit to cost ratio than the other options. To achieve these benefits, a single, coordinating authority is needed.

There are a number of ways a Great Artesian Basin Authority will improve water use efficiency and integrate water and land use. By education, metering and community participation it can standardise licence administration and encourage piping.

The Authority can develop economic optimising models and pricing policy, stabilise bore flows, set regional drawdown targets and manage recharge areas. It can develop regional landuse policy to improve regional productivity, control land degradation, improve equity, manage wildlife and plan future development for industry, tourism and recreation (Eigeland, 1991).

6 CONCLUSION

The formation of a Great Artesian Basin Authority will promote integrated water and land management and maximise net benefits to the community. The increase in water use efficiency by piping and other technology, along with greater public contribution to policy formulation and assessment, will help achieve sustainable and equitable use of the resource for existing and future generations.

REFERENCES

Australian Water Resources Council, (A.W.R.C.)April 1987. Report to Groundwater Committee of Great Artesian Basin Sub-Committee.

Christiansen, G. 1991. Piping the Milroy Bore:A Benefit-Cost Analysis. Soil Conservation Service, May 1991.

Custodio E. 1989. Strict Aquifer Rules verses Unrestricted Groundwater Exploitation., in Groundwater Economics. U.N. Symposium Papers, Barcelona, Spain, 1989, Elsevier Scientific Press, New York.

Eigeland N.M. 1991. Pipe the Bores. Great Artesian Basin Rehabilitation Workshop, Department of Water Resources, Dubbo, October 1991.

Eigeland N.M. 1991. Management Plan for the Great Artesian Basin. Department of Conservation and Land Management New South Wales, (Unpublished).

Habermehl, M.A. 1980. The Great Artesian Basin, Australia. Bureau of Mineral Resources, Journal of Australian Geology & Geophysics 5: 938.

Habermehl M.A. 1986. Proceedings of the International Conference on Groundwater Systems Under Stress, A.W.R.C. Conference Series No13, Canberra, 1987.

Habermehl M.A. 1991. in, Management Plan for the Great Artesian Basin, Eigeland N.M., 1991, Department of Conservation and Land Management,N.S.W.(Unpub.).

Interstate Working Committee on the Great Artesian Basin, 1989. Minutes. Convenor C. Hazel, Queensland Water Resources Commission.

Lange, R.T., Nicolson, A.D. and Nicolson, D.A. 1984. Vegetation Management of Chenopod Rangelands in South Australia. Aust. Rangeland Journal 6(1): 46-54.

Morrisey J.G and O'Connor, R.E.Y. 1988. 28 Years of Station Management. Australian Rangeland 5th Biennial Conference, Longreach 1988.

Pasqual J. and Rocabert I. 1989. Consumption of Groundwater as a Private and Public Good.,in Groundwater Economics. U.N. Symposium Papers, Barcelona, Spain, 1989, Elsevier Scientific Press, New York.
Pittman E.F. 1912. in, Interstate Conference on Artesian Water. 1st. Report. Government Printer, Sydney.
Purvis, J.R. 1986 Nurture The Land:My Philosophies of Pastoral Management in Central Australia. Aust. Rangeland Journal 8(2): 110-117.
Queensland Government, 1954. Artesian Water Supplies in Queensland. Report following First Interim Report (1945) to investigate certain aspects relating to the Great Artesian Basin (Queensland portion) with particular reference to the problem of diminishing supply. Department of the Co-ordinator of Public Works Queensland. Parliamentary Paper A56-1955.
Seidel G. 1980. Application of the GABHYD Groundwater Model of the Great Artesian Basin, Australia. BMR Journal of Geology and Geophysics, 5, 1980, 39-45.

Environmental Management, Geo-Water & Engineering Aspects, Chowdhury & Sivakumar (eds)
© 1993 Balkema, Rotterdam. ISBN 90 5410 099 0

How bad is your RORB modelling?: Sensitivity fitting – An innovation

P.I. Hill, N.S. Fleming & T.M. Daniell
Department of Civil Engineering, University of Adelaide, S.A., Australia

ABSTRACT A Flood Frequency Analysis was undertaken for the Aroona Dam catchment in the northern Flinders Ranges using historical and design data. The aim of the study was to validate the design information contained in Australian Rainfall and Runoff (1987) for arid regions of South Australia. Reverse Routing was employed to produce a useful data set for Flood Frequency Analysis. Design flood magnitudes were predicted using a RORB catchment model. A new model parameter selection method was developed called Sensitivity Fitting. Inadequacies in the design procedures included in Australian Rainfall and Runoff (1987) have been identified.

1.0 INTRODUCTION

Hydrological studies of the arid regions of Australia are frequently faced with a lack of data and inadequate design procedures. Such problems have been encountered by designers associated with recent developments in the arid regions of South Australia.

Several studies have identified possible inadequacies in the method of developing design floods from Australian Rainfall & Runoff (AR&R, 1987). Research was undertaken to validate the existing design recommendations included in AR&R (1987) for this region.

A Flood Frequency Diagram was developed using records of historical events from the Aroona Dam catchment. Reverse Routing was used to generate the data set which allowed a Flood Frequency Analysis to be undertaken.

The design procedures in AR&R (1987) were then tested against the historical Flood Frequency Diagram by generating a design Flood Frequency Diagram. This was undertaken using a RORB catchment model, with AR&R (1987) storms of differing duration and temporal patterns.

Sensitivity Fitting was developed as an extension of traditional RORB fitting procedures to determine the optimum catchment model parameter set. An appreciation of the impacts of parameter selection is offered by this method.

2.0 PREVIOUS STUDIES

The South Australian Department of Road Transport constructed both the Stuart Highway and Hawker-Lyndhurst road in the past decade. The design of related structures such as bridges and culverts required information regarding expected flood flows. This was obtained from AR&R (1977). Road flooding and costly maintenance resulted.

A study was initiated in an attempt to establish more reliable flood estimates. Kemp (1989) attempted this using the best available hydrologic data for the area - that associated with the catchment of Aroona Dam. A RORB computer runoff routing model of the catchment was established and model parameters determined using historical flood data.

The study highlighted a large difference between the newly derived design flows and those used to design the road.

Kinhill Engineers (1989) were commissioned by Electricity Trust of South Australia (ETSA) to determine the magnitude of the 20 year and 100 year ARI flood events and to compare these events with the Aroona Dam spillway capacity. Initial modelling was undertaken using the rainfall data and areal reduction factors recommended by AR&R (1987). The results produced were inconsistent with the observed recent runoff events. This lead to the development of new areal reduction factors.

The highest calculated discharge over the spillway was 1060 m^3/s, corresponding to the 100 year ARI storm of duration 2 hours. The spillway was barely adequate to pass the 100 year ARI flood event.

3.0 CATCHMENT DESCRIPTION

Aroona Dam lies west of Leigh Creek South, in the Flinders Ranges, as shown in Figure 1. It was constructed between January 1953 and September 1955 to serve as a reliable surface water supply to the township. Its current operation and maintenance is the responsibility of ETSA.

The dam has a storage capacity to spillway level of 4860 ML. The dam is 12.8 metres high, concrete lined and has an ogee spillway 97 metres wide. The discharge capacity of the spillway is approximately 1000 m^3/s.

Two sub-catchments, Windy Creek (442 km^2) and Emu Creek (229 km^2), contribute to the dam catchment of total area 693 km^2. The confluence of these two creeks lies just downstream of the Hawker-Lyndhurst road. The catchment is characterised by low ranges with some higher ridges. It has a mixed cover of low open woodland, shrubland and tall open mallee scrub, being used for extensive livestock grazing.

4.0 DATA ACQUISITION

A lack of hydrologic data is common in the arid regions. This is partially due to the difficulties encountered in measuring flows in arid and semi-arid areas. Problems pertinent to Aroona Dam are (Kotwicki, 1986):

1) Events are irregular and are quite often of a short duration,
2) High stream velocities and debris loads are common, resulting in the possible loss of gauging instruments, and
3) The gauging stations are not always stable and flooding may alter the channel cross-section.

The Aroona Dam catchment has five pluviometers and four stream-gauges, but the length of record is quite short. These pluviometers were installed by ETSA in 1986, along with two additional stream gauges on Windy and Emu Creeks. Refer to Figure 1.

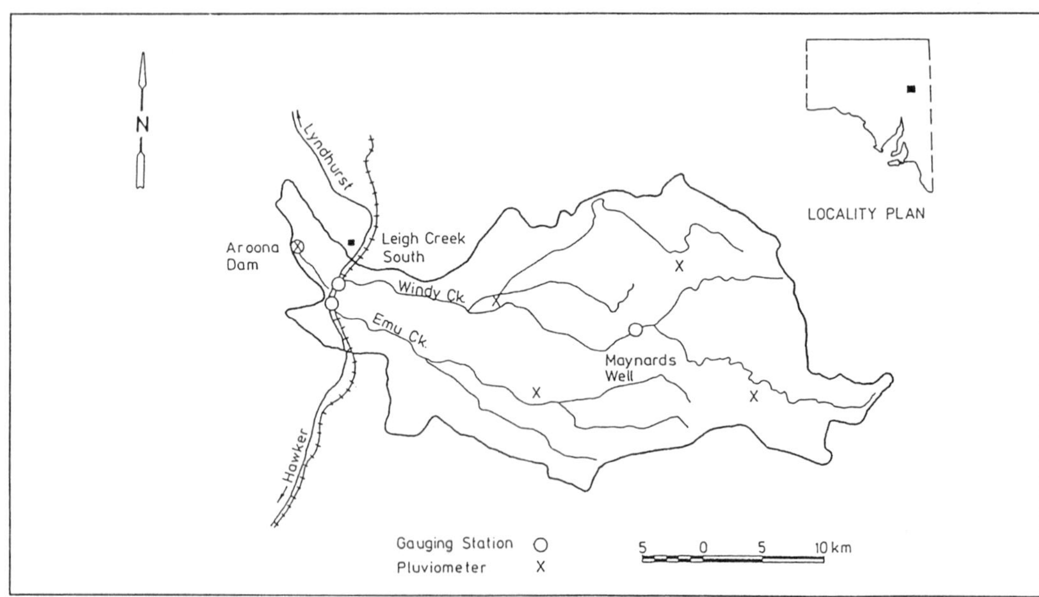

Figure 1 Aroona Dam Locality Map

Rating tables exist for all gauging stations on the catchment, but their accuracy should be viewed with some scepticism, due to the limited number of gaugings and shifting controls.

Data were obtained from the Engineering and Water Supply Department of South Australia in the form of HYDSYS files. HYDSYS is an interactive data archiving system that has been adopted by authorities across Australia. It not only provides rainfall and gauging records - it can also use rating tables to convert gauging records into flows (Heweston & Daniell, 1988).

4.1 *Rainfall Characteristics*

The Bureau of Meteorology estimates the average annual rainfall for the area to be 200 mm per year. The pluviometer records from 1986 for the Aroona Dam catchment and surrounding district indicate that the average rainfall during this period is considerably higher than the estimated mean. Because the record is short, it is difficult to obtain a true indication as it may be biased by the recent large events.

Two main rainfall mechanisms dominate the period of record. In the summer months, long dry periods are occasionally punctuated by storms with high intensities giving rise to large flows. Rainfall events are more frequent in the winter months but are of lower intensity.

5.0 FLOOD FREQUENCY ANALYSIS

The aim of a Flood Frequency Analysis is to analyse historical flood records to produce a Flood Frequency Diagram. This relates the peak flow to the Annual Exceedance Probability (AEP) or the Average Recurrence Interval (ARI). AR&R (1987) suggests that a 'moderate' length of record (10 to 15 years) is required for a Flood Frequency Analysis.

The rainfall pattern and the discontinuous nature of the records required that a partial series be used. A partial flood series consists of all floods above a certain chosen value. The year in which the flow occurs is not considered. Because the base value is arbitrary, the number of events (K) does not have to equal the number of years of record (N). Following McDermott and Pilgrim (1982) and Jayasuriya and Mein (1985), K was chosen to equal N for this study.

Thirteen years of stage height records exist at the dam. During this time, only five events were recorded which caused the dam to spill. This is not a sufficient number of events to produce a meaningful Flood Frequency Analysis. It is therefore more useful to use calculated dam inflows to derive a Flood Frequency Diagram. Although the flow is not measured at the inflow to the dam, it can be calculated by Reverse Routing from the stage readings at the dam.

An alternative method is to rout the known flows just upstream of the confluence to the inflow of the dam using a method such as the Muskingum Method which is outlined in AR&R (1987). This method was not chosen because the stream flow records at these two locations are not continuous and lack a reliable rating curve.

5.1 *Reverse Routing*

The Reverse Routing technique uses the dam outflow and stage records, and makes use of the following form of the Continuity Equation:

$$I_i = Q_i + \frac{dS}{dt}\Big|_i \tag{1}$$

The derivative in equation (1) is expressed in finite difference form. The simplest expression is a two point approximation:

$$\frac{dS}{dt}\Big|_i = \frac{(S_{i+1} - S_{i-1})}{2\,\Delta t} \tag{2}$$

This expression has been demonstrated to give similar results to the more complex higher order formulae (Boyd et al., 1989). It was therefore adopted for the Reverse Routing.

The choice of time-step is very important. If a large time step is chosen, the number of calculation steps is reduced, but the calculated inflow hydrograph may not predict the inflow peak correctly. Too small a time-step has been shown to produce instabilities in the calculated inflow hydrograph (Boyd et al., 1989). A time-step of half an hour was adopted as this adequately defined the inflow hydrograph.

In order to check the accuracy of the Reverse Routing procedure, the calculated inflow hydrograph was compared to the two measured upstream hydrographs at the road. For the five events which produced a discharge from the dam, the calculated inflow hydrograph was also compared to the outflow.

An indication that the Reverse Routing was successful is that the peak of the outflow hydrograph cuts the falling limb of the inflow hydrograph, as shown in Figure 2. This behaviour satisfies the mathematical relationship expressed in equation (2).The dual peak is a result of there being two main tributaries in the catchment.

5.2 *Generation of Flood Frequency Diagram*

The largest thirteen calculated peak inflows were used in the Flood Frequency Analysis. Ten theoretical distributions were fitted for these points using the statistical frequency analysis package, WSO87 (Kopittke, 1976), (the modified version of WSO6). Using goodness of fit tests and a visual examination of the plotted points, it was decided that the Log-Pearson III distribution was the appropriate distribution.

6.0 RORB RUNOFF ROUTING AND CATCHMENT MODELLING

AR&R (1987) was followed for the derivation of design rainfalls and floods for the Aroona Dam catchment. The RORB Runoff Routing Program was used for this study. This was due to its availability, greater flexibility, range of options and the existence of a Aroona Dam catchment model (Kemp, 1989).The package was used to model the runoff characteristics of the catchment in the prediction of the design floods. These were then compared to those from the historical Flood Frequency Analysis.

6.0.1 *The Storage Equation*

RORB's routing procedure is based on a relationship between storage and discharge, of the form :

$$S = 3600.k_r.k_c.Q^m \qquad (3)$$

where S is the storage (m³), ($k_r.k_c$) is a dimensional empirical coefficient, Q is the outflow discharge (m³/s) and exponent m is dimensionless.

This relationship models the reach storage behaviour between nodes. The coefficient k_r is a dimensionless ratio called the relative delay time. The coefficient k_c is an empirical coefficient applicable to the entire catchment and stream network.

The exponent m is a measure of the catchment's nonlinearity. AR&R (1987) recommends the use of m=0.8 where sufficient data does not exist to calibrate the model. The value of m usually adopted lies between 0.7 and 0.9, but AR&R (1987) also notes that for flood routing in major rivers with flood plains and through storage reservoirs, values between 1.0

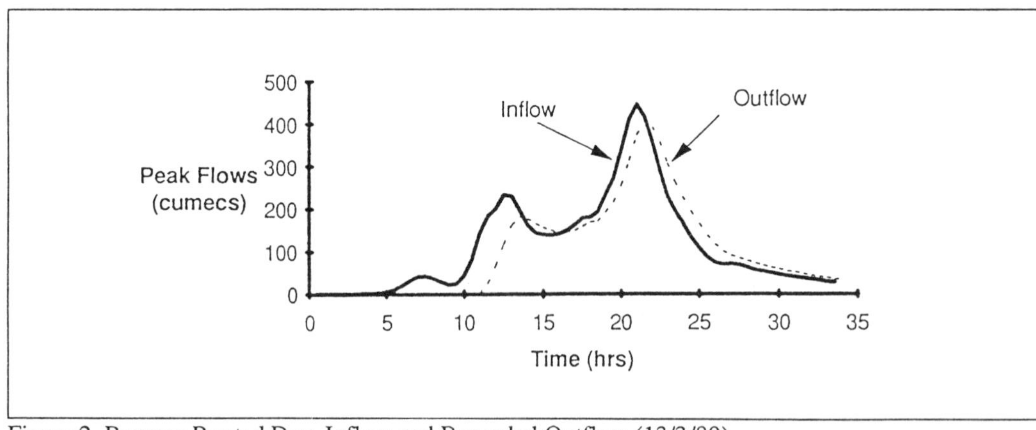

Figure 2 Reverse Routed Dam Inflow and Recorded Outflow (13/3/89)

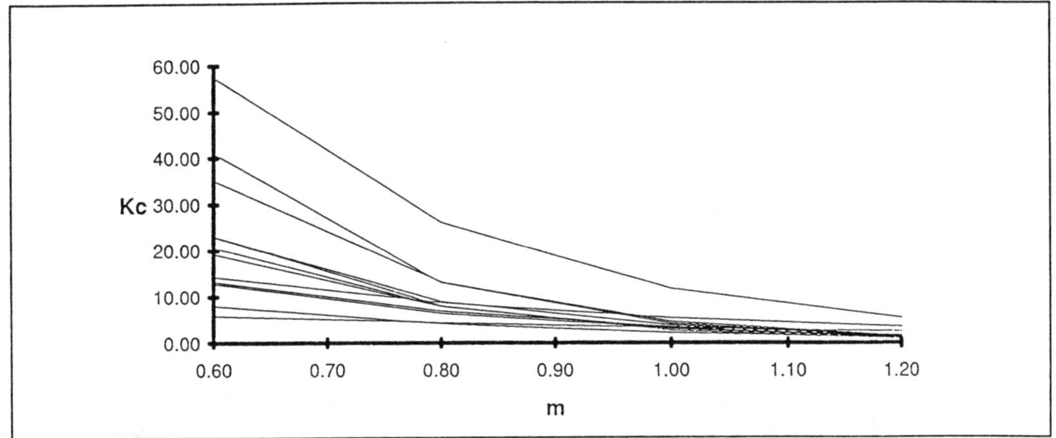

Figure 3 RORB Fit Parameters m and k_c for Aroona Dam Reverse Routed Inflows

and 1.5 (or higher) are not unusual.

6.0.2 *Parameter 'Fitting'*

Fitting with RORB involves a trial and error adjustment of the parameters k_c, m and initial loss (IL) and continuing loss (CL) until the calculated hydrograph matches the observed hydrograph; in flow peak, lag and shape.

The fitted parameters; k_c, m, IL and CL, are strictly only pertinent to the single event from which they were derived. Where data from several floods is available, derived parameter values are examined for trends to aid in the selection of suitable design values.

6.0.3 *Fitting the Aroona Dam Catchment Model*

The RORB model of the catchment, established by Kemp (1989), has 22 sub-areas, 22 model storages, 5 pluviometers and 4 gauging stations.

The model parameter fit runs were undertaken using historical rainfall and dam inflow data.

The thirteen largest inflow events obtained by Reverse Routing were selected for parameter fitting. With the corresponding rainfall records, fitting was undertaken for each of these events resulting in thirteen separate curves of k_c versus m. When these graphs for several events are then superimposed, they ideally indicate a unique pair of k_c and m values that provide a good fit for all events. Five general areas of curve intersection were identified as shown in Figure 3. The trend of decreasing k_c with increasing m (in the range of 0.6 to 1.2) dominated the plot. No satisfactory pair of parameters could be estimated from the plot.

In both cases the IL and CL parameters were observed to be highly variable - indicative of fluctuating natural conditions and a lack of definition of rainfall events. An average of the derived IL and CL values was considered to be most appropriate.

The trial and error fitting left the impression that with increasing m, the sensitivity of the calculated flow to small changes in k_c increased markedly. It was decided to test whether a set of parameters, m and k_c, existed for which the error in calculated flows was a minimum for all events.

Further analysis was undertaken using the results of the RORB fit runs in order to test the sensitivity of the model to a variation in k_c. This method of analysis has been named *Sensitivity Fitting.*

6.1 *Sensitivity Fitting*

Sensitivity Fitting is a method of testing the sensitivity of the model to a variation of k_c. It is a simple extension of the normal RORB parameter fitting process. It was considered reasonable that a range of m should exist for which a change in k_c would result in a minimal change in calculated peak discharge.

The five general areas of curve intersection identified on the k_c versus m plot correspond to m values of 0.78, 0.91, 1.05, 1.11 and 1.17. For each of these values of m, the average value of k_c was estimated.

Nine of these 13 events corresponded to Summer storms. Only these nine events were considered further, as Summer events correspond to the most intense rainfall and highest flood peaks and were therefore of prime interest.

The nine fit models were then re-run with the five new pairs of parameters m and k_c, and the calculated and historical peaks were recorded. Average IL and CL values were used, however other values may have been more appropriate. For these 45 values of flow, the percentage error between actual and calculated flow peak was evaluated. When plotted against the range of m values selected for the test, a 'sensitivity' envelope with a narrow band between an m of 0.95 and 1.05 resulted. These are plotted in Figure 4 and show the wide band of results that were obtained.

To further rationalise this plot in an attempt to select an ideal pair of k_c and m, a curve corresponding to the average value of percentage error in flow peak for each m was plotted. This curve has its lowest point at an m value of 1.05 (and k_c of 3.0).

This method has resolved the problem of choosing a pair of parameters (k_c and m) which are most suitable for all events. It should be emphasised though that these parameters are strictly applicable only to summer events, the records from which they were determined.

It is also important to note the significant errors that would have been encountered with the selection of m=0.8 as is recommended practice in AR&R (1987), should no better information be available. An error of up to 100 percent could be expected in the calculation of peak flows, in addition to the errors associated with the data!

Figure 4 indicates an optimum choice of parameter m for all summer events. It does not however indicate the relationship between flood magnitude and percentage error in the calculated flood peak for a given value of m. Such a relationship was sought, and is shown in Figure 5.

This figure indicates that events having a flood peak less than 50 m^3/s are associated with large errors for all tested values of m. For this catchment, the RORB model was unable to adequately predict flow peaks for small events.

For the chosen value of m=1.05 (from Figure 4), Figure 5 illustrates that the error is less than 25% over a broad range of recorded inflow. A choice of m about 0.8 would result in large errors in predicting floods of magnitude greater than 300 m^3/s. This is significant because it is the large flood events that are of concern for design purposes.

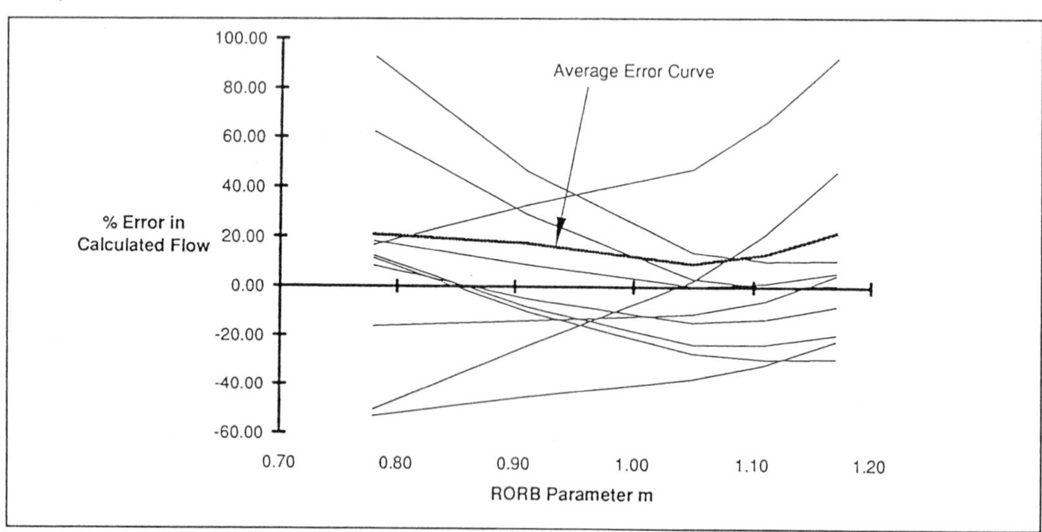

Figure 4 Sensitivity Fitting Envelope

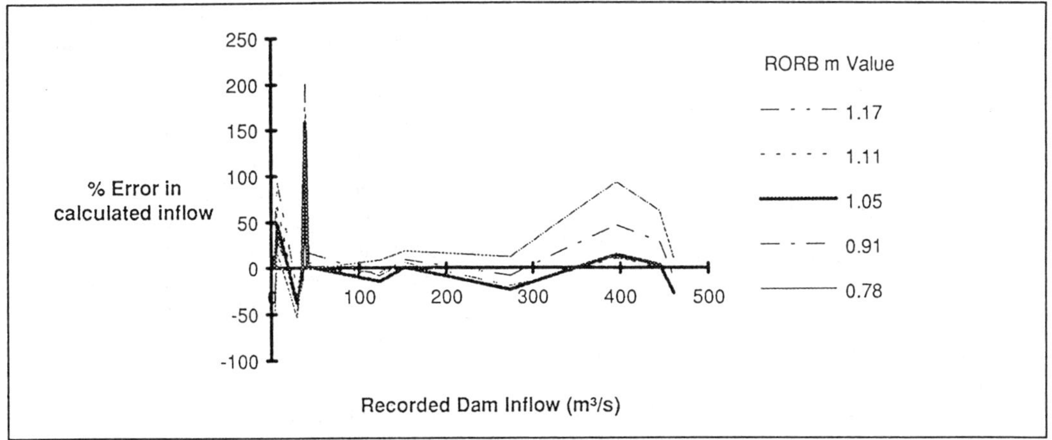

Figure 5 Percentage error in RORB calculated flow versus recorded peak flow

Since the small flow events have been associated with large estimation errors, they were eliminated from the Sensitivity Fitting data set, resulting in the new average curve as shown on Figure 6. This figure suggests a more appropriate choice of m of 1.0, corresponding to a zero average error.

Sensitivity fitting does not eliminate the disparity between the calculated and observed peak flows. It does however attempt to minimise the inherent errors. Particular note should be made of the 40% error which still exists even with the optimum m and k_c.

This new method of choosing the most appropriate model parameters, k_c and m, may be far more desirable than previously recommended methods. It is not expected to be limited to use in the Aroona Dam catchment, and should be equally applicable to non-arid regions.

6.2 *Design Flood Frequency Analysis*

A spreadsheet was formulated to calculate the design rainfalls for a range of ARI, storm durations and temporal patterns. The RORB model as previously established for the 'fit' runs was converted for use in 'design' runs. Thirty-two design storms were modelled, corresponding to ARIs of 2, 5, 10 and 20 years.

The model was run using parameters m=1.05, k_c=3.0, IL=23.8 mm and CL=5.13 mm/hr, these being the parameters derived from the Sensitivity Fitting procedure. The results indicate that the 6 hour storm gave the largest

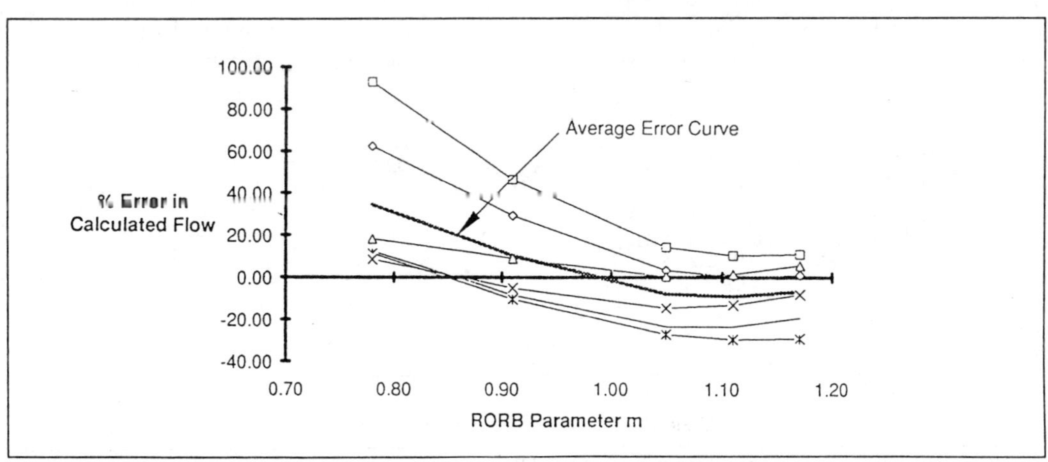

Figure 6 Sensitivity Envelope using Summer event flows exceeding 50 m³/s

design flow; that is 6 hours is the critical duration.

6.3 *Comparison with Historical Flood Frequency Diagram*

The generated design flows were plotted on the Flood Frequency Diagram in order to test the accuracy of the existing AR&R (1987) design methods. Refer to Figure 7.

It is evident that the predicted design flood peaks are less than the fitted Log-Pearson III distribution, by at least 20 percent.

Design temporal patterns in AR&R (1987) are derived from intense burst within longer duration storms. Therefore it may be suggested that to model design storms with zero (or a reduced) initial loss would be suitable, because the initial loss would have been satisfied prior to the burst from which the temporal pattern was generated. A reduced initial loss is also suggested in AR&R(1987) on page 123.

The models were re-run with zero initial loss and all parameters as before. This analysis produced significantly greater flows, as shown in Figure 7.

Significant errors were also identified in the areal reduction factors and Intensity-Frequency-Duration information in AR&R (1987). Areal reduction factors were far too low for this region. Based upon the work of Kemp (1989), Kinhill (1989) and our own studies, higher Areal Reduction Factors were adopted.

A detailed analysis of the historic rainfall data for the area indicated that the IFD design values for this location were low. These design rainfalls together with unmodified Areal Reduction Factors would have resulted in low estimations.

7.0 CONCLUSIONS

A number of important details regarding the nature and consequences of the rainfall common to this region have been established. Furthermore, a set of catchment modelling and data utilisation methods have been employed and developed.

Reverse Routing is a valuable and easy method of using stage and discharge records in developing flood predictions. The Reverse Routing procedures of Boyd et al (1989) have produced dam inflow hydrographs which appear to be correct estimations when compared with

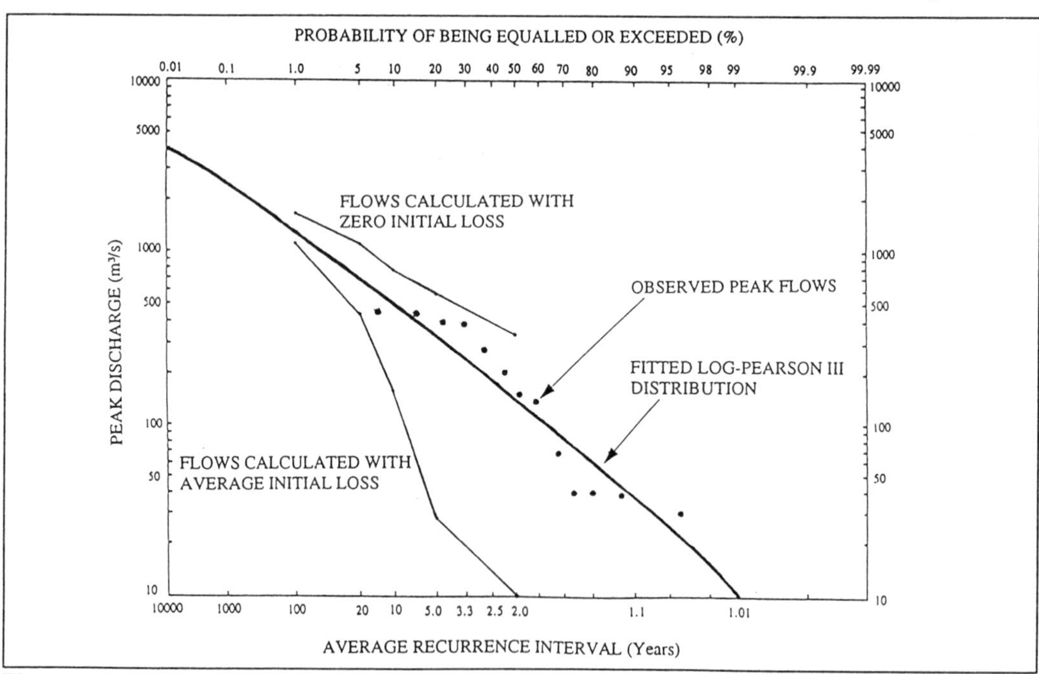

Figure 7 Historical and Design Flood Frequency Diagram

other dam inflow estimates and expected catchment runoff behaviour.

Sensitivity Fitting has proved an important extension to the normal RORB model parameter selection procedures. It provides a method of selecting an m and k_c which best fits the catchment model, in terms of minimising the expected calculated flood error. It is also anticipated that the principles of Sensitivity Fitting are equally applicable to the calibration of catchment runoff routing models other than RORB, as well as to catchments in non-arid regions.

Employing this parameter optimisation method may however still yield calculated flood peaks in significant error (40 percent) to the actual flow magnitude. This study has also concluded that the recommended choice of RORB parameter m of 0.8, would result in large errors in predicted flood magnitude, of the order of 100 percent. It is stated in AR&R (1987) that values of m up to 1.5 are not rare; it should be questioned however, in light of these results, how extreme the errors may be for such a high value of m. For this reason, catchment modelling should always be based upon calibrated parameters, by way of Sensitivity Fitting, should sufficient data exist to undertake such an analysis.

8.0 ACKNOWLEDGEMENTS

The authors gratefully acknowledge the following persons for their invaluable assistance during the period of research :

David Kemp, Department of Road Transport, for initial catchment information and RORB model;

Mark Pivovaroff, Kinhill Engineers Pty Ltd, (Adelaide);

Robin Leaney and Sarah Meanes, Engineering and Water Supply Department for assistance with catchment hydrological data;

Russell Mein, Monash University, for providing a PC version of RORB Runoff Routing software.

9.0 REFERENCES

Boyd M.J., Pilgrim D.H., Knee R.M., Budiawan D., (1989) Reverse Routing to Obtain Upstream Hydrographs, *Hydrology and Water Resources Symposium 1989, Christchurch NZ, 28-30 November 1989* The Institution of Engineers, Australia Preprints of Papers pp 372-6 (National Conference Publication No. 89/19)

Heweston P.F., Daniell T.M., (1988) HYDSYS - The ACT Water Administration Hydrometric Archiving System, *Hydrology and Water Resources Symposium 1988, Canberra 1-3 February 1988.* The Institution of Engineers, Australia Preprints of Papers pp 271-2 (National Conference Publication No. 88/1)

Institution of Engineers Australia, (1987) *Australian Rainfall and Runoff - A Guide to Flood Estimation*, Volumes 1 & 2, Publication of I.E.Aust.

Jayasuriya M.D.A., Mein R.G., (1985) Frequency Analysis using the Partial Series, *Hydrology and Water Resources Symposium 1985, Sydney 14-16 May 1985.* The Institution of Engineers Australia, pp81-85 (National Conference Publication No 85/2)

Kemp D.J., (1989) A Regional Flood Frequency Analysis for the Northern Flinders Ranges, *Hydrology and Water Resources Symposium Christchurch NZ, 28-30 November 1989* The Institution of Engineers, Australia Preprints of Papers pp 131-5 (National Conference Publication No. 89/19)

Kinhill Engineers Pty. Ltd., (1989) *Electricity Trust of South Australia Aroona Dam Spillway Flood Risk Study*, Adelaide

Kopittke R.A., Stewart B.J., Tickle K.S., (1976) *WS06 Frequency Analysis, DEC System 10 Documentation and User Manual*, Irrig. and Water Supply Commission, Queensland

Kotwicki V., (1986) *Floods of Lake Eyre,* Publ. by Engineering and Water Supply Department, Adelaide

McDermott G.E., Pilgrim D.H., (1982) A Design Method for Arid Western New South Wales based on Bankfull Estimates, *Hydrology and Water Resources Symposium 1982,* The Institution of Engineers Australia,

Environmental Management, Geo-Water & Engineering Aspects, Chowdhury & Sivakumar (eds)
© 1993 Balkema, Rotterdam. ISBN 90 5410 099 0

Flood mitigation study for Moonee Ponds Creek – An Urban Flood Management case study

R.J. Keller
Monash University, Vic., Australia

B.J.C. Perera
Victoria University of Technology, Vic., Australia

ABSTRACT: The Moonee Ponds Creek has a catchment area of 145 square kilometres, most of which is fully urbanised. This paper describes a study undertaken to investigate flooding characteristics of the lower urbanised part of the catchment and to identify an appropriate flood mitigation strategy for the 1% flood. The CELLS model was utilised with substantial modifications, and was successfully calibrated against a major flood event which occurred in the catchment in 1933. An appropriate flood mitigation strategy is described and a construction sequence identified to minimise adverse flooding effects during construction.

1 INTRODUCTION

The Moonee Ponds Creek has a catchment area of 145 square kilometres above Footscray Road bridge. In general the land upstream of the Jacana retarding basin is undeveloped but downstream the land is fully developed and urbanised.

Flood levels have been designated for that part of Moonee Ponds Creek between Footscray Road and Mt. Alexander Road based on historical records of a major flood flow in the creek in 1933. However, since this flood, the nature of the creek has been changed by extensive flood mitigation works and the addition of new bridges over the creek. In particular, the flood levels do not take account of the effects of mitigation works undertaken since 1933, nor are they able to take account of the effects of proposed flood mitigation works.

A detailed flood study was undertaken at Monash University for that part of Moonee Ponds Creek between Footscray Roand and Mt. Alexander Road and the results have been presented in Keller et al. (1988). In this paper the emphasis is placed on the development of an appropriate flood mitigation strategy and on the identification of the best construction sequence to minimise adverse flooding effects in the event of a major flood during construction.

Presented first is a brief description of the CELLS modelling approach developed by Weiss (1976) and of the special modifications undertaken for this study. The setup and calibration of the data set are then presented. Investigation of the flood mitigation options is then discussed and the development of a suitable construction sequence described.

2 THE MATHEMATICAL MODEL

2.1 *Review of the CELLS model approach*

The model takes as its starting point the actual topography of the area under consideration. It is assumed that the flood plains, as well as the actual bed of the river and its tributaries, can be divided into a certain number of cells. Each cell communicates with the neighbouring cells and the links between them correspond to an exchange of flow in the physical system.

The division of the given area into cells is not arbitrary but is based as far as possible on natural boundaries, such as roads, dykes, natural bank levées, etc. If the absence of natural obstacles leads to cells of unwieldy size, their further subdivision is no longer based on natural features, but is merely used to allow more precisely for the slope of the plain.

There are two types of cells (excluding dummy cells) available to define the cell layout; they are the river cells and storage cells. In general the river cells can be used to define the river sections and storage cells to define the flood plains. River cells model water surface gradient along the main flow direction but assume horizontal water levels in the lateral direction. However, the storage cells assume horizontal water levels throughout the cells. If it is necessary to model the water surface gradient in these storage cells, then the user is required to subdivide the cell into more storage cells, if river cells cannot be used.

For each cell the continuity equation may be written, equating the change in cell volume to the summation of inflows from all adjacent cells. The discharge between an adjacent cell and the cell under consideration is expressed using an appropriate equation of motion. Typically this may be a river type relationship, a weir type relationship applicable to boundaries over which the flow passes (e.g. roads or levée banks), or an orifice type relationship applicable to boundaries through which the flow passes (e.g. culverts).

The continuity equation is then written for each of N cells in the area to be modelled and, incorporating the discharge equations, this leads to a system of N non-linear ordinary differential equations in which the dependent variables are the N free surface elevations. A full statement of the equations and solution techniques are presented in Cunge et al. (1980).

2.2 *Special enhancements*

Bridge Loss Modelling: Refined modelling of flow through bridges is considered necessary when they create a severe hindrance to the flow and the quality of calibration data is adequate. The original model treats the flow through bridges as weir flow when the bridge is undrowned and as orifice flow when it is drowned. This method does not take into account the head losses due to the presence of the bridge and the roughness of the river reach between the centroids of the adjacent cells of the bridge directly. Furthermore, since generally the bridges are constructed transverse to the main flow direction, the flow through the bridge is better represented by river flow. Hence in this model flows through bridges are represented as follows:

(i) when the water level is below the soffit level of the bridge, the flow is considered as river flow with an equivalent (increased) Manning's n to cater for the head loss due to the presence of the bridge (which accounts for the number of piers, skewness of the bridge, pier shape and size, etc.), and the roughness of the river reach between the centroids of the river cells adjacent to the bridge;

(ii) when the later level is above the soffit level of the bridge (but below the deck level), the flow is considered as orifice flow.

For case (i), the energy loss (ΔH) between the cell centroids is given by:

$$\Delta H = \underbrace{C\alpha_2 \frac{Q^2}{2gA^2}}_{\text{bridge loss}} + \underbrace{\frac{Q^2 n^2 \Delta x}{A^2 R^{4/3}}}_{\text{reach roughness loss}} \qquad (1)$$

where C is the bridge loss coefficient
α_2 accounts for the non-uniform velocity distribution under the bridge
Δx is the distance between cell centroids
A is the bridge gross waterway area (including the piers).

The value of n refers to the roughness of the reach between centroids of the cells adjacent to the bridge.

From Manning's equation

$$\Delta H = \frac{Q^2 n_{eq}^2 \Delta x}{A^2 R^{4/3}} \qquad (2)$$

where n_{eq} is the equivalent value of Manning's

n including the effect of the bridge. Equating the right-hand sides of Eqns. (1) and (2) yields

$$n_{eq} = \sqrt{\frac{R^{4/3}}{\Delta x}\left[\frac{\alpha_2 C}{2g} + \frac{n^2 \Delta x}{R^{4/3}}\right]} \qquad (3)$$

The computation process then utilises Manning's equation, with the n value given by Eqn. (3).

As is apparent in Eqn. (3), it is necessary to know the water level at the bridge to compute n_{eq}. Hence it is required to run the model without the presence of the bridges to get a first order approximation for the water level and then use this water level to compute n_{eq} which, in turn, is used in the model to compute water levels including the presence of the bridges. Although this process is iterative, usually a second order approximation for water levels thus obtained is quite satisfactory.

Non-horizontal Weir Crest Modelling: Flow over levées and roads, and the cross flow between river and flood plain are modelled using weir flow relationships. These weir flow relationships require the specification of the level of weir crest in terms of the levée or road crest, or as the level of the boundary between river/flood plain. In general these weir crests are non-horizontal. However, the original model was restricted as it could handle only horizontal weir crests.

Figure 1 shows a schematic of a non-horizontal weir crest modelled as a series of horizontal crests. For a rectangular weir, the undrowned weir discharge Q is given by

$$Q_c = \frac{2}{3}(\frac{2}{3}g)^{0.5} L H^{1.5} \qquad (4)$$

where L is the crest length
H is the head on the crest.

For the example shown in Fig. 1, the total undrowned weir discharge, Q_{ce}, is given by:

$$Q_{ce} = Q_{c1} + Q_{c2} + Q_{c3} \qquad (5)$$

Substitution of Eqn. (4), written for each horizontal sub-weir, into Eqn. (5) yields:

$$Q_{ce} = \frac{2}{3}(\frac{2}{3}g)^{0.5}[L_1 H_1^{1.5} + L_2 H_2^{1.5} + L_3 H_3^{1.5}] \qquad (6)$$

The total undrowned weir discharge may also be written as:

$$Q_{ce} = \frac{2}{3}(\frac{2}{3}g)^{0.5} L H_e^{1.5} \qquad (7)$$

where $L = L_1 + L_2 + L_3$, and
H_e is the equivalent head.

Equating the right-hand sides of Eqns. (6) and (7) yields:

$$H_e = \{\frac{1}{L}[L_1 H_1^{1.5} + L_2 H_2^{1.5} + L_3 H_3^{1.5}]\}^{2/3} \qquad (8)$$

Since H_2 and H_3 can be expressed geometrically in terms of H_1, H_e can be expressed as a unique function of H_1 with Eqn. (8) and this function and the total undrowned discharge over the weir, Q_{ce}, determined from Eqn. (7), are the inputs to the CELLS model.

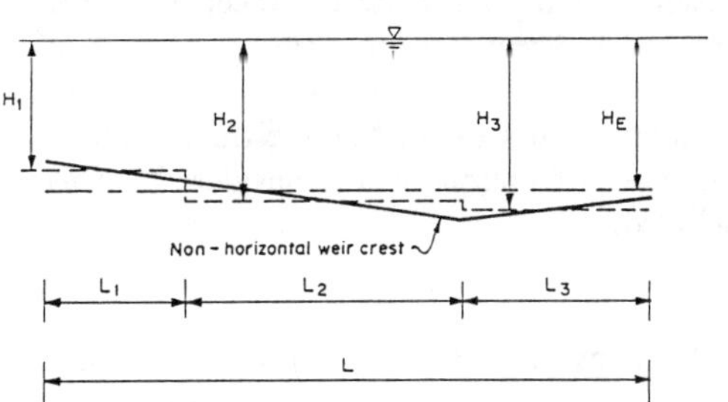

Fig. 1. Schematic of Non-horizontal Weir Crest.

3 MOONEE PONDS CREEK FLOOD STUDY

The model with enhancements described in Section 2 was applied to the study area briefly described in Section 1 subject to periodic extensive flooding.

The study area is relatively long and narrow with a central water course ranging from a small concrete or rock-lined low flow channel placed within a wide grassed trapezoidal channel to a 30m wide trapezoidal tidal channel.

Of major importance to the flood levels in the reach are the sixteen bridges and service crossings. Of these, six are road bridges, four are service crossings, and the remaining six are railway bridges. These obstructions generally consist of multiple spans and piers and cause major inteference to flood flows, especially when the flood level exceeds the bridge soffit level. Detailed descriptions of the bridge sites and of the study area are included in D'Agata (1988) and Withers (1987).

The worst flood to occur in the creek was in 1933 when a discharge of 230 m^3/sec was recorded. Some observed water surface elevations were available for this flood and were used in calibration.

The calibration process required first a careful and informed construction of the data set. The study area was represented by a series of river cells and storage cells. Flows in the river sections were generally represented by river cells whereas in flood plains, the flow situation was represented by river or storage cells as appropriate. Flow over the levées and the cross flow from the creek to flood plain were modelled as weir flow. In some cases weir links were not considered since the levées at these locations are at such a high level that there would not be any flow over them. This reduced the computer execution time required by the CELLS model. Further, care was taken to ensure that the cells were small enough to represent the study area satisfactorily.

In all, 41 cells were used to define the study area. Of these two were dummy cells (cells 1 and 41) used to define the boundary conditions. Twelve of the flood plain cells were storage cells, and the remainder were river cells. The river cells were represented by cross-sections at the upstream and downstream ends and their roughnesses or, where relevant, segment roughnesses. Where a bridge was present, the bridge was considered as a boundary for a river cell and the cross-section details and equivalent roughness were input.

Following the determination of the final arrangement of cells, a series of calibration runs was carried out during which a number of adjustments to the roughness characteristics, discharges, and bridge waterway areas were made.

The latter adjustments were necessitated by the high velocities under flood conditions at some of the constricted bridge sites. Constriction scour then significantly increases the bridge waterway area during the flood passage with a consequent reduction in local water levels. The method developed by Keller (1983) was used to determine the waterway opening under flood conditions.

The calibration process led to the establishment of Manning's n values between 0.015 for lined channel sections and 0.05 for flood plains. These values accord well with field inspections and are considered to be realistic.

To calculate the 1% flood profile under existing conditions, some changes to the data set were required to reflect site changes since the calibration flood. The final cell layout for existing conditions is shown schematically in Figure 2. The 1% design flood hydrograph was determined using the RORB program developed by Laurenson and Mein (1988).

Running the model for the 1% design flood showed significant design problems. Because of relatively low soffit levels, the Racecourse Road, Macaulay Road and Main Lines Railway bridges all created significant upstream ponding. The ponding created significant flooding of the eastern flood plain just upstream of the Main Lines Railway bridge and of the western flood plain between Racecourse Road and Arden Street (see Fig. 2).

In an attempt to mitigate the flood problems a number of options were considered. These included:

(i) construction of flood proof walls along the west bank between Racecourse and Macaulay Roads and on the east bank between Sutton and Mark Streets and between Macaulay Road and Straker Street;

(ii) raising the height of the levée along the west bank between Macaulay Road and Arden Street;

(iii) construction of levées on the west bank between Gravitation Railway bridge and Dynon Road and between Dynon Road and Standard Gauge Main Lines Railway bridge;

(iv) removal of Macaulay Road Bridge.

Additionally these options were considered in conjunction with enlargement of flow areas within the creek.

In the absence of modifications to flow areas within the creek the construction of the flood proof walls was shown to eliminate flooding in the eastern and western flood plains upstream of Arden Street. However, flood levels in the creek both upstream and downstream of Arden Street were consequently increased. These increased levels led to flooding of the eastern flood plain further downstream (cell 37 - see Fig. 2) which was not evident before construction of the walls. This illustrates clearly the adverse effects which may arise from the installation of flood mitigation measures.

Enlargement of flow areas within the creek was shown to create mixed benefits with channel improvements between Racecourse and Macaulay Road bridges being much more effective in reducing water levels within the creek than is the case with enlargement of the flow area under the Main lines Railway bridge.

The program was also used to determine the most appropriate construction sequence assuming the occurrence of the 1% flood during the construction of a range of flood mitigation options. The lowest flood levels were achieved by building flood mitigation measures in the following sequence:

1. Construction of the flood proof wall on the west bank between Racecourse Road and Macaulay Road.
2. Raising height of the existing levée on the west bank between Macaulay Road and Arden Street.
3. Construction of the flood proof wall on the east bank between Sutton and Straker Streets (should be extended to Arden Street).

4 CONCLUSIONS

The CELLS model with significant modifications has been successfully applied to the prediction of flood levels within an urban catchment. Possible flood mitigation options were identified and tested and the most appropriate construction sequence established.

The study has shown that care is necessary in identifying flood mitigation options because of possible detrimental effects elsewhere.

By use of the modified bridge routine and non-horizontal weir crest routine in conjunction with the CELLS model, accurate modelling of a complex urban situation was carried out.

ACKNOWLEDGMENTS

This project was funded under a grant from Melbourne Water (formerly Melbourne and Metropolitan Board of Works).

REFERENCES

Cunge, J.A., F.M. Holly & A. Verwey 1980. Practical aspects of computational river hydraulics. Boston, Pitman.

D'Agata, S. 1988. Moonee Ponds Creek Flood Study - Footscray Road to Mount Alexander Road - Available Data. Melbourne and Metropolitan Board of Works, Feb.

Keller, R.J. 1983. General scour in a long contraction. Proc. XXth Congress of IAHR, Moscow, USSR. II:280-289.

Keller, R.J., B.J.C. Perera & E.M. Laurenson 1988. Moonee Ponds Creek Flood Mitigation

Study. Report for Melbourne and Metropolitan Board of Works.

Laurenson, E.M. & R.G. Mein 1988. RORB Version 4. Runoff Routing Program - User Manual, Monash University, Department of Civil Engineering.

Weiss, H.W. 1976. An integrated approach to mathematical flood plain modelling. Report No. 5/76, Hydrological Research Unit, University of Witwatersrand, Johannesburg, South Africa.

Withers, D. 1987. Flood mitigation report - Moonee Ponds Creek - Footscray Road to Mount Alexander Road - Part 1 - Preliminary Investigations. Melbourne and Metropolitan Board of Works.

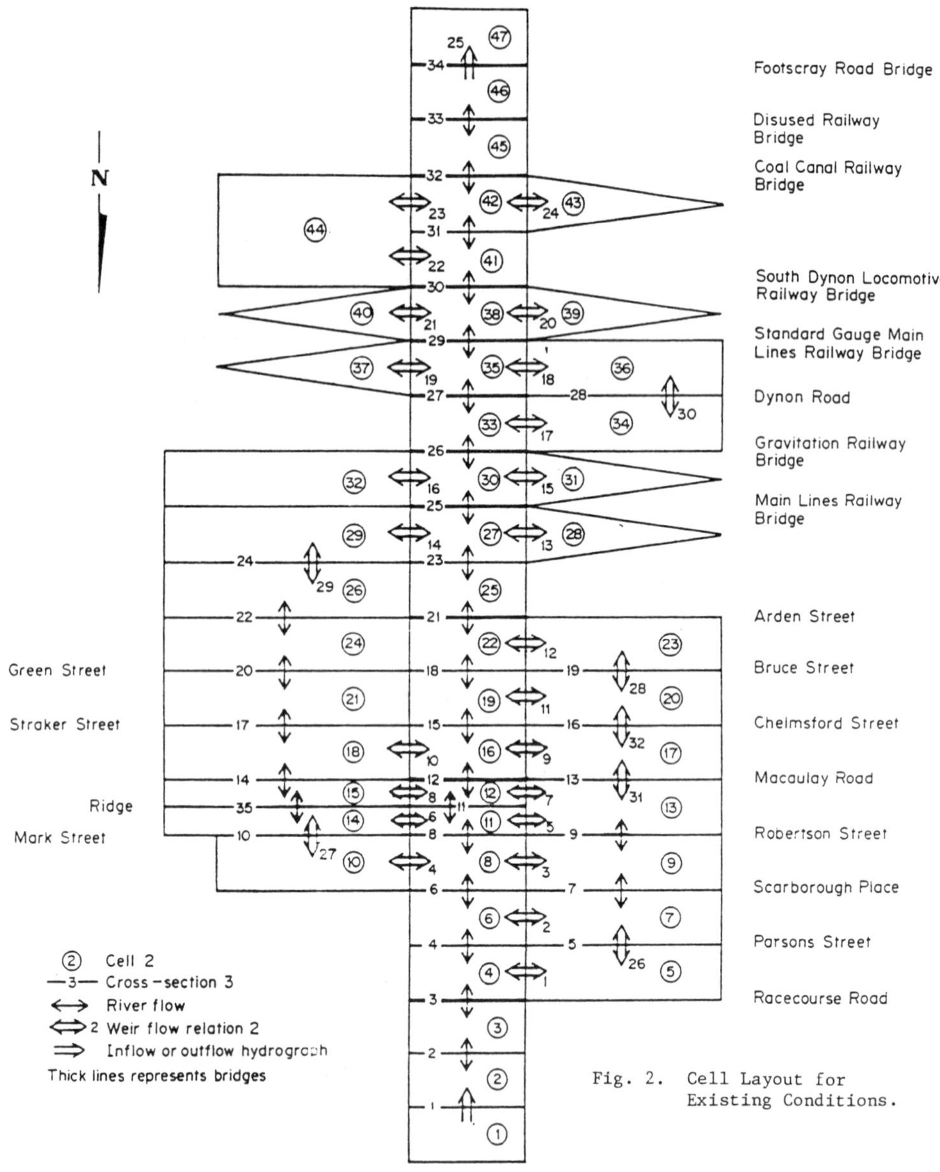

Fig. 2. Cell Layout for Existing Conditions.

Environmental Management, Geo-Water & Engineering Aspects, Chowdhury & Sivakumar (eds)
© 1993 Balkema, Rotterdam. ISBN 90 5410 099 0

The housing market and urban flood control policy

Des Lambley
NSW Department of Local Government, N.S.W., Australia

Ian Cordery
University of New South Wales, N.S.W., Australia

ABSTRACT: It is commonly believed that the value of urban property is depressed when identified as floodprone. Recent investigation indicates that a temporary reduction in dwelling values can result from flooding. However, there is stronger evidence that the sale price of flood liable dwellings in Sydney are persistently lower than those which are flood-free, except when the attention of the community becomes dominated by economic extremes. It is clear that many socio-economic and environmental variables dominate the thinking of individuals causing them to trade certain levels of risk for other, quite subtle advantages offered by a property which sometimes suffers flooding. There is no doubt recurrent flood damage to the nation's housing stock is an economic burden to the owners and also to the community who contribute to disaster relief and reconstruction funds through taxation.

URBANISATION

Urban populations continue to increase and the demand for homes tends to drive residential development onto lands prone to flooding. When a house is flooded as a result of earlier planning decisions, the property owner, the occupier and the community at large suffers socio-economic losses. Parker (1988) discussing the privatisation of the water industry in England and Wales says that "Catchment urbanisation is now probably the major factor in worsening flood hazards, and large and extensively paved urban areas produce increased flows, rapid runoff and sudden floods".

However, neglecting to designate a house as floodprone does not make it any less subject to damage from flooding. The press has stated that to name a house as floodprone would take an average of 30% off its value (Monaghan 1984). Such sensational press and the popularization of this assumption as an election issue, exploited by both Labor and the Liberal candidates during the 1984 election campaign, contributed to a decision by the then Premier, Neville Wran, to direct that the floodplain mapping programme for NSW be terminated and the published maps be withdrawn.

One mitigation measure is to allow market forces to dictate risk acceptability. However, such a policy tends to warn or protect potential owners of floodprone property at the expense of current owners. In addition there are people within a society who do not have the knowledge of risk or who are unable to comprehend the impact of flooding in urban areas. This has the potential to reduce their socio-economic circumstances. Former Prime Minister of Australia, Malcolm Fraser (Fraser 1992) reinforces the view that it is negligent to allow the market place to determine everything. Government, its agencies and particularly the floodplain planning and management industry all have an important role in education, the provision of information and disbursing a duty of care in this regard.

However, evidence exists to demonstrate the "woodenheadedness" of Government (Tuchman 1984). All too frequently government agencies and local government bodies allow the problem to perpetuate by selling land or building public housing estates, or zoning land as residential in known floodprone areas. For example, in New South Wales both the Department of Housing and Landcom have pursued such policies. Planning of this kind gambles with people's lives as it trades the problem and costs downward to the occupants of dwellings in these areas, and who all too often are those with most to lose.

FLOOD IMPACT ON HOUSE VALUES

Naturally, the impact of flooding upon the residential property market should be clearly reflected in a discontinuity in values. However, the impact of flooding upon housing values in the Sydney metropolitan area is not as apparent as popularly believed. This research tends to demonstrate that extraneous variables have a greater impact upon the market and cause any disutility from flooding to be obscured.

The purchase of a house is the largest single investment normally undertaken by a household. Continued maintenance of a house which is subject to frequent flooding has the effect of over-capitalizing the investment. Yet if the restoration of the property does not take place, deterioration is accelerated and normal capital appreciation would diminish. Thus, the life of the property may be shortened, or its physical or social amenity afforded to the owner or occupants could be reduced. This effect may be reflected in its depressed market value.

The value of a flood damaged dwelling may be sustained or improved by appropriate restoration, and its useful life extended. The nature of the repairs will be dictated by the owner's intention to retain or dispose of the asset. Its value is also governed by whether the property is owner occupied or let, the age of the occupants and their inclination to maintain it. In Australia where there is a high level of home ownership, further decision constraints may be imposed by the magnitude or residue of a mortgage, household income, employment history of the occupants and the vitality of the general economy.

RESEARCH IN THE USA

Recent US research has been directed towards distinguishing market values of floodprone and non-floodprone lands on the basis of the 1 in 100 year flood isopleth. Tobin and Newton (1986) argue that floods reduce land values through damage to structures. They say, "The extent of this reduction in utility will be dependent upon the temporal, spatial and hydrological features of the flood hazard. The subsequent recovery of the land (and housing) market will be dependent on various socio-economic criteria found in the metropolitan area along with the prevailing flood conditions".

Other studies by Tobin and Montz (1988) concluded that whilst reductions in the residential utility resulting from flooding can be illustrated, the economic impact is often variable. In a more recent work by Montz and Tobin (1988) it is suggested that areas "...flooded to the greatest depth had a more active housing market, with more houses listed and sold in the post-flood period than the residential areas with less severe flooding. This may be the result of persons unloading hazardous or damaged property".

These US studies have identified that urban flooding can be spatially differentiated by investigating the housing market. There are however, greater difficulties in predicting the length of time it takes for the differences to disappear.

SYDNEY STUDY

Following the example set by the US studies referred to above, this paper examines the hypothesis that properties inundated by floodwater should experience depressed values.

In Sydney, much flooding is brought about by rapid urbanization further upstream in small catchments and in areas inappropriate for housing and development. As a consequence, some people occupying older, downstream areas are now being flooded perhaps for the first time. On the Georges River (the area studied) two major and several smaller floods have been experienced since January 1984, as shown in Table 1.

During April 1988 persistent rain climaxed in widespread flooding in New South Wales. The flood was considered to be a major event in the western suburbs of Sydney, with a recurrence interval of 25 years. From a list of 2650 applications for Disaster Relief Assistance after that flood about 1200 metropolitan properties were identified for further investigation (Disaster Relief Committee, 1989). A search of the Water Board's data base and the property registers in 8 councils indicated that 519 of these properties had changed ownership since January 1984. The date sold and the sale price for each of these properties was extracted for examination.

To minimise any bias, the list was reduced by 39 properties acquired by local government bodies under the Voluntary Purchase Scheme

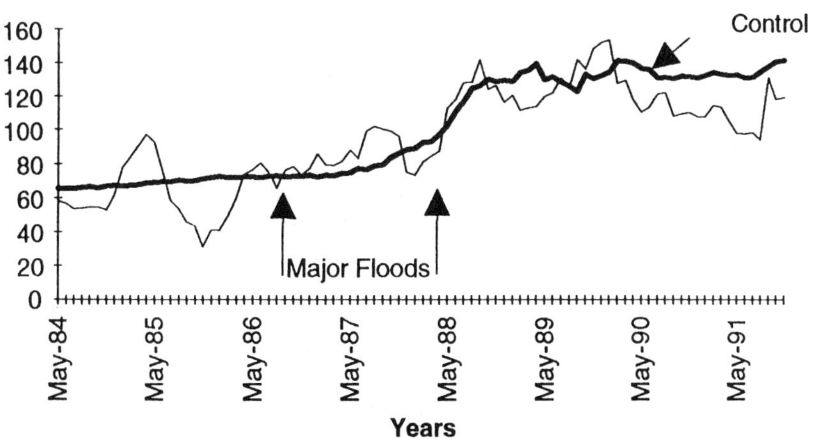

Figure 1 Average Monthly Sale Price

Table 1. Significant flooding in Western Sydney measured at the Liverpool Weir on the Georges River.

Year	Month	Height (Metres)
1984	July	1.38
1984	November	1.65
1986	August	4.6
1987	October	2.81
1988	April	4.6
1988	May	3.5
1990	August	2.4
1991	June	3.2

during this time as payments under this scheme were in may cases generously above the market value.

The Voluntary Purchase Scheme operates to provide for the removal of floodprone houses by local government who then rezone the land parcels as open space. A total of 480 properties remained for analysis. Approximately 40% of the 1200 addresses identified as subject to the April, 1988 flood had been sold at least once since January, 1984.

It is noted that several score of the 1200 addresses from which Disaster Relief Claims were made were identifiable as tenanted Housing Department (public housing) properties. Some of the addresses in the data set were Department of Housing properties which were sold to the tenants under earlier home purchase schemes.

Handmer and Partlett (1988) estimate that the Western Sydney urban floodplain contains some 2800 dwellings subject to inundation by a *1%* flood. The 1986 flood event has been estimated to have damaged some 700 dwellings in this area (Public Works Department 1986). The sample size (480 dwellings) is therefore considered to be significant and statistically acceptable, being about 17% and 68% respectively of these "populations".

The 480 properties extracted occupy land subject to a 25 year recurrence interval flood and most of them are therefore considered to have suffered flooding from major events in both August, 1986 and April-May 1988. Some localized damage would have resulted from a number of smaller floods, and these events would more importantly have maintained in the householders an awareness of the flood problem. A monthly mean sale price was calculated for the 480 properties. While an average of just under 5 properties per month were sold over this period there were some months in which no sales occurred. For control purposes, a monthly mean sale price for all properties sold by the principal estate agency co-operative was obtained for this same period in Western Sydney, the region experiencing the flooding events studied (EAC Multilist, 1992). The large monthly differences both the sale price and the control data were smoothed by applying a 5 month running mean. This is shown at Figure 1.

While previous research sought to demonstrate a disjunction in the sale price of floodprone dwellings, an additional approach used in this study compares the number of

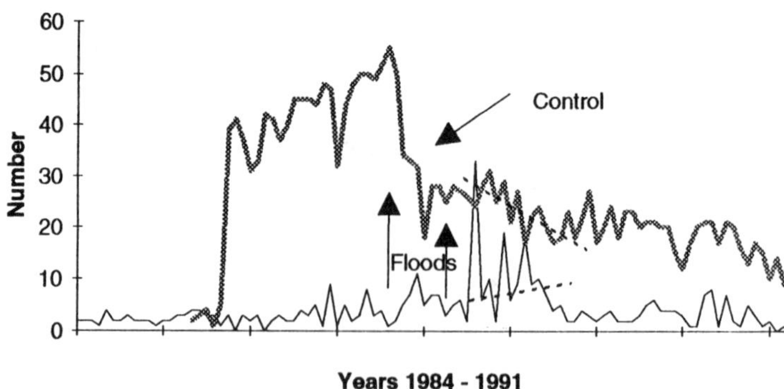

Figure 2 No. Houses Sold/Month

sales of floodprone and control properties each month for the period January 1984 - March 1992. This is shown at Figure 2.

Field studies indicated both the sample and the control groups of properties represent a diversity of type, style and age. Few are likely to be of double brick or stone construction but they are largely typical of the housing type built in strictly working class suburbs some 20 to 30 years ago. Levee banks are not a feature of this region, but house raising, retarding basins, voluntary purchase and channel maintenance am improvements are flood abatement works now being undertaken. However, at the same time upstream development is continuing. All of these dwellings occupied land zoned for residential use.

RESULTS

The month to month inconsistencies between prices of the 480 floodprone properties and the control group shown in Figure 1 probably reflects sampling problems. The variations in prices is probably caused by the sale of one particularly high or low priced property in a small monthly sample of properties sold. Presentation of running totals does not entirely overcome this problem.

Apart from six short periods in which the sale prices of floodprone properties exceeded the control averages, the general trend for 1984-1992 shows that floodprone property values fell slightly, relative to the control group. In 1985 the average floodprone property price was about 100% of the price of control properties and by 1991 this had fallen to about 80%. Although this differentiation is unlikely to operate effectively as a warning to people or be an inhibitor to commercial interests developing flood liable lands, it may indicate a marketplace discrimination.

While there seems to be some evidence that prices fell two or three months after the August 1986 flood for the floodprone properties, a similar fluctuation did not occur after the same level of flooding in April 1988.

The "recent" computer basing of these records and the fact that computer data bases recorded only the last sale in many cases, or a sale date without the sale price, gives some cause for suspicion about the accuracy of property turnover rate. However, the price fall of floodprone properties appears to be consistent and is probably real and not a function of record keeping.

As a comparison, data was processed to show the actual number of sales occurring per month since 1986. This alternative measure eliminates the vagaries of the supply-demand, and other external forces operating to obscure the prices which might be paid for floodprone dwellings. The data shown at Figure 2 tends to support the US contention that perhaps the effects of flooding can be seen in the market-place. Considerable lumpiness is present including the typical low turn-over rate during the Christmas
New Year period.

Figure 2 shows that the number of transactions fell sharply three months after the major flood of 4.6 metres in August 1986. It is considered that this lag is consistent with the conveyancing time and indicates that unsold properties were withdrawn from the market immediately they were subject to flooding. During the following two months there was a large increase in the number of sales of

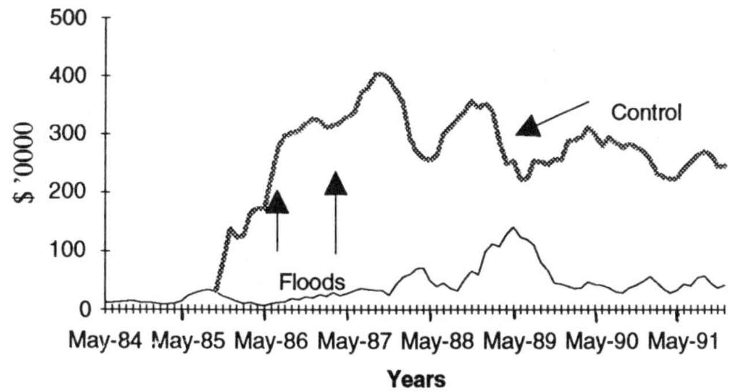

Figure 3 Turnover Comparison

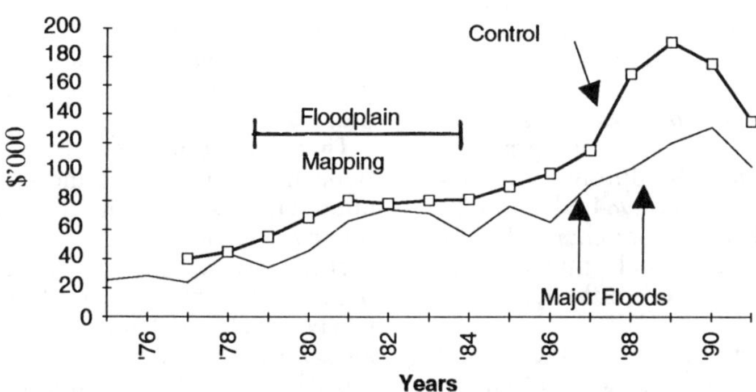

Figure 4 Annual Averages

floodprone properties. This supports a notion there is a popular desire to off load floodprone properties as people seek to escape from the disutility of flooding. This fall in turnover of floodprone properties and the subsequent surge in sales some 4 months later is in sharp contrast to the control property numbers.

In April 1988 another major flood of 4.6 metres was experienced in the Georges River catchment. The number of houses sold in both the floodprone and control data sets show coincidental decline during the month of the flood. This is considered to reflect a down-turn in business activity. A reduction in the number of properties sold at a time 3 months after this flood is similar to that experienced in 1986 but the flooded properties show a larger proportional decline. This feature is suspected to be a continuation of the general economic downturn. However, in 1988 off-loading seems to be a particularly clear feature. During the 15 months following the low in April, the number of floodprone dwellings sold increased while the control data show a steady decline in the number of sales per month. There is a clear, converging trend in sales numbers as shown by the heavy line plotted at Figure 2. This supports the contention that the experience of flooding prompts many people to sell their property after flood damages have been repaired. Examination of the gross value of sales per month of both the control set and the floodprone properties demonstrates the relative investment movements per month and further supports this conclusion. This feature is shown at Figure 3.

Use of a longer set of data, including an earlier period where there were fewer sales per month, provides insight into other possible causes of price fluctuations. Figure 4 shows mean annual sale prices for the floodprone properties and a control group of all Sydney properties. The floodprone properties are from the same sample as shown in Figure 1 but the control group is different.

Annual information is used for two reasons Firstly only annual data was available for any form of control group before 1984. Secondly there were fewer sales of the floodprone properties before 1984 and the number of sale per month was so small as to cause monthly data to fluctuate wildly. Figure 4 shows that from 1979 to 1989 sale prices of floodprone properties fell 2 - 3% per year compared with all-Sydney prices and then regained much of the lost value in 1989. The 1989 gain for floodprone properties probably reflects the effect of very high interest rates and consequent buyer focus on lower priced properties to minimise their loan at the time of house purchase.

The sharp decline in prices of floodprone properties in 1984 and subsequent rise in 1985 is of considerable interest. It is thought that the fall in 1984 was caused by the availability of an interim flood map for the Georges River floodplain at this time and the intense local publicity given to the mapping programme. Figure 4 shows that floodprone property values fell by about 25% in 1984, close to the amount suggested by Monaghan (1984). However, once the flood maps were withdrawn, values recovered to the long term trend.

The volatility of the housing market means that price is a variable which is subject to many influences in addition to the effects of flooding. For example, rises in mortgage interest rates, inflationary trends and diminished wage parity are considered to have had considerable effects on market prices. The declining business activity since about 1987 and the growth of unemployment to about 10% of the workforce are important external reasons which impact upon housing values.

It seems that the marketplace disjunction of housing prices is an inadequate tool to differentiate floodprone and non-floodprone lands in the Sydney metropolitan area. Other unrelated variables have a greater impact upon the market and cause any disutility of flooding to be obscured. However, assessment of the number of dwellings sold confirms the conclusions drawn from some US studies.

POLICY IMPLICATIONS

Inadequate housing market data is a serious impediment to research on these matters. Data for this study had to be synthesized from two primary (Disaster Relief Committee & EAC Multilist Realtors) and several secondary sources. A central register of residential properties affected by floods would seem to be a sensible, helpful management tool. Such a register would identify the turnover histories of flood affected property, the date built, date(s) sold, sale price(s), depth and velocity of floodwaters commonly experienced, and various government moneys which have been paid to the occupants or the owners of those properties over the years. In this way it would be possible to more accurately assess the wisdom of earlier planning decisions in terms of aggregate costs to the community and to pinpoint the focus of attention for mitigation works, house raising assistance and Voluntary Purchase Schemes.

The justification for this data collection and coordination would be that the expenditure of public moneys by way of grants, or relief to owners or occupiers of land, by any sphere of government represents a defacto capitalization of the land.

That residential properties continue to be flooded is testimony to the fact that available information is not being used wisely (or at all for town planning and land zoning purposes particularly when repetitive events have occurred over many years. It may well be that newly developed land parcels should carry a guarantee that the land is flood-free to a specified frequency. In this way closer, legal attention could be given to the matter from a consumer point of view. An ambivalent marketplace seems to encourage housing on floodprone lands in the Sydney metropolitan area and some would say this vindicates government policy which allow such development to take place. The State Government and its Agencies may argue they are on the "horns of a dilemma" in that the demand for housing persists and using land which is liable to infrequent flooding is a risk worth taking.

CONCLUSION

The effects of floods on the housing market in Sydney is not as clear as might have been expected, particularly in dollar terms. However, there is numerical evidence that a disjunction in the marketplace is evident following damage from flooding. The number of dwellings sold decline as people withdraw their properties from sale for one or two months following flooding Then sales rise sharply shortly after major flooding as people attempt to off-load flood liable property. The majority of these sales probably indicate owners who are no longer prepared to risk further losses from flooding.

A market-place disjunction of housing (as measured in dollar terms) is considered to be an inadequate tool to differentiate floodprone and non-floodprone lands in the Sydney metropolitan area. While a disjunction in values may result from flooding, other external variables such as supply and demand, and the national economy all prevail at the time of the sale. These are seen as more important influences on prices. While floods may have an effect upon housing values in Sydney, it seems that this influence is uncertain and therefore highly unpredictable. Although the evidence seems somewhat contradictory and inconclusive, it would appear that residential development on lands subject to irregular flooding in the Sydney metropolitan area is not regarded by some people as being economically unwise.

REFERENCES

Disaster Relief Committee 1989. Personal Communications. Sydney.

EAC Multilist Realtors 1989. Personal Communications. Sydney.

Fraser, Malcolm 1992. The darker side of Westpac's $1.6 bn loss. *The Sun-Herald,* June 7, p.38.

Handmer, J & Partlett, D. 1988. *Flood Warnings and Legal Liability: The Georges River Floods, August 1986.* CRES, ANU, Canberra.

Handmer, J. 1986. *Property Acquisition for Flood Damage Reduction.* CRES, ANU, Canberra.

Monoghan, D. 1984. Flood Zoning the Land: A good idea in theory. *Sydney Morning Herald,* 19 May, p.4.

Montz, B. E. & Tobin, G.A. 1988. The spatial and temporal variability of residential real estate values in response to flooding. *Disasters: The Journal of Disaster Studies and Management,* (12) 4: 345-355.

Parker, Dennis J. 1988. *The Privatisation of the Water Industry in England and Wales and the Implications for Flood Hazard Management.* CRES Paper 1988/13. ANU, Canberra.

Public Works Department (1986). *The Sydney Floods of August 1986: Residential Flood Damage Survey. Vol.l.* Sydney.

Tobin, G.A. & Newton, T.G. 1986. A Theoretical Framework of Flood Induced Changes in Urban land Values. *Water Resources Bulletin* Vol.22, No.l, pp. 67-71. American Water Resources Association.

Tobin, G.A & Montz, B.E. 1988. Catastrophic Flooding and the Response of the Real Estate Market. *The Social Science Journal* Vol.25 No.2, pp.167-177. JAI Press Inc.

Tuchman, Barbara W. 1984. *The March of Folly* Michael Joseph, London.

Water Board 1990. Personal Communications. Sydney.

ACKNOWLEDGEMENT

* Special acknowledgement is given to the Water Board, The Public Works Department, EAC Multilist and the Disaster Relief Committee for their assistance in providing the data upon which this paper is based.

* The origins of this work was part funded by grants from the Australian Water Research Advisory Council and the Public Works Department of NSW for research on "Criteria for selection of design levels for development on flood plains".

Environmental Management, Geo-Water & Engineering Aspects, Chowdhury & Sivakumar (eds)
© 1993 Balkema, Rotterdam. ISBN 90 5410 099 0

A study of ground water hydrology of Bijauli Block, Tehsil Atrauli, District Aligarh, India

Gauhar Mahmood
Department of Civil Engineering, JMI University, New Delhi, India

ABSTRACT: Systematic inventory of 90 dug wells, 20 shallow tubewells and 26 deep tubewells yield relevant hydrogeological data that furnishes valuable information on the groundwater regime of the study area.
In order to study the occurrence and movement of groundwater in several physiographic units of the area, depth to water level, water table contour and water level fluctuation maps have been prepared. In addition, lithological logs of the deep tubewells (70 to 143 metres below ground level) were studied and utilised to prepare a fence diagram to depict sub-surface geology and the aquifer system in the study area.
It is concluded that shallow aquifers (0-50 metres below ground level) are overdrawn as compared to deeper aquifers (51 to 143 metres below ground level) and therefore, in the study area either more deep tubewells are needed or an unlined canal built. Apart from these conclusions, some more useful suggestions were made for engineers and hydrogeologists working on this problem.

INTRODUCTION

The area under study is located in the northern part of the Aligarh district between the latitudes of 27 50° N to 20 05° N and longitudes of 78 20° E to 78 35° E. The area is covered by toposheet Nos. 53 L/4, 53 L/8 and 54 I/5 of The Survey of India (Figure I)

The present study deals with groundwater condition and movement in Bijauli Block, Tehsil Atrauli, district Aligarh, U.P.(India), as related to development of groundwater reservoirs and an understanding of the hydrogeological cycle and the local hydrologic system. Hence a precise evaluation of groundwater resources of the area is an essential prerequisite to proper development and water resource conservation.

OBJECTIVE AND SCOPE

The investigations were carried out with the view of determining the origin, occurrence and movement of groundwater in the study area which is also important in delineating different aquifer horizons and their relationship to groundwater recharge and discharge, with special emphasis on areas recommended for groundwater development.

In this context it is very important to make refined quantitative evaluation of the groundwater resources, starting at the block level and progressing to district level. Such investigation will depict a harmonious hydrogeological picture of the country, encompassing blockwise occurrence behaviour, water quality, quantity and state of development of the water resources with the scope of its future development.

ANALYSIS AND DISCUSSION

Groundwater Condition

The coarser clastics in the alluvial sequence forms the major repository of groundwater in the area. Groundwater occurs under unconfined, semi-confined and confined conditions depending upon the occurrence of aquitard and aquicludes as confining beds. The shallow aquifers are unconfined, whereas the deeper aquifers are semi-confined in nature The rainfall is the main source of groundwater recharge, but the recharge also occurs through irrigation return flow and effluent seepage from the Lower Ganga canal.

Evolution of Aquifers

The evolution of aquifers in a fluvial system depends upon the hydrodynamics of the flow regime, local geology and topography of the terrain leading to the clastic deposits which are typically represented as channel deposists, flood plain and and back swamp deposits.

Channel Deposits

The typical channel deposits of the river Ganga as observed in the area form bottom upwards sequence comprised of coarse sand mixed with gravel to fine sand and silt with a thin clay layer at the top. The top clay and some fine grained sand layers are commonly washed away during the succeeding flood periods and a fresh body of sand in a thinning upward sequence may be deposited annually forming thereby a reasonably thick terrigenous clastic deposit which persists even after the river changes its course. These thick bodies of sand are potential repositories of groundwater aquifers (Fig.II)

Flood Plain Deposits

During flood season when the flood water overflows the bank, medium to fine grained sand bodies of moderate thickness but limited areal extent are deposited across the flood plain. These lenticular bodies of sand form the moderate potential aquifers in comparison to highly potential aquifers of the channel deposits.

Back-swamp Deposit

The flood water further moves down the slope in the low lying areas where it deposits predominantly fine grained suspended materials which have settled out under the influence of gravity and forms the lensoid bodies of sand overlain by still finer clastics i.e. clay, thus there occurs enclaves of sand intercalated within the underlying and overlying thick clay beds; such bodies of sand are poor aquifers. These aquifers are typical representatives of back-swamp deposits,.thereafter the river changes its course and with the passage of time, the position of the channel, flood plain and back-swamp deposits start changing, that is why most of sand or clay are irregular or discontinuous. The above lithological variations are attributable to the mode of deposition by constant shifting nature of the River Ganga. In summary, the fluvial system which has generated various aquifers in the area contains:-

Channel deposits: Thick linear aquifers of indefinite areal extent but the best groundwater reservoirs.

Flood plain deposits: These are lenticular aquifers, limited in thickness and areal extent and only moderate in groundwater potential.

Back-swamp deposits: These are lensoid bodies of sand occurring as enclaves or stringers within thick clay beds and generally are low yield potential aquifers, often with quality problems in the Ganga alluvium system, the complexes of channel, flood plain and ox-bow lake faces may reappear several times in a single well drill. Thus the terregenous clastic deposit of the River Ganga witness the complex hydrodynamic regime which has generated the various aquifers in the Great Ganga plain.

Aquifer Geometry

The fence diagram (Fig. II) reveals the vertical and probable lateral extension of the several lithological units. A perusal of the fence diagram shows that in general, there occurs three to four tier aquifer system in the Bijauli block.

By and large these aquifers appear to merge and behave as a single aquifer system. The fence diagram shows that the granular zones comprising of sand and gravel occupies about 30% to 40% of the total formation encountered down to a depth of 143 meters below ground level. In north western portion of the area, gradually the clay beds pinch out towards the central part of the area and the granular zone attains greater thickness. This trend continues progressively to the south and southeast where the aquifer attains the maximum thickness of 90 metres. In the northeast also the granular predominates the previous beds as sand and gravel are 50% to 60% of the total litho units. The most peculiar situation is shown by the southwest direction where the clay predominates over the sand, the aquifer is lenticular body of sand trapped within thick clay beds, and the granular zone comprises 15 % to 20 % of the total strata down to a depth of 140 meters below ground level. Fine to medium grained, with a rare admixture of coarse sand with minor gravel, generally comprises the aquifer material in the area, mostly sands are of various shades grey and micaceous in nature.

On the basis of geological section descriptive lithology of the borehole and hydrogeological properties, the aquifers can be placed under two distinct categories:

(a) Shallow aquifers at depth less than 50 metres.

(b) deeper aquifers between 51 metres to 143 metres.

Shallow Aquifers

Shallow aquifers within the depth of 50 metres are comprised of mainly fine to medium sand grains. The thickness of the aquifers ranges from 5 to 10 metres and the groundwater is unconfined. These aquifers are generally tapped by open wells, hand pumps and shallow farmer tubewells but because of excessive withdrawal of groundwater from these aquifers, they have become strained. The discharge of these aquifers varies from 30 cu m/h to 50 cu m/h at nominal drawdown of 3.0 to 4.5 metres (Dutt, 1969).

Deeper Aquifers

Deeper aquifers are encountered generally within the depth range of 51 to 143 metres below ground level. The fence diagram reveals that these aquifers are semi-confined to confined in nature. Generally tubewells tapping these granular zones have discharge of 50 cu m/h to 227 cu m/h, with drawdown ranges from 1.97 to 11.72 metres (Raju and Rao, 1965).

Depth to Water Level

Water data for 90 dug wells, evenly spaced at a distance of one kilometre were utilized to prepare the depth to water level maps of the area. The maps were prepared for pre-monsoon (Nov) 1986. In the premonsoon depth to water ranges from 1.7 to 12.0 metres whereas in the post-monsoon period depth to water ranges from 1.5 to 11.5 metres. The area can be divided into six depth to water zones: (i) Less than 2.0 metres, (ii) 2.0 to 4.0 m, (iii) 4.0 to 6.0 m, (iv) 6.0 to 8.0 m, (v) 8.0 to 10.0 m, (vi) 10.0 to 12.0 metres below ground level. The deepest water level, 12.0 m, was recorded in the Bijauli Upland while the shallowest water level (1.7) was recorded at Dinapur in the Ganga Khadir area. A perusal of depth to water level maps show that in the Bijauli upland area the depth to water generally ranges from 5.0 to 10.0 m. This condition encompasses the major portion of the area except for a tract along the bank of the unlined lower Ganga Canal. Consequently, the general water level proximal to the canal has progressively risen. The small narrow area parallel to the Ganga bank is known as the low valley of Ganga or Khadir. Here depth to water level ranges from 1.7 m to 2.4 m. In general, the abovementioned depth to water level found in conformity with the general physiographic units of the area.

Movement of Groundwater

Water level data collected during the pre-monsoon periods were analysed, and the elevation of the standing water levels with reference to the Mean Sea Level were plotted and water table contour maps were prepared with a contour interval of one metre, the water table contour maps were very useful in determining groundwater flow direction, gradients and area of recharge and discharge (Todd 1980).

The elevation of the water table ranges from 179 metres in the northwest to 169 metres in the southeast A perusal of the water table maps show a general direction of groundwater flow is north to south with minor local variation. In the west the groundwater flows from north to south with minor local variation. In the west the groundwater flows from northwest to southeast while in the eastern part the groundwater flows from west to east. In general, the gradient varies from 0.4 m/km to 0.6 m/km and areas with wide contour spacing (Flat Gradient) seems to possess higher hydraulic conductivity than areas with narrow contour spacing (Steep Gradient). In the northeastern part of the area near the River Ganga, the hydraulic gradient is very steep, i.e. 2.0 m/km. This steep gradient is probably caused by two factors viz.

(1) Due to heavy withdrawal of groundwater, or

(2) Due to low permeability of the aquifer material.

It has been observed during the field survey that the concentration of the wells is relatively low in the eastern part from which it could be deduced that the steep gradient in this tract is due to low hydraulic conductivity.

The general slope of the water table is towards the River Ganga, i.e. west to east, which indicates that the River Ganga is an effluent stream. Similarly the River Kali which passes through southern and of the area also is effluent.

The maps show that a groundwater mound has been formed in the southeastern part of the

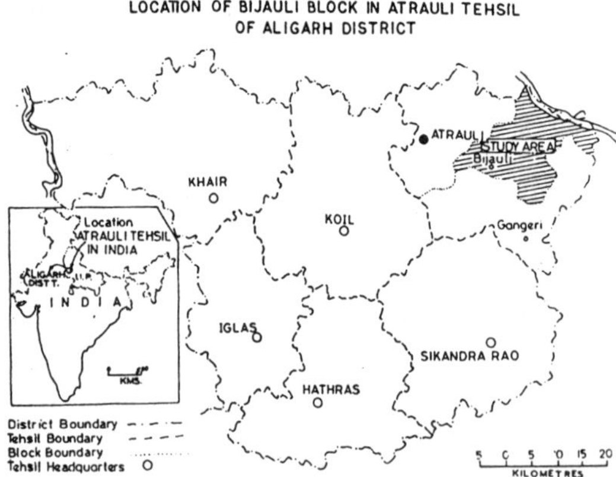

Figure 1

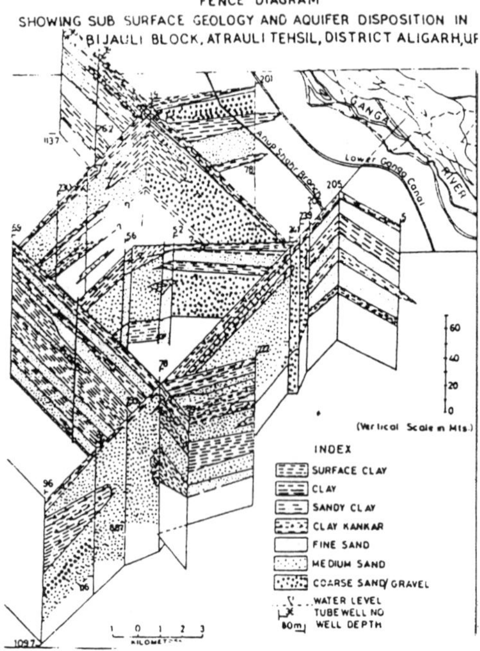

Figure 2

area, proximal to the right bank of the Upper Ganga Canal system due to excessive seepage of the surface water into shallow equifers through unlined bed of the canal and its distributories. This mound then sheds water from its eastern flaks into the River Ganga.

Groundwater Behaviour

Water level in the key observation wells in the area have been utilized in preparing continuous hydrographs with the object of studying response in time and space and also dependence on meterologic controls. Hydrographs of three wells for the period 1982 to 1986 are utilised. A perusal of these hydrographs indicate that the water level fluctuation is cyclic and sinusoidal. The water level is lowest during June and highest during November. It is observed that the water level starts rising by the last week of June. From mid November onwards, there is a sharp decline in water level until January, afterwards the recession in water is slow, indicating natural groundwater discharge through a steady sub-surface outflow, in harmony with regional groundwater movement.

From the above discussion it is seen that the water level rising and declining with respect to the time and rainfall infilteration.

Water Level Fluctuation

The map shows seasonal water level fluctuations in the area. The fluctuations are represented by contours of water level variance in pre-monsoon and post-monsoon for the period of June and November 1986. The area is shown by four water level fluctuation zone on a contour interval of 0.2 metres. The maximum fluctuation of 0.8 metres is recorded only in a few wells and in general the fluctuation ranges between 0.4 m to 0.6 m. This relatively minor fluctuation is attributed to scanty and sporadic rainfall during 1986 monsoon season. It is also evident from the map that the fluctuation in the upland is greater than in the low valley of the Ganga. It appears that these high fluctuation areas are also areas of high relief with minor local variations. As shown by the groundwater movement the upland tracts are groundwater recharge areas, where the groundwater moves towards the Ganga and Kali Rivers or down the regional gradient, i.e. due southeast.

Correlation with Rainfall

In order to study the long range trend of the water levels as a function of rainfall (Fig. VI), hydrographs of wells at Bijauli and Sankara have been selected to provide a groundwater behaviour in the area.

The correlation of the groundwater levels with rainfall were made with available data from 1982 to 1986. The hydrograph of Bijauli block represents the upland area, whereas the hydrograph of the well of Sankara is representative of the upland margin as well as the low valleys of Ganga. A critical study of

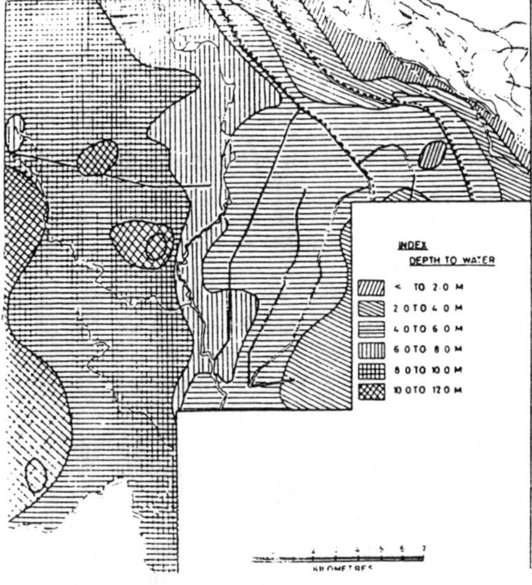

Figure 3

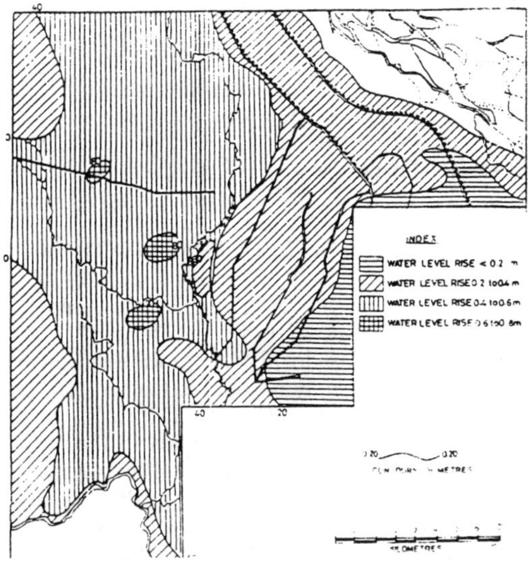

Figure 5

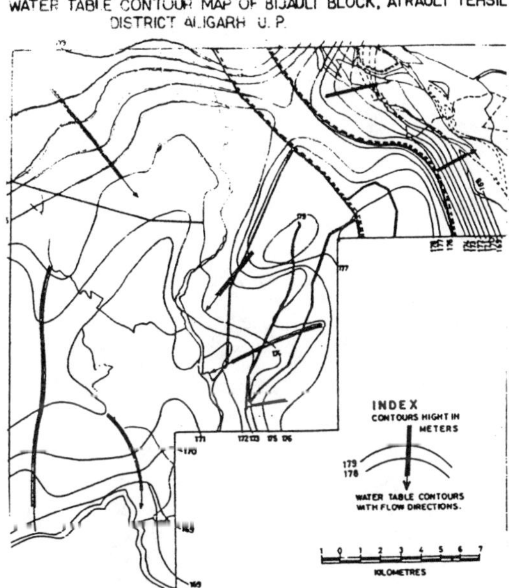

Figure 4

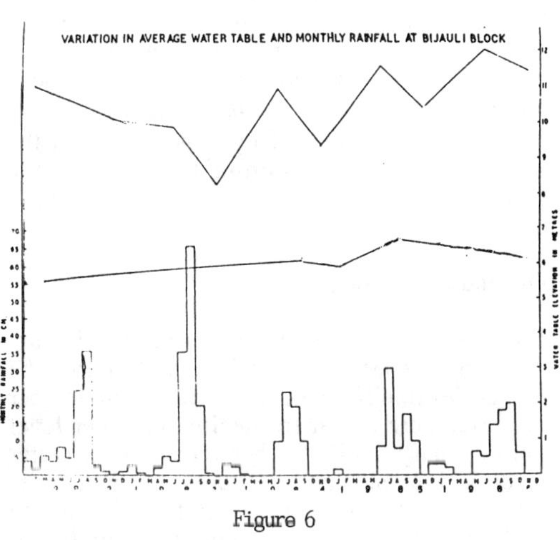

Figure 6

the hydrographs indicate that the response of the water levels to rainfall and drought is reasonably quick. The axcent of the water level is also greatly affected by intensity and duration of rainfall.

The study further shows that the response of rainfall at Bijauli is quicker as compared to Sankara. This is because, in the upland area like Bijauli, the rainfall is the only source of groundwater recharge. However, the maximum decline in the water table was recorded during May and June but the Sankara hydrograph shows a less declining water level than observed in Bijauli. This is because the hydrograph well at Sankara is situated at the upland margin between the lower Ganga canal and the River Ganga. The shallow aquifers below the canal bed are constantly being recharged by the seepage from the lower Ganga canal and, moreover, the movement of groundwater is towards the Ganga river, so an effect of scanty and sporadic rainfall or drought are compensated for excess recharge, although perhaps some of the subdued response to rainfall can be attributed to the clayey character of surfacial materials.

CONCLUSION AND RECOMMENDATION

Groundwater in the area occurs, in part, in shallow, unconfined aquifers. These shallow aquifers extend as deeply as 50 metres below ground level and may be pumped at the rate of 30 cu m/h to 50 cu m/h, with a drawdown of 3.0 m to 4.5 m. Of 250 tubewells in the area, about 235 are installed in the shallow aquifers and as a result of these aquifers being overdrawn, eventually an increasing shortage of irrigation water will have a severe negative effect on harvests of the area. The deeper aquifers occur 50 to 143 meters below ground level and may be pumped at rates of 50 to 227 cu m/h, with a drawdown of 1.97 to 11.72 metres. These deeper aquifers are confined, so the groundwater supply in these aquifers is more stable. It is also recommended from overall analyses of hydrogeology that the unlined canals be commissioned or the groundwater be extracted from the deeper aquifers only.

Groundwater Behaviour

It has been concluded that the water level starts rising by the last week of June and attains the highest level in November: from mid November onwards there is a sharp decline in water level until January, while further recession in water level is slow, indicating of natural groundwater discharge through steady state sub-surface outflow in harmony with regional groundwater movement.

REFERENCES

Bhaweja, B.K. & Karanth, K.R., 1980, Groundwater Recharge Estimation in India Technical Series, H. Bull. 2, C.G.W.B.

Burrard, S.G., 1915, Origin of Indo-Gangetic Trough, Commonly called Himalayas Foredeep, Proc. Roy. Soc., London, 91 A, pp 220-238.

Dutt, D.K., 1969, Hydrogeology of Aligarh District, Uttar Pardesh, Indian Minerals, Vol. 23 (2), pp 1-14.

Hassan, N., Jouhari, P.K. & Hassan F., 1982, Intensive Geohydrological Investigation Block, District, Badayun, U.P., pp 112-114.

Krishnan, M.S., 1982, Geology of India and Burma, Sixth Edition, pp 489-536.

Nevelle, H.R., Gazetter of Aligarh District, Vol, VI, pp 218-222.

Oldham, R.D., 1917, The Structure of Himalayas and Gangetic Plain Mem. Geol. Survey. Ind., V.42 (2), pp 16.

Pascoe , E .H., 1984 , A Manual of India and Burma. Govt. of India Press Calcutta, pp 21-22.

Pathak, D.D., 1985 , Hydrogeology and Ground Water Potential of Uttar Pardesh.

Pathak, D.D., 1986, Ground Water in India and its Scope for Further Development, Current Trend in Geology, Vol. VII-VIII. pp 573-577.

Rao, K.V. & Raju, T.S., Quantitative Studies on Ground Water, Storage, Recharge, Draft and Safe Yield in Atrauli Area, Aligarh District, Uttar Pardesh, Indian Geohydrology, Vol . I (1), pp 46-67.

Environmental Management, Geo-Water & Engineering Aspects, Chowdhury & Sivakumar (eds)
 ISBN 90 5410 099 0

Effect of water harvesting structure on groundwater recharge – A case study

H.K.Mittal
Directorate of Extension Education, Udaipur, Rajasthan, India

Jaspal Singh
College of Technology & Agricultural Engineering, Udaipur, Rajasthan, India

ABSTRACT: Recurring droughts are causing serious damage to the production system of the country. Hence, stress was given to enhance rainwater in small water harvesting structures called anicut with the objective to recharge downstream wells and to provide life saving irrigation to crops. Study reveals that these structures have successfully proved their functional utility. The higher average groundwater recharge was observed which ranged from 1.15 to 1.26 times more in downstream wells as compared to distant or upstream wells.

1 INTRODUCTION

There is considerable scope to utilize the rainfall for productive purposes in the areas underlain by aquifers of good quality water. In view of dwindling water suppplies and declining water table in many areas greater emphasis is now needed on managing rainfall and its utilization for irrigation. The rain water is enhanced by (1)Insitu conservation in soil profile(2)conservation in small reservoirs or anicuts(3) artificial groundwater recharge (Greb et al 1967), (Subbaiha 1991).Keeping in view the above said strategies, three water harvesting structures (anicuts) were constructed in Agad watershed of Udaipur district in Rajasthan State (India) for increasing groundwater recharge to existing wells. The area is characterized by subhumid climate with an average annual rainfall of 700 mm. Furthermore,anicut is stone structure with a weir constructed across a stream or river with more or less perennial flow. It not only reduces erosive velocity of runoff but also prevents the gullies from further erosion.

2 MATERIALS AND METHODS

2.1 Recovery test for estimating transmissivity

To study the effect of anicut on recuperation index i.e. transmissivity(T), one well on each side of anicut was selected and recovery data were noted in terms of residual draw down s'. This recovery method (1935) was used for analysis for recovery data. For small values of 'u' (dimensionless time, $r^2S/4tT$), the relationship used was

$$s' = 2.3Q/4\pi T \times \log t/t', \quad (1)$$

where t is the time since pumping started and t' is the time since pumping was stopped (t=tp + t'), tp being time of pumping, Q is constant rate of pumping and 'T' being transmissivity. The solution of 1 is obtained by plotting s' on arithmetic and t/t' on logarithm scale. Straight lines were fitted for early and late recovery & average periods separately.The slope of the line in each case was equal to $2.3Q/4\pi T$ and also equal to the change in $\Delta s'$ in s' per unit log cycle. Thus transmissivity was calculated as

$$T = 2.3Q/4\pi\Delta s' \quad (2)$$

2.2 Groundwater storage changes

This study deals with assessment of groundwater recharge due to percipitation using water fluctuation method. Water levels of the respective wells of the area were measured and precipitation data from a nearby meteorological station was obtained. Four wells were selected on the downstream side of the anicut and two wells on the upstream side and away from anicut. The groundwater recharge of these wells was estimated and compared to study the effect of anicut construction.The water levels were recorded in monsoon season i.e. July to November, 1990.

2.3 Effect of water harvesting structure on groundwater recharge

The rise of water level can be expressed as

$$h = Pi/Sy \qquad ------ \qquad (3)$$

where h = rise of water level cm, Pi =portion of precipitation that percolates to the water table i.e., recharge to groundwater cm, Sy=specific yield percent.

The rise of water level was obtained from the data,specific yield was calculated from the relation $h = hoe^{-\alpha t}$. This is analogous to Boussinesq's equation $Q = Qoe^{-\alpha t}$ in which 'α' is governed by hydrogeological characteristics of the basin. Analogous to this, α may be considered to give the value of specific yield of a particular hydrogeological formation in the relation $h = hoe^{-\alpha t}$ Here, h= water in the well at any time t, ho = initial water level in the well, t= time interval of water level rise from ho to h. Thus having calculated average specific yield for each well Pi is calculated as Pi = h x Sy. Further recharge in terms of percentage of rainfall at different well locations was also estimated and compared. These recharge values can be suitably adjusted (weighted) for a normal rainfall year also.

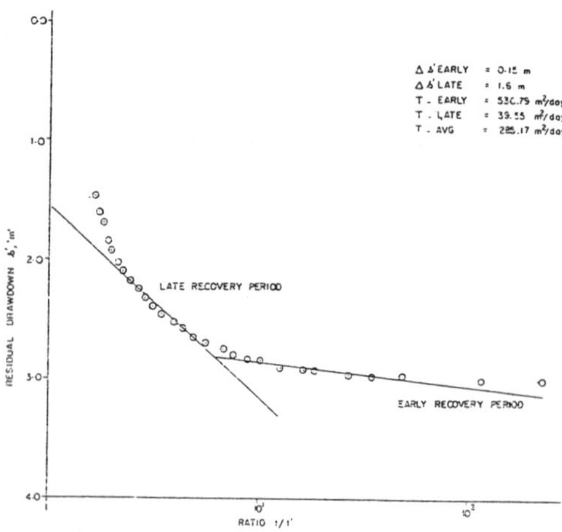

Fig. 1. Recovery in the well located near Anicut in Agad W/S

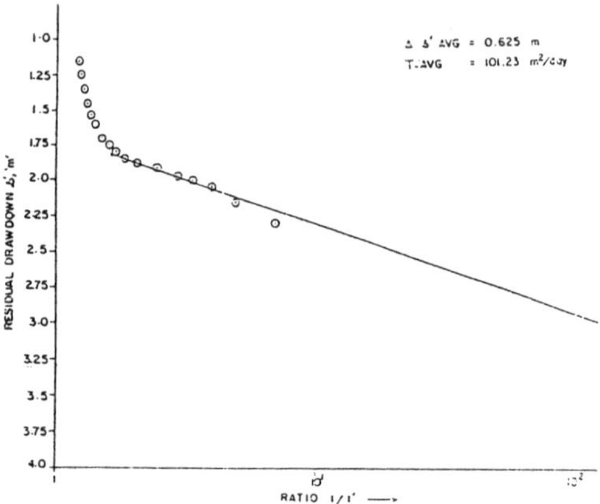

Fig. 2. Recovery in the well located away from Anicut in Agad W/S

3 RESULTS AND DISCUSSION:

From the data collected at the site and from experiments the different aspects of groundwater recharge due to construction of anicut was studied and discussed herein.

3.1 Transmissivity

Fig 1 and 2 shows that average transmissivity of the area was 285.17m2 /day and 101.23 m2/day in wells located near and away from anicut respectively. This shows that transmissivity or recuperation index of the well under recharge of anicut is 2.81 times as compared to the well located on upstream side of anicut.As the geological formation is quartzite the variation in transmissivity can safely be said to be due to construction of anicut, the higher values being obtained in the well located at downstream of the structure.

3.2 Recuperation rate

The recovery rate for well near the anicut was 0.141 m ^{3}/min whereas the well on the upstream side of the anicut has registered the recovery rate of 0.087 m3/min (Table 1).This shows that recuperation rate was 62.06 per cent higher in the well near the anicut as compared with well away from anicut. This confirms the hypothesis that a well downstream of anicut will attain its original static water level in lesser time as compared to upstream well after pumping is stopped.

3.3 Water level fluctuations

In Agad watershed total precipitation during monsoon months was 899.0 mm.Due to this net rise was 5.66 m, 6.08 m,5.9m and 5.65 m in the well no.1 to 4 located near anicut. In well no. 5 & 6 the net rise was 4.7 m and 4.9 m(Fig.3).This shows that the net rise was 22.21 to 29.36 per cent more in wells under influence of anicut as compared to well no.5 and 15.30 per cent to 24.08 per cent more as compared to well No.6. Further, the available water column upto December 1990 was 63.0 to 89.6 per cent more in downstream wells as compared to well No.5 and 35.7 per cent to 57.14 per cent more as compared to well No.6. Thus it can be concluded that higher rise in water column during monsoon period and more availability of water column depths after monsoon period in well no.1 to 4 is due to construction of anicut.

Table 1. Response of wells under the influence of anicut(Quartzite formation)

Particulars	Discharge qp(m^3/day)	Pumping duration tp (min)	Draw_down 'm'	T avg. m^2/day	Recovery rate m^3/min
Well near anicut	345.60	225.0	3.01	285.17	0.141
Well away from anicut	345.60	60.0	2.50	101.23	0.087

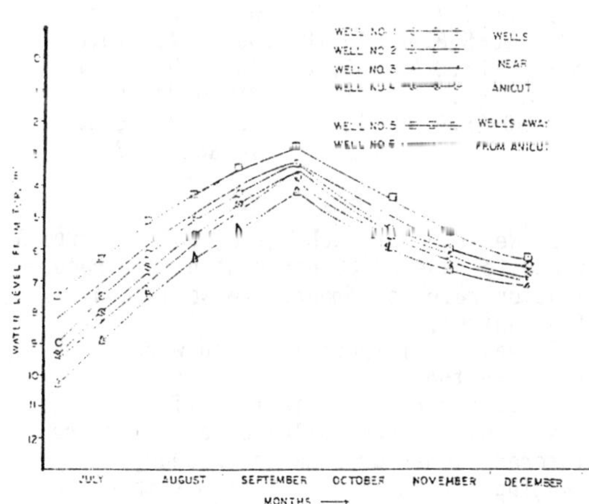

Fig.3. Water level fluctuations (1990) in selected wells of Agad W/S

3.4 Estimation of specific yield

Average values of specific yield in the study area was estimated to be 1.20×10^{-2}, 1.07×10^{-2}, 1.16×10^{-2}, 1.12×10^{-2} in well no.1 to 4, and 1.21×10^{-2} and 1.10×10^{-2} in well no.5 and 6 respectively (table 2 &3). The variation in the values must be because of variation information characteristics i.e. fracture intensity, hardness etc. The values of specific yield were comparable with the values reported by NABARD (1964) for quartzite formation.

Table 2. Estimation of specific yield of wells near anicut

S.No.	ho (m)	h (m)	t (days)	% α	Avg.specific yield
Well No.1					
1.	9.0	7.50	15	1.20	
2.	7.50	6.14	16	1.25	
3.	6.14	5.12	16	1.13	1.20% $=1.2\times10^{-2}$
4.	5.12	4.24	15	1.25	
5.	4.24	3.34	20	1.19	
Well No.2					
1.	10.30	8.90	15	0.97	
2.	8.90	7.30	16	1.23	
3.	7.30	6.30	16	0.92	1.07% $=1.07\times10^{-2}$
4.	6.30	5.40	15	1.02	
5.	5.40	4.22	20	1.23	
Well No.3					
1.	9.5	8.30	15	0.90	
2.	8.30	7.00	16	1.06	
3.	7.00	5.80	16	1.17	1.16% $=1.16\times10^{-2}$
4.	5.80	4.80	15	1.26	
5.	4.80	3.60	20	1.43	
Well No.4					
1.	9.40	8.00	15	1.07	
2.	8.00	6.60	16	1.20	
3.	6.60	5.55	16	1.08	1.12% $=1.12\times10^{-2}$
4.	5.55	4.55	15	1.32	
5.	4.55	3.75	20	0.96	

Table 3. Estimation of specific yield of wells away from anicut

S.No.	ho (m) A	h (m) B	t (days) C	% α D	Avg. specific yield E
Well No.5					
1	7.50	6.30	15	1.16	
2.	6.30	5.12	16	1.29	
3.	5.12	4.30	16	1.16	1.21%
4.	4.30	3.45	15	1.46	$=1.21\times10^{-2}$
5.	3.45	2.80	20	1.04	
Well No.6					
1.	8.20	7.20	15	0.86	
2.	7.20	5.85	16	1.29	
3.	5.85	4.90	16	1.10	1.10%
4.	4.90	4.05	15	1.27	$=1.10\times10^{-2}$
5.	4.05	3.30	20	1.02	

3.5 Effect of water harvesting structure on groundwater recharge

In Agad watershed the maximum recharge was estimated to be 12.34, 12.50, 13.78 and 11.38 per cent in well No.1 to 4 respectively which were located near anicut (Table 4). The well no.5 and 6 located away from anicut registered maximum recharge as 10.42 and 10.83 per cent respectively (Table 4). The maximum recharge per cent for the wells near the anicut was more as compared to wells away from anicut. Further, it can be estimated from these tables that average recharge per cent was registered minimum 1.15 times to maximum 1.21 times in well No. 1 to 4 as compared to well No.5 and 1.16 times to 1.26 times as compared to well no.6 respectively. As higher recharge has been observed in the wells on downstream side of the structure, it can be inferred that it is due to construction of anicut near these wells. Hence, it can be concluded that water harvesting structure has certainly influence on increasing groundwater recharge.

Table 4. Groundwater recharge as per cent of rainfall (near and away from anicut)

Period	Rainfall (cm)	'h' cm	Pi cm	% R
Well No.1				
5.7.90-20.7.90	14.70	150	1.81	12.34
21.7.90-5.8.90	13.70	136	1.63	11.91
6.8.90-21.8.90	13.70	102	1.22	8.93
22.8.90-5.9.90	16.00	88	1.05	6.60
6.9.90-25.9.90	10.00	90	1.08	10.69
		Average		10.09
Well No.2				
5.7.90-20.7.90	14.70	140	1.49	10.19
21.7.90-5.8.90	13.70	160	1.71	12.49
6.8.90-21.8.90	13.70	100	1.07	7.81
22.8.90-5.9.90	16.00	90	0.96	6.01
6.9.90-25.9.90	10.00	118	1.26	12.50
		Average		9.80
WellNo.3				
5.7.90-20.7.90	14.70	120	1.39	9.46
21.7.90-5.8.90	13.70	130	1.50	11.00
6.8.90-21.8.90	13.70	120	1.39	10.16
22.8.90-5.9.90	16.00	100	1.16	7.25
6.9.90-25.9.90	10.10	120	1.39	13.78
		Average		10.33
Well No.4				
5.7.90-20.7.90	14.70	140	1.56	10.66
21.7.90-5.8.90	13.70	140	1.56	11.38
6.8.90-21.8.90	13.70	105	1.17	8.58
22.8.90-5.9.90	16.00	100	1.12	7.00
6.9.90-25.9.90	10.10	80	0.89	8.87
		Average		9.3
Well No.5				
5.7.90-20.7.90	14.70	120	1.45	9.87
21.7.90-5.8.90	13.70	118	1.42	10.42
6.8.90-21.8.90	13.70	82	0.99	7.24
22.8.90-5.9.90	16.00	85	1.02	6.42
6.9.90-25.9.90	10.10	65	0.78	7.78
		Average		8.34
Well No.6				
5.7.90-20.7.90	14.70	100	1.10	7.48
21.7.90-5.8.90	13.70	135	1.48	10.83
6.8.90-21.8.90	13.70	95	1.04	7.62
22.8.90-5.9.90	16.00	85	0.93	5.84
6.9.90-25.9.90	10.10	75	0.82	8.16
		Average		7.98

4 CONCLUSIONS

From the above study following conclusions can be drawn.

1. The value of transmissivity was found higher in the case of a well located downstream of the anicut as compared to upstream or distant wells.

2. Wells under recharge of anicut were found to have 62.06 per cent higher recuperation rate as compared to wells away from anicut.

3. Values of specific yield were found in the range of 1.07×10^{-2} to 1.21×10^{-2} for quartzite formation.

4. Wells under influence of anicut registered higher groundwater recharge ranging from 1.15 to 1.26 times than the wells located away from anicut.

5. Finally, it is concluded that it is

useful to construct such anicuts for protecting crops from the vagaries of monsoon.

REFERENCES

Greb, B.W., D.E.Smika & A.L.Black 1967. Effect of straw mulch rates on soil water storage during summer fallow in great plains.Soil Science Society of America Proceeding. 31:536-559.

Mittal, H.K. 1991.Evaluatory study on resource development in Thakarda-I and Agad watersheds. An Unpublished M.E.Thesis, Coll.of Tech.& Ag.Engg. Udaipur:148-161.

NABARD 1984. Norms for groundwater assessment-based upon report of the groundwater estimation committee: 1-13.

Subbaiah, R. 1991. Decision support for managing rainwater-A case study, Indian Journal of Agricultural Engineering. 1(1):59-65.

Environmental Management, Geo-Water & Engineering Aspects, Chowdhury & Sivakumar (eds)
© 1993 Balkema, Rotterdam. ISBN 90 5410 099 0

Large-scale catchment simulation using the MIKE-SHE model: 2. Modelling the Berriquin irrigation district

J.F. Punthakey, R. Cooke & N.M. Somaratne
NSW Department of Water Resources, Sydney, N.S.W., Australia

R.S. Carr
Lawson and Treloar Pty Ltd, Sydney, N.S.W., Australia

K.K. Watson
University of New South Wales, Kensington, N.S.W., Australia

ABSTRACT: Groundwater recharge from rainfall and from excess irrigation water along with inadequate drainage has resulted in rising watertables and land salinisation problems in the Berriquin Irrigation District (BID). By 1990 approximately 91,300Ha of land in the BID had watertables within 2m of the surface. The MIKE-SHE model, a physically-based, distributed modelling system will be used to assess the impact of irrigation and water logging against a number of on-farm options and proposed structural changes within the irrigation district. The model domain uses a uniform grid spacing of 2,000m in each direction to cover an area of approximately 3,200km^2. The four process components of SHE utilised for modelling the BID were interception/evapotranspiration, overland and channel flow, unsaturated flow, and saturated flow.

1 INTRODUCTION

Irrigation first commenced in the Berriquin Irrigation District (BID) in the 1939-40 season with some 3,600Ha being irrigated; this had increased to 150,000Ha by 1971-72. During the last two decades the irrigated area has fluctuated between 110,000 and 140,000Ha, with the water delivered to farms being approximately 800,000 Ml in most seasons. The study area is shown in Figure 1.

The flat topography of the BID allows water to be supplied through a system of unlined channels which are operated and maintained by the NSW Department of Water Resources (DWR). Water can be drained or discharged from the supply channels into a number of 'escape' creeks or drainage channels which carry the water to major creeks and rivers outside of the BID. Natural surface drainage is generally poor because of the flat terrain and the numerous road and supply channel embankments that traverse the area. Rainfall and excess irrigation water may cause accumulation of water at the soil surface, resulting in waterlogging for long periods particular during wet winters.

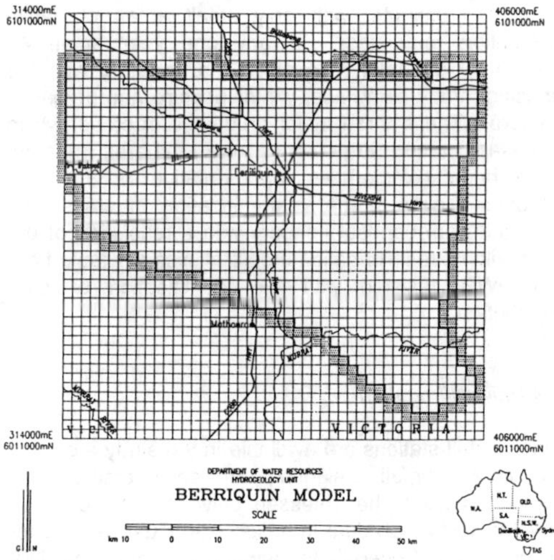

Figure 1: Study area

Watertables in the BID have risen inexorably over the years such that in 1990 some 91,300Ha of land had watertables within two metres of the soil surface. High watertables can have a detrimental affect on agricultural production through salinisation of the land surface, and degradation of the regional water resource through increased seepage on the one hand, or surface runoff on the other, of high salinity water.

To ensure sustainability of irrigation, improved management practices need to be implemented to control rising watertables and any resulting waterlogging, whilst minimising negative impacts on downstream water users. The MIKE-SHE model will be used to evaluate the effects of several possible management strategies that are being considered for the BID. The types of strategies under consideration include the provision of surface and sub-surface drainage, changes in on-farm practices, upgrading of the supply infrastructure, and changes in institutional arrangements to bring about changes in water usage. A detailed description of the MIKE-SHE model is given in (Carr et. al. 1993). The application of the MIKE-SHE model in a water resource management, planning and policy role will also be evaluated in a future study.

The specific objective of this present study is the modelling of the BID using the MIKE-SHE model to analyse surface and sub-surface flow processes in order to estimate the quantities of water and salt that would be discharged from the district under a proposed drainage system.

2 SITE DESCRIPTION

2.1 Site

The BID is the largest of four irrigation districts located north of the Murray river in the vicinity of Deniliquin in south-west New South Wales and covers an area of 320,000Ha. Average annual rainfall is 400mm, while the annual potential evaporation is 1800mm.

2.2 Hydrogeology

The aquifer system modelled includes the upper 30m of the Shepparton formation (Wolley & Williams 1977) which includes sand lenses associated with prior stream deposition. Below 30m to a depth of 95m the formation consists primarily of clays. The Shepparton formation is underlain by the Calivil sands which are - between 30 to 50m in thickness. These sands are in turn underlain by the Olney formation overlying bedrock. Depth to bedrock changes from 325m below ground level in the western part of the irrigation area, becoming progressively shallower in a south-westerly direction until outcropping becomes evident east of Tocumwal.

2.3 Soils

(Smith 1945; Churchwood & Flint 1956) have mapped soil groups covering most of the BID. The survey by (Smith 1945) mapped the soils in the area using 10 broad soil groupings ranging from the heavy clays of the Riverina association, through the sandy clay loams of the Cobram-Katunga association to the Sandmount sands. The survey was carried out on a reconnaissance basis with profiles examined every 0.5 to 1km. Another survey by (Johnston 1950) defined 14 soil groups over a limited area of 5,000Ha in the southern part of the district.

3 ESTIMATION OF DATA FOR THE MIKE-SHE MODEL

The flexibility of SHE (Abbott et. al. 1986a & 1986b) makes it possible, with one modelling system to perform predictions for a wide range of hydrological problems. A total modelling system for the BID is required as the impact of irrigation and waterlogging has to be assessed against a number of on-farm options and structural changes within the irrigation district. In addition to the surface drains disposing of the excess irrigation runoff the entire district has to limit the total salt exported in order to minimise any adverse impact on downstream users.

The deterministic nature of the model and the broad range of hydrologic processes described by it imply the availability of extensive data sets. This presents a planning challenge in relation to data procurement, particularly in relation to future usage of the model, and creative studies to provide adequate input data where field data is limited.

The four process components of SHE utilised for modelling the BID are: interception/evapotranspiration; overland and channel flow; unsaturated flow; and saturated flow. An additional component, FRAME, organises the input of catchment and meteorological data (Danish Hydraulic Institute 1991). The data required for each grid cell is topography, rainfall with climate station description, vegetation types, soil characteristics and depths, and the location of impermeable beds.

3.1 Model Grid and Boundary Conditions

The saturated zone component of the model was developed for the BID using a 2x2 km grid spacing. The groundwater system was conceptualised as a single layer unconfined aquifer of variable thickness, underlain by an impermeable base. Zero flux boundary conditions were specified for the eastern model boundary where elevated bedrock is encountered. The northern boundary follows Billabong creek, and the southern boundary is prescribed by Tuppal creek and the Murray river, both being fixed head boundaries. The western boundary has specified flux boundary conditions where the groundwater flows from the BID into the adjacent Denimein irrigation area.

3.2 Landuse/Vegetation

Within the BID introduced agricultural crops and pastures dominate, with only small areas of native vegetation remaining. In any one year about 40 percent of the total area of BID is irrigated, with the principal crops being rice, winter pasture and perennial pasture. These three major groupings of crops, which represent around 90 percent of the total area irrigated, have been selected as representative of the total irrigation regime. The yearly location of crops was obtained from Landsat imagery with a resolution of 0.5 x 0.5 kilometres. Irrigation crop statistics available from the DWR assisted with the identification of crop distribution.

3.3 Rainfall/Irrigation

Nine rainfall stations are available in the study area and the areal rainfall weighting for each station was completed using the Thiessen polygon method. The DWR collects individual irrigation water usage data for accounting and operational purposes. From such data a sample pattern of irrigation applications for individual crops has been obtained since 1975. For the preliminary simulation the median irrigation application rate only has been used. Sensitivity tests will be undertaken in due course to evaluate the choice of application rate.

3.4 Supply and Drainage Channels

Extensive survey and design information are available for all the channels in the area. Due to the unlined nature of the supply channels, the channel flow component of the model requires data on bed seepage rate, bed layer thickness, and the characteristics of the underlying soils.

3.5 Evapotranspiration

Evapotranspiration was estimated independently of the MIKE-SHE model using a quasi two-layered resistance energy balance model (Punthakey 1988; Punthakey & Gurner 1986). The model is based on a modified Penman-Monteith combination method which incorporates aerodynamic and surface resistance as controls on the evaporation process, and utilises a numerical procedure to evaluate the energy balance of a composite crop-soil surface. The ET model requires hourly or daily averages of solar radiation, air temperature, wet bulb temperature and wind speed. Estimates of emissivity, albedo, crop height, and leaf area index also need to be specified.

The landuse in the BID was divided into three major categories; irrigated rice, annual pasture, and perennial pasture. For each landuse, crop height and leaf area index were specified temporally, for input to the evapotranspiration model. The model was run using daily time steps from July 1985 to June 1990. The estimated evapotranspiration from the resistance energy balance model provided the input for the evapotranspiration component of the model.

3.6 Unsaturated Zone

The unsaturated zone forms a crucial part of the BID and its effective modelling. This zone extends from the ground surface to the water table, with the hydrologic characteristics of the soils in this zone determining the surface infiltration rates and the recharge volumes to the groundwater system. The soil profile in this zone, is in general quite variable and some simplification of the profile found at different locations within the model domain is necessary if the CPU time is to be kept within reasonable bounds. In addition, in the upper part of the profile the root zone has a major impact on the profile as water is removed by evapotranspiration and replenished by rainfall or irrigation.

In order to assemble the data requirements for the unsaturated component of the SHE model, three sources of data were used. In the first instance, a comprehensive data set for three soils in the NW of the BID were available (Bridge 1968). Secondly, soil moisture characteristic curves were estimated using a model (van Genuethen 1980) which generated this characteristic curve from information on the particle size distribution of a soil. Thirdly, drillers logs for 685 bores in the model area were used to describe approximately the soil type for the upper 2m of the soil profile to obtain general estimates of saturated hydraulic conductivity across the area. Although caution must be exercised in relation to the absolute values obtained, the comparative information is valuable in conjunction with accurate local values.

For each soil category specified in the defined area it was necessary to provide detailed data on the soil water retention curve and the hydraulic conductivity-water content relationship together with other relevant soil parameters such as field capacity.

3.7 Estimation of Saturated Hydraulic Conductivity

The aquifer parameters, horizontal hydraulic conductivity K, together with the elevation of the bottom of the aquifer were obtained from drillers logs. For each bore the drillers logs were used to estimate K values using a data base of 5000 bore log descriptions commonly in use by drillers. About 370 bores with complete drillers logs were found within the model domain. The data was triangulated and grid-average values obtained for each of the model cells. Figure 2 shows low K values to the NW and NE part of the model grid. High K values are found in the SE section of the grid which is characterised by sand dunes, and a large zone of high K values indicating permeable sediments in the central portion of the model domain. The permeable sediments follow a East to N-West trend which corresponds well with prior stream activity.

4 SIMULATIONS USING THE MIKE-SHE MODEL

4.1 Water levels

Water tables are rising in both dryland and irrigated areas due to the clearing of deeper rooted native vegetation and their replacement with shallower rooted agricultural crops. Water use by agricultural crops is generally less, consequently more rainfall passes through the root zone to the watertable. In addition, the average irrigation intensity in the BID is 2.4 Ml Ha^{-1}. This is equivalent to 60 percent of the average annual rainfall and further results in rising watertables. The rise in watertables is highest in wet years as the winter rainfall results in increased infiltration of water through the rootzone to the watertable.

In 1980 the BID had 15,000Ha with high watertables, i.e. within 2m of the surface. By July 1990 this had increased to 91,300Ha. Figure 3 shows the areas susceptible to water logging and salinisation due to watertables less then 2m from the surface.

Initial simulations were performed for a 2 year period from July 1989 through to June 1991. The July starting date enabled the model to start with a known equilibrium moisture content for the soil profile.

4.2 Unsaturated Zone

Figure 4 shows the estimated unsaturated zone profile for one soil category (Cobrum-Kutunga) during the two year simulation period. The 2m profile remains wet between July and September due to winter rainfall and low evapotranspiration (ET) rates. As the ET increases the profile tends to dry with the upper 0.15m drying out more rapidly, whereas the lower layers exhibit a pronounced time lag in profile drying.

During the winter months, rainfall results in rapid wetting of the soil profile down to 2m; this presents a high potential for water logging to occur during these months. During the summer, moisture contents range from 0.12 at the surface to close to saturated conditions at 2m depth. Figure 4 also shows a pronounced difference in potential and actual ET particularly in the summer months indicating limiting moisture conditions in the root zone.

5 CONCLUSIONS

The use of a single modelling system, the MIKE-SHE, to model the BID is presented. The data requirements for the major flow components modelled include landuse, rainfall/irrigation, supply and drainage channels, evapotranspiration, unsaturated zone properties, and aquifer parameters. Preliminary simulation results indicate the satisfactory performance of the model particularly in relation to the potential for waterlogging to occur in large areas of the BID during winter months as a result of high rainfall and inadequate drainage (Figure 5).

6 ACKNOWLEDGMENTS

The cooperation and assistance of NSW Department of Water Resources Water Distribution and Regional Staff, and the NSW Department of Agriculture Staff located at Finley and Deniliquin are gratefully acknowledged. Dr S A Prathapar, CSIRO Griffith is thanked for his assistance with the unsaturated zone modelling.

REFERENCES

Abbott, M.B., Bathurst, J.C., Cunge, J.A., O'Connell, P.E. & Rasmussen, J. 1986. An Introduction to the European Hydrological System - Systeme Hydrologique Europeen, 'SHE', 1: History and Philosophy of a physically-based Distributed Modelling System. *J. Hydrol., 87: 45-59.*

Abbott, M.B., Bathurst, J.C., Cunge, J.A., O'Connell, P.E. & Rasmussen, J. 1986. An Introduction to the European Hydrological System - Systeme Hydrologique Europeen, 'SHE', 2: Structure of a physically-based Distributed Modelling System. *J. Hydrol., 87: 61-77.*

Bridge, B.J. 1968: Soil Physical Properties as influenced by Gypsum. *University of Sydney, MSc. Thesis, 112 p.*

Carr, R.S., Punthakey, J.F., Cooke, R. & Storm, B.R. 1993: Large scale catchment simulation using the MIKE-SHE model: 1. Process simulation of an irrigation district. *Int. Conf. on Environ. Management.* Wollongong, Australia.

Churchwood, H.M., & Flint, S.M. 1956: Jenargo Extension of the Berriquin Irrigation District, New South Wales. *CSIRO, Soil and Land Use Series, No. 18.*

Danish Hydraulic Institute 1991. SHE - European Hydrological System Version 4.3 Users Guide and Technical reference. Horsholm, Denmark.

Johnston, E.J. 1950: The morphology and evolution of the soils of 'Pine Lodge' estate, New South Wales. *CSIRO, Bulletin, No. 259.*

Punthakey, J.F. 1988: Actual Evapotranspiration and Energy Balance for Irrigated Horticultural Crops in the Riverland of South Australia. *Dept. of Agric. South Australia, Tech. Paper No. 20. 134 p.*

Punthakey, J.F. & Gurner, E.K. 1986: Predicting Actual Evapotranspiration by Modelling the Crop Energy Balance. *Hydrology and Water Resources Symposium, Institution of Engineers Australia pp 33-37.*

Smith, R. 1945. Soils of the Berriquin Irrigation District, NSW. *CSIRO, Bulletin, No. 189.*

Van Geuchten, M.T. 1980: A closed-form equation for predicting the hydraulic conductivity of unsaturated soils. *Soil Sci Soc. Am. J. 44:892-898.*

Wolley, D.R. & Williams, R.M. 1977: Tertiary stratigraphy and hydrogeology of the eastern part of the Murray Basin, New South Wales. *Hydrogeological Report, No. 1977/6.*

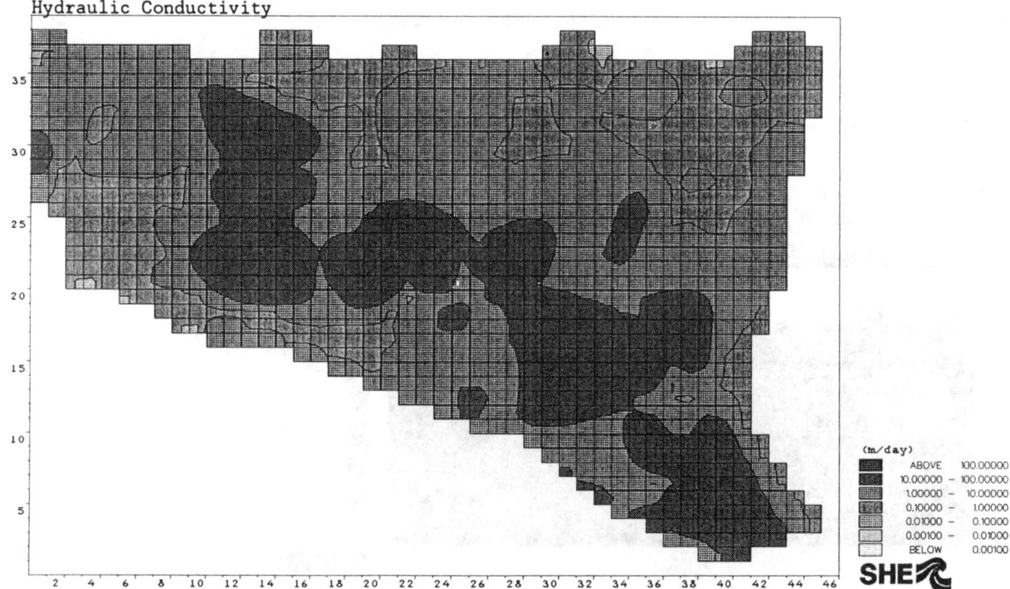

Figure 2: Hydraulic conductivity estimates.

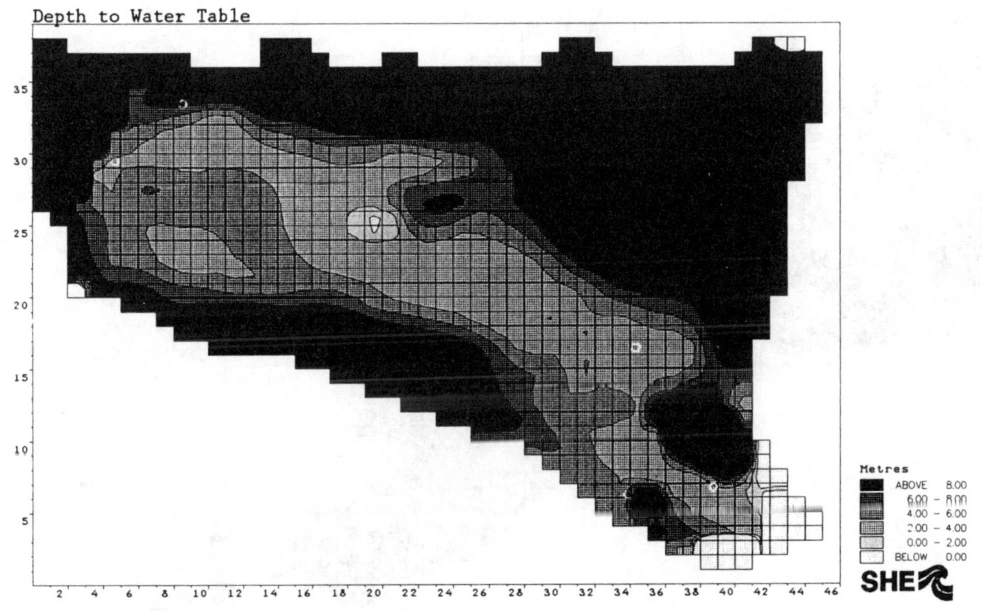

Figure 3: Depth to water table, July 1989.

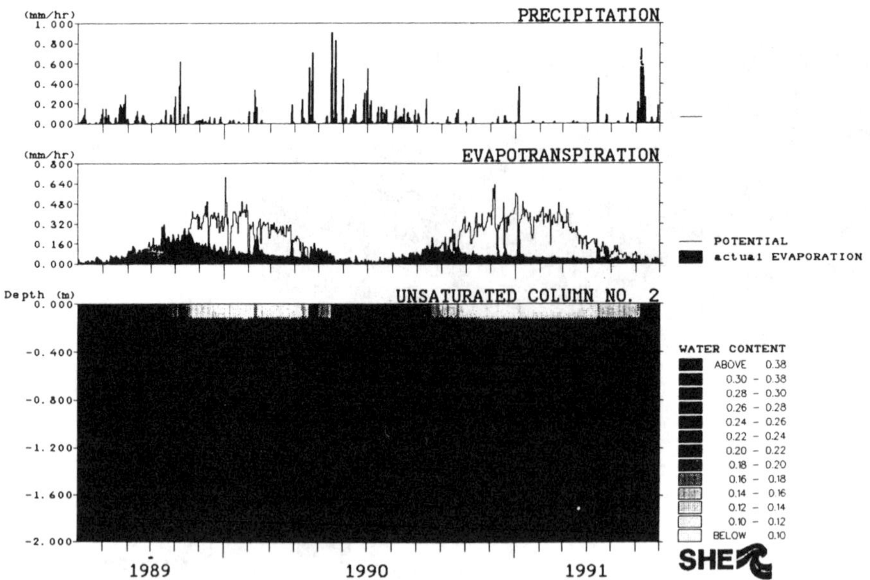

Figure 4: Unsaturated zone and soil surface dynamics.

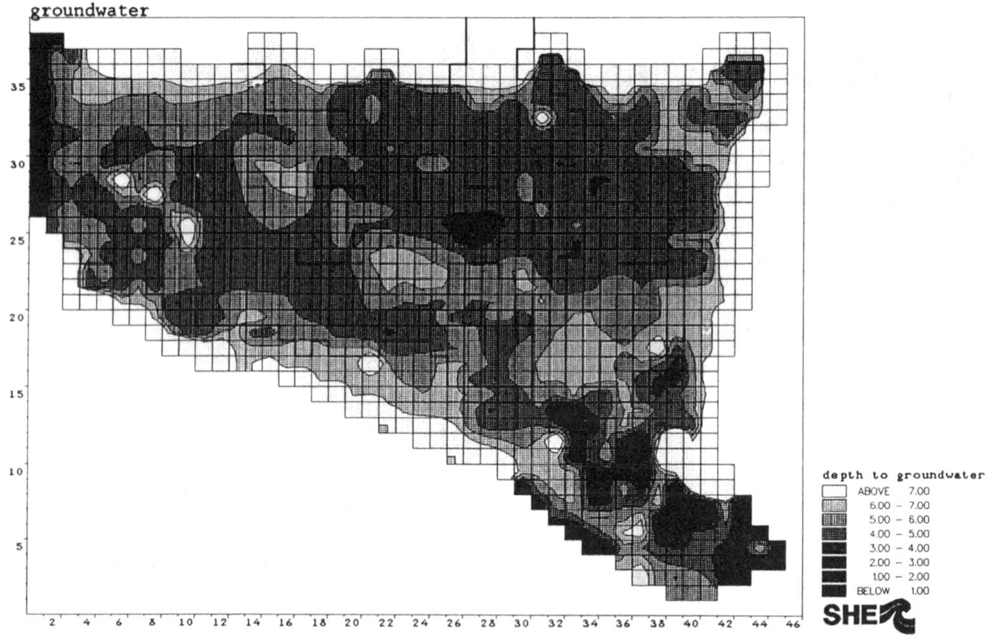

Figure 5: Depth to groundwater table after 2 years.

Environmental Management, Geo-Water & Engineering Aspects, Chowdhury & Sivakumar (eds)
 ISBN 90 5410 099 0

Modelling flood flows in Macquarie Rivulet using FESWMS-2DH

E.H.Rigby
Forbes Rigby Pty Ltd, Wollongong, N.S.W., Australia

ABSTRACT: A finite element based two dimensional hydraulic model (FESWMS-2DH) was established to model complex flood behaviour in the lower reaches of Macquarie Rivulet, above its point of discharge into Lake Illawarra. Using this model it was possible to directly evaluate the impact of proposed excavation and filling on flood behaviour in this sensitive wetland area. It is the author's conclusion that this model represents a significant improvement over other tools available for modelling two dimensional flood flows and may in some hydraulically complex situations provide the only means of solution currently available.

1 INTRODUCTION

Over the last two decades, the increasing power and ready availability of computers has spawned a vast collection of software for modelling the hydraulics of flood flows in channels and streams. In the early stages of development, such software was based on the simplified concept of steady uniform flow such as might be achieved in a laboratory flume.

With time these models evolved to include many of the features of real streams, but most still retained their original one dimensional (1D) structure. Over this same period several pseudo two dimensional models evolved based on '1D' links between nodes or cells to approximate the two dimensional flow domain. It is mostly this class of model that is used today, to model two dimensional flows.

This paper sets out the experiences of the author in establishing and applying a model, which solves the full St Venant equations for flow in two dimensions, to evaluate the impact of various flood plain filling options on flooding in Macquarie Rivulet, a small stream to the south of Sydney, Australia.

2 MACQUARIE RIVULET

The catchment of Macquarie Rivulet is located approximately 100 km south of Sydney, on the narrow coastal strip formed by the Illawarra escarpment to the west and Pacific Ocean to the east.

This 105km^2 catchment is predominantly rural in character although residential development is expanding rapidly in the lower one third of the catchment.

Macquarie Rivulet is typical of the many mountain streams traversing this coastal strip, with headwaters on the escarpment, falling steeply down the escarpment and over the foothills onto a low lying coastal plain. Like many streams in the region, Macquarie Rivulet discharges into a coastal lagoon (Lake Illawarra) before reaching the sea.

Macquarie Rivulet has a stream length of approximately 22.5 km with a fall of the order of 680 metres over this length. Some 450m of this fall occurs in the first kilometre of the stream's length.

Whilst the upper reaches are therefore quite steep with some spectacular waterfalls, the bed grade for most of the streams length is relatively flat (typically about 2m per km).

Where Macquarie Rivulet discharges into Lake Illawarra a substantial delta has formed, creating a complex wetland habitat around the delta and on the foreshores of the adjacent bays.

As with many streams in the Illawarra, conflict arises between the east to west flowing streams and north/south oriented road and rail network. Substantial impediment to flow in major floods is created by road and rail bridges, over Macquarie Rivulet near its outlet to Lake Illawarra.

The combination of these waterway constrictions and the meandering channel downstream of the bridges creates complex two dimensional flows in all events of sufficient magnitude to overtop the natural levees.

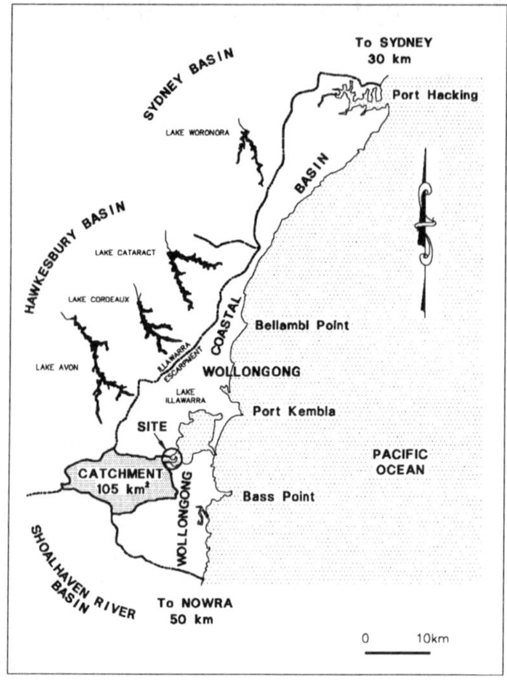

Fig. 1 Location plan

3 PROPOSED DEVELOPMENT

A mixed residential and recreational development has been proposed on land bounded by Macquarie Rivulet to the south, the Illawarra rail line to the west, Tallawarra Power Station ash ponds the north and Lake Illawarra to the east.

The conceptual layout for this development comprises residential development on higher land to the north of the site with a golf course on lower land, part of the northern flood plain of Macquarie Rivulet.

The relationship of the development site to the catchment and to other features of the area is reproduced in Figure 1.

In order to meet State Planning guidelines for development on flood prone land, this scheme required filling part of the northern flood plain to raise residential land above the designated (1% AEP) flood event.

Whilst a standard step (1D) backwater model had initially been used to establish predevelopment flood levels adjacent to the site in the designated flood, such a model, even with overbank flows separately modelled, could not be expected to adequately represent the impact of the excavation, filling and reshaping proposed, on flood flows in the vicinity of the site.

4 MODEL SELECTION

A Finite Element Surface Water Modelling System for two dimensional flows in a horizontal plane (FESWMS-2DH) was chosen to simulate the complex flow field in the lower reaches of Macquarie Rivulet, for a range of flood plain development options.

This model was developed for the US Federal Highways Administration by the Water Resources Division of the US Geological Survey to provide a means of simulating flows where natural processes and man made structures have created complex hydraulic conditions, that are difficult to evaluate using conventional methods. This modelling system is able to simulate complex flow conditions which are essentially two-dimensional in the horizontal plane. The FESWMS-2DH modelling system may be applied to a broad range of flow conditions such as flow at multiple opening bridge crossings, with or without pressure flow and/or overtopping, flow around islands, or flow over an irregular flood plain. Unsteady flow may be modelled by specifying varying boundary conditions at different points in time.

FESWMS-2DH solves the complete St Venant equations, where two equations describe the conservation of momentum and one equation describes the conservation of mass. Frictional losses may be modelled using the Chezy or Manning equation. Turbulence is modelled using the Boussinesq eddy viscosity concept. The value of the kinematic eddy viscosity may be determined as a function of the shear velocity and flow depth. Optionally the effects of wind stress and the Coriolis force may be included. A Galerkin finite element method is used to solve the resulting system of differential equations. The solution methodology is described by Lee, Froehlich, Gilbert, and Wiche (1983) and details of the model as implemented by the USGS are set out in the associated FESWMS User Manual, Froelich (1989).

Studies by Lee et al (1983), Shearman et al (1986) and Lee & Froehlich (1989) have shown that FESWMS-2DH is capable of simulating the significant features of flow over an irregular flood plain and that the model is able to reproduce the water surface elevation and velocity field near bridge openings more completely and more accurately than one-dimensional models such as HEC-2.

Given, in addition to the above capabilities, the models ability to accept input from an existing digital terrain model (DTM) of the study area and it's ability to display the pre and post development flow fields on screen or on a plotter, it was resolved to trial the model in this study.

5 MODEL CONSTRUCTION

Initially a detail survey of the flood plain, creek and adjacent bays was undertaken, to augment existing topographic data.

In conjunction with supplementary data digitised from 1:4000 contoured orthophoto mapping, this data was ingested into Autocad, filtered and exported as 3D faces to the 2D modelling package FESWMS using a small utility program developed by the author.

Particular care was taken to ensure that the shallow depressions and raised areas on the downstream flood plain were correctly represented in the model. Additional (smaller) elements were provided in areas where flow was expected to vary rapidly (viz in the vicinity of both bridge openings).

Consideration was given to the use of curved sided elements to better describe the geometry of the rivulet banks but not adopted as a large proportion of the flow in the event to be modelled occurred above bank level and across the adjacent flood plain.

The triangular and quadrilateral element layout of the model created by the above procedure is reproduced in Fig 2.

Contours of the surface represented by the final element layout were then output and

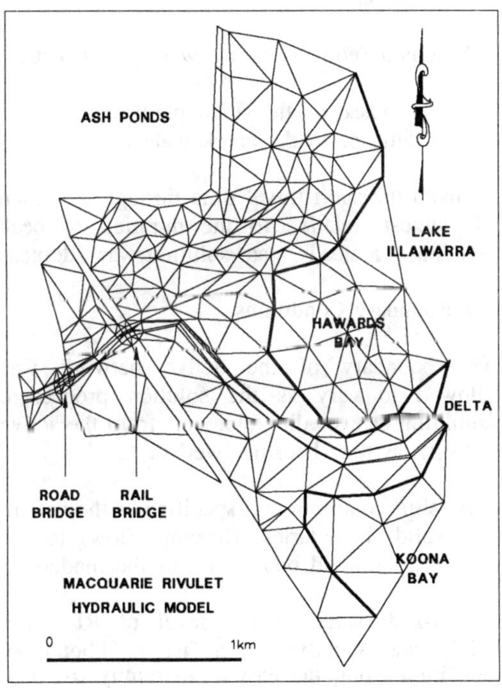

Fig. 2 Finite element network

compared with the original contours from the Autocad DTM to confirm that the surface had been faithfully recreated in the model.

6 MODEL APPLICATION

6.1 Model Parameters

At the time of the study no data was available for calibration of the model in the study area - hence all results were based on subjective assessment of surface friction and eddy loss parameters following guidelines presented by Chow (1959), Arcement & Schneider (1984) and Lee & Froelich (1989).

Bed friction (Mannings 'n') was specified as a function of depth such that for elements,

(a) In The Creek Waterway
for depth < 0.1m, n = 0.100
> 0.5m, n = 0.035
with n varying linearly between these limits

(b) On The Flood Plain
for depth < 0.1m, n = 0.100
> 0.5m, n = 0.055
with n varying linearly between these limits

Kinematic eddy viscosity was specified as,

$2m^2$/sec on the flood plain and
$5m^2$/sec in the creek waterway

Given the short travel time through the reach of interest, relative to the duration of peak flooding - a steady state solution was adopted.

6.2 Boundary Conditions

The boundary of the active network was allowed to vary as the solution proceeded, admitting or excluding elements from the active network as they became 'wet' or 'dry'.

A "slip" condition was specified as the default for solid boundaries (forcing flow to be tangential at solid boundaries of the model).

A fixed water surface level of RL 2.0m AHD was specified as a "natural" boundary condition along the eastern extremity (outflow boundary) of the model where the flood plain meets Lake Illawarra (corresponding to the 1% AEP flood level of Lake Illawarra).

A total flow of $1900m^3$/sec was specified as inflow to the model at the western extremity (inflow open boundary) of the model, upstream of the highway bridge (corresponding to the 1% AEP peak discharge in Macquarie Rivulet - Boyd et al (1989)).

6.3 Options Modelled

Since the final layout of the proposed development was not known, a series of options for filling of the flood plain were investigated.

These options ranged from what might be considered minimal to maximal filling of the triangular parcel of flood plain land, bounded by the rail embankment to the west, the ash pond embankment to the north, and Macquarie Rivulet to the south. The boundary of each of these fill zones is shown on fig 3.

Simulation of these fill options, ranging from Option A (minimal filling) to Option C

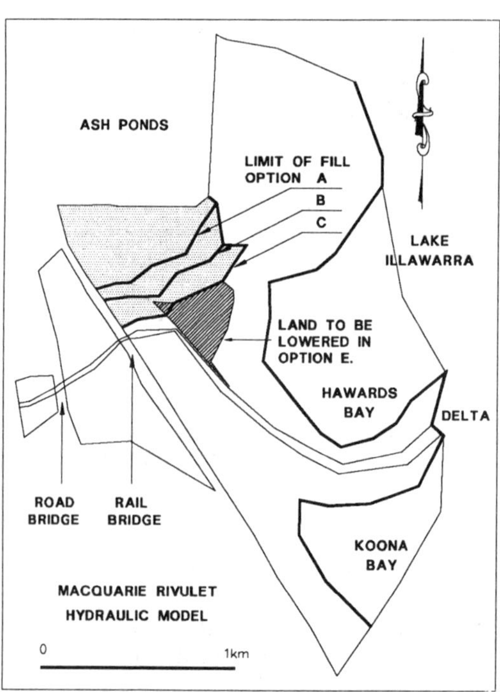

Fig. 3 Earthworks options

(maximal filling), was undertaken by changing the FESWMS "property type" of elements to be filled to zero - thereby excluding these elements from the active network before re-running the model.

In summary this range of fill options had the following impact on flood velocities and levels, relative to existing conditions.

Fill Option	Flood Level Impact	Velocity Impact
A (minimal)	Insignificant change	Insignificant change
B	+300 above rail bridge	Minimal change
C (maximal)	+1000 above rail bridge	Significant change

Since the impact of fill Options B or C were unacceptable and Option A did not provide sufficient flood free land to support the residential development, a further option (E) was investigated wherein filling to the line of Option B was combined with removal of high ground on the northern flood plain downstream of the bridge.

This elevated deposit of levee material adjacent to the bend in the stream appeared a likely candidate for improving the flow across the flood plain in the designated event and a source of fill for the development. Since final levels for the golf course had not been developed at this time, this initial review of Option E was based on the flood plain to the east of the bend, downstream of the rail bridge, being reduced to RL 1.5m AHD. This combined cut to fill (Option E) produced insignificant net change in flood level or velocity in the study area demonstrating that filling to the line of Option B with compensating cut on the flood plain could be undertaken without adversely impacting flood levels or velocity in the outfall reach.

A further option (D) involving filling to create a more gradual expansion of flow out onto the flood plain below the rail bridge was also modelled, but found to impact levels as much as option B. Since this option adversely impacted flood levels and did not accommodate residential development as well as option B, it was not explored further.

7 CONCLUSIONS

- o Whilst all two dimensional flood modelling exercises involve considerable data collection and model establishment time, the model chosen in this study has proven relatively easy to apply and amend.
- o Insights provided by the model into the distribution of flood levels and velocities across the flood plain, and how these change with encroachment of fill onto the floodplain have greatly assisted in the assessment of the impact of earthworks on flooding, flood safety, erosion and sediment transport.
- o With respect to the hydraulics of flood flows in Macquarie Rivulet, the model was able to demonstrate in a clear, quantified manner, the impact that filling would have on flood flows and to confirm that the particular combination of excavation and filling proposed (option E) would not adversely impact flooding.
- o It is the authors conclusion that FESWMS-2DH represents a significant advancement in tools available for modelling two dimensional flows with direct application in areas similar to that outlined in this paper. In more hydraulically complex circumstances involving non steady flows with flow transitions, mixed weir and pressure flows, wind shears and coriolis effects, FESWMS-2DH may provide the only means of solution currently available.

REFERENCES

Arcement G J & Schneider V R 1984, Guide for selecting manning's roughness coefficients for natural channels and flood plains, US Department of Transportation, Federal Highway Administration, Virginia, Report No FHWA-TS-84-204.

Boyd, M et al 1989. A comparison of design flood estimation methods. Proc. Fifth National Local Government Engineering Conference, Sydney, September 1989.

Chow V T, 1959, Open channel hydraulics. McGraw Hill, New York.

Froelich D C, 1989, Finite element surface water modelling system. Two dimensional flow in a horizontal plane. Users Manual, US Department of Transportation, Federal Highway Administration, Virginia, Report No FHWA-RD-88-177.

Gilbert J J, and Froehlich D C, 1987, Simulation of the effect of US highway 90 on Pearl River floods of April 1980 and April 1983 near Slidell, Louisiana: US Geological Survey Water Resources Investigations Report 85-4286, 52 p.

Gilbert J J, and Schuch-Kolben R E, 1987, Effects of proposed highway embankment modifications on water-surface elevations in the lower Pearl River flood plain near Slidell, Louisiana: US Geological Survey Water-Resources Investigations Report 86-4129, 40p.

Lee J K, and Bennett C S, III, 1981, A finite-element model study of the impact of the proposed I-326 crossing on flood stages of the Congaree River near Columbia, South Carolina: US Geological Survey Open-File Report 81-1194, 56p.

Lee J K & Froehlich D C 1989, Two-dimensional finite-element hydraulic modelling of bridge crossings, Research Report, US Department of Transportation, Federal Highway Administration, Virginia, Publication No FHWA-RD-88-146.

Lee J K, Froehlich D C, Gilbert J J, and Wiche G J, 1983 "A Two-dimensional finite-element model study of backwater and flow distribution at the I-10 crossing of the pearl river near Slidell, Louisiana", US Geological Survey Water Resources Investigations Report 82-4119, NSTL Station, Mississippi.

Shearman J O, Kirkby W H, Schneider V R, and Flippo H N, 1986, Bridge waterways analysis model: research report, Federal Highways Administration Report FHWA/RD-86/108, Reston, Virginia.

Environmental Management, Geo-Water & Engineering Aspects, Chowdhury & Sivakumar (eds)
 ISBN 90 5410 099 0

Spatial velocities in aquifers with 2-D behaviour

S.V.K.Sarma
Federal University of Paraiba, Campina Grande, Brazil

A.de S.Iêdo
Federal University of Pernambuco, Recife, Brazil

ABSTRACT: In free aquifers with irregular base and curved phreatic surface,the concept of radius of influence of well would be redundant. This situation arises in aquifers adjacent to rivers where spatial components of velocity are to be evaluated,as zones of influence do not conform with conventional circular form. An alluvial aquifer in Paraiba State, Brazil was chosen for study and the velocity components evaluated. Results were presented in the form of tables and 3-D graphs.Application of results for public water supply was suggested.

1 INTRODUCTION

By the turn of the century, the demand for water will increase many-fold, due to the population explosion on our planet and consequent domestic, industrial and agri-culturally dependent exigencies. Only 30 % of world population now has guaranteed water supply while the remaining 70% depend on wells, lakes, and springs that are not perennial and are liable to get contaminated. Inspite of the dire need for groundwater in Latin America and Carribean states, very little is being done for its preservation.

Presently, many towns are partly or wholly dependent on this resource. Effect of super-exploration of aquifers through battery of wells is felt in different parts of Brazil, more so in semi-arid northeast, resulting in lowering of phreatic levels and contamination of aquifers.

2 TWO-DIMENSIONAL FLOW IN FREE AQUIFERS

Two and three dimensional models should be adopted, when flow is complex as in multi-layered, non-homogeneous and anisotropic aquifers. Freeze and Whitherspoon (1966) were among the first who used numerical modeling for 2-D flows in non-homogeneous aquifers on a regional sacle. In areas of discharge and recharge, flow will not be even near-horizontal. As such, it would not be realistic to assume that the potential head doesn't vary along the depth of aquifer.

In free aquifers, the permeability coefficient K, and effective posority ne , vary in space. Lee and Cherry (1979) studied the hydrodynamics of rivers and aquifers in which vertical gradients are significant. The 3-D flow equation for aquifers would have components of flow velocity, qx, qy & qz in the three directions, hydraulic gradients, Jx, Jy and Jz and permeabilities, Kxx, Kyy and Kzz. With symmetry in permeability tensor, the useful components are three for 3-D flows and two for 2-D flows (Bear, 1972).

3 STUDY AREA IN PARAIBA STATE, BRAZIL

The study area in Paraiba river basin near Pilar is 12 km from Juripiranga (Fig. 1), 800m above mean sea level where rainfall is 800 to 1,500 mm/yr.The government organ, CDRM in 1988 investigated for alternative sources for 150 m^3/h of demand and an alluvium near Pilar was found suitable. Twenty-two soundings, each of 5 cm were drilled and a test well established.

The saturated thickness of aquifer (200 m x 500m) was about 8m ; 40,000 m^2 of study area was chosen. It was subdivided into 2 configurations: quadrilaterals and squares, to know the difference they make in spatial velocity components, the former being made by joining the existing wells,the latter by making the area a near-square. Effects of excessive pumping on 1- and 2-D heads developed in the area were studied. The results obtained were interpreted in the

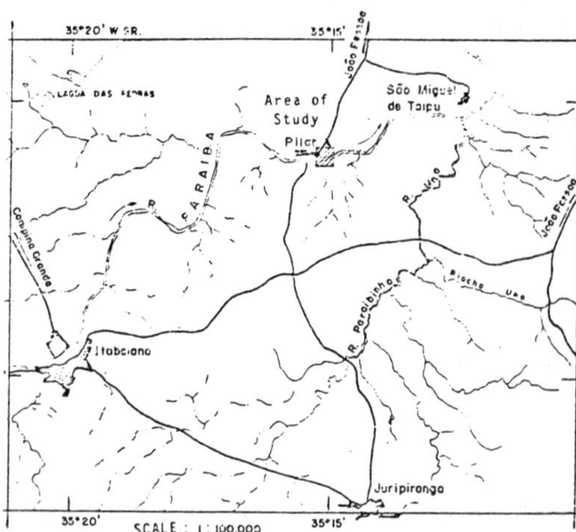

Fig. 1 Area of Study in Pilar,Pb, Brazil

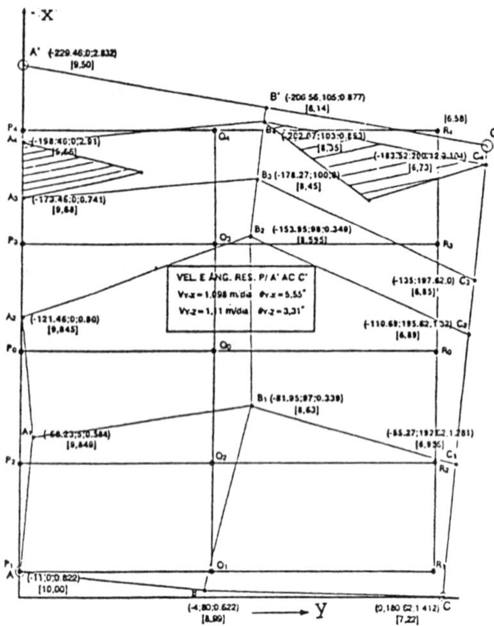

Fig. 2 Configurations adopted for Aquifer

light of sloping phreatic surface and undulating bed to know the implications on contaminant transport in public water supply.

The area between AA' and CC' forming 193 m x 189m (Fig.2) was investigated for modeling. The figures are irregular because of the irregular choice of wells by CDRM. The heads at A, B and C are higher than those at A', B' and C'. The equivalent rectangle is as per Fig.2. The river is an effluent one as the aquifer was losing water to the river at the time of study.

4 CALIBRATION DONE TO TEST THE MODEL

The test well is pumped at 26.6 m^3/h. Being a free aquifer, storage coefficient does not enter into the calculations. However, permeability coefficient K, and transmissivity T, of the aquifer were determined from pumping test results as 0.00025 m/s and 0.00212 m^2/s, adopting test well discharge. The effective or drainable porosity was determined at 0.30. While Kx was taken as equal to Ky, Kz in z-direction was treated 100*Kx. From calibration, the radius of influence was found to be 65m. In the finite difference method adopted, grid interval was taken as dx = dy = 65m/12 = 5.42m. The dynamic levels registered in the three observation wells were respectively 6.31m, 8.32m & 8.40 m.

5 2-D FLOW EQUATION FOR FREE AQUIFERS

The governing equation adopted here-in for regional groundwater flow is as given by Rhuston and Redshaw (1979) which is applicable for free aquifers. However, in free aquifers, head distribuition along an exposed surface defines also the physical boundary at the surface of aquifer. Although a majority of aquifers exhibit horizontal flow, for the present case, modeling needs to be done using 2-D flow equation involving x & y.

The observed data were treated as independent variables that describe temporal and spatial changes within the aquifer. The hydraulic heads, be them one- or two-dimensional, may influence groundwater fluctuations. Whether flow is one-, two- or three dimensional depends on bed level undulations and nature of hydraulic heads. Local rainfall and evaporation losses may also govern the nature of phreatic surface. Observation wells that indicate the flow regime in the given locality, would be useful in investigating contaminent transport which demands information about velocity and flow components. This information is used to know the behavior of the aquifer under various pumping schedules.

Pinder and Gray (1977) suggested ways of calculating velocity and flow components in space, using data on piezometric heads. Their systematic method for calculating heads is a practical one to attain this objective , with known values of porosity and permeability coefficient. While the gradient of h is easily calculable for a regular well configuration, the problem arises when wells are randomly located. With hydraulic heads at corners of a square well configuration known, Finite Difference Method can be used for knowing derivatives, dh/dx & dh/dy. On the otherhand, for an irregular configuration (Fig. 2), one has to resort to other approximations.

6 SPATIAL COMPONENTS OF VELOCITY

When an aquifer system shows significantly high gradients, as in the present case, it is essential to treat spatial velocity components. Pinder and Gray (1977) considered a 3-D potential field for wells randomly located having undulating base. Using their method, one can estimate spatial components of velocity by grouping at a time the heads at four corners of a tetrahedron by way of linear interpolation.

Fig.2 represents quadrilaterals as A'ABB' and rectangles as P1P4R4R1 in two configurations. The CDRM wells helped form sub-quadrilaterals A1ABB1, A2ABB2...To soften the irrigularities, interpolation was done between points AA1 to get such points as P2 with the object of making global rectangles as P1P4R4R1 and sub-rectangles as P1P2Q2Q1. Being regular figures, individual rectangles gave relatively good values of magnitudes and directions of velocities. The velocity componentes and their resultants in and with x-y planes (Tables 1) for the configurations were calculated using 2-D heads and base levels. For want of space, only two cases are presented here.

Three-D figures were plotted with the objective of having perspective vision of base and water surface. While base levels vary as per crystalline formation of the aquifer, the phreatic surface varies as per hydrodynamic laws. Fig. 3 represents heads so developed in perspective view for better visualisation of the phenomenon.

Tables 1 and 2 refer to quadrilaterals A'ABB' and rectangles P'PQQ' showing coordinates, phreatic heads, spatial velocity components and corresponding resultants in x-y (horizontal) and y-z (vertical) planes. Also is shown permeability value Kz, in z-direction, which gave reasonable values for velocities. The negative values are because of choice of axes.

As is evident, velocities along -x axis are less as compared to those along y, once that hydraulic gradients are greater

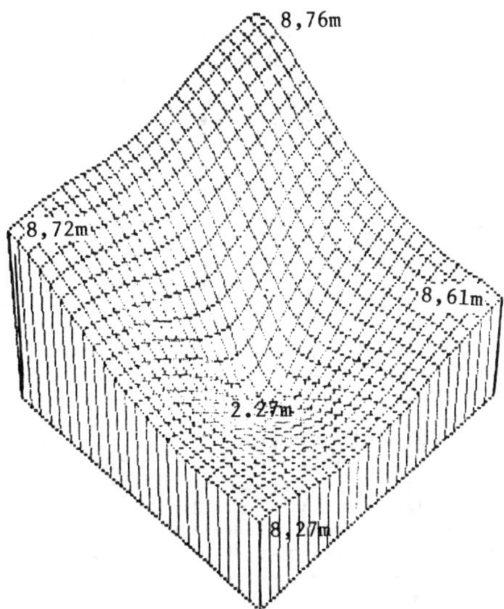

Fig. 3 Perspective View of Heads for pumping rate 2Q

along y, sides of figures being practically the same. A max. velocity Vx, of 1,353m/d was registered with P2P1Q1Q2, with a min. zero. However, values were higher along y, being Vy = 2.76 m/d for P2P1Q1Q2, with a min. 0.01 m/d for Q0Q2R2R0 and Q4Q3R3R4. Higher values for Vy are attributed to greater base inclinations and higher hydrauli gradients. The resultants in quadrilaterals varied from a maximum of 1,38 m/d in B'BCC' with an angle of 6.31° to a minimum of 0,94 m/d with 11.84° in A'ABB'. The maximum and minimum velocities do not necessarily correspond to maximum and minimum angles. The angle 11.84° corresponds to minimum velocity, while the minimum angle 1.32° corresponds to the velocity 1.29 m/d, variation in inclination being 10.52°. The non-compatibility of velocities to angles is due to undulating base of the aquifer.

Spatial velocities Vxyz, with y-x plane are obtained from resultants, Vy-z and Vz. Maximum and minimum velocities of 1.34 m/d at 0.09° in B.'BCC' and minim of 0.94m/d at 1.843° in A'ABB' were registered, the difference between maximum and minimum being 0.402 m/d. The maximum flow in rectangles was 20.27 m^3/ for 2-D flow (P2P1Q1Q2), while in quadrilaterals, it was 12.53 m^3/d (A'ABB).

7 EFFECT OF HIGH PUMPING RATES ON DRAWDOWNS

The hydraulic heads realized in the aquifer due to application of various pumping rates

TABLE 1. Velocity Components and Angles made on and with x-y Plane
Case: Quadrilateral Configuration - A'ABB'

-Vx	Vy	Vz	-x	y	z	H	-x2	y	z	h	-x3	y	z	H	-x4	y	z	H
Resul. Vel in x-y plane= 0.94/Vel.with x-y plane= 0.941:Angle1=11.84/Angle2= 1.84																		
Case: Two-Dimensional																		
0.193	-0.03	00	9.5	00	10	80	8.99	105	8.14									
0.918	229.5	2.21	11	0.2	4	--	206.6	0.255										
Case: Three-Dimensional																		
0.193	-0.03	00	9.5	00	10	80	8.99	105	8.14									
0.918	229.5	2.21	11	0.2	4	--	206.6	0.255										

Note: One-Dimensional case does not arise in quadrilateral configuration

TABLE 2. Velocity Components and Angles made on and with x-y Plane
Case: Rectangular Configuration - P'PRR'

-Vx	Vy	Vz	-x	y	z	H	-x2	y	z	h	-x3	y	z	H	-x4	y	z	H
Result.Vel.in plane x-y=0.89: Vel.with x-y plane xy=0.897: Angle1=16.6: Angle2=7.0																		
Case : One-Dimensional																		
00	00	00	10	00	10	82	8.99	82	8.99									
0.89	198.5	2.65	11	0.22	11	00	198.46	0.45										
Case : Two-Dimensional																		
0.36	00	00	9.5	00	10	82	8.99	82	8.14									
0.85	198.5	3.65	11	0.22	11	00	198.5	0.45										
Case : Three-Dimenssional																		
0.36	0.13	00	9.5	00	10	82	8.99	82	8.14									
0.85	198.5	2.65	11	0.22	11	00	198.5	0.45										

x.y.z.H in meters; Vx.Vy.Vz.Vxy= vels.(m/day): 1.2= angles in and with x-y plane

TABLE 3 Hydraulic Heads developed with 1-D Flow with Pumping Rates of Q,1.5Q & 2Q

Center Line Heads h in the Direction of the River, in meters													
Pumping rate = Q	8.78	8.62	8.44	8.23	8.03	7.44	6.29	7.30	7.66	7.83	7.91	7.94	
=1.5Q	8.78	8.56	8.30	7.99	7.55	6.78	4.77	6.63	7.26	7.57	7.75	7.87	
= 2Q	8.78	8.49	8.16	7.74	7.15	6.06	2.46	5.88	6.85	7.32	7.60	7.80	
Lateral to the River Flow:													
Pumping rate Q	8.63	8.47	8.30	8.10	7.83	7.37	6.29	7.37	7.83	8.10	8.30	8.47	8.63
1.5Q	8.59	8.37	8.14	7.85	7.44	6.71	4.77	6.71	7.44	7.85	8.14	8.37	8.59
2Q	8.56	8.27	7.97	7.58	7.02	5.98	2.46	5.98	7.02	7.58	7.97	8.27	8.56

TABLE 4 Hydraulic Heads developed with 2-D Flow with Pumping Rates of Q,1.5Q & 2Q

Centre Line Heads h in the Direction of the River, in meters													
Pumping rate = Q	8.70	8.56	8.40	8.21	7.94	7.48	6.38	7.41	7.80	8.00	8.10	8.15	
= 1.5Q	8.70	8.48	8.25	7.95	7.52	6.77	4.79	6.65	7.30	7.61	7.77	7.84	
= 2Q	8.70	8.41	8.08	7.67	7.08	5.98	2.27	5.80	6.76	7.20	7.43	7.53	
Lateral to the River Flow:													
Pumping rate Q	8.66	8.53	8.38	8.19	7.92	7.46	6.38	7.44	7.89	8.14	8.33	8.49	8.64
1.5Q	8.60	8.41	8.18	7.89	7.47	6.74	4.79	6.71	7.43	7.84	8.13	8.37	8.58
2Q	8.55	8.29	7.98	7.58	6.99	5.92	2.27	5.89	6.95	7.52	7.92	8.24	8.52

were studied in a 60m x 60m area near well B'. The drawdowns so obtained for different pumping rates, Q are as shown below:

$Q(m^3/h)$	1.1	1.3	1.5	1.7	1.9	2.1	2.3
s(m)	6.1	5.5	4.8	3.9	2.9	1.6	0

With Q referring to test discharge of 26.6 m^3/h, for pumping rates of Q, 1.5Q and 2Q, one-D and two-D heads developed in the aquifer were studied. The one-D case occurs when heads along the river flow vary, but not transversal to it. A net work of 13x13 heads with the well placed at the centre was considered; however, only 12 heads along the river and 13 heads transversal to it were shown below. The test well of CDRM was located at 30m from point B'. Heads varying from 8.77m, 8.73m to 8.72 m, 8.75 m refer to that side distant from river, while values 8.61m, 8.77m to 8.54m, 8.35m refer to the side nearer to it, the head at center of well being 6.38 m.

Tables 3 and 4 show that the head variation along the river is greater than that in the transversal direction in all the three cases. Also, drawdowns in one-D case were less than those for two-D case, till the pumping rate was a little higher than 1.5 Q. However, this tendency is reversed above rates of 1.5Q and higher, which is perhaps due to due undulating base, as also to the the anisotropy in aquifer material. (Figures 4 and 5).

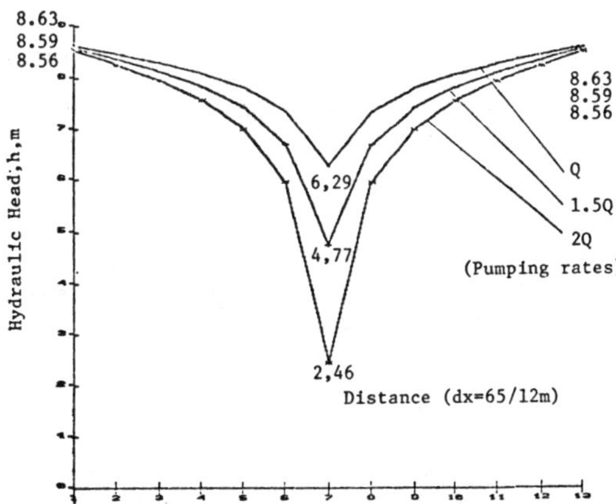

Fig. 4. Two-Dimensional heads developed transversal to River-One-D Case

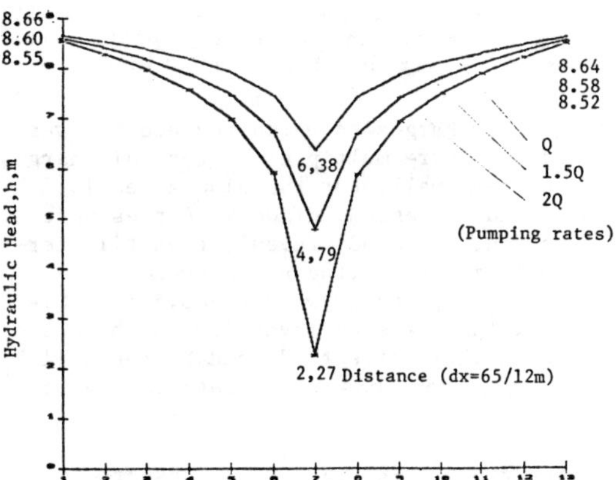

Fig. 5. Two-Dimensional heads developed transversal to River-Two-D Case

8 PROTECTION ZONE AGAINST CONTAMINATION

It is common to make provision for zone of protection for public water supply wells, to identify potential contaminant sources and to separate then. An arbitrarily fixed radius of influence would result in additional land acquisition to protect the contamination zone. On the other hand, subprotection by choosing a lesser radius results in dangerous outcome (Ramos, 1989). Hence if the specified radius does not include the total captation area or is not sufficiently great to permit adeqaute attenuation, the pollutants may destroy the well, resulting in high costs of substitution.

Considerable increase in the area of contribution zone occurs when inclination of phreatic surface diminishes, with pumping rate reduced from 2Q to Q, when the well can capture from an area much more than that with a higher rate of discharge. The equipotential lines are no longer oriented along the river ; the distortion in flownet was compared with the conventional case. The milder the gradient and pumping rate, more concentric are the circles that form the transport zone (Ramos, 1989). A detailed discussion is unwarranted as it is out of scope of this paper.

9 THE IMPORTANCE OF CONTRIBUITION ZONE

If hydraulic gradients before pumping are appreciably large, it is inevitable to protect the well against pollution. The contribution zone defines the surface area of discharge and subsurface regions of flow in the pumped area from which the water is received. This important concept is not understood thoroughly by water supply engineers of old school of thought. The cone of depression or zone of influence of a well defines the recharge area for the well which means that the zone of influence is the same zone of contribution. However, phreatic aquifers, in which water is in continual motion, have appreciable hydraulic gradients which give rise to inclined water surfaces, with or without pumping. In such cases the contribution zone would not be same as the zone of influence. The numerical difference between these two zones depends on the degree of inclination and pumping rate. With inclination, more water would be on upstream than on downstream side of the well. Thus, contaminants that are beyond the cone of influence but are within the zone of contribution, eventually migrate to the well.

The inclination of phreatic surface would cause the water divide or the point of stagnation downstream of well. If the well pertains to public water supply, the contaminants beyond the zone of influence and downstream of divide do not pollute the well. On the otherhand, if the well is meant to remove pollutants, the contaminants downstream of stagnation point are not going to be removed by the well, inspite of their being within the cone of influence.

10 CONCLUSIONS

Analytical models utilized here-in proved to be useful in testing the data of CDRM and in implementing the same for public water supply for Pilar region, PB, Brazil. In aquifers adjacent to rivers that exhibit 2-D heads, Pinder and Gray's (1982) artifice is adequate to find velocity components in and with x-y plane, for regular or irregular configurations.

The discharges to which the aquifer was subjected were multiples of test discharge Q, of test well. For the discharges 1Q, 1.5Q and 2Q tested, drawdown curves were compared. The study reveals that till certain limit (1.5Q), the one-D heads showed higher values than two-D heads. After this limit, there was an inversion in the phenomenon. These effects depend on the soil, and also on base undulations of the aquifer.

The two-D subsurface flow presents an important role in de-contamination of aquifers as also in pumping for public water supply, depending on the case. Enviromental engineers should be aware of such situations and protect the waters from getting polluted. In the case of water wells, the objective is to define the protection zones, while in the case of pollution capture zones are to be clearly defined to avoid undue entry of pollutants.

11 ACKNOWLEDGMENTS

The authors gratefully acknowledge the CNPq, Brasilia, D.F. for the financial help rendered and Civil Engg. Department, UFPB, Campina Grande, Brazil for the moral support given to conduct this study.

REFERENCES

Bear, J. 1972. Dynamics of Fluids in Porous Media, Elsevier, New York.

Freeze, R.A. & Witherspoon, P.A. 1966. Theoretical Analysis of Regional Groundwater Flow: 1. W R R, vol.2, 641-656

Lee, D.R. & J. A. Cherry. 1979. A Field Exercise on Groundwater flow using seepage meters. J.of Geological Education. vol. 27, 6-10.

Pinder, G.F. & W.G. Gray. 1977. F.E.Simulation in Surface and Subsurface Hydrology. New York.Academic Press Inc. 1-295.

Ramos, F. et al. 1989. Engenharia Hidrológica.Coleção ABRH de Recursos Hídricos, Vol.2. Rio de Janeiro, EFRJ. 293-404.

Rushton, K.R. & S.C. Redshaw- 1979. Seepage and Gr.water Flow. Wiley,UK. 339-376.

Environmental Management, Geo-Water & Engineering Aspects, Chowdhury & Sivakumar (eds)
© 1993 Balkema, Rotterdam. ISBN 90 5410 099 0

Fluid flow through a single rock joint

J. Sheng & Y.F. Bu
National Central University, Taiwan

ABSTRACT: A pair of rough surfaces is generated by using a fractal model of surface topography to form a single rock joint. The aperture of the joint which dominates the flow rate is converted as the apparent permeability. The joint is thus treated as a plane flow domain with uniform thickness and varying permeability. A two-dimensional finite element program SEEP is used to calculate the flow rate through the joint. A correction factor is introduced to reduce the flow rate accounting for the tortuosity in the direction normal to the conceptual x, y plane.

1 INTRODUCTION

Underground storage of hazardous waste catches the environmental concern in recent years. It is well recognized that the migration of most pollutants is strongly related to the groundwater condition in the region where the waste is buried (Huyakorn and Pinder, 1983). In most deep underground storage projects, rock mass is inevitablly involved. The groundwater flow in a rock mass is mainly through various types of discontinuities, i.e. fissures, cracks, joints, etc. It is of primary interest to understand the fluid flow behavior through a single joint before dealing with a complex rock mass. Studies on this subject started from the assumption of laminar flow between two perfectly smooth parallel plates and lead to the so-called 'cubic law' (e.g., Tsang and Witherspoon, 1981). When the apertures of a rock joint do not vary dramatically from point to point, the cubic law gives a satisfactory estimation on the flow rate. However, it is not unusual that some rock joints present a rather high degree of roughness with a relatively small average aperture. This geometric feature may intuitively decrease the flow rate obtained from the cubic law. The Reynolds equation defined on an x, y plane was solved by using the finite difference technique and the results showed that the flow rate could be as low as 70% of that predicted by the cubic law (Brown, 1987). The flow velocity field on the x, y plane clearly demonstrated the effect of the varying aperture on the flow behavior. With further examination, however, there is still a missing parameter, the component of actual flow path in the direction normal to the x, y plane. In stead of involving a three-dimensional analysis, a simple correction factor $1/\tau_z$ is proposed in this paper to handle this problem.

2 THEORETICAL BACKGROUND

2.1 Rock joint generation

The rock joints used in this research are generated numerically rather than real models. The advantage of doing so is that the parametric study can easily be conducted without the efforts of tedious surface measurement. The generated joint surfaces have to contain certain characteristic features of natural rock joints. The power spectral character of natural rock joint surfaces was found similar to that of the fractal model of surface topography (Brown and Scholz, 1985; Mandelbrot, 1983). It is thus believed that the fractal model is adequate to describe the character of a natural rock joint for the purpose of this research. The specific numerical technique used in this paper is the recursive subdivision method (Fournier et al., 1982) which was successfully adopted in rock joint simulation with minor

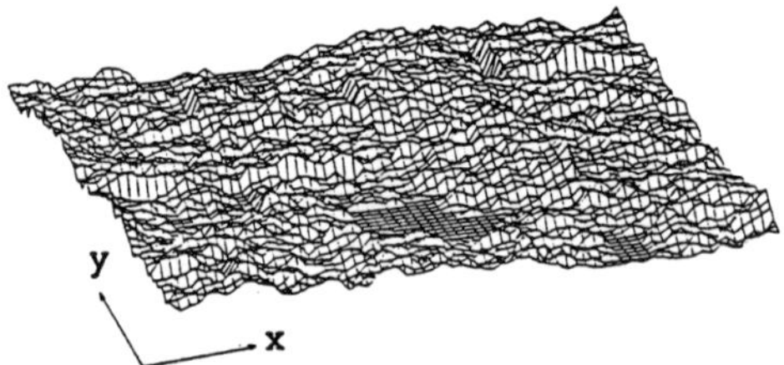

Fig.1 A generated rock joint

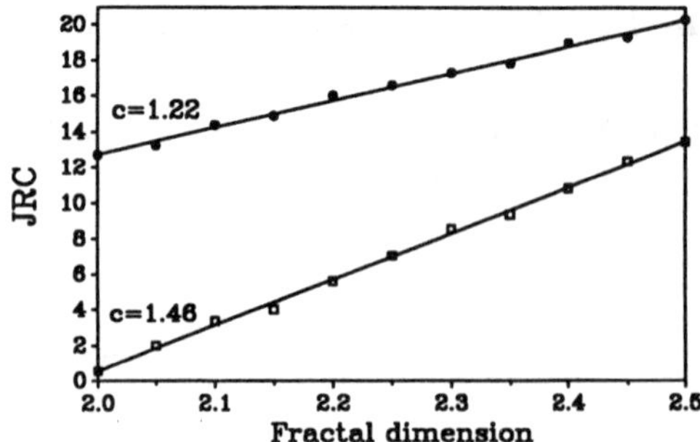

Fig.2 Suitable range of constant C

modification (Brown, 1987). Figure 1 shows a rock joint which is formed by placing together a pair of numerically generated surfaces at a reasonable distance 2σ. σ is the standard deviation of height of the joint surfaces. The flat regions in Figure 1 are the contact areas of these two surfaces, where no flow occurs. These two surfaces have the same fractal dimension (D=2.5), the same standard deviation of height, but different shapes as a result of using different 'seed' in the surface generating program (Fournier et al., 1982). There is a constant C used in the program to be adjusted to fit a specific application. Figure 2 gives a reasonable range of C to generate a surface with a fractal dimension $2.0 \leq D \leq 2.5$ and a JRC value within 20. These constraints ensure the generated surfaces bearing similar features of natural rock surfaces.

2.2 Flow rate estimation

The cubic law is probably the simplest way in estimating the flow rate through a single rock joint (e.g., Tsang and Witherspoon, 1981):

$$Q = -L_y \frac{d^3}{12\mu} \frac{dp}{dx} \tag{1}$$

where Q is the flow rate, L_y is the width of the joint in the direction normal to the fluid pressure gradient dp/dx, d is the separation between the assumed smooth parallel plates, and μ is the fluid viscosity. For a better estimation, it is first assumed that the cubic law holds locally (Walsh, 1981). The equation of continuity, $\nabla \cdot Q = 0$, is then applied through out the flow domain to obtain

$$\nabla \cdot (\frac{d^3}{12\mu} \nabla p) = 0 \tag{2}$$

which is identical to the simplified Reynolds equation for a confined steady state flow of an incompressible fluid (Brown, 1987). The above equation can be solved numerically for a domain of complex geometry such as a single rock joint. In order to use the available finite element program SEEP (Wong and Duncan, 1985), the permeability K of dimension L/T has to be determined. The two-dimensional governing equation solved in SEEP has the form

$$\nabla \cdot (\tilde{K} \nabla h) = 0 \tag{3}$$

where $\tilde{K}$ is the permeability tensor and h is the total head of the fluid. A direct comparison between Eqs.(2) and (3) gives the relation

$$K = d^3 \gamma / 12\mu \tag{4}$$

in which γ the density of the fluid, equals p/h and the constant K is for an isotropic medium. Eq.(4) needs a further explanation: Eq.(1) can be rearranged in the form of Darcy's Law as

$$\begin{aligned} Q &= (\frac{d^2\gamma}{12\mu})(-\frac{dh}{dx})(L_y d) \\ &= KiA \end{aligned} \tag{5}$$

in which K is proportional to d^2 rather than d^3 shown in Eq.(4). Since SEEP is a program for a two-dimensional analysis, the flow domain has to be of constant thickness which is in contrast to a rock joint. In order to overcome this difficulty, the extra term d is incorporated in K which may be called the apparent permeability. Thus each element with a specific thickness (varying aperture d) is converted to an element of unit thickness with a modified permeability as a compensation. Thus the calculated flow rate Q_{seep} has a spatial dimension of volume (L^3) rather than volume per unit thickness (L^3/L).

3 DISCUSSION

3.1 Geometry of joint

Since the rough surface is generated by the recursive subdivision method, its surface topography is intuitively isotropic. First, the joint roughness coefficient JRC is used here as an index to check if it is the same in x and y directions for any generated surface. Most of the surfaces used here are square with a 65 by 65 mesh size. The JRC's along x and y directions of 17 joints are shown in Figure 3a. Same procedure is done for σ and the results is in Figure 3b. Both JRC and σ suggest that the surfaces are indeed isotropic, topographically. Secondly, the flow rates Q_x and Q_y are calculated separately by switching the same boundary conditions between the boundaries along x and y directions. If the joint is isotropic as far as the permeability is concerned, then $Q_x = Q_y$. Surprisingly, $Q_x \simeq Q_y$ holds only when these two surfaces have a separation d_m greater than a value about 4σ. As d_m decreases, the difference between Q_x and Q_y increases. For some cases with $d_m = 1\sigma$, this difference exceeds a factor of 2 as shown in Figure 4. Besides, there is no trend in knowing which direction would get a higher flow rate. A few cases are examined with a mesh size of 129 by 129. It is found that this anisotropic flow behavior is just slightly eased.

There is no intention in this research to find a relation among the fractal dimension D, the standard deviation σ, the coefficient of roughness JRC, and the flow rate Q for the following reasons: (1) The definition of the fractal dimension of a surface is not unique for the time being. Furthermore, a perfectly smooth plane may get a fractal dimension greater than 2 or less than 2 depending on the method used. (2) The JRC is originally developed for the purpose of shear strength analysis of the rock joints. A surface isotropic with respect to JRC may present anisotropic flow behavior. Similar situation exists for σ. (3) The flow rate Q varies dramatically for several joints with a same set of D and σ under a same boundary conditions as shown in Figure 5. All these imply that D, σ, and JRC are inadequate for predicting Q.

3.2 Estimation methods

There are several methods proposed to estimate the flow rate between a pair of rough surfaces. These methods can be grouped into four categories: (1) direct application of the cubic law with different estimations of the representive aperture (e.g., Neuzil and Tracy, 1981), (2) a flow rate reducing factor accounting for the flow path tortuosity or for the contact areas of the rough surfaces, (3) empirical equation based on laboratory data or numerical results (e.g., Louis, 1969), and (4) direct numerical simulation (e.g., Brown, 1987). The aperture terms used in the first cat-

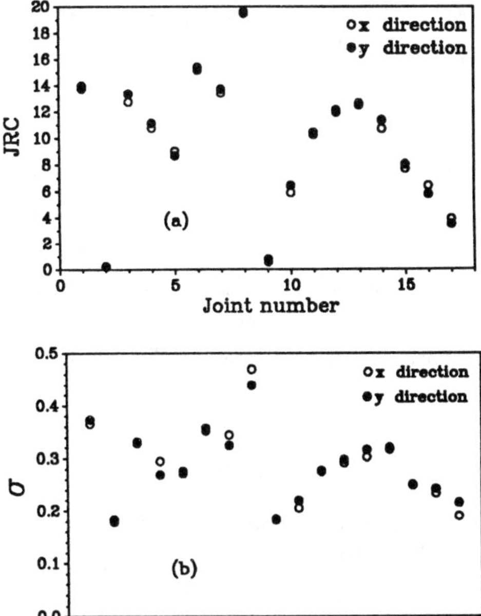

Fig.3 JRC and σ of x and y directions

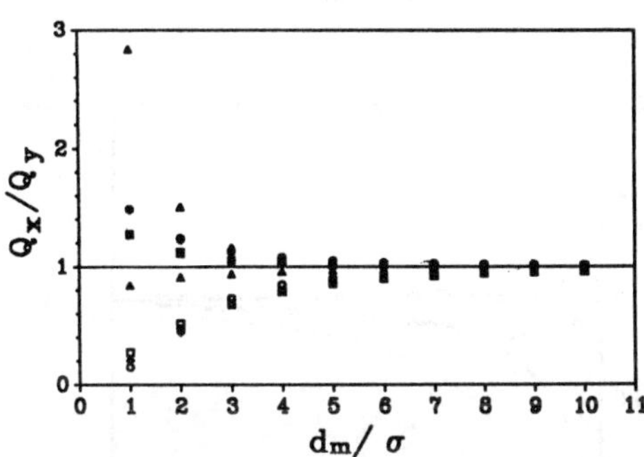

Fig.4 Anisotropic flow behavior of joints with isotropic topography

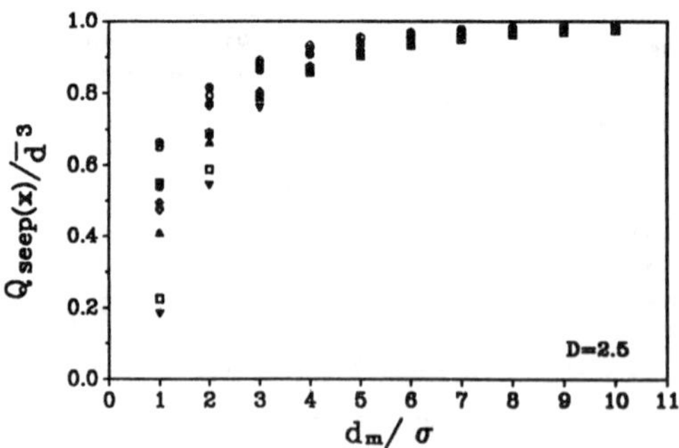

Fig.5 Variation of Q of statistically equivalent joints

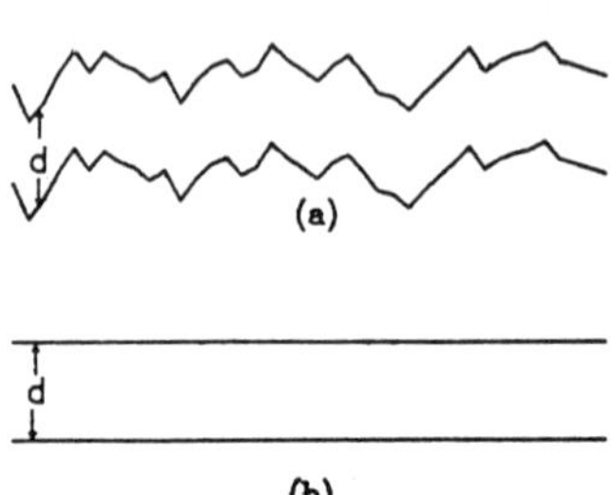

Fig.6 Imaginary rock joints

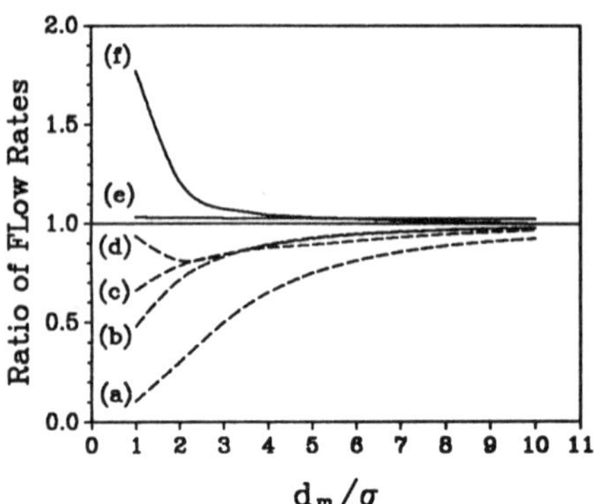

Fig.7 Results of six estimation methods

egory include d_m^3, $\overline{d}^3$, and $\overline{d^3}$. Here a bar accent stands for an arithmetic average. In the second category, a reducing factor $(1-\alpha)/(1+\alpha)$ accounts for the contact areas (Walsh and Brace, 1984) and a factor $1/\tau^2$ takes care of the tortuosity (Walsh, 1981). The coefficients involved in the empirical equations in the third category are not constants (Brown, 1987) and are difficult to give any physical meaning.

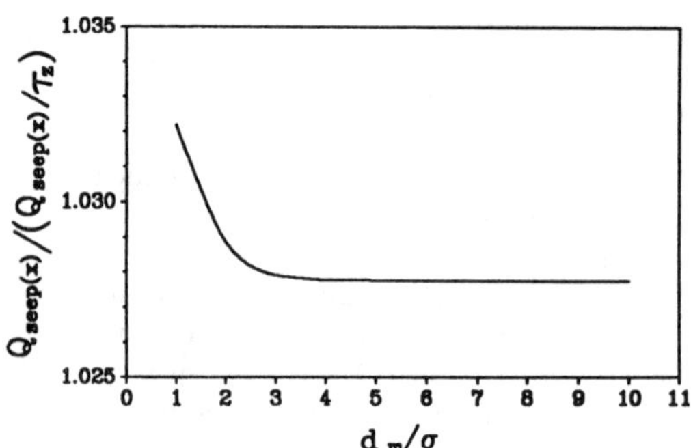

Fig.8 Effect of correction factor $1/\tau_z$

Figures 6a and 6b gives the cross sections of two imaginary rock joints with constant aperture d. Through a two-dimensional numerical simulation, both joints give a same flow rate as long as a same pressure gradient is applied accross the boundaries. However, since the flow path in joint 6a is longer than that of joint 6b, the pressure gradient along joint 6a is less than that of joint 6b, so is the flow rate. Thus, a correction factor $1/\tau_z$ is proposed here to reduce the flow rate calculated by SEEP. The term τ_z is defined as the length ratio of the line connecting the mid-points of the apertures to the base line of the conceptual x, y plane along the direction of the applied pressure gradient. This is different from the idea behind the reducing factor $1/\tau^2$ mentioned before. One should not try to use $1/\tau_z^2$ because the tortuosity on the x, y plane is already considered in a two-dimensional analysis.

Figure 7 gives the results of six estimation methods. The lines are the results of (a) $\overline{d^3}$, (b) $\overline{d}^3$, (c) reducing $(1-\alpha)/(1+\alpha)$ of (b), (d) d_m^3, (e) reducing $1/\tau^z$ of Q_{seep}, and (f) reducing $1/\tau^2$ of (b). A horizontal line at level 1.0 represents the result of SEEP, Q_{seep}. Solid lines above this horizontal line give lower flow rate than Q_{seep} while those below (dashed) give higher. Details of line (e) in Figure 7 is shown in Figure 8. The correction factor $1/\tau_z$ exhibits a minor effect on the flow rate estimation. There is only a 3% reduction at $d_m = 1\sigma$ and about a constant 2.5% for $d_m > 4\sigma$.

4 CONCLUSIONS

Fluid flow through a single rock joint is estimated through a two-dimensional numerical analysis. Although the fractal model of surface topography is adequate in generating the rock joints, the related parameters are insufficient in predicting the flow behavior. The unexpected anisotropic flow behavior may be caused by the relative small mesh size used. A correction factor is proposed accounting for the effect of the tortuosity in the direction normal to the plane in ayalysis. Results from several methods are presented for comparison. Without reliable experimental data for comparison, it is hard to say which method gives the best estimation. Nevertheless, a direct numerical approach avoids most 'averaging' processes and thus keeps most of the details through out the whole analysis. While the theoretical backgrounds are similar among these methods, the numerical simulation is no doubt of high potential in further development.

Acknowledgement. This research is supported by the National Science Council of ROC in Taiwan under the contract No. NSC81-0410-E-008-02.

REFERENCES

Brown, S. R. 1987. Fluid flow through rock joints: The effect of surface roughness. J. Geophys. Res. 92:1337-1347.

Brown, S. R. & C. H. Scholz. Broad bandwidth study of the topography of natural rock surfaces. J. Geophys. Res. 90:12575-12582.

Fournier, A., D. Fussell & L. Carpenter 1982. Computer rendering of stochastic models. Commun. Assoc. Comput. 25:371-384.

Huyakorn, P. S. & G. F. Pinder 1983. Computational methods in subsurface flow. New York: Academic Press.

Louis, C. 1969. A study of groundwater flow in jointed rock and its influence on the stability of rock masses. Rock Mech. Res. Rep. London: Imperial College.

Mandelbrot, B. B. 1983. The fractal geometry of nature. San Francisco:Freeman.

Neuzil, C. E. & J. V. Tracy 1981. Flow through fractures. Water Resour. Res. 17:191-199.

Tsang, Y. W. & P. A. Witherspoon 1981. Hydromechanical behavior of a deformable rock fracture subject to normal stress. J. Geophys. Res. 86:9287-9298.

Walsh, J. B. 1981. Effect of pore pressure and confining pressure on fracture permeability. Int. J. Rock Mech. Min. Sci. Geomech. Abstr. 18:429-435.

Walsh, J. B. & W. F. Brace 1984. The effect of pressure on porosity and the transport properties of rock. J. Geophys. Res. 89:9425-9431.

Wong, K. S. & J. M. Duncan 1985. SEEP:A computer program for seepage analysis of saturated free surface or confined steady flow. Blacksburg: CE of VPI&SU.

Environmental Management, Geo-Water & Engineering Aspects, Chowdhury & Sivakumar (eds)
© 1993 Balkema, Rotterdam. ISBN 90 5410 099 0

Method of fragments – Quick solutions to seepage problems

N. Sivakugan & M.Y.S. Al-Aghbari
Sultan Qaboos University, Sultanate of Oman

ABSTRACT: The seepage problems are traditionally solved by flow nets. This paper describes the method of fragments, a very efficient method for seepage problems. Several different types of seepage problems are solved by flow net method and the method of fragments. The quantity of seepage, exit gradient and uplift force are computed by both methods. Excellent agreement was observed. The method is used in a parametric study, where the optimum position, length and number of sheetpiles are selected.

1 INTRODUCTION

The quantity of seepage and the stability with respect to piping and uplift are the major considerations in the design of concrete dams. While ensuring that the quantity of flow is below acceptable levels, it is necessary to ensure adequate safety against piping and uplift.

Piping is a phenomenon where the seeping water progressively erodes or washes away the soil particles, creating a "pipe" beneath the dam. The passage begins at the downstream side and works backward to meet the free water at the upstream side. Once the tunnel-shaped passage is formed, the downstream side of the dam gets flooded. The safety factor with respect to piping is defined as (Harza 1935):

$$[SF]_{piping} = \frac{\text{critical hydraulic gradient}}{\text{maximum exit gradient}}$$

Since failure due to piping is often catastrophic, very high safety factors, in the order of 3-5, are generally recommended (Harza 1935; Holtz and Kovacs 1981; Scott 1982).

Traditionally, seepage problems are solved by flow nets. The quantity of flow, exit gradient and the uplift force can be computed directly from the flow net. This method was known to be quite satisfactory for all types of seepage problems, and had been used successfully over the past few decades.

The major disadvantage in the flow net method is that it requires substantial effort in drawing a good flow net. In the preliminary design stages, where several alternatives are tried, drawing a separate flow net for every individual configuration becomes a tedious task. Under such circumstances, it is very much desirable to have some methodology to estimate the necessary values quickly with sufficient accuracy.

The "method of fragments" presented in this paper, is a very simple and efficient technique developed by Pavlovsky (1956) in Russia. By this method, the quantity of flow, exit gradient and the uplift can be estimated without drawing the flow nets. The method provides quick but approximate solutions, with accuracy sufficient for all practical purposes. It is a very good tool for the decision makers, who require speedy solutions with little sacrifice in the accuracy.

2 METHOD OF FRAGMENTS

The method of fragments is a quick, but approximate, analytical design method for the solutions of confined flow problems. Many cases may be investigated in little more than the time it usually takes to assemble paper, pencils and erasers for drawing flow nets. The method, developed by Pavlovsky (1956) in Russia, was brought to the attention of the other parts of the world by Harr (1962). The basic assumption in this approach is that the equipotential lines at selected critical

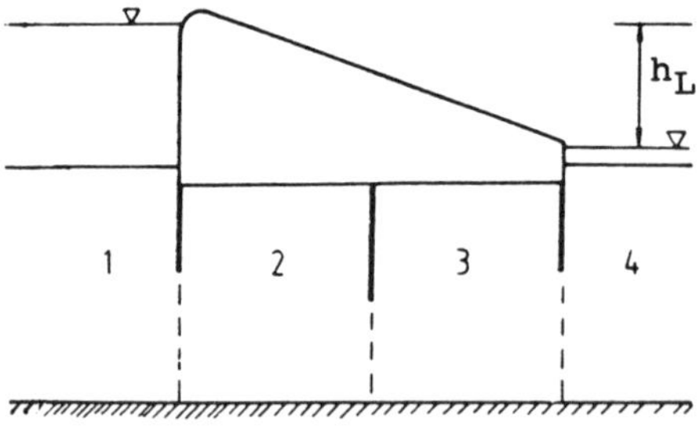

Fig. 1 Fragments in the flow region

points (see Fig. 1) in a flow net are vertical and that they divide the flow region into fragments. The flow region in Fig. 1 is divided into four different fragments, by the vertical equipotential lines extending from the tip of the sheetpiles. A dimensionless quantity, known as form factor (Φ_m) is introduced here. It is defined as:

$$\Phi_m = \frac{\text{No. of equipotential lines in } m^{th} \text{ fragment}}{\text{No. of flow lines}}$$

Therefore, the quantity of seepage can be written as:

$$Q = (k\, h_L)/\Sigma\Phi \qquad (1)$$

where k and h_L are the permeability and the head loss respectively. $\Sigma\Phi$ is given by:

$$\Sigma\Phi = \Phi_1 + \Phi_2 + \cdot\cdot\cdot + \Phi_n \qquad (2)$$

The form factors for different types of fragments, as given by Harr (1977), are presented in Figs. 2 and 3. Since the flow is the same through all fragments,

$$\frac{Q}{k} = \frac{h_m}{\Phi_m} = \frac{h_1}{\Phi_1} = \frac{h_2}{\Phi_2} = \cdot\cdot\cdot = \frac{h_L}{\Sigma\Phi} \qquad (3)$$

From the fact that the form factor is proportional to the head loss in the fragment (Eq. 3), the total head at any point in the flow region can be estimated. The fragment at the downstream side is usually of Type II (Fig. 2). The maximum exit gradient for this fragment, given by Harr (1977) are presented in Fig. 4. It can be sen from Fig. 2 that, the fragments 1, 2, 3 and 4 of Fig. 1 are of types II, VI, VI and II respectively.

For most of the confined flow problems, such as seepage under a concrete dam having sheet piles and a blanket, the quantity of seepage, exit gradient and the uplift can be estimated within minutes once the form factors are determined for each fragment in the flow region.

Fragment type	Illustration	Form factor, Φ (h is the head loss through fragment)
I	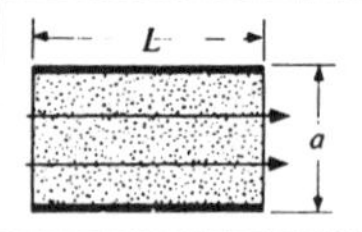	$\Phi = \frac{L}{a}$
II	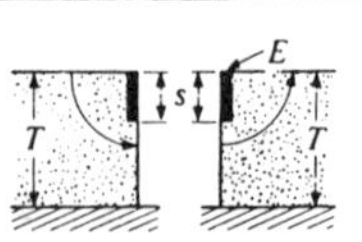	see Fig. 3
III	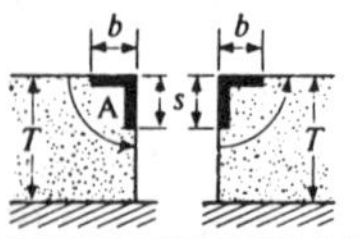	see Fig. 3
IV	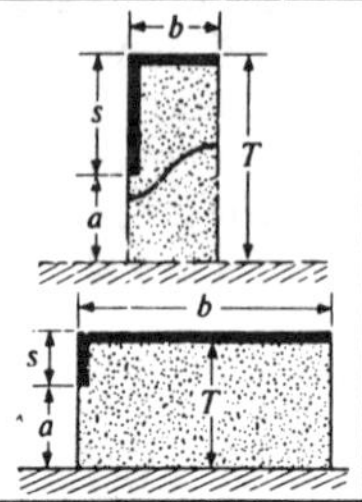	$b \le s$: $\Phi = \ln(1 + b/a)$ $b \ge s$. $\Phi = \ln(1 + s/a) + (b-s)/T$
V	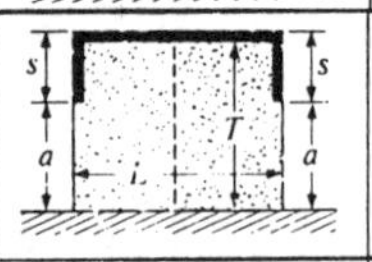	$L \le 2s$: $\Phi = 2\ln(1 + L/2a)$ $L \ge 2s$: $\Phi = 2\ln(1 + s/a) + (L-2s)/T$
VI	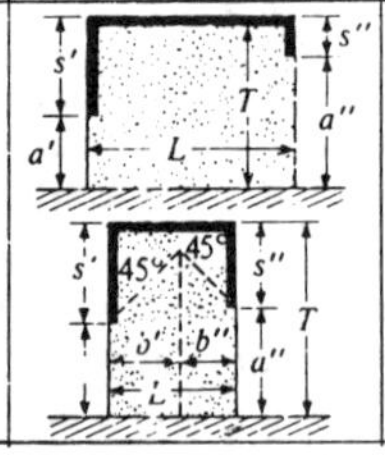	$L \ge s'+s''$: $\Phi = \ln[(1+ s'/a')(1+ s''/a'')] + (L-s'-s'')/T$ $L \le s'+s''$: $\Phi = \ln[(1+ b'/a')(1+ b''/a'')]$ where $b' = (L+s'-s'')/2$ $b'' = (L-s'+s'')/2$

Fig.2 Types of fragments and form factors (after Harr 1977)

3 VALIDATION OF THE MODEL

Nine randomly selected seepage problems are solved here by the conventional flow net method and by the method of fragments. The normalised quantity of flow (Q/kh_L per m width), the maximum exit gradient (i_{exit}) and the uplift force (kN per m width) are computed by both methods.

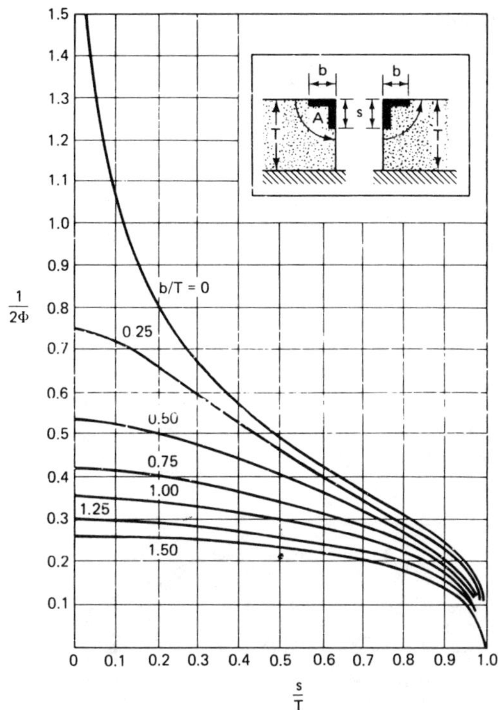

Fig.3 Form factors for types II and III fragments (after Harr 1977)

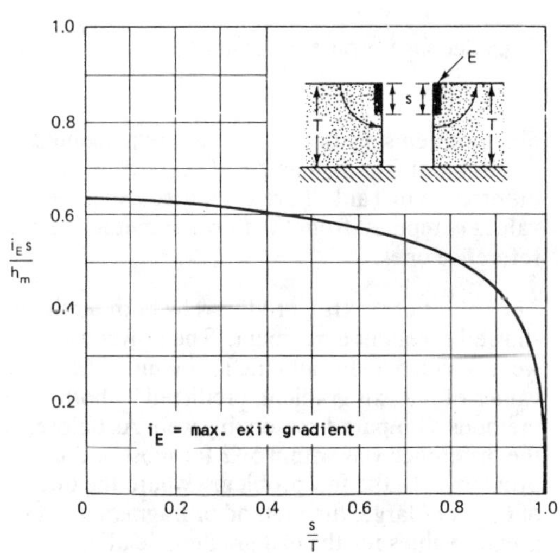

Fig.4 Exit gradient for type II fragments (after Harr 1977)

Problem 1:

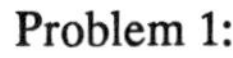

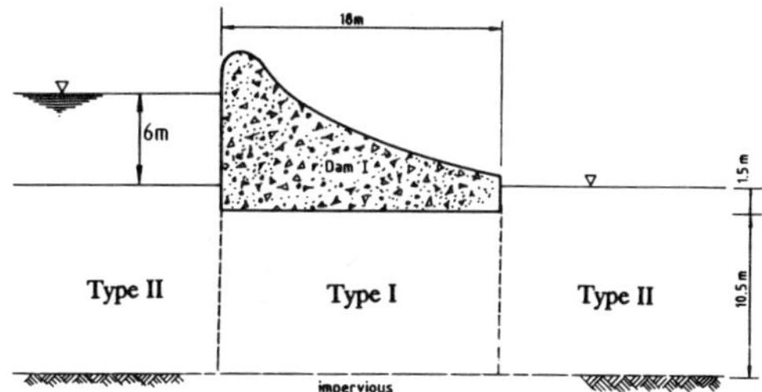

Problem 2:

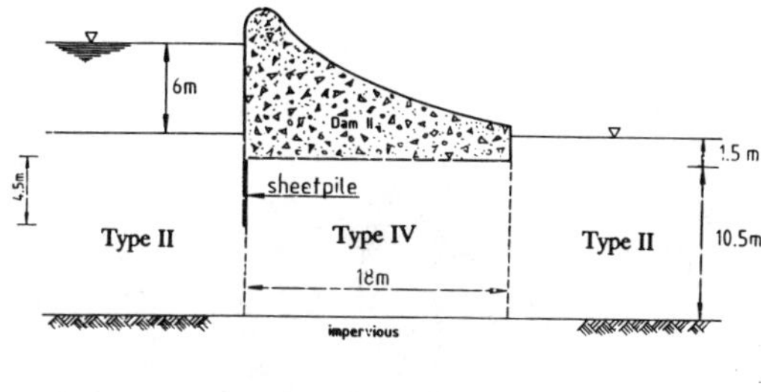

Problem 3:

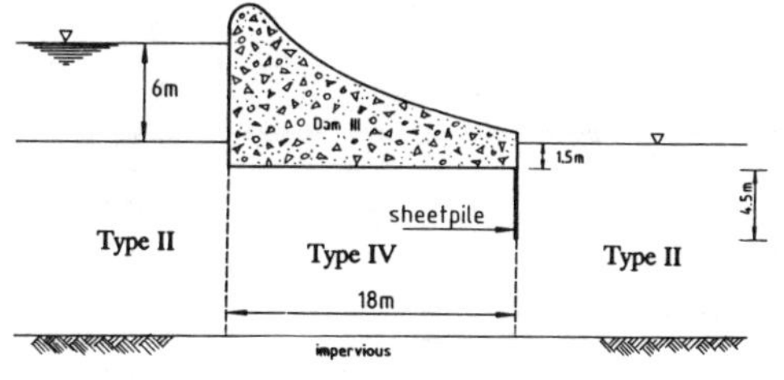

Problem 4:

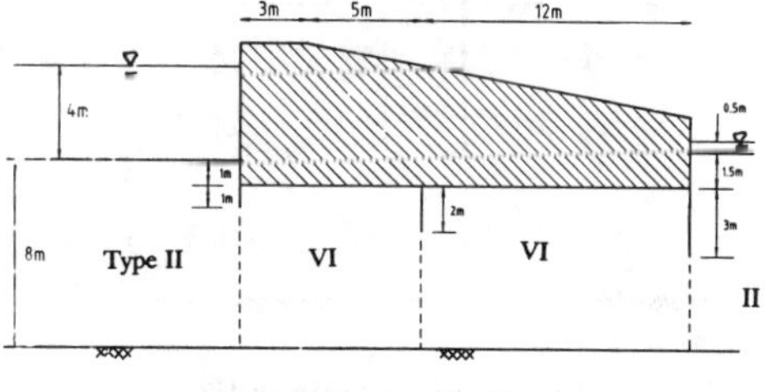

Fig.5 Seepage problems

Problem 5:

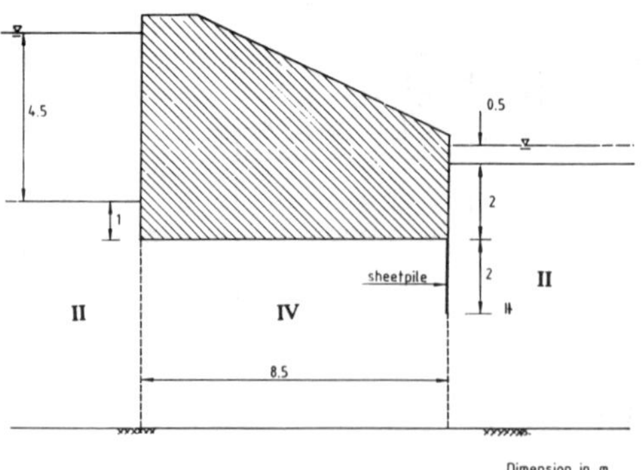

Problem 6:

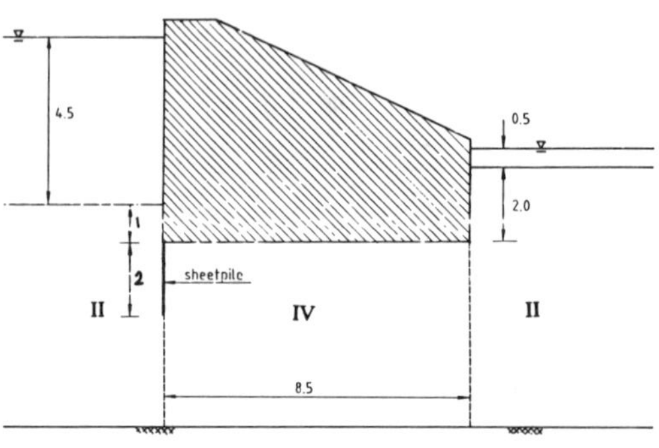

Problem 7:

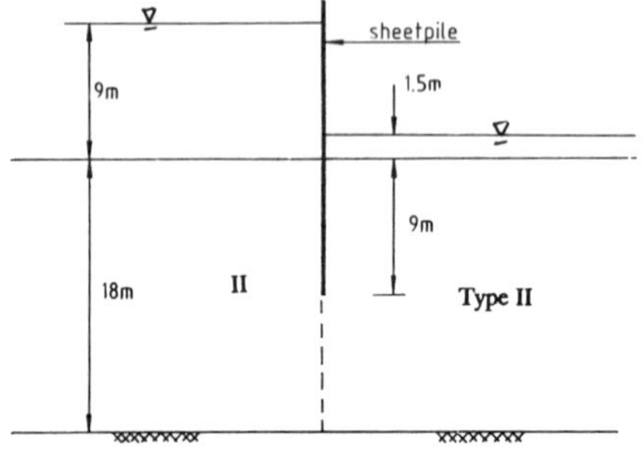

Fig.5 Seepage problems (continued)

The nine problems studied here are shown in Fig. 5, where the types of fragments are also given.

The values of the quantity of seepage (Q/kh_L per m width), maximum exit gradient (i_{exit}) and the uplift force (kN per m width), computed for these

Problem 8:

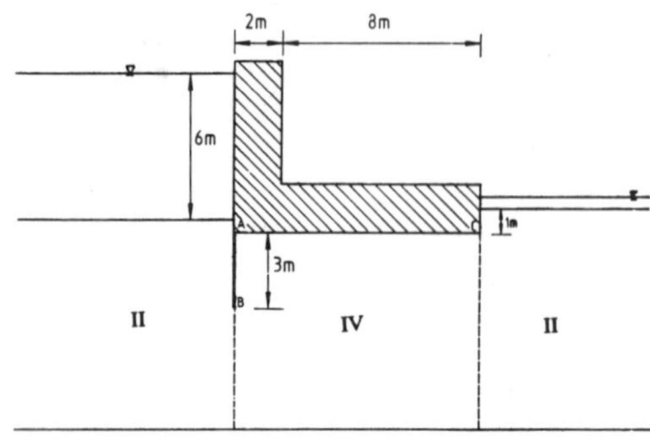

Problem 9:

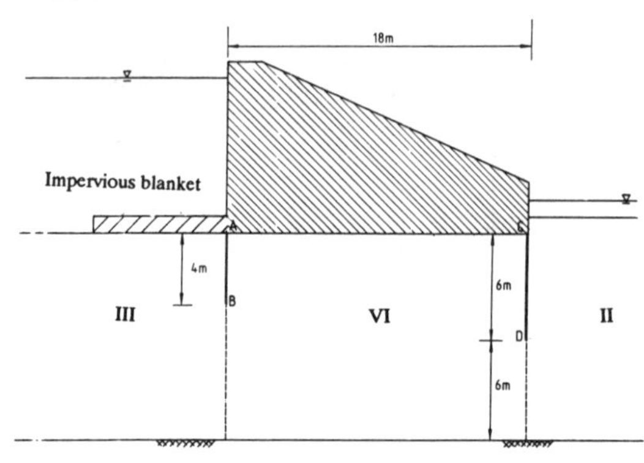

Fig.5 Seepage problems (continued)

nine problems, by both methods, are presented in Table 1 and Figs. 6, 7 and 8. The percentage differences in Table 1 are computed using the values computed from the flow net method as the reference ones.

The values of Q/kh_L, predicted by both methods, showed excellent agreement. The differences were within 5 % for all nine problems. The values of the exit gradient, predicted by both methods, compared reasonably well. As before, the difference was within 5 % for most of the problems. In the few problems, where the differences were large, the method of fragments gives greater values for the exit gradient, leading to conservative solutions to the problems.

In the flow net method, the uplift force was computed using a fairly accurate pressure diagram. In the method of fragments, the pressure was determined at two or three points at the base of the dam and the distribution was assumed to

Table 1. Results from flow net method and the method of fragments

No.	Method of Analysis						% Difference		
	Flow Net			Fragments					
	Q/kh_L per m	i_{exit}	uplift kN/m	Q/kh_L per m	i_{exit}	uplift kN/m	Q/kh_L	i_{exit}	uplift
1	0.33	0.44	796.8	0.36	0.48	794.6	-8.9	-8.0	0.3
2	0.29	0.38	681.1	0.30	0.39	661.2	-3.8	-2.0	3.0
3	0.29	0.19	908.1	0.30	0.18	928.0	-3.8	5.0	-2.0
4	0.21	0.08	679.1	0.20	0.08	722.2	6.9	0.0	-6.0
5	0.25	0.15	327.7	0.28	0.14	323.2	-12.0	5.0	1.0
6	0.25	0.20	266.5	0.28	0.19	269.9	-10.0	2.0	-1.0
7	0.50	0.31	—	0.50	0.50	—	0.0	-59.0	—
8	0.36	0.38	323.4	0.37	0.58	329.6	-0.5	-34.0	-2.0
9	0.22	0.27	889.6	0.24	0.16	969.3	-11.5	38.0	-9.0

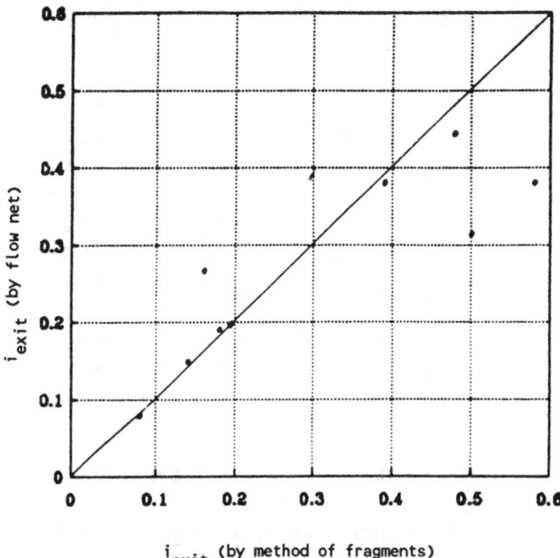

Fig.7 Predicted values of maximum exit gradient

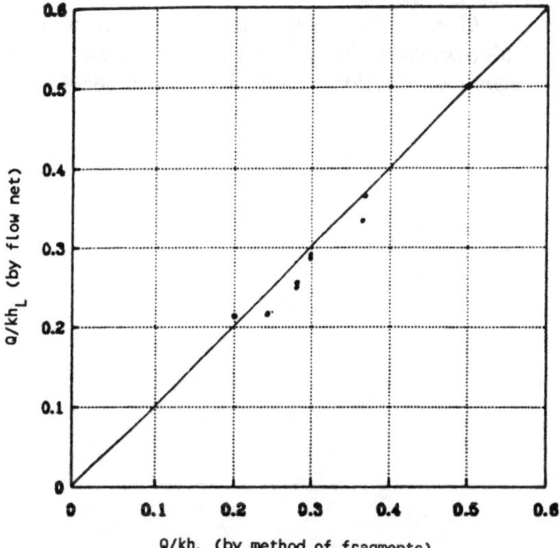

Fig.6 Predicted values of Q/kh_L

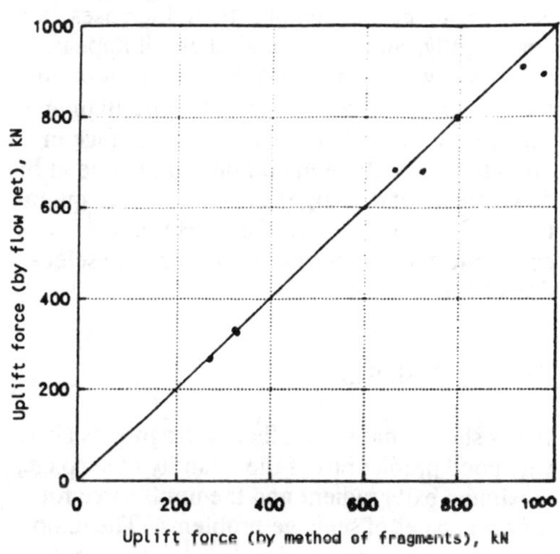

Fig.8 Predicted values of uplift force

be linear. Interestingly, the uplift forces computed by both methods showed very good agreement. The average difference between the values determined from both methods was less than 5%.

4 SEEPAGE BENEATH A CONCRETE DAM

The method of fragments is used in studying the effect of the position and the length of a single sheet pile on the dam shown in Fig. 9.

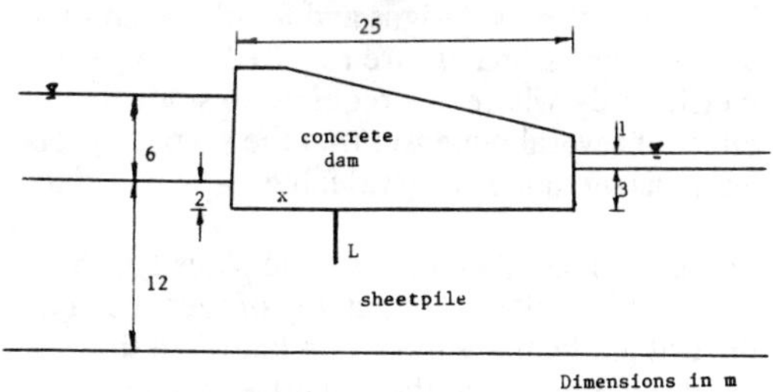

Fig.9 Concrete dam with single sheetpile

Table 2. Computed values for different positions of a single sheetpile

x (m)	Q/kh_L per m	i_{exit}	uplift kN per m
0	0.24	0.13	1402
5	0.26	0.14	1450
10	0.26	0.14	1471
15	0.26	0.14	1492
20	0.26	0.14	1513
25	0.24	0.09	1559

For different values of x and L, the quantity of seepage, exit gradient and the uplift were computed. The computed values are given in Table 2. For a particular length of the sheetpile, the quantity of seepage was about the same for all values of x. However, the exit gradient is reduced by about 40 % when the sheetpile is placed at the downstream end. The uplift force increases, but only slightly, with x. From all these, it appears that the downstream end is the best position for the sheetpile. Several trials with more than one sheetpile showed that there is no advantage in providing a sheetpile in addition to the one at the downstream end. Larger the L, lower the quantity of seepage and exit gradient. Here, the economic considerations may govern the selection of the length L.

5 CONCLUSIONS

It was shown that the method of fragments gives very good predictions of the quantity of seepage, maximum exit gradient and the uplift force for different types of seepage problems. The major advantage of the method is that it gives simple and quick solutions to the seepage problems without sacrificing the accuracy. It also has a very good theoretical basis. The method is very useful in the preliminary designs and feasibility studies, where only "estimates" are required. In a parametric study, where it is required to study the effect of several dimensions in the seepage problem, this method is very valuable.

When a single sheetpile is to be placed at the bottom of the dam, the quantity of seepage and the exit gradient are the minimum when the sheetpile is placed at the downstream end of the dam. However, the uplift force increases, but only slightly, when the sheet pile is moved from the upstream end to the downstream end.

It was also found that providing more than one sheetpile, in addition to the one provided at the downstream end, does not reduce the quantity of seepage or exit gradient significantly. Only the first sheetpile, at the downstream end, proved to be very effective in reducing these quantities.

REFERENCES

Harr, M.E. 1962. *Ground water and seepage.* McGraw-Hill.

Harr, M.E. 1977. *Mechanics of Particulate Media.* McGraw-Hill.

Harza, L.F. 1935. Uplift and seepage under dams on sand, *Transactions*, ASCE, Vol.100.

Holtz, R.D. and Kovacs, W.D. 1981. *An introduction to geotechnical engineering*, Prentice-Hall.

Pavlovsky, N.N. 1956. Collected works, Akad. Nauk USSR, Leningrad.

Scott, G.A. 1982. Piping potential of weak zones under concrete dams. *J. of the geotechnical engineering division*, ASCE, 108(GT3): 488-493.

Environmental Management, Geo-Water & Engineering Aspects, Chowdhury & Sivakumar (eds)
© 1993 Balkema, Rotterdam. ISBN 90 5410 099 0

Sustainable urban drainage

S. Sunjoto
Gadjah Mada University, Indonesia

ABSTRACT: Generally speaking, what is meant by urban drainage is to allow runoff water flowing from drainage areas to a trench or a canal , then to a river and finally to the sea. Consequently the decrease of groundwater recharge in the urban areas is directly proportional to the increase of the pavement and roof area. In this paper, the infiltration well idea will be introduced. It is designed to facilitate the storage and infiltration of precipitation which falls on the roof or pavement. A method of analysis to determine the minimum dimension of infiltration well and to get the maximum quantity of water infiltrated is proposed, using a new formula of calculation derived mathematically. This formula is based on Darcy's law concerning the unsteady state radial flow and that the measurement of the infiltration well is related to the roof or pavement area, intensity of rainfall for certain period, hydraulic conductivity and shape factor of the bottom of the well. The idea of this system is environmentally approached, which is to facilitate groundwater storage by maintaining water conservation as to avoid the deficit of water in the future and to support the principles of sustainable development.

1. INTRODUCTION

1.1. Drainage concept

The drainage concept applied so far can be called "pull and push" which means pulling water from the surface to prevent ponding water , by using drainage networks which pushes water to the river and eventually to the sea. The concept is applied to both problems: surface drainage and subsurface drainage, and it results in the abandonment of water resources and is against environmental principles.

1.2. Water balance in Indonesia

According to the findings by many researchers, the water table level of most big cities in Indonesia is decreasing. In Jakarta, for instance, the capital of the Republic of Indonesia, the surface of confined groundwater decreases between 30 - 50 meters during the last century, and the elevation of unconfined groundwater surface decreases about 0.50 meter per year. According to the Time magazine (November, 1990), the precious resource of world's water is becoming scarcer due to the population growth and development Both have depleted and polluted the world's water supply, raising the risk of starvation, epidemic and even war. The rate of population growth in Java and Madura is so high that the Departement of Public Works (1984) predicted that both island will have a deficit of 50 % in the year 2000 because the water necessity is one and a half times the available water.

1.3. Prevention

To prevent the above problems, general disciplines intend to improve water balance deficit by increasing dependable flow, re-utilisation, terracing and other technical related means. There is in Indonesia a traditional technique to conserve surface water which then becomes groundwater by using the concept of an infiltration well. Sunjoto (1988) proposed a formula to calculate an optimum dimension of the infiltration well.

2. INFILTRATION WELL

The idea is based on the concept of natural water balance as a part of hydrological cycle. When the ground surface is covered fully with vegetation and is permeable, some precipitation will then infiltrate naturally and the rest as runoff flows along the ground surface. But now that the world's population is growing which results in the increasing needs of housing, manufacturing, transportation facilities etc., much of ground area is covered by impervious layers such as roofs, pavements, etc. These layers prohibit water infiltration process and consequently there will be more runoff water than that infiltrated and stored as ground water. It is the infiltration well that will infiltrate precipitation in spite of the impervious layers over the ground surface. This technique allows water to get in to the well which functions as a storage facility, as well as to infiltrate, to become ground water.

3. PARAMETERS OF CALCULATION

An appropriate method of calculation using relevant parameters is needed in order to get a suitable design. In other words, the dimension of an infiltration well depends on:

3.1. Surface coverage area

A surface coverage area is an area where precipitation falls and flows to the infiltration well.

3.2. Intensity

Intensity is a value of the depth of precipitation which is calculated statistically, and gives different value for different geography, dominant duration, and frequency of precipitation.

3.3. Hydraulic conductivity

Sunjoto's formula (1988) is based on the Darcy's law, that is the discharge of the infiltration flow depends on this coeffient.

The three parameters are the main parameters to calculate the dimension of the infiltration well. While other parameters necessary to control the result of the calculation are as follows:

3.4. Dominant duration

Dominant duration of precipitation is the most frequent occurrence of precipitation in a certain region. It is used to determine the value of the intensity by a curve of duration against intensity.

3.5. Time lag

When water gets into the infiltration well, it will infiltrate as much as the flow capacity of the well into the deeper soil. Meanwhile the rest of water will stay in the well for a certain period of time. The time lag of precipitation can be used to control the duration of the previous storage in order to prevent spilling from the well.

3.6. Groundwater surface

The bottom of the well should be at a higher level than the highest surface of the local unconfined groundwater as to obtain a filter layer and maintain the economic interest.

3.7. Wells distribution

The distribution of infiltration wells and digging wells affects the water table level , and "multi well systems" can be applied to answer the existing problem.

4. DERIVATION OF FORMULA

Sunjoto (1988) proposed a mathematical formula based on flow phenomena to mesure the dimension of well infiltration, with the derivation as follows: (See Fig. 1)

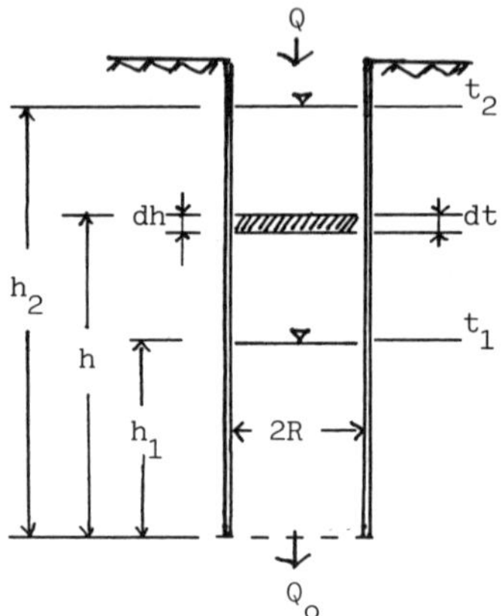

Figure 1. Schema of flow phenomena in the infiltration well.

The infiltration discharge is:

$$Q_o = F.K.h$$

Storage is the difference between roof water discharge and the infiltration discharge:

$$dv = (Q - Q_o) \, dt$$

$$dv = (Q - F \, K \, h) \, dt \qquad (1)$$

The quantity of water equals the depth of water times the cross section area of the well :

$$dv = S\ dh \quad (2)$$

where $S = \pi R^2$

Equation (1) = (2) so:

$$S\ dh = (Q - F K h)\ dt$$

$$dt = \frac{S\ dh}{Q - F K h}$$

$$t = \int \frac{\frac{S\ dh}{F K}}{\frac{Q}{F K} - h}$$

The result of the integration is:

$$T = - \frac{S}{F K} \left(\ln \left(\frac{Q}{F K} - H \right) - \ln \frac{Q}{F K} \right)$$

$$- \frac{F K T}{S} = \ln \left(1 - \frac{F K H}{Q} \right)$$

$$1 - \frac{F K H}{Q} = e^{- \frac{F K T}{S}}$$

with $S = \pi R^2$ so:

$$H = \frac{Q}{F K} \left(1 - \exp \left(- \frac{F K T}{\pi R^2} \right) \right)$$

Where:

H : water depth in the well (m)
F : the shape factor (m), See Table 1.
Q : water discharge from the roof (m^3/s)
T : the flow duration (dominant duration) (sec)
K : hydraulic conductivity (m/s)
R : the radius of the well (m)

The result is that the formula shows an unsteady state radial flow, but if the value of the "e power" is big, then the formula becomes:

$$H = Q/FK \text{ or } Q = F K H$$

and this formula indicates the steady state.

The shape factor depends on the layer of the soils, the form of the well bottom, and the type of wall side of the well either permeable or impermeable.

5. CONSTRUCTION

Infiltration well is like a digging well only of different function. A digging well keeps water to be drawn but an infiltration well is to conserve water. The well should have a brick wall or be of concrete pipe, or as such. The inner space of the well should be left empty as to be able to store water before it infiltrates. The bottom of the well is covered with broken rocks and gravels to prevent the layer from being eroded by the water fall.

The bottom should have a porous layer in order to have minimum dimension. The upper part of the well is covered with concrete slab and soil put above the slab. Here one can grow flowers for better landscaping and sturdier construction.

6. ADVANTAGE

The advantage of these systems are:

1. A quantity of unconfined ground water can be conserved.
2. The surface of the unconfined groundwater stays stable.
3. The dimension of the drainage networks is minimized.
4. The area of ponding water is minimized
5. Groundwater pollution is minimized.
6. Land subsidence is prevented.
7. Salted water intrusion in coast areas is avoided.

7. CONCLUSION

There are a lot of techniques to conserve water. But using an infiltration well for urban areas is considered appropriate. In which case the design is meant to provide a solution to the problems of potential water deficiency.

The calculation procedure is as follows:

1. Calculate water discharge from the roof to the infiltration well using Rational Formula as:

$$Q = C.I.A$$

where:

Q : discharge
I : intensity
A : area of the roof
C : runoff coefficient of the roof

2. Calculate the depth of water in the infiltration well using Sunjoto's formula (1988) as:

Table 1. Value of shape factors

No.	Case	Shape factor, F (m)	References
1	-2R-	$4 \pi R$	Samsioe (1931)* Dachler (1936)* Aravin (1965)
2	-2R-	$2 \pi R$	Samsioe (1931)* Dachler (1936)* Aravin (1965)
3	-2R-	$4 R$	Forchheimer (1930)* Dachler (1936)* Aravin (1965)
4	-2R-	$5.5 R$	Harza (1935)* Taylor (1948)* Hvorslev (1951)*
		$2 \pi R$	Sunjoto (1989)
5	-2R-, L	$\frac{2 \pi L}{\ln(L/R + \sqrt{(L/R)^2 + 1})}$	Dachler (1936)*
		$\frac{2 \pi (L + 2/3 R)}{\ln((L+2R)/R + \sqrt{(L/R)^2 + 1})}$	Sunjoto (1989)
6	-2R-, L	$\frac{2 \pi L}{\ln(L/2R + \sqrt{(L/2R)^2 + 1})}$	Dachler (1936)*
		$\frac{2 \pi (L + 2/3 R)}{\ln((L+2R)/2R + \sqrt{(L/2R)^2 + 1})}$	Sunjoto (1989)
7	-2R-, h_w, h_o	$\frac{\pi (h_o + h_w)}{\ln(h_w/R + \sqrt{(h_w/R)^2 + 1})}$	Sunjoto (1989)
8	-2R-, D	$\frac{2 \pi D}{\ln(2(D+2R)/R + \sqrt{(2D/R)^2 + 1})}$	Sunjoto (1989)

Source : * Olson & Daniel (1981)

$$H = \frac{Q}{F K} \left(1 - \exp \left(- \frac{F K T}{\pi R^2} \right) \right)$$

3. Control the calculated water depth by using time lag data, the unconfined groundwater table level and finally multi well systems if there are many wells.

REFERENCES

Aravin, V.I. & Numerov, S.N. 1965. Theory of fluid flow in undeformable porous media. Transl. from Russian. Jerusalem.

Dachler, R. 1936. Grundwasserstromung. Julius Springer, Wien.

Darcy, H. 1856. Histoire des fontaines publiques de Dijon. Dalmont, Paris.

Forchheimer, P. 1930. Hydraulik, 3rd ed. Teubner, Leipzig.

Harza, L.F. 1935. Transactions, ASCE, Vol. 100, pp. 1352-1385. New York.

Hvorslev, M.J. 1951. Time lag and soil permeability in groundwater observation. Bul. No. 36, Waterways Experiment Station. Vicksburg, Missisippi.

Olson, R.E. & Daniel, D.E. 1981. Measurement of the hydraulic conductivity of fine grained soils, Permeability and groundwater contaminant transport. ASTM, STP 746. Zimmie, T.F. & Riggs, C.O.

Samsioe, A.F. 1931. Zeitschrift fur angewandte mathematik und mechanik. Vol. 11, pp. 124-125.

Sunjoto, S. 1988. Optimation of infiltration well to avoid salted water intrusion in coast areas, Proc. IUC-UGM, Yogyakarta.

Sunjoto, S. 1989. The development of ground water hydraulic model. IUC-UGM, Yogyakarta.

Taylor, D.W. 1948. Fundamentals of soil mechanics. New York.

Environmental Management, Geo-Water & Engineering Aspects, Chowdhury & Sivakumar (eds)
© 1993 Balkema, Rotterdam. ISBN 90 5410 099 0

Areal investigation of surface runoff propensity in a model basin (Turbolo Torrent, Calabria, Italy)

A.Taddei & O.Terranova
Research Institute for Hydraulic-Geologic Protection in Meridional and Insular Italy, Italian National Research Council, Italy

ABSTRACT: The aim of this investigation is to choose a geomorphoclimatic model capable of automatically recognizing the areas in Italy characterized by the greatest surface runoff propensity in cases of heavy rainfall. A recently conceived model (Celico; 1992) chosen from the literature on the subject, was taken as a point of reference. Calibration was carried out by applying the methodology to the Turbolo Torrent model basin. Of the physical factors which most noticeably influence the separation mechanism of surface streaming from infiltration, three were analized: lithology, land use and ground-surface slope. The result of this preliminary investigation can be found in the thematic maps of the three factors, analized on a scale of 1:25.000 and by one final map relating to the surface runoff propensity obtained from the superimposition of all three.

1 INTRODUCTION

The problem of flood prevention is, in Italy, a very real one. It is, consequently, extremely important to find a rapid procedure which allows us to distinguish areas on the basis of the ways in which these contribute to flood formation.

The methodologies proposed in the literature, SHE (Abbott et al.; 1986), SCS Curve Number (Soil Conservation Service; 1972), NERC (1975), often require the kind of information which exists only in those few italian areas that have been the subject of in-depth study. In order, therefore, to model flood formation in basins without direct measurements, we need a method which uses easily-acquirable, large scale physiographical data. To this end, Celico et al. (1992) have proposed a geomorphoclimatic model whose aim is to calculate the average runoff coefficient. Running parallel to this study, the pricipal aim of the present investigation is the delimitation of areas with different surface runoff propensities. The attainment of this objective constitutes the basis for the construction of distributed-parameter rainfall-runoff models and their subsequent calibration.

The methodology in question is currently being experimented in the Turbolo Torrent basin which has been the subject of geological and geomorphological studies for some years now (Terranova et al., 1991), and which is equipped with the appropriate network of rain and stream gauges.

2 GEOGRAPHICAL AND CLIMATIC BACKGROUND

The Turbolo Torrent lies in the north-central part of Calabria (Cosenza, Italy), and is a tributary of the River Crati which it joins in the middle stretch. The torrent originates in the Coastal Chain and develops longitudinally from west to east for a length parallel to the main stream of approximately 11.5 km. The subtended basin is about 29 kmq and is noticeably elongated in shape.

Precipitation within the basin is extremely variable both in space and time. In fact the average annual rainfall is between 900 and 1600 mm, with an overall average of 1220 mm in the basin. The distribution of precipitations in time shows that, on average the wettest months are, in order, December and January and the least wet are July and August. Average annual temperatures vary from 10°C in the montainous areas to 17°C in those lower down. The hottest month is August and the coldest January. These climatic conditions affect the seasonal runoff cycle of the Torrent which, not being recharged by a ground water reservoir, runs dry for long periods in the summer and sometime floods violently during the winter.

3 GEOLOGICAL AND MORPHOLOGICAL BACKGROUND

There are extensive outcrops of marine and continental sedimentary rock, and to a lesser extent, metamorphic rock in the Turbolo Torrent basin. The metamorphic rocks are represented by biotitic-garnet type gneiss which tectonically overhang ophiolitic rocks. These materials, which form the skeleton of the strata buid-up of the Coastal Chain, appear to be intensely fractured and degraded. The metamorphic rocks crop out in the western part of the basin. The sedimentary rocks are for the most part clays, sand and conglomerates from the period between the Miocene and Olocene. The central-western band of the basin is occupied above all by blue-grey clays (Upper Pliocene), while further towards the east pleistocenic sand crops out extensively. After the depositing of these marine sediments, the final phase of continental filling (conglomerate with a sandy-silty matrix) resulted in extensive terracing. From the Pliocene onwards the area was involved in intense tectonic activity, and from the Pleistocene in considerable polyphasic uplifting.

The harsh metamorphic relief is linked at the valley bottom by ancient alluvial fans and by debris from more recent strata.

The hydrographic network is dendritic and subdendritic in the montainous areas and in those areas where Pliocene clays crop out (the western part of the basin), and because of faulting and stratification, it is predominantly rectangular in shape in the sandy areas. The network is sparse with a drainage density of 3.97 km/kmq and a stream density of 11 channels/kmq (Terranova et al., 1991).

4 METHODOLOGY USED

The formation of surface runoff is described via an analysis of those factors for which data are readily available throughout whole Italy. This description uses the Horton-type (Horton, 1933) mechanism of surface runoff production which predominates in our areas for geologic and climatic reasons.

In this model the only losses taken into consideration are those caused by infiltration, as others, like evapotranspiration or storage-detention, are negligible in cases of heavy rainfall.

4.1 *Lithology*

As a detailed knowledge of the lithology of an area can often make up for lack of more direct information, a lithological map with a scale of 1:25.000 was produced. Successively, using granulometric analyses, in-situ infiltration tests and bibliographical research as a basis, various grouping were made between the lithotypes identified. Four classes with different infiltration capacity were defined:

	surface
1) High ($f>10^{-2}$) breccia of alluvian fan	4.7%
2) Medium ($10^{-2}>f>10^{-4}$) metalimestone, conglomerate with a sandy matrix, sand, stratum debris	49.1%
3) Medium to low ($10^{-4}>f>10^{-6}$) gneiss, metabasites, sandy clays, conglomerates with sandy-silty matrices, sandy muddy gravelly alluvium	26.5%
4) Low ($10^{-6}>f$) phyllites and schists, clays and gneiss, clays and silty-clays.	19.7%

4.2 *Land use*

Land use influences infiltration. Ploughing, for example, increases infiltration, at least initially, while woodland intercepts water and models floods. Seven types of land use have been identified using aerial photography studies:

	Surface
1) Heavely wooded areas	12.2%
2) Sparsely wooded	15.5%
3) Olive groves	32.4%
4) Crops	14.5%
5) Uncultivated	21.5%
6) Bare	2.4%
7) Urban	1.5%

4.3 *Ground-surface slope*

The stepness of slopes also influences both the amount of water which collects and the speed at which it flows. Taken from information found in IGMI (Italian Military Geographic Institute) maps, five classes of ground-surface slope were distinguished:

	Surface
1) Gentle (0%-10%)	20.9%
2) Medium to gentle (10%-25%)	28.6%

3) Medium (25%-50%) 31.0%
4) Medium to steep (50%-100%) 16.1%
5) Steep (>100%) 3.4%

5 RESULTS OF THE INVESTIGATION

Thematic maps relating to the three physical factors under examination were made. These maps were produced and superimposed on one another using ILWIS cartographic software. The first table was produced by superimposing the infiltration capacity map on the land use one (table 1).

Table 1. Superimposing of infiltration capacity map on the land use one

	land use:	1	2	3	4	5	6	7
inf.	1	1	2	3	4	5	6	7
cap.	2	8	9	10	11	12	13	14
f:	3	15	16	17	18	19	20	21
	4	22	23	24	25	26	27	28

Of the 28 possible combinations, the 12 with a crop out surface of less than 1.5% were moved to either the class above or the one below, following logical criteria (similarity or adjacency). These particular groupings involved less than 7% of the total area. The 16 classes thus defined were listed in order of increasing surface runoff propensity. When listing, greater weight was given to infiltration capacity rather than to land use, as this former seems to have greater influence on rainfall-runoff transformation (Musgrave & Holtan, 1984). The urban areas were all included in class 16 (table 2).

Table 2. First grouping in classes of increasing surface runoff propensity

Table 1 class	New class	Surface	Table 1 class	New class	Surface
1,2,3, 4,5,6	-> 1	4.5%	17	-> 9	5.5%
8	-> 2	5.2%	18	-> 10	8.3%
9	-> 3	11.1%	19,20	-> 11	4.6%
10	-> 4	20.3%	22,23	-> 12	3.5%
11	-> 5	4.5%	24	-> 13	2.8%
12,13	-> 6	7.2%	25	-> 14	1.6%
15	-> 7	4.9%	26,27	-> 15	11.6%
16	-> 8	2.7%	7,14, 21,28	-> 16	1.5%

From this first superimposition of one map on another we find that land with medium permeability with a covering of either olives or sparse woodland shows noticeably extended out-cropping (20.3% and 11.1% respectively). Uncultivated areas and mediterranean bush with low infiltration levels are also extensive.

The map showing the 16 classes resulting from the superimposition of infiltration capacity on land use was, in turn, superimposed on the map showing ground-surface slope. Further grouping was carried out on the resulting table of 5 columns and 16 rows, arriving at the class subdivisions to be found in table 3.

Table 3. Final grouping in classes of increasing surface runoff propensity

Table 2 class \ Ground-surface slope class	1	2	3	4	5
1	1	1	1	1	2
2	1	1	1	2	2
3	1	1	2	2	3
4	1	2	2	3	3
5	2	2	3	3	4
6	2	3	3	4	4
7	3	3	4	4	5
8	3	4	4	5	5
9	4	4	5	5	6
10	4	5	5	6	6
11	5	5	6	6	7
12	5	6	6	7	7
13	6	6	7	7	8
14	6	7	7	8	8
15	7	7	8	8	8
16	7	8	8	8	8

From the final map we can see how, by following this methodology, the areas characterized by a greater propensity to surface runoff are to be found in central-western part of the basin, an area in which Pliocene clays crop out. The western portion of the basin benefits from the effects produced by vegetation, while the rest of the area is affected by the presence of lithotypes with high infiltration capacity.

6 CONCLUSIONS

The aim of the investigation carried out in the Turbolo Torrent basin was to isolate and calibrate a

TURBOLO TORRENT BASIN – CALABRIA – ITALY
SURFACE RUNOFF PROPENSITY MAP

Research Institute for Hydraulic Geologic Protection in Meridional and Insular Italy
Italian National Research Council (Cosenza – Italy)

GRAPHIC DISPLAY: *Dr. Antonella Taddei*
SUPERVISOR: *Eng. Oreste Terranova*

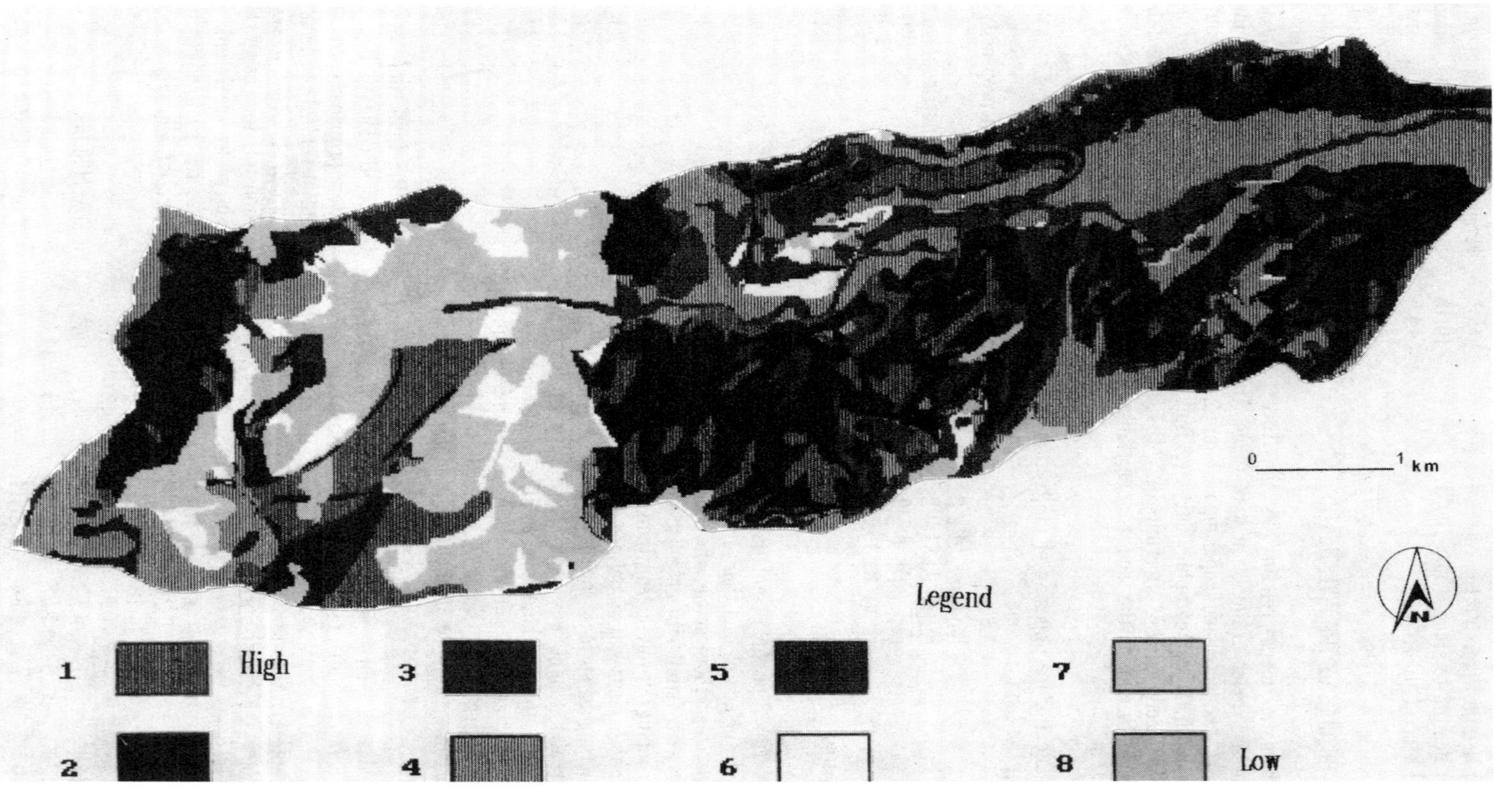

methodology which, using certain important physical parameters, was able to supply information on surface runoff and on flood formation in areas without hydrometrographic equipment. The literature on the subject contains numerous model which analyze surface runoff at basin scale. These models, however, rely on parameters which are rarely available in our country, such as depth of ploughing, soil moisture, slope, etc.. Consequently we decided to use such information as lithology, land use and ground-surface slope. An initial analysis of these factors resulted in a preliminary map on surface runoff propensity which was processed by the ILWIS G.I.S.. The second phase of this investigation forsees a verification of the methodology via a comparison with the hydrological data available on the area under examination. This will, in effect, be a means of calibrating the methodology, and will help to better define the relations which exist between the analyzed physical parameters. It will also help to verify the validity or oterwise of the analysis of a complex process, such as the rainfall-runoff transformation, based only on the study of a few physical factors.

AKNOWLEDGEMENT

The authors wish to thank Prof. P.B. Celico of the Salerno University for the useful discussions.

REFERENCES

Abbott M.B., Bathurst J.C., Cunge J.A., O'Connell P.E. & Rasmussen J. 1986. An introduction to the European Hydrological System - Systeme Hydrologique Europeen, "SHE", 2: structure of a physically-based, distributed modelling system. *Journal of Hydrology*, 87.

Celico P.B., De Innocentis M., Rossi F. & Villani P. 1992. Influenza dei parametri fisici del bacino sul coefficiente di deflusso di piena. *Rapporto 89' linea 1 GNDCI-CNR*, in stampa.

Horton R. E. 1933. The role of infiltration in the hydrologic cycle. *Trans. Am. Geophys. Union*, 14.

Musgrave G. W. & Holtan H.N. 1964. Infiltration. In *Handbook of applied hydrology* (V.T. Chow - Editor).

Natural Environment Research Council 1975. Estimation of flood peaks from catchment characteristics. In *Flood Studies Report*, 1, Londra.

Soil Conservation Service 1972. *National Engineering Handbook, Hydrology. Sect. 4*, Washington D.C.

Taddei A. 1991. Studio geologico del bacino idrografico del Torrente Turbolo (Bacino del Crati, Calabria). *Rapporto interno CNR IRPI (CS)*, 339.

Terranova O., Catalano E., Langellotti M. & Sorriso-Valvo M. 1990. Individuazione dei parametri caratterizzanti i fenomeni idraulici ed erosivi di un picccolo bacino (T. Turbolo - Calabria). *1st workshop on Informatica e Scienze della Terra* Sarnano 1989, De Frede, Napoli.

Environmental Management, Geo-Water & Engineering Aspects, Chowdhury & Sivakumar (eds)
 ISBN 90 5410 099 0

Integrated approach to the exploration of groundwater along the Tuticorin coast, India

R.Thirugnana Sambandam, A.Balasubramanian, R.Chellasamy & V.Radhakrishnan
Department of Geology, VO Chidambaram College, Tuticorin, Tamil Nadu, India

ABSTRACT: An integrated approach involving the hydrogeological, hydrochemical and geoelectrical methods has been attempted to explore the fresh water zones along the coastal aquifers of Tuticorin, Tamil Nadu, India. The study area is comprised of Archaeans, Tertiaries and Recent to Sub-Recent formations. Groundwater occurs mostly under water table conditions. Electrical resistivity studies indicate the lateral and vertical occurrence of saline water and fresh water. Sand and gravel (porosity 30 to 40%) contain fresh water. Clayey formations (porosity as high as 60%) are saturated with saline water. Graphical interpretation of hydrochemical data reveals that the aquifers contain mostly groundwaters of NaCl type and the mechanism controlling the chemistry is found to be saline intrusion. The potable potential zones having < 250 ppm of chloride in groundwaters are existing far away from the coastline. Modelling studies are recommended for appropriate management strategies.

1 INTRODUCTION

The rational use and management of groundwater in coastal areas and islands are of particular importance because of

i) the increasingly high concentration of population in coastal belts all around the world, especially in deltas, major sea ports and tourist resorts,

ii) the limited availability of potable water due to high salinity and

iii) man-made contamination.

Saltwater intrusion in coastal aquifers is a major hazard to the public in the coastal areas all over the world. Prediction and control of saltwater encroachment in a particular area is essential for the management of water resources in coastal aquifers. Water resources evaluation in coastal aquifers should take into account the presence of saltwater and the mixture of fresh and salt waters. No single technique can provide unique results with necessary accuracy and hence an integrated appraoch using hydrogeological, hydrochemical and geoelectrical methods has been employed in the present study to explore the zones of potable fresh groundwater in the coastal aquifers of Tuticorin (Pearl City), Tamil Nadu, India.

2 LOCATION

Tuticorin coast, which encompasses an international sea port, is located between latitudes 8°15' to 9°0' N and longitudes 77° 50' to 78°15' E and covers an areal extent of 2000 sq.km. (Fig.1). This coastal belt falls under the semiarid climate zone and receives a very scanty rainfall with an annual average of 677 mm which is far less than that of the state average, 942 mm. Topographically, this tract has a wide stretch of low land (15m above msl) and a narrow midland (between 15 and 50m). Groundwater is replenished by the river Tambraparni and the small ephemeral stream, Karamaniyar, both of which confluence with the Bay of Bengal.

Tuticorin and its neighbouring towns are inhabited by a population of 900,000 people. Cultivation is the major land use in the western portion besides the presence of a large number of salt pans and salt-based and fishery industries along the coast. Chief soil types are red loamy soil, occuring in the western part of the area and black cotton soil in the northeastern part. Beach sand occurs along the coastline.

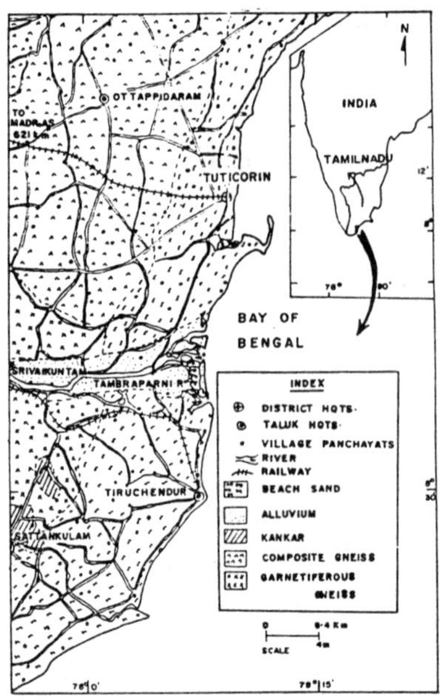

Fig. 1 Location of Tuticorin coast, Tamil Nadu, India

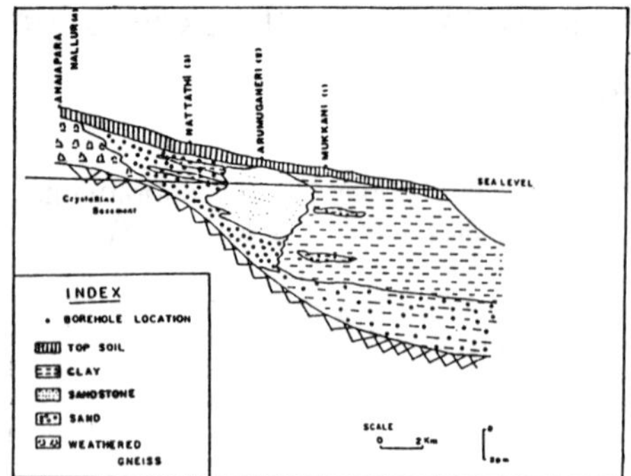

Fig. 2 Subsurface geology of the coastal zone

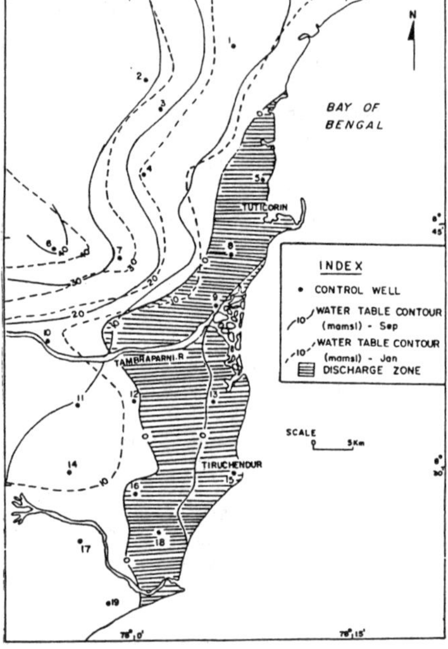

Fig. 3 Groundwater levels and zone of discharge

3 HYDROGEOLOGIC SETTING

Tuticorin coast is covered by the Archean, Tertiary and Recent to Sub-Recent formations (Varadaraj 1989). The Archean complex includes ridges of quartzites, bands of charnockites, calc-granulites and the basement peninsular gneisses. The Tertiaries are made up of a few patches of hard, compact sandstones and shell limestones. They occur as a narrow belt with a thickness varying from 2 to above 20m. The sandy coast is 3 to 20 km in width. A narrow strip of recent alluvium is confined to the course of the river Tambraparni, with a width of about 100-500m on either side and its thickness varies from 5-15m. The subsurface lithology encountered in the area is depicted in Fig. 2, based on the borehole lithologs.

Groundwater occurs under watertable conditions in the fissured crystalline complex and also under watertable and semi-unconfined conditions in the sedimentary aquifers. The average depth to the watertable varies from 1 to 10m below ground level (bgl) during post-monsoon period (Oct.-Dec.) and 5 to 15m bgl in pre-monsoon period (Jun-Sept.). The analysis of longterm mean watertable contours of September and January (Fig. 3) indicates that the water levels go below the mean sea level (msl) during September denoting the drought conditions, poor storage characteristics and possibilities of saltwater encroachment in the aquifers. Based on the grid deviation concept of watertable analysis (Biswas & Chatterjee 1967), the area has been divided into zones of recharge and discharge. Any developmental activities confined to the discharge zones will develop sea water incursion which needs an appropriate management strategy to tackle.

4 GEOELECTRICAL SURVEYS

Electrical resistivity methods are normally employed to delineate the groundwater potential zones and also to demarcate the saltwater - freshwater interface in coastal regions (Arora and Bose 1981; Balasubramanian et al. 1985; Melanchthon et al. 1988). Vertical Electrical Soundings (VES) have been conducted at sixty locations using Wenner configuration with a maximum electrode spacing of 100m using Aquameter - CRM 500 (Computerised Resistivity Meter) model. In order to understand the relationships between the ground water quality, subsurface lithologic units and aquifer resistivity, the field curves have been compared with aquifer details (Fig.4.). In the earlier investigations made in different parts of the world (Zohdy et al. 1974; Stewart 1982; Sathyamoorthy and Banerjee 1985), it has been reported that the zones saturated with saltwater show a very low resistivity of < 10 ohm-m. The results of the present work conform to the earlier studies (Balasubramanian et al. 1985). Iso-apparent resistivity map (Fig. 5) has been prepared with reference to 0, 10, 30, & 50m below msl despite the limitations involved in the concept of depth of penetration. Figure 6 shows the geoelectric sections of the coastal belt. The zones with apparent resistivity of <10 ohm-m have been found to be increasing in areal extent with reference to depth.

The coastal belt can be conveniently classified into four zones based on resistivity values as given in Table 1.

Table 1. Aquifer resistivity and formation charcteristics.

Resistivity in ohm-m	Formation charcteristics
< 10	Impermeable clay deposits or permeable sand formations with highly saline water (Ec: > 6,000 micro-mhos)
10-20	Sand or clay horizon with brackish water (Ec : 3000 micro-mhos)
20-60	Sandstone, shell limestone and weathered crystallines with fresh water (Ec : < 1000 micro-mhos)
> 60	Sedimentaries or crystallines with or without water

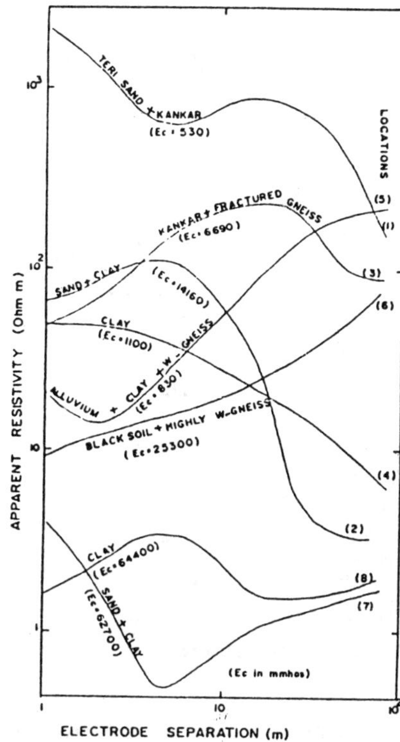

Fig. 4 Electrical sounding curves and aquifer characteristics

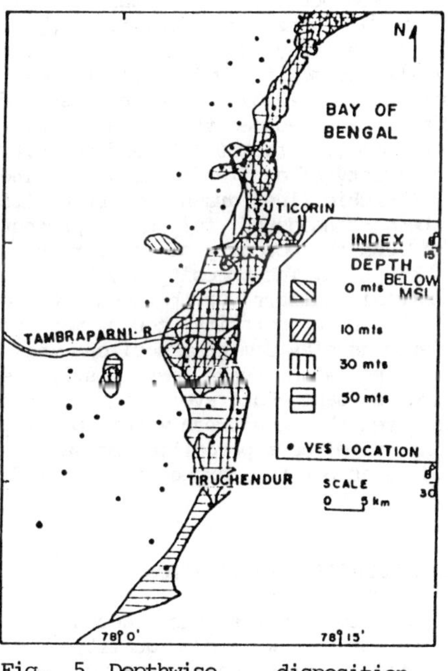

Fig. 5 Depthwise disposition of saltwater zones

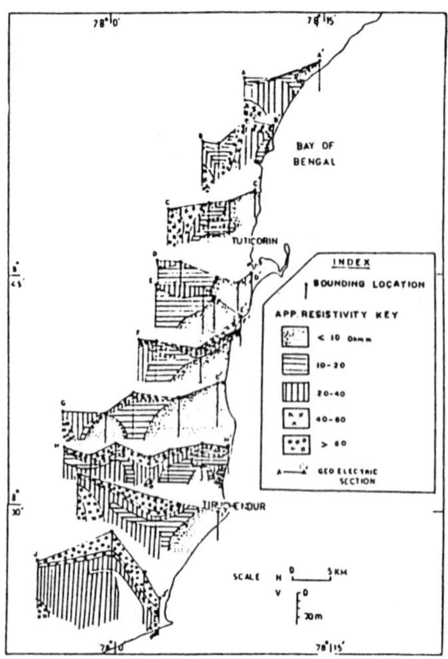

Fig. 6 Geoelectric sections

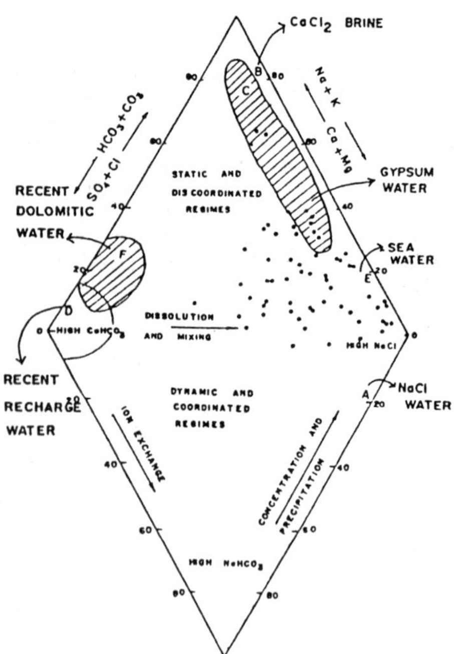

Fig. 7 Hydrochemical facies of groundwaer (After Johnson 1975)

The relationships between hydrologic parameters such as porosity, permeability, transmissivity, etc. to the aquifer resistivities and thickness have been studied by several authors. Worthington (1976) worked out the relationship between porosity (ϕ), water resistivity (ρ_w), aquifer resistivity (ρ_o) and formation factor (FF) by modifying Archie's formula. Jackson, et al. (1978) conducted a set of experiments and deduced the relationships between formation factor (FF) and porosity for a variety of aquifer materials. The best fit Archie lines drawn for FF=n^-1.5 has been considered for our present analysis and the porosity of aquifers have been computed. The porosity values of the coastal aquifers vary between 30 to 60%. The high porosity zones may be considered as clay dominant horizons while the low porosity areas may mostly be occupied by sand and gravel. It can also be inferred that the fresh water bearing sand and gravel have porosities between 30 and 54 % as per the expectation (Todd 1959).

5 HYDROCHEMICAL INVESTIGATIONS

In order to analyse the geochemical patterns of flow and quality of groundwater, water samples have been collected from sixty locations where the resistivity soundings were conducted. The water quality parameters such as pH, electrical conductivity, total dissolved solids(TDS) and temperature of groundwaters have been measured using a digital water analysis kit. The concentration of major ions (Ca, Mg, Na, K, Cl, NO3, SO4, and HCO3) have been analytically determined following ISI methods of water analysis (1964). The hydrochemical facies of groundwater are normally interpreted using the Piper's trilinear plots (Piper 1944) and Durov's diagram(Zaporozec 1972).

Figure 7 shows the hydrogeochemical environments of groundwater of Tuticorin and the facies to which they belong. The characteristics of sea water, dissolution, mixing and the effects of contamination with gypsum have also been inferred from this plot. The chemical data has been plotted over Durov's diagram (Fig. 8) and it has been found that almost all groundwater samples are of NaCl type. This type would be the ultimate product of either prolonged groundwater residence or saline intrusion.

Stuyfzand (1989) classified groundwaters of coastal areas into eight main types based on chloride concentration as shown in Table 2.

Table 2. Main groundwater types based on Chloride.

Main type	ppm
Very Oligohaline	< 5
Oligohaline	5 - 30
Fresh	30-150
Fresh - Brackish	150-300
Brackish	300-1000
Brackish salt	1000-10,000
Salt	10,000-21,000
Hypersaline	> 21,000

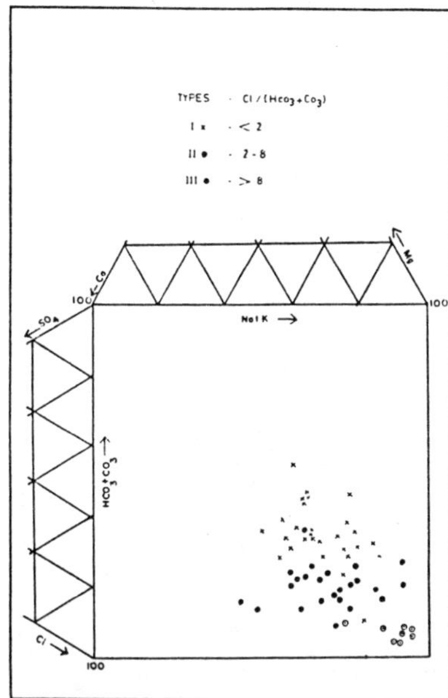

Fig. 8 Durov's plot

The quality of groundwater of this terrain has been assessed based on this criterion (Fig. 9). It could be inferred that the entire area is occupied with fresh-brackish water away from the coast line, brackish to saltwater nearer the coastline and saltwater along the coast.

6 SUMMARY AND CONCLUSIONS

On the basis of the results of an integrated study to the exploration of fresh groundwater in the coastal aquifers of Tuticorin, Chidambaranar district, Tamil Nadu, it has been established that saline

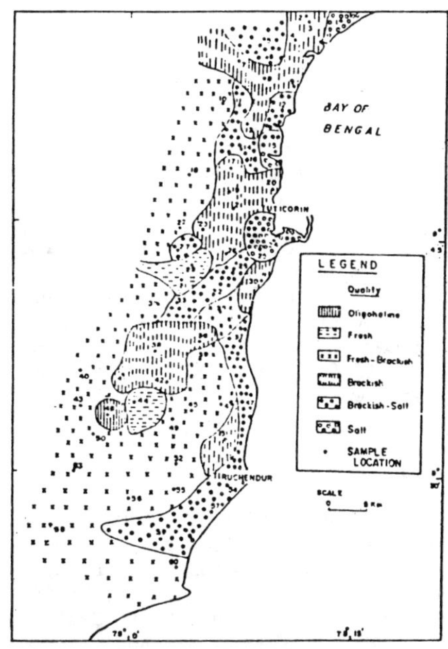

Fig. 9 Quality of groundwater in the coastal belt

water is invariably present at varying depths. Interpretation of electrical resistivity sounding data indicates that the aquifers saturated with brackish to freshwater are characterised by the resistivities of 10-20 ohm-m, those with freshwater posses 20-60 ohm-m. The higher order of resistivities were obtained mostly in the western part (where the weathered crystalline hard rocks are exposed). The low resistivity (<10 ohm-m) is indicative of the presence of clay saturated with saltwater in the sediments of the eastern part. The resistivities of formations confirm the presence of salt water at depths and also the disposition of saltwater-freshwater interface. The interpretation of the hydrogeochemical facies reveals the mechanisms controlling the chemistry of groundwaters and their qualitative characteristics for domestic, industrial and irrigational needs. Only salt tolerant crops could be grown in the brackish water zones. Exploitation of fresh groundwater should be done within he safe yield conditions which needs a detailed study of quantitative resources. Possible water management proposals could be made through numerical modelling of the aquifers.

ACKNOWLEDGEMENT

We gratefully acknowledge the Department of Science and Technology,Government of India, New Delhi for providing necessary funds to carry out this research work. Authors are especially grateful to Shri A.P.C.V. Chockalingam, Secretary, V. O. C. Educational Society, and Prof.S.Duraisamy, Principal, V.O. C. College, Tuticorin for their encouragement. We thank the Chief Engineer, Groundwater(PWD),Govt. of Tamil Nadu, for providing longterm water level data. Sincere thanks are also due to our project staff for their logistic support in the field work.

REFERENCES

Arora,C.L. & Bose,R.N. 1981. Demarcation of fresh and saline-water zones, using electrical methods (Abohar area, Ferozepur dist., Punjab) J. Hydrol. 19: 75-85.

Balasubramanian,A., Sharma,K.K. & Sastri, J.C.V. 1985. Geoelectrical and hydrogeochemical evaluation of coastal aquifers of Tambraparni basin, Tamil Nadu, Geophys. Res. Bull. 23(4):203-209.

Biswas,A.B. & Chatterjee,P.K. 1967. On representation of water table by grid deviation method. Bull. Geol. Soc. Ind. 4:12-14.

ISI 1964. Indian standard methods of sampling and test (physical and chemical) for water used in industry:122.

Jackson,P.D., Smith,D.T. & Stanford,P.N. 1978. Resistivity - porosity -particle shape relationships for marine sands. Geophysics.43(6): 1250-1268.

Johnson,J.H. 1975. Hydrochemistry in groundwater exploration. Groundwater symp. Bulawaya. 1975.

Melanchthon,V.J., Raju,V.N., Ramakrishna, A. & Bhowmick,A.N. 1988. Resistivity survey for mapping fresh water pockets. J. Assoc. of Expl. Geophys.IX(2):71-78.

Piper,A.M. 1944. A graphic procedure in the geochemical interpretation of water analysis. Am. Geophy. Union Trans. 25:914-923.

Sathyamurthy,K. & Banerjee,B. 1985. Resistivity surveys for locating brine pockets in the little Rann of Kutch, Gujarat. J. Assoc. of Expl. Geophys. VI (2):7-13.

Stewart,M.T. 1982. Evaluation of electromaganetic methods for rapid mapping of salt water interface in coastal aquifers. Groundwater. 20(5):538-545.

Stuyfzand,P.J. 1989. A new hydrochemical classification of water types. IAHS Publ.182:89-98.

Todd,D.K. 1959. Groundwater Hydrology, John Wiley & Sons, Inc, New York: 336.

Varadaraj, N. 1989. Groundwater resources and developmental potential of Chidambaranar district, Tamil Nadu, Unpublished GCWB Report. Southern Region, Hyderabad:23.

Worthington, P.F. 1976. Hydrogeophysical equivalence of water salinity porosity and matrix conduction in arenaceous aquifers . Geophys. Divi. NRPL, CSIR Pretoria, S.A.

Zaporozec,A.1972. Graphical interpretation of water quality data. Groundwater. 10(2):32-43.

Zohdy,A.A.R., Eaton,G.P. & Mabey,D.R.1974. Application of surface geophysics to groundwater investigations, in Tech. Wat. Resour. Inv. USGS, Book 2,CD1:116.

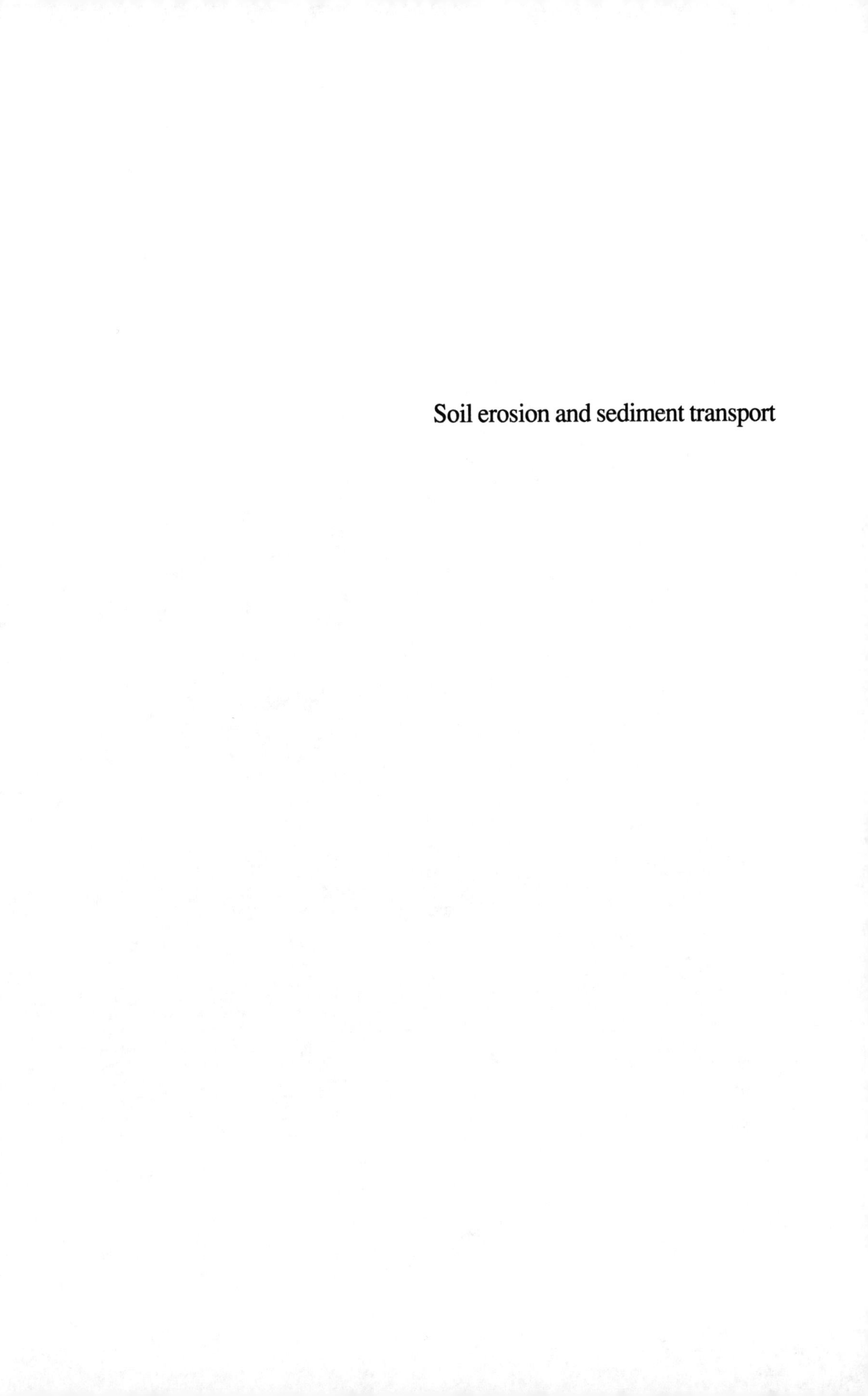

Soil erosion and sediment transport

Environmental Management, Geo-Water & Engineering Aspects, Chowdhury & Sivakumar (eds)
 ISBN 90 5410 099 0

Some erosion characteristics of residual soil slopes in Malaysia

Jamal Mohd Amin & Kamaruddin Abu Taib
Universiti Kebangsaan Malaysia, Malaysia

Izihan Ibrahim
A.I. and Associaties, Malaysia

Ahmad Azizi Dian
Za'aba Consultant, Malaysia

ABSTRACT: Erosion resulted from constructions in residual soil area has posed a problem to engineers. Occurence of erosion failures in slopes such as rills and gullies have been sighted. This paper attempts to study erosion due to surface wash. Surface wash is simulated by Flume test. Flume test on compacted samples are carried out with varying parameters such as tractive stress (T), duration and relative compaction (R.C.)

1 INTRODUCTION

As a growing country, Malaysia is undergoing many construction projects. Projects such as North-South expressway, gas pipeline utilization, housing development, golf course etc. contribute to the advancement of Malaysia. However associated with the developments, problems such as erosion during constructions are taking place.

Generally, process leading to erosion can be divided into three parts. The first is the surface wash where water moves as thin sheets over shallow or deep channel. The third is the subsurface flow where infiltration of surface water into earth contribute to piping due to dispersivity soil. Finally the third one is the rainfall drops that give impact to the soil particle and carries it down.

Runoff produced by rainfall on surfaces of slopes (surface wash) is a major cause of erosion especially in residual soils. The wet climate of Malaysia aggravates this condition resulting in failures of such slopes and substantial erosion. Water flows as thin sheets or in gullies on the slopes while transporting the soil particles downhill. To obtain an understanding of the processes involved, tests are designed in an open-channel flume to observe the behavior of erosion based on the relative compaction of the soil, tractive stress on soil surface due to the flowing water, and the duration of flow.

2 MATERIALS FOR THE STUDY

The soil sample used for this study are residual soil (Grade VI) taken at first 1 feet from the ground. The parent rocks are metasediment argillaceous and arenaceous rocks. The soil samples are taken from a golf construction site located near University Kebangsaan Malaysia campus. Location and geology of the study area are shown in Figure 1.

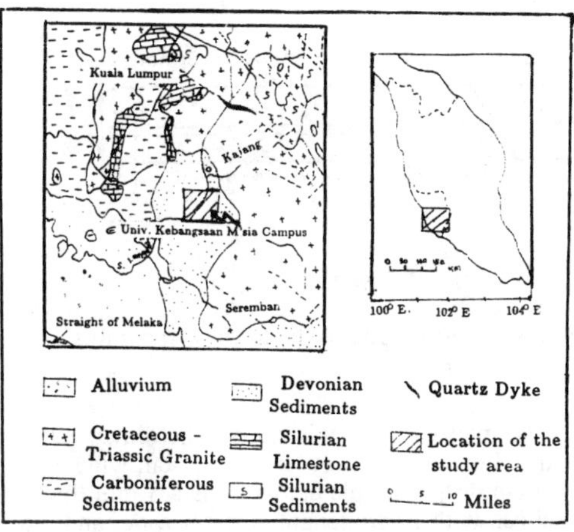

Figure 1: Location and Geology of the Study Area (after Tajri,1984)

The soils are yellowish red in colour. Disturbed samples passing 2.0 mm sieve opening are used for this study. The physical properties of the soil are shown in Table 1.

3 EXPERIMENTAL SET-UP

Figure 2 depicts a schematic of the flume and

Table 1: Some physical properties of the soil sample

Property	Soil Sample
Liquid limit	24
Plastic limit	63
Plasticity index	39
Specific gravity	2.71
Optimum moisture content*	16%
Maximum dry density*	19%
Fine grained soil	60%
Coarse grained soil	40%

* Standard Proctor Compaction

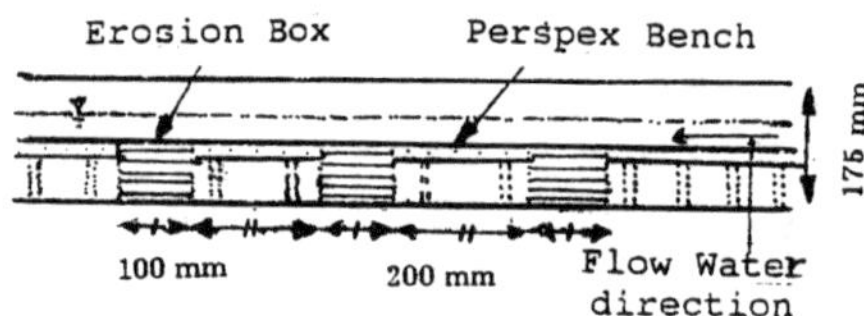

Figure 2 : Schematic of the testing Apparatus

soil samples used for this study. The dimensions of the flume are 480 cm length, 7.5 cm wide and 17.5 cm deep. Sample container is a 3 mm thick steel can with 10 cm length, 7.5 cm wide, and 3.5 cm deep. The flume floor was modified by placing perspex seated in between sample box. For a fixed flowrate, a value of tractive stress T acting on the surface can be derived from the Boys equation given by Chow(1959)(Shaikh et al, 1988):

$$T = \gamma R \sin \alpha$$

where

γ - unit weight of water
R - hydraulic radius of flow
α - slope of bed

The slope of the flume in this set-up is fixed at 0.07 due to limitations of the equipment.

Three samples of residual soils obtained from the site are compacted to a certain relative compaction and weighted accordingly. Once situated in the flume inside cans as shown, a

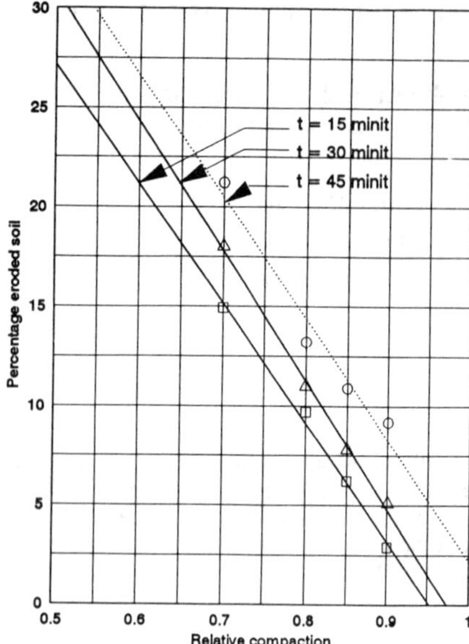

Figure 3: Percent eroded soil vs R.C. at different time for T = 14.6 N/m^2

certain discharge (from tap water) is obtained for a predetermined period of time. The amount of soil left after erosion is the dried weight to estimate the soil loss. In total, 45 samples have been tested by varying the parameters as follows:

R.C. : 0.70, 0.80, 0.85, 0.90
T : 1.18 N/m^2, 1.46 N/m^2, 70 N/m^2
t : 7,15,30, 45 minutes

4 RESULTS AND DISCUSSION

The amount and percentage of eroded soil is measured for each sample tested. From the tabulated results, several relevant plots have been constructed. They are percent erosion versus relative compaction and duration t for fixed tractive stresses T, percent eroded soil versus relative compaction and T for certain durations, percent eroded soil versus duration t and T for various relative compaction.

An example of the first set of plots is depicted in Figure 3. As can be seen, straight lines with negative slopes describe the relationship. This indicates that as the soil is more compact, the loss due to erosion is reduced somewhat linearly. The lines are parallel for durations of 15, 30, and 45 minutes. A densely compacted soil has less void ratio and higher shear strength to resist erosion than less compact soils. However, as the duration of flow increases more soil gets eroded, as expected.

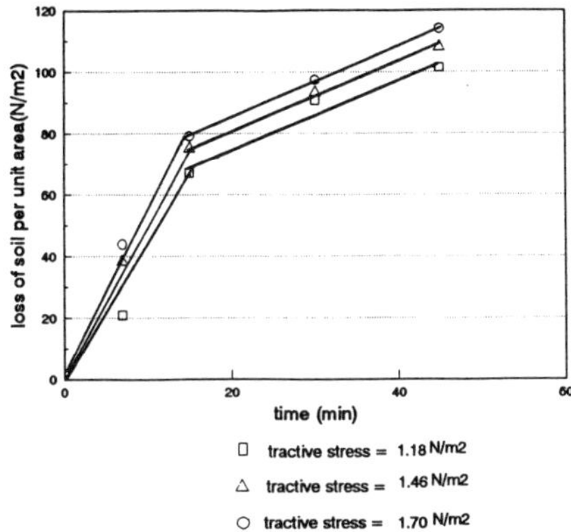

Figure 4: Loss of soil per unit area vs time for relative compaction 0.7

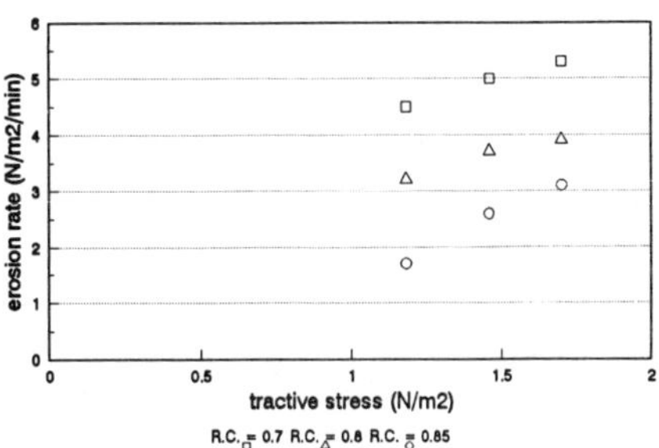

Figure 5: Rate of erosion vs tractive stress for different relative compaction, trend 1

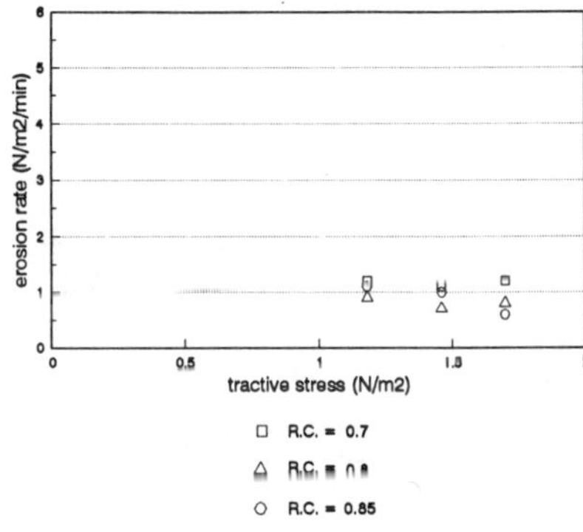

Figure 6: Rate of erosion vs tractive stress for different relative compaction, trend 2

The second set of curves more or less shows the same trend as the first. Here, the amount of soil eroded is inversely proportional to the relative compaction for a fixed tractive stress T. This pattern somewhat prevails for all three durations.

For the third set of plots, one interesting observation can be made. For a certain T and relative compaction, the soil loss per unit area is plotted against duration. As expected, Figure 4 depicts a general trend of increased erosion as time progresses. However this trend is not uniform throughout. In fact, two trends can be seen, where the first, up to 15 minutes, increases at a faster rate than the second. This implies that the rate of erosion is reduced abruptly at a certain threshold of time. Two possible explanations come to mind. The first is concerned with the non-uniform degree of compaction in the soil samples. Apparently, when the sample is compacted layer by layer from necessity, the bottom part is denser than the top. Thus the erosion rate decreases once the top layers are scoured away. A better compaction technique is needed to confirm this observation.

A second reason is that the velocity of the flow is somewhat reduced as scoured hole in the sample gets deeper with time. Assuming that the flowrate is constant, the deeper flow channel reduces the velocity over the sample. It is concluded that this results in a lower tractive stress acting on the soil surface, and hence a lower rate of loss. The fourth group of curves for various tractive stresses confirms this observation.

To conclude all these relationships, the rates of erosion are plotted against T for various relative compaction as the two trends in Figure 4 and Figure 5. As observed and discussed above, trends 1 and 2 are probably caused by the two factors. In Figure 4, the rate increased with T for all three relative compaction. The same however cannot be said for trend 2 (Figure 5). Erosion seems to slowly decrease with T. This is apparent at relative compaction of 0.80 and 0.85.

5 CONCLUSION

Results obtained from the flume tests are very encouraging as their behaviour is mostly not unexpected. In general, the amount of soil eroded from the samples decreases as relative compaction is increased. As the flowrate or discharge of water increases, so does the tractive stress acting on the soil surface, resulting in greater soil loss.

The non-uniformity of compaction in the samples and velocity changes in flow both contribute to the apparent two-trend behaviour in the erosion rate. Finally, the curves for these relationships can be used to estimate the rate of soil loss once the parameters of relative compaction and tractive stress have been determined.

6 REFERENCES

Ahmad Azizi bin Hj. Dian, (1992), "Surface erosion of residual soil near Univ. Kebangsaan M'sia" Thesis presented to Universiti Kebangsaan Malaysia, Bangi, in partial fulfillment of the requirements of Bachelor of Science in Engineering.(in Malay)

Chow,V.T. (1959),Open Channel Hydraulics,Mc Graw Hill,New York N.Y. 168 and 200.

Mohammad Tajri Bin Kamaruzzaman, (1984), "Engineering geological properties for metasediment weathering profile in Bandar baru Bangi, Selangor", Thesis presented to Universiti Kebangsaan Malaysia, Bangi, in partial fulfillment of the requirements of Bachelor of Science in Geology.(in Malay)

Shaikh,A.,Ruff,J.F.,Charlie,W.A. and Abt. S.R (1988), "Erosion rate at dispersive and nondispersive clays" Journal of Geotechnical Engineering, Vol. 114, No. 5, ASCE, pg 589-598.

Environmental Management, Geo-Water & Engineering Aspects, Chowdhury & Sivakumar (eds)
© 1993 Balkema, Rotterdam. ISBN 90 5410 099 0

Introducing computer model ANSWERS and its application in soil and water conservation

S.Amin
Shiraz University, Iran

H.Momtahan
Fars Regional Water Authority, Ministry of Energy, Shiraz, Iran

ABSTRACT: Lumped and distributed parameter models have been used extensively to simulate watersheds responses to rainfall events. In lumped parameter models, the averages of the input data are used, while distributed parameter models use spatially distributed inputs to predict the spatial effect of land use, soil type, fertility, and rainfall.

ANSWERS is an event oriented model which predicts peak flow, total runoff, and sediment yield and concentration from agricultural watersheds for an event. The simulation of 9-4-80 and 6-22-81 storms on an experimental watershed , Hoeppner, (located in Allen County, northeast Indiana, USA) showed that the simulated runoff for both storms was very close to the observed values. The prediction of sedimentation is also demonstrated by ANSWERS for the Upper Black Creek watershed of Allen County. The effect of different land use management for a natural storm on a small agricultural watershed located in College of Agriculture, Shiraz, Iran was also verified.

INTRODUCTION

In the past two decades the development of hydrologic and nonpoint sources simulation models has increased rapidly. Two major types models in agricultural watersheds are lumped and distributed parameter models.

This paper has a brief discussion about lumped and distributed parameter models used in watershed simulation. ANSWERS a distributed parameter model is introduced. IN contrast to lumped parameter models which require 3 to 5 years data from a watershed for calibration the ANSWERS model does not require calibration therefore, it can be easily transferred for simulation of different watersheds.

DIFFERENT TYPE OF MODELS

Many researchers have tried to make simplifications and abstractions to simplify the various aspects of a watershed behavior. According to Biswas (1976), hydrologists have produced symbolic or mathematical models of all types. There are linear and nonlinear models; deterministic and stochastic models, lumped and distributed parameter models, continuous time and discrete time models; and so on. In simulation of a watershed, the lumped and distributed parameter models are used extensively.

LUMPED PARAMETER MODELS

A lumped parameter model does not account for the spatial distribution in the input variables (Clrak, 1973), and consequently, input and output relations are expressed as a function of time but not space. In these models the characteristics of a watershed are "averaged or lumped " together and the processes of the watershed are described mathematically by some ordinary differential equations or system of them (Beasley, 1977; Woolhiser, 1973). Some of the advantages of lumped parameter models are that they are simple and basically more economical to use (Huggins and Monke, 1966). On the other hand , the disadvantages of lumped parameter models are:

1. They do not normally predict contribution from different part of a watershed.

2. they are base on calibrated coefficients which are not easily transferable to the other watersheds.

3. They are not capable of predicting affects of rapidly changing land use patterns.

The degree of the complexity of the lumped parameter models varies greatly. Two of the simplest examples are USLE (Universal Soil Loss Equation) for estimating of soil loss (Wischmeier and Smith, 1978). The more complex lumped parameter model used in agricultural watershed simulation is ARM (Agricultural Runoff Model) which developed in 1979 by Davis and Donigian.

DISTRIBUTED PARAMETER MODELS

In distributed parameter models, the spatial variability of a watershed is considered. These models are, therefore, usually more complex than lumped parameter models. The processes of a watershed are governed by partial differential equations or systems of them. Numerical approaches rather than analytical solutions are needed to solve these equations, procedure which takes time. Huggins and Monke (1966) considered this as the main disadvantage of this kind of model. these model require a geometrical network of points (Clark, 1973; Jayavardena and White, 1979; Foster, 1980). Huggins et al. (1973) listed the advantages of these models as:

1. Increased accuracy due to their inherent capacity to evaluate spatially variable factors within a watershed.

2. the watershed behavior is described in a comprehensive manner.

Distributed parameter models are able to evaluate the effects of different land use and crop management over a watershed with different soils types (Beasley, 1977).

THE ANSWERS MODEL

The ANSWERS (Areal Nonpoint Source Watershed Environment Response Simulation) model was developed by Beasley (1977). This distributed parameter model uses a grid system to define a watershed. The watershed under study is divided into a number of small elements or cells as shown in Figure 1.

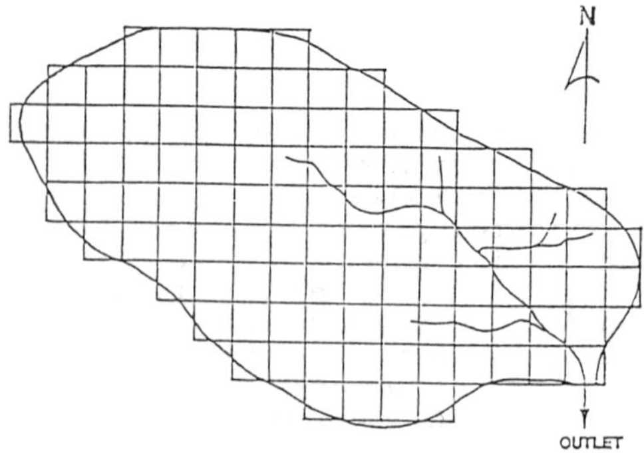

Figure 1. Grid system representing a watershed. (In this particular case, it is the Hoeppner Watershed located in Allen County, Indiana , USA), (From Amin Sichani, 1982).

These cells are independent from each other in soil characteristics, slope, topography, crop coverage and management, etc. The size of each cell must be such that uniformity of all hydrological characteristics of the cell can be assumed. Important hydrological and soil erosion parameters are defined in a predata file for a watershed before a storm event is simulated. The model outputs are a hydrograph of runoff and sediment concentration, total sediment yield from a watershed, and the sediment detachment and deposition from each element of the watershed.

A definite advantage of the ANSWERS model is that it can evaluate rapidly changing land use in a watershed. ANSWERS can evaluate the net erosion and deposition from elements of areas which are sensitive to erosion. These areas can be tested with different soil and crop coverage management practices to determine reductions in erosion or even the cost effectiveness of different conservation practices.

DESCRIPTION OF THE WATERSHEDS

The capability of the ANSWERS model to predict runoff and sediment from small and large sized watersheds located in Allen County, northeast Indiana, USA, is demonstrated. To show the effects of crop coverage management the model is used on a third small agricultural watershed in College of agriculture, Shiraz, Iran (Momtahan, 1989). The properties of these watersheds are outlined in Table 1.

RESULTS AND DISCUSSION

The ANSWERS model was used to simulte the storms of 9-4-80 and 6-22-81 of the Hoeppner watershed. The summary of the response of the watershed is in Table 2. Figure 2 and 3 show the watershed outputs from the watershed for these two events.

Figure 4 shows the observed and predicted hydrologic response and sediment losses from the Upper Black Creek watershed for a storm of 9-4-75. The output of the watershed from this storm is also depicted in Table 2.

The two previous watersheds and their corresponding storms were used to demonstrate the ability of the ANSWERS to closely predict actual response to a natural storm. An understanding of the capability

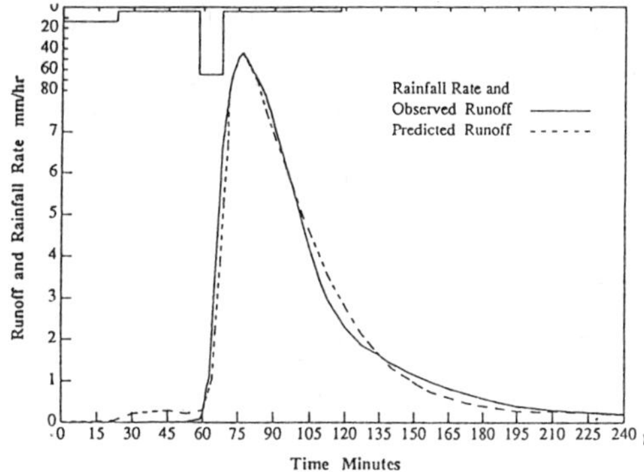

Figure 2. Rainfall and observed versus predicted runoff hydrographs at Hoeppner watershed (storm: 9-4-80).

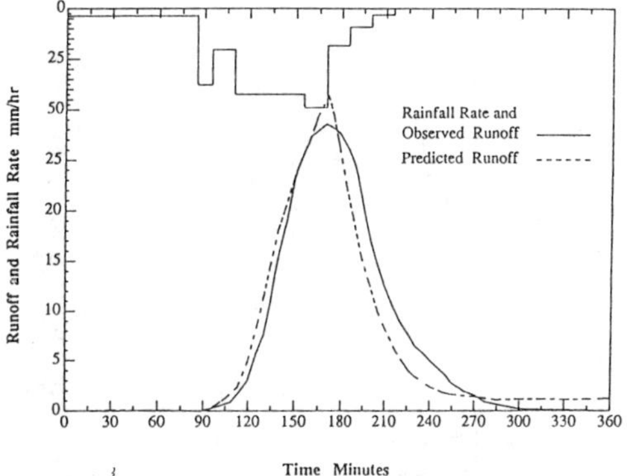

Figure 3. Rainfall and observed versus predicted runoff hydrographs at Hoeppner watershed (storm: 6-22-81).

Table 1. Experimental Watersheds properties

Watershed Name	Area (ha)	Average slope (%)	Soil type	Crop coverage
Hoeppner, USA	4.3	2.7	silt loam	small grain (30%), corn (70%)
Upper Black Creek, USA	714	1.1	loam, silty loam, loam, clay loam, slity clay	small grain (60%), row crop (40%)
Agricultural College, Iran	4.8	2.6	sandy loam clay loam	small grain (60%), fallow (40%)

Table 2. The watersheds responses to the rainfall events used in simulation.

Watershed	Rainfall			Runoff				
	Date	Depth (mm)	Duration (hr)	Observed			Predicted	
				depth (mm)	Peak flow (mm/hr)	Time of peak (min)	Peak flow (mm/hr)	Time of peak (min)
Hoeppner	9-4-80	21	2	6.8	9	78	9	78
	6-22-81	72.5	2.5	31.0	29	171	31.8	171
Upper Black Creek	9-4-75	28.2	34	2.8	0.53	644	0.48	579
A gricultural College, Iran	11-30-86	111.2	24	-	-	-	7.6	560

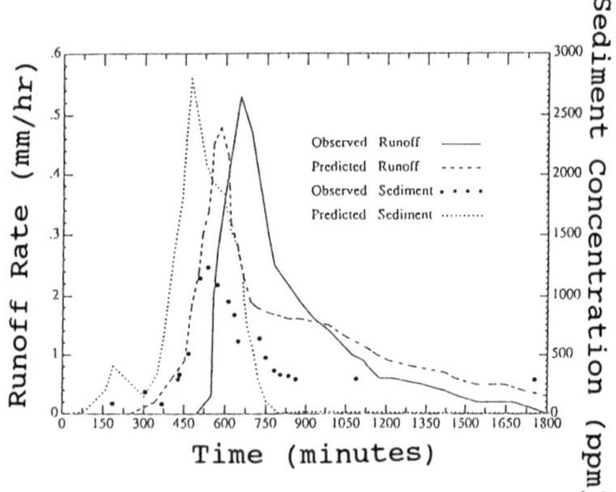

Figure 4. Black Creek watershed response (Storm: 9-4-75).

of the ANSWERS model for planning can be best illustrated with third example. This example is based on the Agricultural College small watershed when subjected to the storm of 11-30-86 (Azar, 9, 1365).

Figure 5 shows ANSWERS prediction of sediment detached through the catchment with fallow condition during that storm. The "contour" lines were created by connecting equal soil detachment points. Figure 6 represents simulation results which analyze the relative benefite of changing the coverage of the watershed surface from fallow to small grain, such as wheat, to reduce sediment yield and its association pollution. The

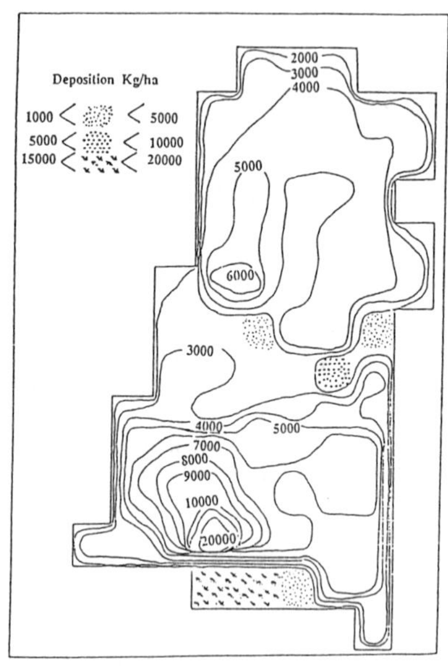

Figure 5. Response of the small Agricultural Watershed of the Agricultural College, Shiraz, Iran (storm:11-30-86), Fallow condition.

predicted sediment yield is less than one-tenth of the previous watershed response.

SUMMARY AND CONCLUSION

The ANSWERS model, a distributed, rate oriented, single event type model which was originally designed

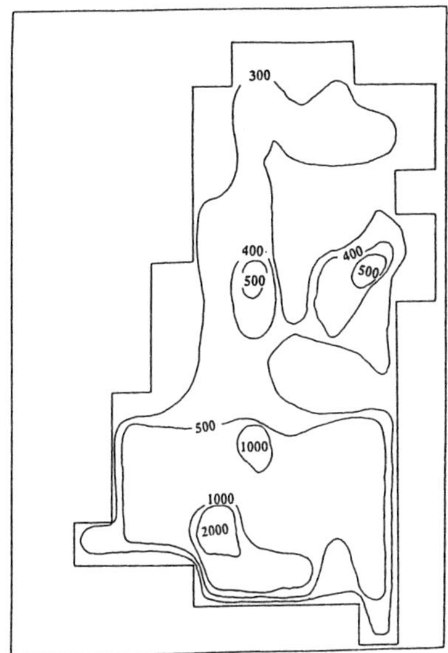

Figure 6. Response of the small Agricultural Watershed of the Agricultural College, Shiraz, Iran (storm:11-30-86), small grain coverage condition.

for agricultural applications, is introduced in this paper. The capability of the model in simulation of hydrologic responses were demonstrated using two storms events on a small agricultural watershed with 4.3 ha area in Allen County, northeast of Indiana, USA. The simulated and observed values of runoff were very close. Prediction of the sedimentation from the watersheds was demostrated using Upper Black Creek watershed with 714 ha area. The simulation results of sedimentation from this site was higher than observed values. However, the values are in reasonable and acceptable agreement.

The model was finally used for estimating the effects of different land management including crop coverage and fallow on a small agricultural watershed in College of Agriculture, Shiraz, Iran. Measurements are not available to directly determine the accuracy of these predictions.

The ANSWERS model has been tested in several climatic conditions in the USA and results were found to be in close agreement with respect to observed values.

REFERENCES

Beasley, D.B. 1977. ANSWERS: A mathematical model for simulating the effect of land use and management on water quality. Ph.D. Thesis, Purdue Univer. W. Lafayette, IN., USA. 266 p.

Biswas, A.K. 1976. System Approach to Water Management. McGraw-Hill Book Co., New York, NY., USA. 429p.

Clark, R.T. 1973. A review of some mathematical models used in hydrology, with observations on their calibration and use. J. Hydrology 19:1-20.

Davis, H.H., Jr., and A.S. Donigian, Jr. 1979. Simulation nutrient movement and transformations with ARM model. Trans. ASAE 22(1):151-154.

Foster, G.R. 1980. Soil erosion modelling: special consideration for nonpoint pollution evaluation on field-size areas. IN: Overcash, M.R. and J.M. Davidson (eds.), Environmental impact of nonpoint source pollution. Ann Arbor Science, Ann Arbor, MI, USA. pp. 123-240.

Huggins, L.F. and E.J. Monke. 1966. The mathematical simulation of the hydrology of small watersheds. Rept. 1, Water Resources Research Center, Purdue University. W. Laf. IN., USA. 130p.

Huggins, L.F., J.R. Burney, P.S. Kunda and E.J. Monke. 1973. Simulation of hydrology of ungaged watersheds. Rept. 38, Water Resources Research Center, Purdue University. W. Laf.IN., USA. 70p.

Jayavardena, A.W. and S.K. White. 1979. A finite element distributed catchment model. II. Application to real catchment. J. Hydrology 42: 231-249.

Momtahan, H. 1989. Application of the ANSWERS model for prediction of runoff and sediment from small agricultural watersheds. M.S. Thesis, Irrigation Engineering Department, Agricultural College, Shiraz University, Shiraz, Iran. 218p. (In Persian)

Wischmeier, W.H. and D.D. Smith. 1978. Predicting rainfall erosion losses. Agricultural Handbook NO. 537. Science and Education Administration. U.S. Department of Agriculture. 58p.

Woolhiser, D.A. 1973. Hydrology and watershed modeling- State of art. Trans. ASAE 16(3):553-559.

Environmental Management, Geo-Water & Engineering Aspects, Chowdhury & Sivakumar (eds)
© 1993 Balkema, Rotterdam. ISBN 90 5410 099 0

Initial unit weight of reservoir deposited sediments

M. Bina
Shahid Chamran University, Iran (Presently: University of New South Wales, N.S.W., Australia)

M. Ghomeshi
Shahid Chamran University, Iran (Presently: University of Wollongong, N.S.W., Australia)

A. Shokrollahi
Khuzestan Water and Power Authority, Iran

ABSTRACT: This study was conducted in order to provide an estimate of deposited sediment unit weight in Dez Reservoir which is located in south-western Iran. 121 samples were taken from the reservoir deposited sediments and analysed for grain size. The initial unit weights of the sediments were estimated using the method of Lara and Pemberton (1965), Miller (1953) and Lane and Koelzer (1943). These estimations were compared with the dry unit weights of the undisturbed samples from the Dez Reservoir. The Lara and Pemberton (1965) method results were the most reliable. In this paper, ultimate unit weight of accumulated sediments, variation of particle size and solid density along the lake and also the reservoir operation type are presented.

1. INTRODUCTION

Sediment sampling from a large reservoir is a tedious job, specially if it is necessary to collect undisturbed samples. However, it does provide valuable information on texture, void ratio, density and other characteristics of the accumulated materials, which are required for the estimation of the volume occupied by the transported sediments, and for reservoir functioning evaluation. This paper is particularly concerned with the Dez Dam, which is the highest dam in Iran and is located in the northern part of Dezful, Khuzestan Province. It is part of a study which has been conducted in order to provide a reliable estimate of deposited soil unit weight in large reservoirs operating in the south-western part of Iran, under semi arid conditions.

2. CLIMATE

Reservoir sediment problems have been observed in all climatic regions and sedimentation processes in a reservoir are quite complex because of the wide variation in many of the influencing factors.

Dez watershed, with an area of 21 720 km^2 is located between 48° 10′ to 50° 21′ east longitude and 31° 34′ to 34° 7′ north latitude. Rainfall is variable from year to year and decreasing from north to south, with an overall annual average of 640 mm. Isohyets of mean annual rainfall on the catchment area are shown in Figure 1.

Range of temperature is also large, from −10°C in the winter up to 45°C in the summer, with little cloudiness and mostly sunshine. As a consequence, potential evaporation is more than 2 500 mm per year.

Runoff tends to be at minimum in late summer and maximum in later winter and early spring (March to May). Annual average discharge at Tale-Zang station which is the nearest station upstream of the dam site, is 230 m^3/s and monthly average discharge variation is demonstrated in Figure 2.

3. RESERVOIR AND DAM

Dez Dam started operation in 1962. It provides flood protection, hydro-electric power and irrigation for 125 000 ha of agricultural lands. Some of the specifications of this double curvature arched dam are indicated in Table 1 and illustrated in Figure 3.

Table 1. Reservoir and dam specifications

1	Dam site upstream catchment area	17 365 km^2
2	Length of reservoir	65 km
3	Area of reservoir	63 km^2
4	Reservoir volume	3.33×10^9 m^3
5	Sediment yield	18.57×10^6 T/y
6	Average daily discharge	230 m^3/s
7	Height of dam	203 m
8	Dam height – length ratio	0.96
9	Type of dam	Double curvature

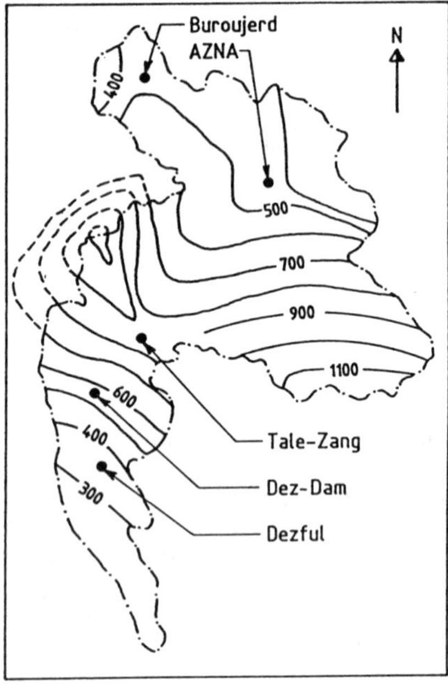

Fig. 1: Mean annual rainfall distribution on catchment area

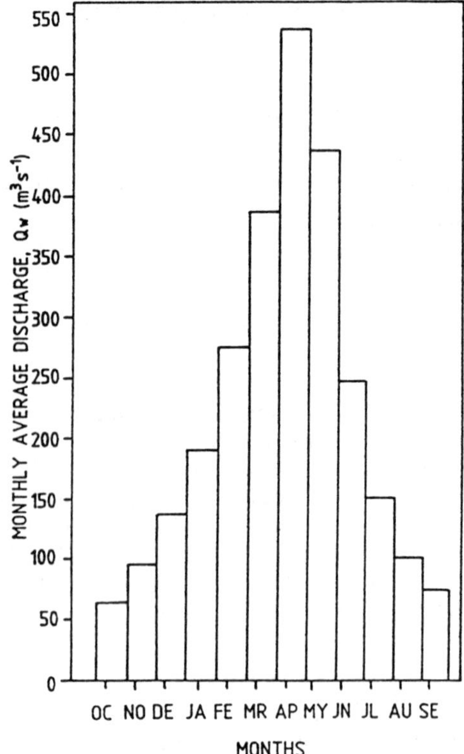

Fig. 2: Monthly average discharge variation at Tale-Zang station

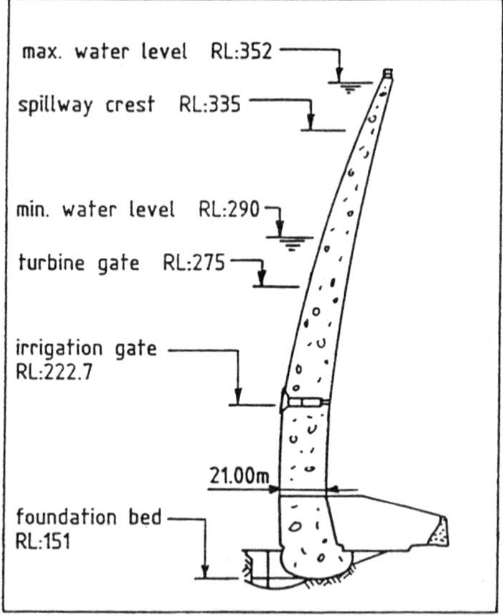

Fig. 3: Dam cross-section

4. ANALYSIS AND DATA

4.1 Sediment size

The size of the incoming sediment particles has a significant effect on the unit weight. Using the American Geophysical Union soil texture classification, the mechanical analysis of the 10 year record of bed load sediments at Tale-Zang station (which is located 10 km

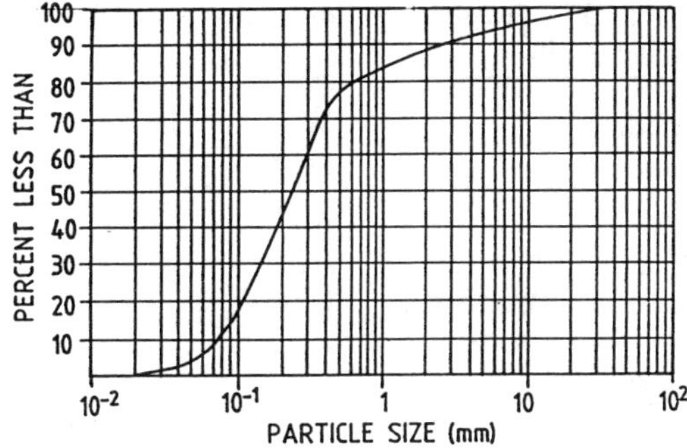

Fig. 4: Bed load particle size distribution at Tale-Zang station

Table 2. Illustration of sediment texture for selected lengths of Dez Reservoir

Lake Length *	Texture	No. of Samples	$\bar{x}$	σ
km			%	%
0–10	Clay	24	36.2	4.14
	Silt	24	63.8	3.04
	Sand	24	00.0	0.00
40–50	Clay	10	31.5	3.69
	Silt	10	68.5	3.13
	Sand	0	00.0	0.00
50–60	Clay	8	13.1	1.64
	Silt	8	74.0	5.45
	Sand	8	12.9	4.20

* Reservoir length measured from dam site towards upstream

from the upstream end of the reservoir) showed that the samples are high in sand, 85% (see Figure 4).

High suspended sediment load of silt causes the accumulated sediment in the reservoir to be high in silt. Texture analysis of 121 samples taken in 1983 from Dez Reservoir deposited sediment showed no sand in the accumulated sediments along the downstream 50 km of the lake. However, it was observed that the sand content increased for the last upstream 10–15 km of the lake. Also, silt content increased from 64% to 74% but clay content decreased moving upstream (see Table 2).

Overall average percentage of particle size distribution for the 121 soil samples can be summarised as 30.5% clay, 65.0% silt and 4.5% sand. The sampling locations are shown in Figure 5.

4.2 Reservoir operation

The reservoir operation is probably the most influential factor on the unit weight. A reservoir subjected to considerable drawdown may undergo greater consolidation, whereas the Dez Reservoir with a nearly stable pool does not allow the sediments to dry out and consolidate as much. This study showed that the Dez Reservoir deposited sediments are mostly just submerged, described as Operation Type 1 in Table 3 below.

Table 3. Reservoir operation type criteria

Operation Type	Operation
1	Sediment always submerged or nearly submerged
2	Normally moderate to considerable reservoir drawdown
3	Normally empty
4	River bed sediment

4.3 Consolidation rate

The deposition of new sediment on top of previously deposited ones changes the unit weight of earlier deposits. Therefore, the consolidation rate of sediments varies with time. This consolidation process affects the average unit weight over the estimated life of the reservoir.

4.4 Initial unit weight and solid density

Lane and Koelzer (1943), Miller (1953), Lara and Pemberton (1965) and Lara and Sanders (1970) have studied the initial unit weight of deposited sediment in reservoirs, considering three influencing factors (reservoir operation, particle size and consolidation rate). Their investigations give tabulations for initial unit weight (W_1) estimation and have been extracted and adapted as per Table 4.

Table 4. Values of W_1 (kg/m^3) and K for ultimate density determination

O*	I**	Sand		Silt		Clay	
		W_{1s}	K_s	W_{1m}	K_m	W_{1c}	K_c
	L&P	1550	0	1120	91.0	416	256.0
1	M	1410	0	1073	91.3	208	256.0
	L&K	1490	0	1041	91.3	481	256.0
	L&P	1550	0	1140	29.0	561	135.0
2	M	1410	0	1217	43.2	-	171.4
	L&K	1490	0	1185	43.2	737	171.4
	L&P	1550	0	1150	00.0	641	000.0
3	M	1410	0	1297	16.0	-	96.3
	L&K	1490	0	1266	16.0	961	96.3
	L&P	1550	0	1170	00.0	961	000.0
4	M	1410	0	1346	00.0	-	-
	L&K	1490	0	1314	00.0	1250	000.0

* Reservoir operation type
** Investigator:
L&P = Lara and Pemberton
M = Miller
L&K = Lane and Koelzer

Initial unit weight can be obtained using Table 4 and Equation (1) below:

$$\overline{W}_1 = W_{1c} \cdot P_c + W_{1m} \cdot P_m + W_{1s} \cdot P_s \quad (1)$$

where $\overline{W}_1$ is initial unit weight (kg/m^3), W_{1c}, W_{1m} and W_{1s} are initial unit weight for clay, silt and sand respectively, taken from Table 4. P_c, P_m and P_s are percentage of clay, silt and sand, respectively, of the inflowing sediment. In this study unit weights were calculated in three different methods then compared with the measured average. The average value of 21 samples measured for dry unit weight, was 0.90 g/cm^3 with a standard deviation of 0.07. As a result, the Lara and Pemberton (1965) method seems reliable for Dez Reservoir sediment initial unit weight estimation (see Table 5).

Solid density of 45 samples averaged to be 2.68 with standard deviation of 0.015.

4.5 Ultimate unit weight

In determining the unit weight after T years of reservoir operation, it is recognised that each year's deposits will have a different compaction time. Miller (1953) developed an approximate solution for determining the average unit weight of all sediment deposited in T years of operation as follows:

Table 5. Measured and estimated initial unit weight

Method	$\overline{W}_1$ g/cm^3	Difference %
Miller	0.75	–16.7
Lane and Koelzer	0.83	–7.8
Lara and Pemberton	0.86	–4.4
Measured from samples *	0.90	0.0

* $N = 21$, $\sigma = 0.07$, $P_C = 0.372$, $P_M = 0.628$, $P_S = 0.000$

$$W_t = \overline{W}_1 + 0.4343\,\overline{K} \left[\frac{T}{T-1} (L_n T) - 1 \right] \quad (2)$$

where W_t is average unit weight after T years of reservoir operation, W_1 is the initial unit weight derived from Equation (1) and K is a constant based on type of reservoir operation and sediment size obtained from Table 4 and equation below:

$$\overline{K} = K_c \cdot P_c + K_m \cdot P_m \quad (3)$$

Using the overall average of measured particle size in Section 4.1, W_1 and K values from Table 4 and Equations (1), (3) and (2), the average unit weight of deposited sediments was estimated as 1.11, 1.15, 1.17 and 1.19 g/cm^3 for 50, 100, 150 and 200 years of operation.

5. CONCLUSIONS

The general conclusions and results can be summarised as below.

1. The silt content of the deposited sediments ranged from 64% to 74% with 74% being at the upstream end of the lake.
2. The clay content of the deposited sediments ranged from 13% to 36% with 13% being at the upstream end of the lake.
3. The sand content of the deposited sediments ranged from 0% to 13% with 0% being along the downstream 50km, then

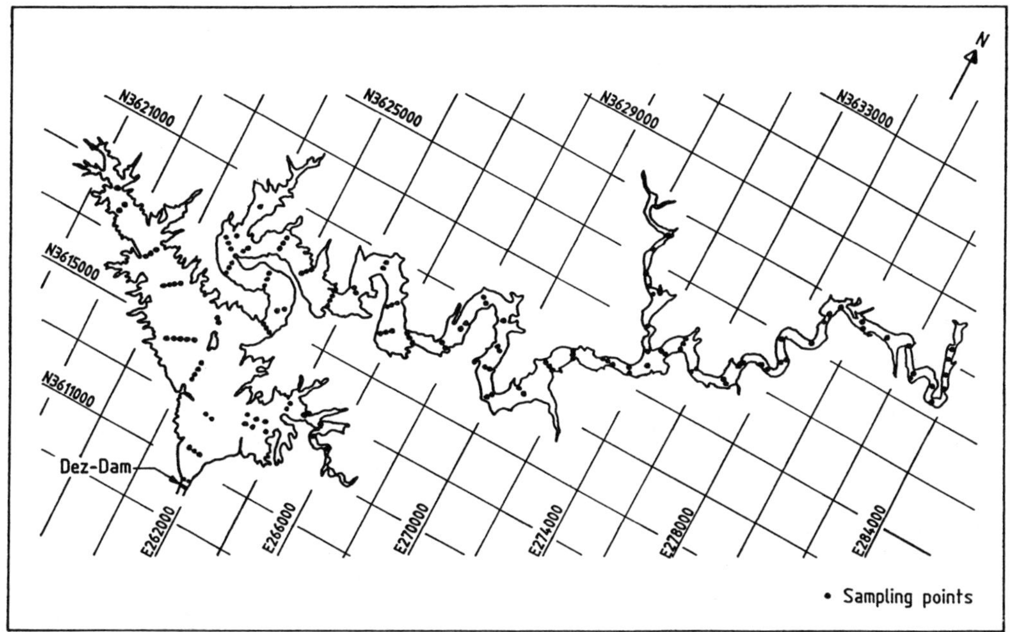

Fig. 5: Dez Reservoir and sampling locations

increasing up to 13% at the upstream end of the lake.

4. Samples with more silt and sand cause the density to be heavier, therefore in volume determination it is recommended to use different unit weights for different areas of accumulated sediments.

5. Lara and Pemberton (1965) method which is also described in USBR (1980), turns out to be the most reliable one for the estimation of unit weight of very low sand content in deposited sediment and a large reservoir of operating type 1.

6. Solid density of deposited sediments ranged from 2.66 to 2.70.

6. ACKNOWLEDGEMENTS

The authors are grateful to the Khuzestan Water and Power Authority for all the assistance and cooperation provided; Dr Rein Nittim for discussion and the Water Research Laboratory, University of New South Wales for the help in producing the manuscript.

7. REFERENCES

Lane, E.W. & V.A. Koelzer 1943. Density of sediments deposited in reservoirs. A study of methods used in measurement and analysis of sediment loads in streams. Interagency Committee on Water Resources; Report 9.

Lara, J.M. & E.L. Pemberton 1965. Initial unit weight of deposited sediments. Proceedings of Federal Interagency Sedimentation Conference, 1963, USDA, Agriculture Research Service, Miscellaneous Publication 970: 818-845.

Lara, J.M. & H.I. Sanders 1970. The 1963-64 Lake Mead survey. US Bureau of Reclamation.

Miller, C.R. 1953. Determination of unit weight of sediment for use in sediment volume computations. US Bureau of Reclamation.

United States Bureau of Reclamation. 1987. Design of small dams. A Water Resources Technical Publication, US Government Printing Office, Denver, Colorado.

Environmental Management, Geo-Water & Engineering Aspects, Chowdhury & Sivakumar (eds)
© 1993 Balkema, Rotterdam. ISBN 90 5410 099 0

Reservoir sedimentation: Modelling and control

G. Boari, B. Molino & G. Valenti
Department of Environmental Engineering and Physics, University of Basilicata, Potenza, Italy
C. D'Anisi
Pugliese Waterworks, Division of Potenza, Italy

ABSTRACT: Sedimentation of man-made reservoirs is one of the major problem for the correct management of hydraulic resources. Material transported by the river current as bed load or suspended load, sensibly reduce the reservoir capacity till it become useless. Experimental field tests was engaged by the D.E.E.P. of the University of Basilicata on the Camastra reservoir where sediment deposition is remarkable, to validate theoretical results.

1. INTRODUCTION

The dynamics of reservoir sedimentation processes can be viewed as part of wide range of problems in which the features of the basin modelization are analyzed through the description of each hydraulic or hydrologic process behaving into the catchment itself.
A good schematization of the sediment erosion transport and deposition phenomena both inside of the river network and into adjacent areas, can be able to give the formulation of distributed models, extremely useful during the design and management phases.
These models require a suitable database in order to calibrate the parameters. A much important aspect is the accuracy of the measurement phases and the detail of the informations available. In fact the output of the model is strictly depending on the space and time step used during the acquisition phase of the informations (hydrologic, geomorphologic, etc.).
The D.E.E.P. of the University of Basilicata (Italy) is monitoring the hillslopes of the Camastra basin in Basilicata, from a long time. The measurements are performed in order to study erosion and transport processes at the hillslope scale and then going up to the basin scale through G.I.S. techniques of analysis. Moreover, recently, experimental measurements of the Camastra reservoir bottom have been performed by laser apparatus getting the coordinates of the measurement point and the respective depth of the water automatically. Then a DTM of the bottom has been derived.
The objectives of the research can be stated as:
- systematic measurement of the bottom lake deriving the relative DTM. The comparison of the DTMs should be able to give an estimation of the sedimentation amount related to average catchment hydrology and extreme events;
- validation of some empirical models for the aggradation process and relative parameters calibration;
- improvement of the DTM as management control tool of the sediment removing techniques: the traditional dredging or sluicing or SEPS.

2. FIELD MEASUREMENT

In order to obtain a real time series of data relative to the soil volumes getting into a reservoir during the ordinary management phases, a field measurement have been performed by a combined use of total laser station and echo sound.
The investigation reservoir is the Camastra lake, placed into the downvalley of the Camastra creek in the Apulo Lucano Apennines (Southern

Italy), about 1 km upstream of the confluence with the Basento river. The water surface has 1.5 sq km as extension with 42 Mcm of capacity. The upward catchment is 390 sq km drained, in the upper part, by two creeks confluent into the Camastra stream.

Litologically the catchment belonging the blue clays and the Gorgolione flysh, with alternance of thin clay layers, lime stones, clayey marns and silty marns.

The measurements achieved into the reservoir have been interested globally the bottom and particularly the delta of the Camastra stream and the area close to the dam.

Besides, much more attention has been performed during the measurements around the exhaustive bottom drain valve. The bathymetry of the lake has been obtained through an alternative technique in which the position of the bathymetric measurement point has been detected by a total laser station device fixed on the boundary. The apparatus used during the measurements sis composed by laser station and echosound radio connected to the laser station.

This apparatus is able to get the distance of the bottom point automatically in terms of polar coordinates and to store them in to the hydrologger (fig.1).

The aparatus behave handy giving easy measurement without fixed alignments. In fact whatever position and course take the echosound, its position is automatically detected and stored on magnetic device by the laser station, giving a high number of the measurement points.

In this case, as first step, some measurement stations have been placed on the shores, deriving the correspondent geographic coordinates with reference to the existing datum points on the dam.

Nine station have been placed around the lake, 1 of them is on the dam, 4 to the right shore and 4 to the left one.

The data have been acquired by a boat, on which the echosound were placed. Longitudinal and transversal courses, with respect to the valley's axis, have been followed by the boat during the measurements, defining a spatial grid 10x10 m.

The points have been detected, along the courses with 2pts per seconds as frequency corresponding to 1 meter as relative distance. The total amount of the measured points is 11321.

Before deploying the apparatus into the field measurements, some tests have been performed in order to verify the accuracy of both echosound and laser station.

The tests have been made both in field and in laboratory deriving the errors for the single component and for the global station. The maximum error in the depth measurements is around 1 cm for depths no greater than 20 m increasing, not sensitively, for depths greater than 20 m. Anyway the error is about 0.5 percent of the depth measured. About the laser station, the error is quasi-linear depending on the distance measured, corresponding to the 10 percent. The composition of the errors, following the classical theory, give rise an average percentage error with order of magnitude close to 2 percent.

Then the measurements have been analyzed and interpolated through classical techniques (Kringing) deriving the DTM of the lake bottom. In figg.2 and 3 are shown the DTM of the Camastra lake and the lell curves while in the figg. 4 and 5 are shown the longitudinal and cross sections of the lake.

3. AN EMPIRICAL MODEL FORECASTING THE SEDIMENTATION INDEX IN BASILICATA'S RESERVOIRS

The importance of the sedimentation processes occurring into the lakes and the complex phenomena causing them, are allowed the studies of the researchers in these field. Muchmore, the researchers have been developed in the reservoirs where experimental data can be obtained and then used for the calibration of the parameters of the models forecasting the sedimentation index (T.I.).

Generally the number of the parameters characterizing the interaction between catchment and reservoirs is really high. That is, taking attention on the variety of the parameters, hydrological, geomorphological, climatic, agricultural, anthropological and so on, each of them is not so easy to estimate.

Some researchers, Brune, Allen, Brown, Dandy (Brune et oth., 1941; Brown, 1944; Brune, 1953; Dendy et oth., 1973), are fixed their attention on the formulation of empirical models based on a few parameters taking mainly into account the geometric characteristics of the reservoirs-catchment system.

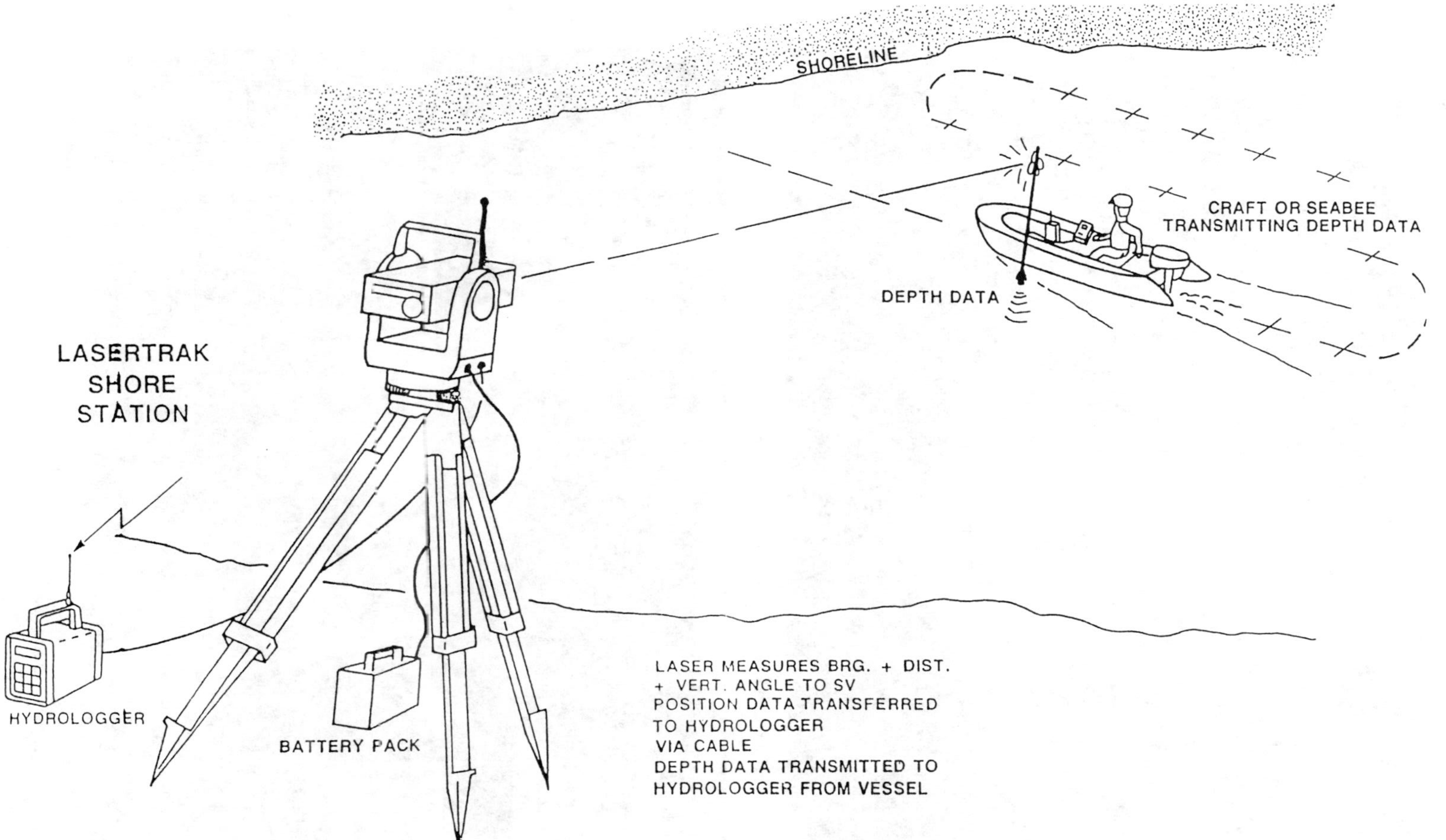

Fig.1 The schemes of the laser aparatus employed for the bathymetric measurements

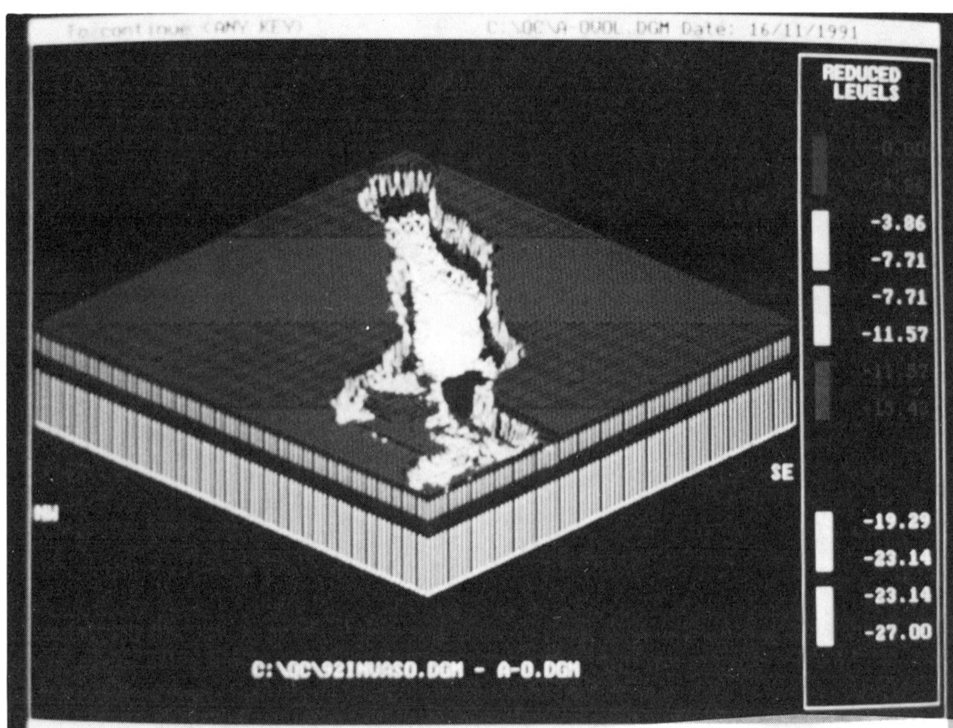

Fig.2 The DTM of the Camastra lake

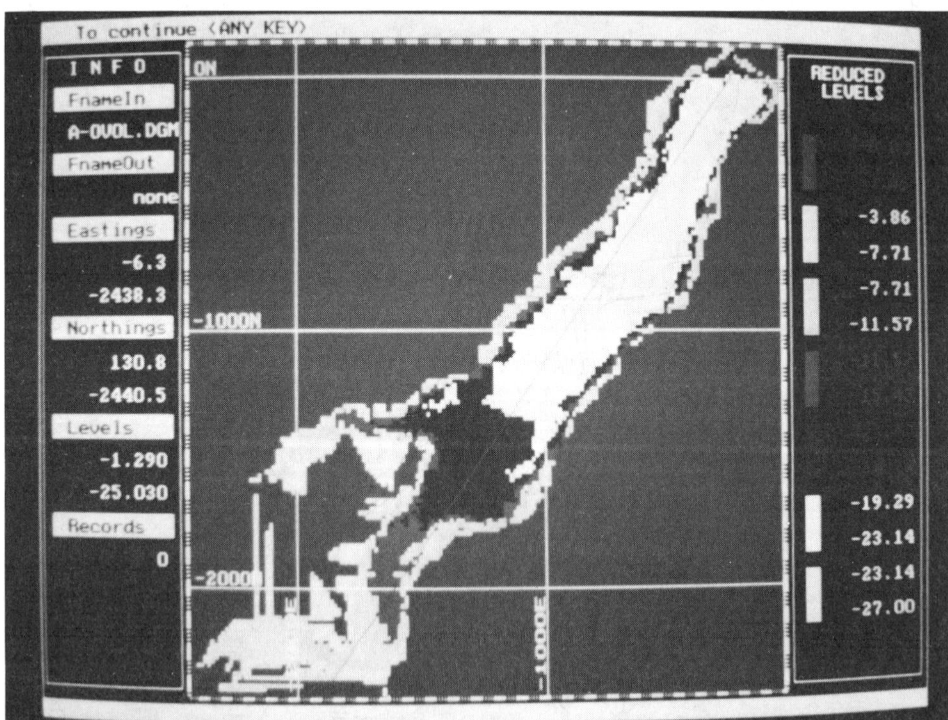

Fig.3 The elevation map of the Camastra lake bottom

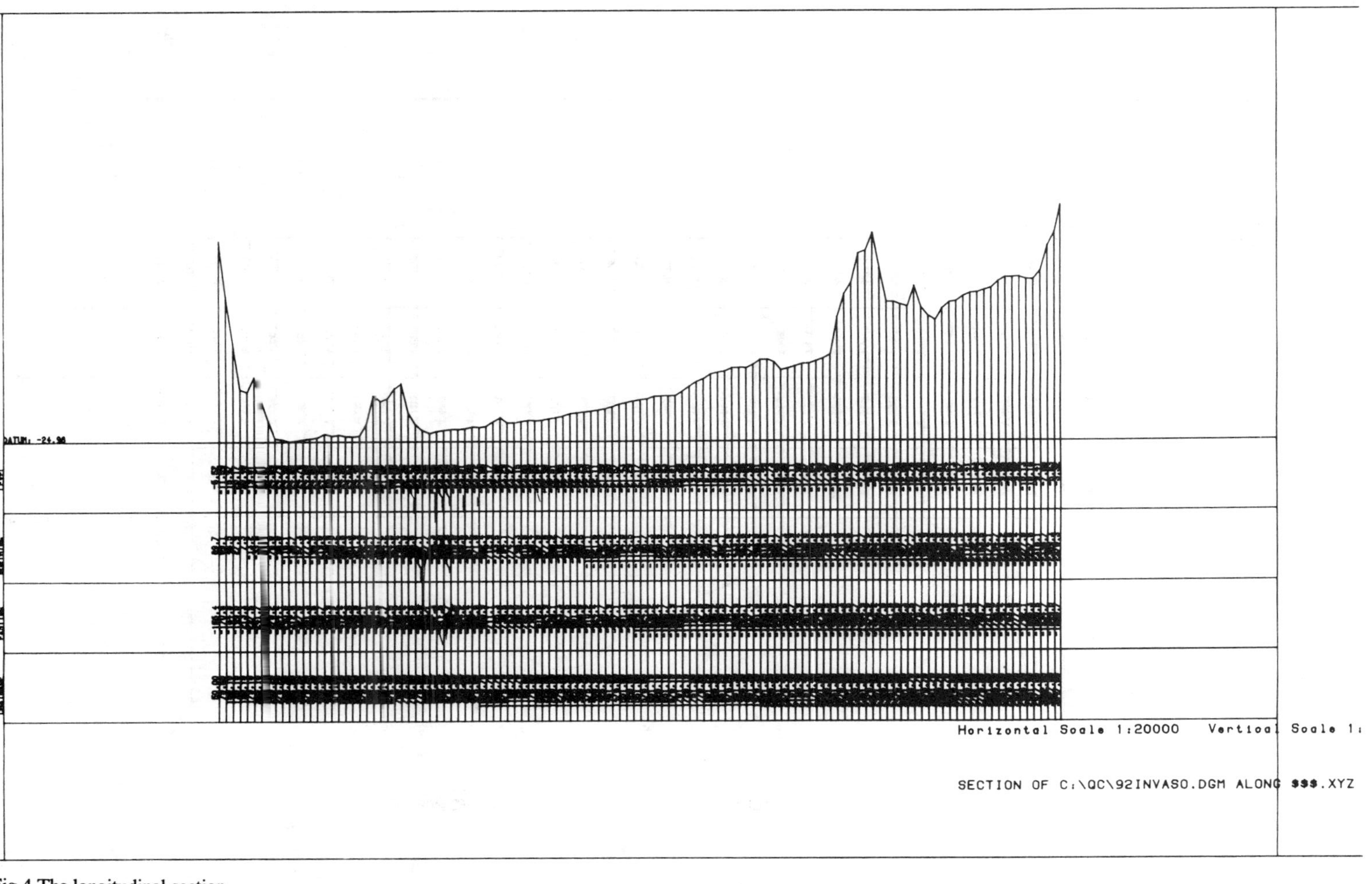

Fig.4 The longitudinal section

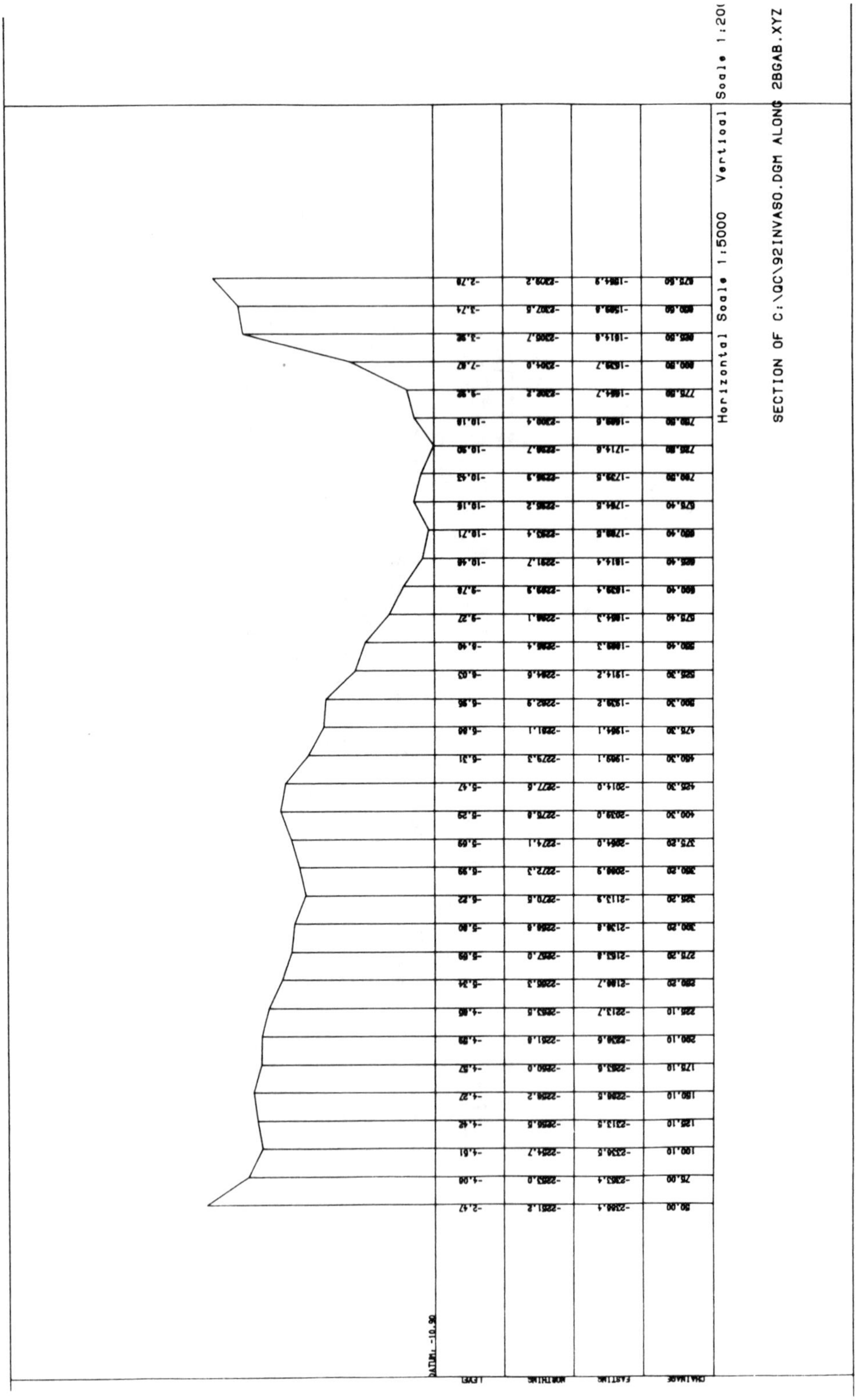

DATUM: -10.90

CHAINAGE	EASTING	NORTHING	LEVEL
50.00	-2388.4	-2251.2	-2.47
75.00	-2363.4	-2253.0	-4.00
100.10	-2336.5	-2254.7	-4.61
125.10	-2313.5	-2256.5	-4.48
150.10	-2288.5	-2258.2	-4.27
175.10	-2263.6	-2260.0	-4.57
200.10	-2238.6	-2261.8	-4.89
225.10	-2213.7	-2263.5	-4.86
250.20	-2188.7	-2265.3	-5.34
275.20	-2163.8	-2267.0	-5.69
300.20	-2138.8	-2268.8	-5.80
325.20	-2113.9	-2270.5	-6.22
350.20	-2088.9	-2272.3	-5.99
375.20	-2064.0	-2274.1	-5.69
400.30	-2039.0	-2275.8	-5.29
425.30	-2014.0	-2277.6	-5.47
450.30	-1989.1	-2279.3	-6.31
475.30	-1964.1	-2281.1	-6.80
500.30	-1939.2	-2282.9	-6.96
525.30	-1914.2	-2284.6	-8.03
550.40	-1889.3	-2286.4	-8.40
575.40	-1864.3	-2288.1	-9.27
600.40	-1839.4	-2289.9	-9.78
625.40	-1814.4	-2291.7	-10.48
650.40	-1789.5	-2293.4	-10.71
675.40	-1764.5	-2295.2	-10.16
700.50	-1739.5	-2296.9	-10.53
725.50	-1714.6	-2298.7	-10.90
750.50	-1689.6	-2300.4	-10.18
775.50	-1664.7	-2302.2	-9.98
800.50	-1639.7	-2304.0	-7.82
825.50	-1614.8	-2305.7	-3.98
850.50	-1589.8	-2307.5	-3.74
875.50	-1564.9	-2309.2	-2.78

Horizontal Scale 1:5000 Vertical Scale 1:20(

SECTION OF C:\QC\92INVASO.DGM ALONG 2BGAB.XYZ

Fig.5 The cross section

Tab.1 Data of investigated reservoirs

SERBATOIO	C ($m^3 10^6$)	S (km^2)	TI ($\frac{m^3}{km^2 year}$)	H (mm)
Montecotugno	530.0	804.0	1870	67.25
Pertusillo	167.0	530	910	62.52
Camastra	42.0	350.0	540	48.31
Basentello	41.7	267.0	230	37.55
S.Giuliano	107.0	1631.0	460	47.05
Rendina	22.8	385.0	630	51.52

Tab. 2 Linear models

INDEPENDENT VARIABLE	DEPENDENT VARIABLE	MODEL	CORRELATION COEFFICIENT
T.I.	H	T.I. = 15.929 H	r = 0.436
T.I.	C	T.I. = 3.82 C	r = 0.678
T.I.	K,H	T.I. = -1757.51+48.329H	r = 0.815
T.I.	H,C	T.I. = 7.82H+2.47C	r = 0.946
T.I.	H,C,S	T.I. = 9.66H+2.51C-0.16S	r = 0.965
T.I.	K,H,C,S	T.I. = -528.53+20.36H+2.05C-0.114S	r = 0.981

About Italian researchers, the studies are mainly allowed on regional systems, in order to characterize the features of the catchment-reservoir system (Bazzoffi, 1985; Gervasoni, 1987) and basin hydrological characteristics (Tamburino et oth., 1990).
Studies carried out by the authors of the present paper, have shown how the empirical models are not generalizable. Therefore the improvements of these models, in conditions far away from the original ones, in which they have been formulated, give rise not so good results.
Moving from these basis, studies about the possibility of the development of a regional empirical model forecasting the sedimentation index, are carrying on by the D.E.E.P. of the University of Basilicata.
As reported in literature, the most important parameters strictly related to the sedimentation process are:
a) initial capacity of the reservoir (C)
b) the amplitude of the upstream catchment (S)
c) the mean of the maximum annual rainfall with duration equal to one day (H).
These parameters, have been employed into a multivariate analysis, in order to derive the optimum model both in the case of simple dependence and double dependence. Finally the analysis have been performed in the case in which all the parameters have been taken in to account.
The data used in the calculations, are relative to six Basilicata's reservoirs (Southern Italy).
In tab.1 are reported the capacity C, the extension, in the mean of the maximum annual rainfall with duration one day and the sedimentation index for each reservoir. Each set of data is referred to a time period greater than 10 years. For the Rendina and S.Giuliano reservoirs, the data have been obtained through bathymetric measurement, while for the other ones, the values have been given by the reservoir managers.
In tab.2 are reported the relationship of the linear

Tab. 3 Non linear models

INDEPENDENT VARIABLE	DEPENDENT VARIABLE	MODEL	CORRELATION COEFFICIENT
T.I.	H	T.I. = 23.157 $e^{(0.063H)}$	r = 0.976
T.I.	C	T.I. = 337.224+2.874 $C^{1.011}$	r = 0.954
T.I.	S	T.I. = 53.14 $S^{0.392}$	r = 0.370

models. From this table it is easy to note how the correlation coefficient increases when more of the variables are taken into account.

The two-parameters model C-H seems to have a good behavior; in tab.3 is shown a synoptical table of forecasting models derived through a one-dimensional-nonlinear interpolation. Also in this case the models depending both on C and H have a good performance.

4. CONCLUSIONS

The reservoir sedimentation phenomena is mainly depending on a few parameters, as reported in literature. These variables are the reservoir capacity, the area of the upstream catchment and the mean of the maximum annual rainfall with duration one day. At the same time, a universal model cannot be derived because the sedimentation process is depending on more local characteristics (hydrological, geomorphological, pedological, etc.).

The investigations, carried out in Basilicata, seem to show an empirical model validate at the regional scale and easily structured. That is, the reservoir sedimentation index (T.I.) is depending on just one parameter: the mean of the maximum annual rainfall with duration one day. The relationship give a good fitting of the real data set. Of course, a multivariate analysis give a rise a good performance for the three parameters linear model. But the exponential model, also having the correlation coefficient less than the correspondent of the three parameters model (r_e=0.976 against r_4=0.981), has the advantage of the one-dependence on an easy estimable parameter (H).

The results shown above are, obviously, temporary and changeable once more data will be available.

ACKNOWLEDGMENT

The authors should be like to give best thanks to the EAAP and much more to the President of the Society, for the support received during the research and for the cooperation in the phases of the data acquisition and field measurement.

REFERENCES

BAZZOFFI P. (1985): Models for the assessment of sedimentation in reservoirs as a tool for correct management and duration estimate in the project phase CEE "Agricultural water management" Arnnem (Nederlands) .

BROWN C.B. (1944): Discussion of "Sedimentation in reservoirs". by B.O. W Taig, Transactions American Society of Civil Engineers, 109, pp. 1080-1086.

BRUNE M.G., ALLEN R.E. (1941): A consideration of factors influencing reservoir sedimentation in the Ohio Valley Region. Transactions, American

Geophysical Union. , 22, pp. 649-655.
BRUNE M.G. (1953): Trap efficiency of reservoirs. Transactions, American Geophysical Union., 34, 3.
DENDY F.E. et alii (1973): Reservoirs sedimentation surveys in the United States. in: man-made Lakes: Their problems and environmental effects. American Geophysical Union - Washington.
GERVASONI S. (1987): Interrimento dei serbatoi artificiali. Dissertazione finale Dottorato di ricerca tra le sedi di Ferrara, Firenze, Parma, Pavia, Perugia, Siena.
TAMBURINO V., BARBAGALLO S., VELLA P. (1990): L'influenza delle precipitazioni e delle caratteristiche dei bacini sull'interimento dei serbatoi artificiali in Sicilia XXII Convegno di Idraulica e Costruzioni Idrauliche -Cosenza .

Environmental Management, Geo-Water & Engineering Aspects, Chowdhury & Sivakumar (eds)
© 1993 Balkema, Rotterdam. ISBN 90 5410 099 0

Control of the surface drainage structures at the hillslope scale

V.A.Copertino, M.Greco, G.Spilotro & G.Vacca
Department of Environmental Engineering and Physics, University of Basilicata, Potenza, Italy

ABSTRACT: The hillslope processes have an important role in the study of the basin features.Both morphological evolution phenomena and erosion patterns at the hillslope scale assume great interest in order to control the sediment yield for the river network and to check the processes of planimetric shape and/or the growth of the drainage structures.

1 .INTRODUCTION

Surface water flow is vitally important in the understanding of stream flow processes, but it is also relevant for the movement of the material on a hillslope.

In the last years, the interest into the study of the interactions between the small scale processes (hillslope scale) and the large scale (basin scale) is growing up considerable.

In fact, the component "hillslope" constrain directly the general problems proper of the basin scale, like sediment supply, reservoirs sedimentation, land changes, etc. That is, allowing the need to study the hillslope's processes in detail.

Many approaches have been developed in the past, in order to obtain a quantitative description of the erosion and transport phenomena behaving on a hillslope, and something more is going on.

As first attempt, some empirical models (Musgrave, 1935; Horton, 1939; Zingg, 1940) based on a few simple parameters, have been developed in order to estimate the amount of the sediment eroded volume. Then, moving step by step, multidepending empirical models have been stated (Musgrave, 1947; Wischmeier, 1960) till to formulate some simulating numeric models, in which some aspects, relative to the sediment supply, are strictly connected to the study of the space and time evolution of the surface drainage structures.

Actually, the advanced terrain data acquisition technologies, both in data management and storage (DTM, G.I.S.), allow the aims of the researchers to perform some more studies into the hillslope field merging classical empirical methodologies and physically based analysis able to give a full description of the hillslope state analogously to the larger scales.

In the present paper, the hillslope's modeling will be presented (empirical and conceptual) applying some of these models to the real set of data relative to an Apulo-Lucano Apennine hillslope (Italy). The DTMs of the hillslope are known, because systematically detected in different time.

2. HYDRAULIC PROCESSES AND SURFACE DRAINAGE STRUCTURES AT THE HILLSLOPE SCALE

Differently from channel flow, proper of the river network, the overland flow is depending on a large number of variables not so easy to define. Beside an analytical description based on the classical hydraulic relations is best with many difficulties.

Generally the flow is unsteady and spatially varied because directly depending on the rainfall and infiltration processes, neither space and time uniform.

This leads the need to define the circles of the studies, consequently the correspondent space

Fig.2.1 - Rills and gullies, morphologic index of no uniform erosion

and time scale, in which some simplifications can be used in order to describe the overland flow. Of course these simplifications should not be enforcing the system significantly.

Connected with rainfall and infiltration variability, some more no negligible phenomena are behaving, like raindrop, leading more mathematical difficulties.

The flow should be laminar or turbulent or both. But some experimental results have shown low Reynolds number values allowing principally a laminar flow regime (Kirkby, 1978). Otherwise, computing the raindrop effect the Reynolds number increase giving a transient stage of flow (quasi-turbulent flow).

Moreover, in the upper part of the hillslope (convex), the flow seems to be laminar tending to became turbulent downward as increasing as the length.

The depth of the flow vary between greater and lower values then the critical one. Sometimes the flow pass from supercritical to sub critic becoming strongly unsteady with roll waves or rain waves features.

Some experiments, carried out in laboratory, have shown how tends the water to flow uniformly space distributed like a film, with not so sensitive depth changes. On the other hand, both partial vegetable cover and no uniform surface roughness lead a lateral concentration into the flow, similar to the shallow-braided channel.

Erosion, transport and deposition processes are associated to the overland flow. That is, characterizing the overland flow process from an engineering point of view. In fact great interest assume the estimation of the soil volume amount coming down from the hillslope and reaching the river network characterizing the soil transport processes at the basin scale.

The problems rising up are related to a wide range of employment both design and management and control (reservoir sedimentation, sediment delivery, etc.).

The recognition of surface drainage structures at the hillslope scale, more o less organized leads a simple qualitative erosion index. In fact rill structures placed on an hillslope are representative of a faster erosion process. Beside the overland flow can be characterized through the morphological features of the drainage net: rills or gullies lead a no uniform surface flow (fig.2.1) while them absence give rise flow features like sheet flow. Anyway rill flow and sheet flow are not incompatible and the passage from each to the other one should be possible passing a certain threshold. This threshold is in someway depending on a lot of parameters constraining the flow and erosion processes, because the threshold overcoming can correspond to different output of the system; for example, a climate change, like increase of rainfall intensity, leads a variation into the erosive flow potential. This variation can raise the flow velocity deepening the surface drainage structures or, on the other hand, increase the flow depth with consequent space uniform erosion.

The last condition should be corresponding to the case in which an extreme event washoff the hillslope surface deleting the drainage network. This event is classified as catastrophic with respect to the drainage network growth (Roth and Siccardi, 1990).

3. CONTROL OF THE HILLSLOPE EROSION AND PATTERN DRAINAGE EVOLUTION

The rill formation have been observed during an overland flow on a steep surface where rainfall give rise a cospiciuos discharge. Meyer and Monke (1965) making laboratory test, during which they used glass spheres to simulate a bottom cohesiveless surface observed that the flow were mainly in rills and that the erosion intensity as increased as the slope and the discharge. For slopes close to 10% (observed values in to New Fork River Site 1) the erosion was advanced with strict deep rills and with downvalley direction. Moreover the erosion, increasing with the runoff, tend to be uniform till a extreme condition at which a concentrated particle flow was generated into the potential rills and them self become fully developed.

Other experiments carried on in field have shown a positive trend between the slope and soil supply while negative one exist between the last one and cover vegetable density.

An important result obtained during simulation, is that high values of sediment concentration are present during the initial phase of runoff while tend to decrease going on.

The observed values detected into New Fork River Site 1, show a high sediment concentration variability starting from 228 mg/lt. after 24 min. of flow to 184 mg/lt. at 35 min., 41 mg/lt. at 82 min. and 36 mg/lt. at 119 min. Similar results have been observed by Lowdermilk and Sundling (1950) with erosion decrease during the flow till to reach a limit condition in which all the fine particles were removed and the surface was as paved. Also Swanson, Dedrick and Weakly (1965) detected a reduction of the soil concentration in time, but they did not recognize any paved surface.

Some more studies were carried on by US Geological Survey on a physical model well reproducing the Badger Wash Basin in Western Colorado. They outlined that:

-during an event, the discharge flowing over the surface is depending on the infiltration rate, then on the soil saturation degree;

-erosion as increase as the runoff.

Moreover, neglecting raindrop effect, in order to have erosion and transport, a sufficient discharge must flow.

An analysis of variance for net erosion on convex, straight and concave segments shows significant differences at the 5% level for erosion between different hillslope segments. Straight segment of slope have greatest erosion and there was no significant differences between convex and concave segments (fig.3.1)

In tab.3.1 are reported some empirical models derived in order to estimate the sediment discharge coming down during an overland flow. At the same time some conceptual models have been developed. These models provide to check the steady or unsteady condition of a surface

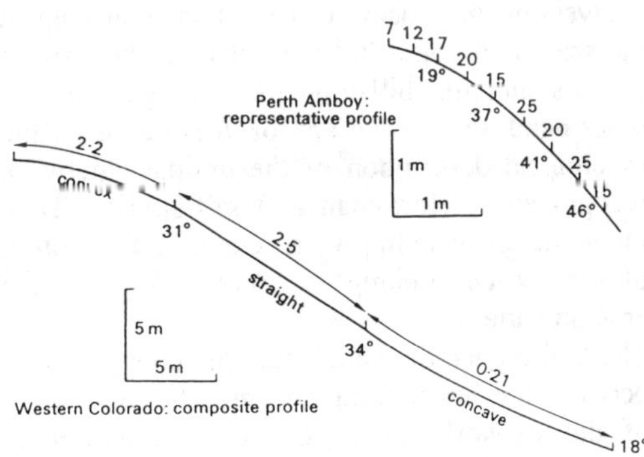

Fig.3.1 - Straight, concave and convex hillslope's segments (Kirkby,1971).

Tab.3.1 - Empirical model for the estimation of the eroded volume from a hillslope.

Relationships	Parameters
$e_r = k_2 \frac{\sin\theta\, x^{0.6}}{\tan^{0.3}\theta}$	θ=angle of the hillslope $k_2 = k_2(\sigma,n)$ σ=rainfall intensity n=roughness coefficient x= distance from the watershed e_r=eroded volume
$e_r = 0.43+0.30\,S+0.043\,S^2$	S= tan θ = slope
$e_r = k_3\, S^{1.3}$	S= tan θ = slope k_3=0 ÷1
$e_r = k_4\, x^{0.5}$	k_4=k_3=0÷1 =empirical coefficient
$M_e = 0.3(\frac{R}{100})^{2.2}$	R= mean annual rainfall M_e=eroded mass
$M_e \propto x^{1.6} \tan^{1.35}\theta$	S= tan θ = slope M_e=eroded mass
e_r=0.04 $q_s \tan^{0.62}\theta$	q_s=mean rainfall intensity

drainage structure recognizing the sediment sources areas.

The models can be derived following two different approaches: upward and downward.

The first one consider a local physically based analysis of the erosion process then going up in the scale defining a linkage between the erosion process and the hillslope morphology. On the other hand, the second approach is based on the topological description of the drainage network recognized as Hortonian and self-similar. Then the model provide to give an external description of the system linking the morphology to the erosion pattern.

The hydrodynamic model state the energetic link between the contributing area and the local slope of the network, explaining it as function of geometric and dynamic characteristics of both soil and flow:

$$AS^{\alpha}=f(\tau_*,d_s,i) \qquad (3.1)$$

in which A is the contributing area into each link of the network, S is the local slope of the link, τ_* is the dimensionaless bottom shear stress, ds is the soil particles diameter, i a value representing the mean rainfall intensity relative to the upstream area and α is depending on the flow regime (turbulent or laminar) and geometry (channel flow or sheet flow). In the case α=2 the leftside of the 3.1 relationship represent the minimum rate expenditure of energy in the link. Starting from the classical hydrodynamic relations, the rightside assignment in the 3.1 can be stated as (Greco et oth., 1992):

$$f(\tau_*, d_s, i) = k\frac{\tau_*^{\,a}\, d_s^{\,b}}{i} \qquad (3.2)$$

in which K is a dimensional coefficient while a and b are depending on the flow regime assumptions.

The morphologic models, derived in terms of topological network characteristics, give out (Flint, 1974):

$$S_u = K_{sa}\, A^{\beta}{}_u \qquad (3.3)$$

in which u is the hortonian order of the link:

$$\beta = -\frac{\log R_s}{\log R_a} \qquad (3.4a)$$

$$K_{sa} = \frac{K_s}{K_a{}^{\beta}} = \frac{S_s\, R_s{}^{s}}{(A_1 S_1)^{\beta}} \qquad (3.4b)$$

with Ra and Rs the area and slope ratios respectively (Schumm, 1956; Strahler, 1952), s is the network magnitude, A_1 and S_1 are respectively the mean area and slope of the first order link.

The combined use leads more useful advises about the drainage pattern configuration and the erosion processes behaving inside the system.

Fig.4.1 - The real hillslope investigated .

4. FIELD INVESTIGATION ON AN APENNINES HILLSLOPE

Real-life observations are necessary, not only to provide correct input information to the modeling attempts, but also to provide verification of the theoretical predictions.
In order to improve both empirical and conceptual models at the hillslope scale, a real hillslope in Italy was monitoring systematically through a laser apparatus (Copertino et al. 1992).
The place investigated is in the downvalley of the Camastra creek, in the Apulo Lucano Apennines, upstream the confluence with Basento River.
Being the site close to the Pontefontanelle reservoir, the rainfall data can be obtained by the raingauge placed on the dam.
The site is geologically characterized by flysh facies belonging to the Gorgoglione flysh.
The litological structure is constituted by thin layers of limestone, clayey marns, silty marns and clays.
Just a part of the area has been choose for the monitoring. The hillslope has about 4000 sqm as extension with an average slope close to 50°. The site has a very high degree of erosion with an organized surface drainage structures look liking classical features as rills and gullies (fig.4.1).
The measurements are performed by laser apparatus, obtaining the surface in terms of polar coordinates with respect to the fixed base station.
The data are collected systematically after a set of rainfalls and are stored in a CDU. Then the data are downloaded to a pc in order to perform some analysis in order to derive the Digital Terrain Model (DTM) of the system.

5. DIGITAL TERRAIN MODEL

The DTM is a simply defined as an ordered array of numbers representing the spatial distribution of terrain attributes (Moore et al., 1991).
In hydrological view point, the DTM is an important source of general information about the landscape.
In the last years more interest the analysis of the DTM have been assumed, not so much for the amount of the informations that can be possible to obtain from them, but for the handle of system.

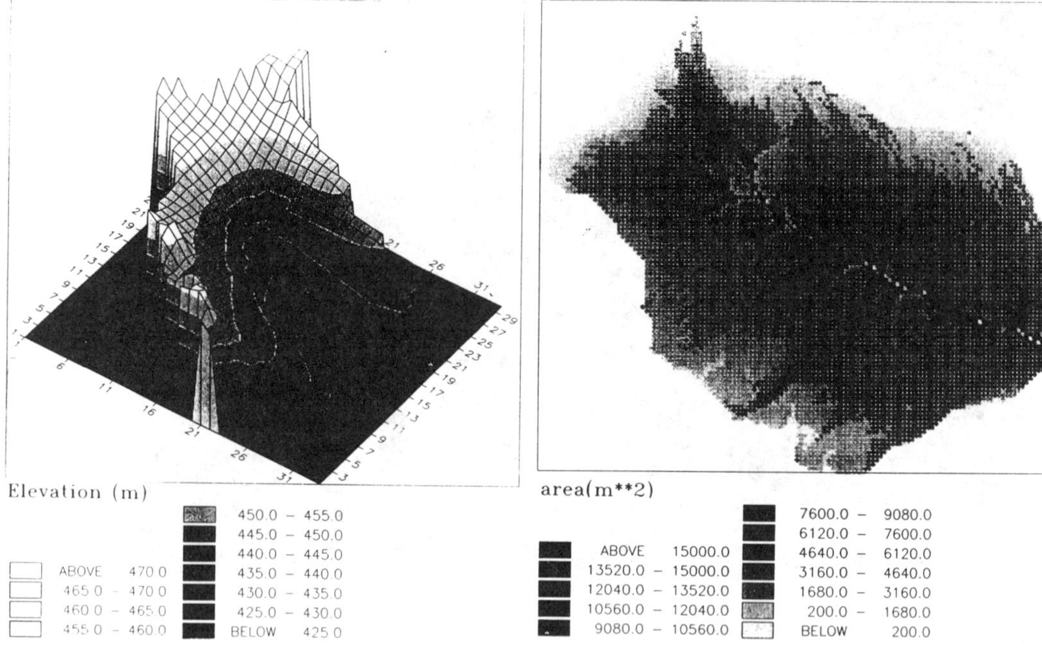

Fig. 5.1 - An example of DTM (Quinn, 1991)

Fig.5.2 - Drainage area map (Quinn, 1991)

Generally, the DTM is a data digital support collected by map.

This means that the level of the information is strictly equal to the one of the cartography, but the employment of data in this form let the customer make a readily acquisition and storage.

The research in the field of the DTM's data analysis is growing day by day in the hydrology giving some more opportunity about the accuracy of the model's application (figg. 5.1-5.2).

Missing the analytical interpretation of the elevation data, we like to state the very high importance that this "tool" can give into the hydrological research showing the primary topographic attributes obtaining by DTM analysis in tab.5.1 (Speight, 1980).

The primary topographic attributes are suitable parameters able to describe or characterize the spatial variability of some specific processes occurring in the landscape.

DTM analysis have been performed on the digital elevation model obtained by the field measurements described above.

The real set of data has been collected following a grid with very small size (1x1 m), with a high accuracy of the measurements (± 0.005 m).

The analysis have been carried on in order to derive the amount of the data defining the input of the empirical and conceptual models.

From the hillslope's DTM (fig.5.3) the surface drainage network has been derived in terms of links (fig.5.4) and corresponding contributing area and local slope.

About the surface drainage pattern some more calculations have been developed characterizing the network by topological parameters belonging the Horton-Strahler laws.

In the tab.5.2 are shown the sorted network and the relative bifurcation, order, area and slope ratios.

Tab.5.1 - Primary topographic attribute.

Attribute	Definition	Hydrologic significance
Altitude	Elevation	Climate, vegetation type, potential energy
Upslope height	Mean height of upslope area	Potential energy
Aspect	Slope azimuth	Solar irradiation
Slope	Gradient	Overland and subsurface flow velocity and runoff rate
Upslope slope	Mean slope of upslope area	Runoff velocity
Dispersal slope	Mean slope of dispersal area	Rate of soil drainage
Catchment slope*	Average slope over the catchment	Time of concentration
Upslope area	Catchment area above a short length of contour	Runoff volume, steady-state runoff rate
Dispersal area	Area downslope from a short length of contour	Soil drainage rate
Catchment area*	Area draining to catchment outlet	Runoff volume
Specific catchment area	Upslope area per unit width of contour	Runoff volume, steady-state runoff rate
Flow path length	Maximum distance of water flow to a point in the catchment	Erosion rates, sediment yield, time of concentration
Upslope length	Mean length of flow paths to a point in the catchment	Flow acceleration, erosion rates
Dispersal length	Distance from a point in the catchment to the outlet	Impedance of soil drainage
Catchment length*	Distance from highest point to outlet	Overland flow attenuation
Profile curvature	Slope profile curvature	Flow acceleration, erosion/deposition rate
Plan curvature	Contour curvature	Converging/diverging flow, soil water content

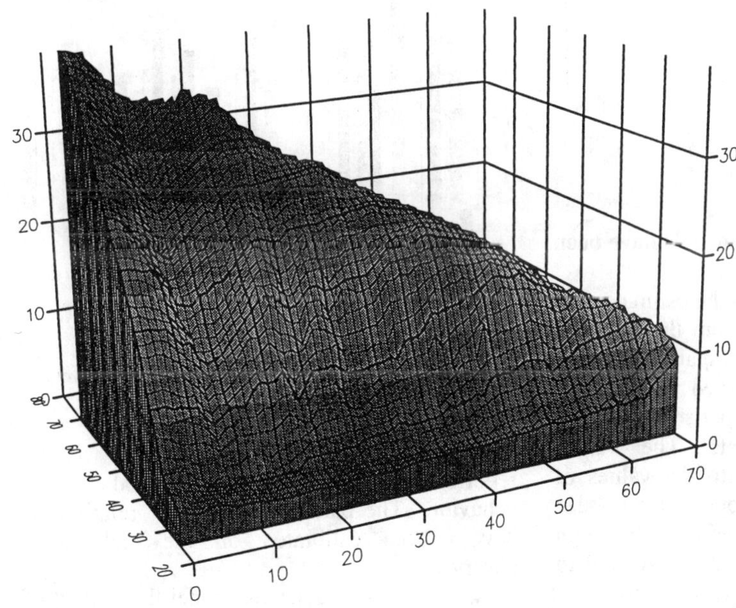

Fig.5.3 - The DTM of the monitored hillslope

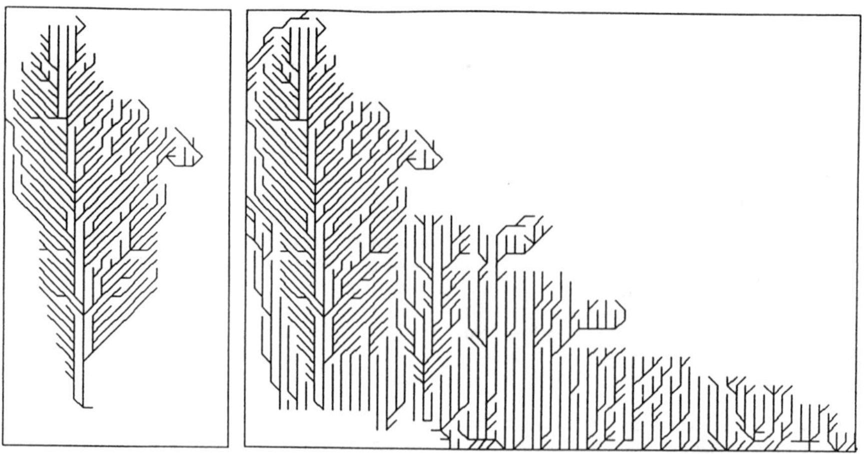

Fig.5.4 - The observed surface drainage structure

Tab.5.2 - The topological parameters of the monitored drainage structure.

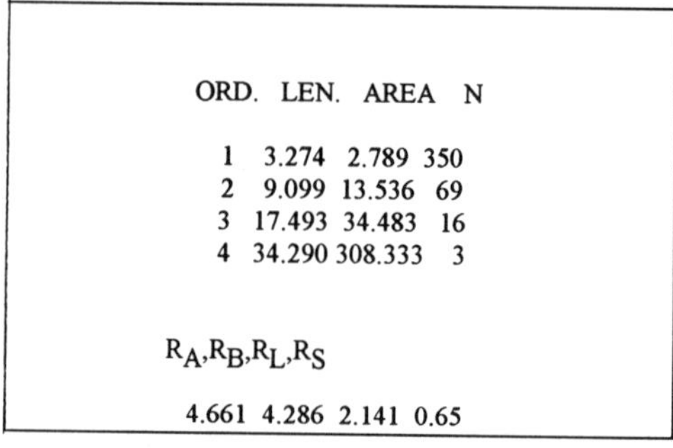

ORD.	LEN.	AREA	N
1	3.274	2.789	350
2	9.099	13.536	69
3	17.493	34.483	16
4	34.290	308.333	3

R_A	R_B	R_L	R_S
4.661	4.286	2.141	0.65

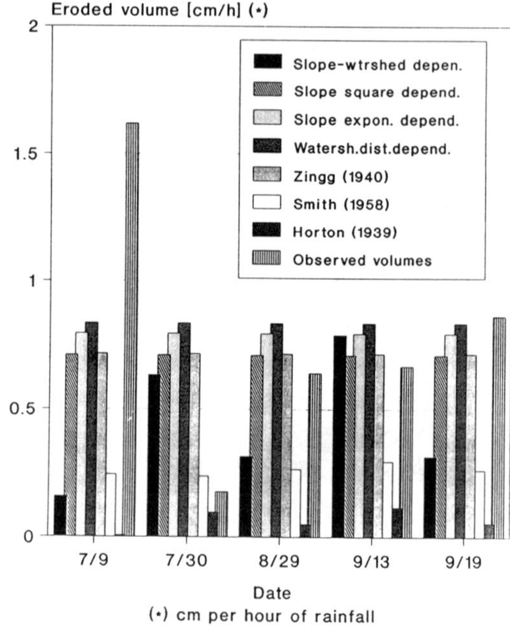

Fig.6.1 - Comparison between the observed eroded volume and the empirical calculated values

6. MODEL IMPROVEMENT

Both empirical and conceptual models have been applied on a real set of data.
The empirical model relate to the estimation of the eroded volume washoff from the hillslope during the storm have been compared with the observed values. The observed values have been calculated by the comparison between two consecutive measurements. The values shown in fig.6.1 are relative to the values of observed rainfall intensity, slope, grain size and length of the hillslope. Perhaps the high variability of the numerical results is related to the strict number of variables involved into the relationships.
On the other hand, the conceptual models, reported above, are shown a very good behavior. The parameters of the relationship have been calibrated on the real values observed.
Only for the hydrodynamic model the value of the bottom shear stress has been assumed equal to the theoretical value of the Shield (τ_*

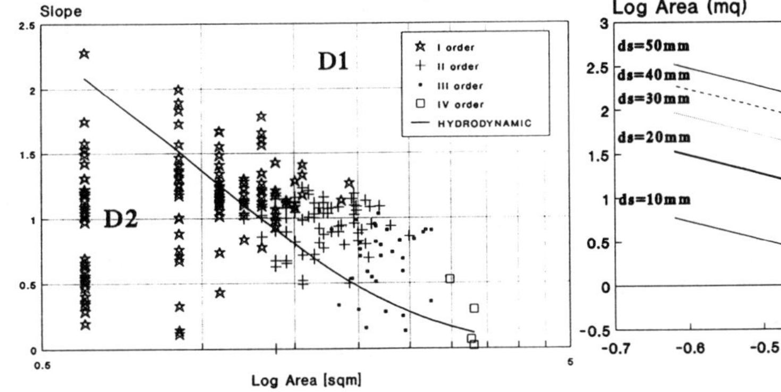

Fig.6.2 - The hydrodynamic relationship

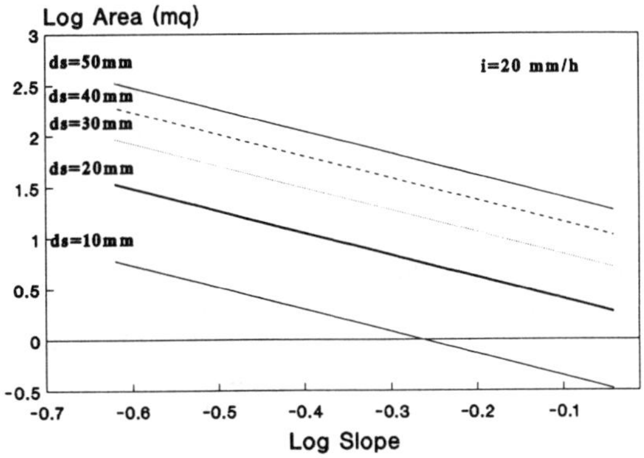

Fig.6.4a - Critical curves for different values of diameter .

=0.047). This allows that the hydrodynamical curve is representing the "critical condition" for the network area-slope ratio with respect to the sediment transport condition (fig.6.2). In fact the set of the points below the critical curve (D_2) correspond to a steady areas with respect to the sediment transport while the points above (D_1) are representative of unsteady sediment transport condition.

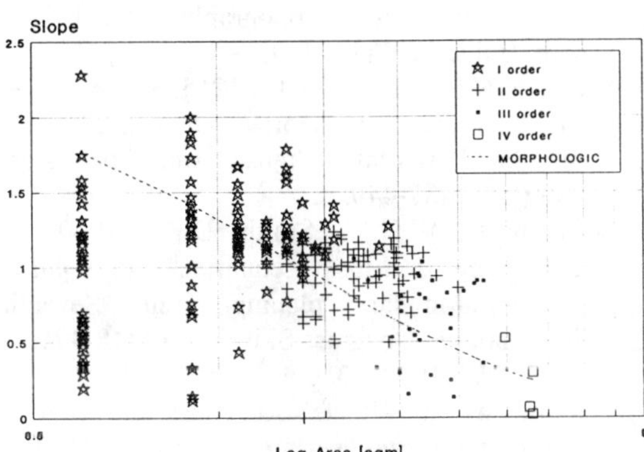

Fig.6.3 - The morphologic relationship

That is, the area corresponding could be recognized as "sources area for the sediment supply".
The morphological curve is describing the actual state of the drainage network as observed by the topological ratios. In the fig.6.3 the curve represent the average condition of the system outlined an average unsteady state of the system respect to the critical condition.

This means that the network tend to grow in order to reach a steady state in which the local slope and the contributing area are in equilibrium with the boundary conditions imposed.

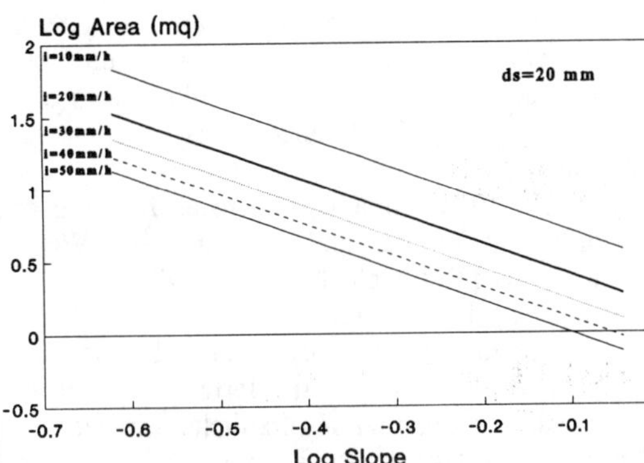

Fig.6.4b - Critical curves for different values of rainfall intensity.

Obviously the critical curve could change varying the values of the parameters. For example switching different values of the rainfall intensity, with a fixed grain size, or changing the characteristic soil diameter, with fixed rainfall intensity value, the critical curve move up or down in the plane, defining different equilibrium condition (fig.6.4a-6.4b).

7. CONCLUSIONS

A systematic monitoring of real hillslope in Italy,

has been performed in order to employ some empirical and conceptual models.
The results show the objective limitations related to the use of empirical models both for the differences of the site investigated and for the strict number of parameters involved into the relationships.
On the other hand, the conceptual models, hydrodynamically and topologically based, show a good performance in the phase of pattern drainage control. Moreover the hydrodynamic model seems to be able to switch on steady and unsteady areas respect to the sediment transport phenomena, while the morphological one is representative of the actual state of the structure. A combined use should be proposed in order to perform some more analysis on the surface drainage structures at the hillslope scale.

REFERENCES

Copertino V.A., Greco M., Molino B., 1992, Systematic monitoring of an erodible surface at the hillslope scale, International Conference Hydrocomp'92, Budapest.

Flint J.J., 1974, Stream gradient as function of order, magnitude and discharge, Water Resources Research, 10 (5), 969-973.

Greco M., La Barbera P. e Roth G., 1992, Sull'interazione tra morfologia di bacino ed i processi di erosione e trasporto, Convegno di Idraulica e Costruzioni Idrauliche, Firenze.

Kirkby M J.1971. Hillslope process-response models based on the continuity equation in Slopes: form and process. Institute of British Geographers special Pubblications,3, London: Institute of British Geographers.

Kirkby M.J., 1978, Hillslope hydrology, A Wiley-Interscience Publication.

Lowdermilk W.C, Sundling H.L., 1950, Erosion pavement formation and significance, Trans. Am. Geophys. Union, 31, 96-100.

Meyer L.D., Monke E.J., 1965, Mechanism of soil erosion by rainfall and overland flow, Trans. Am. Soc. Agr. Engrs., 8, 572-577 and 580.

Moore I.D., Grayson R.B. Landson A.R., 1991, Digital terrain modelling: a review of hydrological, geomorphological and biological applications, Hydrlogical Processes, 5, 3-30.

Musgrave G.W, 1935, The infiltration capacity of soils in relation to the control of surface runoff and erosion, J. Am. Soc. Agronomy, 27, 336-345.

Musgrave G.W., 1947, Quantitative evaluation of factors in water erosion - a first approximation. Journ. of Soil and Water Conserv.,2, 133-8.

Quinn P., Beven K., Chevallier P, Planchon O., 1991 The prediction of hillslope flow paths for distributed hydrological modelling using Digital Terain Models, Hydrological Processes, 5, 59-80.

Roth G. e Siccardi F., 1990, Un criterio di stabilità d'insieme per lo sviluppo di una rete di drenaggio alla scala di versante, XXII Convegno di Idraulica e Costruzioni Idrauliche, Cosenza.

Schumm, S.A., 1956, Evolution of drainage system and slopes in badlands at Perth Amboy, New Jersey, Geol. Soc. Am. Bull.,67, 597-646.

Speight J.G., 1980, The role of topography in controlling throughflow generation: a discussion, Earth Surface Proceses and landforms, 5, 187-191.

Strahler, A.N., 1952, Hypsometric (area-altitude) analysis of erosional topography, Geol. Soc. Am. Bull., 63, 1117- 1142.

Swanson N.P., Dedrick A.R., 1965, Soil particles and agregates transported in runoff from simulated rainfall, Trans. Am. Soc. Agr. Engrs., 8, 437-440.

Wischmeier W.H., Smith D.D., 1960, A universal soil-loss equation to guide conservation farm planing, Trans. Seventh International Congress Soil Sci., 1, 418-425.

Zingg A.W, 1940, Degree and length of land slope as it affects soil loss in runoff. Agricultural Engineering, 21, 59-64. Horton R.E., 1939, The analysis of runoff plot experiments. Trans. of Amer. Geoph. Union, 20, 693.

Environmental Management, Geo-Water & Engineering Aspects, Chowdhury & Sivakumar (eds)
© 1993 Balkema, Rotterdam. ISBN 90 5410 099 0

Tests on a new method of suspended load estimation

M. Habibi & M. Sivakumar
Department of Civil & Mining Engineering, University of Wollongong, N.S.W., Australia

ABSTRACT: This paper introduces the theoretical basis and fundamental equations of a recently developed method of sediment transport prediction and tests its applicability to natural alluvial channels. The new theory is based on turbulence energy concepts and, in particular, the idea that "suspended sediment concentration at any depth above the stream bed is proportional to the total turbulent energy production at the same level". The proposed equations are simple and suitable for engineering application. They involve a small number of easily measurable well-known hydraulic and sediment parameters. The applicability of the new sediment discharge formula has been tested under a wide range of hydraulic and sediment parameters. The recently published field data of Nakato(1990) and Voogt et al (1991) has been used for this purpose. The data have been collected from a typical natural river in Sacramento, California and two tidal channels in the southwestern part of the Netherlands. They cover a wide range of flow velocities and sediment sizes. The test results have shown that, despite its simple structure, the newly developed method has good predictability.

1 INTRODUCTION

Sediment transport problems have attracted the attention of the engineers and scientists for decades. Since DuBoys(1879), who introduced the first bed load discharge formula, a great number of theories have appeared in the literature to relate the rate of sediment transport by a river to hydraulic properties of flow and physical characteristics of sediment particles. The applicability of these methods have been tested against measured sediment data from laboratory flumes as well as from natural rivers. The tests have generally shown a large discrepancy between the predicted transport rates and measured sediment discharges(ASCE Task Committee, 1975; Yang, 1972; White et al., 1975; Bechteler and Vetter, 1989; and Nakato, 1990). As a result, none of the existing theories have gained universality. Nevertheless some of the available procedures, such as Einstein(1950) bed load function and Samaga et al(1986a and b), lack simplicity and often make use of various correction factors to be read from a series of tables and graphs. This procedure not only makes the computation cumbersome and unsuitable for engineering application, but results in loss of accuracy at each stage of the calculations. Hence investigation towards development of more accurate and simpler formulae is still continuing all over the world.

Based on the generally accepted principles of turbulence and fluid mechanics, the authors have developed a simple and a powerful procedure for the estimation of concentration and transport rate of sediment materials for the river systems. A brief introduction of the proposed theory is presented herein. Since the usefulness of a sediment transport equation depends on its applicability to natural rivers, the developed relationships are tested against data from three natural channels.

2 THEORETICAL BASIS OF THE NEW THEORY

Flow expends energy for transportation of sediment particles. Several investigators including Bagnold(1966), Engelund and Hansen(1967), Ackers and White(1973), Yang(1973, 1979 and 1984), Wiuff(1985), and Celik and Rodi(1991) tried successfully to relate the transport rates of sediment materials to some form of flow energy. However no attempt has been made so far to relate the local sediment concentration, i.e. concentration at any level y above the channel bed, to the flow energy. Habibi and Sivakumar(1992a) concentrated their study on the possibility of deriving an expression for local concentration based on energy concepts. They argued that the concentration of suspended solids at any depth y is directly proportional to the total turbulence energy production of the flow at the same level.

The theoretical support for this idea is obtainable from the definition of a suspension which is regarded as a mode of sediment motion in which the submerged weights of the particles are supported by the turbulence fluctuations of the flow velocity (Einstein, 1964). Thus it is postulated that a part of the turbulence energy production is used to keep sediment materials in suspension and prevent from settling. This particular part is assumed to be proportional to the total turbulence energy production and is equivalent to the work performed on the suspended particles to counteract the gravitational forces responsible for their settling.

3 DERIVATION OF THE SUSPENDED LOAD EQUATION

Expressions for the total turbulence energy production of the flow at any local depth y above the channel bed and for the work of the gravitational forces on suspended solids at the same level were derived. Using these two expressions and a proportionality parameter β, Habibi and Sivakumar(1992a) developed the following concentration equation.

$$C_y = \beta \frac{S\, u_*}{\Delta\, k\, \omega} \left(\frac{D-y}{y}\right) \qquad (1)$$

in which $\Delta = (\rho_s/\rho) - 1$. This equation is dimensionally homogeneous and expresses the volumetric concentration of suspended particles C_y at any depth y as a function of important hydraulic parameters of flow and sediment characteristics including energy or water surface slope S, total flow depth D, local depth y, mass density of the fluid ρ, von Karman constant k, shear velocity u_*, mean fall velocity of suspended particles ω and mass density of sediments ρ_s. The expression is not valid for a very thin layer close to the bed where suspension is impossible due to large particle sizes.

Transport rate of the suspended particles per unit of channel width, q_s, was then obtained from the depth integration of the product of local concentration, C_y, and flow velocity, u, as (Habibi and Sivakumar, 1992a) :

$$q_s = \beta \frac{S\, D\, V^2}{2\, \Delta\, \omega} \left[1 + \left(\frac{u_*}{k\, V}\right)^2 \right] \qquad (2)$$

in which V denotes the average cross sectional flow velocity.

For the development of Eq. (2) the proportionality parameter β was assumed to be independent of the local depth y and a logarithmic velocity distribution was assumed for the turbulent channel flow. Discussion about functional relationship of β, whether it is a constant or a function of other sediment flow parameters, is one of the most important aspects of the new development and research is continuing in this area. For the present study, β is assumed to be constant and is equal to 2%. Similar concepts have been used by Bagnold(1966) and Celik and Rodi(1991) in the evaluation of suspended load efficiency. For the computation of the fall velocity, ω, median diameter of the bed material mixtures, d_{50}, is used in the following applications.

4 APPLICABILITY OF THE SUSPENDED LOAD FORMULA

Four independent sets of data from sediment transport in three natural channels will be used to test the applicability and accuracy of the proposed expression of suspended load transport as given in Eq. (2). The data include 149 individual measurements which covers a wide range of flow velocities (from 0.52 m/s to 2.28 m/s) and sediment sizes (d_{50} from 0.18 mm to 6.30 mm).

For the purposes of the analysis, a discrepancy ratio r which is defined as the ratio of computed sediment discharges to the measured values, i.e., $r = (q_{s,comp} / q_{s,meas})$ is used. The closer the calculated r values approaches unity, the more accurate will be the predictions.

It should be noted that, due to complexity of sediment transport processes and difficulties and inaccuracies in measurement of flow and sediment parameters, particularly in the case of natural rivers, sediment discharges in general can not be predicted precisely. In sediment transport problems an equation which is able to predict transport rates within a range of two times smaller or larger than the measured values, i.e. $0.5 < r < 2.0$, for more than 60% of the applied cases, may be considered reasonable.

4.1 *Test against data from Sacramento river*

Nakato(1990) published 29 sets of data from suspended sediment transport in Sacramento river, California. The data were collected within the years 1977 to 1979 from two gauging stations along the river, one in Butte City and the other near Colusa. Tables 1 and 2 show the reported data which cover a wide range of water discharges (Q from 142 m^3/s to 2243 m^3/s), mean flow velocities (V from 0.52 m/s to 1.79 m/s) and sediment sizes (d_{50} from 0.33 mm to 6.30 mm). In these tables Q_s refers to measured suspended load and does not include wash load, A refers to cross sectional area of the flow, P is the wetted perimeter, B is the surface width and T is the water temprature.

Using Eq. (2) and data from Tables 1 and 2 with Δ=1.65 and k=0.4, the transport rates of suspended bed materials were predicted. Columns 10 in Tables 1 and 2 show the calculated values of discrepancy ratios r. Based on these values about 69% and 62% of the predicted transport rates fall in the range of 0.5~2.0 times the measured sediment discharges for gauging stations at Butte City and Colusa respectively. Considering the wide range of hydraulic and sediment parameters and the difficult task of measurements and the errors involved in field data collection, the accuracy obtained is acceptable.

The authors have applied five existing theories of sediment transport to the same data and have shown that the proposed method is one of the best available procedures in terms of predictability(Habibi and Sivakumar, 1992b). The discrepancy ratio r is shown in Fig. 1 for the Sacramento river data. The line r=1 represents the line of perfect agreement between calculated and measured sediment discharges.

4.2 *Tests against data from tidal alluvial channels in Netherlands*

Recently Voogt et al(1991) collected 120 sets of data from two tidal channels in Netherlands, namely Krammer and Scheldt Estuary tidal channels. The measurements represent transportation of fine sediments(d_{50} about 0.25 mm) under high velocity (V up to 2.28 m/s) conditions. Because of the very small sizes of the bed materials and high values of flow velocities, it is assumed that the sediment materials mostly travel in suspension and the bed load component is negligible. This assumption can be supported through comparisons of computed values of bed and suspended load transport rates using a number of existing theories. Hence the reported measured total sediment discharges can be

Table 1. Sediment flow data from Sacramento river at Butte City, California (After Nakato, 1990).

Data point (1)	Q m^3/s (2)	Q_s kg/s (3)	A m^2 (4)	P m (5)	B m (6)	S $(*10^4)$ (7)	T ^oC (8)	d_{50} mm (9)	r (10)
1	632.8	57.0	566.5	151.1	149.5	2.25	10.0	0.47	0.93
2	294.0	5.0	358.1	145.5	144.6	2.04	9.5	0.39	4.00
3	2024.4	168.0	1144.2	154.3	154.0	2.82	10.0	0.43	2.18
4	918.4	157.0	712.6	152.7	149.5	2.24	11.0	0.43	0.61
5	2242.8	233.0	1293.0	165.7	160.1	2.88	11.0	0.43	1.74
6	470.4	10.0	492.1	149.1	147.3	1.83	13.0	3.80	0.64
7	218.96	1.0	351.6	146.3	145.5	1.11	18.5	6.00	0.95
8	228.76	1.1	348.8	146.1	145.2	1.25	18.5	5.70	1.09
9	178.08	0.4	314.4	145.8	145.2	1.06	13.5	6.30	1.63
10	205.24	0.8	328.4	145.3	144.6	1.22	8.0	0.33	9.81
11	1052.8	250.0	768.4	153.3	149.8	2.45	7.5	0.40	0.57
12	150.1	0.3	289.3	144.6	144.0	0.99	10.0	0.40	10.00
13	1170.4	166.0	863.3	155.8	151.9	2.22	9.0	0.40	0.84
14	319.2	1.8	406.5	146.6	145.2	1.46	13.5	3.15	1.72
15	262.4	1.9	368.4	145.2	144.0	1.35	13.5	0.41	4.87
16	245.3	1.3	350.7	145.2	144.3	1.40	18.0	5.10	1.23

Table 2. Sediment flow data from Sacramento river at Colusa, California (After Nakato, 1990).

Data point (1)	Q m^3/s (2)	Q_s kg/s (3)	A m^2 (4)	P m (5)	B m (6)	S $(*10^4)$ (7)	T ^oC (8)	d_{50} mm (9)	r (10)
1	164.9	2.6	235.3	82.5	81.4	1.60	11.0	0.45	2.48
2	257.3	22.0	311.6	86.4	85.1	1.83	14.5	0.38	0.71
3	142.0	3.1	213.0	81.5	80.5	1.64	14.5	0.39	2.15
4	380.2	27.3	434.4	92.1	89.7	1.44	9.5	0.42	0.67
5	1122.9	74.5	910.7	117.0	112.8	1.78	10.5	0.62	0.87
6	1079.7	162.0	887.5	116.1	111.6	1.78	10.0	0.60	0.39
7	1117.2	90.5	910.7	117.0	112.8	1.78	11.5	0.88	0.55
8	616.0	43.4	609.3	98.0	94.5	1.47	14.0	2.00	0.23
9	206.9	11.3	261.4	82.6	81.4	1.80	9.0	0.35	1.22
10	686	56.6	634.4	102.6	99.4	1.69	7.5	0.40	0.90
11	1038.8	12.6	856.8	109.9	104.9	1.72	9.0	0.48	0.57
12	310.8	11.8	398.2	87.9	85.4	1.21	11.5	0.38	1.03
13	280.0	60.6	360.0	86.8	84.8	1.35	14.5	0.37	0.20

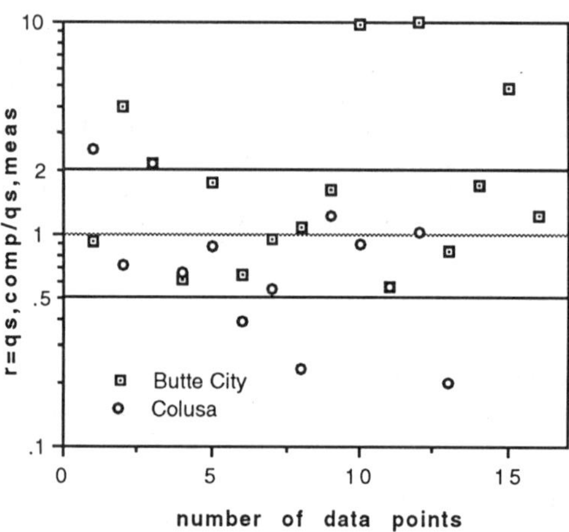

Fig. 1. Comparison of computed sediment discharges with measured transport rates for Sacramento river.

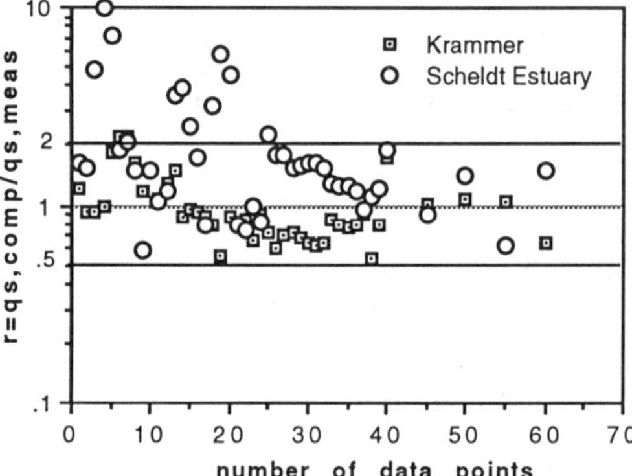

Fig. 2. Comparison of computed sediment discharges with measured transport rates for Netherlands tidal channels.

directly compared with predicted transport rates of suspended materials.

Using the proposed expression for the estimation of suspended load, i.e. Eq. (2), and collected data of Voogt et al(1991) with Δ=1.65 and k=0.4, the transport rates of suspended bed materials were predicted and discrepancy ratios r were calculated. Graphical comparisons of computed r values are presented in Fig. 2. As no slope measurements have been reported by Voogt et al, Chezy formula was used for the estimation of water surface slope S.

Based on the calculated discrepancy ratios, about 97% and 81% of the predicted transport rates fall in the range of 0.5-2.0 times of the measured sediment discharges for Krammer and Scheldt Estuary tidal channels respectively (Table 3). This indicates a very good predictability for the new theory in application to the streams carrying fine sands under high flow velocity conditions.

In another investigation the authors have applied five methods of sediment transport estimation to the same 149 sets of data to campare the predictability of the new theory

Table 3. Statistical analysis of the test results.

Channel name	No. of data points	Discrepancy ratio r: % within 0.5-2.0	Mean value	Standard deviation
Sacramento river at Butte City	16	69	2.67	2.97
Sacramento river at Colusa	13	62	0.92	0.66
Krammer tidal channel	60	97	1.03	0.39
Scheldt Estuary tidal channel	60	81	1.95	1.34
Total data points	149	85	1.57	1.30

Table 4. Summary of the test results for the whole 149 data sets.

Investigator	r: % within 0.5-2.0	Mean value	Standard deviation
Yang(1973)	79	1.70	1.42
Rijn(1984)	77	1.60	1.73
Wiuff(1985)	10	7.80	7.33
Samaga et al(1986b)	74	1.07	0.80
Celik and Rodi(1991)	18	0.56	0.50
Habibi and Sivakumar (1992)	85	1.57	1.30

with a number of existing formulae (Habibi and Sivakumar, 1992b). The selected equations included those of Yang(1973), Rijn(1984), Wiuff(1985), Samaga et al(1986b) and the recently published approach of Celik and Rodi(1991). Summary of the comparison and statistical analysis of the calculated discrepancy ratios for these theories are shown in Table 4 and Figure 3. Based on this study Habibi-Sivakumar's approach, with a score of 85% of estimated transport rates within the range 0.5-2.0 times of the measured sediment loads, has shown the best predictability.

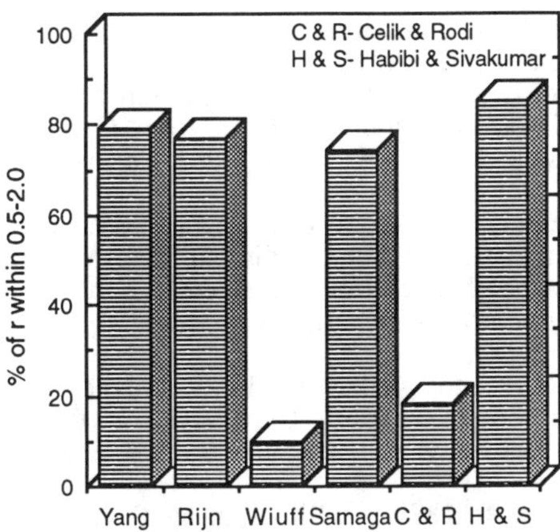

Fig. 3. Applicability of the selected formulae.

5 SUMMARY AND CONCLUSION

Theoretical basis and fundamental equations of a new theory, which enables the computation of local sediment concentration and transport rate of suspended bed materials, are presented. The proposed theory was applied to 149 sets of data from three natural channels. A summary of the test results is shown in Table 4. Considering all the data points about 85% of the predicted discharges by the developed equation are within 0.5~2.0 times of the actual measurements. This can be considered as very good in terms of the accuracy for a sediment transport estimation model in river systems. In particular, the new theory has shown excellent predictability for sediment-laden flows with high velocity and fine bed materials.

ACKNOWLEDGMENT

This work is part of the activities of the Water Engineering and Geomechanics Research Program of the Department of Civil and Mining Engineering of the University of Wollongong and the support of the University is acknowledged. The first author wishes to thank the Iranian Government for providing a postgraduate research scholarship.

REFERENCES

Ackers, P. and White, W. R. 1973. Sediment Transport: New Approach and Analysis: *J. Hydr. Engrg.*, ASCE, 99: 2041-2060.

ASCE Task Committee. 1975. *Sedimentation Engineering*, ASCE.

Bagnold, R.A. 1966. An Approach to the Sediment Transport Problem from General Physics: *U.S. Geol. Survey,* Prof. Paper 422-I, 37 p.

Bechteler, W. and Vetter, M. 1989. The Computation of Total Sediment Transport in View of Changed Input Parameters. *Proc. Int. Symp. on Sediment Transport Modeling*. S.Y. Wang(ed.).

Celik, I. and Rodi, W. 1991. Suspended Sediment Transport Capacity for Open Channel Flow: *J. Hydr. Engrg.,* ASCE, 117: 191-204.

Du Boys, P. 1879. Le Rhone et les Rivieres a Lit Affouillable: *Annales des Ponts et Chaussees,* Series 5: 18: 141-195.

Einstein, H. A. 1950. The Bed-Load Function for Sediment Transportation in Open Channel Flows: *USDA, Tech. Bull.* No. 1026.

Einstein, H. A. 1964. River Sedimentation: Section 17-II: 17.35-17.67, Handbook of Applied Hydrology, Ven Te Chow(ed.), McGraw Hill Book Co., Inc., New York, NY.

Engelund, F. and Hansen, E. 1967. A Monograph on Sediment Transport in Alluvial Streams, Technical University of Denmark, Hydraulic Laboratory.

Habibi, M. and Sivakumar, M. 1992a. New Formulation of Suspended Sediment Transport: *J. Hydr. Engrg.*, ASCE(Submitted).

Habibi, M. and Sivakumar, M. 1992b. Review of Selected Methods of Sediment Transport Estimation: *11th Australasian Fluid Mechanics Conference*, Hobart, Tasmania, Australia, 14-18 December, 1992.

Nakato, T. 1990. Tests of Selected sediment Transport Formulas: *J. Hydr. Engrg.,* ASCE, 116: 362-379.

Rijn, L. C. van. 1984. Sediment Transport, Part II: Suspended Load Transport: *J. Hydr. Engrg.,* ASCE, 110: 1613-1641.

Samaga, R. B., Ranga Raju, G. K., and Garde, J. R. 1986a. Bed Load Transport of Sediment Mixtures: *J. Hydr. Engrg.,* ASCE, 112: 1003-1018.

Samaga, R. B., Ranga Raju, G. K., and Garde, J. R. 1986b. Suspended Load Transport of Sediment Mixtures: *J. Hydr. Engrg.*, ASCE, 112: 1019-1035.

Voogt, L., Rijn, L. C. van, and Den Berg, J. H. van. 1991. Sediment Transport of Fine Sands at High Velocities: *J. Hydr. Engrg.,* ASCE, 117: 869-891.

White, W. R., Milli, H., and Crabbe, A. D. (1975). "Sediment Transport Theories: a Review," Proc. Inst. of Civil Engineers, London, England, Part 2, Vol. 59, pp. 265-292.

Wiuff, R. 1985. Transport of Suspended Materials in Open Submerged Streams: *J. Hydr. Engrg.,* ASCE, 111: 774-792.

Yang, C. T. 1972. Unit Stream Power and Sediment Transport: *J. Hydr. Engrg* ., ASCE, 98: 1805-1826.

Yang, C. T. 1973. Incipient Motion and Sediment Transport: *J. Hydr. Engrg* ., ASCE, 99: 1679-1704.

Yang, C. T. 1979. Unit Stream Power Equation for Total Load: *J. of Hydrology*, 40: 123-138.

Yang, C. T. 1984. Unit Stream Power Equation for Gravel: *J. Hydr. Engrg* ., ASCE, 110: 1783-1797.

Environmental Management, Geo-Water & Engineering Aspects, Chowdhury & Sivakumar (eds)
© 1993 Balkema, Rotterdam. ISBN 90 5410 099 0

Geo-water maps used to manage leaching into rivers

E.J. Heidecker
The University of Queensland, Qld, Australia

ABSTRACT: Dispersible sodic clays are leaching into the Burdekin River, affecting engineering, health, tourism, and parks along the river. Geo–water maps indicate key management areas for protective terraces and mulch gardens that provide clay–bonding humus.

1 CLAY LEACHING, A GROWING PROBLEM

Sodic–clay leaching is best known along the Murray–Darling River system where Knight et al (1989) have recognised combined geological and water controls. The impact of dispersed sodic clays along the Burdekin River in north Queensland is likely to be wider than that of simple salinisation. Extensive sheet erosion progresses to micro–karst, tunnel, and then gully erosion as shown in Fig.1 which is at "A" in Figs. 2 and 3.

Certain dispersed clays remain suspended in brackish and sea waters. Thus clay leachates generated along the Burdekin River are now reaching productive delta lands and estuaries. Ultimately coastal fisheries and the Great Barrier Reef are likely to be affected.

1.1 *Need for geo–water maps*

Fig. 2 shows from the air that devastated area "A" lies at a coincidence of complex geology and water impacts. Drainage has been impacted by vehicles and tourists visiting Big Bend recreation area to the north on the Burdekin River. Geo–water maps such as Fig. 3 are needed to resolve such complex geological, water, and impact relationships.

1.2 *Use of geo–water maps*

Knight et al (1989) have used many types of geological and water maps to recognise geological controls in a large area of sodic erosion at the head waters of the Murray Darling system. Geo–water map Fig. 3 integrates geological, water, and impact elements (roads, tracks, and tourist parks.) In this integrated view area "A" is at an impacted river crossing below and adjacent to palaeo–drainage channels blocked by basalt so that salt and sodium have accumulated.

Once control relationships are recognised it is possible to identify similar sites elsewhere which should receive special management attention if subjected to tourist and vehicular impacts. "B" in Fig. 3 illustrates such a site. In that area steep banks down into Fletcher Creek overlie a Pleistocene palaeo–channel where tourist visitations to Dalrymple National Park are likely to concentrate.

CONCLUSIONS

Geo–water maps can identify areas for preventative management before impact triggers destructive leaching. Thus there are opportunities for long–term management measures. Humus landscaping with retentive terraces and humus–producing groves and gardens is an attractive long–term self–sustainable measure under way at Dalrymple in Fig. 3. Forward management here demonstrates use of geo–water maps and the principles of soil stabilisation with clay–bonding humus (Brady 1984).

REFERENCES

Brady, N.C. 1984. *The nature and properties of soils*, 264–260.

Knight, M.L., Saunders, B.J., Williams, R.M. & Hillier, J. 1989. Geologically induced salinity at Yelarbon, Border Rivers area, New South Wales, Queensland. *BMR Journal of Australian Geology & Geophysics* 11:355–361.

Fig. 1 A leached area at "A in Figs 2 & 3. Micro–karst sink holes in the foreground lead down to tunnels in dispersible sodic clay. Further leaching and erosion has opened these tunnels into gullies in the middle distance.

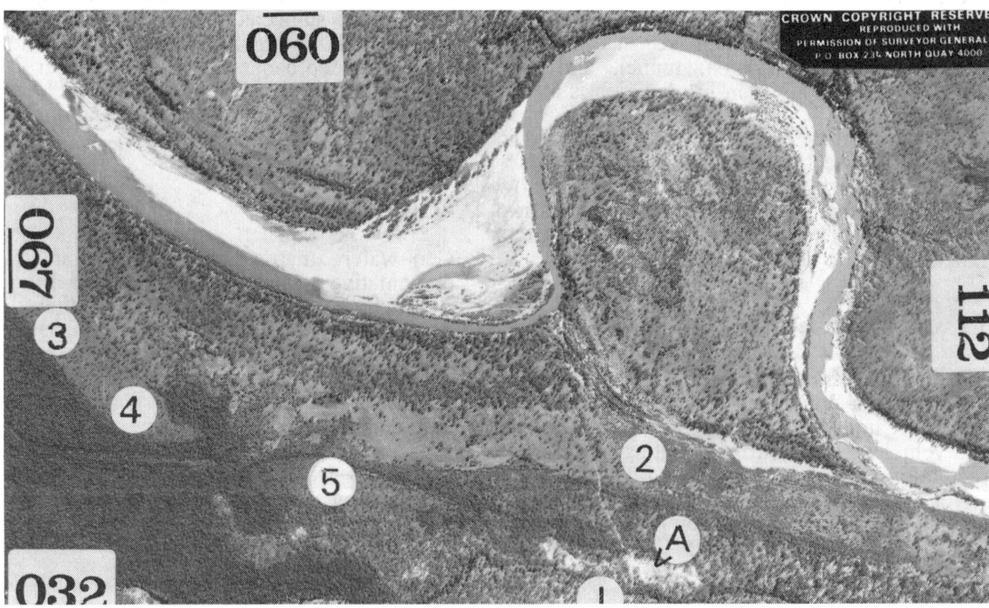

Fig. 2 An aerial photograph of the Big Bend recreation area and Burdekin River in Fig. 3. "A" is an area of sodic–clay leaching shown in fig. 1. This leached area is near a creek crossing at 1 and adjacent to a palaeo–channel covered by basalt at 2 and on a palaeo–channel continuing the line of an older channel–filling basalt flow 3, 4, 5. Grid numbers refer to Fig. 3, the geo–water map.

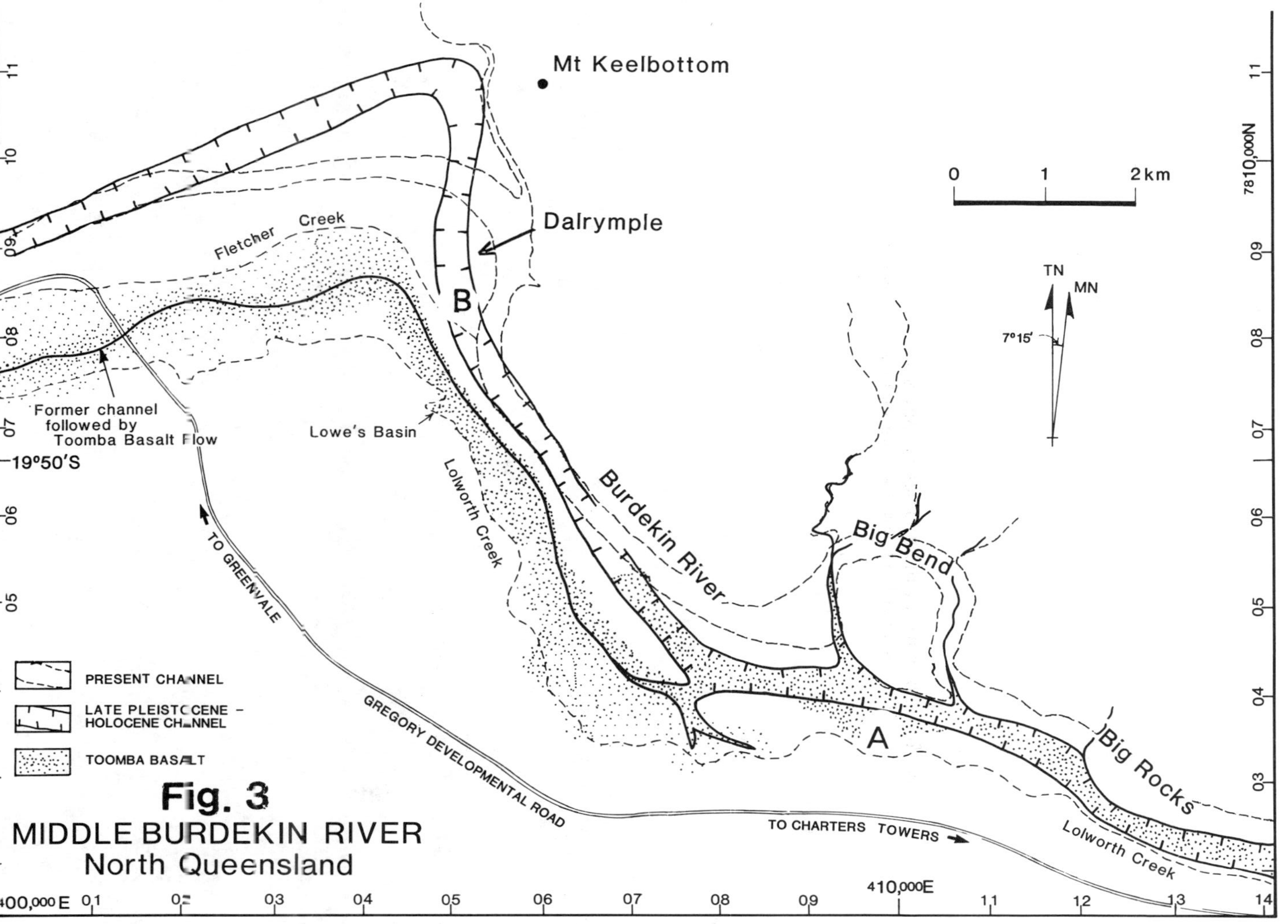

Fig. 3
MIDDLE BURDEKIN RIVER
North Queensland

Environmental Management, Geo-Water & Engineering Aspects, Chowdhury & Sivakumar (eds)
© 1993 Balkema, Rotterdam. ISBN 90 5410 099 0

A model to predict the stable natural channel

G.A.Jenkins
Queensland University of Technology, Qld, Australia

R.J.Keller
Monash University, Vic., Australia

ABSTRACT: Natural channels tend to adjust their average values of widths, depths, bed slopes and meander sizes to accommodate the imposed sequences of flow and sediment loads. The mechanism of adjustment depends on the bed material and erodability of the banks. This paper describes a numerical model which has been developed to define the stable geometric properties of a natural stream, due to the imposition of a dominant steady water discharge and sediment load. The theoretical aspects of the model are described, as well as an application of the model to a section of a natural stream in Victoria.

1 INTRODUCTION

With the introduction of European settlement in Australia, urbanisation and deforestation have considerably altered the amount and duration of both water and sediment flows in the natural streams. In both the urban and agricultural environment, rivers and streams have generally been seen as serving the role of drainage channels for excess rainfall. Alterations to the natural rivers have usually been made in order to protect farm land from being encroached, or to protect from the adverse effects of flooding. Due to these biased views of the benefits of natural rivers, severe degradation of the natural river systems has taken place in Australia.

The effect on the natural rivers has included the loss of the complex aquatic environments which exist in these ecosystems. The loss of the large number of plant and animal species which live in the natural river environment has accelerated the degradation of the physical structure of the river systems.

An example of the physical degradation of natural channels occurs when the downstream conditions of a river are altered, such as draining a swamp to provide farm land. As the natural channel is a result of an equilibrium condition for the imposed regime of water flow and sediment load, the river will adjust itself to satisfy the new hydraulic conditions. This results in the formation of a headward erosion, formed by the action of the channel adjusting its bed slope to satisfy the new downstream control.

In a natural river, the geometric properties of the channel vary throughout its course, depending on the physical properties of the material which forms the bed and banks of the channel. Complex interactions take place in the formation of a natural river system, due to the imposition of a varying sequence of water flow and sediment discharge. Due to the difficulty in modelling these interactions, it is usual to define the steady water and sediment discharges which have been dominant in the formation of the natural system. The bankfull discharge is generally taken as this dominant discharge, Chang (1988). The formation of the channel can then be modelled using the hydraulic processes of a steady dominant discharge and sediment load.

2 THEORETICAL BACKGROUND

The mathematical model which has been developed as part of this study, has two major components which define the stable profile for an alluvial channel. The first component is the derivation of the water surface elevation, velocity and friction slope at each cross-section from the equations of steady gradually varied flow in open channels. The second is the determination of the cross-section shape so that the sediment carrying capacity equals the imposed steady sediment load at all locations along the channel.

In the definition of steady gradually varied flow in the natural channel, it is assumed that the flow is one-dimensional along the longitudinal axis. No allowance has been made for secondary currents in the transverse direction, which may be important in the development of meandering channels.

The channel is represented by a series of cross-sections defined along the longitudinal axis. A typical natural river cross-section is shown in Figure 1. Generally the cross-section can be divided into moveable and non-moveable sections. The moveable section represents the alluvial bed and banks which will be scoured due to the hydraulic action in the channel. The non-moveable sections represent parts of the cross-section which are protected against scour

In the model developed, each cross-section is divided into three sub-sections; a left and right non-moveable section, and a central moveable section which generally represents the main channel part of the cross-section.

The water surface elevation, total energy level and average velocity can be calculated at each cross-section from the equation of gradually varied flow, Equation 1. For any particular cross-section, the sediment carrying capacity under these hydraulic conditions, may not equal the dominant sediment load. In this case, the channel can not be said to be in an equilibrium, or stable condition.

In a natural channel, where the sediment carrying capacity of a cross-section is in excess of the dominant sediment load, the moveable parts of the cross-section will be scoured until both the sediment carrying capacity and sediment load are equal. The opposite effect where the sediment carrying capacity of the cross-section is less than the sediment load will cause silting at the cross-section. Therefore, in the model developed, this action is represented by the moveable parts of the channel being either lowered or raised respectively, so that both the gradually varied flow equations and the sediment carrying capacity equations satisfy the

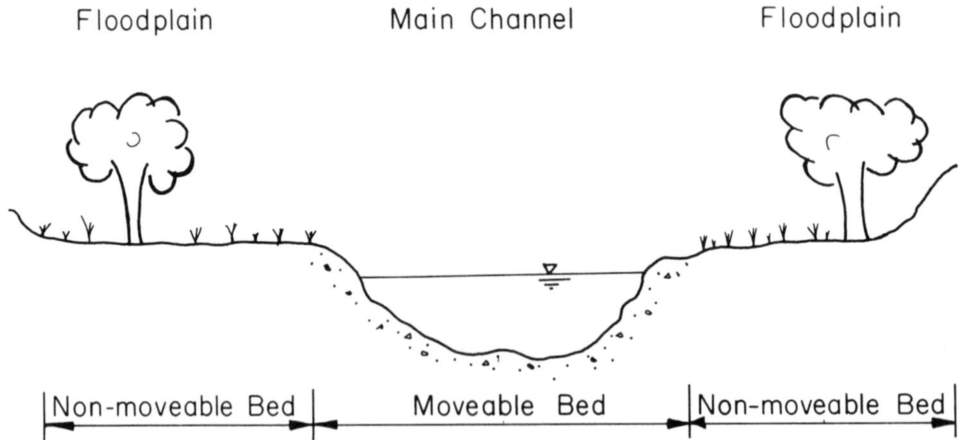

Fig. 1 - Typical Cross-section in a Natural Channel

imposed dominant water and sediment flow conditions.

The model uses the Manning's roughness coefficient to define the roughness of each sub-section. No allowance has been made in the roughness coefficients for the effects of form roughness due to the formation of dunes, ripples etc, in the alluvial bed. If it is felt that these roughness effects are important in the prediction of the stable channel, then the roughness coefficients chosen should be adjusted manually.

The description of the steady gradually varied flow conditions are given by the integration of the equation of motion,

$$\frac{dE}{dx} = S_o - S_f \tag{1}$$

where; E - Specific energy
x - Longitudinal distance along channel
S_O - Bed slope
S_f - Friction slope

The solution to Equation 1 is found using the standard step method, described in detail by Henderson (1966). In this analysis of the stable natural channel, it has been assumed that subcritical flow conditions exist.

In predicting the sediment discharge in the channel, the model developed allows the user to choose from one of the following sediment transport equations. These equations have been chosen from the many available because they are used widely by practising river engineers. The model could be simply modified to permit the use of alternative bed load equations.

Du Boys Formula, Chang (1980)

$$q_s = \psi_D \tau_o (\tau_o - \tau_c) \tag{2}$$

where; q_S - Sediment discharge

ψ_D - $\dfrac{7.011 \times 10^{-6}}{d_{50}^{3/4}}$

τ_o - Bed shear stress
τ_c - Critical shear stress, $= 0.5985 + 0.9097 d_{50}$
d_{50} - Median size of the bed sediment,

Einstein-Brown Formula, ASCE (1975)

$$\begin{aligned} 0.465\Phi &= e^{-0.391\psi} && \text{if } \psi > 5.5 \\ \Phi &= 40\left(\frac{1}{\psi}\right)^3 && \text{if } \psi \le 5.5 \end{aligned} \tag{3}$$

where;

$$\Phi = \frac{q_s}{\gamma_s F \sqrt{g\left(\dfrac{\gamma_s}{\gamma} - 1\right)}}$$

$$\psi = \frac{(\gamma_s - \gamma) d_{50}}{\tau_o}$$

$$F = \sqrt{\frac{2}{3} + \frac{36\ ^2\nu}{g d_{50}^3 \left(\dfrac{\gamma_s}{\gamma} - 1\right)}} - \sqrt{\frac{36\ ^2\nu}{g d_{50}^3 \left(\dfrac{\gamma_s}{\gamma} - 1\right)}}$$

in which;
γ_s - Sediment specific weight
γ - Specific weight of water
g - Gravitational acceleration
ν - Kinematic viscosity of water

Engelund-Hansen Formula, (ASCE, 1975)

$$q_s = 0.05 \gamma_s V^2 \sqrt{\frac{d_{50}}{g\left(\dfrac{\gamma_s}{\gamma} - 1\right)}} \left[\frac{\tau_o}{(\gamma_s - \gamma) d_{50}}\right]^{3/2} \tag{4}$$

where; V - Mean velocity of flow

The bed shear stress at any point across a cross-section is given by,

$$\tau_o = \gamma\, h S_f \cos\phi \tag{5}$$

where; h - Local flow depth

ϕ - Transverse boundary slope

The equations shown above have been derived to calculate the sediment discharge in an open channel. Natural channels are generally irregular in cross-section, and in order to predict the stable channel profile, it is necessary to define a total sediment carrying capacity for the cross-section.

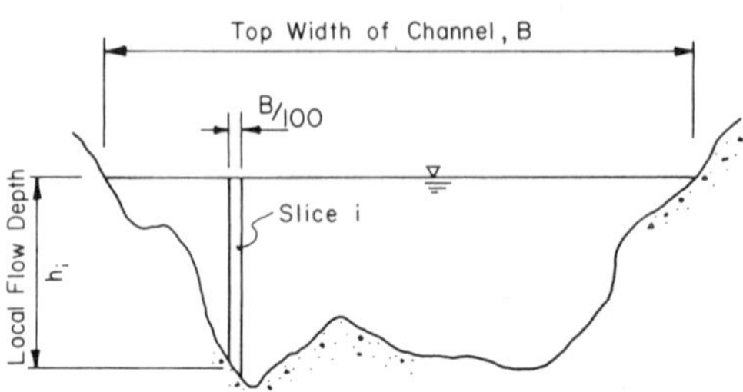

Fig. 2 - Local Flow depth for calculation of Bed Shear Stress

This can be achieved using the equations described above, by slicing the wetted part of the cross-section into thin slices of equal width, as shown in Figure 2. Each slice has local depth h, and the bed shear stress can be calculated from Equation 5 above. The sediment discharge for the slice can then be calculated from one of the bed load equations above, using the bed shear stress calculated. This sediment discharge can then be taken as a unit rate for the slice. The total sediment carrying capacity for the cross-section can be found by summing the discharge rates calculated for each slice. In the model developed, the top water surface is divided into 100 slices, in the process of calculating the sediment carrying capacity.

For a stable channel in alluvial material, no slope can be greater than the angle of repose for the sediment. Any transverse slope greater than the angle of repose of the material will not be in a stable state. This is modelled by setting all transverse slopes, within the moveable part of the cross-section and which are greater than the angle of repose, to the angle of repose. The angle of repose for the bed material is determined from the chart shown in Figure 3, taken from Simons et al (1982). In this derivation, it is assumed that the angle of repose of a bed material is only dependent on the median particle size.

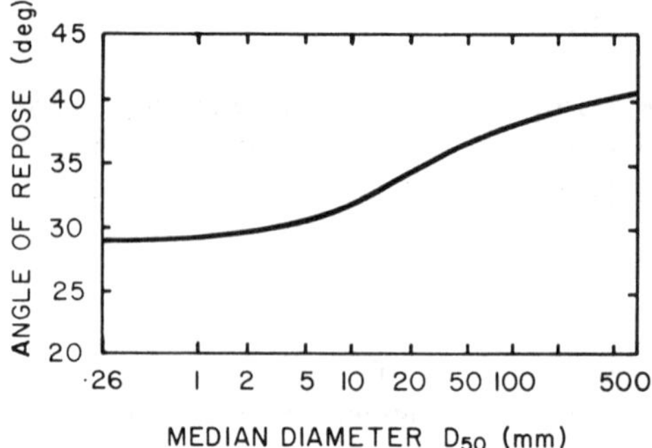

Fig. 3 - Angle of Repose for Sediment (from Simons et al (1982))

From the parameters of dominant water discharge, sediment load and bed material properties, the model computes each stable cross-section by solving the equations of continuity of mass, conservation of energy and sediment carrying capacity. An explicit solution to these equations is not possible and an iterative procedure must be employed.

It has been assumed that the flow in the channel is subcritical. Therefore, as the flow is controlled by the downstream conditions, computations begin with the downstream cross-section. The three equations of continuity of mass, conservation of energy and sediment carrying capacity are then solved at each cross-section in turn.

At each cross-section the water surface elevation, mean velocity and friction slope are found iteratively, using the standard step method, Henderson (1966). From these hydraulic conditions, the sediment carrying

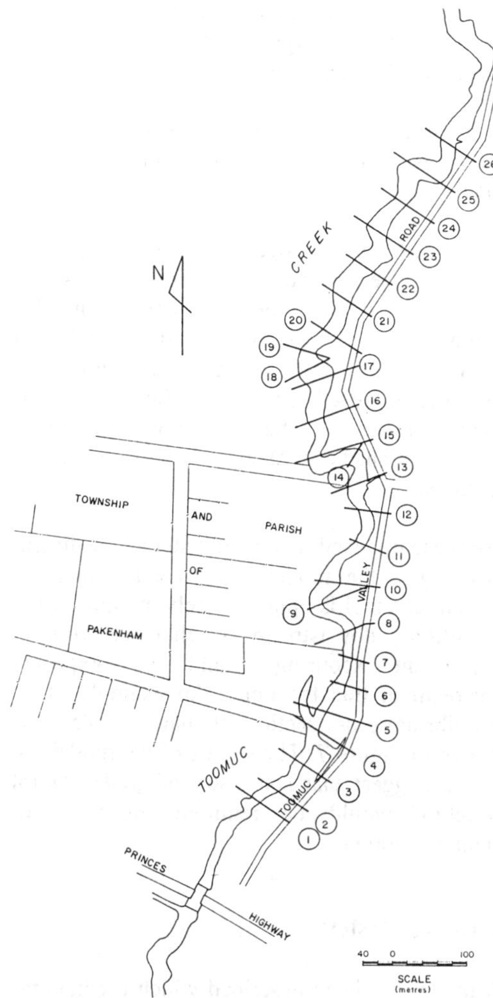

Fig. 4 - Toomuc Creek Site Plan

capacity of the cross-section is determined. If the sediment carrying capacity is not equal to the sediment load, the moveable part of the cross-section is adjusted up or down. The method employed uses a set increment of depth adjustment, in which the elevation of each moveable point on the cross-section is adjusted, at each iterative cycle.

The hydraulic conditions at the cross-section are then recalculated using the standard step method and the process repeated. The solution uses a form of increment halving, which is continued until the sediment carrying capacity of the cross-section equals the imposed sediment load. At this stage the computations proceed to the next cross-section upstream, and the process is repeated until the last cross-section is reached.

3 NUMERICAL EXAMPLE

A numerical example has been prepared to demonstrate the application of the model to the prediction of a stable channel profile for a natural channel. The example prepared is for a section of Toomuc Creek, located in South East Victoria, shown in Figure 4.

The bankfull discharge for this section of creek was estimated as 6.94 m^3/sec, assuming an ARI of one year from a partial series flood frequency analysis, I E Aust. (1987). A stream gauging station is located along the creek in the region of the study section.

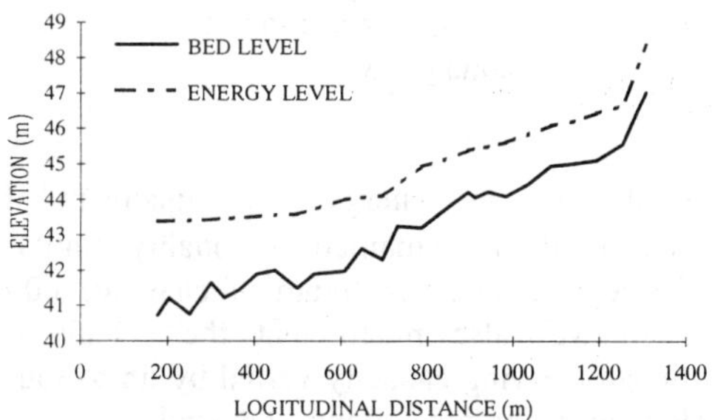

Fig. 5 - Backwater Analysis at Bankfull Discharge

The model was applied to the study section to determine the existing hydraulic conditions at the site. A backwater analysis was undertaken , in which a downstream control level of RL 43.38 m AHD was assumed. The resulting longitudinal section demonstrated that there is a distinct change in the energy gradient around chainage 600 *m*, see Figure 5.

A sensitivity analysis was undertaken to determine the effect on the hydraulic conditions of changing the downstream control level. It was noted that regardless of the downstream

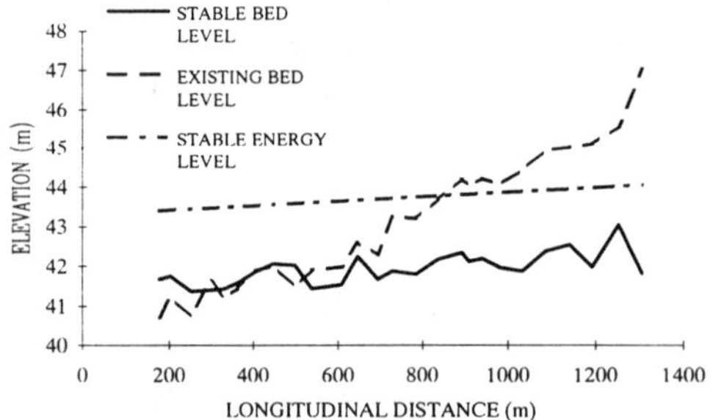

Fig. 6 - Stable Channel Analysis

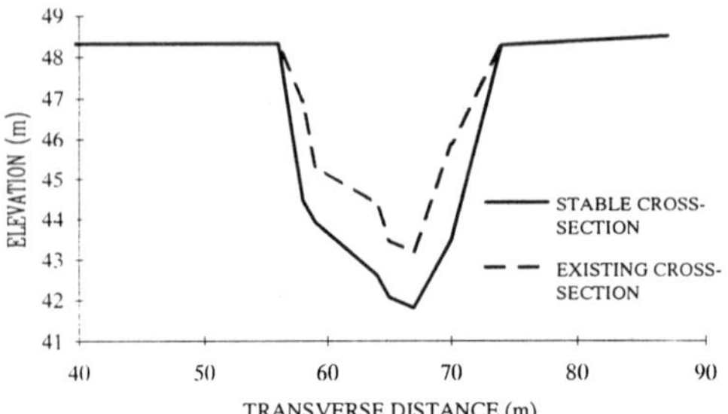

Fig. 7 - Stable Cross-section at Chainage 785 *m*

control level, the energy slope upstream of chainage 700 *m* remained reasonably similar and steeper than downstream of chainage 600 *m*. It was also noted that the calculated sediment carrying capacity varied by up to four orders of magnitude along the study section. Du Boys Formula was used in all of the runs undertaken in this study.

A stable channel analysis of the study section was then undertaken to gain some understanding of the effect of the various parameters involved. It was observed that the energy gradient for the stable channel is constant along the study reach, see Figure 6. This is in agreement with the theory of minimum stream power for a channel reach, which has been defined as,

> *The equilibrium geometry of an alluvial channel reach of equal discharge is so adjusted that the power expenditure is a minimum subject to the given constraints. Minimum power expenditure is equivalent to uniform energy gradient along the channel.* Chang (1988),

Furthermore, the stable energy gradient increases with an increase in dominant sediment load.

A detailed site investigation for the study reach has not been undertaken for this study. Therefore, it has been assumed in this demonstration of the model that the sediment particle size is 0.5 *mm* at each cross-section, and that the imposed constant sediment load is 0.004 m^3/sec. The Manning's roughness coefficient has been taken as 0.035 for all cross-sections.

For these imposed dominant water and sediment discharges, the model has predicted that the present channel is approximately at equilibrium conditions, downstream of chainage 700 *m*. Significant scouring could be expected upstream of this location. An example of an equilibrium cross-section, at chainage 785 *m*, is shown in Figure 7. The results of the model run would suggest that some sort of grade control structure would be required, upstream of chainage 700 *m*.

4 CONCLUSION

A model has been described which predicts the stable profile for natural channels. The model assumes steady water and sediment discharges as being dominant in the formation of the stable channel profiles. The natural channel is represented by a series of cross-sections, which can be subdivided into three regions which are either moveable or non-moveable, representing the mobile and fixed parts of the channel respectively.

In the derivation of the stable channel, the model satisfies the conditions that under gradually varied flow, each cross-section must have a sediment carrying capacity equal to the dominant sediment load. The solution procedure uses the standard step method, Henderson (1966), to solve the gradually varied flow equations and an iterative technique to satisfy the sediment carrying condition at each cross-section in turn.

It has been assumed in the derivation of the model that the properties of the channel may vary along the study reach and that the existing channel geometry will play an important role in the development of the stable geometry of the channel

The model has been applied to a section of creek in South Eastern Victoria to demonstrate qualitatively, its use in predicting a stable natural geometry. To achieve quantitative results from this study a more detailed site investigation would be required to determine the dominant sediment load applied to the study section.

ACKNOWLEDGMENTS

This work was sponsored by the Land and Water Research and Development Corporation, the Victorian Department of Water Resources, Melbourne Water and the Victorian Rural Water Commission.

REFERENCES

ASCE 1975. *Sedimentation Engineering.* ASCE Manuals and Reports on Engineering Practice - No. 54, New York.

Chang, Howard H. 1980. Stable Alluvial Canal Design. *ASCE, Journal of the Hydraulics Division, Vol. 106, No. HY5.*

Chang, Howard H. 1988. *Fluvial Processes in River Engineering.* John Wiley & Sons, New York.

Henderson, F. M. 1966. *Open Channel Flow.* Macmillan, New York.

I E Aust. 1987. *Australian Rainfall and Runoff, Volume 1.* Editor D.H. Pilgrim, ACT

Simons, Li and Associates 1982. *Engineering Analysis of Fluvial Systems.* Simons, Li and Associates, Inc., Fort Collins, Colorado.

Environmental Management, Geo-Water & Engineering Aspects, Chowdhury & Sivakumar (eds)
© 1993 Balkema, Rotterdam. ISBN 90 5410 099 0

Prediction of local scour depth at bridge abutments

Ricky T.F. Kwan
Sinclair Knight, Sydney, N.S.W., Australia

ABSTRACT: This paper presents an overall picture of the mechanics of flow and erosion at bridge abutments and similar groyne-like structures, along with the hydraulic factors that a design engineer has to consider when estimating the maximum local scour depth. A scour depth predictor relationship is proposed based on laboratory data.

1 INTRODUCTION

Scour is the removal of bed material from a stream bed by the erosive action of flowing water. One of the concerns to the bridge engineer or river engineer is the problem of scour at the base of bridge piers or abutments. Unless the bridge foundations can be founded on rock scouring may undermine and result in the damage or failure of the bridge. Brice et al (1978) reported that damage to bridges and highways in the United States caused by scour amounted to about $100 million per major event. Similarly Sutherland (1986) reported that approximately 30% of all major bridge failures in New Zealand resulted from abutment scour.

One problem which often confounds the designer is the diversity of equations and opinions that have been put forward for the estimation of scour at bridge piers and abutments (Raudkivi and Sutherland 1981). This is attributed, in part, to the large number of interacting variables affecting scour and the different conditions under which various relationships are derived. Relationships derived from one set of conditions may therefore be inappropriate for another.

To date our understanding of the scour process is still incomplete, particularly for abutment scour. Recent research conducted at the University of Auckland in New Zealand has, however, made it possible to explain many of the earlier inconsistencies and better account for the factors affecting scour. Some of these results are discussed below for local scour at bridge abutments. Pier results are only referred to when necessary.

2 LOCAL SCOUR AT BRIDGE ABUTMENTS

Local scour occurs as a result of the presence of the abutment and is characterised by the formation of scour holes at the base of the structure.

Most researchers report that the scour process is initiated by flow acceleration around the abutment (Wong 1982, Tey 1984). Rajaratnam (1983) found in his experiments on a planar bed that this effect increased the local bed shear stress by up to five times that of the mean upstream value. Scour therefore occurs where the combined effect of increased bed shear stress and turbulent agitation exceeds the critical value for sediment entrainment. This usually commences near the upstream corner of the abutment where the maximum scour depth is also usually found (Ahmad 1953, Wong 1982).

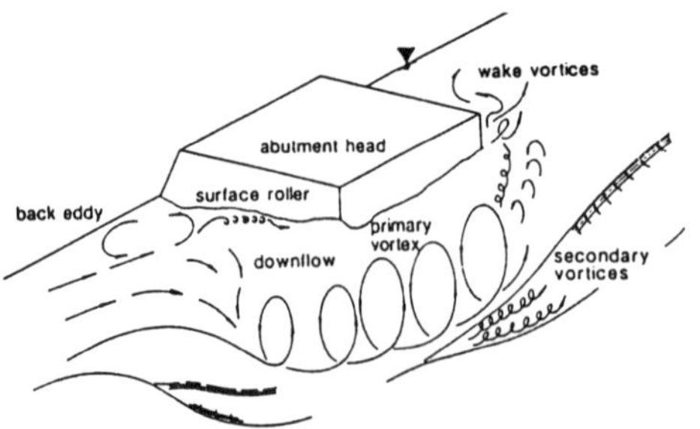

Figure 1: Abutment flow structures.

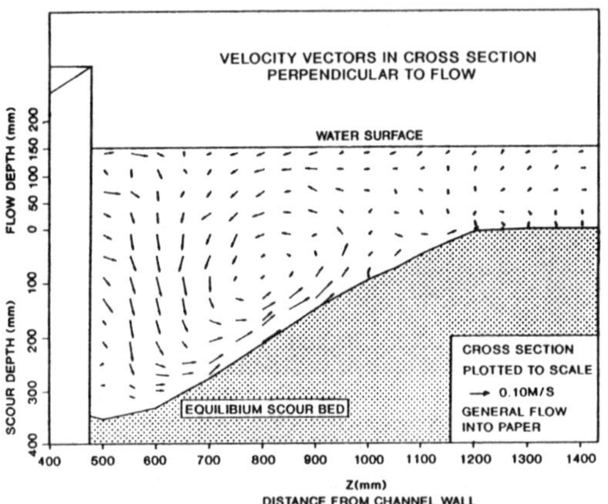

Figure 2: The primary vortex at a wingwall abutment.

In recent years much debate has centred on the nature of the primary mechanism responsible for the development of the scour hole. While many researchers infer that it is due to a vortex structure, others have disputed the presence of any vortex system significant enough to cause scour, Morton (1984,1986). There is conclusive evidence now that once scour has been initiated a major vortex does form - the primary vortex, and that it is the primary vortex which is the dominant mechanism for scour, Figures 1 and 2.

Kwan (1988) showed that the downflow associated with the primary vortex was strong near the upstream corner of the abutment, with a peak downward velocity 75% that of the approach flow. Kwan concluded that it was the downflow impinging on the bed of the scour hole which was the main scour agent. He also found in his experiments that the primary vortex was similar to a Rankine vortex and confined mostly within the scour hole below the original bed level.

Kwan (1988) also found that the mean and maximum velocities at the scour hole at equilibrium are, in general, approximately 0.75 and 1.4 times that of the mean upstream approach velocity, respectively.

Local scour may occur with or without general sediment transport. The former is referred to as clear water scour and the latter as live-bed scour. For local scour at an abutment under clear water conditions a static equilibrium scour depth is reached asymptotically over a relatively long period of time, Tey (1984), Kwan (1984).

For local scour under live-bed conditions however, a dynamic state of equilibrium is reached over a relatively short period. Dynamic equilibrium occurs when over a period of time the sediment inflow from upstream equals the sediment outflow from the scour hole. This is then followed by aperiodic oscillations about a mean equilibrium depth of scour with the passage of bed forms through the scour hole, Chiew (1984), Kandasamy (1989).

It is therefore important to distinguish between clear water scour and live-bed scour in the interpretation of scour data. Much of the confusion and inconsistencies in earlier papers published were in fact partly due to not differentiating between these two types of scour.

3 THE PARAMETERS OF LOCAL SCOUR

The complex relationship between the depth of local scour and the parameters affecting it can be summarised as follows:

$$d_s = f\ (\text{flow, fluid, abutment, sediment, time}) \qquad (1)$$

It can further be simplified to the following using dimensional analysis:

$$d_s/L = K_s K_\theta K_d K_\sigma K_U K_y \qquad (2)$$

$$d_s/y = K_s K_\theta K_d K_\sigma K_U K_L \qquad (3)$$

where d_s is the maximum scour depth, and the dimensionless K factors are used to account for the effects of abutment shape S, abutment alignment θ, median sediment size d, geometric standard deviation σ, flow velocity U, flow depth y, and abutment length L.

In the subsequent sections current understanding on the relative importance and influence of these parameters on local scour is discussed.

3.1 Effect of abutment shape

It is well established that the shape of an abutment, represented by the shape factor K_s, is important in determining the local scour depth. Different shapes divert flows in a different manner and one would naturally expect streamlined shapes to create less disturbance and hence produce less scour than blunt shapes. The shape factor can be determined in the laboratory by varying the shape of an abutment (while maintaining the other variables constant) and then measuring and comparing the resulting equilibrium scour depths.

Table 1: Shape factor K_s for various abutment shapes.

Abutment shape	K_s
Vertical plate or narrow vertical wall	1.0
Vertical wall with semi circular end	0.9
45° wing wall	0.8
1H:1V spillthrough	0.7
1.5H:1V spillthrough	0.6

Values of K_s for five different abutment shapes are given in Table 1, based on averaged values obtained by Field (1971), Laursen (1970), Gill (1972), Wong (1982), Tey (1984), Kwan (1984,1988), Kandasamy (1985, 1989), Dongol (1990) and Melville (1991).

The vertical plate or narrow vertical wall produced the maximum scour and is used as the reference shape, i.e. $K_s = 1$. Thus the use of a 1.5H: 1V spillthrough abutment is seen to reduce scour by up to 40% as indicated by its K_s factor of 0.6.

3.2 Effect of abutment alignment

The abutment alignment is defined by the angle θ in Figure 3. It has generally been found that the scour depth increases with an increase in θ. Using the perpendicularly aligned abutment (θ = 90°) as a reference, abutments pointed upstream are seen to produce larger scour depths (up to 10% increase for angles between 150-180°). By contrast the scour depth is reduced as θ is decreased, and for the range of values investigated, the maximum decrease is 17% when the abutment is pointed downstream at 30° (Kwan 1988).

There is evidence from clear water experiments that in some cases the maximum scour depth may actually decrease for θ > 90°. This is usually the case when the flow is relatively shallow and the abutment length relatively short

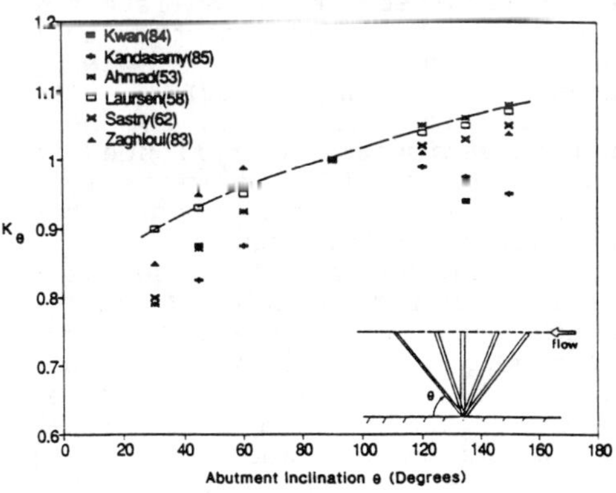

Figure 3: Effect of abutment alignment.

such that the scour hole extends to the upstream bank (Kwan 1984, Kandasamy 1985). For most practical purposes it is suggested that the curve in Figure 3 is appropriate, where $K_\theta = 1$ for $\theta = 90°$.

3.3 Effect of sediment size and grading

There are few data available which show the influence of sediment size on local scour depth at abutments. Pier data, however, indicate that under clear water conditions scour initially increases with sediment size and then decreases slightly with further increments in sediment size (Ettema 1976, Chiew 1984).

Chiew concluded that under both clear water and live-bed conditions the influence of sediment size is insignificant if $L/d_{50} > 50$. If the equivalent $L/d_{50} > 50$ is applicable to abutments then for most practical purposes the scour depth at an abutment may be assumed to be independent of sediment size. This is because $L/d_{50} < 50$ is unlikely to occur in the field or even small scale laboratory experiments. It is therefore suggested that K_d in equations (1) and (2) be taken as 1 until further information becomes available.

By contrast the variation of particle size distribution or sediment grading σ has generally been found to have a pronounced effect on local scour depth. Non-uniform sediment ($\sigma > 1.5$) have been reported to consistently produce lower scour depths than uniform sediments (Ramu 1973, Ahmad 1973, Wong 1982). The reduction in scour depth for non-uniform sediments is attributed to the formation of an armour layer which protects the bed thus preventing further scour.

The influence of sediment grading is indicated in Figure 4 using data by Wong (1982) for clear water conditions. In this figure the ordinate K_σ is plotted as the ratio of the scour depth for a sediment grading of σ to that for a uniform sediment ($\sigma = 1.5$), without the effects of the other variables.

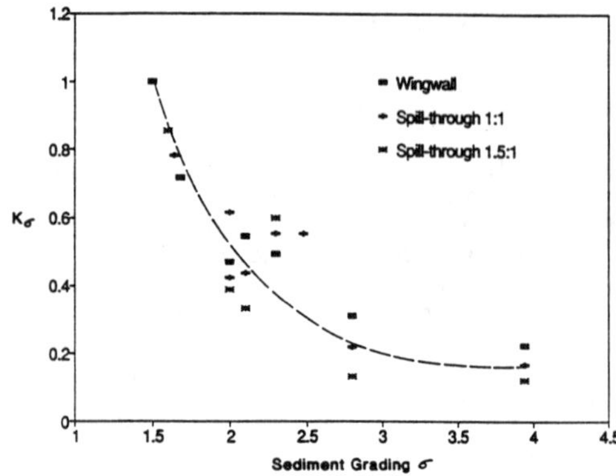

Figure 4: Effect of sediment grading.

These results show a definite trend as discussed above, and for a sediment mixture with a grading $\sigma = 3$ the scour depth is only about 20% that of the value for a uniform sediment.

At higher values of U/U_c however the effect of armouring is expected to become less significant. This is due to the fact that at velocities exceeding the critical armour velocity the armour layer breaks and the non-uniform sediment then effectively behaves like a uniform sediment under live-bed conditions (Baker 1986, Chin 1985). For this reason K_σ in equations (2) and (3) may be assumed to be unity, as the armour layer is likely to become unstable under flooding conditions.

3.4 Effect of approach flow velocity

The influence of approach flow velocity may be expressed in terms of the velocity ratio U/U_c or the shear velocity ratio u_*/u_{*c}, where U_c is the threshold velocity for sediment entrainment and u_{*c} is the corresponding shear velocity based on Shields criterion. Thus $U/U_c < 1$ for clear water scour and $U/U_c > 1$ for live bed scour.

It has been well established that clear water scour occurs only if $U/U_c > 0.5$. Thereafter it increases almost linearly with increasing velocity to reach a maximum scour depth at $U/U_c = 1$

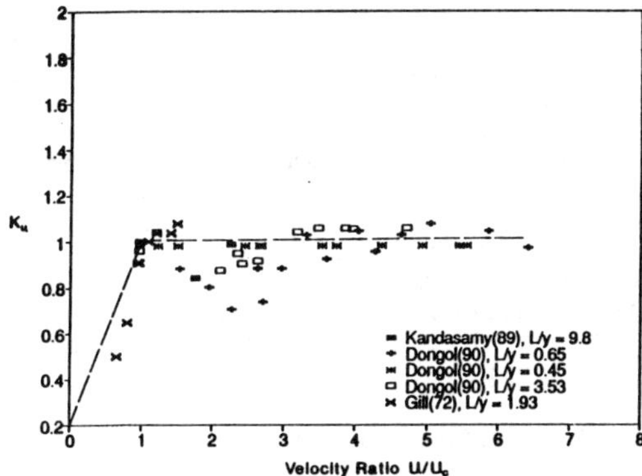

Figure 5: Effect of flow velocity on scour depth.

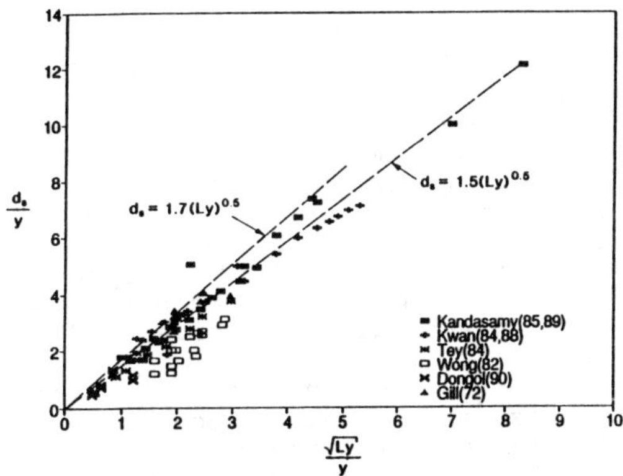

Figure 6: Effect of abutment length and flow depth.

(Liu et al 1961, Field 1971, Tey 1984).

Relatively few data are available for abutment scour under live-bed conditions. A comprehensive set was collected by Kandasamy (1989) and Dongol(1990). These results are plotted in Figure 5 together with data from Gill (1972), where the ordinate K_U is the ratio of the scour depth at a particular velocity ratio U/U_c relative to that at the threshold condition ($U/U_c = 1$), without the effects of the other variables. These results showed that the equilibrium scour depth first decreases on exceeding $U/U_c = 1$ but thereafter increases again at higher velocities to about the threshold depth. It is apparent that at higher flow conditions up to $U/U_c = 6$ the maximum scour depth is still only approximately equal to the corresponding scour depth in the vicinity of $U/U_c = 1$. The same trend has been established for pier scour (Chee 1982, Chiew 1984). Thus in practice a value of $K_u = 1$ may be adopted.

3.5 Effect of flow depth and abutment length

The abutment length and flow depth have been reported to be two of the most significant parameters affecting scour. Many relationships incorporating these two parameters have been proposed, as summarised in Raudkivi and Sutherland (1981).

Recent research indicates that the scour depth is related to the abutment length and flow depth as follows (Kwan 1988, Kandasamy 1989, Kwan and Melville 1992):

$$d_S \ \alpha \ (Ly)^{0.5} \qquad (4)$$

Kwan (1988) derived the above function on the basis of measurements of the flow field at an abutment and based on a physical representation of the scour hole. There is therefore a physical basis for the form of equation (4).

Figure 6 is a plot of equation (4) and comprises data from Gill (1972), Wong (1982), Tey (1984), Kwan (1984, 1988), Kandasamy (1985,1989), and Dongol (1990). All the data were taken at threshold conditions, i.e U/Uc = 1 and using a uniform sediment with a median particle size of 0.9 mm. In this plot the effect of abutment shape has also been incorporated using the shape factors in Table 1, i.e. Figure 6 applies to a vertical wall abutment with $K_s = 1$. The plot therefore demonstrates only the effect of abutment length and flow depth without the effect of any of the other variables in equations (2) and (3).

It can be seen that equation (4) gives a reasonably good fit to the data in Figure 6 and that the constant of proportionality is approximately 1.5. Equation (4) is therefore:

$$d_S = 1.5 \ (Ly)^{0.5} \qquad (5)$$

where 70 > L/y > 0.25.

Data from Tey (1984), Kwan (1984, 1988) and Kandasamy(1985,1989), also showed that scour depth increases at a decreasing rate with both flow depth and abutment length. This suggests that the effect of abutment length and flow depth on scour depth should become less significant for relatively long abutments and relatively short abutments, respectively. In this context a relatively long abutment is one with a large L/y value and a relatively short abutment is one with a small L/y value.

Various values of L/y at which the scour depth may become independent of either the abutment length or flow depth have been suggested, (Kwan 1988, Kandasamy 1989, Melville 1991). However it is not evident from the range of data plotted in Figure 6.

Considering the wide range of values covered, equation (5) is considered to be appropriate for most practical purposes.

4 PREDICTION OF MAXIMUM SCOUR DEPTH

Based on results established in section 3 the depth of local scour at a bridge abutment, represented by equations (1), (2) and (3), may be rewritten as follows:

$$d_s = 1.5K_sK_\theta K_dK_\sigma K_U(yL)^{0.5} \quad (6)$$

where 70 > L/y > 0.25.

For reasons given in the preceding sections equation (6) has been derived for a vertical wall abutment (K_s = 1), an abutment perpendicular to the approach flow direction (K_θ = 1, θ = 90°), assuming no effect from sediment size (K_d = 1), no effect from sediment grading (K_σ = 1) and at threshold velocity conditions (K_U = 1, U/U_c = 1).

For a different abutment shape and alignment, the shape factor K_s is given by Table 1, and the abutment alignment factor K_θ by Figure 3. As discussed the factors K_d, K_σ, and K_U representing sediment size, sediment grading and flow velocity may be taken as unity for design purposes.

Thus equation (6) is relatively easy to apply and only the abutment length, flow depth, abutment shape and abutment alignment need to be known in the estimation of local scour depth. It may also be appropriate to use a slightly higher constant in equation (6) if a higher degree of conservativeness is desired. From Figure 6 it appears that the following would be conservative:

$$d_s = 1.7K_sK_\theta K_dK_\sigma K_U(yL)^{0.5} \quad (7)$$

for 25 > L/y > 0.25. For values of L/y > 25, Figure 6 indicates that equation (7) may be too conservative and equation (6) is suggested to be more appropriate.

It should be noted that the scour depth predictor equations (6) and (7) have been derived solely from laboratory data. They have yet to be confirmed in the field, due to the absence of prototype scour data and the obvious difficulties in collecting them. They are however shown to conform well with laboratory data and are consistent with our understanding of the physical processes of scour.

5 CONCLUSIONS

Present understanding on local scour at bridge abutments and the parameters affecting it have been discussed. A relationship for the prediction of the maximum local scour depth at a bridge abutment is proposed. This is shown to be $d_s = 1.5K_sK_\theta K_dK_\sigma K_U(yL)^{0.5}$, where 70 > L/y > 0.25, and the K factors are as discussed in the preceding sections.

6 REFERENCES

Ahmad M (1953), "Experiments on Design and Behaviour of Spur Dikes", Proc Inter Hydr Convention, Minnesota.

Ahmad M (1973), Discussion on paper by Gill (1972).

Brice J.C and Blodgett J.C (1978), Countermeasures for Hydraulic Problems and Bridges", Vols 1,2, Fed Hwy Adm, Dept of Transp, Washington.

Baker R.E (1986),"Local Scour at Bridge Piers in Non-Uniform

Sediment", Masters Thesis, Auckland University.

Chee R.K.W (1982),"Live Bed Scour at Bridge Sites", Masters Thesis, Auckland University.

Chiew Y.M (1984), "Local Scour at Bridge Piers", Phd Thesis, Auckland University.

Chin C.O (1985), "Stream Bed Armouring", Phd Thesis, Auckland University.

Dongol (1990), Unpublished Data, Auckland University.

Ettema R (1980),"Scour at Bridge Piers", PhD Thesis, Auckland University.

Field W.G (1971), "Flood Protection at Highway Bridges", Eng Bulletin CE3, Newcastle University.

Gill M.A (1972), "Erosion of Sand Beds Around Spur Dikes", ASCE Jour Hydr Div, Vol 98(9).

Kandasamy J.K (1985), "Local Scour at Skewed Abutments", Masters Thesis, Auckland University.

Kandasamy J.K (1989),"Abutment Scour", PhD Thesis, Auckland University.

Kwan T.F (1984), "Study of Abutment Scour", Masters Thesis, Auckland University.

Kwan T.F (1988), "A Study of Abutment Scour", PhD Thesis, Auckland University.

Kwan T.F and Melville B.W (1992), "Local Scour and Flow Measurements at Bridge Abutments", Jour Hydr Research. In press.

Laursen E.M (1970), discussion on paper by Shen et al(1969), "Local Scour Around Bridge Piers", ASCE Jour of Hydr Div(95).

Liu H.K, Chang F.M and Skinner M.M (1961), "Effect of Bridge Constriction on Scour and Backwater", Eng Resr Centre, Colorado State University, Report No CER 60 HKL22.

Melville B.W (1991), Private Communication.

Melville B.W and Sutherland A.J (1988), "Design Method for Local Scour at Bridge Piers", Jour Hydr Engrg, ASCE, 114(10).

Morton B.R (1984), "The Generation and Decay of Vorticity", Geophys Astrophys Fluid Dynamics (28).

Morton B.R and Evans J.L (1986), "Horseshoe Vortices and Bridge Pier Erosion", 9th Australasian Fluid Mechs Conference.

Rajaratnam N and Nwachukwu B.A (1983), "Flow near Groyne-Like Structures", Jour Hydr Engrg 109(3).

Ramu K.L.V (1973), discussion on paper by Gill M.A (1972).

Raudkivi A.J and Sutherland A.J (1981),"Scour at Bridge Crossings", RRU Bulletin 54, NRBNZ.

Sutherland A.J (1986),"Report on Bridge Failure", RRU, NRBNZ.

Tey C.B (1984),"Local Scour at Bridge Abutments", Masters Thesis, Auckland University.

Wong W.H (1982), "Scour at Bridge Abutments", Masters Thesis, Auckland University.

Environmental Management, Geo-Water & Engineering Aspects, Chowdhury & Sivakumar (eds)
 ISBN 90 5410 099 0

Study of riverbank erosion and design of bank protection works, Mekong River bridge project

Ricky T.F. Kwan & Neil Mayo
Sinclair Knight, Sydney, N.S.W., Australia

ABSTRACT: The design considerations involved in the design of the river bank protection works for the Mekong River bridge are discussed. These included aspects of flooding, river geomorphology, river scour and the processes of bank erosion in the river. A rock-filled wire basket system was proposed as the most effective and economic means of bank protection for the bridge.

1 INTRODUCTION

The Mekong River (4200 km) is the longest in South East Asia and one of the world's greatest and most famous rivers. Millions of people along its banks rely on the river for transport, water and agriculture on the fertile floodplains.

However for political, financial and technical reasons there are no bridges spanning the river from the China-Burma border to the South China Sea, a distance of more than 2400 kilometres. Now, as part of Australia's aid to the region, a bridge will finally be constructed, at the Thailand-Laos border. Its completion is expected to greatly improve access and aid in the social and economic development of the two countries.

Design of the bridge was carried out by the Australian engineering consultants Maunsell and Sinclair Knight, in a joint venture. Construction of the bridge is currently underway and is expected to be completed in 1994.

This paper focusses on one element of the the design, namely the riverbank protection works for the bridge. This was an important element in the overall design to ensure the long term security of the structure. In the following sections a brief description of the study area is provided, followed by a review and analysis of the data available. The various hydrologic and geomorphological factors influencing the design of the protection works, including the flood levels, flood velocities, possible changes to the river course, lateral migration rates and river bed scour are assessed. The local processes of bank erosion currently operating in the river are then identified. Finally an appropriate method of riverbank protection is proposed.

2 THE STUDY AREA

The Mekong River rises in Tibet and drains through Tibet, southern China, Burma, Thailand, Laos, Cambodia and Vietnam before reaching the South China Sea (Figure 1).

The bridge site is located at the town of Nong Khai, Thailand, and the village of Tha Naleng, about 30 km downstream of Vientianne, the capital of Laos. The river at this location is about 640 m wide. From Vientianne to Thakhet (400 km) the river flows through deposits of alluvium along a highland plain. The riverbanks are generally composite in nature, consisting of layers of non-cohesive and cohesive materials. The upper reaches near

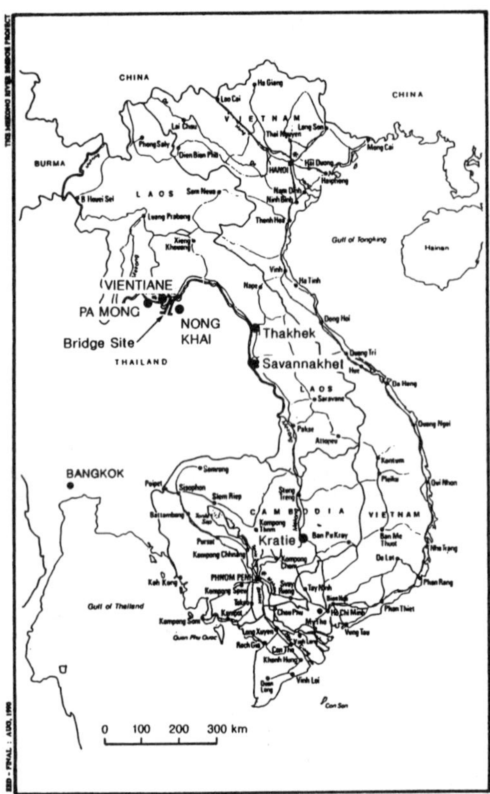

Figure 1: Locality map.

Vientianne are characterised by a series of relatively gentle meanders, with radius of curvature to width ratios ranging from 5 to 10. By contrast, the lower downstream 240 km near Thakhek is relatively straight. Overall, the average water surface gradient along this part of the river is 8-10 cm/km.

Further upstream and downstream, the river flows along a narrow and steep valley, flanked by mountains and hills on either side. This includes the river from the Burmese border down to Pa Mong (800 km), and from Savannakhet to Kratie (600 km). The river in these regions is characterised by numerous rapids and waterfalls. The water surface gradients are also relatively steep at about 20 cm/km.

3 REVIEW AND ANALYSIS OF DATA

Excellent data were available from a number of sources and the following data at the bridge site were reviewed and analysed:

1937-1990 hydrological and flood level records, 1956-1960 Aerial Photos (flown for Canadian Aid) at 1:40,000, 1967-1968 Maps (prepared from 1967 aerial photos by US Army Corp of Engineers) at 1:10,000, 1973 Pa Mong Study maps prepared from 1967 aerial photos, 1977 Aerial Photographs enlarged to 1:5000, 1989 Hydrographic Surveys (Finnish Aid) at 1:20,000, 1991 Survey of bankline, and 1991 Geotechnical data.

A frequency analysis of the flood levels was carried out for the 53 years of records. The results indicated that the flood level for the 1 in 100 year flood event was RL 167.95m in the vicinity of the bridge, which corresponded to a discharge of 22,500 m^3/s. For the 1 in 500 year flood a flood level of RL 168.59 and a discharge of 24,000 m^3/s was established.

These analyses formed the basis of the design level for the top of the riverbank works, as well as the minimum bridge elevation. The top of the riverbank works was eventually set at RL 168 m.

The effectiveness of any riverbank works was first assessed from a geomorphological perspective. The possibility of a major shift in the river course, lateral bend migration rates, and the extent of river bed scour anticipated during the life of the structure were some of the factors investigated. The local processes of bank erosion were then identified prior to the formulation of an appropriate method of bank protection.

3.1 Major shift in river course

Several historical channels are evident on the inside of two bends within 10 km upstream of the bridge site, as shown in Figure 2. It was possible that these channels would act as floodways during heavy flooding, which raised the concern that a new shortened channel could form

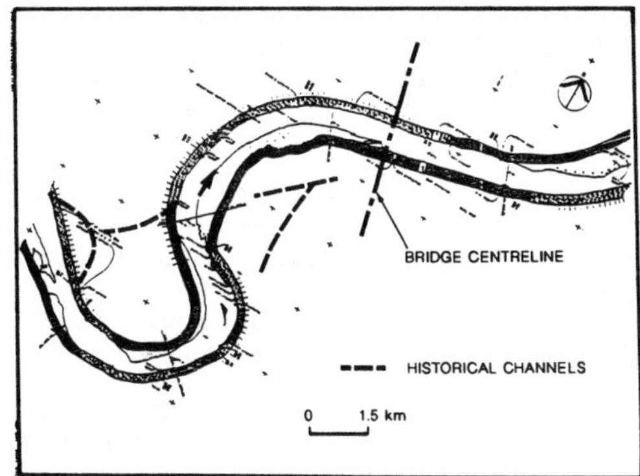

Figure 2: Historical channels.

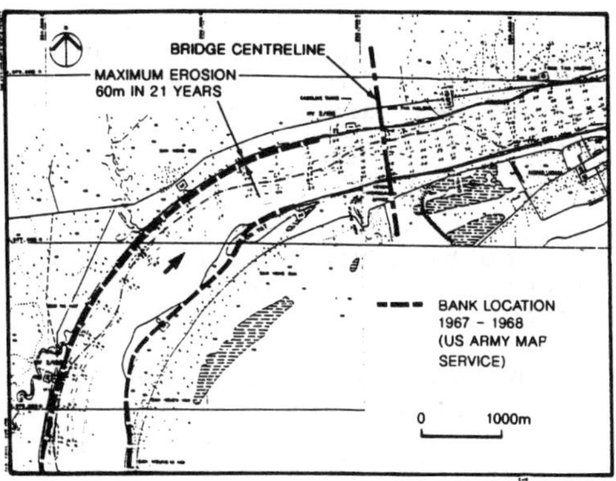

Figure 3: Rates of bank retreat

within the lifespan of the bridge and threaten to outflank the bridge.

An investigation of the topographical and geological features in the area was carried out in order to assess the possibility of this change occuring. However, it was found that the many natural highlands, man-made levees and road embankments existing in the area precluded many short-cuts for the river. The relatively small overbank floodways and flow velocities (upper limit of 3 m/s in a major flood) were also considered to be inadequate to pose such a threat.

Further investigations also revealed that parts of the river bed were stabilised by the presence of natural rock outcrops. Another mitigating factor was that the area, being densely populated and regularly maintained by the Thai and Laos authorities, was considered to be too closely monitored for any such changes to be left unnoticed and uncontrolled. It was therefore concluded that a major shift in the river course could not occur.

3.2 Lateral bend erosion

It is apparent from Figure 2 that the river upstream of the site has shifted laterally by erosion of the concave (outer) bank and deposition of the convex (inner) bend. These historical rates of erosion were quantified using the aerial photographs and maps listed in section 3. The maps provided a history of bank erosion at the bridge site over the last 21 years. The results are shown in Figure 3.

The maximum erosion was found to occur on the apex of the bend about 1.5 km upstream of the bridge site. The extent of bank retreat was found to be about 60 metres over this period, which equated to an erosion rate of about 3 m/year. At the bridge site at the downstream end of the bend the rate of erosion was found to average 1 metre per year. These rates were generally substantiated by the landowners fronting the river.

Based on these estimates a movement of 100 metres could be expected at the bridge site over the next 100 years. Erosion at the outer bend upstream was also expected to increase as the bend radius decreased, with the bend expected to tend to translate downstream. Overall the results indicated that a translation of 300 metres at the site over the next 100 years was not unrealistic.

Riverbank protection works were therefore considered essential in controlling the rate of bank retreat. While it would have been desirable to protect the entire bend, in order to minimise the risks of erosion at the bridge site, it was not considered to be economically viable as it involved protecting over 6 km of the riverbank. The gradual, rather than catastrophic, nature of the erosion was also such that it would provide ample time for any remedial work to be carried out should there be a threat to the bridge. It was therefore proposed that the bank protection works be limited to the vicinity of the bridge.

Based on these considerations and in order to protect several water supply pipelines in the area, it was concluded that the protection of the riverbank to a distance 300 metres upstream and 100 metres downstream of the bridge centreline was adequate.

3.3 River bed scour

The anticipated river bed scour level at the bridge site was established from hydrographic surveys of the area and measurements of changes in bed level at other parts of the river. These results indicated that the maximum likely scour level was at approximately 5 metres below the existing bed level or RL 149.5 m. Drilling investigations also showed that the bed at this level consisted of weathered siltstone which would be highly resistant to erosion. The maximum bed scour level was therefore assumed to be at RL 149.5 m.

4 PROCESSES OF RIVERBANK EROSION

The cross-sectional profiles of the riverbank at the bridge site were found to be similar and generally steep and almost vertical, particularly at the upper 2-3 metres. The banks were in general 6-7 metres high and consisted of clay to a depth of 6 m, silts for the next 1.5 m and sands down to a depth of 18 m.

Bank failure mechanisms were identified to be a combination of the following:

- Saturation of the riverbank during flooding and rapid drawdown of the river level relative to the groundwater table, leading to slumping of the banks. This rapid drawdown was characteristic of the Mekong. A maximum difference of 3 m between the river and groundwater table water levels was found to be fairly representative.
- Vertical tension cracks at the upper zone, which weaken the stability of the bank. A maximum tension crack depth of up to 3 m was found to be not uncommon, particularly during the dry season.
- Transient or permanent bed scour at the bank toe which lead to undermining, steepening and increase in riverbank height, thus initiating progressive collapse of the upper bank. This was a primary process of bank retreat considered to be critical to the effectiveness of the riverbank works.
- Seepage through lenses of sand and coarser material sandwiched between layers of cohesive banks, leading to piping failures, and
- Wave action from wind and boats.

These factors were incorporated in the detailed analyses and design of the riverbank protection works.

5 DESIGN OF RIVERBANK PROTECTION

The detailed design of the riverbank protection works was based on the factors discussed in the preceding sections.

Bank stability under various loading conditions, including the effects of rapid drawdown of the river level, tension cracks, construction loads, and seismic loads was analysed using the computer programs XSLOPE and STABL5P. Overall the riverbank was designed to have a minimum factor of safety of 1.3 under static loads and 1.1 under earthquake conditions.

Protection of the bank toe against basal scour was considered to be of fundamental importance to the stability of the protection works. It was essential that the protective apron had the ability to readjust its geometry to fill and pave any scoured profile, thus preventing further scour, without affecting the overall stability of the riverbank works. The two major types of protection considered to have the flexibility to accomodate these changes were

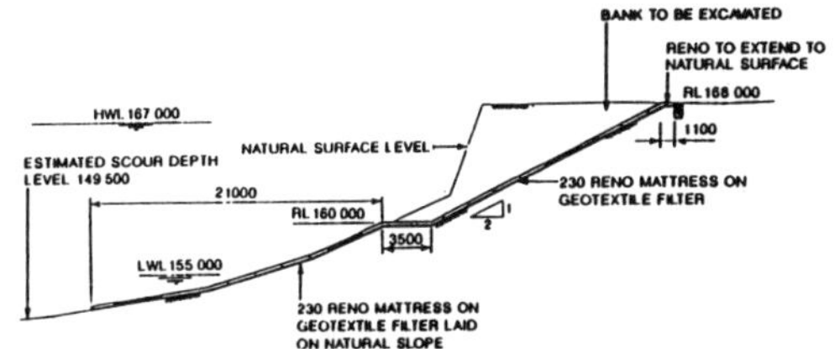

Figure 4: Typical section of design adopted.

1) rip-rap protection and 2) a wired rock basket (reno mattress) system. Both options were considered in detail, including aspects of reliability, durability and cost.

A reno mattress system was finally selected on the basis of a number of factors. They were: past success in use in the Mekong River; durability at least equal to that for rip-rap, if not better; lower maintenance costs; and initial capital costs (A$950,000) of only half that for a rip-rap system.

Other factors in favour of reno mattress construction were that the rock volume and rock sizes required were much smaller, which meant that the rocks were more readily available. While 4000 m^3 of rock was required for the reno mattresses 16000 m^3 of rock, or four times more volume, was required for the rip rap system. Similarly the size of rock required for the reno mattresses was only up to 120 mm in comparison to a size of up to 500 mm for rip rap. A typical section of the adopted reno mattress design is shown in Figure 4.

The riverbank was designed at side slopes of 1:2 down to a berm level of RL 160 m. The top of the protection works was set at RL 168 m, as noted earlier. The berm, which was 3.5 m wide, was provided to facilitate ease of construction and maintenance. The apron length below the berm was also sized to ensure that the reno mattress would rotate and protect any scour profile down to a maximum river bed scour level of RL 149.5 m.

Overall the bank protection works were provided for a distance 300 m upstream and 100 m downstream of the bridge centreline, as mentioned earlier. During final design a 100 m long section under the bridge, consisting of 2 m high reinforced concrete piles and infill panels, at the edge of the river bank, was also provided for enhancement and use as a bridge viewing area.

6 CONCLUSIONS

A study of the factors influencing the design of the riverbank protection works for the Mekong River bridge project has been summarised. The major parameters affecting the design were found to include lateral migration of the river, rapid drawdown of the river level relative to the groundwater table, and scour at the bank toe. A reno mattress system was designed on the basis of greater reliability and durability and considerable savings in costs.

7 ACKNOWLEDGEMENTS

Input from the design teams of Sinclair Knight and Maunsell during the course of the project is gratefully acknowledged.

8 REFERENCES

Maunsell-Sinclair Knight, "The Mekong River Bridge Project", Phase 3, Design reports, Jul 1991 and Jan 1992.

Ian Drummond and Associates, "Mekong River Bridge Project: Hydraulics, Hydrology, Bank Protection", Proof Check, July 1991.

Maunsell-Sinclair Knight, "The Mekong River Bridge Project", Phase 2, Feasibility Study Update, Engrg and Econ. Data, Final Report, August 1990.

Environmental Management, Geo-Water & Engineering Aspects, Chowdhury & Sivakumar (eds)
© 1993 Balkema, Rotterdam. ISBN 90 5410 099 0

Tendency of erosion in Loess Plateau and silt transport in Yellow River, Huanghe

Jian Li & Suqing Liu
Chengdu Institute of Mountain Disaster and Environment, Chinese Academy of Sciences, People's Republic of China

ABSTRACT: The soil erosion in Loess Plateau is quite serious. The annual modulus of soil erosion is 10,000-20,000 tons/sq km year. The maximum content of silt in gullies reaches 1100-1200 kg/cubic metre. As a result, the yellow River (Huang He) is famous for its high concentration of fine material in the water. The annual volume of silt of Yellow River is 1600 million tons most of which comes from the erosion of Loess Plateau. It brings a very bad result of rapid silting the river bed of Yellow River and, at the same time, it forms a large piece of new land in the area of delta near the sea .

1 INTRODUCTION

Both Loess Plateau and Yellow River are well known in the world. The former one is famous by its large area and thickness of Loess. The latter one is a surprise to all who have a chance to see with so high concentration of fine materials in its water which come mainly from the former and has changed its colour of water into yellow.

The problems of soil erosion in Loess Plateau and the troubles caused by silt in Yellow River have already been a major argument of science and practice from very beginning in China. It is because the process of erosion in Loess Plateau not only causes reduction in agriculture land itself, but also a series of troubles with rapid silt in the middle and lower reaches of yellow River.

In the period of thousands of years, the research work looking for a countermeasure is being continued in China. A series of projects have been proposed to build a new ecosystem and a number of measures have been taken against the erosion in the Plateau and improving the situation of Yellow River after the foundation of the People's Republic of China. These include the work of soil and water conservation with planting trees and grasses on a large scale, terracing the sloping land, reforestation and building of reservoirs and so on.

Many observation stations have been established. Many typical basins have been brought under control successfully. The problems concerning mountains, rivers, farmlands, forests, pastures and roads have been tackled in a comprehensive manner. So the achievements have been obtained in this aspect.

Along with growth of population, environmental pollution and degradation of ecosystem, new problems with erosion appear in many places of Loess Plateau. The human activities of cultivation on steep slopes, opencasting of mining, destroying forest and vegetation cover, overgrassing and construction of roads and factories are accelerating the erosion process. Naturally, the raising of river bed of Yellow River is still a troublesome business in its middle and lower reaches at the present time.

2. THE EROSION IN LOESS PLATEAU

The Loess Plateau is located in middle reaches of Yellow River within the provinces of Shanxi, Shaanxi, Gansu and Ningxiia Huizu Zizhiqu (Autonomous Region). The area of the Loess Plateau is 580,000 sq km and the thickness is 10-200 metres. The precipitation in the region ranges 350-500 mm per year. It belongs to the semi-arid region. The total population is 72.5 millions (1981). It is about 32 person/sq km and the maximum is 1000 person/sq km in the plain of Weihe River, Shaanxi Province.

Owing to the specific structure of Loess Plateau and social and economic conditions of the region, the soil erosion of the plateau has

Fig. 1. The Yellow River Basin

Table 1. The average modulus of soil erosion in different Provinces of Yellow River Basin (Bureau of Yellow River Management, 1990)

Province	Number of Tributaries	Modulus of Soil Erosion (Tons/sq km year)
Qinghai	2	1330
Gansu	7	4390
Ningxia	8	3452
Inner Mongolia	11	4602
Shaanxi	13	6153
Shanxi	15	4503
Henan	5	1471
Shangdong	2	577

been continuing for thousands of years. The observation has shown that the modulus of erosion in average for Loess Plateau is 10, 000-20,000 T/sq km year (Sun Jianxuan 1985). The most serious erosion is in the main part of Loess Plateau located in Shaanxi and Shanxi provinces where the maximum value of erosion modulus was estimated in Yulin prefecture, Shaanxi Province. It reaches as high as 30, 000 T/sq km year.

The main reason of serious erosion is rapid growth of population and the unreasonable activities of human beings. According to the data, the total population for Loess Plateau was 35 .5 millions in 1957, and then it became 72.5 millions in 1981. It means that the population increased 37 millions in the period of 24 years. The increasing rate is 2.2 % (Tian Houmu 1985). The rapid growth of population has led to the destruction of the ecosystem balance. In the Loess area, it has long been customary for local people to get food by increasing the cultivation of slopes including the very steep slopes (more than 25 degrees). It is calculated that in Loess area of Shaanxi Province, the area of cultivated slopes with gradient of more than 25 degrees is about 1.5 million hectares. In many places the cultivated fields amount to 1 ha/person and even 2-2.5 ha/person. The erosion from this kind of cultivated land is 60 times more than the value comes from the natural grass land and it is 30 times more than the value comes from the natural surface with sparse vegetation cover. At the same time the action of cutting trees and bushes for fuel and overgrassing with cattle and sheep have become so widespread that the bared slopes can be seen nearly everywhere in the Loess Plateau at the present time.

The mining action for coal and construction of roads or highways is another reason of serious erosion in this region. This action is getting stronger during the last 20 years. It is estimated that the amount of eroded material by the action mentioned above in the Basin of

Wuding River has increased 48 million tons in the period of 1950-1984.

For the whole area of Shanxi Province, the total volume increased by over 60 million tons per year due to activities such as mining and construction.

Based on the investigation by means of comparison with the airphotos taken in 1967 and 1979, the gully erosion developed very quickly. The rate of erosion for the head of gullies is 3.52 m/year, and the maximum value reaches 15.7 m/year. According to the historical material of county annuals of Guyuan County, Ningxia Huizu, Autonomous Region, the largest piece of flat surface left on the plateau from ancient time named Dongzhiyan was 34 km wide in Tong Dynasty (AD 6I8) and it is 18 km wide now. At the most narrow place, there is 0.5 km left only. It is clear Dongzhiyan will soon be divided into two parts with much smaller flat surface in the near future. This kind of example of changing from large flat surface to ridges and gullies can be found in thousands in the Loess Plateau nowadays.

3. THE SILT AND TRANSPORT OF YELLOW RIVER

Yellow River in China is famous enough by its high concentration of silt in the water. The annual volume of silt measured in Huayankou Gauge Station in Henan Province is 1,600 million tons which comes from the erosion of Loess Plateau. The content of silt in the water in average is 34.6 kg/cubic m in the period of flood. The maximum concentration of silt in the water for tributaries reaches 1000 kg/cubic m. There are 250 million tons of silt going to the field with the water in irrigation systems along the river in Henan and Shandong Provinces.

Another 250 million tons is silted in bed between the Huayuankou Gauge station, Henan Province and Lijin Gauge station near the mouth of Yellow River in Shandong Province, which has caused the river bed to rise about 10 cm per year. The remaining 1100 million tons of silt is to be transported to the sea to form the delta of Yellow River. Based on this situation, the river bed of Yellow River grows higher and higher day by day and it is always 5-10 metres above the ground of both side. The Yellow River becomes very unstable. The direct result would be the bursting of dikes by big floods, and the course of Yellow River would be changed after-wards.

It causes terrible catastrophes and brings great damages in the lives and properties of local people. So the Yellow River has vividly been named as Hanging River, Mud River, Hazardous River and Yellow River itself by the yellow colour of the water.

According to the historical data, Yellow River had overflowed 7 times and changed its course once during one thousand years before Qin Dynasty (BC 1500-BC 221). But in the period of Qin Dynasty to Western Han Dynasty (BC 221-AD 24) Yellow River overflowed 4 times and burst 7 times and twice changed its course.

From Tang Dynasty (AD 618) till 1949, Yellow River became intolerable. It overflowed 404 times, burst 1100 times and the course was changed 1546 times. Actually, from Yan Dynasty (AD 1271) till 1949 (the People's Republic of China was established), the violence of Yellow River went to extreme. It overflowed and burst nearly every year and it changed its course of main river once a year, even twice a year (Tian Honmou, 1985). The changes of the course of Yellow River alternate southwards and northwards with maximum distance of 2000 Kilometres, intruding into Jiangsu Province in the south and reaching Hebei Province in the North. It is worth to point out that Yellow River has been brought under control for nearly half century. During this period, the river bed has risen 5-10 metres above the ground. In some places it is already 12-13 metres. This is a serious latent danger of bursting the dikes and changing its course.

Another obvious result is rapid silt in the delta and forming new land into the sea. It is estimated that the total area of new land formed by Yellow River during 1855-1984 is 2530.4 sq km. But the area of 606.8 sq km was formed during the last 10 years (1974-1984). Table 2 shows the speed of land forming in different periods (Report of Dongying City, Shansong Province).

It is clear enough from Table 2 that the tendency of soil erosion is becoming stronger and the silt of Yellow River is still a troublesome problem in China.

CONCLUSIONS

The erosion in Loess Plateau is very serious. It is the main reason of troubles with Yellow River in the middle and lower reaches. Prohibiting erosion processes by human activities should be an effective and successful approach to solving this problem. Most importantly:

1) An installation of a bureau of the executive authorities within the government to have an

Table 2. Silt progress and land formation in the delta of Yellow River at different stages

Periods	1855-1934	1934-1959	1959-1974	1974-1984	1855-1984
Number of years of river flow	77	17	15	10	119*
Distance of silt into the sea (km)	14.26	0.07	1.35	4.24	18.47
Speed of silt (km/yr)	0.19	0.00	0.09	0.42	0.16
Land forming (sq km)	1725.6	9.2	188.8	606.8	2530.4
Speed of land forming (sq km/yr)	22.41	0.54	12.59	60.68	21.26

* There were 10 years without water to the area of delta

administrative supervision function and to control an observance of laws ensuring the counter-measures against soil erosion in Loess Plateau.

2) Detailed administrative rules of environment protection and law of water and soil conservancy must be developed in the places where the erosion is most intensive and the rules must be set into action by the local government.

3) An appropriate method to be found to describe characteristics of Yellow River and suitable method for local people to change the course of Yellow River artificially in the near future must be thought and should be carried out as soon as possible.

REFERENCE

Sun Jianxuan,, Yang Hongye, Zhang Bingyi and Qiao Jiming, 1985. The problem of soil degradation and erosion while harnessing in works of soil and water conservation is in need of immediate solution. Bulletin of Soil and Water Conservation. No. 3, 12-14.

Tian Uoumou 1985, The increase of population in Plateau and its effect on soil loss. Bulletin of Soil and Conservation. No.3. 7-11.

Bureau of management for Yellow River, 1990.. The benefit of the soil and Water conservation in the Yellow River Basin. Soil and Water Conservation in China. No.5 20-24.

Environmental Management, Geo-Water & Engineering Aspects, Chowdhury & Sivakumar (eds)
© 1993 Balkema, Rotterdam. ISBN 90 5410 099 0

Application of the ANSWERS model for prediction of runoff and sediment from a small agricultural watershed of Iran

H. Momtahan
Fars Regional Water Authority, Ministry of Energy, Shiraz, Iran

S. Amin
Shiraz University, Iran

ABSTRACT: In order to make use of a model to predict results of watersheds in Iran, the model should be tested with conditions common to Iran. A watershed with an area of 4.83 hectares in southern Iran in the Bajgah valley located northwest of Shiraz was chosen. The ANSWERS model was used to simulate runoff and sediment from watershed using four different rainfall events. The simulated runoff values correlated well with the observed data. Correlation coefficients of the observed and predicted runoff of the four storm events ranged from 0.86 to .99 with average of 0.93. Therefore, the ANSWERS model can be used for prediction of runoff in small agricultural watersheds of Iran. However, for the types of storms common to the south part of Iran the erosion component of the ANSWERS model was not able to estimate sediment concentrations of the runoff accurately. Therefore, further research for development of a sediment component is needed considering the nature of the erosion of the watersheds of the southern part of Iran.

INTRODUCTION

One of the most important resources in the production of food is soil. It is estimated that 9.6×10^5 metric tons of soil are washed from Iranian watersheds each year (Iran agricultural Ministry, 1981). Because soil erosion is a selective process with respect to particle size, the higher fraction of plant nutrients, organic matter, and fine colloidal particles are lost from the soil and transported to the standing waters by runoff flow.

Estimates of soil erosion and sediment yield from watersheds are needed to select the best management practices to control sediment yield and improve water quality. Mathematical models are used to estimate runoff and sediment movement from various agricultural watersheds. The ANSWERS model (Beasley et al. 1980) is a distributed parameter model that uses a grid system to define a watershed area for simulation. The watershed under study is divided into number of small elements or cells. These cells are independent from each other in soil characteristics, slope, topography, crop coverage, and management. The size of each cell must be such that uniformity of all hydrological characteristics of the cell can be assumed (Huggins and Monke, 1966). Important hydrological and soil erosion parameters are defined in a predata file for a watershed before a storm event is simulated.

The ANSWERS model is an event-oriented model which predicts the complete hydrograph of a runoff event and gives the total volume of runoff at the end of the simulation of a single event. The model is capable of predicting the impact of alternative land management practices on sediment yield from an agricultural watershed. It calculates the total sediment yield and the concentration of the sediment corresponding to the time increments of hydrograph. The ANSWERS model

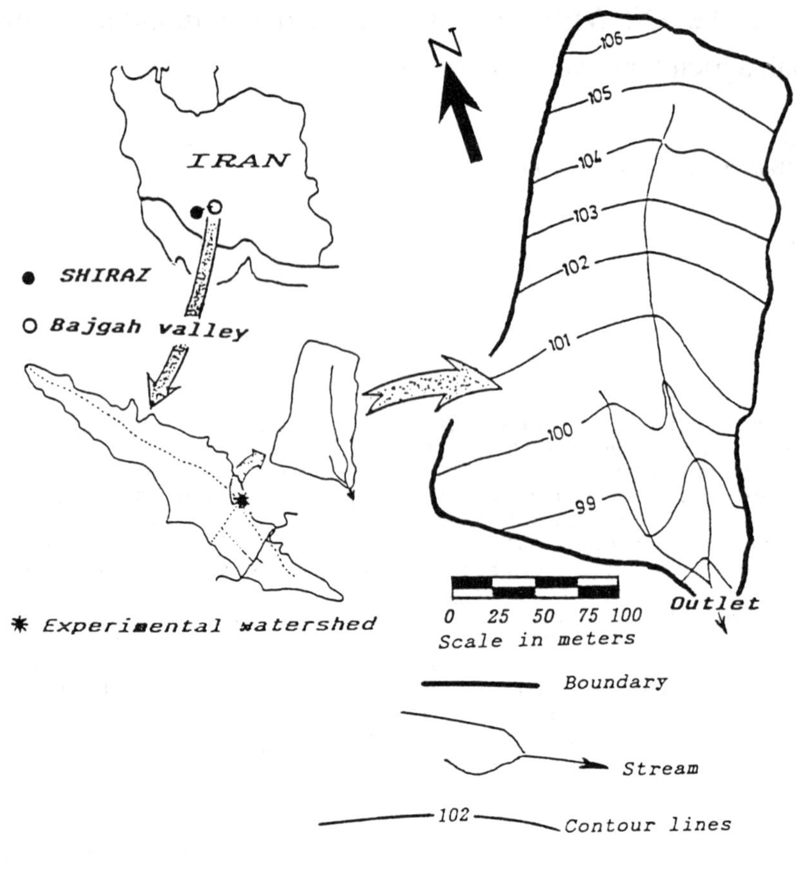

Figure 1. Location of the experimental watershed in Bajgah valley in the north of Shiraz and its contour map.

Table 1. The results of the simulation of the four rainfall of 1988 on the watershed under study.

Date	Rainfall (mm)	Duration (hr)	Runoff depth (mm)	Maximum runoff (mm/hr)	Average soil loss kg/ha	R^2 *
2/18	26.4	7	0.7	0.3	80	0.94
2/25	49.5	15.7	6.9	2.5	893	0.93
3/05	16.5	3.7	1	0.5	84	0.99
3/24	20.8	5	1.53	0.8	196	0.98

* Coefficient of determination for observed and predicted runoff.

also shows the total erosion or deposition for each element of the watershed under consideration. Therefore, a land planner can easily find the critical points in a watershed. The model has been applied, tested, and verified on different watersheds for various climatic conditions in locations including Illinois, Iowa, Ohio, Oklahoma, Pennsylvania, Texas, Virginia, and Ontario, Canada (

Dillaha, 1981). The ANSWERS model has been enhanced to simulate phosphorus transport in surface runoff (Amin Sichani , 1982), heavily forested watersheds (Thomas, 1984), and distribution of transported sediment particles from agricultural watersheds (Beasley and Huggins, 1982). The ANSWERS model was applied in different watersheds with high intensity rainfall events of the USA, therefore, for application of the model in Iran which in most part the rainfall amount and intensities are low the model should be tested and verified for runoff and sediment prediction.

EXPERIMENTAL WATERSHED

The ANSWERS model was applied using data from a small agricultural watershed located at the Agricultural College, Shiraz, Iran (Momtahan, 1989). This small agricultural watershed has an area of 4.83 hectares and short time of concentration of 15 minutes due to slope steepness and small size. The watershed has an average slope of 2.6 percent with minimum slope around 0.2 percent and a maximum slope of around 4.8 percent. A topographic map of this watershed is shown in Figure 1. This small agricultural watershed was used because it is representative of the major portion of the watersheds of the Fars Province in southern Iran. Precipitation occurs primarily during November to May each year with greatest frequency during February. Average Rainfall of the valley is 400 mm/year and rainfall events fit the B distributed form. The experimental watershed contains two soil types; Kuye-Asatid, (Loamy-skeletal over fracmental, carbonatic, mesic, Fluentic, Xerorthents), sandy loam lies in the upper portion of the watershed and Ramjerdi, (Fine, mixed, mesic, Fluentic, Xerochrepts), clay loam in the lower portion. A 15 cm (6 inches) Parshall flume was installed at the outlet of the watershed to measure the runoff flow rate from the site. As the flow was measured, water samples were taken for laboratory measurement of the suspended solid materials. Four different storms in 1988 were selected to simulate the response of the watershed. A grid system of 25 meters square was adapted on the watershed to simulate the four rainfall events. The rainfall data used were from the Agricultural College weather station. The storm events used in this study are representative of those received in this region of Iran. Information concerning soil textures, total porosity, soil types, and the area of each soil type in the watershed was collected from Solhi (1988), (see Table 1). Soil samples were taken intermittently during the course of the study in order to update the antecedent soil moisture information, ASM, needed in the simulation. Representative plots of each soil type were completely saturated in order to measure the field capacity of the soils. This was determined by taking samples over two days. Infiltration capacity of the soils were determined by the double ring infiltrometer method. The soil erodibility factor, K, was determined using particle size distribution of the soils of the watershed under study; and the graphs of the Agricultural Handbook 537 (Wischmeier and Smith, 1978),

Channel sizes and specifications were determined through direct measurements using the topographic map of the watershed. Other informations were extracted from the ANSWERS user's manual (Beasley and Huggins, 1982) according to the given tables. At the time of simulation, the upper part of the watershed was planted in dray farming wheat, while the lower part was fallow.

RESULTS AND DISCUSSION

Recommendation of the ANSWERS model for simulation of runoff and sediment from agricultural watersheds of Iran needs the application of the model on a small experimental watershed. Therefore, the ANSWERS model was used to simulate four storm events during the months of February and March 1988. A summary of all the storm events with the correlation coefficients of the linear regression analysis between observed and simulated run-

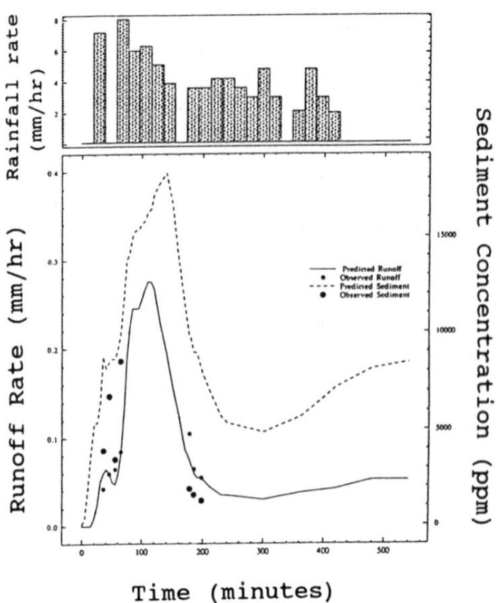

Figure 2. Rainfall rate, predicted and observed runoff and sediment from the experimental watershed, storm of 2/18/88.

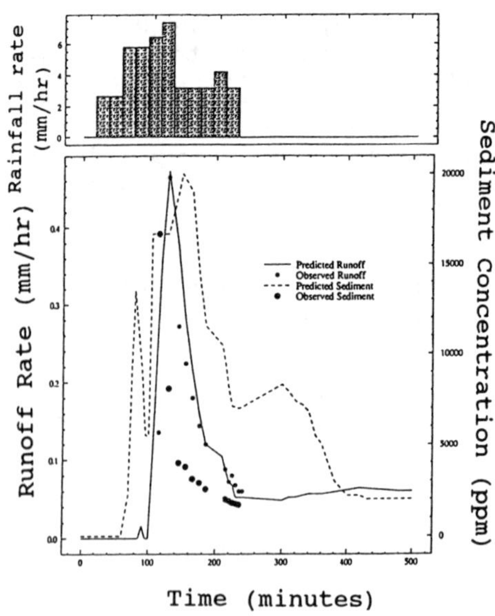

Figure 4. Rainfall rate, predicted and observed runoff and sediment from the experimental watershed, storm of 3/05/88.

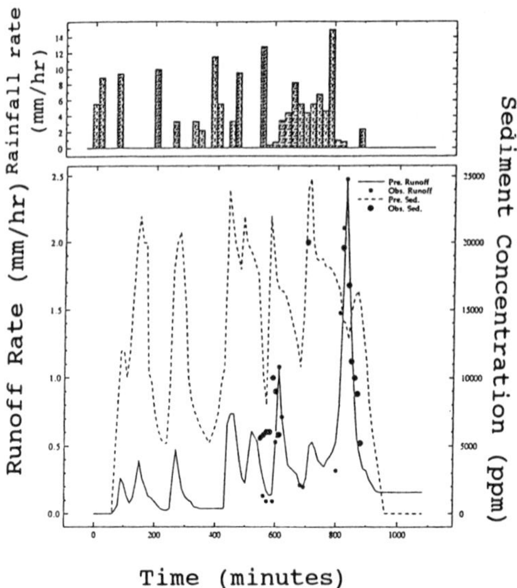

Figure 3. Rainfall rate, predicted and observed runoff and sediment from the experimental watershed, storm of 2/25/88.

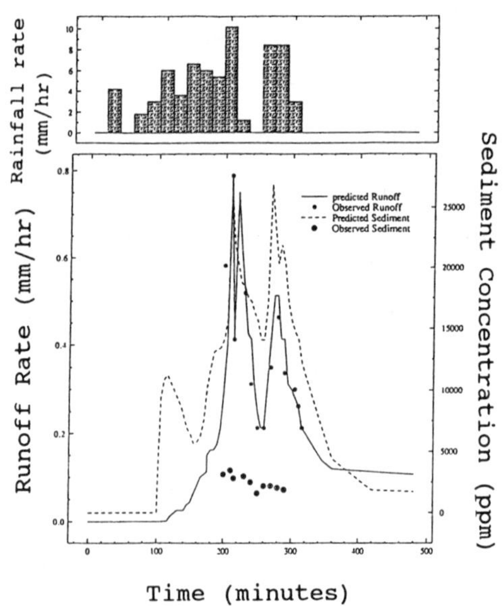

Figure 5. Rainfall rate, predicted and observed runoff and sediment from the experimental watershed, storm of 3/24/88.

off are given in Table 1. Figures 2 to 5 also shows the close agreement between observed and predicted runoff by the ANSWERS model for these storms.

REFERENCES

Amin Sichani, S. 1982. Modelling of phosphorus transport in surface surface runoff from agricultural watersheds. Ph.D. Thesis. Purdue Univ. West Lafayette, IN. USA. 157 pp.

Beasley. D.B. and L.F. Huggins. 1982. ANSWERS user's manual. EPA-905/9-82-001.U.S. Environmental Protection Agency, Great Lakes National Program Office. Chicago, IL. 54 p.

Beasley. D.B., L.F. Huggins and E.J. Monke. 1980. ANSWERS: A model for watershed planning. TRANSACTIONS of the ASAE 23(4): 938-944.

Dillaha, T.A.,III. 1981. Modelling the particle size distribution of eroded sediments during shallow overland flow. Ph.D. Thesis. Purdue University, W. Laf. IN. USA. 189 pp.

Foster, G.R. 1980. soil erosion modelling: Special consideration for nonpoint pollution evaluation of field size areas. In: Overcash, M.R. and J.M. Davidson (eds.), Environmental impact of nonpoint source pollution, Ann Arbor, MI. pp. 213-240.

Foster, G.R., L.J. Lane, J.D. Nowlin, J.M. Laflen, and R.A. Young 1980. A model to estimate sediment yield from field size areas: Development of model. In: Knisel, W.G. (ed.), CREAMS. A field scaled model for Chemicals Runoff, and Erosion from Agricultural Management Systems. Conservation Res. Rept. No. 26, USDA-SEA. pp. 36-64.

Huggins, L.F. and E.J. Monke. 1966. The mathematical simulation of the hydrology of small watersheds. Technical Report No. 1. Water Resources Research Center. Purdue Univer. W. Laf. IN. USA. 130 pp.

Momtahan. H. 1989. Application of the ANSWERS model for prediction of runoff and sediment from small agricultural watersheds of Iran. M.S. Thesis. Irrigation Engineering Department, Agricultural College, Shiraz Univ. Shiraz, Iran. 218 pp.

Purschwitz, M.A. 1981. Using ANSWERS on Indiana and Texas watersheds with four element sizes. M.S. Thesis. Purdue University, West Lafayette, IN. USA. 95 pp.

Soil Conservation Office .1981. An exploratory analysis of the economical and social effects of watershed management. Iran Agricultural Ministry. (In persian).

Solhi, M. 1988. Soil genesis, morphology, physiochemical properties and classification of Bajgah region soils (Fars Province). M.S. hesis. Soil Science Department, Agricultural College, Shiraz Univ. Shiraz, Iran. 140pp. (In Persian).

Thomas, D.L .1984. A physically based, distributed parameter, forest hydrology model. Ph.D. Thesis. Purdue Univ. W. Laf., IN. USA. 260 pp.

Wischmeier, W.H. and D.D. Smith. 1978. Predicting rainfall erosion losses. Agricultural Handbook No. 537. Science and Education Administration, U.S. Department of Agriculture. 58 pp.

Environmental Management, Geo-Water & Engineering Aspects, Chowdhury & Sivakumar (eds)
© 1993 Balkema, Rotterdam. ISBN 90 5410 099 0

Incipient motion criteria below circular jets

M. Shafai-Bajestan
Shahid Chamran University, Ahwaz, Iran

ABSTRACT: A total of 146 tests using four sizes of uniform and three sizes of graded sediments were conducted to develop a relationship for incipient motion of sediment particles at pipe outlet. For visual observation of sediment movement the Half-Jet Technique was used. This technique allows a large number of tests to be conducted in a relatively short period of time. In this paper the experimental set-up, procedures and the criteria for threshold conditions are discussed.

1. INTRODUCTION

In many practical problems in the field of hydraulic engineering, flowing water emerges from circular pipe on a boundary consisting of alluvial materials. The removal of the bed material through the action of high local velocity produces a scour hole which may result in the failure of the structure. Currently, the common recommended methods to minimize the extent of scour are: to pre-excavate a plunge pool below the pipe outlet deep enough to dissipate the excess kinetic energy of the jet, or to place sediment-material large enough to be stable at pipe outlet, or to have a combination of a low deep plunge pool covered with riprap materials. For the design of any of the above methods a relationship at the point of incipient motion is needed. This relationship should relate the flow conditions and pipe diameter to the sediment properties when the particle is about to move. It should be simple and practical so the engineers would like to use it and of course accurate enough for the design purposes.

Although such relationships have been developed for open channel flows such as Shieds diagram, Isbach equation [1], or Neill equation [2] and many others (for more information see [3]), very little information on incipient motion below pipe outlet is available. It is the purpose of this paper is to develop such relationship. To accomplish this, a general relationship at the point of incipient motion based on stability analysis of a sediment particle has been developed (See [3]). Then the necessary coefficients are supplied through an extensive experimental program. The experimental set-up tests procedure and the final equations are presented in this paper.

2 GENERAL RELATIONSHIP

The general relationship at incipient motion below a circular jet has been developed by the author in the following form:

$$\frac{V}{[g(G_s-1)D_s]^{0.5}} = C_4\left(\frac{P_e}{D}\right) \qquad (1)$$

In which V is the pipe velocity, D is the pipe diameter, g is the acceleration of gravity, G_s is the specific gravity, D_s is the representative of the sediment size, C_4 is a coefficient which can be determine from the experimental data, and P_e is the diffusion length or the length which the jet penetrates through the tailwater. To determine the values of P_e and C_4, two cases are discussed as follow:

A - Vertical pipe: When the jet impinges the bed vertically, as shown in Fig 1, the value of P_e is equal to the vertical distance from the water surface to the bed. For example Poreh and Hefez [4] conducted experimental study to find the conditions of particle movement below the vertical jet. They used air in their experimental

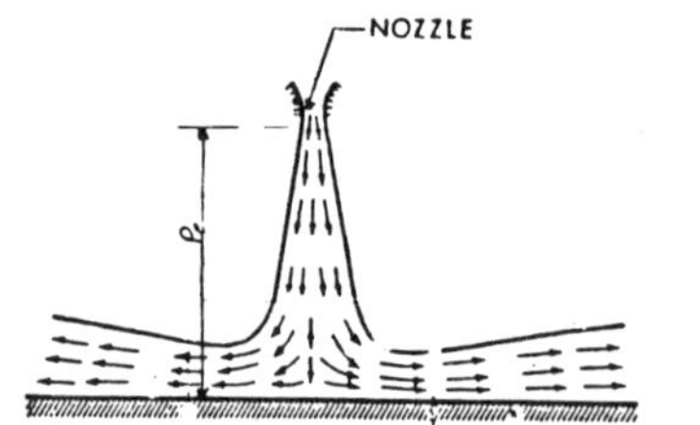

Fig. 1. Definition sketch for vertical jet

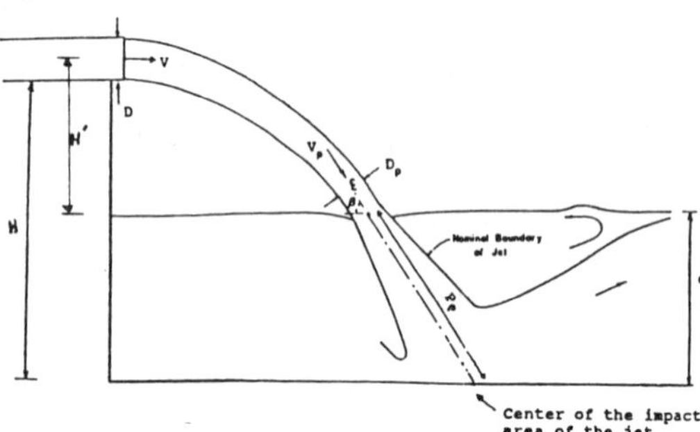

Fig. 2. Definition sketch for horizontal jet

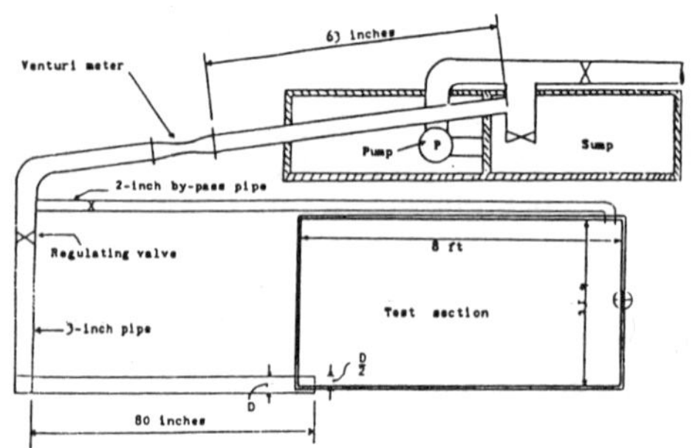

Fig. 3. Plan view of the experimental set-up

study and developed an expression which if rewritten in the form Eq. 1, the value of C_4 would be in the order of 0.28. Note that in Poreh and Hefez, the value of P_e should be taken as the vertical distance from the pipe to the bed. (Since they penetrated air into the air).

B - Horizontal Pipe: Fig.2 shows a sketch at which the flowing water emerges from a horizontal pipe on a river bed: the flow depth in the river is shown as d_t. The important variables should be measured at the point of incipient motion. Incipient motion here is defined when the first displacement of the bed material is observed. Visual observation of the movement of the bed material below a pipe outlet in three dimensions is difficult due to the large amount of turbulence, air bubbles and fine sediment stirred up by the impact of the jet. The indirect way of determining the incipient motion, in which to measure the scour depth, and extrapolate for the no-scour conditions, requires a long time and a very exhaustive procedure to obtain an approximately steady scour hole. To overcome these difficulties, and to make visual observation of incipient motion possible, a semicircular half jet nozzle was used. Its centre placed in line with the inside of the glass basin wall so that half of the pipe cross section was located inside the basin. Using this technique a larger number of tests could be done in a relatively short period of time.

The experimental facilities utilized through this study consisted of a basin 8 ft in length, 4 ft in width and 5 ft in height. The basin wall on the side of jet was made of plexi-glass for visual observation. Near the end of the cantilevered pipe where the flow entered the basin, a transition from full circular to half circular jet was inserted so that the jet issuing from the pipe had a more uniform velocity distribution with the flat side in contact with the plexi-glass wall. The cantilevered pipe was long enough, more than 20D, to ensure uniformity of the flow at the nozzle. Because of contraction of the flow at the nozzle, the effects of the upstream flow were further minimized. Figure 3 shows the plan view of the experimental set-up for this study.

The general procedure followed for completion of each test was that once the cantilevered pipe was adjusted to the desired elevation and the bed material were placed, the pump was started and the basin was filled to a proper elevation (a tailwater deep enough to ensure dissipation of enough of the jet energy that movement of the material would not occur). Then keeping a constant discharge, the tailwater was lowered gradually (by opening the tailgate) and the movement of the bed material was carefully observed and/or videotaped. The tailwater depth was recorded when the first transport of the bed material (incipient motion) was observed. The range of the variables which have been investigated in this study are as follows:

$V \rightarrow$ 0.61-4.3 m/s

$Q/(gD^5)^{0.5} \rightarrow$ 0.4-5 7

H/D	→	4-24
D_{50}	→	6.9-51.0 mm
$\sigma = (D_{84}/D_{16})^{0.5}$	→	1.2-3.0

in which Q is the flow discharge, H is the height of pipe above the bed surface and is the standard deviation of the sediment material.

3. GENERAL OBSERVATIONS

A - Path of the Jet: The important observation in all tests was the path of the iet through the tailwater. It was found that the jet tended to follow a straight line inclined from the water surface to the bed surface with an angle of ß to the tail-water surface, rather than a free fall trajectory. This observation *reveals,* that in the analysis of sediment movement below a pipe outlet, penetration length (the length of the path of the jet after it strikes the pool) should be used instead of the tail-water depth as it has been in the past.

B - Bed Material Movement: Usually, as the tailwater was lowered, a level was reached at which a few rocks on the bed surface continuously bounce and move back and forth with no displacement. By decreasing the tailwater level a bit more, the first rock movement was observed. This was considered to be incipient motion. Generally, the tailwater level causing incipient motion was higher for graded material than for uniform material of the same D_{50}. This is because the graded material contains a smaller particle size than uniform material of the same D_{50}. This observation indicates that incipient motion is governed by the smaller particle size in the mixture.

When the tailwater was gradually reduced, it eventually reached a level where the sediment particles of uniform size started to wash away from the impact area (incipient failure).

The situation for graded sizes of sediments was quite different. As the tailwater was reduced, only a small fraction of the mixture washed away from the top of the bed layer. The larger material size stayed in the scour hole protecting the finer rock size of the mixture. Likewise, the finer material protected the coarser material from flow penetration under and around the particle. Thus at any tailwater level some of the particles could be washed away while most of the material remained stable and in place. The incipient failure occurred only when the tailwater was so low that the driving force was capable of entraining and transporting the larger sizes.

4. CRITERIA OF INCIPIENT MOTION

A total of 146 tests were conducted with the procedures explained in the previous sections. The data from these tests were used to supply the coefficient C_4 of Eq. 1. The linear regressional analysis was used and the following equation was developed:

$$\frac{V}{[g(G_s-1)D_{30}]^{0.5}} = 0.382\left(\frac{P_e}{D}\right) \quad (3)$$

$r^2 = 91\%$

in which D_{30} is the sediment size which 30% by weight is finer. This equation shows that D_{30} is the representative particle size for incipient motion.

Fig 4 shows a plot of Eq. 3 along with the experimental results.

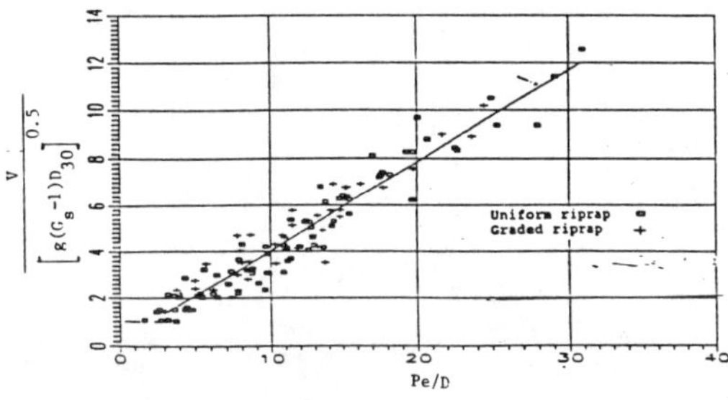

Fig. 4. Proposed equation along with the experimental data

REFERENCES:

[1] Isbach, S.V. 1936. Construction of Dams by Depositing Rock in Running water. 2nd Congress on Large Dams, paper C2, pp 123-136.

[2] Neill, C.R. 1968. Note on Initial Movement of Coarse Uniform Bed-material. J. of Hydraulic Research, IAHR, Vol. 6, No.2, pp 143-176.

[3] Shafai-Ba;estan, M. 1989. Riprap Criteria at

Pipe Outlet. Ph.D Dissertation. Dept. Of Civil Engineering, CSU, Colorado.
[4] Poreh, M. and Hefez, E. 1967. Initial Scour and Sediment Motion Due to an Impinging Jet. Proceedings of 12th Congress of IAHR, Vol.3, Ft. Collins, pp. 9-16.

NOTATIONS

C_4 coefficient
D pipe diameter
D_s representative size of sediment
D_{16}, D_{30}, D_{50}, D_{84} - particle size which 16%, 30%, 50% and 84% by weight is finer.
d_t tailwater depth
g acceleration of gravity
G_s specific gravity of sediment
H height of pipe above bed
H' vertical distance from pipe centre to tailwater surface
P_e diffusion length
Q flow discharge
V flow velocity
ß angle as shown in Fig 2
σ standard deviation of sediment

Environmental Management, Geo-Water & Engineering Aspects, Chowdhury & Sivakumar (eds)
© 1993 Balkema, Rotterdam. ISBN 90 5410 099 0

The effect of soil tillage on rainfall induced aggregate breakdown

N. M. Somaratne
Hydrogeology, Department of Water Resources, Parramatta, N.S.W., Australia

K. R. J. Smettem
CSIRO Division of Soils, Davies Laboratory, Townsville, Qld, Australia

G. Stanger
School of Earth Sciences, The Flinders University of South Australia, S.A., Australia

ABSTRACT: Effects of soil tillage on high-energy raindrop-induced aggregate breakdown was investigated. The comparison was made on direct drilled and conventionally cultivated soils by using simulated rain to create a surface seal on bare soil and by using unsaturated disc permeameter measurements to characterise the surface hydraulic properties. A significant reduction of hydraulic conductivity(k_0) at -20 mm pressure potential was observed in both tillage treatments after the surface aggregate stability was disrupted by raindrop impact. Such a difference was not observed in sorptivity(s_0). Two types of sealing, "macro-sealing" and "matrix-sealing", are introduced. Derived flow-weighted mean pore radii using s_0 and k_0 data suggest that the reduction of hydraulic conductivity is primarily due to the structural breakdown of soil aggregates under raindrop impact with subsequent macro-sealing this occurring irrespective of the tillage treatment employed.

1 INTRODUCTION

Apart from the loss of crop production, soil erosion from agricultural lands exerts detrimental effects on the environment by the washing away of productive lands and the silting up of water ways and reservoirs. The runoff from eroding lands may carry agricultural chemicals and fertilisers with them thus causing further problems in environment management.

Cultivation for weed control and for the provision of a favourable soil environment for root growth increases the risk of erosion. The awareness of this problem has led to adoption of direct drilling which minimises the soil disturbance compared with conventional cultivation methods. The main benefit of direct drilling is considered to be reduced water erosion resulting from reduced surface runoff through improved infiltration.

Infiltration is greatly influenced by the degree of surface sealing which results from raindrop impact on bare soil. Raindrop induced surface seal (or crust) is formed due to a variety of reasons. The process of seal formation is related to the breakdown of aggregates by raindrop impact and slaking, the movement of clay and silt particles into pores causing clogging and the deposition of suspended particles on the soil surface (Onofiok and Singer, 1984; Chen et al, 1980; Arshad and Mermut, 1988; Mualem and Assouline, 1989).

The objective of this study was to characterise the unsaturated hydraulic properties of raindrop induced aggregate breakdown on two tillage treatments by using disc permeameters. Since the effect of tillage is time dependent, measurements were made during the seed bed preparation period on freshly tilled soil and during the post harvesting period.

2 THEORY

The theoretical background of the disc permeameters and the method of their application have been discussed in detail by Perroux and White (1988) and Smettem and Clothier (1989). However, a brief description is provided here in order to illustrate the relation between the relevant soil parameters. Wooding (1968) showed that when water is applied at a supply potential Ψ_0, the steady state volumetric flow rate q is,

$$q = \pi r_0^2 (k_0 - k_n) + 4 r_0^2 \phi \qquad (1)$$

where r_0 is the disc radius, k_0 is the hydraulic conductivity at pressure potential Ψ_0, k_n is the hydraulic conductivity of the initial soil water potential Ψ_n and ϕ is the matrix flux potential defined by

$$\phi = k_0 \lambda_c \qquad (2)$$

where λ_c is the macroscopic capillary length, which is related to sorptivity s_0 and hydraulic conductivity k_0 by

$$\lambda_c = \frac{b s_0^2}{(\theta_0 - \theta_n) k_0} \qquad (3)$$

θ_n is the initial moisture content corresponding to Ψ_n and θ_0 is the equilibrium moisture content corresponding to Ψ_0. b is a constant with a mean value for field soils of 0.55 (White and Sully, 1987). Equations (1), (2) and (3) give

$$k_0 = \frac{q}{\pi r_0^2} - \frac{4 b s_0^2}{\pi r_0^2 (\theta_0 - \theta_n)} \qquad (4)$$

The sorptivity dominates the early time infiltration rate (Philip, 1969) and therefore,

$$\frac{Q}{\pi r_0^2} = s_0 \, t^{1/2} \qquad (5)$$

where Q is the cumulative infiltration and t is the time from the commencement. From λ_c, Philip (1985) derived a characteristic mean pore radius, λ_m using capillary theory,

$$\lambda_m = \frac{\sigma}{\rho_w g \lambda_c} \approx \frac{7.4}{\lambda_c} \qquad (6)$$

where σ is the surface tension, g is the acceleration due to gravity and ρ_w is the density of water.

3 MATERIAL AND METHODS

3.1 Site and soil

The study was conducted in conjunction with the CSIRO tillage and crop rotation trial near Kapunda, South Australia. The soil is a red brown earth comprising of 100 to 300 mm of hard setting loam over a heavy clay B-horizon (Hignett, 1989). The research at this site was commenced in 1983 by the CSIRO Division of Soils.

3.2 Methods

Measurements for this study were made during the 1990 crop cycle in direct drilled (DD) and conventionally cultivated (CC) plots reported by Smettem and Ross

(1992). The CC plots consisted of three cultivations between the opening Autumn rain and sowing time using 150 mm wide shares. Direct drilling involved no tillage prior to seeding with a 10-row SIRODRILL until 1986, after which a McKay modified sowing point of 20 mm width on a 10-row seed drill was used. Initial soil hydraulic properties of the two tillage treatments were characterised by insitu measurements using disc permeameters at -20 mm pressure potential. At this near-saturated pressure potential, the influence of some large structural pores on soil hydraulic properties is excluded. Thereafter, steel target frames of area 1 m^2 were embedded in the ground for rainfall simulation.

All vegetation remnants and loose materials were removed using a portable vacuum cleaner hence exposing the soil surface. A 35 mm hr^{-1} intensity storm of 30 minutes duration was simulated over the target area using the micro-processor controlled rainfall simulator described by Pederson and DuBois (1986). After rain simulation, the target area was covered with polythene sheets to protect the soil surface from rapid drying, which would otherwise cause cracking of the surface layer. Once the soil water content had returned to their pre-rainfall simulation value, measurements were repeated at - 20 mm pressure potential at the soil surface in order to characterise the post-rain hydraulic properties. Measurements were replicated by six at pre-seeding and four at post-harvest.

4 RESULTS AND DISCUSSION

The in situ measurements of unsaturated soil hydraulic properties and derived flow

Table 1. Summary of hydraulic properties at pre-seeding (May-June) and post-harvest (Nov-Dec). Means followed by the same letter are not significantly different at P=0.05.

	Pre - seeding			Post - harvest		
	Pre - rain		Post - rain	Pre - rain		Post - rain
	DD	CC	Combined DD & CC	DD	CC	Combined DD & CC
s_0 (mm $hr^{-1/2}$)	18.4 (1.8)^a	11.2 (0.8)^b	9.5 (0.8)^b	24.0 (1.1)^a	25.0 (1.0)^a	22.5 (1.0)^a
k_0 (mm hr^{-1})	35.6 (2.0)^a	15.3 (4.3)^a	4.5 (0.4)^b	30.5 (3.1)^a	24.9 (1.3)^a	8.6 (0.7)^b
λ_c (mm)	16.3	17.3	39.3	28.1	37.3	87.5
λ_m (µm)	452	426	187	263	198	84
θ_0 (cm^3 cm^{-3})	0.43 (0.01)	0.37 (0.02)	0.41 (0.14)	0.38 (0.02)	0.38 (0.01)	0.39(0.03)
θ_n (cm^3 cm^{-3})	0.11	0.11	0.13	0.01	0.01	0.02
ρ_b (g cm^{-3})	1.32 (0.01)	1.18 (0.01)		1.32 (0.01)	1.28 (0.01)	

weighted mean pore radii during seed bed preparation and post harvesting periods are presented in Table 1.

4.1 Sorptivity

The sorptivity reflects the surface area available for adsorption. The sorptivity values before rain simulation during pre-seeding period is significantly higher under DD than CC. The difference in s_0 may be attributed to large pores created as a result of tillage in the CC plots which prevents initial adsorption even though the θ_n was same in DD and CC. However post-rain s_0 value in combined DD and CC were reduced to 9.5 ± 0.8 (n=6) mm $hr^{-1/2}$.

Such a difference was not observed in post-harvest s_0 data (Table 1). Before rain simulation, s_0 under DD was about the same as that of CC. The post-rain s_0 value in combined DD and CC was about the same as in the case of pre-rain soils.

In general, post harvest s_0 is greater than that of seed bed preparation period. This is mainly due to the variation in θ_n. Post harvest θ_n was 0.01 ± 0.002 cm^3 cm^{-3} in DD and CC plots whereas in seed bed preparation period, θ_n was 0.11 ± 0.03 cm^3 cm^{-3} in both tillage treatments.

4.2 Unsaturated hydraulic conductivity

There was a marked difference in hydraulic conductivity at -20 mm pressure potential between DD and CC before rain simulation during the seed bed preparation period (Table 1). This observation was expected as tillage-created large pores in CC soil may not be contributing to water flow at a negative pressure potential of -20 mm. These large pores may be effective only under ponded conditions, once the surface aggregate disrupted, hydraulic conductivity was significantly reduced irrespective of tillage treatments employed.

Hydraulic conductivity and bulk density (ρ_b) in the DD treatments were similar for both measurement times but were significantly different in CC. In CC treatments, observed lower hydraulic conductivity and bulk density at pre-seeding, presumably due to loosening of soil and disrupting of pore continuity due to the effect of tillage. However this effect was diminished at post harvest due primarily to the reconsolidation of tilled soil in CC.

Time to incipient ponding was recorded during the post harvest rainfall simulation. The observed ponding time in DD was 6.7 ± 0.8 min and in CC 7.2 ± 1.2 min. This suggest that there is no significant difference in aggregate stability between DD and CC treatments during the post harvest period. This conclusion is in good agreement with the observed seal hydraulic properties discussed previously.

4.3 Flow weighted mean pore radius

Flow weighted mean pore radius was derived using s_0 and k_0 data. We observed a reduction of mean pore radius in both tillage systems after simulated rainfall (Table 1). The reduction during seed bed preparation period was from 452 μm in DD and 426 μm in CC to the 187 μm. During the post harvest period mean pore radius reduced as a result of aggregate disruption from 263 μm in DD and 198 μm in CC to the 84 μm. The range of above pore classes are of mesopore type (10-1000 μm) as designated by Wilson and Luxmoore (1988). They indicated that mesopores may contribute to rapid infiltration than macropores (>1000 μm) because it is often insufficient to initiate channelling in macropores while being sufficient to fill the mesopores and initiate preferential flow. Smettem and Collis-George (1985) also explained that at or near saturation, preferential flows dominates the infiltration process.

Our observations suggests that the reduction of hydraulic conductivity with seal development was primarily due to the loss of conducting pores of mesopore class. Two types of crusting or sealing are apparent. The "macro-sealing" where sealing of macro pores and meso pores occur at the initial period of rainfall under aggregate disruption and "matrix-sealing" which follows the macro-sealing stage under continued rainfall and the sealing of soil matrix occur leading to inhomogeneous dense layer. The initial sharp reduction of hydraulic conductivity as observed in this study was due to macro-sealing rather than matrix-sealing.

Onofiok and Singer (1984) described two stages of sealing, "early crust" stage that occurred just after ponding and "late crust" stage observed after 60 minutes of rainfall simulation. Their scanning electron microscopy study on three California soils, Columbia fine sandy loam, Wyo clay loam and Yolo silt loam showed that early crust had more micropores compared to the uncrusted sample. They explained this is mainly due to aggregate breakdown and pore plugging by individual particles. This observation is similar to our macro-sealing stage where large conducting pore sizes were reduced at the early stage of the seal development.

At the late crust stage of Onofiok and Singer (1984), they observed 0.1 mm thick upper layer of high porosity but below this considerably reduced porosity of lower layer of high concentration of clay and silt size particles in Yolo and Wyo soils. In Columbia late crust they observed an upper compact surface layer of 0.05 mm thick and a lower more open layer with particles uniformly reduced in size. Tarchitzky et al (1984) observed reduction of porosity in Sandy, Sandy Loam and Clayey soils by 34.3%, 33.2% and 29.1% as a result of formation of surface seal under simulated rain. In this study, application of rainfall was insufficient to create such a well established seal or matrix sealing,which would cause infiltration and runoff to maintain steady state under a constant rate of rainfall.

5 CONCLUSIONS

The result of this study indicates that seven years of tillage practices on this soil do not substantially affect soil hydraulic properties following high energy rainfall on bare soil. The sharp reduction of hydraulic conductivity with high energy raindrop impact was due to the macro-sealing of conducting pores.

6 ACKNOWLEDGMENT

Dr. A.D.Rovira and Mr. D.K.Roget are thanked for allowing access to the field trial at Kapunda. Professor K.K.Watson is thanked for reviewing the manuscript.

REFERENCES

Arshad, M.A. & A.R.Mermut 1988. Micromorphological and physico-chemical characteristics of soil crust types in Northern Alberta, Canada. Soil Sci. Soc. Am. J. 52:724-729.

Chen, Y., J.Tarchitzky, J.Brouwer, J.Morin & J.Banin 1988. Scanning electron microscope observations on soil crusts and their formation. Soil Sci. 130:49-55.

Hignett, C.T. 1989. Physical measurements on a red-brown earth at Kapunda. CSIRO Division of Soils, Report No. 107.

Mualem, Y. & A.Assouline 1989.

Modelling soil seal as a non-uniform layer. Water Resources Research, Vol. 25:2101-2108.
Pederson, R.N. & B.DuBois 1986. Microprocessr controlled, variable intensity, variable timing, rainfall simulator. Conference in Agricultural Engineering, Adelaide.
Perroux, K.M. & I.White 1988. Design for disc permeameters. Soil Sci. Soc. Am. J. 52:1205-1215.
Philip, J.R. 1969. Theory of infiltration. Adv. Hydrosci., 5:215-296.
Philip, J.R. 1985. Reply to "Comment on 'Steady infiltration from spherical cavities'". Soil Sci. Soc. Am. J., 49:788-789.
Smettem, K.R.J. & B.E.Clothier 1989. Measuring unsaturated sorptivity and hydraulic conductivity using multiple disc permeameters. Journal of Soil Sci.,40:563-568.
Smettem, K.R.J. & N.Collis-George 1985. The influence of cylindrical macropores on steady state infiltration in a soil under pasture. Journal of Hydrology, 79:107-114.
Smettem, K.R.J. & P.J.Ross 1992. Measurements and prediction of water movement in a field soil: The matrix-macropore dichotomy. Hydrological Processes, 6:1-10.
Tarchitzky, J. A.Banin, J.Morin & Y.Chen 1984. Nature, formation and effects of soil crusts formed by water drop impact. Geoderma, 33:135-155.
Wilson, G.V. & R.J.Luxmoore 1988. Infiltration, macroporosity and meso porosity distributions on two forested watersheds. Soil Sci. Soc. Am. J., 52:329-335.
White, I. & M.J.Sully 1987. Macroscopic and microscopic capillary length and time scales from field infiltration. Water Resources Research, 23:1514- 1522.
Wooding, R.A. 1968. Steady infiltration from a shallow circular pond. Water Resources Research 4:1259-1273.

Mining impacts and solutions

Environmental Management, Geo-Water & Engineering Aspects, Chowdhury & Sivakumar (eds)
© 1993 Balkema, Rotterdam. ISBN 90 5410 099 0

Coal mining and water quality with illustrations from Britain

F.G. Bell & A. Kerr
Department of Geology and Applied Geology, University of Natal, Durban, South Africa

ABSTRACT: Coal mining can result in a variety of waste waters and effluent emissions. The main pollutants are suspended solids, dissolved salts, acidity and ferruginous discharges. They are produced at various stages during the extraction, processing, handling and storage of coal, and even after mine closure. Examples are given of pollution associated with two abandoned mines in Scotland.

1 INTRODUCTION

Different types of waste waters and process effluents are produced as a result of coal mining. These may arise due to the extraction process, by the subsequent preparation of coal from the disposal of colliery spoil or from coal stockpiles. The strata from which the groundwater involved is derived, the mineralogical character of the coal and the colliery spoil, and the washing processes employed all affect the type of effluent produced. Problems also result from past mining operations, for example, there are 700 disused spoil heaps in the South Wales coalfield which cover an area of 54 km^2 amounting to about 2.5% of the total area.

Generally the major pollutants associated with coal mining are suspended solids, dissolved salts (especially chlorides), acidity and iron compounds. Colliery discharges have little oxygen demand, the BOD normally being very low. Elevated levels of suspended matter are associated with most coal mining effluents, with occasionally high values being recorded. Although not all minewaters are highly mineralized, a high level of mineralization is typical of many coal mining discharges and is reflected in the high values of electrical conductivity. Highly mineralized minewaters usually contain high concentrations of sodium and potassium salts and minewaters which do not contain sulphate may contain high levels of strontium and barium. For example, barium rich waters were pumped from the workings in the Harvey and Beaumont seams at Eccles colliery, Backworth, Northumberland. In the former, the barium content averaged 4500 mg/l, in the latter 1450 mg/l. The low pH values of many minewaters are commonly associated with highly ferruginous discharges. In some streams the pH value is less than 4.0, the iron concentration is greater than several hundred milligrams per litre and the sulphates exceed one thousand milligrams per litre.

2 EFFLUENTS ARISING FROM COAL MINING

The most common pollutant of a receiving stream associated with coal mining is the increased concentration of suspended solids. In particular the blanketing effect of coal slurry particles on the bed of a river is unacceptable both in terms of appearance and its influence on the flora and fauna in the stream. The turbidity of some streams into which colliery waste is discharged may restrict the penetration of light into the water and so have adverse effects on plant life therein. Indeed the effects of particles on aquatic plants, both in suspension and after settlement, have long been recognized, those resulting from the change in light and unsuitability of substrate for plant colonization being most notable. Reductions in both abundance and diversity of invertebrates occur in reaches of streams where fine particles are deposited. However, where suspended solid concentrations are low or siltation sporadic, community structure may remain unaffected. Where suspended solids concentrations averaged 110 mg/l and

Table 1. Composition of mine drainage waters (Courtesy of British Coal)

Column number	1	2	3	4	5	6
Approximate percentage of waters in each class	55%	25%	10%	7%	7%	2%
Quality	Hard alkaline		Moderately saline	Alkaline and ferruginous	Acidic and ferruginous	Highly saline
pH value	7.8	6.8	8.2	6.9	2.9	7.5
Alkalinity, mg/l $CaCO_3$	260	850	240	340	Nil	190
Calcium, mg/l	75	28	90	190	125	2560
Magnesium, mg/l	90	17	40	130	90	720
Dissolved iron, mg/l	0.1	0.5	0.1	25	122	0.6
Suspended iron, mg/l	0.1	2	0.1	21	0.1	0.2
Manganese, mg/l	0.1	0.1	0.1	6	7	0.9
Chloride, mg/l	180	200	3400	42	50	30800
Sulphate, mg/l	170	210	250	1720	1250	350

occasionally exceeded 2000 mg/l in a river receiving minewater, Edwards (1981) reported no change in the composition of the fauna although a few species were virtually eliminated and there was a 90% reduction in overall abundance. In general, good fisheries are unlikely to be found where average concentrations of suspended solids exceed about 80 mg/l.

With respect to disposal of spoil, solids are lost mainly from areas of current tipping, particularly during periods of heavy rainfall. Losses of solids from stable spoil heaps with a vegetation cover are not severe. However, in the nineteenth and early twentieth centuries, as there was no major demand for fine coal, much was tipped. Spoil disturbance, coupled with the use of inefficient washing procedures in the recovery of this fine coal, is commonly a cause of high suspended solids concentrated in nearby rivers.

Drainage water from coal mines in Britain can be classified as hard, alkaline, moderately saline, highly saline, alkaline and ferruginous, and acidic and ferruginous (Best and Aikman, 1983; Table 1). The ferruginous discharges constitute about 8% of the total, of which 7% are regarded as alkaline and ferruginous. The high level of dissolved solids which is often present in minewaters represents the most intractable water pollution problem connected with coal mining. This is because dissolved salts are not readily susceptible to treatment or removal. The only treatment normally possible is dilution and dispersion in a receiving water course which may involve transporting the discharge considerable distances. In some situations this option is not available. The range of dissolved salts encountered in minewater is variable, with electrical conductivity values up to 335 000 µS/cm and chloride levels of 60 000 mg/l being recorded (Woodward and Selby, 1981). Some average values for various coal mining effluents associated with the Nottinghamshire coalfield are given in Table 2.

The acidity of minewater discharges may affect the aquatic life in a receiving stream but the precipitation of metal hydroxides as the acidity is neutralized by the alkalinity of the stream is more serious. For example, acidic minewaters often contain high concentrations of dissolved metals, particularly iron in the form of ferrous sulphate, The latter is oxidized in streams and precipitated as hydrated ferric oxides. Where the pH exceeds

Table 2. Average quality characteristics of coal mining effluents in the Nottinghamshire coalfield (Courtesy of Severn-Trent Water Authority)

Type of Effluent	BOD (ATU) (mg/l)	Suspended Solids (mg/l)	Chloride (mg/l)	Electrical Conductivity (μS/cm)	Minimum pH	Other potential contaminants
Minewaters	2.1	57	4,900	14,000	3.5	Iron, barium, nickel, aluminium, sodium, sulphate
Drainage from coal stocking sites	2.6	128	600	2,200	2.2	Iron, zinc
Spoil tip drainage	3.1	317	1,600	4,100	2.7	Iron, zinc
Coal preparation plant discharges	2.1	39	1,500	4,200	3.2	Oil froth flotation chemicals
Slurry lagoon discharges	2.4	493	2,000	6,100	3.8	

5 and soluble metals are not toxic, the impact on the benthic fauna is similar to that of coal siltation with a decline in the diversity of faunal species. Acidic ferruginous minewater also may contain high concentrations of aluminium which precipitates as hydroxide as the pH value rises on entering a receiving stream, giving a milky appearance to the water. Concentrations of heavy metals may be high in some acid waters and may exert a toxic effect, however, their ecological significance is masked by the effects of acidity.

The leaching of the oxidation products of iron pyrites (i.e. sulphuric acid and the sulphates and hydroxides of both ferrous and ferric iron) can cause major pollution problems. Whilst this acid and ferruginous drainage is occasionally associated with spoil heaps, modern compaction techniques and the fine nature of the spoil usually limits oxygen supply and the infiltration of water, thus inhibiting pyritic decomposition. Significant concentrations of sulphates sometimes occur in low volume seepages from older, more permeable, spoil heaps. Acidic drainages from such sources occasionally contain low concentrations (a few mg/l) of copper, nickel and zinc but other heavy metals are rarely found in concentrations greater than 0.1 mg/l.

Waters from coal stockpiles are variable in quality, with major acid flushes occurring in periods of heavy rainfall after long periods of dry warm weather. Nonetheless acid minewaters are frequently associated with drainage from coal storage sites, particularly where the content of sulphur in coal exceeds 1%.

3 SALTS IN MINE DRAINAGE WATERS AND THEIR CONTROL

The character of drainage from coal mines varies from area to area and from coal seam to coal seam. Hence mine drainage waters are liable to vary in both quality, and also in quantity, sometimes unpredictably, as the mine workings develop. Although not all minewaters are ferruginous, and in fact some are of the highest quality and can be used for potable supply, they are commonly high in iron and sulphates, low in pH and high in acidity. Ferruginous discharges can give rise to disastrous conditions in receiving streams and can affect many kilometres of otherwise good quality water.

The principal groups of salts in mine discharge waters are chlorides and sulphates. The former occur in the waters lying in the confined aquifers between coal seams in most coalfields in Britain, South Wales being a notable exception. These salts are released into the workings by mining operations. In general, the salinity increases with depth below the surface and with distance from the outcrop or incrop. The concentration of ions in minewaters conform to established ratios which are remarkably consistent throughout British coalfields. The more saline waters contain significant concentrations of barium, strontium, ammonium and manganese ions. Dissolved sulphates occur only in trace concentrations in waters of confined aquifers.

Mine drainage waters at the point of discharge almost invariably contain sulphates which are either present in the waters lying in the more shallow unconfined aquifers, or are generated in the workings by the action of atmospheric oxygen on pyrite. The physical changes such as delamination, bedding plane separation and fissuring of rock masses caused by mining permit air to penetrate a much larger surface area than the immediate boundaries of the working faces and associated roadways. These changes also alter the hydrogeological conditions within a coalfield and allow wider movement of groundwater through the rock masses than existed prior to mining. These two factors mean that groundwater comes in contact with a large surface area of rock which is exposed to atmospheric oxidation and hence the groundwater becomes contaminated. Pyrite may occur in high concentrations, particularly in the upper part of a coal seam or in associated black shales. The primary oxidation products of pyrite are ferrous and ferric sulphates, and sulphuric acid. The three principal reactions involved in the production of ferric sulphates and sulphuric acid from pyrites can be written as follows:

$$2FeS_2 + 7O_2 + 2H_2O \quad - 2FeSO_4 + 2H_2SO_4 \qquad (1)$$
$$4FeSO_4 + 2H_2SO_4 + O_2 - 2Fe_2(SO_4)_3 + 2H_2O \qquad (2)$$
$$3Fe_2(SO_4)_3 + 12H_2O \quad - 2HFe_3(SO_4)_2(OH)_6 + 5H_2SO_4 \qquad (3)$$

Reactions (1) and (3) are chemical reactions but due to the low rate of chemical oxidation of $FeSO_4$ at pH values below 4.0, compared with the high rate in the presence of Thiobacillus ferroxidons, reaction (2) is mainly bacteriological. Reactions (1) and (2) require aerobic conditions while reaction (3), the most acid producing, is essentially a hydrolysis reaction that can proceed without air. Sulphates and sulphuric acid react with clay and carbonate minerals to form secondary products including manganese and aluminium sulphates. Further reactions with these minerals and incoming waters give rise to tertiary products such as calcium and magnesium sulphates. Generally the stratal water is sufficiently alkaline to ensure that only the tertiary products appear in the discharge at the surface (columns 1, 2, 3 and 6 of Table 1). Exceptionally both primary and secondary products may appear in waters from intermediate depths (column 4 of Table 1). The primary oxidation products tend to predominate in very shallow workings liable to leaching by meteoric water (column 5 of Table 1).

The oxidation of pyrite can be eliminated by flooding a mine with water. However, it is obviously not possible to flood an active mine. It is also not possible to flood abandoned mine workings which are above the water table and which are drained by gravity. If abandoned mine work workings below the water table are flooded, then advantage may be taken of the alkaline nature of many of the stratal waters to neutralize acidic ferruginous waters. The subsequent mixtures are usually neutral and free from suspended solids but they may contain significant concentrations of ferrous iron (up to 150 mg/l)

If adequate control of contamination by the pyrite oxidation products (ferrous, ferric, manganese, and aluminium salts) at source is not possible, then chemical treatment may be used to achieve any required quality although the costs are considerable. In terms of quality limits these should be about 5 to 10 mg/l of iron, manganese and aluminium together in the suspended state, with less than 1 mg/l of any of the three in solution. Any traces of heavy metals present in the raw water are reduced to negligible concentrations by the treatment methods involved.

Waters containing temporary acidity may be treated by cascade aeration and sedimentation in lagoons. The general form of the aeration reaction is:-

$$2\,Fe(HCO_3)_2 + 0 \quad - \quad Fe_2O_3.2\,H_2O + CO_2$$

The sludges accumulated in the lagoons have a dry solid content of 5 to 10% which increases to 25 to 35% after drainage. The dry solids can contain over 40% of iron.

Water containing permanent acidity require the use of an alkali such as lime. The basic reactions are:-

$$FeSO_4 + Ca(OH)_2 \quad - \quad FeO.H_2O + CaSO_4$$
$$Fe_2(SO_4)_3 + 3\,Ca(OH)_2 \quad - \quad Fe_2O_3\,.3\,H_2O + 3\,CaSO_4$$

Other alkalis such as sodium hydroxide may be used, but are more expensive and do not produce such dense sludges as do lime. The sludges recovered from alkali neutralization followed by sedimentation and consolidation are of relatively low density (2 to 5% dry solids), but drain fairly well. The drained sludge does not re-imbibe rain water, does not re-dissolve, and obtains a density of 20 to 30% dry solids after a few weeks exposure. The iron content of the solids is ususally in the range 10 to 20% and contains considerable impurities such as calcium sulphate and carbonate.

The prevention of pollution of groundwater by coal mining effluent is of particular importance. Movement of pollutants through strata is often very slow and is difficult to detect. Hence effective remedial action is either impractical or prohibitively expensive. Because of this there are few successful recoveries of polluted aquifers. In coalfields of the east Midlands and south Staffordshire colliery spoil is often tipped on top of the Sherwood Sandstone, which is the second most important aquifer in Britain. Although modern tipping techniques may render spoil impervious, surface water run-off can leach out soluble salts, especially chloride. This may result in the loss of up to 1 tonne of chloride per hectare of exposed spoil heap per annum under average rainfall conditions (spoil heaps may extent to many hundreds of hectares in extent). The run-off from these spoil heaps may discharge directly into drainage ditches or to land around the periphery of the heap and infiltrate into the aquifer below or run into streams. Figure 1 shows the isopleths for chlorine ion concentration in the groundwater of the Sherwood Sandstone in the concealed part of the Nottinghamshire coalfield. It also indicates the locations of the collieries. Accordingly it demonstrates the relationship between elevated chlorine ion level in the groundwater and mining activity. Such pollution of a aquifer can be alleviated by lining the beds of influent streams which flow across the aquifer or by providing pipelines to convey mine discharge to less sensitive water courses which do not flow across the aquifer.

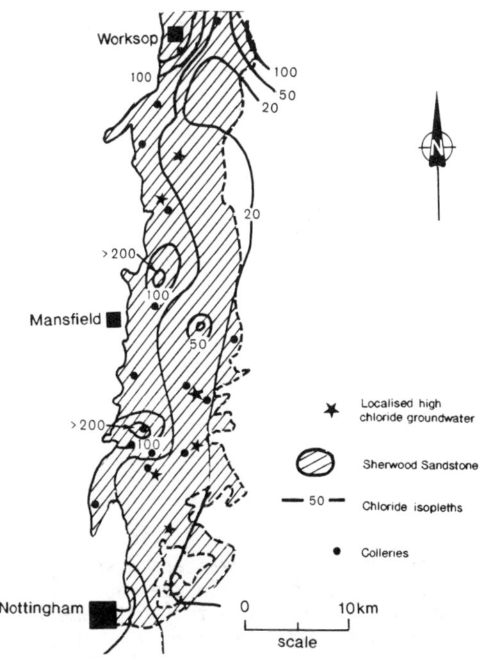

Fig. 1 Chloride isopleths for groundwater in the Sherwood Sandstone of the concealed Nottinghamshire coalfield.

4 USE OF MINE DRAINAGE WATER

It has been estimated that some 30 Ml/day of mine drainage water was being used in the 1980's by water authorities and industry in England and Wales, and that a further 100 Ml/day was suitable for use by industry without treatment. Around 400 Ml/day would be suitable after treatment. Most of the surplus mine drainage water was available in Scotland (76 Ml/day), north east England (130 Ml/day) South Wales (200 Ml/day) and west Yorkshire (54 Ml/day). However, the quantities of surplus water available at individual mines is usually small (e.g. 2 Ml/day) and the distances between sources and potential users are often considerable. These disadvantages, together with the general uncertainty as to the continued quality and availability of mine drainage waters suggest that no appreciable increase in utilization of these waters is likely.

The great majority of the surplus water is discharged to surface water courses. Approximately 25% is discharged to estuaries or coastal waters which are more tolerant to low quality, especially saline, discharges than are the inland water courses. Of the remaining about three quarters of the discharge to inland water courses are made as trade effluents under consents granted by the relevant water authorities. The rest consists of mine drainage discharges exempt under existing legislation. Also as no one is responsible for discharges from abandoned mines, water authorities cannot impose consent conditions. Typical trade effluent consent conditions in Britain for discharges of surplus water from coal mines to inland water courses lie within the following ranges (Rae, 1978):

pH value	greater than 5 (or 6) and less than 9
Suspended solids	30 to 200 mg/l
Iron (total) as Fe	5 to 30 mg/l
Oil	5 mg/l

Table 3. Examples of iron loading along River Ore

Location	Ave flow l/s	Ave Iron mg/l	Ave loading kg/day	Receiving stream	Origin
1 Inchgall	0	-	-	Ore	Coal outcrop
2 Glencraig	1	-	-	Fitty burn	Fissure?
3 W Colquhally	22	8	15	Fitty burn	Drainage adit
4 Minto pit.	38	4	13	Ore	Drainage adit
5 Cardenden gas works	8	2	1	Den burn	Coal outcrop
6 Cardenden Sewage Works upstream	4	11	4	Ore	Seepage
7 New Garden No 1	10	13	11	Ore	Pit shaft?
8 New Garden No 2	5	11	5	Ore	Borehole
9 Cluny	16	49	68	Ore	Pit shaft/ borehole
10 Blairenbathie	36	5	15	Ore	Pit shaft

The effects of minewaters discharged into major rivers in Britain are negligible. For instance, it is estimated that some 75 such discharges into the Trent contribute less than 18 mg/l of chloride into the river at Nottingham. The effects on smaller rivers and streams draining active coalfields may be more noticeable and can account for over one half the dry weather flow. Locally coal mine surplus waters can have a predominant effect on small streams and may account for the total flow in dry weather. However, the total length of streams seriously affected by coal mine discharges is only a small proportion of the total length in a hydrological region.

In general the surplus waters discharged from coal mines are an acceptable and sometimes a welcome addition to the flow in rivers and streams. The number and severity of problems caused by these discharges has decreased in recent years, mainly as a result of investment in new plant, the introduction of new technologies and more effective monitoring and control systems.

5 THE PROBLEM OF ABANDONED MINES

A major source of acid minewater results from the closure of mines. When a mine is abandoned and removal of water by pumping ceases, the water level rebounds and groundwater reoccupies the strata. Groundwater then eventually drains to the surface from old drainage adits, river bank mine mouths, faults, springs and shafts which intercept strata in which water is under artesian pressure. However, it may take a number of years before this happens. Old adits are often unmapped and unknown, and even currently discharging ones are not always immediately evident. An examination of the catchment data is often the only way that such discharges come to light. Discharges from old adits and mine mouths are usually gravity flows. Some of these are close to surface waters and so ferruginous water can drain directly into them.

5.1 Case history 1: the Fife coalfield

The Fife coalfield in Scotland has suffered more from ferruginous minewater discharge, especially from abandoned mines, than possibly any other in Britain. One of the worst areas affected by such discharge occurred in the south of the county in a roughly triangular area, bounded by Cardenden in the north, Fordell in the east and Dumfermline in the west (Fig. 2). The geological structure of the area consists of a syncline plunging to the north east with the Lower Limestone Group outcropping in the south of the syncline and the Passage Group in the north. The majority of the area is occupied by the Limestone Coal Group in which more than 20 coal seams were worked. Most of the surface water flows from west to east through the River Ore into the River Leven. A low watershed separates this basin from that of Keithing Burn which drains into the Firth of Forth. Both these water courses were seriously affected by ferruginous discharges. The number of gravitational discharges from abandoned mines are shown in Figure 2 and the associated discharges are listed in Table 3. The effects of these discharges on the receiving streams varied according to their size and the ferruginous loading of the discharge. For instance, in

February 1977 a discharge started to flow from an old adit near Fordell. Initially the rate was 380 l/s (3.3 Ml/day) but within a year this had fallen to a steady 280 l/s (2.4 Ml/day). The heavy iron loading which discharged into Keithing Burn was responsible for staining the floor and banks of the stream an ochre colour. Then ferruginous waters appeared at successively higher reaches along the River Ore and the surrounding farmland was saturated. At Cardenden the foundations of some buildings were flooded by rising groundwater levels. The average annual flow in the Ore is 1.37 m^3/s (11.9 Ml/day) with an average iron loading of 0.6 t/day 2 km downstream of Cardenden. This ferruginous loading is too great for the stream to cope with by dilution.

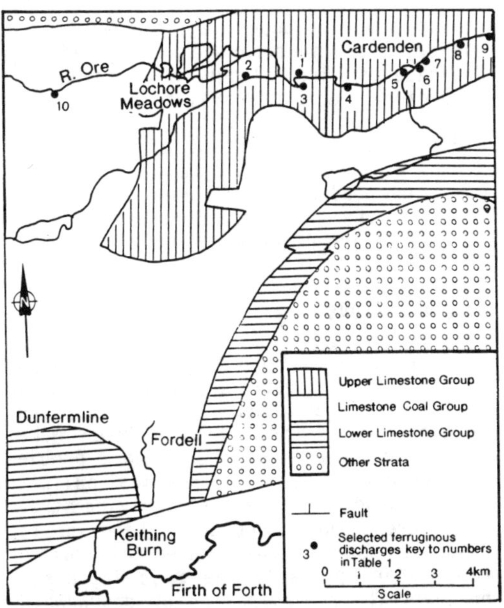

Fig. 2 Location map of places referred to in Table 3 along River Ore, Fifeshire coalfield.

5.2 Case history 2: west Ayrshire

Best and Aikman (1983) described minewater from Kames colliery at Muirkirk in Ayrshire, Scotland. The mine was closed in 1969 and the area was landscaped. However, the pumping main used for draining the mine was left in position so that the mine workings filled with water. The water overflowed from the pumping main, making its way into Garpel Water, a tributary of the River Ayr. The minewater was characterized by the absence of dissolved oxygen, with a pH value of just over 6 and contained ferrous iron in solution (Table 4). As the discharge flowed into Garpel Water and mixed with river water, the pH value increased and iron was oxidized and precipitated on the river bed. The process affected the quality of Garpel Water and the River Ayr for about 10 km. Five years after the mine closed the rate of discharge varied between 0.022 m^3/s and 0.047 m^3/s (0.2 to 0.4 Ml/day) and subsequently fell to an annual rate which ranged between 0.026 and 0.03 m^3/s (0.24 to 0.27 Ml/day). Initially the concentration of iron in the discharge was nearly 70 mg/l but over the years the concentration decreased to a steady 25 mg/l. The iron oxide precipitate blanketed the river bed.

Best and Aikman (1983) showed that the relationship between oxygen saturation, pH value and precipitation of iron for minewater from Kames Colliery was as indicated in Table 5. They further demonstrated that the rate of oxidation in minewaters with pH values between 6.2 to 6.5, meant that 90% of the iron was oxidized within 8 hours. Their tests on minewater from Kames Colliery indicated that 50% of the iron settled in 1 hour for 100% oxygen saturation, in 2 hours for 84% saturation and in 3.8 hours for 37% saturation. This indicates that iron can be removed quickly from minewater by aeration. Accordingly Best and Aikman designed a treatment plant which comprised an aeration cascade which fed the minewater into a biological filter, from the base of which the aerated water passed into a settling lagoon in which ferric hydroxide

Table 4. Chemical analysis of minewater discharged from Kames Colliery

	Average	Range
Total iron (mg/l)	26	18-32
Soluble iron (mg/l)	26	18-32
Dissolved oxygen (% sat)	nil	-
pH	6.35	6.2-6.6
Conductivity (µS/cm)	1350	1600-1500

Table 5. Effect of aeration on Kames minewater (After Best and Aikman, 1983)

Dissolved oxygen (% saturation)	pH value	Insoluble iron (% of total
30	6.4	17
87	6.4	44
100	7.0	64
104	7.2	89

precipitate was deposited. The wet sludge was dried out in separate drying beds. The drainage of the sludge to a handleable consistency took approximately one month. Such a treatment plant was capable of reducing the iron content of the 125 m^3/h discharge by some 75% from the original 28 mg/l, and produced an effluent which did not lead to the deterioration of the water quality in the River Ayr.

6 CONCLUSIONS

Major pollutants typically associated with coal mining are suspended solids, dissolved salts, acidity and iron compounds. Abandoned mines and old spoil heaps make a large contribution to the problem. Suspended solids act as a physical pollutant which degrades the beds of receiving streams and eventually reduces biodiversity. Dissolves salts, which are difficult to remove, are probably the most intractable problem. The contamination of groundwaters is a frequent hazard as the remediation of aquifers is rarely successful. Oxidation of pyrite leads to increased acidity and ferruginous compounds, and the resultant chemical treatment may require considerable costs. Treatment methods vary, and include cascade aeration and sedimentation or the addition of neutralizing chemicals. In England and Wales, some mine drainage waters, with and without treatment, are used to augment local water supplies to industry.

REFERENCES

Best, G.A. and Aikman, D.T. 1983. The treatment of ferruginous groundwater from an abandoned colliery. Water Pollution Control, 82:557-566.

Edwards, R.W. 1981. The impact of coal mining on river ecology; Proceedings Symposium on Mining and Water Pollution, Nottingham, Institution of Water Engineers and Scientists, 31-38.

Rae, G. 1978. Mine Drainage from Coalfields in England and Wales. A Summary of its Distribution and Relationship to Water Resources. Central Water Planning Unit, Technical Note No 24, National Coal Board, London.

Woodward, G.M. and Selby, K. 1981. The effect of coal mining on water quality. Proceedings Symposium on Mining and Water Pollution, Nottingham, Instituion of Water Engineers and Scientists, 11-19.

Environmental Management, Geo-Water & Engineering Aspects, Chowdhury & Sivakumar (eds)
© 1993 Balkema, Rotterdam. ISBN 90 5410 099 0

Environmental scenario of Indian mining industry

B.B.Dhar
Central Mining Research Station, Dhanbad, India

ABSTRACT : Indian mining industry is an age old industry in the country. After independence, India inherited a sick exploited mining industry particularly Coal sector. This was beset with environmental problems; many of them beyond the control. However, with more thrust in energy sector and general awakening in the country, environmental protection and mitigation has become an important issue.

Today mineral industries, in India, have to undergo strict environmental laws and regulations before a project can be financed.

In this paper, an attempt has been made to highlight some of the environmental issues associated with mining industry. Also, reference has been made to environmental laws associated with industry in India. This paper is expected to help an investor to assess the environmental obligations and issues prevalent in the country for any investment in mineral sector.

INTRODUCTION

In the process of development, mining is one of the core industries which is contributing knowingly or unknowingly towards the pollution of environment. In the process, Man some how seems to have ignored or perhaps forgotten the importance and the need for clean and pure environment required for his own survival. But at the same time, mining industry,among various others,is one which assures the raw materials,and is indispensable for a country. Its development in the modern sense of technology in the world has a number of effects on the environment which may be positive or negative.

To strike a balance between the development and its impact on mine workers and the surrounding environment , an environment protection element should be introduced at the conceptual stage of the mining project itself. This is the modern concept of sustainable development.

Any country in general, and developing countries like India in particular need a long term proper planning for each of the mining projects. For this, an Environmental Management Plan (EMP) is the most integral part of the project. Development of an integrated EMP, today, is the result of increased public awareness of the harmful environment and social effects of development. The impact assessment method and impact apprisal must consider,in assessment (Dhar 1991), the following :

1. Impact identification,
2. Impact measurement,
3. Impact interpretation,
4. Communication of information on impacts to decision makers and the public,and
5. Impact monitoring.

In environmental management of mining operations in India or elsewhere, the prime responsibility lies with those associated with mine planning, operations, and research and development organisations and enforcement agencies. Large scale openpit mining, in recent years in India and abroad, has brought into focus the need for environmental impact of mining on land, water and air. This awakening is well established in developed countries, but is still in infancy as far as developing countries are concerned. Fortunately, India has taken a lead in this group of countries and cannot be considered too far behind to the developed countries as far as the environmental assessment is concerned.

THE INDIAN SCENARIO

To briefly review the existing Indian mining scenario from environmental angle is rather difficult. However it may not be out of place to point out what has been done or achieved as far as mining industry is concerned. The author had the privilege of being associated with the first committee formed in 1979 by the then Department of Science & Technology, Government of India, to assess the need for environmental laws and regulations in mining. This was an interdisciplinary team of experts to frame a broad set of directives and guidelines for the protection of environment as a result of mining. These guidelines have since been accepted by the Government of India, as a result of which the Environmental Appraisal Committee - for " Mining Projects" by the Department of Environment & Forests was created. This Committee has been responsible for ensuring the environmental requirements in mining and allied activities based on its impact for all major mining projects in the country.Unfortunately, this was not sufficient because it was not obligatory on small mines to follow the criteria. As a resu- lt of this, in 1986-87, Mines and Metals (Regulation and Development (MMRD) Act 1957 was modified and a provision was made for environmental protection. These new laws and regulations are since then being followed vigorously in this country. Thus a new dimension was added to the status of environment and ecology for its protection and management. Dhar (1990) has given a detailed account of environmental management scenario in India in his book, a reference to which has been made in one of the recent reports of UNEP (1991) as well.

MINING AND ENVIRONMENT :

The impacts of mining operations as summarised by Dhar (1990,1992) can broadly be classified into the following :

a) Impacts on natural resources in terms of

- air,
- land (Pollution and degradation),
- water(surface and subsurface),
- biological regime (including flora and fauna),
- noise and vibrations,
- accelerated deforestations, causing,and
- climatic changes.

b) Socio-economic impacts - may vary from country to country. In developing countries,this impact is more significant because of the complex natural situations where one has to look into

- need of the people as a result of population explosion and
- capacity to develop (poverty)

c) The environmental costs associated with such activities, namely

i) Social Costs -that include :

- Unemployment of outstees,
- Disruption of social fabric of outstees,
- Exploitation of simple folk,
- Increase in cost of essential commodities,
- Higher incidence of crimes and new vices,
- Higher incidence of respiratory borne diseases,
- Transportation and communication development,
- Availability of medical,education & other facilities to limited population, and
- Some employment.

ii) Economic Costs - that include :

- Mineral availability,
- Land acquisition,
- Rehabitilation,
- Employment generation,
- Infrastructure buildup,
- Machinery equipment.

iii) Environmental Costs - that include :

- Losses of forest and /or wildlife habitat,
- Water pollution and related diseases,
- Air Pollution and related diseases,
- Land Pollution, and
- Loss of land fertility due to erosion.

ENVIRONMENTAL PLANNING :

Therefore, for any activity, the above impacts have to be asessed and analysed. The analysis is generally based on the following approaches :

- Qualitative,
- Quantitative,
- Combination of Qualitative and Quantitative,
- Resource Management Approach, and
- Simulation and mathematical modelling techniques.

To accomplish meaningful results as a result of analysis of impacts, the requirement is proper environmental planning right at the conceptual stage of the mining project. The term used today, in India, is known as EMP (Environmntal Management Plan) and the important issues that are taken in to consideration, after base line data, has been generated and impacts assessed (broadly) are :

- Deforestation,
- Flora and Fauna,
- Land degradation and subsidece
- Surface and ground water pollution,
- Air pollution,
- Noise pollution and vibration,
- Human displacement, and
- Compensatory afforestation.

Based on this,a comprehensive plan is developed for a proper mining method, and the plan of reclamation and for the total life of the project including environmental costs associated in mitigation and control.

Once an environmental clearance is given based on the above dsigned EMP, the project is normally considered for funding.

MINERAL LAWS AND ASSOCIATED ENVIRONMENTAL REGULATIONS FOR MINING

No industrial activity can be seen in isolation with respect to its impact on environment and ecology. Mining, therefore, is no exception.

Mining and environment have got three issues in today's concept of development that are complimentary to each other, and no single issue can produce the desired results (Dhar 1991). These are :

1. The policies of the government,
2. The regulations, and laws of the country, and
3. The enforcement mechanism.

Policies will be defined by the various governments/countries based on broad guidelines for development. Regulations will be either the modifications of regulations developed by developed countries, or developing countries, with issue based on their own actions, keeping in mind the complex nature of the problems faced by an individual country. Enforcement is essential because without that, no law or regulation can take shape unless it is properly implemented. There are countries where policies and regulations are existing, but there is no implementation or enforcement, and result being large scale environmental degradation. Since these issues are complex and vary from country to country no universal policy can be adopted. India, in recent years, has come up with well defined laws and regulations for the development and exploitation of the mineral resources. A sustainable development is perhaps possible, if the pressure on natural resources is maintained within reasonable levels. The Indian scenario about mining and environmental laws is summarised below for ready reference.

i) Laws and agreement pertaining to exploration in mining:

There are national laws and policies, under which mineral exploration and exploitation are carried out in India as follows:

1. Mines Act, 1952,
2. Mines Rules, 1955,
3. Mines and Metals (Regulation & Development Act),1957,
4. Coal Mines Regulation,1957,
5. Mineral Conservation and, Development Rules,1958
6. Land Acquisition (Mines)Act, 1885,
7. Petroleum Concession Rules, 1949,
8. Payment of Wages (Mines)Rules, 1956,
9. Mineral Concession Rules,1960, and
10. Metalliferous Mines Regulations, 1968.

Beside these, mineral development has to be got cleared by the Department of Environment and Forests, Government of India in case of large projects for environmental clearance and small mines by Indian Bureau of Mines, which is again government of India organisation, before mining begins under pollution rules and acts.

Only the major acts are enlisted above. The years mentioned reveal the time of their enactment but each of them has been amended several times so far.

ii) Legal requirements for environmental protection

The following constitutional provisions and Acts and the Rules framed there under cover the entire gamut of the environment and ecology in India.

I. Constitution of India provides :

- Article 48 A : Protection and improvement of Environment and safeguarding and forests wildlife
- The state shall endeavor to protect and improve the environment and to safeguards the forests and wildlife of the country
- Article 51 A : It should be the duty of every citizen of India, to protect and improve the natural environment including forests, lakes, rivers and wildlife to have compassion for living creatures.

II.The following acts are of direct relevance to environment management of mining operations :

- Wildlife Protection Act, 1972 (amended 1980),
- The Water (Prevention and Control of Pollution) Act, 1974, with amendments in 1988,
- The Air (Prevention and Control of Pollution) Act, 1981 with amendments in 1987,
- Forest (Conservation) Act 1980 with amendments in 1988,
- The Environment (Protection) Act, 1986 (is the most recent addition),
- The Mines and Minerals (Regulation and Development) Act,1957 with amendments in 1986 and effective from February 1987.

The ministry of Environment and Forests serves as a focal point for the planning,promotion and coordination of various environmental programmes. The main bodies implementing these acts are the central and state pollution control boards. All major projects require clearance from central and state pollution control boards.

iii) Environmental Trends

Awareness about environmental issues is growing among the people living in and around the mining areas. There is no as such anti-mining group in India. Central and State Pollution Control Boards are taking proper care for the protection of environment.

Environmental Management Plan (EMP) is required to be formulated on the basis of rapid or exhaustive EIA comprising all the major environmental attributes, which present the base line.

There is no reclamation act existing till date. Environmental clearance for major projects is granted by the Ministry of Environment & Forests through its Environment Assessment Committee on mining projects, and rest of the small mining projects being assessed by IBM (Indian Bureau of Mines).

There has been growing expertise regarding ability to predetermine environment related preobligations and Environment Management Cells are being created in each of the mining organisations for the purpose.

Regional centres of Ministry of Environment are being built up to keep the vigilance whether the commitment to that ministry is followed and fulfilled,in respect of legal requirements for healthy environment.

CONCLUSION

It may therefore be concluded that the above summary should help an individual or a group to assess for themselves the legal requirements that are necessary for any investment in mineral sector in India These laws and regulations have to be followed, may it be in coal or non-coal mining sector. Even the small scale mining operations have to fulfill the legal requirements and a satisfactory mine plan and EMP provided. Expertise by now is well developed in the country and the mining companies find no difficulty in the preparation of either investment or environmental mine plans.

REFERENCES

Dhar,B.B.,Jamal,A.,Srivastava,B.K., Ratan,S.,1991.Environmental Impact Assessment of Opencast Coal Mine - A Quantitative approach Minetech CMPDIL,Ranchi, India. Vol.12 No.6 pp 17-24.

Dhar,B.B.,(Ed) 1990. Environmental Management of Mining Operations Ashish Publication House, 8/81 Punjabi Bagh, New Delhi pp 1-390.

Dhar,B.B., Rao,Srinivas Ch., Jamal, A.,1990. Environmental Impact Assessment and Analysis due to Mining.Proceedings of the Second Asia Pacific Mining Conference, Jakarta pp 115-126.

Dhar,B.B., 1992. Environment and Sustainable Development in Developing Countries with special reference to Mining Industry Third Asia Pacific Mining Conference; Minerals, Energy and the Environment : The onlook for the rest of the decade, Manila, Philippines 18-21 March pp 459-473.

Dhar,B.B., 1991 Perspectives for developing countries on Environmental Policies, Regulations and Enforcement; Documentation III-IE Int.Round Table on Mining and Environment Berlin ;Sponsered UNDTCD and DSE (German Foundation for International Development) June 25-29.

UNEP Report Series No.5 (1991) on " Environmental Aspects of Selected Non-ferrous Metals Ore Mining" - A technical guide pp 85.

Environmental Management, Geo-Water & Engineering Aspects, Chowdhury & Sivakumar (eds)
© 1993 Balkema, Rotterdam. ISBN 90 5410 099 0

Legislation and management of mine tailings storages in Western Australia

H.Jones, I.H.Lewis & C.F.Swindells
Department of Minerals and Energy, W.A., Australia

ABSTRACT: Western Australia has a large number of mine tailings storages in a wide range of geographical, geological, and environmental settings. Management aspects of these storages, particularly from a legislative perspective, must reflect the wide range in physical and other factors that may influence the performance of the tailings storage during both the operating and post-operating phases. Interim guidelines on safe design and operating standards have been prepared by the Department of Minerals and Energy to manage the design, operation, and rehabilitation of mine tailings storages in Western Australia. The guidelines reflect the differing design and operating standards that may exist as a result of the range in specific safety and environmental considerations at an individual site. The legislative framework for the management of mine tailings storages is outlined, and the goals for post-operating rehabilitation are presented. Case histories are used to illustrate both the legislative process as well as the operating standards that are required to be maintained during both operating and post-operating phases of tailings storages.

INTRODUCTION

There are currently over 200 open pit mines and some 45 underground mines operating in Western Australia. Processing of various types of ore from these operations results in the annual disposal of significant volumes of mostly fine grained waste products in the form of tailings.

The rapid growth in the number of open pit and underground gold mines in Western Australia in the 1980s led to the current numbers of operating mines. Many of these mines have exploited medium to large sized low-grade gold orebodies, where mine profitability is closely related to processing techniques and the rate of production. The rapid increase in the quantity of treated gold ore during the period 1980 to 1991 (Figure 1) has been mirrored by a comparable increase in the required storage volume for mine tailings.

These developments within the mining industry have coincided with a period of increasing community awareness of environmental matters. In particular, the level of public interest in relation to the safety and environmental impact of mining operations in Western Australia has been significantly heightened.

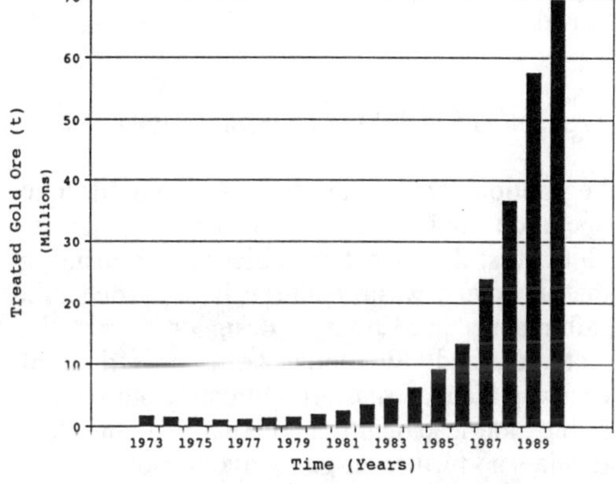

Figure 1. Annual ore production.

The Department of Minerals and Energy (formerly the Department of Mines) Western Australia, has placed a major emphasis on the environmental aspects of mining operations in recent years, with particular attention being paid to industry performance in respect of tailings management. Following an independent review of the current tailings practices commissioned by the Department in 1985, the following were identified as key issues:

- the wide variation in the standards of design, operation, and rehabilitation of tailings storages that were present within the industry;
- the lack of a unified approach to the regulation and licencing of tailings storages, with a number of Government Departments and instrumentalities involved in the regulatory process;
- a legacy of poorly managed and abandoned tailings storages from previous periods of mining activity in Western Australia; many of the historical examples have yielded unsatisfactory environmental impacts and major rehabilitation problems;
- the need for a clear regulatory framework to cover the design, operation, and rehabilitation of tailings storages.

The development of the current regulatory framework arose from the initial review and presents an attempt to provide a technically based, yet practically workable, approach to regulating mine tailings storages in Western Australia.

TAILINGS STORAGE LEGISLATION

Legislation governing tailings dam design, operation and rehabilitation has two major objectives: the short term objective of ensuring that unwanted waste material is deposited in a safe manner in a structure designed to take the material, and the long term objective of ensuring the materials are ultimately stored in a stable, non-leaching, rehabilitated structure that requires no further on-going maintenance. This means that the legislation must address a number of potentially conflicting requirements. For example; placement of tailings in an old worked out pit may be a cheap way of disposing of the material in the short term but in the post-operational phase, the unconsolidated material in such a pit is a significant public safety hazard. Another example of conflicting objectives is the method of building the dam wall out of deposited tailing. While this has proved a very safe and economical method during the construction of the tailing dump, many of those dumps are now major environmental problems due to wind erosion of the unprotected dry tailings.

The specific operation and stability requirements to meet safe short, medium, and long-term conditions vary from site to site, and require careful consideration of a range of technical factors.

The difficulties associated with specifying various short, medium, and long-term stability requirements may compound when environmental considerations are taken into account. A structure may be designed and operated safely from engineering and other technical considerations but may still result in an unacceptably high environmental impact.

Technical and environmental conflicts may be further complicated when viewed in conjunction with the options available for regulatory framework. The key to improved industry performance is a balance between rigidly policed minimum criteria and wider measures aimed at increasing industry awareness of the issues to be considered, and education in respect of the inadequate level of current-day industry performance.

The need for legislation in respect of tailings dam safety has been recognized almost universally. The International Commission on Large Dams (ICOLD) established the ICOLD Committee on Mine and Industrial Tailings Dams in 1984. This committee developed a set of guidelines to cover all safety aspects of tailings dams (ICOLD, 1989) largely in response to the increasing numbers of large tailings dams that were being constructed around the world.

Overseas approaches to tailings storage legislation have largely been based on direct regulation of the various aspects of tailings storage including technical considerations. A good example is the South African legislation in respect of dam safety proclaimed in 1986. These regulations prescribe conditions and requirements for the management of dam safety and include tailings storages on mine sites or works covered by the South African Mines and Works Act (1956).

A similar legislative approach has been adopted in Finland (the Safety Code for Tailings and Other Waste Dams, 1986 [Saarela, 1989]) as well as in other states of Australia, eg the Queensland Water Resources Act, 1989, the Victorian Water Act, 1989, and the New South Wales Dam Safety Act, 1978.

Table 1: Hazard Potential And Size Classification, South African Dam Safety Legislation

Size class	Maximum wall height in metres
Small	More than 5 but less than 12 m;
Medium	equal to or more than 12 but less than 30 m;
Large	equal to or more than 30 m.

Hazard potential rating	Potential loss of life	Potential economic loss
Low	None	Minimal
Significant	Not more than ten ..	Significant
High	More than ten	Great

A feature common to many of the legislative approaches has been the classification of storages and dams in relation to both size and potential safety hazards. An example of the South African classification is shown in Table 1.

The main thrust of this legislation is directed to ensuring no catastrophic embankment failure occurs during the operational life of the dam. The Western Australian legislation addresses these safety aspects during the operational phase of the tailings dam and in addition requires the post-mining rehabilitation of the completed dam to be addressed.

FACTORS INFLUENCING TAILINGS STORAGE LEGISLATION AND MANAGEMENT

The purpose in the introduction of legislation and procedures to regulate tailings storage is to ensure that all aspects of design, operation, and rehabilitation are adequately addressed by mine operators to minimise the potential for long-term environmental problems.

Studies of the performance of tailings storages (eg Ivanov et al., 1989) have shown that operating failures may commonly result from:

- breaks in service and pipeline facilities;
- substandard construction control, particularly associated with the primary embankment;
- design errors arising from an inappropriate or under-designed method of tailings storage.

These common types of operating failure all involve consideration of technical matters. Clearly any legislation to control tailings storage must ensure that adequate levels of technical input are adopted by mine operators. The initial review of tailings storage performance in Western Australia clearly identified an inadequate recognition of technical issues in the design of tailings storages.

The majority of the State's mining projects occur within crystalline basement rocks in which groundwater resources are usually found within superficial unconfined aquifers, or localised within fractured rock aquifers. Both sources of groundwater, used for human or livestock consumption, are susceptible to contamination. Extensive resources of hypersaline groundwater are also present and used in extractive processing.

Despite the generally arid nature of much of Western Australia, an active flora and fauna is sustained in even the harshest of climatic conditions. Uncontrolled leakage of even acceptable quality water from say a tailings storage, could yield significant adverse environmental impacts on plants finely tuned for harsh conditions. Conversely, in other areas characterised by hypersaline groundwaters and saline soils, the release of similar acceptable quality process water may have little or no discernible impact on prevailing groundwater quality and/or natural ecosystems.

In developing a legislative framework, it is important to recognize that the intended outcome of the legislation is an appropriate course of action being followed by the operator of the tailings storage. When this is linked to the technical considerations that are so fundamental to the safe operation of a tailings storage, and the manner in which these may vary from site to site, then the level of prescription in the adopted legislation requires careful consideration.

TAILINGS STORAGE LEGISLATION PROPOSED FOR WESTERN AUSTRALIA

In developing a legislative framework for Western Australia, a number of factors have been considered, including:

- the preference for a single authority to assess, licence, and monitor tailings storages;

Table 2. Hazard ratings - mine tailings storages (after ANCOLD, 1986 and Queensland Water Resources Commission, 1991)

Type of effect	Hazard category High	Hazard category Significant	Hazard category Low
Uncontrolled Releases or Seepage			
Loss of human life	Location such that contamination of a water supply likely to be used for human consumption and consumption of the contaminated water is expected	Location less critical but contamination of a water supply likely to be used for human consumption and consumption of the contaminated water is possible but not expected	No contamination of a water supply likely to be used for human consumption expected
Loss of stock	Location such that contamination of a currently used stock water supply and consumption of the contaminated water is expected	Location less critical but contamination of a currently used stock water supply and consumption of the contaminated water is possible but not expected	No contamination of a stock water supply expected
Environmental damage	Location such that damage to an environmental feature of significant value is expected	The significance of the environmental feature is less or damage is possible but not expected	No environmental features of significance or no damage expected
Embankment Failure			
Loss of human life	Loss of life expected because of community or other significant developments	No loss of life expected, but the possibility recognized. No urban development and no more than a small number of habitable structures downstream	No loss of life expected
Direct economic loss	Excessive economic loss such as serious damage to communities, industrial, commercial or agricultural facilities, important utilities, mine infrastructure, the storage itself or other storages downstream	Appreciable economic loss, such as damage to secondary roads, minor railways, relatively important public utilities, mine infrastructure, the storage itself or other storages downstream	No significant economic loss, such as limited damage to agricultural land, minor roads, mine infrastructure, etc
Indirect Economic Loss	Storage essential for services and repairs not practicable	Repairs to storage practicable	Repairs to storage practicable. Indirect losses not significant

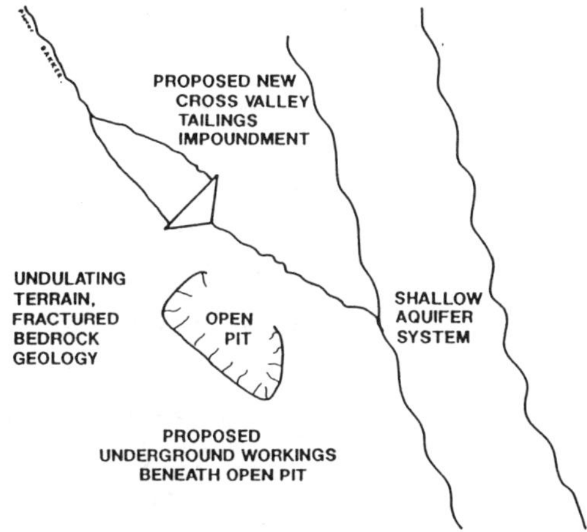

Figure 2. A proposed new tailings storage

- the requirement for an approach flexible enough to accommodate the variation in physical conditions and technical considerations that will occur from site to site;
- a strategy that would accommodate future changes in technical knowledge and/or community expectations.

The present strategy was developed in conjunction with other Government Departments, consultants involved in designing tailings storages, and the mining industry.

The intention in the development of the strategy has been to provide a common approach to the safe design, construction, operation and rehabilitation of tailings storages, and to provide a systematic method of classifying their adequacy under normal and worst case operating conditions.

An important element in the overall approach has been the development of a set of interim guidelines that have been based on a hazard rating system. By using a hazard rating system, differences in design philosophy, degree of technical input and the type and nature of tailings storage construction are clearly recognized.

The hazard rating (Table 2) is derived by considering:

- the potential impact on the environment in the vicinity of the tailings storage that would result from either a controlled or uncontrolled escape of material or seepages;
- the potential impact in terms of safety and economics arising from failure of the storage embankment.

The hazard rating given to an individual tailings storage is not an assessment of the risk of failure of the storage embankment, but the potential impact in the event of escape of material, seepage, or embankment failure.

The hazard rating is the basis not only for classifying and registering an individual tailings storage, but also defines the on-going design and operating standards required to provide acceptable levels of safety, minimal environmental impact, and adequate rehabilitation of an individual tailings storage. The design and operating standards are shown in Table 3. Clearly more stringent and technically based standards and procedures are required for those storages with significant or high hazards.

PRACTICAL EXAMPLES

To illustrate the administration of the guidelines two hypothetical practical examples have been selected. The examples are derived from recent projects that have been reviewed by the Department of Minerals and Energy and are typical of the range in tailings storages in operation in Western Australia.

(a) Proposed New Tailings Storage

A proposed new tailings storage is shown in Figure 2. The storage is located on a cross valley position in close proximity to a currently operating open pit and a proposed underground mine to be developed below the pit. The site for the tailings storage encompasses part of a creek system that recharges shallow aquifers supplying local stock water supply.

On the basis of size, environmental impact and safety, the dam is classified as posing a high hazard. As a consequence, the following will be required:

- a proposal be submitted to the Department of Minerals and Energy describing the intended development and supported by a detailed design report prepared by a geotechnical/ engineering specialist detailing the logic of site selection; flood studies; hydrological and hydraulic analyses; water balance studies; details of the geotechnical investigations undertaken; information on construction materials and proposed construction methods;

TABLE 3: Design and operating requirements - mine tailings storages

HAZARD CATEGORIES		
HIGH	SIGNIFICANT	LOW
. DETAILED DESIGN REPORT PREPARED BY GEOTECHNICAL/ ENGINEERING SPECIALIST	. DESIGN REPORT PREPARED BY GEOTECHNICAL/ ENGINEERING SPECIALIST	. COMPLETION OF A NOTICE OF INTENT OR WORKS APPROVAL DOCUMENT FOR NEW OR EXISTING TAILINGS DAM
. COMPLETION OF THE TAILINGS STORAGE DATA SHEET	. COMPLETION OF THE TAILINGS STORAGE DATA SHEET	. COMPLETION OF THE TAILINGS STORAGE DATA SHEET
. CONSTRUCTION SUPERVISED BY GEOTECHNICAL/ ENGINEERING SPECIALIST	. BRIEF CONSTRUCTION REPORT AND AS BUILT DRAWINGS	. INSPECTION REPORT COMPLETED BY A GEOTECHNICAL/ ENGINEERING SPECIALIST AT REHABILITATION PHASE OR EVERY 5 YEARS
. SUBMISSION OF CONSTRUCTION CONTROL AND TESTING RECORDS	. TWO-YEARLY INSPECTION REPORTS BY GEOTECHNICAL/ ENGINEERING SPECIALIST (ALSO TO BE COMPLETED AT COMMENCEMENT OF REHABILITATION)	. ROUTINE MONTHLY TECHNICAL INSPECTION BY SITE PERSONNEL
. DETAILED CONSTRUCTION REPORT PRESENTING VALIDATION OF DESIGN	. ROUTINE FORTNIGHTLY INSPECTION BY SITE PERSONNEL	. PROVISION OF AN EMERGENCY ACTION PLAN
. ANNUAL INSPECTION REPORT BY GEOTECHNICAL/ ENGINEERING SPECIALIST (ALSO TO BE COMPLETED AT COMMENCEMENT OF REHABILITATION)	. PROVISION OF AN EMERGENCY ACTION PLAN	
. ROUTINE WEEKLY INSPECTION BY SITE PERSONNEL		
. PROVISION OF AN EMERGENCY ACTION PLAN		

details of slope stability and seepage analyses; descriptions of operating procedures proposed; instrumentation and monitoring proposals; dambreak studies and proposed emergency action plans; and rehabilitation proposals;

- after assessment by the Department on mining, technical, geotechnical, and environmental grounds as necessary, the proposal is either recommended or rejected; rejected proposals are returned to the proponents for additional study, while recommended proposals are passed to other Government agencies (eg Water Authority of WA, Environmental Protection Authority) as required on a site by site basis

In a project such as this particular example, the following specific aspects would probably require particular attention:

- the storage facility would need to have the embankment construction supervised by a geotechnical/engineering specialist aware of the design concepts; construction control and testing records would have to be submitted along with a construction report presenting a validation of the design;
- it is likely that, due to the location within a recharge area, foundation preparation would be required to minimise seepage potential;
- tailings deposition techniques to maximise water reclaim and tailings densities, and to minimise water ponding would be essential;
- a comprehensive groundwater monitoring/ recovery network would have to be installed, with monitoring of water level and chemistry from the network and the water supply on a frequent, regular timetable with

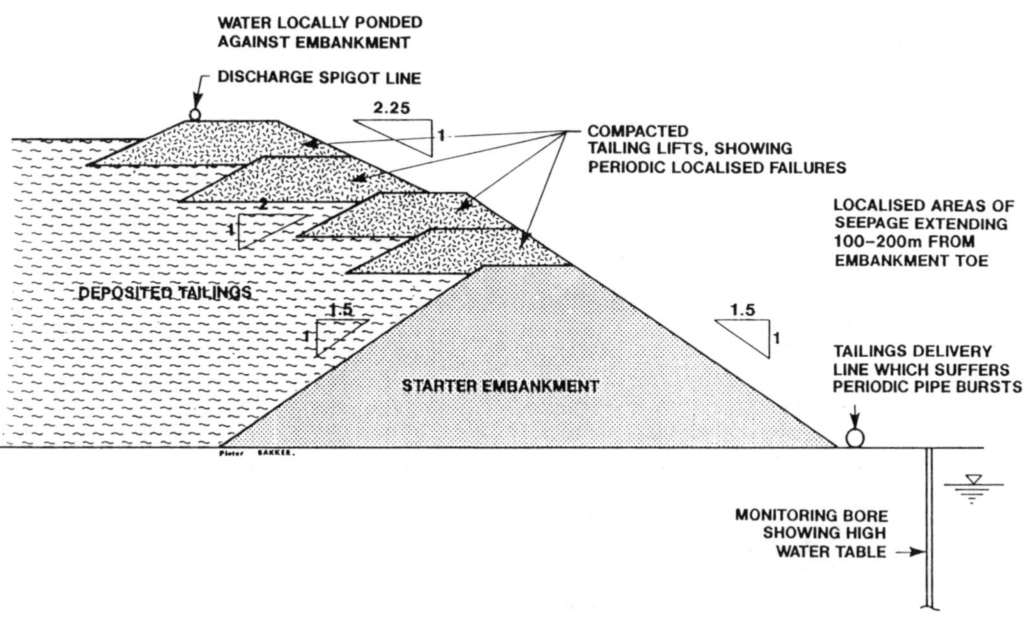

Figure 3. An existing tailings storage

periodic reporting of results;

- a containment bund around the perimeter of the open pit, sized as a result of dambreak studies, to prevent uncontrolled losses of material from entering the open pit and potentially endangering the underground operations;
- identification and sealing of all exploration boreholes and old workings beneath the area of the impoundment;
- prevention of the location of any new underground workings beneath the general area of the tailings dam.

(b) Existing Tailings Storage

An existing tailings storage (Figure 3) is within 2 years of its final completion, with the final addition to the storage being constructed. The storage, constructed with upstream construction methods using tailings, has had a history of problems including:

- pipe breakages leading to uncontrolled tailings leakage;
- extensive areas of high salinity seepage in the surrounding near-surface soil layers, and a rise in the groundwater table in the vicinity of the storage leading to vegetation deaths and distress;
- poor tailings management leading to areas of ponded water close to the embankment walls within the storage;
- localised failures of the embankment.

Because of its size and assessed potential environmental impact, this storage would be classified as a significant hazard site. As a result of the mandatory two yearly technical review and report, the condition of the dam would be defined by the geotechnical/ engineering specialist or consultant. A similar report would be prepared at the commencement of final rehabilitation.

Successful rehabilitation would require some alterations to the present management practices, specifically:

- the installation of a recovery and pumpback system to control and contain surface discharges;
- the installation of a suitable subsurface recovery system to control the development of the groundwater mound beneath the dam; such a recovery system may need to be operated after the completion of mining operations, until the seepage losses could be demonstrated to be decreasing by a steady

reduction in the mound beneath the dam;

- improvement in tailings management practices to ensure that tailings are placed uniformly and in thin layers around the full perimeter of the storage area, to force the pond of stored water away from the embankment walls.

For long-term rehabilitation and decommissioning of the system, the Department of Minerals and Energy requires final outer walls to be 20 degrees or less and covered with an waste rock layer and suitable drainage control to ensure that there is no erosion.

Once adequate strengths have developed within the tailings, the top surface of the tailings would be rehabilitated. Where saline process water has been used, the top surface of the tailings is covered with at least 500 mm of suitable waste rock. Where potable process waters were used, the top surface of the storage can be ripped, seeded, fertilised, and revegetated to prevent erosion.

Topsoil is spread on all external surfaces and where necessary, additional seed and fertiliser applied to establish a self-regenerating cover compatible with the surroundings.

The decant and any underdrainage systems would be decommissioned, sumps refilled and a surface drainage system developed incorporating rock-filled drop structures as necessary to shed rainfall runoff from all external surfaces without causing erosion.

CONCLUSION

The recent performance of mine tailings storages in Western Australia has highlighted the need for a clear regulatory framework within which to legislate the design, operation and rehabilitation of storages. Interim guidelines have been developed with which to regulate industry performance. The guidelines are based on the use of a hazard rating system linked to design and operating standards to achieve acceptable levels of safety, minimal environmental impact and rehabilitation. Initial implementation of the guidelines has been effective with generally positive industry response. The effectiveness of the guidelines will be reviewed over the next twelve months, prior to their finalisation.

ACKNOWLEDGEMENTS

This paper is published with the permission of the Director, Geological Survey of Western Australia.

REFERENCES

ICOLD, 1989. Tailings Dam Safety, Bulletin 74 International Commission on Large Dams, Paris, France.

Ivanov, P.L., Kolpachkova, A.B., & Krunkov, G.T. 1989. Methods to estimate and reduce negative impacts of tailings dams on the environment. Proc. 12 Int. Conf. on Soil Mech. and Fndtn Engng; Rio de Janeiro, p.1877-1880.

Saarela, J. 1989. Dam safety code for tailings and other waste dams in Finland. Proc. 12 Int. Conf. on Soil Mech. and Fndtn Engng, Rio de Janeiro, p.1907-1910.

Environmental Management, Geo-Water & Engineering Aspects, Chowdhury & Sivakumar (eds)
© 1993 Balkema, Rotterdam. ISBN 90 5410 099 0

The effects of sand mining on water quality in a coastal unconfined aquifer in New South Wales, Australia

Stephen R.Jones
D.J.Douglas & Partners Pty Ltd, Newcastle, N.S.W., Australia

Michael J.Thom
D.J.Douglas & Partners Pty Ltd, Sydney, N.S.W., Australia

Francis William Down
Hunter Water Corporation, Newcastle, N.S.W., Australia

ABSTRACT: Sand mining commenced at Tomago, just north of the major industrial city of Newcastle in New South Wales, in 1972. The unconfined aquifer at Tomago in which mining has taken place is also the source of a significant proportion of water supplied to Newcastle and the surrounding districts in summer. As a consequence, the effects of mining on groundwater quality have been closely monitored since mining commenced, indicating that the quality of water does decrease with mining but not to a point where the water becomes untreatable for use as domestic supplies. The main effect is a potential increase in cost to remove additional soluble iron from the water. Research is continuing on the overall effects of mining, remining and deep mining, particularly whether the high iron water passes into the undisturbed cemented sand or whether it is precipitated when these conditions are encountered.

This paper outlines the chemical effects of sand mining on groundwater quality by analysis of monitoring data. The compounding effects observed with second mining and deep mining are discussed. Possible mechanisms of iron solubilisation are presented.

1 INTRODUCTION

This paper briefly reviews the available hydrogeological characteristics of the Tomago Sandbeds, and the observed effects on the aquifer resulting from the sometimes conflicting utilisation of the natural resources.

The area known as the Tomago Sandbeds lies on the northern side of Newcastle Harbour and forms part of a large coastal aquifer system which also includes the North Stockton Sandbeds and the Anna Bay Sandbeds (see Fig 1).

The Tomago Sandbeds is broadly defined by the pump station layout in Fig 1 and comprises an inner barrier dune system deposited during the late Pleistocene age. The area extends over 30 km in length, trending north-east, and is 2 to 5 km in width. The average depth of the aquifer in the developed area is 15 to 20 metres.

The topography of the sandbeds is low to moderate relief, with ground surface levels generally in the range RL 6m (AHD) to RL 10m, with isolated dune systems up to RL 40m: the most notable being Duckhole Dune in the vicinity of the RAAF Radar Station.

The area contains a number of major natural resources including: groundwater which is harvested as one of three main water sources for the Newcastle area; and heavy mineral sands including rutile, zircon, ilmenite and monazite which are extracted predominantly by dredging.

The Hunter Water Corporation (HWC) are trustees of the sandbeds and have been responsible for development of the area as a major source of potable water since 1939. The sandbeds supply up to 30% of Newcastle's water needs during peak demand periods.

Mineral sands extraction has been carried out since 1972 by RZM Pty Ltd within a series of mine leases administered by the Department of Minerals and Energy. Mining takes place under a series of controls with extensive monitoring, which enables the integrity of the groundwater resource to be maintained while allowing extraction of the valuable mineral resource.

The controls are designed to minimise the impact of mining on the aquifer, particularly in terms of water quality and aquifer recharge/yield characteristics.

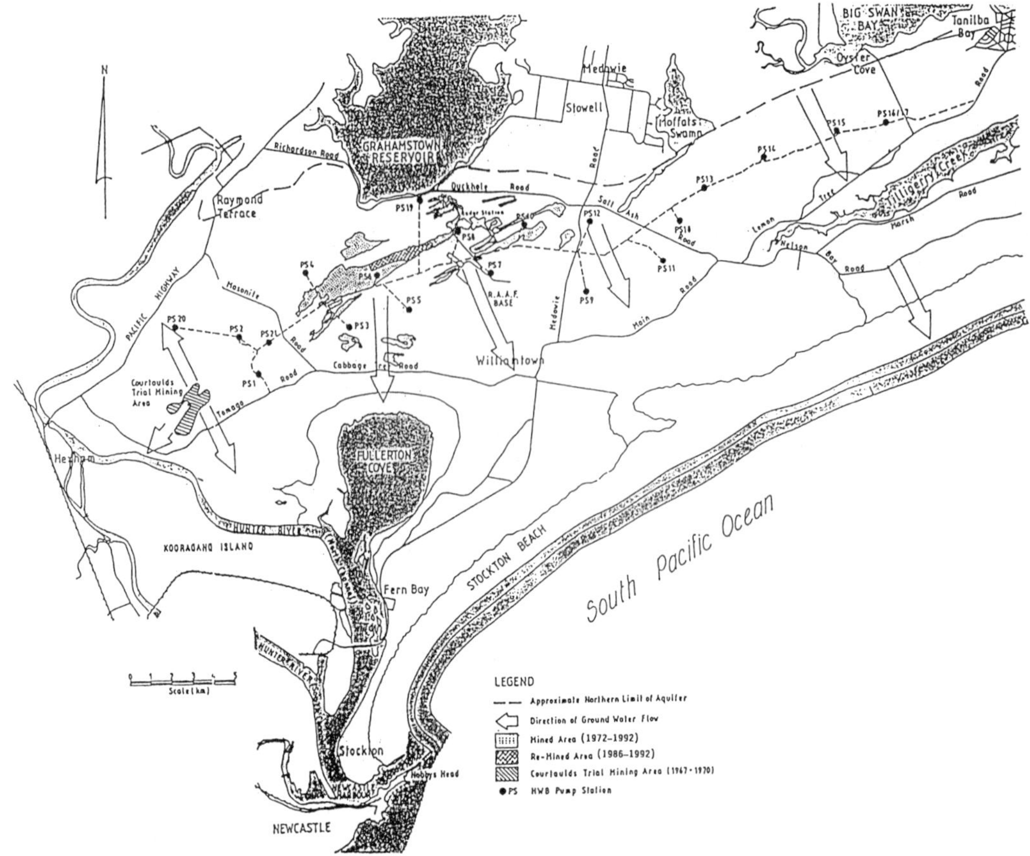

Fig.1 Location of Tomago Sandbeds

To date, approximately 700 ha of the Sandbeds has been mined, of which about 90 ha has been remined. The remined area includes about 20 ha of deep mining to the base of the aquifer.

2 STRATIGRAPHY

The sandbeds may be described in summary as consisting of a medium grained quartzose sand deposit containing occasional clay and indurated sand layers.

From a study of several hundred soil samples (Ref 1) taken throughout the sandbeds, it was found that a convenient division of the strata could be made based on colour as follows:

- upper light coloured sands (ULS)
- dark coloured sand (DS), occasionally cemented or indurated
- lower light coloured sand (LLS)
- grey coloured sand (GS)

The ULS, DS and LLS are of similar particle size with an average D_{10} size ranging from 0.15 mm to 0.16 mm. Mining to date has been generally confined to these units. The lower GS strata has substantially more fines than the above with an average D_{10} size of 0.08 mm.

The sandbeds are underlain by the Tomago Coal Measures and varying thicknesses of clay deposited within the Tomago Basin prior to the deposition of sand, which contains the heavy minerals.

3 WATER QUALITY EFFECTS

The groundwater quality within the mined zone, and surrounding control points, is regularly sampled from an array of more than 250 monitoring locations. Each monitoring location has two or more spear-points at different depths within the aquifer. Test results are stored, collated and analysed by Douglas & Partners using a purpose-written database system.

Monitoring of up to 13 chemical constituents has shown that the main parameter of concern is soluble iron, which must be less than 0.3 mg/l in water that leaves the treatment works in order to meet the National Health and Medical Research Council requirements. Current HWC treatment plant and techniques require that the influent iron concentration should not exceed 5 mg/l. This is achieved by mixing "good" and "bad" quality waters from the various pumping stations.

Four of the monitored parameters are discussion below, namely iron, suplhate, chloride and pH.

3.1 Iron

The natural levels of soluble iron vary widely across the sandbeds and the response in iron levels to mining also varies widely between individual monitoring points. Analysis of the data across the catchment indicates the following general trends in iron concentrations:

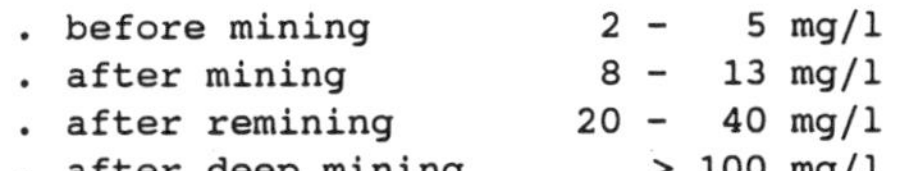

- before mining 2 - 5 mg/l
- after mining 8 - 13 mg/l
- after remining 20 - 40 mg/l
- after deep mining > 100 mg/l

A typical response to both first mining and remining at a single monitoring point is shown in Fig 2 (SK6780). The effect is best understood, however, by observing overall trends.

The ratio of median values before and after mining are plotted for each monitoring site against depth in Fig 3. A clear post-mining increase is observed, occurring predominantly within the tail-ings zone and 1 to 2 m below mining level (pond bottom). This effect appears to persist for many years.

If the median iron concentration of the most recent three readings are compared to the premining median and the ratio of these is plotted against years since mining (Fig 4), there is no clear trend indicating a general return to premining levels. A significant number of sites still indicate iron concentration above premining levels, 15-18 years after mining.

Various mechanisms have been postulated to explain the rises in iron levels following mining (Table 1) and include factors such as soil chemistry, microbiology, leaching of organic acids, and changes in water level (both climatic

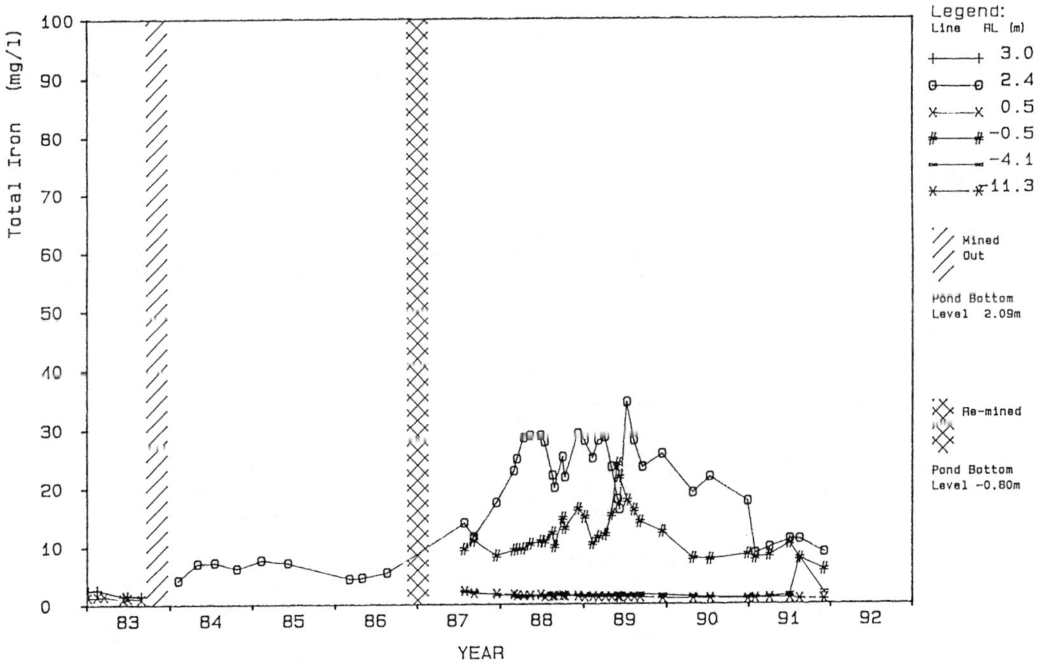

Fig.2 Iron concentrations at spearpoint SK 6780

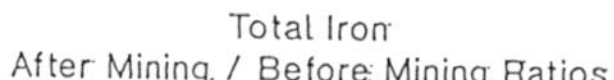

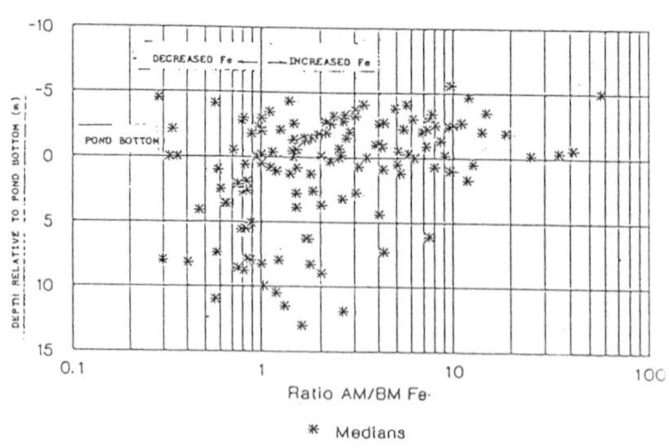

Fig.3 Median iron ratios in groundwater

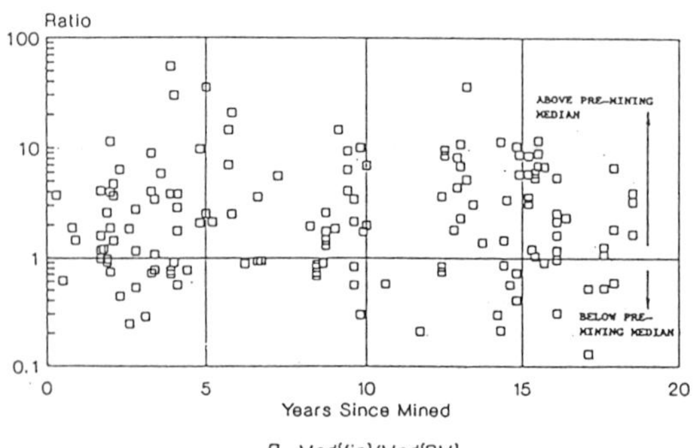

Fig.4 Median iron ratios related to time since mining

Table 1. Possible iron solubilisation

1.	Vadoze zone leaching	Geochemical mechanism supported by various laboratory experiments.
2.	Pond oxygen reaction	Detailed mechanism never postulated but first identified by Soil Mechanics Ltd (ref 1). Suggested mechanisms involve chemical/ biological transformation of iron sulphides to soluble sulphates involving oxygen introduced through the mining pond.
3.	"Slime" layer reaction	Originally observed at Courtaulds area, detailed mechanism never postulated but abundance of iron, organics and anaerobic conditions suggested as significant. Complicated by past removal of fines at Tomago and use of flucculants.
4.	Aquifer base reaction	Occurs where mining strikes weathered bedrock (clay).
5.	Ground controlled reactions	Historically observed by HWC in areas of pumping, always very localised and thought to be related to Eh-pH conditions controlled by local soil chemistry.
6.	Salinity controlled reactions	Occasional reactions which appear to relate to chlorides for unknown reasons. Sulphate often absent.

and pumping induced). Due to the complexity and variability of these reactions, none have been fully explained and research is continuing by various organisations.

It is clear that the range of chemical responses observed at Tomago cannot be explained by one single mechanism and at any given site a combination of mechanisms is in play. Regardless of the cause, the result is potentially greater treatment costs to the HWC and loss of flexibility in mixing low and high quality waters. Annual water extraction fees paid by RZM to HWC are designed to cover the additional costs.

The effect of deep mining to the aquifer base is demonstrated at SK3472 (Fig 5) where spearpoints at all depths show a substantial increase in soluble iron. The reactions caused by both remining and deep mining are thought to principally involve mechanism 2 (Table 1).

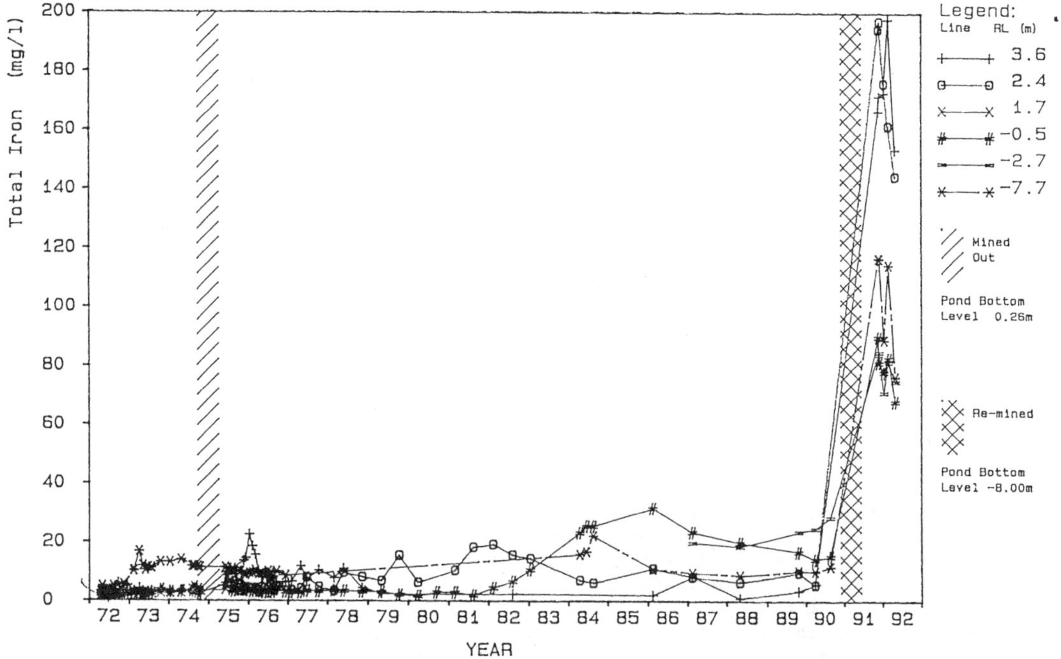

Fig.5 Iron concentrations at SK 3472

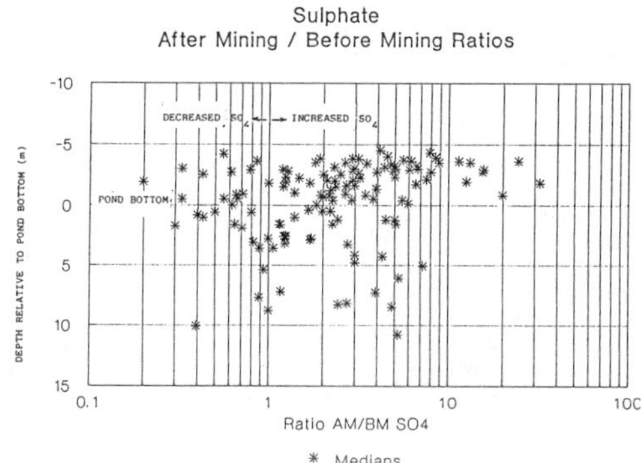

Fig.6 Sulphate concentrations

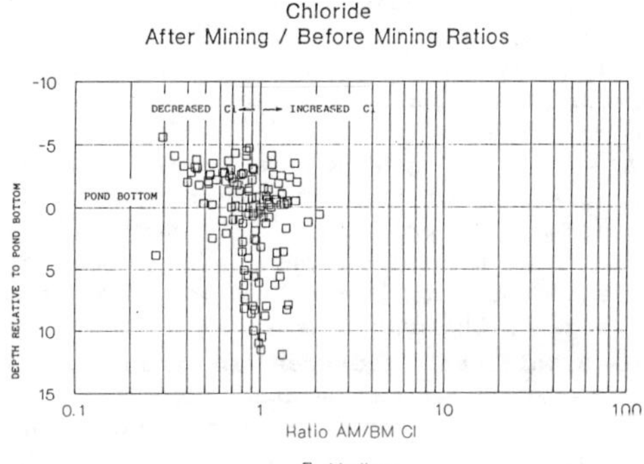

Fig.7 Chloride concentration

3.2 Sulphate

Rises in sulphate (Fig 6) after mining have been previously identified, linked to soluble iron, much of which is in the form of iron sulphate. The median sulphate ratios reflect this connection, showing a definite trend to increase following mining. The increases predominantly occur within the tailings zone, as with iron.

3.3 Chloride

Median chloride ratios (Fig 7) show no apparent strong trends, except perhaps a tendency for increasing chloride with depth after mining. This may be attributed to the influx of fresh water into the top of the aquifer reducing chloride concentration, aided by high permeability in the tailings zone after mining.

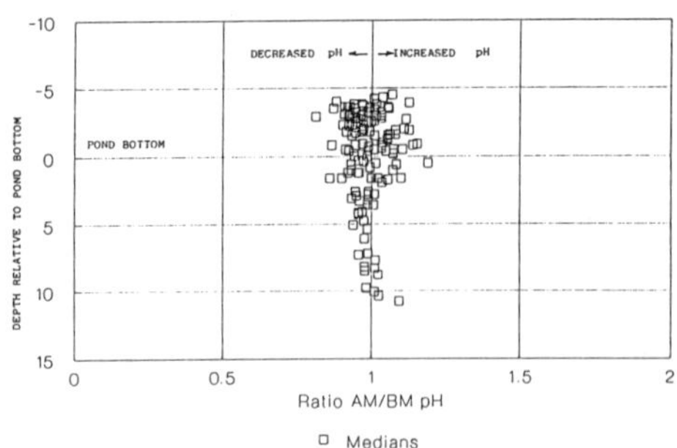

Fig.8 pH ratio

3.4 pH

Median pH ratios (Fig 8) generally show little variation due to mining. At individual spearpoints, a definite trend of pH with depth is apparent, with the near surface water being relatively acidic, and approaching neutrality (pH=7) with increasing depth.

4 MOBILITY OF SOLUBLE IRON

Past efforts to assess whether the soluble iron moved with the groundwater flow, or merely moved in and out of solution in response to ground conditions, had proven inconclusive.

A series of spearpoints were installed down gradient of the deep remining area as mining was in progress. The spearpoints were in ground mined once and placed at distances of 0 m, 10 m, 20 m and 40 m from the edge of the deep remining.

A significant increase in iron levels was observed within months of mining at successive spearpoints and successive depths. This may be partly due to the fact that the soluble iron is moving into mined ground (relatively homogenous, permeable tailings). It has not been conclusively demonstrated (Thom, 1992) whether soluble iron moves significant distances into unmined ground.

REFERENCES

Soil Mechanics Limited (in association with Ercon), 1970: Tomago aquifer study, Newcastle, NSW, Australia. Unpublished.

Viswanathan, M N, 1984:Recharge characteristics of an unconfined aquifer from the rainfall/water table relationship. Jnl of Hydrology, 70, pp 233-250.

D J Douglas & Partners Pty Ltd, 1991. Annual review of mining at Tomago. Unpublished report No 9000-13 to the HWC

D J Douglas & Partners Pty Ltd, 1987: Annual review of mining at Tomago, Unpublished report No 9000-3 to the HWC.

Ercon Australia, 1985: Mineral sands mining at Tomago, NSW. Annual review of mining and monitoring. Unpublished report No 5884/11 to the HWC.

Viswanathan, M N, 1983: Optimum yield of an unconsolidated coastal aquifer, Int Conf on Groundwater and Man, 1983.

Thom, Michael J, 1992: The geochemical and geophysical characteristics of an unconfined coastal aquifer at Tomago, NSW. M.App.Sc Thesis, Centre for Hydrogeology & Groundwater Management, UNSW, unpublished.

Environmental Management, Geo-Water & Engineering Aspects, Chowdhury & Sivakumar (eds)
© 1993 Balkema, Rotterdam. ISBN 90 5410 099 0

Mining coal in water bearing strata in Western Australia

I. Misich & G. Hammond
Western Collieries Ltd, W.A., Australia

A.W. Evans & O.W. Jones
Curtin University of Technology, W.A., Australia

R. MacPherson
Griffin Coal Mining Company Pty Ltd, W.A., Australia

ABSTRACT : Coal extraction in the weak, water saturated Collie Basin sediments is carried out by both open cut and underground methods. Groundwater has historically played a major role in the planning of all operations in the basin and was often the cause of pit closure in the past. The two remaining mining companies, Western Collieries Ltd (wholly owned by Wesfarmers) and the Griffin Coal Mining Company Pty Ltd have developed effective site specific dewatering techniques to give stable and safe mining conditions and a more cost efficient operation.

Recent improvements in extraction techniques by Western Collieries Ltd (WCL) in underground mining have created new groundwater problems as strata collapses above these mines. A research project relating to subsidence prediction in the Collie Basin - jointly funded by WCL and the Minerals and Energy Institute of Western Australia (MERIWA) - though only partially completed, has identified a method by which the potential groundwater inflow can be estimated.

1 INTRODUCTION

Collie Basin coal is ideal for power station feed having a low sulphur and ash content and being non-coking. It is also used in the mineral sands industry, cement works, alumina refining and nickel smelting.

In 1953 the Muja Mine commenced as a combined open cut and underground operation but in 1965 the underground mine was abandoned when an uncontrolled water flow entered the Hebe Seam via a drill hole. The open cut operations were then upgraded to their current status, which now includes another small deposit. WCL have maintained both open cut and underground mining since operations began in 1952.

Groundwater has historically been and will always be the major factor influencing the performance of all mines in the Collie Basin. Current water extraction from the basin in the process of mining is in the order of 50Ml per day.

In open cut mining, groundwater causes problems with :

. Materials handling (spoil)
. Reduced wall stability
. Increased sump and pumping requirements
. Wet blasting
. Increased potential for floor heave and leakage through the pit floor
. Reduced stability of backfill dumps
. Reduced trafficability

In underground mines, groundwater related problems are :

. Reduced roof stability
. Materials handling (equipment)
. Reduced floor stability/trafficking
. Increased sump and pumping requirements

Each mining company has adopted a different approach to these groundwater problems, basically due to site specific characteristics of each mine area. Both approaches are successful and are continuously appraised and improved where necessary.

In 1991 Western Collieries Ltd produced 2.8M tonnes of coal and pumped out over 17M tonnes of water in the process. Griffin mined 2.6M tonnes and extracted 4.2M tonnes of water.

2 LOCATION AND GEOLOGY

The Collie Basin is a geological anomaly in that

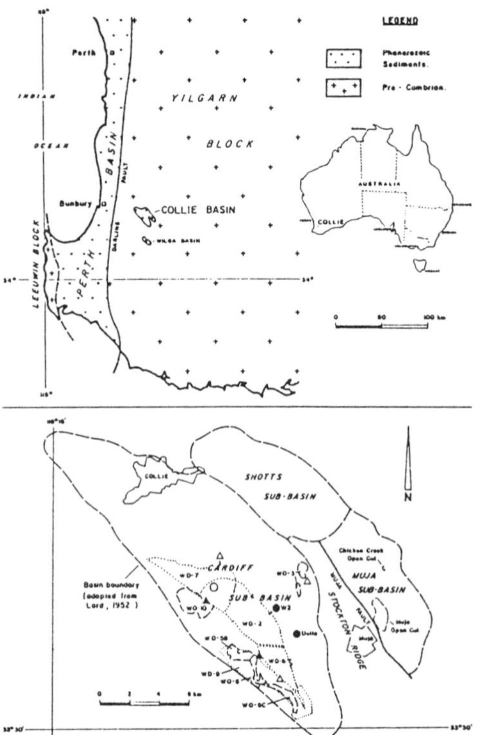

Figure 1. Collie Basin, Location & regional geological setting.

it is a small, isolated Permian Coal Basin preserved within the Archaean basement rocks of the Yilgarn Block (Lord, 1952). The Basin is situated approximately 150km south of Perth in Western Australia and 27km east of the Darling Fault, which forms the western margin of the Yilgarn Block (Figure 1).

The basin is 180 - 230m above sea level and has dimensions of 26km by 15km. It is elongated in a NW-SE direction, parallel to the strike of the major basinal faults, and contains the only commercially exploited coal in the state of Western Australia, with a total annual production of approximately 5.4Mt. The Collie Basin is a bi-lobate, probably fault-controlled basin with a maximum depth of approximately 1,500m (Low, 1958).

Coal seams are generally consistent in their form and thickness across the basin. Faulting within the majority of the currently mined reserves can be considered minimal with only a few localised exceptions where the mine planning strategies are dominated by the presence of faults.

Fully saturated sandstone aquifers separate the coal seams, which act as aquitards, effectively isolating one aquifer from the next.

3 GENERALISED MINING/GROUNDWATER STRATEGIES

Both companies employ shovel/truck methods in their open cut mines. The philosophy behind Griffin mining strategies is to dewater-in-advance all interburden and overburden. This system works well in the "Muja" Sub-basin and allows Griffin to use conventional coal mine blasting and benching techniques for removal of waste material.

WCL only partially dewater the sediments by sumping into the high wall and allowing the water to drain naturally to a sump. As a result, the majority of the operations are carried out on top of the coal seams rather than on the soft interburden and overburden material. Until recently no blasting of overburden and interburden was necessary and ripping has been minimal. Both mining companies use bulldozers in selected areas to doze down batters to loading equipment. Consequently the mining conditions in each mining area, though basically very similar, do vary significantly in some aspects.

Underground mining, until the last decade, has had the philosophy of maintaining stable roof to prevent water inflow into the mines. As a result, apart from localised 'pressure points', there was no planned large scale dewatering.

A project to increase coal recovery was initiated and jointly funded by the National Energy Research Development and Demonstration Programme (NERDDP), WCL and the State Energy Commission (SECWA). The project showed that Collie coal could be successfully mined by the Wongawilli method and techniques used for dewatering the relevant aquifers are being used currently. [Hebblewhite & Humphreys, 1988]

The aquifers are dewatered/depressurised by a combination of conventional production bores and vertical in-pit drainage holes. Water make within the mine is then channelled to sumpage areas and pumped to the surface via sump bores.

It has become increasingly obvious that the cost of pumping water from higher aquifers as the mine deepens would be very high. Consequently a research program was set up to assess the effect of 'total extraction' on the superincumbent strata. The aim of the research was to be able to predict - ahead of mining - the source and volume of water for panels of various dimensions to enable mining, dewatering and pumping strategies to be defined.

4 DEWATERING STRATEGIES

4.1 Muja Open Cut Operations

Nine major seams ranging up to 12m thick (Hebe Seam) are mined together with several minor seams from the Muja coal measures. The seams occur to a maximum depth of 235m below the original ground surface with the current pit floor about 195m deep. The sequence occurs as a small basin some 4km long and 1km wide striking SE to NW and results in an artesian basin configuration with a finite groundwater volume.

Historically it has been the old underground workings which have played the major role in groundwater management. The large open bords are used as collector sumps of groundwater as it flows into the mine via drainage bores.

As mining progresses to the north the available groundwater volume decreases resulting in decreasing abstraction requirements with time. At present in excess of 10,500kL/day is pumped from the Muja Mine to SECWA's Muja Power Station.

There are three main sources of groundwater flow into the Muja Open Cut, all of which require different dewatering strategies, they are :

Basal Nakina Sands. These sands occur in remnant Cretaceous drainages and have been intersected along the western perimeter of the open cut. The sands are predominantly medium to coarse grained and occur as thin layers interbedded with pebble bands, which can yield significant volumes (300 - 600kL/day) of groundwater. Currently these sediments are being dewatered by a conventional borefield with additional drains, low angled bores and sumps located where necessary. At present abstraction from bores into the Nakina Formation totals about 750kL/day

Above Hebe Aquifers. There are six main aquifers identified within the mined sequence which consist mainly of clayey, poorly cemented sandstone, clayey sandstone and minor siltstone and shale interbeds. Thicker shale and coal horizons form aquitards which retard vertical movement of groundwater between aquifers, although faulting and wash outs, do allow vertical connection in localised areas. These aquifers are both unconfined and

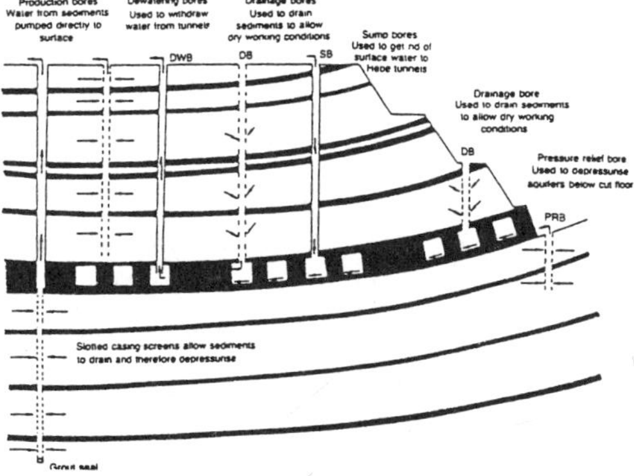

Figure 2. Schematic diagram of production, dewatering, drainage, sump and pressure relief bores.

confined and exhibit intergranular permeability and require dewatering to facilitate mining. An abstraction target of 1,340kL/day has been set currently to ensure that complete dewatering of these aquifers are maintained in advance of mining. Historically, abstraction has been significantly larger than current requirements with up to 8000kL/day being pumped. This target is achieved by a number of methods.

These are :

1. Above Ate seam drainage bores (ADS). These bores basically allow drainage of the shallow aquifers into the more permiable aquifers and are aimed at enhancing 'dewatering' conditions in the early stages of mining block development.

2. Drainage bores (DB). These bores drain groundwater from above-Hebe aquifers into the abandoned Hebe underground workings which is used as a sump. This water is then pumped via pipeline and booster to SECWA for use in the power station. Currently, more than 500kL/day is pumped from these underground workings.

3. Production bores (PB). Production bores dewater and depressurise both above Hebe and/or below Hebe aquifers and pump this discharge via pipelines, booster pumps and large dams to SECWA. In recent years these bores have been located at the lowest point of the basin and on the highwall of developing blocks and are thus shallow and of shorter life than the previous production bores.

Below Hebe Aquifers

To the depth of interest (up to 400m below ground surface and 200m below Hebe), twelve aquifers (numbered 1 to 12) occur below the Hebe seam. These sandstones are well

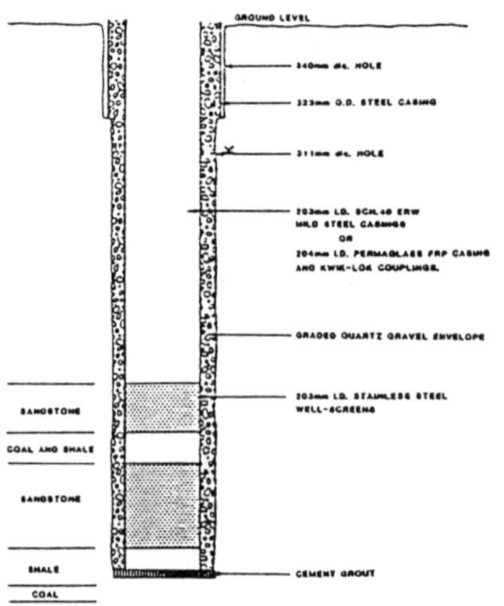

Production bores, standard designs and specifications.

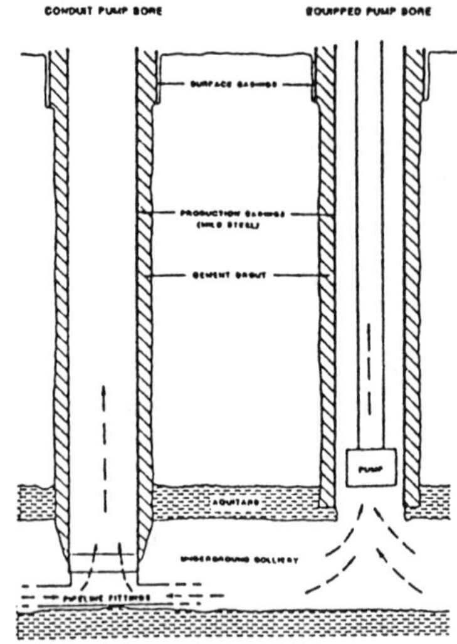

Sump bores, typical design and construction.

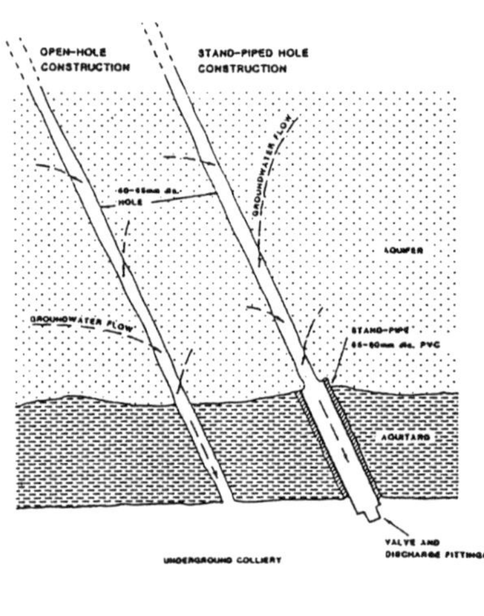

U/G drainage bores, typical design and specifications.

Figure 3.

cemented and much more indurated than the above Hebe aquifers and exhibit intergranules and fracture permeabilities. To prevent heave and leakage in the floor of the open cut, the upper two aquifers require dewatering and the lower 10 require partial depressurisation. The current abstraction target of 6,100kL/day is achieved by the development of pressure relief bores drilled from the lower mining benches or pit floor.

Additional water inflow from surface runoff and seepage is controlled by surface drains and sumps and can produce in excess of 500kL/day.

Currently there is a transition between the present approach of using the underground workings as a reservoir to the more traditional production bore approach as open cut mining of the Hebe underground workings reduces the available storage capacity and the last of the

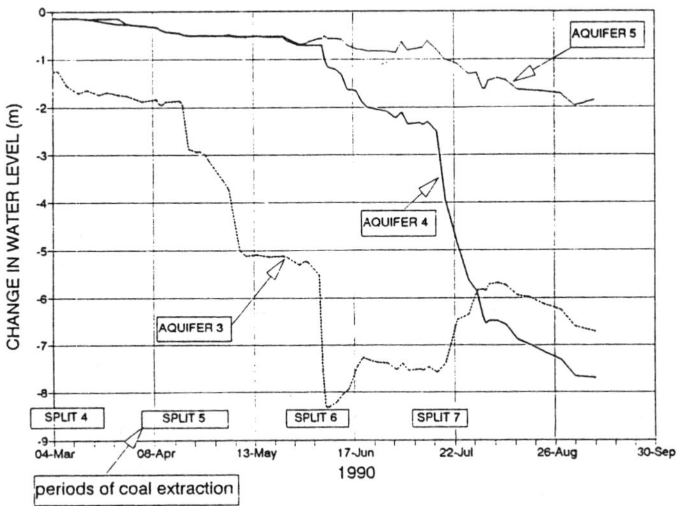

Figure 4. Red panel aquifer levels.

tunnels are mined out (Figure 2).

4.2 *WCL Open Cut Operations*

At WCL no dewatering is attempted in advance of the mining face.

Dewatering is by means of natural drainage at the mining face which results in a relatively steep cone of depression behind the advancing highwall and only partial dewatering of the overburden/interburden material. Water produced at the face gravitates to lower sumpage areas and is pumped out of the immediate mining area by portable diesel operated pumps to larger more permanent sumps equipped with pontoon mounted electric pumps which discharge water to SECWA facilities or the local river system after passing through settling ponds.

Piezometers in the WO5B area have shown relatively low drawdown rates of less than 1m per year with the mining face over 200m away. As mining approaches to within about 150m of piezometers the drawdown rate is observed to increase dramatically to about several metres per month.

At WO5B and WO5F Mines sumps are made in advance and the area is allowed to drain for some time (at least six months) before mining recommences in that area. This has proved particularly useful in very wet horizons where clay content is too high to allow effective borehole drainage.

Volumes averaging 5,500kL/d have been pumped out of the WO5B area in recent years.

4.3 *WCL Underground Operations*

During the last decade WCL underground operations have been dewatered by a combination of conventional production bores, sump bores and underground drainage holes (Figure 3).

Conventional production bores are used to dewater specific aquifers in advance of mining and provide pressure relief for development roadways. Bores are installed at least two years before mining is planned to commence.

Underground drainage holes are generally installed in development roadways surrounding panels which will be mined by 'total extraction' methods. Usually the underground holes dewater and/or depressurise the first two aquifers immediately overlying the coal seam being mined (viz the Wyvern Seam at WD2 and WD6 Collieries) in about six months.

Such holes can also be used to provide localised pressure relief in the roof or floor when required.

Water produced from the underground drainage holes is piped or allowed to flow over the floor to designated sumps where it is pumped to the surface through sump bores. Water due to seepage and the occasional roof fall is removed from the mine in the same way.

4.4 *Prediction of Aquifer/Aquitard Rupture and Water Inflows and Dewatering/Mining Strategies*

Total extraction results in large scale collapse

and fracture of super incumbent strata which ruptures overlying aquifer systems. A research project into the subsidence characteristics of the Collie Basin sediments has been set up, partly to predict the effects of total extraction on aquifers.

The prediction of aquifer/aquitard integrity was based on the hypothesis that the water from upper aquifers could not breach through separating aquitard layers unless the bending strength of the impermeable layer was exceeded and became fractured. (Bending moments and stresses being part of the subsidence process above total extraction panels.

It was not considered important or necessary to consider horizontal changes in permeability, rather to concentrate on vertical fracturing through aquitards which would provide direct access into the mining horizon.

Following a comprehensive research study into the subsidence characteristics of the Collie Basin by WCL, it was concluded that aquitard integrity was dependent on the following parameters:

(i) Effective mining/extraction width and height

(ii) Height of the aquitard above the mining horizon

(iii) Mechanical/bending properties of the aquitard.

A series of nomograms have been developed by Misich in 1990 (see Misich, Evans & Jones, 1991) which predict, for any combination of the above parameters, the likely occurrence of aquitard rupture. Using this information, water inflows can be estimated by using Darcy's Law :

$$Q = KiA, \quad \text{where :}$$

i = the head of water in the aquifer
K = the vertical hydraulic conductivity of fractured aquitard
A = the area of the fractured zone in the aquitard at the base of a particular aquifer.

'Back analysis' of one Wongawilli extraction panel (*viz.* Red Panel) estimated the aquitard at the base of Aquifer 3 to have a vertical hydraulic conductivity of 1m/d over the fracture zone. The aquitard at the base of Aquifer 4 (55m above) had an initial vertical hydraulic conductivity of 0.1m/d which increased to 0.25m/d with increases in extraction area. The aquitard at the base of Aquifer 5 (95m above) had an initial vertical hydraulic conductivity of 0.05m/d which increased to 0.15m/d.

Inflows into Red Panel reached a maximum of about 11,000kL/d following fracturing of the aquitard at the base of Aquifer 5. Figure 4 illustrates the impact of mining on groundwater levels in each of the aquifers.

Parameters derived from Red Panel have been used to predict inflows of over 20,000 kL/d in future proposed 'total extraction' panels.

Such high inflow predictions have resulted in a modification of mine plans, schedules and extraction sequences in an effort to minimise the amount of water that has to be pumped in the short and long term.

Research is continuing in this field to further fine-tune existing nomograms and to better assess the effects of narrow, sequential (panel/pillar) extraction systems. The results from this study will play a very important role in the future of underground mining in the Collie Basin.

5 CONCLUSIONS

Both Griffin and Western Collieries Ltd. have adopted different approaches to open cut mining in water saturated sediments in the Collie Basin.

Conventional borefield dewatering strategies in the Muja pit have greatly improved mining conditions, however development costs are high and blasting is now required to maintain productivity.

In WCL's open cuts, groundwater is drained into advance sumps. This has minimal set-up cost and has ensured that relatively simple digging procedures can be maintained. However, materials handling and trafficability can be difficult in some horizons. Dump stability can also be a problem when mining wet clayey horizons.

Each of these approaches are successful in their own "micro environment", however the cost benefits and disadvantages for dewatering strategies in future mining areas will need to be carefully evaluated.

Underground mining which historically has been limited to bord and pillar workings with only 35-40% extraction may still prove to be the most economic method due to the expense of pumping large volumes of water from total extraction panels. Current research into the subsidence characteristics of the strata is an important element in assessing the most economic method of coal extraction at Collie.

ACKNOWLEDGEMENTS

The authors wish to thank Western Collieries

LTD and Griffin Mining companies for their support. Thanks are also extended to consultant Mr I.Brunner for his assistance at various stages. The views expressed in this report are not necessarily those of the mining companies'.

REFERENCES

Hebblewhite B & Humphreys D. (1988). Underground mining research in the Collie coalfield, western Australia. NERDDP Project No.664.

Lord, H. (1952) Collie mineral field. West Australian Geological Survey, Bulletin No. 105.

Low, G.H. (1952) Collie mineral field. West Australian Geological Survey, Bulletin No. 105.

Misich I, Evans A, & Jones O. (1991) Control and strata investigations to allow total extraction of coal by underground methods in the Collie basin (Western Australia). Proc. 4th Int. Mine Water Congress, Ljublana, Slovenia Sept. 1991.

Environmental Management, Geo-Water & Engineering Aspects, Chowdhury & Sivakumar (eds)
© 1993 Balkema, Rotterdam. ISBN 90 5410 099 0

Experimental laboratory studies of the effect of dewatering on rock stability at depth in the Collie Coal Basin

H.R. Nikraz & M. Press
School of Civil Engineering, Curtin University of Technology, W.A., Australia

ABSTRACT: Deep mining in the Collie Basin has suffered from a high level of water flow and sediment in-rushes throughout its century long history. A major source of these problems is the very extensive system of weak, saturated, sandstone aquifers. As a result, past underground operations had been limited to room and pillar extraction achieving 30-40 percent recovery. The trial introduction of the Wongowilli method of short-wall extraction, increasing recovery to 60 - 70 percent, required an enhanced understanding of strata mechanics to enable confident application of engineering design. Controlled strata deformation was required for safe underground operations, and to limit surface subsidence. The correct modelling of the mechanical properties under conditions prevailing during dewatering was essential to these operations.

This paper describes the research carried out to examine the variation in the mechanical characteristics of saturated sandstone aquifers in the Collie Coal Basin under conditions anticipated during dewatering.

1 INTRODUCTION

Many of the Collie underground mines throughout this century reported problems due to "heavy" types of water often causing temporary mine closures. One of the most serious water-related problems of recent times was the forced permanent closure of the Hebe Colliery in 1965. This was caused by one of the working places exposing a major water-bearing aquifer via a borehole which rapidly flooded the mine. It was never re-opened as an underground mine. An interesting and comprehensive account of the water problems and mine de-watering techniques in the Collie Basin has been given by Jones (1985).

Due to the water problems, underground mining at Collie had been restricted to first workings only, hence the very low overall recovery achievable (around 35% at best). The problem of the water is compounded by the extremely weak nature of the strata overlying the coal seams, particularly the sandstone aquifers themselves. This combination of large quantities of overlying water in very weak strata prevented any extensive form of secondary workings or total extraction for fear of uncontrolled goaf flooding, slurry inrushes and uncontrollable surface subsidence.

In order to increase the recovery to approximately 70%, the Wongawilli method of short-wall mining has been introduced. Extensive aquifer de-watering was carried out to enable this mining method to be applied. The porous and weak nature of the aquifers provides a potential sources of subsidence (due to pore closure), and strata failure (due to increasing the effective stress), as a result of pore pressure reduction upon de-watering. The proposed development of multiple seam extraction below areas sensitive to surface subsidence has increased the need to establish the strata mechanical properties. This will assist in confident application of rock mechanics principles for predictive modelling of strata behaviour.

The importance of a knowledge of the effects of induced stress changes as a result of de-watering a weakly cemented sandstone aquifer has been previously referred to by Kawecki et al., 1988. The two aspects that are of primary interest are (1) deformations produced by changes in the state of stresses of the aquifer as a result of de-watering and (2) an erosional effect due to the flow of water induced by de-watering, which would tend to exacerbate any detrimental effect of pore pressure reduction.

Although seepage and drainage of both soils and hypothetical porous media have been investigated theoretically and experimentally by geotechnical engineers, drainage of poorly-consolidated rock has been neglected. There is little reference in the literature to the basic properties of rock deformation caused by migration of rock particles. Furthermore, most failure theories and experimental work are related to non-porous and low-porosity rock types, while major problems during aquifer de-watering relate to relatively weak formations such as loose sand and/or poorly consolidated sandstone as is the case in the Collie Basin.

In an analytical attempt, author (Nikraz, 1991) has shown that the migration of rock materials in poorly cemented rock are to be expected, when seepage velocity V_s exceeds the critical velocity V_{sc}:

$$V_{cs} = 5.11 \times 10^{-11} \tan \phi \, n^{4.1} \, d^2 \frac{\gamma s - \gamma f}{\gamma f}$$

where

V_{cs} = Critical seepage velocity (m/s)

ϕ = angle of internal friction (degree)

n = porosity (%)

d = average grain size diameter (mm)

γs = specific weight of particle (kg/m^3)

γf = specific weight of fluid (kg/m^3)

However, Nikraz (1991) concluded that the critical seepage velocity calculated in this way is considerably smaller than the actual critical value, because in the layer the grain is compelled to move in a complicated conduit including rising sections and not on a plain surface. The only conclusion which can be drawn from equation (1) is that the critical seepage veloicity is a function of porosity, average particle diameter and friction assuming the specific weight of both water and particle are constant. Thus the investigation of erosional effect due to flow of water induced by dewatering can only be determined by experiments.

This paper contains a description of the equipment commissioned, test techniques, results, analysis and interpretation of the data obtained.

2 GEOLOGY

The Collie Basin is comprised of two unequal lobes in part separated by a fault controlled, basement high known as Stockton Ridge. The basin itself consists of three sub-basins, the Cardiff, Shotts and Muja. The Cardiff sub-basin is the deepest and contain at best 1300m of permian sediments. The Shotts and Muja sub-basins contain 600m and 800m of permian sediments, respectively.

Underground mining occurs in the Cardiff sub-basin which contains Coal measures up to 15m thick. This research based on sediment samples from the Cardiff sub-basin.

The groundwater is contained within the Collie Coal Measures, which constitute a multi-layer of sandstone beds separated by confining beds of shale, mudstone and coal beds. About 75% of the succession are sandstone of which 70% are sufficiently permeable to be regarded as good aquifers. Coal seams and shale bands within the Collie Basin act as aquitards and aquicludes respectively for high permeability zones within sandstone units, which in turn lead to semi-confined or confined aquifer zones.

3 TRIAXIAL EQUIPMENT

The scope of the research required unique triaxial testing of rock. As no commercial system was available, a system with the appropriate capabilities was designed.

An automated data capture system utilising transducers and dynamic recording were designed and commissioned. The overall system was designed to withstand a maximum predicted hydraulic pressure of 14 MPa.

While the equipment developed for use in this and in other associated programmes of research was similar in principle to that described by Bishop and Hankel (1962), a number of refinements had to be introduced because of the markedly higher strength and stiffness characteristics of soft rock which required the use of significantly higher testing pressures.

The system developed for this study consisted of six integrated units :

1. A triaxial cell
2. A confining pressure system
3. A pore pressure system
4. A stiff load frame and servo-controlled loading ram
5. Monitoring equipment for applied loads, water flows, and specimen deformation, and
6. Data capture and display system.

Table 1 - Summary of triaxial and associated tests

CATEGORY	TYPE	RELEVANT PARAMETERS
Modified Triaxial Tests	Pore Pressure Reduction & Water Flow Induction	Permeability, volume change and % weight particles washed out
	Creep	Volume change
Associated Tests	Strength & Deformability	Saturated uniaxial Compressive strength dry and saturated tensile strength, saturated tensile strength after modified triaxial tests, and deformability modulus
	Bulk Properties	Porosity, void ratio Saturated moisture content; and dry density
	Textures	% gravel content, % sand content, % clay content, median diameter sorting and uniformity coefficient
	Ultrasonic Wave Velocity	Compressional wave velocity
	This section	-
	Packer Test	In situ permeability

Full details of the equipment design may be found in Nikraz (1991).

4 MODIFIED TRIAXIAL TEST

The test was designed to simulate in situ strata conditions as it is dewatered. the design approach was influenced by observed in situ deformation which ranged from roof failures to significant strata stabilisation.

The test parameters considered most significant were :

(a) The triaxial and confining stresses
(b) Pore pressure reduction rate
(c) Water head applied to induce flow through the specimen
(d) Test duration

These were standardised for the tests which consisted of six stages, namely :

1. Specimen mounting in triaxial cell
2. Specimen saturation
3. Load application
4. Pore pressure reduction
5. Axial water flow induction
6. Time dependent testing

Only a brief description of test procedures will be presented herein but full details may be found in Nikraz (1991).

Each specimen was photographed prior to testing. The specimens were then placed into the triaxial cell and saturated. Specimen permeability was determined upon completion of the saturation. Any water passing through the specimen was collected for analysis.

The test stresses were increased incrementally to their predetermined values. In the absence of in situ stress measurement or indications of high lateral stress, hydrostatic stress conditions were assumed. Thus, the confining stresses were based on the depth pressure, ie

$$\sigma_1 = \sigma_2 = \sigma_3 = 10^{-6}\ \rho.g.h.\ (MPa) \qquad (2)$$

where σ_1, σ_2 and σ_3 are the major and minor principle stresses

ρ = average saturated strata density (kg/m^3)
g = gravity constant (m/s^2)
h = depth of cover (m)

Once the desired stress conditions had been attained, pore pressure reduction was achieved by opening the fine metering valve to the atmosphere, with the top pore pressure

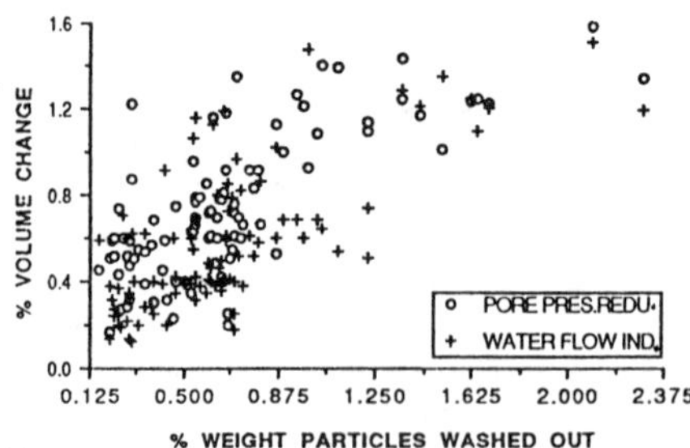

Figure 1. Percentage of volume change vs percentage of weight of particles washed out

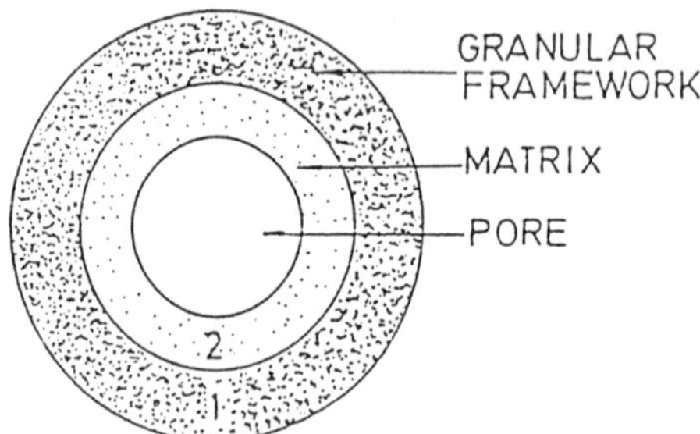

Figure 2 - Conceptual model of cylindrical capillary.

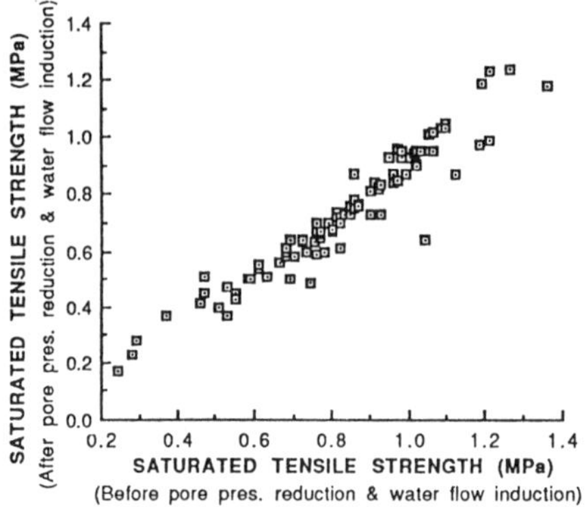

Figure 3 - Saturated tensile strength after pore pressure reduction and water flow induction vs saturated tensile strength.

inlet closed. Any particles washed out with the water were collected.

If no failure occurred, the final volume change associated with pore pressure reduction was noted once the predetermined stress conditions hasd been re-attained.

A head of distilled water was then applied to the top of the specimen for 20 minutes using the specimen saturation system to stimulate the highest predicted insitu head. The water flowing out of the specimen was sampled and analysed for specimen deterioration. The permeability at the start and upon completion of the test was determined, and the final volume change evlauated.

Upon completion of the water flow test, some specimens were continued to be monitored for up to another 35 hours under the same stress conditions, except that no water flow was induced. This provided an indication as observed after 20 minutes, compared to that attained after extended exposure to changes in stress.

Upon removal from the cell, the specimen was again photographed to illustrate the extent and location of any deterioration. Any rock particles remaining on the bottom pedestal and in-line filter were also collected.

In addition, an extensive set of laboratory index tests were performed to supplement the modified triaxial test results (see Table 1). The methodology used to determine the geotechnical properties may be found elsewhere (Nikraz 1991).

5 RESULTS ANALYSIS AND INTERPRETATION

Thirteen specimens failed in total and failure type ranged from a single classic triaxial failure plan, to complete collapse into a elastic, soil-like consistency. Analysis of the parameters monitored suggested that porosity, coefficient of permeability, saturated compression wave velocity, uniformity coefficient, and tensile strength, might be used as indicators of potential failure (Nikraz, 1991; Nikraz and Press; 1992).

Degradation was observed on most specimens. This was quantified by the weight of particles washed out from the specimen, and any that were collected from the bottom pedestal and inline filter upon completion of the tests. Due to the small weight of particles collected, in most cases it was not possible to further quantify the degradation by evaluation of the different size fractions.

Although the origin of the larger particles was readily identified by the cavities at the specimen surface, the origin of the clay sized fraction was more difficult to determine. Migration of the fines through the specimen, or from near the specimen surface could not readily be distinguished.

Most deterioration was at the specimen base, and in some cases extended significantly (1 - 2 cm) up the specimen. this was not classified as failure, although further tests may indicate a need for a change in failure classification.

The location of the majority of the deterioration suggested the washing out of the fines that contributed to the cementing matrix. The possibility of fines migration through the specimen was supported by the deterioration of some specimens at the top. The matrix fines could realistically only have been carried in the direction of water flow. Whether these passed along the entire length of the specimens during testing could not be determined without some means of distinguishing the fines origin.

The percentage weight of particles washed out increased with increasing volume change (see Figure 1). Whether this was as a result of the increased porosity facilitating migration or whether the increased volume change was due to the loss of particle was not clear. The latter was less likely as the weight included the larger particles that remained on the pedestal. Hence, the volume change was attributed primarily to a closure of pore spaces, which was enhanced by a weakening of the specimen, due to loss of fines from the cementing matrix.

A simple conceptual model which explains the permeability behaviour as a result of the pore pressure reduction and the water flow induction, is one in which the quartz grains form a rigid framework which supports the externally applied stress. The pore water channels are surrounded, at least in part, by the matrix materials (clay and various accessory materials). Figure 2 schematically illustrates such a structure. The outer ring represents the quartz skeleton. The pore space, clay and accessory minerals, which are distributed between the quartz grains, are simply shown here as a single pore surrounded by the clay material.

In this model, any increase in confining pressure or reduction in pore water pressure produces a strain in region 1 which in turn induces a stress in region 2 and thus results in a strain at the pore boundary. Consequently, a small radius of pore may result. Furthermore, the seepage force as a result of pore pressure reduction or the water flow induction may result in washing the matrix materials in region 2 away and, as a consequence, a bigger pore radius may be obtained.

As indicated in Table 1, saturated tensile

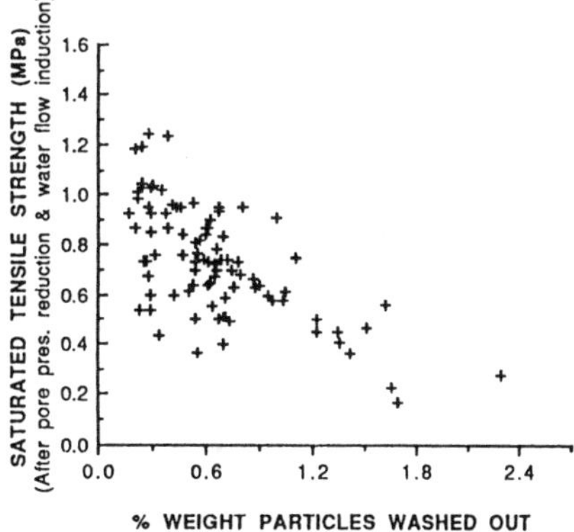

Figure 4 - Saturated tensile strength after pore pressure reduction and water flow induction vs percentage of weight of particles washed out.

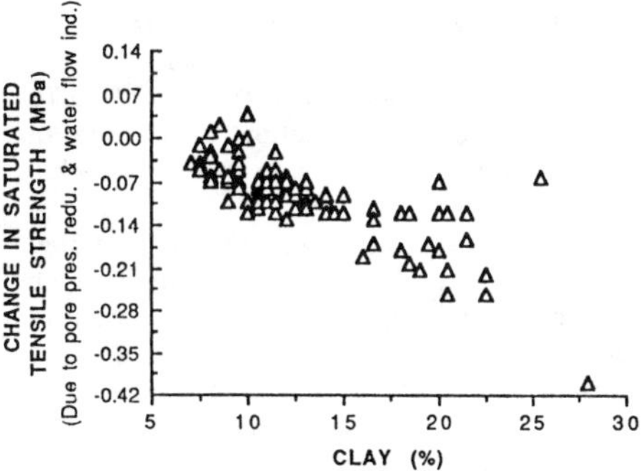

Figure 5 - change in saturated tensile strength due to pore pressure reduction and water flow induction vs percentage clay-sized fraction.

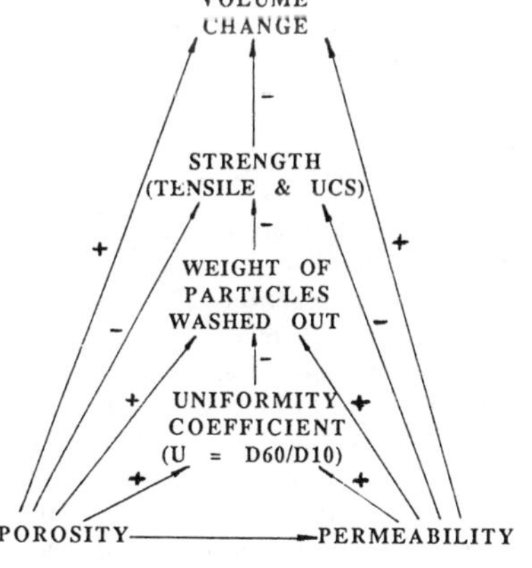

Figure 6 - Effect of causal factors on volume change.

strength was also measured after triaxial tests. Figure 3 shows the relationship between the saturated tensile strength and saturated tensile strength of those specimens that had not failed under triaxial test (pore pressure reduction and water flow induction). There appears to be a weakening of rocks as a result of pore pressure reduction and water flow induction action. On average there is a reduction of rocks as a result of pore pressure reduction and water flow induction action. On average there is a reduction of 12% in saturated tensile strength after the triaxial test compared to the saturated tensile strength before the test. Similarly, a plot of saturated tensile strength at the end of triaxial test against the weight of particles washed out could suggest increased washing out with decreased saturated tensile strength after the triaxial test (see Figure 4). Thus, this relationship could be used to positively support the concept of weakening the rock as a result of particle migration through the specimen.

Moreover, correlation coefficient of 0.76 was observed in the correlation of change in tensile strength with the percentage of clay content (see Figure 5).

This shows that the reduction in saturated tensile strength increases as the percentage clay content increases, which must be used to further support the concept of weakening of the rock as a result of mainly fine particles migration through the specimen.

Finally, the inter-relationship between the volume change and associated parameters can be summed up in the form of a schematic "cause and effect" diagram (after Miller and Kahn, 1962) in which some prior knowledge of the factors exist (see Figure 6). In such a system, the effects of one variable on another are indicated by an arrow : positive correlations are shown by a plus sign, negative ones by a minus sign.

6 CONCLUSIONS

An appropriate triaxial test procedure has been established for the purpose of simulating the in situ reaction of saturated strata to dewatering operations.

Failure of the sandstone was seen to be possible during simulated de-watering. Fourteen percent of the specimens failed and the failure type ranged from the classical single shear plane failure associated with standard triaxial testing, to collapse of the specimen into a clastic, soil-like consistency. Most of the failures could readily be attributed to the pore pressure reduction. Permeability, porosity, compressional wave velocity, uniformity coefficient and tensile strength values have suggested the possibility of their use as indicators of potential failure under given stress conditions.

The rock degradation was observed and quantified by the weight of particles washed out from the specimen. Degradation appeared to depend mainly upon specimen porosity, permeability and uniformity coefficient. For given stress conditions, the possibility of degradation increased with increasing porosity, permeability and uniformity coefficient. The possibility of fines migration through the specimen was supported by the deterioration of some specimens at their upper surface.

The strengthening effect of de-watering was indicated by the saturated tensile strenth test results. Approximately 12 percent reduction in saturated tensile strength recorded due to simulated de-watering. This reduction tended to increase with increase in specimen clay content. Moreover, saturated tensile strength measured at the end of triaxial tests tended to decrease with increase in particles washed out, permeability variation and volume change.

Permeability tended to decrease with pore pressure reduction and increased during water flow induction. Change in permeability as a result of water flow induction correlated well with clay content. This suggested that the washing out of clay and that of silt sized particles attributed to the formation of water conduits through the specimen. A simple conceptual model which explains the permeability behaviour as a result of the pore pressure reduction and the water flow induction has been proposed.

7 REFERENCES

Bishop A W and Henkel, D O, (1962) "The Measurement of Soil Properties in the Triaxial Test", second edition, Edward Arnold, London, p227.

Cistin, J (1965), "Problems of deformation of some Mechanic Filters" (in Czechoslovakian) Vodohospodarski Chasopis, No. 2.

Creager, W P, Juston, J D and Hinds, J (1945), "Engineering for Dams", John Wiley, New York.

Istomina, V S (1957), "Filtration Stability of Soils", Gostroizdat, Moscow, Leningrad.

Jones I O (1986), "Drainage and de-watering of Coal Mines". In Martin C H, Ed, Australian

Coal Mining Practice, pp 468-486.

Kawecki M, Evans A W and Nikraz H, (1988), "The Influence of Pore Pressure Variation on Sandstone Behaviour, and its Contribution to Sandstone and Strata Behaviour Interpretation in the Collie Basin". Western Australian Mining and Petroleum Research Institute - Project 74, Final Report, June 1988.

Lubotohkov, E A (1965), "Graphical and Analytical Methods for Determination of the Properties of Non Cohesive Soils Characterizing suffusion (in Russian). Izvestia VN11G, No. 78.

Miller R C and Kahn J S (1962), "Statistical Analysis in the Geological Sciences". John Wiley and Sons Inc New York, p 483.

Nikraz H R (1991), "Laboratory Evaluation of the Geotechnical Design Characteristics of the Sandstone Aquifers in the Collie Basin, PhD thesis, Curtin University of Technology, p. 317.

Nikraz H R and Press M (1992), "Geotechnical Evaluation of Sandstone Aquifers in the Collie Coal Basin for optimisation of Dewatering Operations. Western Australian Conference on Mining Geomechanics", Western Australian School of Mines, Kalgoorlie, Western Australia, June 1992.

US Army Corps of Engineers (1955), "Drainage and Erosion Control Subsurface Drainage Facilities for Airfields", Engineering Manual, Military Construction, Washington.

Environmental Management, Geo-Water & Engineering Aspects, Chowdhury & Sivakumar (eds)
© 1993 Balkema, Rotterdam. ISBN 90 5410 099 0

Delayed water irruptions: Events and circumstances

O. Sammarco
Italian Bureau of Mines, Ministry of Industry, Italy

ABSTRACT: Those particular irruptions of water or of water and mud that occur with a delay with respect to the onset of the causes that determine them are considered. The circumstances and events responsible for these irruptions, classified on the basis of the initiatory cause, as well as the ensuing effects are analysed. For each type of delayed inrush some occurrences are cited, illustrating the evolution of the events that concurred in causing them, and others that were avoided in time. Controls are proposed by which it is possible to identify the responsible factors in order to eliminate them or counteract them efficiently with specific operations which are also proposed.

1 INTRODUCTION

In reality water irruptions cannot be rigorously subdivided into instantaneous and delayed ones. On the basis of the immediate consequences, each inrush must be considered instantaneous since its effects are manifested in an impulsive way. However, on the basis of the time that elapses between the initiatory cause and the inrush, the latter must be considered instantaneous if the cause is substantially simultaneous with the irruption, and delayed if a sufficiently long period of time elapses between the first cause and the violent effect. Hence, adopting the latter criterion, an irruption due to the yielding of a pillar between a flooded inactive mine and an active one must be considered instantaneous if the yielding depended on the blasting and delayed if it occured by spontaneous and progressive weakening.

Usually the irruptions considered to be delayed on the basis of the second criterion are the ones hardest to foresee as the causes that determine them are negligible at the outset and therefore hard to detect. For this reason it is necessary to study the dynamics of these inrushes in order to better identify circumstances and events that could make prediction less arduous.

The irruptions in question, classified on the basis of the causes that originate them, are subdivided here in delayed irruptions due to:

1 Blockage of drainage holes and pathways along which the water flows into underground.
2 Inflows, even small ones, of water into cavities.
3 Introduction of water in materials that are to be kept dry.

In the following, reference will be made to each of the three types of irruption cited by referring only to the initiatory cause.

2 BLOCKAGE OF DOWNFLOWS

Underground workings may be located near overlying natural water bodies or flooded mines. In such circumstances, if it were possible to conduct the water away from those flooded cavities, through pre-existing fractures or fractures produced by the workings or through specially drilled holes, the hydrostatic pressure could be reduced and the workings could come closer to those cavities.

If, in case of closer workings, the downflow paths later became less conductive or even plugged, the hydrostatic pressure around the workings could increase until it causes them to be flooded. Flooding would occur in the case of pillars yielding due to insufficient resistance to withstand the forces corresponding to the restoration of the earlier hydrostatic pressures or in the case of detachment of rocks around the workings such as to bring the workings into communication with flooded cavities.

The factors which govern the mechanism of such inrushes are:

1 The geometry of the flooded cavities, fractures, joints (Singh 1986) and faults.

2 The hydraulic supply system of these water bodies.

3 The nature and state of the rock around the workings.

4 The ways by which the downflows from the flooded cavities towards the workings are obstructed.

The ways in which downflows can become obstructed are various and complex. The following causes, which may also occur contemporaneously, can determine obstruction:

1 Swelling of clays and cavings inside the flooded cavity so as to impede the regular downflow of the water, or along the drainage pathways.

2 Deposition of materials, admitted into the cavity by the water that reaches it and/or deriving from erosion in the cavity itself and along the drainage ways (Routhier 1963).

3 Precipitations, at times scaling, which depend on the particular chemicophysical conditions (Dagan 1989).

In this regard cases which occurred in underground mines in southern Tuscany are discussed.

The Merse abandoned mine was again flooded after the holes that had allowed it to drain became plugged. The quick emptying of it by means of drainage holes, made near the first holes, and by pumping with a submersible pump lowered into one of its old flooded shafts, probably avoided the flooding of the Campiano mine, situated below the Merse mine and hydraulically connected to it by pathways along the Boccheggiano fault (Sammarco 1986). These passages proved to be completely clogged, but they could be reopened as a result of the increase in hydrostatic pressure upstream from them.

The inrushes of water and mud that occurred, fortunately in the absence of personnel, at the face of the gallery at the -200 m level of the Selvena mercury mine on 17 April 1963 and 27 June 1973 are probably attributable to partial obstruction of downflows. An increase in the water pressure at the face (from 90 to 130 MPa) was encountered before the 1963 irruption and a decrease in the flow rate of the drained water (from 15 to 10 m^{-3}/s) before the 1973 irruption.

Values to be kept check on in order to foresee such inrushes are therefore: the hydrostatic pressure around the workings, measured in all the cavities if they are separate; and the flow-rate at which the water flows through the drainage holes.

In order to avoid any foreseen irruptions one could, by means of boreholes drilled from secure positions, empty the water bodies that have shown themselves to be insidious and/or adequately cement the rocks in the discontinuities around the workings in order to increase their resistance. The latter intervention might precede the digging itself if it is desired to guarantee the stability in advance (Kipko 1984 and Spichsk 1985).

3 WATER INFLOW IN CAVITIES

A cavity which is located above or at the same level as mine workings or as other zones to be protected could be flooded if water were to flow in, even in a discontinuous way and at slow rates.

If this were to happen, the water accumulated in the cavity could flow out violently, even long after the creation of the cavity itself and the start of its flooding, in the case that:

1 A dike of the flooded cavity gives way because it is no longer sufficient to withstand the hydrostatic pressure reached with the increase of the water level or because it is weakened as a result of water infiltrations or thinned by slippages in the submerged part.

2 A wall or an unsubmerged side of the cavity gives way.

3 The roof or part of it collapses, in the case of underground cavities.

In the first case, through the openning created by the collapse of the dike, there would be a water irruption with an energy equal to the potential energy of the mass of water in the flooded cavity with respect to the elevation at which the liquid stops, which may be written as:

$$e \iiint_V h dV \qquad (1)$$

where V is the volume of the portion of the flooded cavity that would dry as a result of the irruption, e is the specific weight of the liquid and h is the difference in elevation between the initial position and that where the elementary particle weighing $e dv$ stops.

In the other two cases, the water overflowing the dike would pour with and energy that depends on: the dimensions of the mass of rock detached; its position with respect to the free surface of the liquid; the difference in elevation between this surface and the level of the overflow; the height from which the water falls after flowing over the dike.

In particular the kinetic energy

$$M_r v_r^2 / 2 \qquad (2)$$

attained by a rock of mass M_r detached from the roof of a flooded cavity which reaches, at the moment of the impact with the water, the velocity v_r, will substantially determine:

1 The formation of a wave which, if sufficiently high, will overflow from the basin.

2 Sprays and the formation of wave motions and whirlpools which will extinguish themselves without participating in the irruption.

3 Oscillations of the water level

4 The rise of the water level in the cavity from the initial elevation to that of the regime that will stabilize after the event, if the volume of the detached mass is greater than that of the water that will overflow.

The water will pour towards the base of the dike where will transfer an energy of

$$K M_r v_r^2 / 2 + g M_w h \qquad (3)$$

where K is a coefficient of reduction, having indicated with $(1-K)M_r v_r^2/2$ the share of kinetic energy of the rock that does not feed the irruption.

An example of irruption determined by the landslide of a side of a flooded valley over three years after the flooding had begun, exceptional for the dimensions of the masses involved, for the amount of energy released, for the rapidity with which the scenario was modified and for the damage produced, is represented by the destructive wave created 9 October 1963 in Valle del Vaiont: an enormous mass of rock, detached from the slopes of Monte Toc, slid into the Vaiont lake, creating a gigantic wave which struck the shores, spilled over the top of the lake dam and raced into the valley below (Selli and Trevisan 1964).

In Figure 1, where the longitudinal profile and the right slope of Vaiont valley are represented, the trend of the wave and the body of the landslide are shown in section.

The events relating to this landslide, the biggest ever occurred in Europe in historical times, took place as follows. A mass of rock-about 300 million m^3, average thickness 157 m and 1.9 km² in area, composed of a succession of strata inclined towards the lake, mainly calcareous and ranging in age between the Malm and the Upper Cretaceous - detached itself above the left shore of the lake and plunged down in just 45 s, reaching a maximum speed of 17 m/s, then partially climbed up the other shore, pushing the water along in front of it. This resulted in the formation of a huge wave that struck the shore opposite that of the landslide, submerging it for over 200 m above the initial level (700.4 m a.s.l.).

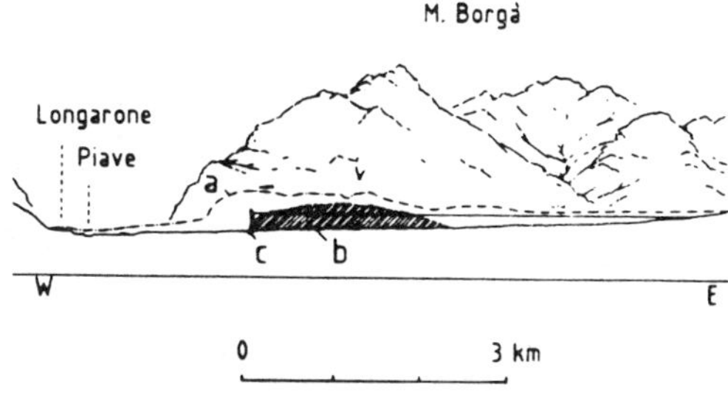

Fig. 1 Right slope of Vaiont valley: profile of the wave (a), section of the body of the landslide (b) and dam (c).

The wave then swept back onto the landslide, towards the side opposite that of the dam and mainly towards it. The dam resisted the impact of the wave but was overswept by it, so that a mass of 25 million m^3 of water irrupted into the Vaiont torrent and the Piave valley, denuding rocks, felling trees and constructions and causing over 1700 deaths.

On the basis of the data supplied by Ciabatti (1964) I have calculated the following approximate values relative to single phases of the event:

1 Energy produced because of the detachment, understood as variation of the potential energy that the mass of debris underwent in descending from the initial position to the final one:

2,000.00 x 10^6MJ

2 Energy transmitted by the landslide to the wave that spilled over the dam:

1.00 x 10^6MJ

3 Kinetic energy acquired by the wave during the fall, assumed free, from the top to the bottom of the dam:

64.00 x 10^6MJ

4 Maximum kinetic power with which the wave irrupted at the bottom of the dam:

3.80 x 10^6MW

It thus emerges: that the energy transmitted by the landslide to the overflow water was 2000 times less than the energy released by the landslide itself; that the energy unleashed by the irruption was over 60 times more than the energy of the spill wave that directly caused it; and that the power of the irruption was incredibly great.

The most significant circumstances that contributed to the catastrophe, the dynamics of which it was possible to reconstruct only after the event, were:

1 The inclination towards the lake of the strata composing the slope that gave way.

2 The rather plastic state of the stratum on which the slide took place.

3 The deep incision, created due to the erosion caused by the Vaiont torrent, which, gradually deepening, reached the abovesaid stratum and made the equilibrium of the slope unstable.

4 The steepness of the flanks on the valley.

5 The existence in the slope of a ground water table, strongly dependent on the level of the water in the lake and on precipitation, which soaked and lubricated the slide surfaces, reduced the internal friction and hence the cohesion of the rock affected by it and exerted greater hydrodynamic pressures on the rock the greater the slope of the ground water table and therefore the lower the level of the lake was.

6 The oscillations of the level of the lake.

The Vaiont landslide was preceded by: displacements of benchmarks, more accentuated during oscillations in the level of the lake and after precipitations, up to 3.22 m in three years and with velocities up to 0.04 m/day; the appearance of fractures and, in the last days prior to the disaster, sinking and inclination of trees; detachments of disarticulated masses and slide movements in concomitance with variations of the lake level. These signals were underestimated, however, because they were considered forewarnings only of small landslides that would only have helped give more stability to the slopes, as had normally occurred for analogous flooded cavities.

In the light of what happened, where analogies might be suspected with the cited case, in-depth controls should be made, especially with continuous monitoring of the parameters of the motion of benchmarks adequately distributed over the largest areas possible.

This would help identify the position, the geometry and the kinetic characteristics of any unstable masses and foresee the effects of their detachment in order to prevent damaging ones.

4 INTRODUCTION OF WATER IN WASTE FILLINGS

There could be a cavity, apparently dry but in reality subject to water infiltrations, filled with sandy, silty or clayey material.

Such material could violently invade nearby places communicating with the cavity if enough water accumulated to fluidify it and allow immediate runoff towards the outside. The seepage of water into cavities represents the initiatory cause and can determine the following events which govern the inrush mechanism.

1 Variations of the mechanical characteristics of the backfilling.

2 Reduction of the frictional resistence between the wall of the cavity and the backfill in correspondence to any wet zones around the cavity itself.

3 Increases in pressure around clayey sectors if they are prevented from swelling.

4 Weighing down of the waste filling due to the water introduced and resulting increase of the pressures, also in correspondence to the cited communications of the cavity with the exterior.

5 Formation of voids in the waste filling itself because of losses of water and/or removal of solid particles, so that if the backfill overlying the voids gave way it could greatly stress the underlying material and favour its expulsion from the cavity.

6 Water seepage through dikes composed of backfilling located in correspondance to the abovesaid communications. These infiltrations, especially since they cause the removal of solid material and/or landslips, cause the dikes to weaken.

Not all of these events are necessary for the inrush to occur, but once they begin they progress over time and concur in causing it, either through gradual or impulsive variations, as in the case of the fall of the overlying waste filling onto the one settled below.

The energy that the backfill conveys to the exterior when it exits from the cavity depends on the amount of waste filling mobilized and its initial altimetric distribution, and the energy consumed before the mud reaches the outside. Other parameters being equal, the destructive effect of an irruption will be greater the shorter the duration of the outflow and in particular the lower the viscosity of the mud, upon which the duration also depends.

An irruption of the type described occurred in the Campiano mine in Tuscany on the night of May 26, 1990 (Sammarco 1990). In September 1989 a raise, no longer utilized and considered completely dry, was backfilled with loose dry material. This material completely filled the raise and at its base, in the underlying gallery, to form a heap with a slope of about 50° to the horizontal. Eight months after backfilling, without any forewarning, the raise was almost totally emptied in a spontaneous outburst. In the form of mud, the material contained in it irrupted violently, invading the gallery opposite the raise and section of gallery leading into it for a total of 110 m. After the event, mud filled the raise only for 10 m and declining slightly from its base displayed an average height of 1.50 m in the flooded zone.

The energy released by the irruption, calculated on the basis of the geometric data and the specific weight of the mud was 520.00 MJ.

The force of the irruption is deducible from the effects discovered: breaking of pipes placed 3 m above the floor of the flooded gallery; displacement of a power loader, weighing 0.3 MN and parked 40 m from the base of the raise, for 25 m.

Without a doubt this irruption was triggered by water infiltrations. Indeed, mud came out of the raise, although dry material had been put into it. Furthermore, it was seen that after the event, when the raise was connected up to the ventilation circuit, the air passing through it underwent an increase in absolute humidity: with thermohygrometric measurements in the ventilation air at the bottom and top of the raise repeated several times, it was seen that very weak water immissions occurred in the raise at a variable flowrate between 10 and 40 mN/s, not directly detectable because of the immediate evaporation (Figure 2). As a result the slowly accumulating water gradually reduced the internal resistance of the material and the frictional resistance between the material and the walls of the raise until it made the self-supporting capacity of the material and the frictional resistance between the material and the walls of the raise until it made the self-supporting capacity of the waste filling and the reactions around it, probably with the concurrence of other unfavourable factors, no longer sufficient to offset the weight taken on by the material in the raise.

After this experience the necessity emerges of subordinating any filling of a cavity with loose material to the results of controls, also indirect ones (Sammarco 1991), which assure the permanent absence of water in the cavity itself. In the case of voids already filled or which must be filled with loose material for which there is no certainty of the absence of infiltrations, in correspondence to the communications with the exterior, structures able to offset the greater pressures that can be hypothesized and to resist any impulsive stresses must be realized.

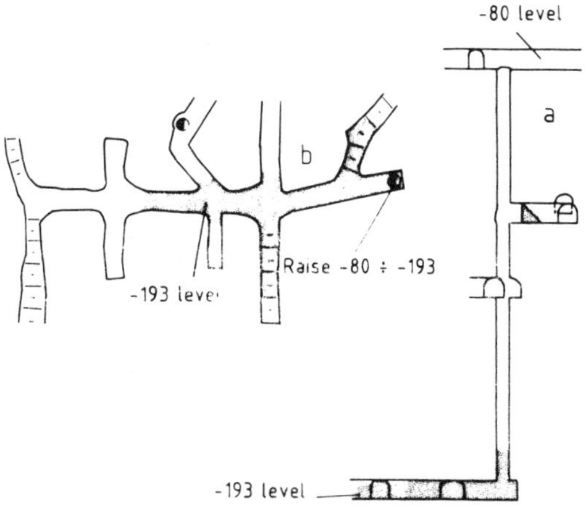

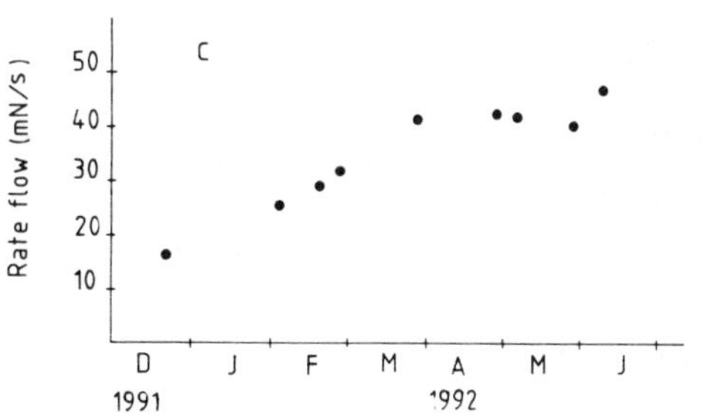

Fig. 2 Sudden, violent emptying of a raise of the Campiano mine: section of the raise (a), invaded zone (b) and trend of the water infiltration flow rate in the raise (c).

These prevention criteria are now in force in the mine where the described irruption took place. In order to obtain adequate pillars experimentation is also being done in this mine with the blasting of the rock around communicating galleries with spaced and delayed fan shots so that the firing will produce blocks which, wedging themselves together, create arch effects without hindering any removal of the water from the cavity.

5 CONCLUSIONS

Delayed irruptions are very insidious

because they occur as a result of situations which are hard to recognize and whose evolution is difficult to foresee without resorting to in-depth controls which unfortunately are not always possible. Detailed knowledge, also obtained through indirect investigations, is needed of places, circumstances and events which might occur to find the initiatory cause of a possible irruption, of the factors that would favour it and the elements which make it possible to evaluate the risk. The greater the risk, the greater the precautions to be taken must be. These precautions consist in acting ahead of time to avoid situations that cause or favour irruptions. In the possibility of preventing irruptions it is necesary to curb them by setting up systems capable of nullifying their destructive effects or attenuating them also by minimizing the damages of the flooding.

The author thanks Enel-Dpt/Deputy Direction for Geothermal Activities of Pisa and Rimin S.P.A. for their kind and active collaboration.

REFERENCES

Ciabatti, M. 1964. La dinamica della frana del Vaiont. Giornale di Geologia, vol. XXXII: 139-154.

Dagan, G. 1989. Flow and transport in porous formation. Berlin: Springer-Verlag.

Kipko, E. Ja, Palozov, Ju. A. and Spichak Ju. N. 1984. Hydrosealing and consolidation of geological faults during tunnel driving. International Journal of Mine Water, Vol. 3, No 3: 35-41.

Kipko, E. Ja. 1985. New developments in the field of grouting using integrated method. Proc. Second International Mine Water Congress, Vol. 1. Granada, Spain: 365-376.

Routhier, R. 1963. Les gisements métallifère, Tome I. Paris: Massom.

Sammarco, O. 1986. Spontaneous inrushes of water in underground mines. International Journal of Mine Water, Vol. 5, No 2: 29-41.

Sammarco, O. 1990. Sull'improvviso e spontaneo svuotamento del fornello 7 nell'unità mineraria Campiano. Italian Bureau of Mines, Internal Report.

Sammarco, O. 1991. Mine water inrushes foreseeble with specific humidity control of the ventilation air. Proc. Fourth International Mine Water Congress, Vol. 1 , Ljubljana, Slovenia-Pörtschach, Austria: 305-311.

Selli, R. and Trevisan, L. 1964. Caratteri ed interpretazioni della frana del Vaiont. Giornale di Geologia, Vol. XXXII: 7-67.

Singh, R.N. 1986. Mine inundations. International Journal of Mine Water, Vol. 5, No 2: 1-28.

Environmental Management, Geo-Water & Engineering Aspects, Chowdhury & Sivakumar (eds)

Restoration of opencast mine sites – A case study

R.N.Singh
Department of Civil and Mining Engineering, University of Wollongong, N.S.W., Australia

ABSTRACT: Opencast mining is an efficient method of coal mining under a variety of conditions. However, environmental consequences of opencast mining should be considered carefully. Restoration and rehabilitation of opencast mine sites is becoming increasingly important. Moreover, the performance of such sites must be monitored in order to improve the efficiency of compaction and restoration methods. In this paper, a case history of backfill compaction studies is briefly presented.

1 INTRODUCTION

Opencast mining is an important sector of coal production which has a vital role in the recovery of shallow low cost coal reserves. In particular, this is an efficient method of coal mining under the following conditions;

i. Large reserves of prime quality coal in seams which are too shallow to be developed by underground mining methods.
ii. Coal reserves sterilised in the seams which have been mined in the past, where the pillars of coal were left *in situ* to support the surface features and provide safety to the working areas. Modern techniques of opencast mining now enable these pillars to be extracted safely and economically, along with small coal reserves previously left unworked because of the lack of market for the coal.
iii Seams less than 0.6m in thickness, which can not be economically worked by underground mining methods. Modern opencast mining techniques permit exploitation of 120-150 mm thick seam in a profitable manner in a multipleseam environment.
iv Coal reserves from an area of industrial dereliction which can be cleared as a part of the opencast mining scheme. The land is restored to a useful purpose with the cost met by the mining operations. In many major mining countries, there were many derelict post mining areas which scarred the landscape for a long period of time. These areas have now been worked by the opencast mining method and restored to productive agricultural land, national parks, nature reserves and recreational areas
v. Production costs per tonne of opencast coal are significantly lower than those of deep mined coal. Thus, the profits earned by the opencast operations are substantial and are of great benefit to the industry and to the national economy.
vi. A substantial amount of opencast coal mined is of high quality, or can represent the major source of special coals with particular characteristics. Examples of these are anthracite, high quality coking coals and low sulphur, low phosphorus and low chlorine coals.
vii Low cost of production makes post -mining environment improvement possible.

Millions of tonnes of coal reserves which could be sterilised by major civil engineering projects such as a major road construction or the development of industrial or residential sites, may now be commissioned once the coal has been extracted by the opencast mining method. The site may require special preparation

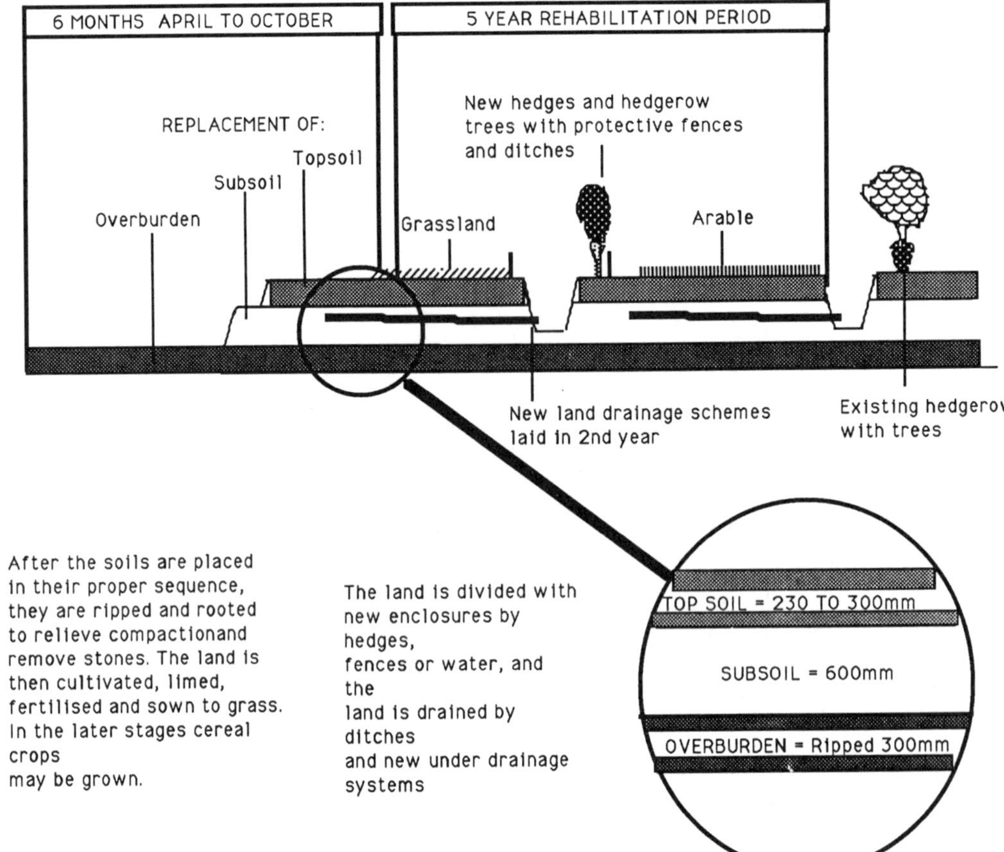

Figure 1. Restoration and rehabilitation scheme for an opencast mine site.

such as suitable compaction prior to the development of the site.

2 ENVIRONMENTAL CONSEQUENCES OF OPENCAST MINES

In the past few decades, environmental concerns have grown from a minor problem to a major obstacle facing the Surface Coal Mining Industry. The problem is accentuated as new mines are being developed in close proximity to external features such as housing areas, roads and industrial developments. The removal of overburden to extract coal deposits not only alters the nature of the near surface strata but also has significant bearing on the ecological balance of the area. Environmental effects of surface mining may be categorised as those experienced during the mining operation, and those experienced during post mining period. Examples of the first category are (a) blast vibrations, (b) air pollution, (c) noise pollution and (d) water pollution. The post mining effects may be exemplified as (a) land degradation and (b) water pollution.

In order to mitigate the surface environmental effects, mining operations are carried out in a planned manner followed by progressive site restoration and rehabilitation.

In the past, most of the restored opencast land was returned to agricultural use, however, in recent years, more consideration is being given to restore the opencast sites for urban construction, including road development, residential sites and industrial buildings. One of the major problems concerning civil engineering construction on the restored opencast sites is the spatial movements of opencast backfill. Until recently, in the U.K., the compaction of backfill followed the Department of Transport's specifications for highway work; however, these specifications are not directly relevant to deep opencast backfills.

3. DEVELOPMENT OF RESTORED OPENCAST MINE SITES:

The post-mining development of opencast mine sites may consist of one or a combination of the following;

- Agriculture, (predominant after-use).
- Recreation and Leisure (Golf Course, Water Sport Centres, Wetlands).
- Development, (Buildings, Roads or other Civil Engineering structures. Industrial or Residential Developments. Forestry).

By far the majority of sites are returned to an agricultural use by a well established procedure, (BJ 1984). Opencast reclamation where possible has been advantageous to the local community as a method of restoring industrially derelict land or as a means of providing recreational facilities.

3.1 Agriculture restoration

The restoration of opencast sites is a major item discussed in the planning permission for any site. When restoration for agricultural use is envisaged, the advice of the Agricultural Department is sought before, during and after sites are developed, on matters such as site restoration and after-care. The site restoration is usually carried out by a contractor. Fig. 1 is a diagrammatic representation of an opencast site restoration and rehabilitation scheme routinely carried out before the site is returned to its owner for agricultural use.

Where possible, progressive restoration takes place during the life of the site, rather than leaving the whole restoration programme until coal extraction has been completed. Once the overburden has been placed to the agreed contours, it is rooted to a minimum depth of 300 mm. The sub-soil is usually spread in two layers to allow free drainage through the layers and into the overburden. Finally, the top soil is placed, cultivated to its full depth, and rooted into the sub-soil. Soil spreading is

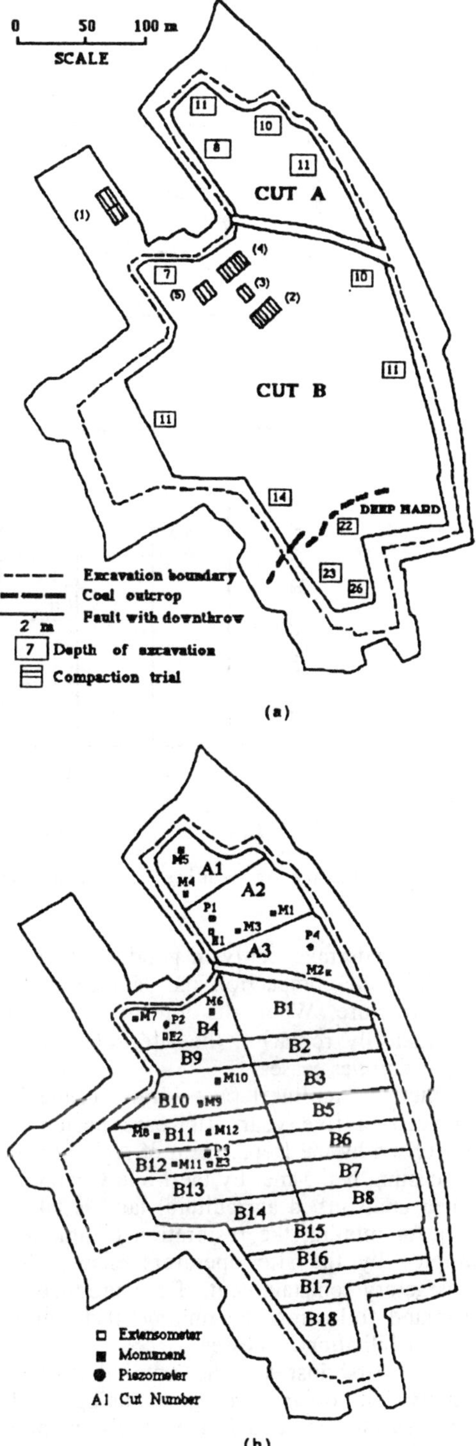

Figure 2. Plan showing lokation of site of investigation.

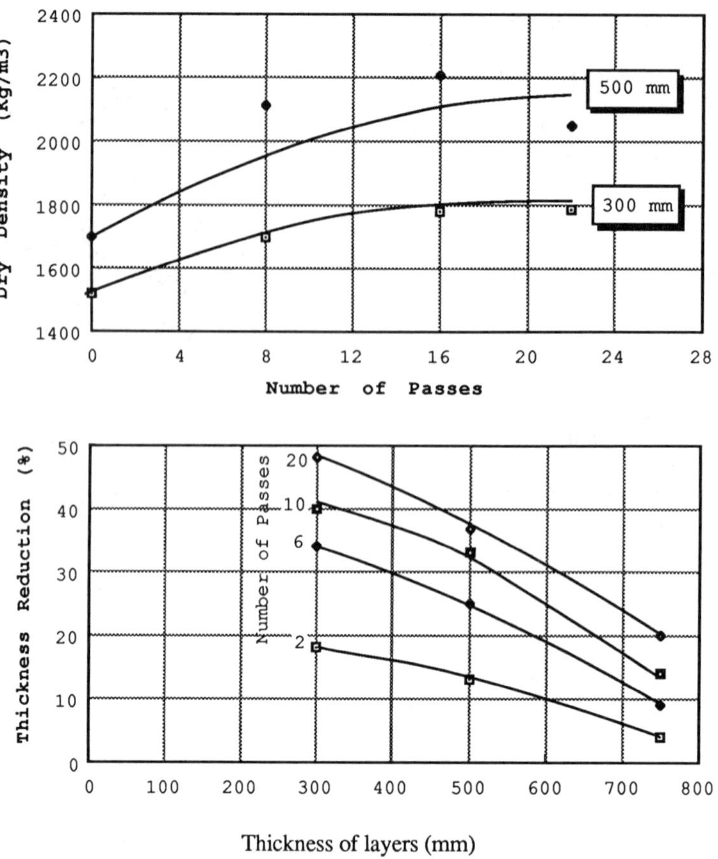

Figure 3. Results of compaction trials, number of passes vs density.

usually undertaken only in good weather to minimise the compaction and damage to the soil structure. When the topsoil has been satisfactorily replaced, the site contractor's responsibilities ceases.

Where agricultural post-mining usage is intended, a five year after-care period is supervised by the Department of Agriculture, the costing being met by the Mine Operator. When undisturbed agricultural land is taken over for mining, the restoration techniques adopted by the site operators ensure the satisfactory reinstatement of the productive farmland following the mining, restoration and rehabilitation processes.

In the first instance the restored land is usually laid down with grass using seed mixtures chosen to suit the environmental conditions, as well as the designated use - grazing, silage, or hay production. Restored soils are frequently acidic and deficient in nitrogen and phosphorus and require generous quantities of these fertilisers and, when available, organic manure.

Careful attention is given to the drainage during the five years of after-care. Open ditches and water courses are created to clear the surface water . An under-drainage system is developed and completed during the second year, following which ploughing and re-seeding takes place. Tree and hedge planting are an important part of restoration schemes. Advance planting often takes place on site boundaries even before work has started on site. Native trees and shrubs are planted and every effort is made to assist the early re-establishment of wild life on the site.

The factors affecting settlement of backfilled sites include material properties, depth of fill, surface infiltration, groundwater recovery, site features and mining methods. Discussion of these factors is outside the scope of the present paper due to space restrictions. The same applies to methods of increasing backfill stability. The remaining part of the paper is, therefore devoted to a case history of backfill compaction studies.

4 A CASE STUDY OF BACKFILL COMPACTION STUDIES AT AN OPENCAST COAL SITE INTENDED FOR INDUSTRIAL DEVELOPMENT

4. 1 Introduction

Investigations were carried out at one opencast coal site designated as Site G in the Midlands in the U.K., where compaction of backfill was included as part of the site operation in the working contract. The site was planned for industrial development and the backfill was instrumented with the aim of supplying information on the behaviour of a mudstone backfill compacted by vibrating rollers according to the Department of Transport specifications. Instrumentation of the backfill comprised extensometers, piezometers and tilt stations.The results of compaction trials necessary for the determination of the loose layer thickness, moisture content and compacting efforts are discussed. The instrumentation scheme was aimed at monitoring the settlement of the backfill, creep, collapse and the recovery of the groundwater within the fill mass.

4.2 Mine Geology

The site is under-lain by strata of Coal Measures age. Two seams are being worked, the Deep Hard Seam and the underlying Piper seam with the Deep Hard being present only in the south of the site. The dip is generally undulating east, and varies from 1 in 13 in the west to about 1 in 50 at the centre of the site. Local dips exceed these values in the proximity of the faults illustrated in Fig. 2(a). Above the Deep Hard, the strata are mainly grey siltstones, which are occasionally sandy and reach a maximum thickness of about 13 m. Between the Deep Hard and Piper seams, the strata has a maximum thickness of 16 m comprising mudstone with seatearths, which at other localities changes into grey siltstone. The siltstone becomes progressively sandy towards the Piper Seam. Underlying the siltstone is a persistent sandstone with an average thickness of 8.9 m which is underlain by siltstone.

4.3 Working Method

The average excavation depth was 13 m with a maximum depth of 27 m. The total volume of the excavation including coal was estimated at around 3,000,365 m^3 of which 315,120 m^3 was planned to be excavated from cut A, the remainder from cut B (Fig. 2(b)). The site was worked with rippers (Cat. D9G's and Kom. 355 A's) and scrapers (Cat. 631C's and Cat. 631D's).

4.3.1 Backfilling and Compaction

(i) Compaction Trials

Following the commencement of excavation work on the site, during the period of June-August 1986, compaction trials were carried out to determine suitable layer thickness, moisture content and degree of compaction required for backfilling of the workings.

(a) Trial Procedure

Five trial areas within the actual site were chosen (Fig.2a). Prior to trials, these areas were prepared by stripping all topsoil, levelling the underlying layer and subsequently compacting it by eight passes of a tamping roller with a weight of 7300 kg per m width of roll. At each area, trial strips 15 m long and 5m wide were prepared and in the first three trials material was evenly spread into each strip in different loose layer thicknesses and different moisture contents. The material was then compacted and the aim was to establish the relationship between passes. In-situ tests to determine bulk and dry density were conducted intermittently between passes. The latter trials were carried out in order to determine the reduction in layer thickness as a result of compaction of the weathered siltstone.

(b) Trial Results

The first three trials were conducted on the weathered siltstone samples which constituted the largest proportion of the material used in

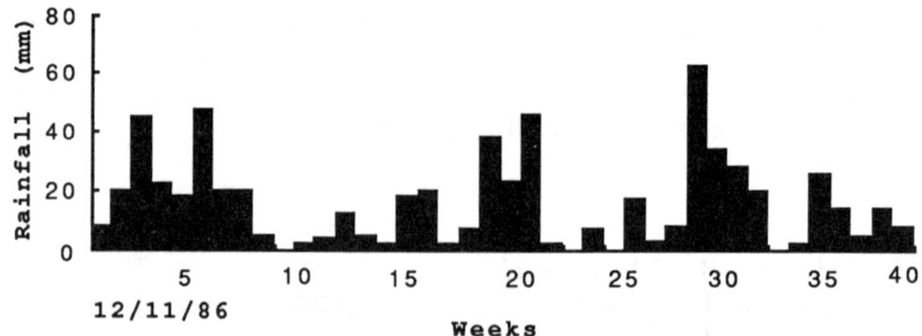

Figure 4. Rainfall record at the site.

the backfilling of Cut A. Trial dry density values were plotted against number of passes for the Bomag BW6 and the Cat. 825 B tamping roller (Fig. 3). As seen in the figure, there is a considerable scatter of results, and no significant additional compaction was obtained when the number of passes was increased to 22 in Trial 3. This can be explained by the fact that the Bomag BW6 possesses an operating weight of 3.5 t/m run, which is below the specification of 5.0 t/m run. The trials suggested that the Bomag BW6 would not be suitable in lifts greater than 500 mm. However, there is a well demonstrated increase in dry density with number of passes when Cat. 825B tamping roller (7.3 t/m run) was employed. The weathered sandstone was utilised in Trial 4, and the results of the dry density values are plotted against the number of passes of Cat. 825B tamping roller in Fig. 3. A significant increase in dry density was obtained with an increase in number of passes. Trial 5 was conducted to investigate the reduction in layer thickness in weathered sandstone following compaction using the Cat. 825B tamping roller (Fig3). On the evaluation of compaction trial results, the recommendation for layer thickness and number of passes for both tamping and vibrating rollers were given by the consulting company. It was also suggested that material with a point load strength less than 1 MPa should be placed in the top two metres. Material stronger than that should be placed towards the base of the excavation. Suitable backfill material was defined as follows:

Moisture contents of optimum $\pm$ 3% and dry density not less than 95% of maximum dry density (BS 1377, 1975).

(ii) Compaction Procedure

When the deposition and compaction of material was started in Cut A, the compaction trial results were not available, therefore Cut A was backfilled and compacted in accordance with the specification given by the mine operator. Cut B was compacted at the loose layer thickness, moisture content and number of passes of specified equipment as determined from the compaction trails. In-situ and laboratory tests were conducted regularly to determine the suitability of compacted material. The moisture content of the

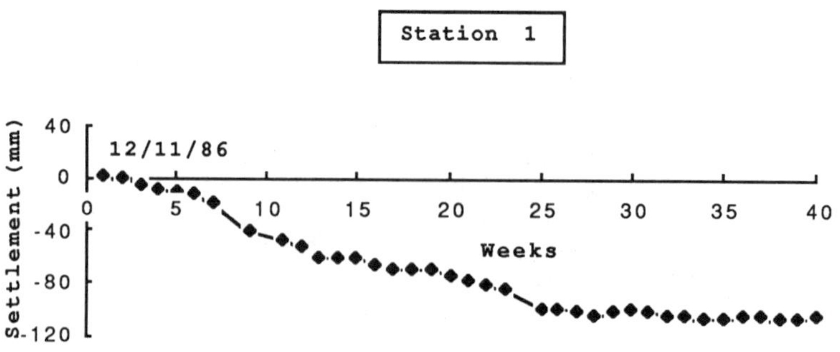

Figure 5a. Settlement monitoring results from station 1.

compacted layers were between 7 and 11 percent and were usually drier than optimum. Dry density values were normally between 90 and 95 percent of the standard Proctor maximum dry density. Occasionally water spraying of material was required to increase moisture content and dry density values.

4.3.2 Instrumentation of Backfill

Immediately following backfilling, the installation of monitoring equipment, including surface monuments, magnetic extensometers and standpipe piezometers, was initiated with the aim of monitoring settlement of the backfill and determining the depth of groundwater within the fill. Eleven surface monuments were installed before September 1987 and by completion of the site operation this has risen to over twenty. The surface monument consisted of a 1 m^3 concrete block on which a metal pin with a rounded top is attached The levels of these stations relative to permanent ground markers outside the zone of influence were determined at weekly intervals. In the same time period, three magnetic extensometers and four standpipe piezometers were installed. The number of extensometers and piezometers eventually increased to ten.

4.4 Backfill Monitoring Results

Following completion of the backfilling and compaction operation in Cut A, five settlement points were installed in November 1986. A period of about two months elapsed between the completion of backfilling and the installation of the stations.

Fig. 4 shows the rainfall record during the period of investigation. The settlement reading results over a forty -week period are presented in Fig. 5. During this period all stations had shown a steady settlement of less than 20 mm with the exception of station 1 where a settlement of more than 100 mm was recorded. This was probably due to higher voids within the backfill in this part of the site. In order to discover the extent of this high settlement zone, a number of additional settlement points consisting of metal bars were installed around Station 1. These points were levelled, but due to insufficient data being available no realistic conclusions can be drawn.

Results from magnetic Extensometer System 1 in Fig. 6 show how movement varies with depth within the fill. It is of interest that during the period between the 12th and 20th week the rate of settlement rapidly increased with time. This was almost certainly a direct effect of the heavy rainfall which occurred during this period (Fig. 4).

Instrumentation of backfill in Cut B was started in March 1987 with the installation of surface monuments. Although the stations have been monitored for only 30 weeks, generally they indicate higher settlement values than in Cut A. Cut B4 was not instrumented and monitored for four months following completion of backfilling. The surface monuments installed in this cut produce varying settlement values. Station 6 realised a maximum settlement of more than 40 mm over thirty five weeks, whereas station 7 located near the site boundary did not show any settlement over the same period. The 35

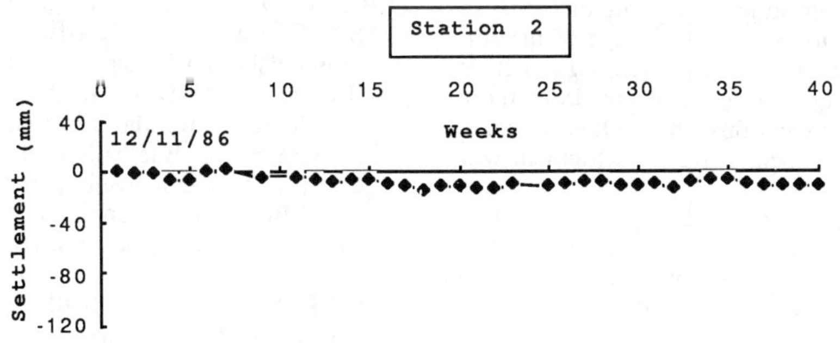

Figure 5b. Settlement monitoring results from station 2.

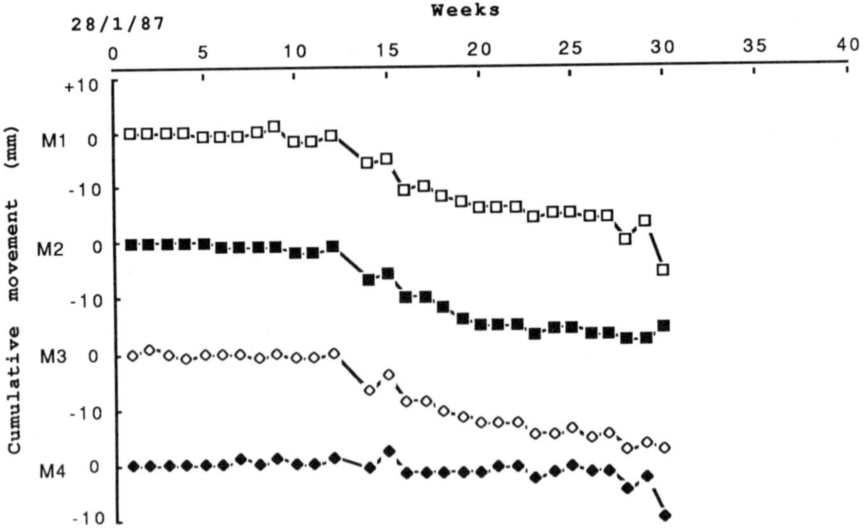

Figure 6. Extensometer no 1 results (Magnet M1=0.63m, M2=1.77m, M3=4.51m, M4=7.57m and datum=10.27m)

cm water rise recorded on Piezometer 2, located close to station 6, may have been the reason for such a high settlement. The cumulative movement of various layers in this cut was also monitored by means of magnetic extensometers.

5. CONCLUSIONS

Backfill material properties and the mining method are the most dominant factors affecting backfill behaviour. Ground preparation technique controls the size and shape of the material produced so that ripper prepared ground is generally smaller and rounder, capable of being handled by scrapers, compared to the much larger sizes produced by blasting. The particle size is, in turn, related to the initial bulkage of the fill, such that the larger the particle sizes within the fill the greater the bulkage. Potentially this leads to a more pronounced breakdown of size and consequently greater settlement will take place.

A case history has been presented in this paper with details of compaction trials and of the monitored settlement with time. Some of the extensometer results are also presented.

6. ACKNOWLEDGEMENTS

The research work reported in this paper was carried out by Dr S. Reed, Dr I. Egretli and Mr F.I. Francis as part of their post-graduate theses from 1985 to 1988, which are gratefully acknowledged. Thanks are due to the British Coal Opencast Executive for providing financial assistance in support of the project. Thanks are also due to various engineers, geologists, managers and site staff within the Opencast Executive for providing continuous help without which the project would not have been successful.

7. REFERENCES

BRENT-JONES (1984), Flagstaff Opencast Site,Geotechnical Certification report, (unpublished), Chesterfield, 32p.

BUIST D. S. AND W.G. DUTCH (1984), "Deflection of M1 Motorway due to settlement of Backfilled opencast site, Strelley, Nottinghamshire, England, Proc. Third International Conference on Ground Movements and Structures, Uwist, Cardiff, 1984.

CHARLES, J.A. (1984), "Settlement of Fills", International Conference on Ground Movements and their Effects on Structures,

P.B. Attwell and R.K. Taylor, Editors, Surrey University Press, pp 26-45

CHARLES, J.A., NAISMITH, W.A. AND BURFORD, D (1977), "Settlement of Backfill at Horsley Restored Opencast Coal Mining Site, Building Research Establishment Current Paper, pp 46-77

CONDON, F.I. (1986), "Restoration of Opencast Mine Sites for Future Development, M Phil Thesis, University of Nottingham (Unpublished)

Environmental Management, Geo-Water & Engineering Aspects, Chowdhury & Sivakumar (eds)
© 1993 Balkema, Rotterdam. ISBN 90 5410 099 0

The assessment of the long-term erosional stability of engineered structures of a proposed mine rehabilitation

G.Willgoose
Department of Civil Engineering and Surveying, University of Newcastle, N.S.W., Australia

S.Riley
The Office of the Supervising Scientist, Alligator Rivers Region Research Institute, Jabiru, N.T., Australia

ABSTRACT: Ranger Uranium Mine (RUM) is contemplating a mine rehabilitation strategy that may involve the permanent storage of mine tailings in the existing ground tailings dam. The Guidelines of the Code of Practice on Management of Radioactive Wastes from the Mining and Milling of Radioactive Ores requires a structural life of containment structures of at least 1000 years. A landscape evolution model using runoff and erosion physics to simulate the changing form of a catchment with time, developed by the first author, is used to assess the long term stability of the proposed containment structures. The model was calibrated using hydrologic and erosion data from field experiments conducted by the Office of Supervising Scientist on the waste rock dumps and natural surfaces at Ranger. The simulated evolution with time of the engineered landforms was studied. The sensitivity to settlement was examined. The implications of the methodology to the design, and the long term stability, of rehabilitation structures at other mine sites are discussed.

1 INTRODUCTION

The Ranger Uranium Mine (RUM) is located 260 km east of Darwin and is surrounded by the World Heritage listed Kakadu National Park (Figure 1). By circa 2012 when RUM orebodies 1 and 3 have been mined out, more than 100 million tonnes of tailings, subeconomic grade ore (BOGUM) and waste rock will require containment. One rehabilitation proposal incorporates the tailings and other waste material into a landform of 4 km^2 that will rise about 17m above the surrounding area (the "above-grade option").

The short-term threat from the rehabilitated landform is the input of sediment to the local fluvial system from the erosion of the waste rock and BOGUM material. In the longer term radioactive materials and heavy metals from the tailings pile may be eroded and dispersed into the local riverine systems thus posing a potential hazard to humans and the ecosystem.

This paper outlines the method used to predict the landforms that will exist after 1000 years of the exposure to the climate, runoff and erosion and presents preliminary results. A computer model developed by the first author that can simulate the evolution of landscapes over time (SIBERIA), is calibrated to existing data and used to predict the future landform.

1.1 SIBERIA - Long term landscape evolution model

SIBERIA (Willgoose, et. al., 1989) is a computer model developed to study the erosional development of catchments and their channel networks. A crucial component of this model is that it explicitly incorporates the interaction between the hillslopes and the growing channel or gully network based on physically observable mechanisms. Elevations changes – both hillslope and channel – are simulated by a mass transport continuity equation applied over geologic time. Mass transport includes fluvial sediment transport, such as modelled by the Einstein–Brown equation, and mass movement mechanisms such as creep, rainsplash or landslide. An explicit differentiation between the processes that act on the hillslopes and in the channels is made. The growth of the channel network is governed by a physically based threshold mechanism, where if a function (called the channel initiation function) is greater than some predetermined threshold then channel head advance occurs. The channel initiation function is primarily dependent on the discharge and slope at that point, and the channel initiation threshold is dependent on the resistance of the catchment to channelisation. The elevations on the hillslopes and the growing channel interact through the different transport processes in each regime and the preferred drainage to the channels that results. The interaction of these processes produces the long–term hill and valley form of the catchment.

The first component of the model, elevation changes within the catchment, are simulated by a mass transport continuity equation applied over geologic time. If more material enters a region than leaves it then the elevations rise, and vice versa. The model averages the mass transport processes in time so that the elevations (and the channel network) are indicative of the average with time of the full range

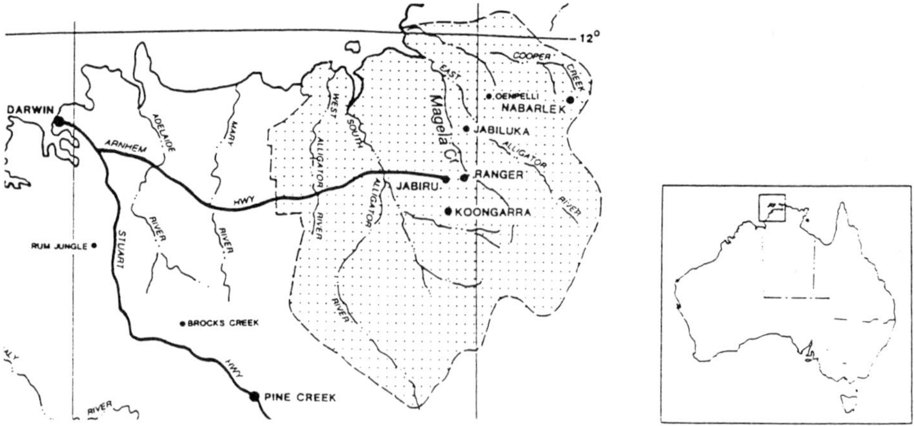

Figure 1. Location map for the Ranger Uranium Mine.

of erosion events; the elevations simulated are average elevations with time. The form of the elevation equation used here (a subset of the total capabilities of SIBERIA) is

$$\frac{\partial z}{\partial t} = \frac{\nabla \cdot q_s}{\rho_s (1-n)} + D \cdot \nabla^2 \cdot z \tag{1}$$

where z is the elevation, t the time, q_s the sediment transport per unit width (mass/time), $\rho_s(1\text{-}n)$ is the bulk density of the sediment and D is the diffusivity of diffusive transport (rainsplash, creep).

The model's second component simulates the advance of the gully heads into the surrounding hillslopes and the dynamics of gully networks on the basis of physical processes. The detailed behaviour of these equations and their implications for catchment geomorphology are discussed elsewhere (Willgoose, et. al. 1991a, b, c, d).

The slope in the fluvial sediment transport equation is determined directly from the catchment elevations and the direction of steepest downhill drainage. The discharge relationship, dependent on area and slope, can be formulated to reflect the processes that occur in the field. However, it is important to note that if the sediment transport equation is to model the long term average sediment transport equation then the discharge per unit width, q, should be interpreted as the mean annual peak discharge, analogous to the idea of a dominant discharge (Willgoose, et. al., 1989), so that

$$Q = \beta_3 A^{m_3} \tag{2}$$

where q is the discharge per unit width, β_3 the runoff rate constant, A is the area per unit width and m_3, n_3 coefficients that are fitted to the data. This empirical relationship accounts for runoff routing effects within the catchment and the spatial correlation of rainfall (Leopold, et. al. 1964; Huang and Willgoose, 1992).

1.2 DISTFW - Hydrology model

The runoff is the most important determinant of soil erosion. To simulate the runoff a digital terrain map (DTM) based rainfall-runoff model (DISTFW) was calibrated to the field data. This model was then used to calculate the runoff required by SIBERIA. The hydrology model used to fit to the rainfall simulator and natural rainfall plots is based on the 1-D kinematic wave flood routing model described by Field and Williams (1987) called the Generalized Kinematic Catchment Model (GKCM). This model has been extended to use DTM data on a square grid: hereafter this new extended model will be called the Distributed Field-Williams Model (DISTFW).

1.3 Erosion model

The overland flow erosion model, one commonly used by geomorphologists and soil scientists (Moore, 1992), is of the general form

$$q_s = \beta'_1 q^{m_1} S^{n_1} (\tau - \tau_c) \tag{3}$$

where q_s is the sediment discharge/unit width, q the discharge/unit width, S the local slope, τ the bottom shear stress for the flow and τ_c a shear stress threshold. The parameters β_1, m_1 and n_1 are fixed by the flow geometry and erosion physics. This equation parameterises the total load; i.e. the sum of both the suspended and bed loads. For instance, for flow in rills the parameters are approximately $m_1 \approx 1.3$ and $n_1 \approx 2.2$ (Moore and Burch, 1986; Willgoose, et. al., 1989). Exact values for these parameters depend on the rill geometry. The parameter β_1 gives the rate of sediment transport and is primarily a function of sediment grain size.

Riley (1992) attempted to identify the shear stress threshold for the material from the waste rock dump

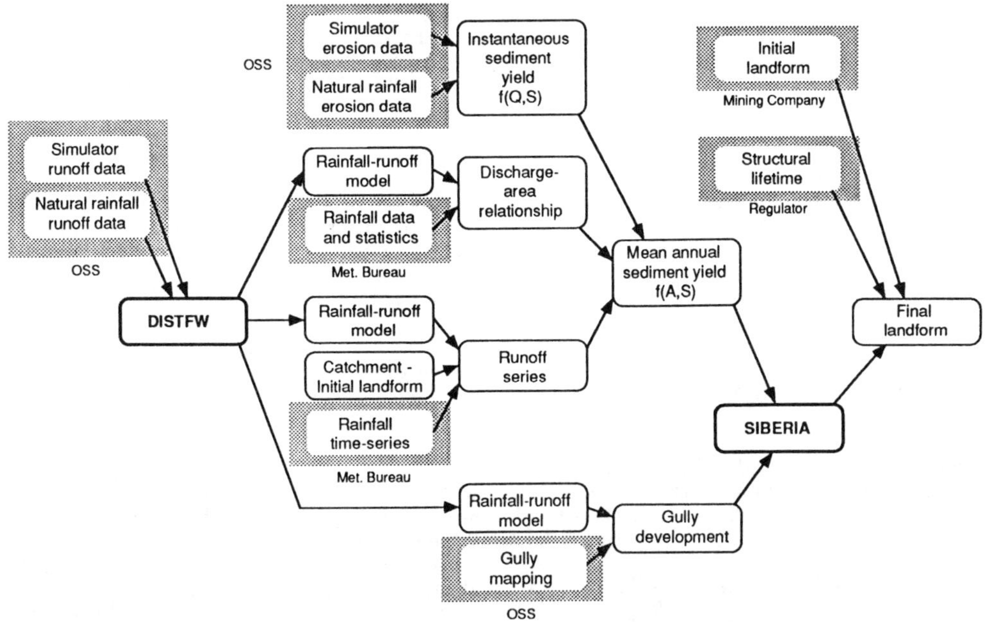

Figure 2. Schemetic of the calibration procedure (greyed regions are data requirements).

using a small flume and concluded that the value was so small that he was unable to reliably estimate it. For smaller areas the overland erosion is dominated by rainsplash (or rain–flow) erosion rather than the fluvial erosion of Equation (3). Rainsplash is generally modelled by an additive Fickian diffusion term where the diffusivity, D, is a function of the applied rainfall energy (in turn a function of the energy of the individual raindrops and the rainfall rate) so that $D = D' R$ where R is the rainfall rate. The total erosion rate is more conveniently expressed in terms of the concentration measured

$$\begin{aligned} c &= \beta_1 q^{m_1-1} S^{n_1} + \frac{D S}{q} \\ &= \beta_1 q^{m_1-1} S^{n_1} + \frac{D' R S}{q} \end{aligned} \tag{4}$$

As the discharge or the area of the plot decreases then the second, diffusive, term will begin to dominate. For large areas the diffusive term is relatively less important. That the diffusive term is additive implies that the processes that cause diffusive transport and those that cause fluvial transport do not interact; i.e. a higher or lower level of diffusive transport does not of itself change the rate of fluvial transport.

2 HYDROLOGY MODEL CALIBRATION

Runoff data from natural rainfall events and rainfall simulation experiments (Riley and East, 1990; Riley 1992) sites on the waste rock dump at RUM were used for the calibration of the rainfall-runoff model. Reliable events were selected for several sites and the model parameters adjusted by trial and error to give a good fit. The broad range of hydrographs available in the data (single and double peaked hydrographs for a number of closely spaced sites) exercised all components of the model.

2.1 *Calibration*

Runoff data from some sites were not used for calibration so that they could be used for independent verification of the calibrated model parameters. A typical calibrated hydrograph is shown in Figure 3. These calibrated parameters were then used to simulate runoff events for independent, verification, storms and the resulting model validation was satisfactory, supporting the validity of the calibrated model. An estimate of Mannings n based on grain size data of surface material from the waste rock of RUM and using empirical relationship between grain size and Mannings n (Henderson, 1966) was in good agreement with the calibrated value.

3 EROSION MODEL CALIBRATION

Sediment transport data for natural rainfall runoff events and rainfall simulator experiments were used for the calibration of the sediment transport model.

The largest (area≈100m^2) simulator catchments were used to calibrate the fluvial sediment transport

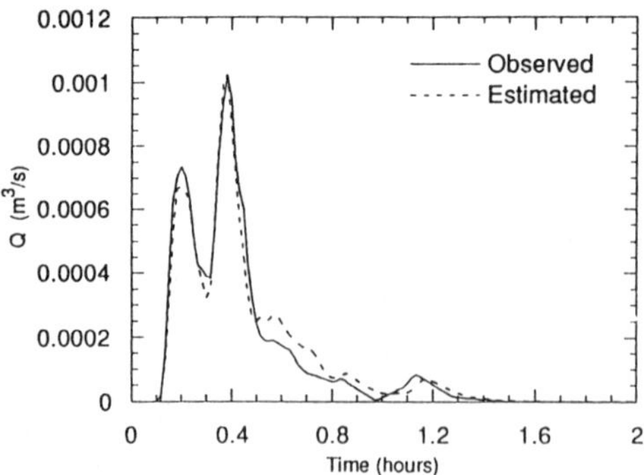

Figure 3. Calibration for CWT2 for the rainfall event of 10/1/91.

equation as they were least dominated by rainsplash and rainflow effects. Multiple regression was then used to directly fit the exponents on discharge and slope. The multiple regression yielded (r^2=0.64)

$$c = 3.59\, q^{0.68}\, S^{0.69} + \frac{0.178\, R\, S}{q} \qquad (5)$$

The exponent on discharge is consistent with other field data. The exponent on the slope in equation (7), however, is somewhat less than that expected for a rilled surface (about 1.5-2). It is possible that the batter (hence higher slope) plots had a coarser lag layer (compared with the cap rock plots) so reducing the transport rate on the higher slope surfaces. The exponent of 1.5–2 is derived using the assumption that the material grading properties do not change with discharge or slope.

The rainsplash diffusivity calibrated above was checked against data from paired small plots (area≈1m^2) one of which was covered to protect it from direct rainslash impact. The difference in transport rates between the plots, the rainsplash transport, was in satisfactory agreement with calibrated diffusivity.

4 DETERMINATION OF PARAMETERS FOR SIBERIA

The schematic of Figure 2 outlines the parameter estimation process for SIBERIA. The procedure can be divided into two stages. The first stage defines how the mean peak discharge varies with area; the spatial dependence of the runoff hydrology. The rainfall-runoff model was used to simulate the variation of discharge with area for rainfall data of Jabiru. The second stage defines how the erosion varies with time; the mean annual sediment yield. The erosion that occurs in runoff events was averaged to give the mean annual sediment yield. A synthetic runoff time-series was simulated, using the rainfall-runoff model and observed pluviograph records for Jabiru. The erosion model was then used to simulate an erosion time-series. The resulting erosion series was then averaged over the simulated record and the average sediment transport rate related to the area and thus the mean peak discharge estimated by the hydrology model.

4.1 Scale dependence of the hydrology

One of the most important parameters in SIBERIA is the relationship between the mean peak discharge and catchment area; it influences both the rate of erosion and gully incision. This parameter was determined from the rainfall-runoff model by using a DTM of the waste rock dump to generate a design 1 in 2 year storm for each point on the DTM. These discharges were related to the catchment area by the fitted relationship

$$q_2 = \alpha\ A^{0.88} \text{ for } r^2 = 0.99 \qquad (6)$$

where q_2 is the 1 in 2 year discharge/30m width (the resolution of the DTM) and A is the area draining through a node (with effective width 30m). The coefficient on the proportionality in this relationship always appears in conjunction with the average erosion rate so is determined in the calculations of the next section.

4.2 Long term erosion rate and timescales for the simulation

Generating a runoff series of sufficient accuracy to determine the average sediment transport rate using the DTM based DISTFW rainfall-runoff model requires considerable computational power. To speedup these calculations a multistage procedure was adopted. The DTM of the waste rock dump was used to generate a hydrograph using a measured rainfall event and the calibrated parameters. A lower resolution rainfall-runoff model for the catchment was then fitted to this hydrograph and this lower resolution model used to generate the 20 year runoff record based on the Jabiru 30 minute pluviograph data at 5 minute resolution. This runoff time series was used to simulate an erosion time series using the adopted erosion model in equation (7). This erosion series was then averaged and the average rate of erosion used as input to SIBERIA.

5 ASSESSMENT OF PROPOSED RUM LANDFORMS

Erosion of the rehabilitation landform was simulated for the equivalent of 1000 years using the parameters calibrated in the previous sections (Figure 4). There is significant erosion both on the caprock layer of the waste rock dump and around the outside batters. The waste rock dump is almost 20m high and the maximum valley depth at 1000 years on the cap rock is 7.7m with the maximum deposition being 6.1m and on the batters erosion is in the range 3-7m (Figure 5). Depths of erosion at 500 years are about

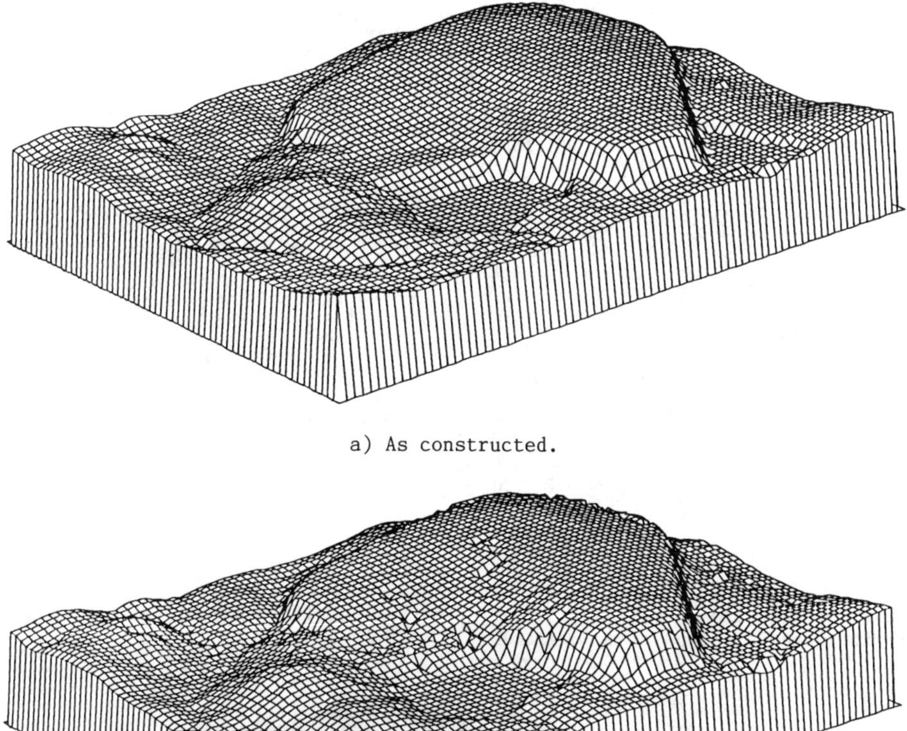

a) As constructed.

b) 1000 years after construction.

Figure 4. Erosion on the proposed containment structure (displayed on 60m grid).

75% that of the 1000 year levels. The erosion on the cap rock layer is localised to valleys while the batter erosion is more widespread. It is almost possible to define the extent of the waste rock dump from the peaks of erosion around the outside of the batters.

Simulations imposing random settlements arising from spoil settling (Figure 6) suggested that though the exact spatial pattern of erosion varied depending upon the initial pattern of settlement.

None of these simulations involve the use of gully erosion capabilities of SIBERIA and solely simulate the erosion as a result of sheet erosion mechanisms.

6 DISCUSSION AND CONCLUSIONS

This paper has demonstrated how a model that simulates the evolution of landscapes (SIBERIA) and their hydrology (DISTFW) can be used to assess the long-term stability of an engineered landform. Using the procedure outlined in this paper it has been shown to be feasible to calibrate SIBERIA and DISTFW using field data, predict minesite runoff and erosion rates and simulate how the minesite geomorphology may evolve in future times.

A caveat is that SIBERIA, at this time, cannot explicitly model the development of soil profiles and vegetation changes that may occur over the 1000 years for which the RUM landforms were assessed. SIBERIA can, however, model the sensitivity to such changes if there is some knowledge about how these might change. Field work to provide this type of data is currently in progress.

Despite this caveat the digital terrain basis of SIBERIA and DISTFW facilitates their integration into existing digital terrain based mine-planning packages so that erosion and runoff assessment can be be done routinely as a part of the design of proposed rehabilitation landforms at the mine planning stage. The first author is currently involved

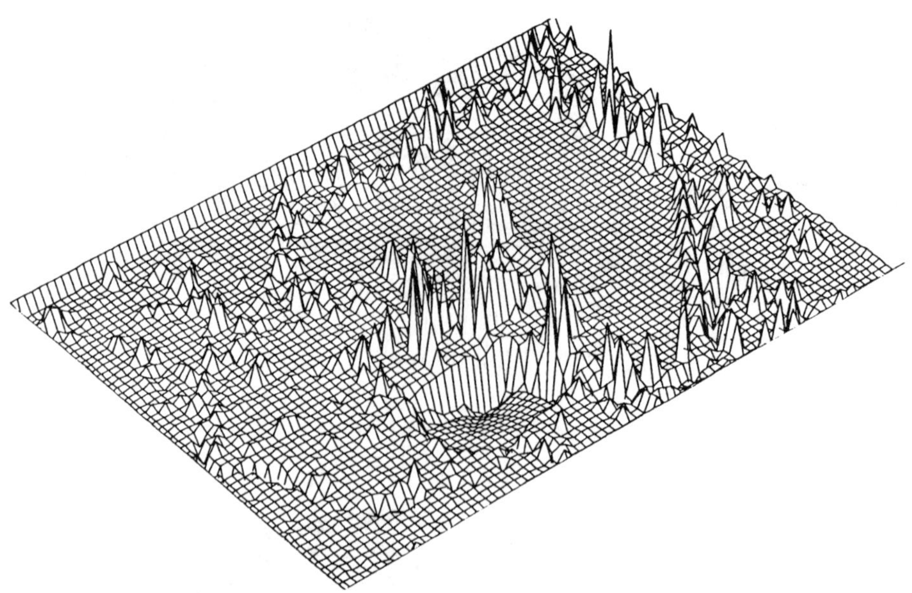

Figure 5. Elevation differences between initial and 1000 year elevations showing erosion upwards, deposition downwards.

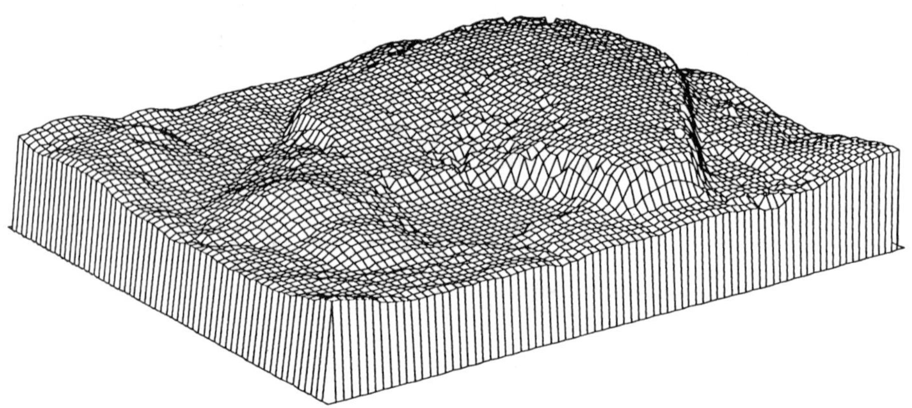

Figure 6. Erosion on the proposed containment structure with spatially random 1m settlements (displayed on 60m grid).

in such project for the Queensland Coal Association.

7 ACKNOWLEDGMENTS

The assistance of field workers of the Geomorphology Group of The Office of the Supervising Scientist, Alligator Rivers Research Institute, Jabiru is appreciated. Computational work was carried out using resources of the School of Engineering, The University of Newcastle, NSW.

8 REFERENCES

Field W G and Williams B J. 1987. "A generalized kinematic catchment model", *Water Resources Research*, 23(8), pp 1693-1696.

Henderson F M. 1966. *Open channel flow*, MacMillan, New York.

Huang H Q and Willgoose G R. 1992. *Numerical analyses of relations between basin hydrology, geomorphology and scale*, Research Report No 075.04.1992, The University of Newcastle, Department of Civil Engineering and Surveying.

Leopold L B, Wolman M G and Miller J P. 1964. *Fluvial processes in geomorphology*, Freeman, London.

Moore I D and Burch G J. 1986. "Sediment transport capacity of sheet and rill flow : Application of unit stream power theory", *Water Resources Research*, 22, (8), pp 1350 - 1360.

Riley S J. 1992. *Small scale studies of the erodability of Ranger waste rock dump*, Internal Report 64, Supervising Scientist for the Alligator Rivers Region, Sydney.

Riley S J and East T J. 1990. *Investigation of the erosional stability of waste rock dumps under simulated rainfall: A proposal*, Internal Report of The Office of the Supervising Scientist.

Willgoose G R, Bras R L and Rodriguez-Iturbe I. 1989. *A physically based channel network and catchment evolution model*, TR 322, Ralph M. Parsons Laboratory, Dept. of Civil Engineering, MIT, Boston, MA.

Willgoose G R, Bras R L and Rodriguez-Iturbe I. 1991a. "Physically based coupled network growth and hillslope evolution model: 1 Theory", *Water Resources Research*, 27(7), pp 1671-1684.

Willgoose G R, Bras R L and Rodriguez-Iturbe I. 1991b. "A physically based coupled network growth and hillslope evolution model: 2 Applications", *Water Resources Research*, 27(7), pp 1685-1696.

Willgoose G R, Bras R L and Rodriguez-Iturbe I. 1991c. "A physical explanation of an observed link area-slope relationship", *Water Resources Research*, 27(7), pp 1697-1702.

Willgoose G R, Bras R L and Rodriguez-Iturbe I. 1991d. "Results from a new model of river basin evolution", *Earth Surface Processes and Landforms*, 16, pp 237-254

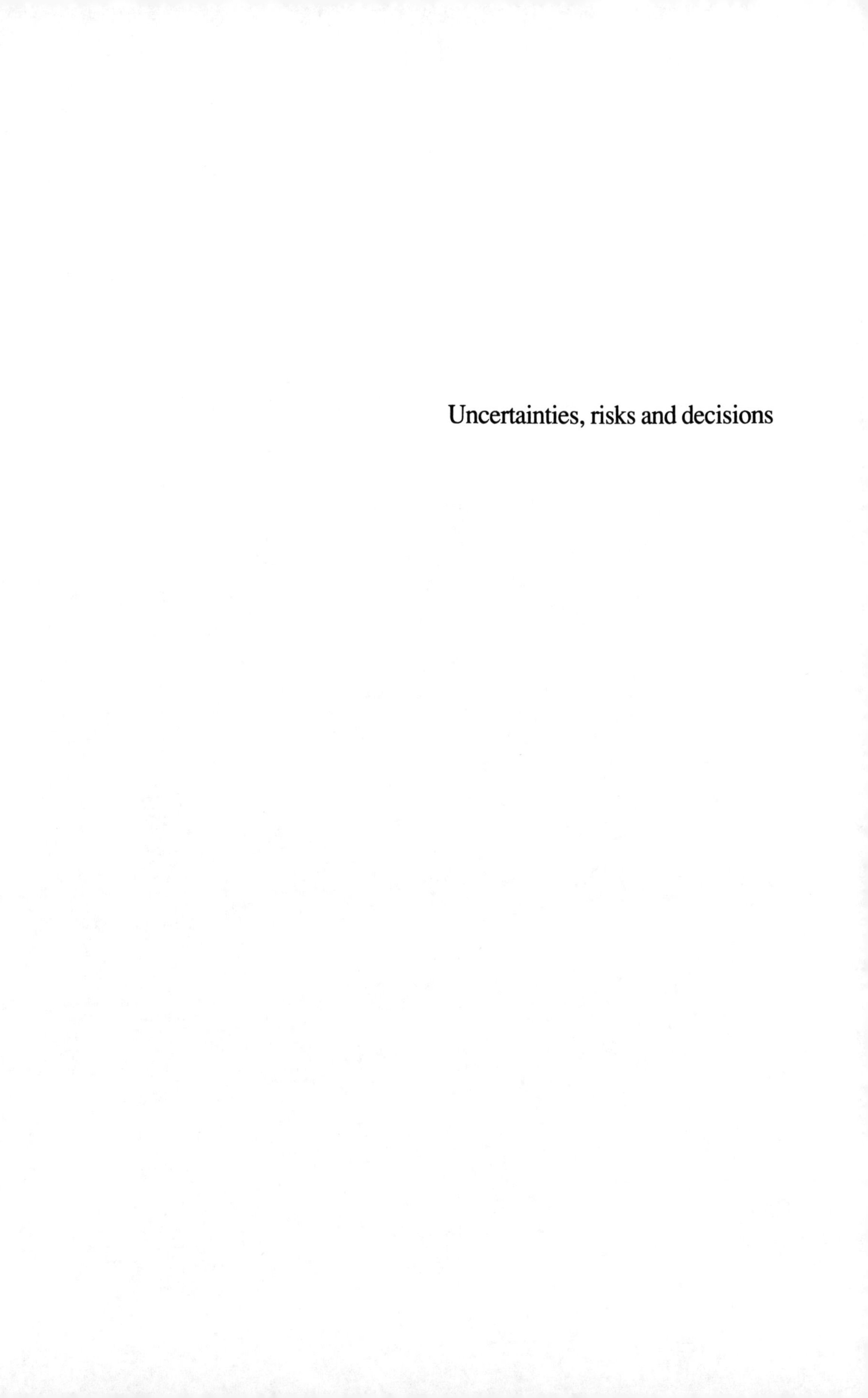

Uncertainties, risks and decisions

Environmental Management, Geo-Water & Engineering Aspects, Chowdhury & Sivakumar (eds)
© 1993 Balkema, Rotterdam. ISBN 90 5410 099 0

Observational approach for landslides – The probabilistic context

R.N.Chowdhury
Department of Civil & Mining Engineering, University of Wollongong, N.S.W., Australia

S.Zhang
Runge Mining, Brisbane, Qld, Australia

ABSTRACT: In this paper the use of progressive failure analysis within a probabilistic framework is considered in the context of an observational approach for slope stability and landslides. Progressive action is often a dominant feature during the development of landslides. Therefore, the probability of catastrophic or total failure changes with time. Several approaches can be used for updating estimates of the probability of sliding (or of reliability). A new approach, highlighted in this paper, relates the probability of sliding to the proportion of slip surface over which localised failure has been observed. The proposed approach can be very useful for assessment of hazard or risk as well as for decision-making in relation to slope stability problems.

INTRODUCTION

In many geotechnical engineering projects decisions have to be taken in the face of uncertainties about the ground conditions. These uncertainties include geological details, geotechnical properties and pore water pressures. Prediction of the behaviour of sloping areas subject to landsliding can be extremely difficult, and conventional analytical approaches have to be supplemented by observation followed by revision and updating of initial calculations. Decision-making involves the use of judgement based on experience and there is considerable scope for further research to develop more powerful aids to these processes.

The important role of 'the calculated risk' in geotechnical and earthwork engineering has, of course, been recognised for a considerable time (Casagrande, 1965). Moreover, observational approaches have been important in the development of modern methods of implementing geotechnical engineering projects. Various important features of 'the observational approach in applied soil mechanics' as developed by Terzaghi in a systematic and rational way, have been presented by Peck (1969) who discussed both the advantages and limitations of such an approach.

The most important advantage of an observational approach is that knowledge of soil properties and of other ground conditions can be updated and that design can be modified in the light of geotechnical performance. However, exercise of engineering judgement is required at every step and analytical approaches as well as experience play an important part in these processes. Research and innovation are required to develop better aids to engineering judgement.

In this paper a new probabilistic approach is highlighted.

WHY A PROBABILISTIC APPROACH ?

The very fact that, as far back as 1964, Casagrande chose to discuss the role of "the calculated risk" in earthwork and geotechnical engineering demonstrates awareness of probabilistic considerations in decision-making. It is more remarkable that, in his discussion of Casagrande's paper, Sherman (1965) referred to the use of probabilistic analysis in a formal sense and discussed the relationships between computed factor of safety and computed probability of failure for three different probability distributions. Traditionally, however, probabilistic considerations were incorporated only subjectively in the exercise of engineering judgement and, even to-day, some engineers are uncomfortable with the use of formal probability.

In the last few decades, considerable work has been done to develop probabilistic approaches specifically suited to analysis of geotechnical engineering problems. It is now widely recognised that these methods enable systematic analysis of uncertainties of different kinds. Acceptance of probabilistic approaches is, therefore, growing although it is by no means universal.

The general impression among geotechnical engineers is that the computed 'probability of failure' is simply an alternative to the computed 'factor of safety' or, at best, an additional index of safety. The fact that probabilistic methods of analysis may be used in many ways including some completely new ones, is the most important reason for working within a probabilistic framework. Moreover, it should be emphasised that, as aids to engineering judgement and decision-making, probabilistic and deterministic approaches should always be regarded as complementary. According to Whitman (1984) "........... Bayesian updating may be viewed as a formalisation of the observational approach advocated by Terzaghi and Peck". Yet, Bayesian updating is just one way in which the estimated probability of success or failure of an earth mass or a geotechnical structure can be reassessed in the light of observation and experience.

PROBABILITY OF SLIDING BY PROGRESSIVE FAILURE

Failure of a slope along a potential slip surface can be considered in probabilistic terms (Chowdhury Tang and Sidi, 1987). The probability of overall sliding by a process of progressive failure may be denoted by p_{pf}. The probability of 'simultaneous failure' when the shear strength over the whole of the slip surface is at its peak value may be denoted by p_p and when the shear strength over the whole of the slip surface is at its residual value by p_r.

For a saturated clay slope under undrained conditions, it has been found that in general p_{pf} is greater than p_p and smaller than p_r, i.e.,

$$p < p_{pf} < p_r \tag{1}$$

The quantities p_p and p_r can be evaluated considering any limit equilibrium model and correspond respectively to estimated factors of safety F_p and F_r based on assumptions of either peak or residual shear strength being operative all along a slip surface.

On the other hand, a new conceptual model is required to evaluate the quantity p_{pf}. There is no counterpart to this quantity in a conventional deterministic framework. However, geotechnical engineers intuitively expect that the real factor of safety will be lower than that based on peak strength as a consequence of progressive failure. The extent to which the factor of safety decreases is related to the residual factor, i.e., the proportion of the slip surface over which the shear strength has decreased to a residual value. In contrast the quantity p_{pf} is calculated, on the basis of an appropriate model, without making any assumptions about the extent of decrease of shear strength to a residual value or to any value below the peak strength. The modification of p_{pf} to incorporate observational data concerning the slope is, however, feasible and has been demonstrated for saturated clay slopes under undrained conditions (Chowdhury 1992, Chowdhury and Zhang 1993).

Table 1. Geometrical and other parameters of natural slope with slip surface.

Slope Inclination	$i = 30°$
Potential Slip Surface Inclination	$\beta = 30°$
and its Depth Below Ground Level	= 4m
Unit Weight of soil	γ = 17 kN/m^3
Pore Water Pressure	= 0

A DEVELOPING LANDSLIDE

Consider a slope which has been instrumented as part of an observational approach and for which an initial estimate of the progressive failure probability p_{pf} has been made. The observations may reveal that there is no local failure along any part of the slip surface. As long as monitoring of the slope leads to that conclusion, it is conservative and safe to retain the initial estimate of progressive failure probability p_{pf}. In fact, one would be justified in using a Bayesian updating approach to revise the calculated value downwards. Even if this is not done, decisions concerning slope management based on initial analyses need not be changed.

Consider, however, the situation of a developing landslide or a slope which has started showing signs of movement during the life of the project. Observations may reveal that some proportion of the slip surface has suffered significant relative deformation. In other words, monitoring may have led to the conclusion that 'local failure' has occurred over some proportion of the slip surface. Under these circumstances the progressive failure probability must be updated. Denote the updated probability of progressive failure when a proportion x of the slip surface has suffered local failure by p_{pfx}. This is the probability of a 'slide' or of 'overall failure' or of 'catastrophic failure'. The use of one or other of these terms may be preferred in the context of the particular landslide - prone environment.

Since a slide has not occurred, it is obvious that :

$$P_{pf} < P_{pfx} < 1, \quad x < 1 \tag{2}$$

It is important to note that the probability of progressive failure of the whole slope will increase as x increases and that a slide is imminent if and when x and p_{pfx} approach unity.

$$x \rightarrow\rightarrow 1, \quad p_{pfx} \rightarrow\rightarrow 1 \tag{3}$$

To an engineer concerned with the management of

Table 2. Statistical parameters of shear strength along the slip surface in a natural slope.

$c_p = c_r = 0$,	$\tan\bar{\phi} = \tan 37°$,	$\tan\bar{\phi}_r = \tan 33°$
	$V_p = 20\%$,	$V_r = 15\%$
	$\Delta_p = 15\%$,	$\Delta_r = 10\%$
	$\theta_p = 1.52m$,	$\theta_r = 6.1m$

a slope or sloping area subject to instability, the following are important.

(a) the estimated magnitude of p_{pfx} at any given time

(b) the rate at which p_{pfx} is expected to increase at any given time.

An engineer may wish to change the management strategy for the landslide if ppfx increases to some 'critical' value selected or based on experience or engineering judgement. Similarly the management strategy may have to change if p_{pfx} is expected to increase rapidly based on analysis and simulation.

In this next section an illustrative example of a natural slope is briefly discussed. Long-term stability is considered in terms of effective shear strength parameters. Previous publications have been concerned only with saturated clay slopes under undrained conditions.

RISK ASSESSMENT FOR NATURAL SLOPE - A CASE STUDY

A natural slope with an inclination of 30° is considered in which the potential slip surface, at a depth of 4 metres, is parallel to the slope. The soil mass has a unit weight of 17 kN/m^3 and, for simplicity, pore water pressures are ignored in this analysis (The model can handle any spatial distribution of pore water pressures). The basic data are summarised in Table 1.

The statistical parameters of shear strength are summarised in Table 2. These include the mean values of peak and residual shear strength (effective cohesion is assumed to be zero) and their coefficients of variation. Natural variability is denoted by V_p and V_r for peak and residual shear strength respectively and the corresponding components due to statistical estimation error are denoted by Δp and Δr. The correlation distances are denoted by θ_p for peak shear strength and θ_r by residual shear strength. Moreover, correlation between peak and residual shear strength may be varied. In this paper either a zero correlation or a perfect correlation are assumed. The calculated

Table 3. Failure probabilities for natural slope

ρ_{pr}	p_p	p_r	p_{pf}	No of slices	slip surface length
0	0.07	0.19	0.11	40	20m
1	0.07	0.19	0.13	40	20m
0	0.065	.17	0.10	40	40m
1	0.66	0.17	0.12	40	40m

failure probabilities are shown in Table 3 for two slip surface lengths, either 20 m or 40 m.

A typical shape for the failure probability p_{pfx} is shown in Fig. 1. It should be noted that the starting point of the curve in Fig. 1 is the calculated p_{pf} value is shown in Table 3.

DISCUSSION

The use of a probabilistic model of progressive failure has now been demonstrated for a natural slope. It is clear that the probability of progressive failure p_{pfx} is higher than the probability of failure based on peak shear strength and lower than that based on residual shear strength.

The probability of failure decreases somewhat with increase in length of the slip surface.

Based on observed 'local failure' along the slip surface, the updated probability increases very sharply. From a magnitude of about 10% the updated value increases to about 90% as 'local failure' is assumed over a relatively small proportion (about 5%) of the slip surface. This clearly indicates that the occurrence of a slide is very likely.

In previous work (e.g., Chowdhury, 1992) concerning saturated clay slopes under undrained conditions, various shapes of the curve of p_{pfx} have been obtained. In some cases the value of the failure probability changed little with initial increases in the value of x and a steep increase occurred only after x had assumed a relatively high value, say 70%. The results in the present study concerning a natural slope are quite different. One would have initiatively expected such a result considering that there is obviously a high correlation between different segments of the potential sliding mass.

It has been shown that the progressive failure model works well and gives consistent results. It is clear that observational data can be used more meaningful in decision-making concerning landslides if simulations of progressive failure probability are available.

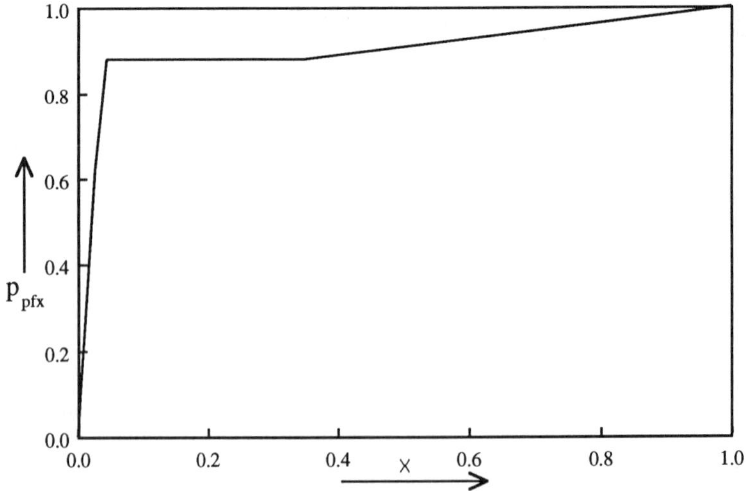

Fig. 1 Updated progressive failure probability-variation with 'failed' proportion of slip surface

ACKNOWLEDGEMENTS

The work was carried out with funding from the Australian Research Council and the senior author would like to acknowledge the ARC and the University of Wollongong for research support. The work was carried out as part of the activities of the Water Engineering and Geomechanics Research Program which has been established with the senior author as Coordinator. The second author worked as a Senior Research Assistant on the ARC project.

REFERENCES

Casagrande, A. 1965. The role of the "calculated risk" in earthwork and foundation engineering, Journal of Soil Mechanics and Foundations Division, A.S.C.E., Vol. 91, NoSM4 reprinted in Terzaghi Lectures (1963-1972) ASCE, 72-111.

Chowdhury, R.N., Tang W.H. and Sidi, I, 1987. Reliability model of progressive slope failure, Geotechnique 37: No. 4, 467-481.

Chowdhury, R.N. 1992. Simulation of risk of progressive slope failure, Canadian Geotechnical Journal, 29: No. 1, 94-102.

Chowdhury, R.N. and Zhang, S. 1993. Modelling the risk of progressive failure - New Approach. Reliability Engineering and System Safety 39: No. 3, in press.

Peck, R.B. 1969. Advantages and limitations of the observational method in applied soil mechanics, Geotechnique 19: No. 2, 171-87.

Sherman, I. 1965. Discussion on ' The role of the "calculated risk" in earthwork and foundation engineering' reprinted in Terzaghi Lectures 1963-1972, A.S.C.E. 119-122.

Whitman, R.V. 1984. Evaluating Calculated Risk in Geotechnical Engineering, Journal of Geotechnical Engineering Division, A.S.C.E. 110: No. 2, 314-357.

Environmental Management, Geo-Water & Engineering Aspects, Chowdhury & Sivakumar (eds)
© 1993 Balkema, Rotterdam. ISBN 90 5410 099 0

Error minimization in environmental risk analysis

E. Dembicki & T. Chi
Gdansk Technical University, Poland

ABSTRACT: In the problem of environmental risk analysis, different studies (Suokas & Kakko, 1989, Santamarina & Chameau, 1987) show that Engineering Knowledge (EK) plays an important role in the scope of uncertaintainty in non-frequency probability, which is caused by incomplete knowledge and the use of vague concepts in its representation. Thus, the use of an appropriate mathematical apparatus to express and aggregate multiple estimates of EK with the smallest residual is necessary. For this problem, we present an approach based on the ideas of fuzzy sets, optimization and Bayesian theory. It can assist researchers in environmental risk analysis.

1 INTRODUCTION

In geological problems, evaluation of the properties of geological materials must take into account the uncertainties which occur in the process of exploration and evaluation of the results obtained from different tests. In the process of risk analysis, distinguishing between the process of exploration and that of evaluation is necessary. Underlying these processes are two ideas: the idea of differences and the idea of similarities between observation attributes. In the explorative process, observation elements are recognised by their differences based on quantative techniques used for measuring the attribute in real values like metres, newtons, etc.

Fundamental to the quantative techniques based on a two-valued logic (0 - false, 1 - true) is the notion that an item x either is or is not a member of any set E:

$$x \in E \quad \text{or} \quad x \notin E \tag{1}$$

The membership functions are as follows:

$$\mu_x(x) = \begin{cases} 1 & \text{for } x \in E \\ 0 & \text{for } x \notin E \end{cases} \tag{2}$$

where:

$$X = \{(x_i, \mu_x(x), \quad x_i \ \forall E, \quad i = 1, 2, \dots n\} \tag{3}$$

In solving any geotechnical problem, a mathematical model must usually be formulated under the assumption that all of the relations describing soil properties, configurations of the structure built in soil and boundary conditions are unlikely to be determined. Analysis must include uncertainties which are considered by repeated experiments in a frequency sense, i.e. related to occurrence or non-occurrence of any event. It can be defined as function:

$$P : Z \Rightarrow [0,1] \tag{4}$$

where Z is the set of all subsets of the sample space. In this phase, the probability theory is used as a convenient base to face so-called objective uncertainty.

In the evaluated process, observation elements are recognised by their similarity. For this case, an expert estimate based on multivalued logic is used for describing similarity level of the observation elements. Such an estimate is derived from psychological experiments and has been traditionally used by experts for describing the knowledge base. In this paper, it is formalised as a fuzzy subset in interval [0,1] independent on the semantics of evaluated attribute but dependent on the state of mind of the individual involved in the problem. In this case, Bayesian method was not invented to take care of the characteristic of man: develop expert estimates, correct them in the process of adaptation and, at the same time, assess its own behaviour

This paper suggests a constructive way to determine the degree of membership (a number between 0,1) in order to account for the systematic deviation of human behaviour from the expected evaluated values.

2 EVALUATION OF EXPLORATIVE DATA

As a part of the process of the environmental risk analysis it is necessary to evaluate the observation data in geological problems. The assessment of any possibility values usually requires an engineering judgment to evaluate the influence of various factors, fuzziness/partial ignorance and achieve the necessary synthesis of the available information. Therefore, any of the relations under consideration can be determined only within certain degree of truth.

In the evaluation process, we are generally able to define, in semantic measures, a geotechnical medium as one of fuzzy sets in which any geotechnical property should be considered from the point of view of its truthfulness. Thus, any property x of a soil medium should not be a real number in the usual sense but a fuzzy number of fuzzy subset A:

$$x = \{ x \in A \mid \mu_A(x),\ \mu_A(x) \in T\} \qquad (5)$$

$$T = [0,1] \qquad (6)$$

where T is called the truth space represented by real numbers from 0 to 1 and $\mu_A(x)$ is called the membership function that represents numerically the degree to which an element x belongs to A. In semantic sense, if $\mu_A(x)=1$ then it characterises precisely, in this case, the evaluation property. If $\mu_A(x)<1$ then it provides only a fuzzy description of the evaluated property. For instance, in evaluating the angle of internal friction Φ, an expert may give an interval estimate of, say, 32° to 36°. Also, words may be used to convey certain concepts, for example "about 33°". All this can be modelled as a fuzzy subset as follows:

$$\Phi=\{32 \mid 0.4,\ 33 \mid 0.9,\ 34 \mid 0.8,\ 35 \mid 0.5\} \qquad (7)$$

Now, the question arises how to quantify the human error or the level of fuzziness/partial ignorance underlying the membership form:

$$\mu_A(x)=\{0.4,\ 0.9,\ 0.8,\ 0.5\} \qquad (8)$$

Generally, in the process of evaluation based on similarities of estimative elements, we can be aware of the abilities of classifying evaluated elements in the true sense of their similarity. Thus we should consider any two elements from the point of view of their similarity. It is assumed here that the evalution of the degree of their similarity is possible.

We will now define a subset B by the following function $M_B(x_i,x_j)$:

$$M_B(x_i,x_j)=\frac{\mu_x(x_i)}{\mu_x(x_j)} \qquad (9)$$

where M_B denotes the level that estimates the element x_i when it is compared with element x_j. This is represented as the degree of similarity that exists between the element $x_i \in X$ and $x_j \in X$. This function satisfies the equations:

$$M_B(x_i,x_j)=1/M_B(x_j,x_i),\quad x_i \ \forall\ X,\quad x_j\ \forall\ X \qquad (10)$$

$$M_B(x_i, x_i)=1 \tag{11}$$

Equation (11) expresses the fact that an element is similar to itself to the maximum degree.

3 MINIMIZATION OF HUMAN ERROR

It is well known that, in order to account for systematic deviations of human behaviour in which the mutual exclusiveness property needed in the probability theory is lost, the modelling of partial ignorance/ fuzziness in a systematic way is necessary. In this case, an expert is defined as an individual who tries to minimize losses. This could be interpreted as an effort to minimize the error involved in the estimative values of the degree of similarity of the observation elements.

Let $\phi_1, \phi_2, \ldots, \phi_n$ be the numbers of a fuzzy set Φ. We are interested in evaluating the membership values $\mu_\Phi(\phi_1), \mu_\Phi(\phi_2), \ldots, \mu_\Phi(\phi_n)$ of the above members ϕ_i, $i=1,2,\ldots n$.

Assume that $m_{ij}=M_\Phi(\phi_i, \phi_j)$ denotes the number that estimates the relative membership of element ϕ_i when it is compared with element ϕ_j. Obviously:

$$m_{ij}=\mu_\Phi(\phi_i)/\mu_\Phi(\phi_j) \tag{12}$$

$$m_{ji}=1/m_{ij} \tag{13}$$

$$m_{ii}=1 \tag{14}$$

where $\mu_\Phi(\phi_i)$, $i=1,2,\ldots,n$ are unknown membership values. When a fuzzy set Φ contains n elements then this method requires the estimation of the n(n-1)/2 comparisons. For example, for n=6 we have 15 comparisons as follows:

$$\begin{matrix} * & m_{12} & m_{13} & m_{14} & m_{15} & m_{16} \\ & * & m_{23} & m_{24} & m_{25} & m_{26} \\ & & * & m_{34} & m_{35} & m_{36} \\ & & & * & m_{45} & m_{46} \\ & & & & * & m_{56} \end{matrix} \tag{15}$$

Let us now consider the case in which it is possible to have a perfect judgement. It is:

$$m^*_{i,j}=\mu_\Phi(\phi_i)/\mu_\Phi(\phi_j) \tag{16}$$

where $\mu_\Phi(\phi_i)$, $\mu_\Phi(\phi_j)$ denote the actual values of the evaluation of ϕ_i and ϕ_j.

Thus, in the non-perfect cases, we have:

$$m_{i.j}-\mu_\Phi(\phi_i)/\mu_\Phi(\phi_j)=\varepsilon_i \tag{17}$$

where ε_1 denotes the deviation of m_{ij} from the perfect judgement. Obviously, if $\varepsilon_1=0$ then m_{ij} was perfectly evaluated: $m_{ij}=m^*_{ij}$. In that case, we obtain $m_{ij}-\mu_\Phi(\phi_i)/\mu_\Phi(\phi_j)$ = $m_{ij}\mu_\Phi(\phi_j)-\mu_\Phi(\phi_i)=0$, $j=1,2,..n$, i.e:

$$-\mu_\Phi(\phi_i)+m_{ij}\mu_\Phi(\phi_j)=0, \quad i,j=1,2,\ldots,n, \quad j>i \tag{18}$$

where the $\mu_\Phi(\phi_i)$ is a membership function that has a positive value in the unit interval. It holds:

$$\sum_{i=1}^{n}\mu_\Phi(\phi_i)=1 \tag{19}$$

From (18) and (19), we have:

$$Ry=E \tag{20}$$

$$Ay=1 \tag{21}$$

where:

$$y=\{\mu_\Phi(\phi_1), \mu_\Phi(\phi_2), \ldots \mu_\Phi(\phi_n)\}^T \tag{22}$$

* y is the n x 1 vector of all elements to be evaluated,
* E is the (n(n-1)/2 x 1 vector of the possible errors,
* A is the 1 x n vector in which all the elements are equal to 1.
* R is the (n(n-1)/2 x n matrix describing the relationship between elements ϕ_i, ϕ_j, $i,j=1,2,\ldots,n$. It is formulated, for the case n=6, as follows:

$$R=\begin{bmatrix} -1 & m_{12} & 0 & 0 & 0 & 0 \\ -1 & 0 & m_{13} & 0 & 0 & 0 \\ -1 & 0 & 0 & m_{14} & 0 & 0 \\ -1 & 0 & 0 & 0 & m_{15} & 0 \\ -1 & 0 & 0 & 0 & 0 & m_{16} \\ 0 & -1 & m_{23} & 0 & 0 & 0 \\ 0 & -1 & 0 & m_{24} & 0 & 0 \\ 0 & -1 & 0 & 0 & m_{25} & 0 \\ 0 & -1 & 0 & 0 & 0 & m_{26} \\ 0 & 0 & -1 & m_{34} & 0 & 0 \\ 0 & 0 & -1 & 0 & m_{35} & 0 \\ 0 & 0 & -1 & 0 & 0 & m_{36} \\ 0 & 0 & 0 & -1 & m_{45} & 0 \\ 0 & 0 & 0 & -1 & 0 & m_{46} \\ 0 & 0 & 0 & 0 & -1 & m_{56} \end{bmatrix} \quad (23)$$

Generally, the best value y* must be chosen so as to minimize a given function f(y):

$$f(y)=(E-Ry)^T(E-Ry) \quad (24)$$

subject to:

$$Ay=1 \quad (25)$$

By incorporating equation (25) to the function (24) through the Lagrange multiplier λ, we obtain:

$$f(y)=(E-Ry)^T(E-Ry)+\lambda^T(Ay-1) \quad (26)$$

The estimates y* are found by taking partial derivatives of function (26) with respect to y* and setting these derivatives to zero, i.e.:

$$\delta f(y)/\delta(y)=0 \quad (27)$$

which yields:

$$y^*=(R^TR)^{-1}(R^TE-A^T\lambda) \quad (28)$$

where the vector λ is to be determined in such a way that the function (25) is satisfied. This is a typical linear least-square problem (see Song, 1981).

In the other case in which the state of partial ignorance should be represented by means of a uniformly distributed probability measure, it can be justified on the basis of the maximum entropy principle, which is an assumption rejected in this paper. Thus, our problem is formulated as the constrained optimization problem, as follows:

maximize:

$$En=-\sum_{i=1}^{n}\mu_\Phi(\phi_i)\ \ln(\mu_\Phi(\phi_i)) \quad (29)$$

subject to:

$$-\mu_\Phi(\phi_1)+m_{12}\mu_\Phi(\phi_2)\le\varepsilon_1 \quad (30)$$

$$-\mu_\Phi(\phi_1)+m_{13}\mu_\Phi(\phi_3)\le\varepsilon_2 \quad (31)$$

....

$$-\mu_\Phi(\phi_1)+m_{1n}\mu_\Phi(\phi_n)\le\varepsilon_{n-1} \quad (32)$$

$$-\mu_\Phi(\phi_2)+m_{23}\mu_\Phi(\phi_3)\le\varepsilon_n \quad (33)$$

....

$$-\mu_\Phi(\phi_2)+m_{2n}\mu_\Phi(\phi_n)\le\varepsilon_{2n-3} \quad (34)$$

....

$$-\mu_\Phi(\phi_{n-1})+m_{n-1,n}\mu_\Phi(\phi_n)\le\varepsilon_{n(n-1)/2} \quad (35)$$

$$1-\sum_1^n\mu_\Phi(\phi_i)\le\varepsilon_{n(n-1)/2+1},\quad i=1,2,\ldots,n \quad (36)$$

$$\mu_\Phi(\phi_i)>0,\quad i=1,2,\ldots,n \quad (37)$$

In particular, we now examine an illustrative example based on the following geotechnical problem:

4 EXAMPLE

Let us assume that a set of n=4 values of the angle of internal friction, ϕ_i, i=1,2,3,4 obtained from one penetration test are given, in degrees, as follows: ϕ = { 43, 45, 47, 49 }. In evaluating these data an expert may use the words: "value of ϕ is about 49°". This can be formulated as a fuzzy subset as follows:

$$\Phi=\{43|\mu_\Phi(43),\ 45|\mu_\Phi(45),\ 47|\mu_\Phi(47),\ 49|\mu_\Phi(49)\}$$

On the other hand, similar relation between these data is given as:

*	m_{12}=0.8,	m_{13}=0.7,	m_{14}=0.55
	*	m_{23}=1.2,	m_{24}=1.5
		*	m_{34}=0.9

When the model (Equations 28-36) is applied using Monte-Carlo simulation with error vector, ε is given as $\varepsilon_1=\varepsilon_2=\varepsilon_3=\varepsilon_4=\varepsilon_5=\varepsilon_6=0.0001$, $\varepsilon_7=0.005$. The simulation program, written in TURBO.C, solves the above constrained optimization problem yielding:

$E=1.25541$,

$\mu_{\Phi}(\phi_i)=\{0.1059,\ 0.1715,\ 0.2774,\ 0.4489\}$

5 CONCLUSIONS

In this paper, elements are recognised for mere evaluation by their similarity. The subjective uncertainty is defined based on D. Dubois' and H. Prade's idea (1990): "subjective uncertainty cannot be but a compromise between the ideally optimal Bayesian theory and the limited precision and vagueness of the available, often subjective, knowledge". This is a possibility measure from an error minimization point of view. This paper shows the simplicity of the introduction of subjective opinion in the evaluation of uncertainty into the mathematical model for the risk analysis in which the membership function takes on fuzzy values. This is seen as a numerical model of the state of knowledge. It is a useful tool for the complex problem of environmental risk analysis. However, the likelihood ratio test for correctness of fit to a completely specified hypothetical distribution regarding the exact experiment on the basis of fuzzy and random information must be handled. This investigation is based on the notion of the probability of a fuzzy event proposed by Zadeh (1968): $P(K)=\int_X \mu_K(x)dP(x)$, $\forall x \in X$, where X denotes a finite set of fuzzy events and K denotes a fuzzy event on X.

Other implications and practical conclusions resulting from further research in this area will be presented elsewhere.

6 REFERENCES

Dubois, D. & Prade, H. 1990. Coping with uncertain knowledge. Computers and Artificial Intelligence. 2:115-143.

Santamarina, J.C. & Chameau, J.C. 1987. Expert system for geotechnical engineers. J. of Computing in Civil Engineering. 4:241-253.

Soong, T.T. 1981. Probabilistic modelling and analysis in science and engineering. New York.

Suokas, J. & Kakko, R. 1989. On the problem and future of safety and risk analysis. Journal of Hazardous Materials. 21:105-124.

Zadeh, L.A. 1968. Probability measure of fuzzy events. Journal of Math. Anal. Appl. 23:421-427.

Environmental Management, Geo-Water & Engineering Aspects, Chowdhury & Sivakumar (eds)
© 1993 Balkema, Rotterdam. ISBN 90 5410 099 0

Decision analysis for optimized urban risk management

Waldemar Hachich
Geoexpert and Escola Politécnica, University of São Paulo, Brazil

ABSTRACT: Risk management is a major challenge, especially in densely populated areas threatened by hydrological and geotechnical risks. A method is proposed for optimal resource allocation within a given budget, based on Decision Analysis and on a criterion that privileges human life.

1 INTRODUCTION

Municipalities of large cities, especially in the so-called third world, are faced with an ever growing need for housing. Since supply typically lags far behind demand, slums develop as inhabitants of lesser means try to solve their housing problems by themselves. In São Paulo, Brazil, for example, in 1989 about 800,000 people lived in 1600 such areas. A preliminary analysis showed about 240 of these areas posed high geotechnical or hydrological risks.

Risk reduction is therefore a major concern in such circumstances. This paper sets forth a Decision Analysis framework for rational allocation of the funds available for a risk management program, and discusses the use of alternative objective functions (decision criteria).

2 PROBABILISTIC MODELS IN PERSPECTIVE

About three decades ago it was realized that no direct relationship could be established between safety factors and safety itself. Similar safety in different projects could only be achieved through different safety factors, primarily because safety factors do not explicitly take uncertainties into account. Accordingly, the main objective of probabilistic analyses was then the introduction of new safety indicators (Langejan 1965, Wu and Kraft 1967). The inductive branch of the conceptual probabilistic model, that is, Statistical Inference and Decision Making, was then seen almost as a by-product (Höeg and Murarka 1974).

Probabilistic models offered rational safety indicators, such as the reliability index and the probability of failure, which:

- kept a close relationship with safety itself;
- provided a uniform measure of safety, regardless of the failure modes.

Difficulties in the determination of the applicable distributions, however, led to the now common idea that failure probabilities should — in principle at least — be regarded as <u>nominal</u> values.

On the other hand, it is nowadays recognized that most of the benefits to be gained from probabilistic models stem from their potential for increasingly rational decision-making, opening possibilities for:

- planning, on the basis of a common safety measure;
- developing rational, consistent and reproducible procedures for optimal resource allocation (through Decision Analysis), for example in a risk management program.

This latter challenge is exactly the one faced by engineers who are expected to devise the most effective way of spending the ever shrinking funds for risk management in slum areas.

3 CONCEPTUAL RISK MANAGEMENT FRAMEWORK

The steps involved in implementing a risk management program for slum areas are not fundamentally different from those in other decision-making contexts. Some of them, however, are specific to the

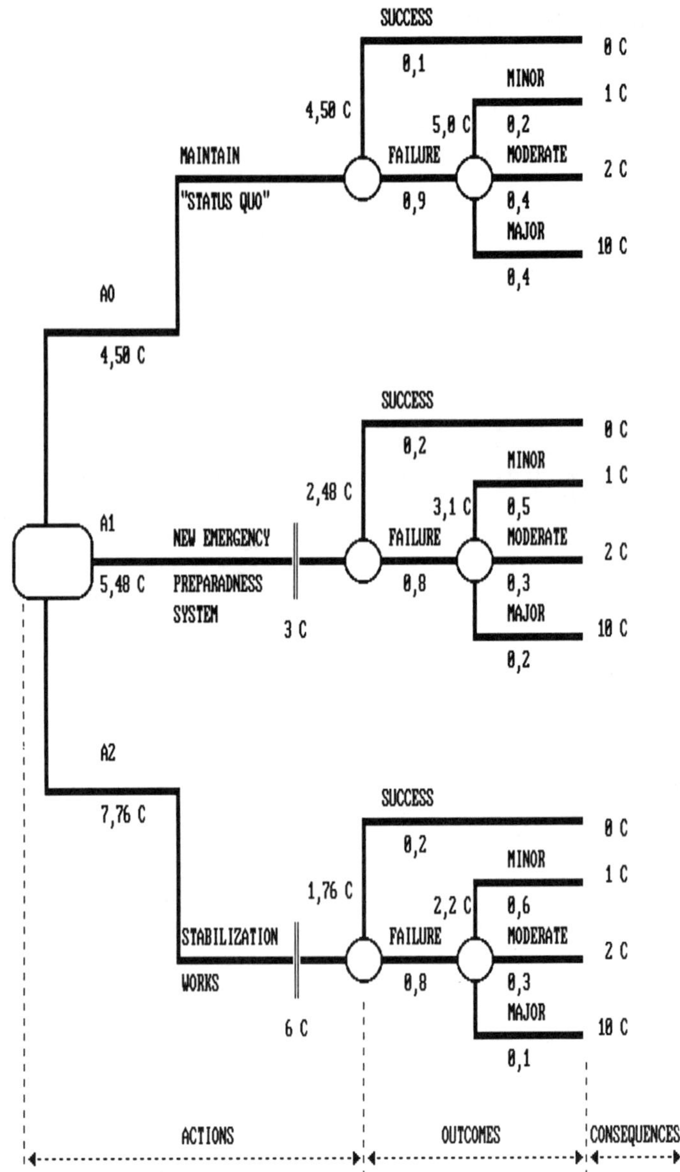

Fig. 1 Decision tree: monetary consequences

technical aspects of geotechnical or hydrological risks. The proposed method requires:

- gathering of topographical, geological, pluviometric, hydro-geological and geotechnical information;
- definition of failure mechanisms (for example, for geotechnical risks, type, volumes, setup time, event duration, physical extension); in some cases more than one mechanism may be present in a certain location;
- definition and cost estimate of possible actions for risk mitigation;
- precise definition of relevant events (outcomes of different actions);
- inventory of areas potentially affected by failure;
- evaluation of consequences;
- gathering of all possible actions, outcomes and consequences in a decision tree;

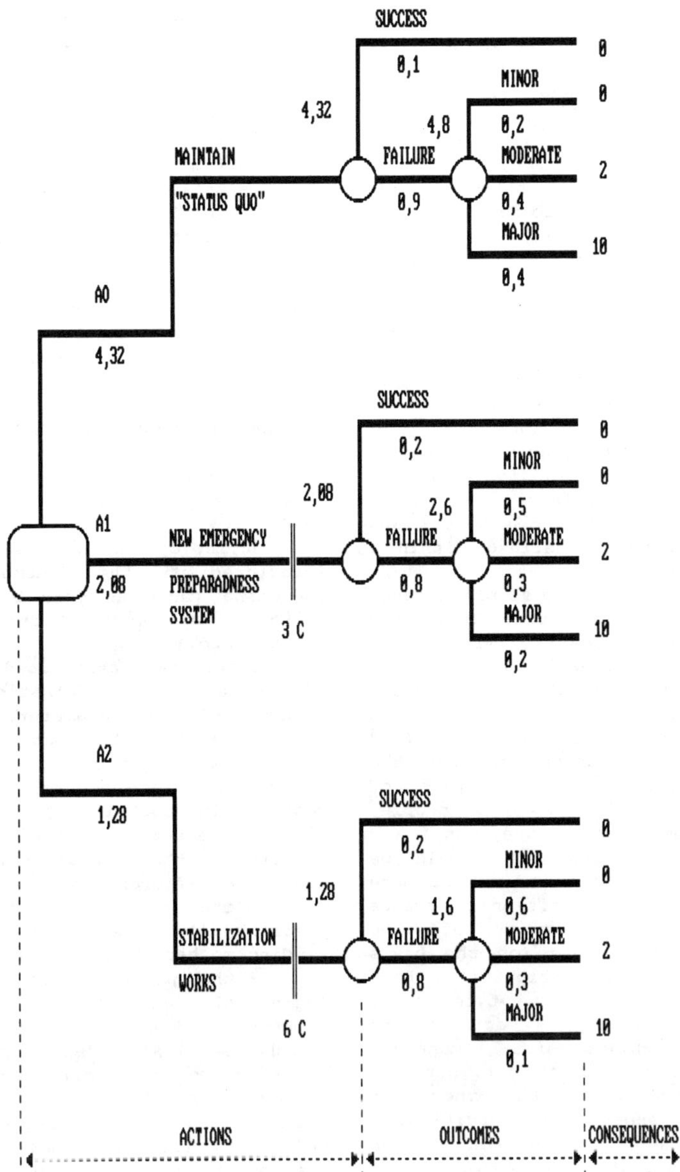

Fig. 2 Decision tree: lives lost

- computation of probabilities of the outcomes;
- choice of decision criterion;
- choice of most effective action.

Some of these steps are not discussed here because they are no different from what is usually done by engineers in a risk management team. This is particularly true of the first three steps, which require specific technical knowledge.

Suffice it to say that as far as definition of possible actions, one of them, the "do nothing" alternative (A0, see Fig. 1), should always be retained as the background for the analysis, since it is against this "status quo" that benefits derived from all other actions should be weighed. Each possible alternative should be listed and a preliminary evaluation of its applicability to the several sites should be carried out. In examining this checklist, the engineer should exercise judgment so as to reject actions that are clearly unsuited for the situation at hand (either for technical or for economical reasons), and which would otherwise just clutter the analysis.

In defining outcomes (i.e., in visualizing failure scenarios), it should be kept in mind that later in the analysis it will be necessary to compute the probabilities and the consequences associated with each of those relevant events. It is therefore mandatory that a clear-cut, as quantifiable as possible, definition of those events be set forth. For example, slope failure probabilities are quite often reported with no mention of the time frame of the event. On a geological time scale the probability of failure is inescapably unity. However, this event, its probability and its consequences are completely void of practical significance. Success and failure should both be defined within a certain time span.

While "success" is a simple event, failure is a compound event. There are several types and degrees of failure. Each such event should, therefore, be described in terms of its characteristics (e.g., volume, velocity, extension of affected areas, etc.) as well as quantified attributes which adequately measure the extent of damage (e.g., millions of dollars, lives lost, displaced people, acres rendered improper for use, etc.).

Evaluation of consequences should be straightforward — albeit tedious — once there is a clear-cut definition of events and a thorough inventory of the potentially affected areas (on the basis of the aforementioned attributes). There are essentially three types of consequences: monetary, social and environmental. While for monetary consequences the choice of attribute is obvious, it is not so for the other two; the help from experts in social and environmental aspects may be required for a proper assessment of consequences.

The decision tree (Fig. 1) is the tool for logical (and chronological) set up of actions, outcomes and consequences. Even in circumstances where it may become unfeasible to perform a complete decision analysis, the decision tree is still an extremely valuable tool to help visualize the problem: experience shows that the required structuring of information lends itself to exhaustive analysis of alternatives, preventing overlooking of certain combinations action-outcome-consequences (or, at least, forcing a justification for any such omission).

The choice of the most effective action is mechanically performed through the process of averaging-out-and-folding-back prescribed by Decision Analysis (Raiffa 1970). However, the rational paradigm of Decision Theory involves four steps before this process can be started:

- problem structuring;
- evaluation of consequences;
- probability assessment;
- establishment of a decision criterion;

of which only the first two have been discussed so far. The other two are exactly the remaining two steps in the proposed risk management framework: the next two sections are devoted to them.

4 COMPUTATION OF PROBABILITIES OF OUTCOMES

Uncertainties are usually classified in three categories (Benjamin, and Cornell 1970):

- intrinsic, the natural uncertainties associated with the randomness of natural phenomena (such as mean annual rainfall, patterns of sedimentation in fluvial deltas, etc.);
- statistical, associated with the lack of data or information for the determination of parameters of models; it usually can be reduced at additional costs (larger sampling programs);
- model, the uncertainties caused by possible inadequacies of the adopted model to describe actual behaviour; it should be pointed out that, in general, there are at least two different models at play:
  - the physical model adopted to describe the phenomenon, such as the sliding mechanism of a natural slope;
  - the probabilistic model (normal, log-normal, exponential, Poisson, etc.) adopted to describe the random variables of the physical model (for example, the distribution of soil strength or of maximum rainfall).

Although model uncertainty can also be reduced by increased sampling, seldom are the additional costs warranted by the benefits on a single-project scale. Model uncertainty is usually not taken explicitly into account. All probabilities are, therefore, conditional on the validity of the adopted model. Experience and judgment should guide the engineers in choosing realistic models for the phenomena they try to explain.

In a broad-scope program such as the one described herein, site-specific data are usually scarce, so that statistical and intrinsic uncertainties can hardly be isolated from each other. Target areas are in general too unsuitable for engineering purposes to have ever been thoroughly

Table 1 Optimal actions according to different decision criteria

Action	Expected cost	Expected number of lives lost	Expected number of lives saved (relative to A0)	Cost per life saved
A0	4.50	4.32	0	—
A1	5.48	2.08	2.24	2.45
A2	7.76	1.28	3.04	2.55

investigated and the urge for remedial or preventive measures usually precludes any additional investigation. The probabilities associated with some of these intermingled uncertainties remain more than anything else a matter of expert judgment.

In principle probabilities can be assessed from historical records, from expert opinions, or through analytical procedures. The probabilities of hydrological events can usually be derived from historical records. Expert opinions (by themselves or in conjunction with the other two approaches, as in McCann, Franzini and Shah 1984), despite their inevitable subjectivity, remain the most effective way of dealing with geotechnical uncertainties, for the reasons discussed above.

5 THE CRUCIAL PROBLEM: CHOOSING A DECISION CRITERION

Philosophically it can be argued that the best decision criterion is the maximization of expected utility for society. However, such a definition leaves a lot to be desired in practical terms. What does one mean by utility? And what does society really mean?

In situations that involve exclusively monetary consequences and in which the decision rests with risk-neutral people or organizations (such as government or large corporations), it has been shown that utility is linearly related to monetary value (Raiffa 1970). The decision criterion then becomes minimization of expected cost (Fig. 1). Risk neutrality is assumed in this paper.

Whenever consequences have to be expressed in terms of multiple attributes, the specification of a utility function becomes extremely delicate, since it requires explicit tradeoffs among the several preferences (Kenney and Raiffa 1976). Questions such as: "how many lives is society willing to trade for x monetary units?" would then inevitably surface.

There is, of course, an alternative approach that avoids this disturbing question when the problem is that of establishing the best action within a prescribed budget. Then, instead of discussing the value of human life, one evaluates the cost per life saved for each remedial action, and the decision criterion becomes the minimization of the expected cost per life saved.

Such a criterion enables the engineer responsible for risk management to recommend the action that has the largest potential for minimizing the risk of loss of human lives. This is arguably the best way to allocate the available funds.

Social consequences (lost lives) replace, in Fig. 2, the monetary consequences of Fig. 1. Table 1 combines data from both figures in a form that invites a decision based on the cost per life saved. This hypothetical example was conceived so as to emphasize the fact that the "optimal" decision may be completely different depending on the adopted criterion (minimum expected cost, minimum expected number of lives lost, minimum cost per life saved).

It might be argued that such a criterion overlooks environmental consequences, which are nowadays a major concern. However, it should be kept in mind that this approach is tentatively suggested to tackle a problem in which minimization of human death and suffering can be regarded as a sound environmental policy.

Another point that deserves mention is the fact that the method is passive in that it simply suggests what to do with the available funds. The question of actively striving for more funds is not addressed herein, since it belongs to the realm of political action.

6 CONCLUSION

The proposed procedure rests upon a sound conceptual basis, proven in other application areas, and should, therefore, entice better answers to the challenge of managing hydrological and geotechnical risks in urban areas.

There has not been an opportunity for practical application of the proposed method yet. In 1988 a slide in one of those areas claimed dozens of lives. In the aftermath of this tragedy, most of the decisions made during the past two years were based essentially on expert opinions and a heuristic algorithm, because doing something to reduce risks was far more relevant than the quest for an optimal solution.

Nevertheless, this circumstance may even prove to have been fortunate. A Doctoral thesis is under development at the University of São Paulo, in which all the data and decisions will be reviewed on the basis of the proposed methodology. Comparison of the decisions actually made with the would-be decisions dictated by this model may turn out to be a quite enlightening experience.

REFERENCES

Benjamin, J.R. and Cornell, C.A. 1970. Probability, Statistics, and decision for civil engineers. McGraw-Hill Book Company.

Höeg, K. and Murarka, R.P. 1974. Probabilistic analysis and design of a retaining wall. Journal of the Geotechnical Engineering Division, ASCE, Vol. 100, n. GT3.

Kenney, R.L. and Raiffa, H. 1976. Decision with multiple objectives: preferences and value tradeoffs. John Wiley & Sons.

Langejan, A. 1965. Some aspects of the safety factor in Soil Mechanics, considered as a problem of probability. Proceedings of the Sixth International Conference on Soil Mechanics and Foundation Engineering, Montréal.

McCann Jr., M.W., Franzini, J.B. and Shah, H.C. 1984. Preliminary safety evaluation of existing dams. Technical Reports ns. 69 and 70. The John A. Blume Earthquake Engineering Center, Stanford University.

Raiffa, H. 1970. Decision Analysis. Addison-Wesley Publishing Company.

Wu, T.H. and Kraft Jr., L.M. 1967. The probability of foundation safety. Journal of the Soil Mechanics and Foundations Division, ASCE, Vol. 93, n. SM5.

Environmental Management, Geo-Water & Engineering Aspects, Chowdhury & Sivakumar (eds)
© 1993 Balkema, Rotterdam. ISBN 90 5410 099 0

Progressive reliability model of opencut soil slopes

R.A. Halatchev & D.A. Gabeva-Halatcheva
University of Mining & Geology, Sofia, Bulgaria

ABSTRACT: A new progressive reliability model of soil slopes in opencuts is suggested, which is developed in conformity with the classical treatments of the theory of reliability. It includes the use of an average and local factors of safety as criteria of the stability state of the slope and a vertical slice of the slide mass respectively, which are changing in time on the basis of a long-term shear strength model. The reliability is assessed with regard to the slope as a system of functioning elements, i. e. vertical slices and to the particular elements. Expressions of the statistical characteristics of the reliability parameters are worked out and an index of reliability is suggested. A probable prognostication of the progressive failure mechanism of the slope also is included in the modelling. The applicability of the developed approach is demonstrated on a hypothetical opencut slope.

1 INTRODUCTION

The approach for developing methods for probable assessment of the opencut slopes stability takes a worthy place in the contemporary problems of the Geomechanical Engineering. It is a natural result of the development of the knowledge in this scientific field and corresponds to the character of the processes passed in these engineering works. Its actuality undoubtedly is due to the possibilities for an effective management of the slope stability state with an aim to decreasing the negative influence of the mining activity on the environment.

In parallel with the creation of slope risk models successfull attempts are made for developing realiability models (Vanmarcke, 1977, Chowdhury et al., 1987, Halatchev, 1991). In this way qualitatively a new trend is set up in the probabilistic slope modelling, which has its own specific features.

The present paper suggests a reliability model of soil slopes in the opencuts. As a version a probabilistic model of a slope progressive failure is also developed. The soil is characterized with forces of cohesion and friction. The time aspect of the shear strength variability is taken into account by an adoption of a deterministic model of a long-term shear strength in the massive. It is achieved through modelling the cohesion as a random process with a nonstationary character of variability. The frictional coefficient is treated as a random quantity.

2 DETERMINISTIC STABILITY ANALYSIS

2.1 *Technique for slope stability analysis*

The Swedish slicing method and the method of algebraic summation of forces (Ilin et al., 1985) are used widely in the stability analysis of soil slopes. They are based on the principle of static limit equilibrium and use a discretion of the slide mass with vertical slices (Fig. 1). Their formular apparatus is a suitable tool for back analysis. The analytical expression of the factor of safety for these methods is the following:

$$F_s = \left[\sum_{i=1}^{n} N_i' \operatorname{tg} \phi_i + \sum_{i=1}^{n} l_i c_i\right] / \sum_{i=1}^{n} T_i, \quad (1)$$

where: $\operatorname{tg} \phi_i$, c_i are frictional coefficient and cohesion in an effective form; n — number of the slices.

From theoretical point of view F_s as a criterion of the slope state represents an average quantity since it is assessed at every point of the soil mass as a relation of the shear strength of the soil to the induced shear stress. Therefore F_s is a variable quantity along the whole failure surface, which is confirmed by the investigations of Chugh (1988). Having proceeded from this position, we introduce a local factor of safety F_{l_i} as a criterion of the i-th slice state of the slide mass:

$$F_{l_i} = (N_i' \operatorname{tg} \phi_i + l_i c_i) / T_i. \qquad (2)$$

After some transformations of (1) and (2) the following relationship between the average and local factors of safety can be written:

$$F_s = \sum_{i=1}^{n} F_{l_i} \delta_i, \qquad (3)$$

where: δ_i is a coefficient, $\left(\delta_i = T_i / \sum_{i=1}^{n} T_i\right)$.

It is obvious that δ_i carries out the role of a coefficient of proportionality, which balances the influence of all F_{l_i} in the common assessment of F_s.

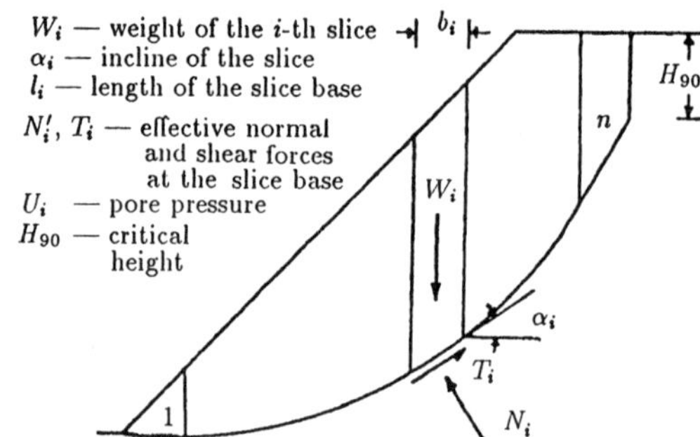

Fig. 1 Slope geometry and forces acting in a single slice.

2.2 *Long-term shear strength of soil*

At the investigation of the long-term stability of soil slopes the emphases is put on the determination of the shear strength, which varies in the process of the soil creep. There are some investigations on the problem of long–term shear strength of clays in the opencut massive, which made Ilin, Galperin & Streltsov (1985) to make a general summary of the following approaches for a determination of this strength:

First approach: It proceeds from the situation that the shear strength tends to zero during the increase of the load duration (Fig. 2a, curve 1). The strength, which corresponds to the term of mining work existence, is accepted as an ultimate strength τ_u;

Second approach: A standpoint is adopted that the strength of the clay is increased in the creep process (Fig. 2a, curve 2) and there is a tendency for reaching some value τ_{fn} at a long test, which is bigger then the instantaneous strength τ_0. Such a variation of the strength at the stage of a fixed creep is explained with the equilibration of the failure process and establishment of new connections between the clay particles;

Third approach: It is based on the assumption that during a long time action of the load the shear strength is on decrease in time to some value, which is constant for every soil type and it pulls up asymptotically to the so called limit of the long–term shear strength τ_∞ (Fig. 2a, curve 3).

The last approach is adopted as leading in the research practice, because the quantity τ_∞ is suitable for the assessment of the long-term slope stability. Experimentally this approach got a confirmation. For example, Bjerrum (Ilin et al., 1985) determined that the intensity of the shear strength fall from τ_p (instantaneous strength of the clay till the moment of the failure surface formation) to τ_d (shear strength at the process of the failure surface formation) depends on the intensity of the normal stress σ_n, which contributed to the failure surface creation (Fig. 2b). He found that τ_∞ as a measure of the ultimate soil strength almost corresponds to the strength τ_d ($\tau_\infty = \tau_d$) for the plastic clays and exceeds considerably τ_d ($\tau_\infty > \tau_d$) for the overconsolidated clays.

Skempton (1964) approved that the clays have a peak (τ_p) and residual (τ_r) shear strength and the development of the progressive failure along the whole length of the failure surface is realized till the moment of the fall of the average shear strength to a residual value, which is nearly equal to zero.

The long–term shear strength in accordance with the theory of Vialov & Zaretskii can be assessed by the formula (Ilin et al., 1985):

$$c(t) = c_0 - (c_0 - c_\infty) t_p / (T_p + t_p), \qquad (4)$$

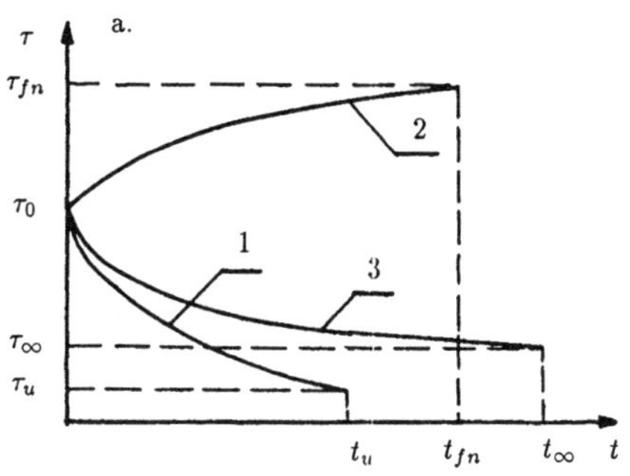

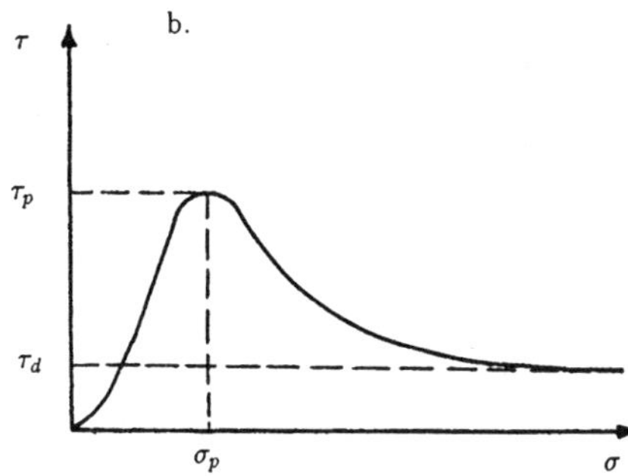

Fig. 2 Relationships of the shear strength and: *a* — time, *b* — the stress σ.

where: c_0 is the instantaneous shear strength of the soil;

c_∞ — limit of the long–term shear strength;

t_p — moment of the soil failure or time of the slope stability functioning;

T_p — parameter with a time dimension.

Regarding to the frictional coefficient it is found out that it has a slight variability in the process of a long–term functioning of the slope and can be treated as a constant quantity.

3 TERMINOLOGY OF THE RELIABILITY MODEL

The development of the reliability model of soil slopes needs the definitions of some basic terms, used in the theory of reliability (Avduevskii et al., 1986), to be concretized. In this way it is possible to be elucidated the connection between the stability and reliability of the slopes:

Failure of the element: limit equilibrium setting in the formed failure surface at the base of a slice as a result of which its working capacity is disturbed;

Failure of the system: limit equilibrium setting in the formed failure surface of the slope massive at which its working capacity is disturbed;

Working capacity of the element: a state of the slice, at which the value of the stability criterion, characterizing its ability to fulfil the destined functions, corresponds to the normal requirements. The definition of the working capacity of the system is analogous and it is valid for the state of the whole slope;

Reliability of the element, respective to the system: the property of the element to conserve its exploitation qualities in time, respective to the system.

It is understood from the given definitions that the element is identified with a slice of the potential slide mass as it is indivisible into constituent parts for the adopted level of the slope stability modelling. In relation to the system there is an identification with the potential slide mass and it represents as aggregate of mutually connected and acted elements, which form a united whole.

4 RELIABILITY ANALYSIS OF SOIL SLOPES

4.1 *Reliability of element of the system*

The reliability af an element should be assessed with the probability that during its period of functioning a failure could not occur. This probability can be represented as:

$$P(t) = \text{Prob}\left[F_{l_i}(S) > 1\right], \quad S \in [0, t]. \quad (5)$$

The local factor of safety in (5) is considered as a function of time, i.e. a realization of a random process. Indeed it is because $F_{l_i}(t)$ is a continuous random quantity in every momemt of time as it is a function of physical parameters of the the soil massive, which are also continuous random quantities. For the i-th slice of the slide mass $F_{l_i}(t)$ can be represented as:

$$F_{l_i}(t) = a_i f_i + b_i c_i(t), \quad (6)$$

where: a_i and b_i are coefficients, ($a_i = N_i'/T_i$, $b_i = l_i/T_i$).

We assume that all parameters in (6) are deterministic quantities with the exception of f_i (random quantity) and $c_i(t)$ (random process). The statistical characteristics of F_{l_i} will be:

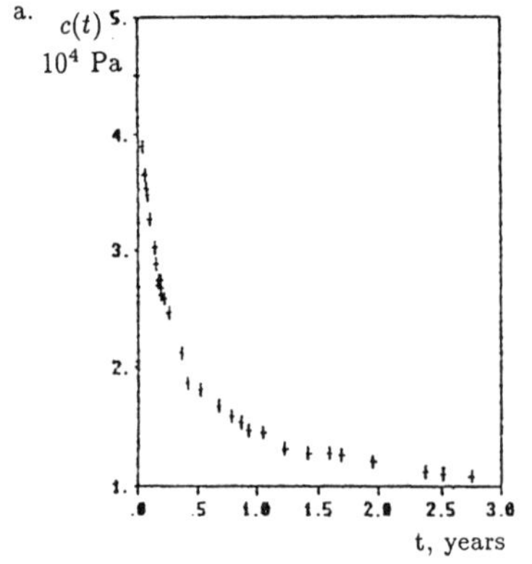

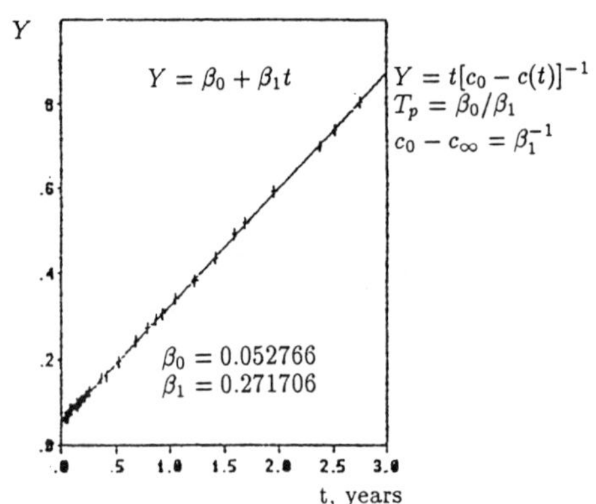

Fig. 3 (a) Correlation field of experimental data and (b) regression model of the same data after a transformation.

— Mathematical expectation:

$$E\left[F_{l_i}(t)\right] = a_i E(f_i) + b_i E[c_i(t)], \quad (7)$$

where: $E(f_i)$ is the mathematical expectation of f_i;

$E[c_i(t)]$ — mathematical expectation of $c_i(t)$.

The mathematical expectation of $c_i(t)$ should be assessed with respect to the model (4). This is however comparatively a complex problem in technical relation and because of that we represent this model as a nonstationary process of the kind—

$$c_i(t) = n_i(t) + m_i(t), \quad (8)$$

where: $n_i(t)$ is a deterministic monotonous function of time;

$m_i(t)$ — normal stationary process with $E[m_i(t)] = 0$.

Therefore the expression of the mathematical expectation of $c_i(t)$ should be the following:

$$E[c_i(t)] = n_i(t), \quad (9)$$

— Dispersion:

$$D[F_{l_i}(t)] = a_i^2 D(f_i) + b_i^2 D[c_i(t)] + 2a_i b_i \mathrm{Cov}[f_i,\, c_i(t)], \quad (10)$$

where: $D(f_i)$ is the dispersion of f_i;

$D[c_i(t)]$ — dispersion of the process $c_i(t)$:

$$D[c_i(t)] = D[m_i(t)], \quad (11)$$

$D[m_i(t)]$ — dispersion of the process $m_i(t)$;

$\mathrm{Cov}[f_i,\, c_i(t)]$ — common covariation of f_i and $c_i(t)$:

$$\mathrm{Cov}[f_i,\, c_i(t)] = E[f_i - E(f_i)][c_i(t) - E[c_i(t)]], \quad (12)$$

— Correlation function:

$$K_{F_{l_i}}(t_1,\, t_2) = a_i^2 D(f_i) + a_i b_i \{\mathrm{Cov}[f_i,\, c_i(t_1)] + \mathrm{Cov}[f_i,\, c_i(t_2)]\} + b_i^2 \mathrm{Cov}[c_i(t_1),\, c_i(t_2)], \quad (13)$$

where: $\mathrm{Cov}[c_i(t_1),\, c_i(t_2)]$ is the autocorrelation function of the shear strength process $c_i(t)$.

— Normalized correlation function:

$$R_{F_{l_i}}(t_1,\, t_2) = K_{F_{l_i}}(t_1,\, t_2) D\left[F_{l_i}(t)\right]^{-1} \quad (14)$$

— Common correlation function:

$$K_{Fl_i(t),F_{l_j}(t)}(t_1,\, t_2) = a_i a_j \mathrm{Cov}[f_i,\, f_j] + a_i b_j \mathrm{Cov}[f_i,\, c_j(t_2)] + a_j b_i \mathrm{Cov}[c_i(t_1),\, f_j] + b_i b_j \mathrm{Cov}[c_i(t_1),\, c_j(t_2)]. \quad (15)$$

The expression (15) allows the correlation function of two detached processes $F_{l_i}(t)$ and $F_{l_j}(t)$ to be assessed, which characterize the state of two slices of the slide mass.

With aim to be reached a completeness of the reliability model we suggest the use of a reliability index:

$$\beta_i(t) = E[F_{l_i}(t)] D[F_{l_i}(t)]^{-0,5}. \quad (16)$$

4.2 *Reliability of the system*

The reliability of the system on the analogy of the slice reliability is assessed with the probability that during its period of exploitation a failure couldn't occur:

$$P(t) = \mathrm{Prob}[F_s(S) > 1],\ S \in [0,\, t]. \quad (17)$$

The average factor of safety should be treated as a random process for considerations presented in the model of the element relia-

Table 1. Results for the slices reliability

Slice No.	Coefficients a	b	Index β	$E[F(t=0)]$ l	$D[F(t=0)]$ l	Probability %
1	40.0000	29.0600	9.15	139.25	231.8105E-00	100.00
2	33.3333	8.0920	9.85	43.50	19.5036E-00	100.00
3	22.2222	3.2599	10.42	19.39	3.4669E-00	100.00
4	9.0909	0.9803	10.83	6.34	0.3429E-00	100.00
5	7.1429	0.6192	11.15	4.31	0.1491E-00	100.00
6	4.0000	0.3064	11.33	2.23	3.8743E-02	100.00
7	2.8571	0.2059	11.41	1.53	1.8062E-02	99.99
8	2.5000	0.1705	11.49	1.30	1.2784E-02	99.58
9	2.0000	0.1378	11.47	1.04	8.2974E-03	67.00
10	1.6667	0.1199	11.42	0.89	6.1327E-03	8.08
11	1.4286	0.1192	11.20	0.84	5.6219E-03	1.70
12	1.0000	0.1368	10.51	0.83	6.2058E-03	1.58
13	0.7407	0.2057	9.73	1.08	1.2396E-02	76.42

bility. Having used the assumptions in that model for the character of the strength parameters variation we determine the following expressions of the characteristics of F_s:

— Mathematical expectation:

$$E[F_s(t)] = \sum_{i=1}^{n} E[F_{l_i}(t)]\delta_i, \qquad (18)$$

— Dispersion:

$$D[F_s(t)] = \sum_{i=1}^{n}\sum_{j=1}^{n} \delta_i\delta_j \mathrm{Cov}[F_{l_i}(t),\, F_{l_j}(t)], \qquad (19)$$

where: $\mathrm{Cov}[F_{l_i}(t),\, F_{l_j}(t)]$ is a common covariation of $F_{l_i}(t)$ and $F_{l_j}(t)$:

$$\mathrm{Cov}[F_{l_i}(t),\, F_{l_j}(t)] = a_i a_j \mathrm{Cov}[f_i,\, f_j] + a_i b_j \mathrm{Cov}[f_i,\, c_j(t)] + a_j b_i \mathrm{Cov}[c_i(t),\, f_j] + b_i b_j \mathrm{Cov}[c_i(t),\, c_j(t)], \qquad (20)$$

— Correlation function:

$$K_{F_s(t)}(t_1,\, t_2) = \sum_{i=1}^{n}\sum_{j=1}^{n} \delta_i\delta_j K_{F_{l_i}(t), F_{l_j}(t)}(t_1,\, t_2), \qquad (21)$$

— Normalized correlation function:

$$R_{F_s}(t_1,\, t_2) = K_{F_s}(t_1,\, t_2) D[F_s(t)]^{-1}. \qquad (22)$$

5 PROBABILISTIC MODEL OF PROGRESSIVE FAILURE

An original stochastic interpretation of the progressive failure is given by Chowdhury et al. (1987), who used for this aim the conditional probability theorem. Here we suggest a solution keeping in common lines to their approach because of the circumstance that the progressive failure modelling of the slopes is impossible without taking into account the convention at the realization of the events having a relation with the dynamics of the slide process.

The consideration of the slide mechanism on the basis of the long-term strength model of a saturated cohesive soil under drained conditions assumes the use of two levels of this strength. For the frictional coefficient these are the peak f^p and residual f^r values. As regards the cohesion, which varies in time, the peak and residual values are inherent for the moment of the slope construction ($t = 0$). For the functioning slope the shear strength can vary within the borders of the intermediate $c^m(t)$ and residual $c^r(t)$ values till the moment of reaching the limit of the long-term strength c_∞. It is sensibly here for the residual strength to be accepted the limit of the long-term shear strength for the plastic clays in accordance with the opinion of Bjerrum, while for the hard overconsolidated clays it is necessury to be used the minimum shear strength c_d ($c_d < c_\infty$). The slope stability at the beginning of the deformation process, which precedes the slope failure, should be assessed by the formula:

$$F_s(t) = \sum_{i=1}^{k} F_{l_i}^{r}(t)\delta_i + \sum_{i=1}^{z} F_{l_i}^{m}(t)\delta_i, \qquad (23)$$

where: $F_{l_i}^{r}(t)$ and $F_{l_i}^{m}(t)$ are local factors of safety, evaluated with strength parameters $(f^r,\, c^r(t))$ and $(f^p,\, c^m(t))$ respectively

Table 2. Results for the slope reliability

Period, years	$E[F(t)]$ s	$D[F(t)]$ s	Reliability %
0.0	1.4943	2.1912E-02	
0.1	1.2077	2.5015E-03	99.99792
0.2	1.0666	2.4991E-03	88.89713
0.3	0.9825	2.4741E-03	29.58566
0.4	0.9268	2.3346E-03	3.47511
0.5	0.8871	2.4343E-03	0.24152
1.0	0.7884	2.3747E-03	0.00000
1.5	0.7479	2.5348E-03	0.00000
2.0	0.7259	2.4549E-03	0.00000

k, z — number respectively of the cut end uncut slices ($k + z = n$).

From formula (23) it follows that "k of n" slices of the slope are unstable, i.e. the requirement for a limit equilibrium is broken at the bases of these slices and a section of the failure surface is formed. The rest "z" slices are stable, i.e. they work and support the whole loading of the slope. It is understood that this formula reflects the most common case of the slope state. If we use the classical treatments of the theory of reliability, it follows that the considered system cant't be classified as a system with serially connected elements (the failure of a slice in the most common case doesn't lead to a failure of the slope) or with parallel connected elements (theoretically the failure of the system at $F_s(t) < 1$ doesn't mean that all $F_l(t) < 1$). This system can be identified with the failsafe system, for which at the reaching a limit equilibrium at some elements it gets a redistribution of the stresses and this system continues to be operable (Augusti et al., 1984).

The slide mechanism is considered as a progressive development of the failure surface from slice or group of slices to slice. In relation to the events E_{j-1} (failure of all slices from the 1st to and including the jth slice) and A_j (failure of the jth slice), the probability, that failure along the surface $(1, j-1)$ is followed by the failure of the surface (j) at a given moment, is denoted by $P\left(A_j/E_{j-1}\right)$. Using the conditional probability theorem:

$$P(A_j/E_{j-1}) = \frac{P\left[F_{s_{j-1}}(t) < 1 \cap F_{l_j}(t) < 1\right]}{P\left[F_{s_{j-1}}(t) < 1\right]}, \tag{24}$$

where: $F_{s_{j-1}}(t)$ and $F_{l_j}(t)$ are factors of safety, associated with the events E_{j-1} and A_j:

$$F_{s_{j-1}}(t) = \sum_{i=1}^{j-1} F^r_{l_i}(t)\delta_i + \sum_{i=j}^{n} F^m_{l_i}(t)\delta_i, \tag{23a}$$

$$F_{l_j}(t) = a_j f^p_j + b_j c^m_j(t). \tag{6a}$$

The calculation of of $P(A_j/E_{j-1}$ requires the obtainment of assessments for the statistical characteristics of $F_{s_{j-1}}(t)$:

— Mathematical Expectation:

$$E\left[F_{s_{j-1}}(t)\right] = \sum_{i=1}^{j-1} E\left[F^r_{l_i}(t)\right]\delta_i + \sum_{i=j}^{n} E\left[F^m_{l_i}(t)\right]\delta_i, \tag{25}$$

where: $E[F^r_{l_i}(t)]$ and $E[F^m_{l_i}(t)]$ are expectations of $F^r_{l_i}(t)$ and $F^m_{l_i}(t)$ respectively:

$$E[F^r_{l_i}(t)] = a_i E[f^r_i] + b_i E[c^r_i(t)], \tag{7a}$$

Table 3. Results for the progressive failure of the slope

Number of cut slices	$E[F(t)]$ s $t = 0.2$	$D[F(t)]$ s $t = 0.2$	$P[E]$ %	$P[E$ and $A]$ %	$P[A/E]$ %	Reliability %
0	1.0666	2.4991E-03	9.14820			88.89713
1	1.0378	2.5142E-03	22.59489	0.00000	0.00000	73.83014
1 - 2	1.0074	2.4812E-03	44.12426	0.00000	0.00000	51.18319
1 - 3	0.9755	2.4058E-03	69.11880	0.00000	0.00000	26.69292
1 - 4	0.9422	2.3013E-03	88.60602	0.00000	0.00000	9.10092
1 - 5	0.9074	2.1827E-03	97.62334	0.00000	0.00000	1.71061
1 - 6	0.8711	2.0742E-03	99.76636	0.52923	0.53047	0.14714
1 - 7	0.8333	1.9981E-03	99.99037	64.86053	64.86678	0.00510
1 - 8	0.7944	1.9653E-03	99.99982	99.99980	99.99998	0.00008
1 - 9	0.7542	1.9950E-03	100.00000	100.00000	100.00000	0.00000
1 - 10	0.7125	2.0966E-03	100.00000	100.00000	100.00000	0.00000
1 - 11	0.6702	2.2773E-03	100.00000	100.00000	100.00000	0.00000
1 - 12	0.6249	2.5543E-03	100.00000	100.00000	100.00000	0.00000

$$E[F_{l_i}^m(t)] = a_i E[f_i^p] + b_i E[c_i^m(t)], \qquad (7b)$$

— Dispersion:

$$D[F_{s_{j-1}}(t) = \sum_{i=1}^{j-1}\sum_{q=1}^{j-1} \delta_i\delta_q \text{Cov}[F_{l_i}^r(t),\ F_{l_q}^r(t)]$$

$$+ 2\sum_{i=1}^{j-1}\sum_{q=j}^{n} \delta_i\delta_q \text{Cov}[F_{l_i}^r(t),\ F_{l_q}^m(t)]$$

$$+ \sum_{i=j}^{n}\sum_{q=j}^{n} \delta_i\delta_q \text{Cov}[F_{l_i}^m(t),\ F_{l_q}^m(t)], \qquad (26)$$

The assessments of the characteristics of $F_{l_j}(t)$ can be obtained using (7) and (10).

It is assumed for this progressive reliability model that the slope soil is perfectly brittle and the shear strength drops to the residual value at the place of the failure realization.

6 PRACTICAL EXAMPLE

The stability of a hypothetical slope is investigated. The data for the physical properties of the clays built the slope are borrowed from (Ilin et al., 1985). The assessments of the mean and deviation for f^p are 0.2126, 0.0611, for f^r — 0.1773, 0.0108, for c_0 — $4.5.10^4$ *Pa*, $1.0735.10^4$ *Pa*. These parameters have a normal distribution. The data of the process $c(t)$ are obtained through a back analysis of fallen slopes. Fig. 3a illustrates their variability in time, which is linear approximated in the least squares method by a transformation of the model (4) (Fig. 3b). The following estimates are obtained for T_p — 0.1942 years, c_∞ — $0.8196.10^4$ *Pa*.

The verification of the hypothesis for a zero mathematical expectation of the normal stationary process of c(t) with the t-test confirmed its plausibility at a confidence probability 0.95.

The procedures on the calculation of the slope reliability are made by the program SLOPREL, written in QBasic.

The results of the reliability of the slices are given in Table 1. They show that the local factor of safety varies within large borders at $t = 0$: from 0.83 to 139.25. This tendency in the variability of the reliability index and probability of slices failure is analogous.

The results for the slope reliability are given in Table 2. The reliability is assessed for some periods of the slope functioning. The results in Table 3 are concerned with the slope reliability at a progressive failure. Here the value of c_∞ is used as an estimate for the long-term residual shear strength.

A problem arises at the calculation of the slope reliability. This is the formation of a sample in a section of the process c(t) for assessing covariances and autocorrelation functions. In the present case only one realization is known, not an ensemble of realizations. That doesn't allow a sample of values to be formed for every section of the process. For solving this problem we suggest a method for experimental values projection in a process section.

From the analysis of Fig. 4a it is possible to be fixed two parts of the process: nonstationary and stationary. This division can be made by criteria for trend. Having used such a criterion we fixed the border of these parts at $t = 1.25$ years. For the stationary part ac-

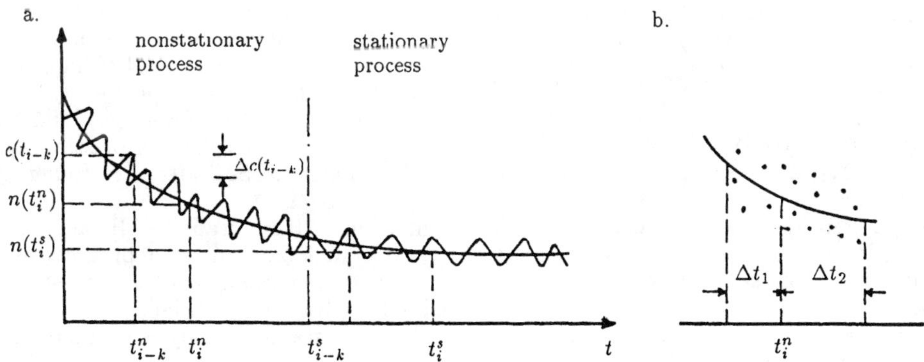

Fig. 4 Structure of the long-term shear strength process.

cording to the trend analysis is known that the values of the process are independent in between and they can be treated as realizations of a random quantity. Hence the formation of a sample in the section t_i can be realized through projection of experimental values in it irrespective of their situation within the borders of this part. Analytically it may be presented in the following way:

$$c^j(t_i^s) = c(t_{i-k}^s), \qquad (27)$$

where: $c^j(t_i^s)$ is a projected value in the section t_i^s; $c(t_{i-k}^s)$ — a value of the section t_{i-k}^s being projected in the section t_i^s.

For the formation of a sample with a given size an equal quantity of experimental values from the left and right of the section can be taken as they are nearest by time.

Regarding the nonstationary part, which has a different speed of variation, the projection of values in the section t_i^n should be realized in this way:

$$c^j(t_i^n) = n(t_i^n) + \Delta c(t_{i-k}^n), \qquad (28)$$

where: $n(t_i^n)$ is the deterministic component of the process in t_i^n; $\Delta c(t_{i-k}^n)$ — residual in the section t_{i-k}^n:

$$\Delta c(t_{i-k}^n) = c(t_{i-k}^n) - n(t_{i-k}^n), \qquad (29)$$

For this part of the process the interval division into two equal subintervals like the stationary process is incorrect as the process has a different speed of variation. Having taken into account the circumstance that the left subinterval is characterized with a bigger speed of variation and therefore the experimental values have weaker connection with the given section as derivates of time in contrast to these ones of the right subintervals, we suggest the projection of a given number of values in the section t_i^n (Fig. 4b) to be made on the basis of the following equality:

$$|n'(t_i^n - \Delta t_1)|\Delta t_1 = |n'(t_i^n + \Delta t_2)|\Delta t_2, \qquad (30)$$

where: $n'(t_i^n - \Delta t_1)$, $n'(t_i^n + \Delta t_2)$ are the process speeds respectively in time $(t_i^n - \Delta t_1)$ and $(t_i^n + \Delta t_2)$; Δt_1, Δt_2 — length of the left and right subintervals.

Using a step-by-step assignment of Δt_1 it is possible to assess Δt_2 so that the necessary quantity of data must fall into the whole interval.

The formation of samples in the suggested method doesn't change the stationary character of variation respectively of the residuals for the nonstationary part and of the values of the long-term strength for the stationary part.

Another problem, which also arises, is the determination of the standart deviation of c_∞ using the expression $c_0 - c_\infty = 1/\beta_1$ (Fig. 3b). Having used the rules of the interval mathematics (Alefeld et all., 1987) we obtain the variation interval of c_∞:

$$[c_\infty^l, c_\infty^u] = [c_0^l - 1/\beta_1^l, c_0^u - 1/\beta_1^u], \qquad (31)$$

where: l and u mean the low and upper interval limits.

Here the confidence intervals assessment is used for determining the intervals of c_0 and β_1 with a confidence probability 0.95. After calculating the interval of c_∞ its standart deviation (0.1527) was assessed with the help of the 3σ rule.

7 CONCLUSION

The developed reliability approach takes into account the time aspect of the soil slope stability in the opencuts through the use of a long-term shear strength model of a cohesive soil under drained conditions. The results obtained from its application on a hypothetical slope confirm its applicability and the correctness of the modelling.

REFERENCES

Alefeld, G., Herzberger, J., 1987. Introduction to interval computations. 1st ed., p. 356 (Mir: Moscow), (in Russian).

Avduevskii, V.S. et al., 1986. Reliability and effectiveness in technics: Manual, 1st ed., (1), p. 223 (Mashinostroenie: Moscow), (in Russian).

Augusti, G., Baratta, A. and Casciati, F., 1984. Probabilistic methods in structural engineering, 1st ed., p. 584 (Chapman and Hall: London — New York).

Bjerrum, L., 1967. Progressive failure in slopes of overconsolidated plastic clay and clay shales. *J. Soil Mech. & Found. Div.*, (93) 5: 3–49.

Chowdhury, R.N., Tang, W.H. and Sidi, I., 1987. Reliability model of progressive slope failure, *Geotechnique*, (37) 4:467–481

Chugh, A.K., 1988. Variable factor of safety in slope stability analysis by limit equilibrium method, IE(I) Journal — CI, (69): 149–155.

Ilin, A.I., Galperin, A.M. and Streltsov, V.I.,

1985. Control of long-term stability of opencut slopes, 1st ed., p. 248 (Nedra: Moscow), (in Russian).
Halatchev, R.A., 1991. Reliability model of spoil banks in the deep open-pit mines, *J. Minno Delo*, No. 1: 27–29, (in Bulgarian).
Skempton, A.W., 1964. Long-term stability of clay slopes. *Geotechnique*, (14) 2: 77–102.
Vanmarcke, E.H., 1977. Reliability of Earth slopes. *J. Geotech. Engng Div. ASCE*, (103) 11: 1227–1246.

Environmental Management, Geo-Water & Engineering Aspects, Chowdhury & Sivakumar (eds)
© 1993 Balkema, Rotterdam. ISBN 90 5410 099 0

Uncertainties in water quality management

N.B.Harmancioglu & N.Alpaslan
Dokuz Eylul University, Turkey

ABSTRACT: The paper presented aims to analyze the diverse sources of uncertainties underlying water quality assessment and management activities. Basically three types of uncertainties are identified: (a) *natural uncertainties*, as those that result from the random nature of water quality phenomena; (b) *man-made uncertainties* that represent the impact of society; and (c) *technological uncertainties* that stem from man's failure to understand and predict the behavior of water quality processes. Such uncertainties are discussed with respect to both their occurrences and their eventual consequences in water quality management. Possible means of evaluating and handling of such uncertainties are proposed for an engineering-based management of water quality problems.

1 INTRODUCTION

The ultimate goal of water quality management lies in two areas: public health and water use. Public health is directly affected by the quality of water supplied for various uses. Water use refers to the utilization of water resources to meet a specific demand and is also a basic objective of water resources development. In this respect, water availability is significantly affected by its quality. Essentially, the two goals of water quality management are interrelated since the consequent impact of water use is on public health.

The above two issues, public health and water use, have technical and economical implications which necessitate efficient and cost-effective programs to control and improve the quality of water resources.The development of such programs, however, is subject to significant uncertainties in both the nature of water quality phenomena and the technological means of interpreting this highly complex issue. Thus, water quality assessment and management constitute a problem of decision making under numerous uncertainties whose sources are often diffuse and complex.

The paper presented aims to analyze the diverse sources of uncertainties underlying water quality assessment and management activities. Basically three types of uncertainties are identified: (a) *natural uncertainties*, as those that result from the random nature of water quality phenomena; (b)*man-made uncertainties* that represent the impact of society; and (c) *technological uncertainties* that stem from man's failure to understand and predict the behavior of water quality processes.

In essence, an engineering approach to control and management of water quality requires the evaluation of associated risk factors as well as the expected benefits. There has been a great deal of advances in assessing various types of uncertainties and the resulting risks for water resources decision-making and management. Such developments relate particularly to water quantity; yet, there still remains a lot to be achieved with respect to risk and reliability aspects of water quality problems. The confusion in water quality risk assessment basically stems from the uncertainties underlying water quality phenomena. Consequently, an efficient risk analysis requires assessment of underlying uncertainties which lead to various risk factors in water quality management.

The emphasis is placed here on water quality monitoring as it is man's only means of being informed about water quality. Monitoring constitutes the link between the actual process and our understanding, interpretation, and assessment of the highly complex phenomena. Adequately accomplished monitoring can increase our knowledge on water quality processes and hence reduce the uncertainties; whereas, results of poor monitoring practices may lead to erroneous interpretations and decisions. Thus, water quality monitoring may be considered as the most significant remedial action with respect to evaluation of the two main goals of water quality management.

In the following sections, types of uncertainties in water quality management are evaluated on practical examples of current management problems. Next, the relation between such uncertainties and accruing risk factors are discussed to finally stress the role of water quality monitoring as the crucial step between assessment of uncertainties and identification of risk factors in water quality management. The basic idea conveyed by the analysis is that best management policies can be achieved only by giving proper credit to uncertainties underlying water quality phenomena.

2 UNCERTAINTIES IN WATER QUALITY AS SOURCES OF RISK

2.1 Uncertainties in the nature of water quality

Water quality occurs as a result of two fundamental mechanisms: (a) the natural hydrologic cycle; (b) man-made effects, which are often referred to as the "impact" of society. Both of these mechanisms, particularly the first one, are affected by the laws of chance so that water quality has to be recognized as a stochastic or random process by nature (Sanders et al., 1983). These two causal factors have different aspects as sources of uncertainty in the nature of water quality so that they are discussed separately in these two sections.

Water may be considered "pure" only in its vapor state. Impurities start to accumulate as soon as it condenses. Then, during different phases of the hydrologic cycle, the quality of water reflects a dynamic character varying as a function of time, space, or both. For example, dissolving of minerals from the soil, occurrence of certain microbiological activities, or the variation of water quantity itself all constitute natural interferences in the quality of water. These effects eventually lead to variations in the physical,chemical, and biological properties of water so that water quality, by nature, becomes a highly complex and uncertain process.

It follows from the above that a multitude of natural factors contribute to water quality. Ortolano (1984) states that, among these natural interferences, "ecological diversity and stability and weather and climate" are highly complicated and "poorly understood". Consequently, it becomes very difficult to predict the quality of water at any point in time and/or space due to the inherent uncertainties. For example, the answer to the question "what is the quality of water?" has to include the outcomes of several variables so that one has to deal with a "vector" of variables instead of a "single" discharge variable in case of water quantity. Sanders et al. (1983) point out that "several hundred variables have already been identified that may be of interest to different users in a comprehensive description of water quality processes".

The random nature of water quality processes has been recognized to be highly significant in water quality assessment and management. Researchers emphasize that one must have an understanding of how water quality processes occur in nature to properly assess and manage them (Sanders et al., 1983; Ward, 1989). However, our understanding is subject to a risk factor produced by the uncertainties inherent in the nature of water quality. The only means of reducing these uncertainties is to collect data on water quality processes.

2.2 Uncertainties related to impact of society

Impact by man also causes random fluctuations in water quality processes. Furthermore, it leads to continuous changes in these processes to make "their properties both inconsistent...and nonhomogeneous" (Sanders et al., 1983). Thus, impact effects complicate water quality phenomena, adding further uncertainties which contribute to a risk factor in our understanding of water quality.

Man-made interferences occur in the form of point or nonpoint pollution sources. Some researchers and practitioners claim that, as a result of intensive research and management efforts, point sources of pollution are fairly well recognized and taken care of (Schad, 1984; Alm, 1984). This may be true to a certain extent for the most conventional pollutants. However, the rapid developments in industry still shed doubts on our understanding of point pollutants.

An important result of industrial development is the change in by-products, coupled with changes also in raw material inputs. For most of these new synthetic products, health hazards have not been identified. Sors (1982) point out that 4 million chemicals have been synthesized within the last decade, and at present this figure has significantly increased. Such chemicals, recently considered to be "hazardous wastes", inevitably appear in both industrial wastewaters and solid wastes. The treatment and disposal of hazardous wastes are generally difficult and expensive. Their illegal or haphazard disposal into surface waters directly results in water pollution with hazardous health effects. On the other hand,improperly managed land disposal of such wastes indirectly produces the same effect as the leachate from these areas reaches surface waters. It is highly difficult to assess the quality of water with respect to these new hazardous products because: first, there are no agencies or equipment, even in developed countries, to produce up-to-date information on how these pollutants are reflected in the quality of water; and second, it may as well be the case that the industrial producer himself is not aware of what specific constituents of the by-products are discharged into surface waters. Furthermore, the problem becomes more complicated if one considers the possible synergistic or antagonistic effects of the new products, which are also unknown when the pollutants themselves are unidentified. For example, several organic compounds detected in drinking water supplies are suspected to be carcinogens. Some of these compounds are synthetically produced and some are formed during the disinfection process. Unfortunately, a few of these have been identified and tested for their hazardous effects on health(Ortolano, 1984).

The magnitude of uncertainties with respect to nonpoint source pollution (NSP) is much greater than it is in the case of point pollution since the former has received relatively less attention in management practices. Schad (1984) claims that NSP is still one of the principal problems related to water quality. This is because the major efforts within the last two decades have been devoted to building of treatment plants for the control of point pollution. Vigon (1985) also makes similar remarks, emphasizing that water quality assessment and management can never be successful unless due consideration is given to NSP. He describes the problem associated with NSP to be the relatively poor identification of nonpoint sources and their effects, coupled with an uncertain approach to their control and management. Indeed, assessment of NSP is subject to uncertainties with respect to both the producing sources of pollution and the required means of control. Currently, the majority of the problem is attributed to agricultural activities, where the use of various types of fertilizers, insecticides and herbicides with different poisonous constituents leads to uncertainties in the occurrence of NSP.

The preceding discussion shows that numerous aspects of point and nonpoint source of pollution contribute to the impact of related uncertainties in water quality phenomena to eventually produce a certain risk in our understanding and assessment of water quality.

2.3 Technological uncertainties

Technological uncertainties stem from man's failure to understand and predict the behavior of water quality processes, which eventually leads to failures in the assessment and management phases. The technological tools for evaluating water quality cover data collection (monitoring) activities, data analysis techniques, model development, establishment of standards, and finally making decisions as to the quality of water. This last procedure is the link between assessment and planning/management phases, the latter covering the design and application of required control measures. The planning and management phase has its own technological uncertainties like, for example, the limitations and uncertainties in certain cleanup technologies or treatment efficiencies. However, if the assessment phase is poorly achieved, the second phase starts with wrong decisions from the very beginning.

Allison et al. (1988) argue that there is uncertainty in our understanding of the cause-effect relationships in water quality so that the decision-making process becomes fairly difficult. They claim that we fail to recognize causes that produce desired or adverse outcomes. The major reason for this uncertainty lies in pollution measurements such that there is "a lack of commonly agreed on means of measuring contamination". Furthermore, data accuracy appears to be the most significant factor in the identification of causes and eventually in the decision making process.

Data collection and acquisition activities, or water quality monitoring practices, are man's only means of being informed about water quality. Thus, water quality monitoring is the most crucial activity on man's side with respect to the assessment of water quality and the associated risk factors. Considering this significant role of monitoring in water quality assessment and risk evaluation, uncertainties that pertain to monitoring practices will be handled here separately in section 2.4.

Current literature provides a vast amount of work on water quality data analyses and modeling efforts. An attempt to cite these could not possibly exhaust all studies. Therefore it will suffice to state here that the majority of researchers complain about the "messy" character of water quality data. Available records are often inadequate in amount with lots of missing and irregularly observed values, lacking also required levels of reliability and accuracy. These are all problems which create uncertainties in our understanding of water quality processes and which are directly associated with monitoring practices.

With respect to water quality modeling, Beck (1987) provides a comprehensive review of relevant uncertainties, which result from model structures (selection of probability distribution functions and models), estimated model parameters, prediction errors, and experimental design "to reduce the critical uncertainties associated with a model". According to Beck, significant progress has been achieved to analyze the above uncertainties; yet there are still further issues to be investigated.

Alpaslan and Harmancioglu (1991) point out that current modeling practices have become more sophisticated and detailed; yet such a trend does not necessarily indicate increased efficiency and accuracy. Problems in model development are related to conceptual difficulties (i.e., inability to understand the basic concepts underlying real phenomena), lack of sufficient and reliable data for model calibration and verification, improper selection of model structure, and, above all, failure to recognize objectives of modeling.

An important source of uncertainty in the technological sense relates to the establishment of standards and assessment of compliance. Two measures are used in evaluating the quality of surface waters: water quality criteria and standards. The former is specified by those characteristics of water that do not produce adverse effects on human health and the natural environment. They are often expressed as the limiting levels of concentrations and intensities of key quality variables, determined for intended uses of water. Standards are based on water quality criteria but are stated as legal rules and regulations by governmental, federal or state authorities to assure compliance. It is desirable to keep the standards at the level of criteria for an effective control of adverse water quality characteristics.

Basically, the determination of standards entails uncertainty when they are associated with water quality criteria. This is because the latter are subject to continuous change as new information about the nature and behavior of water quality is obtained.

Furthermore, new industrial products and their constituents, which may have hazardous effects when discharged, can easily escape the standard simply because they have not yet been recognized in criteria or in standards. Under all these changing conditions, the assessment of compliance by preset standards is subject to uncertainty and consequently to a risk factor. The problem may be that we don't recognize the correct indicators of pollution to develop proper standards. Or, it may be the case that we fail to select among available standards in evaluating water quality characteristics.

Another difficulty associated with compliance is that it is based on sampling. In this context, it is highly affected by sampling errors and the resulting uncertainties. For example, the actual quality of water at a time and space point may be bad, but it may go unnoticed if it is not sampled at that time. Or, if a sample taken at a certain time shows that the quality is good, it is assumed until the next sampling that it will remain good. In each case, our decisions carry a risk of failing to observe the actual quality of water (Warn, 1988). This risk factor depends significantly on the selection of sampling frequencies and is therefore associated with monitoring activities.

2.4 Uncertainties associated with water quality monitoring

The role of monitoring has an important consequence: we can assess various uncertainties (and risks) realistically, provided that monitoring is efficiently realized. Otherwise, if monitoring practices are underlined by specific uncertainties of their own, these will be superposed on what already exists in water quality processes. Then, the assessment of water quality and the associated risks will be set on incorrect grounds to begin with. No matter how well the underlying processes for water quality are recognized, resulting assumptions cannot be verified when reliable and sufficient data do not exist (Ortolano, 1984).

In essence, monitoring is recognized to be one of the principal problem areas in water quality assessment. Schad (1984) claims that "we are not really sure of the cost-effectiveness of some of the programs accomplished to date because of the lack of adequate monitoring of water quality in our streams, lakes, and estuaries". According to Ward (1989), our understanding of environmental processes and problems evolve quite rapidly, whereas monitoring systems develop at a slower pace, often becoming out of date with respect to recently emerging issues and purposes of water quality assessment. On the other hand, the decision-making process in water quality management is highly sensitive to the reliability and accuracy of available data.

At present, monitoring activities have indeed become sophisticated with new methods and technologies. However, when it comes to utilizing collected data, no matter how numerous they may be, one often finds that available samples fail to meet specific data requirements foreseen for the solution of a certain problem. In this case, monitoring practices are described to be unsatisfactory. The problem seems to be that we are good at collecting data but poor in extracting and evaluating the information they convey, basically because of our failure to define properly the objectives of monitoring. Ward et al. (1986) refer to this situation as the "data-rich but information-poor syndrome" in water quality monitoring.

It follows then that there are still problems and failures associated with current monitoring practices, which are consequently reflected in the collected data. First of all, there are several monitoring systems especially in developed countries, but they are not coordinated into an overall information system (Sanders et al., 1983). The design of water quality monitoring networks is still a controversial issue since it is not yet clear what is expected from monitoring. Once the design is inadequately realized, the resulting data are also inadequate. Consequently, there are uncertainties with respect to design of monitoring systems, to begin with. Next, there are operational difficulties such as equipment failures, measurement inaccuracies and uncertainties of sampling and laboratory techniques which lead to unreliable and inadequate data. Physical conditions like variations in pollutant loading, streamflow levels, dilution effects, travel times and variations of water quality across the river section, contribute to uncertainties in water quality sampling.

Some researchers claim that the current measurement techniques have become so advanced that we can now identify more variation and consequently more uncertainties in the behavior of water quality processes (Sors, 1982; Beck and Finney, 1987). Therefore, it is no longer sufficient to measure the average or the extreme values of water quality variables. This is both an advantage and a disadvantage. On one hand, we gain more insight into water quality processes; on the other hand, we are faced with new uncertainties as to the kinds and levels of constituents, which require further analyses.

More specifically, there are four basic problems that need to be solved for an efficient monitoring system. These problems pertain to the selection of sampling sites, sampling frequencies, variables to be sampled, and the duration of sampling. The majority of the uncertainties associated with monitoring stem from the selection of these four factors and, therefore, represent the risk factors in monitoring. For example, the failure to pinpoint representative sampling sites can lead to an incorrect assessment of water quality conditions over a region. The selection of temporal frequencies is an equally significant issue, where a compromise has to be made between the costs of frequent sampling and the risk of infrequent sampling as the failure to detect changes in quality conditions. Selection of variables to be sampled is another source of uncertainty that affects water quality assessment. For example,

Chapman et al. (1982) refer to EPA's list of "priority of pollutants" comprising 129 toxic chemicals and point out to the difficulty of developing sampling programs to include all of these pollutants. If sampled variables are not representative of ambient water quality conditions, the decisions based on irrelevant data may be misleading. Furthermore, the natural and impact-induced inputs that determine water quality change in time so that previously selected variables may no longer be sufficient to indicate any sign of pollution. This is especially true for the discharge of new constituents of industrial by-products, which are often unidentified even by the producers. Currently, the emphasis is on biomonitoring to describe the effects of micropollutants on health and ecosystem conditions. Biomonitoring is an essential part of monitoring, yet it is highly complicated due to difficulties in measuring micropollutants. Furthermore, the possible effects of micropollutants on human health and the overall ecosystem are often uncertain.

On the other hand, there are still uncertainties on how to determine the duration of sampling programs. The question then is "how long is long enough" for a proper assessment of water quality conditions.

In view of the preceding considerations, it may be stated that there are still difficulties associated with water quality monitoring activities. These problems eventually lead to uncertainties and therefore risks in the assessment of water quality.

3 RELATIONSHIP BETWEEN UNCERTAINTIES AND RISK FACTORS

It is common practice in engineering to define safety factors upon which all decisions are based. In design and operation of all projects, e.g. highways, bridges, high-rise buildings or even dams, safety factors are defined to minimize the eventual risks to people. To assure safety in water quality management, the same approach has to be used. However, the basic problem is that it is highly difficult to define such safety factors in water quality related problems due to numerous uncertainties involved. What's more, we are not 100% sure what "safety" or "safety measures" are in case of water quality. This is essentially why water quality management is highly complex and difficult.

The above problem then puts significant demands on risk assessment in water quality management. Risk factors pertaining to each decision regarding water quality must be clearly identified. Furthermore,they have to be defined in quantitative terms to achieve reliable results in management.However, risk identification is again one of the major problem areas related to water quality, as reflected by the prevailing confusion in risk and reliability analyses.

One possible way of approaching this problem is to identify first the uncertainties underlying the problem faced. Such uncertainties may be one or more of the three types discussed in section 2. In essence, these uncertainties are the sources of basic risk factors to be considered in water quality assessment and management. Accordingly, with respect to their sources, one may define, as in Fig.1, *natural risks* as those that stem from the uncertainties in the nature of water quality, *impact risks* to result from uncertainties due to the impact of society, and *technological risks* that relate to uncertainties in the technologies applied. Such risk factors exist as *true risk* despite our knowledge or lack of knowledge about them (Chow, 1979). The quantifiable part of these factors is referred to as "risk" (natural, impact, or technological); whereas, the unquantifiable part still remains as uncertainty to represent our lack of knowledge about any *risk factor*. In this case, uncertainty refers to the inevitable errors in man's efforts to understand and predict the behavior of water quality processes.

An important step in risk assessment covers the quantification of risk factors specified. The quantifiable uncertainty, or risk, is basically a statistical concept, representing the probability of failure of achieving a specified goal. In case of water quality, the risk factors identified with respect to their sources have to be expressed as probabilities of failing to accomplish specified objectives. In reality, the random nature of water quality processes justifies the use of probabilistic measures. Recently, the increase in measurement capabilities permits us to observe much better the variability and the uncertainty in the behavior of water quality processes. Consequently, the temporally varying character of water quality variables can be expressed by probability distributions. If water quality is described in probabilistic terms, then the accruing risk factors can also be quantified by probabilistic measures.

The quantification of risk factors covers then the use of probabilistic measures commonly used for engineering-based risk and reliability analyses. They are applicable to the case of water quality by a proper selection of relevant risk factors.

On the other hand, assessment of natural, impact and technological risk factors is an information-based procedure; that is, it requires reliable and sufficient data on water quality variables. This requirement puts significant demands on monitoring networks, as shown in Fig.1 where monitoring appears as the link between uncertainty and the accruing risk factors. Unfortunately, the current status of both the design and the operation of monitoring systems is characterized by significant uncertainties to result in risks of achieving what is expected from monitoring. These uncertainties and risks are mistakenly attributed to water quality itself, since it is evaluated inevitably within the framework of monitoring practices. Thus, the determination of assessment risks depends highly on monitoring, so that uncertainties in the latter need to be cleared out first to properly address the risks in the former.

At present, the only reasonable approach for minimizing uncertainties in monitoring is to specify first the objectives of monitoring since each objective of water quality assessment and management has different data requirements. Such an approach helps to decrease the uncertainties on what to monitor,where, when and how long. In practice, however, the definition of objectives is not an easy task since it requires the consideration of several factors, including social, legal, economic, political, administrative, and operational aspects of monitoring goals and practices.

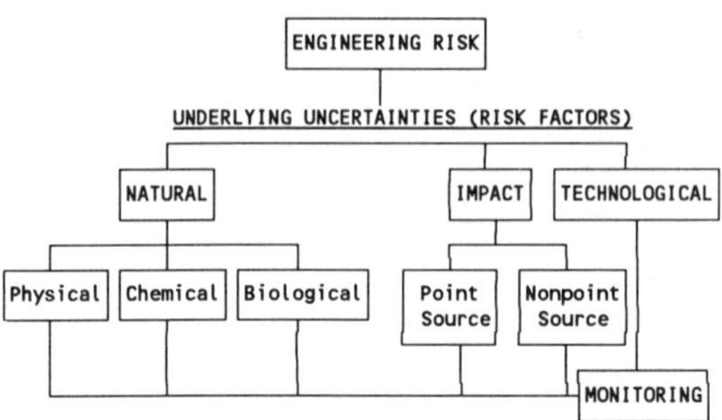

Figure 1. Uncertainties (risk factors) underlying risk components.

Eventually, the lack of a clear and precise statement of objectives results in uncertainties in the initial design of water quality monitoring networks.

Despite its own uncertainties, monitoring appears to be the only effective means of dealing with water quality related uncertainties as there is no other way of being informed about the phenomenon. We need to monitor not only the prevailing conditions of water quality but also the results of every decision we make. Accordingly, the major emphasis must be put on minimizing uncertainties in water quality monitoring networks to effectively deal with the three basic types of uncertainties.

4 CONCLUSION

The paper presented reviews the diverse sources of uncertainties underlying water quality assessment and management activities. *Natural*, *man-made*, and *technological uncertainties* are discussed with respect to both their occurrences and their eventual consequences in water quality management.

The above major types of uncertainties also constitute the natural, impact and technological risk factors to be assessed within the decision-making process. Water quality management requires a reliable risk assessment procedure which can be realized by identifying the nature of underlying uncertainties. Next, by quantifying such risk factors, the best management policies can be obtained.

The emphasis is placed here on water quality monitoring as it is the only means of reducing uncertainties and effectively quantifying risk factors. Since monitoring has its own uncertainties, major efforts must be spent on minimizing these uncertainties so that water quality related problems can be effectively dealt with.

REFERENCES

Allison, R.C., R. Durand & G.Schwebach 1988. Risk: a key water policy issue.*J. of Water Poll. Cont. Fed.* vol 60, 7:1211-1213.

Alm, A.L. 1984. Meeting water quality goals. In: Schad, T.M.(ed.) Options for reaching water quality goals, *Symposium Proceedings, 20th Annual Conference of AWRA,* Washington, D.C., 9-12.

Alpaslan, N. & N. Harmancioglu 1991. Environmental decision models. ECOWARM, *Proc. of the "European Conference on Advances in Water Resources Technology"*, Athens, A.A. Balkema Publ., 339-347.

Beck, M.B. 1987. Water quality modelling: a review of the analysis of uncertainty. *Water Resour. Res.* 23(8), 1393-1442.

Beck, M.B. & B.A. Finney 1987. Operational water quality management: Problem context and evaluation of a model for river quality. *Water Resour. Res.* 23 (11), 2030-2042.

Chapman, P.M., G.P. Romberg & G.A. Vigers 1982. Design of monitoring studies for priority pollutants. *J. of Water Poll. Cont. Fed.* vol.54, 3:292-297.

Chow, V.T. 1979. Risk and reliability analysis applied to water resources in practice. In: McBean, E.A., K.W. Hipel & T.E. Unny (eds) Reliability in Water Resources Management, Fort Collins, *Water Resour. Publ.*, 243-252.

Ortolano, L. 1984.*Environmental planning and decision making.*John Wiley & Sons,New York.

Sanders, T.G., R.C. Ward, J.C. Loftis, T.D. Steele, D.D. Adrian & V. Yevjevich 1983. *Design of networks for monitoring water quality.*Littleton CO:Water Res. Publ.,328p.

Schad, T.M. 1984. Introduction In: Schad, T.M.(ed) Options for reaching water quality goals: Symp. Proc., *20th annual conference of AWRA*, Washington, D.C., p.1.

Sors, A.I. 1982. Risk assessment and its use in management a state-of-art review. In: Evaluation and risk assessment of chemicals:*Proceedings of a seminar, 1980, WHO Interim Document* 6, 236-294.

Vigon, B.W. 1985. The status of nonpoint source pollution: its nature, extent and control. *Water Resour. Bull.* 21(2),179-184.

Ward, R.C., J.C. Loftis & G.B. McBride 1986. The data-rich but information-poor syndrome in water quality monitoring.*Env.Man.*, 10, 291-297.

Ward, R.C. 1989. Water quality monitoring-a systems approach to design. In: Ward, R.C., J.C. Loftis & G.B. McBride (eds) *Proc., Int. Symp. on the Design of Water Quality Information Systems,* Fort Collins, CSU Information Series no.61, 37-46.

Warn, A.E. 1988. Auditing the quality of effluent discharges. In: Workshop on Statistical Methods for the Assessment of Point Source Pollution, 12-14 September, *Canada Centre for Inland Waters,* Burlington, Ontario, Canada.

Environmental Management, Geo-Water & Engineering Aspects, Chowdhury & Sivakumar (eds)
© 1993 Balkema, Rotterdam. ISBN 90 5410 099 0

Uncertainty in modelling water resource management system for mines

C.V. McQuade & A.D.S. Gillies
Mining Engineering Department, University of Queensland, Qld, Australia

A.H. Lee
Science Faculty, Northern Territory University, N.T., Australia

ABSTRACT: Monte Carlo simulation is used to investigate uncertainty in mine water resource management systems using a mine site in the Northern Territory, Australia. Uncertainty in model parameters are included in a basic water balance model used to simulate monthly pond volume, the output is reviewed for suitability in presenting risk associated with model design. As most proposed mines have little or no relevant historical data for design purposes the uncertainty in system performance is significant.

1 INTRODUCTION

Mining operations use considerable quantities of water in the beneficiation of ore. Pulles (1992) estimated that water use in gold mining operations comprised more than one per cent of the annual water use in South Africa. A mine water resource management system (WRMS) is an integral component of a mining operation, comprising water supply, dewatering, water disposal and flood management. Water is also a major pathway for the transport of potential contaminants from the mine control area into the wider environment. It is for this latter reason that governments are tightening legislation relating to mine WRMS.

Failure of a mine WRMS can be defined by the inability of the system to meet operational water needs or by exceeding government guidelines. These guidelines may require a "no release" system or a release no more frequent than a given annual exceedence probability. The causes of mine WRMS failure are limited and may include insufficient or too much rainfall-runoff water yield, pump failure, insufficient pump capacity, a sudden change in production rate, a change in government requirements (such as the inclusion of rehabilitation guidelines). However, the primary cause of system failure is likely to be the inability of the system to manage water volumes. Too much water may result in exceedance of government guidelines, too little water may result in a mill or mine shut down.

Mine WRMS are typically designed using empirical or physically based models relying on historically relevant data or by simulating expected system performance using stochastically generated data. These models need to be sensitive to the mine physical characteristics, which include:

1. A developing open pit.
2. Waste rock dumps which vary in size and shape during the mining operation.
3. Infrastructure areas which typically have minimal infiltration.
5. Tailings disposal system.
6. Water runoff control areas associated with water supply or environmental management.
7. Water supply requirements which tend to be reasonably constant throughout the mine life, primarily related to the processing rate of ore.
8. Rehabilitation requirements.

Table 1. Characteristics of the Simulated Rainfall

Month	Median (mm)	3rd Decile (mm)	7th Decile (mm)
Oct	25	16	36
Nov	141	112	167
Dec	219	190	252
Jan	336	301	378
Feb	258	198	331
Mar	238	182	205
Apr	43	24	73

Table 2. Pan/lake Coefficients

Month	Vardavous (1989)	Manton Dam	Water Balance	Estimate	SE
Sep	0.66	0.83	0.64	0.64	0.08
Oct	0.66	0.88	0.79	0.79	0.07
Nov	0.75	0.91	na	0.75	0.15
Dec	0.84	1.10	na	0.85	0.15
Jan	0.92	1.13	na	0.90	0.15
Feb	0.92	1.09	na	0.90	0.15
Mar	0.95	1.06	na	0.95	0.15
Apr	0.77	1.01	na	0.75	0.15
May	0.70	0.94	0.50	0.70	0.20
Jun	0.70	0.85	0.75	0.75	0.07
Jul	0.66	0.79	0.82	0.80	0.10
Aug	0.64	0.75	0.71	0.70	0.15

Uncertainty is inherent in system design which raises the question: what is the reliability of the system in meeting the water supply requirements and in meeting government control guidelines? Such questions are of considerable concern in municipal water supply systems throughout the developed world and continue to receive considerable research effort (Singh, 1987; McTeran and Kaplan, 1990). However, these question do not appear to have been addressed for mining operations.

The reliability of simulation results produced by mine WRMS models is a function of the uncertainties associated with nature, data, model parameters and model structure (Schilling and Fuchs, 1986). Natural uncertainties refer to random, temporal and areal fluctuations inherent in natural processes. Data uncertainty is associated with measurement inaccuracy and errors, adequacy of the parameter data and data handling. The uncertainty in selection of the appropriate model parameter values to be used for a simulation comprises the model parameter uncertainty. Model structures are at best simplifications of the physical system and form the last uncertainty.

Current design techniques tend to use simple models which are (Hollingsworth, Dames and Moore, 1992) based upon a monthly time step, as the rainfall runoff relationship becomes more linear with increased time step. Rainfall is often pseudo randomly extracted from the single realisation of the site or regional record. Model parameters are estimated from regional studies or engineering best guess. The estimation of the expected values (means) of the model parameters fails to give an insight on the possible variation in modelling output. Parameter sensitivity analysis appears to be the most likely tool to examine possible output variability and is the minimum recommended requirement for such an exercise (Pilgrim, 1987).

Output from sensitivity analysis is not generally in a form suitable for examination of system performance risk. The approach based upon Monte Carlo simulation may be used to investigate the distributions and variabilities of the outcomes (Fedra, 1983).

This paper summarises a preliminary investigation into the use of stochastic modelling techniques applied to a mine at its proposal stage and operational stage which has five years of rainfall-runoff data. The mine is located within the climate zone Summer Rainfall-Tropical (Lee and Gaffney, 1986).

2 METHODOLOGY

A basic empirical water balance model is proposed for the mine's main water supply pond having a maximum surface area of 19ha. The pond has two catchment types comprising a concrete/asphalt area (50ha) and an ore stockpile area (40ha). For the proposed stage runoff and evaporation coefficients are estimated from engineering and site experience. Based on five years of rainfall, evaporation, pump distribution and catchment area data a rainfall-runoff relationship is derived for use within the model. The model has the form:

$$PondVol_{t+1} = PondVol_t + Runoff - Evap - Milluse \quad (1)$$

where Milluse is 10,000m^3 from January to May and 50,000m^3 for the remaining months.

Daily rainfall data were collected at station 1 (located approximately 3km north west of the supply pond) from 1971, these data were filtered and cumulated into monthly totals. Oenpelli station, approximately 20kms to the north east, was used to extend the data set to 1910 by fitting monthly relationships using least squares.

To generate simulation data the individual months from October to April were fitted to different gamma distributions, the remaining months were assumed to have nil effective rainfall. We also attempted various other distributions, but the gamma model seems to provide the best fit. Two hundred years of monthly rainfall data were simulated using the gamma distribution parameters and zero inserted for the months May to September. This method is preferred to the traditional time series technique used by McQuade, Lee and Gillies (1991) to model Oenpelli rainfall since the month of October is now incorporated into the model where it could not be satisfactorily normalised as required in time series modelling. The median, 3rd and 7th decile values for the simulated data are included in table 1.

Troutman (1982) suggested that the uncertainty in the effective rainfall distribution of the catchment can have a major impact on the success of any hydrologic modelling exercise. A shorter period record available from station 2 (located within 500m of the pond) was compared to the record from station 1 (Figure 1). From this comparison it was considered that station 1 is appropriate for the water balance modelling.

Runoff model parameters for the proposed stage were based upon catchment type (concrete/asphalt and ore dump), catchment area and rainfall and were estimated from experience. The evaporation coefficient was assumed to have a constant value of 0.7, with a standard deviation 0.2 and these values were applied to the average monthly pan evaporation

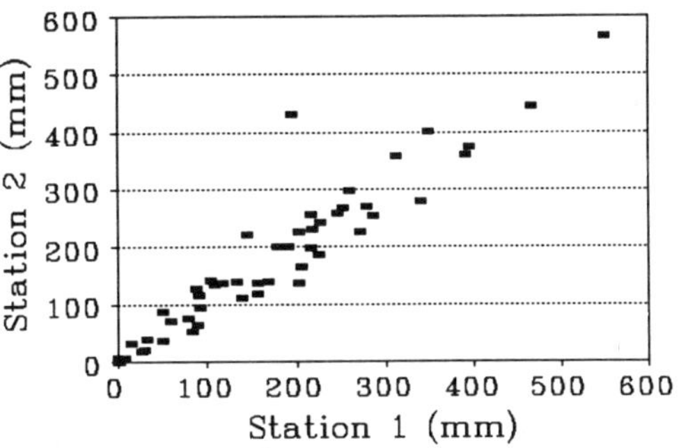

Figure 1. Station 2 Rainfall vs Station 1 Rainfall

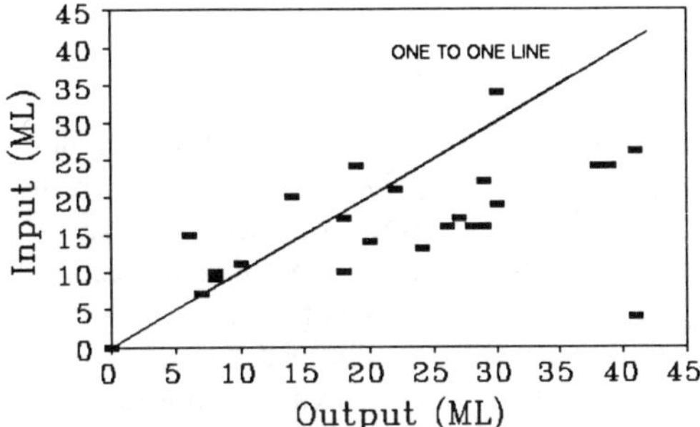

Figure 2. Tank Input vs Output

data for the site.

For the operating stage water transfer volumes and estimated evaporation volumes were extracted from the pond volume for each month to derive rainfall-runoff volume data. Evaporation data were based on evaporation pan readings at station 1 and expected pan/lake coefficients estimated from water balance calculation on a pond 1km to the west. Water transfers were measured using turbine water meters and recorded weekly.

Chiew and McMahon (1992) suggested that pan data are not an appropriate estimate for potential evaporation and should be used with caution. It is therefore expected that considerable error may be associated with the use of pan information. Pan/lake coefficients sighted as applicable to the area are listed in table 2. Considerable variation is evident and the last column lists the selected coefficients and its standard error, assuming that the coefficient

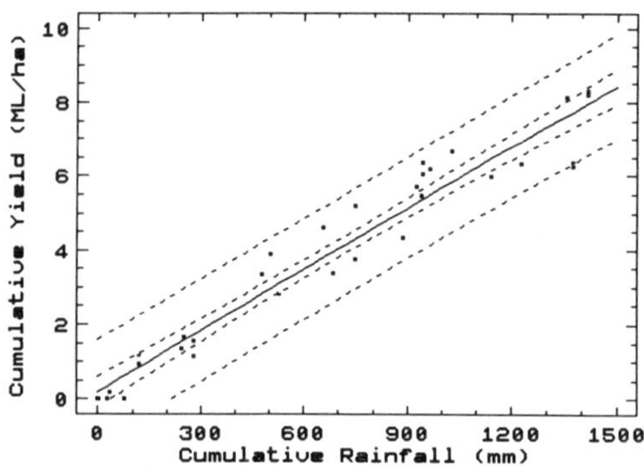

Figure 3. Cumulative Yield vs Cumulative Rainfall

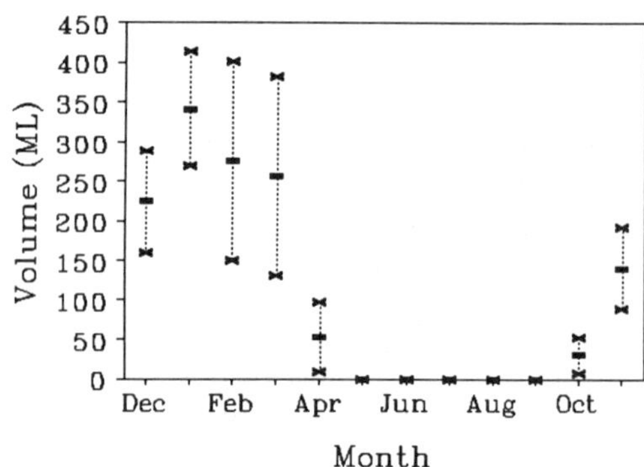

Figure 4. Rainfall-Runoff with Coefficient of 1.0

distribution is Gaussian.

Quoted accuracy for turbine meters installed to specification is ± 2%. However, an examination of the pond water balance data indicated possible higher errors from this source. To check the accuracy of water transfer data, weekly water volumes pumped into a staging tank were plotted against water volumes pumped out of the tank (Figure 2). The water meter recording tank output had been checked twice over the five year period and found to operate within specification. It is clear that the assumption of water meter accuracy may not always be correct.

Several rainfall-runoff relationships were evaluated for fit. The best fit was found using cumulative pond yield data and cumulative rainfall, the relationship is plotted in figure 3 with the 95% prediction limits, yield=0.192453+0.00548766×rainfall, and standard error from the fit is 0.654787; yield is in units of ML/ha.

The model cumulates pond water volume at the end of each month. In December water is either added or disposed such that the volume is 50ML at the end of the month. This is the only flexibility permitted in the operation of the water management system for this basic model. In the actual system there are alternative water storage and water may be added or subtracted to the pond in each month to meet operational objectives and government guidelines. This effectively resets the system in December each year. For the proposed stage the runoff volume for each rainfall value is randomly selected using the appropriate runoff coefficient and its standard error. For the operational stage the expected value and standard error are selected from the rainfall-runoff relationship and used to randomly select the monthly runoff. Evaporation for the operational stage is also randomly selected from the appropriate monthly estimates and standard errors to depict variability of the process.

3. RESULTS AND DISCUSSION

To gauge the natural uncertainty associated with rainfall the generated rainfall is converted to runoff volumes assuming a runoff coefficient of 1.0 for the entire catchment (Figure 4).

The squares represent the expected values and the crosses the standard error for each month. The maximum error of approximately 120ML occurs in the months of February and March. The modelling output for the proposed stage is plotted in figure 5 and the output for the operational stage is plotted in figure 6. In both figures the full squares represent the expected pond water volume at the end of the month and the crosses represent the cumulative error.

The estimated values increase with wet season rainfall till May and then begin to

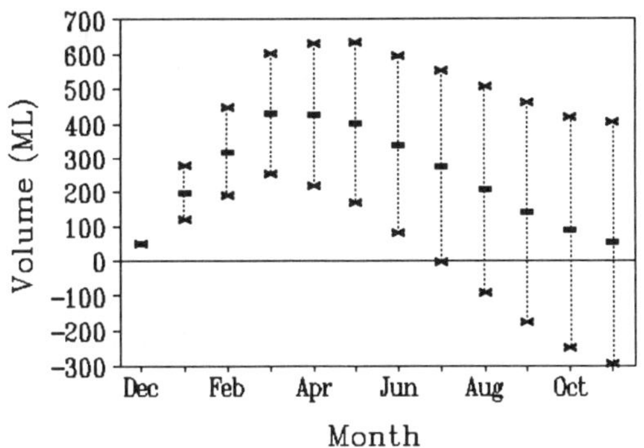

Figure 5. Pond Volume - Proposed Stage

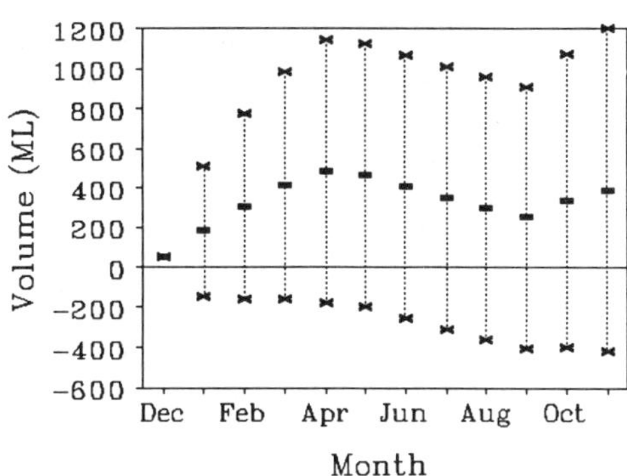

Figure 6. Storage Pond Water Volume

decrease due to the extraction of mill water and evaporation loss. The expected values for both stages are comparable to September at which point for the proposed stage the volume continues to decrease while in the operational case volume increase reflects the commencement of the wet season. The good agreement of the expected values in each case would not normally be expected. The original runoff coefficients used in the design of the mine WRMS were unavailable so we estimated them from specific site experience. Site experience has shown that runoff coefficients for the ore stockpile range between 0.01 and 0.3, such low coefficients would not normally be expected where surface infiltration rates are approximately 3mm/hr.

Uncertainty in pond volumes over the rainfall range is considerable, varying in April from 250ML for the proposed stage to 600ML for the operational stage. The smaller error for the proposed compared to the operational stage was not expected. It occurs for two main reasons, firstly the standard error of the runoff coefficients are estimated and possibly should be larger. Secondly, the error for the proposed stage is separately calculated for the two catchments and then combined, as the runoff from the ore stockpile is small the absolute error from this catchment is small. However the operational stage error is calculated across the combined catchments although the majority of the error is associated with the concrete/asphalt catchment. The application of rainfall-runoff relationships for each catchment type may reduce the overall error in the operational case.

The addition of the evaporation error (approximately 5ML/month) is not significant in the overall error of the pond volume in either case. For practical purposes evaporation could be ignored or considered a constant loss term related to the pond surface area. Evaporation from the mine's tailings dam which is operated as a source of recycle water for the mill could not be ignored as the maximum surface area is $1km^2$ and its catchment area is $1.05km^2$. The uncertainty in the rainfall-runoff relationship represents the major source of error in this basic water balance, which concurs with the findings of Schilling and Fuchs (1986).

The simulation model output may be used to estimate the frequency of exceeding operational criteria if we divide the number of times the system exceeds the criteria by the total months of simulation. Such information may be valuable in the selection of the optimum mine WRMS design.

4 SUMMARY AND CONCLUSIONS

The inclusion of random fluctuations in model parameters allows the presentation of total model error in a single output. This information can then be used to aid the design process and quantify the expected confidence in meeting operational and government guidelines. Our basic model suggests that a high degree of confidence in meeting a requirement of zero release may be expensive to meet, however the output do provide the necessary data for the assessment.

Cordery and Cloke (1991) have suggested that the reduction in uncertainties warrants the collection of quality data which inturn will allow the more accurate determination of the risk or reliability. The five years of data collected in this case does not reflect this statement. If mining operations undertake data collection for future WRMS design modification or to demonstrate compliance with government guidelines then the evidence presented here would indicate priority should be given to collect quality data and not just data.

REFERENCES

Chiew, F.H.S. and McMahon, T.A. 1992. An Australian comparison of Penman's potential evapotranspiration estimates and Class A evaporation pan data. *Aust. J. Soil Res.* 30-:101-112.

Cordery, I. and Cloke, P.S. 1991. An over view of the value of collecting streamflow data, *International Hydrology and Water Resources Symposium:* Perth 2-4 October 1991, Institution of Engineers, Australia.

Fedra, K. 1983. A Monte Carlo approach to estimation and prediction, In; Uncertainty and Forecasting of Water Quality. Ed. Beck and Straten, Springer-Verlag: New York:259-289.

Hollingsworth, Dames and Moore. 1992. *McArthur River Project, Draft Environmental Impact Statement, for Mount Isa Mines Limited*, Job No. 003-293-696.

Lee, D.M. and Gaffney, D.O. 1986. *District rainfall deciles - Australia.* Aust. Govern. Pub. Service, Canberra.

McQuade, C.V., Lee, A.H. and Gillies, A.S. 1991. Rainfall analysis for mine water management system design and operation. *4th International Mine Water Association Congress:* Ljubljana (Slovenia)-Pörtschach (Austria) Sept. 25-30 1991, pp295-303.

McTeran , W.F. and Kaplan, E. 1990. *Risk Assessment for Groundwater Pollution Control.* Amer. Soc. Civil Eng.: New York.

Pilgrim, D.H. 1987. *Australian rainfall and runoff.* The Institution of Engineers: Canberra.

Pulles, W. 1992. Water pollution: its management and control in the South African gold mining industry. *J. Mine Ventilation Soc. South Africa* February 1992:18-36.

Schilling, W, and Fuchs, L. 1986. Errors in stormwater modelling - A quantitative assessment. *ASCE J. Hydraulic Engineering* 112(2):111-123.

Singh, V.P. 1987. *Application of Frequency and Risk in Water Resources.* Reidel: New York.

Troutman, B. 1982. An analysis of input in precipitation-runoff models using regression with errors in the independent variables. *Water Res. Res.* 18(4):947-964.

Vardavas, I.M. 1989. A simple computer model for terrestrial and solar radiation transfer. The Dept. of the Supervising Scientist, *Technical Memorandum No. k26.* Aust. Govern. Pub.

Environmental Management, Geo-Water & Engineering Aspects, Chowdhury & Sivakumar (eds)
© 1993 Balkema, Rotterdam. ISBN 90 5410 099 0

Risk maps for rockfall prone areas: Environmental/human aspects and remediation projects

F. Oboni & V.T.G. Angelillo
Oboni and Associates, Inc., Lausanne, Switzerland

ABSTRACT: On September 24, 1990 several blocks of weathered conglomerate rock rolled down a very steep hill situated just above a developed zone of the north shore of Lake Geneva causing severe damage to houses but miraculously no casualties.

The accident was immediately reverberated by the media resulting in a general concern for this kind of phenomenon in the area. Remarkably, the long-established residents of the community were conscious of the existence of these kinds of events and were ready to assume the risks of living in those beautiful surroundings, whereas new residents felt menaced and insecure after the rockfall and sought legal advice for claiming protective works to be undertaken under state financing. Emergency procedures were immediately organized by the community authorities to temporarily avoid recurrence of rockfalls on the same site and protect the population. Long term stabilization was submitted for public approval and voluntary financing by the potential victims, since Swiss law does not recognize anyone as responsible for damage caused by natural disasters. This paper presents the methodology that we, as the consultants, developed in order to compile natural hazard risk maps for the entire territory of the community along with the solution used to stabilize the cliff. The method used is derived from seismic hazard analysis procedures, and based on probabilistic approaches that combine subjective probability assessment and analytical techniques to define the risks generated by natural hazards.

1. INTRODUCTION

Rockfalls are quite common phenomena in the european Alps; however, the existence of these phenomena has not inhibited human colonization of these areas nor has it stopped Alpine people from building very dense and intricate road networks. Constructions were actually made possible because observation over long periods of time allowed intuitive definition of the most hazardous areas.

In many cases, adequate passive and active protection measures were taken so that villages, roads and in some cases railroads were sheltered. The protective works for this kind of slope can range from ablation of dangerous masses to the setting of berms and other energy absorbing or deflecting devices up to the construction of protection galleries. The design of these works can be effectively achieved when correct data type and quantity are available and all relevant hazards are duly taken into account.

In order to determine the necessity of such protective works or to define a reasonable development strategy in areas affected or prone to rockfalls, the first step is to prepare a Risk Map for Rockfall Prone Areas (RMRPA).

2. RISK MAPS FOR ROCKFALL PRONE AREAS (RMRPA)

When elaborating a RMRPA, suitable methodologies have to be developed, eventually based on subjective probabilities theory because of the scarcity of available data.

It is absolutely necessary that the used methodology permits dialogue among the specialists in charge of different aspects as well as a comparison of the risk level among different sectors. Only then can a rational approach to the risk mitigation be undertaken leading to proportional distribution of both the budget and attention to the different endangered structures.

Outcrops are present in a great diversity of morphological conditions. The particular methodology chosen for this case enables comparing different areas in distinct geological formations, even if various scientists are implied in the job.

This methodology is very similar to those generally used for seismic risk analyses since it uses the same probabilistic techniques, with limited statistical support.

We would like to mention that statistics and probability differ one from the other in the sense that the first condenses data relative to a population or a sampling of a population, whereas the second solves problems that involve stochastic variables or uncertain parameters.

Only rarely are statistics available in the applied earth sciences. More often, it is probabilistic methods which are used in order to deal with "pseudo" stochastic variables characterized by evaluated statistical moments, generally of the first and second order, and only exceptionally of the third order.

When stochastic events are associated with geological features, discrete probabilistic distribution of a parameter can be defined by working with subjective probabilities, i.e. by utilizing judgement, experience and good sense.

2.1 *Terminology*

It is useful to define certain terms that will be used in this paper.

a. Sources are defined as portions of the area studied where stones or blocks can be freed from an outcrop by unidentified natural causes.
b. Projectiles are stones or blocks of stochastic volume v_i, detached from a source, moving downhill under stochastic course and velocity. Generally $v^0<v_i<v^u$ with v^0 and v^u defining respectively the lower threshold value of the volume considered to be potentially damageable, and the higher threshold value of the potential volume dictated by the geology.
c. Fragmentation is the phenomenon where, during its course, a projectile is separated into smaller elements by shocks against obstacles. The projectile of volume v_i can fragment into smaller v_{ip} elements. If the $v_{ip}<v^0$, then the projectile is annihilated during its course.
d. Targets are structures or valuable objects that can intercept a projectile during its course, before its annihilation. Typical targets are houses, roads, electric lines, cultivated fields, forest, equipment and machinery, etc.
e. Fall height is the difference in level between the source and the considered potential target.
f. Potential energy is the product of the projectile's mass and its fall height.
g. Attenuation is the reduction of the kinetic energy of a projectile due to natural or artificial obstacles intercepted during the course.
h. Vulnerability is the susceptibility of the target to damage.
i. Cost of a hit is the cost resulting from the impact of a projectile on a target. Vulnerability plays the role of amplifier of the cost function.
j. Risk, R, is the product of a probability of occurrence of a hit and the related cost.

$$R=p_x \cdot C_x \tag{1}$$

2.2 *Methodology*

Risk evaluation depends on complex data such as:

- Geometry and layout of sources.
- Frequency of phenomena (rarely available).
- Expected trajectories and attenuation.

All these data are characterized by uncertainties mainly due to the incomplete nature of the available or potentially available data.

The following steps have to be taken to obtain a rational RMRPA:

- Definition of the sources, i.e. location of the outcrops, tectonic model, water, and roots.
- Geometry of the sources, i.e. cracks and fractures model, joints, dip.
- Volume of projectiles, fall height linked to each source.
- Frequency of the phenomena for each source (based on historical data if available).
- Potential trajectories, volume-distance relationship, and fragmentation based on observation of ancient phenomena or based on judgement.
- Attenuation as a function of existing obstacles (based on observation of ancient cases, judgement, etc.).

In order to define an RMRPA for a given area we outline four phases of analysis.

The first phase is centered on defining the potential hazardous areas including delineating the potentially endangered targets such as buildings, pilons, roads, etc. in order to determine whether further analysis is pertinent.

The second phase consists of superimposing on a preliminary map the geology and the man-made environment. The compilation of these documents implies a complete geological survey.

The third phase consists of preparing final cartographic documents of the zones where the man-made environment is endangered by the considered hazard. This implies a probabilistic analysis of each source-target couple, as detailed in the next chapter.

The fourth phase consists of the diagnostic inventory of the endangered sectors for each area delineated in the third phase. The description of this phase goes beyond the scope of this paper. The most dangerous sources should generally be monitored or rehabilitated.

2.3 *General RMRPA probabilistic model*

The probability of occurrence of phenomenon Z overstepping a threshold value z can be evaluated as follows:

$$p(Z>z|t) \leq \mu \cdot t \quad (2)$$

where $\mu \cdot t$ is equal to the number of phenomena larger than the evaluated threshold value during observation time t.

If the probability level is low, then:

$$\mu(Z) \cdot t \approx p(Z>z|t) = p_z \quad (3)$$

where $\mu(Z)$ = frequency of overstepping

and, finally, the risk is evaluated as $R = p_z \cdot C_z$, where C_z is the resulting cost of the phenomenon.

If this concept is applied to rockfalls, and if the sources endangering a potential target are independent, then $\mu(Z)$ is evaluated as the sum of the $\mu_n(Z)$ resulting from each one of the n sources on the same target, i.e.

$$\mu(Z) = \Sigma_n \mu_n(Z) \quad (4)$$

where $\mu_n(Z)$ =

$$\alpha(v^0) \cdot \Sigma_i \Sigma_j p(V=v_i|v^0,v^u) \cdot p(S=s_j|v_i) \cdot p(C>c|v_i,s_j)$$

and

$\alpha(v^0)$ = frequency of a phenomenon starting from a generic source, overstepping the evaluation threshold value.

The three conditional probabilities are defined as follows:

a. $p(V=v_i|v^0,v^u)$
Probability that the phenomenon has a volume $v^0<v_i<v^u$. Probability that a phenomenon of a given volume happens. Note that the maximum volume is defined by physical limits sometimes easily detectable on the sources and that the lower significant volume v^0 is arbitrarily fixed by the expert, eventually as a function of the vulnerability and the type of targets, as well as the local topography.

b. $p(S=s_j|v_i)$
Probability that the volume v_i hits a target before fragmentation. Probability of occurrence of a trajectory run by a non-fragmented projectile, function of the geometry of the source, of the constitution of the projectile and the slope.

c. $p(C>c|v_i,s_j)$
Probability that the cost of the consequences over-run a threshold value c given v_i,s_j.

In order to cope with the stochastic or "pseudo stochastic" character of the data, subjective probabilities techniques are used. These are based on the decision tree concept, which takes into account the uncertainties in the evaluation of conditional probabilities by optimally using experience and judgement of qualified experts. Decision trees are prepared for each single potential source detected within the region to be mapped (Power et al, 1981, Kulkarni et al, 1984, Youngs et al, 1985).

The volume distribution for each source is based on in-situ observations and geological analysis. The number of defined branches should cover satisfactorily the probabilistic distribution of the projectiles' volume. Generally three to five branches are sufficient to cover the range, each one receiving a subjective probability so that the sum of the probabilities is equal to unity, following the rule for mutually exclusive phenomena.

After this first step, starting from each "volume branch", the decision tree presents branches corresponding to "hit" classes S. Here, three classes are generally sufficient for an acceptable accuracy of the description of the distribution:

- Crossing hit, the worst, where the projectile crosses through the target like an artillery bullet.
- Penetrating hit, where the projectile penetrates the structure but stops within it.
- Absorbed hit, in the case where the projectile is stopped by the target without penetration.

As done with the volumes, the probabilities of each group of the three classes sum up to unity.

The last level of the decision tree is used to determine the cost of the consequences of a hit. Here very simple solutions can be adopted, using for example a demonetarized cost factor instead of the real cost which can prove extremely intricate and tedious to define. Generally four classes are used:

- Total damage Cost Factor=3
- Major damage Cost Factor=2
- Minor damage Cost Factor=1
- Negligeable damage Cost Factor=0

The probability of occurrence is obtained by successive products along the different branches and the risk is obtained by multiplication of the probability with the cost.

At this point simple statistical rules can evaluate first and second moments of the risk for each source-target couple as well as the probability of overstepping a predefined cost limit.

Once all the sources have been evaluated, it is possible to class them by increasing average risk, and, by selecting appropriate critical thresholds it is possible to create groups.

Three groups of sources are generally defined:

R1: Sources that require an immediate detailed diagnostic inventory and risk mitigation program.

R2: Sources that should be checked and could be classified as R1 afterwards.

R3: Sources that do not require any kind of specific checking and are declassified.

Classes R1 and R2 are generally equipped with monitoring devices.

3. EXAMPLE

On September 24, 1990 several blocks of weathered conglomerate rock rolled down a very steep hill situated just above a developed zone of the north shore of Lake Geneva causing severe damage to houses but miraculously no casualties.

The accident was immediately reverberated by the media resulting in a general concern for this kind of phenomenon in the area. Remarkably, the long-established residents of the community were conscious of the existence of these kinds of events and were ready to accept the risks of living in those beautiful surroundings, whereas new residents felt menaced and insecure after the rockfall and sought legal advice for claiming protective works to be undertaken under state financing.

Emergency procedures were immediately organized by the community authorities to temporarily avoid recurrence of rockfalls on the same site and protect the population. Long-term stabilization was submitted for public approval and voluntary financing by the potential victims, since Swiss law does not recognize anyone as responsible for damage caused by natural disasters nor does it foresee the financing of the rehabilitation of known natural hazards with public funds.

A two year struggle to find a suitable solution ensued between the community authorities, their legal and technical consultants and the residents.

During this time, the local authorities investigated the possibility of forcing the residents to finance the rehabilitation program: it became evident that such a procedure could, in fact, be implemented by law, but that the duration of the procedure would be totally incompatible with the evolution of the risks involved.

Finally, a new and different approach suggested by our office, consisting of organizing information meetings, guided tours of the potential sources and a turnkey rehabilitation program based on a lumpsum contract proved to be successful.

A better understanding of the actual situation as well as the suppression of the monetary loss fear convinced most of the involved residents to spontaneously participate in the financing. This first group asked the authorities to begin the work as soon as possible and undertook the task to convince the rest of the population to adhere to the program.

The rehabilitation works are described in the next section.

Residents who were not directly affected by the event of September 1992 but who live in other exposed sectors of the community had asked the authorities to provide a global survey on the risks generated by similar phenomena elsewhere.

The study focused on a surface area of 4.7km^2, with a total of 17.6km of conglomerate rock cliffs. At the moment of the study, the number of selected R1 and R2 sources were defined as 5 and 9, respectively from a total of over 60 analyzed sources.

The final result is an RMRPA with classification based on the existing environment at the time of the study. The map should be periodically updated to include important modifications of the buildings/road/lines distribution.

4. PROTECTIVE STRUCTURES

Passive measures such as defining specific construction policies which enhance the safety of the building and residents without noticeably increasing the costs should always be stressed. These construction policies are specific to the project and take into account all the pertinent parameters including the design of the active protective structures, in order to provide a rational, well balanced global design.

4.1 *Complementary data*

Once an RMRPA has been developed for a general area, the sets of complementary data that should be made available in order to design safe and economical protective structures and to develop construction policies are the following:

a. Detailed data on ancient rockfalls.
These data will be used in order to perform computer simulations of rockfalls leading to the evaluation of probable trajectories, running distances, etc., and allowing mathematical modeling of the protective works to be undertaken.

b. Monitoring systems.

- Measures of movement of chosen major rock elements:

A monitoring system should be implemented in order to measure the existence of a divergence of some major rock elements from their initial (today) position. This monitoring system should be designed in such a way as to serve as a long term surveillance and alert system.

- Measure of movement of superficial terrain.

A monitoring system should be implemented to check the stability of the superficial layers of the slopes in order to correctly understand their probable behaviour and thereby prevent local failures due to incorrect construction techniques or wrong implantation.

4.2 *Preliminary design considerations and protective structures*

In this section we will review some of the typical protective structures that are used in rockfall prone areas.

a. Protective nets at cliff surface.
These devices are generally used to protect one individual structure or a group of structures by preventing the initiation of a rockfall. The travelling distance and the velocity of a detached block are dramatically decreased by reducing the freedom of movement. These devices can be implemented on a globally stable cliff, allowing only superficial phenomena, with small to medium sized potentially falling elements. The slopes on which these nets can be installed range from gentle to vertical, with the typical height at 20m.

b. Protective nets in planar screens.
These structures are used to perform global protection at long or short distances. They are designed and built in order to stop falling rocks. The performances of these structures depend primarily on their position along the trajectory of the falling rocks, on the energy dissipation capabilities, and, of course, on the resistance of the constitutive elements as well as the foundations. Efficacity is reduced if the falling rocks are flying above 3-4m from the ground surface.

c. Deflecting and frontal berms.
These structures are becoming more frequent nowadays, particularly because of the enhanced

capabilities of modern earth moving equipment which perform satisfactorily in difficult topographical conditions. These structures can be designed as global or individual protective measures. They are designed to stop or deflect falling rocks just uphill of the protected area, provided that enough space is available between the protective device and the area to protect (minimum 10m for small structures).

d. Composite caissons.
This technique consists of creating platforms in very steep areas by using local materials, i.e. earth and timber. The platforms are then used as shock absorbing berms, for building roads, or as construction platforms, and are even sometimes used in the Alps as foundation for light timber chalets. These structures are very quickly colonized by the local flora so that in a few seasons they disappear under vegetation.

e. Anchoring/Bolting.
Anchoring and bolting with passive or prestressed elements is generally an expensive operation because of the scaffolding that is often necessary. When designing this kind of work one should give due regard to corrosion of the tendons and their heads. This solution often is the only solution when the price of the surrounding terrain is extremely high and any loss of surface has to be proscribed.

f. Mass suppression and stability enhancement.
Mass suppression is often a very economical solution, provided that enough surfaces are available at the crest and at the toe of the outcrop, where the chipped rocks are generally stocked. Stability enhancement can be obtained by casting of concrete elements which increase the rotational stability.

4.3 *Rehabilitation of the event*

The rehabilitation program undertaken after the accident and after completion of a diagnostic inventory, was based on a combination of protective nets in planar screens, mass suppression, drainage and consolidation at the toe of the outcrops, and construction of concrete shoulders to increase the rotational stability of selected elements. At the foot of the cliff, a stock of chipped materials was created then covered with topsoil and seeded with appropriate grasses.

5. CONCLUSIONS

This paper presented the global approach used to define risks level when dealing with rockfall prone areas.

Risk mitigation programs can be effectively and economically designed when a specific risk map (RMRPA) is elaborated following a probabilistic methodology that takes advantage of the subjective probabilities techniques to fill the informational gaps and copes with the stochastic aspects of a number of parameters.

An RMRPA is not a static document, since it has to follow the building development of the considered area. The RMRPA is an instrument that classifies the dangerous sources and helps to detemine the appropriate monitoring or rehabilitation programs to be implemented.

An actual example has been presented to describe an application of methodology and interest focused on the human aspects that lead to important decisions.

Finally, the preliminary considerations leading to the design of protective structures have been presented and the rehabilitation project briefly described, keeping in mind the operational and environmental aspects of these structures.

REFERENCES

Kulkarni, R.B. et al 1984. Assessment of confidence intervals for results of seismic hazard analysis. Proceedings of the Eight World Conference on Earthquake Engineering, San Francisco, California.

Power, M.S. et al 1981. Seismic exposure analysis for the WNP-2 and WNP-1/4 site. Appendix 2.5K to Amendment No.18 Final Safety Analysis Report WNP-2, for Washington Public Power Supply System, Richland, Washington.

Youngs, R.R et al 1985. Seismic hazard assess ment of the Hanford region, Washington State. Proceedings of the DOE Natural Phenomena Hazards Mitigation Conference, Las Vegas, Nevada.

Environmental Management, Geo-Water & Engineering Aspects, Chowdhury & Sivakumar (eds)
1993 Balkema, Rotterdam. ISBN 90 5410 099 0

Reliability analysis of a rock slope with a wedge

V. Ravi
Central Building Research Institute, Roorkee, India

ABSTRACT : In this paper, reliability based stability analysis of a rock slope is presented, where the failure occurs along a wedge. The short solution given by Hoek and Bray for the limit equilibrium analysis has been taken as the basis. Cohesion and frictional coefficients of the intersecting planes are assumed to be correlated random variables The study has been carried out for three different probability distributions viz. normal lognormal and logistic. The analysis has been applied for hypothetical examples taken from Hoek & Bray. Further, a parametric study of probability of failure was carried out for different values of correlation coefficient assumed. The influence of pore water pressure on the probability of failure was also studied. A computer program RELWEG was developed incorporating the reliability analysis into the short solution. The study of the effect of a tension crack, an external force and a tensioned cable are outside the scope of this paper.

1 INTRODUCTION.

Stability analysis of rock slopes assumes importance because of so many factors such as socio-economic development, human settlements etc. In rock slopes failure occurs mainly in four modes viz. (i) planar failure (ii) circular failure (iii) wedge failure and (iv) toppling failure. Out of these modes circular failure occurs rarely whereas the other modes are fairly common. Application of probabilistic methods for the stability analysis of rock slopes gathered momentum in recent years. Uncertainty in geotechnical parameters arises due to the variation in relevant properties of the rock mass. External loads due to excavation, seismicity etc. affect the slope instability. The factor of safety obtained by conventional limit equilibrium approaches does not take into account various uncertanties associated with the parameters and hence the slope instability. The probabilistic approaches consider systematically all sorts of uncertainties associated with the parameters and solve the problems faced by the engineers in decision making.

The following is a brief review of literature regarding the application of probabilistic methods to rock slopes. Shuk [1970] expressed the probability of failure of the slope as a function of the geometry of a slope and included it in the design of slopes. Later, McMahon [1971, 1974] developed methods to design rock slopes combining probabilistic concepts with economic aspects of the slope. Madhav and Ramakrishna [1980] presented probabilistic analysis of the stability of rock slopes with two conjugate sets of joints by considering the inherent variability of the inclination of the joint sets. They concluded that the variability in the inclination of shear joints has a predominant effect on the stability of a slope. Chowdhury [1986] determined the risk of rock slope failure along planar discontinuities by considering the shear strength parameters as random variables. Also, a risk model for the occurrence of multiple or succesive failures was presented. In an improved model, Chowdhury [1987] calculated the probability of slip along planar discontinuities considering both geometrical and shear strength parameters as random variables. Stimpson et.al. [1987] developed a probabilistic limit equilibrium analysis for planar shear slope failure to predict the tension crack locations and hence positions of successive failure blocks. Scavia et.al. [1990] analysed the stability of block toppling failure in rock slopes probabilistically. Ravi and Sridevi [1991] developed a reliability model for a rock slope with two conjugate sets of joints by considering shear strength parameters and pore water pressure as independent random variables following normal and lognormal distributions. It was concluded that the variability of pore water pressure and effective angle of friction have significant impact on the stability of slope. Sridevi and Ravi [1991] presented a probabilistic limit equilibrium analysis for a planar failure in a rock slope with a tension crack in its upper surface. C and Tan ϕ are considered as independent random variables following normal distribution. Sensitivity analyses of probability of failure have been presented by varying the geometric parameters and the deterministic geotechnical parameters.

In this paper, reliability analysis of a rock slope is presented where the failure occurs along a wedge. This is a kind of failure in which structural features upon which sliding can occur strike across the slope crest and where sliding takes place along the line of intersection of two such planes. The analysis presented here has been illustrated for hypothetical examples taken from Hoek and Bray [1974]. Sensitivity analysis of probability of failure has been carried out for the examples.

2 RELIABILITY ANALYSIS.

The main assumptions in the analysis are as follows.

(i) Cohesion parameters of the two intersecting planes are assumed to be correlated random variables.

(ii) Frictional coefficients of the two intersecting planes are also assumed to be correlated random variables.

(iii) However, for the sake of simplicity, the cohesion and frictional parameters for a given plane are assumed to be independent.

(iv) Effect of pore water pressure has been considered in the analysis but it has not been considered as a random variable.

(v) The influence of a tension crack, an external load and tension bolts on the stability of the slope has not been considered in this study.

(vi) The factor of safety has been assumed to follow normal, lognormal and logistic probability distributions.

The short solution for the rapid computation of the factor of safety given by Hoek and Bray [1974] has been taken as the basis for the limit equilibrium analysis. It takes into account cases such as whether the slope overhangs the crest of the slope or not. Also it makes allowance for various cases such as (i) wedge has contact on both the intersecting planes (ii) on only one of the intersecting planes. (iii) it loses contact on both the planes and floats as a result of water pressure. The deterministic limit equilibrium analysis has been supplemented with the reliability analysis. This is accomplished through the use of Rosenblueth's Point Estimate method. It is an approximate numerical integration technique. In this method the expected value and other moments are found out by adding several terms (i.e. four only if the basic random variables are two and eight only if the basic random variables are three etc.) These terms constitute the magnitude of the factor of safety calculated at values of the variables one standard deviation either side of their mean, the correlation coefficients between basic variables come as multiplying factors to the terms. This method is very convenient to use and has proved to be successful. It does not require the calculation of derivatives and hence can be used regardless of how complex the expression for F is. Once the statistical parameters of F are calculated, the analysis has been carried out by assuming (i) normal, (ii) lognormal and (iii) logistic distributions for the factor of safety. A user friedly computer program has been developed in FORTRAN 77 incorporating the short solution and the reliability analysis described above. It includes an initial check as whether a wedge has formed or not for the given set of data and carries out the analysis in three cases viz. (i) wedge has contact on both the intersecting planes (ii) it has contact on one of the planes (iii) it loses contact on both the planes and floats as a result of water pressure acting on both the intersecting planes. The probability of failure for diffferent distributions is calculated by subroutines provided in the program. The factor of safety defined by the ratio of the total force resisting sliding and the total force tending to induce sliding is given in three cases as follows.

If $n_1 > 0$ and $n_2 > 0$ the wedge has contact on both the planes and the factor of safety is given by

$$F = \left[n_1 \mathrm{Tan}\phi_1 + n_2 \mathrm{Tan}\phi_2 + |p| C_1 + C_2\right] \sqrt{k} / |Zi| \tag{1}$$

If $n_2 < 0$ and $m_1 > 0$ contact is there on plane 1 only and the F is given by

$$F = \frac{m_1 \mathrm{Tan}\,\phi_1 + |p|\, C_1}{[Z^2\{1-a_z^2\} + kU_2^2 + 2\{ra_z - b_z\}ZU_2]^{1/2}} \tag{2}$$

If $n_1 < 0$, $m_2 > 0$ contact is there on plane 2 only and the F is given by

$$F = \frac{m_2 \mathrm{Tan}\,\phi_2 + C_2}{[Z^2 b_y^2 + k\, p^2\, U_1^2 + (rb_z - a_z)\, pZU_1]^{1/2}} \tag{3}$$

where,

$a_x = \mathrm{Sin}\,\varphi_1 \quad \mathrm{Sin}\,(\alpha_1 - \alpha_2)$

$a_y = \mathrm{Sin}\,\varphi_1 \quad \mathrm{Cos}\,(\alpha_1 - \alpha_2)$

$a_z = \mathrm{Cos}\,\varphi_1$

$f_x = \mathrm{Sin}\,\varphi_4 \quad \mathrm{Sin}\,(\alpha_4 - \alpha_2)$

$f_y = \mathrm{Sin}\,\varphi_4 \quad \mathrm{Cos}\,(\alpha_4 - \alpha_2)$

$f_z = \mathrm{Cos}\,\varphi_4$

$b_y = \mathrm{Sin}\,\varphi_2$

$b_z = \mathrm{Cos}\,\varphi_2$

$i = a_x\, b_y$

$g_z = f_x\, a_y - f_y\, a_x$

$q = b_y(f_z\, a_x - f_x\, a_z) + b_z\, g_z$

$r = a_y\, b_y + a_z\, b_z$

$k = 1 - r^2$

$Z = \gamma\, H\, q / 3\, g_z$

$p = -\, b_y\, f_x / g_z$

$$n_1 = \left[(Z/k)\,(a_z - rb_z) - p\, U_1\right] p/|p|$$

$$n_2 = \left[(Z/k)\,(b_z - r\, a_z) - U_2\right]$$

$$m_1 = (Z\, a_z - r\, U_2 - pU_1)\, p/|p|$$

$$m_2 = (Z\, b_z - r\, p\, U_1 - U_2)$$

If $m_1 < 0$ and $m_2 > 0$, contact is lost on both the planes and the wedge floats as a result of water pressure acting on both the intersecting planes and in this case factor of safety falls to zero.

Evaluation of the probabilities associated with normal density function requires numerical integration. Lack of closed form solution to determine the cumulative normal distribution hinders the development of complete reliability analysis. To overcome this difficulty, a distribution capable of providing similar results as given by normal distribution has to be used. Logistic distribution fits this category giving results without significant loss of accuracy.

Its density function is given by

$$f_X(x) = \frac{\pi \exp\{-\pi\,(x - \mu)/\sqrt{3}\,\sigma\}}{\sqrt{3}\,\sigma\,[1 + \exp\{-\pi(x - \mu)\}]^2} \tag{4}$$

where, μ and σ are the mean and the standard deviation of the logistic variable X. A brief account of the logistic distribution and its applications is given in Barbosa et al (1989).

3 NUMERICAL EXAMPLES.

Example 1 :

The data for this example taken from Hoek and Bray (1974) is as follows :

$\varphi_1 = 47^\circ$ $\varphi_2 = 70^\circ$, $\varphi_3 = 10^\circ$, $\varphi_4 = 65^\circ$

$\alpha_1 = 52^\circ$, $\alpha_2 = 18^\circ$, $\alpha_3 = 45^\circ$, $\alpha_4 = 45^\circ$

$\gamma = 25$ KN/m^3, H = 20 m, $U_1 = U_2 = 30$KN/m^2

$\bar{C}_1 = 25$ KN/m^2, $\bar{C}_2 = 0$,

$\overline{\mathrm{Tan}\,\phi_1} = 0.577$, $\overline{\mathrm{Tan}\,\phi_2} = 0.7002$

The results in the form of probability of failure are presented in Table 1 showing the effect of the presence of pore water pressure.

Example 2 :

The data for this example taken from Hoek and Bray (1974) is as follows :

$\varphi_1 = 45^o$, $\varphi_2 = 70^o$, $\varphi_3 = 12^o$, $\varphi_4 = 65^o$

$\alpha_1 = 105^o$, $\alpha_2 = 235^o$, $\alpha_3 = 195^o$, $\alpha_4 = 185^o$

$\gamma = 25.63$ KN/m^3, H = 39.65 m,

$U_1 = 20$ KN/m^2, $U_2 = 10$ KN/m^2

$\bar{C}_1 = 24.4$ KN/m^2, $\bar{C}_2 = 48.9$ KN/m^2,

$\overline{\text{Tan}\ \phi_1} = 0.364$, $\overline{\text{Tan}\ \phi_2} = 0.5773$

The results in the form of probability of failure are presented in Table 2 showing the influence of the correlation coefficient between C_1, C_2 and Tan ϕ_1, Tan ϕ_2 by varying it between 0 and 1.

Example 3 :

The data for this example is as follows :

$\varphi_1 = 95^o$, $\varphi_2 = 70^o$, $\varphi_3 = 92^o$, $\varphi_4 = 85^o$

$\alpha_1 = 105^o$, $\alpha_2 = 235^o$, $\alpha_3 = 195^o$, $\alpha_4 = 185^o$

$\gamma = 15$ KN/m^3, H = 39.65 m,

$\bar{C}_1 = 24.4$ KN/m^2, $\bar{C}_2 = 48.9$ KN/m^2,

$\overline{\text{Tan}\ \phi_1} = 0.364$, $\overline{\text{Tan}\ \phi_2} = 0.577$

U_1 and U_2 are varied from 0 to 40 KN/m^2 with various combinations. The results in the form of probability of failure are presented in Table 3 showing the effect of pore water pressure. For all the three examples, a coefficient of variation of 20% is assumed for the basic random variables.

4 RESULTS AND DISCUSSION.

The results of example 1 show that when $U_1 = U_2 = 30$ KN/m^2, lognormal distribution yielded more probability of failure compared to logistic or normal distributions. When $U_1 = U_2 = 0$, probability of failure decreased sharply for all distributions conforming to physical understanding. In example 2, the sensitivity analysis of probability of failure is performed for variations in correlation between C_1 and C_2 and between Tan ϕ_1 and Tan ϕ_2. Probability of failure increased steadily for all distributions with the increase in correlation coefficient. This is in agreement with our intuition. In example 3 a parametric study was conducted by varying U_1 and U_2. Steady increment in the probability of failure

Table. 1 : Results of Example 1

U_1 U_2 (KN/m^2)	$\bar{F}$	$\tilde{F}$	Normal P_f	Logistic P_f	Lognormal P_f
30 30	0.63	0.35	0.854	0.872	0.875
0 0	1.15	0.48	0.377	0.362	0.442

Table 2 : Results of Example 2

r	$\bar{F}$	$\tilde{F}$	Normal P_f	Logistic P_f	Lognormal P_f
0	1.58	0.165	0.0002	0.0017	7×10^{-6}
0.25	1.58	0.184	0.0008	0.0033	5×10^{-5}
0.5	1.58	0.201	0.002	0.005	0.0002
0.75	1.58	0.212	0.004	0.0077	0.0005
1.0	1.58	0.231	0.006	0.01	0.001

Table. 3 : Results of Example 3

U_1 U_2 (KN/m^2)	$\bar{F}$	$\tilde{F}$	Normal P_f	Logistic P_f	Lognormal P_f
20 10	1.44	0.251	0.0397	0.0399	0.022
30 20	1.33	0.242	0.0887	0.0797	0.071
40 30	1.198	0.231	0.196	0.1745	0.197
10 20	1.722	0.305	0.0089	0.0135	0.001
20 30	0.453	0.089	1.0	0.9999	0.9999
0 0	2.08	0.313	0.0002	0.0018	64×10^{-8}
10 10	1.87	0.307	0.0023	0.0058	87×10^{-6}
15 15	1.34	0.25	0.087	0.078	0.068
20 20	1.29	0.247	0.118	0.104	0.1037
25 25	1.24	0.245	0.162	0.143	0.1567

was observed for all distributions thereby showing the predominant influence of the pore water pressure.

5 CONCLUSIONS.

The short solution for the deterministic limit equilibrium analysis has been supplemented with reliability analysis. Three numerical examples were solved using the analysis presented. Rosenblueth's Point Estimate Method has been used for the releiability analysis. C_1 and C_2 are assumed to be correlated random variables and so also the Tan ϕ_1 and Tan ϕ_2. This is to enable proper simulation of wedge failure when the wedge has contact on both the intersecting planes. But C_1, Tan ϕ_1 and C_2, Tan ϕ_2 are assumed to be independent. The program is capable of tackling three cases viz. (i) when the wedge has contact on both the planes, (ii) when it has contact on only one of the planes and (iii) when it loses contact on both the planes. In fact, all these cases, have been encountered in the three examples solved in the paper. In example 2, only case (i) is encountered. Hence, a parametric study was conducted by varying correlation coefficient. In the remaining examples, a mixture of all the three cases were noticed. When U_1 or U_2 is greater than 40 KN/m^2, the contact is lost on both the planes and the wedge floats as a result of water pressure on both the intersecting planes. Correlation coefficient between the shear strength parameters of the intersecting planes has been found to have an influence on the probability of failure. As it is increased, probability of failure was found to be increasing for all the distributions. Pore water pressure has a tremendous effect on the probability failure. For values of U_1 more than U_2 steady increase in probability of failure was observed. Also it is found to have a predominent effect on the type of wedge failure, though no definite pattern has been noticed. That is, different combinations of U_1 and U_2 gave rise to one of the three alternatives cited above.

As expected logistic distribution gave similar values of probability of failure as given by normal distribution. The consideration of effect of tension crack and external loads is outside the scope of the paper.

ACKNOWLEDGEMENTS.

The author is thankful to the Director, Central Building Research Institute, ROORKEE for permitting us to send this paper to the Journal. Thanks are due to Dr. R. K. Bhandari for his encouragement. The excellent word processing of Shri Nirmal Kumar Rashtriya Commercial Institute, Roorkee is well acknowledged.

REFERENCES.

Barbosa, M.R., Morris, D.V. and Sarma, S.K. (1989) Factor of safety and probability of failure of rock fill embankments, *Geotechnique*, Vol. 39, No. 3, pp 471-483.

Chowdhury, R. N. Geomechanics risk model for multiple failures along rock discontinuties , *Int. J. Rock Mech. Min. Sci. and Geomech. Abstr.*, **23**. 337-346 (1986).

Chowdhury, R. N. Risk of slip along discontinuities in a heterogeneous medium , *Mining Science and Technology*, Elsevier Science Publishers, **4**, 241- 255 (1987).

Hoek, E. and Bray, J. W. *Rock slope engineering*, Inst. Min. Metall.,London (1974).

Madhav, M. R. and Ramakrishna, K. S. A probabilistic analysis of stability of rock slopes, *ISL*, New Delhi, 267-275 (1980).

McMahon, B. K. Statistical methods for design of rock slopes, *Proceedings First Australia - New Zealand Conference Geomechanics*, 1, 314-321 (1971).

McMahon, B K. Design of rock slopes against sliding on preexisting fractures, *Proceedings Third Congress International Society of rock mechanics*, Denver, Colorado, USA, 2, 803 - 808 (1974).

Ravi, V., and Sridevi, B. Reliability Model for the Planar Failure in Rock Slopes with two sets of Conjugate Joints, *First Romanian Symposium on Landslides, Held in September 1991 at Piatra Neamt, Romania.*

Scavia C., Barla, G. and Bernaudo V. Probabilistic stability analysis of block toppling failure in rock slopes. ,*Int. J. Rock Mech. Min. Sci. and*

Geomech. Abstr., **27**,465 - 478 (1990).
Shuk, T. Optimisation of slopes designed in rock , *Proceedings Second Congress, International Society of Rock Mechanics*, Belgrade, **3**, Sect. 7-2 (1970).
Sridevi B. and Ravi V. Probabilistic stability analysis of a rock slope with a tension crack (Under review by *International Journal for rock mechanics and Mining Sciences,* U.K.).
Stimpson. B, Barron. K, and Kosar. K. Multiple block plane shear slope failure, *Canadian Geotechnical Journal,* **24**, 479-489 (1987).

Environmental Management, Geo-Water & Engineering Aspects, Chowdhury & Sivakumar (eds)
© 1993 Balkema, Rotterdam. ISBN 90 5410 099 0

The probabilistic analysis for stability of rock slope

G.L.Xu
China University of Geoscience, Wuhan, People's Republic of China

ABSTRACT: The probabilistic analysis for stability of rock slope is becoming more common in recent years, which provides a practice and comprehensive method in this area. The models of joint geometries, ie. orientation, trace length, spacing and roughness, are established by joint investigation and Monte-Carlo technique is used to simulate rock mass structure patterns and probabilistic stability model. Strength parameters of rock mass, such as friction and cohesion value, are evaluated by Hoek-Brown empirical criterion, Jennings apparent strength and Lajtai method. The sensitivity analysis of parameters is also presented here.

1 INTRODUCTION

Probabilistic analysis for stability of rock slopes has become more common and familiar in this decade. The reason is, quite simply, that it can consider uncertainties of calculating parameters, boundary and stress condition. Hence, it seems more reasonable for evaluating stability of rock slope.

The principal difficulties, such as the structure pattern, friction and cohesion values, which inhibit the stochastic model from application, are imperative to be overcome. Here, the aim is to try to provide a practical and comprehensive method for rock slope stability analysis. Four sections are discussed in the following. Section one is the structure model of rock mass on the basis of field investigation and statistics. Section two is the stochastic principle and method of slope stability analysis. Calculating parameters are estimated in section three and sensitivity analysis of parameters in section four.

2 STRUCTURE PATTERN OF ROCK MASS

2.1 Joint geometry

Joint can be described by geometric characteristics, ie. shape, size, orientation, spacing and roughness etc.

Generally, joint shape suggests irregular. However, elliptical joints have been observed (Bankwitz, 1966; Kulander et al., 1979; Barton, 1983), also, circular joints have been produced through hydraulic fracturing in the laboratory (Cleary, 1984). To simplify, we can convert irregular shape into equivalent disc, that is,

$$D = 2\sqrt{S/\pi} \tag{1}$$

where D is the diameter of equivalent joint, S is the area of any shape joint.

Joint size can be represented by trace length formed by the intersection of a joint with an outcrop surface. Trace length is estimated either by scanline method (Hudson, 1979) or by sampling window (Kulatilake et al, 1984; Xu, 1989). Given the stochastic character of trace length, a number of distributions have been proposed: exponential, uniform, lognormal, normal, hyperbolic. According to fractal geometry theory, joint size seems to have fractal characteristic proved by Kojima et al (1987) and Xu (1992), which hints it follows an exponential distribution. And the relationship between trace length and diameter has been discussed by Warburton (1980) and by fracture mechanism, that is,

$$E(l / D) = \pi / 4 \tag{2}$$

Orientation of a joint in space is described by dip and dip angle. The common distributions of orientation are uniform and normal.

Spacing, the distance between two joints

measured along a line perpendicular to the joints (ISRM, 1978) is used to describe location. Poisson process and exponential spacing form are supported by extensive field observation (Snow, 1968; Priest and Hudson, 1976; Einstein et al., 1980). Using fractal dimension, the same result has been shown by Xu (1992).

The wall roughness of a joint, which can be described by joint roughness coefficient (JRC) and is quite difficult to be estimated, is a potentially important component of its shear strength. However, it is a very easy way to evaluate JRC with fractal dimension under the assumption that a wall roughness profile is similar to Koch snow curve. The fractal dimension and JRC are respectively,

$$D_f = \frac{\ln 4}{\ln [2 (1+\cos \tan^{-1} 2h / L)]} \quad (3)$$

$$JRC = 85.2671 (D_f - 1)^{0.5679} \quad (4)$$

$$r = 0.99$$

where h is the mean height of primary roughness, L is the mean base length of primary roughness and r is the relative coefficient (Xie et al, 1992).

2.2 Structure pattern of rock mass

To simulate structure pattern of rock mass by computer is just an inverse process of joint geometry investigation in the field. Structure pattern of rock mass or joint system model describes the joint characteristics as an entity. Here, Baecher Disk Model is introduced to establish joint system model with program DNWSP, which can output very important data, such as RQD and persistence coefficient.

Fig. I is an existing joint system network of graystone in Wangji Minging Pit, Hubei Province.

Fig l. An example of joint network

3 PROBABILISTIC ANALYSIS MODEL

Stability factor K, which denotes the stability index of a slope, is a function of a series of parameters, ie.,

$$K = g (x_1, x_2, \cdots, x_m) \quad (5)$$

Therefore, Stability factor K is also a stochastic variable. If the probabilistic density function is f(K), the failure probability P_f is defined as,

$$P_f (K < 1) = \int_0^1 f(K)\, dK \quad (6)$$

Since f(K) is usually difficult to be expressed in analytical expression, it should be estimated by stochastic simulation.

3.1 Stochastic variables

For standard normal distribution N(0,1), the equation of random sampling is,

$$\left. \begin{aligned} U_1 &= \sqrt{-2\ln r_1} \cos 2\pi r_2 \\ U_2 &= \sqrt{-2\ln r_1} \sin 2\pi r_2 \end{aligned} \right\} \quad (7)$$

where r_1 and r_2 are independent random numbers in range [0,1], U_1 and U_2 are independent stochastic variable samples which follow N(0,1).

For other distribution forms, the equations of random samplings of normal, lognormal and exponential, respectively, are,

$$\left. \begin{aligned} X_1 &= \mu + \sigma\sqrt{-2\ln r_1} \cos 2\pi r_2 \\ X_2 &= \mu + \sigma\sqrt{-2\ln r_1} \sin 2\pi r_2 \end{aligned} \right\} \quad (8)$$

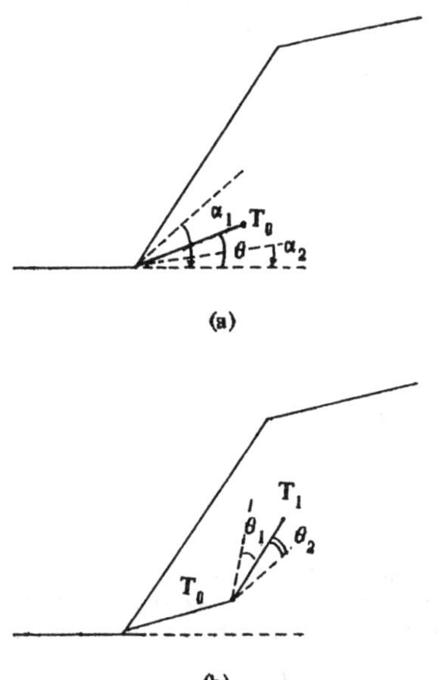

Fig 2

$$\left.\begin{aligned} X_1 &= \exp(\mu + X_1\sigma) \\ X_2' &= \exp(\mu + X_2\sigma) \end{aligned}\right\} \qquad (9)$$

$$X = -\frac{1}{\lambda}\ln r \qquad (10)$$

And uniform random number among [0,1] can be represented by pseudo-random number with mathematics method.

3.2 Failure probabilistic model

Firstly, the potential sliding path has to be determined based on the structure pattern of rock mass. The initial segment T_0 (Fig.2a) is given by,

$$\theta = \alpha_2 + (\alpha_1 - \alpha_2)\, r^2 \qquad (11)$$

The deviation angle $\Delta\theta$ of successive segment T_i (Fig.2b) is,

$$\Delta\theta = \frac{\pi}{4}\left[1 + (1 - r^2)\, r^{(i+r)}\right] \qquad (12)$$

where α_1 and α_2 are the maximum angle of initial segment T_0 along clockwise and anti-clockwise, r is random number.

The trace lengths and spacings are determined stochasticly with joint geometry model. And repeat the above procedures until the path intersects the top surface. Now, an irregular sliding path is produced randomly.

Secondly, force condition analysis is made as shown in Fig.3. The sliding force acted on the ith slice is,

$$\begin{aligned} E_i = {} & W_i \sin\alpha_i + E_{i-1}\cos(\alpha_{i-1} - \alpha_i) \\ & - [W_i \cos\alpha_i \tan\phi_i + c_i L_i \\ & + E_{i-1}\sin(\alpha_{i-1} - \alpha_i)\tan\phi_i] \end{aligned} \qquad (13)$$

where W_i is the weight of ith slice, L_i is the ith sliding length, ϕ, is the friction angle of the ith sliding face, c_i is the cohesion value of ith sliding face and α_i is the angle between sliding face and horizontal line.

Thirdly, the failure probability of rock slope can be calculated if the calculating number is sufficient (at least 200 times), that is,

$$P_f = \frac{\sum_{i=1}^{m} P_i\ (P_i < 1)}{n} \qquad (14)$$

where n is the total number of calculating and m is the number which $P_i < 1$ occurs.

Compared with all of the potential sliding path and its failure probability, a maximum one of that path, which is called a critical path, can be determined. Additionally, if the sets of joints are greater than 4, this rock mass can be treated as homogenous material and the sliding surface can be considered as curved.

4 STRENGTH PARAMETERS ESTIMATING

The shear strength of rock mass or joint is the key problem in failure probability analysis. Because of the test cost, we usually cannot get mean, variation and distribution of shear strength. However, from the world's experience, the most of distribution of those values are of normal form. Variation can be estimated from the variation of rock mass classification grading. Therefore, mean values of shear strength are focused here.

4.1 Hoek-Brown empirical criterion

The Hoek-Brown failure criterion has been defined by following equation (Hoek and Brown, 1980),

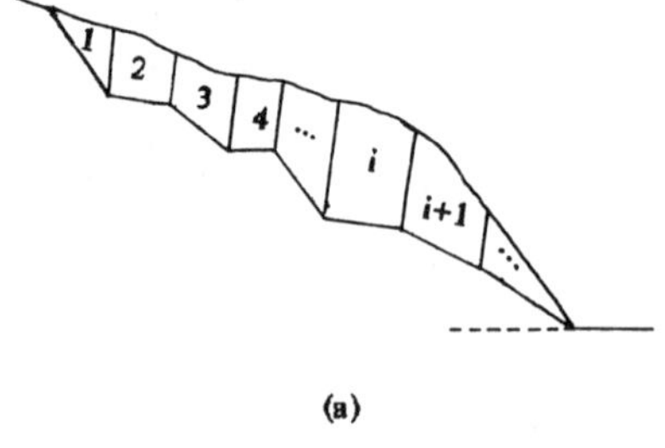

(a)

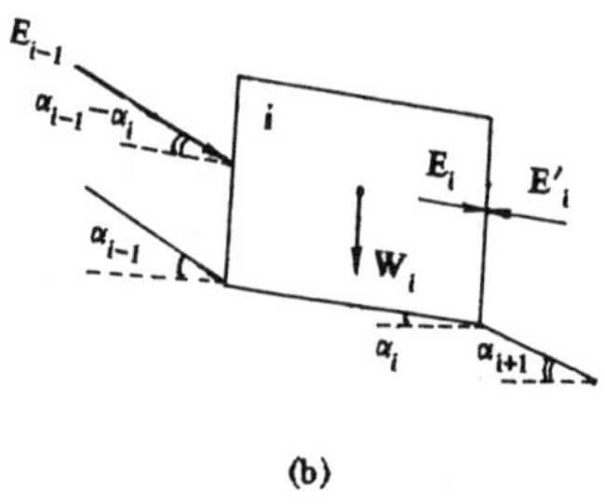

(b)

Fig 3

$$\sigma_1 = \sigma_3 + \sqrt{m\,\sigma_c \sigma_3 + s\,\sigma_c^2} \qquad (15)$$

where σ_1 and σ_3 are the major and minor principal effective stress, σ_c is the uniaxial compressive strength of the intact rock, m and s are material constants.

In order to provide a basis for linking their criterion to measurements or observations which can be carried out in the field, Hoek and Brown suggested a set of relations between the rock mass rating (RMR) from Bieniawski's rock mass classification and the constants m and s. Following Priest and Brown (1983), the relations were presented in the form of the following equations:

For disturbed rock masses,

$$\left.\begin{aligned} m &= m_i \exp\left(\frac{RMR - 100}{14}\right) \\ s &= \exp\left(\frac{RMR - 100}{16}\right) \end{aligned}\right\} \qquad (16)$$

For undisturbed or interlocking rock masses,

$$\left.\begin{aligned} m &= m_i \exp\left(\frac{RMR - 100}{28}\right) \\ s &= \exp\left(\frac{RMR - 100}{9}\right) \end{aligned}\right\} \qquad (17)$$

where m_i is the value of m for the intact rock, determined from the results of triaxial tests.

Under specified normal stress σ_n, the friction and cohesion values are derived from equation (15) (Hoek, 1990).

$$\begin{aligned} h &= 1 + \frac{16(m\,\sigma_n + s\,\sigma_c)}{3\,m^2\,\sigma_c} \\ \theta &= \frac{1}{3}\left(90 + \arctan\frac{1}{\sqrt{h^3 - 1}}\right) \\ \phi &= \arctan\frac{1}{\sqrt{4\,h\cos^2\theta - 1}} \\ \tau &= (\cot\phi - \cos\phi)\frac{m\,\sigma_c}{8} \\ C &= \tau - \sigma_n \tan\phi \end{aligned} \qquad (18)$$

4.2 Lajtai equation

Joint persistence can be used to estimate the strength of a rock mass against sliding along a given plane, which can be expressed as,

$$\left.\begin{aligned} C_a &= (1 - k)\,C_r + k\,C_j \\ \tan\phi_a &= (1 - k)\tan\phi_r + k\tan\phi_j \end{aligned}\right\} \qquad (19)$$

where C_a and $\tan\phi_a$ are called Jenning's equivalent friction and cohesion parameters, r and j denote the intact rock and joint.

Although these equations are very simple to be handled, there are several shortcomings: (1) failure surfaces are restricted in joint plane, (2) shear failure does not typically occur for the usually low value of σ_n, and (3) small variations of persistence produce large variations of resistance.

When the σ_n is low (usually engineering stress within this range), the minor principal stress reaches the tensile strength easily and tensile failure occurs (Lajtai, 1969), that is,

$$\tau = \sqrt{\sigma_i(\sigma_i + \sigma_n)} \qquad (20)$$

Therefore, the comprehensive shear strength of a given rock mass is,

$$\tau = (1 - k)\sqrt{\sigma_i(\sigma_i + \sigma_n)} + C_k k(C_j + \sigma_n \tan\phi_j) \qquad (21)$$

If the joint shear strength is expressed with Barton equation, (21) can be rewritten as,

$$\tau = (1 - k)\sqrt{\sigma_i(\sigma_i + \sigma_n)} + C_k k\,\sigma_n \tan\left(\phi_r + JRC\lg\frac{JCS}{\sigma_n}\right) \qquad (22)$$

where JCS is the joint compressive strength and C_k, is the joint wall strength coefficient between (0,1).

5 SENSITIVITY ANALYSIS

In order to improve the stability of rock slope and optimal slope design, some measures must be made. And sensitivity analysis on some factors can provide the database or guideline for those measures.

There are many factors, such as angle, height. groundwater level, friction, cohesive value and seismic force, affecting the stability of rock slope. Sensitivity analysis is carried out to arrange the importance and regulation of these factors. An illustrative example is omitted due to limitation on space.

6 CONCLUSIONS

Because of the uncertainties of calculating parameters, failure probability analysis with Monte-Carlo technique seems more reasonable in rock slope stability analysis. A critical path of a given slope can be searched out randomly and sensitivity analysis can also be fulfilled conveniently. Two empirical criteria, Hoek-Brown criterion and Lajtai equation, are introduced to estimate the mean values of shear strength, which are the main variables in analysis. It has been shown that the above thinking is quite good for practical use.

REFERENCES

Bieniawski, Z.T. 1989. Engineering rock mass classifications. John Wiley & Sons: 50-177.

Glynn, E. F. et al. 1978. The probabilistic model for shear resistance of jointed rock. 19th U.S. Symp. on Rock Mech, Nevada: 66-76.

Hock, E.1990. Estimating Mohr-Coulomb friction and cohesion values from Hoek-Brown failure criterion. Int. J. Rock Mech. Min. Sci,27. 227-229.

Hudson, J.A. and Priest, S.D. 1979. Discontinuities and rock mass geometry. Int. J. Rock Mech. Min. Sci., 16:339-362.

Kulatilake, P.H.S.W. and Wu, T.H. 1980. Estimation of mean trace length of discontinuities. Rock Mech. and Rock Engrg., 17:215-232.

Xie, H.P. et al. 1992. Fractal estimation of joint roughness coefficients. Congr. of Fractured and Jointed Rock Mass. Univ. of California at Berkeley.

Xu, G.L. 1992. Fractal analysis for joint geometry characterizations. Hydrology & Engineering Geology (to be published, in Chinese).

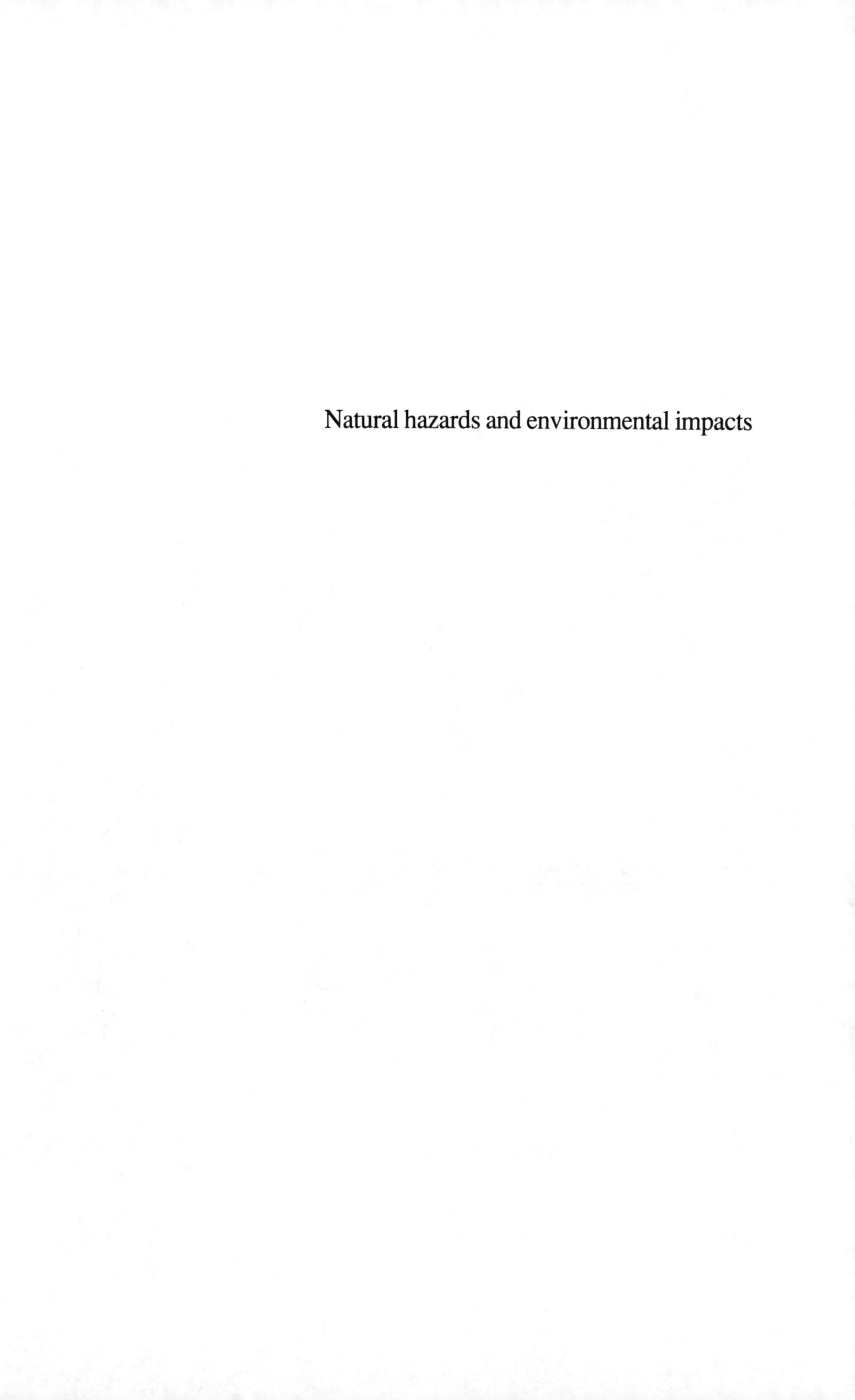

Natural hazards and environmental impacts

Environmental Management, Geo-Water & Engineering Aspects, Chowdhury & Sivakumar (eds)
© 1993 Balkema, Rotterdam. ISBN 90 5410 099 0

Some aspects of environmental impacts of dam reservoirs in Himalayan region

R. Anbalagan, Bhawani Singh & Sanjeev Sharma
University of Roorkee, India

ABSTRACT: Construction of dams leads to creation of an artificial lake eco-system, modifying the fluvial eco-system causing environmental imbalances both on the up-stream and downstream areas of the dam. Recently there has been a sudden spurt in the construction activities of dams in Himalaya which is characterised by a fragile mountain eco-system. The paper deals with some specific and vital environmental impacts of man-made lakes in the Himalayan region.

1. INTRODUCTION

The Himalaya accounts for a major share of unused hydroelectric resource as only 12% of the estimated potential has been used so far in India. As such the harnessing of surface water resource through multi-purpose dams in Himalaya has recieved high priority in the recent years. A number of dams have been constructed in the recent past and some are under construction (Fig. 1). However, the environmental impacts of these dams on the fragile eco-system of Himalaya have been a subject of major debate among the environmentalists. The cry is for stopping construction of high dams. The preservation of eco-wealth in developing countries is essential for stability of developed countries. So the concern is world wide. The most imporotant impact is the creation of a lake eco-system modifying the fluvial eco-system and thereby causing imbalances both on the upstream and downstream of the dam. Some of the impacts pertaining to the dams of the Himalayan region are discussed below.

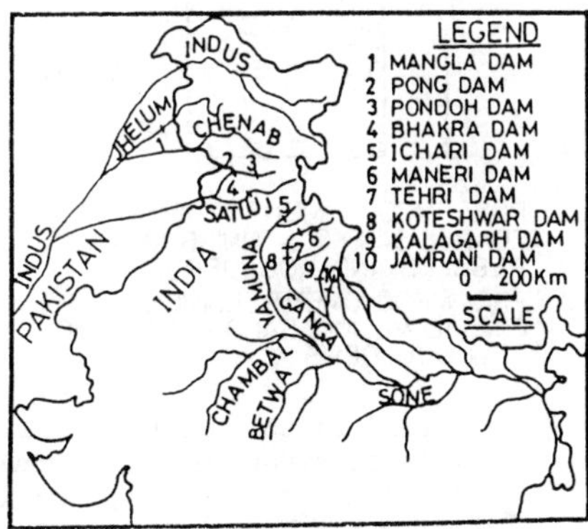

Fig. 1 Location map of dams in Himalaya

2. IMPACT ON RIVER BED MORPHOLOGY

The sudden transformation of the river eco-system into a lake eco-system brings in drastic changes in the erosion and sedimentation pattern of the area. As the river from the catchment enters the reservoir lake, the sudden drop in its velocity leads to the deposition of the traction load at the mouth of the lake, while the suspended load may be carried further deeper into the reservoir. Thus the sedimentation in the peripheral area of the reservoir progressively gives rise to a delta (Fig. 2). As the size of the delta increases there would be flattening of the river bed gradient. The locally flattened river bed gradient would cause back-water effects resulting in the deposition of more and more sediments on the river bed above the reservoir leading to aggradation conditions (Fig. 2).

An examination of the grain size distribution of the sediments of pre-impoundment and post-impoundment on the river bed above the Gobind Sagar Reservoir

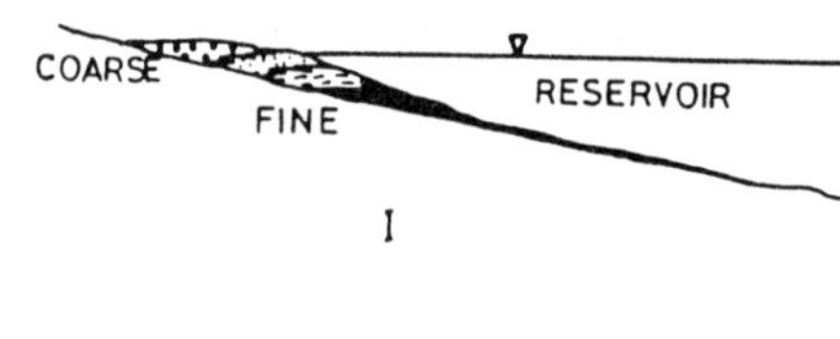

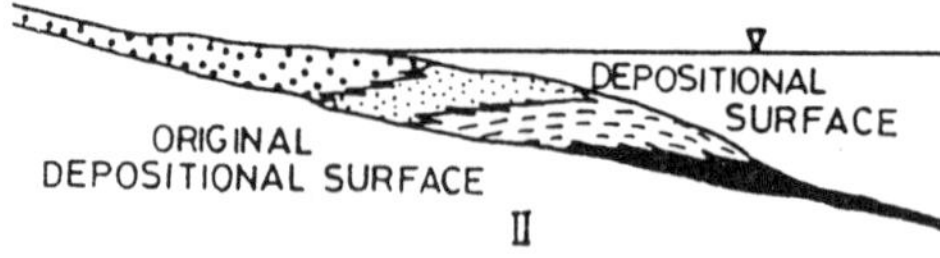

Fig. 2 The sedimentation close to the mouth of the reservoir gives rise to prograding delta

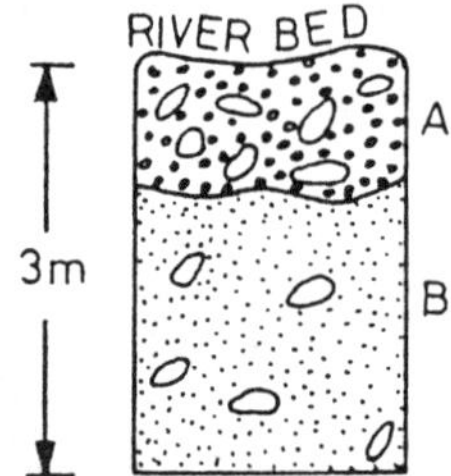

Fig. 3 Variation in the nature of sediment below river bed as observed in a pit about 2 km upstream of Gobind Sagar Reservoir

of Bhakra dam, in Punjab, shows that the post-impoundment sediments are much coarser than the sediments of pre-impoundment period indicating the aggradation conditions due to the growth of the delta (Fig. 3). The life of the reservoir of 130 m high Jamrani dam in Kumaun Himalaya calculated on the basis of the changed environment of deposition indicated about 500 years against 100 years calculated on the observed sediment load under fluvial conditions (Anbalagan, 1986).

In case of high dams characterised by deaper and wider reservoirs such as 225 m high Bhakra dam and 260.5 m high Tehri dam, delta formation would take place far off from the dam and would act as an effective natural barrier for the traction load due to locally flattened gradients. But in case of smaller dams characterised by shallow and shorter reservoirs, the river bed delta may not function effectively as a barrier due to considerably shorter distance between the dam and the delta. The 59.25 m high Ichari dam took a little over two years for its reservoir to get filled with sediments upto the crest level of the spillway (Mohan and Tyagi, 1982). It emphasizes the fact that the high dams (> 100 m) are more favourable in the Himalayan environment form the point of view of longer life expectancy of the reservoir.

3. IMPACT ON RESERVOIR RIM STABILITY

The reservoir operation is associated with fluctuation of water level between Maximum Reservoir Level (MRL) and Dead Storage Level (DSL). When the water is at MRL, the valley slopes close to the rim would have two characteristic zones a dry zone above the water level and a water submerged zone. In the submerged zone, though the weight of the rocks gets reduced due to the uplift pressure of the water, the lateral thrust of the standing water prevents the sliding tendency of the slopes. If there is a sudden significant draw-down it would cause 3 different zones - a dry zone, a water charged zone and a water submerged zone. In the water charged zone, the weight of the slope material is considerably increased, while the shear strength gets reduced causing favourable conditions for slope instability. On the other hand if the reservoir operation is carefully planed so as to avoid sudden draw-down conditions, the stability conditions of the rim may not get changed rapidly. The debris due to slides, unlike a fluvial environment, are not carried away by flowing river and as such lie close to the toe of the slide (Fig. 4). With the increased accumulation of the debris, the general gradient of the slope would get flattened. The subsequent compaction of the debris due to the lateral thrust of the reservoir water and vegetation growth over them would result in the stabilization of the debris mass so as to provide an effective toe supporot to the unstable slopes above. An excellent example of this phenomena can be quoted from the Gobind Sagar Lake of Bhakra dam, where a known active slide for many years on the right bank prior to impoundment had remarkably attained stability in less than 5 years mainly because of the toe support of the stabilised debris material subsequent to the reservoir impoundment.

4. IMPACT ON THE SEISMIC STATUS OF THE REGION

The seismic zoning map of India prepared by Bureau of Indian Standards places the Himalaya, which is a young, folded, overthrust mountain system, on high order zones of IV and V. The large scale concentration of tectonic stresses are often released in the form of earthquakes in this region. Though the construction of a dam and impounding water behind it causes impact on the seismic status of the region, it is not precisely known the extent of changes brought about by water impoundment. The changes in the seismic status of the area due to reservoir filling are more influenced by the geologic and tectonic setting of the area. Huang Nain (1982) indicates three important conditions, in case of high dams, for the Reservoir Induced Seismic (RIS) activities.

a) the availability or otherwise of channels of deeper infiltrations,

b) the existence of deeper tectonic conditions, such as stress build-up in the pre-existing fault planes or weak zones, and

c) physical and textural properties of litho units.

In Himalaya, the important dams such as Bhakra, Kalagarh, Pong and Pondoh in India as well as Mangla in Pakistan, all located in the Outer Himalaya, did not indicate any significant change in the seismic status after impoundment. In case of Kalagarh, Pong and Pondoh dams, there was a slight increase in the number of seismic activities, just after reservoir filling, but gradually declined subsequently (Srivastava and Dube, 1982). A perusal of the geological setting of these reservoirs indicates that they are characterised by alternating sequence of soft sedimentary rocks such as claystone, mudstone, siltstone and sandstone of Siwalik Group, in contrast to brittle rocks, such as trap rocks, exposed in the vicinity of dams like Koyna, which showed steep increase in RIS after impounding. It indicates that the occurrence of RIS is closely related to the brittleness of rocks in terrains where the tectonic stresses are close to critical conditions

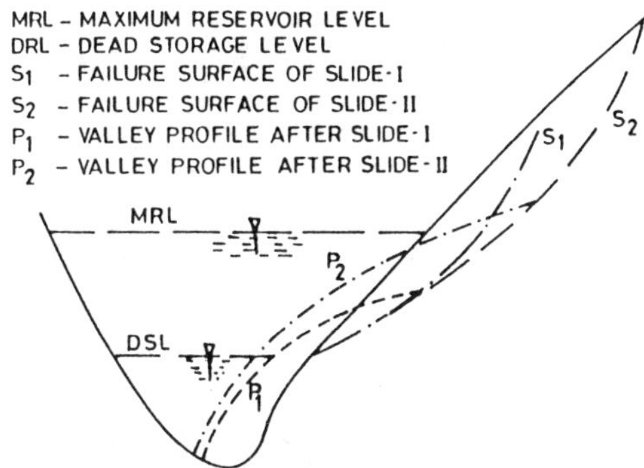

Fig. 4 Influence of reservoir on stabilization of surficial landslides.

of failure. The failure of brittle rocks are associated with sudden vibratioins and the attendant seismic shocks, in contrast to softer rocks, which do not cause significant vibrations on failure.

However, in Lesser and Higher Himalaya, the rocks are brittle and slopes unstable. Thus the risk due to natural earthquakes and RIS is quite high. Hence the earth and rockfill dams are more preferable in such terrains. Moreover the RIS can be reduced by gradually increasing the rate of reservoir filling.

5. IMPACT ON AQUATIC COMMUNITIES

Aquatic communities are differently affected by the change from fluvial to lacustrine environment. In the initial stages, the presence of abundance of organic matters accelerates the microbiological activities often leading to the explosion of some species and dwindling of some other species. In Gobind Sagar Lake of Bhakra dam, the 'gid' fish (labeodero) population showed a marked increase, but the 'gugli', (Schyzothorex) became rare, while the 'masheer' (Torputitora) and 'kunni' (Labeodero diecdhylus) declined considerably. Further the unlined canals fed by the Gobind Sagar Lake faced an unprecedented algal growth leading to choking of canals with the aquatic weeds (Valdiya 1987).

6. IMPACT ON LAND USE

The inundation of a large tract of land due to reservoir filling in Himalaya mostly include the most valued forests

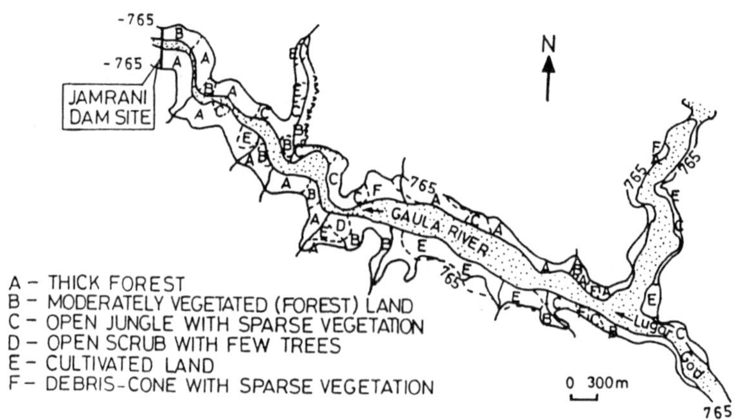

Fig.5 Vegetation pattern map of the area of submergence of Jamrani dam reservoir

including the reserved forests and fertile agricultural terraces. The quantum of various categories of land to be submerged by the projects has to be evaluated in detail for a better appraisal of the environmental impacts. The 4.5 sq km of water spread of Jamrani Dam Reservoir area has been mapped for the various categories of vegetation to be submerged (Fig. 5). The submergence include about 0.46 sq km of agricultural land and 1.6 sq km of reserved forest encompassing.

a) Thickly vegeted forest - 0.65 sq km
b) Moderately vegetated forest - 0.44 sq km
c) Open Jungle with sparse vegetation - 0.67 sq km

The Jamrani Project plan outlay contains compensatory afforestation cost to compensate the loss of reserved forest by bringing in an equal area, if not more, of non-forest land into protected afforestation. Similarly, it should be mandatory that the future project plans incorporate eco-development plans so as to minimise environmental degradation due to construction of projects in Himalaya. Bio-engineering methods of surficial slope stabilization should be used extensively as it is the cheapest method.

7. IMPACT ON SOCIO-ECONOMIC STATUS OF REGION

Demand for energy is ever growing often exponentially in many countries. In India, Super Thermal Power Stations are being constructed to meet this demand but the emission of sulphur dioxide and carbon monoxide is causing depletion of ozone layer and carbon dioxide is producing green house effect. On the other hand hydel generation is pollution-free and is based on renewable resource at far cheaper cost but its construction is very slow with high initial investment.

The most tragic part of the dam construction is the uprooting of the human communities from their ancestral dwellings causing emotional tumult, social insecurities and economic distress. In Himalaya, the gigantic Tehri dam reservoir would submerge the Tehri town and 23 villages, in addition to partially submerging 72 villages. A recent survey indicates that about 70,000 people would be ousted because of submergence. The Pong dam rendered 16,000 families homeless (Valdiya, 1987).

The Jamrani dam would partially submerge 13 villages, resulting in displacement of 1190 people. The rehabilitation of the uprooted people, especially in Himalaya, is a complex and and sensitive problem since the relocates are subjected to new life style in a different physiographic setting and perhaps different climatic conditions. As such, it should be handled with utmost care so as tnot to cause disrespect to the sentiments of the hill people. However, these projects during construction, generate a significant employment potential both skilled and unskilled. Moreover the dam construction in hill areas is invariably associated with better communication facilities such as construction of roads and bridges, promoting the cause of development of the hill region. In the completed projects of Himalaya, the relocates were given priority in employment. Further the hydro-electric generation from the projects, boosted the industrial development and agricultural production of the region. For example, the Punjab State which ranked just average in agricultural production and per capita income till the construction of Bhakra dam, had pushed down all other states of India to achieve a per capita income of twice the national average during its rapid pace of economic growth due to 'green revolution' in the post-Bhakra era. If the reservoir lake and its surrounding are developed in a planned way to create a hill resort with all modern facilities it will help to boost the tourism and the revenue earnings.

8. CONCLUSIONS AND RECOMMENDATIONS

The construction of a dam and impounding of water behind it causes a sudden transformation of a river eco-system into a lake eco-system. Though these changes in Himalaya are associated with some adverse impacts on the environment of the region, a careful evaluation of all the impacts reveals that they are overwhelmingly positive. The formation of delta at the mouth of the reservoir causes aggradation conditions in the catchment which reduces the erosion of the area. In addition, a systematic silt control programme through check dams has to be implemented in the catchments of the Himalayan reservoirs for their efficient functioning. In the Himalayan environment, characterised by steeper gradients and high sediment yield, the reservoir sedimentation pattern favours high dams (100 m) as they will have a comparaitively longer life. The modified lacustrine conditions will increase increase stability of slopes around the reservoir.

Though the RIS is dependent on the geologic and tectonic setting of the area, it is a cause of concern only for the future reservoirs especially in the initial stages of their functioning. A regulated reservoir filling and releasing of water, at a considerably slow pace may gradually dissipate the accumulated energy of tectonic stresses. The level of reservoir filling could be raised in increments every year in the initial stages say 1/2, 2/3, 3/4 and then full capacity of the reservoir, as it may help to avoid a sudden increase in RIS. The geologic and tectonic setting of Lesser and Higher Himalaya favour rockfill dams, especially if the height is > 100 m, as they have better stability in an earthquake prone environment. Though the dam construction causes economic distress and social insecurity to the people to be displaced due to the dam, the problem can be solved satisfactorily if it is handled sympathetically preserving local culture and fulfilling the requirements of the people of the area. This programme may form a part of affordable and sustainable economic development of hilly areas.

There is no doubt that the water resources development projects, in-spite of some adverse environmental impacts have contributed substantially to the growth of the Indian economy.

REFERENCES

Anbalagan, R. 1986, Geotechnical study and Environmental Appraisal of a Water Resources Development Project in Kumaun Himalaya, U.P., Ph.D. Thesis, Kumaun University, Nainital, India.

Anbalagan, R. 1990, Appraisal of environmental impacts of the construction of Jamrani dam in Nainital Distt., Kumaun, Himalaya, India, Jour. Geol. Soc. Ind. v.36, 307-316.

Mohan, J. and Tyagi, S.S. 1982, Siltation of reservoirs in Ganga Yamuna Valley, 4th Cong. Int. Assoc. Engg. Geol., New Delhi, v. 7, 3-10.

Naian, H. 1982, Engineering Geological analysis method for the forecast of reservoir induced seismicity. 4th Cong. Int. Assoc. Engg. Geol. New Delhi, v. 7, 141-146.

Srivastava, H.N., Dube, R.K., 1982, Seismicity studies of some important dams in India. 4th Cong. Int. Assoc. Engg. Geol. New Delhi, v.8, 219-228.

Environmental Management, Geo-Water & Engineering Aspects, Chowdhury & Sivakumar (eds)
 ISBN 90 5410 099 0

Environmental impact of Roodbar landslide in Iran

S.A.Anvar, L.Behpoor & A.Ghahramani
Shiraz University, Iran

ABSTRACT: The huge Roodbar Landslide happened during the M7.3 Manjil-Roodbar Earthquake in 1990. The earthquake produced the initial slip and the seepage of water and oil from the broken pipes initiated the slide. Unfortunately, no remedial steps were taken for improving the condition of slide except for rerouting the broken petroleum line. Continuous seepage of water into the slide has produced several ponds on the slide. Furthermore, the discontinuation of olive tree preservation on the slide has caused surface weathering and crack formation and the destruction of more than 50000 olive trees in this scenic region has had considerable impact on the environmental conditions. It is recommended to design lifelines in earthquake prone regions and on the slopes with special precautions.

Fig. 1 Roodbar landslide

1 INTRODUCTION

The devastating Manjil-Roodbar earthquake of magnitude M7.3, occurred on June 21,1990 and shook an area of about 600,000 square kilometers in the Northwest of Iran resulting in more than 30,000 loss of life , structural damages inflicted upon more than 100,000 homes, and leaving hundred thousands homeless. In addition, the earthquake resulted in liquefaction in vast areas and several major landslides which caused damages to buildings, environment and lifelines

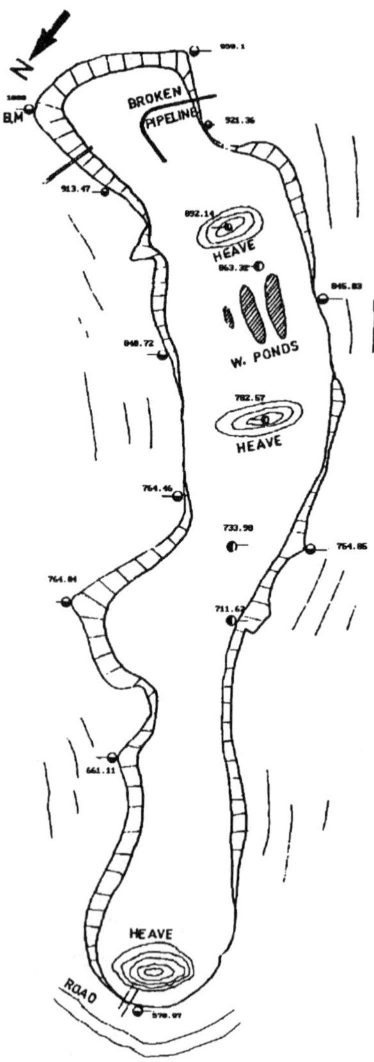

B. M. ELEVATION = 1000 m.

ELEVATION OF NATURAL GROUND

ELEVATION AFTER SLIDING

Fig. 2 Slide geometry

in the region.

Roodbar landslide, one of the biggest landslides yet reported in Iran in recent years, 2 km long, 300 m average width (ranging from 200 to 500 meters), and 10 m average depth (ranging form 5 to 30 meters) happened in the region. This slide has moved more than 5 million cubic meter of soil, with a movement at some points reaching 400 meters.

A brief description of this landslide together with the results of soil properties measurements and stability analysis were reported by the authors (1992). In the present paper, the geometry of the landslide measured by land survey is presented.The mechanism of the slide is discussed and the effect of oil on strength reduction of soil is studied and it is postulated that the penetration of oil from a petroleum lifeline, broken during the earthquake, could be one of the contributing factors in triggering the slide 48 hours after the main shock of earthquake. The impact on environment is discussed and recommendations are offered for designing lifelines in earthquake prone zones and on the slopes.

2 MECHANISM AND CHARACTERISTICS OF THE SLIDE

According to measurements and analysis carried out previously by Anvar, Behpoor and Ghahramani (1991) and recent investigations, the mechanism and characteristics of the slide can be summarized as follow. An overview of the slide area is shown in Fig.1 and the slide gometry based on new field survey of the Roodbar slide is presented in Fig. 2.

1. The initial average slope of the ground of about 20 degrees has been reduced to about 15 degrees in upper part after the slide occurrence, the movement being up to 400 meters.

2. By considering ground topography, measurements of soil properties, and using finite element and slope formulas it was shown that the slope had a factor of safety of about 2.6 prior to earthquake.

3. The acceleration of earthquake was reported to be 0.65 g at the station 65 km away from the area. Using this horizontal acceleration, calculations showed that the safety factor during the earthquake was reduced to 1. Although the slide did not occur during the earthquake, the earthquake acceleration could produce an initial slip.

4. After the earthquake, using the residual strength the slide had a factor of safety of about 1.8. Thus after initial slip which occurred during earthquake and its after shocks, the slope was still stable.

5. The petroleum pipeline located in the upper part of the slide was broken during the earthquake (most probably as the consequence of the initial slip). Petroleum flew to the body of the

slide and continued to flow for 48 hours before the start of the slide.

6. 48 hours after the earthquake, the reduction in soil residual shear strength due to petroleum and water penetrations resulted in dropping of factor of safety below unity. Consequently the slide was triggered and continued for almost a month before coming to rest, during which local heaves and water ponds were created.

7. The slide destroyed several houses and irrigation ponds and about 50,000 olive trees and resulted in a considerable damage to the environment (see Fig. 1). It is to be noted that the petroleum line was repaired and rerouted within the first month, (see Fig. 3).

Fig. 4 shows the variations of safety factor in time sequence.

Fig. 3 Rerouted petroleum line

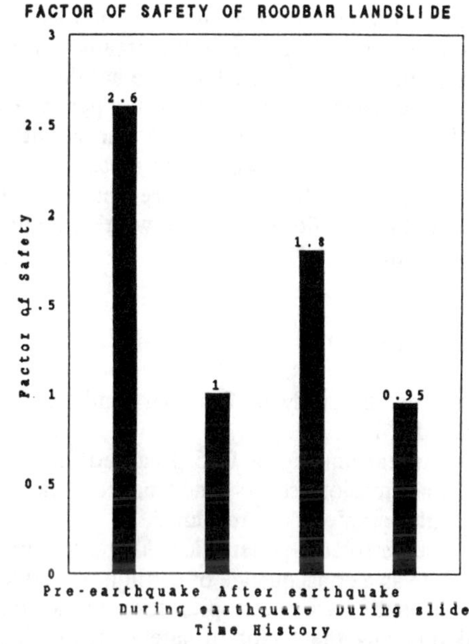

Fig. 4 Time sequence of factor of safety

3 DISCUSSION

The Roodbar earthquake and consequent landslide showed clearly that the design of lifelines, especially gas and petroleum lines, in areas prone to earthquake and on slopes need special attention. Gas line rapture may cause explosion with subsequent damages to the environment and even loss of life. The breakage of gas line due to landslides, at least, can shut off the natural gas supply to urban areas and industrial consumers causing severe discomfort. Behpoor and Ghahramani (1988) reported a gas line rapture in Iran together with the stability measures performed to prevent future sliding in new gas line location.

Breakage of petroleum lines can cause landslides and severe damage to the environment. In the case of Roodbar, the petroleum line rupture resulted in the penetration of the oil in the fissures and subsequent reduction of the angle of internal friction and cohesion of the soil. In Roodbar before the earthquake soil parameters were c = 140 kpa and angle of internal friction ϕ = 27.5 degrees. After the initial shook of the earthquake the strength parameters were reduced to residual, c= 100 kpa and ϕ = 20 degrees. However the slope had still sufficient factor of safety to be stable after the earthquake. The infiltration of petroleum and water reduced the angle of internal friction and cohesion with subsequent triggering of the slide. The soil samples which were taken from the slide area were tested. The hydrometer test shows the existence of petroleum (see Fig. 5) To further study this phenomenon, the soil samples taken from the slope were mixed with different amount of oil and strength parameters were measured. Fig. 6 shows the results of this study. It is clear that a small amount of petroleum would considerably reduce the strength parameters. It should be noted that petroleum does not ordinarily penetrate deep into the soil due to high viscosity, however the existence of tension cracks due to either weathering or earthquake as in the case of Roodbar (see Fig. 7) would ease the penetration of petroleum and subsequent reduction in soil strength, thus weak zones are produced which may trigger the sliding of the slope. It is also important to note that although the gas pipelines are equipped with automatic shut-off valves, the petroleum pipelines are not normally equipped with such devices, therefore petroleum could leak for several hours before being shut off from pumping stations.

4 ENVIRONMENTAL IMPACT OF ROODBAR LANDSLIDE

The Roodbar area is located in the South of Caspian Sea endowed with forests, beautiful landscape, rivers, fruitful agricultural fields and olive trees on the slopes. These have made an attractive environment being used also as a resort area by Iranians during summer months. The income of many residents in the area is basically dependent on olive trees. The slide has destroyed more than 50,000 olive trees and the penetration of oil has affected many others. As the conditions are similar in many neighboring areas adequate measures should be taken to prevent landslides and lifelines should be designed with earthquake considerations.

5 CONCLUSIONS

Based on the study of Roodbar landslide it is concluded that:

1. The earthquake of 0.65 g caused an initial slip and tension cracks and reduced the soil strength parameters to residual.
2. The petroleum penetration due to rupture of pipeline as a consequence of earthquake reduced soil strength parameters producing weak zones and reducing the factor of safety of the slope. This triggered the landslide 2 days after the earthquake.
3. The environmental impact and the economical loss to the area was considerable.
4. The slide destroyed some houses and water pools and about 50,000 olive trees with considerable damage to the environment.

6 RECOMMENDATIONS

Considering that lifelines are of fundamental importance to economic growth in developing countries, the following recommendations are made.

1. Special attention should be given to design of lifelines in earthquake zones and on the slopes. The lifelines should be equipped with its own independently operated communication lines as is practiced in South of Iran.
2. On slopes in earthquake zones, the petroleum and other lifelines should be equipped with automatic shut-off valves to be activated in case of rupture.

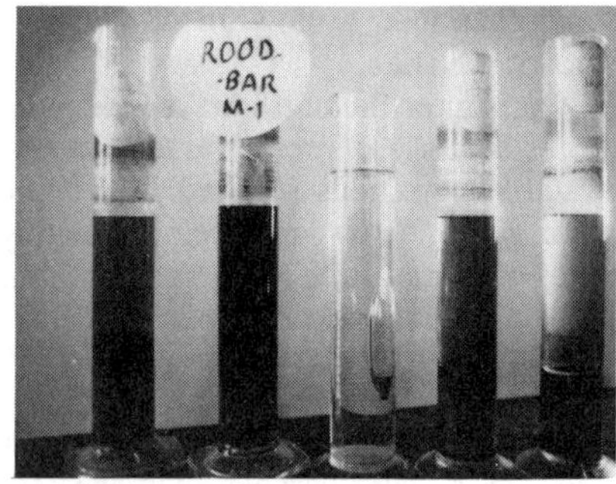

Fig. 5 Hydrometer showing oil content

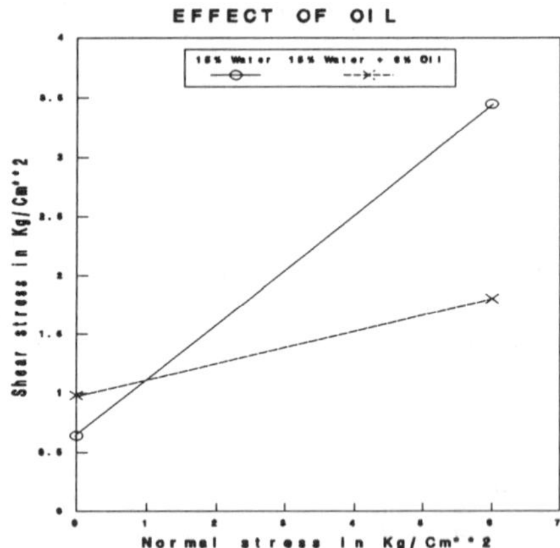

Fig. 6 Shear strength with oil

ACKNOWLEDGEMENT

The authors would like to acknowledge the support of the School of Engineering of Shiraz University and the Power and Water authority of The Guilan Province. The efforts of Mr. Tortiz and Mr. Rezanejhad is acknowledged.

Fig. 7 Tension cracks on the slide

REFERENCES

Anvar, S.A., L. Behpoor and A. Ghahramani 1992. Landslide in recent Roodbar earthquake in Iran. Sixth international symposium on landslides. New Zealand: Balkema.

Behpoor, L. and A. Ghahramani 1988. Landslide and stabilization at gas pipeline station 201 in Iran. Fifth international symposium on landslides. Switzerland: Balkema.

Environmental Management, Geo-Water & Engineering Aspects, Chowdhury & Sivakumar (eds)
 ISBN 90 5410 099 0

The magnitude and nature of 'noise' in world sea-level records

E.A. Bryant
Department of Geography, University of Wollongong, N.S.W., Australia

ABSTRACT: While average world sea-level is rising at a uniform rate of 1-1.5 mm yr $^{-1}$, regional rates can vary by an order of magnitude. Over time scales of several years these rates can be 10-100 times greater because sea-level is affected at this scale by highly changeable meteorological and oceanographic variables. The inherent "noise" level in world sea-level records is 35 mm. Much of this is expressed as fluctuations on the order of 20-100 mm with a frequency of 3-5 years. This latter "noise" is highly coherent at tide gauges around the globe and appears unrelated to resonance or wave excitation in oceans. It is suggested that this variability reflects changes in the world hydrological budget linked to the Southern Oscillation. This latter phenomenon relates to the strength of trade winds in the tropical Pacific Ocean and can generate significant drying followed by flooding over continental landmasses.

1 INTRODUCTION

The outcome of the present debate on global sea-level in a greenhouse-warmed world is dependent upon not only the magnitude of future warming but also the nature of global sea-level behaviour. To date, that debate has often naïvely assumed that any sea-level response will be of equal magnitude globally and that it will reflect melting of ice-caps and thermal expansion of the upper layers of the ocean (Titus, 1986). Our present knowledge of sea-level behaviour based upon long term historic records from various parts of the globe is however at odds with this view (Emery and Aubrey, 1991). Analysis of long-term historic records from various parts of the globe has found that regional climatic change in precipitation, barometric pressure and temperature induce quite divergent responses in sea-level (Bryant, 1988). A fundamental characteristic of global sea-level is its large spatial and temporal variability over distances as little as 200-300 km and timescales as short as a few days. This brings into question the appropriateness of the currently perceived global rise in sea-level of 1.0-1.5 mm yr^{-1} (Barnett, 1983; Gornitz and Lebedeff, 1987). It also indicates that the likely outcome of any climatic change, of which greenhouse-induced warming is but one part, will be regional discrepancies in the magnitude of sea-level rises and falls.

There are other aspects about the nature of sea-level that have been poorly considered both in terms of studies on sea-level and future projections. While the variation in sea-level appears to be climatically induced, the timing and magnitude of fluctuations in sea-level records appears to be coherent over great distances. The aim of this paper is to examine a) the magnitude and distribution of the background variability present in world sea-level records, b) the role of climate in producing this variability, c) the degree of coherence globally present in this background "noise" and d) the likely causes of this coherence, in particular the role of the global hydrological cycle. This latter information is necessary if engineers need to evaluate the effectiveness of large engineering projects proposed to mitigate future global sea-level rise.

2 THE MAGNITUDE AND DISTRIBUTION OF "NOISE" IN SEA-LEVEL TRENDS

The present eustatic rate of sea-level rise ranges from 1.0-1.5 mm yr^{-1} (Barnett, 1983; Gornitz and Lebedeff, 1987). These values suffer from a number of limitations including high spatial and temporal variability and gaps in world coverage, especially in the centre of oceans and in the southern hemisphere (Bryant, 1988). More seriously, such values have not been evaluated relative to the base level of variability inherent in world sea-level records. Such a determination is crucial, because rates of sea-level rise can only be judged as significant when they exceed this base level. In order to determine the global variability in annual sea-level records, 219 stations with ≥15 years of measurement between 1969-1979 were extracted from sea-level records compiled by the Permanent Service for Mean Sea Level, Bidston Observatory, UK. A linear trend calculated using Pearson product moment regression analysis was subtracted from each record and the standard deviation of residuals calculated for each record. A histogram of these standard deviations is

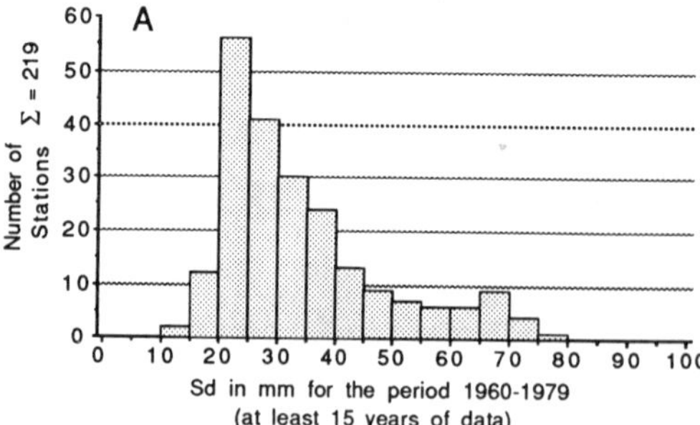

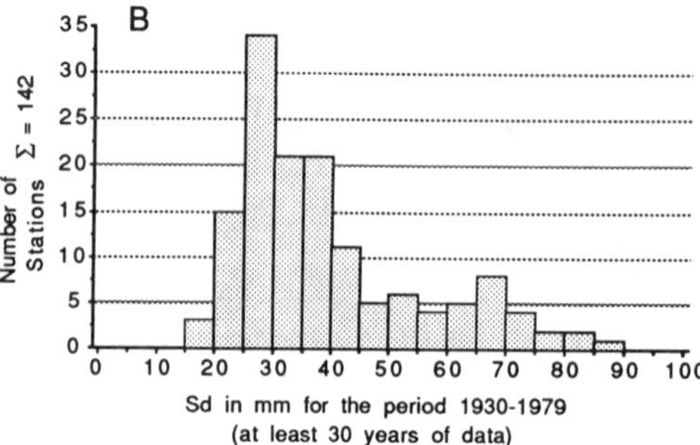

Figure 1 Histograms of the standard deviation of sea-level records a) 1960-79 b) 1930-1979

presented in Figure 1a, while a progressive record of these values eastwards around the globe is plotted in Figure 2.

The first standard deviation generally characterises the magnitude of variability within any time series. For sea-levels this value ranges from 13.0 to 93.8 mm with a mean of 35.0 mm equivalent to a variance in global sea-level of 1225 mm^2. Figure 2 illustrates that there are preferred regions where the inter-annual variability in sea-level records is exceptionally high. Generally most sea-level records have standard deviations between 20-40 mm. The Baltic and Yellow Seas, central Pacific and Gulf of Mexico contain a high proportion of stations where standard deviations exceed 50 mm. These regions are ones where climatically induced sea-level fluctuations such as storm surges or seiching are prominent. The high values in the central Pacific region are mainly due to changes induced by fluctuations in the Southern Oscillation characterised by changes in the intensity of trade winds.

The value 35.0 mm defines globally the average variability against which annual increments in eustatic sea-level rise must be referenced before any sea-level rise can be judged as significant. For instance, presently accepted rates of sea-level rise of 1.0-1.5 mm yr^{-1} (Barnett, 1983; Gornitz and Lebedeff, 1987) must be measured for periods of 23-35 years before they can be considered as trends and not part of the underlying variability in sea-level. Enhanced rates of sea-level rise of 20-30 cm in the next 50 years can only be credibly defined on a

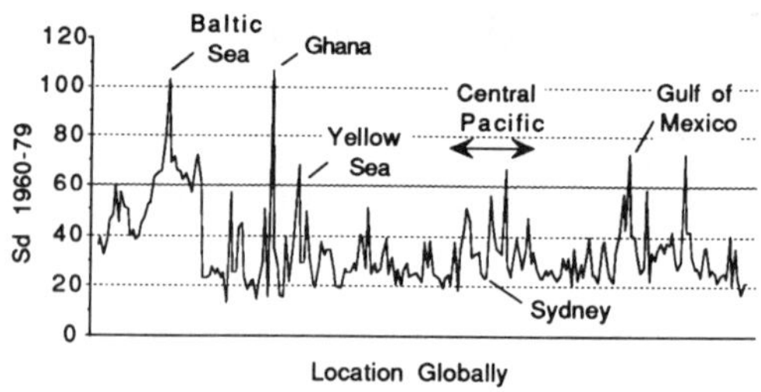

Figure 2 Standard deviations of sea-level records with at least 15 years data 1960-1979 plotted eastwards around the globe beginning in Scandinavia

global scale with at least a decade of data. Trends for individual stations may require even longer periods of measurement before they can be dubbed as significant. There are two caveats that must be put on a base level of 35 mm. Firstly, it is doubtful if the stations used in this analysis are completely representative of the globe. The data set excludes about 60% of the world's coastline and includes only a few values from the centres of oceans. Secondly, this value may not be representative of long-term variability in sea-level records. An analysis of 142 stations extracted from the same data base for a longer period, 1930-1979, produced an average standard deviation of 39.3 mm (Figure 1b). The concept of an eustatic rise in sea-level induced by Greenhouse warming must incorporate the possibility that the inherent variability in global sea-level can also change.

3 CLIMATIC REASONS FOR SEA-LEVEL VARIABILITY

Tectonic factors cannot be used to account for the spatial variation evident in Figure 2 because the scale over which this forcing operates is generally much longer. Climatic factors, especially atmospheric-oceanographic coupling, offer much wider scope for explaining this noise in sea-level records and have been applied to this end quite satisfactorily over timescales ranging from days to decades (Barnett, 1984; Bryant, 1988). Regionally these climatically induced changes can exceed the 1-1.5 mm yr^{-1} eustatic rate by a factor of 10-100 times. On annual timescales river discharge is important in the East China Sea and accounts for 7-21 per cent of the variance in sea-level along the eastern United States coastline (Meade and Emery, 1971; Emery and Aubrey, 1991). More significant are the persistent inter-annual fluctuations in sea-level associated with El Niño-Southern Oscillation (ENSO) events in the equatorial Pacific region (Wyrtki, 1982). Generally tropical air movement in the Pacific is dominated by strong easterlies which blow warm surface water towards the western side of the Pacific, piling it up to heights of 20 cm or more. The easterlies oscillate (the Southern Oscillation) in strength every three to five years. Their failure leads to an El Niño-Southern

Oscillation event that causes the propagation of a Kelvin wave eastwards across the Pacific in the space of two months. During the 1982/83 event, sea-levels fluctuated between 30-40 cm across the equatorial Pacific (Harrison and Cane, 1984). These sea-level fluctuations are not restricted to the tropics. They can spread along the west coast of the Americas over several months (Enfield and Allen, 1980). The 1982/83 El Niño-Southern Oscillation event raised sea-levels 35 cm above average as far north as the Oregon coast (Komar, 1986). Long term fluctuations in sea-level can also reflect changes in the intensity of winds, shifts in the location and intensity of pressure cells or long term changes in rainfall over the coastal sector of an ocean (Bryant, 1991). Finally a study of monthly sea-level, atmospheric pressure, temperature and precipitation at 91 stations around the world for the period 1933-1980 found that 50%, 60%, 24% and 34% of the variance respectively in these variables was inter-related (Bryant, 1988).

4 THE NATURE OF COHERENCE

The above observations imply a degree of coherence in not only the behaviour of sea-level globally but also in the behaviour of the climatic forcing variables. This coherence has been recognised previously. For instance Hamon et al. (1975) identified a strong spatial coherence in sea-level at most stations along the east coast of Australia. Across the Pacific Ocean sea-levels generally are inversely correlated in response to the Southern Oscillation (Wyrtki, 1982). Pariwono et al. (1986), extended this further and showed excellent correspondence between sea-levels in South Australia with those in Southern California. Thus while sea-level records evidence a high degree of noise, it appears that this noise is synchronous over long distances. As an illustration of this, annual sea-level records for the period 1930-1974 for London, Trieste (Venice), New York, Tokyo (Aburatsubo) and Sydney, were plotted (Figure 3) and analysed statistically to determine the degree of cross-correlation. The stations represent both northern and southern hemispheres. They also represent open ocean and enclosed sea environments which are not in resonance with each other. On a gross scale it is evident that some of the major fluctuations in these sea-level records occur contemporaneously. For instance, there is a prominent dip in all records around 1949-50 and a peak around 1952-53. Many of the fluctuations in the figure also correspond to those appearing in a composite of South Australian and southern Californian records (Pariwono et al., 1986).

The records were first detrended using Pearson product moment regression analysis. Cross-correlation was then performed at yearly lags to discern any precession in the peaks of fluctuations. This analysis tended to indicate that fluctuations in the Sydney record preceded ones in the London and Venice records by 1 year which themselves preceded those in New York and Tokyo by an additional year.

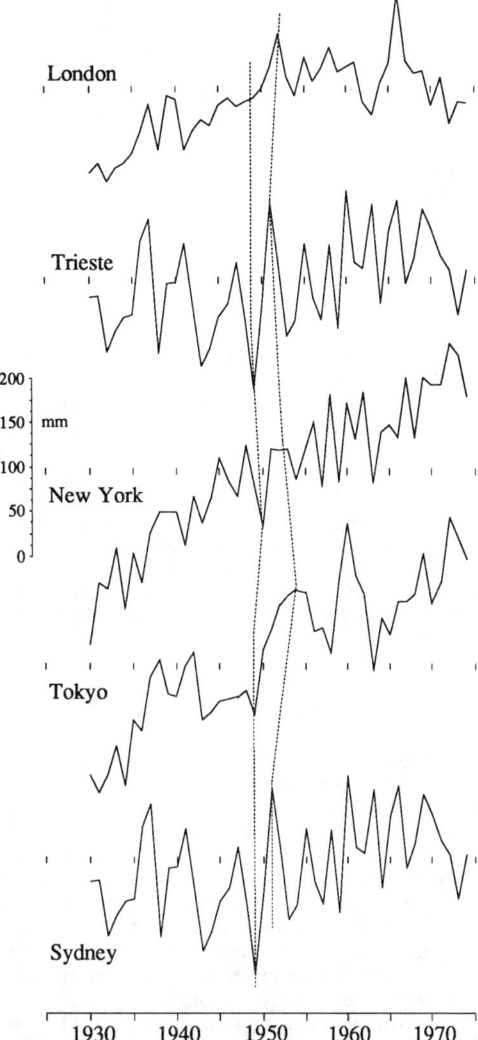

Figure 3 Annual sea-level records between 1930-74 for five major tide gauges

It should be pointed out that these are the best fit lags and that there are times when fluctuations in sea-level appear contemporaneously between records. The time series were then shifted to reflect these best fit lags and analysed using Principal Component Analysis (PCA). PCA has the capacity to extract the structure of coherence from data sets by grouping together similar variables (in this case residuals of lagged sea-level records) into groups or components (Ebdon, 1985). The magnitude of this coherence for variance groupings of records can then be assessed by calculating factor scores. Two coherent groupings are evident above the noise level of the data. The factor scores for each grouping together with the percentage of the variance explained for each city are

Table 1. Principal Component Analysis of lagged sea-level records of 5 major tide gauges 1930-1974

City	lag in years	% Variance in Group 1	% Variance in Group 2
New York	1	16.4	64.5
Tokyo	1	45.4	12.3
Sydney	-1	45.2	28.4
London	0	57.5	0.4
Venice	0	31.3	7.7
All gauges		39.1	22.6

presented in Table 1. The largest grouping accounts for 39.1% of the total variance in the 5 detrended sea-level records. This group includes all records except New York. It indicates that 45% of the inter-annual variability in the Tokyo and Sydney records and 57.5% of that in the London record are coherent when lagged slightly relative to each other. The second grouping indicates 28.4% of the variability in the Sydney record is correlated with 64.5% of that in New York. Allowing for the fact that the records have been shifted relative to each other, this result implies that sea-level at New York rises after it does at Sydney.

5 REASONS FOR THIS COHERENCE

Pariwono et al. (1986) attributed this high coherence in sea-level to pressure patterns associated with the Southern Oscillation; however a more fundamental factor must be involved because the results include two European gauges not normally evidencing pressure responses to the Southern Oscillation. If atmospheric pressure is not implicated, then this coherence must relate to changes in either temperature or rainfall. There is little doubt that both air and sea surface temperature fluctuations over 2-3 year periods are coherent globally (Folland et al., 1984). However the magnitude of these oscillations (0.5-1.0°C) appears unlikely to be sufficient to affect sea-levels through thermal expansion to the degree observed in Figure 3. Rainfall is the more plausible cause; yet this factor cannot operate solely at the local level to produce such a coherent signature globally. Rainfall can only be invoked at this broad scale as part of the global hydrological cycle. The Southern Oscillation has a major effect on the global hydrological cycle because it is associated with widespread, alternating droughts and floods over Africa, Australia, India and southeast Asia (Bryant, 1991). Oscillations between drought and flood occur at intervals of 3-5 years, the same cycle as fluctuations in sea-level. During its dry phase (El Niño-Southern Oscillation events) when trade winds in the tropical Pacific Ocean are reduced, sea-levels drop in the west Pacific and rise in the east. At the same time drought dominates land masses with water draining from continents and accumulating in the oceans. During wet phases (La Niña events) the trade winds strengthen leading to enhanced monsoonal rainfall conditions in Australia, India, southeast Asia and Africa. Rain accumulates over these continental land masses in drainage systems, lakes or watertables. During these periods evaporation from oceans supplying this rainfall lowers global sea-levels. The area of landmass involved is on the order of 16 x 10^6 km^2 equivalent to 10-11% of the surface area of the oceans. Typical rainfall amounts to 200-400 mm in the transition from a dry ENSO phase to a wet La Niña phase over the space of 2-3 months. This volume of rainfall averaged over the oceans represents a change in sea-level of 22-45 mm. This approximates the standard deviation inherent in global sea-levels and begins to equal the magnitude of many of the inter-annual variations evident in Figure 3. While the Southern Oscillation may be responsible for these rainfall variations, the delays in the exchange of water between continents and oceans results in coherent but lagged changes in sea-level measured at tide gauges distributed around the globe.

6 CONCLUSIONS

The inherent variability in world sea-level records is around 35 mm. This value must be exceeded before rises in average global sea-level can be viewed as representing a significant trend. If the sparse coverage of tide gauges is symbolic of world sea-level, then the sea-level rise of 100-150 mm over the past century is meaningful. Much of the variability in tide gauge records regionally is due to climatic-oceanographic forcing; however a significant percentage of this "noise" at inter-annual timespans is coherent across the globe. Because many places globally are affected by drought at the same time, water drains from the continents and accumulates in the oceans forcing world sea-level up. Heavy rainfall also tends to occur at the same time in many places. Water is then taken from the oceans and temporarily stored in lakes, rivers and watertables on continents. As a consequence world sea-level drops. This exchange has implications for large scale projects designed by engineers to mitigate against future sea-level rises. For instance a shift to more frequent ENSO events leading to droughts over Africa, India, Southeast Asia and Australia will not only exacerbate these rises, but will also require longer time periods for the diversion of water from oceans to continents to be effective.

REFERENCES

Barnett, T.P. 1983. Recent changes in sea level and their possible causes. *Climatic Change* 5: 15-38.

Barnett, T.P. 1984. The estimation of "Global" sea level change: a problem of uniqueness. *Journal Geophysical Research* 89: 7980-7988.

Bryant, E.A. 1988. Sea-level variability and its impact within the greenhouse scenario. In G.I. Pearman (ed.), *Greenhouse: planning for climate change*. C.S.I.R.O., Melbourne, 135-146.

Bryant, E.A. 1991. *Natural Hazards*. Cambridge, Cambridge Univ. Press.

Ebdon, D. 1985. *Statistics in Geography*. 2nd ed. Oxford, Basil Blackwell.

Emery, K.O. and D.G. Aubrey 1991. *Sea Levels, Land Levels, and Tide Gauges*. New York, Springer-Verlag.

Enfield, D.B. and J.S. Allen 1980. On the structure and dynamics of monthly mean sea level anomalies along the Pacific coast of North and South America. *Journal of Physical Oceanography* 10: 557-78.

Folland, C.K., D.E. Parker and F.E. Kates 1984. Worldwide marine temperature fluctuations 1856-1981. *Nature* 310: 670-73.

Gornitz, V. and S. Lebedeff 1987. Global sea-level changes during the past century. In D. Nummedal, O.H Pilkey and J.D. Howard (eds.), *Sea-level change and coastal evolution*. Society of Economic Paleontologists and Mineralogists Special Publication 41: 3-16.

Hamon, B.V., J.S. Godfrey and M.A. Greig 1975. Relation between mean sea level, current and wind stress on the east coast of Australia. *Australian Journal of Marine and Freshwater Research* 26: 389-403.

Harrison, D.E. and M.A. Cane 1984. Changes in the Pacific during the 1982-83 event. *Oceanus* 27: 21-8.

Komar, P.D. 1986. The 1982-83 El Niño and erosion on the coast of Oregon. *Shore and Beach* 54: 3-12.

Meade, R.H. and K.O. Emery 1971. Sea level as affected by river runoff, Eastern United States. *Science* 173: 425-28.

Pariwono, J.I., J.A.T. Bye and G.W. Lennon 1986. Long-period variations of sea-level in Australia. *Geophysical Journal Royal Astronomical Society* 87: 43-54.

Titus, J.G. 1986 The causes and effects of sea level rise. In J.G. Titus (ed.), *Effects of changes in stratospheric ozone and global climate v. 1: An overview*. United States Environmental Protection Agency, 219-248.

Wyrtki, K. 1982. The Southern Oscillation, ocean-atmosphere interaction and El Niño. *Journal Marine Technology Society* 16: 3-10.

Environmental Management, Geo-Water & Engineering Aspects, Chowdhury & Sivakumar (eds)
 ISBN 90 5410 099 0

Earthquake analysis for embankments

R.N.Chowdury & D.W.Xu
Department of Civil and Mining Engineering, The University of Wollongong, N.S.W., Australia

ABSTRACT: In this paper a new method is outlined for the analysis of embankments and earth dams under earthquake conditions. One main part of this method is the estimation of a critical seismic coefficient. The other part consists of a formulation for dynamic response analysis in which the critical seismic coefficient is one of the important parameters. The significant innovation in this method is the facility for updating the critical seismic coefficient as the slope weakens during an earthquake. The deformation of the slope can be estimated as a function of time after the inception of an earthquake. The conditions for catastrophic failure can be studied using this method, provided that reliable data on soil properties are available.

1 INTRODUCTION

Design of earthquake-resistant earth structures is a major concern of geotechnical engineers. The assessment of the capability of an earth structure to resist dynamic forces without loss of integrity is not an easy task. Important factors include the following:

(a) The time history of ground motion and the magnitude of the earthquake
(b) The geometry and design of the earth structure including the geotechnical properties of the soils such as shear strength parameters, pore water pressure development during dynamic loading, strain-softening characteristics
(c) Dynamic soil properties such as natural frequency and damping ratio
(d) Whether the earth structure is a dam or is likely to be a water-retaning structure in the future

A simple approach for earthquake analysis is the pseudo-static approach. It is well known and widely used but has significant limitations mainly because of the difficulty in selecting an appropriate value of the pseudo-static seismic coefficient. Analyses based on widely accepted values of this coefficient have often given misleading results and computed factors of safety well above unity have been computed for some earth structures which have failed. Therefore, more realistic and sophisticated methods of analysis have been developed in recent years (Newmark, 1965; Seed, 1979).

2 EMBANKMENT DISPLACEMENTS

While failures of some types of earth structures have occurred during or soon after earthquakes, complete failures of embankments of clay have rarely been reported. According to Seed (1979) all reported cases of slope failure have involved sandy soils. However, deformation under earthquake loads can lead to damage even if complete failure does not occur and, therefore, analyses which enable estimation or prediction of deformation are very valuable. In this connection the method developed by Newmark (1965) has proved to be popular. In this method a sliding block model is used to simulate the acceleration, velocity and displacement of a slope during an earthquake. Permanent displacements occur as a consequence of the inertia forces exceeding the yield resistance along a potential slip surface within the slope. Modifications or refinements of the Newmark method have been suggested by Seed and Martin (1966) and Ambraseys and Sarma (1967) among several others. Provided the yield resistance of soil can be estimated reliably and that shear strength does not change significantly during an earthquake, Newmark's approach gives a reliable estimate of embankment displacements.

Estimates of embankment deformations considering the decrease of effective peak acceleration with increasing depth of the slip surface have also been made by several research workers. The results for earthquakes of magnitudes 6.5 and 8.25 respectively have been summarised by Seed (1979) in a tabular form. These calculations are based on the assumpations that (a) no strength loss occurs, or (b) the strength loss is limited to 15%. These results may not be applicable directly to water-retaining embankments espectively if they involve saturated sand or brittle clayey soil. The yield acceleration coefficients are considered to be in the range of K_c = 0.05 - 0.20 in the studies for which the results have been tabulated by Seed (1979). It is interesting to note that for 6.5 magnitude earthquakes the estimated

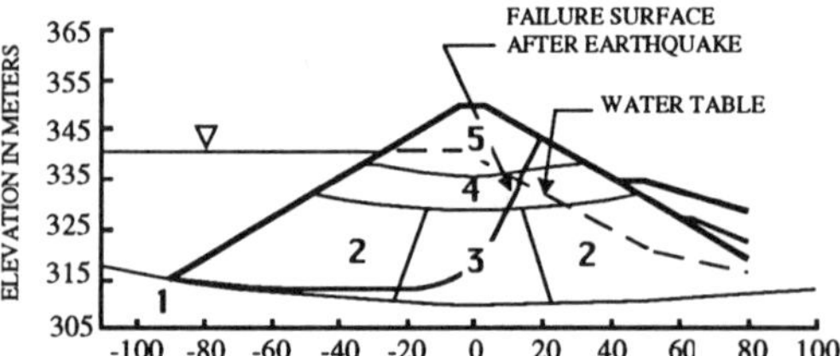

Fig.1 Cross-section of Lower San Fernando dam

displacements do not exceed about 1.2 m and for 8.25 magnitude earthquakes the estimated displacements do not exceed about 5.2 m. According to Seed (1979), displacements up to about 1m may be considered as acceptable. This type of performance can generally be expected from an embankment designed to withstand an inertia force of 0.1 to 0.15g without yielding, provided that little or no excees pore pressures are developed in the soil and that shear strength loss during an earthquake is less than 15%. In other words, seismic coefficients of 0.1g and 0.15g should be used in pseudo-static analyses for 6.5 and 8.25 magnitude earthquakes respectively, and a factor of safety of 1.15 ensured in both cases.

It would seem that some types of clay soil, dry sands and dense seturated sands do not lose significant resistance to deformation as a result of earthquake loading. In contrast, however, loose to medium dense cohesionless soils may have significant strength loss during an earthquake. In particular, under conditions of saturation, high excess pore water pressures will generally develop in such soils during dynamic loading. Local excess pore pressures will, of course, redistribute within the earth structure during or after the earthquake.

The main aim of the present paper is to demonstrate how the steady increase in pore pressure during an earthquake can be simulated and incorporated in a seismic response analysis. Such an analysis can, therefore, facilitate a realistic estimation of embankment performance not only for insensitive soils but also for soils which may suffer significant decrease of shear strength during an earthquake.

3 SEISMIC RESPONSE USING STRESS DEFORMATION STUDIES

Considering the limitations of the pseudo-static approach and the difficulties of evaluating the earthquake response of saturated cohesionless soils, Seed and his co-workers (Seed, 1979) developed a comprehensive procedure involving both static and dynamic finite element analyses combined with laboratory tests to determine pore pressure development under cyclic loading. This procedure, known as the 'Seed-Lee-Idriss analysis procedure', has been used to 'explain' the embankment slides that occurred as a result of the San Fernando Earthquake of 1971 (Seed *et. al*, 1976). It would be interesting to have case studies with analyses based on this method, made before the occurrence of a failure caused by an earthquake. However, no such analyses have been reported in the literature.

A limitation of the Seed-Lee-Idriss procedure is that the change in the factor of safety or in K_c, the critical seismic coefficient (also called 'the yield seismic coefficient') during an earthquake cannot be studied. For example, the factor of safety after earthquake of the upstream slope of the Lower San Fernando Dam was evaluated after zones of liquefaction failure had been identified independently based on laboratory tests under stress conditions evaluated from finite element studies. However, the variation in the factor of safety from the value before the earthquake to the post -earthquake conditions was not simulated. The need for a consistent, relatively simple and comprehensive approach is, therefore, obvious. A dynamic response analysis coupled to a limit equilibrium approach which enables simulation of changes to excess pore pressure and factor of safety during an earthquake would be useful in this regard. Previous extensions to Newmark's method do not go far enough to achieve that aim.

4 THE PROPOSED METHOD

The method developed by the authors (Chowdhury-Xu method) consists of the following elements:
(a) A comprehensive limit equilibrium procedure which can handle slip surafces of circular and arbitrary shape under different loading conditions. This method should enable determination of factor of safety as well as critical seismic coefficient or critical acceleration. The method should also enable the determination of critical slip surfaces, i.e., slip surfaces with either the minimum factor of safety or the minimum value of K_c, the critical seismic coefficient
(b) A method for simulating the acceleration-time history of earthquakes
(c) A dynamic response analysis model
(d) A method for evaluating principal stress

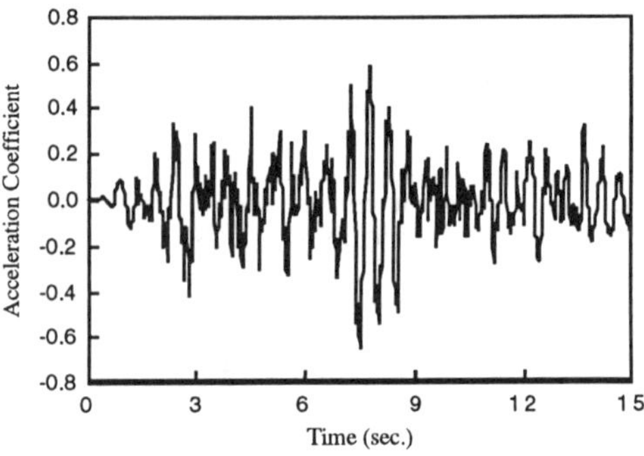

Fig. 2 Time history of earthquake acceleration

increments at any time during an earthquake and hence the shear stress increment

In addition to data on shear strength parameters, initial pore water pressures and slope geometry, results of pore pressure development under cyclic loading conditions should be available. From the latter and the shear stress history during the earthquake, excess pore water pressures, developed at any time during the earthquake, may be estimated as part of the proposed new method.

4.1 Some detail of the proposed new method

For the motion of a typical vertical slice of a potential sliding mass, the following equation has been developed

$$\ddot{X}_i(t) + 2\nu\omega\,\dot{X}_i(t) + \omega^2 X_i(t) = g\,\frac{\cos(\alpha_i - \theta - \phi_i')}{\cos\phi_i'}\,[K(t) - K_c] \qquad (1)$$

in which α_i is the inclination of the base of the i-th slice to the horizontal, ω is the natural frequency, ν is damping ratio, g is gravitational acceleration, θ is the inclination to the horizontal of K(t), the seismic acceleration coefficient at time t, K_c is the critical seismic coefficient (i.e., the coefficient which leads to a factor of safety of unity) and ϕ_i' is the interal friction angle of the soil. The acceleration, velocity and displacement of the slice are denoted respectively by $\ddot{X}_i$, $\dot{X}_i$ and X_i. The damping ratio ν and natural frequancy ω are related to the mass of the slice M_i, stiffness coefficient K_s and damping coefficient C as follows:

$$C = 2\,\nu\,\omega\,M_i \qquad (2)$$

$$\omega^2 = K_s / M_i \qquad (3)$$

It is interesting to note that Equation (1) appears to be independent of cohesion c'. However, this is not so since cohesion is included in the critical seimic coefficient K_c.

Table 1 Soil properties used in analysis

Material No.	c_u (KN/m ^{2})	ϕ_u	γ (KN/m ^{3})
1	57.5	20°	17.3
2	57.5	20°	16.0
3	103	0	19.0
4	0	33°	20.0
5	103	0	19.0

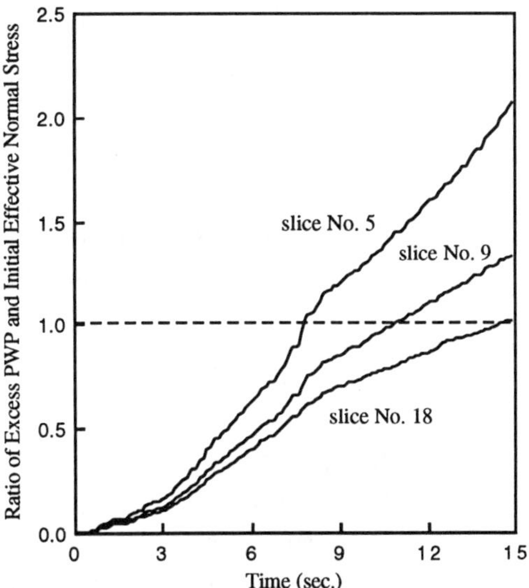

Fig. 3 Variation of excess pore water pressure with time for $A_1 = 0.67$ and $A_5 = 6.0$

The method of solution of the differential equation, Eq. (1), will not be discussed in this paper because of space limitation. However, it is important to note that K_c is considered to be a variable during earthquake motion. This is the most important new element in the proposed response analysis procedure. The inclusion of natural ferquency and damping ratio is also novel since the relevant terms have been excluded from all previously published limit equilibrium methods and from Newmark-type models.

The change in K_c during an earthquake is a consequence of decreased shear strength. Only the potential decrease in shear strength due to generation of excess pore water pressure is considered in this paper.

4.2 Excess pore water pressure

Much has been written about the generation of excess pore water pressures under static loading and Skempton's pore pressure coefficients A and B are widely used for estimation of transient pore pressure under static loading conditions. Under dynamic loading the generation of pore water pressures has been studied mainly from laboratory tests. The influences of initial stress condtions, the stress level of cyclic loading and the number of cycles of loading have also been explored by many researchers. No simple equation relating the increase of pore water pressure to the increase of shear stresses during earthquakes is available. However, Sarma (1988) has proposed the determination of a dynamic pore pressure parameter A_n similar to the Skempton coefficient A for static loading. Only the determination of A_n is of interest here because the manner in which Sarma (1988) proposed to use the parameter is not suitable for estimating the change in K_c as a function of time or of the increase in displacement attributable to the change in K_c.

The following relationshipe have been proposed relating excess pore water pressure Δu to increments of either the major principal stress $\Delta\sigma_1$ (triaxial loading conditions) or the shear stress $\Delta\tau$ (simple shear conditions) during dynamic loading

$$\frac{\Delta u}{\sigma'_{3c}} = A_n \frac{\Delta\sigma_1}{\sigma'_{3c}} \quad (4)$$

$$\frac{\Delta u}{\sigma'_{v0}} = A_n \frac{\Delta\tau}{\sigma'_{v0}} \quad (5)$$

where σ'_{3c} is the minor principal effective consolidation stress and σ'_{v0} is the effective vertical overburden stress.

By re-ploting published data concerning the results of cyclic loading tests on different sands, linear relationships were found to exist between $A_n^{1/2}$ and log(n) (where n is the number of cycles) for different values of the ratio $\frac{\Delta\tau}{\sigma'_{v0}}$ or $\frac{\Delta\sigma_1}{2\sigma'_{3c}}$. The following equation was proposed for A_n:

$$A_n^{1/2} = A_1^{1/2} + \beta \log n \quad (6)$$

where, A_1 is the parameter for one cycle of loading and β is a constant.

By using Sarma's suggested definition of A_n the variation of A_n during the time history of an earthquake may be determined if a concept such as that of 'the number equivalent uniform stress cycles' is invoked. If results of pore pressure development from more sophosticated tests (such as shaking table tests) are available, the variation of A_n during an earthquake may be determined even more realistically.

During an earthquake, the excess pore pressures are primarly generated as a consequence of the increments in shear stress due to dynamic loading. Therefore, at any time t, the cumulative excess pore water pressure may be related to the time-average of the maximum shear stress increment as follows:

$$\Delta u(t) = A_n(t)\,\{[\Delta\sigma_1(t) - \Delta\sigma_3(t)]_{average}\} \quad (7)$$

It is impotant to understand that A_n is a parameter relevant to the cumulative excess pore water pressure and not to the development of pore water pressure between time t and (t $+\Delta t$). The increments in principal stresses with reference to the initial principal stresses fluctuate dramatically with time due to the nature of earthquake motion. Therefore, the best approach for calculating the cumulative excess pore water pressure is to use the averaging process as proposed above in Eq. (7) for the increment of shear stress at any time. The time-average of the maximum shear stress is, therefore, always calculated from the start of the earthquake.

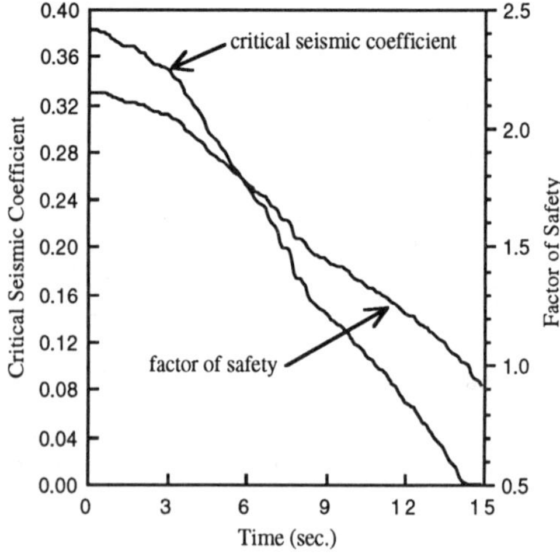

Fig.4 Variation of factor of safety and critical seismic coefficient with time for $A_1 = 0.67$ and $A_5 = 6.0$

Table 2 Average, maximum and minimum displacements

$A_n = A_5$	Ave. Disp. (cm)	Max. Disp. (cm) (slice No.)	Min. Disp. (cm) (slice No.)
1.67	5.28	6.64 (19)	4.35 (20)
3	9.04	11.07 (19)	7.45 (20)
5	19.82	24.25 (19)	16.32 (20)
6	37.01	45.45 (19)	30.59 (20)

5 TYPICAL RESULTS

The method proposed above was used to carry out a number of studies of the Lower San Fernando Dam which has previously been studied by Seed and his co-workers (Seed, et. al., 1976). The embankment cross-section is shown in Fig. 1 along with one of the observed slip surfaces. The simulated earthquake acceleration-time history closely resembling that used by Seed and his co-workers is shown in Fig. 2. The soil properties used for determination of the values of factor of safety and critical seismic coefficient are shown in Table 1.

The variation of generated excess pore water pressures with time for three typical slices is shown in Fig.3 based on the assumed values of dynamic pore water pressure coefficient, $A_1 = 0.67$ and $A_5 = 6$. This means that the number of equivalent uniform cycles is considered to be n = 5 and that A_n has a value varying from 0.67 to 6 during the 15 second period of earthquake motion; the value is considered at be constant ($A_1 = 0.67$) over the first 3 seconds.

The variation of the factor of safety with time and the variation of the critical seismic coefficient with time are shown in Fig. 4 for $A_1 = 0.67$ and $A_5 = 6.0$.

For different values of A_5, the estimated maximum and minimum displacements are shown in Table 2. The number of total slices is 20 and slices 19 and 20 are located near the top (crest) of the slope.

6 DISCUSSION OF RESULTS

From the results presented in Table 2 it is clear that displacements are quite small at a value of $A_5 = 1.67$. The displacements do increase with an increase in the value of A_5 but , even with $A_5 = 6$, the maximum displacement is only about 46cm. With such a high value of A_5 the critical seismic coefficent does eventually reach zero almost towards the end of the earthquake. Therefore, a slide could be expected to occur only if material behaviour was known to correspond to this high value ofA_5. However, values of A_5 in the range (1.35-2.58) have been interpreted by Sarma (1988) on the basis of available data concerning the upper and lower San Fernando dams. It is obvious that failure is not prediced on the basis of undrained shear strength parameters used for static and pseudo-static analyses of these dams before the earthquake. In a separate paper, it will be shown that, under earthquake loading, failure is prediced with a realistic value of A_5 if effective-shear strength parameters are used for the analysis.

7 CONCLUSION

A new approach for earthquake analysis of embankments has been outlined. This approach has special relevance to water-retaining earth structures such as earth dams in which there are zones of sand or any other soils which can develop excess pore water pressures during cyclic loading. The method works well and only typical results have been presented here. These results confirm that failure of the Lower San Fernando dam during the San Fernando earthquake could not have been predicted on the basis of undrained shear strength parameters (for the sands) which were used for static and pseudo-static analysis during the design stages. It has been shown that the magnitudes of critical seismic coefficient and the factor of safety of an embankment decrease significantly during an earthquake and may reach critical values if the soil shear strength decreases sufficiently. This can happen in soils such as saturated sands which develop significant pore pressures during dynamic loading.

ACKNOWLEGEMENTS

The work has been undertaken within the Water Engineering and Geomechnics Research Program established by the University of Wollongong as part of its Research Management Strategy in 1988 with the first author as Coordinator.The support of the University of Wollongong for research activities is acknowledged.

REFERENCES

Ambraseys N.N. and Sarma S.K. 1967. *The response of earth dams to strong earthquakes*, Geotechnique, Vol. **17**, 181-213.

Newmark, N.M. 1965. *Effect of earthquakes on dams and embankments*, Geotechnique,Vol.**15**, No.2, 139-160.

Seed, H.B. and Martin, G.R. 1966. *The seismic coefficient in earth dam design*, Journal SMFD, Proc. ASCE, Vol. **92**, No. SM3, May, 25-58.

Seed, H.B. 1979. *Considerations in the earthquake-resistant design of earth and rockfill dams*, Geotechniques, Vol. **29**, No. 3, 215-263.

Environmental Management, Geo-Water & Engineering Aspects, Chowdhury & Sivakumar (eds)
© 1993 Balkema, Rotterdam. ISBN 90 5410 099 0

Warning and danger index of dam safety monitoring

Li Zhenzhao & Zhou Xuyin
Wuhan University of Hydraulic & Electric Engineering, Hubei, People's Republic of China

ABSTRACT: This paper is intended to discuss and analyse the determination of safety monitoring indexes for dams. It is presented that dam monitoring indexes can be classified as warning value and danger value; the former is a bound to judge whether the observed value is normal; the latter is a bound to judge whether the structure reflected by the observed values is in safe state. The warning value should be based on mathematical model of monitored quantity, and the stability and strength is a prerequisite for danger value. A method to determine monitoring indexes is presented and some examples are introduced in this paper.

1 FOREWORD

It will exert a great influence on natural environment to build a dam. A dam in the state of safe operation, will play a positive role in power generation, water supply, flood control, irrigation, breeding and improvement of environment. In the case of failure, however, it can bring about disaster and destruction of environment. It is the main task of dam monitoring to obtain the knowledge on the actual state of a dam, and to find the latent problems and take preventive measure in good time. To judge whether the behaviour of a dam is normal and safe, the critical values of monitored quantities should be determined so as to obtain the scientific criterions of safety control.

To ensure the safety of a dam, a great deal of calculation and demonstration has been carried out during the design stage, through which, the type, size, material, structure, flood carrying capacity and foundation treatment of the dam have been determined. The principal means taken in design for the safe operation of a dam is stipulating characteristic stage (such as normal storage water level, design flood level and check flood level etc.) and corresponding load conditions (such as earthquake intensity, wave pressure and elevation of silting-up). The stability and strength of the dam should be taken into consideration in design to meet with relevant safety requirements in the case of stipulated load combination.

Because the design theory is not so mature to exactly evaluate the actual operation state of a dam, many factors in construction may be different from those expected in design, and behaviour of the dam in use may change. The possibility that the abnormal and unsafe phenomena appear can not be ruled out even the dam is used with a water level within the range of design. On the contrary, the dam is probably in normal state when it is used in a water level beyond the range given by design. All those mentioned above show that the safety of a dam cann't be controlled only by water level. We can also judge the safety of a dam according to actual behaviour of the dam, which is a more accurate and more effective method than that by design.

According to the observed data for several dams in China, we studied the problem of safety monitoring and provided some monitoring indexes for these dams. This paper is to present some consideration and methods we have taken in our researching work.

2 SELECTION OF MONITORING INDEXES FOR DAMS

2.1 Selection of objects of monitoring indexes

Usually, there are a large number of items and points to be observed, so we should select the dominating and typical observed items and points as the objects of monitoring indexes to carry out safety monitoring in time and more effectively.

Generally, the observed items such as seepage quantity, uplift pressure, pore water pressure, deformation and stress can be taken as principal minitoring quantities.

Following factors should be taken into consideration to select typical points from a group of survey points for a monitoring quantity.

1. The typical survey points should be selected respectively for different types of dam blocks (such as nonoverflow block, overflow block and intake block etc.).

2. Among the dam blocks in same type, the typical survey points should be located at those blocks which are higher or in complex geologic conditions.

3. It had better to select those with a large variation range of observation value or a widely varying tendency as typical survey points.

A minitoring object is usually a single survey point. It is easy to use a single survey point to construct a mathematical model and provide monitoring indexes. But the effect of monitoring operation with single point is limited. If possible, it had better to use a group of connected survey points to provide monitoring indexes.

For a dam or an independently working dam block, after selecting monitoring indexes for various items, we ought to comprehensively analyse the dam or dam block to find out whether the problem is a local one or a general one, and evaluate the behaviour of the dam (or dam block) as a whole.

2.2 Expression of monitoring indexes

Monitoring indexes can be classified into warning value and danger value. The warning value is a bound of observed value between normality and abnormality. It can be determined on the basis of physical relation between observed values and loads, variation range and regularity of previous observed values. The warning value can be subclassified into ordinary warning value and serious warning value.

The danger value is a safety limit of structure reflected by observed values, which can be determined according to safety requirements. The danger value can be subclassified into ordinary danger value, in the case of permissible state, and serious danger value in the case of limiting state.

Monitoring indexes are usually expressed as a function connected with operation conditions of a dam (such as water level of upstream and downstream, temperature, precipitation and age etc.) and the positions of survey points. In this way, the function can be used in various conditions for daily monitoring operation. In the case of the most disadvantageous combination of (loads) conditions, the calculating result can be taken as monitoring limit.

A monitoring index should be expressed as a confidence interval in any conditions owing to the random errors of observed values for dams.

The general expression for monitoring indexes can be written as

$$[y]^{\pm} = f[x_1, x_2, \ldots, x_n] \pm \delta \qquad (1)$$

where $[y]^{\pm}$ are the bounds of monitored indexes; $[y]^{+}$ is the upper bound; $[y]^{-}$ is the lower bound; $x_1, x_2, \ldots, x_n$ are the main factors that affect the monitored quantity y (such as operation conditions and loads), including positional variates of survey points when a group of points are used in monitoring operation and time variates when time-variant effect should be taken into consideration; δ is half breadth of confidence belt.

y_i is the monitored quantity y observed at i-th time, if

$$y_i \in ([y]^{+}, [y]^{-}) \qquad (2)$$

then, y_i can be thought normal or safe, otherwise, y_i probably be abnormal or unsafe.

3 DETERMINATION OF WARNING VALUE FOR MONITORED QUANTITY

The warning value for monitored quantity is a bound of observed values between normality and abnormality, for which the monitoring expression can be obtained on the basis of its mathematical model. To ensure the reasonableness and validity of the warning value, the mathematical model should be tested, treated and improved if necessary, and an appropriate confidence belt should be selected to set up a practical monitoring expression.

3.1 Mathematical model of monitored quantity

Monitored quantity y always varies with various environmental factors (including time factor and positional factor of survey points) x_i (i=1,2,...,n). y can be taken as conditional random variates conditioned upon factors $x_1, x_2, \ldots, x_n$. The statistic relation between the monitored quantity and all factors can be written as

$$y = f(x_1, x_2, \ldots, x_n) + \varepsilon \qquad (3)$$

where $f(x_1,x_2, \ldots,x_n)$ represents the variation caused by $x_1,x_2, \ldots, x_n$; ε represents random error, which is a random variate in normal distribution $N(0,\sigma^2)$.

The mathematical model can be written as

$$\hat{y} = f(x_1,x_2, \ldots ,x_n) \quad (4)$$

where $\hat{y}$ is an estimation for mathematical expectation $E(y \mid x_1,x_2,\ldots,x_n)$ of conditional random variate y.

3.2 Requirements of mathematical model for determining a warning value

For ensuring a good quality of mathematical model, on which the warning value based, and comprehensively and accurately reflecting the relations among the monitored quantities and the factors connected, a physical analysis should be carried out, enough and appropriate factors and a reasonable structure of the model have to be selected, the parameters of the model should be calculated in proper mathematical method, in addition, the residual error series of the model

$$\eta_j = y_j - \hat{y}_j ; \quad j = 1,2,\ldots,m \quad (5)$$

should be tested.If the mean of η_j approaches zero, the distribution of η_j is a normal random distribution.If the changing regularity no longer exists, we use $\hat{y}$ to estimate the mathematical expectation of random variate y, use s^2 of η_j to estimate the variance σ^2 of y. s is standard deviation of residual error series of expression (4), which is refered to as "residual standard deviation", and can be obtained from variance analysis of expression (4). m is the number of observed value y.

3.3 Test and improvement of mathematical model

3.3.1 Rejection of gross errors

Before constructing a mathematical model, we ought to identify and reject the gross errors of observed data. Some gross errors can be identified by rationalization analysis, or distinguished and rejected with a multistep method of construction of model.

3.3.2 Check and rejection of linear terms in residual error series

The existence of linear terms in residual error series in a model means that the linear time effect is not involved in the model, which will affect the warning value. The residual graphic analytical method and linear regression analysis method can be used to check whether the linear terms exist.

Using m η_i and t_i, we establish a regression equation as

$$\eta_i = a_0 + a_1 t_i \quad (6)$$

If the linear relation of expression (6) and the mathematical model is significant under significance level α, then linear terms exist in residual errors. We add expression (6) to expression (4), thus

$$\hat{y} = f(x_1,x_2, \ldots ,x_n) + a_0 + a_1 t_i \quad (7)$$

The linear terms in expression (7) is included in $\hat{y}$, so the linear terms of residual errors have been rejected.

3.3.3 Check and rejection of periodic terms in residual error series

The method of trigonometric polynomial regression is often used to establish multivariate wisestep regression equation, in which,trigonometric function of time t_i with different periods are regarded as factors. For example, following formula can be obtained

$$\eta_i = b_0 + \Sigma \ [a_k \cos(2\pi\, kt/365) + b_k \sin(2\pi\, kt/365)] \quad (8)$$

where q is the maximum degree of harmonic wave, we can take q=2 or larger. a_k and b_k are regression coefficients; b_0 is a regression constant.

Under significance level α, if the equation is significant,then the periodic terms represented by factors selected in equation must be included in residual errors. Adding the right-hand side of expression(8) to the right-hand side of expression (4), we have model expression $\hat{y}$ without periodic terms of residual errors.

3.3.4 Test of normality of residual error series

Using χ^2-testing goodness of fit,we can test whether the residual data are in normal distribution.

3.4 Determination of warning value for monitored quantity

If we have constructed the mathematical

model of monitored quantity and have made necessary test and improvement, and the residual error series is a random series in normal distribution and in which no regular component exists, then the upper and the lower bounds of warning values for monitored quantity can be written as

$$[y] = \hat{y} \pm ks \tag{9}$$

where $\hat{y}$ is the value of mathematical model of monitored quantity; s is residual standard deviation of the expression of mathematical model; k is coefficient determined by sample number n, it is between about 2.1 to 3.3.

Taking account of the objective existence of random errors of monitored quantity, we take the breadth of confidence belt as 2ks in expression (9).

4 DETERMINATION OF DANGER VALUES FOR MONITORED QUANTITY

The danger values of monitored quantity are limitations to indicate whether a dam is in safe operation state. The both stability and necessary strength are the principal factors reflecting the dam safety. The method to determine stability and strength is stipulated definitely in design codes of various countries. The method of reliability analysis will be a new developing direction in the safety design of dams. Therefore both aspects have to be taken into consideration in the determination of danger values.

4.1 The foundamental formula necessary for dam safety

The requirements of strength and stability for dams are given by the current design codes of dam of China in the method of safety coefficient. The formulas are

$$\sigma_0 / [k] = [\sigma] > \sigma \tag{10}$$

$$R / [k] > S \tag{11}$$

where σ_0 is limiting strength of dam material or foundation rock; $[\sigma]$ is permissible stress of dam material or foundation rock; σ is actual stress of dam material or foundation rock; [k] is safety coefficient of strength or stability stipulated; R is resistance such as resistance to sliding; s is effect of loads such as sliding force.

If the strength and stability conform to formula (10) and (11), then the dam is in safe state. If they don't conform to both formulas, then the dam is in an unsafe state. But it only means that the calculated safety coefficient doesn't reach the stipulated safety coefficient, but doesn't indicate that the dam is beyond the limiting state of safety.

Both resistance R and load effect S are regarded as random variates in reliability theory. The equation of limiting state can be written as

$$Z = R - S \tag{12}$$

If $Z>0$, then the dam is safe. If $Z<0$, then the dam ceases to be effective. If $Z=0$, the dam is in limiting state. The safety requirements of a dam can be expressed as

$$(R - S) > 0 \tag{13}$$

With the result of reliability analysis, more accurate safety bound and probability of reliability can be obtained. It become easier to determine the critical value of monitored quantity.

There is a common requirement in formula (10), (11) and (13), that the load effect should be smaller than resistance R under given conditions. Because the resistance is less variable, the load and its effect are the main factors to be controlled for dam safety. Monitored quantity should be connected with load effect, and the monitoring danger value has to be predicated on the safety requirements. According to the difference between the method of current design codes and the method of reliability, we think that the monitoring danger value given by the former method is "ordinary danger value". the monitorting danger value given by the latter is "serious danger value".

4.2 Determination of danger boundary of monitored quantity

4.2.1 Danger value of stress

We can regard $[\sigma]$ of formula (10) as "ordinary danger value" of stress. The "serious danger value" of stress can be given by formula (13). In formula (13), uplift pressure adopted can be determined by the mathematical model constructed with observed values.

4.2.2 Danger value of uplift pressure

Using formula (11) and (13), we can calculate the danger value of uplift pressure of dam foundation. The danger value of uplift pressure for different reservoir levels are different under given conditions. For convenience of daily control, we can establish an equation for danger values of uplift pressure varying with different water level, or we can grade the danger values of uplift pressure according to different water levels.

4.2.3 Danger value of deformation

The deformation of a dam cann't be reflected directly in formula (10), (11) and (13), but it is closely related with strength and stability. Its controlling value can be determined by two steps of calculation. The first step is calculating one or more kinds of dominating unfavourable loads combination which just meet with the requirements of strength and stability. The second step is calculating the correspoinding deformation under this load combination, which can be taken as the danger values of deformation.

4.2.4 Confidence interval of danger value

Usually, mathematical model of observed values will be used to determine the danger value of monitored quantity. And also the monitored objects are observed values with random errors. Similarly to the boundary of warning value, a confidence interval has to be considered to determine the boundary of danger value. But in practice, only upper bound of the confidence interval or the mid-value is used for safety monitoring, which can be expressed as

$$[y] = [y]_0 \tag{14}$$

$$[y]^+ = [y]_0 + ks \tag{15}$$

where $[y]^+$, $[y]_0$ are upper bound and mid-value of the danger values for monitored quantities; $[y]$ is danger value of monitored quantities; s is surplus standard deviation of mathematical model of monitored quantities; k is coefficient, which is between 2.1 to 3.3 according to sample number n.

5 EXAMPLES OF MONITORED QUANTITIES

5.1 Warning value of monitored quantity based on statistical model

Based on 198 observed values of horizontal displacement for survey point No.1 and No. 9 on the dam crest of Taipingshao gravity dam in China, these values were obtained from Jan.1982 to Dec.1986, we constructed statistical models respectively as follows

$$\hat{y}_1 = -2.92 - 4.886*10^{-2}\ln(t+1)/2 -1.2513*10^{-2}T_{10} -2.6503*10^{-2}T_{150} + 0.1451H \tag{16}$$

$$\hat{y}_9 = 39.51 - 0.1578\ln(t+1)/2 -8.3413*10^{-2}T_{10} - 0.1043T_{30} -2.0656*10^{-2}T_{120} + 0.1348H \tag{17}$$

where, $\hat{y}_1$ and $\hat{y}_9$ are respectively regression calculation value of horizontal displacement for survey point No.1 and No.9; t is the order number of date numbered from July 1st in 1981; H is the reservoir water level on the day that the observation operation was carried out; T_i is the mean temperature of i days before that day.

The warning value for survey point No.1 and No.9 are

$$[y] = \hat{y} \pm 3s \tag{18}$$

where s is residual standard deviation of the equation, which can be taken as 0.133 mm for $\hat{y}_1$, 0.308 mm for $\hat{y}_9$.

Testing 24 values observed in 1987, we found that observed values y_9 for survey point No.9 are all within the limitations of warning values

$$\hat{y}_9 + 0.92 > y_9 > \hat{y}_9 - 0.92$$

which indicates the observed value y_9 are normal. Testing observed value y_1 of survey point No.1, however, we often found

$$y_1 < \hat{y}_1 - 0.40$$

which shows that the observed values of survey point No.1 exceed the bound of warning value. The reason is that soils and rocks were filled on the downstream face at which the survey point No.1 located in the autumn of 1986,which caused a horizontal displacement towards upstream.

5.2 Warning value of monitored quantity based on hybrid model

The hybrid model of horizontal displacement for dam block No.5 of Gutianxi dam in China was established according to water pressure component derived by deterministic calculation and temperature component derived by statistical calculation. The residual standard deviation of the equation is 0.53mm, regression value $\hat{y}_5$ can be written as

$$\hat{y}_5 = 1.21 + 1.14*10^{-4}H^3 - 1.043*10^{-3}H^2 - 0.3343T_{90} + 0.2851T_{80} \quad (19)$$

Warning value are: $[y_5]^+ = \hat{y}_5 + 1.6$ and $[y_5]^- = \hat{y}_5 - 1.6$. According to the test, the observed value y_5 during the period from Jan. 1980 to July 1984 are all between $[y_5]^+$ and $[y_5]^-$. So the horizontal displacement of survey point No.5 is normal.

6 PRELIMINARY CONCLUSIONS

From all those described above,we can draw some preliminary conclusions as follows

1. The monitoring indexes of a dam are scientific criterions to recognize the state of the dam,which plays an important role in safety control . To use the monitoring indexes in safety controlling, we must take account of the actual operation state of a dam on the basis of safety design.

2. The principal moniotring indexes of a dam are uplift pressure (or pore water pressure),seepage quantity,deformation and stress. The dominating and typical survey points should be selected to set up the monitoring indexes. A group of survey points should be used as far as possible to establish the equation of monitoring indexes, and a synthetic analysis of the monitoring indexes of multi-term is necessary.

3. Monitoring indexes can be classified into warning value and danger value. The former is the boundary which can be used to judge whether the observed values are normal, the latter is a boundary used for judgeing whether the state of the structure reflected by observed values is normal.

4. It is better to express the monitoring indexes as a mathematical expression related with operation conditions of the dam,age and even positions of survey points to make daily monitoring control easier in arbitrary conditions. We can also calculate the results in most unfavorable condition and use it as limiting value for monitoring control.

5. It is better to give a confidence interval for monitoring indexes to fit with the existence of random errors in model and observed values. The bandwidth of confidence interval can be determined in proper method according to residual standard deviation and the size of the sample.

6. The warning value can be derived based on mathematical model of monitored value and confidence interval. The residual errors of mathematical model are required to be in normal distribution without any effect of regularity. So the model should be tested firstly and improved if unsatisfactory.

7. The stability and strength are the perequisites to the determination of a warning value. The method of safety coefficient and the method of reliability analysis are the principal methods to determine a warning value. The mathematical model of monitored quantity is often used to infer the load effect according to equilibrium equations of stability and strength. The danger value can be determined by two steps. Firstly we infer the dominating load combination which just meets with the requirements of dam safety. Secondly,we calculate the deformation value under the load combination and take it as the danger value. The expression of danger value ought to contain confidence interval.

REFERENCES

Li Zhenzhao, 1989. Measured data analysis for concrete dams, Beijing, Press of hydraulic and electric engineering.

Li Zhenzhao, Li Shuru. 1989. Deformation performance analysis of the Gutian-1 hydropower station dam. Hangzhou. Large Dam & safety 7-8: 21-33.

Li Zhenzhao, Zhang shuli, Jin Chunsan. 1989. Report on statistical analysis for deformation data measured by the laser surving system with vacuum tube in the Taipingshao dam. Changchun. Observation technology 2: 43-66.

Singh, V.P., & Zhenzhao Li.1992. Modeling deformation of concrete dams. Geomechnics & water engineering in environmental management, R.N. Chowdhury (Ed), Balkema.

Environmental Management, Geo-Water & Engineering Aspects, Chowdhury & Sivakumar (eds)
© 1993 Balkema, Rotterdam. ISBN 90 5410 099 0

Earthquake hazard in Australia and rational management strategies

Robert E. Melchers
Department of Civil Engineering and Surveying, The University of Newcastle, N.S.W., Australia

ABSTRACT: Because the Australian continent lies within a socalled intra–plate zone, the risk of damage to buildings and structures and the risk of death and injury due to earthquakes is relatively low for Australia as a whole. Nevertheless, the Newcastle experience points to the need to re–examine existing design and construction standards and in particular the details used in construction. A brief overview of these matters is given. There is a danger, however, that new regulations may adversely affect the continued existence of existing building stock. The paper argues that for buildings which have a projected life which is short relative to the occurrence rate of ground shaking resulting from (an) identified earthquake generating feature, the risk of defined degrees of damage is low and might be ignored under Australian conditions. This is relevant for structurally adequate existing buildings and for new buildings constructed in the first few years after the occurrence of major ground shaking.

KEYWORDS: Earthquake; Risk Management; Intra–Plate; Codes; Seismic Hazard; Existing Construction; Buildings.

1 INTRODUCTION

It is generally accepted that the Australian continent is not one with a high degree of earthquake activity but, as the recent earthquake in the Newcastle region has reminded, Australia is not entirely free of earthquakes either. This should not be unexpected: surveys of world–wide seismic activity show that virtually no substantial part of the world is seismically inactive. The regions most severely affected by major earthquake activity are those at the borders of crustal plates. Fortunately, Australia is well within such a plate. Earthquake activity within Australia is known as "intra–plate" activity. Unfortunately, the scientific understanding of intra–plate seismicity is not as well developed as that of plate boundary seismicity.

What is perhaps not so well known is that, with the exception of its west–coast belt, much of the USA is also part of an intra–plate region. Similarities between Australia and the central and eastern US (CEUS) have been commented upon repeatedly by seismologists (e.g., Johnston, 1989). This is fortunate for Australia since much effort has been made in the USA to increase understanding of intra–plate earthquake causes and activity. Significant observations and inferences now exist. A very brief review of these will be given in the next section since they have clear implications also for Australia. Further, their importance in relation to seismic hazard analysis and for risk management will be indicated, particularly in relation to the degree of risk and amount of earthquake design measures required for existing structures.

2 MAIN FEATURES OF INTRA–PLATE SEISMICITY

A comprehensive overview of the seismicity of Australia is available (Gaull *et al.*, 1990), including a critical examination of what this means for risk assessment and risk amelioration for buildings and structures (Melchers,1992). Some of the more important factors are:

1. The rate of occurrence of continental, intra–plate events is far lower than that of earthquake events in other earthquake zones: according to Johnston (1989): "*Over the past 200 years, the stable continental plate interior (SCI) regions of the earth experienced 2 independent $M > 8$ earthquakes, roughly 7 – 10 events of $M = 7.0 – 7.9$ and about 50 events of $M = 6.0 – 6.9$. Within the same time frame, the plate boundaries of the earth have produced approximately 9 earthquakes exceeding $m > 9.0$, (some) 140 events of $M > 8.$, over 2000 magnitude 7's and over 34000 magnitude 6's. Stated differently, for every SCI earthquake of at least magnitude 6, plate boundary zones produced more than 600*

such events. Therefore, the seismicity of stable continental interiors is between two and three orders of magnitude below that of plate boundaries".

2. Contrary to conventional wisdom, the magnitude (M) of an earthquake is not, in itself, of particular interest to engineers and others working in damage containment and risk assessment. The factor of prime importance is the degree of ground shaking which may be expected at a particular site and the probability of occurrence of that degree of shaking, whatever the source. The conventional measure of ground shaking is the modified Mercalli (MM) scale. A better measure, however, if it can be obtained, is that of ground acceleration as this allows the computation of earthquake induced forces (e.g., Dowrick, 1987).

3. The degree of ground shaking at a site is governed by the distance the epicentre of an earthquake is removed from the site and the degree of attenuation of ground shaking with distance. For intra–plate events, the attenuation rate appears to be considerably less than that for plate boundary events (e.g., Coppersmith and Youngs, 1989). As a result, the area affected by an intra–plate earthquake event of a given degree of ground shaking as a result of given magnitude M is considerably larger for intra–plate events. Hence the risk at any one location of being affected by a given intensity of ground shaking due to an intra–plate event occuring somewhere in the region is greater than suggested by the figures given by Johnston (1989) but is likely to change his orders of magnitude by only about one, at most.

4. There is evidence, mainly from China, but also from Europe and the CEUS that the time between earthquake events in intra–plate regions is of the order of hundreds of years if not more. Also, regions seismically active may change, relatively quickly (in the geological time scale), to seismically inactive and vice versa (e.g., Coppersmith and Youngs, 1989). This suggests that the available earthquake records for Australia can represent only an incomplete picture of the seismicity of the country as a whole

5. The Charleston and other experience in the CEUS suggests that the time between intra–plate events sourced by the one seismic feature appears to be of the order of hundreds of years, if not longer (Coppersmith and Youngs, 1989). This inference is comparable with estimates which have been made for the Newcastle region, for which it appears not unlikely that one feature is responsible (Melchers, 1990). More generally, where no feature has been identified or seismic activity has not yet become a matter of interest, similar orders of return periods may be expected (Coppersmith and Youngs, 1989). Despite the possible non–stationary nature of seismic activity, it is conventional to assume stationarity and to derive or estimate the mean occurrence rate of ground shaking from whatever information is available.

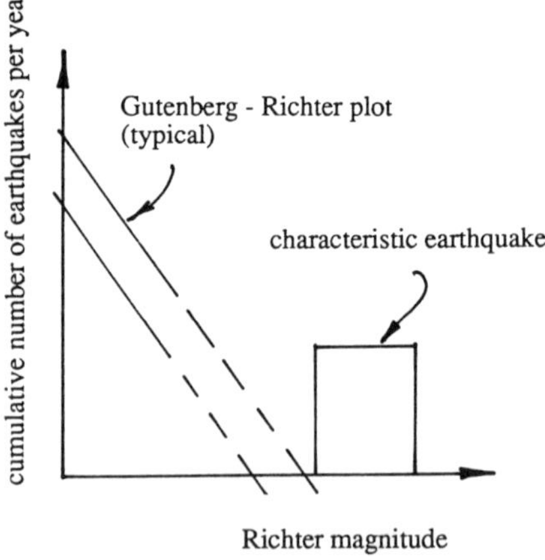

Figure 1. Gutenberg–Richter Plot and Characteristic Earthquake Description

6. It is accepted by earth scientists that intra–plate activity is the result of in–plane forces (usually compressive) in the upper crust (which is, typically, some 30 km thick). The upper crust is "separated" from the lower crust by the "brittle–ductile transition", which is the result of the depth dependent temperature gradient affecting the strength of crustal rock (e.g., Kearey and Vine, 1990). Earthquakes also originate in the lower crust but at a comparatively much lower rate. The rate of compressive crustal deformation is very small. Typical strain rates of $10^{-9} - 10^{-12}$/year have been recorded (Johnston, 1989). If the crust is considered to be largely elastic and to have a creep rate lower than the applied strain rate (a reasonable assumption), it follows from solid mechanics principles that there will be a gradual increase of crustal stresses with time. Earthquakes in the intra–plate crust are considered to occur when the local stress state is such as to cause failure of the local rock strength. This has been speculated to occur at faults and other local zones of weakness and at stress concentrations (e.g., Talwani, 1989). However, the precise mechanisms are still subject to speculation.

3 PROBABILISTIC MODELLING

3.1 General Remarks

There appears to be an increasing amount of opinion which supports the notion that seismic

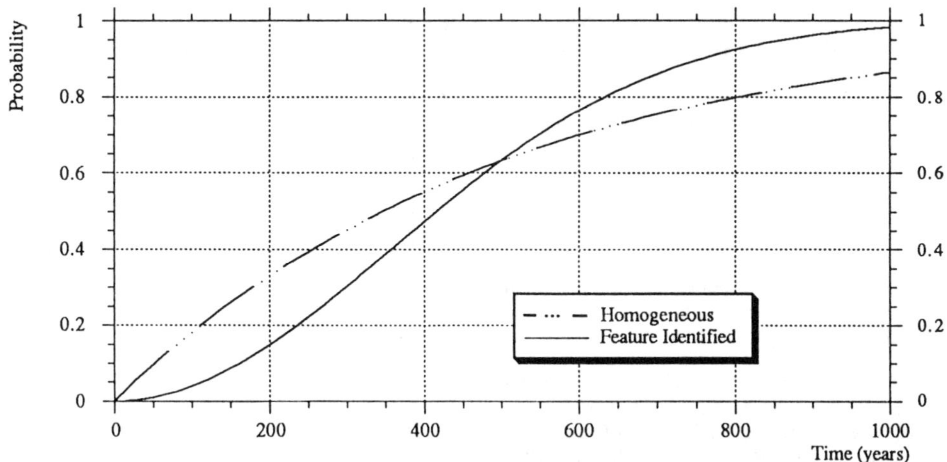

Figure 2. Probability of Exceedance as a Function of Time

activity is correlated in time for considerable periods and that "individual" and rather sudden earthquakes are but part of a much longer sequence in time of seismic activity (e.g., Heaton, 1990). This suggests that, particularly for intra–plate earthquake activity at a given site, it is appropriate to consider the effects of earthquake activity over a wide area. And these effects should be considered as a stochastic or time dependent random process at that site.

From time to time the site–specific stochastic process will upcross (i.e., exceed) one or more given levels of nominal activity. It follows from stochastic process theory that provided the nominal levels used are sufficiently high, the upcrossing event may be considered to be independent of all other such events, although clusters may need to be considered as the one event (cf., Leadbetter, Lindgren and Rootzen, 1982).

3.2 *Plate Boundaries*

For subduction type plate boundary earthquake events the assumption is commonly made that each large magnitude event can be considered to be independent of all other such events. Such a model is reasonable if seismic activity is considered to be a semi–continuous process of many small disturbances with occasional larger ones and rare extreme events. Seismic activity at a site can then be considered to be a stochastic process, even for the one earthquake generating feature. This allows the occurrence in time of earthquakes of magnitude above a given (high) level to be modelled by means of a Poisson distribution (Cornell, 1968). Although a major assumption, it is adequate for engineering purposes (Cornell and Winterstein, 1988).

It has been argued that the Poisson model is deficient where one earthquake generating feature has been identified and it shows regularity of seismic events at relatively short intervals rather than randomness. The approach most commonly adopted for this case is that of the socalled "characteristic" earthquake (e.g., Youngs and Coppersmith, 1985). On the Gutenberg–Richter plot for earthquake magnitude–frequency relationship, the characteristic earthquake plots as a rectangular (but strictly uncertain) region of magnitude and recurrence (see Figure 1). However, Cornell and Winterstein (1988) claim that within the limits of accuracy normally available, the characteristic model is superior to the Poisson model only if the events show strongly regular characteristic time behaviour.

3.3 *Intra–Plate Regions*

For intra–plate earthquakes the situation is not much different. If there is an identified earthquake generating feature semi–continuously seismically active, the analysis is essentially the same as that for subduction/plate boundary events but, of course, the recurrence interval between significant events will be much longer in general. The assumption of a Poisson process to describe the occurrence of large magnitude events appears reasonable.

In the case where the earthquake generating features are not well–defined, there is a large degree of uncertainty about the distribution of epicentres which may contribute to ground shaking at a particular site. In view of the observations earlier, it appears entirely appropriate, therefore, to assume equally likely (but of course not equal) contributions from all surrounding points on the earth's crust. This is the usual assumption for individual zones as used in seismic risk mapping (e.g., Gaull *et al.*, 1990). In what follows this will be termed the "homogeneous" case. For this case, too, it is reasonable to adopt a Poisson process model for event exceedances.

The average rate of occurrence (ν) of ground shaking represents the probability p_a of the occurrence of an event per year. Conversely, the well-known "return period" T_r is given by:

$$T_r = 1/\nu \tag{1}$$

and represents the mean time between events. Here an event is defined as the occurrence of ground shaking greater than a particular level, a, say.

Let it be assumed that it is possible from available information to estimate the average rate of a particular level of ground shaking. For simplicity let this level be one at which serious structural damage is likely to occur. For Australian conditions and masonry construction not specifically designed to resist earthquake shaking, this might correspond to a Modified Mercalli level MM = a = 6.

From well-established probability theory it follows that if the upcrossings are Poissonian, the expected time between upcrossings is given by the Exponential distribution. The probability of ground shaking of a given magnitude a or greater following a previous such event is then given by (e.g., Melchers, 1987):

$$P\begin{pmatrix}\text{ground shaking}\\ > a\end{pmatrix} = 1 - e^{-\nu t} \tag{2}$$

where t is the time period of interest.

For the "homogeneous" case it would be expected that the average rate of upcrossing (i.e., the occurrence rate) will be roughly constant with time, since a number of earthquake features contribute to it. Hence it follows that expression (2) can be used immediately. It has been graphed in Figure 2 for an assumed return period T_r = 500 years (i.e., = ν = 0.002/year).

For the situation where a dominant earthquake generating feature has been identified and the seismicity is governed by an increasing strain rate in the crust, the situation is different. Figure 3 shows a schematic stress history. The stress level rises slowly with time, with step-wise stress-relief as a result of each significant earthquake. It follows that the upcrossing rate varies with time. The average upcrossing rate does not represent the true situation particularly well, since it implies equal likelihood of earthquake occurrence (and hence its effect) at any point in time. Hence a more appropriate model is to assume that the upcrossing rate varies with time. A simple model is to assume it varies linearly with time as:

$$\nu(t) = c \cdot t \tag{3}$$

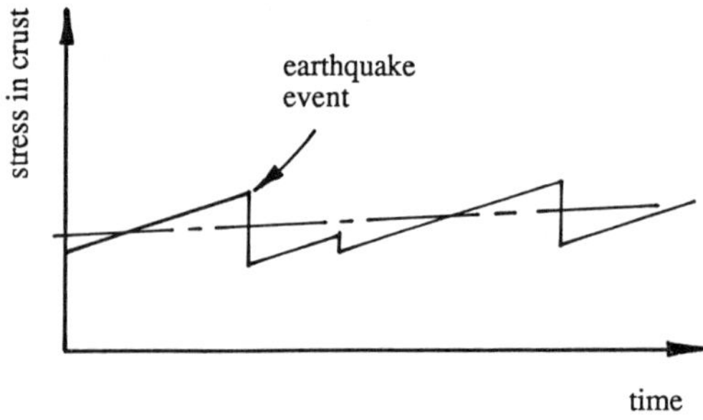

Figure 3. Time-Dependence of Stress State in Upper Crust (Schematic)

where

$\nu(t)$ = instantaneous upcrossing rate

$c = \text{constant} = 2\nu_m/T_r = 2\nu_m^2 = 2/T_r^2$

obtained by assuming

$$\nu_m = \int_0^{T_r} ct\, dt$$

Relationship (3) with T_r = 500 years has been plotted in Figure 2 and shown as the "feature identified" case. It is seen that in the period immediately after the occurrence of an earthquake, the probability of a given level of ground shaking being reached is considerably lower than given by the conventional curve for the "homogeneous" case.

4 APPLICATION TO BUILDING CODES

4.1 New Construction

In principle, the rational setting of requirements in building regulations must consider the probability of a given level of ground shaking occurring at a particular site, the probability of damage of various types and intensities (including allowance for the potential loss of human life – a difficult topic which will not be addressed herein) and the cost of measures to ameliorate earthquake induced ground shaking effects. It is well-established that for new construction the cost of amelioration measures is relatively small, at least for Australian conditions. Typical estimates range from almost zero cost for well-designed structural steel and reinforced concrete structures to up to about 5% of structural cost for buildings involving masonry construction. Even in the latter case the costs may be less, as the probability of building or structural failure under earthquake shaking conditions depends on a number of factors, including:

Table 1. Probability of Exceedance as a Function of Time

	PROBABILITY		
TIME (YEARS)	HOMOGEN. CASE	FEATURE IDENTIFIED	RATIO
50	0.10	0.01	10
100	0.18	0.04	4.5
150	0.26	0.086	3

1. degree of structural earthquake resistance built into the structure or provided later
2. the amount of attention given to connection details during design
3. the care taken in construction to ensure the connection details are properly made
4. the layout of the building in plan and vertically, and
5. the condition of the building at the time of the ground shaking.

4.2 Existing Construction

The situation for existing buildings and structures is much more difficult. For these experience indicates that the cost of installing earthquake damage control measures is high. Experience along the west coast of the USA and elsewhere shows that tough local government building regulation requirements on seismic upgrading can work, but also that many owners simply opt out because of the expense, demolishing the building rather than upgrading it. This can have severe implications for the city–scape and perhaps the psychology of inhabitants, particularly where heritage buildings are involved. Certainly the heritage issue was a prominent one in Newcastle (Melchers, 1990).

It has not been customary to differentiate to any great extent between existing construction and new construction in terms of the design standards which are expected to be met for seismic resistant design. For regions which are particularly earthquake prone, this is quite rational, but, as will be argued below, there is a case for differentiation under certain conditions within intra–plate earthquake regions.

4.3 Relative Risk of Ground Shaking

The curves in Figure 2 may be used in conjunction with cost–benefit analysis ideas and other criteria for the setting of realistic requirements in building regulations, particularly where existing buildings are of interest. As noted, it is these for which the potential cost implications of seismic upgrading are significant.

Consider the situation where the local average recurrence rate of earthquake shaking of magnitude MM = a > 6 (say) has been estimated as 500 years (cf., Newcastle, say). If a dominant intra–plate earthquake generating feature has been identified, the lower curve in Figure 2 indicates the risk of such shaking occurring with the elapse of time. The upper curve indicates the risk of a ground shaking occurring due to any earthquake feature at all (i.e., the unidentified, homogeneous case).

Consider now a building which expert opinion suggests may "fail" (however defined) should ground shaking greater than MM = 6 occur. If the building is designed on other criteria to last 50 years and it was in existence at time t = 0 and survived or was adequately repaired to its previous condition (or built shortly thereafter), then the probability of occurrence of earthquake ground shaking of at least MM = 6 within the 50 year period can be read from Figure 2 at t = 50. It is seen to be about 0.01 in the case earthquake "feature identified" and about 0.1 for the "homogeneous" case. Similar differences in risk can be obtained for other points in time, see Table 1.

It follows directly that a structure which is considered to be adequate for a low level of earthquake induced ground shaking may be acceptable provided: (i) it was in existence or constructed shortly after the previous occurrence of ground shaking; and (ii) its expected life is relatively short compared with the mean return period of the considered level of ground shaking.

Relative to AS2121–1979, the Australian Standard for earthquake design of structures, a change in risk level of a factor of 10 at t = 50 years corresponds, roughly, to a change of one zone, i.e., from zone A to 1 or from zone zero to zone A.

It is evident from Figure 2 that for a 50 year lifetime structure built (say) 50 years after the occurrence of (the MM = 6 level of) ground shaking, the risk ratio is that at 100 years, namely about 4.5 and significantly less than if the structure had been built at t = 0 years. In terms of AS2121, this corresponds to rather less than a change of one seismic zone. This illustrates that for structures whose lifetime extends closer towards the mean return period, seismic resistance measures become more and more warranted.

5 IMPLICATIONS FOR EXISTING BUILDINGS

The above arguments have no great interest in

regions where earthquake generating features have not been identified and the best current descriptions of understanding are subsumed under the "homogeneous" case. Nor are they of interest where the recurrence rate of earthquake ground shaking is high, as it is in most marginal regions, since, although the earthquake generating feature is usually well–identified, the difference between the risk curves in Figure 2 becomes much smaller with increasing occurrence rates.

The discussion does have an interest in intra–plate regions where the ground shaking is due largely to an identified earthquake generating feature(s). Under these conditions, the above indicates that for buildings and structures having relatively short expected lifespans compared with the mean recurrence rate of ground shaking, upgrading of seismic resistance may not be warranted. This is particularly the case for buildings which survived the previous ground shaking and which are still considered adequate under previous (perhaps non–seismic) design conditions.

6 CONCLUSION

For buildings and structures having short projected lives relative to the mean return period of damaging ground shaking due to an identified earthquake generating feature(s), it has been argued that seismic design requirements might be relaxed. This is likely to be the case only in intra–plate earthquake regions.

7 REFERENCES

Coppersmith, K.J. and Youngs, R.R., (1989), Issues Regarding Earthquake Source Characterisation and Seismic Hazard Analysis Within Passive Margins and Stable Continental Interiors, (in) S. Gregersen and P.W. Basham (Eds), Earthquakes at North–Atlantic Passive Margins: Neotectonics and Post–Glacial Rebounds, Nato ASI Series C, Dordrecht, Kluwer Acad. Pub., pp. 601–631.

Cornell, C.A., (1968), Engineering Seismic Hazard Analysis, Bull. Seism. Soc. America, 58(5), 1583–1606.

Cornell, C.A. and Winterstein, S.R., (1988) Temporal and Magnitude Dependence in Earthquake Recurrence Models, Bull. Seism. Soc. America, 78(4), 1522–1537.

Dowrick, D.J., (1987), Earthquake Resistant Design, Second Edn., Wiley–Interscience, Chichester.

Gaull, B.A., Michael–Leiba, M.O. and Rynn, J.M.W., (1990), Probabilistic Earthquake Risk Maps of Australia, Aust. J. of Earth Sci., 37( ), 169–187.

Heaton, T.H., (1990), Evidence For and Implications of Self–Healing Pulses of Slip in Earthquake Rupture, Physics Earth Planetary Interiors, 64( ), 1–20.

Johnston, A.C., (1989), The Seismicity of "Stable Continental Interiors", (in) S. Gregersen and P.W. Basham (Eds), Earthquakes at North–Atlantic Passive Margins: Neotectonics and Post–Glacial Rebounds, Nato ASI Series C, Dordrecht, Kluwer Acad. Pub., pp. 299–327.

Kearey, P. and Vine, F.J., (1990), Global Tectonics, London, Blackwell Scientific Publications.

Leadbetter, M.R., Lindgren, G. and Rootzen, H., (1982), Extremes and Related Properties of Random Sequences and Processes, Springer–Verlag, New York.

Melchers, R.E., (1987), Structural Reliability Analysis and Prediction, Ellis Horwood/John Wiley & Sons, Chichester.

Melchers, R.E., (1990), (Ed.), Newcastle Earthquake Study, I.E. Aust.

Melchers, R.E., (1992), On Earthquake Design Standards for Australia, Trans. I.E. Aust., CE41(1), 63–71.

Melchers, R.E. and Page, A.W., (1992), The Newcastle Earthquake, Proc. Instn. Civil Engrs., 94(2), 143–156.

Talwani, P., (1989), Characteristic Features of Intraplate Earthquakes and Models Proposed to Explain Them, (in) S. Gregersen and P.W. Basham (Eds), Earthquakes at North–Atlantic Passive Margins: Neotectonics and Post–Glacial Rebounds, Nato ASI Series C, Dordrecht, Kluwer Acad. Pub., pp. 563–579.

Youngs, R.R. and Coppersmith, K.J., (1985), Implications of Fault Slip Rates and Earthquake Recurrence Models to Probabilistic Seismic Hazard Estimates, Bull. Seism. Soc. America, 75(4), 939–964.

Environmental Management, Geo-Water & Engineering Aspects, Chowdhury & Sivakumar (eds)
© 1993 Balkema, Rotterdam. ISBN 90 5410 099 0

Response of landfill liner and collection systems during earthquake loading conditions

S. Singh
Santa Clara University, Calif., USA

B.J. Murphy
PRA Group Inc., Hayward, Calif., USA

ABSTRACT: The results of the first phase of a study on the response of landfill liner and leachate collection systems to dynamic loads are presented. The report discusses the possible mechanisms which may cause rupture or cracking in the system. The effects of static and dynamic loadings on settlement, slope stability and foundation failure are described and the application of the finite element method for gaining further insight into the behavior of the liner and collection systems is recommended.

1 INTRODUCTION

In California, it is very likely that a solid waste landfill will be subjected to dynamic loading conditions due to an earthquake event. In some cases the severity of the dynamic loading can be such that it causes serious damage to the clay liner and leachate collection system. The system can develop rupture or cracks leading to an escape of leachate and other hazardous materials. It is therefore important that the clay liner systems are designed to resist the dynamic loads. Much of the attention in recent years have been directed towards the development of a clay liner having very low permeability and resistance to the deteriorating action of leachate. A noticeable gap exists in relation to the study of the integrity of clay liner systems under excessive static or dynamic loads. Accordingly, efforts were made to develop an understanding of this problem. This report presents results of the first phase of the studies made to date.

A review of the existing studies of landfills, and hazardous or solid waste, has surprisingly indicated a total lack or neglect of the evaluation of the adverse effects on the integrity of the liner and leachate collection systems due to the earthquake loading conditions. Only one study made by Ertech on the interface shear strength of a composite multilayered clay liner system has considered earthquake effects. The earthquake effects were incorporated as a seismic coefficient in the slope stability analyses in a conventional psuedostatic approach. Clearly, a systematic approach to the study of the problem of the seismic response of liner systems is warranted, especially for fills located in California. Accordingly, a systematic study was undertaken at Santa Clara University. The first few steps in the study efforts were to (i) identify, (ii) define and (iii) evaluate the static and dynamic loading conditions which are detrimental to the integrity of the liner system.

2 STATIC LOADING

The static loading due to the overburden of the waste can cause differential settlements in a weak or soft foundation leading thereby, to the development of differential strains in the clay liner system. It is therefore important for weak or soft foundations, or wherever the potential for differential settlements exist, that the severity of the development of excessive differential strain in the compacted clay liner and the HDP geomembrane system be evaluated. The lateral spreading of weak foundations, for example for fill on recent bay mud or at a site in New Jersey which experienced foundation failure, should be evaluated. Accordingly, the static bearing capacity and settlement of the foundation soils are important considerations but these topics will not be discussed further in this paper.

Failure of the toe of the landfill can cause downdrag on the liner system. If a liner system is contiguous from the side wall to the base of a fill, settlements in the fill can cause significant downdrag on the liner system and may even cause ripping off of the HDP and other units of the liner system. An evaluation of the mechanism of the downdrag force must be considered in the design against rupture or cracking of the liner system.

3 DYNAMIC LOADING

The most severe dynamic loading conditions would result where a landfill is located directly above a fault/fault zone. A recent study by Sohn (1988) on the propagation of crack in an earth embankment subjected to fault movements has indicated that the crack development in earth embankments occurs in two ways: shear failure deep inside the embankment and tension failure near the surface. Apparently due to the inherent capacity of a municipal refuse fill to undergo large strains without failure, the tension cracks near the surface of the landfill are not likely to result in the rupture of the fill surface. Excessive settlement due to slumping could create tension stresses along the upper half of the slope of the fill. Open surface cracks can therefore develop as a result of the combined effect of the two aforementioned phenomena. Sohn's study which involved centrifuge and finite element analyses, indicated that the surface cracks disappear deep in the embankment. The study, however, showed that in the case of a strike-slip fault movement, a shear rupture develops from the base level of the embankment and propagates upward continuously in the transverse direction. Since the liner system is located at the base level of a landfill, the shear rupture can have significant adverse effect on the integrity of the liner system. Accordingly, an evaluation of the potential for the development of an open leakage channel must be investigated. Such an evaluation can follow an approach similar to what is given in the ASCE's Guidelines for the Seismic Design of Oil and Gas Pipeline Systems, especially for buried pipelines crossing a fault/fault zone.

When the landfill is located in a seismically active area like California, but is not situated directly over a fault/fault zone, the effect of ground shaking can manifest itself on the liner system in two ways: (i) slope instability with its sliding surface passing through the liner system, and (ii) dynamic shear stresses resulting from the shear wave propagation up through the liner system. Flexural stresses induced by seismic waves traveling along the earth's surface should also be addressed.

Seismic stability of landfills has recently been addressed (Singh and Murphy, 1990). The approach is essentially based on soil mechanics principles. Conventional psuedostatic and deformational analysis approaches after Nemark (1965) and Makdisi & Seed (1978) have been used to estimate displacement along a potential failure surface. There are, however, two important considerations. Because of the relatively low to very low shear strengths along the interfaces of the various members of a typical composite liner system, the non-circular shape of the sliding surface and the widely varying deformations at which the strength is mobilized along the interface of the composite members and through the material of the landfill, the conventional approach should be applied with caution. Also, as pointed out by Seed et.al, 1990, the three dimensional aspects of the problem of the landfill stability cannot be ignored because in some cases, for example in the case of the Kettleman fill failure, the three dimensional analysis of the failure block gave a lower factor of safety than the two dimensional approach. A rigorous approach should include finite element modeling, in which case a proper characterization of the material properties, especially of the fill material, can pose problems. Because of the complex and heterogenous nature of the material, there still does not exist sufficient data on the dynamic strength properties of the fill material. However, the dynamic strength properties of different composite members of a liner and the collection system can be reasonably estimated from laboratory testing.

There are two most important aspects of the problem of the dynamic stability of a landfill. One is the mechanism of failure and the other is the nature of failure. The assessments, especially those based on finite element analyses, can provide insights into the potential mode/mechanism of failure. At this stage, it appears that the block sliding mechanism is the most appropriate mode of failure. Seed et.al. 1990, used this mechanism to analyze the failure of the Kettlement Hills Landfill. They also considered multiple block sliding analyses to allow for differential movement of the sliding mass. These analyses were for the static loading case. In the case of the dynamic loading due to the earthquake shaking, the initial mode/mechanism of failure could be the sliding block type, but because of the rather very low shear strength parameters along the interfaces especially when there are wet or saturated conditions, for example right at the surface of the clay liner where there will be a strong potential for high pore pressure generation, the inertia forces in the sliding block of the fill material can cause excessive drag forces on the sides and the base of the fill. These excessive forces are likely to be concentrated at sharp transition points/planes near the bends or separation points of the sliding block. An analysis which incorporates these situations can be significant in providing a more realistic picture of the mechanism of failure as well as the nature of the failure.

It is important to gather field performance data on the behavior of clay liner and collection systems during earthquake loading conditions. It is highly desirable that the new fills should have provision for proper instrumentation. Any sharp change in the leachate discharge collection soon after an earthquake event can be significant and can relate to the failure by

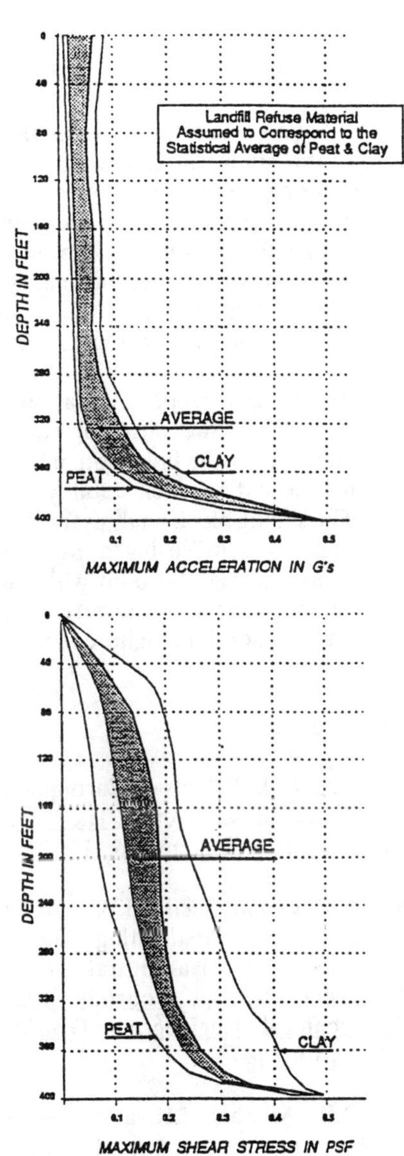

Fig - Results of Response Analysis by SHAKE for 400 ft. High Landfill

cracking of the liner system.

The figure below presents the results of a SHAKE analysis for a 400 ft. high refuse fill. The dynamic properties of the fill material were assumed to be the average of that of peat and clay. It is important to note that the maximum shear stress and acceleration levels remain quite high at the liner location, thus the potential for slip surface through the liner system.

4 CONCLUSIONS

On the basis of the results from the first phase of this study, the following important conclusions can be made:

1 Bearing capacity and the settlements of the foundation soil can adversely affect the integrity of the liner system.

2 The potential for a slip surface through the liner system is very high during earthquake loading. Accordingly, proper estimation of the strength of the different composite members of the liner system is crucial for the sliding block stability analysis.

3 The sliding block analysis for dynamic slope stability, using a psuedostatic approach, may be used with caution by allowing for the adverse effects of three dimensional failure if it exists.

REFERENCES

Newmark, N.M., Effects of Earthquakes on Dams and Embankments", The Rank line Lecture Geotechnique, No., 15, 1965.

Makdisi, F.I. and Seed, H.B., "Simplified Procedure for Estimating Dam and Embankment Earthquake-Induced Deformations", Journal of the Geotechnical Engineering Division of ASCE, July, 1978.

Seed, R.B., Mitchell J.K. and Seed, H.B., "Slope Stability Failure Investigation: Landfill Unit B-19, Phase I-A, Kettleman Hills, California", Report No. UCB/GT/80- 01, July, 1988.

Singh, S., and Murphy, B.J., "Evaluation of the Stability of Sanitary Landfills, in, 'Geotechnics of Waste Fills' -- Theory and Practice", ASTM STP; 1070, 1990.

Sohn, J. "Crack Propagation in Earth Embankment subjected to Fault Movement", Ph.D., thesis in Civil Engineering, University of California, Davis, 1988.

Environmental Management, Geo-Water & Engineering Aspects, Chowdhury & Sivakumar (eds)
© 1993 Balkema, Rotterdam. ISBN 90 5410 099 0

Assessment of dam break models for downstream potential hazard analysis

K.C.Tai
Department of Water Resources, Vic., Australia

V.Fazio
Department of Electrical Engineering, Monash University, Clayton, Vic., Australiaa

ABSTRACT: Dam break models provide an estimate of the potential downstream hazard of dam failure. This paper presents the advantages and disadvantages of different dam break models for estimating purposes. The potential hazard analysis is presented under two consideration: the qualitative and the quantitative. For the latter, the full dam break model as developed by the United States National Weather Service should be used.

1 INTRODUCTION

From a technical viewpoint, the best approach to deal with a dam that is, or is likely to be, hazardous to life or property, is to carry out a structural integrity analysis and a downstream hazard analysis. The former will identify the cause of the concern, centred on the likely mechanism of failure and the risk or the probability of failure of the hazardous dam. The latter analysis will identify the likely injury, property damage and economic loss to downstream community and development.

The structural integrity analysis requires reviewing and monitoring relevant dam data related to the design, construction, performance and present condition of the dam (ANCOLD 1992). The downstream hazard analysis can be assessed through the use of dam break models. The assessment could focus on two considerations of hazard: the qualitative and the quantitative aspects of hazard posed by a hazardous dam.

However, dam owners may be reluctant to undertake the above mentioned approach for the following reasons:

* the cost of the exercise is seen to be high or unwarranted; and

* the dam owners do not see themselves as being accountable and responsible for dam safety, because it is seen to be a government responsibility.

However, in Victoria, the recent introduction in dam safety legislation and a variety of liabilites for damages caused by the failure of a dam under the *Water Act 1989* means that dam safety is the responsibilty of the dam owners. The role of government is to explore and publicise appropriate ways in which dam owners can manage their dams and assess their liabilities. This development has provided the incentive for the authors to make an assessment of dam break models to estimate downstream potential hazard.

For downstream hazard analysis, it means asking the question: what are the potential consequences downstream, if a potentially hazardous dam were to breach or fail in a certain manner? The "what/if" approach is amenable to the use of dam break models to simulate the potential downstream consequences of a dam breach upstream.

The benefits of this modified approach are:

* such information will impact upon the extent of responsibility for dam owners, including their duty and standard of care, and their minimisation of negligence and liability for potential compensation claims should their dam fail; and

* this awareness by the dam owners will impact upon their respective dam safety management and dam safety programs, in response to the different levels of potential hazard identified and assessed.

2. THE STUDY APPROACH

The study approach requires an assessment of the usefulness of different dam break models that have been developed in recent years, in analysing downstream potential hazard.

The authors' attention has been drawn to three recently developed dam break models: the US National Weather Service Simplified Dam Break (SMPBRK) Model, the New South Wales Dam Safety "Generalised Guidelines" Model, and the US National Weather Service sophisticated and advanced Dam Break (DAMBRK) Model. The first two models are in fact simplified versions of the full DAMBRK Model.

To assess the usefulness of these models to the approach, the authors have divided the study into two parts.

The first part deals with the qualitative consideration of downstream potential hazard analysis:

(a) the need to ascertain that downstream potential hazard to life or property is the result of the dam's location; and

(b) that the level of downstream potential hazard posed by the potentially hazardous dam can be categorised qualitatively into high, significant or low category.

The second part deals with the quantitative consideration of downstream potential hazard analysis:

(a) the need to ascertain the level of potential risks to different parties in terms of personal injury, property damage and economic loss; and

(b) the need to estimate the magnitude of the costs of the potential hazard to different parties (for example, the dam owner, the local community downstream of the dam, and the general community which has to pay taxes to repair damaged infrastructure and utilities).

3 THE DAM BREACH MODELS

A brief description is given for each of the three dam break computer models.

3.1 The DAMBRK Model

The DAMBRK Model is an unsteady flow dynamic routing model which develops an outflow hydrograph due to spillway and/or dam failure flows. The hydrograph is routed through the downstream river valley.

DAMBRK was developed in 1977, and is used widely in the United States by most Federal and State agencies and around the world for dam safety and design, real-time hazard warning and hazard mitigation planning (Fread 1987).

The DAMBRK Model has wide applicability, for example:

* routing of a flood from a dam breach, the flood generated being much larger in magnitude than any flood resulting from runoff from rainfall and/or snow-melt;
* routing of any specified hydrograph through reservoirs, rivers, canals, or estuaries as part of general engineering studies of waterways; and
* routing of mud/debris flow hydrographs or rainfall/snowmelt flood hydrographs.

The DAMBRK Model can operate with various levels of input data from rough estimates to complete data specification.

The Model can also deal with complex problems arising from the downstream propagation of the floodwave, e.g., bridge/embankment flow constrictions, tributary inflows, river sinuosity, levees, channel storage and tidal effects.

The Model has been calibrated against five historical cases of dam failures in the US. The calibration parameters were observed downstream peak discharges, peak stages and flood peak travel times.

Calibration of the model against the Teton Dam, which failed in 1976, produced a variation of computed and observed values, of the above-mentioned parameters, of less than 5%. Calibration of the model against the Buffalo Creek "coal waste" dam, a high tailings dam, indicated an average difference of about 9% between the computed and observed peak flows. The dam breached in 1972, resulting in a loss of 118 lives and a property damage of over US$50 million.

3.2 The Simplified Dam Break (SMPDBK) Model

The SMPDBK Model can be used with an inexpensive micro-computer to produce within minutes, the predicted dam break floodwave peak flows, depths and travel times at selected points downstream of the breached dam (Wetmore & Fread, undated).

This is a simplified procedure compared to DAMBRK and can operate on a minimum of data, time, computer facilities and technical expertise.

The SMPDBK Model can be used as (a) a forecasting tool in a dam failure emergency with short warning response time, little available data, and inaccessible large computer facilities; and (b) as a tool for "pre-event" dam failure analysis, by emergency management personnel preparing disaster contingency plans, when the use of the advanced DAMBRK Model is precluded by limited resources.

The model is designed for interactive use by allowing the user to enter as much or as little data as are available. The model uses automatic preprogrammed defaults when data are not available.

The model is capable of producing only approximate flood forecasts by first computing the peak outflow at the breached dam, based on the reservoir size, and the temporal and geometrical description of the breach. The computed floodwave and channel properties are then used together with routing curves to determine (a) the attenuation of the peak flow as it moves downstream; (b) the maximum depth reached by the peak flow; and (c) the time required for the peak to reach each forecast point at these downstream points.

In addition, if the threshold depth for flooding to occur at that forecast point is entered as an input data, the model can compute the time at which that depth is reached, as well as when the flood wave recedes below that depth. This can be used to calculate the time duration of flooding for flood evacuation and flood fortification.

The SMPDBK Model retains the critical deterministic components of the DAMBRK Model, while reducing the need for extensive numerical calculations through the approximation of the downstream channel/valley as a prism. The model utilises dimensionless peak flow routing graphs developed by using the DAMBRK Model.

However, the performance of the SMPDBK Model will be affected by the progression of the floodwave downstream. The Model loses its accuracy if there are significant backwater effects created by downstream dams or bridge embankments, by temporary off-channel storage, or by other downstream channel complexities, such as levee overtopping, flow volume losses, downstream dams, weirs and lakes. When these constraints are present, the DAMBRK Model should be used, rather the SMPDBK Model.

In several theoretical comparison runs between the DAMBRK Model and the SMPDBK Model made by the NWS, the average difference between predicted peak flows and peak travel times was found to be 10 to 20%, with difference in depth of less than 0.3 m.

The error in the evaluation of hypothetical dam failure is due to errors resulting from the simplification of the SMPDBK Model, since the DAMBRK Model is considered to be more accurate than the SMPDBK Model.

3.3 The New South Wales Dam Safety Committee (NSW-DSC) "General Guidelines" Model

The NSW-DSC has used a "General Guidelines" Model to determine whether a full dam break analysis is required to establish the hazard potential of existing small dams (DSC-NSW 1988).

The guidelines have been developed by running the NWS DAMBRK Model several times against real cases of local dam failures. The guidelines are not meant to be an exhaustive analysis of cases of small dam failures, but they do provide a quick and simple guide to gauge the possible effects of flooding from a small dam breach.

The General Guidelines limit small dams to a maximum height of 50 m and a maximum storage capacity of 5000 ML.

The failure modes for small dams are assumed to be two: (a) failure through runoff from a large catchment into a relatively small size dam, with the inflow flood being rated as the Probable Maximum Flood (PMF) based on a three hour representative storm; and (b) failure through a large body of stored water breaching the dam without the PMF inflow.

As a guide, inflow floods causing dam failure under case (a) conditions are for dams up to 10 m in height and 5000 ML in capacity or dams up to 20 to 30 m in height and 500 ML in capacity with relatively large catchments. However, under case (b) conditions, dams greater than 30 m in height and 500 ML in capacity, the outflow peak from a full supply water storage is the result of dam breach under "sunny day" conditions.

Since one of the most common causes of failure of small dams is due to inadequate spillway capacity, the General Guidelines have adopted a representative spillway width of 40 m in the study.

The model can only be used when "typical" conditions are fulfilled. Typical conditions include gradual expansion of the regular flood plain; gradual variation in the slope of the channel; either a narrow or a wide valley; and the fit of the slope of the peak flow attenuation curve which needs to be either steep, mild or flat. When the typical conditions are not met by specific dam characteristics, geography (which is limited to the eastern third of New South Wales), and stream morphology, the model cannot be used. (The model may be applied to Victoria, if the relationship between peak PMF inflow and catchment area for prescribed dams in Victoria is established.)

Pending extensive field testing and requirement for feedback from actual and hypothetical failure cases for small dams, the General Guidelines Model offers no statements of accuracy.

4 RESULTS OF ASSESSMENT AND DISCUSSION

4.1 Qualitative requirements

For establishing the usefulness of the models to handle the qualitative requirements of the approach to handle downstream potential hazard analysis, the authors have investigated and found that it is useful to apply:

* the "Generalised Guidelines" Model to small dams, served by large catchments, where the dam break mode is dominated by the PMF inflow rather than the full supply pool of water stored behind the dam; and

* the Simplified Dam Break (SMPDBK) Model to small and large dams where the failure mode is influenced by dam breach under "sunny day" conditions, where the flood peak results from the release of a large static body of water, and where catchment size is small and the resulting PMF inflow is insignificant in comparison to the large body of water kept behind the dam.

There are certain trade-offs in advantages and disadvantages when using the two simplified dam break models. The acceptance of the trade-offs are influenced however by the objectives of the exercise.

If the objectives are to establish qualitatively whether a hazard to life or property is the result of the location of a potentially hazardous dam and whether the level of potential hazard is high, significant or low, the simplified dam break models are the most appropriate models to use.

The trade-offs are between minimal data requirement, speed and ease of use of the models on the one hand, and on the other hand, approximate results but with the attendant reduction in accuracy judged to be acceptable, despite the simplified assumptions to get rid of downstream channel complexities.

4.2 Quantitative requirements

For establishing the usefulness of the models to the quantitative requirements of the downstream potential hazard analysis, the authors have found that:

* it is not appropriate to use the "Generalised Guidelines" Model for small dams to produce quantitative results as they are not as reliable as those results obtained from DAMBRK analysis of the hazard;

* it is only under stringent conditions that the Simplified Dam Break (SMPDBK) Model should be used to analyse breaches from small and large dams to produce approximate quantitative results that are credible;

* appropriate dam break models may be used to ascertain the magnitude of potential risks to different parties, conditional upon the different modes of dam failure that can be expected or are likely to occur; but

* it is almost impossible to use dam break models to estimate the absolute risk or absolute probability of failure of a hazardous dam without consideration of the other aspect of hazard analysis - the analysis of the structural safety aspect of hazardous dam.

Despite the above assessment, the National Weather Service DAMBRK Model should always be used when serious reservations exist about the safety of the dam in question, especially when the risks and liabilities of likely dam failure need to placed on the appropriate responsible parties.

However, there are still trade-offs available in using the full DAMBRK Model. The trade-offs are between the requirements for significant amount of accurate data input and the high cost associated with the gathering of such vital data, especially from field survey of channel complexities downstream of the hazardous dam, on the one hand, and on the other hand, the ability of DAMBRK to undertake extensive numerical computation to deal with downstream channel complexities. Downstream complexities include significant cross-channel changes, dead or off-channel storage areas; changes in roughness coefficients; lateral inflows; local losses such as sever contractions and/or expansions (e.g., bridge openings); and effects of channel sinuosity and flood plains (Fread 1978; 1981).

As the authors of the DAMBRK Model (Wetmore & Fread, undated) have written:

Ideally, to obtain the most reliable estimate of probable flood elevations and travel times, the sophisticated NWS DAMBRK Model would be used in long-term disaster preparedness study where sufficient computer resources are available. However, for short term studies with limited resources, the SMPDBK Model would be most helpful in defining approximate peak stages, discharges, and travel times.

5 CONCLUSIONS

The paper presents an approach to analysing downstream potential hazard through an assessment of dam break models, in order to promote dam safety awareness in Victoria.

The paper indicates that potential hazard to life or property can be evaluated on a qualitative and quantitative basis. The paper shows the appropriate ways in which the qualitative and quantitative requirements can be satisfied, so that the approach can be operated effectively.

ACKNOWLEDGMENTS

The authors express their gratitude to Jim Keary of the Department of Water Resources, Victoria, for his valuable comments and suggestions, and to Ms Carol Roberts for her caretaking efforts in editing the paper.

DISCLAIMER

The opinions expressed by the authors in this paper do not necessarily reflect the policy of the Department of Water Resources, Victoria

REFERENCES

Australian National Committee on Large Dams (ANCOLD) 1992. Guidelines on Dam Safety Management. Draft Report by ANCOLD Working Group on Dam Safety Management.

Dams Safety Committee of New South Wales 1988. General guidelines for determining flood conditions resulting from the failure of small dams. Department of Water Resources, June.

Fread, D.L. 1978. National Weather Service operational dynamic wave model. Hydrologic Research Laboratory, W/OH3, NWS, NOAA, Silver Spring, MD 20910. Reprinted April 1987.

Fread, D.L. 1981. Some limitations of dam-breach flood routing model. Presented at the ASCE Fall Convention, St. Louis, Missouri, Oct. 26 - 30.

Fread, D.L. 1987. NWS dam breach models for micro-computers. Presented at the ASCE Annual Conference of Irrigation and Drainage Division. Portland, Oregon, July 28-31.

Wetmore, J.N. & Fread, D.L. Undated. The NWS Simplified Dam Break Flood Forecasting Model for desk-top and hand-held microcomputers. Hydrologic Research Laboratory, Office of Hydrology, National Weather Service, NOAA, 8060, 13th Street, Silver Spring, Maryland.

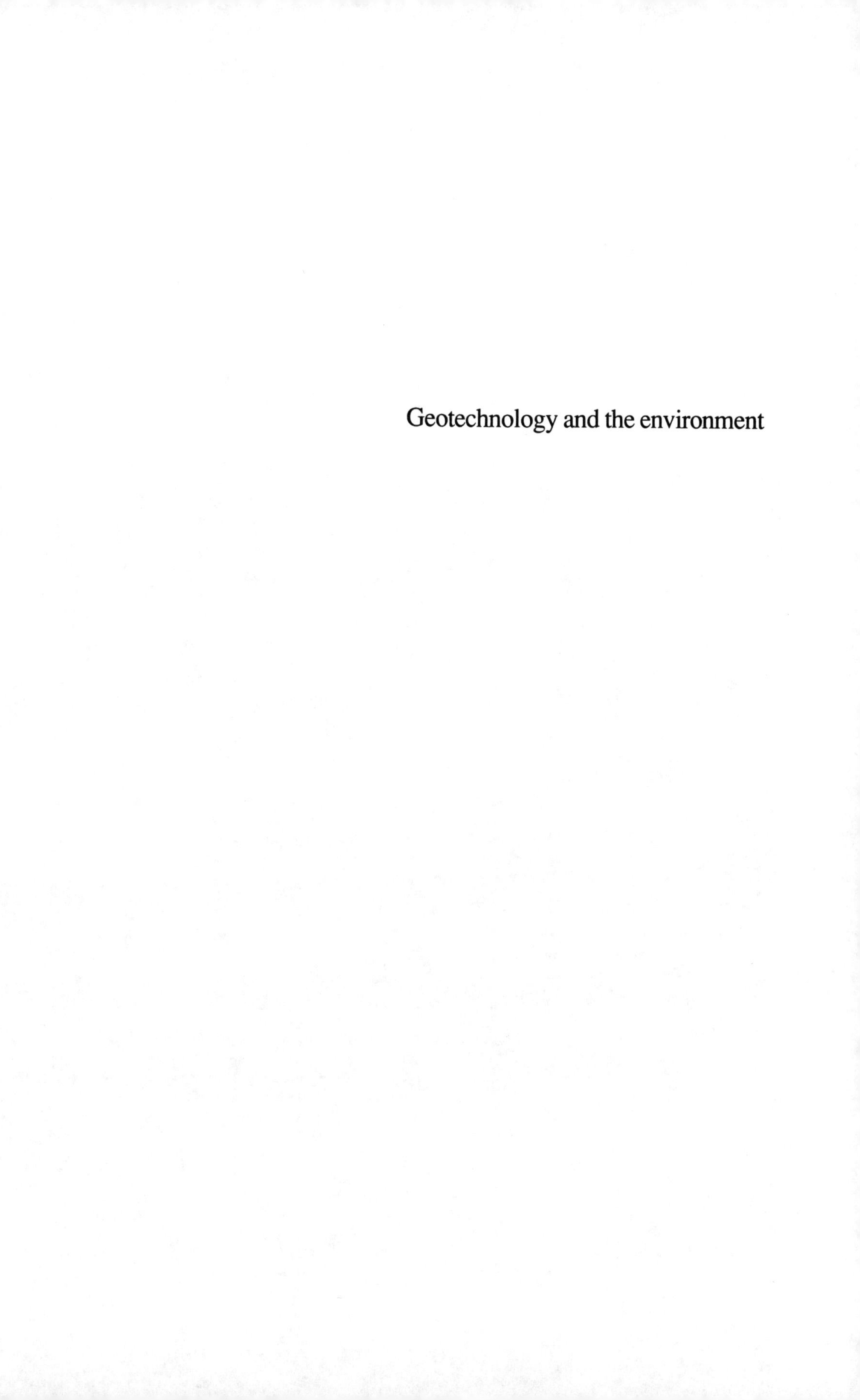

Geotechnology and the environment

Environmental Management, Geo-Water & Engineering Aspects, Chowdhury & Sivakumar (eds)
© 1993 Balkema, Rotterdam. ISBN 90 5410 099 0

Reinforced earth in coastal protection

R.M.Arenicz & P.J.Farris
The University of Wollongong, N.S.W., Australia

ABSTRACT: Fundamentals of reinforced earth and two of its first applications in coastal protection are presented following a review of coastal processes and conventional protection methods.

1 INTRODUCTION

Coastal protection is one of the important civil engineering problems which has to be dealt with in almost all countries in the world. Over the centuries, various methods and techniques have been developed to protect coastal land and man-made structures situated on the coastline from erosion caused by the variations in sea water level (tides), wave action, wind action, and associated littoral drift processes.

A coast can be naturally protected - headlands, reefs, rocky shores, dunes, river sediments can and do form a natural protection barrier against erosion of land. In other places however, man-made protection is needed. The techniques and structures used depend on location, protection requirements and economic viability of the proposed engineering undertaking.

To be successful, the choice and design of a suitable man-made coastal protection has to be based on an understanding of the natural coastal defense mechanisms against tides, littoral processes and wave action (including wind induced waves). This enables the designer to predict and assess the effects of protective structures on the natural processes in progress - hence the most appropriate technique can be selected. The designer's basic choice is, essentially, between:

(a) structures which harden the shoreline making it more resistant to erosion;

(b) stabilising the shoreline by altering prevailing coastal processes;

(c) artificial creation, reconstruction or maintenance of beaches and sand dunes.

In recent years, a new composite material - reinforced earth - has been used successfully in shoreline protection. The aim of this paper is to present this new development in relation to the coastal erosion processes and other traditional structures used in coastal protection.

2 COASTAL PROCESSES

2.1 Tides

Knowledge of the tidal process parameters is needed in design of marine structures and in planning the construction of these structures. The following data is normally required:

- times and levels of high and low water on any day;
- mean high and low water spring tides, mean high and low water neap tides, mean tide level at parts around the coast;
- height of the tide at a given time, and time at which the tide reaches a given height.

2.2 Wave action

Ocean waves are examples of periodic progressive waves, mainly generated by the action of wind and water. Once formed, ocean waves can travel for vast distances, spreading in area and reducing in height, but maintaining wavelength and period. As waves approach a shoreline, their height and wavelength are modified by the process of refraction and shoaling before breaking on the shore. During this process (Chadwick and Morfelt, 1986), deep water waves first become transitional (when wavelength L is approximately equal to half of the water depth d), then shallow waves (when $d/L < 0.4$). As the wave breaks, sand is being transported up the beach and, on return of the wave, its backwash movement drags the sand with it. The combined effect of these two actions depends on the wave frequency - short waves tend to move sand seawards, while long waves tend to move it landward (Brunn, 1985).

Since the forces exerted by waves on a marine structure can be destructive, the structure has to be designed to resist and withstand the highest wave expected at the structure. The wave conditions (non breaking, breaking, or broken) at the structure depend critically on the water level. Consequently, a design water depth must be established in determining wave forces on a structure.

2.3 Wind action

The actual water level is affected by wind, especially under storm and hurricane conditions. The wind causes tangential stresses at the water surface, thus inducing a surface current. The end result is a piling up of water at the leeward side and a lowering of water level at the windward side, with a return flow along the bottom. Such an effect, termed "storm surge", has been defined (Brunn, 1985) as the rise of normal water level due to the action of wind stresses on the water surface. The magnitude of storm surge depends on wind velocity and direction, on the distance of wind blow, and on water depth.

2.4 Design water depth

When the wind action is taken into consideration, the total mean water level used in design can be calculated from the following basic expression (Brunn, 1985):

$$dT = d_o + A_s + S_o + S_p + S_x + S_y + \alpha H_b$$

where:

- dT - total mean water level for design;
- d_o - mean low water level from hydrographic charts;
- A_s - astronomical tide from tide tables;
- S_o - initial rise in tide caused by storm approaching continental shelf prior to arrival of winds;
- S_p - component of storm surge due to reduction ΔP in atmospheric pressure ($S_p = 1.14\Delta P$);
- S_x - wind tide due to direct wind stress component perpendicular to coast;
- S_y - component of storm surge due to wind driven current parallel to coastline caused by wind stress component parallel to coast;
- α - empirical constant of value between 0.1 and 0.2 depending on bottom slope;
- H_b - height of braking wave.

2.5 Littoral drift process

Coastal stability can be adversely affected by the littoral drift - a term used for natural movement of material along the shore. Its direction and quantity depend on a number of factors like wind and wave actions, associated currents, and density of transported material. Due to the complexity of its physical processes, predicting the littoral drift is a difficult task and a number of theories have been put forward to determine dynamics of the drift. In general, the quantity of the littoral drift material would depend on wave power and its angle of attack (Chadwick and Morfelt, 1986).

3 TRADITIONAL PROTECTION STRUCTURES

The following represent a brief review of traditional, man-made structures used to protect coastlines from erosion. The choice of a particular coastal protection would be based on an assessment of the erosion situation, function required and economical considerations.

3.1 Seawalls

Seawalls are structures whose function is to protect the land and property behind them from damage by heavy wave action. Relying on their own substantial weight to hold them in the position enables seawalls to serve a secondary purpose of retaining land fill behind them.

Seawalls can be either permeable or impermeable, smooth or rough and may have any face shape (or combination of shapes). Traditionally, seawalls are built from concrete, rock, rock-fill timber cribbing or gabions, and cellural steel sheet pile.

The barrier that a seawall forms parallel with the shoreline may have an undesirable effect of depleting the supply of material available for littoral drift as well as increasing the rate of erosion of the adjacent seabed due to wave reflection. Such an erosion may result in undercutting of wall foundations unless the wall is founded on solid rocks or a very substantial toe protection is provided. Such a protection can consist of rock blankets laid on the ground seaward of the toe or sheet pile cut off walls in front of the toe (Chadwick and Morfelt, 1986). Regarding the erosion rate, it can be reduced by building the wall with face curved and/or inclined at low angle to the horizontal. Such a structure will, however, be significantly more expensive than the vertical one.

The forces exerted on a seawall by wave action can be considered to be composed of three parts: the static pressure forces, the dynamic pressure forces and the shock forces. The shock forces arise due to breaking waves trapping pockets of air which are rapidly compressed, resulting in high localised forces exerted on the structure. Thus, seawalls must have a high structural strength and their construction material must be able to withstand the shock forces.

3.2 Revetments

Revetments are structures designed to protect the shore from wave and current action. They rest upon, and are supported by, the land behind them which is usually at or near its natural angle of repose. Revetments can be either smooth or rough in surface texture, permeable or impermeable and are usually constructed with a sloping face. As structures, they are lighter than seawalls and are not suited for heavy wave attack. The slope and roughness of revetments reduce the seaward erosion near the structure.

There are three basic, commonly used, types of revetments:

- concrete revetments (rigid);
- riprap revetments (flexible);
- interlocking concrete block revetments (flexible).

The rigid concrete revetments provide for excellent shoreline protection but the site must be dewatered to allow concrete to be placed. The flexible revetments offer almost as good protection as the rigid ones, with the added advantage of being able to withstand minor settlements without structural failure. They also allow for relief of the hydrostatic uplift pressure through underlying filter and bedding layers.

3.3 Bulkheads

Bulkheads are basically sheet pile structures used to retain fill and prevent the land behind them from sliding into the water. They are well suited to berthing due to their vertical or near vertical form. Although the bulkhead's vertical nature results in sever scour, they are constructed to withstand such high forces. A protection from undercutting is also required, either through placing a riprap at the toe or by providing an adequate toe penetration.

Bulkheads can be built using a variety of materials: timber, steel, aluminium, concrete, and asbestos cement. Regarding construction methods and type of structural support, bulkheads are classified into two categories: cantilevered and anchored. The former are held up by the strength of the sheet pile alone, driven deeply into the foundation soil. The latter rely on anchoring systems to help them remain upright against the forces from the landward side.

3.4 Groins

These are protective structures designed to alter longshore drift in such a way so as to build or stabilise a beach, by reducing the rate of littoral transport and by trapping littoral material. For groins to work properly, a detailed understanding of the functional design of groins, as well as of the littoral process in a given location, is essential.

Groins are narrow structures, varying in length from less than 30m to 100m or more, extended from the point landward of predicted shoreline recession into the water. Depending of their construction and dimensions, groins can be classified as either permeable or impermeable, high or low, long or short, and fixed or adjustable.

Groins can by built as individual structures or as an interacting system of groins. The added advantage of the latter system is in interaction between groins causing the downdrift side of each groin to benefit from accretion on the updrift side of the adjacent downdrift groin, hence resulting in widening of the beach.

For a groin system to work properly, it is necessary to have the correct spacing between groins. If the spacing is too wide, excessive recession on the downdrift sides would occur. Insufficient spacing would produce a net offshore transport of sand thus increasing erosion.

3.5 Artificial beach replenishment

The beach erosion caused by the littoral drift can often be accelerated by the placing of a shore protection structure behind the beach. In such a case, it may be necessary to take the remedial measure of stockpiling (with periodic replenishment) a suitable beach material at the updrift section of the problem area. Under suitable conditions, such an artificial nourishment of beach area may be economically preferable to groins and other littoral drift altering structures. It also gives an advantage of having the deficiency of natural sand supply remedied without damaging the shore beyond the problem area.

3.6 Breakwaters

A breakwater is a structure designed to protect a designated area from wave action. Breakwaters can be either located offshore or connected to the shore.

Offshore breakwaters are generally sited so as to provide shelter to a harbour entrance or to create a littoral reservoir. Due to their location in deep water, the offshore breakwaters can control a wide part of the littoral zone and they are usually much more effective than groins in interception of littoral material.

A shore-connected breakwater forms a littoral barrier in the zone between the seaward end of its shorearm and the limit of wave uprush until the capacity of the structure to impound drifting sand is reached and natural bypassing of the littoral material is resumed.

4. REINFORCED EARTH IN COASTAL PROTECTION

4.1 Fundamentals

Reinforced earth in its contemporary form was invented in 1966 by the French engineer Henri Vidal. Basically, it is a mass of soil strengthened by the inclusion of reinforcement material able to withstand tensile stresses which otherwise might cause internal failure within the soil. The concept of reinforced earth is analogous to that of reinforced concrete: in both cases the mechanical properties of the original material are improved by the inclusion of reinforcement resisting the tensile stress.

To date, the most established and successful application of reinforced earth has been primarily as a material used in the construction of various types of retaining walls, embankments, bridge abutments and roads, where it has been proven to be a very reliable and economically superior (Jones, 1985) alternative to other, more traditional materials or techniques of enhancing soil strength.

A reinforced earth structure consists of three basic components:

- soil fill, usually cohesionless;
- reinforcement, typically in the form of horizontal metal strips, either smooth or ribbed, attached to:
- facing (or skin) elements, commonly made of either metal or concrete, forming the outside surface of structures such as retaining walls.

In some cases, cohesive soils can be used as fill and, increasingly often in recent years, polymer fabrics and grids are used as reinforcement instead of metal strips. Also, some of the applications might not require the use of facing elements - e.g. road pavements or foundation soils.

Due to its versatility and low cost, the application of reinforced earth in various other engineering structures has also been considered, with some promising results from the prototype trials. In particular, the first large scale reinforced earth seawalls were constructed during the last twenty years in France, U.K., U.S.A., Canada and the Solomon Islands, proving to be a viable alternative to other conventional structures regarding cost effectiveness and performance, thus paving the way towards a wider application of rein-forced earth in coastal protection. The two case studies presented below are intended to illustrate the most recent developments in this area: first prototypes of reinforced earth seawalls constructed under water.

4.2 Honiara Copra Wharf

In 1986, a reinforced earth wharf 85 m long and 3-6 m high was built in Honiara (Solomon Islands) to replace the existing old and decaying structure. The wharf was designed to withstand an external loading combined of 15 kPa surcharge, up to 50 T bollard forces and 0.15 g earthquake acceleration (Boyd and Ryan, 1988).

To form the wharf facing, heavy precast concrete rectangular panels 3x2.25x0.55m were used. Each panel has vertical slotted keyways at its back allowing strip reinforcement of the fill to be connected to the panel and then to be slid into its proper position (Boyd and Ryan, 1988).

The fill reinforcement consisted of horizontal steel strips 60x9mm in cross-section and 6m long anchored in the slotted keyways at the back of the facing panels. For this purpose, each strip has its end twisted at right angles to fit into the vertical slot in the panel (Boyd and Ryan, 1988).

As a backfill material, a sandy river gravel was used, with the 75 μm fraction not exceeding 5%, coefficient of uniformity of 6.5 and the internal friction angle of at least 35 degree regardless of saturation. The fill was placed in 0.75 m thick layers vibrated to required density and leveled with poker vibration (Boyd and Ryan, 1988).

Construction of the wharf was carried out under water, with the use of divers and floating equipment.

4.3 Bluffers Park Prototype

The Bluffers Park Underwater Wall was constructed in 1986 on the shore of the Great Lake near Toronto (Canada). The height and length of the wall were 4.8 m and 36 m, respectively (Farris, 1991).

As the facing elements, heavy precast concrete rectangular panels were used, typically 4x2x0.3m in size, with the reinforcement strips attached to the panels prior to the actual construction of the wall. This was due to the lack of visibility under water which made it impossible to attach the strips to the panels during the construction process (Farris, 1991).

The reinforcement strips were horizontal, made of steel, 50x6 mm in cross-section and 6.2 m long. They were attached to the facing panels through articulated connections with a horizontal bolt fitting into a slotted vertical recess in the back of the panel. The strip ends were twisted at right angles to fit into the vertical slots in the panels. The connections allowed accommodation of the differential settlement between the fill and facing panels (Farris, 1991).

Uniform fine beach sand excavated from the construction site was also used as a backfill material.

The construction of the wall was carried out without divers, with the use of basic equipment: crane for panel installation, clamshell for excavation and backfilling, and concrete mixer (Farris, 1991).

5 CONCLUSIONS

Since the mid-seventies, reinforced earth has successfully been used in construction of coastal seawalls in several countries (France, U.K., U.S.A.), with the structures built under dry conditions of low tide using largely conventional construction procedures.

In the mid-eighties a further significant development took place: reinforced earth has been used for the first time to build underwater seawalls in Canada and the Solomon Islands. Although design of these two prototypes varied in a number of structural details, as well as in construction techniques, both structures were of similar height, with large and heavy rectangular precast concrete panels used as facing and with reinforcement anchorage allowing accommodation of the differential settlement between strips and facing panels. Also, both structures were built in sheltered, calm water environment (no heavy wave attack).

The settlement of reinforced earth fill being placed under the water table can be significant and should be monitored during and after construction. Connections between facing panels and reinforcement have to allow for the settlement of strips embedded in the fill, otherwise the tensile forces exerted on reinforcement might be significantly increased.

Based on the successful construction and behaviour of the prototypes, reinforced earth structures offer a viable alternative to conventional shore protection methods by means of mass concrete or sheet-pile seawall structures. However, from Canadian experience, alignment techniques for placing heavy facing panels in their correct position under water have to be refined for future constructions. Also, there is a need for further research to develop adequate design and construction procedures counteracting the adverse effects of waves on stability and compaction of reinforced earth fill if such structures are to be built outside protected waters.

6 REFERENCES

Boyd, M.S. & Ryan, M.J. 1988. A reinforced earth solution for marine structures. Proceedings of the 2nd Australasian Port, Harbour and Offshore Engineering Conference, Vol 2., Brisbane, 70-74.

Brunn, P. 1985. Design and construction of mounds for breakwaters and coastal protection. New York: Elsevier.

Chadwick, A. & Morfelt, J. 1986. Hydraulics in civil engineering. London: Allen & Unwin.

Farris, P.J. 1991. Coastal protection methods with particular reference to reinforced earth. BE thesis: University of Wollongong, Australia.

Jones, C.J.F.P. 1985. Earth Reinforcement and Soil Structures. London: Butterworth.

Environmental Management, Geo-Water & Engineering Aspects, Chowdhury & Sivakumar (eds)
© 1993 Balkema, Rotterdam. ISBN 90 5410 099 0

Behavior of the waterfront quaywall constructed by steel pipe piles in soft soil slope

W.P.Hong
Chung-Ang University, Seoul, Korea

J.P.Ahn
Cho-Sun University, Kwangju, Korea

ABSTRACT : The horizontal displacement of a waterfront quaywall constructed by placing steel-pipe-piles in soft soil at close intervals in a row was investigated and an analytical approach was attempted to analyze the behavior of the waterfront quaywall. For estimating lateral force acting on piles for quaywall in soils undergoing lateral movement, the concept of the passive pile was applied. The pile deflection was dependent considerably on the soil modulus above potential sliding surface, which was related to the slope-stability.

1. INTRODUCTION

Recently a waterfront quaywall has been constructed in South Korea by placing steel-pipe-piles in soft soil at close intervals in a row. Spaces between piles were opened except the upper part of the piles which was connected partially each other. When the soil moves laterally through spaces between piles, the piles can resist to the soil movement due to soil arching action.

This type of quaywalls is treated as a kind of sheet-pile quaywalls. The mechanism of earth pressure and the behavior of quaywalls, however, are quite different from those of the conventional sheet-piles quaywalls.

A landing pier was constructed in front of this quaywall for vessels to anchor. However, the landing pier had some trobles resulted mainly because of the backfill behind the quaywall. The quaywall was subjected to unsymmetrical surcharges due to the backfill above soft soil. The soft soil has undergone lateral movement, which is related to the stabilities of both quaywalls and slope under platform of the landing pier. Consequently, such lateral movements may induce severe damages to the landing pier with wharf facilities.

In several wharf facilities, also severe damages have been reported by Heussink & Wenz (1969), Peck & Raamot(1964) and Uriel et al. (1976). Some remedial designs were provided to control such trouble (Tschebotarioff 1973).

The wharf facility is very sensitive to the lateral displacement of landing pier. Therefore, in order to design the steel-pipe-pile quaywalls, the behavior of quaywells should be analyzed accurately. However, the behavior of the quaywalls is much related to the mechanism of the lateral force acting on the steel-pipe-pile quaywalls due to backfill.

The lateral force acting on steel-pipe-piles for quaywalls results from interaction between soils and piles in the soil undergoing lateral movement. Accurate estimation of the lateral force was possible by considering the mechanism of plastic deformation of the soil between piles (Ito & Matsui 1975, Matsui, Hong & Ito 1982).

In this paper, first an investigation is performed on the behavior of the waterfront quaywall constructed by steel-pipe-piles in soft soil slope undergoing lateral movement due to unsymmetrical surcharge on the slope. Then, an analytical approach is presented to analyze the behavior of the waterfront quaywall.

2. PREVIOUS WORK ON PASSIVE PILES

When landing piers and bridge abutments are constructed on soft soils, piles are used to support the loads from upper structures. The loads resulted by backfills, however, works on the soft soils as unsymmetrical surcharges with consequent development of lateral soil deformation. Such lateral soil movements, which is generally unfavorable deformation, induce severe damages on upper structures. The piles are subjected to secondary lateral force resulted from the interaction between soils and piles. The piles subjected to the lateral force due to lateral soil movement is called by the passive pile (De Beer 1977).

A number of cases has been reported about movements of landing pier(Tschebotarioff 1973) and bridge abutments(Nicu, Antes & Kessler 1971). Marche & Lacroix(1972) have presented the property of lateral flow of soft soil based

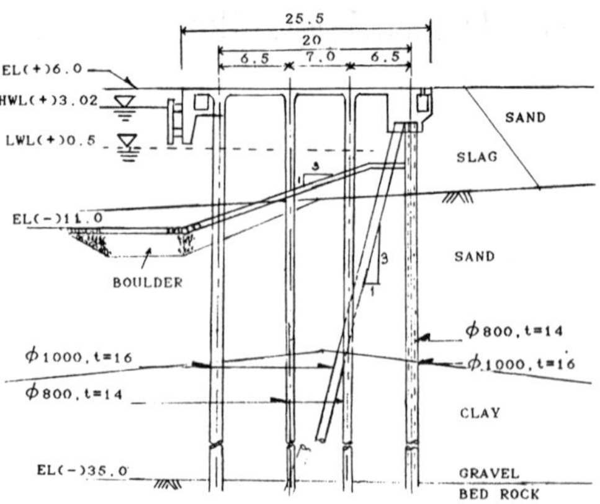

Fig.1 Cross section and plan of landing pier

on investigation on severe records. Tschebotarioff(1971), De Beer & Wallays(1972), etc. also have contributed on the studies about the passive piles. Marche(1973) has described the relation between the lateral flow of soil and the safety factor of slope in case of embankment on soft soils, Franke (1977) has recommended a method to decide the required safety factor of slope on the soil undergoing lateral movement by relating the consistency index of soil to the slope-stability. Based on investigation for embankment cases, Tschebotarioff(1973) has presented the fill height to induce the shearing deformation or failure in soil.

The behavior of passive piles in soft soil has been analyzed by use of several analytical methods such as conventional methods (Tschebotarioff 1971, De Beer & Wallays 1972), the theory of subgrade reaction (Marche 1973), theory of elasticity(Poulos 1973) and finite element method(Moser 1973). Ito, Matsui & Hong(1979) have described a basic design method for the stability analysis of the slope with landing pier.

3 BEHAVIOR OF A WATERFRONT QUAYWALL

3.1 Waterfront Structures

Fig.1 shows a case where some trouble resulted because of backfill behind a waterfront quaywall of a wharf in South Korea. The waterfront quaywall was constructed by steel-pipe-piles in soft soil slopes behind a pile -supported landing pier.

The cross section and plan of the landing pier, the waterfront quaywall and the soil slope are shown in Fig.1. The wharf was constituted by three blocks of a 122.2 m long reinforced-concrete platform supported by steel-pipe-piles with diameters of 1000 mm and 800 mm. That is, expansion joint was provided every 122.2 m. The depth of water was planned at EL(-)11.0 m and EL(-)7.5 m so that vessels of 20,000 DWT and 5,000 DWT burden capacity could anchor, respectively.

Five rows of piles were driven to support the concrete platform by EL(-)35.00 m, where bed rock was located. In four of the five rows the piles were vertical and in one row the piles had a 3:1 batter. The last row of piles on the inboard side of the platform was driven at close interval to retain the backfill. This row of piles is used for a steel-pipe-pile waterfront quaywall. The pile heads of the batter row were connected with those of the last pile row. In the first row and batter row piles had 1000 mm in diameter, 16 mm in thickness and 4 m in interval between piles. In the second and third rows piles had 800 mm in diameter, 14 mm in thickness and 4 m in interval between piles.

In the last row, piles with 1000 mm in diameter and 16 mm in thickness were dirven in 2 m interval between piles. Additionly piles with 800 mm in diameter and 14 mm in thickness were driven between the piles with 1000 mm in diameter. Upper part of these piles were connected partially to form a wall. The spaces between rows of piles were 6.5 m, except 7.0 m between the second and third rows. The width of the crane rail was 20 m on the 25.5 m-wide concrete platform

3.2 Soil property

Under the concrete platform of the wharf a slope of 2 horizontal to 1 vertical was formed by armour stones. The toe of the slope was dredged to provide the required depth of water. Then, a geotextile was placed and hydraulically filled with sand dredged out from a borrow area in the vicinity. Slag was used to fill partially behind the waterfront quaywall as shown in Fig.1. Backfilling was completed by sand fill on the geotextile placed on the backfill slag surface.

The original soil below the fill strata was formed by 15 m thick silty sand strata up to EL(-)21.6 m and below the silty sand strata silty clay with 11.2 m thickness was formed. And lower strata was constituted by fine sand, sandy gravel, boulder and weathered rocks.

The upper sand had N-value about 6 and the silty clay had an unconfined compressive strength of 0.9 t/m^2 to 1.0 t/m^2.

3.3 Lateral Displacement

The first cracks were observed on the reinforced concrete platform during backfilling behind the waterfront quaywall after construction of the platform of the landing pier.

14 measuring points were provided on the platform of landing pier to observe the seaward lateral movement of wharf as given in Fig.2.

Fig.2 shows that the lateral movements at the site of M3 and M4 measuring points were severe.

Because expansion joint was provided at the site of those measuring points, the stiffness of the platform at this site was lower than that at other sites.

Fig.3 illustrates the variation of the seaward movement of the wharf at the site of M4 measuring point. The behavior of lateral movement can be divided into three steps. For first step during 100 days the horizontal displacement show constant value about 25 mm because backfill was not reached yet the site behind M4 measuring point. For the second step during 150 days, the horizontal displacement increased up to 160 mm with the velocity of 27 mm/month(0.9 mm/day). For the third step, the horizontal displacement had a velocity about 1.9 mm/month and was converged to 199 mm.

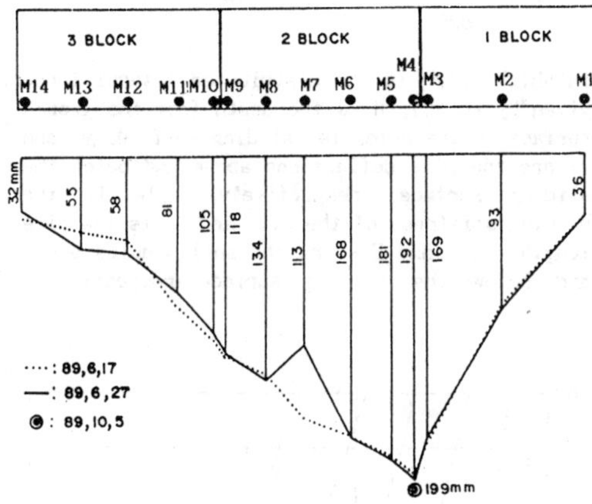

Fig.2 Seaward lateral displacement of wharf

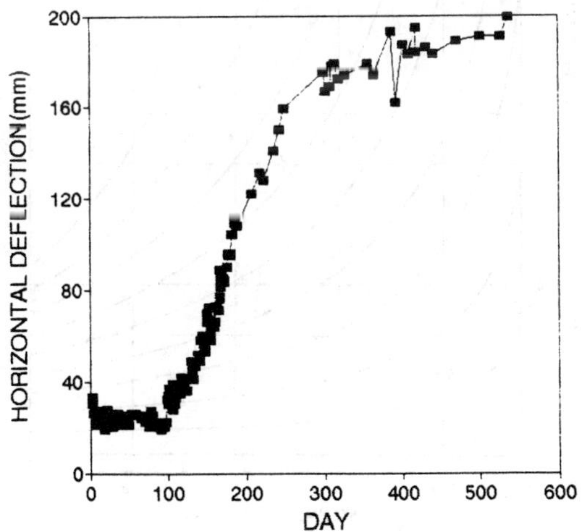

Fig.3 lateral displacement at M4 mearing point

4. ANALYSIS OF BEHAVIOR OF THE WATERFRONT QUAYWALL

4.1 Remarks

The steel pipe piles for waterfront quaywall may be expected to have a preventive effect against the lateral soil movement resulted by backfills, while the piles are subjected to lateral force above the potential sliding surface in soil slope under the platform of landing pier. Thus, it is very important that an accurate estimation of the lateral force is an important key-factor to analyze the behavior of the steel-pipe-pile quaywall, because its effects on the pile- and slope- stabilities are contrary. That is, overestimating the lateral force leads to conservative pile-stability results, but the slope-stability will be unconservative, while in the case of its underestimation, the contrary occurs.

In this chapter, the behavior of quaywall investigated in the previous chapter is analyzed by considering the stabilities of both piles and slopes. In order to facilitate this study, the following assumptions are made:

1) The modulus of elasticity of steel pile is 2.1×10^6 kg/cm^2.

2) The allowable bending and shearing stresses of steel piles are 1400 kg/cm^2 and 800 kg/cm^2, respectively.

3) The soil modulus recommended by Poulos (1971) is used. That is, the soil modulus in sand strata below the potential sliding surface is 175 t/m^2, which is the average recommended soil modulus. And the soil modulus in clay strata below the potential sliding surface is 40 times the undrained shear strength of the clay.

4) The fixity-conditions of pile head and tip are, respectively, unrotated and hinged ones.

5) The piles are subjected to only the lateral forces resulted by soil movements above sliding surfaces in soils.

4.2 Stability analysis of slope containing piles

Potential unstable slopes can be made stable by the addition of the effective resisting force of piles only when the piles have enough stiffness against the lateral force. That is, to control slope failure by use of piles, not only the slope but also the piles should be stable. Therefore, it is necessary to carry out two kinds of analyses: one for the pile-stability and the other for the slope-stability (Ito, Matsui & Hong 1979).

The lateral force p(z) acting on a row of piles per unit depth of soil can be estimated by Eq.(1), which has been derived by considering the interval between piles and also assuming that the Mohr-Coulomb's plastic condition occurs in the surrounding ground just around the piles(Ito and Matsui 1975, Matsui, Hong & Ito 1982).

$$p(z)/d = K_{p1}\ c + K_{p2}\ \sigma_H(z) \cdots\cdots\cdots\cdots (1)$$

in which z is the depth from ground surface, d is the pile diameter, c is the cohesion of soil, K_{p1} and K_{p2} are the coefficients of lateral force(Hong 1986) and $\sigma_H(z)$ is the earth pressure acting on the outboard side of the pile row.

Fig.4 is available to estimate K_{p1} and K_{p2}, which are dependent on the angle of internal friction of soil ϕ and interval ratio D_2/D_1 (D_1 and D_2 ($=D_1-d$), respectively, are the center-to-center and clear intervals between piles). As for the earth pressure $\sigma_H(z)$, it is reasonable to assume that it is equal to the active earth pressure because the sliding mass moves beyond the row of piles.

The lateral force used in design of piles is the value within the critical lateral force obtained by Eq.(1). The lateral force will be mobilized from zero to the critical value.

Experimental results by Matsui, Hong & Ito(1982) show that the lateral force increases almost linearly with the displacement of soil within the values given by the theoretical equation. Thus, the mobilized lateral force $p_m(z)$ may be explained by Eq.(2).

$$p_m(z) = \alpha_m\ p(z) \cdots\cdots\cdots\cdots\cdots\cdots (2)$$

in which α_m is defined as the mobilization factor of lateral force. The coefficient α_m varies from zero to unity. p(z) is given by Eq.(1).

As for the pile-stability, an analytical method of piles subjected to horizontal loads can be applied, considering that the lateral force acts on piles through the soil mass above the sliding surface. Assuming that the piles shall be subjected to the lateral reaction from soil in proportion to the pile deflection resulted by the distributed lateral force above sliding surface, the following differential equation are used for the pile-stability analysis.

$$\left.\begin{aligned} E_pI_p \frac{d^4y_1}{dz^4} &= p_m(z) - E_{s1}y_1 \quad (0 \leq z \leq H) \\ E_pI_p \frac{d^4y_2}{dz^4} &= E_{s2}y_2 \quad (H < z \leq L_p) \end{aligned}\right\} \cdots (3)$$

in which $p_m(z)$ is the distributed lateral force given by Eq.(2), H is the depth from the ground surface to the potential sliding surface, y_1 and y_2 are the pile deflections above and below the sliding surface, respectively, E_pI_p is the bending stiffness of the pile and L_p is the pile length. E_{s1} and E_{s2} are the soil modulus above and below the sliding surface, respectively.

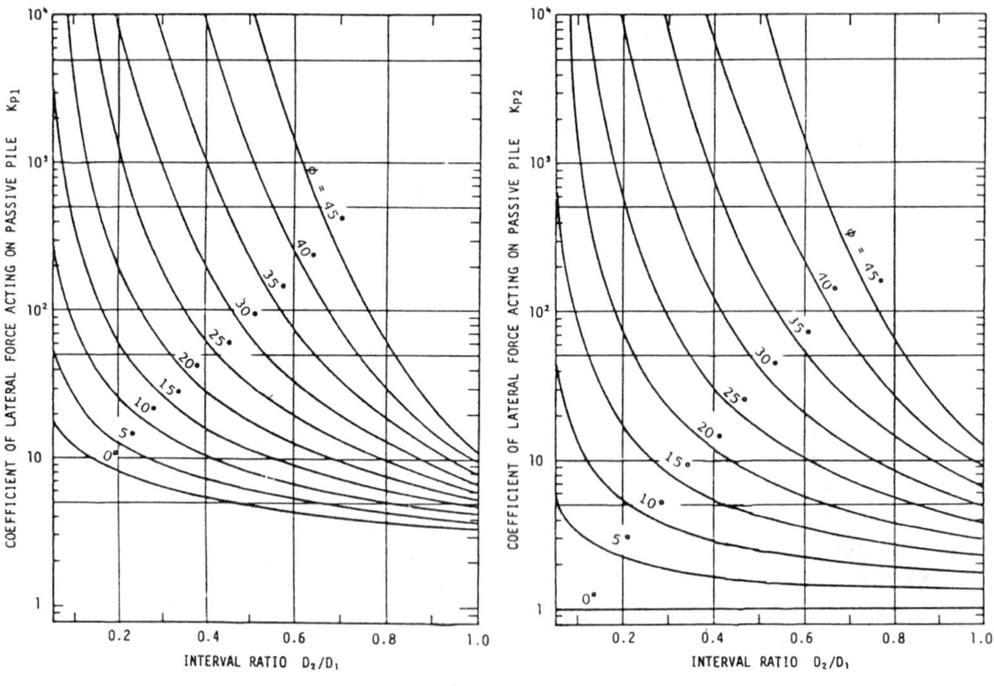

Fig.4 Coefficients of later force, K_{p1} & K_{p2}

When the plastic deformation above the sliidng surface is considerable, E_{s1} may be given by zero since the soil reaction above the sliidng surface is not abailable.

These differential equations can be solved under both the fixity condition at head and tip of piles and the continuity condition of pile at the sliding surface of slope.

The pile-stability can be investigated by comparing the allowable bending stress with the maximum induced stress.

Using the circular arc analysis, the slope-stability can be analyzed by a comparison between the resisting and deriving moments.

The resisiting moment may be obtained as the sum of both the resisting moments due to the shearing resistance along the potential sliding surface of slope and the reaction force of piles, respectively.

4.3 Safety factor of slope

In order to investigate the slope-stability for the wharf shown in Fig.1, the soil properties and strata were decided as shown in Fig.5 by site investigations and soil tests at the site of M4 measurring point, where the lateral displacement of platform was most severe.

The minimum safety factor of slope under the landing pier was 0.87 without the preventive effect of piles against slope failure along the potential sliding surface. The safety factor of slope, however, was increased to 1.23 by considering the preventive effect of the four rows of vertical piles except a row of batter.

Now the slope can have the slope-stability against sliding. However, if design surcharge of 5.33 t/m^2 is loaded on the waterfront storage yards behind the quaywall, the safety factor of slope will be decreased to 1.06, which is not enough to provide stability of the slope. Therefore, some countermeasures should be provided in order to prevent additional lateral movement of the wharf for the case when any manufactured goods is loaded on the ground surface of the waterfront storage yards.

4.4 Behavior of piles

The piles under landing pier moved horizontally because of the lateral force acting on the piles in soils undergoing lateral movements, while the resistance of piles was contributed on increment of the slope-stability.

Fig.6 shows the deflection behavior of piles estimated according to the pile-stability analysis described above. Fig.6(a) illustrates the pile deflection in case when the soil resistance above sliiding surface of slope was neglected; that is E_{s1} is assumed as zero. The

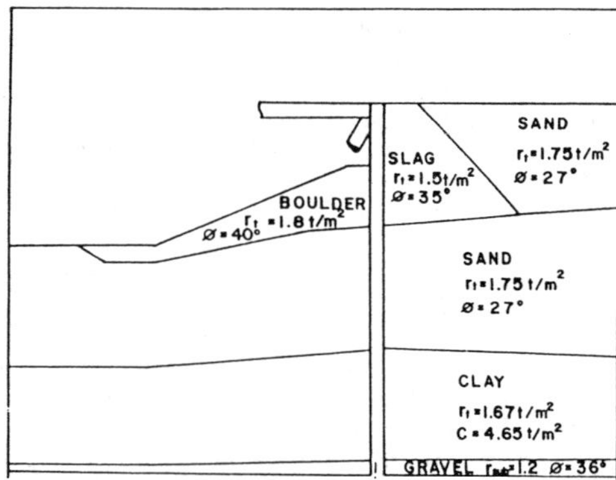

Fig.5 Soil properties and strata of slope

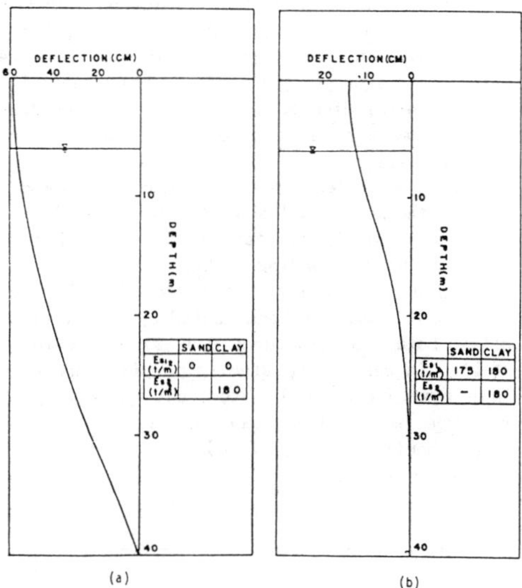

Fig.6 Deflection behavior of piles

maximum deflection at pile head in Fig.6(a) represents 580 mm.

In case when the soil modulus above sliding surface is assumed as same to the soil modulus below sliding surface the pile deflection at the pile head shows 110 mm as shown in Fig.6(b).

For various values of the soil modulus E_{s1}, the variation of the maximum deflextion at pile head could be given in Fig.7.

Consequently, it may be known that the pile deflection is dependent considerably on the soil modulus above sliding surface. Especially, when the soil modulus E_{s1} decreases below 40 t/m^2, the maximum pile deflection increases consederably, while the behavior of the maximum pile deflection converges under E_{s1} over 100 t/m^2.

Generally the soil modulus E_{s1}, however, is dependent on the slope stability. That is, the higher soil modulus can be produced under the higher slope-stability.

The measured maximum deflection, which was 199 mm at pile head as shown in Fig.3, is same to the value estimated in case of Esl about 65 t/m^2, which can be produced under low slope-stability condition.

5. CONCLUSIONS

The behavior of horizontal displacement of a wharf, in which a landing pier was placed in front of a steel-pipe-pile waterfront quaywall, was investigated. And also an analytical approach was attempted to analyze the behavior of the waterfront quaywall.

The slope under the landing pier could be stable by the preventive effect of four rows of piles to support the reinforced concrete platform. However, in case when design surcharge is placed on the ground surface at the waterfront storage yards behind the quaywall, the horizontal displacement of the landing pier may increase more because of decrease of the slope-stability.

The pile deflection was dependent considerably on the soil modulus above potential sliding surface, which is related to the stability of the slope under landing pier.

When a landing pier is placed in front of waterfornt quaywalls constructed by steel-pipe-piles, both the structures and slope should be stable. In case where steel-pipe piles are used in a row to construct waterfront quaywalls, it is suggested in this paper that both piles and slope should be designed under the criterion to satisfy both stabilities.

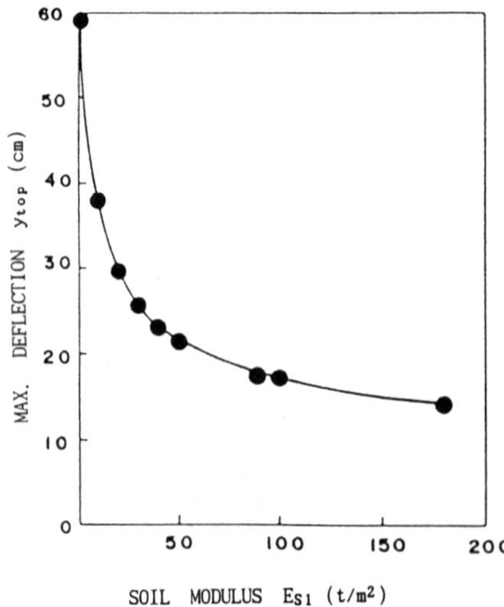

Fig.7 Relation between maximum pile deflection and soil modulus above sliding surface

ACKNOWLEDGEMENT

The study presented here was supported by the Korea Science and Engineering Foundation under Grant No. 881-1306-012-2. Grateful appreciation is expressed to the foundation for the support of this study.

REFERENCES

De Beer, E.E. & M. Wallays. 1972. Forces induced in piles by unsymmetrical surcharges on the soil around the pile. Proc. 5th ECSMFE 1:325-332.

De Beer, E.E. 1977. Piles subjected to static lateral loads, State-of-the-Art Report. proc. 9th ICSMFE Specialty Session 10.1-14.

Franke, E.1977. German Recommendation on passive piles. Proc. 9th ICSMFE Specialty Session 10. 193-194.

Heussink, H. & K. P. Wenz 1969. Storage yard foundations on soft cohesive soils. Proc. 7th ICSMFE 2:149-155.

Hong, W.P. 1986. Design method of piles to stabilize landslides. Proc. Inter. Sym. Envir. Geot. 1:441-453.

Ito,T., T.Matusi & W.P.Hong. 1979. Design Method for the stability analysis of the slope with landing pier. Soils and Foundations.19(4):43-57.

Marche,R. & Y.Lacroix. 1972.Stabilite des culees de ponts establies sur des pieux traversant une couche molle.Canadian Geotechnical Journal 9(1):1-24.

Marche,R. 1973. Discussion, Specialty Session 5. Proc. 8th ICSMFE 4(3):247-252.

Matsui, T., W.P. Hong & T.Ito.1982. Earth pressures on piles in a row due to lateral soil movements. Soils and Foundations 22(2):71-81.

Moser,M.A. 1973. Lateral pressure of calyey soils on structures. Proc.8th ICSMFE Specialty Session 5.4(3):252-253.

Nicu, N.D., D.R. Antes & R.S. Kessler 1971.Field measurements on instrumented piles under an overpass abutment. Highway Research Record 354:90-102.

Peck, R.B. & T. Raamot 1964. Foundation behavior of iron ore storage yards. Jour. SMFD ASCE 90(SM3):85-123

Poulos,H.G.1973.Analysis of piles in soil undergoing lateral movement. Jour.SMFD ASCE 99(SM5):391-405.

Uriel,S. et al. 1976. Behavior of precast piles under lateral pressures at a shipbuilding site Proc. 6th ECSMFE 1:585-590.

Stermac, A.G., M.Devata & K.G.Selby 1968.Unusual movements of abutments supported on end-bearing piles. Canadian Geotechnical Journal 5(2):69-79.

Tschebotarioff, G. P. 1971. Discussion. Highway Research Record 354:99-101.

Tschebotarioff,G.P. 1973. Foundations Retaining and Earth Structures. pp.559-566.Tokyo:McGraw-Hill, Kogakusha.

Environmental Management, Geo-Water & Engineering Aspects, Chowdhury & Sivakumar (eds)
 ISBN 90 5410 099 0

Use of pozzolanic fuel ash in problematic soil modification

B. Indraratna & R. N. Chowdhury
Department of Civil & Mining Engineering, University of Wollongong, N.S.W., Australia

N. Kuganenthira
Department of Civil Engineering, John Hopkins University, Ma., USA

ABSTRACT: This paper is concerned with the influence of thermal plant waste (predominantly fly ash) on the stabilization of an erodible residual soil commonly found in Thailand. The effect of blending pozzolanic ash on the rate of erosion, dispersivity, strength and frictional properties, compaction and consolidation characteristics are discussed. It is found that the addition of a small quantity of waste ash inhibits erosion, and improves the strength and deformation characteristics. The long-term properties are related to the pozzolanic nature of this ash which contributes to self-hardening of the stabilized soil.

1. INTRODUCTION

Problems associated with highly erodible soils have been reported in many parts of the world, including Australia. In the northern parts of Thailand, dispersive soils are commonly encountered, and several irrigation dams constructed in this region have suffered severe internal and surface erosion. Prominent erosional features include gully marks, channels, internal cavities and tunnels within the soil mass. Rapid erosion of soil responsible for the failure of natural slopes is initiated by the dispersion of clay particles, and subsequent 'wash-out' along existing discontinuities by seepage water. Cole et al., (1977) reported that erodible soils obtained in the vicinity of irrigation dam sites in Thailand contained predominantly quartz, with a total clay fraction less than 50%. The soil analyzed in the present study were obtained from clayey formations of the Khorat plateau, where the average exchangeable sodium percentage is greater than 60% with a corresponding dispersion of about 40%. These residual soils can be categorized as yellowish brown silty clay. Field observations revealed extensive soil erosion and gully patterns which indicate the highly unstable nature of these soils. Apparent cloudiness of stagnant water ponds also reflect the dispersive nature of the underlying clay layers.

In the past, chemical methods such as reservoir water treatment, cement and lime treatment have been used (Chandra and James, 1984; Joshi and Nagaraj, 1987) to inhibit soil dispersion and erosion. However, chemical stabilization techniques are relatively expensive, in contrast to the possibility of using waste materials in achieving the same objectives. Pulverized fuel ash or fly ash, a by-product of coal combustion in power plants, has been considered as an effective construction fill (Indraratna et al., 1991). Due to its low specific gravity, it could be used easily handled with conventional construction equipment and effectively compacted (Toth et al., 1988). As certain coal ashes can provide an adequate array of multi-valent cations capable of flocculating of fine clay particles, it can be anticipated that even the most dispersive clays may be stabilized by blending with fly ash. Moreover, lignite fly ash is a pozzolanic material that encourages self-hardening with time. Consequently, a sufficient amount of pozzolanic fly ash mixed with a given erodible soil should result in a stiffer soil matrix which is resistant to erosion.

In Australasia, large quantities of mine waste and power plant waste are disposed as open stockpiles or pumped to artificial lagoons. There is no doubt that the disposal of coal ash is also a genuine concern in terms of land usage and environmental pollution. In particular, the possibility of fly ash in polluting the atmosphere is a serious problem with regard to open untreated dumps. Therefore, large scale utilization of fly ash for practical engineering projects must be encouraged.

2. PROPERTIES OF FLY ASH AND ITS USES

Ash is produced in large quantities in many parts of the world where thermal power plants utilize coal or lignite as fuel. For instance in New South Wales, more than nine million tonnes of fly ash are produced annually. The properties of fly ash vary from one place to another, depending upon the quality of coal, type of furnace and combustion process among other factors. Therefore, ash properties can vary within the same boiler in response to the varying electrical energy demand. In a boiler, two types of waste ash are found, namely the bottom ash and fly ash. The heavier bottom ash collects in the bottom part of the boiler, whereas the relatively light fly ash collects mainly on the sides of the boiler. In comparison with bottom ash, fly ash has a very fine texture, and possesses a tendency to 'fly' when disturbed. This paper will highlight the use of fly ash in soil modification.

In the past fly ash has been widely used in concrete industry as an additive to cement. Although waste coal ash has some history of being used as a geotechnical fill in England and North America (e.g. Joshi et al., 1981; Gray and Lin, 1972), only recently it has become popular in geotechnical practice in Australasia. Fly ash together with lime or cement has been employed successfully to stabilize soft Bangkok clay (Balasubramaniam et al., 1990). Injection of lime and fly ash slurries for in-situ soil stabilization has also been a recent development in Canada. Joshi and Nagaraj (1987) have reported that cementation during pozzolanic activity in a fly ash based backfill is likely to reduce long-term lateral pressures against retaining walls.

The usefulness of a waste ash in geotechnical practice depends upon two main properties, firstly its low specific gravity and secondly its potential pozzolanic reactivity. The low specific gravity of fly ash makes it an ideal embankment fill over alluvium and other soft compressible deposits. If properly compacted, higher embankments can be raised using fly ash with less ground settlements. The pozzolanic nature enables self-hardening with time. Therefore, fly ash mixed with a natural soil produces a stronger and stiffer material for construction purposes. The principal constituents that control the behaviour of fly ash are silica (SiO_2), Alumina (Al_2O_3) and lime (CaO). In sufficient moisture, fly ash can provide an array of divalent and trivalent cations that can contribute to the immediate flocculation of clay particles. Consequently, the cohesion of the clay matrix can increase considerably enhancing the short-term stability of otherwise dispersive soils. This paper will elucidate on the effect of a pozzolanic fly ash on the erodibility of a tropical residual soil.

3. FLY ASH BLENDED SOIL MATRIX

Combinations of ash-soil mixtures were investigated to evaluate the effect of fly ash on the behaviour of dispersive soils. A range of compacted specimens were prepared and tested to obtain compaction, consolidation and triaxial characteristics before and after mixing the natural dispersive soil with fly ash.

3.1 Physical Properties

Grain size distribution of natural soil samples revealed a clay size content of about 40%, based on wet sieving and hydrometer tests. The natural moisture content varied from 3% to 5%, while the liquid and plastic limits were estimated at 30% and 15%, respectively. On the basis of Atterberg limits, this residual clay can be classified as a low plastic clay, within the range of other dispersive soils (Sherard et al, 1976). In order to study the effect of fly ash content on the soil plasticity, samples with varying amounts of fly ash were analysed. The influence of ash on the Atterberg limits of the erodible soil is illustrated in Fig. 1. As the ash content is increased, a significant reduction in

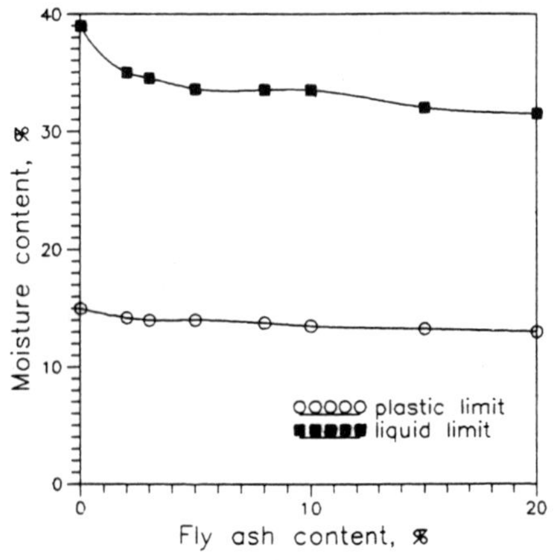

Fig. 1. Effect of fly ash on soil plasticity

liquid and plastic limits is observed. However, increasing the ash content beyond 3% shows only a marginal effect.

3.2 Dispersivity

The extent of soil dispersivity could be assessed on the basis of crumb test interpretation guide (Sherard et al, 1976):

Grade 1: Possible slaking of crumb without any cloudiness.
Grade 2: A bare hint of cloudiness near the surface of crumb.
Grade 3: Recognizable cloud of colloids in suspension.
Grade 4: A considerable reaction with colloidal clouds completely covering the bottom of the beaker.

The natural soil could be classified under the dispersive grade 3 - 4. The degree of cloudiness was obvious after 10 minutes. Subsequently, natural soil samples were mixed with various ash contents from 2% to 10%. The cloudiness diminished with increasing ash content, and above 5% fly ash content, the blended soils could not be regarded as dispersive.

3.3 Erodibility

Pinhole tests were conducted to evaluate the erodibility of the natural and blended soil samples. Cylindrical test specimens (38mm in length) with a 1.0mm diameter hole along the axis were prepared, and subjected to hydraulic heads of 50, 180 and 380 mm. The time-dependent flow through the pinhole was measured and related to the enlargement of the pinhole due to erosion. In order to study the effect of fly ash on the rate of erosion, the natural soil was mixed with fly ash (3% to 20% by weight), and tested under the same conditions. The results are summarized in Fig. 2. The natural soil falls in the D1 or D2 region, which is assigned for highly dispersive soils. With 3% fly ash, some improvement in stability is observed (range D2 - D4), whereas with 5% and 8% fly ash contents, a dramatic reduction in the flow rate is encountered. The specimens containing higher fly ash quantities indicate non-dispersive characteristics (ND1), as illustrated in Fig. 2b. However, increasing the fly ash content beyond 8% caused a greater flow rate. This increased erodibility is attributed to the reduction in overall cohesion due to excessive ash. It may be concluded that 5-8% fly ash seems to be sufficient in stabilizing this highly erodible soil as shown in Fig. 3.

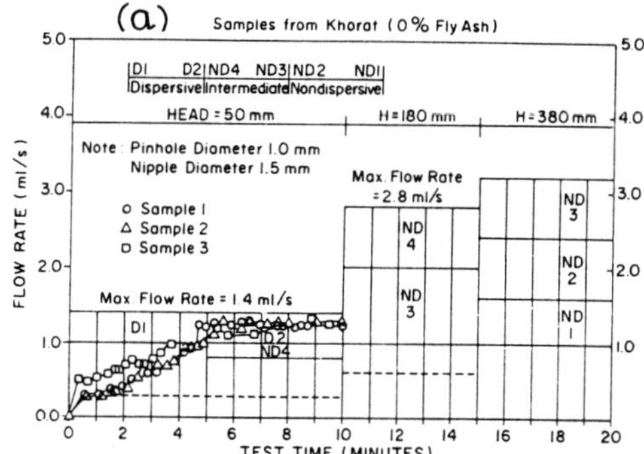

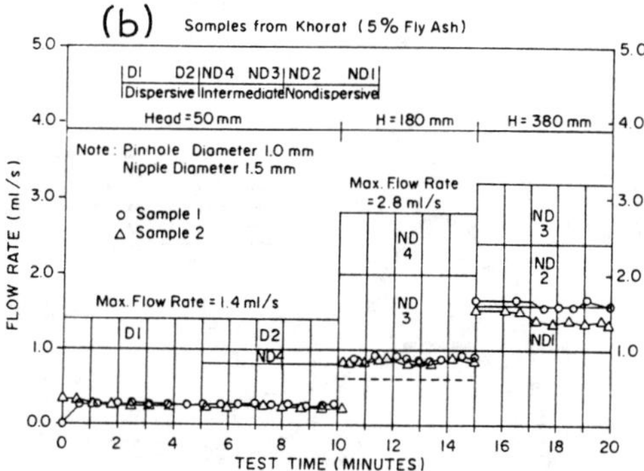

Fig. 2. Rate of Erosion measured in pinhole apparatus: (a) natural soil and (b) 5% fly ash.

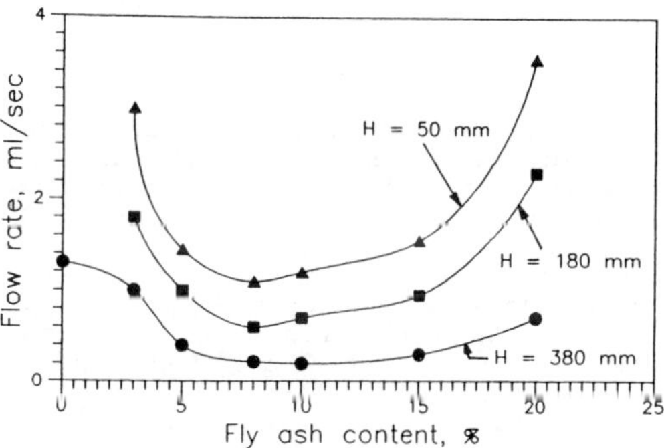

Fig. 3. Effect of fly ash on soil erodibility

3.4 Strength and Frictional Characteristics

A series of blended samples (diameter 52.5 mm and height 105 mm) were tested to failure in unconfined compression. Selected stress-strain responses after

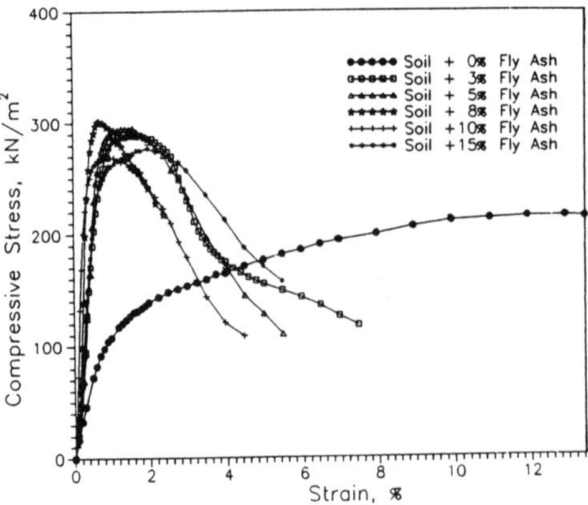

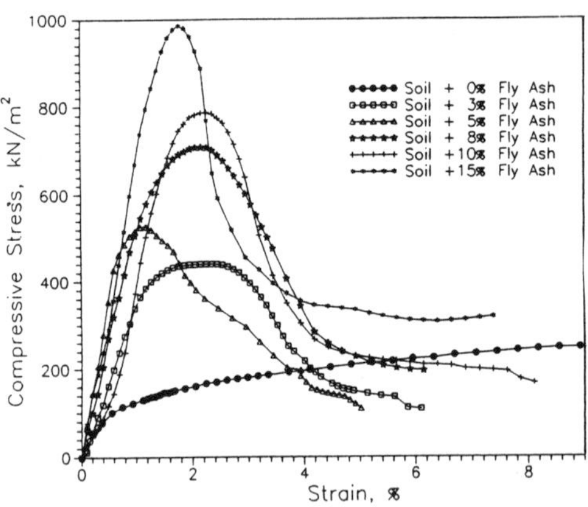

Fig. 4. Uniaxial deformation response after curing for (a) one day and (b) two weeks

1 day and 2 weeks curing are illustrated in Fig. 4. After one day of curing, a significant increase in strength is observed, but the difference due to increasing ash contents is not marked until two weeks of curing. Indraratna et al (1991) have shown that the preliminary increase in strength is due to the initial flocculation of clay particles, whereas the long-term strength is a function of the pozzolanic hardening of the blended soil. It is of importance to note that the increase in strength is also associated with increased brittleness (Fig. 4b). As a result, excessive fly ash contents (greater than 8%) may not be desirable if the stabilized soil is utilized as am embankment or structural fill.

In order to evaluate the influence of fly ash on the immediate strength of treated dispersive soil, CIU triaxial tests were conducted on compacted samples after one day curing. The cell pressure was raised to the required effective consolidation pressure (50 to 150 kPa), and the blended samples were subjected to isotropic consolidation for one day. The effective stress paths (q-p plots) determined from the triaxial compression tests are shown in Fig. 5, which indicate a lightly overconsolidated behaviour of the compacted samples. The degree of overconsolidation seems to increase with the increasing ash content, although at higher consolidation pressures, a normally consolidated response is still maintained. The shear strength envelopes are also plotted for the natural soil as well as for the blended samples. It is obvious that the slope of the envelopes (related to the angle of friction) increases with the increasing ash content as expected.

The addition of fly ash to this residual soil promotes immediate particle flocculation, and in the long-term, the effect on strength is even more encouraging due to internal cementation (Koo, 1990). Consequently, the soil matrix becomes more stable, and the degree of erosion becomes less on saturation. Beyond 5% fly ash content, a significant increase in shear strength is observed, reflecting the favourable effect of this waste material on the behaviour of erodible soils. On the basis of pinhole tests, it was discussed earlier that excessive ash quantities (beyond 8%) leads to greater erosion. In fact, this observation is supported by Fig. 5d, which indicates a lack of cohesion ($q' = 0$) at a fly ash content of 15%.

3.5 Compaction of blended soils

The construction of earth embankments requires favourable compaction properties of a soil. In other words, sufficient dry densities must be achieved by conventional compaction machinery. Natural soil and blended soil specimens were compacted according to the standard Proctor method, and the effect of fly ash content on dry densities was evaluated accordingly. Figure 6 illustrates the standard Proctor curves for fly ash contents varying from 3% to 20%. The maximum dry density and optimum water content of the natural soil were determined as 19 kN/m^3 and 11%, respectively. As the fly ash content is increased to 8% by weight, the mean dry density increases to a maximum of about 20.5%. Further increase in fly ash does not affect the overall dry density, but decreases the optimum moisture content. This suggests that with available field equipment a fly ash blended soil can be compacted to an acceptable degree at a reduced moisture content, in comparison with conventional soils.

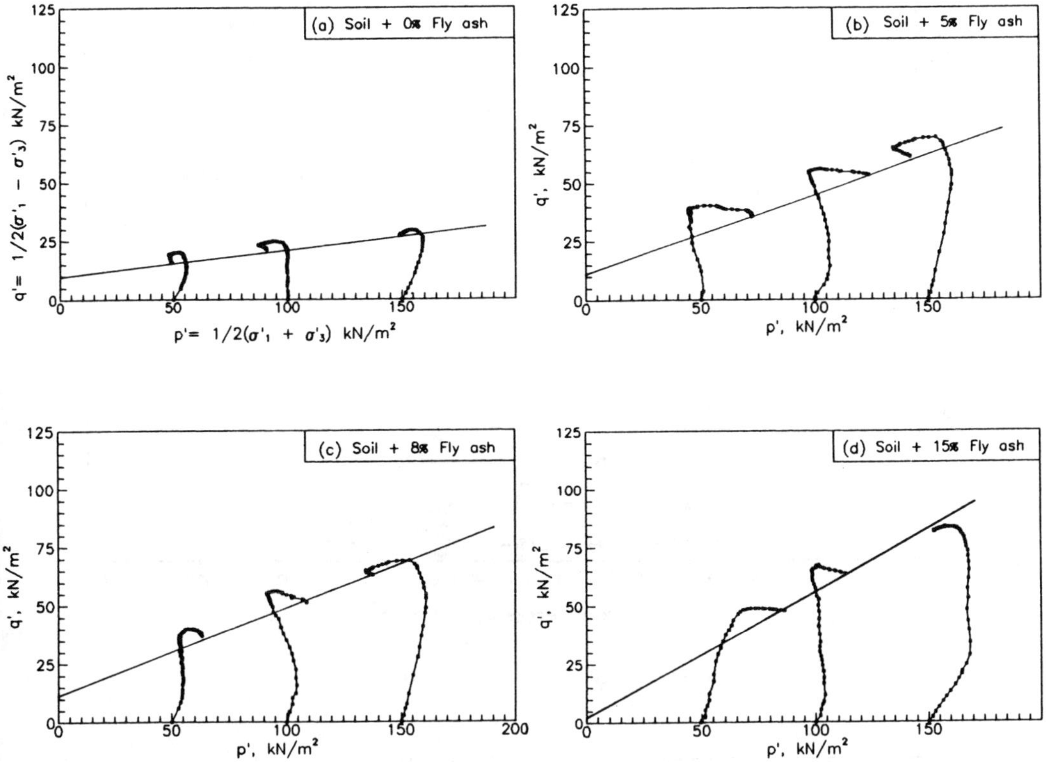

Fig. 5 Triaxial behaviour of natural and blended soil samples

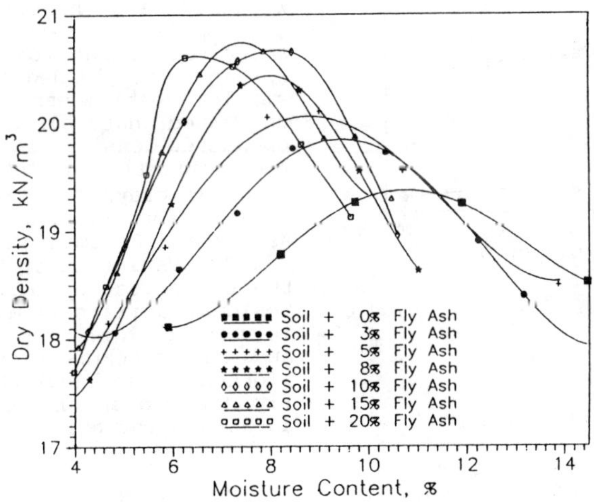

Fig. 6. Effect of fly ash on Proctor compaction

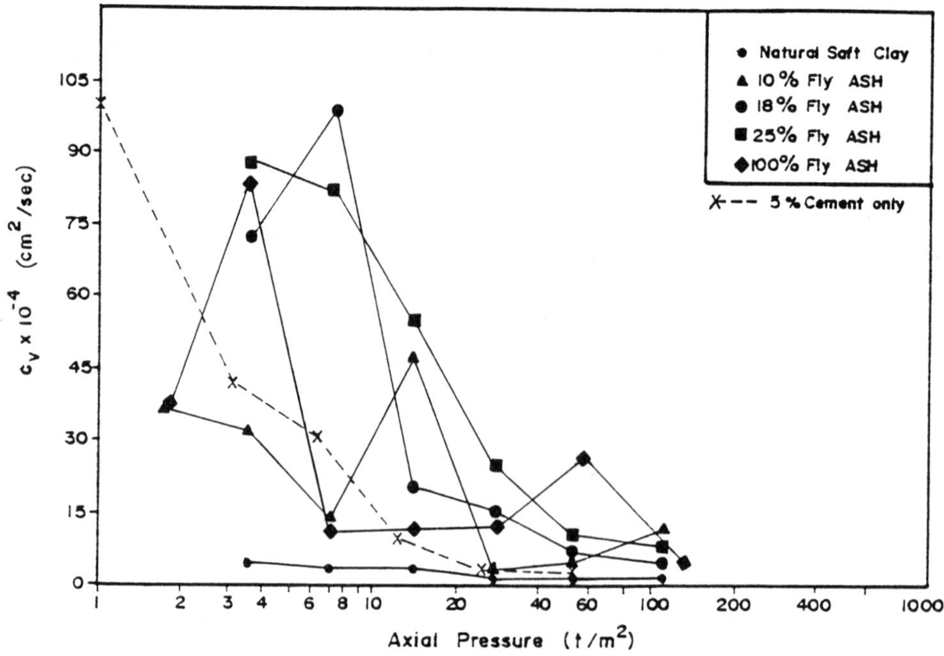

Fig. 7 Consolidation of fly ash blended soft soil

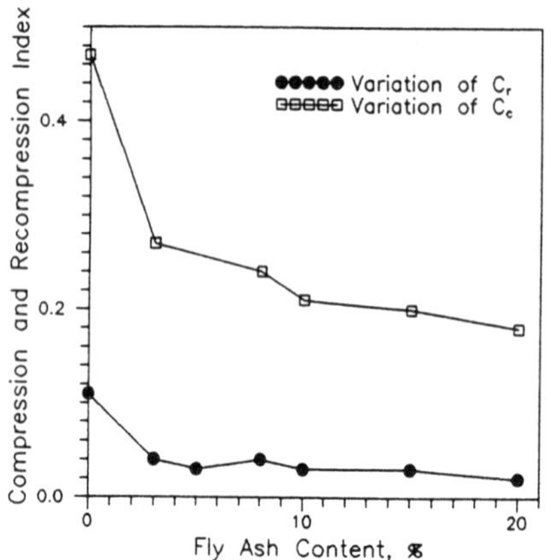

Fig. 8. Influence of fly ash on soil compression

3.6 Consolidation of blended soil

The coefficient of consolidation of fly ash is considerably higher than a natural clay soil under similar dry densities. With sufficient curing and subsequent pozzolanic reactivity, the stiffness of compacted fly ash increases further. Figure 7 illustrates the effect of increasing fly ash content on the coefficient of consolidation of a blended soil containing 5% cement (Khan, 1990). The benefits of increasing the fly ash content is most evident at low axial surcharge pressures. At elevated stress levels, the cementitious bonds are weakened, and the effect of fly ash is not marked. Therefore, it may be anticipated that for a given surcharge loading, the blended matrix will produce smaller settlements in contrast to the untreated natural soil at the same compacted density. In particular, it has been found that mixing fly ash with soft alluvial clays has tremendous benefits in construction practice in Southeast Asia (Balasubramaniam et al., 1990).

The variation of fly ash content on the consolidation behaviour of blended residual soil is summarized in Fig. 8, in which both the compression index (C_c) and the recompression index (C_r) are plotted. It is clear that increasing the fly ash content beyond 10% has little advantage for this particular soil.

4. CONCLUSIONS

Fly ash can be recommended as an effective stabilizing agent for the improvement of erodible and dispersive soils. Use of fly ash in the stabilization of erodible soil terrains can contribute positively in finding some solution to the ever increasing disposal problems of power plant waste. Pozzolanic lignite ash provides an economically attractive alternative to chemical additives such as cement and lime commonly used in soil modification. The chemical constituents of coal ash is responsible for immediate flocculation of dispersed clay particles via cation exchange under sufficient moisture contents. The pozzolanic reactivity (formation of cementitious compounds) of fly ash can ensure long-term stability of erodible and dispersive soils, if blended with appropriate proportions. Nevertheless, excessive amounts of fly ash say beyond 8% by weight has an adverse effect of reducing the overall cohesion of the blended matrix. Therefore, on the basis of this study, a fly ash content of about 5% by weight is recommended to inhibit erosion of unstable silty soils with a quartz content greater than 50%.

REFERENCES

Balasubramaniam, A.S., Buensuesco, B.R., Phien-wej, N. and Bergado, D.T. 1990. Engineering behaviour of lime stabilized soft Bangkok clay. 10th Southeast Asian Geotechnical Conference, Taiwan, pp. 23-28.

Chandra, S. and James, C.G.L. 1984. Improvement of dispersive soils by using different additives. Indian Geotechnical Journal, Vol. 14, No. 3, pp. 202-216.

Cole, B.A., Ratanasen, C., Maikland, P., Ligging, T.B. and Charapun, S. 1977. Dispersive clay in irrigation dams in Thailand. ASTM Special Technical Publication 623, Sherard, J.L. and Decker, R.S. (eds.), pp. 25-41.

Gray, D.H. and Lin, Y.K. 1972. Engineering properties of compacted fly ash. ASCE J. of Soil Mechanics and Foundation Engineering Division, Vol. 98, pp. 360-380.

Indraratna, B., Nutalaya, P., Koo, K.S. and Kuganenthira, N. 1991. Engineering behaviour of a low carbon pozzolanic fly ash and its potential as a construction fill. Canadian Geotechnical Journal, Vol. 28, pp. 542-555.

Joshi, R.C., Natt, G.S. and Wright, P.J. 1981. Soil improvement by lime-fly ash slurry injection. Proc. Tenth Int. Conf. on Soil Mech. & Foundation Engg., Stockholm, 12/27, pp. 707-712.

Joshi, R.C. and Nagaraj, T.S. 1987. Fly ash utilization for soil improvement. Environmental geotechnics and problematic soils and rocks, Balasubramaniam et al. (eds.), Balkema Rotterdam, pp. 15-24.

Khan, M.J. 1990. Effect of fly ash with lime and cement additives on the engineering behaviour of soft Bangkok clay. M.Sc. Thesis No. GT-89-16, Asian Institute of Technology, Bangkok, Thailand.

Kuganenthira, N. 1990. Engineering behaviour of fly ash and its effectiveness in stabilizing dispersive soils. M.Sc. Thesis No. GT-89-7, Asian Institute of Technology, Bangkok, Thailand.

Sherard, J.L., Dunnigan, L.P. and Decker, R.S. 1976. Identification and nature of dispersive soils. ASCE, Geotechnical Division, Vol. 102, GT4 pp. 69-87.

Toth, P.S., Chan, H.T. and Cragg, C.B. 1988. Coal ash as structural fill with special reference to Ontario experience. Canadian Geotechnical Journal, Vol. 25, pp. 694-704.

Environmental Management, Geo-Water & Engineering Aspects, Chowdhury & Sivakumar (eds)
© 1993 Balkema, Rotterdam. ISBN 90 5410 099 0

Evaluation of foundation seepage at Doroodzan earth dam

M.Javan & M.R.Farjood
Shiraz University and Fars Water Authority, Iran

ABSTRACT: An evaluation of foundation seepage at Doroodzan earth dam in Fars Province, Southern Iran is presented. The evaluation is based on the analysis of data from tip and tube-type piezometers collected during the period of 1972-1985. It was found that there is lateral seepage from the left abutment towards the dam foundation. Most of the tube-type and tip-type piezometers installed at foundation are in good condition and are functioning properly. At normal water surface in reservoir, foundation seepage is calculated to be 0.340 CMS. The predicted value of foundation seepage at design stage was 0.2 CMS. At normal water level, about 0.045 CMS of seepage water is removed by pressure relief wells. The cut-off wall is effective in reducing seepage with an efficacy of 80 to 83 percent.

1 INTRODUCTION

Adequate control of seepage at foundation is one of the most important aspects of dam safety. Dam failures have caused many catastrophies and about 23% of the failures are associated with foundation problems ICOLD (1973) and ASCE/USCOLD (1975). Furthermore 10% of all dam failures have been caused by foundation seepage (Clevenger, 1973). Statistical interpretation of dam failures is currently under study by the International Commission of Large Dams (ICOLD) Ad Hoc Committee. Some preliminary results on 142 failures have been obtained. According to these preliminary results, 12 of the dams had percolation problems in their foundations (Serafim et al, 1989).

A certain amount of foundation seepage occurs at all dams but this seepage must be controlled and kept within allowable safe limit. Periodic evaluations including measurement of leakage and piezometric heads are necessary (Ley, 1973). Recently with developments in computers and electronics, real time systems have been developed for data collection in dam instrumentation and managemen (Walz,1989). In the computerized dam surveillance Expert Systems Approach the point at which seepage is observed can be determined at various views of the dam. Depending on the location of the point and data collected, the system provides diagnostic messages and recommendations (Grime et al, 1988).

The purpose of this study was to analyze the piezometric data collected during the 14 years for the purpose of foundation seepage evaluation and control.

Doroodzan dam is located on the Kor river about 100 kilometers North West of Shiraz, Fars Province, South of Iran. The dam was constructed on a 27 meter thick layer of gravel, sand and silt and was completed in 1972. The reservoir has a total capacity of 993 million cubic meters. The dam and reservoir were constructed for irrigation and municipal water supply but other benefits like hydroelectric power and flood control are also derived from the reservoir.

2 SEEPAGE CONTROL CONSIDERATIONS

Doroodzan dam shown in Figs. 1 and 2 is an earthfill structure. The crest of the dam is 700 meters long at an elevation of 1683.5 meters. The width is 6 meters at the top and 375 meters at river bed level. The height of the dam is 56.5 meters from the river bed and the depth of water behind the dam at normal water surface is 49.5 meters. The spillway is of overflow ogee type with crest length of 150 meters designed for a maximum flood of 3100 CMS.

Due to the pervious nature of the foundation material, the following seepage control methods have been used.

1. Cutoff wall of slurry trench type extending to a depth of 27 meters below river bed.

2. Eleven pressure relief wells with diameter of 356 mm penetrating to a depth of 41 meters.

3. A blanket 3 meters thick extending 400 meters upstream.

4. A rock toe drain.

These seepage controls are also shown in Fig. 1.

Thirteen tip-type piezometers were located in the embankment as indicated by the notation, E in Fig. 1. However, the data recorded by these

TUBE TYPE PIEZOMETERS

Design-ation	Distance along Dam Axis	Distance from Dam Axis	Bottom Elevation	Top Elevation
C-1	0+435	R-15.0	1612.0	No Data
C-2	0+595	R-15.0	1612.0	1678.70
C-3	0+405	R-65.0	1585.0	1659.83
C-4	0+495	R-65.0	1585.0	1659.40
C-5	0+585	R-65.0	1585.0	1659.54
C-6	0+360	R-215.5	1585.0	1627.22
C-7	0+448	R-215.5	1585.0	No Data
C-8	0+536	R-215.5	1585.0	No Data
C-9	0+580	R-215.5	1585.0	No Data

TIP TYPE PIEZOMETERS (Embankment)

Design-ation	Distance along Dam Axis	Distance from Dam Axis	Bottom Elevation
E-27	0+421	L-65	1647.00
E-28	0+421	L-56	1636.08
E-29	0+421	L-35	1646.70
E-30	0+421	L-31	1636.20
E-31	0+421	L-10	1646.45
E-32	0+421	L-6	1636.30
E-33	0+421	R-10	1646.25
E-34	0+421	R-14	1636.47
E-35	0+421	R-35	1646.00
E-36	0+421	R-39	1636.60
E-37	0+421	R-65	1645.70
E-38	0+421	R-69	1636.80
E-39	0+421	R-99	1636.90

TIP TYPE PIEZOMETER (Foundation)

Design-ation	Distance along Dam Axis	Distance from Dam Axis	Bottom Elevation
F-1	0+340	L-10.5	1617.00
F-2	0+395	L-10.5	1612.00
F-3	0+450	L-18.0	1612.00
F-4	0+510	L-18.0	1612.08
F-5	0+570	L-18.0	1611.33
F-6	0+625	L-18.0	1616.88
F-7	0+340	L-3.0	1617.20
F-8	0+390	L-3.0	1612.20
F-9	0+440	L-16.0	1597.20
F-10	0+450	L-16.0	1612.00
F-11	0+510	L-16.0	1611.93
F-12	0+525	L-16.0	1596.44
F-13	0+570	L-16.0	1611.94
F-14	0+525	L-16.0	1616.86
F-15	0+440	L-60.0	1596.00
F-16	0+520	L-60.0	1597.01

Fig. 1 Specifications of different type piezometers and location of embankment piezometers

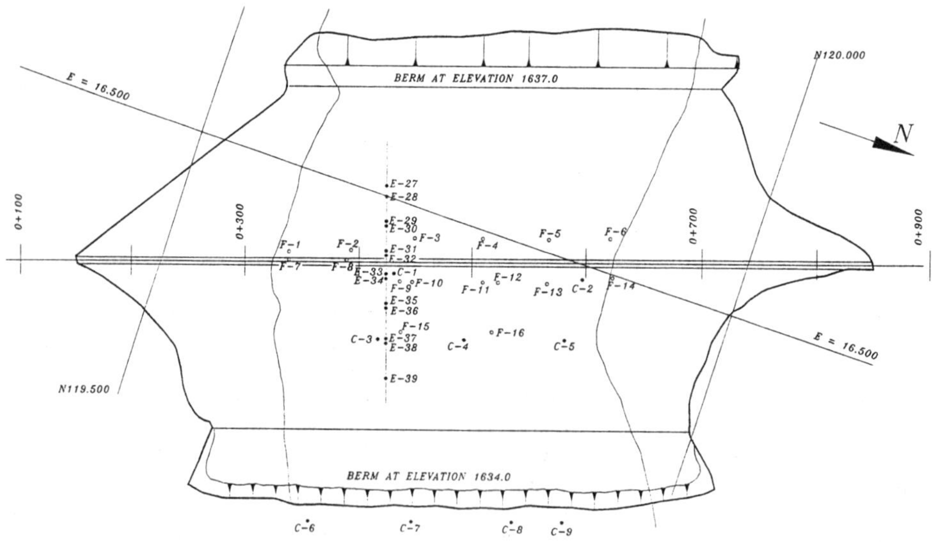

Fig. 2 Plan view of the dam, showing locations of the piezometers

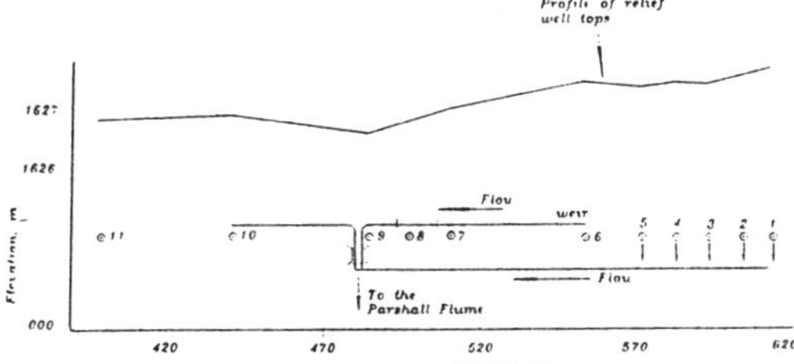

Fig. 3 Locations of relief wells and their corresponding elevations

instruments were not used in the analysis presented in this paper.

To measure pore pressure at foundation level, 16 tip-type and 9 tube-type piezometers were installed in the dam. Figures 1 and 2 show the location and specification of each piezometer.

The foundation tip-type piezometers are indicated by the notation F and the Casasgrande by the notation C in the Figures.

The flow from eleven relief wells were measured individually using V-notches and volumetric methods. Fig. 3 shows the location of relief wells and weirs; there was no flow from relief well No. 11. Daily discharge and reservoir surface level measurements began in September 1984 and continued for a period of 13 months.

There were some data collected from the tip-type piezometers for the period, 1972 to 1985. These data were processed and analyzed for foundation seepage evaluation. No data was available on the tube type piezometers therefore, in November 1984, data collection on piezometers C1,C2,C3,C4 and C7 began for the first time. Measurements of piezometric head at C5 and C6 was not possible due to clogging problems and piezometers C8 and C9 were never found.

A 600 mm Parshall Flume, installed at the time of construction, measures the total seepage flow through the dam, including relief wells, chimney drain, horizontal drain and seepage flow from the left abutment. Data from this total flow measurement system enables verification of the seepage calculations based on piezometer readings.

3 ANALYSIS OF DATA

Analysis of the data from the tip-type piezometers showed that 12 of the 16 installed were functioning properly;piezometers F3,F5,F7 and F8

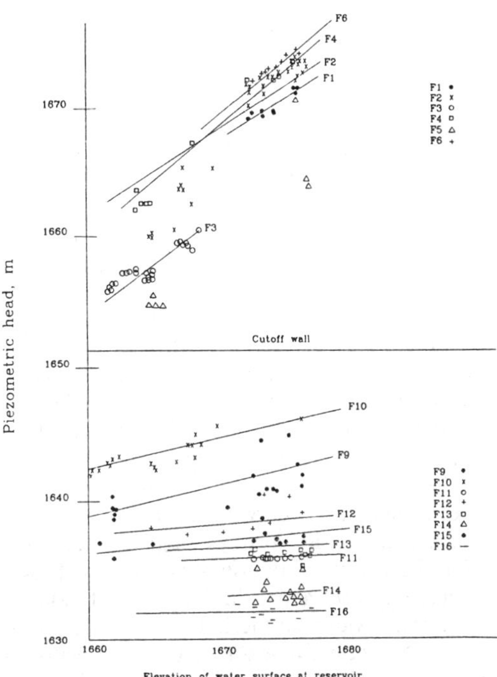

Fig. 4 Variation of piezometric pressure with reservoir level fluctuations

were malfunctioning. This had already been confirmed during a study made on the possibilities of increasing reservoir storage (Taleghani-Daftari, 1979).

The piezometric heads versus water surface level in the reservoir is shown in Fig. 4.

The large pressure head difference between upstream and downstream of the cutoff wall shows the effectiveness of this wall in reducing uplift pressure. Data from F6 and F14 piezometers were used to determine efficacy of the cutoff wall. For the highest piezometric head difference in 1973, the efficacy was calculated to be 83.5% while for the lowest difference in head during 1974 the efficacy was 80%. The cut off wall efficacy, E is calculated as follows.

$$E=\frac{P_1-P_2}{P_0} \qquad (1)$$

where P1 = piezometric head at F6, upstream of cutoff wall, (m), P2 = piezometric head at F14, downstream of cutoff wall, (m) and P0 = depth of water in reservoir, (m).

$$Efficacy=\frac{1674.2-1632.7}{49.7}=83.5\% \quad \text{(in 1973)}$$

$$Efficacy=\frac{1672.4-1635.4}{46.2}=80\% \quad \text{(in 1974)}$$

It can be observed from Fig. 4 that the variations of piezometric head with reservoir water level is approximately linear. The gradients of the pressure versus reservoir head plots are greater for locations upstream of the cut-off wall. This is due to the fact that pore pressure is relieved in the downstream section. the effect is reduced as one moves upstream. A comparison between the gradients of F10 and F16 piezometers reveals this effect. Similar observations were made on Nerskogen dam in Norway (Tueit et al, 1985).

Fig. 5 shows the variation of piezometric head at foundation level along the longitudinal dam axis at normal water surface. This indicates, quite clearly, the effectiveness of the cut-off walls and the relief wells. This is an analysis similar to the one used for the Hitakura dam in Japan (Sugimura, 1985).

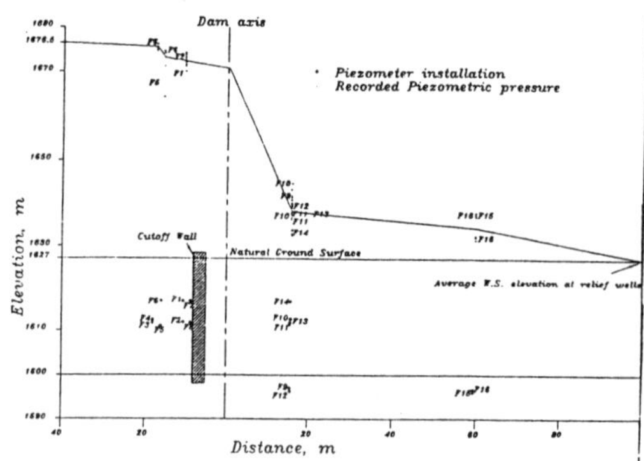

Fig. 5 Piezometric pressure variations along longitudinal dam axis at normal reservoir level

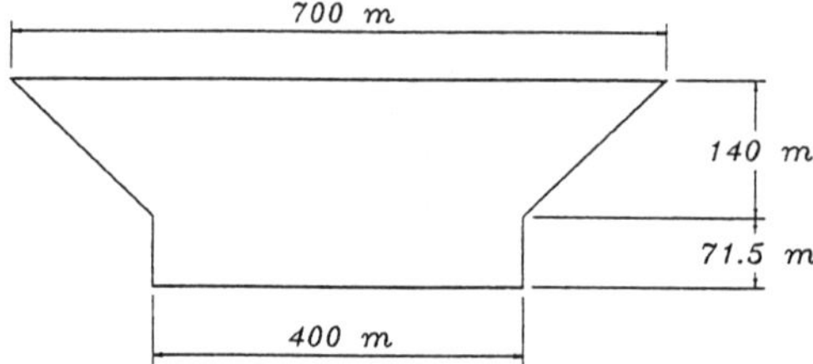

Fig. 6 Area through which foundation seepage occurs

4 SEEPAGE CALCULATIONS

The weighted permeability, K in a 14.8 m thick layer in which piezometers F9 and C1 are installed is 1.47×10^{-4}cm/s. The area, A, under the foundation, through which seepage occurs is considered to be 105600 m^2 as shown in Fig. 6. distance between the cut-off wall and relief well axis is 211.5 m (i.e. 114 + 71.5). Since normal water surface was not attained during this study, the foundation seepage at this water level was calculated based on linear interpolation of piezometric heads of piezometer F9. The foundation seepage rates for different water levels calculated using Darcy equation are shown in Table 1.

It is interesting to compare these values with the predicted value of foundation seepage at the time of design, which was 0.2 m^3/s (Personal Communication). The excess is probably due to the abutment seepage.

Table 1. The foundation seepage rates at different water levels

Reservoir water level,m	Seepage rate ,CMS
1660.78	0.175
1664.22	0.211
1676.50 (N.W.S)	0.340

5 FLOW DIRECTION

Piezometers C1,C3 and F9 in Fig. 7 have approximately the same alignment in a direction perpendicular to the dam axis. As it is seen from this figure, the piezometric head is decreasing in the upward direction. This shows that the direction of flow is upward towards the foundation. Fig. 8 shows the direction of flow at elevation of 1612 m. The difference in piezometric heads C1 and C2 indicates that water is flowing from left abutment along dam axis toward the right abutment. This might confirm the seepage of reservoir water through the limestone formation of the left abutment. In a memorandum of 1969 the designers believed that near the abutments

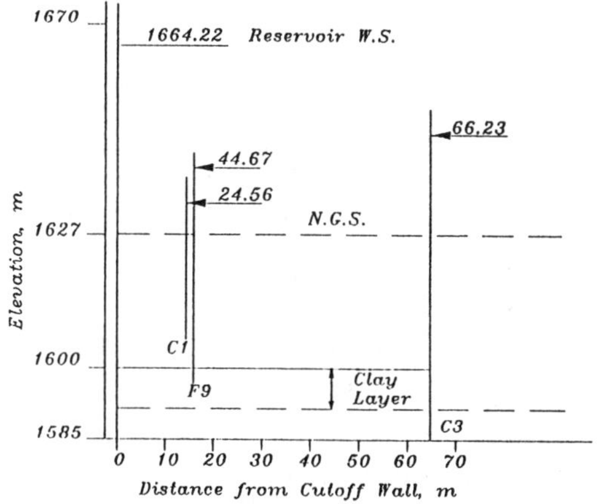

Fig. 7 Heads of C1,F9 and C3 piezometers at different depths of foundation for reservoir W.S. elevation of 1664.22 m.

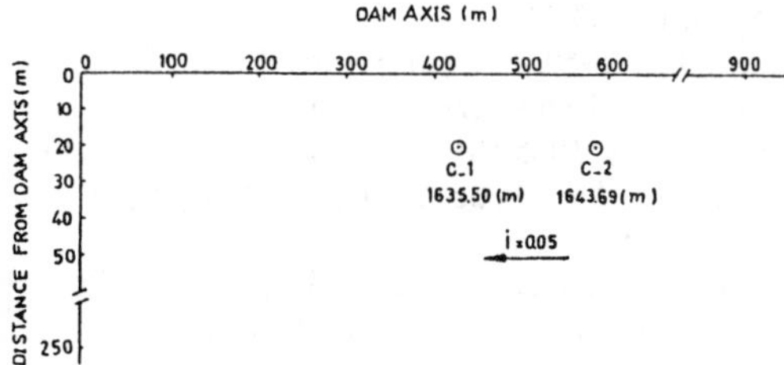

Fig. 8 Flow direction at elevation of 1612 m.

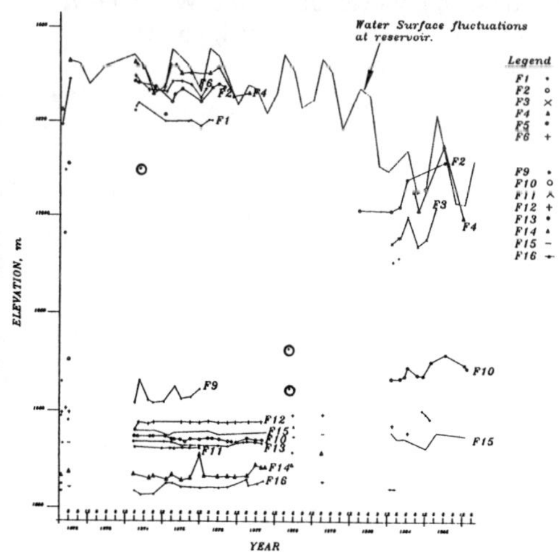

Fig. 9 Changes in relief well discharge with water surface level in the reservoir

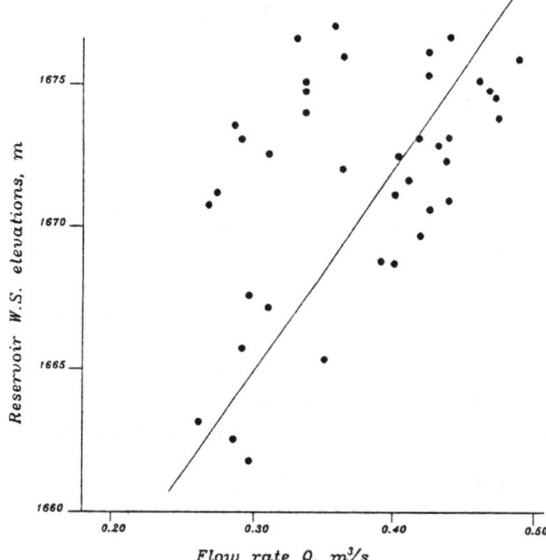

Fig. 10 Variation of discahrge through the Parshall Flume (measured total seepage rate) with water surface levels at the reservoir

greater quantities of seepage could be anticipated (Sidor, 1969). The following reasons have been given for the higher than normal seepage of the abutments.

1. The limestone abutments were very pervious as indicated by the large grout takes.

2. The existence of compressible soft clays.

The relationship between total relief discharge and reservoir water level is shown in Fig. 9. In a study made on Hitokura dam, this relation was found to be linear (Sugimura, 1985), but in this study this type of relationship was not found. This might be due to the effect of temperature on the viscosity and consequently in Fig. 9 the circles indicate readings taken for periods of low temperature during winter. The relief wells reduce pore pressures of the toe area down to level of 1626.0 m. This elevation corresponds to the elevation of the invert of horizontal drains. The total discharge of relief wells is measured to be 0.045 CMS at reservoir water level of 1664.0 m. The remaining seepage water finds its way downstream under the foundation.

6 TOTAL SEEPAGE

The total seepage discharge through the dam including embankment, foundation and left abutment is measured by a 600 mm concrete Parshall Flume. The discharges for the period of 1973 to 1985 are shown in Fig. 10. There are discrepancies in observations due to errors in the available data. One error might be due to the inclusion of runoff water in the measurements. In a more precise analysis runoff water from precipitation must have been accounted for. Such precautions had been taken into account in the seepage study of Bell Canyon dam (Connel et al, 1985). However, according to Fig 10, there is an increase in the total discharge with water level rise in the lake. The total discharge from Fig. 10 at reservoir water level of 1664.0 m is found to be .290 CMS. Excluding the discharge from relief wells (0.045 CMS). The seepage from the left abutment is estimated to be 0.245 CMS.

7 CONCLUSIONS

Based on the data collected from foundation piezometers which were operational during this study, the following conclusions were obtained.

1. Seepage through the foundation was calculated to be 0.34 CMS at normal water surface in the reservoir. This value had been predicted to be 0.2 CMS at design stage.
2. About 0.045 CMS of seepage water is removed by relief wells. These wells reduce uplift pressure down to a level of 1627.0 m.
3. There is seepage from left abutment towards foundation along cut-off wall.
4. At reservoir water level of 1664.0 m an average flow of 0.290 CMS passes through the Parshall Flume of which 0.245 CMS is the seepage water from left abutment.
5. Most of the tube-type and tip-type foundation piezometers are in good condition except C5,C6,C8,C9,F3,F5,F7 and F8.

The following recommendations are made:

1. Discharge through relief wells should be measured regularly at two weeks interval. Additional discharge measurements should be taken during floods and after earthquakes.
2. A more detailed subsurface investigation on left abutment is needed before grouting.
3. Since most of the tube-type piezometers are operational, data from these piezometers should be collected more frequently during the year.

REFERENCES

ASCE/USCOLD 1975. Lessons from dam incidents, U.S.A. ASCE. N. Y.

Clevenger, W. A. 1973. When is foundation seepage unsafe. Proc. ASCE, Engineering Foundation Conference, Pacific Grove, Ca. pp. 570-583.

Connel, D. H. et al 1985. Grouting solves complex problems at Bell Canyon dam. Proc. 15th ICOLD Congress, Lausanne, Suisse.

Grime, D., T. Phillips and M. Waage 1988. The Kelly system: On-line expertise. Hydro Review, August pp. 36-46.

Junstin, J.B, and N. C. Courtney 1966. Design reports on Doroodzan Multipurpose Projects.

Ley, J.,E. 1973. Foundation of existing dams seepage control. Proc. ASCE, Engineering Foundation Conf.,Pacific Grove.pp. 584- 608.

Serafim, J. L. and J. M. Coutinho-Rodrigues 1989. Statistics of dam failures: a preliminary report. Int. Water Power and dam construction, April, pp. 30-34.

Sidor, J.R. 1969. Memorandum, Justin, J. B. consulting Engineers, Philadelphia, U.S.A.

Sugimura, Y. 1985 Remedial measure for uplift at Hitokura dam. Proc. 15th ICOLD Congress, Lausanne, Vol. 3.

Taleghani-Daftari Consulting Engineering Co. 1979. Final report on Doroodzan storage.

Tueit, T. et al, 1985. Foundation treatment at Nerskogen dam. Proc. 15th ICOLD Congress,Lausanne, Suisse, Vol.3.

Walz, A. H. 1989 Automated data management for dam safety evaluation. Int. Water Power and Dam Construction, April, pp. 23-25.

Environmental Management, Geo-Water & Engineering Aspects, Chowdhury & Sivakumar (eds)
© 1993 Balkema, Rotterdam. ISBN 90 5410 099 0

Road construction in hilly areas versus environment degradation

Prabhat Kumar & N.S. Bhal
Structural Engineering Division, Central Building Research Institute, Roorkee, India

ABSTRACT : The cycle of gradual environment degradation which commences at slope cutting and may end up at a natural disaster is described in this paper. This form of cycle is already in operation as substantiated by the statistical estimates available in the published literature. Special efforts are needed to slow down and ultimately reverse this cycle of gradual destruction. This paper contains a proposal to construct debris icelands along the hill roads to quickly, economically and harmlessly dispose debris produced by slope cutting. A large number of construction strategies are possible to implement the proposed concept. One such implementation strategy is also described herein.

1. INTRODUCTION

The hilly regions in India are important from economic as well as strategic considerations. Strategically, the frontiers of the country have to be defended by the armed forces against any enemy aggression. Economically, hilly regions contribute significantly to national wealth. Most of the mighty Indian rivers originate in the hills. These have great irrigation and hydropower generation potential. A large number of schemes to utilise this potential are in operation. Besides, hilly regions are being developed for promoting tourism and sports. Several shrines of religious importance are also located in hills and are visited by large number of people. It is not surprising that developmental activities at great pace are taking place in the hilly regions. The population explosion adds to this pressure for making more land available for cultivation.

Most of these activities involve deforestation and cutting of hill slopes. The latter operation produces debris which ought to be disposed off quickly, economically and harmlessly. The debris disposal methods which have so far been practiced do it quickly and, perhaps, economically but definitely not harmlessly. Consequently, it has led to serious environmental degradation and ground failure problems. This paper presents an analysis of the cycle of events which commence with debris disposal and may end up at serious natural disasters. The possibilities have to be explored so that hill area development could continue but without causing environment degradation. The proposals put forward herein are conceptual because a field trial has not been done. Also, there may be many more ways to achieve the above objective.

2. CYCLE OF GRADUAL DESTRUCTION

Fig. 1 shows a lush green valley which is full of natural wealth. But soon with the advent of developmental activities, what remain of it is a network of roads and barren land (Fig. 2). The transition is caused primarily by the debris which is produced in slope cutting. This debris is usually allowed to roll down the hill slopes. On its way this debris tramples the vegetation, kills it and buries it under its weight. A large part of this debris reaches the streams in valley and is swept away. The deforestation may also be done directly at a bigger scale for several other reasons.

The soil on the hill slopes without vegetation cover is subjected to wind and rain erosion. The eroded soil and the debris add to the sediment load of the streams in the valley. The sediment gets deposited in the reservoirs behind dams, in the river bed and, also in the rich and fertile farmland of the plains

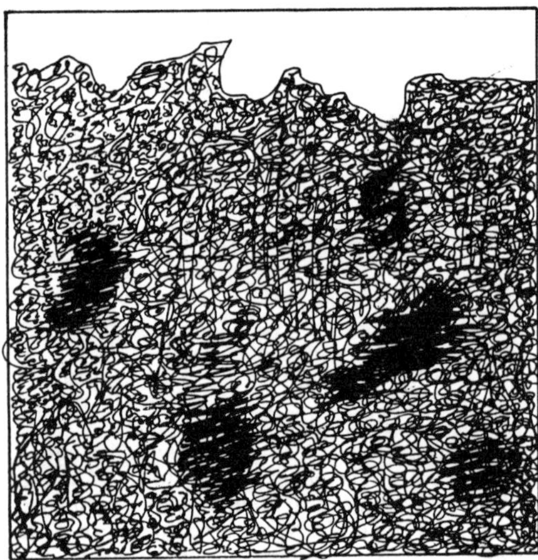

Fig. 1 A virgin lush green valley

Fig. 2 Valley subjected to gradual destruction

after flood waters recede. Most of the rain water, the flow of which was slowed down by the vegetation now reaches the stream straightaway. The floods become more frequent and the flood peaks become higher leading to more soil erosion. Higher flood peaks tend to lubricate, saturate and soften more hill slopes.

The gradually altered hill slopes without the reinforcement and protection of vegetation become open to ground failures. These failures accelerate the cycle of erosion, sedimentation and more frequent floods with higher peaks. Thus, this cycle becomes more and more destructive with each passing year. Sometimes, the debris of ground failures form landslide dams across a stream, the sudden bursting of which causes flash floods and devastation. Some such major events in the Himalayan region are listed by Bhandari, 1987.

3. EVIDENCES OF CYCLE OF DESTRUCTION

Some statistical data is available in the published literature to describe magnitude of the problem caused by slope cutting in the Himalayan region and to suggest that the cycle of gradual destruction is well and truely operational. The exclusive features of himalayan region and hurdles against environment regeneration have been described by Kumar and Bhal (1992).

It is estimated (Valdiya, 1985) that there is a network of 44000 kilometer long roads in Himalayas. Through this construction 2650E06 cubic meters volume debris was produced. In addition, each year every kilometer of road is estimated to produce about 550 cubic meter of debris through landslide and rockfall.

The slopes without vegetation cover could not be expected to hold soil cover together and widespread erosion is a natural consequence (section 2). It is estimated that (Bhandari, 1987) nature takes about 1000 years to produce a few centimeters of top soil but destablizing forces of nature in the Himalaya wipe away millions of cubic meter of top soil in just a second. The present rate of erosion in catchment area of Himalayan rivers is estimated to be upwards of 1 mm per year (Valdiya, 1985). The amount of sediment in the rivers is estimated at 16.5 hectare-meter per hundred square kilometer of the catchment area per year (Valdiya, 1985). Satellite photographs taken in 1974 reveal that eroded debris carried by Himalayan rivers has created a new land mass about 50,000 square kilometer in area extending to about 700 square kilometer into the sea (Bhandari, 1987).

These figures are alarming enough. Lampe (1982) voices his concern as follows - "All efforts should focus on human needs and securing of a means of existence. In nearly all parts of the world the ecological balance has already been destroyed through uncontrolled growth of population. It would be in vain to try to restore this balance in

Fig. 3 Hill roads and debris icelands

near terms. Yet this must be ultimate aim,using reasonable measure to correct as far as possible the already existing damage, and to prevent this new destruction".

These words are very important in that these suggest that the developmental process must go on. The strategy should be to try to slow down the cycle of gradual destruction and to do nothing to accelerate it any further.

4. SOLUTION STRATEGY

Theoretically, it is possible to construct hill roads in such a way that cut and fill are exactly equal; so that no debris is produced at all. But in actual practice, this shall require high level planning and control during execution. It is usually not possible to achieve this target because of hostile working conditions and other resource constraints, particularly, the time. However, a compromise solution is always possible which is schematically illustrated in Fig. 3. The advantages of this scheme are thought to be as follows,

1. The debris produced neither destroys vegetation nor it falls into the stream.
2. The debris icelands may be utilized for emergency stops, parking areas and recreation centres.
3. The debris deposited in the icelands is always available for any recycling at a latter time.
4. The weight of the debris icelands add to the stability of ground on which it rests. This statement is based on the general observation regarding slope stability.

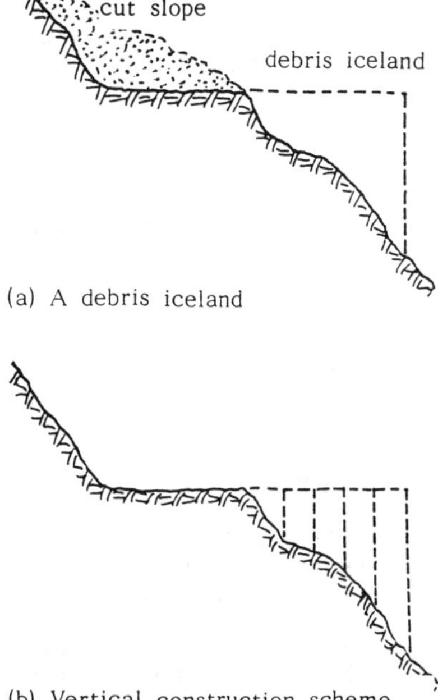

(a) A debris iceland

(b) Vertical construction scheme

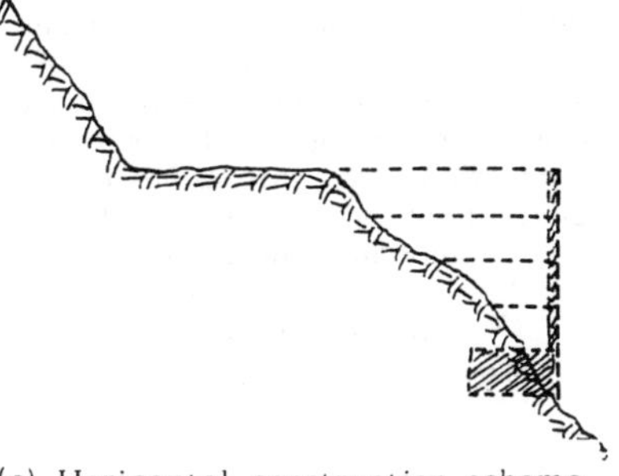

(c) Horizontal construction scheme

Fig. 4 Schemes of debris iceland construction

This scheme ensures that the rolling of debris into the valley is reduced to accidental spills only. It will be necessary to ensure that the freedom of allowing accidental spills is not misused.

5. CONSTRUCTION STRATEGY

In the construction of such icelands, economy is the biggest consideration. Although, these icelands are expected to help in environment protection without stopping the developmental activities, the cost of which may well offset additional cost of iceland construction : It is unlikely to be seen by the planners and builders in that perspective. Therefore, if the proposed strategy is acceptable, to make the construction of debris icelands economically viable is an open challange to engineering profession. The actual construction strategy shall depend on the following factors,

1. Accessibility of the site below cut slope.
2. Terrain characteristics
3. Weather conditions

6. A CONSTRUCTION PROPOSAL

The debris iceland (Fig. 4a) is essentially a gravity retaining wall which requires bearing for its toe for stability.

A scheme of vertical staged construction in given in Fig. 5 which may involve some pile driving and anchoring. On the other hand, the scheme of horizontal stages is simpler. The precast ferrocement units may be used as formwork for various lifts. This formwork is meant to be left in its place to become an integral part of the iceland. The principle to be used in the construction of various lifts is that the resultant of all forces must not fall outside the middle third region at all stages. The stability of the lower most lift is of vital importance and may have to be secured by anchoring, Fig. 6.

The above schemes of construction are schematic only. In actial construction some modifications may have to be introduced in view of practical difficulties.

REFERENCES

The opinions expressed in this paper are of these authors.

Bhandari, R.K., 1987
Himalaya and Strategy for development, Indian Geotechnical Journal, 17(1):1-78.

Kumar, P. and Bhal, N.S., 1992a.
Mitigation of ground failures in third world contries with special reference to India, 2nd US-Asia Conference on Engineering for Mitigating Natural Hazard Damage, Indonesia Disaster Mitigation Centre, Jakarta, June 1992.

Kumar, P. and Bhal, N.S., 1992b.
Eqrthquake hazard impact reduction in countries like India, 7th International Conference on Earthquake Prognosis, Asian Institute of Technology, Bangkok, Thailand, September 1992.

Lampe, K.J., 1992.
Rural development in mountain areas:why progress is so difficult, Mountain Research and Development, 3(2):125-129.

Valdia, K.S., 1985.
Accelerated erosion and landslide prone zones in the Central Himalayan Region, Proceedings of Seminar on Environmental Regeneration in Himalaya-Concept and Strategies, J.S. Singh (ed.), 12-37.

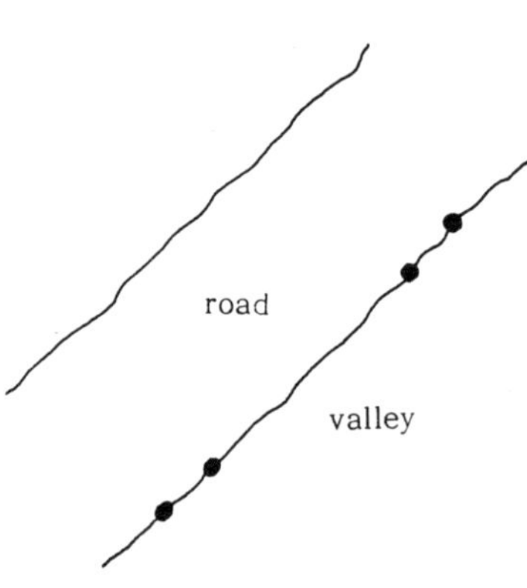

(a) Erect foundation blocks

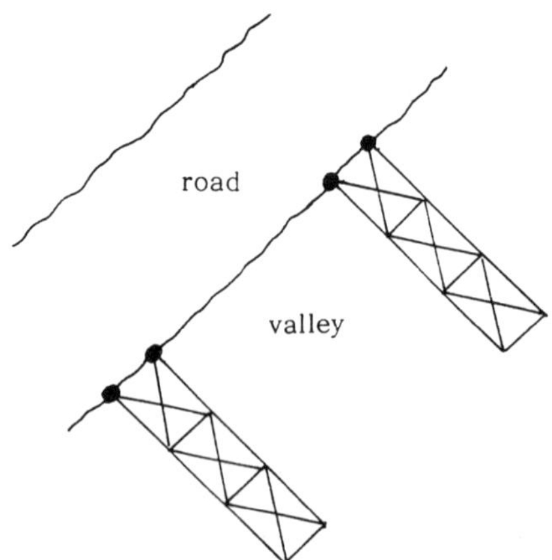

(b) Construct trussed cantilever beam using road as work plateform

Fig. 5 (a) and (b) Details of vertical construction

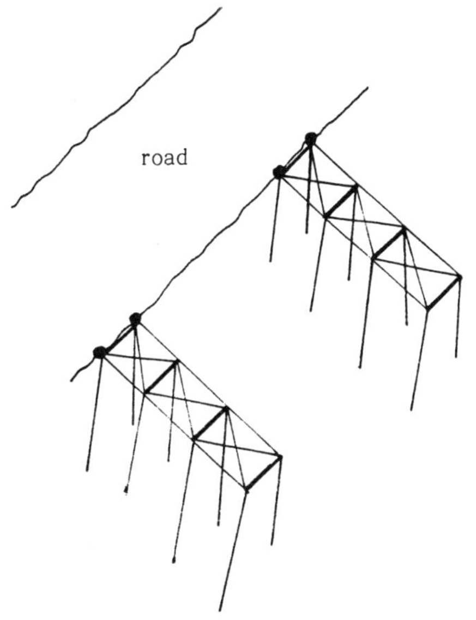

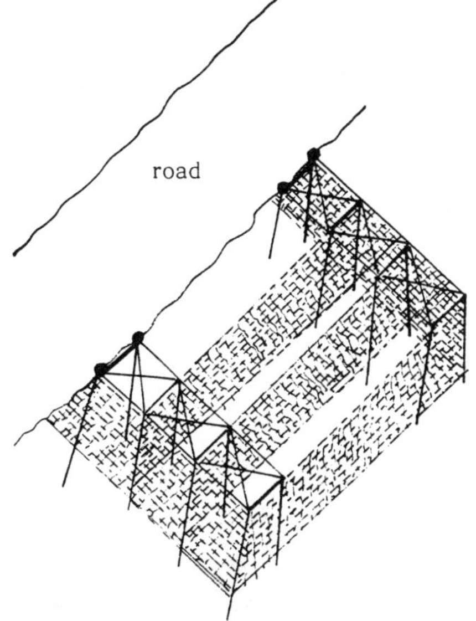

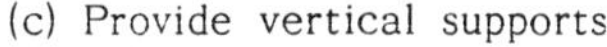
(c) Provide vertical supports

(d) Make enclosures of wire mesh

(e) Fill enclosures with debris

(f) Extend construction by using already constructed iceland as work plateform. Extend trussed cantilever beam and repeat steps c, d and e.

Fig. 5 (c), (d), (e) and (f) Details of vertical construction

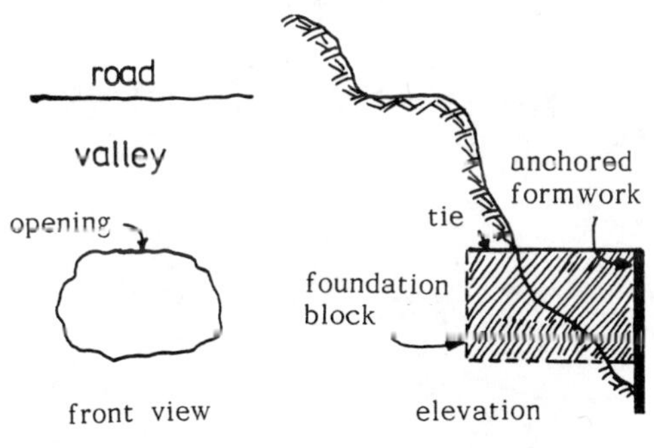

(a) Build foundation of grouted debris

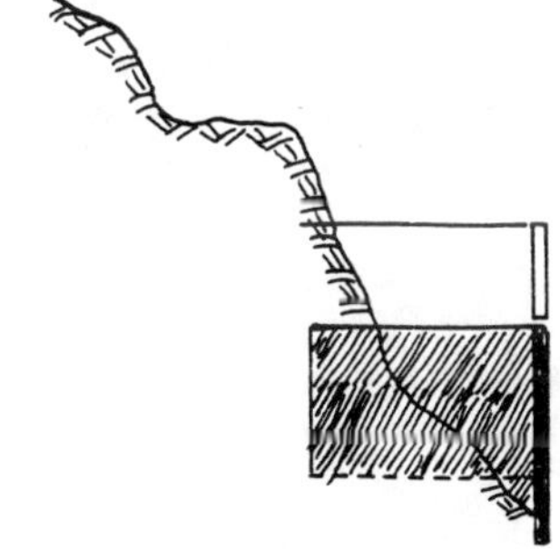

(b) Erect interlocking ferrocement formwork to make an enclosure

(c) Fill with debris and continue with next lift

Fig. 6 Detailed horizontal construction of debris iceland

Environmental Management, Geo-Water & Engineering Aspects, Chowdhury & Sivakumar (eds)
© 1993 Balkema, Rotterdam. ISBN 90 5410 099 0

A simplified method for estimating the land subsidence due to dewatering

C.J. Lee, J. Sheng & C.Y. Chang
National Central University, Taiwan

ABSTRACT: A simplified method for estimating the land subsidence due to dewatering is proposed in this paper. The vertical strain of the standard soil under various groundwater conditions is constructed at first. An influence factor is defined as the ratio of the vertical strain of a given soil and that of the standard soil. The relations of the influence factor and soil properties are also shown with figures. Hence the subsidence can be easily found by examining a series of figures. The subsidence calculated with this method show a satisfactory agreement with the field data.

1 INTRODUCTION

Along the southwestern coastal area of Taiwan, a large amount of fresh water is needed for the thousands of fishery farms every day. The only economic way for the fishery farmers to get enough fresh water is to take the groundwater locally by drilling wells. Following the expansion of fishery farms, wells increase rapidly. Water tables are permanently and substantially lowered due to over-withdrawal of the groundwater. The lowest groundwater level of 20m below sea level is found in some area. The overuse of groundwater results in land subsidence. The historical data of elevations of several bench markers in Pingtung coastal area shows that the maximum land subsidence reaches 2.5m. The regional land subsidence causes many serious environmental problems (Lin, 1986). Terzaghi's one-dimensional consolidation theory is widely used to estimate the land subsidence. However, since both the saturated and unsaturated deformation behavior are unavoidablly involved in this problem, a more delicated analysis is required. In this paper, the one-dimensional finite strain consolidation theory (Gibson, 1981) is employed to model the behavior of land subsidence due to dewatering. The design engineers and the administrators are more concerned about what the total settlement would be than how the consolidation proceeds. A quick, convenient and reliable method for estimating the total settlement of different types of soils and various groundwater conditions is presented.

2 STRESS STATES IN UNSATURATED ZONE

The one-dimensional finite strain consolidation equation for saturated and unsaturated soils is derived under the similar assumptions in Terzaghi's theory with an addition that the air phase in all the unsaturated pores are connected to the atmosphere (Chen, 1989). This implies that the pore air pressure is equal to the atmospheric pressure, i.e., $u_a = 0$. The water in the soil voids is gradually squeezed out and is partly displaced by the air during consolidation. The effective stress, σ', at a depth h below the ground surface is

$$\sigma' = \sigma - Su_w \tag{1}$$

and

$$u_w = \gamma_w(h - Z_i) \tag{2}$$

where σ = the total stress, S = the degree of saturation, u_w = the pore water pressure, γ_w = the unit weight of water, and Z_i = the distance between the groundwater table and the ground surface. The total stress σ in Eq.(1) can be claculated by the equation below:

$$\sigma = \frac{(G_s + Se)\gamma_w}{1+e}h \tag{3}$$

when the void ratio e, S, and the specific gravity G_s of soil are given. Eq.1 may not be absolutely justifiable for all the cases, nevertheless, it is reasonable for a first attempt to describe the stress states in the unsaturated zone during dewatering.
The degree of saturation of a soil element at a distance $(Z_i - h)$ above the groundwater table can be obtained with the empirical water retention equation (van Genuchten, 1980):

$$\frac{S - S_r}{1 - S_r} = [1 + (\alpha(Z_i - h)^n]^{\frac{1}{n} - 1} \qquad (4)$$

in which S_r is the residual degree of saturation, and α and n are material constants. The stress states at a given depth can be calculated after S is determined by Eq.(4). A normally consolidated soil exhibits a linear $e - logp$ relationship in which the slope is defined as the compression index, C_c. For both C_c and one pair of (σ', e) are given, the initial void ratio e_0 at any depth (or node) of a soil deposit can be uniquely determined after the corresponding effective stresses are calculated. When there is a change of the groundwater table ΔZ, the pore water pressure becomes

$$u_w = \gamma_w(h - Z_i - \Delta Z)$$

The stress states at any depth after dewatering are determined by Eqs.(1), (3) and (4). The final void ratios are calculated with the same procedure described above. Therefore, changes of the void ratio at various depths (or nodes) can be uniquely determined.

3 PARAMETRIC STUDY AND PARAMETER REDUCTION

A computer program AAA is coded to solve for the total settlement of a soil deposit due to dewatering based on the above analysis. There are totally nine parameters involved in the settlement calculation. Before developing a simplified method, it would be wise to conduct a comprehensive parametric study in order to understand the sensitivity of these parameters in estimating the land subsidence.

The specific gravity of most soils falls between 2.6 and 2.9. The calculated results show that only a 4% variation of the total settlement corresponds to G_s. Several sets of representive values of α, n, and S_r for common soils (Kool et al., 1985) are used to investigate their effects upon the total settlement. Results indicate that 3.6% variation of the total settlement corresponds to those representive values. After a simple statistic analysis, $\alpha = 1.5$, $n = 3.5$, $S_r = 0.2$, and $G_s = 2.7$ are used as the fundamental properties of the 'standard soil' and are kept as constants through out the whole analysis. The definition of the 'standard soil' will be introduced latter. The remaining five parameters play important roles for the estimation of settlement. They should be treated carefully in developing the simplified method.

4 DEVELOPMENT OF SIMPLIFIED METHOD

The concept of 'standard soil' is the fundamental base for the development of the simplified method proposed here. The consolidation behavior of the standard soil is extensively analysed through a series of numerical simulations. The simulated results are used to construct a reference base for any other types of soils. The following properties of the standard soil are carefully chosen for the simulation: G_s =2.7, $(C_c)_s$ =0.1, e_s =1.0, σ'_s = 1.0ton/m^2, and the parameters of water retention curve (n=3.5, α=1.5, and S_r=0.2). Note that the subscript s represents a parameter of the standard soil.

4.1 Vertical strain and effective pressure

For a soil deposit of thickness H, the vertical strain due to consolidation is defined as

$$\epsilon_v = \frac{\Delta H}{H} = \frac{C_c}{1 + e_0} \log(\frac{\sigma'_{vo} + \Delta\sigma'}{\sigma'_{vo}}) \qquad (5)$$

in which ΔH is the total settlement, e_0 is the initial void ratio, σ'_{vo} is the initial effective overburden pressure, and $\Delta\sigma'$ is the change of effective overburden pressure.
The effective overburden pressure σ'_{vo} at a specific depth h and the change of the effective overburden pressure $\Delta\sigma'$ can be calculated by the following equations:

$$\sigma'_{vo} = \int_0^{Z_i} \gamma dh + (\gamma_{sat} - \gamma_w)(h - Z_i)$$
$$= \bar{\gamma} Z_i + (\gamma_{sat} - \gamma_w)(h - Z_i)$$
$$= (\gamma_{sat} - \gamma_w)h + (\bar{\gamma} - \gamma_{sat} + \gamma_w)Z_i$$

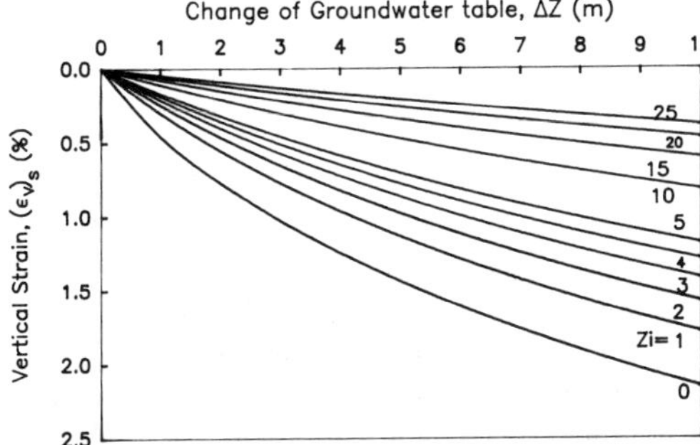

Fig.1 Vertical strain and changes of groundwater table with various initial groundwater tables for standard soil

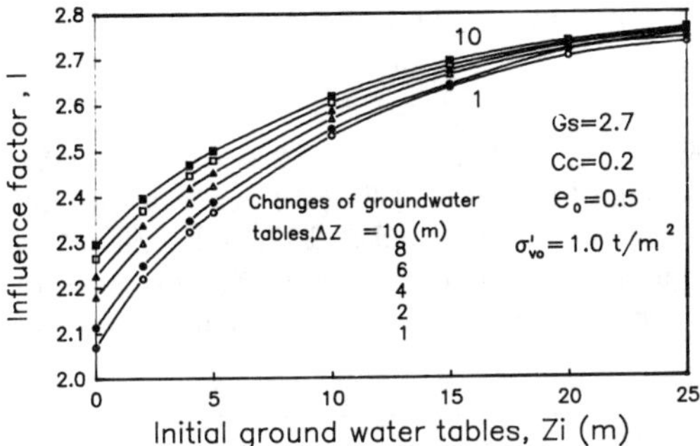

Fig.2 Influence factor vs. initial groundwater table for different changes of groundwater tables

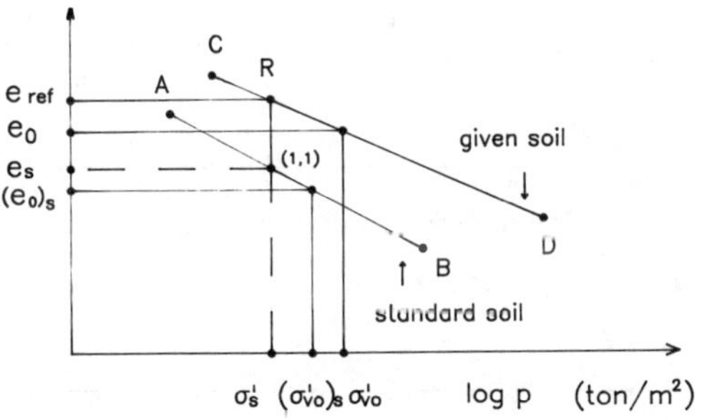

Fig.3 Graphic procedure for determining e_{ref}

$$= m + pZ_i \qquad (6-a)$$

and

$$\Delta\sigma' = \gamma_w \Delta Z \qquad (6-b)$$

where $\bar{\gamma}$ is the average unit weight of soil above the groundwater table, $m = (\gamma_{sat} - \gamma_w)h$, and $p = (\bar{\gamma} - \gamma_{sat} + \gamma_w)$. Hence Eq.(5) can be rewritten as:

$$\epsilon_v = \frac{C_c}{1+e_o}\log\left(\frac{m + pZ_i + \Delta Z\gamma_w}{m + pZ_i}\right) \qquad (7)$$

The vertical strain $(\epsilon_v)_s$ under various combinations of Z_i and ΔZ is calculated as a reference solution in the program AAA. Figure 1 shows that the relations between $(\epsilon_v)_s$ and Z_i at different $\Delta Z's$ for the standard soil.

4.2 Influence factor

A soil with properties different from those of the standard soil gives different vertical strain even under a same groundwater condition (including any changes). An influence factor, I, is thus defined as the ratio of the vertical strain ϵ_v for a given soil and $(\epsilon_v)_s$ under a specific groundwater condition:

$$I = \frac{\epsilon_v}{(\epsilon_v)_s} = \frac{\frac{C_c}{1+e_o}\log\left(\frac{m+pZ_i+\Delta Z\gamma_w}{m+pZ_i}\right)}{\frac{(C_c)_s}{1+(e_o)_s}\log\left(\frac{m_s+p_sZ_i+\Delta Z\gamma_w}{m_s+p_sZ_i}\right)} \qquad (8)$$

Here again, the subscript s stands for a parameter of the standard soil. Figure 2 gives an example of the influence factor of a specific soil under various groundwater conditions.

4.3 Reference void ratio

The linear $e - \log p$ relation for the standard soil is given in Figure 3 as a straight line AB. A standard point on AB is selected where $e_s = 1.0$ and $\sigma'_s = 1.0 ton/m^2$ as shown in the same figure. Assume that CD line is the virgin compression curve of a given soil. This line can be uniquely determined if its C_c and a point of (σ', e) is known. This point is generally the initial in situ state at a specific sampling depth, (σ'_{vo}, e_o). As a matter of fact, any other point with known state qualifies. Hence a point R with $\sigma'_s = 1.0 ton/m^2$ and e_{ref} is chosen as a reference state. Here e_{ref} is the reference void ratio of this given soil under an overburden pressure $\sigma'_s = 1.0 ton/m^2$. Note that point R has a same σ'_s as the standard point on line AB does. This reference point R was, is, or will be on the line CD for this given soil. The reference void ratio e_{ref} can be calculated by the following equation:

$$e_{ref} = e_o - C_c(log\sigma'_{ref} - log\sigma'_{vo}) = e_o + C_c log\sigma'_{vo} \qquad (9)$$

The equation (9) can be rewritten as

$$log\sigma'_{vo} = log(m{+}pZ_i) = (e_o{-}e_{ref})/C_c \quad (10-a)$$

Similarly, $(\sigma_{vo})s'$ can be expressed as

$$\begin{aligned} log(\sigma_{vo})'_s &= log(m_s + p_s Z_i) \\ &= [1-(e_o)_s]/(C_c)_s \quad (10-b) \end{aligned}$$

4.4 Hyperbolic model

The hyperbolic relations between the influence factor I and (1) the initial groundwater table Z_i and (2) the change of groundwater table ΔZ are derived in this section. Substituting Eqs.(10-a) and (10-b) into Eq.(8), the influence factor I is described as

$$\begin{aligned} I &= \frac{\frac{C_c}{1+e_o}\left[\log(m+pZ_i+\Delta Z\gamma_w) - \frac{(e_o-e_{ref})}{C_c}\right]}{\frac{(C_c)_s}{1+(e_o)_s}\left[\log(m_s+p_sZ_i+\Delta Z\gamma_w) - \frac{(1-(e_o)_s)}{(C_c)_s}\right]} \\ &= \frac{A_o\log(m+pZ_i+\Delta Z\gamma_w) - B_o}{A_s\log(m_s+p_sZ_i+\Delta Z\gamma_w) - B_s} \quad (11) \end{aligned}$$

where $A_o = C_c/1+e_o$, $A_s = (C_c)_s/1+e_s$, $B_o = (e_o - e_{ref})/1+e_o$, and $B_s = (1-(e_o)_s)/1+(e_o)_s$. Now it is clearly shown in Eq.(11) that I is a function of Z_i and ΔZ. Keeping ΔZ as a constant, the factor I becomes

$$\begin{aligned} I &= \frac{A_o logC_o(1+(pZ_i/C_o)) - B_o}{A_s logC_s(1+(p_sZ_i/C_s)) - B_s} \quad (12) \\ &= \frac{(A_o logC_o - B_o) + A_o log(1+(pZ_i/C_o))}{(A_s logC_s - B_s) + A_s log(1+(p_sZ_i/C_s))} \end{aligned}$$

in which $C_o = m + \Delta Z\gamma_w$ and $C_s = m_s + \Delta Z\gamma_w$.

Taking the Taylor's expansion of $log[1+(pZ_i/C_o)]$ and $log[1+(p_sZ_i/C_s)]$ and neglecting the high-order terms, Eq.(12) can be modified as

$$\begin{aligned} I &= \frac{(A_o logC_o - B_o) + (A_o p/C_o)Z_i}{(A_s logC_s - B_s) + (A_s p_s/C_s)Z_i} \\ &= \frac{D_o + E_oZ_i}{D_s + E_sZ_i} \quad (13) \end{aligned}$$

in which $D_o = A_o logC_o - B_o$, $D_s = A_s logC_s - B_s$, $E_o = A_o p/C_o$, and $E_s = A_s p_s/C_s$. Eq.(13) can be rearranged as

$$\begin{aligned} \frac{Z_i}{I} &= \frac{E_s}{E_o}Z_i + \frac{D_s}{E_o} + \frac{D_o/E_o}{I} \\ &= bZ_i + a \quad (14-a) \end{aligned}$$

or

$$I = \frac{Z_i}{a + bZ_i} \quad (14-b)$$

in which $b = (E_s/E_o)$ and $a = (D_s/E_o)/(D_o/(E_oI))$. Transforming Figure 2 into a Z_i/I against Z_i plot ends up a series of parallel straight lines with a constant slope b as shown in Figure 4. So this constant b is a property of the given soil and is independent of Z_i and ΔZ. Figure 5 is provided to obtain the constant b for soils with different C_c and e_{ref}. The value of $1/a$ is the tangents of those curves in Figure 2 at $Z_i = 0$. From the mathematical definition of a, it is clear that a is propotional to the factor I. Since I is a function of Z_i and ΔZ, a depends only on ΔZ at $Z_i = 0$ for a given soil. Their relation can be expressed as

$$a = a_{coeff}\Delta Z + a_{const} \quad (15)$$

From Figure 2 (for $C_c = 0.2$ and $e_{ref} = 0.5$), through Figure 4 and Eq.(15), one can get only a point in Figures 6 and 7. To complete these two figures, one should have a series figures similar to Figure 2 with different combinations of e_{ref} and

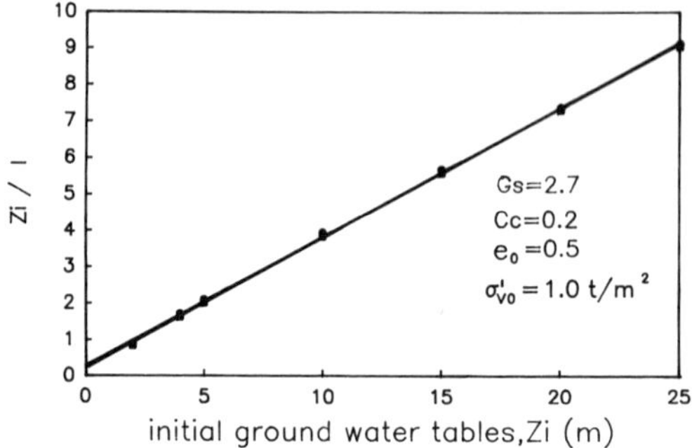

Fig.4 $\frac{Zi}{I}$ against Zi curve

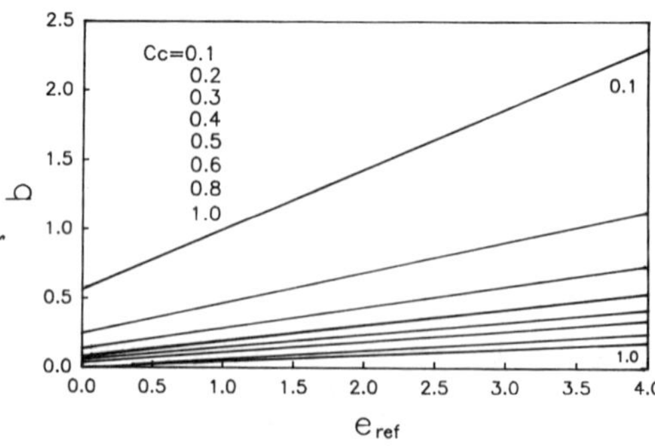

Fig.5 Variation of b with Cc and e_{ref}

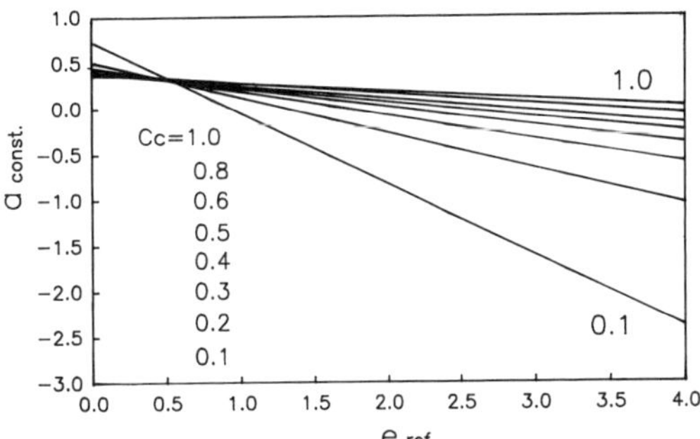

Fig.6 Variation of a_{const} with e_{ref} and Cc

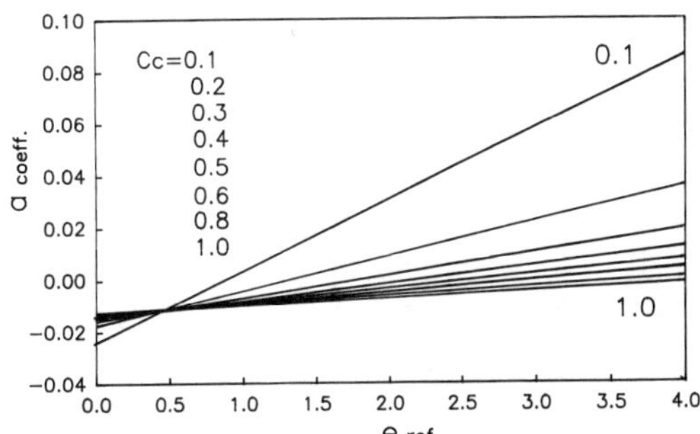

Fig.7 Variation of a_{coeff} with e_{ref} and Cc

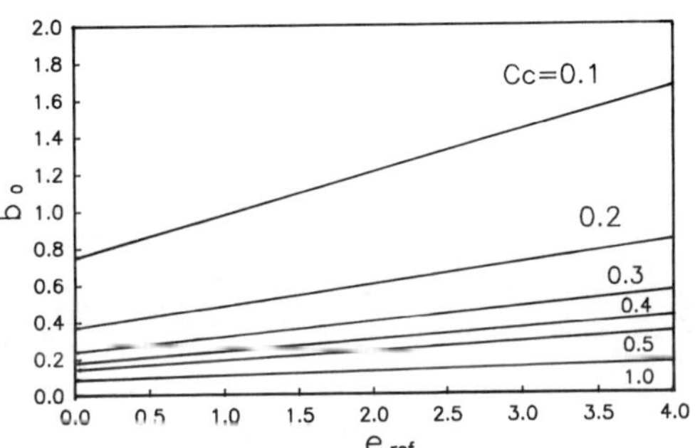

Fig.8 Variation of b_0 with e_{ref} and Cc

C_c. Through a similar procedure, the relation of the influence factor I and ΔZ can be obtained as

$$I = \frac{\Delta Z}{a_o + b_o \Delta Z} \tag{16}$$

Note that the above equation is specifically for the case of $Z_i = 0$. The reason for the case of $Z_i = 0$ so special can be explained by Eq.(14-b) in which I becomes undefined mathematically

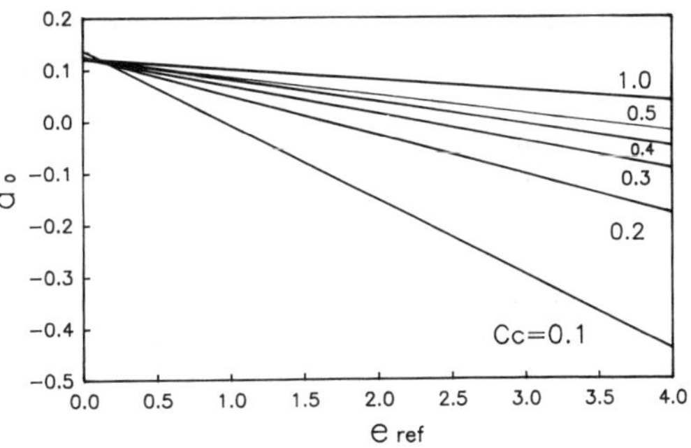

Fig.9 Variation of a_0 with e_{ref} and Cc

Table 1. Comparison of measured and computed subsidence

year	1975	1981	1984	1988
Z_i(m)	11.02	14.63	18.33	20.14
ΔZ(m)		3.61	7.31	9.12
measured(cm)		39.00	69.00	88.00
computed(cm)		41.37	78.00	94.64
simplified(cm)		43.26	80.12	95.68
Terzaghi's(cm)		66.83	127.05	155.07

when $Z_i = 0$. This is why those curves in Figure 2 intersect the $I-axis$ at different points. These intersections are determined by Eq.(16). Figures 8 and 9 give the values of a_o and b_o for different types of soils.

The procedures of this simplified method are summarized below.

a) Determine the parameters C_c, e_o, and σ'_{vo} of the virgin compression curve of a given soil.

b) Determine e_{ref} from Eq.(9) or Figure 3.

c) Based on the given values of Z_i, ΔZ, and H, find b, a_{const}, and a_{coeff} from Figures 5, 6, and 7 respectively.

d) If $Z_i = 0$, find a_o and b_o from Figures 8 and 9.

e) Calculate a From Eq.(15)

f) Calculate I from Eq.(14-b), or from Eq.(16) if $Z_i = 0$.

g) Find $(\epsilon_v)_s$ from Figure 1.

h). Calculate ϵ_v and finally the total settlement.

5 COMPARISON OF RESULTS

The west of Yunlin county is identified as a land subsidence area. One of the monitoring records of groundwater tables and ground levels is selected as an example for comparison. Parameters are determined based on the records obtained from the site near the monitoring station, including $H = 50m$, $C_c = 0.48$, $e_o = 0.71$, and $\sigma'_{vo} = 34.5ton/m^2$. Table one gives the comparison among the measured settlement and other estimated results. The discrepency between the results of the simplified method and those of the computer simulation fall within 5%. The estimated results show a satisfactory comparison with the field measurements. This validates the accuracy and the applicability of the simplified method.

6 CONCLUSIONS

The one-dimensional finite strain consolidation theory is employed to model the behavior of land subsidence. The simplified method with the help of a series of figures are proposed. This method is simple, quick, convenient, and reliable. It is capable of predicting subsidence due to dewatering for a wide variety of soils.

REFERENCES

Chen, Peng-Teng 1989. Theory and application of one-dimensional large strain consolidation of unsaturated soil. Master thesis of National Central University.

Gibson,R.E. & R.L. Schiffman 1981. The theory of one-dimensional consolidation of saturated clays II, Finite nonlinear consolidation of thick homogeneous layers. Can. Geotech. J. 18: 280-293.

Kool, J.B.,J.C. Parker, & M. T. van Genuchten 1985. Determing soil hydraulic properties from one-step experiments by parameter estimation: I. theory and numerical studies. Soil Sci. Soc. Am. J. 49:1348-1354.

Lin, Yung-Teh 1986. Information on land subsidence in coastal areas of Taiwan. Report of Council of Agriculture, R.O.C

van Genucnten, M. T. 1980 A closed-form equation for predicting the hydraulic conductivity of unsaturated soils. Soil Sci. Soc. of Am. J. 44: 892 -898.

Environmental Management, Geo-Water & Engineering Aspects, Chowdhury & Sivakumar (eds)
© 1993 Balkema, Rotterdam. ISBN 90 5410 099 0

Geosynthetic systems in environmental management

Michael A. Sadlier
Geosynthetic Consultants Australia Pty Ltd, Melbourne, Vic., Australia

Barry R. Christopher
Technical Services, Polyfelt, Inc. USA

Abstract. Liner systems would appear on first appearance to be dominated by considerations relating to the liner materials (geomembrane and/or soil) and their proper installation, but in fact there are many other considerations relating to such items as leachate or gas control and management that critically affect the proper function of the system. This paper discusses the systems approach to the use of geosynthetics in liner systems as typified in the latest US EPA regulations and guidance documents and seeks to highlight the interactive nature of an effective system design.

1. INTRODUCTION

In broad terms the components of an effective containment system include:

- suitable foundation support.
- synthetic and/or clay liners
- liner protection layers.
- leachate collection and detection systems.
- hydrostatic pressure relief and gas venting.
- cover and capping systems.

In evaluating the requirements for each of the components, an evaluation of their compatibility with respect to the other components is essential to a successful design. Special consideration should be given to interface compatibility for side slope stability, the durability of each component and how the deterioration of any component would effect other components, and the influence of construction on these relationships. Some of these elements are complex in their behavior and have led to advances in the testing and evaluation of geosynthetic materials for liner systems. Each component system will be discussed in this paper.

2.0 THE SYSTEMS STRATEGY

The basis of a systems approach is the combination of several individual components which may in themselves be single or multi-functional to achieve a level of performance that is normally considerably greater than that of the individual component. Likewise, the improper performance of any component must be evaluated to limit its influence on the performance of other components and the entire system. Particularly for critical environmental work a systems approach lends itself well to development of enhanced security as the components combine to reduce the risk and consequence of a breakdown of any of the critical sealing elements. A well designed system will develop on a synergy of individual component strengths that works to minimise the impact of individual component weaknesses.

In this way we see for example in landfills the provision of primary and secondary liners often of different but complementary materials which are in turn complemented by leachate detection, collection and drainage layers. This system guards against the possibility of leachate escape and at the same time reduces the likely consequence of a leak by significantly reducing the head driving the leachate. This consequence of leakage is also reduced by combining the ability of synthetic and clay liners to respectively restrict flow and migration of contaminants. Thus the realistic system approach recognizes the potential for a leak in any liner and focuses on the purpose of the containment which is to prevent contaminants from entering the surrounding environments.

Other geosynthetics such as geotextiles and geonets are conventionally used in each component of the containment system due to the potential for significant advantages over the classical granular soil layers they replace including:

- Improved performance if properly selected.
- Potential material cost savings, especially if granular materials must be imported.
- Construction cost and time savings
- Substantial savings in cap thickness reducing landfill settlement due to weight and more importantly increasing air space for disposal (providing the potential for substantial cost savings beyond any material cost issues).

As with the graded granular soil layers that it will replace, the geosynthetic must be carefully selected based on the properties required to perform adequately its intended function in the specific application. This is an excellent example of the Design by Function approach. This paper reviews the use of geosynthetics in each component of the containment system and provides design and material selection guidance. The influence of each component on the over all function of the containment system is also reviewed.

3.0 SITE STABILIZATION FOR LINER CONSTRUCTION

Good liner installation starts with an important element in the containment system, a good stable foundation including any slopes or walls.

Construction of an clay liner requires a sound base in order to achieve proper compaction of the clay layers as the liner is built up. Placement of a geomembrane or clay and synthetic composite requires a smooth, relatively dry, compacted surface as a working platform for proper installation and welding of seams.

A geotextile is often placed below seams to provide a clean, dry working surface. Even if properly prepared, a good foundation condition is quite often difficult to maintain, considering the influence of weather on construction. For example, ruts in the foundation which often occur due to construction traffic, can lead to an over stressed conditions if the liner spans these voids. Wet, soft or otherwise unsuitable foundations will require stabilization for proper liner construction. In these cases, a geotextile with a granular support layer may be used to expedite construction and provide an increased structural supporting element for the liner installation. If properly designed, the support layer may have a dual function as a gas vent layer. Some care is required with slopes and friction angles when placing a geotextile directly under a synthetic liner. This is discussed in detail by Hausmann et al.

4.0 SYNTHETIC LINER SELECTION

Synthetic liners have been improving in many aspects in recent years and can now offer excellent chemical and weathering resistance with positive installation and quality assurance techniques that have lead to an increasing level of confidence.

A variety of synthetic liners such as butyl rubber, polyvinyl chloride (PVC), chlorosulphonated polyethylene (CSPE), chlorinated polyethylene (CPE), Ethylene Propylene Diene Terpolymer (EPDM) and elasticized polyolefins have been used over the years in pursuit of an ideal combination of chemical resistance, weathering performance, flexibility and weldability. Many of these involve the use of additives or polymer modifiers which help performance in some respects but detract in others. For instance modification of the basic polyethylene structure (CPE, CSPE) to improve weldability and flexibility has had a detrimental effect on the weathering and chemical resistance. Plasticisers used to provide flexibility in other polymers may dissipate or leach out over time or even encourage liner consumption by animals.

Figure 1 shows the relative usage of different liner types in the U.S.A. and Canada based on IFAI 1992 projections.

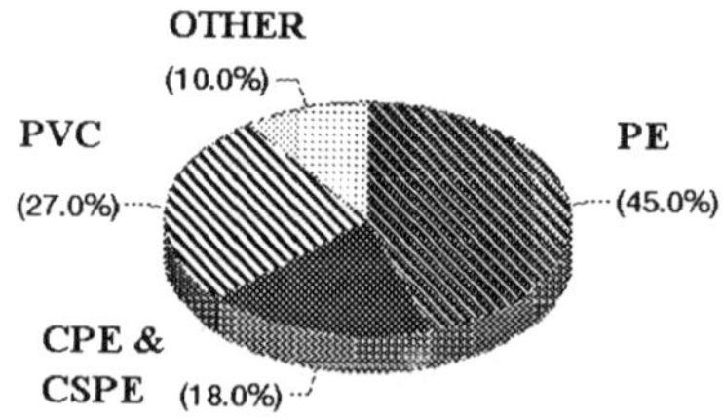

Figure 1.

A relatively recent innovation in terms of lining technology, High Density Polyethylene (HDPE) has established itself quickly as the liner of choice for both solid and liquid waste containment. Quality HDPE liners are based on pipe grade polymer resins with around 2% carbon black and no other additives which can dissipate over time. They are produced in wide seamless (6 m plus) sheets and continuing development in welding technology has made their field installation reliable and efficient. A potential for brittle cracking around weld zones has been all but eliminated by the development of more controlled extrusion welding techniques and wedge welding methods that are very efficient and provide an easy avenue for quality control testing. HDPE provides excellent weathering and chemical resistance and flexibility suitable for most applications. A recent innovation is Very Low Density Polyethylene (VLDPE) a less crystalline form of polyethylene polymer which can be used when greater flexibility is desired and weathering and chemical resistance are less important.

5.0 GEOMEMBRANE PROTECTION

Impermeable geomembrane liners used in containment systems are relatively thin and can be damaged easily, both during installation of the liner system, as well as, after completion of construction. Adequate mechanical protection must be provided to resist construction equipment loads and loads imparted by the waste.

Experience has shown that a layer of sand or a heavy weight (>200 g/mm^2) needlepunched nonwoven geotextiles can play an important role in successful geomembrane installations and long-term performance by acting as a cushion to prevent puncture damage to the geomembrane. As such, the geotextile cushion layer can be placed below geomembranes to resist puncture and wear due to abrasion caused by sharp-edged rocks in the subgrade and above the geomembrane to resist puncture caused either by drainage aggregate or direct contact with waste materials and similar benefits can be gained during intermittent covers and final closures.

Polyethylene geomembranes are the materials of choice for waste containment and mine heap leaching applications due to their high chemical resistance, biological inertness and seamability. However, due to the semi-crystalline thermoplastic nature of the sheet material, polyethylene sheets can be damaged very easily by impact or puncture. Nonwoven geotextiles will effectively protect polyethylene geomembranes and, in fact, can result in cost savings by allowing for a thinner polyethylene sheet design.

Figure 2 illustrates the protection provided by geotextile cushion layers based on results of puncture tests on HDPE membranes of different thicknesses with a 400 g/sqm polypropylene continuous filament needlepunched geotextile. This testing was reported by Puhringer and used a modified CBR plunger method with an 90^0 angle pointed tip and clearly shows the effectiveness of geotextiles used in this way.

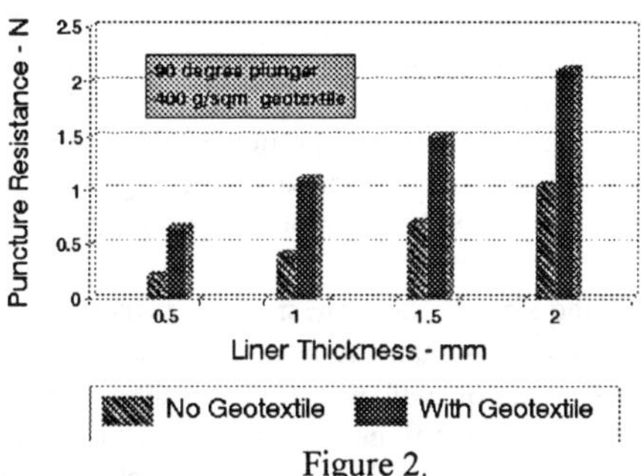

Figure 2.

6.0 LEACHATE COLLECTION AND DETECTION SYSTEMS

A primary leachate collection system is required to both remove leachate from the system so that it can be treated and minimizing the head on the liner system. A secondary system removes any leachate that penetrates the primary liner such that it does not build up head and leak through the secondary liner. The system will consist of a drainage layer and an adequate filter. Drainage element requirements include a layer (either geonet or gravel) that is adequately free flowing to remove all leachate based on an evaluation of inflow-outflow requirements, sufficiently open such that any fines entering the layer can be flushed out, and an adequate slope to maintain a sufficient gradient for flow (typically greater than 4%). Geonets offer several advantages over gravel. They reduce the potential of damage to the membrane during construction and long-term loading and they occupy significantly less space for an equal flow capacity.

An adequate filter will be required to prevent fines from entering and clogging the leachate collection/detection layer. Geotextiles are commonly used as filters for leachate collection systems. providing protection to geonets, sand or gravel layers.

Filtration is the most common application of geotextiles. However, adequate performance depends on proper selection of the geotextile in relation to its compatibility with the material that is to be filtered. Designing with geotextiles for filtration is essentially the same as designing graded granular filters which geotextiles have come to replace. The geotextile is similar to a soil in that it has voids (pores) and particles (filaments and fibers). However, with geotextiles, the geometric relations between filaments and voids are more complex than in soil because of the shapes and compressibility of the filaments. In geotextiles, we generally try to measure the pore size directly instead of, as with soils, using the particle size to estimate the pore size. Since the pore size is directly measured, relatively simple relationships between pore sizes and the particle sizes of the soil to be retained can be developed. Looking at particle retention, three simple filtration concepts are:

1. If the size of the largest pore in the geotextile filter is smaller than the larger particles of soil (or waste), the soil will not pass the filter. As with graded granular filters, the larger particles of soil will form a filter bridge over the hole, which in turn, filters the smaller particles of the soil, in turn, retaining the soil and preventing piping.

2. If the smaller openings in the geotextiles are sufficiently large such that the smaller particles of soil are able to pass through the filter, then the geotextile will not "clog".

3. A large number of openings should be present in the geotextile so that proper flow can be maintained even if some of the openings later become clogged.

These simple concepts and analogies with soil filter design criteria are used to establish design criteria for geotextiles. Specifically, the criteria are:

- The geotextile must retain the soil (retention criteria)
- while allowing water to pass (permeability criteria)
- over the life of the structure (clogging resistance criteria).

To perform effectively, the geotextile must also survive the installation process (survivability criteria). For the same reasons that heavier and denser geotextiles provide better puncture resistance, only geotextiles with a weight/unit area of 200 g/m2 or greater should be considered. For a detailed discussion of the development and background along with recommended geotextile filtration criteria, see Christopher and Holtz, 1985 and 1988. Other considerations for design of leachate collection filters are related to chemical and biological compatibility with the leachate. The main consideration relates to the potential for chemical precipitants or microbes to be trapped in the pore space of the filter and reduce flow. Also there is a potential for microbes, once trapped, to grow and clog the filter. A recent study was performed by Koerner and Koerner (1991) for the U.S. EPA to evaluate the clogging potential of various filter material. Their study found that for the filter to be effective, the dissolved solids and microorganisms must be permitted to pass through the filter. Their study was performed under extreme biological conditions. The results indicated that very open structures (as found in needle-punched nonwoven geotextiles) offered the highest biological clogging resistance (with flow reductions over a 6 month period of less than an order of magnitude for most cases) and performed substantially better than sand filters in all cases. Tests performed with tightly woven geotextiles and heat bonded geotextiles found substantially reduced flow rates (on the order of several magnitudes) .

7.0 HYDROSTATIC PRESSURE RELIEF AND GAS VENTING

Relatively thick needlepunched nonwoven geotextiles have a three dimensional fiber structure and high percentage of air voids which allows multidirectional free flow of liquids and gases in the plane of the fabric. As such, they can be used as a drainage layer in the following situations:

- Between the two membranes of a double lining system, in order to drain any leakage that may occur through the primary (top) liner.

- Between membrane and soil, or between membrane and waste material to drain any slope, waste seepage or groundwater. (Moisture between a geomembrane and underlying soil can significantly decrease the interface friction angle.)

- Beneath a geomembrane liner in a liquid containment system to divert gases from beneath the system that can accumulate due to organics in the underlying soils.

- Beneath intermittent and final landfill cover systems over waste disposal to act as gas transmission media and divert gases to collection systems.

Water Transport Design. Design and selection of the geotextile will depend on the magnitude of anticipated seepage, the normal stress anticipated over the geotextile, and the geotextile characteristics including thickness, weight per unit area, and fiber type (continuous or staple filament).

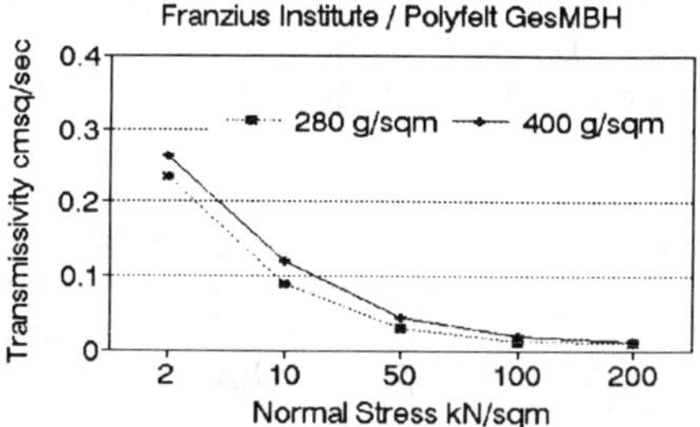

Figure 3 (after Franzius Inst.)

Figure 3 gives the transmissivity response curves of 280 and 400 g/sqm polypropylene continuous filament needlepunched nonwoven geotextiles as a function of normal stress. Using Darcy's formula, in-situ transmissivity can now be checked to determine if it is sufficient for the desired water discharge quantity in question:

$$Q = T \times b \times i$$

where, Q = Water quantity
b = Width of geotextile
T = Transmissivity
i = Hydraulic gradient

Gradient i is dependent on the water pressure present and on the landfill configuration. In the least complicated situation, drainage along a slope of inclination angle ß, i = sin ß. If greater transmissivity is required than provided by heavyweight needlepunched nonwoven geotextiles, special layered systems such as geonets must be installed.

Gas Transport Design. In this application, the same design considerations as for water transmission can be used. Because of the high porosity of needlepunched nonwoven geotextiles (90% air voids in an uncompressed state), gas transmissivity is usually sufficient, even under very high applied normal loading. Under comparable conditions, gas transmissivity is approximately two orders of magnitude greater than water transmissivity (Koerner, et al., 1984). Therefore, it is sufficient to base the design on water transmissivity, since even with a high water content in the geotextile there will be sufficient gas transmissivity available for the venting of gases.

8.0 COVER AND CAPPING SYSTEMS

Cover and capping systems are also integral to the long-term performance of the containment system. For a detailed discussion see USEPA 1989. In addition to the liner materials to prevent infiltration of precipitation into the contained mass, other geosynthetics including geonet and geotextiles form an integral part of the final cover. Geonets again provide drainage layers while geotextiles may be used for stabilization of the waste mass to allow cap construction, transmission of gas below the liner, hydrostatic pressure relief and frictional improvement between clay layers and the liner and filtration of soil layers adjacent to the drainage system.

9.0 INTERFACE COMPATIBILITY CONSIDERATIONS FOR SIDE SLOPE STABILITY

An evaluation must be made of the interface compatibility between layers of the multilayer systems to prevent side slope slippage and to avoid significant downdrag on the liner system. The minimum friction tends to be between the geomembrane and cohesive soil layers, especially for smooth sheet type geomembranes. These values are often low due to the desire to place soils wet of optimum to achieve low hydraulic conductivity levels. The friction value can be reduced further by moisture accumulation at the interface between the geomembrane and the underlying soil as a result of thermal differentials, freeze thaw activities, or seepage from consolidation of the wet cohesive soil layer and/or the landfill.

Geotextiles with lateral drainage capacity can be placed at the geomembrane soil interface to disperse moisture accumulation, dissipate pore pressure development from seepage forces and maintain uniform friction resistance. Drain lines should be placed at the base of side slopes (e.g. in the anchor trench) to allow for removal of any moisture accumulation.

The actual geotextile/geomembrane, geonet/geotextile, geonet/geomembrane and geotextile/soil friction resistance will depend on the specific materials. Due to the complexities of this situation and the potential risks involved in such designs, it is recommended that the designer perform direct shear tests to accurately evaluate anticipated interface conditions. Designers are referred to Koerner (1990) for typical values as a first order review of design requirements.

CONCLUDING REMARKS

Geosynthetics have been successfully used in conjunction with natural sealing materials for landfills, heap leach and reservoir systems since the early 1970's. Use of these materials improve the designer's confidence in the performance of a system as well as provide cost savings through reductions in material costs as well as decreased volumes for the liner system design which in turn increases the volume of containment systems. To provide confidence in performance, the designer should consider the specifics of the application and as with any good engineering design, carefully select the materials to meet the specifics of the application requirements. By taking a systems approach to evaluation and selection of materials, the designer can be assured of the synergistic behavior of the containment system components.

REFERENCES

Christopher, B.R., "Geotextiles in Landfill Closures," Geotextiles and Geomembranes, Special Issue on Landfill Closures - Geosynthetics, Interface Friction and New Developments, Elsevier, 1991.

Christopher, B.R. and Holtz, R.D., Geotextile Engineering Manual, Prepared for Federal Highway Administration, Washington, D.C., DTFH61-83-C-00150, 1985.

Christopher, B.R. and Holtz, R.D., Geotextile Design and Construction Guidelines, Prepared for Federal Highway Administration, Washington, D.D., HI-89-050, 1989.

Franzius Institute, Hamburg. Geotextile Testing carried out for Polyfelt GesmBH, August 1981.

Hausman M.R., Beckingsale C.and Sadlier M.A 'Geomembranes, Geotextiles and Slope Stability' ANZ Geomechanics Conference, Christchurch February, 1992

Koerner, R.M., Bove, J.A. and Martin, J.P., "Water and Air Transmissivity of Geotextiles," International Journal of Geotextiles and Geomembranes, Volume 1, Number 1984.

Koerner, R.M. and Koerner, G.R., "Leachate Flow Rate Behavior Through Geotextile and Soil Filters and Possible Remediation Methods," Proceedings of the 5th GRI Seminar on Geosynthetics in Filtration, Drainage and Erosion Control, Geosynthetics Research Institute, Philadelphia, PA.,1991.

Hakonson, T.E., "Evaluation of Geologic Materials to Limit Biological Intrusion into Low-Level Radioactive Waste Disposal Sites," Los Alamos National Laboratory Report No. LA-10286, 1986.

Polyfelt, Inc., Geotextiles Design and Practice Manual, Evergreen, Alabama, 1987.

Puhringer H. 'Geotextile Geomembrane Composite Pyramid Testing' ASTM Symposium on Geosynthetic Testing for Waste Containment Applications, Las Vegas 1990.

Sadlier M.A. 'Geosynthetic Containments in Environmental Protection' Proceedings of Geotropika '92 Conference, Johore Bahru, Malaysia, April 1992

United States Environmental Protection Agency, 1987, Geosynthetic Design Guidance for Hazardous Waste Landfills and Surface Impoundments, EPA-600/2-87-097, Hazardous Waste Engineering Research Laboratory, Cincinnati, 1987.

United States Environmental Protection Agency, "Technical Guidance Document: Final Covers on Hazardous Waste Landfills and Surface Impoundments," EPA-530/SW-89-047, Environmental Protection Agency, Cincinnati, OH,1989

Environmental Management, Geo-Water & Engineering Aspects, Chowdhury & Sivakumar (eds)
© 1993 Balkema, Rotterdam. ISBN 90 5410 099 0

Author index